A Synonymized Checklist of the Vascular Flora
of the United States, Canada, and Greenland

A Synonymized Checklist of the Vascular Flora of the United States, Canada, and Greenland

VOLUME II THE BIOTA OF NORTH AMERICA

by John T. Kartesz and Rosemarie Kartesz
Biota of North American Program

in confederation with
Anne H. Lindsey and C. Ritchie Bell
North Carolina Botanical Garden

THE UNIVERSITY OF NORTH CAROLINA PRESS
CHAPEL HILL

To Earl L. Core and Lincoln Constance
who have devoted their lives
to a better understanding of our flora

© 1980 The University of North Carolina Press
All rights reserved
Manufactured in the United States of America
ISBN 0-8078-1422-9
Library of Congress Catalog Card Number 80-263

Library of Congress Cataloging in Publication Data

Kartesz, John T
    A synonymized checklist of the vascular flora of
the United States, Canada, and Greenland.

    "Volume II, The biota of North America."
    Includes index.
    1. Botany—United States—Nomenclature.
2. Botany—Canada—Nomenclature. 3. Botany—
Greenland—Nomenclature. I. Kartesz, Rosemarie,
joint author. II. Title.
QK110.K37   582.097'01'4   80-263
ISBN 0-8078-1422-9

Contents

Foreword

A LIST IS A REFERENCE POINT; it can be a beginning or an ending, a preface or a summary. A list can be stultifyingly rigid and static or realistically flexible and vital. Depending upon how it is used, a list can cause stagnation in a field or provide a challenge that leads to new ideas, new discoveries, and new knowledge. Indeed, a few very special lists, catalogs, and enumerations, by a few very talented and dedicated botanists, have marked the milestones of the floristic history of the world over the past three centuries. *Species Plantarum*, although it contained limited descriptive and reference material, was essentially an orderly listing or enumeration of the approximately 6,000 species of plants then known to Linnaeus. The list obviously did not remain static, but grew over the years into the extensive floristic works which now cover the 300,000 or more species of vascular plants in the world.

Just as John Clayton's "catalog" of plants native to Virginia provided a foundation for the early floristic work in North America, so may the Kartesz checklist provide the foundation and incentive for the current cooperative taxonomic work necessary to produce the unified treatment of the rich and varied flora of North America, which Muhlenberg first proposed in the years 1790-1792 with the comments, "Let each of our American botanists do something and the wealth of American work would soon be recognized.... If then one of our younger associates ... would combine the different floras into one, how pleasant it would be for the botanical world."

C. Ritchie Bell
Chapel Hill, North Carolina
26 October 1979

Introduction

THE VASCULAR FLORA OF NORTH AMERICA represents the second most diversified group of higher organisms of the continental biota; it is surpassed in numbers of species only by the enormous class Insecta. For more than a century American taxonomists have pondered and probed the colossal task of attempting to coalesce the vast resources of floristic information into a functional whole. Unfortunately, many perplexing problems and the sheer magnitude of the task have so far thwarted the efforts of some dedicated and widely acclaimed systematists in their genuine attempts to generate a realistic treatment of the continental flora.

Aside from the traditional complexities of political and financial conflicts which unfortunately often overshadow, delay, or even destroy such worthy national and international enterprises, the tremendous effort and time required to synthesize the primary source material to produce even a fundamental checklist is in itself a formidable obstacle to progress. Botanical journals dealing directly or indirectly with North American floristics number well into the hundreds. In addition, there are hundreds of floras, manuals, guides, checklists, theses, and minor or local floristic treatments which often seem to contribute as much to the taxonomic problems of the area as they do to our knowledge of the flora. Also, a relatively high number of introduced escapes and waifs not previously reported for North America are added to the flora each year and augment the problem of realistically defining the limits of our naturalized flora.

If, indeed, the development of an acceptable and authoritative continental checklist or flora involved simply a mechanical assembly or compilation of data from existing literature, the task, although extensive, would be relatively simple. However, differences in taxonomic judgment, opinion, and fundamental theory, coupled with historical misinterpretation of data, basic errors in orthography and documentation, and numerous misidentifications, have compounded the problem and created a high degree of taxonomic confusion. To untangle completely the collective nomenclatural web of truth, error, and synonymy which has accrued to the flora of North America will require years of detailed taxonomic research by individual specialists. Even in the Northeast, where vascular plant systematics has been an active interest for well over two centuries and where the amount of botanical exploration is unparalleled anywhere in the Western Hemisphere, many nomenclatural and taxonomic uncertainties still prevail. In reality, the system can never be totally free of error, confusion, and the inevitable differences of valid taxonomic opinion. But these problems should not be used as excuses to delay or abandon efforts toward a current *Flora of North America*.

It must be recognized that this first attempt to produce a comprehensive checklist of the flora of such a large area will neither be perfect nor receive universal acceptance. Any attempt to develop a continental checklist now can only represent a provisional best effort toward an accurate account of our present floristic knowledge and must be taken only as an endeavor to standardize and assign meaning to the massive taxonomic data which has accumulated. By no means does this checklist complete the task, but it should provide a more solid foundation for the preparation of an atlas, and ultimately a guide, to the flora of North America.

Alphonse de Candolle closed the introduction to his "Lois de la Nomenclature botanique," presented to the Botanical Congress of 1867 in Paris, with the following comments which are still pertinent:

> There will come a time when all the plant-forms in existence will have been described; when herbaria will contain indubitable material of them; when botanists will have made, unmade, often remade, raised or lowered, and redefined several hundred thousand groups from classes to mere varieties, and when synonyms will have become much more numerous than accepted groups. Then science will have need of a great revision of its formulae.... Then perhaps there will arise something wholly different from Linnaean nomenclature, something so designed as to give certain and definite names to certain and definite groups.
>
> That is the secret of the future, a future still very far off.
>
> In the meantime let us perfect the system of binomial nomenclature established by Linnaeus. Let us try to adapt it better to the continual, necessary changes in science ... to drive out small abuses, small negligences; and if possible, to come to agreement

among ourselves on disputed points. Thus we shall prepare the way for the better progress of taxonomy.[1]

Developmental Design

The intention of this checklist is to provide an enumeration which is both comprehensive and flexible, to account as accurately and completely as possible for the multiplicity of pertinent names of all known native and naturalized vascular plant taxa in the regions covered, and to be open to such periodic revisions of the computerized data as will be necessary to maintain the vitality of the list as a basis for further work.

The work was begun in 1973 with an intensive literature search, especially for the period from 1900 to the present. This search included a survey of all pertinent regional manuals, state floras, species checklists, botanical journals and monographs, the Gray Card Index and other plant indices, rare species literature, foreign floras and other works available which directly or indirectly relate to North American plants. Individual reference cards were prepared for each plant name encountered; these cards included the names of families, genera, species, subspecies, varieties, subvarieties, cultivars, hybrids, nothomorphs, and forms. These references were then compiled into an expanded data base which now includes entries for nearly 350,000 scientific plant names. A complete reference was kept for the botanical source or sources of each scientific name, the associated vernacular name, the known state distribution for each taxon, pertinent synonymy, source of synonymy, and notations of orthographic variations. Neither space nor costs will permit the inclusion of the approximately 3,000 primary references which document the data base, but a list of the major floristic references has been prepared for separate publication.

Due to the size of the data base, it is neither physically nor fiscally possible to include all of the information in a single volume. Thus it seemed appropriate to publish, as funds become available, the basic material in a series of three interrelated volumes of realistic size, cost, and content. This synonymized checklist is the first of the three publications and attempts to account for as many plant names as possible that are in current use, i.e., those scientific names treated in regional manuals, state floras, monographs, and most journal articles in current use for all regions of North America north of

---

1. Quoted from a "free translation" by C. A. Weatherby in a Symposium on Botanical Nomenclature, II: Botanical Nomenclature Since 1867, *American Journal of Botany* 36:5-7 (1949).

Mexico, plus Hawaii, Greenland, Puerto Rico, and the U.S. Virgin Islands. It is anticipated that the data base can, in a few years, be expanded to include Mexico.

Format

For convenience this checklist is divided into three sections: Pteridophyta (includes the current Psilophyta, Lepidophyta, Calamophyta, and Filicophyta), Gymnospermae (includes the current Cycadophyta and Coniferophyta), and Angiospermae. Within each section all names of taxa are arranged sequentially and alphabetically by family, genus, species, subspecies, and variety. Family names follow contemporary treatments and are synonymized in the index for ease of reference. Family names for the Angiospermae follow the convenient treatment by Windler and Windler (Phytologia 26(2): 103-15; 1973), but Gymnospermae family names generally follow those of Little's 1979 U.S. Forest Service *Checklist of United States Trees (Native and Naturalized)*. The families of the lower vascular groups included in the Pteridophyta are as delineated by Crabbe, Jermy, and Mickel (Fern Gaz. 11:141-62; 1975). Hybrid names are included but unnamed hybrids (those known only by formula crosses) are omitted; only the name representing the most recent taxonomic status of known or presumed hybrids is given. The names of forms, cultivars, and other infraspecific taxa below the varietal level are not included. A few names of uncertain application or status are indicated by a question mark before the name.

Excluded Taxa

Although all names and combinations encountered in the literature search were cataloged and included in the data base, a few of the names not in current use and of questionable synonymous application are not included in the checklist, but seem to be more appropriately left with the less used monographic synonyms in the data base. These omitted names are primarily those from such superseded works as *Flora of Utah and Nevada* (1922) by Tidestrom, *Flora of the Prairies and Plains of Central North America* (1932) by Rydberg, and *Ferns of the Southeastern States* (1938) by Small. In addition, a few anomalous names and the names of a number of cultivated plants not appropriate to the checklist have also been excluded, as indicated below and on pages xxxviii-xli.

Geographic Scope

The geographic scope of this checklist includes all of the continental United States and Canada including St. Pierre and Miquelon, Greenland, Hawaii, and the Caribbean islands of

Puerto Rico and the U.S. Virgin Islands. It is regrettable that the extremely diversified and interesting flora of Mexico could not realistically be included in the present treatment, but through cooperative international efforts plans are currently under way to expand the present data base to include the vascular flora of Mexico.

For the determination of occurrence of a given taxon within the range of this checklist, we have relied primarily on actual published and documented specimen citations. If, for example, the literature indicates that plants of a particular taxon are known either as bonafide native local populations or, if not native, as established populations or persisting without cultivation, we consider their presence documented. Conversely, if a taxon is merely included without discrimination in a list or mentioned in the literature with no actual account of its establishment indicated, it has not been included in this checklist. Indeed, most of the several hundred names in this latter category in our data base are clearly best omitted from the checklist without further note, in spite of the fact that they have appeared in some local checklists, guides or other taxonomic publications relating to the flora of our area. Most of these plants appear to be cultivated, and no actual specimens, specimen citations, or published indications of status have been found, either as established populations or as mere waifs. Accordingly, these taxa should not be considered as components of our native or naturalized flora until such status is warranted by appropriate documentation. However, it did seem appropriate to list (on pages xxxix-xli) a few of the excluded taxa in order to emphasize or perhaps question their exclusion. Hopefully, many of the problems or questions of plant occurrence and distribution within our area will either be solved or more finely focused after the completion of the data base for an atlas of the vascular flora of all of North America. This atlas is currently in progress. As with all aspects of this checklist, the authors would appreciate any comments and corrections regarding inclusion or exclusion of names that would make the checklist more accurate and valuable.

Synonymy   For most taxa in the checklist a high degree of functional synonymy is accomplished with only "manual level" synonyms. However, for certain genera such as *Salix*, *Lupinus*, *Crataegus*, *Rubus*, *Antennaria*, and *Galium*, a more complete assessment of synonymy is necessary to provide an accurate understanding of the elements in these confusing and controversial genera. Despite the more detailed treatment of a few genera, it must be kept in mind that the general intent of the checklist is primarily to account only for those names which appear in the current standard floristic treatments of the areas concerned and thus to provide functional or "manual level" rather than detailed or "monographic level" synonymy, which would add nearly 300,000 more names and would more than triple the size and cost of the checklist! All synonyms are clearly indicated in the list by the abbreviation SY before the indented name.

In order to further reduce the time and cost of the computerized manuscript preparation and the physical size of the checklist, synonyms are presented in an abbreviated form whenever such abbreviation will not lead to confusion or impair the usefulness of the checklist. However, it should be noted that the names concerned with such abbreviations may, in some cases, require an ending different than the ending of the full name immediately above to which the abbreviation refers.

Author Citation

A most diligent attempt was made to standardize author citations in conformity with commonly accepted form: the citations for generic names, for example, follow those in *Index Nominum Genericorum*. Since numerous author names have historically been either misapplied, misspelled, or abbreviated incorrectly or unsystematically, it has been impossible so far to substantiate, clarify, and standardize a small proportion of the citations. In all such instances, assistance is earnestly invited and suggestions for appropriate alteration appreciated.

Specialist Review

Once the checklist was completed in draft form, copies of the various portions of the manuscript (i.e., a group of families, a single family, a subfamily, or in some special cases a single difficult genus or section) were submitted to taxonomic specialists for review. In all, over 250 taxonomic specialists generously provided many helpful comments and suggestions in the early stages of the project and many also provided excellent critical reviews of their pertinent portions of the final manuscript. The reviewers' comments and questions were answered, as far as possible, and their corrections, additions, and deletions were incorporated in the computerized draft version of the checklist. With few exceptions, all the comments, suggestions, and changes received from specialists were also incorporated into the revised checklist manuscript. However, in a few instances it seemed more appropriate to allow the original draft to stand unchanged or to augment an already reviewed manuscript in order to provide the most appropriate or recent taxonomic concept of a group or to comply with the basic format and established goals of the checklist. Thus the senior author ultimately assumes the full responsiblity for any nomenclatural errors or questionable differences of taxonomic opinion or judgment in this checklist.

Uniformity and Anomalous Names

One major disadvantage of local or regional floristic or monographic work is its potential for incompleteness and for lack of uniformity of treatment of the botanical elements of the total flora or even of the taxa within a given family, genus, or species. Such problems concerning uniformity of taxonomic treatment can be especially obvious and difficult to resolve when caused by differences in the taxonomic concepts of various contemporary specialists who publish on a given taxon or in a given geographic area. Thus, although the apparent inconsistencies in the use of infraspecific categories in the checklist may be criticized by some, and although a few names may be considered anomalies by others, such lack of uniformity is only a usual consequence of our current taxonomic methodology (which often seems to include a bit of taxonomic license at the infraspecific levels!), and the result of an enumeration of our plant taxa that has attempted to be as objective as possible.

However, it is inevitable that a few names cannot be treated within the general format and established synonymy of a taxonomic enumeration of this scope; these names represent individual anomalous situations which require special explanation and listing. A few such anomalies are those names or new combinations that sometimes appear in various floristic treatments or other botanical publications prior to their actual valid publication under the rules of nomenclature.

Most of the changes needed to validate these anomalous names very likely represent logical taxonomic or geographic extensions of prior work by specialists who may, indeed, have already solved any taxonomic or nomenclatural problems involved. It therefore seemed appropriate to eschew the temptation of arbitrarily and mechanically making and publishing these nomenclatural changes in a wholesale manner in order to allow the various specialists an opportunity to evaluate the anomalies individually and then to produce whatever remedial publications might be indicated. In order to prevent further misapplication, confusion, and error concerning the few anomalous names encountered to date, they have been excluded from the main body of the checklist, but are included for cross-referencing on pages xxxviii-xxxix, with a brief comment about the reason for each apparent anomaly.

Acknowledgments

THE AUTHORS ARE extremely grateful for the overwhelming support and cooperation from the hundreds of botanists and specialists who have contributed so generously in various ways and in varying degrees toward the development of this work. Through their laborious and painstaking efforts and friendly cooperation over the past few years, they have helped make possible the first significant concrete step toward the ultimate production of a "Flora" for our continent. The strong endorsement implied by such active participation in this work by members of the botanical community, both collectively and individually, is especially encouraging and deeply appreciated, as is the official endorsement of this checklist by the Flora of North America Steering Committee. In order to recognize fully the individual work of each of the many reviewers, these tremendously valuable contributions are listed by plant family in a separate section following the introduction. The unending assistance of Dr. Harold St. John, of the Bernice P. Bishop Museum, with the many complex plant taxa of the Hawaiian Islands deserves special mention and special thanks.

We extend our most sincere appreciation to our confederates in the production of this publication: Dr. C. Ritchie Bell, Director of the North Carolina Botanical Garden, and Dr. Anne H. Lindsey, Research Associate, both at The University of North Carolina at Chapel Hill, who have labored endlessly at developing the necessary data input methods, data organization, and format structure. They have assisted with the final editing and proofing of the manuscript and have provided many invaluable taxonomic suggestions which have added enormously to the quality and utility of this work. Indeed, it is doubtful that the transition from the data cards and extensive typescript of the authors to the camera-ready copy and actual publication of the checklist could have been completed without their expert advice and constant support at every phase of the project.

The preparation of the extensive data base behind this checklist would not have been accomplished without ready access to extensive library materials in both public and private collections. Thus, we are deeply indebted to the following libraries and institutions which have provided the majority of the essential reference materials necessary for our work: Hunt Institute for Botanical Documentation, Pittsburgh; Carnegie Museum, Pittsburgh; New York Botanical Garden, Bronx; Library of the Ohio State University, Columbus; West Virginia University Main Library, Morgantown; the John N. Couch Botanical Library of The University of North Carolina, Chapel Hill; Library of the University of California at Berkeley; Library of Energy Impact Associates, Pittsburgh; and the Hillman Library, Pittsburgh. Many other libraries often were the source of special references and we express our sincere thanks to the staff at each of these institutions. In addition, the personal collections of botanical references were made freely available to the senior author by a number of botanists, especially C. Ritchie Bell, The University of North Carolina at Chapel Hill; Ronald Stuckey, Ohio State University; Larry Morse, New York Botanical Garden; Jesse Clovis, West Virginia University; Earl L. Core, West Virginia University; and David Boufford, Carnegie Museum. We appreciate their cooperation. We also wish to acknowledge with thanks the generous loans by Elbert L. Little and Lyman Benson of their draft manuscripts.

The ultimate production of the actual camera-ready copy for the checklist was only possible through the continued assistance (and understanding!) of a number of very capable and sympathetic members of the staff of The University of North Carolina Computer Center. Thus we owe much to Don N. Mitchell, William R. Jarrell, Charles H. Dunham, William M. Yeager, John Cole, and Kar Wang Lau, without whose computer expertise and active personal cooperation our task would have been far more difficult. The critical but tedious and exacting task of accurately entering the thousands of scientific names and authorities into the computer system and then editing the tapes was very ably done by Anne Lindsey, with assistance from Patricia Cooley Thorburn and Sharon Morgan at The University of North Carolina Department of Botany. The typographic accuracy of the checklist is the result of their truly dedicated and most appreciated efforts.

As with any major research project done without grant or contract funds, the thousands of hours and thousands of dollars expended in the preparation of the primary data base have been totally the authors' personal contribution. Thus it is with profound thanks that we acknowledge the critical fiscal contributions toward the actual computerization of the first portion of the data base and its associated manuscript preparation, in the form of loans or appropriate budget increases by The University of North Carolina at Chapel Hill, its School of Arts and Sciences, and its Department of Botany, and by gifts and loans from the private Botanical Garden Foundation, Inc., which is associated with the North Carolina Botanical Garden.

We extend our thanks to Thomas Duncan at the University of California at Berkeley for his continuing assistance in promoting an interest in this BONAP-NCBG project, aid with the specialist reviews, and encouragement throughout the project. We also thank Larry E. Morse at the New York Botanical Garden, who promoted and endorsed the work of BONAP and who has provided helpful technical advice and editorial comments. The senior author especially wishes to thank Drs. Earl L. Core, Jesse F. Clovis, and Roy B. Clarkson of the Department of Botany at West Virginia University for their continued encouragement, advice, and moral support, which have been so valuable, so appreciated, and of such great assistance in the completion of the extensive data base.

Reviewers and Cooperating Specialists

UNLESS OTHERWISE NOTED (by the appropriate generic name in parentheses following the reviewer's institution name), the many helpful reviews of the material were general and concerned all of the family or genus indicated. However, a number of exceptionally detailed and comprehensive reviews of large or difficult groups were provided which deserve special mention and thanks. Reviews in this category are each marked with an asterisk in the following list:

PTERIDOPHYTA

PTERIDOPHYTA
John T. Mickel
    The New York Botanical
    Garden
    Bronx, New York

PTERIDOPHYTA
Alan Smith*
    University of California at
    Berkeley
    Berkeley, California

PTERIDOPHYTA
Warren H. Wagner, Jr.
    University of Michigan
    Ann Arbor, Michigan

ADIANTACEAE
Rolla M. Tryon
    Gray Herbarium
    Harvard University
    Cambridge, Massachusetts
    (Notholaena)

CYATHEACEAE
Rolla M. Tryon
    Gray Herbarium
    Harvard University
    Cambridge, Massachusetts

LYCOPODIACEAE
Joseph M. Beitel
    The University of Michigan
    Ann Arbor, Michigan

PARKERIACEAE
Robert M. Lloyd
    Ohio University
    Athens, Ohio

GYMNOSPERMAE

GYMNOSPERMAE
Elbert L. Little, Jr.
    U. S. D. A. - Forest Service
    Washington, D.C.

CUPRESSACEAE
Thomas Zanoni
The University of Oklahoma
Norman, Oklahoma
(Juniperus)

EPHEDRACEAE
Lyman Benson
Pomona College
Claremont, California

ANGIOSPERMAE

ACANTHACEAE
Martha Meagher*
Leisure City, Florida

ACERACEAE
Thomas J. Delendick
Brooklyn Botanic Garden
Brooklyn, New York

AGAVACEAE
Howard S. Gentry
Desert Botanical Garden
Phoenix, Arizona

AGAVACEAE
Susan Verhoek
Lebanon Valley College
Annville, Pennsylvania
(Manfreda)

ALISMACEAE
Ernest O. Beal
Western Kentucky University
Bowling Green, Kentucky
(Sagittaria)

ALISMACEAE
Jean Wooton
University of Southern
Mississippi
Hattiesburg, Mississippi

AMARANTHACEAE
James Mears
The Academy of Natural
Sciences
Philadelphia, Pennsylvania
(Gomphrenoideae)

AMARANTHACEAE
Clyde F. Reed
Reed Library and Herbarium
Baltimore, Maryland

AMARANTHACEAE
Jonathan D. Sauer
University of California at
Los Angeles
Los Angeles, California
(Amaranthus)

ANACARDIACEAE
William T. Gillis
Michigan State University
East Lansing, Michigan

ANACARDIACEAE
David A. Young
University of Illinois at
Urbana-Champaign
Urbana, Illinois

ANNONACEAE
Robert Kral
Vanderbilt University
Nashville, Tennessee
(Asimina)
(Deeringothamnus)

APIACEAE
Lincoln Constance*
University of California at
Berkeley
Berkeley, California

AQUIFOLIACEAE
Robert B. Clark
Meredith, New Hampshire

AQUIFOLIACEAE
Ross C. Clark
Erskine College
Due West, South Carolina

AQUIFOLIACEAE
Theodore R. Dudley
U.S. National Arboretum
Washington, D.C.

ARACEAE
Normand Cornellier
Montreal Botanic Garden
Montreal, Quebec

ARALIACEAE
Harold St. John
Bernice P. Bishop Museum
Honolulu, Hawaii

ARECACEAE
Sidney F. Glassman
University of Illinois at
Chicago Circle
Chicago, Illinois

ARECACEAE
Harold E. Moore, Jr.
Cornell University
Ithaca, New York

ARISTOLOCHIACEAE
Howard W. Pfeifer
University of Connecticut
Storrs, Connecticut
(Aristolochia)

ASCLEPIADACEAE
Donald J. Drapalik
Georgia Southern College
Statesboro, Georgia

ASTERACEAE
Dennis M. Anderson
Ohio Department of Natural
Resources
Columbus, Ohio
(Silphium)

ASTERACEAE
Loran C. Anderson
Florida State University
Tallahassee, Florida
(Bigelowia)
(Chrysothamnus)

ASTERACEAE
Theodore M. Barkley
Kansas State University
Manhattan, Kansas
(Senecio)

ASTERACEAE
Alan A. Beetle
The University of Wyoming
Laramie, Wyoming
(Artemisia)

ASTERACEAE
Mark W. Bierner
The University of Tennessee
Knoxville, Tennessee
(Amblyolepis)
(Balduina)
(Dugaldia)
(Gaillardia)
(Helenium)
(Hymenoxys)
(Plummera)

ASTERACEAE
Judith M. Canne
University Guelph
Guelph, Ontario
(Galinsoga)

ASTERACEAE
Gerald D. Carr*
University of Hawaii at Manoa
Honolulu, Hawaii
(Calycadenia)
(Dubautia)
(Osmadenia)

ASTERACEAE
W. Dennis Clark
Arizona State University
Tempe, Arizona
(Hazardia)

ASTERACEAE
Jerry A. Clonts
Anderson College
Anderson, South Carolina
(Elephantopus)

ASTERACEAE
James Coleman
Universidad Estadual Paulista
Sao Paulo, Brazil
(Verbesina)

ASTERACEAE
Lincoln Constance
University of California at
Berkeley
Berkeley, California
(Eriophyllum)

ASTERACEAE
W. S. Davis
University of Louisville
Louisville, Kentucky
(Lacamothrix)
(Malacothrix)

ASTERACEAE
D. C. D. DeJong
University of Cincinnati
Cincinnati, Ohio
(Astranthium)

ASTERACEAE
George W. Douglas*
Douglas Ecological
Consultants, Ltd.
Victoria, British Columbia
(Antennaria)
(Arnica)

ASTERACEAE
Robert I. Ediger
California State University at
Chico
Chico, California
(Senecio)

ASTERACEAE
James Estes
The University of Oklahoma
Norman, Oklahoma
(Artemisia)

ASTERACEAE
T. Richard Fisher*
Bowling Green State
University
Bowling Green, Ohio
(Heliopsis)
(Silphium)

ASTERACEAE
Robert Gardner
Baylor University
Waco, Texas
(Lipochaeta)

ASTERACEAE
Leslie Gottlieb
University of California at
Davis
Davis, California
(Stephanomeria)

ASTERACEAE
Vernon L. Harms*
University of Saskatchewan
Saskatoon, Saskatchewan
(Heterotheca)
(Petasites)

ASTERACEAE
Charles B. Heiser, Jr.
Indiana University
Bloomington, Indiana
(Helianthus)

ASTERACEAE
Robert Jansen
The Ohio State University
Columbus, Ohio
(Spilanthes)

ASTERACEAE
Miles F. Johnson
Virginia Commonwealth
University
Richmond, Virginia
(Ageratum)
(Eupatorium)

ASTERACEAE
R. Roy Johnson
Grand Canyon National Park
Grand Canyon, Arizona
(Porophyllum)

ASTERACEAE
Almut G. Jones*
University of Illinois at
Urbana-Champaign
Urbana, Illinois
(Aster)
(Machaeranthera)
(Xylorhiza)

ASTERACEAE
Samuel Jones
The University of Georgia
Athens, Georgia
(Vernonieae)

ASTERACEAE
Rudy G. Koch
University of Wisconsin
Superior, Wisconsin
(Bidens)

ASTERACEAE
Donald Kyhos
University of Hawaii at Manoa
Honolulu, Hawaii
(Chaenactis)

ASTERACEAE
Timothy Lowrey
University of California at
Berkeley
Berkeley, California
(Tetramolopium)

ASTERACEAE
Ronald McGregor
The University of Kansas
Lawrence, Kansas
(Echinacea)

ASTERACEAE
Bassett Maguire
The New York Botanical
Garden
Bronx, New York
(Arnica)

ASTERACEAE
James Mears
The Academy of Natural
Sciences
Philadelphia, Pennsylvania
(Parthenium)

ASTERACEAE
L. Maynard Moe
California State College at
Bakersfield
Bakersfield, California
(Matricaria)

ASTERACEAE
John Mooring
The University of Santa Clara
Santa Clara, California
(Eriophyllum)

ASTERACEAE
Robert T. Neher
La Verne College
La Verne, California
(Tagetes)

ASTERACEAE
Guy Nesom
University of North Carolina
Chapel Hill, North Carolina
(Chaptalia)
(Conyza)
(Erigeron)

ASTERACEAE
Wesley E. Niles
University of Nevada at
Las Vegas
Las Vegas, Nevada
(Perityle)

ASTERACEAE
David K. Northington
Texas Tech University
Lubbock, Texas
(Pyrrhopappus)

ASTERACEAE
Robert Ornduff
University of California at
Berkeley
Berkeley, California
(Lasthenia)

ASTERACEAE
Gerald B. Ownbey*
University of Minnesota
St. Paul, Minnesota
(Cirsium)

ASTERACEAE
James C. Parks
Millersville, State College
Millersville, Pennsylvania
(Melanthera)

ASTERACEAE
Dennis R. Parnell
California State University,
Hayward
Hayward, California
(Corethrogyne)

ASTERACEAE
Donald J. Pinkava
Arizona State University
Tempe, Arizona

ASTERACEAE
Richard W. Pippen
Western Michigan University
Kalamazoo, Michigan
(Cacalia)

ASTERACEAE
A. Michael Powell
Sul Ross State University
Alpine, Texas

ASTERACEAE
Edward L. Richards
Arkansas State University
State University, Arkansas
(Ratibida)

ASTERACEAE
John C. Semple*
University of Waterloo
Waterloo, Ontario
(Borrichia)
(Chrysopsis)
(Heterotheca)
(Pityopsis)
(Xanthisma)

ASTERACEAE
Beryl B. Simpson
University of Texas at Austin
Austin, Texas
(Acourtia)

ASTERACEAE
Edwin B. Smith
University of Arkansas
Fayetteville, Arkansas
(Coreopsis)
(Haplopappus)

ASTERACEAE
G. Ledyard Stebbins
University of California at
Davis
Davis, California
(Antennaria)
(Crepis)

ASTERACEAE
Warren P. Stoutamire
The University of Akron
Akron, Ohio
(Gaillardia)

ASTERACEAE
John L. Strother
University of California at
Berkeley
Berkeley, California

ASTERACEAE
Tod F. Stuessy
The Ohio State University
Columbus, Ohio

ASTERACEAE
Constance Taylor*
Southeastern Oklahoma State
University
Durant, Oklahoma
(Euthamia)
(Solidago)

ASTERACEAE
Andrew M. Torres
University of Kansas
Lawrence, Kansas

ASTERACEAE
Ronald J. Tyrl*
Oklahoma State University
Stillwater, Oklahoma
(Achillea)

ASTERACEAE
Lowell Urbatsch
Louisiana State University
Baton Rouge, Louisiana
(Ericameria)
(Haplopappus)
(Tetragonotheca)

ASTERACEAE
Gene Van Horn
The University of Tennessee at
Chattanooga
Chattanooga, Tennessee
(Pentachaeta)

ASTERACEAE
Thomas J. Watson, Jr.
University of Montana
Missoula, Montana
(Xylorhiza)

ASTERACEAE
Ronald R. Weedon
Chadron State College
Chadron, Nebraska
(Bidens)

ASTERACEAE
James R. Wells
Cranbrook Institute of Science
Bloomfield Hills, Michigan
(Polymnia)

ASTERACEAE
Dieter H. Wilken
Colorado State University
Fort Collins, Colorado
(Hulsea)

BALSAMINACEAE
Robert Ornduff
University of California at
Berkeley
Berkeley, California

BETULACEAE
Knud E. Clausen
U. S. D. A. Forest Service
Carbondale, Illinois
(Betula)

BETULACEAE
John Furlow
Capital University
Columbus, Ohio
(Alnus)

BIGNONIACEAE
Alwyn H. Gentry
Missouri Botanical Garden
St. Louis, Missouri

BIXACEAE
Richard Keating
Southern Illinois University at
Edwardsville
Edwardsville, Illinois

BORAGINACEAE
Larry C. Higgins
West Texas State University
Canyon, Texas

BORAGINACEAE
Francia Hommersand
University of North Carolina
Chapel Hill, North Carolina
(Amsinckia)

BORAGINACEAE
Alfred Richardson
Texas Southmost College
Brownsville, Texas
(Tiquilia)

BRASSICACEAE
Gerald A. Mulligan
Biosystematics Research
Institute
Ottawa, Ontario

BRASSICACEAE
Reed C. Rollins*
Gray Herbarium
Harvard University
Cambridge, Massachusetts

BRASSICACEAE
Elizabeth Anne Shaw
Gray Herbarium
Harvard University
Cambridge, Massachusetts

BRASSICACEAE
Ronald Stuckey
The Ohio State University
Columbus, Ohio
(Armoracia)
(Nasturtium)
(Rorippa)

BROMELIACEAE
Lyman B. Smith
Smithsonian Institution
Washington, D.C.

BUDDLEJACEAE
Harold St. John
Bernice P. Bishop Museum
Honolulu, Hawaii

BURSERACEAE
Duncan Porter
Virginia Polytechnical Institute
and State University
Blacksburg, Virginia

CACTACEAE
  Lyman Benson*
    Pomona College
    Claremont, California

CALLITRICHACEAE
  Ernest O. Beal
    Western Kentucky University
    Bowling Green, Kentucky

CAMPANULACEAE
  Janet Hohn
    Fish and Wildlife Service
    Portland, Oregon
    (Githopsis)

CAMPANULACEAE
  Stanwyn Shetler
    Smithsonian Institution
    Washington, D.C.
    (Campanula)

CAPPARIDACEAE
  Hugh H. Iltis
    University of Wisconsin
    Madison, Wisconsin

CAPRIFOLIACEAE
  Theodore R. Dudley
    U. S. National Arboretum
    Washington, D.C.

CAPRIFOLIACEAE
  Peter J. Salamun
    The University of Wisconsin
    Milwaukee, Wisconsin

CARYOPHYLLACEAE
  Garrett E. Crow
    University of New Hampshire
    Durham, New Hampshire
    (Sagina)

CARYOPHYLLACEAE
  Arthur Kruckeberg
    University of Washington
    Seattle, Washington
    (Lychnis)
    (Silene)

CARYOPHYLLACEAE
  John McNeill*
    Canadian Department of
    Agriculture
    Ottawa, Ontario

CELASTRACEAE
  Barry A. Prigge
    Rancho Santa Ana Botanic
    Garden
    Claremont, California

CERATOPHYLLACEAE
  Richard M. Lowden
    Universidad Catolica Madre
    Y Maestra
    Republica Dominicana
    (Ceratophyllum)

CERATOPHYLLACEAE
  Jean Wooten
    University of Southern
    Mississippi
    Hattiesburg, Mississippi

CHENOPODIACEAE
  H. Lamont Arnold
    College of Eastern Utah
    Price, Utah
    (Salsola)

CHENOPODIACEAE
  Peter W. Ball
    University of Toronto
    Mississauga, Ontario
    (Arthrocnemum)
    (Salicornia)
    (Sarcocornia)

CHENOPODIACEAE
  I. John Bassett
    Plant Research Institute
    Central Experimental Farm
    Ottawa, Ontario
    (Atriplex)

CHENOPODIACEAE
  Will H. Blackwell, Jr.
    Miami University
    Oxford, Ohio

CHENOPODIACEAE
  John McNeill
    Canadian Department of
    Agriculture
    Ottawa, Ontario
    (Atriplex)

CHENOPODIACEAE
  Clyde F. Reed
    Reed Library and Herbarium
    Baltimore, Maryland

CHENOPODIACEAE
  Howard Stutz
    Brigham Young University
    Provo, Utah
    (Atriplex)

CISTACEAE
  Larry E. Morse
    The New York Botanical
    Garden
    Bronx, New York
    (Hudsonia)

CISTACEAE
  Robert L. Wilbur
    Duke University
    Durham, North Carolina

CLUSIACEAE
  Preston Adams
    DePauw University
    Greencastle, Indiana

COMMELINACEAE
  David Dunn
    University of Missouri at
    Columbia
    Columbia, Missouri

COMMELINACEAE
  D. T. MacRoberts*
    Louisiana State University in
    Shreveport
    Shreveport, Louisiana

CONVOLVULACEAE
  Daniel F. Austin
    Florida Atlantic University
    Boca Raton, Florida

CONVOLVULACEAE
  Jew-Ming Chao
    Board of Chosen Freeholders
    of the County of Burlington
    Mount Holly, New Jersey

CRASSULACEAE
  Robert T. Clausen*
    Cornell University
    Ithaca, New York

CRASSULACEAE
  Reid Moran
    Natural History Museum
    San Diego, California

CROSSOSOMATACEAE
Larry DeBuhr
Cottey College
Nevada, Missouri

CROSSOSOMATACEAE
Barry A. Prigge
Rancho Santa Ana Botanic
Garden
Claremont, California

CUCURBITACEAE
Thomas W. Whitaker
U. S. D. A.
La Jolla, California

CYPERACEAE
Earl L. Core
West Virginia University
Morgantown, West Virginia
(Scleria)

CYPERACEAE
John E. Fairey, III
Clemson University
Clemson, South Carolina
(Scleria)

CYPERACEAE
Frederick J. Hermann
University of Wisconsin at
Madison
Madison, Wisconsin
(Carex)

CYPERACEAE
Robert Kral
Vanderbilt University
Nashville, Tennessee
(Bulbostylis)
(Fimbristylis)
(Fuirena)

CYPERACEAE
Alfred E. Schuyler
The Academy of Natural
Sciences
Philadelphia, Pennsylvania
(Scirpus)

CYPERACEAE
Henry K. Svenson
Osterville, Massachusetts
(Eleocharis)

CYPERACEAE
John W. Thieret
Northern Kentucky University
Highland Heights, Kentucky
(Cyperus)

CYRILLACEAE
Joab Thomas
North Carolina State
University
Raleigh, North Carolina

DIAPENSIACEAE
Richard B. Primack
Boston University
Boston, Massachusetts

DIAPENSIACEAE
Wesley N. Tiffney
University of Massachusetts
Nantucket, Massachusetts

DROSERACEAE
Donald E. Schnell
Statesville, North Carolina

EBENACEAE
Frank White
University of Oxford
Oxford, Great Britain

ERICACEAE
Theodore R. Dudley
U.S. National Arboretum
Washington, D.C.
(Rhododendron)

ERICACEAE
John E. Ebinger
Eastern Illinois University
Charleston, Illinois
(Kalmia)

ERICACEAE
James C. Hickman
University of California at
Berkeley
Berkeley, California
(Menziesia)

ERICACEAE
Martha Prince
Locust Valley, New York
(Rhododendron)

ERICACEAE
H. T. Skinner
U.S. National Arboretum
Washington, D.C.
(Rhododendron)

ERICACEAE
S. P. Vander Kloet
Archbold Biological Station
Lake Placid, Florida
(Vaccinium)

ERICACEAE
Gary D. Wallace
Department of Arboreta and
Botanic Gardens
Arcadia, California
(Montropoideae)

ERICACEAE
Philip V. Wells
The University of Kansas
Lawrence, Kansas
(Arctostaphylos)

ERICACEAE
Robert L. Wilbur*
Duke University
Durham, North Carolina

ERIOCAULACEAE
Robert Kral
Vanderbilt University
Nashville, Tennessee

ERIOCAULACEAE
Harold N. Moldenke
Plainfield, New Jersey

EUPHORBIACEAE
Derek Burch*
University of Florida
Institute of Food and
Agricultural Sciences
Fort Lauderdale, Florida

EUPHORBIACEAE
Grady L. Webster
University of California,
Davis
Davis, California

FABACEAE
Rupert Barneby
  The New York Botanical
  Garden
  Bronx, New York
  (Astragalus)
  (Cassia)
  (Dalea)
  (Parryella)
  (Psorothamnus)
  (Rhynchosia)
  (Sophora)
  (Sphinctospermum)

FABACEAE
Andre F. Clewell
  Florida State University
  Tallahassee, Florida
  (Kummerowia)
  (Lespedeza)

FABACEAE
Wilbur H. Duncan
  The University of Georgia
  Athens, Georgia
  (Galactia)

FABACEAE
David Dunn*
  University of Missouri at
  Columbia
  Columbia, Missouri
  (Lupinus)

FABACEAE
John M. Gillett*
  National Museum of Natural
  Sciences
  Ottawa, Ontario
  (Trifolium)

FABACEAE
Charles R. Gunn
  U. S. D. A., Beltsville
  Agricultural Research Center
  Beltsville, Maryland
  (Vicia)

FABACEAE
Duane Isely
  Iowa State University
  Ames, Iowa

FABACEAE
Richard E. Riefner, Jr.
  Towson State University
  Towson, Maryland
  (Crotalaria)

FABACEAE
Bernice Schubert*
  The Arnold Arboretum
  Harvard University
  Cambridge, Massachusetts
  (Desmodium)

FABACEAE
Steven G. Skinner
  Towson State University
  Towson, Maryland
  (Crotalaria)

FABACEAE
Stanley L. Welsh*
  Brigham Young University
  (Hedysarum)
  (Oxytropis)

FABACEAE
Donald R. Windler
  Towson State University
  Towson, Maryland
  (Crotalaria)

FAGACEAE
Wilbur H. Duncan
  The University of Georgia
  Athens, Georgia

FAGACEAE
Roland L. Guthrie
  West Virginia University
  Morgantown, West Virginia
  (Quercus)

FAGACEAE
John M. Tucker
  University of California at
  Davis
  Davis, California
  (Quercus)

FLACOURTIACEAE
Richard Keating
  Southern Illinois University at
  Edwardsville
  Edwardsville, Illinois

FLANKENIACEAE
Molly Whalen
  University of Texas at Austin
  Austin, Texas

GARRYACEAE
Gerald Dahling
  Macalester College
  St. Paul, Minnesota

GENTIANACEAE
C. Rose Broome
  University of Maryland
  College Park, Maryland
  (Centaurium)

GENTIANACEAE
John M. Gillett*
  National Museum of Natural
  Sciences
  Ottawa, Ontario

GENTIANACEAE
Robert Ornduff
  University of California at
  Berkeley
  Berkeley, California
  (Nymphoides)

GENTIANACEAE
James D. Perry
  University of North Carolina at
  Asheville
  Asheville, North Carolina
  (Sabatia)

GENTIANACEAE
Douglas M. Post
  San Francisco State University
  San Francisco, California
  (Frasera)

GENTIANACEAE
James Pringle*
  Royal Botanical Gardens
  Hamilton, Ontario

GERANIACEAE
Peter F. Yeo
  University Botanic Garden
  Cambridge, Great Britain

GESNERIACEAE
Harold St. John
Bernice P. Bishop Museum
Honolulu, Hawaii

GOODENIACEAE
Harold St. John
Bernice P. Bishop Museum
Honolulu, Hawaii

HALORAGIDACEAE
Susan Aiken
University of Minnesota
St. Paul, Minnesota
(Myriophyllum)

HALORAGIDACEAE
Harold St. John
Bernice P. Bishop Museum
Honolulu, Hawaii

HALORAGIDACEAE
Jean Wooten
University of Southern
Mississippi
Hattiesburg, Mississippi

HIPPOCASTANACEAE
James W. Hardin
North Carolina State
University
Raleigh, North Carolina

HYDROCHARITACEAE
Harold St. John
Bernice P. Bishop Museum
Honolulu, Hawaii

HYDROPHYLLACEAE
Lincoln Constance*
University of California at
Berkeley
Berkeley, California

IRIDACEAE
Peter Goldblatt
Missouri Botanical Garden
St. Louis, Missouri

IRIDACEAE
Douglas Henderson
University of Idaho
Moscow, Idaho
(Sisyrinchium)

IRIDACEAE
Homer Metcalf*
Montana State University
Bozeman, Montana

JUGLANDACEAE
Wayne E. Manning
Bucknell University
Lewisburg, Pennsylvania

JUNCACEAE
John E. Ebinger
Eastern Illinois University
Charleston, Illinois
(Luzula)

JUNCACEAE
Frederick J. Hermann
University of Wisconsin
Madison, Wisconsin
(Juncus)

LAMIACEAE
Peter W. Ball
University of Toronto
Mississauga, Ontario

LAMIACEAE
John R. Bozeman
Department of Natural
Resources
Brunswick, Georgia
(Dicerandra)

LAMIACEAE
Philip Cantino
Gray Herbarium
Harvard University
Cambridge, Massachusetts
(Physostegia)

LAMIACEAE
Joseph L. Collins
Tennessee Valley Authority
Norris, Tennessee
(Scutellaria)

LAMIACEAE
Robert S. Irving
University of Nebraska
Omaha, Nebraska

LAMIACEAE
Harlan Lewis
University of California at
Los Angeles
Los Angeles, California
(Trichostema)

LAMIACEAE
Roger W. Sanders
The Ohio State University
Columbus, Ohio

LAMIACEAE
Rainer W. Scora
University of California at
Riverside
Riverside, California
(Monarda)

LAMIACEAE
Arthur O. Tucker
Delaware State College
Dover, Delaware
(Mentha)

LEMNACEAE
Ernest O. Beal
Western Kentucky University
Bowling Green, Kentucky

LENTIBULARIACEAE
Robert K. Godfrey
The Florida State University
Tallahassee, Florida
(Utricularia)

LENTIBULARIACEAE
Donald E. Schnell
Statesville, North Carolina

LILIACEAE
Wilbur Duncan
University of Georgia
Athens, Georgia
(Yucca)

LILIACEAE
Walter S. Flory*
Wake Forest University
Winston-Salem, North Carolina
(Cooperia)   (Leucojum)
(Crinum)   (Lycoris)
(Furcraea)   (Narcissus)
(Galanthus)   (Pancratium)
(Habranthus) (Zephyranthes)
(Hippeastrum)

LILIACEAE
John D. Freeman
Auburn University
Auburn, Alabama
(*Trillium*)

LILIACEAE
James C. Hickman
University of California at
Berkeley
Berkeley, California
(*Veratrum*)

LILIACEAE
Lee W. Lenz
Rancho Santa Ana Botanic
Garden
Claremont, California
(*Triteleia*)

LILIACEAE
Roger M. Macfarlane
Palo Alto, California
(*Fritillaria*)

LILIACEAE
Louis V. Mingrone*
Bloomsburg State College
Bloomsburg, Pennsylvania
(*Allium*)

LILIACEAE
Sherman Preece
University of Montana
Missoula, Montana
(*Zigadenus*)

LILIACEAE
Harry L. Sherman
Mississippi University for
Women
Columbus, Mississippi
(*Hastingsia*)
(*Schoenolirion*)

LILIACEAE
Frederick Utech
Carnegie Institute
Pittsburgh, Pennsylvania

LILIACEAE
Robert L. Wilbur
Duke University
Durham, North Carolina
(*Uvularia*)

LINACEAE
Claude M. Rogers*
Wayne State University
Detroit, Michigan

LOASACEAE
Henry J. Thompson
University of California at
Los Angeles
Los Angeles, California

LOGANIACEAE
Harold St. John
Bernice P. Bishop Museum
Honolulu, Hawaii

LORANTHACEAE
Delbert Wiens
University of Utah
Salt Lake City, Utah

LYTHRACEAE
Shirley Graham
Kent State University
Kent, Ohio

MAGNOLIACEAE
James W. Hardin
North Carolina State
University
Raleigh, North Carolina

MALPIGHIACEAE
Bruce MacBryde
Fish and Wildlife Service
Washington, D.C.
(*Galphimia*)

MALVACEAE
Paul A. Fryxell*
Texas A & M University
College Station, Texas

MARTYNIACEAE
Peter Bretting
Indiana University
Bloomington, Indiana

MARTYNIACEAE
Richard H. Hevly
Northern Arizona University
Flaggstaff, Arizona

MELASTOMATACEAE
Robert Kral
Vanderbilt University
Nashville, Tennessee
(*Rhexia*)

MELASTOMATACEAE
Thomas Morley
University of Minnesota
St. Paul, Minnesota
(*Mouriri*)

MELASTOMATACEAE
John Wurdack
Smithsonian Institution
Washington, D.C.

MYRTACEAE
Rogers McVaugh
University of Michigan
Ann Arbor, Michigan

MYRTACEAE
Harold St. John
Bernice P. Bishop Museum
Honolulu, Hawaii

NAJADACEAE
Robert R. Haynes
University of Alabama
University, Alabama

NYCTAGINACEAE
Leo A. Galloway
Missouri Western State College
St. Joseph, Missouri

NYCTAGINACEAE
Clyde F. Reed
Reed Library and Herbarium
Baltimore, Maryland

NYCTAGINACEAE
Richard W. Spellenberg
New Mexico State University
Las Cruces, New Mexico

NYMPHAEACEAE
Ernest O. Beal
Western Kentucky University
Bowling Green, Kentucky

NYSSACEAE
Robert H. Eyde
Smithsonian Institution
Washington, D.C.

OLEACEAE
Elbert L. Little
U. S. D. A. Forest Service
Washington, D.C.
(Fraxinus)

ONAGRACEAE
David E. Boufford
Carnegie Institute
Pittsburgh, Pennsylvania
(Circaea)

ONAGRACEAE
Harlan Lewis
University of California at
Los Angeles
Los Angeles, California
(Clarkia)
(Gayophytum)

ONAGRACEAE
Howard Towner
Loyola Marymount University
Los Angeles, California
(Calylophus)

ORCHIDACEAE
Robert Dressler
Smithsonian Tropical Research
Institute
Miami, Florida
(Encyclia)
(Epidendrum)

ORCHIDACEAE
Carlyle A. Luer*
The Marie Selby Botanical
Gardens
Sarasota, Florida

ORCHIDACEAE
Loyal A. Mehrhoff, III
University of North Carolina
Chapel Hill, North Carolina
(Calypso)
(Platanthera)

OROBANCHACEAE
Turner Collins
Evangel College
Springfield, Missouri
(Orobanche)

OXALIDACEAE
Melinda Denton
University of Washington
Seattle, Washington

PANDANACEAE
Benjamin Stone
University of Malaya
Kuala Lumpur, Malaysia

PAPAVERACEAE
Robert W. Kiger
Hunt Institute for Botanical
Documentation
Pittsburgh, Pennsylvania
(Papaver)

PAPAVERACEAE
Gerald B. Ownbey*
University of Minnesota
St. Paul, Minnesota
(Argemone)
(Corydalis)
(Dicentra)

PIPERACEAE
Harold St. John
Bernice P. Bishop Museum
Honolulu, Hawaii

PITTOSPORACEAE
Harold St. John
Bernice P. Bishop Museum
Honolulu, Hawaii

PLANTAGINACEAE
I. John Bassett
Plant Research Institute
Central Experimental Farm
Ottawa, Ontario

PLUMBAGINACEAE
James L. Luteyn
Universidad Catolica
Quito, Ecuador
(Limonium)
(Plumbago)

POACEAE
Bernard R. Baum
Biosystematics Research
Institute
Ottawa, Ontario

POACEAE
Alan A. Beetle
University of Wyoming
Laramie, Wyoming

POACEAE
Jere Brunken
The Ohio State University
Columbus, Ohio

POACEAE
George Church
Brown University
Providence, Rhode Island

POACEAE
Beecher Crampton
University of California at
Davis
Davis, California

POACEAE
James de Wet
University of Illinois
Urbana, Illinois
(Bothriochloa)   (Setaria)
(Cynodon)   (Sorghum)
(Dichanthium)   (Triticum)
(Echinochloa)   (Zea)
(Eleusine)   (Zizania)
(Pennisetum)

POACEAE
Douglas R. Dewey
Utah State University
Logan, Utah
(Triticeae)

POACEAE
Leroy H. Harvey
Smithsonian Institution
Washington, D.C.
(Eragrostis)

POACEAE
M. J. Harvey
Dalhousie University
Halifax, Nova Scotia

POACEAE
Arthur H. Holmgren
Utah State University
Logan, Utah

POACEAE
Stephen D. Koch
Rama de Botanica
Colegio de Postgraduados
Chapingo, Mexico

POACEAE
Vladimir J. Krajina
The University of British
Columbia
Vancouver, British Columbia
(Festuca)
(Vulpia)

POACEAE
Charles C. Laing
Ohio Northern University
Ada, Ohio
(Ammophila)

POACEAE
Grant L. Pyrah
Southwest Missouri State
University
Springfield, Missouri
(Agroelymus)
(Agrohordeum)
(Agropyron)
(Agrositanion)

POACEAE
John R. Reeder*
Charlotte Reeder*
University of Arizona
Tucson, Arizona

POACEAE
Rhonda Riggins
California Polytechnic State
University
San Luis Obispo, California
(Sporobolus)

POACEAE
Blaine D. Rogers
Columbia College
Columbia, California
(Puccinellia)

POACEAE
H. Scholz*
Meseum Berlin-Dahlem
Berlin, Germany
(Bromus)

POACEAE
Richard W. Spellenberg
New Mexico State University
Las Cruces, New Mexico

POACEAE
Edward E. Terrell
Beltsville Agricultural Research
Center
Beltsville, Maryland
(Aegilops)   (Luziola)
(Arrhenatherum) (Rhynchelytrum)
(Dactylis)   (Sorghum)
(Hydrochloa)   (Triticum)
(Lolium)

POACEAE
F. Douglas Wilson
Western Cotton Research
Laboratory
Phoenix, Arizona
(Sitanion)

POACEAE
John T. Witherspoon
Springfield, Missouri
(Eragrostis)

PODOSTEMACEAE
Jean Wooten
University of Southern
Mississippi
Hattiesburg, Mississippi

POLEMONIACEAE
Alva Day*
California Academy of Sciences
San Francisco, California
(Allophyllum)   (Langloisia)
(Collomia)   (Linanthus)
(Gilia)   (Loeselia)
(Gymnosteris)   (Microsteris)
(Ipomopsis)   (Navarretia)

POLEMONIACEAE
Edgar T. Wherry
Unitarian Universalist House
Philadelphia, Pennsylvania
(Phlox)

POLEMONIACEAE
Dieter H. Wilken*
Colorado State University
Fort Collins, Colorado

POLYGALACEAE
Robert R. Smith
Hartwick College
Oneonta, New York

POLYGONACEAE
James H. Horton
Western Carolina University
Cullowhee, North Carolina
(Polygonella)

POLYGONACEAE
Richard S. Mitchell
New York State Museum and
Science Service
Albany, New York
(Polygonum)
(Rumex)

POLYGONACEAE
James L. Reveal*
University of Maryland
College Park, Maryland
(Centrostegia)   (Lastarriaea)
(Chorizanthe)   (Mucronea)
(Dedeckera)   (Nemacaulis)
(Eriogonum)   (Oxytheca)
(Gilmania)   (Pterostegia)
(Goodmania)   (Stenogonum)
(Hollisteria)

PONTEDERIACEAE
Richard M. Lowden
Universidad Catolica Madre
Y Maestra
Republica Dominicana
(Pontederia)

PONTEDERIACEAE
Jean Wooten
University of Southern
Mississippi
Hattiesburg, Mississippi

PORTULACACEAE
Janet Hohn
Fish and Wildlife Service
Portland, Oregon
(Lewisia)

PORTULACACEAE
Walter Lewis
Washington University
St. Louis, Missouri
(Claytonia)
(Montia)

PORTULACACEAE
John McNeill*
Canadian Department of
Agriculture
Ottawa, Ontario

PORTULACACEAE
James F. Matthews
The University of North
Carolina at Charlotte
Charlotte, North Carolina
(Portulaca)

PORTULACACEAE
Stewart Ware
College of William and Mary
Williamsburg, Virginia
(Talinum)

POTAMOGETONACEAE
Robert R. Haynes
University of Alabama
University, Alabama
(Potamogeton)

PRIMULACEAE
Thomas Cooperrider
Kent State University
Kent, Ohio
(Lysimachia)

RANUNCULACEAE
Lyman Benson*
Pomona College
Claremont, California
(Ranunculus)

RANUNCULACEAE
Carl S. Keener
The Pennsylvania State
University
University Park, Pennsylvania

RHAMNACEAE
Marshall C. Johnston
The University of Texas at
Austin
Austin, Texas

ROSACEAE
Mr. and Mrs. Hannibal A. Davis*
Freeport, Florida
(Rubus)

ROSACEAE
Walter Lewis
Washington University
St. Louis, Missouri
(Rosa)

ROSACEAE
Ghillean T. Prance
The New York Botanical
Garden
Bronx, New York
(Chrysobalanus)
(Hirtella)
(Licania)

ROSACEAE
Kenneth R. Robertson
Illinois Institute of Natural
Resources
Urbana, Illinois

RUBIACEAE
Lauramae Dempster*
University of California at
Berkeley
Berkeley, California
(Galium)

RUBIACEAE
Cheryl A. Lawson*
Duncan, Oklahoma
(Galium)

RUBIACEAE
Walter Lewis
Washington University
St. Louis, Missouri
(Diodia)
(Hedyotis)
(Richardia)

RUTACEAE
Virginia L. Bailey
Detroit Institute of
Technology
Detroit, Michigan
(Ptelea)

RUTACEAE
Duncan Porter
Virginia Polytechnic Institute
and State University
Blacksburg, Virginia

RUTACEAE
Harold St. John
Bernice P. Bishop Museum
Honolulu, Hawaii

RUTACEAE
Rainer W. Scora
University of California at
Riverside
Riverside, California
(Citrus)
(Fortunella)

RUTACEAE
Benjamin Stone
University of Malaya
Kuala Lumpur, Malaysia
(Pelea)
(Platydesma)
(Zanthoxylum)

SALICACEAE
George W. Argus*
National Museum of Natural
Sciences
Ottawa, Ontario

SALICACEAE
James E. Eckenwalder*
University of Toronto
Toronto, Ontario
(Populus)

SAPINDACEAE
Harold St. John
Bernice P. Bishop Museum
Honolulu, Hawaii

SAPOTACEAE
Harold St. John
Bernice P. Bishop Museum
Honolulu, Hawaii

SARRACENIACEAE
C. Ritchie Bell
University of North Carolina
Chapel Hill, North Carolina

SARRACENIACEAE
Frederick W. Case, II
Saginaw, Michigan

SARRACENIACEAE
Donald E. Schnell
Statesville, North Carolina

SAXIFRAGACEAE
John G. Packer
The University of Alberta
Edmonton, Alberta
(Chrysosplenium)

SCROPHULARIACEAE
Gerald D. Carr
University of Hawaii at Manoa
Honolulu, Hawaii
(Pedicularis)

SCROPHULARIACEAE
Thomas Cooperrider
Kent State University
Kent, Ohio
(Chelone)
(Lindernia)

SCROPHULARIACEAE
Wayne J. Elisens
The University of Texas at
Austin
Austin, Texas
(Maurandya)

SCROPHULARIACEAE
Wayland Ezell*
St. Cloud State University
St. Cloud, Minnesota
(Mimulus)

SCROPHULARIACEAE
Lawrence R. Heckard
University of California at
Berkeley
Berkeley, California
(Castilleja)
(Cordylanthus)
(Orthocarpus)

SCROPHULARIACEAE
Noel Holmgren
The New York Botanical
Garden
Bronx, New York

SCROPHULARIACEAE
Lytton Musselman
Old Dominion University
Norfolk, Virginia

SCROPHULARIACEAE
Richard Shaw
Utah State University
Logan, Utah
(Scrophularia)

SCROPHULARIACEAE
Richard M. Straw
Klamath Falls, Oregon
(Chionophila)
(Keckiella)
(Nothochelone)
(Penstemon)

SCROPHULARIACEAE
John W. Thieret
Northern Kentucky University
Highland Heights, Kentucky

SCROPHULARIACEAE
Robert K. Vickery, Jr.
The University of Utah
Salt Lake City, Utah
(Mimulus)

SCROPHULARIACEAE
Peter F. Yeo
University Botanic Garden
Cambridge, Great Britain
(Euphrasia)

SMILACACEAE
Wilbur Duncan
The University of Georgia
Athens, Georgia
(Smilax)

SOLANACEAE
John E. Averett
University of Missouri
St. Louis, Missouri

SOLANACEAE
William G. D'Arcy*
Missouri Botanical Garden
St. Louis, Missouri

SOLANACEAE
Charles B. Heiser, Jr.
Indiana University
Bloomington, Indiana
(Capsicum)
(Solanum)

SPARGANIACEAE
Ernest O. Beal
Western Kentucky University
Bowling Green, Kentucky

STYRACACEAE
James L. Reveal
University of Maryland
College Park, Maryland
(Halesia)

TAMARICACEAE
Bernard R. Baum
Biosystematics Research
Institute
Ottawa, Ontario
(Tamarix)

THYMELAEACEAE
Lorin I. Nevling
Chicago Field Museum of
Natural History
Chicago, Illinois
(Daphne)
(Daphnopsis)
(Dirca)
(Edgeworthia)
(Passerina)

THYMELAEACEAE
Harold St. John
Bernice P. Bishop Museum
Honolulu, Hawaii
(Wikstroemia)

TYPHACEAE
Ernest O. Beal
Western Kentucky University
Bowling Green, Kentucky

URTICACEAE
I. John Bassett
Plant Research Institute
Central Experimental Farm
Ottawa, Canada

VALERIANACEAE
Frederick G. Meyer
U.S. National Arboretum
Washington, D.C.
(*Valeriana*)

VALERIANACEAE
Donna M. E. Ware
College of William and Mary
Williamsburg, Virginia
(*Valerianella*)

VERBENACEAE
Harold N. Moldenke*
Plainfield, New Jersey

VERBENACEAE
Ray Umber
University of Wyoming
Laramie, Wyoming
(*Glanularia*)

VIOLACEAE
Norman H. Russell
Central State University
Edmond, Oklahoma

VITACEAE
Wilbur H. Duncan
The University of Georgia
Athens, Georgia

XYRIDACEAE
Robert Kral
Vanderbilt University
Nashville, Tennessee

ZANNICHELLIACEAE
Ernest O. Beal
Western Kentucky University
Bowling Green, Kentucky

ZINGIBERACEAE
Rosemary M. Smith
Royal Botanic Garden
Edinburgh, Scotland

ZYGOPHYLLACEAE
Duncan Porter
Virginia Polytechnic Institute
and State University
Blacksburg, Virginia

Excluded Names

Excluded Anomalous Names[a]

PTERIDOPHYTA

CYATHEACEAE
Cyathea wilsonii (Hook.) Proctor
(*Cnemidaria horrida* X *Cyathea arborea*)
SY = *Hemitelia w.* Hook.
(an unnamed intergeneric hybrid not appropriately referable to either genus, per se)

ANGIOSPERMAE

APIACEAE
Ligusticum mutellinoides (Crantz) Willar ssp. *alpinum* (Ledeb.) Thellung (based on a misidentification of *Podistera macounii*)
Ptilimnium macrospermum Ahles (nomen nudum)

ASTERACEAE
Chrysothamnus bolanderi (Gray) Greene
SY = *C. nauseosus* ssp. *b.* (Gray) Hall & Clements
(an intergeneric hybrid between *Happlopappus macronema* and *Chrysothamnus nauseosus*; thus, not appropriately referable to either genus per se)
Chrysothamnus nauseosus (Pallas) Britt. ssp. *viscosus* Keck
(an intergeneric hybrid between *Haplopappus* sp. and *Chrysothamnus nauseosus* not referable to either genus per se)

Lipochaeta porophylla Deg. & Deg. (nomen nudum)
Prenanthes alba L. ssp. *pallida* Milstead (not validly published)

BRASSICACEAE
Draba incisa (nomen nudum)

CYPERACEAE
Cyperus X *weatherbianus* Fern. (an intergeneric hybrid between *Dulichium arundinaceum* X *Rhynchospora capitellata*; thus, not appropriately referable to the genus *Cyperus*)
Scirpus caespitosus ssp. *austriacus* (Pallas) Aschers. & Graebn. (nomen nudum)
Scirpus microcarpus var *bissellii* (Fern.) House
SY = *S. sylvaticus* var. *b.* Fern. (appears to be a hybrid between *S. expansus* and *S. microcarpus* but is unnamed and thus excluded from the checklist)
Scirpus peckii Britt.
SY = *S. polyphyllus* var. *macrostachys* Boeckl. (a name applied to a series of hybrids between *S. polyphyllus* X *S. atrovirens* and *S. atrocinctus* X *S. hattorianus*)

EUPHORBIACEAE
Phyllanthus epiphyllanthus L.
SY = *Xylophylla e.* (L.) Britt. (based on a misidentification of *P. angustifolius* (Sw.) Sw., a Jamaican taxon)

FABACEAE
Astragalus stocksii Welsh (later homonym)
Tephrosia intermedia Graham
SY = *Cracca intermedia* Small (name apparently not validly published)

LAMIACEAE
Agastache neomexicana ssp. *havardii* (Gray) Lint (nomen nudum)

LILIACEAE
Trillium erectum L. var. *declinatum* Millsp. (nomen nudum)

LOASACEAE
Mentzelia cronquistii
M. marginata
M. multiflora ssp. *leucopetala*
M. paradoxica
M. sanjuanensis
(The above five names proposed by Thompson and Zavortink are included here

---

a. This list contains a total of 29 names. Numerous other names might well be added to this list. However, rather than perpetuate such nomenclatural anomalies and basic errors, we have chosen to ignore most of these names and address only the above few which are more widely circulated or perhaps botanically significant.

rather than in the body of the Checklist since they have already appeared in print (in 1976, as "ined.") but, as of this date, apparently have not yet been validly published.)

LORANTHACEAE
Phoradendron bolleanum ssp. *hawksworthii* Wiens (not validly published; when published, the taxon will be treated at the species level)

ORCHIDACEAE
Orchis purpurella Steph. & Steph. (based on a misidentification of *Dactylorhiza maculata*)

POACEAE
X *Agroelymus dorei* Bowden (a putative intergeneric hybrid between *Agropyron trachycaulum* var. *glauca* (Pease & Moore) Malte and *Hystrix patula (Elymus hystrix)*; thus, the intergeneric name X *Agroelymus* is not applicable)
Festuca vivipara (L.) Sm. var. *hirsuta* (Lange) Schol. and var. *vivipara* (SY = *F. supina* Schur) are incorrectly attributed to Greenland. In reality, the taxon there is actually *F.* X *viviparoidea*. *F. vivipara* is a European taxon.

POLEMONIACEAE
Gilia nubigena W. A. Anderson ?(nomen nudum)

RUBIACEAE
Houstonia betterwickiae Terrell (Perhaps a recognizable taxon, but the genus *Houstonia* is here treated as *Hedyotis*; therefore, a new combination is necessary)

VERBENACEAE
Lantana scorta Moldenke (invalid name)

Taxa Excluded from the Checklist Pending Evidence Confirming Their Native or Naturalized Occurrence in Our Flora[a]

PTERIDOPHYTA

ASPLENIACEAE
Asplenium fragile Presl
Cochlidium minus (Jenman) Maxon

CYATHEACEAE
Cnemidaria sessilifolia (Jenman) Domin
SY = *Cyathea s.* Jenman
Nephelea pubescens (Mett.) R. Tryon

GLEICHENIACEAE
Gleichenia furcata Spreng.

GRAMMITIDACEAE
Grammitis limbata Fee

LYCOPODIACEAE
Lycopodium volubile Forst.

GYMNOSPERMAE

GINKOACEAE
Ginko biloba L.

PINACEAE
Pinus thunbergiana Franco
SY = *P. thunbergii* Parl.

ANGIOSPERMAE

ACERACEAE
Acer diabolicum Blume ex K. Koch

ARECACEAE
Calyptronoma occidentalis (Sw.) H. Moore
SY = *Calyptrogyne o.* (Sw.) Maza
Phoenix dactylifera L.

ASCLEPIADACEAE
Asclepias mexicana Cav.
SY = *A. galioides* H. B. K.

ASTERACEAE
Ageratum domingense Spreng.
Helenium apterum (Blake) Bierner
H. laciniatum Gray

BEGONIACEAE
Begonia tuberhybrida Voss

BERBERIDACEAE
X *Mahoberberis neubertii* (Hort. ex Lem.) C. K. Schneid

BETULACEAE
Alnus incana (L.) Moench ssp. *incana*

BIGNONIACEAE
Catalpa bungei C. A. Mey.

BRASSICACEAE
Cardamine palustris Peterm.
SY = *C. pratensis* L. var. *palustris* Wimmer & Grab.
Draba lanceolata Royle
Lesquerella argentea (Schauer) S. Wats.
L. berlanderi S. Wats.
L. palmeri S. Wats.

BURSERACEAE
Bursera fagaroides H. B. K.) Engl. var. *f.*
B. odorata Brandeg.

CAPPARIDACEAE
Capparis spinosa L.

CAPRIFOLIACEAE
Viburnum X *burkwoodii* Burkwood & Skipw.
V. X *rytidophyllum* Lemoine

CARYOPHYLLACEAE
Sagina maritima Don

---

a. This list contains a total of 150 taxa.

CLUSIACEAE
  Clusia flava Jacq.

COMBRETACEAE
  Terminalia myriocarpa

CONVOLVULACEAE
  Cuscuta japonica Choisy
  Operculina ventricosa (Bertero) Peter

CORNACEAE
  Cornus alba L.
  C. mass L.

CYPERACEAE
  Carex stellata Mackenzie

ERICACEAE
  Arbutus xalapensis H. B. K.
  Pyrola rotundifolia L. var. *r.*
  Rhododendron japonicum (Gray) Sur.

FABACEAE
  Acacia catechu (L. f.) Willd.
  Astragalus anemophilus Greene
  A. carminis Barneby
  Dalea cliffortiana Willd.
      SY = *Parosela dalea* (L.) Britt.
  Leucaena greggii S. Wats.
  Psorothamnus emoryi (Gray) Rydb.
      var. *arenarius* (Brandeg.)
      Barneby

FAGACEAE
  Quercus suber L.
  Q. variabilis Blume

IRIDACEAE
  Iris hoogiana Dykes
  I. xiphium L.
  Watsonia fulgens (Andr.) Pers.
      SY = *W. angusta* Ker-Gawl.

LAMIACEAE
  Mentha arvensis L. ssp. *a.*
  M. longifolia (L.) Huds.
  Stachys agraria Cham. & Schlecht.

LILIACEAE
  Lilium candidum L.
  L. concolor Salisb.
  L. longiflorum Thunb.

  L. pensylvanicum Ker-Gawl.
      SY = *L. dauricum* Ker-Gawl.
  L. regale E. H. Wilson
  Scilla bifolia L.

MALPIGHIACEAE
  Galphimia elegans Baill.

MYRSINACEAE
  Ardisia crispa (Thunb.) A. DC.

ONAGRACEAE
  Camissonia dentata (Cav.) Reiche
      SY = *Oenothera contorta* Dougl.
      ex Hook. var. *epilobioides*
      (Greene) Munz
      SY = *O. d.* Cav.
  Epilobium alsinifolium Vill.
  E. billardierianum Ser. ssp. *cinereum*
      (A. Rich.) Raven & Engelhorn
      SY = *E. c.* A. Rich.
  Gaura hexandra Ortega ssp. *h.*
      SY = *G. tripetala* Cav.
  Oenothera elata H. B. K.
      SY = *O. simsiana* Ser.

PANDANACEAE
  Pandanus veitchii Hort.

PASSIFLORACEAE
  Passiflora ligularis Juss.

POACEAE
  Agropyron cristatum (L.) Gaertn.

  A. junceum (L.) Beauv.
  A. pungens (Pers.) Roemer &
      Schultes
  A. semicostatum (Steud.) Nees ex
      Boiss.
  Avena nuda L.
  Brachypodium caespitosum (Host)
      Roemer & Schultes
  Chloris berroi Arech
  C. truncata R. Br.
  Cortaderia rudiuscula Stapf
  Cynodon arcuatus J. Presl
  C. barberi Rang & Tag.
  C. magennisii Hurcombe
  Digitaria diagonalis (Nees) Stapf
  D. pentzii Stent
  D. swazilandensis Stent
  Dinebra retroflexa (Vahl) Panzer

Elymus dasystachys Trin. & Ledeb.
E. giganteus Vahl
E. interior Hulten
E. sabulosus Bieb.
E. salmonis C.L. Hitchc.
Eragrostis alba Presl
E. chloromelas Steud.
E. falcata (Gaud.) Gaud.
E. obtusa Munro ex Stapf
Eremochloa ciliaris (L.) Merr.
Erioneuron avenaceum (H. B. K.)
    Tateoka
Eulalia fulva (R. Br.) Kunth
Festuca geniculata (L.) Cav.
Flaveria oppositifolia (DC.) Rydb.
Hilaria cenchroides H. B. K.
Hordeum bulbosum L.
H. hexastichon L.
H. nodosum L.
H. spontaneum K. Koch
Lolium remotum Schrank
Melica ciliata L.
Oplismenus compositus (L.) Beauv.
Oryzopsis coerulescens (Desf.) Hack.
Paspalum nicorae Parodi
Pennisetum macrostachyum (Brongn.)
    Trin.
Phalaris commutatum Roemer &
    Schultes
Saccharum bengalense Retz.
Setaria poiretiana (Schultes)
    Kunth
Sporobolus elongatus R. Br.
Stipa elegantissima Labill.
S. pennata L.
S. splendens Trin.
S. tenacissima L.
Tripsacum laxum Nash
Trisetum sibiricum Rupr. ssp.
    sibiricum
Zea mays L. Var. *everta* (Sturtev)
    Bailey
Z. m. var. *japonica* (Van Houtte)
    Wood
Z. m. var. *tunicata* Larr. ex St. Hil.
Zizania latifolia (Griseb.) Turcz.

POLYGONACEAE
  Eriogonum hastatum Wiggins

PROTEACEAE
  Grevillea banksii R. Br.

ROSACEAE
  Amelanchier sanguinea var.
    grandiflora (Wieg.) Rehd.
  Malus X *magdeburgensis* Hartw.
  Prunus serrulata Lindl.
  P. triloba Lindl.

SCROPHULARIACEAE
  Marrubium candidissimum sensu
    auctt. non L.
  M. incanum Desr.
  Origanum majorana L.
    SY = *Majorana hortensis* (Moench)
  Symphytum X *uplandicum* Nym.
    SY = *S. peregrinum* Ledeb.

SOLANACEAE
  Bouchetia anomala (Miers) Britt. &
    Rusby
  Capsicum baccatum L.
  Cestrum daphnoides Griseb.
  C. laurifolium L'Her.
  C. salicifolium Dunal
  Solanum pterocaulon Dunal
  S. robustum Wendl.

ULMACEAE
  Trema orientalis (Blume)
  T. zeylanica (L.) J. Benn.
  Ulmus davidiana var. *japonica* (Rehd.)
    Nakai
    SY = *U. j.* (Rehd.) Sarg.

VERBENACEAE
  Clerodendrum japonicum var.
    pleniflorum (Schauer)
    Maheshwari
  Stachytarpheta indica Vahl

ZYGOPHYLLACEAE
  Guaiacum unijugum Brandeg.
  Larrea divaricata Cav.

Summary of Vascular Plant Taxa Treated

Family	Genera	Species	Subspecies	Varieties	Hybrids	Synonyms
PTERIDOPHYTA						
1. Adiantaceae	21	136	1	33	2	73
2. Aspleniaceae	32	265	4	48	34	258
3. Azollaceae	1	3	0	0	0	0
4. Blechnaceae	5	22	0	2	0	17
5. Cyatheaceae	6	16	0	0	0	13
6. Davalliaceae	3	11	0	0	1	2
7. Dennstaedtiaceae	10	26	1	8	0	17
8. Equisetaceae	1	11	0	5	4	30
9. Gleicheniaceae	3	7	0	4	0	5
10. Grammitidaceae	4	27	0	3	0	22
11. Hymenophyllaceae	2	46	3	1	0	8
12. Isoetaceae	1	23	3	13	0	22
13. Lophosoriaceae	1	1	0	0	0	1
14. Lycopodiaceae	1	40	0	17	5	136
15. Marattiaceae	3	6	0	0	0	0
16. Marsileaceae	2	10	0	0	0	2
17. Ophioglossaceae	2	28	1	26	0	34
18. Osmundaceae	1	3	0	5	1	1
19. Parkeriaceae	1	3	0	0	0	2
20. Polypodiaceae	2	36	0	7	0	24
21. Psilotaceae	1	2	0	0	0	0
22. Salviniaceae	1	2	0	0	0	1
23. Schizaeaceae	3	15	0	0	0	3
24. Selaginellaceae	1	50	3	5	1	19
Subtotals	108	789	16	175	48	690

Family	Genera	Species	Subspecies	Varieties	Hybrids	Synonyms
GYMNOSPERMAE						
25. Cupressaceae	6	30	0	17	1	40
26. Cycadaceae	1	6	0	0	0	3
27. Ephedraceae	1	10	0	8	2	5
28. Pinaceae	6	68	0	20	7	46
29. Taxaceae	2	5	0	0	0	1
30. Taxodiaceae	4	6	0	0	0	3
Subtotals	20	125	0	45	10	98
ANGIOSPERMAE						
31. Acanthaceae	24	98	4	26	0	139
32. Aceraceae	1	19	0	18	2	62
33. Adoxaceae	1	1	0	0	0	0
34. Agavaceae	11	96	0	43	0	52
35. Aizoaceae	16	29	0	0	0	8
36. Alismaceae	4	35	0	18	0	53
37. Amaranthaceae	18	108	4	35	0	106
38. Anacardiaceae	11	34	8	12	1	44
39. Annonaceae	6	20	0	0	1	13
40. Apiaceae	83	391	28	63	1	190
41. Apocynaceae	27	78	2	26	1	49
42. Aponogetonaceae	1	1	0	0	0	0
43. Aquifoliaceae	2	31	0	4	1	18
44. Araceae	21	41	7	5	0	19
45. Araliaceae	11	49	2	112	0	4
46. Arecaceae	16	59	0	12	0	20
47. Aristolochiaceae	3	35	0	7	0	28
48. Asclepiadaceae	15	145	16	2	1	128
49. Asteraceae	346	2687	342	1205	101	3105
50. Balanophoraceae	1	1	0	0	0	0
51. Balsaminaceae	1	10	0	0	0	6
52. Basellaceae	2	5	0	0	0	6
53. Bataceae	1	1	0	0	0	0
54. Begoniaceae	2	3	0	2	0	1
55. Berberidaceae	10	32	0	4	2	24
56. Betulaceae	5	33	8	16	17	78
57. Bignoniaceae	18	28	2	5	0	22
58. Bixaceae	3	5	0	0	0	0
59. Bombacaceae	4	4	0	0	0	0
60. Boraginaceae	34	384	6	137	0	271
61. Brassicaceae	94	634	95	377	4	492
62. Bromeliaceae	10	42	0	5	0	16

Family	Genera	Species	Subspecies	Varieties	Hybrids	Synonyms
63. Brunelliaceae	1	1	0	0	0	0
64. Burmanniaceae	4	7	0	2	0	2
65. Burseraceae	3	5	0	1	0	3
66. Butomaceae	2	2	0	0	0	0
67. Buxaceae	3	8	0	0	0	4
68. Cactaceae	19	174	0	200	0	242
69. Callitrichaceae	1	13	2	2	0	11
70. Calycanthaceae	1	2	0	2	0	3
71. Calyceraceae	1	1	0	0	0	0
72. Campanulaceae	23	290	8	184	1	60
73. Canellaceae	2	2	0	0	0	1
74. Cannaceae	1	8	0	0	0	0
75. Capparidaceae	9	43	10	4	0	41
76. Caprifoliaceae	8	80	5	64	2	81
77. Caricaceae	1	1	0	0	0	0
78. Caryophyllaceae	35	326	76	193	0	338
79. Casuarinaceae	1	4	0	0	0	1
80. Celastraceae	13	32	0	6	0	9
81. Ceratophyllaceae	1	3	0	0	0	3
82. Chenopodiaceae	25	187	12	57	2	163
83. Chloranthaceae	1	1	0	0	0	0
84. Cistaceae	4	35	0	16	0	41
85. Clethraceae	1	2	0	0	0	2
86. Clusiaceae	6	68	2	7	1	49
87. Combretaceae	5	6	0	2	0	0
88. Commelinaceae	13	49	0	16	0	41
89. Connaraceae	1	1	0	0	0	0
90. Convolvulaceae	18	198	35	55	1	174
91. Cornaceae	1	14	7	2	4	33
92. Corynocarpaceae	1	1	0	0	0	0
93. Crassulaceae	12	108	44	0	0	107
94. Crossosomataceae	3	12	0	4	0	7
95. Cucurbitaceae	26	76	1	14	0	24
96. Cunoniaceae	1	1	0	0	0	0
97. Cyperaceae	26	959	35	276	75	623
98. Cyrillaceae	2	2	0	2	0	3
99. Datiscaceae	1	1	0	0	0	0
100. Diapensiaceae	4	4	2	4	0	6
101. Dilleniaceae	1	2	0	0	0	1
102. Dioscoreaceae	2	13	0	2	0	4
103. Dipsacaceae	6	12	0	0	0	5
104. Droseraceae	2	8	0	5	2	4
105. Ebenaceae	1	7	0	0	0	14
106. Elaeagnaceae	3	10	0	1	0	6

Family	Genera	Species	Subspecies	Varieties	Hybrids	Synonyms
107. Elaeocarpaceae	3	4	0	0	0	0
108. Elatinaceae	2	10	0	0	0	3
109. Empetraceae	3	4	4	0	0	9
110. Epacridaceae	1	2	0	3	0	0
111. Ericaceae	44	219	63	53	9	318
112. Eriocaulaceae	3	16	0	7	0	5
113. Erythroxylaceae	1	4	0	0	0	1
114. Euphorbiaceae	47	358	18	121	1	354
115. Fabaceae	142	1521	175	667	26	1667
116. Fagaceae	5	88	3	39	95	104
117. Flacourtiaceae	9	23	0	2	0	6
118. Flagellariaceae	1	1	0	0	0	0
119. Fouquieriaceae	1	1	0	0	0	0
120. Frankeniaceae	1	5	0	2	0	0
121. Garryaceae	1	8	6	0	0	5
122. Gentianaceae	17	117	17	16	4	200
123. Geraniaceae	3	62	9	12	1	32
124. Gesneriaceae	5	174	1	56	0	4
125. Goodeniaceae	1	12	0	19	1	0
126. Haemodoraceae	4	4	0	0	0	4
127. Haloragidaceae	5	24	0	3	0	17
128. Hamamelidaceae	3	5	0	0	0	6
129. Hernandiaceae	1	1	0	0	0	0
130. Hippocastanaceae	1	7	0	2	6	21
131. Hippuridaceae	1	3	0	0	0	0
132. Hydrocharitaceae	10	24	0	0	0	17
133. Hydrophyllaceae	17	238	17	63	0	112
134. Icacinaceae	2	2	0	0	0	0
135. Illiciaceae	1	2	0	0	0	0
136. Iridaceae	18	109	10	32	9	158
137. Juglandaceae	2	21	0	7	9	54
138. Juncaceae	2	130	17	88	4	116
139. Krameriaceae	1	6	0	2	0	4
140. Lamiaceae	70	482	73	195	11	309
141. Lardizabalaceae	1	1	0	0	0	0
142. Lauraceae	13	35	0	7	0	22
143. Leitneriaceae	1	1	0	0	0	0
144. Lemnaceae	4	17	0	0	0	14
145. Lennoaceae	2	2	0	0	0	0
146. Lentibulariaceae	2	31	0	2	0	22

Family	Genera	Species	Subspecies	Varieties	Hybrids	Synonyms
147. Liliaceae	76	487	55	161	4	336
148. Limnanthaceae	2	10	5	8	0	7
149. Linaceae	4	48	0	19	0	65
150. Loasaceae	4	71	2	19	0	24
151. Loganiaceae	8	47	2	71	0	4
152. Loranthaceae	8	49	10	6	0	36
153. Lythraceae	11	36	0	8	0	23
154. Magnoliaceae	2	11	0	2	0	8
155. Malpighiaceae	11	38	0	0	0	12
156. Malvaceae	41	266	64	47	0	203
157. Marantaceae	3	5	0	0	0	1
158. Marcgraviaceae	1	2	0	0	0	0
159. Martyniaceae	4	9	0	0	0	8
160. Mayacaceae	1	2	0	0	0	0
161. Melastomataceae	19	65	0	3	0	26
162. Meliaceae	5	9	0	0	0	3
163. Menispermaceae	5	11	0	0	0	1
164. Menyanthaceae	3	6	0	2	0	5
165. Moraceae	17	31	1	8	0	19
166. Moringaceae	1	1	0	0	0	1
167. Musaceae	2	2	0	0	1	4
168. Myoporaceae	2	3	2	5	0	0
169. Myricaceae	2	11	0	3	1	12
170. Myrsinaceae	8	35	0	8	0	13
171. Myrtaceae	17	80	6	22	0	51
172. Najadaceae	1	8	0	4	0	11
173. Nyctaginaceae	16	119	6	22	0	103
174. Nymphaeaceae	5	18	8	5	0	52
175. Nyssaceae	1	3	0	2	0	7
176. Ochnaceae	2	4	0	0	0	0
177. Olacaceae	2	5	0	0	0	1
178. Oleaceae	12	65	0	25	1	46
179. Onagraceae	13	252	151	7	3	519
180. Orchidaceae	88	285	0	52	11	294
181. Orobanchaceae	4	23	16	10	0	43
182. Oxalidaceae	1	30	8	8	0	44
183. Paeoniaceae	1	4	0	0	0	1
184. Pandanaceae	2	2	0	6	0	5
185. Papaveraceae	23	93	35	15	0	100
186. Passifloraceae	1	26	0	7	0	2
187. Pedaliaceae	2	2	0	0	0	1
188. Phytolaccaceae	8	14	0	2	0	3
189. Piperaceae	3	88	0	39	0	13

Family	Genera	Species	Subspecies	Varieties	Hybrids	Synonyms
190. Pittosporaceae	2	25	0	28	0	21
191. Plantaginaceae	2	43	7	36	0	40
192. Platanaceae	1	4	0	0	0	2
193. Plumbaginaceae	3	11	6	0	0	33
194. Poaceae	231	1490	61	411	43	1446
195. Podocarpaceae	1	1	0	0	0	1
196. Podostemaceae	1	1	0	0	0	1
197. Polemoniaceae	14	283	217	17	2	263
198. Polygalaceae	3	64	0	25	0	34
199. Polygonaceae	24	446	23	295	7	531
200. Pontederiaceae	5	11	0	4	0	8
201. Portulacaceae	8	108	11	39	0	83
202. Potamogetonaceae	3	43	0	5	11	59
203. Primulaceae	10	90	33	31	3	89
204. Proteaceae	1	1	0	0	0	0
205. Punicaceae	1	1	0	0	0	0
206. Rafflesiaceae	1	1	0	0	0	0
207. Ranunculaceae	24	323	66	212	4	340
208. Resedaceae	2	5	0	0	0	0
209. Rhamnaceae	14	123	17	75	10	53
210. Rhizophoraceae	3	3	0	0	0	1
211. Rosaceae	62	870	112	253	45	1183
212. Roxburghiaceae	1	1	0	0	0	0
213. Rubiaceae	60	317	70	160	1	278
214. Rutaceae	20	130	8	90	1	32
215. Sabiaceae	1	2	0	0	0	0
216. Salicaceae	2	117	20	31	44	669
217. Santalaceae	8	16	3	10	0	7
218. Sapindaceae	17	38	0	28	0	12
219. Sapotaceae	9	39	0	11	0	42
220. Sarraceniaceae	2	9	6	0	17	13
221. Saururaceae	2	2	0	2	0	0
222. Saxifragaceae	36	285	43	187	4	271
223. Scheuchzeriaceae	3	8	1	0	0	5
224. Schisandraceae	1	1	0	0	0	1
225. Scrophulariaceae	69	838	168	248	10	771
226. Simaroubaceae	9	13	0	0	0	6
227. Smilacaceae	1	25	0	6	0	27
228. Solanaceae	33	199	7	99	1	114
229. Sparganiaceae	1	9	0	0	0	10
230. Staphyleaceae	2	3	0	0	0	0
231. Sterculiaceae	11	26	5	3	0	26
232. Styracaceae	2	10	0	11	0	8
233. Symplocaceae	1	4	0	0	0	3

Family	Genera	Species	Subspecies	Varieties	Hybrids	Synonyms
234. Taccaceae	1	1	0	0	0	0
235. Tamaricaceae	1	8	0	0	0	3
236. Theaceae	7	12	0	2	0	10
237. Theophrastaceae	1	6	0	0	0	1
238. Thymeliaceae	6	36	3	13	0	3
239. Tiliaceae	3	16	0	3	1	28
240. Trapaceae	1	1	0	0	0	0
241. Tropaeolaceae	1	1	0	0	0	0
242. Turneraceae	2	7	0	7	0	4
243. Typhaceae	1	4	0	0	1	1
244. Ulmaceae	5	24	0	9	0	25
245. Urticaceae	13	72	2	37	0	22
246. Valerianaceae	4	35	12	2	0	41
247. Verbenaceae	23	135	0	72	11	82
248. Violaceae	4	107	27	49	49	94
249. Vitaceae	4	40	0	9	0	42
250. Xyridaceae	1	22	0	5	0	14
251. Zannichelliaceae	3	3	0	0	0	7
252. Zingiberaceae	6	12	0	1	0	18
253. Zosteraceae	2	5	0	12	0	5
254. Zygophyllaceae	7	20	0	1	0	7
Subtotals	2,793	20,674	2,475	7,788	682	19,641
TOTAL	2,921	21,588	2,491	8,008	740	20,429

NAMES TREATED:	56,431	as tabulated above
	27	nothomorphs not in above tabulation
	56,458	in body of checklist
	179	excluded taxa
	304	family synonyms (index only)
TOTAL NAMES TREATED:	56,941	

Pteridophyta

ADIANTACEAE

Acrostichum L.

    aureum L.
    danaeifolium Langsd. & Fisch.
        SY=?A. excelsum Maxon

Adiantopsis Fee

    paupercula (Kunze) Fee
        SY=Cheilanthes p. (Kunze) Mett.

Adiantum L.

    capillus-veneris L.
        var. capillus-veneris
        var. modestum Fern.
        var. protrusum Fern.
    concinnum Humb. & Bonpl. ex Willd.
    cristatum L.
    fragile Sw.
    hispidulum Sw.
    jordanii C. Muell.
    latifolium Lam.
    macrophyllum Sw.
    melanoleucum Willd.
    obliquum Willd.
    pedatum L.
        var. aleuticum Rupr.
           SY=A. p. ssp. a. (Rupr.) Calder &
              Taylor
        var. pedatum
        var. subpumilum W.H. Wagner &
           Boydston
    petiolatum Desv.
    pulverulentum L.
    raddianum Presl
        SY=A. cuneatum Langsd. & Fisch.
    rigidulum Mett.
    tenerum Sw.
    tetraphyllum Humb. & Bonpl. ex Willd.
    X tracyi C.C. Hall ex W.H. Wagner [jordanii
        X pedatum]
    tricholepis Fee
    villosum L.
    wilsonii Hook.

Ananthacorus Underwood & Maxon

    angustifolius (Sw.) Underwood & Maxon

Anetium (Kunze) Splitg.

    citrifolium (L.) Splitg.

Anopteris (Prantl) Diels

    hexagona (L.) C. Christens.
        SY=Onychium strictum Kunze

Aspidotis Nutt. ex Copeland

    californica (Hook.) Nutt. ex Copeland
        SY=Cheilanthes c. (Hook.) Mett.
    carlotta-halliae (Wagner & Gilbert)
        Lellinger
        SY=Cheilanthes c-h. Wagner &
              Gilbert
    densa (Brack.) Lellinger
        SY=Cheilanthes siliquosa Maxon
        SY=Cryptogramma d. (Brack.) Diels

        SY=Onychium d. Brack.
        SY=Pellaea d. (Brack.) Hook.

Bommeria Fourn.

    hispida (Mett.) Underwood
        SY=Gymnopteris h. (Mett.)
           Underwood

Cheilanthes Sw.

    aemula Maxon
    alabamensis (Buckl.) Kunze
    arizonica (Maxon) Mickel
        SY=C. pyrmidalis Fee var. a.
           (Maxon) Broun
    clevelandii D.C. Eat.
    cooperae D.C. Eat.
    covillei Maxon
    eatonii Baker
        SY=C. castanea Maxon
    feei T. Moore
    fendleri Hook.
    fibrillosa Davenport ex Underwood
    gracillima D.C. Eat.
        var. aberrans M.E. Jones
        var. gracillima
    horridula Maxon
    intertexta (Maxon) Maxon
    kaulfussii Kunze
    lanosa (Michx.) D.C. Eat.
        SY=C. vestita (Spreng.) Sw.
    lendigera (Cav.) Sw.
    leucopoda Link
    lindheimeri Hook.
    microphylla (Sw.) Sw.
    notholaenoides (Desv.) Maxon ex Weatherby
    pringlei Davenport
    pyramidalis Fee
    X tarischii Davenport
    tomentosa Link
    villosa Davenport ex Maxon
    viscida Davenport
    wootonii Maxon
    wrightii Hook.

Coniogramme Fee

    pilosa (Brack.) Hieron.

Cryptogramma R. Br.

    crispa (L.) R. Br. ex Hook.
      ssp. acrostichoides (R. Br.) Hulten
        var. acrostichoides (R. Br.) C.B.
           Clarke
           SY=C. a. R. Br.
        var. sitchensis (Rupr.) C. Christens.
    stelleri (Gmel.) Prantl

Doryopteris J. Sm.

    decipiens (Hook.) J. Sm.
    decora Brack.
    pedata (L.) Fee
    tryonii Deg. & Deg.

Eriosorus Fee

    hispidulus (Kunze) Vareschi
        SY=Psilogramme h. (Kunze) Kuhn
        SY=P. portoricensis Maxon

Hecistopteris J. Sm.

  pumila (Spreng.) J. Sm.

Hemionitis L.

  palmata L.

Notholaena R. Br.

  aliena Maxon
          SY=Cheilanthes a. (Maxon) Mickel
  aschenborniana Klotzsch
          SY=Cheilanthes a. (Klotzsch) Mett.
  aurea (Poir.) Desv.
          SY=Cheilanthes bonariensis
              (Willd.) Proctor
  californica D.C. Eat.
          SY=Aleuritopteris cretacea ssp.
              nigrescens (Ewan) Munz
          SY=Cheilanthes deserti Mickel
          SY=N. californica ssp. nigrescens
              Ewan
          SY=N. candida (Mart. & Gal.) Hook.
              var. accessita Jepson
  candida (Mart. & Gal.) Hook.
          var. copelandii (C.C. Hall) R. Tryon
          SY=Cheilanthes candida var.
              copelandii (C.C. Hall)
              Mickel
          SY=N. copelandii C.C. Hall
  cochisensis Goodding
          SY=Cheilanthes c. (Goodding)
              Mickel
          SY=C. sinuata (Lag.) Domin var. c.
              (Goodding) Munz
          SY=N. s. (Lag.) Kaulfuss var. c.
              (Goodding) Weatherby
  dealbata (Pursh) Kunze
          SY=Cheilanthes d. Pursh
          SY=Pellaea d. (Pursh) Prantl
  fendleri Kunze
          SY=Cheilanthes cancellata Mickel
  grayi Davenport
          SY=Cheilanthes g. (Davenport)
              Domin
  greggii (Mett. ex Kuhn) Maxon
          SY=Cheilanthes g. (Mett. ex Kuhn)
              Mickel
  integerrima (Hook.) Hevly
          SY=Cheilanthes i. (Hook.) Mickel
          SY=N. sinuata (Lag.) Kaulfuss var.
              i. Hook.
  jonesii Maxon
          SY=Cheilanthes j. (Maxon) Munz
          SY=Pellaea j. (Maxon) Morton
  lemmonii D.C. Eat.
          SY=Cheilanthes l. (D.C. Eat.)
              Domin
  limitanea Maxon
          var. limitanea
          SY=Cheilanthes l. (Maxon) Mickel
          SY=Pellaea l. (Maxon) Morton
          var. mexicana (Maxon) Broun
          SY=Cheilanthes l. var. m. (Maxon)
              Mickel
          SY=Pellaea l. var. m. (Maxon)
              Morton
  neglecta Maxon
          SY=Cheilanthes n. (Maxon) Mickel
  newberryi D.C. Eat.
          SY=Cheilanthes n. (D.C. Eat.)
              Domin
  parryi D.C. Eat.
          SY=Cheilanthes p. (D.C. Eat.)
              Domin
  parvifolia R. Tryon

          SY=Cheilanthes p. (R. Tryon)
              Mickel
          SY=Pellaea microphylla Mett. ex
              Kuhn
  schaffneri (Fourn.) Underwood ex Davenport
          var. nealleyi (Seaton ex Coult.)
              Weatherby
          SY=Cheilanthes n. (Seaton ex
              Coulter) Domin
          var. schaffneri
          SY=Cheilanthes n. var. mexicana
              (Davenport) Mickel
          SY=N. n. var. m. Davenport
          SY=N. s. var. m. (Davenport)
              Davenport
  sinuata (Lag. ex Sw.) Kaulfuss
          SY=Cheilanthes s. (Lag. ex Sw.)
              Domin
  standleyi Maxon
          SY=Cheilanthes hookeri (Kummel.)
              Domin
  trichomanoides (L.) Desv.
          SY=Cheilanthes t. (L.) Mett.

Pellaea Link

  andromedifolia (Kaulfuss) Fee
          SY=P. a. var. pubescens Baker
          SY=P. a. var. rubens D.C. Eat.
  atropurpurea (L.) Link
  brachyptera (T. Moore) Baker
  breweri D.C. Eat.
  bridgesii Hook.
  glabella Mett. ex Kuhn
          var. glabella
          SY=P. atropurpurea (L.) Link var.
              bushii Mackenzie
          var. nana (L.C. Rich.) Cody
          var. occidentalis (E. Nels.) Butters
          SY=P. o. (E. Nels.) Rydb.
          var. simplex Butters
          SY=P. atropurpurea (L.) Link var.
              s. (Butters) Morton
          SY=P. suksdorfiana Butters
  intermedia Mett. ex Kuhn
  mucronata (D.C. Eat.) D.C. Eat.
          var. californica (Lemmon) Munz &
              Johnston
          SY=P. compacta (Davenport) Maxon
          var. mucronata
          SY=P. longimucronata Hook.
  ovata (Desv.) Weatherby
  sagittata (Cav.) Link
          var. cordata (Cav.) A. Tryon
          SY=P. cardiomorpha Weatherby
          SY=P. cordata (Cav.) Sm.
  ternifolia (Cav.) Link
  truncata Goodding
          SY=P. longimucronata sensu auctt.
              non Hook.
  viridis (Forsk.) Prantl
  wrightiana Hook.
          SY=P. ternifolia (Cav.) Link var.
              w. (Hook.) A. Tryon

Pityrogramma Link

  calomelanos (L.) Link
  chrysophylla (Sw.) Link
  pallida (Weatherby) Alston & Grant
          SY=P. triangularis (Kaulfuss)
              Maxon var. p. Weatherby
  sulphurea (Sw.) Maxon
  tartarea (Cav.) Maxon
  triangularis (Kaulfuss) Maxon
          var. maxonii Weatherby
          var. semipallida J.T. Howell

Pityrogramma (CONT.)
            var. triangularis
            var. viridis Hoover
            var. viscosa (D.C. Eat.) Weatherby
                  SY=P. v. (D.C. Eat.) Maxon
      trifoliata (L.) R. Tryon
            SY=Trismeria t. (L.) Diels

Polytaenium Desv.

      cajennense (Desv.) Benedict
      feei (Schaffn. ex Fee) Maxon
      lineatum (Sw.) J. Sm.

Pteris L.

      altissima Poir.
      arborea L.
      biaurita L.
      crassipes J. Agardh
      cretica L.
            var. albolineata Hook.
            var. cretica
      deflexa Link
      ensiformis Burm. f.
            var. ensiformis
            var. victoriae Baker
      excelsa Gaud.
      grandifolia L.
      hillebrandii Copeland
      irregularis Kaulfuss
      lidgatei (Baker) Christ
            SY=Schizostege l. (Baker) Hbd.
      longifolia L.
            var. bahamensis (J. Agardh) Hieron.
            var. longifolia
      multifida Poir.
      multilata L.
      pungens Willd.
      quadriaurita Retz.
      tripartita Sw.
      vittata L.

Vittaria Sm.

      graminifolia Kaulfuss
            SY=V. filifolia Fee
      lineata (L.) Sw.
      remota Fee
      rigida Kaulfuss

ASPLENIACEAE

Arachniodes Blume

      carvifolia (Kunze) Ching
            SY=Rumohra c. (Kunze) Ching

Asplenium L.

      abscissum Willd.
      acuminatum Hook. & Arn.
      adiantum-nigrum L.
      auritum Sw.
      X biscayneanum (D.C. Eat.) A.A. Eat.
        [dentatum X myriophyllum]
      bradleyi D.C. Eat.
            SY=A. X stotleri Wherry
      caudatum Forst. f.
      X clermontiae Syme [ruta-muraria X
        trichomanes]
      contiguum Kaulfuss
      cristatum Lam.
      cuneatum Lam.

X curtissii Underwood [abscissum X
  verecundum]
dalhousiae Hook.
      SY=Ceterach d. (Hook.) C.
            Christens.
dentatum L.
      SY=A. trichomanes-dentatum L.
X ebenoides R.R. Scott [platyneuron X
  rhizophyllum]
      SY=X Asplenosorus e. (R.R. Scott)
            Wherry
enatum Brack.
      SY=Asplenium gemmiferum Schrad.
            var. enatum (Brack.) Deg. &
            Deg.
      SY=A. g. var. flexuosum (Schrad.)
            Sims
exiguum Bedd.
      SY=Asplenium fontanum (L.) Bernh.
      SY=A. glenniei Baker
falcatum Lam.
      SY=Asplenium nitidulum Hbd.
feei Kunze ex Fee
      SY=Asplenium sanguinolentum Kunze
formosum Willd.
X gravesii Maxon [bradleyi X pinnatifidum]
      SY=X Asplenosorus g. (Maxon)
            Mickel
X herb-wagneri Taylor & Mohlenbrock
  [pinnatifidum X trichomanes]
      SY=X Asplenosorus h-w. (Taylor &
            Mohlenbrock) Mickel
heterochroum Kunze
heteroresiliens W.H. Wagner
horridum Kaulfuss
X inexpectatum (E.L. Braun ex Friesner)
  Morton [rhizophyllum X septentrionale]
      SY=X Asplenosorus i. E.L. Braun ex
            Friesner
insiticium Brack.
juglandifolium Lam.
      SY=Asplenium integerrimum Spreng.
kaulfussii Schlecht.
X kentuckiense T.N. McCoy [pinnatifidum X
  platyneuron]
      SY=Asplenium k. T.N. McCoy
      SY=X Asplenosorus k. (T.N. McCoy)
            Mickel
laetum Sw.
leucostegioides Baker
lobulatum Mett.
macraei Hook. & Grev.
      SY=Asplenium sphenolobum Zenker
      SY=A. s. var. diplaziosorum
            Hieron.
X mixtum Maxon [platyneuron X trichomanes]
      SY=A. virginicum Maxon
monanthes L.
montanum Willd.
myriophyllum (Sw.) Presl
      SY=Asplenium verecundum Chapman ex
            Underwood
nidus L.
normale D. Don
      SY=Asplenium pavonicum Brack.
obtusifolium L.
palmeri Maxon
pinnatifidum Nutt.
      SY=X Asplenosorus p. (Nutt.)
            Mickel
platyneuron (L.) Oakes ex D.C. Eat.
      var. bacculum-rubrum (Featherman)
            Fern.
      var. incisum (Howe ex Peck) B.L.
            Robins.
      var. platyneuron
      var. proliferum D.C. Eat.

Asplenium (CONT.)
   plenum E. St. John ex Small
   praemorsum Sw.
         SY=A. furcatum Thunb.
         SY=A. rhipidoneuron B.L. Robins.
   pseudoerectum Hieron.
   pumilum Sw.
      var. anthriscifolium (Jacq.) Wherry
         SY=Asplenium a. Jacq.
      var. pumilum
   radicans L.
         SY=Asplenium alloeopteron Kunze
         SY=A. crytopteron Kunze
   resiliens Kunze
   rhizophyllum L.
         SY=Camptosorus r. (L.) Link
   rhomboideum Brack.
         SY=Asplenium fragile Presl var.
             insularis Morton
   rutaceum (Willd.) Mett.
   ruta-muraria L.
      var. cryptolepis (Fern.) Wherry
         SY=Asplenium c. Fern.
      var. ohionis (Fern.) Wherry
         SY=Asplenium cryptolepis Fern.
            var. o. Fern.
      var. subtenuifolium C. Christens.
   salicifolium L.
   schizophyllum C. Christens.
         SY=Asplenium nephelephyllum
             Copeland
   scolopendrium L.
         SY=Phyllitis s. (L.) Newman
         SY=?P. s. var. americana Fern.
         SY=P. s. var. emarginata Fern.
   septentrionale (L.) Hoffmann
   serra Langsd. & Fisch.
   serratum L.
   sintenisii Hieron.
   trichomanes L.
   X trudellii Wherry [montanum X
     pinnatifidum]
         SY=Asplenium pinnatifidum Nutt.
            var. t. (Wherry) Clute
         SY=X Asplenosorus t. (Wherry)
            Mickel
   unilaterale Lam.
   uniseriale Raddi
   varians Hook. & Grev.
   vespertinum Maxon
   viride Huds.
   X wherryi D. M. Sm. [bradleyi X montanum]

Athyrium Roth

   distentifolium Tausch ex Opiz
      var. americanum (Butters) Boivin
         SY=A. alpestre (Hoope) Rylands ex
            T. Moore var. americanum
            Butters
         SY=A. alpestre var. gaspense Fern.
         SY=A. americanum (Butters) Maxon
         SY=A. distentifolium ssp.
            americanum (Butters) Hulten
         SY=A. alpestre (Hoppe) Rylands ex
            T. Moore pro parte
   filix-femina (L.) Roth
      var. angustum (Willd.) Lawson
         SY=A. a. (Willd.) Presl
         SY=A. a. var. rubellum (Gilbert)
            Butters
         SY=A. f-f. ssp. a. (Willd.)
            Clausen
         SY=A. f-f. var. michauxii
            (Spreng.) Farw.
         SY=A. f-f. var. r. Gilbert
      var. asplenioides (Michx.) Farw.

SY=A. a. (Michx.) A.A. Eat.
   var. cyclosorum Rupr.
      SY=A. alpestre (Hoppe) Rylands ex
        T. Moore var. c. (Rupr.) T.
        Moore
      SY=A. angustum (Willd.) Presl var.
        boreale Jennings
      SY=A. angustum var. elatius (Link)
        Butters
      SY=A. f-f. ssp. cyclosorum (Rupr.)
        C. Christens.
      SY=A. f-f. var. californicum
        Butters
      SY=A. f-f. var. sitchense (Rupr.)
        Ledeb.
   microphyllum (J. Sm.) Alston
      SY=Asplenium multisectum Brack.
      SY=Asplenium vexans Heller
      SY=Athyrium poiretianum (Gaud.)
        Presl

Bolbitis Schott

   aliena (Sw.) Alston
      SY=Leptochilus a. (Sw.) C.
        Christens.
   nicotianifolia (Sw.) Alston
      SY=Leptochilus n. (Sw.) C.
        Christens.
   pergamentacea (Maxon) Ching
      SY=Leptochilus p. Maxon
   portoricensis (Spreng.) Hennipman
      SY=B. cladorrhizans (Spreng.)
        Ching
      SY=Leptochilus c. (Spreng.) Maxon

Ctenitis (C. Christens.) C. Christens.

   hirta (Sw.) Ching
      SY=Dryopteris h. (Sw.) Kuntze
      SY=Thelypteris h. (Sw.) Copeland
   honolulensis (Hook.) Copeland
   latifrons (Brack.) Copeland
   nemorosa (Willd.) Copeland
      SY=Dryopteris n. (Willd.) Urban
   rubiginosa (Brack.) Copeland
   sloanei (Poepp.) Morton
      SY=C. ampla sensu auctt. non
        (Hook. & Benth.) Kuntze
      SY=Dryopteris a. sensu auctt.
   squamigera (Hook. & Arn.) Copeland
   subincisa (Willd.) Copeland
      SY=Dryopteris s. (Willd.) Urban
   submarginalis (Langsd. & Fisch.) Copeland
      SY=Dryopteris s. (Langsd. &
        Fisch.) C. Christens.

Cyclopeltis J. Sm.

   semicordata (Sw.) J. Sm.

Cyrtomium Presl

   caryotideum (Wallich) Presl
   falcatum (L.f.) Presl
      SY=Polystichum f. (L.f.) Diels
   fortunei J. Sm.

Cystopteris Bernh.

   bulbifera (L.) Bernh.
      SY=Filix b. (L.) Underwood
   douglasii Hook.
   fragilis (L.) Bernh.
      var. fragilis
        SY=C. dickieana Sims
        SY=C. f. ssp. d. (Sims) Hyl.

Cystopteris (CONT.)
        SY=C. f. var. angustata Lawson
        SY=Filix f. (L.) Underwood
    var. huteri (Hausman) Luerss.
    var. mackayi Lawson
    var. tenuifolia (Clute) Broun
  laurentiana (Weatherby) Blasdell
        SY=C. fragilis (L.) Bernh. var. l.
            Weatherby
  montana (Lam.) Bernh.
        SY=Filix m. (Lam.) Underwood
  protrusa (Weatherby) Blasdell
        SY=C. fragilis (L.) Bernh. var. p.
            Weatherby
  tennesseensis Shaver
        SY=C. fragilis (L.) Bernh. var.
            simulans (Weatherby)
            McGregor
        SY=C. f. var. t. (Shaver) McGregor

Deparia Hook. & Grev.

  acrostichoides (Sw.) M. Kato ined.
        SY=Athyrium thelypterioides
            (Michx.) Desv.
        SY=Diplazium a (Sw.) Butters
  fenzliana (Luerss.) M. Kato
        SY=Asplenium f. Luerss.
        SY=Athyrium f. (Luerss.) Deg. &
            Deg.
        SY=Diplazium f. (Luerss.) C.
            Christens.
  japonica (Thunb.) M. Kato
        SY=Athyrium j. (Thunb.) Copeland
        SY=Diplazium j. (Thunb.) Bedd.
  kaalaanum (Copeland) comb. ined.
        SY=Athyrium k. Copeland
        SY=Diplazium k. (Copeland) C.
            Christens.
  marginalis (Hbd.) M. Kato
        SY=Diplazium m. (Hbd.) C.
            Christens.
  mauiana (Copeland) M. Kato
        SY=Athyrium m. Copeland
        SY=Diplazium m. (Copeland) C.
            Christens.
  prolifera (Kaulfuss) Hook. & Grev.
        SY=Athyrium p. (Kaulfuss) C.
            Christens.

Diellia Brack.

  erecta Brack.
  falcata Brack.
  laciniata (Hbd.) Diels
  mannii (D.C. Eat.) Robins.
  unisora W.H. Wagner

Diplazium Sw.

  centripetale (Baker) Maxon
  cristatum (Desr.) Alston
        SY=D. arboreum (Willd.) Presl
  esculentum (Retz.) Sw.
        SY=Athyrium e. (Retz.) Copeland
  expansum Willd.
  grandifolium (Sw.) Sw.
  l'herminierii Hieron.
  lonchophyllum Kunze
  meyenianum Presl
        SY=D. arnottii Brack.
  molokaiense Robins.
  plantaginifolium (L.) Urban
  pycnocarpon (Spreng.) Broun
        SY=Asplenium p. Spreng.
        SY=Athyrium p. (Spreng.) Tidestrom
  sandwichianum (Presl) Diels

        SY=Athyrium s. Presl
  striatum (L.) Presl
  unilobum (Poir.) Hieron.
        var. hymenodes (Mett.) Maxon
        var. unilobum

Dryopteris Adans.

  acutidens C. Christens.
  X algonquinensis Britt. [fragrans X
    marginalis]
  arguta (Kaulfuss) Watt
  X atropalustris Small [celsa X cristata]
  X australis (Wherry) Small [celsa X
    ludoviciana]
        SY=D. clintoniana (D.C. Eat.)
            Dowell var. a. Wherry
        SY=D. cristata (L.) Gray var. a.
            (Wherry) Blomquist & Correll
  X benedictii (Farw.) Wherry [clintoniana X
    spinulosa]
  X bootii (Tuckerman) Underwood [cristata X
    intermedia]
  X burgessii Boivin [clintoniana X
    marginalis]
  campyloptera (Kunze) Clarkson
        SY=D. austriaca (Jacq.) Woynar ex
            Schinz & Thellung
        SY=D. dilatata (Hoffmann) Gray
            ssp. americana (Fisch.)
            Hulten
        SY=D. spinulosa (O.F. Muell.) Watt
            var. americana (Fisch.)
            Fern.
  celsa (Wm. Palmer) Knowlt., Palmer &
    Pollard ex Small
        SY=D. goldiana (Hook.) Gray ssp.
            c. Wm. Palmer
  cinnamomea (Cav.) C. Christens.
  clintoniana (D.C. Eat.) Dowell
        SY=D. cristata (L.) Gray var.
            clintoniana (D.C. Eat.)
            Underwood
  crinalis (Hook. & Arn.) C. Christens.
  cristata (L.) Gray
  X dowellii (Farw.) Wherry [clintoniana X
    intermedia]
  expansa (Presl) Fraser - Jenkins & Jermy
        SY=D. assimilis S. Walker
        SY=D. dilatata sensu auctt. non
            (Hoffmann) Gray
        SY=D. spinulosa (O.F. Muell.) Watt
            var. d. sensu auctt. non
            (Hoffmann) Underwood
  filix-mas (L.) Schott
  fragrans (L.) Schott
        var. fragrans
        var. remotiuscula Komarov
  fusco-atra (Hbd.) Robinson
  glabra (Brack.) Kuntze
        SY=D. nuda Underwood
  goldiana (Hook.) Gray
  hawaiiensis (Hbd.) Robins.
  intermedia (Willd.) Gray
        SY=D. austriaca (Jacq.) Woynar ex
            Schinz & Thellung var. i.
            (Willd.) Morton
        SY=D. spinulosa (O.F. Muell.) Watt
            var. concordiana (Davenport)
            Eastman
        SY=D. s. var. i. (Willd.)
            Underwood
  X leedsii Wherry [celsa X marginalis]
  ludoviciana (Kunze) Small
  marginalis (L.) Gray
  X neowherryi W.H. Wagner [goldiana X
    marginalis]

Dryopteris (CONT.)
  parallelogramma (Kunze) Alston
          SY=D. paleacea (Sw.) C. Christens.
  parvula Robins.
  patula (Sw.) Underwood
      var. rossii C. Christens.
  X pittsfordensis Slosson [marginalis X
      spinulosa]
  X poyseri Wherry [goldiana X spinulosa]
  sandwicensis (Hook. & Arn.) C. Christens.
  X separabilis Small [celsa X intermedia]
  X slossonae (Hahne) Wherry [cristata X
      marginalis]
  spinulosa (O.F. Muell.) Watt
          SY=D. austriaca (Jacq.) Woynar ex
                  Schinz & Thellung var. s.
                  (O.F. Muell.) Fiori
          SY=D. carthusiana (Vill.) H.P.
                  Fuchs
  X triploidea Wherry [intermedia X
      spinulosa]
          SY=D. intermedia (Willd.) Gray
                  var. fructuosa (Gilbert)
                  Wherry
          SY=D. spinulosa (O.F. Muell.) Watt
                  var. f. (Gilbert) Trudell
  X uliginosa (A. Braun ex Dowell) Druce
      [cristata X spinulosa]
  unidentata (Hook. & Arn.) C. Christens.

Elaphoglossum Schott

  aemulum (Kaulfuss) Brack.
          SY=E. conforme (Sw.) Schott
          SY=E. gorgoneum (Kaulfuss) Brack.
  alatum Gaud.
      var. alatum
      var. crassicaule (Copeland) Anders. &
          Crosby
      var. fauriei (Copeland) Anders. &
          Crosby
      var. parvisquamatum (Skottsberg)
          Anders. & Crosby
  apodum (Kaulfuss) Schott
  crassifolium (Gaud.) Anders. & Crosby
          SY=E. reticulatum (Kaulfuss) Gaud.
  crinitum (L.) Christ
          SY=Hymenodium c. (L.) Fee
  decoratum (Kunze) T. Moore
  erinaceum (Fee) T. Moore
  firmum (Mett.) Urban
  glabellum J. Sm.
  herminieri (Bory & Fee) T. Moore
  hirtum (Sw.) C. Christens.
      var. micans (Mett.) C. Christens.
  longifolium (Presl) J. Sm.
  martinicense (Desv.) T. Moore
          SY=E. underwoodianum Maxon
  maxonii Underwood ex Morton
          SY=E. pteropus sensu auctt. non C.
                  Christens.
  pellucidum Gaud.
          SY=E. micradenium (Fee) T. Moore
  petiolatum (Sw.) Urban
      var. dussii (Underwood ex Maxon)
          Proctor
          SY=E. d. Underwood ex Maxon
  piloselloides (Presl) T. Moore
  rigidum (Aubl.) Urban
          SY=E. flaccidum (Fee) T. Moore
  serpens Maxon & Morton
  wawrae (Luerss.) C. Christens.

Fadyenia Hook.

  hookeri (Sweet) Maxon

Gymnocarpium Newman

  dryopteris (L.) Newman
      var. disjunctum (Rupr.) Ching
          SY=Dryopteris disjuncta (Rupr.)
                  Morton
      var. dryopteris
          SY=Dryopteris d. (L.) Britt.
          SY=D. linnaeana C. Christens.
          SY=Phegopteris d. (L.) Fee
  X heterosporum W.H. Wagner [dryopteris X
      robertianum]
  robertianum (Hoffmann) Newman
          SY=Dryopteris r. (Hoffmann) C.
                  Christens.
          SY=G. dryopteris (L.) Newman var.
                  pumilum (DC.) Boivin
          SY=Phegopteris r. (Hoffmann) A.
                  Braun ex Aschers.

Hemidictyum Presl

  marginatum (L.) Presl

Hypoderris R. Br.

  brownii J. Sm.

Lastreopsis Ching

  effusa (Sw.) Tindale
          SY=Ctenitis e. (Sw.) Copeland
          SY=Dryopteris e. (Sw.) Urban

Lomagramma J. Sm.

  guianensis (Aubl.) Ching
          SY=Leptochilus g. (Aubl.) C.
                  Christens.

Lomariopsis Fee

  amydrophlebia (Slosson ex Maxon) Holttum
          SY=Stenochlaena a. Slosson ex
                  Maxon
  kunzeana (Presl ex Underwood) Holttum
          SY=Stenochlaena k. Presl ex
                  Underwood
  sorbifolia (L.) Fee
          SY=Stenochlaena s. (L.) J. Sm.

Matteuccia Todaro

  struthiopteris (L.) Todaro
          SY=M. pensylvanica (Willd.)
                  Raymond
          SY=M. s. var. p. (Willd.) Morton
          SY=Onoclea s. (L.) Hoffmann pro
                  parte
          SY=O. s. var. p. (Willd.) Boivin
          SY=Pteretis nodulosa (Michx.)
                  Nieuwl.
          SY=P. p. (Willd.) Fern.

Maxonia C. Christens.

  apiifolia (Sw.) C. Christens.

Onoclea L.

  sensibilis L.

Peltapteris Link

  peltata (Sw.) Morton
          SY=Rhipidopteris p. (Sw.) Schott

Phanerophlebia Presl

    auriculata Underwood
          SY=Cyrtomium a. (Underwood) Morton
    umbonata Underwood

Polybotrya Humb. & Bonpl. ex Willd.

    cervina (L.) Kaulfuss

Polystichopsis (J. Sm.) Holttum

    chaerophylloides (Poir.) Morton
          SY=Dryopteris c. (Poir.) C.
             Christens.

Polystichum Roth

    acrostichoides (Michx.) Schott
        var. acrostichoides
          SY=P. a. var. schweinitzii (Beck)
             Small
        var. lonchitoides Brooks
    aleuticum C. Christens.
    andersonii Hopkins
          SY=P. braunii (Spenner) Fee ssp.
             a. (Hopkins) Calder &
             Taylor
    braunii (Spenner) Fee
      ssp. braunii
      ssp. purshii (Fern.) Calder & Taylor
          SY=P. b. var. p. Fern.
    californicum (D.C. Eat.) Diels
    dudleyi Maxon
    echinatum (Gmel.) C. Christens.
          SY=P. triangulum sensu auctt. non
             Fee
    X hagenahii Cody [acrostichoides X
      lonchitis]
          SY=P. X marginale (W.R. McColl)
             Cody
    haleakalense Brack.
    hillebrandii Carruthers
    imbricans (D.C. Eat.) D.H. Wagner
      ssp. curtum (Ewan) D.H. Wagner
          SY=P. munitum (Kaulfuss) Presl
             ssp. c. Ewan
      ssp. imbricans
          SY=P. munitum (Kaulfuss) Presl
             ssp. i. (D.C. Eat.) Munz
          SY=P. m. var. i. (D.C. Eat.) Maxon
    kruckebergii W.H. Wagner
    lemmonii Underwood
          SY=P. mohrioides sensu auctt. non
             (Bory) Presl
          SY=P. mohrioides (Bory) Presl var.
             l. (Underwood) Fern.
    lonchitis (L.) Roth
    microchlamys (Christ) Matsumura
    munitum (Kaulfuss) Presl
          SY=P. m. var. incisoserratum (D.C.
             Eat.) Underwood
    muricatum (L.) Fee
    polystichiforme (Fee) Maxon
    rhizophyllum (Sw.) Presl
    scopulinum (D.C. Eat.) Maxon
          SY=P. mohrioides (Bory) Presl var.
             s. (D.C. Eat.) Fern.
    setigerum (Presl) Presl
          SY=P. alaskense Maxon
          SY=P. braunii (Spenner) Fee ssp.
             a. (Maxon) Calder & Taylor
          SY=P. b. var. a. (Maxon) Hulten
    tsus-simense (Hook.) J. Sm.

Pseudophegopteris Ching

keraudreniana (Gaud.) Holttum

Tectaria Cav.

    X amesiana A.A. Eat. [coriandrifolia X
      lobata]
    cicutaria (L.) Copeland
    coriandrifolia (Sw.) Underwood
    gaudichaudii (Mett.) Maxon
    heracleifolia (Willd.) Underwood
    incisa Cav.
          SY=T. martinicensis (Spreng.)
             Copeland
    lobata (Poir.) Morton
          SY=T. minima Underwood
    plantaginea (Jacq.) Maxon
    trifoliata (L.) Cav.

Thelypteris Schmidel

    acuminata (Houtt.) Morton
    angustifolia (Willd.) Proctor
          SY=Dryopteris a. (Willd.) Urban
    asplenioides (Sw.) Proctor
    augescens (Link) Munz & Johnston
          SY=Dryopteris a. (Link) C.
             Christens.
          SY=Lastrea a. (Link) J. Sm.
    balbisii (Spreng.) Ching
          SY=Dryopteris sprengelii
             (Kaulfuss) Kuntze
          SY=T. s. (Kaulfuss) Proctor
    boydiae (Eat.) K. Iwats.
    brittoniae (Slosson ex Maxon) Alain
          SY=Dryopteris b. Slosson ex Maxon
    cordata (Fee) Proctor
          SY=Dryopteris c. (Fee) Urban
    cyatheoides (Kaulfuss) Fosberg
          SY=Cyclosorus c. (Kaulfuss) Farw.
          SY=Dryopteris c. (Kaulfuss) Kuntze
    decussata (L.) Proctor
          SY=Dryopteris d. (L.) Urban
    deltoidea (Sw.) Proctor
          SY=Dryopteris d. (Sw.) Kuntze
    dentata (Forsk.) E. St. John
          SY=Cyclosorus d. (Forsk.) Ching
          SY=Dryopteris d. (Forsk.) C.
             Christens.
    forsteri Morton
    germaniana (Fee) Proctor
          SY=Dryopteris g. (Fee) C.
             Christens.
    globulifera (Brack.) C.F. Reed
          SY=Dryopteris g. (Brack.) Kuntze
          SY=Lastrea g. Brack.
    grandis A.R. Sm.
    guadalupensis (Wikstr.) Proctor
          SY=Dryopteris domingensis
             (Spreng.) Maxon
          SY=D. g. (Wikstr.) C. Christens.
    hastata (Fee) Proctor
          SY=Dryopteris h. (Fee) Urban
    hawaiiensis C.F. Reed
          SY=Cyclosorus sandwicensis
             (Brack.) Copeland
          SY=Dryopteris stegnogrammoides
             (Baker) C. Christens.
    hexagonoptera (Michx.) Weatherby
          SY=Dryopteris h. (Michx.) C.
             Christens.
          SY=Phegopteris h. (Michx.) Fee
    hudsoniana (Brack.) C.F. Reed
    interrupta (Willd.) K. Iwats.
          SY=Cyclosorus gongylodes (Schkuhr)
             Link
          SY=Dryopteris g. (Schkuhr) Kuntze
          SY=T. g. (Schkuhr) Small

Thelypteris (CONT.)
          SY=T. totta (Thunb.) Schelpe
     X invisa (Sw.) Proctor [grandis X serra]
          SY=Dryopteris oligophylla Maxon
          SY=T. o. (Maxon) Proctor
     kunthii (Desv.) Morton
          SY=Dryopteris normalis C.
               Christens.
          SY=D. saxatilis (R. St. John)
               Broun
          SY=T. macrorhizoma E. St. John
          SY=T. n. (C. Christens.) Moxley
          SY=T. s. R. St. John
          SY=T. unca R. St. John
     leptocladia (Fee) Proctor
          SY=Dryopteris l. (Fee) Maxon
     limbosperma (All.) H.P. Fuchs
          SY=Dryopteris oreopteris (Ehrh.)
               Maxon
     linkiana (Presl) R. Tryon
          SY=Dryopteris l. (Presl) Maxon
     nephrodioides (Klotzsch) Proctor
     nevadensis (Baker) Clute ex Morton
          SY=Dryopteris n. (Baker) Underwood
          SY=D. oregana C. Christens.
          SY=Lastrea o. (C. Christens.)
               Copeland
     noveboracensis (L.) Nieuwl.
          SY=Dryopteris n. (L.) Gray
     opposita (Vahl) Ching
          SY=Dryopteris o. (Vahl) Urban
     opulenta (Kaulfuss) Fosberg
     ovata R. St. John
          var. lindheimeri (C. Christens.) A.R.
               Sm.
          SY=Dryopteris augescens (Link)
               Munz & Johnston var. l. (C.
               Christens.) Broun
          SY=T. X l. (C. Christens.) Wherry
          var. ovata
          SY=T. normalis (C. Christens.)
               Moxley var. harperi (C.
               Christens.) Wherry
     palustris Schott
          var. haleana Fern.
          SY=Dryopteris thelypteris (L.) Sw.
               var. h. (Fern.) Broun &
               Weatherby
          var. palustris
          SY=Dryopteris thelypteris (L.) Sw.
          var. pubescens (Lawson) Fern.
          SY=Dryopteris thelypteris (L.) Sw.
               var. pubescens (Lawson) A.R.
               Prince ex Weatherby
     parasitica (L.) Fosberg
          SY=Dryopteris p. (L.) Kuntze
     patens (Sw.) Small
          var. patens
          SY=Dryopteris p. (Sw.) Kuntze
          var. scabriuscula (Presl) A.R. Sm.
     pennata (Poir.) Morton
          SY=Dryopteris megalodus (Schkuhr)
               Urban
          SY=T. m. (Schkuhr) Proctor
     phegopteris (L.) Slosson
          SY=Dryopteris p. (L.) C.
               Christens.
          SY=Phegopteris connectilis
               (Michx.) Watt
          SY=P. polypodioides Fee
     piedrensis (C. Christens.) Morton
          SY=Dryopteris p. C. Christens.
     pilosa (Mart. & Gal.) Crawford
          SY=Leptogramma p. (Mart. & Gal.)
               Underwood
          SY=L. p. var. alabamensis
               (Crawford) Wherry

     poiteana (Bory) Proctor
          SY=Dryopteris p. (Bory) Urban
     puberula (Baker) Morton
          SY=Dryopteris feei C. Christens.
          SY=T. augescens (Link) Munz &
               Johnston var. p. (Baker)
               Munz & Johnston
     quadrangularis (Fee) Schelpe
          var. inconstans (C. Christens.) A.R.
               Sm.
          var. versicolor (R. St. John) A.R.
               Sm.
          SY=Cyclosorus q. (Fee)
               Tardieu-Blot pro parte
          SY=Dryopteris macilenta (E. St.
               John) Correll
          SY=D. v. (R. St. John) Broun
          SY=T. m. E. St. John
          SY=T. v. R. St. John
     reptans (J.F. Gmel.) Morton
          var. reptans
          SY=Dryopteris r. (J.F. Gmel.) C.
               Christens.
          SY=D. r. var. conformis C.
               Christens.
          SY=Goniopteris r. (J.F. Gmel.)
               Presl
          var. tenera (Fee) Proctor
     resinifera (Desv.) Proctor
          SY=Dryopteris r. (Desv.) Weatherby
     reticulata (L.) Proctor
          SY=Dryopteris r. (L.) Urban
          SY=Meniscium r. (L.) Sw.
     rudis (Kunze) Proctor
     sancta (L.) Proctor
          var. portoricensis (Kuhn) Morton
          var. sancta
          SY=Dryopteris s. (L.) Kuntze
     sclerophylla (Poepp. ex Spreng.) Morton
          SY=Dryopteris s. (Peopp. ex
               Spreng.) C. Christens.
          SY=Goniopteris s. (Peopp. ex
               Spreng.) Wherry
     serra (Sw.) R. St. John
          SY=Dryopteris s. (Sw.) Kuntze
     serrata (Cav.) Alston
          SY=Dryopteris s. (Cav.) C.
               Christens.
          SY=Meniscium s. Cav.
     simulata (Davenport) Nieuwl.
          SY=Dryopteris s. Davenport
     tetragona (Sw.) Small
          SY=Dryopteris subtetragona (Link)
               Maxon
          SY=Goniopteris t. (Sw.) Presl
     torresiana (Gaud.) Alston
          SY=Dryopteris setigera sensu
               auctt. non Blume

Woodsia R. Br.

     X abbeae Butters [ilvensis X scopulina]
          SY=W. confusa T.M.C. Taylor
     alpina (Bolton) S. F. Gray
          SY=W. a. var. bellii (Lawson)
               Morton
          SY=W. glabella R. Br. var. b.
               Lawson
     glabella R. Br.
     X gracilis (Lawson) Butters [alpina X
          ilvensis]
          SY=W. ilvensis (L.) R. Br. var. g.
               Lawson
     ilvensis (L.) R. Br.
     X maxonii R. Tryon [oregana var.
          cathcartiana X scopulina]
     mexicana Fee

Woodsia (CONT.)
  obtusa (Spreng.) Torr.
  oregana D.C. Eat.
      var. cathcartiana (B.L. Robins.)
        Morton
        SY=W. c. B.L. Robins.
     var. oregana
     var. squammosa Boivin
  plummeraæ Lemmon
  scopulina D.C. Eat.
      SY=W. appalachiana T.M.C. Taylor
      SY=W. oregana D.C. Eat. var.
        lyallii (Hook.) Boivin
      SY=W. s. var. a. (T.M.C. Taylor)
        Morton
  X tryonis Boivin [glabella X ilvensis]

AZOLLACEAE

Azolla Lam.

  caroliniana Willd.
  filiculoides Lam.
  mexicana Presl

BLECHNACEAE

Blechnum L.

  divergens (Kunze) Mett.
      SY=Struthiopteris exaltata (Fee)
        Broadh.
  fragile (Liebm.) Morton & Lellinger
      SY=B. polypodioides (Sw.) Kuhn
      SY=Struthiopteris p. (Sw.) Trev.
  lineatum (Sw.) C. Christens.
      SY=Struthiopteris l. (Sw.) Broadh.
  occidentale L.
  orientale L.
  polypodioides Raddi
      SY=B. unilaterale Sw.
  serrulatum L.C. Rich.
      SY=B. indicum sensu auctt.
  spicant (L.) Sm.
      SY=B. s. ssp. nipponicum sensu
        auctt. non (Kunze) Love &
        Love
      SY=B. s. var. elongata (Hook.)
        Boivin
      SY=Lomaria s. (L.) Desv.
      SY=Struthiopteris s. (L.) Weiss
  underwoodiana (Broadh.) C. Christens.
      SY=Struthiopteris u. Broadh.

Doodia R. Br.

  kunthiana Gaud.
  lyonii Deg.

Sadleria Kaulfuss

  cyatheoides Kaulfuss
  ?fauriei Copeland
  pallida Hook. & Arn.
      SY=S. hillebrandii Robins.
  ?rigida Copeland
  souleyetiana (Gaud.) T. Moore
  squarrosa (Gaud.) T. Moore
      SY=S. polystichoides (Brack.)
        Heller
      SY=S. unisora (Baker) Robins.

Stenochlaena J. Sm.

  tenuifolia (Desv.) T. Moore

Woodwardia Sm.

  areolata (L.) T. Moore
      SY=Lorinseria a. (L.) Presl
  fimbriata Sm.
      SY=W. chamissoi Brack.
  radicans (L.) Sm.
  virginica (L.) Sm.
      SY=Anchistea v. (L.) Presl

CYATHEACEAE

Alsophila R. Br.

  brooksii (Maxon) R. Tryon
      SY=Cyathea b. Maxon
  bryophila R. Tryon
  dryopteroides (Maxon) R. Tryon
      SY=Cyathea d. Maxon

Cibotium Kaulfuss

  chamissoi Kaulfuss
  glaucum (Sm.) Hook. & Arn.
  hawaiiense Nakai & Ogura
  st.-johnii Krajina
  splendens (Gaud.) Krajina ex Skottsberg

Cnemidaria Presl

  horrida (L.) Presl
      SY=Cyathea h. (L.) Sm.
      SY=Hemitelia h. (L.) R. Br.

Cyathea Sm.

  andina (Karst.) Domin
      SY=C. escuquensis (Karst.) Domin
      SY=Hemitelia e. Karst.
  arborea (L.) Sm.
  furfuracea Baker
      SY=C. brittoniana Maxon
  parvula (Jenman) Domin
      SY=Alsophila aquilina Christ
      SY=C. a. (Christ) Domin

Nephelea R. Tryon

  portoricensis (Spreng. ex Kuhn) R. Tryon
      SY=Cyathea p. Spreng. ex Kuhn

Trichipteris Presl

  borinquena (Maxon) R. Tryon
      SY=Alsophila b. Maxon
      SY=Cyathea b. (Maxon) Domin
  procera (Willd.) R. Tryon
      SY=Alsophila p. (Willd.) Desv.

DAVALLIACEAE

Nephrolepis Schott

  X averyi Nauman [biserrata X exaltata]
  biserrata (Sw.) Schott
  cordifolia (L.) Presl
  exaltata (L.) Schott

Nephrolepis (CONT.)
  falcata (Cav.) C. Christens.
        SY=N. biserrata (Sw.) Schott var.
          furcans Hort. ex Bailey
  hirsutula (Forst.) Presl
  multiflora (Roxb.) Jarrett ex Morton
  occidentalis Kunze
  pectinata (Willd.) Schott
  rivularis (Vahl) Mett. ex Krug

Oleandra Cav.

  articulata (Sw.) Presl

Rumohra Raddi

  adiantiformis (Forst. f.) Ching
        SY=Polystichum a. (Forst. f.) J.
          Sm.

DENNSTAEDTIACEAE

Dennstaedtia Bernh.

  bipinnata (Cav.) Maxon
        SY=D. adiantoides (Humb. & Bonpl.
          ex Willd.) T. Moore
  cicutaria (Sw.) T. Moore
        SY=D. rubiginosa (Kaulfuss) T.
          Moore
  globulifera (Poir.) Hieron.
  obtusifolia (Willd.) T. Moore
        SY=D. ordinata (Kaulfuss) T. Moore
  punctilobula (Michx.) T. Moore

Histiopteris (J. Agardh) J. Sm.

  incisa (Thunb.) J. Sm.

Hypolepis Bernh.

  nigrescens Hook.
  punctata (Thunb.) Mett.
  repens (L.) Presl
  tenerrima Maxon

Lindsaea Dry.

  ensifolia Sw.
  lancea (L.) Bedd.
  portoricensis Desv.
  quadrangularis Raddi
    ssp. antillensis Kramer
        SY=L. montana Fee
  repens (Bory) Thwaites
    var. macraeana (Hook. & Arn.) C.
        Christens.
  stricta (Sw.) Dry.

Lonchitis L.

  hirsuta L.
        SY=Anisosorus h. (L.) Underwood &
          Maxon

Microlepia Presl

  setosa (Sm.) Alston
        SY=M. hirta (Kaulfuss) Presl
        SY=M. strigosa pro parte
        SY=M. strigosa (Thunb.) Presl var.
         h. (Kaulfuss) Hbd.
  speluncae (L.) T. Moore

Odontosoria Fee

  aculeata (L.) J. Sm.
  uncinella (Kunze) Fee

Pteridium Gleditsch ex Scop.

  aquilinum (L.) Kuhn
    var. arachnoideum (Kaulfuss) Brade
      SY=P. a. (Kaulfuss) Maxon
    var. caudatum (L.) Sadebeck
      SY=P. c. (L.) Maxon
    var. champlainense Boivin
    var. decompositum (Gaud.) R. Tryon
    var. latiusculum (Desv.) Underwood ex
      Heller
      SY=P. l. (Desv.) Hieron.
    var. pseudocaudatum (Clute) Heller
      SY=P. latiusculum (Desv.) Hieron.
        var. p. (Clute) Maxon
    var. pubescens Underwood
      SY=P. a. ssp. lanuginosum (Bong.)
        Hulten
      SY=P. a. var. l. (Bong.) Fern.

Saccoloma Kaulfuss

  domingense (Spreng.) C. Christens.
        SY=Orthiopteris d. (Spreng.)
          Copeland
  inaequale (Kunze) Mett.
        SY=Orthiopteris i. (Kunze)
          Copeland

Sphenomeris Maxon

  chinensis (L.) Maxon
        SY=S. chusana (L.) Copeland
  clavata (L.) Maxon

EQUISETACEAE

Equisetum L.

  arvense L.
        SY=E. a. var. alpestre Wahlenb.
        SY=E. arvense var. boreale (Bong.)
         Rupr.
        SY=E. arvense var. riparium Farw.
        SY=E. calderi Boivin
  X ferrissii Clute [hyemale var. affine X
  laevigatum]
        SY=E. hyemale L. var. intermedium
         A. A. Eat.
        SY=E. i. (A.A. Eat.) Rydb.
  fluviatile L.
        SY=E. f. var. limosum (L.) Gilbert
        SY=E. l. L.
  hyemale L.
    var. affine (Engelm.) A.A. Eat.
      SY=E. a. Engelm.
      SY=E. h. ssp. a. (Engelm.) Calder
        & Taylor
      SY=E. h. var. californicum Milde
      SY=E. h. var. elatum (Engelm.)
        Morton
      SY=E. h. var. pseudohyemale
        (Farw.) Morton
      SY=E. h. var. robustum (A. Braun)
        A.A. Eat.
      SY=E. praealtum Raf.
      SY=E. r. A. Braun
  laevigatum A. Braun
      SY=E. funstonii A.A. Eat.

Equisetum (CONT.)
          SY=E. kansanum Schaffn.
          SY=E. l. ssp. f. (A.A. Eat.)
                Hartman
  X litorale Kuhlewein ex Rupr. [arvense X
    fluviatile]
  X nelsonii (A.A. Eat.) Schaffn. [laevigatum
    X variegatum]
          SY=E. variegatum Schleich. ex
                Weber & C. Mohr var. n. A.A.
                Eat.
  palustre L.
          SY=E. p. var. americanum Victorin
          SY=E. p. var. simplicissimum A.
                Braun
  pratense Ehrh.
  ramosissimum Desf.
  scirpoides Michx.
  sylvaticum L.
          SY=E. s. var. multiramosum (Fern.)
                Wherry
          SY=E. s. var. pauciramosum Milde
  telmateia Ehrh.
      var. braunii (Milde) Milde
          SY=E. braunii Milde
          SY=E. t. ssp. b. (Milde) Hauke
      var. telmateia
          SY=E. maximum Lam.
  X trachyodon A. Braun [hyemale var. affine
    X variegatum]
          SY=E. variegatum Schleich. ex
                Weber & C. Mohr var. jesupii
                A.A. Eat.
  variegatum Schleich. ex Weber & C. Mohr
      var. alaskanum A.A. Eat.
          SY=E. v. ssp. a. (A.A. Eat.)
                Hulten
      var. variegatum
          SY=E. v. var. anceps Milde

GLEICHENIACEAE

Dicranopteris Bernh.

  emarginata (T. Moore) W.J. Robins.
      var. emarginata
      var. inaequalis Deg. & Deg.
  flexuosa (Schrad.) Underwood
          SY=Gleichenia f. (Schrad.) Mett.
  linearis (Burm.) Underwood
      var. linearis
      var. maxima (Christ ex Hochr.) Deg. &
        Deg.
  pectinata (Willd.) Underwood
          SY=Gleichenia p. (Willd.) Presl

Diplopterygium (Diels) Nakai

  pinnatum (Kunze) Nakai
          SY=Hicriopteris p. (Kunze) Ching

Sticherus Presl

  bifidus (Willd.) Ching
          SY=Dicranopteris b. (Willd.) Maxon
          SY=Gleichenia b. (Willd.) Spreng.
  owhyhensis (Hook.) Ching

GRAMMITIDACEAE

Adenophorus Gaud.

  abietinus (D.C. Eat.) K.A. Wilson
  baldwinii (Baker) Copeland
          SY=Polypodium knudsenii Hieron.
  haalilioanus (Brack.) K.A. Wilson
  hymenophylloides (Kaulfuss) Hook. & Grev.
          SY=Amphoradenium h. (Kaulfuss)
                Copeland
          SY=Polypodium h. Kaulfuss
  oahuensis (Copeland) Bishop
  periens Bishop
  pinnatifidus Gaud.
  tamariscinus (Kaulfuss) Hook. & Grev.
      var. epigaeus Bishop
      var. montanus (Hbd.) Bishop
      var. tamariscinus
          SY=Amphoradenium t. (Kaulfuss)
                Copeland
          SY=Polypodium t. Kaulfuss
  tripinnatifidus Gaud.

Cochlidium Kaulfuss

  jungens Bishop
  seminudum (Willd.) Maxon
  serrulatum (Sw.) Bishop
          SY=Grammitis s. (Sw.) Sw.
          SY=Polypodium duale Maxon

Grammitis Sw.

  asplenifolia (L.) Proctor
          SY=Polypodium a. L.
  baldwinii (Baker) Copeland
          SY=Polypodium knudsenii Hieron.
  hartii (Jenman) Proctor
          SY=Xiphopteris h. (Jenman)
                Copeland
  hessii (Maxon) Alain
          SY=Polypodium h. Maxon
  hookeri (Brack.) Copeland
          SY=Polypodium h. Brack.
  mollissima (Fee) Proctor
          SY=Polypodium m. Fee
  myosuroides (Sw.) Sw.
          SY=Polypodium m. Sw.
  nimbata (Jenman) Proctor
  sectifrons (Kunze ex Mett.) Seymour
          SY=Polypodium s. Kunze ex Mett.
  suspensa (L.) Proctor
          SY=G. jubaeformis (Kaulfuss)
                Proctor
          SY=Polypodium j. Kaulfuss
  taenifolia (Jenman) Proctor
          SY=Polypodium t. Jenman
  taxifolia (L.) Proctor
          SY=Polypodium t. L.
  tenella Kaulfuss
          SY=Polypodium pseudogrammitis
                Gaud.
  trifurcata (L.) Copeland
          SY=Polypodium t. L.

Xiphopteris Kaulfuss

  ?saffordii (Maxon) Copeland
          SY=Polypodium s. Maxon

HYMENOPHYLLACEAE

Hymenophyllum Sm.

  asplenioides Sw.
  axillare Sw.
  contortum Bosch

Hymenophyllum (CONT.)
        SY=H. crispum sensu auctt. non
            H.B.K.
  elegantulum Bosch
      var. petiolulatum Morton
  fucoides Sw.
  hirsutum (L.) Sw.
        SY=H. ciliatum Sw.
  lanatum Fee
  lanceolatum Hook. & Arn.
  lineare Sw.
  macrothecum Fee
  microcarpum Desv.
  obtusum Hook. & Arn.
  polyanthos (Sw.) Sw.
  protrusum Hook.
  recurvum Gaud.
  sieberi (Presl) Bosch
  tunbridgense (L.) Sm.
  undulatum (Sw.) Sw.
  wrightii Bosch
        SY=Mecodium w. (Bosch) Copeland

Trichomanes L.

  alatum Sw.
  angustifrons (Fee) W. Boer
        SY=T. pusillum sensu auctt. non
            Sw.
  bauerianum Endl.
        SY=T. baldwinii D.C. Eat.
  boschianum Sturm
  capillaceum L.
  crispum L.
  cyrtotheca Hbd.
  davallioides Gaud.
  draytonianum Brack.
  holopterum Kunze
  hookeri Presl
  hymenophylloides Bosch
  kapplerianum Sturm
  krausii Hook. & Grev.
  lineolatum (Bosch) Hook.
  membranaceum L.
        SY=Lecanium m. (L.) Presl
  minutum Blume
        SY=T. saxifragoides Presl
  ovale (Fourn.) W. Boer
  petersii Gray
  pinnatum Hedw.
  polypodioides L.
  punctatum Poir.
    ssp. floridanum W. Boer
    ssp. punctatum
    ssp. sphenoides (Kunze) W. Boer
        SY=T. s. Kunze
  pusillum Sw.
  radicans Sw.
  rigidum Sw.
  robustum Fourn.
  scandens L.

ISOETACEAE

Isoetes L.

  bolanderi Engelm.
      var. bolanderi
      var. pygmaea (Engelm.) Clute
  butleri Engelm.
  eatonii Dodge
  echinospora Durieu
    ssp. asiatica (Makino) Love
        SY=I. e. var. a. Makino

  ssp. maritima (Underwood) A. Love
      var. m. (Underwood) Eat.
        SY=I. muricata Durieu ssp.
            maritima (Underwood) Hulten
  ssp. muricata (Durieu) Love & Love
    var. braunii (Durieu) Engelm.
        SY=I. b. Durieu
        SY=I. gravesii A.A. Eat.
        SY=I. muricata Durieu var. b.
            (Durieu) C.F. Reed
        SY=I. setacea Lam. pro parte
    var. hesperia (C.F. Reed) A. Love
        SY=I. muricata Durieu var. h. C.F.
            Reed
    var. muricata
        SY=I. e. var. m. (Durieu) Engelm.
        SY=I. e. var. robusta Engelm.
        SY=I. e. var. savilei Boivin
        SY=I. flettii (A.A. Eat.) N.E.
            Pfeiffer
        SY=I. muricata Durieu
        SY=I. setacea Lam. ssp. m.
            (Durieu) Holub
  engelmannii A. Braun
      var. caroliniana A.A. Eat.
      var. engelmannii
        SY=I. e. var. valida Engelm.
      var. georgiana Engelm.
  flaccida Shuttlw. ex A. Braun
  foveolata A.A. Eat.
  howellii Engelm.
  lacustris L.
        SY=I. l. var. paupercula Engelm.
        SY=I. occidentalis Henderson
        SY=I. p. (Engelm.) A.A. Eat.
        SY=I. piperi A.A. Eat.
  lithophylla N.E. Pfeiffer
  louisianensis Thieret
  macrospora Durieu
  melanopoda Gay & Durieu
  melanospora Engelm.
  nuttallii A. Braun ex Engelm.
  orcuttii A.A. Eat.
  piedmontana (N.E. Pfeiffer) C.F. Reed
  riparia Engelm. ex A. Braun
      var. amesii (A.A. Eat.) Proctor
        SY=I. saccharata Engelm. var.
            amesii A.A. Eat.
      var. canadensis Engelm.
        SY=I. dodgei A.A. Eat.
        SY=I. r. var. robbinsii (A.A.
            Eat.) Proctor
      var. palmeri (A.A. Eat.) Proctor
      var. reticulata (A.A. Eat.) Proctor
      var. riparia
  saccharata Engelm.
  tegetiformans Rury
  truncata (A.A. Eat.) Clute
  tuckermanii A. Braun
  virginica N.E. Pfeiffer

LOPHOSORIACEAE

Lophosoria Presl

  quadripinnata (Gmel.) C. Christens.
        SY=Alsophila q. (Gmel.) C.
            Christens.

LYCOPODIACEAE

Lycopodium L.

    alopecuroides L.
            SY=Lepidotis a. (L.) Rothm.
            SY=Lycopodium inundatum L. var. a.
                (L.) Tuckerman
    alpinum L.
            SY=Diphasium a. (L.) Rothm.
            SY=Diphasiastrum a. (L.) Holub
    annotinum L.
        var. acrifolium Fern.
        var. alpestre Hartman
            SY=Lycopodium annotinum ssp.
                alpestre (Hartman) Love &
                Love
        var. annotinum
        var. pungens (La Pylaie) Desv.
            SY=Lycopodium a. ssp. p. (La
                Pylaie) Hulten
            SY=L. dubium Zoega
    appressum (Chapman) Lloyd & Underwood
            SY=Lycopodium adpressum (Chapman)
                Lloyd & Underwood
            SY=L. alopecuroides L. ssp.
                appressum (Chapman) Clute
            SY=L. alopecuroides var. adpressum
                Chapman
            SY=L. chapmanii Underwood ex Maxon
                pro parte
            SY=L. inundatum L. var. appressum
                Chapman
            SY=L. i. var. bigelovii Tuckerman
    aqualupianum Spring
            SY=Urostachys a. (Spring) Herter
    X buttersii Abbe [lucidulum X selago]
    carolinianum L.
            SY=Lycopodiella c. (L.) Pichi
                Sermolli
    cernuum L.
        var. cernuum
            SY=Lepidotis c. (L.) Beauv.
            SY=Lycopodiella c. (L.) Pichi
                Sermolli
            SY=Palhinhaea c. (L.) Franco &
                Vasc.
        var. crassifolium Spring
    clavatum L.
        var. brevispicatum Peck
        var. clavatum
            SY=Lycopodium c. var. laurentianum
                Victorin
            SY=L. c. var. subremotum Victorin
            SY=L. c. var. tristachyum Hook.
        var. integerrimum Spring
        var. monostachyon Grev. & Hook.
            SY=Lycopodium c. ssp. megastachyon
                (Grev. & Hook.) Seland.
            SY=L. c. var. m. Fern. & Bissell
            SY=L. lagopus (Laestad.) Zinserl.
                ex Kuzen
    complanatum L.
            SY=Diphasiastrum c. (L.) Holub
            SY=Diphasium anceps (Wallr.) Love
                & Love
            SY=Diphasium c. (L.) Rothm.
            SY=Diphasium wallrothii H.P. Fuchs
            SY=Lycopodium a. Wallr.
            SY=L. c. ssp. a. (Wallr.) Aschers.
            SY=L. c. var. canadense Victorin
    X copelandii Eig. [alopecuroides X
        appressum]
            SY=Lycopodium chapmanii Underwood
                ex Maxon
            SY=L. inundatum L. var. elongatum
                Chapman
    curvatum Sw.
    dendroideum Michx.

            SY=Lycopodium obscurum L. var. d.
                (Michx.) D.C. Eat.
            SY=L. o. var. hybridum Farw.
    dichotomum Jacq.
            SY=Huperza dichotoma (Jacq.)
                Trevisan
            SY=Urostachys d. (Jacq.) Herter
    digitatum A. Braun
            SY=Diphasiastrum d. (A. Braun)
                Holub
            SY=Lycopodium complanatum L. var.
                flabelliforme Fern.
            SY=L. f. (Fern.) Blanch.
            SY=L. f. var. ambiguum Victorin
    erubescens Brack.
            SY=Urostachys e. (Brack.) Herter
                ex Nessel
    funiforme Bory
            SY=Huperzia f. (Bory) Trevisan
            SY=Urostachys f. (Bory) Herter
    X habereri House [digitatum X tristachyum]
            SY=Diphasiastrum X h. (House)
                Holub
            SY=Lycopodium complanatum L. var.
                h. (House) Boivin
            SY=L. tristachyum Pursh var. h.
                (House) Victorin
            SY=L. t. var. laurentianum
                Victorin
    haleakalae Brack.
            SY=Urostachys h. (Brack.) Herter
    inundatum L.
            SY=Lepidotis i. (L.) C. Borner
            SY=Lycopodiella i. (L.) Holub
    X issleri (Rouy) Domin [alpinum X
        complanatum]
            SY=Diphasiastrum X i. (Rouy) Holub
            SY=Diphasium i. (Rouy) Holub
            SY=Lycopodium alpinum L. ssp. i.
                (Rouy) Chassagne
            SY=L. a. var. decipiens Syme ex
                Druce
            SY=L. complanatum L. ssp. i.
                (Rouy) Domin
            SY=L. c. var. pseudo-alpinum Farw.
    linifolium L.
            SY=Huperzia l. (L.) Trevisan
            SY=Urostachys l. (L.) Herter
    lucidulum Michx.
        var. lucidulum
            SY=Huperzia l. (Michx.) Trevisan
            SY=Lycopodium reflexum Sw. non
                Lam.
            SY=Urostachys l. (Michx.) Herter
        var. tryonii Mohlenbrock
            SY=Lycopodium l. var. occidentale
                (Clute) L.R. Wilson
    mannii (Hbd.) Skottsberg
    meridionale Underwood & Lloyd
    nutans Brack.
            SY=Urostachys n. (Brack.) Herter
                ex Nessel
    obscurum L.
        var. isophyllum Hickey
        var. obscurum
    phyllanthum Hook. & Arn.
            SY=Urostachys p. (Hook. & Arn.)
                Herter
    polytrichoides Kaulfuss
            SY=Huperzia p. (Kaulfuss) Trevisan
            SY=Urostachys p. (Kaulfuss) Herter
                ex Nessel
    porophilum Lloyd & Underwood
            SY=Huperzia selago (L.) Bernh. ex
                Schrank & Mart. ssp.
                lucidula var. p. (Lloyd &
                Underwood) Love &Love

Lycopodium   (CONT.)
        SY=Lycopodium l. Michx. var. p.
            (Lloyd & Underwood) Robins.
            & Fern.
        SY=L. s. L. var. p. (Lloyd &
            Underwood) Clute
        SY=Urostachys lucidulus (Michx.)
            Herter var. p. (Lloyd &
            Underwood) Nessel non Lam.
        SY=U. p. (Lloyd & Underwood)
            Herter
   portoricense Underwood & Lloyd
        SY=Urostachys p. (Underwood &
            Lloyd) Herter
   prostratum Harper
        SY=Lycopodium alopecuroides L.
            var. pinnatum Chapman
        SY=L. inundatum L. var. pinnatum
            Chapman
        SY=L. pinnatum (Chapman) Lloyd &
            Underwood
   reflexum Lam.
        SY=Huperzia r. (Lam.) Trevisan
        SY=Urostachys r. (Lam.) Herter
   sabinifolium Willd.
        SY=Diphasiastrum s. (Willd.) Holub
        SY=Lycopodium armatum Desv.
        SY=L. s. var. patens Victorin
        SY=L. s. var. sharonense Blake
        SY=L. sabinifolium var.
            superfertile Victorin
   selago L.
     var. appressum Desv.
        SY=Huperzia s. (L.) Bernh. ex
            Schrank & Mart. ssp. arctica
            (Grossh.) Love & Love
        SY=H. s. var. densa Trevisan
        SY=Lycopodium appressum Desv.
        SY=L. arcticum Grossh.
        SY=L. s. ssp. appressum (Desv.)
            Hulten
        SY=L. s. ssp. arcticum (Grossh.)
            Tolm.
     var. miyoshianum (Makino) Makino
        SY=Huperzia s. (L.) Bernh. ex
            Shrank & Mart. ssp.
            chinensis (Christ) Love &
            Love
        SY=H. s. var. m. (Makino) Taylor &
            MacBryde
        SY=Lycopodium c. Christ
        SY=L. m. Makino
        SY=L. s. ssp. c. (Christ) Hulten
        SY=L. s. ssp. m. (Makino) Calder &
            Taylor
        SY=L. s. var. c. (Christ) Taylor &
            MacBryde
        SY=Urostachys c. (Christ) Herter
            ex Nessel
        SY=U. m. (Makino) Herter ex Nessel
     var. selago
        SY=Huperzia s. (L.) Bernh. ex
            Schrank & Mart.
        SY=H. s. var. patens (Beauv.)
            Trevisan
        SY=Lycopodium s. ssp. p. (Beauv.)
            Calder & Taylor
        SY=L. s. var. p. (Beauv.) Desv.
        SY=Urostachys s. (L.) Herter
   serratum Thunb.
        SY=Huperzia selago (L.) Bernh. ex
            Schrank & Mart. ssp. serrata
            (Thunb.) Love & Love
        SY=H. serrata (Thunb.) Trevisan
        SY=H. sulcinervia (Spring)
            Trevisan
        SY=Lycopodium helleri Herter

        SY=L. sulcinervium Spring
        SY=Urostachys h. (Herter) Herter
            ex Nessel
        SY=U. serratus (Thunb.) Herter
        SY=U. sulcinervius (Spring) Herter
            ex Nessel
   sintenisii (Herter) Maxon
        SY=Urostachys s. Herter
   sitchense Rupr.
        SY=Diphasiastrum s. (Rupr.) Holub
        SY=Diphasium s. (Rupr.) Love &
            Love
        SY=Lycopodium sabinifolium Willd.
            ssp. sitchense (Rupr.)
            Calder & Taylor
        SY=L. sabinifolium var. sitchense
            (Rupr.) Fern.
   taxifolium Sw.
        SY=Phlegmariurus t. (Sw.) Love &
            Love
        SY=Urostachys t. (Sw.) Herter
   tenuicaule Underwood & Lloyd
        SY=Urostachys tenuicaulis
            (Underwood & Lloyd) Herter
   tristachyum Pursh
        SY=Diphasiastrum t. (Pursh) Holub
        SY=Diphasium chamicyparissus (A.
            Braun) Love & Love
        SY=Diphasium complanatum (L.)
            Rothm. ssp. chamicyparissus
            (A. Braun) Kukk.
        SY=Diphasium tristachyum (Pursh)
            Rothm.
        SY=Lycopodium chamicyparissus A.
            Braun
        SY=L. complanatum L. ssp.
            chamicyparissus (A. Braun)
            Nyman
        SY=L. complanatum var.
            chamicyparissus (A. Braun)
            Doell
        SY=L. complanatum var.
            patentifolium Spring
   venustulum Gaud.
   verticillatum L. f.
        SY=Huperzia v. (L. f.) Trevisan
        SY=Lycopodium setaceum Lam.
        SY=Urostachys v. (L. f.) Herter
   wilsonii Underwood & Lloyd
        SY=Urostachys w. (Underwood &
            Lloyd) Herter
   X zeilleri (Rouy) Victorin [complanatum X
    tristachyum]
        SY=Diphasiastrum X z. (Rouy) Holub
        SY=Diphasium z. (Rouy) Damboldt
        SY=Lycopodium complanatum L. var.
            elongatum Victorin
        SY=L. c. var. gartonis Boivin
        SY=L. tristachyum Pursh var.
            boreale Victorin

MARATTIACEAE

Angiopteris Hoffmann

   evecta (Forst.) Hoffmann

Danaea Sm.

   elliptica Sm.
   jenmanii Underwood
   nodosa (L.) Sm.
   urbanii Maxon

Marattia Sw.

    douglasii (Presl) Baker

MARSILEACEAE

Marsilea L.

    berteroi A. Braun
    macropoda Engelm. ex A. Braun
    mexicana A. Braun
    polycarpa Hook. & Grev.
    quadrifolia L.
    tenuifolia Engelm. ex Kunze
    uncinata A. Braun
    vestita Hook. & Grev.
          SY=M. mucronata A. Braun
          SY=M. oligospora Goodding
    villosa Kaulfuss

Pilularia L.

    americana A. Braun

OPHIOGLOSSACEAE

Botrychium Sw.

    alabamense Maxon
    biternatum (Savaiter) Underwood
        SY=B. dissectum Spreng. var.
            tenuifolium (Underwood)
            Farw.
    boreale (Fries) Milde
        SY=B. b. var. obtusilobum (Rupr.)
            Broun
        SY=B. b. ssp. o. (Rupr.) Clausen
        SY=B. pinnatum St. John
    dissectum Spreng.
        SY=B. d. var. obliquum (Muhl.)
            Clute
        SY=B. d. var. oblongifolium
            (Graves) Broun
        SY=B. obliquum Muhl.
        SY=B. obliquum var. elongatum
            Gilbert & Haberer
    dusenii (Christ) Alston
    jenmanii Underwood
    lanceolatum (Gmel.) Angstr.
      var. angustisegmentum Pease & Moore
        SY=B. a. (Pease & Moore) Fern.
      var. lanceolatum
        SY=B. l. ssp. typicum Clausen
    lunaria (L.) Sw.
      var. lunaria
      var. minganense (Victorin) Dole
        SY=B. l. ssp. m. (Victorin) Calder
           & Taylor
        SY=B. l. ssp. occidentalis Love,
           Love & Kapoor
        SY=B. m. Victorin
      var. onondagense (Underwood) House
        SY=B. o. Underwood
    lunarioides (Michx.) Sw.
    matricariifolium (A. Braun ex Doll) A.
      Braun ex Koch
      var. hesperium (Maxon & Clausen)
        Broun
        SY=B. m. ssp. h. Maxon & Clausen
      var. matricariifolium
    multifidum (Gmel.) Rupr.

      var. californicum (Underwood) Broun
        SY=B. c. Underwood
        SY=B. m. ssp. c. (Underwood)
            Clausen
      var. coulteri (Underwood) Broun
        SY=B. m. ssp. c. (Underwood)
            Clausen
        SY=B. silaifolium Presl var. c.
            (Underwood) Jepson
      var. intermedium (D.C. Eat.) Farw.
        SY=B. m. ssp. silaifolium (Presl)
            Clausen
        SY=B. m. var. s. (Presl) Broun
        SY=B. s. Presl
      var. multifidum
        SY=B. matricariae (Schrank)
            Spreng.
      var. robustum (Rupr.) C. Christens.
    oneidense (Gilbert) House
        SY=B. dissectum Spreng. var. o.
            (Gilbert) Clute
        SY=B. multifidum (Gmel.) Rupr.
            var. o. (Gilbert) Farw.
    pumicola Coville
    simplex E. Hitchc.
      var. compositum (Lasch) Milde
      var. laxifolium (Clausen) Fern.
      var. simplex
        SY=B. s. ssp. typicum Clausen
      var. tenebrosum (A.A. Eat.) Clausen
        SY=B. t. A.A. Eat.
    subbifoliatum Brack.
    ternatum (Thunb.) Sw.
    virginianum (L.) Sw.
      var. europaeum Angstr.
        SY=B. v. ssp. e. (Angstr.) Jav.
      var. virginianum

Ophioglossum L.

    concinnum Brack.
    crotalophoroides Walt.
      var. crotalophoroides
      var. nanum Osten ex de Lichtenstein
    dendroneuron E. St. John
    engelmannii Prantl
    lusitanicum L.
        SY=O. californicum Prantl
        SY=O. l. var. c. (Prantl) Broun
    nudicaul; L.f.
      var. minus Clausen
      var. nudicaule
      var. tenerum (Mett. ex Prantl)
        Clausen
        SY=O. mononeuron E. St. John
        SY=O. t. Mett. ex Prantl.
    palmatum L.
        SY=Cheiroglossa p. (L.) Presl
    pendulum L.
      ssp. falcatum (Presl) Clausen
    petiolatum Hook.
    reticulatum L.
    vulgatum L.
      var. pseudopodum (Blake) Farw.
      var. pycnostichum Fern.
      var. vulgatum
        SY=O. v. var. alaskanum (E.G.
           Britt.) C. Christens.

OSMUNDACEAE

Osmunda L.

    cinnamomea L.

Osmunda (CONT.)
        var. cinnamomea
            SY=O. c. var. frondosa Gray
        var. glandulosa Waters
        var. imbricata (Kunze) Milde
    claytoniana L.
    regalis L.
        var. regalis
        var. spectabilis (Willd.) Gray
    X ruggii R. Tryon [claytoniana X regalis
        var. spectabilis]

PARKERIACEAE

Ceratopteris Brongn.

    pteridoides (Hook.) Hieron.
    richardii Brongn.
            SY=C. deltoidea Benedict
    thalictroides (L.) Brongn.
            SY=C. siliquosa (L.) Copeland

POLYPODIACEAE

Neurodium Fee

    lanceolatum (L.) Fee
            SY=Paltonium l. (L.) Presl

Polypodium L.

    amorphum Suksdorf
            SY=P. montense F.A. Lang
    angustifolium Sw.
            SY=Campyloneurum a. (Sw.) Fee
    astrolepis Liebm.
    attenuatum Humb. & Bonpl. ex Willd.
    aureum L.
            SY=Phlebodium a. (L.) J. Sm.
    australe Fee
    californicum Kaulfuss
        var. californicum
            SY=Polypodium vulgare L. var.
                intermedium (Hook. & Arn.)
                Fern.
        var. kaulfussii D.C. Eat.
    chnoodes Spreng.
    costatum Kunze
            SY=Campyloneurum c. (Kunze) Presl
    crassifolium L.
    decumanum Willd.
    dispersum A.M. Evans
    dissimile L.
    erythrolepis Weatherby
    glycyrrhiza D.C. Eat.
            SY=Polypodium vulgare L. ssp.
                occidentale (Hook.) Hulten
            SY=?P. v. var. commune Milde
            SY=P. v. var. o. Hook.
    hesperium Maxon
            SY=Polypodium vulgare L. ssp.
                columbianum (Gilbert) Hulten
            SY=P. v. var. c. Gilbert
            SY=P. v. var. h. (Maxon) A. Nels.
                & J.F. Macbr.
    heterophyllum L.
            SY=Microgramma h. (L.) Wherry
    latum (T. Moore) Sodiro
            SY=Campyloneurum l. T. Moore
            SY=P. phyllitidis L. var. l. (T.
                Moore) Proctor

    loriceum L.
    lycopodioides L.
    pectinatum L.
    pellucidum Kaulfuss
        var. pellucidum
        var. vulcanicum Skottsberg
    phyllitidis L.
            SY=Campyloneurum p. (L.) Presl
    piloselloides L.
    plumula Humb. & Bonpl. ex Willd.
    polypodioides (L.) Watt
        var. michauxianum Weatherby
        var. polypodioides
            SY=Marginaria p. (L.) Tidestrom
    ptilodon Kunze
        var. caespitosum (Jenman) A.M. Evans
    scolopendria Burm. f.
            SY=Polypodium phymatodes L.
            SY=Phymatodes s. (Burm. f.) Ching
    scouleri Hook. & Grev.
    spectrum Kaulfuss
    squamatum L.
    thunbergiana (Kaulfuss) C. Christens.
            SY=Pleopeltis t. Kaulfuss
            SY=Polypodium lineare Thunb.
    thyssanolepis A. Braun ex Klotzsch
    triseriale Sw.
            SY=Goniophlebium t. (Sw.) Wherry
    virginianum L.
            SY=Polypodium vulgare sensu Amer.
                auctt. pro parte non L.
            SY=P. vulgare var. virginianum
                (L.) Eat.

PSILOTACEAE

Psilotum Sw.

    complanatum Sw.
    nudum (L.) Beauv.

SALVINIACEAE

Salvinia Seguier

    minima Baker
            SY=S. rotundifolia sensu auctt.
                non Willd.
    natans (L.) All.

SCHIZAEACEAE

Anemia Sw.

    adiantifolia (L.) Sw.
    cicutaria Kunze
    hirsuta (L.) Sw.
    hirta (L.) Sw.
    mexicana Klotzsch
    portoricensis Maxon
    wrightii Baker

Lygodium Sw.

    japonicum (Thunb.) Sw.
    microphyllum (Cav.) R. Br.
    palmatum (Bernh.) Sw.

Schizaea Sm.

    germanii (Fee) Prantl
            SY=Actinostachys g. Fee
    pennula Sw.
            SY=Actinostachys p. (Sw.) Hook.
    poeppigiana Sturm
            SY=Lophidium p. (Sturm) Underwood
    pusilla Pursh
    robusta Baker

SELAGINELLACEAE

Selaginella Beauv.

    apoda (L.) Fern.
            SY=S. apus Spring
    arbuscula (Kaulfuss) Spring
    arenicola Underwood
        ssp. acanthonota (Underwood) R. Tryon
            SY=S. a. Underwood
        ssp. arenicola
        ssp. riddellii (Van Eselt.) R. Tryon
            SY=S. a. var. r. (Van Eselt.)
                Waterfall
            SY=S. r. Van Eselt.
    arizonica Maxon
    armata Baker
            SY=S. plagiochila sensu Krug non
                Baker
    asprella Maxon
    bigelovii Underwood
    braunii Baker
            SY=S. plana (Desv.) Hieron.
    cinerascens A.A. Eat.
    cordifolia (Desv.) Spring
    deflexa Brack.
    densa Rydb.
        var. densa
            SY=S. engelmannii Hieron.
        var. scopulorum (Maxon) R. Tryon
            SY=S. engelmanii Hieron. var. s.
                (Maxon) C.F. Reed
            SY=S. s. Maxon
        var. standleyi (Maxon) R. Tryon
            SY=S. engelmannii Hieron. var. s.
                (Maxon) C.F. Reed
            SY=S. s. Maxon
    douglasii (Hook. & Grev.) Spring
    eatonii Hieron.
    eclipes Buck
    eremophila Maxon
    flabellata (L.) Spring
    hansenii Hieron.
    kraussiana (Kunze) A. Braun
    krugii Hieron.
    laxifolia Baker
    lepidophylla (Hook. & Grev.) Spring
    leucobryoides Maxon
    ludoviciana A. Braun
    menziesii (Hook. & Grev.) Spring
    mutica D.C. Eat. ex Underwood
        var. limitanea Weatherby
            SY=S. m. var. texana Weatherby
        var. mutica
    X neomexicana Maxon [mutica X rupincola]
    oregana D.C. Eat.
    ovifolia Baker
    parvula Hbd.
    peruviana (Milde) Hieron.
            SY=S. sheldonii Maxon
    pilifera A. Braun
            SY=S. p. var. pringlei (Baker)
                Morton

    plumosa (L.) Presl
            SY=S. stolonifera (Sw.) Spring
    rupestris (L.) Spring
    rupincola Underwood
    selaginoides (L.) Link
    sibirica (Milde) Hieron.
    stipitata Spring
    subcaulescens Baker
            SY=S. sintenisii Hieron.
    substipitata Spring
            SY=S. portoricensis A. Braun
    tenella (Beauv.) Spring
            SY=S. albonitens Spring
    tortipila A. Braun
    uncinata (Desv. ex Poir.) Spring
    underwoodii Hieron.
            SY=S. u. var. dolichotricha
                Weatherby
    utahensis Flowers
    viridissima Weatherby
    wallacei Hieron.
    watsonii Underwood
    weatherbiana R. Tryon
    willdenovii (Desv. ex Poir.) Baker
    wrightii Hieron.

Gymnospermae

CUPRESSACEAE

Callitris Vent.

    hugelii (Carr.) Franco
    robusta R. Br. ex Mirbel

Calocedrus Kurz

    decurrens (Torr.) Florin
        SY=Heyderia d. (Torr.) K. Koch
        SY=Libocedrus d. Torr.

Chamaecyparis Spach

    lawsoniana (A. Murr.) Parl.
    nootkatensis (D. Don) Spach
    thyoides (L.) B.S.P.
        SY=C. t. var. henryae (Li) Little

Cupressus L.

    arizonica Greene
        var. arizonica
        var. glabra (Sudworth) Little
          SY=C. g. Sudworth
        var. nevadensis (Abrams) Little
          SY=C. macnabiana A. Murr. var. n.
              (Abrams) Abrams
          SY=C. n. Abrams
        var. stephensonii (C.B. Wolf) Little
          SY=C. s. C.B. Wolf
    bakeri Jepson
        SY=C. b. ssp. matthewsii C.B. Wolf
    goveniana Gord.
        var. abramsiana (C.B. Wolf) Little
          SY=C. a. C.B. Wolf
        var. goveniana
        var. pigmaea Lemmon
          SY=C. p. (Lemmon) Sarg.
    guadalupensis S. Wats
        var. forbesii (Jepson) Little
          SY=C. f. Jepson
          SY=C. g. ssp. f. (Jepson)
              Beauchamp
    macnabiana A. Murr.
    macrocarpa Hartw. ex Gord.
    sargentii Jepson

Juniperus L.

    ashei Buchh.
    californica Carr.
        SY=J. c. var. siskiyouensis
            Henderson
        SY=Sabina c. (Carr.) Antoine
    communis L.
        SY=J. canadensis Burgsd.
        SY=J. communis ssp. alpina
          (Neilr.) Celak.
        SY=J. communis ssp. depressa
          (Pursh) Franco
        SY=J. communis ssp. nana (Willd.)
          Syme
        SY=J. communis var. d. Pursh
        SY=J. communis var. jackii Rehd.
        SY=J. communis var. megistocarpa
          Fern. & St. John
        SY=J. communis var. montana Ait.
        SY=J. communis var. saxatilis
          Pallas
        SY=J. sibirica Burgsd.

deppeana Steud.
    var. deppeana
        SY=J. d. var. sperryi Correll
        SY=J. mexicana Schiede & Deppe
    var. pachyphloea (Torr.) Martinez
erythrocarpa Cory
        SY=J. texensis Van Melle
X fassettii Boivin [horizontalis X
  scopulorum]
        SY=J. scopulorum Sarg. var. patens
          Fassett
flaccida Schlecht.
horizontalis Moench
    var. douglasii Hort.
    var. horizontalis
    var. variegata Beissn.
monosperma (Engelm.) Sarg.
    var. gracilis Martinez
    var. monosperma
occidentalis Hook.
    var. australis (Vasek) A. Holmgren &
        N. Holmgren
        SY=J. o. ssp. a. Vasek
    var. occidentalis
        SY=Sabina o. (Hook.) Heller
osteosperma (Torr.) Little
        SY=J. utahensis (Engelm.) Lemmon
        SY=Sabina u. (Engelm.) Antoine
pinchotii Sudworth
        SY=J. monosperma (Engelm.) Sarg.
          var. p. (Sudworth) Van Melle
scopulorum Sarg.
        SY=Sabina s. (Sarg.) Rydb.
silicicola (Small) Bailey
        SY=J. lucayana Britt.
        SY=Sabina s. Small
virginiana L.
        SY=J. v. var. crebra Fern. &
          Grisc.
        SY=Sabina v. (L.) Antoine

Thuja L.

    occidentalis L.
    orientalis L.
        SY=Biota o. (L.) Endl.
        SY=Platycladus o. (L.) Franco
    plicata Donn ex D. Don

CYCADACEAE

Zamia L.

    angustifolia Jacq.
    debilis Ait.
        SY=Z. media Jacq.
    integrifolia Ait.
    latifoliolata Preneloup
    portoricensis Urban
    pumila L.
        SY=Z. silvicola Small
        SY=Z. umbrosa Small

EPHEDRACEAE

Ephedra L.

Ephedra (CONT.)
   antisyphilitica Berl. ex C.A. Mey.
         SY=E. a. var. brachycarpa Cory
   X arenicola Cutler [cutleri X torreyana]
   californica S. Wats.
      var. californica
      var. funerea (Coville & Morton) L.
        Benson
         SY=E. f. Coville & Morton
   coryi E.L. Reed
   cutleri Peebles
   fasciculata
      var. clokeyi (Cutler) Clokey
         SY=E. c. Cutler
      var. fasciculata A. Nels.
   X intermixta Cutler [torreyana X trifurca]
   nevadensis S. Wats.
      var. aspera (Engelm. ex S. Wats.) L.
        Benson
         SY=E. a. Engelm. ex S. Wats.
         SY=E. reedii Cory
      var. nevadensis
   pedunculata Engelm. ex S. Wats.
   torreyana S. Wats.
   trifurca Torr.
   viridis Coville
      var. viridis
      var. viscida (Cutler) L. Benson

PINACEAE

Abies P. Mill.

   amabilis (Dougl. ex Loud.) Dougl. ex Forbes
   balsamea (L.) P. Mill.
        SY=Pinus b. L.
   bracteata (D. Don) D. Don ex Poit.
        SY=A. venusta (Dougl.) K. Koch
   concolor (Gord. & Glend.) Lindl. ex
     Hildebr.
      var. concolor
      var. lowiana (Gord. & Glend.) Lemmon
   fraseri (Pursh) Poir.
        SY=Pinus f. Pursh
   grandis (Dougl. ex D. Don) Lindl.
   lasiocarpa (Hook.) Nutt.
      var. arizonica (Merriam) Lemmon
        SY=A. a. Merriam
      var. lasiocarpa
        SY=A. balsamea (L.) P. Mill. ssp.
         l. (Hook.) Boivin
        SY=A. b. var. fallax (Engelm.)
         Boivin
   magnifica A. Murr.
   X phanerolepis (Fern.) Liu [balsamea X
     fraseri]
        SY=A. balsamea (L.) P. Mill. var.
         p. Fern.
   procera Rehd.
        SY=A. nobilis (Dougl.) Lindl.
   X shastensis Lemmon emend. Liu [magnifica X
     procera]
        SY=A. magnifica A. Murr. var. s.
         Lemmon

Larix P. Mill.

   decidua P. Mill.
   laricina (Du Roi) K. Koch
        SY=L. l. var. alaskensis (W.
         Wight) Raup
   lyallii Parl.
   occidentalis Nutt.

Picea A. Dietr.

   abies (L.) Karst.
   breweriana S. Wats.
   engelmannii Parry ex Engelm.
        SY=P. glauca (Moench) Voss ssp. e.
         (Parry ex Engelm.) T.M.C.
         Taylor
        SY=P. g. var. e. (Parry ex
         Engelm.) Boivin
   glauca (Moench) Voss
        SY=P. canadensis (P. Mill.) B.S.P.
        SY=P. g. var. albertiana (S. Br.)
         Sarg.
        SY=P. g. var. porsildii Raup
   X lutzii Little [glauca X mariana]
   mariana (P. Mill.) B.S.P.
      var. mariana
      var. semiprostrata (Peck) Teeri
   pungens Engelm.
   rubens Sarg.
        SY=P. australis Small
   sitchensis (Bong.) Carr.

Pinus L.

   albicaulis Engelm.
   aristata Engelm.
      var. aristata
      var. longaeve (D.K. Bailey) Little
        SY=P. l. D.K. Bailey
   attenuata Lemmon
   X attenuradiata Stockwell & Righter
     [attenuata X radiata]
   balfouriana Grev. & Balf.
   banksiana Lamb.
        SY=P. divaricata (Ait.)
         Dum.-Cours.
        SY=P. d. var. X muscii Boivin
   cembroides Zucc.
   clausa (Chapm. ex Engelm.) Vasey ex Sarg.
   contorta Dougl. ex Loud.
      var. bolanderi (Parl.) Vasey
      var. contorta
      var. latifolia Engelm.
        SY=P. c. ssp. l. (Engelm.)
         Critchfield
        SY=P. divaricata (Ait.)
         Dum.-Cours. var.
         hendersonii (Lemmon) Boivin
        SY=P. d. var. l. (Engelm.) Boivin
      var. murrayana (Grev. & Balf.)
        Engelm.
        SY=P. c. ssp. m. (Grev. & Balf.)
         Critchfield
        SY=P. m. Grev. & Balf.
   coulteri D. Don
   discolor D.K. Bailey
        SY=P. cembroides Zucc. var.
         bicolor Little
   echinata P. Mill.
   edulis Engelm.
   elliottii Engelm.
      var. densa Little & Dorman
        SY=P. d. (Little & Dorman) Gaussen
      var. elliottii
        SY=P. caribaea sensu Small non
         Morelet
        SY=P. heterophylla (Ell.) Sudworth
   engelmannii Carr.
        SY=P. latifolia Sarg.
   flexilis James
   glabra Walt.
   jeffreyi Grev. & Balf.
   lambertiana Dougl.
   leiophylla Schiede & Deppe
      var. chichuahuana (Engelm.) Shaw

Pinus (CONT.)
    monophylla Torr. & Frem.
    monticola Dougl. ex D. Don
            SY=P. strobus L. var. m. (Dougl.
                ex D. Don) Nutt.
    mugo Turra
    muricata D. Don
            SY=P. remorata Mason
    X murraybanksiana Righter & Stockwell
      [banksiana X contorta]
    nigra Arnold
            SY=P. n. var. austriaca (Hoess.)
                Aschers. & Graebn.
    palustris P. Mill.
            SY=P. australis Michx. f.
    ponderosa Dougl. ex P. & C. Lawson
        var. arizonica (Engelm.) Shaw
            SY=P. a. Engelm.
        var. ponderosa
            SY=P. beardsleyi A. Murr.
            SY=P. benthamiana Hartw.
            SY=P. brachyptera Engelm.
        var. scopulorum Engelm.
    pungens Lamb.
    quadrifolia Parl. ex Sudworth
            SY=P. juarezensis Lanner
    radiata D. Don
            SY=P. r. var. binata (S. Wats.)
                Lemmon
    remota (Little) D.K. Bailey
            SY=P. cembroides Zucc. var. r.
                Little
    resinosa Ait.
    rigida P. Mill.
    X rigitaeda [rigida X taeda]
    sabiniana Dougl.
    serotina Dougl.
    X sondereggeri H.H. Chapman [palustris X
      taeda]
    strobiformis Engelm.
            SY=P. ayacahuite C.A. Ehrenb.
            SY=P. flexilis James var. reflexa
                Engelm.
            SY=P. r. (Engelm.) Engelm.
    strobus L.
            SY=Strobus s. (L.) Small
    sylvestris L.
    taeda L.
    torreyana Parry ex Carr.
    virginiana P. Mill.
    washoensis Mason & Stockwell

Pseudotsuga Carr.

    macrocarpa (Vasey) Mayr
    menziesii (Mirbel) Franco
        var. glauca (Beissn.) Franco
            SY=P. taxifolia (Lamb.) Britt.
                    var. g. (Beissn.) Sudworth
        var. menziesii
            SY=P. taxifolia (Lamb.) Britt.

Tsuga Carr.

    canadensis (L.) Carr.
    caroliniana Engelm.
    heterophylla (Raf.) Sarg.
    mertensiana (Bong.) Carr.

TAXACEAE

Taxus L.

    brevifolia Nutt.

canadensis Marsh.
floridana Nutt. ex Chapman

Torreya Arn.

    californica Torr.
    taxifolia Arn.
            SY=Tumion t. (Arn.) Greene

TAXODIACEAE

Cryptomeria D. Don

    japonica (L. f.) D. Don

Sequoia Endl.

    sempervirens (Lamb. ex D. Don) Endl.

Sequoiadendron Buchh.

    giganteum (Lindl.) Buchh.
            SY=Sequoia g. (Lindl.) Dcne.
            SY=Sequoia wellingtonia Seem.

Taxodium L.C. Rich.

    ascendens Brongn.
            SY=T. distichum (L.) L.C. Rich.
                var. nutans (Ait.) Sweet
    distichum (L.) L.C. Rich.
    mucronatum Ten.

Angiospermae

ACANTHACEAE

Andrographis Wallich ex Nees

    echinoides Nees

Anisacanthus Nees

    insignis Gray
          SY=A. i. var. linearis Hagen
          SY=Drejera puberula Torr.
    thurberi (Torr.) Gray
          SY=Drejera t. Torr.
    wrightii (Torr.) Gray
          SY=A. junceus (Torr.) Hemsl.
          SY=Drejera j. Torr.
          SY=D. w. Torr.

Asystasia Blume

    gangetica (L.) T. Anders.
          SY=A. coromandeliana Nees
          SY=Justicia g. L.

Barleria L.

    cristata L.
    lupulina Lindl.
    nitida Jacq.
    prionitis L.

Berginia Harvey ex Benth. & Hook.

    virgata Harvey ex Benth. & Hook.

Blechum P. Br. ex Juss.

    brownei Juss.
          SY=B. blechnum (L.) Millsp.
          SY=B. pyramidatum (Lam.) Urban

Carlowrightia Gray

    arizonica Gray
    linearifolia (Torr.) Gray
          SY=Schaueria l. Torr.
    mexicana Henrickson & Daniel
    parviflora (Buckl.) Wasshausen
          SY=Dianthera p. (Buckl.) Gray
          SY=Drejera p. (Buckl.) Gray
          SY=Justicia p. (Buckl.) Gray
    serphyllifolia Gray
    texana Henrickson & Daniel
    torreyana Wasshausen
          SY=C. pubens Gray
          SY=Croftia parvifolia (Torr.) Gray
          SY=Dianthera parvifolia (Torr.)
              Gray
          SY=Schaueria parvifolia Torr.

Dicliptera Juss.

    assurgens (L.) Juss.
        var. assurgens
          SY=Diapedium a. (L.) Kuntze
          SY=Justicia a. L.
        var. vahliana (Nees) Gomez
          SY=Dicliptera v. Nees
    brachiata (Pursh) Spreng.
          SY=Diapedium b. (Pursh) Kuntze
          SY=Dicliptera b. var. ruthii Fern.
          SY=Dicliptera b. var. attenuata
              Gray
          SY=Dicliptera b. var. glandulosa
              (Scheele) Fern.
          SY=Dicliptera g. Scheele
          SY=Justicia b. Pursh
    krugii Urban
          SY=Diapedium k. (Urban) Britt.
    pseudoverticillaris Gray
    resupinata (Vahl) Juss.
          SY=Diapedium r. (Vahl) Kuntze
          SY=Justicia r. Vahl
    torryi Gray
          SY=Diapedium t. (Gray) Woot. &
              Standl.
    viridiflora (Nees) R.W. Long
          SY=Rhytiglossa v. Nees

Dyschoriste Nees

    angusta (Gray) Small
          SY=Calophanes a. Gray
          SY=D. oblongifolia (Michx.) Kuntze
              var. a. (Gray) R.W. Long
    crenulata Kobuski
    decumbens (Gray) Kuntze
          SY=Calophanes d. Gray
    humistrata (Michx.) Kuntze
          SY=Apassalus h. (Michx.) Kobuski
          SY=Calophanes h. Shuttlw. ex Nees
          SY=Ruellia h. Michx.
    linearis (Torr. & Gray) Kuntze
          SY=Calophanes l. (Torr. & Gray)
              Gray
          SY=Dipteracanthus l. Torr. & Gray
    oblongifolia (Michx.) Kuntze
          SY=Ruellia o. Michx.

Elytraria L.C. Rich. ex Michx.

    bromoides Oerst.
          SY=Tubiflora acuminata Small
    caroliniensis (J.F. Gmel.) Pers.
        var. angustifolia (Fern.) Blake
          SY=E. virgata Michx. var. a. Fern.
          SY=Tubiflora a. (Fern.) Small
        var. caroliniensis
          SY=E. virgata Michx.
          SY=Tubiflora c. J.F. Gmel.
    imbricata (Vahl) Pers.
          SY=Justicia i. Vahl

Eranthemum L.

    pulchellum Andr.
          SY=E. nervosum (Vahl) R. Br.

Hygrophila R. Br.

    brasiliensis (Spreng.) Lindau
          SY=H. portoricensis Nees
          SY=Ruellia b. Spreng.
    lacustris (Schlecht. & Cham.) Nees
          SY=Ruellia l. Schlecht. & Cham.

Jacobinia Moric.

    candicans (Nees) Benth. & Hook.
          SY=Adhatoda c. Nees
    ovata Gray

Justicia Houst. ex L.

    americana (L.) Vahl

Justicia (CONT.)
          SY=Dianthera a. L.
          SY=D. a. var. subcoriacea (Fern.)
                Shinners
          SY=J. a. var. s. Fern.
          SY=J. mortuifluminis Fern.
          SY=J. umbratilis Fern.
   angusta (Chapman) Small
          SY=Dianthera a. (Chapman) Small
          SY=D. ovata (Walt.) Lindau var. a.
                Chapman
          SY=J. o. var. a. (Chapman) R.W.
                Long
   borinquensis Britt.
   brandegeana Wasshausen & L.B. Sm.
          SY=Beloperone guttata Brandeg.
   californica (Benth.) D. Gibson
          SY=Beloperone c. Benth.
   carthaginensis Jacq.
   comata (L.) Lam.
          SY=Dianthera c. L.
          SY=Stethoma c. (L.) Britt.
   cooleyi Monachino & Leonard
   crassifolia (Chapman) Small
          SY=Dianthera c. Chapman
   culebritae Urban
   martinsoniana Howard
          SY=J. verticillaris (Nees) Urban
          SY=Rhytiglossa v. Nees
          SY=Stethoma v. (Nees) Britt.
   mirabiloides Lam.
          SY=Drejerella m. (Lam.) Lindau
   ovata (Walt.) Lindau
          var. lanceolata (Chapman) R.W. Long
          SY=Dianthera l. (Chapman) Small
          SY=D. o. var. l. Chapman
          SY=J. l. (Chapman) Small
          var. ovata
          SY=Dianthera humilis (Michx.)
                Engelm. & Gray
          SY=D. o. Walt.
          SY=J. h. Michx.
   pectoralis Jacq.
          SY=Stethoma p. (Jacq.) Raf.
   periplocifolia Jacq.
   runyonii Small
   warnockii B.L. Turner
   wrightii Gray

Nelsonia R. Br.

   brunellioides (Lam.) Kuntze
          SY=Justicia b. Lam.

Odontonema Nees

   cuspidatum (Nees) Kuntze
          SY=Thyrsacanthus c. Nees
   nitidum (Jacq.) Kuntze
          SY=Justicia n. Jacq.
   strictum (Nees) Kuntze
          SY=Thyrsacanthus s. Nees

Oplonia Raf.

   spinosa (Jacq.) Raf.
          SY=Anthacanthus s. (Jacq.) Nees
          SY=Justicia s. Jacq.

Ruellia L.

   brittoniana Leonard ex Fern.
          SY=R. malacosperma Small
          SY=R. tweediana sensu auctt. non
                Griseb.
   caroliniensis (J.F. Gmel.) Steud.
      ssp. caroliniensis

      var. caroliniensis
          SY=Pattersonia c. J.F. Gmel.
          SY=R. c. var. cheloniformis Fern.
          SY=R. caroliniensis var. dentata
                (Nees) Fern.
          SY=R. caroliniensis var.
                membranacea Fern.
          SY=R. caroliniensis var. nanella
                Fern.
          SY=R. caroliniensis var.
                parviflora (Nees) Blake
          SY=R. caroliniensis var. salicina
                Fern.
          SY=R. caroliniensis var. semicalva
                Fern.
          SY=R. hybrida Pursh
          SY=R. p. (Nees) Britt.
      var. serrulata Tharp & Barkl.
      var. succulenta (Small) R.W. Long
          SY=R. s. Small
      ssp. ciliosa (Pursh) R.W. Long
         var. ciliosa
          SY=R. ciliosa Pursh
          SY=R. ciliosa var. cinerascens
                Fern.
          SY=R. humilis sensu Small non
                Nutt.
         var. heteromorpha (Fern.) R.W. Long
          SY=R. h. Fern.
          SY=R. hybrida sensu Small non
                Pursh
   coccinea (L.) Vahl
          SY=Barleria c. L.
   corzoi Tharp & Barkl.
   davisiorum Tharp & Barkl.
   drummondiana (Nees) Gray
          SY=Dipteracanthus d. Nees
   drushelii Tharp & Barkl.
      var. drushelii
          SY=R. nudiflora (Engelm. & Gray)
                Urban var. humilis (Nees)
                Leonard
      var. macrocarpa Tharp & Barkl.
   humilis Nutt.
      var. calvescens Fern.
      var. depauperata Tharp & Barkl.
      var. expansa Fern.
      var. frondosa Fern.
      var. humilis
      var. longiflora (Gray) Fern.
          SY=R. ciliosa Pursh var. l. Gray
   lorentziana Griseb.
   malacosperma Greenm.
   metzae Tharp
          SY=R. m. var. marshii Tharp &
                Barkl.
          SY=R. muelleri Tharp & Barkl.
   noctiflora (Nees) Gray
   nudiflora (Engelm. & Gray) Urban
      var. glabrata Leonard
      var. hispidula Shinners
      var. nudiflora
          SY=Dipteracanthus n. Engelm. &
                Gray
          SY=R. tuberosa sensu Gray
   occidentalis (Gray) Tharp & Barkl.
          SY=R. nudiflora (Engelm. & Gray)
                Urban var. o. (Gray) Leonard
          SY=R. strictopaniculata Tharp &
                Barkl.
          SY=R. tuberosa L. var. o. Gray
   parryi Gray
   pedunculata Torr. ex Gray
      ssp. pedunculata
      ssp. pinetorum (Fern.) R.W. Long
          SY=R. p. Fern.
   purshiana Fern.

Ruellia (CONT.)
  runyonii Tharp & Barkl.
        var. berlandieri Tharp & Barkl.
        var. runyonii
  strepens L.
          SY=Dipteracanthus s. (L.) Nees
          SY=D. micranthus Engelm. & Gray
          SY=R. s. var. cleistantha Gray
          SY=R. s. var. m. (Engelm. & Gray)
                 Britt.
  tuberosa L.
  yucatana (Leonard) Tharp & Barkl.
          SY=R. nudiflora (Engelm. & Gray)
                 Urban var. y. Leonard

Siphonoglossa Oerst.

  dipteracantha (Nees) Heller
          SY=Adhatoda d. Nees
          SY=Monechma pilosella Nees
          SY=S. p. (Nees) Torr.
  greggii Greenm. & Thompson
  longiflora (Torr.) Gray
          SY=Adhatoda l. Torr.
  sessilis (Jacq.) D. Gibson
          SY=Justicia s. Jacq.

Stenandrium Nees

  barbatum Torr. & Gray
          SY=Gerardia b. (Torr. & Gray)
                 Blake
  dulce (Cav.) Nees
        var. dulce
          SY=Adhatoda d. Cav.
          SY=Gerardia d. (Cav.) Nees
        var. floridana Gray
          SY=Gerardia dulce (Cav.) Nees var.
                 f. (Gray) Blake
          SY=G. f. (Gray) Small
          SY=S. f. (Gray) Small
  fascicularis (Benth.) Wasshausen
          SY=Crossandra f. Benth.
  tuberosum (L.) Urban
          SY=Gerardia portoricensis Britt. &
                 Wilson
          SY=G. t. L.

Teliostachya Nees

  alopecuroidea (Vahl) Nees
          SY=Lepidagathis a. (Vahl) R. Br.
          SY=Ruellia a. Vahl

Tetramerium Nees

  hispidum Nees
          SY=T. nervosum Nees var. h. (Nees)
                 Torr.
  platystegium Torr.

Thunbergia Retz.

  alata Bojer ex Sims
  fragrans Roxb.
          SY=T. volubilis Pers.
  grandiflora Roxb.

Yeatesia Small

  viridiflora (Nees) Small
          SY=Gatesia laetevirens (Buckl.)
                 Gray
          SY=Justicia l. Buckl.
          SY=Rhytiglossa v. Nees
          SY=Y. l. (Buckl.) Small

ACERACEAE

Acer L.

  campestre L.
  circinatum Pursh
  floridanum (Chapman) Pax
          SY=A. barbatum Michx.
          SY=A. b. var. longii (Fern.) Fern.
          SY=A. b. var. villipes (Rehd.)
                 Ashe
          SY=A. f. var. l. Fern.
          SY=A. f. var. v. Rehd.
          SY=A. saccharinum Wangenh var. f.
                 Chapman
          SY=A. saccharum Marsh. ssp. f.
                 (Chapman) Desmarais
          SY=A. saccharum var. f. (Chapman)
                 Small & Heller
          SY=Saccharodendron b. (Michx.)
                 Nieuwl.
          SY=S. f. (Chapman) Nieuwl.
  X freemanii E. Murr. [rubrum X saccharinum]
  ginnala Maxim.
  glabrum Torr.
        var. diffusum (Greene) Smiley
          SY=A. d. Greene
          SY=A. g. ssp. d. (Greene) E. Murr.
        var. douglasii (Hook.) Dippel
          SY=A. d. Hook.
          SY=A. g. ssp. d. (Hook.) Wesmael
        var. glabrum
          SY=A. g. var. tripartitum (Nutt.)
                 Pax
          SY=A. g. var. typicum (Wesmael)
                 Keller
          SY=A. tripartitum Nutt.
        var. greenei Keller
        var. neomexicanum (Greene) Kearney &
                 Peebles
          SY=A. g. ssp. n. (Greene) E. Murr.
          SY=A. n. Greene
        var. torreyi (Greene) Smiley
          SY=A. g. ssp. t. (Greene) E. Murr.
          SY=A. t. Greene
  grandidentatum Nutt. ex Torr. & Gray
        var. grandidentatum
          SY=A. saccharum Marsh. ssp. g.
                 (Nutt. ex Torr. & Gray)
                 Desmarais
          SY=A. s. var. g. (Nutt. ex Torr. &
                 Gray) Sudworth
        var. sinuosum (Rehd.) Little
          SY=A. g. var. brachypterum (Woot.
                 & Standl.) Palmer
          SY=A. saccharum Marsh. ssp. b.
                 (Woot. & Standl.) E. Murr.
          SY=A. s. var. sinuosum (Rehd.)
                 Sarg.
          SY=A. sinuosum Rehd.
  leucoderme Small
          SY=A. nigrum Michx. f. var. l.
                 (Small) Fosberg
          SY=A. saccharum Marsh. ssp. l.
                 (Small) Desmarais
          SY=A. s. var. l. (Small) Rehd.
          SY=Saccharodendron l. (Small)
                 Nieuwl.
  macrophyllum Pursh
  negundo L.
        var. californicum (Torr. & Gray) Gray
          SY=A. n. ssp. c. (Torr. & Gray)
                 Wesmael
          SY=A. n. var. arizonicum Sarg.
          SY=Negundo c. Torr. & Gray
        var. interius (Britt.) Sarg.

Acer (CONT.)
          SY=A. i. Britt.
          SY=A. n. ssp. i. (Britt.) Love &
               Love
          SY=Negundo i. (Britt.) Rydb.
      var. negundo
          SY=Negundo n. (L.) Karst.
      var. texanum Pax
      var. violaceum J. Miller
  nigrum Michx. f.
          SY=A. n. var. palmeri Sarg.
          SY=A. saccharum Marsh. ssp. n.
               (Michx. f.) Desmarais
          SY=A. s. var. n. (Michx. f.) Small
          SY=Saccharodendron n. (Michx. f.)
               Small
  palmatum Thunb.
  pensylvanicum L.
  platanoides L.
          SY=A. p. var. schwedleri Nichols.
  pseudoplatanus L.
  rubrum L.
      var. drummondii (Hook. & Arn. ex
           Nutt.) Sarg.
          SY=A. d. Hook. & Arn. ex Nutt.
          SY=A. r. ssp. d. (Hook. & Arn. ex
               Nutt.) E. Murr.
          SY=Rufacer d. (Hook. & Arn. ex
               Nutt.) Small
      var. rubrum
          SY=A. r. var. tomentosum Tausch
          SY=A. stenocarpum Britt.
          SY=Rufacer r. (L.) Small
      var. trilobum Torr. & Gray ex K. Koch
          SY=A. carolinianum Walt.
          SY=A. r. var. tridens Wood
          SY=Rufacer c. (Walt.) Small
  saccharinum L.
          SY=A. dasycarpum Ehrh.
          SY=A. s. var. laciniatum Pax
          SY=A. s. var. wieri Rehd.
          SY=Argentacer s. (L.) Small
  saccharum Marsh.
      var. saccharum
          SY=Acer nigrum Michx. f. var.
               glaucum (F. Schmidt) Fosberg
          SY=A. n. var. saccharophorum (K.
               Koch) Clausen
          SY=A. saccharum var. g. (F.
               Schmidt) Sarg.
          SY=Saccharodendron saccharum
               (Marsh.) Moldenke
      var. schneckii Rehd.
          SY=Acer saccharum ssp. schneckii
               (Rehd.) Desmarais
          SY=A. saccharum var. rugellii
               (Pax) Rehd.
  X senecaense Slavin [leucoderme X
      saccharum]
  spicatum Lam.
  tataricum L.

ADOXACEAE

Adoxa L.

  moschatellina L.

AGAVACEAE

Agave L.

  americana L.
      var. americana
      var. marginata Trel.
      var. medio-picta Trel.
      var. striata Trel.
  applanata K. Koch ex Jacobi
  arizonica Gentry & J.H. Weber
  chrysantha Peebles
          SY=A. palmeri Engelm. var. c.
               (Peebles) Little
  decipiens Baker
  desertii Engelm.
          SY=A. consociata Trel.
  eggersiana Trel.
  glomeruliflora (Engelm.) Berger
          SY=A. chisosensis C.H. Muller
  gracilipes Trel.
  havardiana Trel.
  lecheguilla Torr.
          SY=A. lophantha Schiede var.
               poselgeri (Salm-Dyck) Berger
  lophantha Schiede
  mckelveyana Gentry
  missionum Trel.
  murpheyi F. Gibson
  neglecta Small
  neomexicana Woot. & Standl.
  palmeri Engelm.
  parryi Engelm.
          var. couesii (Engelm.) Kearney &
               Peebles
          var. huachucensis (Baker) Little
          var. parryi
  parviflora Torr.
  scabra Salm-Dyck
          SY=A. asperrima Jacobi
  schottii Engelm.
          var. schottii
          var. treleasei (Toumey) Kearney &
               Peebles
  shawii Engelm.
  sisalana Perrine ex Engelm.
  toumeyana Trel.
          var. bella Breitung
          var. toumeyana
  utahensis Engelm.
          var. eborispina (Hester) Breitung
          var. kaibabensis (McKelvey) Breitung
               SY=A. k. McKelvey
          var. nevadensis Engelm. ex Greenm. &
               Roush
          var. utahensis
  weberi Cels

Cordyline Comm. ex Juss.

  fruticosa (L.) Chev.
          SY=C. terminalis (L.) Kunth
          SY=C. t. var. ti (Schott) Baker

Dasylirion Zucc.

  heteracanthium I.M. Johnston
  leiophyllum Engelm.
  texanum Scheele
  wheeleri S. Wats.

Dracaena Vand. ex L.

  aurea Mann
  hawaiiensis (Deg.) Fosberg

Furcraea Vent.

  macrophylla Baker
  selloa K. Koch
  tuberosa (P. Mill.) Ait. f.

Hesperaloe Engelm.

    funifer (K. Koch) Trel.
    parviflora (Torr.) Coult.
        var. engelmannii (Krauskopf) Trel.
        var. parviflora

Manfreda Salisb.

    longiflora (Rose) Verhoek-Williams
        SY=Polianthes runyonii Shinners
        SY=Runyonia l. Rose
    maculosa (Hook.) Rose
        SY=Agave m. Hook. non Regel
        SY=Polianthes m. (Hook.) Shinners
    sileri Verhoek-Williams
    variegata (Jacobi) Rose
        SY=Polianthes v. (Jacobi) Shinners
    virginica (L.) Rose
        SY=Agave lata Shinners
        SY=A. v. L.
        SY=M. tigrina (Engelm.) Small
        SY=Polianthes l. (Shinners)
            Shinners
        SY=P. v. (L.) Shinners

Nolina Michx.

    arenicola Correll
    atopocarpa Bartlett
    bigelovii (Torr.) S. Wats.
    brittoniana Nash
    erumpens (Torr.) S. Wats.
    georgiana Michx.
    interrata Gentry
    lindheimeriana (Scheele) S. Wats.
    micrantha I.M. Johnston
    microcarpa S. Wats.
    parryi S. Wats.
        SY=N. bigelovii (Torr.) S. Wats.
            var. p. (S. Wats.) L. Benson
    texana S. Wats.
        var. compacta (Trel.) I.M. Johnston
        SY=N. affinis Trel.
        var. texana
    wolfii (Munz) Munz
        SY=N. parryi S. Wats. ssp. w. Munz

Pleomele Salisb.

    aurea (H. Mann) N.E. Br.
    fernaldii St.John
    forbesii Deg.

Sansevieria Thunb.

    cylindrica Bojer ex Hook.
    hyacinthoides (L.) Druce
        SY=Cordyline guineensis (L.)
            Britt.
        SY=S. thrysiflora Thunb.
    longiflora Sims
    metalica Gerome & Labroy
    trifasciata Prain

Yucca L.

    aloifolia L.
    angustissima Engelm. ex Trel.
        var. angustissima
        var. avia Reveal
        var. kanabensis (McKelvey) Reveal
        SY=Y. k. McKelvey
        var. toftiae (Welsh) Reveal
        SY=Y. t. Welsh
    arizonica McKelvey
    arkansana Trel.

    baccata Torr.
        var. baccata
        var. vespertina McKelvey
    baileyi Woot. & Standl.
        var. baileyi
        SY=Y. standleyi McKelvey
        var. intermedia (McKelvey) Reveal
        SY=Y. b. var. navajoa (J.M.
            Webber) J.M. Webber
        SY=Y. i. McKelvey
        SY=Y. n. J.M. Webber
    brevifolia Engelm.
        var. brevifolia
        SY=Y. b. var. jaegeriana McKelvey
        var. herbertii (J.M. Webber) Munz
    campestris McKelvey
        SY=Y. intermedia McKelvey var.
            ramosa McKelvey
    carnerosana (Trel.) McKelvey
    constricta Buckl.
    elata Engelm.
        var. elata
        var. utahensis (McKelvey) Reveal
        SY=Y. u. McKelvey
        var. verdiensis (McKelvey) Reveal
        SY=Y. v. McKelvey
    faxoniana Sarg.
    filamentosa L.
    flaccida Haw.
        SY=Y. concava Haw.
        SY=Y. filamentosa L. var. c.
            (Haw.) Baker
        SY=Y. filamentosa var. smalliana
            (Fern.) Ahles
        SY=Y. s. Fern.
    glauca Nutt. ex Fraser
        var. glauca
        SY=Y. angustifolia Pursh
        var. gurneyi McKelvey
        var. mollis Engelm.
    gloriosa L.
    harrimaniae Trel.
        var. harrimaniae
        SY=Y. gilbertiana (Trel.) Rydb.
        var. neomexicana (Woot. & Standl.)
            Reveal
        SY=Y. n. Woot. & Standl.
    louisianensis Trel.
        SY=Y. arkansana Trel. var.
            paniculata McKelvey
        SY=Y. freemanii Shinners
    necopina Shinners
    recurvifolia Salisb.
    reverchonii Trel.
    rostrata Engelm. ex Trel.
    rupicola Scheele
        SY=?Y. pallida McKelvey
    schidigera Roezl ex Ortgies
        SY=Y. mohavensis Sarg.
    schottii Engelm.
    tenuistyla Trel.
    thompsoniana Trel.
    thornberi McKelvey
        SY=Y. baccata Torr. var.
            brevifolia (Schott) Benson &
            Darrow
    torreyi Shafer
        SY=Y. macrocarpa (Torr.) Coville
    treculeana Carr.
        var. canaliculata Trel.
        var. treculeana
    whipplei Torr.
        var. intermedia (Haines) J.M. Webber
        SY=Y. w. ssp. caespitosa (M.E.
            Jones) Haines
        SY=Y. w. ssp. i. Haines
        var. parishii M.E. Jones

Yucca   (CONT.)
            SY=Y. w. ssp. p. (M.E. Jones)
                Haines
        var. percursa (Haines) J.M. Webber
            SY=Y. w. ssp. p. Haines
        var. whipplei
            SY=Hesperoyucca w. (Torr.) Baker
            SY=Y. newberryi McKelvey

AIZOACEAE

Aptenia N.E. Br.

    cordifolia (L. f.) N.E. Br.
            SY=Mesembryanthemum c. L. f.

Carpobrotus N.E. Br.

    aequilaterus (Haw.) N.E. Br.
            SY=Mesembryanthemum chilense
                Molina
    edulis (L.) N.E. Br.
            SY=Mesembryanthemum e. L.

Cypselea Turp.

    humifusa Turp.

Disphyma N.E. Br.

    crassifolium (L.) L. Bolus

Drosanthemum Schwant.

    floribundum (Haw.) Schwant.
    speciosum (Haw.) Schwant.

Galenia L.

    secunda (L. f.) Sond.

Geocarpon Mackenzie

    minimum Mackenzie

Gisekia L.

    pharnacioides L.

Glinus L.

    lotoides L.
            SY=Mollugo l. (L.) C.B. Clarke
    radiatus (Ruiz & Pavon) Rohrb.

Herrea Schwant.

    elongata (Haw.) L. Bolus

Lampranthus N.E. Br.

    coccineus (Haw.) N.E. Br.

Mesembryanthemum L.

    crystallinum L.
            SY=Gasoul c. (L.) Rothm.
    nodiflorum L.
            SY=Gasoul n. (L.) Rothm.

Mollugo L.

    cerviana (L.) Ser.
    gracillima Anderss.

nudicaulis Lam.
verticillata L.
        SY=M. berteriana Ser.

Sesuvium L.

    crithmoides Welw.
    erectum Correll
    maritimum (Walt.) B.S.P.
    portulacastrum (L.) L.
    sessile Pers.
    trianthemoides Correll
    verrucosum Raf.

Tetragonia L.

    tetragonioides (Pallas) Kuntze
            SY=T. expansa Murr.

Trianthema L.

    portulacastrum L.

ALISMACEAE

Alisma L.

    gramineum J.G. Gmel.
        var. angustissimum Hendricks
        var. gramineum
        var. graminifolia (Wahlenb.)
            Hendricks
            SY=A. geyeri Torr.
            SY=A. gramineum var. geyeri
                (Torr.) Lam.

    lanceolatum With.
    plantago-aquatica L.
        var. americana Schultes & Schultes
            SY=A. brevipes Greene
            SY=A. p-a. ssp. b. (Greene)
                Samuelsson
            SY=A. p-a. var. b. (Greene)
                Victorin
            SY=A. triviale Pursh

        var. parviflorum (Pursh) Torr.
            SY=A. subcordatum Raf.
        var. plantago-aquatica

Echinodorus L.C. Rich.

    cordifolius (L.) Griseb.
            SY=E. radicans (Nutt.) Engelm.
    ?ranunculoides (L.) Engelm.
    rostratus (Nutt.) Engelm.
            SY=E. berteroi (Spreng.) Fassett
                pro parte
            SY=E. b. var. lanceolatus
                (Engelm.) Fassett
    tenellus (Mart.) Buch.
        var. parvulus (Engelm.) Fassett
            SY=E. p. Engelm.
            SY=E. t. var. latifolius (Seub.)
                Fassett
            SY=Helianthium p. (Engelm.) Britt.
            SY=H. t. (Mart.) Britt. pro parte

Machaerocarpus Small

    californicus (Torr. ex Benth.) Small
            SY=Damasonium c. Torr. ex Benth.

Sagittaria L.

   ambigua J.G. Sm.
   australis (J.G. Sm.) Small
         SY=S. engelmanniana J.G. Sm. ssp.
            longirostra (Micheli) Bogin
            sensu auctt. non Micheli
         SY=S. l. sensu auctt. non
            (Micheli) J.G. Sm.
   brevirostra Mackenzie & Bush
         SY=S. engelmanniana J.G. Sm. ssp.
            b. (Mackenzie & Bush) Bogin
   calycina Engelm.
      var. calycicna
         SY=Lophotocarpus californicus J.G.
            Sm.
         SY=L. calycinus (Engelm.) J.G. Sm.
         SY=L. depauperatus J.G. Sm.
         SY=S. montevidensis Cham. &
            Schlecht. ssp. calycina
            (Engelm.) Bogin
      var. spongiosa Engelm.
         SY=Lophotocarpus spathulatus J.G.
            Sm.
         SY=L. spongiosus (Engelm.) J.G.
            Sm.
         SY=S. montevidensis Cham. &
            Schlecht. ssp. spongiosa
            (Engelm.) Bogin
         SY=S. m. var. spongiosa (Engelm.)
            Boivin
         SY=S. spathulata (J.G. Sm.) Buch.
   cristata Engelm.
         SY=S. graminea Michx. var. c.
            (Engelm.) Bogin
   cuneata Sheldon
         SY=S. arifolia Nutt. ex J.G. Sm.
   engelmanniana J.G. Sm.
   falcata Pursh
         SY=S. lancifolia L. ssp. media
            (Micheli) Bogin
         SY=S. l. var. m. Micheli
   fasciculata E.O. Beal
   graminea Michx.
      var. chapmanii J.G. Sm.
         SY=S. c. (J.G. Sm.) C. Mohr
      var. graminea
         SY=S. cycloptera (J.G. Sm.) C.
            Mohr
         SY=S. eatonii J.G. Sm.
      var. macrocarpa (J.G. Sm.) Bogin
         SY=S. m. J.G. Sm.
      var. weatherbiana (Fern.) Bogin
         SY=S. w. Fern.
   greggii J.G. Sm.
   guyanensis H.B.K.
   intermedia Micheli
   isoetiformis J.G. Sm.
   kurziana Gleuck
         SY=S. subulata (L.) Buch. var. k.
            (Gleuck) Bogin
   lancifolia L.
         SY=S. angustifolia Lindl.
   latifolia Willd.
      var. latifolia
         SY=S. engelmanniana J.G. Sm. ssp.
            longirostra (Micheli) Bogin
         SY=S. longirostra (Micheli) J.G.
            Sm.
         SY=S. ornithorhyncha Small
         SY=S. planipes Fern.
         SY=S. viscosa C. Mohr pro parte
      var. obtusa (Muhl. ex Willd.) Wieg.
         SY=S. esculenta T.J. Howell
      var. pubescens (Muhl.) J.G. Sm.
         SY=S. p. Muhl.
         SY=S. viscosa C. Mohr pro parte

   longiloba Engelm. ex Torr.
   montevidensis Cham. & Schlecht.
   papillosa Buch.
   platyphylla (Engelm.) J.G. Sm.
         SY=S. graminea Michx. var.
            platyphylla Engelm.
         SY=S. mohrii J.G. Sm.
   rigida Pursh
   sagittifolia L.
   sanfordii Greene
   stagnorum Small
         SY=S. subulata (L.) Buch. var.
            gracillima (S. Wats.) J.G.
            Sm.
   subulata (L.) Buch.
         SY=S. filiformis J.G. Sm.
         SY=S. lorata (Chapman) Small
         SY=S. s. var. natans (Michx.) J.G.
            Sm.
   teres S. Wats.
         SY=S. graminea Michx. var. t. (S.
            Wats.) Bogin

AMARANTHACEAE

Achyranthes L.

   aspera L.
         SY=Centrostachys a. (L.) Standl.
   bidentata Blume
   indica (L.) P. Mill.
         SY=Centrostachys i. (L.) Standl.
   mutica Gray
   nelsonii St. John
   splendens Mart. ex Moq.
      var. reflexa Hbd.
      var. rotundata Hbd.
      var. splendens

Aerva Forsk.

   sericea Moq.

Alternanthera Forsk.

   bettzickiana (Regel) Nichols.
         SY=Achyranthes b. (Regel) Standl.
         SY=Achyranthes ficoidea (L.) Lam.
            var. b. (Regel) Baker
   brasiliana (L.) Kuntze
         SY=Achyranthes b. (L.) Standl.
         SY=Alternanthera dentata (Moench)
            Scheygrond
         SY=Alternanthera jacquinii
            (Schrad.) Griseb. ex Kuntze
   caracasana H.B.K.
         SY=Achyranthes peploides (Willd.
            ex Roemer & Schultes) Britt.
         SY=Alternanthera p. (Willd. ex
            Roemer & Schultes) Urban
   crucis (Moq.) Boldingh
         SY=Achyranthes portoricensis
            (Kuntze) Standl.
         SY=Alternanthera p. Kuntze
   echinocephela (Hook. f.) Christopherson
         SY=Alternanthera menziesii St.
            John
   flavescens H.B.K.
         SY=Achyranthes ramosissima sensu
            auctt. non (Mart.) Standl.
         SY=Alternanthera r. sensu auctt.
            non (Mart.) Chod.
   maritima (Mart.) St. Hil.
         SY=Achyranthes m. (Mart.) Standl.

Alternanthera (CONT.)
  palmeri S. Wats.
  paronychioides St. Hil.
        var. amazonica Huber
        var. paronychioides
              SY=Achyranthes p. sensu auctt. non
                    (L.) Lam.
              SY=Alternanthera polygonoides
                    sensu auctt. non (L.) R. Br.
                    ex Sweet
  philoxeroides (Mart.) Griseb.
              SY=Achyranthes p. (Mart.) Standl.
  pungens H.B.K.
              SY=Achyranthes leiantha (Seub.)
                    Standl.
              SY=Achyranthes repens L.
              SY=Alternanthera achyrantha (L.)
                    R. Br. ex Sweet
              SY=Alternanthera r. (L.) Link
  sessilis (L.) R. Br. ex DC.
              SY=Achyranthes s. (L.) Desf. ex
                    Steud.
  tenella Colla
    ssp. flavogrisea (Urban) Mears &
          Veldkamp
    ssp. tenella
              SY=Achyranthes ficoidea sensu
                    auctt. non (L.) Lam.
              SY=Alternanthera f. sensu auctt.
                    non (L.) R. Br. ex Roemer &
                    Schultes

Amaranthus L.

  acanthochiton (Torr.) Sauer
              SY=Acanthochiton wrightii Torr.
  acutilobus Uline & Bray
              SY=Euxolus emarginatus A. Braun &
                    Bouche
  albus L.
              SY=Amaranthus a. var. pubescens
                    (Uline & Bray) Fern.
              SY=A. graecizans sensu auctt. non
                    L.
              SY=A. g. var. p. Uline & Bray
              SY=A. p. (Uline & Bray) Rydb.
  arenicola I.M. Johnston
              SY=Amaranthus torreyi sensu auctt.
                    non (Gray) Benth. ex Wats.
  australis (Gray) Sauer
              SY=Acnida alabamensis Standl.
              SY=Acnida cuspidata Bert. ex
                    Spreng.
  berlandieri (Moq.) Uline & Bray
  bigelovii Uline & Bray
              SY=Amaranthus torreyi sensu auctt.
                    non (Gray) Benth. ex S.
                    Wats.
  blitoides S. Wats.
              SY=Amaranthus graecizans sensu
                    auctt. non L.
  brownii Christoph. & Caum
  californicus (Moq.) S. Wats.
              SY=Amaranthus albomarginatus Uline
                    & Bray
              SY=A. microphyllus Shinners
  cannabinus (L.) Sauer
              SY=Acnida c. L.
  caudatus L.
  chihuahuensis S. Wats.
  crassipes Schlecht.
  crispus (Lesp. & Thev.) N. Terracc.
  cruentus L.
              SY=Amaranthus hybridus L. var. c.
                    (L.) Moq.
              SY=A. h. ssp. c. (L.) Thellung
              SY=A. paniculatus L.

deflexus L.
dubius Mart. ex Thellung
fimbriatus (Torr.) Benth. ex S. Wats.
        var. denticulatus (Torr.) Uline &
              Bray
        var. fimbriatus
floridanus (S. Wats) Sauer
              SY=Acnida f. S. Wats.
gracilis Desf.
              SY=Amaranthus viridus sensu auctt.
                    non L.
greggii S. Wats.
              SY=Amaranthus annectens Blake
              SY=A. myrianthus Standl.
hybridus L.
              SY=Amaranthus chlorostachys Willd.
              SY=A. incurvatus Tim. ex Gren. &
                    Godr.
              SY=A. patulus Bertol.
hypochondriacus L.
              SY=Amaranthus hybridus L. ssp.
                    hypochondriacus (L.)
                    Thellung
              SY=A. leucocarpus S. Wats.
lineatus R. Br.
lividus L.
              SY=Amaranthus ascendens Loisel.
              SY=A. blitum L.
              SY=A. l. var. polygonoides (Moq.)
                    Thellung
              SY=A. viridus sensu auctt. non L.
muricatus Gillies ex Moq.
obcordatus (Gray) Standl.
palmeri S. Wats.
polygonoides L.
powellii S. Wats.
              SY=Amaranthus bouchonii Thellung
              SY=A. bracteosus Uline & Bray
              SY=A. retroflexus L. var. p. (S.
                    Wats.) Boivin
pringlei S. Wats.
pumilus Raf.
retroflexus L.
              SY=Amaranthus r. var. salicifolius
                    I.M. Johnston
rudis Sauer
              SY=Acnida tamariscina sensu auctt.
                    non (Nutt.) Wood
              SY=Amaranthus t. sensu auctt. non
                    Nutt.
scleropoides Uline & Bray
spinosus L.
tricolor L.
              SY=Amaranthus gangeticus L.
tuberculatus (Moq.) Sauer
              SY=Acnida altissima (Riddell) Moq.
                    ex Standl.
              SY=Acnida a. var. prostrata (Uline
                    & Bray) Fern.
              SY=Acnida a. var. subnuda (S.
                    Wats.) Fern.
              SY=Acnida concatenata (Moq.) Small
              SY=Acnida s. (S. Wats.) Standl.
              SY=Acnida tamariscina (Nutt.) Wood
                    var. p. Uline & Bray
              SY=Acnida t. Moq.
              SY=Amaranthus a. Riddell
              SY=Amaranthus ambigens Standl.
              SY=Amaranthus t. var. p. (Uline &
                    Bray) B.L. Robins.
              SY=Amaranthus t. var. s. S. Wats.
viscidulus Greene
watsonii Standl.
              SY=Amaranthus torreyi sensu auctt.
                    non (Gray) Benth. ex S.
                    Wats.
wrightii S. Wats.

Caraxeron Vaillant ex Raf.

    vermicularis (L.) Raf.
          SY=Cruzeta v. (L.) Maza
          SY=Lithophila v. (L.) Uline ex
              Millsp.
          SY=Philoxerus v. (L.) Sm.

Celosia L.

    argentea L.
    cristata L.
          SY=C. argentea L. var. c. (L.)
              Kuntze
    nitida Vahl
    palmeri S. Wats.
    virgata Jacq.

Chamissoa H.B.K.

    altissima (Jacq.) H.B.K.

Charpentiera Gaud.

    densiflora Sohmer
    elliptica (Hbd.) Heller
    obovata Gaud.
    ovata Gaud.
          var. niuensis Sohmer
          var. ovata
    tomentosa Sohmer
          var. maakuaensis Sohmer
          var. tomentosa

Dicraurus Hook. f.

    leptocladus Hook. f.

Froelichia Moench

    arizonica Thornb. ex Standl.
    braunii Standl.
    drummondii Moq.
    floridana (Nutt.) Moq.
          var. campestris (Small) Fern.
            SY=F. c. Small
          var. floridana
    gracilis (Hook.) Moq.
    interrupta (L.) Moq.

Gomphrena L.

    caespitosa Torr.
          SY=G. viridus Woot. & Standl.
    globosa L.
    haageana Klotzsch
    nealleyi Coult. & Fish.
    nitida Rothrock
    serrata L.
          SY=G. celosioides sensu auctt. non
            Mart.
          SY=G. decumbens Jacq.
          SY=G. dispersa Standl.
    sonorae Torr.

Gossypianthus Moq.

    lanuginosus (Poir.) Moq.
          var. lanuginosus
            SY=Guilleminea l. (Poir.) Hook. f.
          var. rigidiflorus (Hook.) Mears ined.
            SY=Gossypianthus r. Hook.
            SY=Guilleminea l. (Poir.) Hook. f.
               var. r. (Hook.) Mears
          var. sheldonii Uline & Bray
            SY=Gossypianthus s. (Uline & Bray)
               Small

            SY=Guilleminea l. (Poir.) Hook. f.
               var. s. (Uline & Bray) Mears
          var. tenuiflorus (Hook.) Mears ined.
            SY=Gossypianthus t. Hook.
            SY=Guilleminea l. (Poir.) Hook. f.
               var. t. (Hook.) Mears

Guilleminea H.B.K.

    densa (Willd.) Moq.
          var. aggregata Uline & Bray
          var. densa
            SY=Brayulinea d. (Willd.) Small

Hermbstaedtia Reichenb.

    elegans Moq.

Iresine P. Br.

    angustifolia Euph.
    argentata (Mart.) D. Dietr.
    diffusa Humb. & Bonpl. ex Willd.
          SY=I. celosia L.
    flavescens Humb. & Bonpl. ex Willd.
          SY=I. keyensis Millsp.
    herbstii Hook.
    heterophylla Standl.
    lindenii Van Houtte
    palmeri (S. Wats.) Standl.
    rhizomatosa Standl.

Lithophila Sw.

    muscoides Sw.

Nototrichium (Gray) Hbd.

    humile Hbd.
          var. humile
          var. parvifolium Deg. & Sherff
          var. subrhomboideum Sherff
    sandwicense (Gray) Hbd.
          var. decipiens Sherff
          var. forbesii Sherff
          var. kolekolense Sherff
          var. leptopodum Deg. & Sherff
          var. longispicatum Hbd.
          var. niihauense St. John
          var. olokeleanum Sherff
          var. sandwicense
            SY=N. s. var. dubium Sherff
            SY=N. s. var. helleri Sherff
            SY=N. s. var. kavaiense (Gray)
               Hbd.
            SY=N. s. var. lanaiense Sherff
            SY=N. s. var. lanceolatum Sherff
            SY=N. s. var. latifolium Sherff
            SY=N. s. var. macrophyllum Sherff
            SY=N. s. var. mauiense Deg. &
               Sherff
            SY=N. s. var. pulchelloides Deg. &
               Sherff
            SY=N. s. var. pulchellum Sherff
            SY=N. s. var. subcordatum Deg. &
               Sherff
            SY=N. s. var. syringifolium Sherff
    viride Hbd.
          var. oblongifolium Sherff
          var. subtruncatum Sherff
          var. viride

Tidestromia Standl.

    gemmata I.M. Johnston
    lanuginosa (Nutt.) Standl.
          var. carnosa (Steyermark) Cory

Tidestromia (CONT.)
    var. lanuginosa
        SY=Cladothrix l. Nutt.
  oblongifolia (S. Wats.) Standl.
    ssp. cryptantha (S. Wats.) Wiggins
    ssp. oblongifolia
  suffruticosa (Torr.) Standl.

ANACARDIACEAE

Anacardium L.

  occidentale L.

Comocladia P. Br.

  cuneata Britt.
  dentata Jacq.
  dodonaea (L.) Urban
  glabra (Schultes) Spreng.

Cotinus P. Mill.

  coggygria Scop.
  obovatus Raf.
        SY=C. americanus Nutt.

Mangifera L.

  indica L.

Metopium P. Br.

  toxiferum (L.) Krug & Urban

Malosma Nutt.

  laurina (Nutt. ex Torr. & Gray) Nutt. ex
    Abrams
        SY=Rhus l. Nutt. ex Torr. & Gray

Pistacia L.

  texana Swingle

Rhus L.

  aromatica Ait.
    var. arenaria (Greene) Fern.
      SY=R. arenaria (Greene) G.N. Jones
      SY=R. trilobata Nutt. var. a.
        (Greene) Barkl.
    var. aromatica
      SY=R. a. var. illinoensis (Greene)
        Rehd.
      SY=Schmaltzia crenata (P. Mill.)
        Greene
    var. serotina (Greene) Rehd.
      SY=R. trilobata Nutt. var. s.
        (Greene) Barkl.
  copallina L.
    var. copallina
    var. latifolia Engl.
    var. leucantha (Jacq.) DC.
      SY=R. l. Jacq.
      SY=R. obtusifolia (Small) Small
  glabra L.
      SY=R. ashei (Small) Greene
      SY=R. borealis (Britt.) Greene
      SY=R. calophylla Greene
      SY=R. g. var. b. Britt.
      SY=R. g. var. laciniata Carr.
      SY=R. g. var. occidentalis Torr.
  integrifolia (Nutt.) Benth. & Hook. f. ex

Brewer & S. Wats.
kearneyi Barkl.
lanceolata (Gray) Britt.
      SY=R. copallima L. var. l. Gray
michauxii Sarg.
microphylla Engelm. ex Gray
ovata S. Wats.
      SY=R. o. var. traskiae Barkl. pro
        parte
X pulvinata Greene [glabra X typhina]
sandwicensis Gray
      SY=R. chinensis P. Mill. var. s.
        (Gray) Deg. & Greenw.
trilobata (Nutt.) Gray
    var. anisophylla (Greene) Jepson
      SY=Schmaltzia a. Greene
      SY=S. t. var. a. (Greene) Barkl.
    var. pilosissima Engl.
      SY=R. aromatica Ait. var. p.
        (Engl.) Shinners
      SY=R. a. var. mollis Ashe
      SY=R. t. var. malacophylla
        (Greene) Munz
    var. quinata (Greene) Jepson
    var. racemulosa (Greene) Barkl.
      SY=Schmaltzia r. Greene
    var. simplicifolia (Greene) Barkl.
    var. trilobata
      SY=R. aromatica Ait. var.
        flabelliformis Shinners
      SY=R. a. var. t. (Nutt.) Gray
      SY=Schmaltzia t. (Nutt.) Small
typhina L.
      SY=R. hirta (L.) Sudworth
      SY=R. t. var. laciniata Wood
virens Lindheimer ex Gray
  ssp. choriophylla (Woot. & Standl.)
  Young
      SY=R. c. Woot. & Standl.
  ssp. virens

Schinus L.

  longifolius (Lindl.) Speg.
  molle L.
  terebinthifolius Raddi

Spondias L.

  mombin L.
  purpurea L.

Toxicodendron P. Mill.

  diversilobum (Torr. & Gray) Greene
      SY=Rhus d. Torr. & Gray
      SY=T. radicans L. ssp. d. (Torr. &
        Gray) Thorne
  radicans (L.) Kuntze
    ssp. divaricatum (Greene) Gillis
    ssp. eximium (Greene) Gillis
      SY=Rhus toxicodendron L. var. e.
        (Greene) McNair
    ssp. negundo (Greene) Gillis
    ssp. pubens (Engelm. ex S. Wats.) Gillis
    ssp. radicans
      SY=Rhus r. L.
      SY=R. r. var. littoralis (Mearns)
        Deam
      SY=R. r. var. malacotrichocarpa
        (A.H. Moore) Fern.
    ssp. verrucosum (Scheele) Gillis
  rydbergii (Small ex Rydb.) Greene
      SY=Rhus radicans L. var. rydbergii
        (Small ex Rydb.) Rehd.
      SY=R. radicans var. vulgaris
        (Michx.) DC.

Toxicodendron   (CONT.)
                SY=R. toxicodendron L. var. v.
                        Michx.
                SY=T. desertorum Lunell
                SY=T. radicans (L.) Kuntze var.
                        rydbergii (Small ex Rydb.)
                        Erskine
        toxicarium (Salisb.) Gillis
                SY=Rhus toxicodendron L.
                SY=T. toxicodendron (L.) Britt.
        vernix (L.) Kuntze
                SY=Rhus v. L.

ANNONACEAE

Annona L.

        glabra L.
        montana Macfad.
        muricata L.
        reticulata L.
        squamosa L.

Asimina Adans.

        angustifolia Raf.
                SY=A. longifolia Kral
                SY=A. l. var. spatulata Kral
                SY=Pityothamnus a. (Raf.) Small
        incana (Bartr.) Exell
                SY=A. speciosa Nash
                SY=Pityothamnus i. Bartr.
        X nashii Kral [angustifolia X incana]
        obovata (Willd.) Nash
                SY=Pityothamnus o. (Willd.) Small
        parviflora (Michx.) Dunal
        pygmaea (Bartr.) Dunal
                SY=Pityothamnus p. (Bartr.) Small
        reticulata Shuttlw. ex Chapman
                SY=Pityothamnus r. (Shuttlw. ex
                        Chapman) Small
        tetramera Small
                SY=Pityothamnus t. (Small) Small
        triloba (L.) Dunal

Deeringothamnus Small

        pulchellus Small
                SY=Asimina p. (Small) Zimmerman
        rugelii (B.L. Robins.) Small
                SY=Asimina r. B.L. Robins.

Guatteria Ruiz & Pavon

        blainii (Griseb.) Urban
                SY=Cananga b. (Griseb.) Britt.
        caribaea Urban
                SY=Cananga c. (Urban) Britt.

Oxandra A. Rich.

        lanceolata (Sw.) Baill.
        laurifolia (Sw.) A. Rich.

Rollinia St. Hil.

        mucosa (Jacq.) Baill.

APIACEAE

Aegopodium L.

        podagraria L.
                var. podagraria
                var. variegatum Bailey

Aethusa L.

        cynapium L.

Aletes Coult. & Rose

        acaulis (Torr.) Coult. & Rose
        anisatus (Gray) Theobald & Tseng
                SY=Cymopterus a. Gray
                SY=Pteryxia a. (Gray) Mathias &
                        Constance
        filifolius Mathias, Constance, & Theobald
        humilis Coult. & Rose
        macdougallii Coult. & Rose
                ssp. breviradiatus Theobald & Tseng
                ssp. macdougallii
        sessiliflorus Theobald & Tseng

Ammi L.

        majus L.
        visnaga (L.) Lam.

Ammoselinum Torr. & Gray

        butleri (Engelm.) Coult. & Rose
        giganteum Coult. & Rose
        popei Torr. & Gray

Anethum L.

        graveolens L.

Angelica L.

        ampla A. Nels.
        archangelica L.
                ssp. archangelica
                ssp. litoralis (Fries) Thellung
                        SY=A. a. ssp. norvegica (Rupr.)
                                Nordh.
        arguta Nutt.
        atropurpurea L.
                SY=A. a. var. occidentalis Fassett
        breweri Gray
        callii Mathias & Constance
        canbyi Coult. & Rose
        dawsonii S. Wats.
        dentata (Chapman) Coult. & Rose
        genuflexa Nutt.
        grayi Coult. & Rose
        hendersonii Coult. & Rose
        kingii (S. Wats.) Coult. & Rose
        laurentiana Fern.
        lineariloba Gray
        lucida L.
                SY=Coelopleurum actiifolium
                        (Michx.) Coult. & Rose
                SY=C. gmelinii (DC.) Ledeb.
                SY=C. l. (L.) Fern.
        pinnata S. Wats.
        roseana Henderson
        scabrida Clokey & Mathias
        sylvestris L.
        tomentosa S. Wats.
        triquinata Michx.
                SY=A. curtisii Buckl.
        venenosa (Greenway) Fern.
                SY=A. villosa (Walt.) B.S.P.
        wheeleri S. Wats.

Anthriscus Pers. emend. Hoffmann

Anthriscus   (CONT.)
   caucalis Bieb.
            SY=A. neglecta Boiss. & Reut. var.
               scandix (Scop.) Hyl.
            SY=A. scandicina (Weber) Mansf.
   cerefolium (L.) Hoffmann
            SY=Cerefolium c. (L.) Schinz &
               Thellung
   neglecta Boiss. & Reut.
   sylvestris (L.) Hoffmann
            SY=Chaerophyllum s. L.

Apiastrum Nutt.

   angustifolium Nutt.

Apium L.

   graveolens L.
         var. dulce (P. Mill.) Pers.
            SY=Celeri g. (L.) Britt. pro parte
   nodiflorum (L.) Lag.
            SY=Ciclospermum n. (L.) W.D.J.
               Koch
   prostratum Labill.
   repens (Jacq.) Lag.

Berula W.D.J. Koch

   erecta (Huds.) Coville
            SY=B. e. var. incisa (Torr.)
               Cronq.
            SY=B. i. (Torr.) G.N. Jones
            SY=B. pusilla (Nutt.) Fern.
            SY=Siella e. (Huds.) M. Pimen.

Bifora Hoffmann

   americana (DC.) Benth. & Hook.
   radians Bieb.

Bowlesia Ruiz & Pavon

   incana Ruiz & Pavon
            SY=B. septentrionalis Coult. &
               Rose

Bunium L.

   bulbocastanum L.
            SY=Carum b. (L.) W.D.J. Koch

Bupleurum L.

   americanum Coult. & Rose
            SY=B. triradiatum J.E. Adams ssp.
               arcticum (Regel) Hulten
   fontanesii Guss. ex Caruel
            SY=B. odontites L.
   lancifolium Hornem.
            SY=B. subovatum Link
   rotundifolium L.

Carum L.

   carvi L.

Caucalis L.

   platycarpos L.

Centella L.

   asiatica (L.) Urban
   erecta (L. f.) Fern.
            SY=C. repanda (Pers.) Small

Chaerophyllum L.

   bulbosum L.
   procumbens (L.) Crantz
         var. procumbens
         var. shortii Torr. & Gray
            SY=C. s. (Torr. & Gray) Bush
   tainturieri Hook.
         var. dasycarpum S. Wats.
         var. tainturieri
            SY=C. floridanum (Coult. & Rose)
               Bush
            SY=C. procumbens (L.) Crantz var.
               t. (Hook.) Coult. & Rose
            SY=C. t. var. f. Coult. & Rose
            SY=C. texanum Coult. & Rose
   temulentum L.

Ciclospermum Lag.

   leptophyllum (Pers.) Sprague
            SY=Apium l. (Pers.) F. Muell. ex
               Benth. & F. Muell.
            SY=A. tenuifolium (Moench)
               Thellung ex Hegi
            SY=C. ammi (L.) Britt.

Cicuta L.

   bolanderi S. Wats.
   bulbifera L.
   douglasii (DC.) Coult. & Rose
            SY=C. maculata L. var. californica
               (Gray) Boivin
            SY=C. occidentalis Greene
   mackenzieana Raup
   maculata L.
            SY=C. m. var. angustifolia Hook.
   mexicana Coult. & Rose
            SY=C. curtissii Coult. & Rose
            SY=C. maculata L. var. c. (Coult.
               & Rose) Fern.
   victorinii Fern.
            SY=C. maculata L. var. v. (Fern.)
               Boivin

Cnidium Cusson ex Juss.

   cnidiifolium (Turcz.) Schischkin
            SY=Conioselinum c. (Turcz.)
               Porsild
   monnieri (L.) Cusson
            SY=Selinum m. L.

Conioselinum Hoffmann

   chinense (L.) B.S.P.
            SY=C. pumilum Rose
   gmelinii (Cham. & Schlecht.) Steud.
            SY=C. chinense (L.) B.S.P. var.
               pacificum (S. Wats.) Boivin
            SY=C. p. (S. Wats.) Coult. & Rose
   mexicanum Coult. & Rose
   scopulorum (Gray) Coult. & Rose

Conium L.

   maculatum L.

Coriandrum L.

   sativum L.

Cryptotaenia DC.

   canadensis (L.) DC.
            SY=Deringa c. (L.) Kuntze

Cuminum L.

   cyminum L.

Cymopterus Raf.

   aboriginum M.E. Jones
   acaulis (Pursh) Raf.
   basalticus M.E. Jones
   bipinnatus S. Wats.
   bulbosus A. Nels.
   cinerarius Gray
   corrugatus M.E. Jones
        SY=Rhysopterus plurijugus Coult. &
           Rose
   coulteri (M.E. Jones) Mathias
   deserticola Brandeg.
   duchesnensis M.E. Jones
   fendleri Gray
   gilmanii Morton
   glaucus Nutt.
   globosus (S. Wats.) S. Wats.
   higginsii Welsh
   humboldtensis M.E. Jones
   ibapensis M.E. Jones
        SY=C. watsonii (Coult. & Rose)
           M.E. Jones
   jonesii Coult. & Rose
   longipes S. Wats.
   macrorhizus Buckl.
   megacephalus M.E. Jones
   minimus Mathias
   montanus (Nutt.) Torr. & Gray
        SY=Phellopterus m. Nutt.
   multinervatus (Coult. & Rose) Tidestrom
   newberryi (S. Wats.) M.E. Jones
   nivalis S. Wats.
   panamintensis Coult. & Rose
      var. acutifolius (Coult. & Rose) Munz
      var. panamintensis
   planosus (Osterhout) Mathias
   purpurascens (Gray) M.E. Jones
   purpureus S. Wats.
   ripleyi Barneby
   rosei M.E. Jones

Cynosciadium DC.

   digitatum DC.

Daucosma Engelm. & Gray

   laciniata Engelm. & Gray

Daucus L.

   carota L.
   pusillus Michx.

Erigenia Nutt.

   bulbosa (Michx.) Nutt.

Eryngium L.

   alismifolium Greene
   amethystinum L.
   aquaticum L.
      var. aquaticum
        SY=E. virginianum Lam.
      var. ravenelii (Gray) Mathias &
        Constance
        SY=E. a. var. floridanum (Coult. &
          Rose) Mathias & Constance
        SY=E. f. Coult. & Rose
        SY=E. r. Gray
   aristulatum Jepson

      var. aristulatum
      var. parishii (Coult. & Rose) Mathias
        & Constance
   armatum (S. Wats.) Coult. & Rose
   aromaticum Baldw.
   articulatum Hook.
   baldwinii Spreng.
   campestre L.
   cuneifolium Small
   diffusum Torr.
   divaricatum Hook. & Arn.
   foetidum L.
   heterophyllum Engelm.
        SY=E. wrightii Gray
   hookeri Walp.
   integrifolium Walt.
        SY=E. ludovicianum Morong
   leavenworthii Torr. & Gray
   lemmonii Coult. & Rose
   maritimum L.
   nasturtiifolium Juss. ex Delar. f.
   petiolatum Hook.
   phyteumae Delar. f.
   pinnatisectum Jepson
   planum L.
   prostratum Nutt.
        SY=E. p. var. disjunctum Fern.
   racemosum Jepson
   sparganophyllum Hemsl.
   vaseyi Coult. & Rose
      var. castrense (Jepson) Hoover ex
        Mathias & Constance
       SY=E. c. Jepson
      var. globosum (Jepson) Hoover ex
        Mathias & Constance
        SY=E. spinosepalum Mathias
      var. vaseyi
        SY=E. v. var. vallicola (Jepson)
          Munz
   yuccifolium Michx.
      var. synchaetum Gray
        SY=E. s. (Gray) Rose
      var. yuccifolium

Eurytaenia Torr. & Gray

   hinckleyi Mathias & Constance
   texana Torr. & Gray

Falcaria Bernh.

   vulgaris Bernh.
        SY=F. sioides (Wibel) Aschers.

Foeniculum P. Mill.

   vulgare P. Mill.
        SY=F. foeniculum (L.) Karst.

Glehnia F. Schmidt

   littoralis F. Schmidt
      ssp. leiocarpa (Mathias) Hulten
        SY=G. leiocarpa Mathias
        SY=G. littoralis var. leiocarpa
          (Mathias) Boivin

Harbouria Coult. & Rose

   trachypleura (Gray) Coult. & Rose

Heracleum L.

   lanatum Michx.
        SY=H. maximum Bartr.
        SY=H. sphondylium L. ssp. montanum
          (Schleich. ex Gaudin) Briq.

Heracleum (CONT.)
   mantegazzianum Sommier & Levl.
   sphondylium L.
      ssp. sibiricum (L.) Simonkai
         SY=H. s. L.
         SY=H. s. var. angustifolium Jacq.
             pro parte

Hydrocotyle L.

   americana L.
   bonariensis Comm. ex Lam.
   bowlesioides Mathias & Constance
         SY=H. sibthorpioides Lam. var.
            oedipoda Deg. & Greenw.
   hirsuta Sw.
   moschata Forst. f.
   pusilla A. Rich.
   ranunculoides L. f.
   sibthorpioides Lam.
         SY=H. rotundifolia Lam.
   umbellata L.
   verticillata Thunb.
      var. fetherstoniana (Jennings)
        Mathias
      var. triradiata (A. Rich.) Fern.
        SY=H. australis Coult. & Rose
        SY=H. canbyi Coult. & Rose
      var. verticillata

Levisticum Hill

   officinale W.D.J. Koch
         SY=Hipposelinum levisticum (L.)
            Britt. & Rose
         SY=L. paludapifolium (Lam.)
            Aschers.

Ligusticum L.

   apiifolium (Nutt.) Gray
   calderi Mathias & Constance
   californicum Coult. & Rose
   canadense (L.) Britt.
   canbyi Coult. & Rose
   filicinum S. Wats.
   grayi Coult. & Rose
   porteri Coult. & Rose
      var. brevilobum (Rydb.) Mathias &
        Constance
      var. porteri
   scoticum L.
      ssp. hultenii (Fern.) Calder & Taylor
        SY=L. h. Fern.
        SY=L. s. var. h. (Fern.) Boivin
      ssp. scoticum
   tenuifolium S. Wats.
        SY=L. filicinum S. Wats. var. t.
         (S. Wats.) Mathias &
         Constance
   verticillatum (Geyer) Coult. & Rose

Lilaeopsis Greene

   attenuata (Hook. & Arn.) Fern.
        SY=L. carolinensis Coult. & Rose
   chinensis (L.) Kuntze
        SY=L. lineata (Michx.) Greene
   masonii Mathias & Constance
   occidentalis Coult. & Rose
   recurva A.W. Hill

Limnosciadium Mathias & Constance

   pinnatum (DC.) Mathias & Constance
        SY=Cynosciadium p. DC.
   pumilum (Engelm. & Gray) Mathias &

   Constance

Lomatium Raf.

   ambiguum (Nutt.) Coult. & Rose
   bicolor (S. Wats.) Coult. & Rose
        SY=L. leptocarpum Coult. & Rose
   bradshawii (Rose) Mathias & Constance
   brandegei (Coult. & Rose) J.F. Macbr.
   californicum (Nutt.) Mathias & Constance
   canbyi (Coult. & Rose) J.F. Macbr.
   caruifolium (Hook. & Arn.) Coult. & Rose
      var. caruifolium
      var. denticulatum (Jepson) Jepson
        SY=Leptotaenia anomala Coult. &
         Rose
        SY=Leptotaenia humilis Coult. &
         Rose
        SY=Lomatium h. (Coult. & Rose)
         Hoover ex Mathias &
         Constance
   ciliolatum Jepson
      var. ciliolatum
      var. hooveri Mathias & Constance
   columbianum Mathias & Constance
   concinnum (Osterhout) Mathias
   congdonii Coult. & Rose
   cous (S. Wats.) Coult. & Rose
        SY=Lomatium circumdatum (S. Wats.)
         Coult. & Rose
        SY=L. montanum Coult. & Rose
   cusickii (S. Wats.) Coult. & Rose
   cuspidatum Mathias & Constance
   dasycarpum (Torr. & Gray) Coult. & Rose
      ssp. dasycarpum
      ssp. tomentosum (Benth.) Theobald
        SY=Lomatium t. (Benth.) Coult. &
         Rose
   dissectum (Nutt. ex Torr. & Gray) Mathias &
     Constance
      var. dissectum
      var. multifidum (Nutt. ex Torr. &
        Gray) Mathias & Constance
        SY=Leptotaenia m. Nutt. ex Torr. & Gray
        SY=Lomatium d. var. eatonii
         (Coult. & Rose) Cronq.
   donnellii (Coult. & Rose) Coult. & Rose
   eastwoodae (Coult. & Rose) J.F. Macbr.
   engelmannii Mathias
   farinosum (Geyer) Coult. & Rose
      var. farinosum
      var. hambleniae (Mathias & Constance)
        Schlessman
        SY=Lomatium h. Mathias & Constance
   foeniculaceum (Nutt.) Coult. & Rose
      ssp. daucifolium (Torr. & Gray) Theobald
        SY=Cogswellia d. (Torr. & Gray)
         M.E. Jones
        SY=Lomatium d. (Torr. & Gray)
         Coult. & Rose
        SY=L. f. var. d. (Torr. & Gray)
         Cronq.
      ssp. fimbriatum Theobald
      ssp. foeniculaceum
        SY=Cogswellia f. (Nutt.) Coult. &
         Rose
        SY=C. villosa (Raf.) Schultes
      ssp. inyoense (Mathias & Constance)
        Theobald
        SY=Lomatium i. Mathias & Constance
      ssp. macdougalii (Coult. & Rose)
        Theobald
        SY=Lomatium f. var. m. (Coult. &
         Rose) Cronq.
        SY=L. m. Coult. & Rose
   geyeri (S. Wats.) Coult. & Rose
   gormanii (T.J. Howell) Coult. & Rose

Lomatium (CONT.)
  grayi (Coult. & Rose) Coult. & Rose
      var. depauperatum (M.E. Jones)
        Mathias
      var. grayi
  greenmanii Mathias
  hallii (S. Wats.) Coult. & Rose
      SY=Lomatium nelsonianum J.F.
        Macbr.
  hendersonii (Coult. & Rose) Coult. & Rose
  howellii (S. Wats.) Jepson
  idahoense Mathias & Constance
  insulare (Eastw.) Munz
  junceum Barneby & N. Holmgren
  juniperinum (M.E. Jones) Coult. & Rose
  laevigatum (Nutt.) Coult. & Rose
  latilobum (Rydb.) Mathias
  lucidum (Nutt.) Jepson
  macrocarpum (Hook. & Arn.) Coult. & Rose
      SY=Cogswellia m. (Hook. & Arn.)
        M.E. Jones
      SY=Lomatium m. var. ellipticum
        (Torr. & Gray) Jepson
  marginatum (Benth.) Coult. & Rose
      var. marginatum
        SY=Cogswellia m. (Benth.) M.E.
          Jones
      var. purpureum Jepson
  martindalei (Coult. & Rose) Coult. & Rose
      SY=Lomatium angustatum (Coult. &
        Rose) St. John
      SY=L. a. var. flavum G.N. Jones
      SY=L. m. var. a. (Coult. & Rose)
        Coult. & Rose
      SY=L. m. var. f. (G.N. Jones)
        Cronq.
  megarrhizum (A. Nels.) Mathias
      SY=Neoparrya m. (A. Nels.) W.A.
        Weber
  minimum (Mathias) Mathias
  minus (Rose) Mathias & Constance
  mohavense (Coult. & Rose) Coult. & Rose
    ssp. longilobum Theobald
    ssp. mohavense
  nevadense (S. Wats.) Coult. & Rose
      var. nevadense
      var. parishii (Coult. & Rose) Jepson
        SY=Lomatium n. var. pseudorientale
          (M.E. Jones) Munz
  nudicaule (Pursh) Coult. & Rose
      SY=Cogswellia n. (Pursh) M.E.
        Jones
  nuttallii (Gray) J.F. Macbr.
      var. alpinum (S. Wats.) Mathias
        SY=Cynomarathrum a. (S. Wats.)
          Coult. & Rose
      var. nuttallii
        SY=Cynomarathrum n. (Gray) Coult.
          & Rose
  oreganum Coult. & Rose
  orientale Coult. & Rose
      SY=Cogswellia o. (Coult. & Rose)
        M.E. Jones
  parryi (S. Wats.) J.F. Macbr.
  parvifolium (Hook. & Arn.) Jepson
      SY=Lomatium p. var. pallidum
        (Coult. & Rose) Jepson
  peckianum Mathias & Constance
  piperi Coult. & Rose
  plummerae (Coult. & Rose) Coult. & Rose
      var. plummerae
      var. sonnei (Coult. & Rose) Jepson
        SY=Lomatium p. var. austiniae
          (Coult. & Rose) Jepson
  quintuplex Schlessman & Constance
  ravenii Mathias & Constance
  repostum (Jepson) Mathias

  rigidum (M.E. Jones) Jepson
  rollinsii Mathias & Constance
  salmoniflorum (Coult. & Rose) Mathias &
    Constance
  sandbergii (Coult. & Rose) Coult. & Rose
      SY=Peucedanum s. Coult. & Rose
  scabrum (Coult. & Rose) Mathias
  serpentinum (M.E. Jones) Mathias
  simplex (Nutt.) J.F. Macbr.
      var. leptophyllum (Hook.) Mathias
      var. simplex
        SY=Lomatium platycarpum Coult. &
          Rose
  stebbinsii Schlessman & Constance
  suksdorfii (S. Wats.) Coult. & Rose
      var. suksdorfii
      var. thompsonii Mathias
        SY=Lomatium t. (Mathias) Cronq.
  torreyi (Coult. & Rose) Coult. & Rose
  tracyi Mathias & Constance
  triternatum (Pursh) Coult. & Rose
      var. anomalum (M.E. Jones ex Coult. &
        Rose) Mathias
        SY=Lomatium a. M.E. Jones ex
          Coult. & Rose
      var. brevifolium (Coult. & Rose)
        Mathias
      var. macrocarpum (Coult. & Rose)
        Mathias
      var. triternatum
  tuberosum Hoover
  utriculatum (Nutt. ex Torr. & Gray) Coult.
    & Rose
      SY=L. vaseyi (Coult. & Rose)
        Coult. & Rose
  vaginatum Coult. & Rose
      SY=Lomatium utriculatum (Nutt. ex
        Torr. & Gray) Coult. & Rose
        var. papillatum (Henderson)
        Mathias
  watsonii Coult. & Rose
      SY=Lomatium frenchii Mathias &
        Constance

Musineon Raf.

  divaricatum (Pursh) Nutt. ex Torr. & Gray
      var. divaricatum
      var. hookeri Torr. & Gray
        SY=M. h. (Torr. & Gray) Nutt.
        SY=M. trachyspermum Nutt.
  lineare (Rydb.) Mathias
  tenuifolium Nutt.
      SY=M. divaricatum (Pursh) Nutt. ex
        Torr. & Gray var. t. Nutt.
  vaginatum Rydb.

Myrrhis P. Mill.

  odorata (L.) Scop.

Neoparrya Mathias

  lithophila Mathias

Oenanthe L.

  aquatica (L.) Poir.
  laciniata (Blume) Zoll.
  pimpinelloides L.
  sarmentosa Presl ex DC.

Oreonana Jepson

  clementis (M.E. Jones) Jepson
  purpurascens Shevock & Constance
  vestita (S. Wats.) Jepson

Oreoxis Raf.

  alpina (Gray) Coult. & Rose
    ssp. alpina
    ssp. puberulenta W. A. Weber
  bakeri Coult. & Rose
  humilis Raf.

Orogenia S. Wats.

  fusiformis S. Wats.
  linearifolia S. Wats.

Osmorhiza Raf.

  bipatriata Constance & Shan
  brachypoda Torr.
  chilensis Hook. & Arn.
        SY=O. divaricata (Britt.) Nutt. ex
            Torr. & Gray
        SY=O. nuda Torr.
        SY=Washingtonia d. Britt.
  claytonii (Michx.) C.B. Clarke
        SY=O. aristata (Thunb.) Makino &
            Yabe var. brevistylis (DC.)
            Boivin
        SY=Washingtonia c. (Michx.) Britt.
  depauperata Phil.
        SY=O. chilensis Hook. & Arn. var.
            cupressimontana (Boivin)
            Boivin
        SY=O. obtusa (Coult. & Rose) Fern.
        SY=Washingtonia o. Coult. & Rose
  longistylis (Torr.) DC.
        SY=O. aristata (Thunb.) Makino &
            Yabe var. l. (Torr.) Boivin
        SY=O. l. var. brachycoma Blake
        SY=O. l. var. villicaulis Fern.
        SY=Washingtonia l. (Torr.) Britt.
  occidentalis (Nutt. ex Torr. & Gray) Torr.
  purpurea (Coult. & Rose) Suksdorf
        SY=O. chilensis Hook. & Arn. var.
            p. (Coult. & Rose) Boivin

Oxypolis Raf.

  canbyi (Coult. & Rose) Fern.
  fendleri (Gray) Heller
  filiformis (Walt.) Britt.
  greenmanii Mathias & Constance
  occidentalis Coult. & Rose
  rigidior (L.) Raf.
        SY=O. r. var. ambigua (Nutt.) B.L.
            Robins.
        SY=O. r. var. longifolia (Pursh)
            Britt.
        SY=O. turgida Small
  ternata (Nutt.) Heller

Pastinaca L.

  sativa L.
        var. pratensis Pers.
        var. sativa

Perideridia Reichenb.

  americana (Nutt. ex DC.) Reichenb.
        SY=Eulophus a. Nutt. ex DC.
  bacigalupii Chuang & Constance
  bolanderi (Gray) A. Nels. & J.F. Macbr.
    ssp. bolanderi
    ssp. involucrata Chuang & Constance
  californica (Torr.) A. Nels. & J.F. Macbr.
  erythrorhiza (Piper) Chuang & Constance
  gairdneri (Hook. & Arn.) Mathias
    ssp. borealis Chuang & Constance

    ssp. gairdneri
        SY=Atenia montana (Blank.) Rydb.
  howellii (Coult. & Rose) Mathias
  kelloggii (Gray) Mathias
  lemmonii (Coult. & Rose) Chuang & Constance
  leptocarpa Chuang & Constance
  oregana (S. Wats.) Mathias
  parishii (Coult. & Rose) A. Nels. & J.F.
      Macbr.
    ssp. latifolia (Gray) Chuang & Contance
    ssp. parishii
  pringlei (Coult. & Rose) A. Nels. & J.F.
      Macbr.

Petroselinum J. Hill

  crispum (P. Mill.) Nyman ex A.W. Hill
        SY=Apium petroselinum L.

Peucedanum L.

  kauaiense Hbd.
  ostruthium (L.) W.D.J. Koch
        SY=Imperatoria o. L.
  palustre (L.) Moench
  sandwicense Hbd.
        var. hiroei Deg. & Deg.
        var. sandwicense

Phlojodicarpus Turcz. ex Ledeb.

  ?villosus Turcz.

Pimpinella L.

  anisum L.
  major (L.) Huds.
  saxifraga L.
    ssp. nigra (P. Mill.) Gaudin
    ssp. saxifraga

Podistera S. Wats.

  eastwoodae (Coult. & Rose) Mathias &
      Constance
  macounii (Coult. & Rose) Mathias &
      Constance
  nevadensis (Gray) S. Wats.
  yukonensis Mathias & Constance

Polytaenia DC.

  nuttallii DC.
        SY=Pleiotaenia n. (DC.) Coult. &
            Rose
  texana (Coult. & Rose) Mathias & Constance
        SY=Pleiotaenia nuttallii (DC.)
            Coult. & Rose var. t.
            (Coult. & Rose) Coult. &
            Rose
        SY=Polytaenia n. DC. var. t.
            Coult. & Rose

Pseudocymopterus Coult. & Rose

  longiradiatus Mathias, Constance & Theobald
  montanus (Gray) Coult. & Rose
        SY=Cymopterus lemmonii (Coult. &
            Rose) Dorn
        SY=C. m. (Nutt.) Torr. & Gray

Pseudotaenidia Mackenzie

  montana Mackenzie

Pteryxia Nutt. ex Coult. & Rose

Pteryxia (CONT.)
    davidsonii (Coult. & Rose) Mathias &
        Constance
                SY=Aletes d. Coult. & Rose
    hendersonii Coult. & Rose
                SY=Cymopterus h. (Coult. & Rose)
                    Cronq.
                SY=Cymopterus longilobus (Rydb.)
                    W.A. Weber
    petraea (M.E. Jones) Coult. & Rose
                SY=Cymopterus p. M.E. Jones
    terebinthina (Hook.) Coult. & Rose
        var. calcarea (M.E. Jones) Mathias
                SY=Cymopteris t. (Hook.) Torr. &
                    Gray var. albiflora (Torr.
                    & Gray) M.E. Jones
                SY=C. t. var. c. (M.E. Jones)
                    Cronq.
        var. californica (Coult. & Rose)
            Mathias
                SY=Cymopterus t. (Hook.) Torr. &
                    Gray var. c. (Coult. & Rose)
                    Jepson
        var. foeniculacea (Torr. & Gray)
            Mathias
                SY=Cymopterus t. (Hook.) Torr. &
                    Gray var. f. (Torr. & Gray)
                    Cronq.
        var. terebinthina
                SY=Cymopterus t. (Hook.) Torr. &
                    Gray

Ptilimnium Raf.

    capillaceum (Michx.) Raf.
    costatum (Ell.) Raf.
    fluvatile (Rose) Mathias
                SY=Harperella f. Rose
                SY=H. vivipara Rose
                SY=P. v. (Rose) Mathias
    nodosum (Rose) Mathias
                SY=Harperella n. Rose
    nuttallii (DC.) Britt.
    X texense Coult. & Rose [capillaceum X
        nuttallii]

Sanicula L.

    arctopoides Hook. & Arn.
                SY=S. crassicaulis Poepp. ex DC.
                    var. howellii (Coult. &
                    Rose) Mathias
    arguta Greene ex Coult. & Rose
                SY=S. simulans Hoover
    bipinnata Hook. & Arn.
    bipinnatifida Dougl. ex Hook.
                SY=S. b. var. flava Jepson
    canadensis L.
                SY=S. c. var. floridana (Bickn.)
                    H. Wolff
                SY=S. c. var. grandis Fern.
                SY=S. c. var. typica H. Wolff
                SY=S. f. Bickn.
    crassicaulis Poepp. ex DC.
        var. crassicaulis
        var. tripartita (Suksdorf) H. Wolff
    graveolens Poepp. ex DC.
                SY=S. g. var. septentrionalis
                    (Greene) Mathias
                SY=S. nevadensis S. Wats.
    gregaria Bickn.
    hoffmannii (Munz) Bell
    kauaiensis St. John
    laciniata Hook. & Arn.
    marilandica L.
                SY=S. m. var. petiolulata Fern.
    maritima Kellogg ex S. Wats.

    peckiana J.F. Macbr.
    purpurea St. John & Hosaka
    sandwicensis Gray
                SY=S. s. var. lobata H. Wolff
    saxatilis Greene
    smallii Bickn.
    tracyi Shan & Constance
    trifoliata Bickn.
    tuberosa Torr.

Scandix L.

    pecten-veneris L.

Seseli L.

    libanotis (L.) W.D.J. Koch

Sium L.

    carsonii Dur.
    floridanum Small
    suave Walt.
                SY=S. cicutifolium Schrank

Spermolepis Raf.

    divaricata (Walt.) Britt.
    echinata (Nutt.) Heller
                SY=Apium e. (Nutt.) Benth. & Hook.
                    ex S. Wats.
    hawaiiensis H. Wolff
    inermis (Nutt.) Mathias & Constance
                SY=S. patens (Nutt.) B.L. Robins.

Sphenosciadium Gray

    capitellatum Gray

Synelcosciadium Boiss.

    carmelii (Labill.) Boiss.

Taenidia (Torr. & Gray) Drude

    integerrima (L.) Drude

Tauschia Schlecht.

    arguta (Torr. & Gray) J.F. Macbr.
    glauca (Coult. & Rose) Mathias & Constance
    hartwegii (Gray) J.F. Macbr.
    hooveri Mathias & Constance
    howellii (Coult. & Rose) J.F. Macbr.
    kelloggii (Gray) J.F. Macbr.
    parishii (Coult. & Rose) J.F. Macbr.
    stricklandii (Coult. & Rose) Mathias &
        Constance
    tenuissima (Geyer ex Hook.) Mathias &
        Constance
                SY=Leibergia orogenioides Coult. &
                    Rose
                SY=Lomatium o. (Coult. & Rose)
                    Mathias
    texana Gray

Thaspium Nutt.

    barbinode (Michx.) Nutt.
                SY=T. b. var. angustifolium Coult.
                    & Rose
                SY=T. chapmanii (Coult. & Rose)
                    Small
    pinnatifidum (Buckl.) Gray
    trifoliatum (L.) Gray
        var. flavum Blake
                SY=T. aureum Nutt.

Thaspium (CONT.)
    SY=T. t. var. a. (Nutt.) Britt.
    var. trifoliatum
      SY=T. aureum Nutt. var. t. (L.)
        Coult. & Rose

Tilingia Regel & Tiling

    ajanense (Regel & Tiling) Drude
      SY=Cnidium a. (Regel & Tiling)
        Drude

Tordylium L.

    apulum L.

Torilis Adans.

    arvensis (Huds.) Link
      ssp. purpurea (Ten.) Hayek
        SY=T. heterophylla Guss.
    japonica (Houtt.) DC.
    leptophylla (L.) Reichenb. f.
    nodosa (L.) Gaertn.
    scabra (Thunb.) DC.

Trepocarpus Nutt. ex DC.

    aethusae Nutt. ex DC.

Turgenia Hoffmann

    latifolia (L.) Hoffmann
      SY=Caucalis l. L.

Yabea K.-Pol.

    microcarpa (Hook. & Arn.) K.-Pol.
      SY=Caucalis m. Hook. & Arn.

Zizia W.D.J. Koch

    aptera (Gray) Fern.
      SY=Z. a. var. occidentalis Fern.
      SY=Z. cordata (Walt.) DC.
    aurea (L.) W.D.J. Koch
    latifolia Small
    trifoliata (Michx.) Fern.
      SY=Z. arenicola Rose
      SY=Z. bebbii (Coult. & Rose)
        Britt.

APOCYNACEAE

Allamanda L.

    cathartica L.

Alyxia Banks ex R. Br.

    oliviformis Gaud.
      SY=A. o. var. lanceolata Hbd.
      SY=A. o. var. myrtillifolia Gray
        ex Hbd.
      SY=A. o. var. ovata Hbd.

Amsonia Walt.

    arenaria Standl.
    brevifolia Gray
      SY=A. tomentosa Torr. & Frem. ex
        Frem. var. stenophylla
        Kearney & Peebles
    ciliata Walt.

      var. ciliata
      var. filifolia Woods.
        SY=A. c. var. tenuifolia (Raf.)
          Woods.
      var. texana (Gray) Coult.
    eastwoodiana Rydb.
    glaberrima Woods.
    grandiflora Alexander
    hirtella Standl.
      SY=A. pogonosepala Woods.
      SY=A. standleyi Woods.
    hubrichtii Woods.
    illustris Woods.
    jonesii Woods.
    longiflora Torr.
    ludoviciana Vail
    palmeri Gray
      SY=A. kearneyana Woods.
    peeblesii Woods.
    repens Shinners
    rigida Shuttlw.
    salpignantha Woods.
    tabernaemontana Walt.
      var. salicifolia (Pursh) Woods.
        SY=A. s. Pursh
        SY=A. t. var. gattingeri Woods.
      var. tabernaemontana
        SY=A. amsonia (L.) Britt.
    tharpii Woods.
    tomentosa Torr. & Frem. ex Frem.

Anechites Griseb.

    nerium (Aubl.) Urban

Angadenia Miers ex Woods.

    berterii (A. DC.) Miers
      SY=Rhabdadenia corallicola Small

Apocynum L.

    androsaemifolium L.
      ssp. androsaemifolium
        var. androsaemifolium
          SY=A. a. var. glabrum Macoun ex
            Woods.
        var. griseum (Greene) Beg. & Bel.
        var. incanum A. DC.
          SY=A. scopulorum Greene ex Rydb.
      ssp. pumilum (Gray) Boivin
        var. pumilum Gray
        var. tomentellum (Greene) Boivin
        var. woodsonii Boivin
          SY=A. a. var. intermedium Woods.
    cannabinum L.
      var. cannabinum
        SY=A. c. var. greeneanum (Beg. &
          Bel.) Woods.
        SY=A. c. var. nemorale (G.S.
          Mill.) Fern.
        SY=A. c. var. pubescens (Mitchell
          ex R. Br.) A. DC.
        SY=A. p. Mitchell ex R. Br.
      var. glaberrimum A. DC.
        SY=A. album Greene
      var. hypericifolium Gray
        SY=A. c. var. suksdorfii (Greene)
          Beg. & Bel.
        SY=A. h. Ait.
        SY=A. sibiricum Jacq.
        SY=A. sibiricum var. cordigerum
          (Greene) Fern.
        SY=A. sibiricum var. farwellii
          (Greene) Fern.
        SY=A. sibiricum var. salignum
          (Greene) Fern.

APOCYNACEAE   39

Apocynum   (CONT.)
                SY=A. suksdorfii Greene
                SY=A. suksdorfii (Greene) Beg. &
                      Bel.
                SY=A. suksdorfii var.
                      angustifolium (Woot.) Woods.
        X medium Greene [androsaemifolium ssp.
          androsaemifolium var. incanum X
          cannabinum var. hypericifolium]
                SY=A. jonesii Woods.
                SY=A. m. var. floribundum (Greene)
                      Woods.
                SY=A. m. var. leuconeuron (Greene)
                      Woods.
                SY=A. m. var. lividum (Greene)
                      Woods.
                SY=A. m. var. sarniense (Greene)
                      Woods.
                SY=A. m. var. vestitum (Greene)
                      Woods.
                SY=A. milleri Britt.

Carissa L.

    macrocarpa (Ecklon) DC.

Catharanthus G. Don

    roseus (L.) G. Don
                SY=Vinca r. L.

Cycladenia Benth.

    humilis Benth.
          var. humilis
          var. jonesii (Eastw.) Welsh & Atwood
                SY=C. j. Eastw.
          var. tomentosa (Gray) Gray
          var. venusta (Eastw.) Woods. ex Munz

Echites P. Br.

    umbellata Jacq.
                SY=E. echites (L.) Britt.

Forsteronia G.F.W. Mey.

    corymbosa (Jacq.) G.F.W. Mey.
    portoricensis Woods.

Funtumia Stapf

    elastica (Preuss) Stapf

Haplophyton A. DC.

    crooksii L. Benson
                SY=H. cimicidum sensu auctt. non
                      A. DC.

Macrosiphonia Muell.-Arg.

    brachysiphon (Torr.) Gray
    hypoleuca (Benth.) Muell.-Arg.
    macrosiphon (Torr.) Heller

Nerium L.

    oleander L.

Ochrosia Juss.

    compta K. Schum.
    forbesii St. John
    haleakalae St. John
    hamakuaensis St. John
    holei St. John

                SY=O. sandwicensis sensu auctt.
                      non DC.
    kauaiensis St. John
    kilaueaensis St. John
    konaensis St. John
    kondoi St. John
    lamoureuxii St. John
    microcalyx St. John

Plumeria L.

    alba L.
    obtusa L.
          var. obtusa
                SY=P. krugii Urban
                SY=P. portoricensis Urban
          var. sericifolia (C. Wright) Woods.
    rubra L.

Prestonia R. Br.

    agglutinata (Jacq.) Woods.
                SY=Echites a. Jacq.

Pteralyxia K. Schum.

    caumiana Deg.
    kauaiensis Caum
    macrocarpa (Hbd.) K. Schum.

Rauvolfia L.

    degeneri Sherff
    forbesii Sherff
    helleri Sherff
    mauiensis Sherff
    molokaiensis Sherff
          var. molokaiensis
          var. parvifolia Deg. & Sherff
    nitida Jacq.
    remotiflora Deg. & Sherff
    tetraphylla L.
    remotiflora Deg. & Sherff
    sandwicensis A. DC.
          var. sandwicensis
          var. subacuminata Sherff
    viridis Roemer & Schultes
                SY=R. lamarckii A. DC.

Rhabdadenia Muell.-Arg.

    biflora (Jacq.) Muell.-Arg.

Tabernaemontana L.

    citrifolia L.
                SY=T. oppositifolia (Spreng.)
                      Urban
    divaricata (L.) R. Br. ex Roemer & Schultes
                SY=T. coronaria (Jacq.) Willd.

Thevetia Adans.

    peruviana (Pers.) K. Schum.
                SY=Cerbera thevetia L.

Trachelospermum Lem.

    difforme (Walt.) Gray

Urechites Muell.-Arg.

    lutea (L.) Britt.
          var. lutea
          var. sericea R.W. Long
                SY=U. pinetorum Small

Vallesia Ruiz & Pavon

    antillana Woods.
    glabra (Cav.) Link

Vinca L.

    herbacea Waldst. & Kit.
    major L.
    minor L.

APONOGETONACEAE

Aponogeton L. f.

    distachyus L. f.

AQUIFOLIACEAE

Ilex L.

    ambigua (Michx.) Torr.
          SY=I. caroliniana (Walt.) Trel.
    amelanchier M.A. Curtis
    anomala Hook. & Arn.
    aquifolium L.
    X attenuata Ashe [cassine X opaca]
    beadlei Ashe
          SY=I. montana Torr. & Gray var. b.
             (Ashe) Fern.
          SY=I. m. var. mollis (Gray) Britt.
    buswellii Small
    cassine L.
    collinus Alexander
          SY=Nemopanthus c. (Alexander) Ross
             Clark
    cookii Britt. & Wilson
    coriacea (Pursh) Chapman
    cumulicola Small
          SY=I. opaca Ait. var. arenicola
             (Ashe) Ashe
    curtissii (Fern.) Small
          SY=I. decidua Walt. var. c. Fern.
    cuthbertii Small
    decidua Walt.
    glabra (L.) Gray
    guianensis (Aubl.) Kuntze
    hypaneura Loes.
    krugiana Loes.
    laevigata (Pursh) Gray
    longipes Chapman ex Trel.
          SY=I. decidua Walt. var. l.
             (Chapman ex Trel.) Ahles
          SY=I. l. var. hirsuta Lundell
    macfadyenii (Walp.) Rehd.
    montana Torr. & Gray
          SY=I. ambigua (Michx.) Torr. var.
             m. (Torr. & Gray) Ahles
          SY=I. monticola Gray
    myrtifolia Walt.
          SY=I. cassine L. var. m. (Walt.)
             Sarg.
    nitida (Vahl) Maxim.
    opaca Ait.
    sideroxyloides (Sw.) Griseb.
        var. occidentalis (Macfad.) Loes.
        var. sideroxyloides
    sintenisii (Urban) Britt.
    urbaniana Loes.
        var. riedalei (Loes.) Edwin
          SY=I. r. Loes.

        var. urbaniana
    verticillata (L.) Gray
          SY=I. bronxensis Britt.
          SY=I. fastigiata Bickn.
          SY=I. v. var. cyclophylla B.L.
             Robins.
          SY=I. v. var. f. (Bickn.) Fern.
          SY=I. v. var. padifolia (Willd.)
             Torr. & Gray ex S. Wats.
          SY=I. v. var. tenuifolia (Torr.)
             S. Wats.
    vomitoria Ait.

Nemopanthus Raf.

    mucronata (L.) Trel.

ARACEAE

Acorus L.

    americanus (Raf.) Raf.
          SY=A. calamus sensu auctt. non L.
          SY=A. c. var. a. (Raf.) H.D.
             Wulff.

Alocasia (Schott) G. Don

    macrorrhiza (L.) G. Don

Anthurium Schott

    acaule (Jacq.) Schott
    cordifolium Kunth
          SY=A. cordatum (Willd.) G. Don
    dominicense Schott
    scandens (Aubl.) Engl.
    selloum K. Koch

Arisaema Mart.

    acuminatum Small
    dracontium (L.) Schott
          SY=Muricauda d. (L.) Small
    quinatum (Nutt.) Schott
    triphyllum (L.) Schott
      ssp. pusillum (Peck) Huttleston
          SY=A. p. (Peck) Nash
      ssp. stewardsonii (Britt.) Huttleston
          SY=A. atrorubens (Ait.) Blume var.
             s. (Britt.) Stevens
          SY=A. s. Britt.
      ssp. triphyllum
          SY=A. atrorubens (Ait.) Blume

Arum L.

    italicum P. Mill.

Caladium Vent.

    bicolor (Ait.) Vent.
          SY=Cyrtospadix b. (Ait.) Britt. &
             Wilson

Calla L.

    palustris L.

Colocasia Schott

    esculenta (L.) Schott
        var. antiquorum (Schott) Hubbard &
        Rehd.

Colocasia (CONT.)
        var. aquatilis Hassk.
        var. esculenta

Dieffenbachia Schott

    seguine (Jacq.) Schott

Dracontium L.

    polyphyllum L.

Lysichiton Schott

    americanum Hulten & St. John
            SY=L. camtschatcense pro parte non
            (L.) Schott

Montrichardia Crueger

    arborescens (L.) Schott

Orontium L.

    aquaticum L.

Peltandra Raf.

    virginica (L.) Schott
        ssp. luteospadix (Fern.) Blackwell &
            Blackwell
                SY=P. glauca (Ell.) Feay
                SY=P. l. Fern.
                SY=P. sagittifolia (Michx.) Morong
        ssp. virginica
                SY=P. tharpii Barkl.

Philodendron Schott

    fragrantissima (Hook.) Kunth
    giganteum Schott
    hederaceum (Jacq.) Schott
    krebsii Schott
    lingulatum (L.) K. Koch
    marginatum Urban
    nechodomii Britt.
    scandens K. Koch & Sello
        ssp. oxycardium (Schott) Bunting
                SY=P. o. Schott
        ssp. scandens
                SY=P. micans (Klotzsch) K. Koch

Pinellia Ten.

    ternata (Thunb.) Breitenbach

Pistia L.

    stratiotes L.

Symplocarpus Salisb. ex Nutt.

    foetidus (L.) Nutt.
            SY=Spathyema f. (L.) Raf.

Syngonium Schott

    podophyllum Schott
        var. albolineatum Engelm.
        var. podophyllum

Xanthosoma Schott

    atrovirens K. Koch & Bouche
    brasiliense (Desf.) Engl.
    helleborifolium (Jacq.) Schott
    hoffmannii Schott

jacquinii Schott
nigrum (Vell.) Stellfeld
        SY=X. violaceum Schott
sagittifolium (L.) Schott

Zantedeschia Spreng.

    aethiopica (L.) Spreng.
            SY=Calla a. L.

ARALIACEAE

Acanthopanax (Dcne. & Planch.) Miq.

    sieboldianus Makino

Aralia L.

    californica S. Wats.
    hispida Vent.
    humilis Cav.
    nudicaulis L.
    racemosa L.
        ssp. bicrenata (Woot. & Standl.) Welsh &
            Atwood
                SY=A. b. Woot. & Standl.
        ssp. racemosa
    spinosa L.

Cheirodendron Nutt. ex Seem.

    dominii Krajina
    fauriei Hochr.
            var. fauriei
            var. macdanielsii Sherff
    helleri Sherff
            var. helleri
            var. microcarpum Sherff
            var. multiflorum (Skottsberg) Sherff
            var. sodalium Sherff
    kauaiense Krajina
            var. forbesii Sherff
            var. kauaiense
            var. keakuense Krajina
    platyphyllum (Hook. & Arn.) Seem.
    trigynum (Gaud.) Heller
            var. acuminatum Skottsberg
            var. confertiflorum Sherff
            var. degeneri Sherff
            var. fosbergii Sherff
            var. halawanum Sherff
            var. hillebrandii Sherff
            var. ilicoides Sherff
            var. mauiens Levl.
            var. molokaiense Sherff
            var. oblongum Sherff
            var. osteostigma Sherff
            var. rockii Sherff
            var. skottsbergii Sherff
            var. subcordatum Sherff
            var. trigynum

Dendropanax Dcne. & Planch.

    arboreus (L.) Dcne. & Planch.
    laurifolius (E. March.) R.C. Schneid.

Didymopanax Dcne. & Planch.

    gleasonii Britt. & Wilson
    morototonii (Aubl.) Dcne. & Planch.

Hedera L.

Hedera   (CONT.)
  helix L.

Munroidendron Sherff

   racemosum (Forbes) Sherff
       var. forbesii Sherff
       var. macdanielsii Sherff
       var. racemosum

Oplopanax Miq.

   horridus (Sm.) Miq.
         SY=Echinopanax h. (Sm.) Dcne.

Panax L.

   quinquefolius L.
   trifolius L.

Reynoldsia Gray

   degeneri Sherff
   hillebrandii Sherff
   huehuensis Sherff
       var. brevipes Sherff
       var. huehuensis
       var. intermedia Sherff
   mauiensis Sherff
       var. macrocarpa Deg. & Sherff
       var. mauiensis
   oblonga Sherff
   sandwicensis Gray
       var. molokaiensis Sherff
       var. sandwicensis
         SY=R. hosakana Sherff
         SY=R. s. var. intercedens Sherff
   venusta Sherff
       var. lanaiensis Sherff
       var. venusta

Tetraplasandra Gray

   bisattenuata Sherff
   gymnocarpa (Hbd.) Sherff
       var. gymnocarpa
       var. leptocarpa Sherff
       var. megalocarpa Sherff
       var. pupukeensis (Deg.) Sherff
   hawaiensis Gray
       var. awiniensis Sherff
       var. gracilis Sherff
       var. hawaiensis
       var. microcarpa Sherff
   kaalae (Hbd.) H.A.T. Harms
       var. kaalae
       var. multiplex Sherff
   kahanana Deg. & Sherff
   kavaiensis (Mann) Sherff
       var. dipyrena (Mann) Sherff
       var. grandis Sherff
       var. intercedens Sherff
       var. kavaiensis
       var. koloana Sherff
       var. nahikuensis Sherff
       var. occidua Deg. & Sherff
       var. robustior Sherff
   kohalae Skottsberg
   lanaiensis Rock
   lihuensis Sherff
       var. gracilipes Sherff
       var. lihuensis
   lydgatei (Hbd.) H.A.T. Harms
       var. brachypoda Sherff
       var. coriacea Sherff
       var. forbesii Sherff
       var. leptorhachis Deg. & Sherff

       var. lydgatei
   meiandra (Hbd.) H.A.T. Harms
       var. bisobtusa Sherff
       var. bryanii Sherff
       var. degeneri Sherff
       var. hillebrandii Sherff
       var. hiloensis Sherff
       var. leptomera Sherff
       var. longipedunculata Skottsberg
       var. makalehana Sherff
       var. mauiensis Sherff
       var. meiandra
       var. molokaiensis Skottsberg
       var. occidentalis Skottsberg
       var. olowaluana Sherff
       var. ovalis Sherff
       var. polyantha Sherff
       var. polystigmata Sherff
       var. prolifica Sherff
       var. prolificoides Sherff
       var. ramosior Sherff
       var. rhynchocarpa Sherff
       var. rhynchocarpoides Deg. & Sherff
       var. rockii Sherff
       var. simulans Sherff
       var. skottsbergii Sherff
       var. tenuistylis Sherff
       var. tenuistyloides Deg. & Sherff
  micrantha Sherff
  munroi Sherff
  oahuensis (Gray) H.A.T. Harms
       var. eradiata Sherff
       var. fauriei Sherff
       var. hailiensis (Deg. & Sherff) Deg.
         & Sherff
       var. longipes Sherff
       var. oahuensis
       var. pseudolongipes Deg. & Sherff
       var. pseudorhachis Sherff
       var. subglobosa Deg., Deg. & Sherff
       var. venulosior Sherff
  pupukeensis Skottsberg
       var. decipiens Deg. & Sherff
       var. megalopoda Sherff
       var. nitida Deg. & Sherff
       var. pupukeensis
       var. venosa Deg. & Sherff
  sherffii Deg. & Deg.
  turbans Sherff
  waialealae Rock
       var. acrior Sherff
       var. kohuana Deg. & Sherff
       var. pluricostata Deg. & Sherff
       var. subsessilis Sherff
       var. urceolata Sherff
       var. wahiawensis Sherff
       var. waialealae
  waianensis Deg.
       var. palehuana Sherff
       var. waianensis
  waimeae Wawra
       var. angustior Sherff
       var. waimeae

ARECACEAE

Acoelorraphe H. Wendl.

   wrightii (Griseb. & H. Wendl.) H. Wendl. ex
   Becc.
        SY=Paurotis w. (Griseb. & H.
          Wendl.) Britt.

Acrocomia Mart.

Acrocomia (CONT.)
    media O.F. Cook

Aiphanes Willd.

    acanthophylla (Mart.) Burret
                SY=Bactris a. Mart.

Calyptronoma Griseb.

    rivalis (O.F. Cook) Bailey
                SY=Calyptrogyne r. (O.F. Cook)
                    Leon

Coccothrinax Sarg.

    alta (O.F. Cook) Becc.
    argentata (Jacq.) Bailey
    discreta Bailey
    eggersiana Becc.
            var. eggersiana
            var. sanctae-crucis Becc.
    sancti-thomae Becc.

Cocos L.

    nucifera L.

Gaussia H. Wendl.

    attenuata (O.F. Cook) Becc.

Prestoea Hook. f.

    montana (Graham) Nichols.
            SY=Euterpe globosa sensu auctt.
                non Gaertn.
            SY=E. m. Graham

Pritchardia Seem. & H. Wendl.

    affinis Becc.
            var. affinis
            var. gracilis Becc.
            var. halophila Becc.
            var. rhopalocarpa Becc.
    arecina Becc.
    aylmer-robinsonii St. John
    beccariana Rock
            var. beccariana
            var. giffardiana Becc.
    brevicalyx Becc. & Rock
    donata Caum
    elliptica Rock & Caum
    eriophora Becc.
    eriostachya Becc.
    forbesiana Rock
    gaudichaudii (Mart.) H. Wendl.
    glabrata Becc. & Rock
    hardyi Rock
    hillebrandii Becc.
    insignis Becc.
    kaalae Rock
            var. kaalae
            var. minima Caum
    kahanae Rock & Caum
    kahukuensis Caum
    kamapuaana Caum
    lanaiensis Becc. & Rock
    lanigera Becc.
    lowreyana Rock
            var. lowreyana
            var. turbinata Rock
    macdanielsii Caum
    macrocarpa Linden ex Andre
    martii (Gaud.) H. Wendl.
    martioides Rock & Caum

    minor Becc.
    montis-kea Rock
    munroi Rock
    remota Becc.
    rockiana Becc.
    viscosa Rock
    weissichiana Rock

Pseudophoenix H. Wendl. ex Sarg.

    sargentii H. Wendl. ex Sarg.
            SY=P. vinifera sensu auctt. non
                (Mart.) Becc.

Rhapidophyllum H. Wendl. & Drude

    hystrix (Pursh) H. Wendl. & Drude

Roystonea O.F. Cook

    borinquena O.F. Cook
    elata (Bartr.) F. Harper
            SY=R. regia (H.B.K.) O.F. Cook

Sabal Adans.

    causiarum (O.F. Cook) Becc.
    etonia Swingle ex Nash
    mexicana Mart.
            SY=S. texana (O.F. Cook) Becc.
    minor (Jacq.) Pers.
            SY=S. deeringiana Small
            SY=S. louisiana (Darby) Bomhard
    palmetto (Walt.) Lodd. ex Schultes &
    Schultes
            SY=S. jamesiana Small

Serenoa Hook. f.

    repens (Bartr.) Small

Thrinax Sw.

    morrisii H. Wendl.
            SY=T. keyensis Sarg.
            SY=T. microcarpa Sarg.
            SY=T. ponceana O.F. Cook
            SY=T. praeceps O.F. Cook
    radiata Lodd. ex Schultes & Schultes
            SY=T. floridana Sarg.
            SY=T. parviflora sensu auctt. non
                Sw.

Washingtonia H. Wendl.

    filifera (Linden ex Andre) H. Wendl.
    robusta H. Wendl.
            SY=W. filifera (Linden ex Andre)
                H. Wendl. var. r. (H.
                Wendl.) Parish
            SY=W. gracilis Parish
            SY=W. sonorae S. Wats.

ARISTOLOCHIACEAE

Aristolochia L.

    anguicida Jacq.
    bilabiata L.
            SY=A. oblongata Jacq.
    bilobata L.
    californica Torr.
            SY=Isotrema c. (Torr.) Huber
    clematitis L.

Aristolochia (CONT.)
  cordiflora Mutis ex H.B.K.
  coryi I.M. Johnston
      SY=A. brevipes sensu auctt. non
        Benth.
      SY=A. b. var. acuminata S. Wats.
  erecta L.
      SY=A. longiflora Engelm. & Gray
  grandiflora Vahl
  littoralis Parodi
  macrophylla Lam.
      SY=A. durior Hill
  maxima Jacq.
  odoratissima L.
  peltata L.
  pentandra Jacq.
      SY=A. marshii Standl.
  reticulata Jacq.
  serpentaria L.
      SY=A. convolvulacea Small
      SY=A. hastata Nutt.
      SY=A. nashii Kearney
      SY=A. s. var. h. (Nutt.) Duchartre
      SY=A. s. var. n. (Kearney) Ahles
  tomentosa Sims
      SY=Isotrema t. (Sims) Huber
  trilobata L.
  watsonii Woot. & Standl.
      SY=A. porphyrophylla Pfeifer
  wrightii Seem.

Asarum L.

  canadense L.
      var. acuminatum Ashe
      var. ambiguum (Bickn.) Farw.
      var. canadense
        SY=A. rubrocinctum Peattie
      var. reflexum (Bickn.) B.L. Robins.
        SY=A. r. Bickn.
  caudatum Lindl.
      SY=A. c. var. virdiflorum M.E.
        Peck
  hartwegii S. Wats.
  lemmonii S. Wats.

Hexastylis Raf.

  arifolia (Michx.) Small
      var. arifolia
        SY=Asarum a. Michx.
      var. callifolia (Small) Blomquist
        SY=H. c. Small
      var. ruthii (Ashe) Blomquist
        SY=Asarum r. Ashe
        SY=H. r. (Ashe) Small
  contracta Blomquist
  heterophylla (Ashe) Small
      SY=Asarum h. Ashe
  lewisii (Fern.) Blomquist & Oosting
      SY=Asarum l. Fern.
  minus (Ashe) Blomquist
      SY=Asarum m. Ashe
  naniflora Blomquist
  shuttleworthii (Britten & Baker) Small
      SY=Asarum s. Britten & Baker
  speciosa Harper
  virginica (L.) Small
      SY=Asarum memmingeri Ashe
      SY=A. v. L.
      SY=H. m. (Ashe) Small

ASCLEPIADACEAE

Araujia Brot.

  sericifera Brot.

Asclepias L.

  albicans S. Wats.
  amplexicaulis Sm.
  angustifolia Schweig.
  arenaria Torr.
  asperula (Dcne.) Woods.
    ssp. asperula
      SY=A. capricornu Woods. ssp.
        occidentalis Woods.
    ssp. capricornu (Woods.) Woods.
      SY=A. a. var. decumbens (Nutt.)
        Shinners
      SY=A. c. Woods.
      SY=A. d. (Nutt.) Dcne.
      SY=Asclepiodora d. (Nutt.) Gray
  auriculata (Engelm. ex Torr.) Holz.
      SY=Acerates a. Engelm. ex Torr.
  brachystephana Engelm. ex Torr.
  californica Greene
    ssp. californica
    ssp. greenei Woods.
  cinerea Walt.
  connivens Baldw. ex Ell.
      SY=Anantherix c. (Baldw. ex Ell.)
        Feay
  cordifolia (Benth.) Jepson
  cryptoceras S. Wats.
    ssp. cryptoceras
    ssp. davisii (Woods.) Woods.
      SY=Asclepias c. var. d. (Woods.)
        W.H. Baker
      SY=A. d. Woods.
  curassavica L.
  curtissii Gray
      SY=Asclepias aceratoides Nash
      SY=Oxypteryx c. (Gray) Small
  cutleri Woods.
  elata Benth.
  emoryi (Greene) Vail ex Small
  engelmanniana Woods.
  eriocarpa Benth.
      SY=Asclepias e. var. microcarpa
        Munz & Johnston
      SY=A. fremontii Torr.
  erosa Torr.
  exaltata L.
      SY=Asclepias bicknellii Vail
      SY=A. phytolaccoides Pursh
  fascicularis Dcne.
  feayi Chapman ex Gray
      SY=Asclepiodella f. (Chapman ex
        Gray) Small
  X giffordii Eastw. [eriocarpa X speciosa]
  hallii Gray
  hirtella (Pennell) Woods.
      SY=Acerates h. Pennell
  humistrata Walt.
  hypoleuca (Gray) Woods.
  incarnata L.
    ssp. incarnata
    ssp. pulchra (Ehrh. ex Willd.) Woods.
      SY=Asclepias i. var. neoscotica
        Fern.
      SY=A. i. var. p. (Ehrh. ex Willd.)
        Pers.
      SY=A. p. Ehrh. ex Willd.
  involucrata Englem. ex Torr.
  labriformis M.E. Jones
  lanceolata Walt.
      SY=Asclepias l. var. paupercula
        (Michx.) Fern.
  latifolia Raf.

Asclepias (CONT.)
  lemmonii Gray
  linaria Cav.
  linearis Scheele
  longifolia Michx.
        SY=Acerates delticola Small
        SY=Acerates floridana (Lam.) A.S.
            Hitchc.
        SY=Acerates l. (Michx.) Ell.
  macrosperma Eastw.
  macrotis Torr.
  meadii Torr. ex Gray
  michauxii Dcne.
  nivea L.
  nummularia Torr.
  nyctaginifolia Gray
  obovata Ell.
        SY=Asclepias viridiflora Raf. var.
            o. (Ell.) Torr.
  oenotheroides Cham. & Schlecht.
        SY=Asclepias brevicornu Scheele
  otarioides Fourn.
        SY=Acerates lanuginosa (Nutt.)
            Dcne.
        SY=Asclepias l. Nutt.
  ovalifolia Dcne.
  pedicellata Walt.
        SY=Podostigma p. (Walt.) Vail
  perennis Walt.
  prostrata Blackwell
  pumila (Gray) Vail
  purpurascens L.
  quadrifolia Jacq.
  quinquedentata Gray
  rubra L.
        SY=Asclepias laurifolia Michx.
  rusbyi (Vail) Woods.
        SY=Asclepias engelmanniana Woods.
          var. r. (Vail) Kearney
  ruthiae Maguire
        SY=Asclepias eastwoodiana Barneby
  scaposa Vail
  solanoana Woods.
        SY=Solanoa purpurascens (Gray)
            Greene
  speciosa Torr.
  sperryi Woods.
  stenophylla Gray
        SY=Acerates angustifolia (Nutt.)
            Dcne.
        SY=Polyotus a. Nutt.
  subulata Dcne.
  subverticillata (Gray) Vail
  sullivantii Engelm. ex Gray
  syriaca L.
        SY=Asclepias intermedia Vail
        SY=A. kansana Vail
        SY=A. s. var. k. (Vail) Palmer &
            Steyermark
  texana Heller
  tomentosa Ell.
  tuberosa L.
    ssp. interior Woods.
        SY=Asclepias t. var. i. (Woods.)
            Shinners
    ssp. rolfsii (Britt. ex Vail) Woods.
        SY=Asclepias r. Britt. ex Vail
    ssp. terminalis Woods.
    ssp. tuberosa
  uncialis Greene
  variegata L.
        SY=Biventraria v. (L.) Small
  verticillata L.
  vestita Hook. & Arn.
    ssp. parishii (Jepson) Woods.
        SY=Asclepias v. var. p. Jepson
    ssp. vestita

  viridiflora Raf.
        SY=Acerates v. (Raf.) Eat.
        SY=Acerates v. var. ivesii Britt.
        SY=Acerates v. var. linearis Gray
        SY=Asclepias v. var. lanceolata
            (Ives) Torr.
        SY=Asclepias v. var. linearis
            (Gray) Fern.
  viridis Walt.
        SY=Asclepiodora v. (Walt.) Gray
  viridula Chapman
  welshii N. Holmgren & P. Holmgren

Calotropis R. Br.

  procera (Ait.) R. Br.
        SY=Asclepias p. Ait.

Cryptostegia R. Br.

  grandiflora (Roxb.) R. Br.
        SY=Nerium g. Roxb.
  madagascariensis Bojer ex Dcne.

Cynanchum L.

  anegadense (Britt.) Alain
        SY=Metastelma a. Britt.
  angustifolium Pers.
        SY=C. palustre (Pursh) Heller
        SY=Lyonia p. (Pursh) Small
  arizonicum (Gray) Shinners
        SY=Metastelma a. Gray
  barbigerum (Scheele) Shinners
      var. barbigerum
      var. breviflorum Shinners
  blodgettii (Gray) Shinners
        SY=Metastelma b. Gray
  cheesmanii Woods.
        SY=Metastelma decipiens Schlechter
  cubense (Griseb.) Woods.
        SY=Metalepis c. Griseb.
  ephedroides (Griseb.) Alain
        SY=Metastelma e. (Griseb.)
            Schlechter
  grisebachianum (Schlechter) Alain
        SY=Metastelma decaisneanum
            Schlechter
  laeve (Michx.) Pers.
        SY=Ampelamus albidus (Nutt.)
            Britt.
        SY=Gonolobus l. Michx.
  lineare (Bello) Alain
        SY=Metastelma l. Bello
  maccartii Shinners
  medium R. Br.
        SY=Vincetoxicum m. (R. Br.) Dcne.
  monense (Britt.) Alain
        SY=Metastelma m. Britt.
  nigrum (L.) Pers.
        SY=Vincetoxicum n. (L.) Moench
  northropiae (Schlechter) Alain
        SY=Epicion n. (Schlechter) Small
  parviflorum (R. Br.) Alain
        SY=Metastelma p. R. Br.
  scoparium Nutt.
        SY=Amphistelma s. (Nutt.) Small
  sinaloense (Brandeg.) Woods.
        SY=Mellichampia s. (Brandeg.)
            Kearney & Peebles
  unifarium (Scheele) Woods.
        SY=C. palmeri (S. Wats.) Blake
  utahense (Engelm.) Woods.
        SY=Astephanus u. Engelm.
  vincetoxicum (L.) Pers.
  wigginsii Shinners
        SY=Basistelma angustifolium

Cynanchum (CONT.)
                 (Torr.) Bartlett

Gomphocarpus R. Br.

   physocarpus E. Mey.

Gonolobus Michx.

   stephanotrichus Griseb.
         SY=Vincetoxicum s. (Griseb.)
           Britt.

Hoya R. Br.

   carnosa (L. f.) R. Br.

Marsdenia R. Br.

   elliptica Dcne.

Matelea Aubl.

   alabamensis (Vail) Woods.
         SY=Cyclodon a. (Vail) Small
   arizonica (Gray) Shinners
         SY=Gonolobus a. (Gray) Woods.
         SY=Lachnostma a. Gray
   balbisii (Dcne.) Woods.
         SY=Pherotrichis b. (Dcne.) Gray
   baldwyniana (Sweet) Woods.
         SY=Gonolobus b. Sweet
         SY=Odontostephana b. (Sweet)
           Alexander
         SY=Vincetoxicum b. (Sweet) Britt.
   biflora (Raf.) Woods.
         SY=Gonolobus b. Raf.
   brevicoronata (B.L. Robins.) Woods.
         SY=Gonolobus parviflorus (Torr.)
           Woods. var. b. B.L. Robins.
   carolinensis (Jacq.) Woods.
         SY=Gonolobus c. (Jacq.) Roemer &
           Schultes
         SY=Odontostephana c. (Jacq.)
           Alexander
         SY=Vincetoxicum c. (Jacq.) Britt.
         SY=V. hirsutum (Michx.) Britt.
   cordifolia (Gray) Woods.
         SY=Rothrockia c. Gray
   cynanchoides (Engelm.) Woods.
         SY=Gonolobus c. Engelm.
   decipiens (Alexander) Woods.
         SY=Gonolobus d. (Alexander) Perry
         SY=Odontostephana d. Alexander
   edwardsensis Correll
   flavidula (Chapman) Woods.
         SY=Odontostephana f. (Chapman)
           Alexander
   floridana (Vail) Woods.
         SY=Odontostephana f. (Vail)
           Alexander
   gonocarpa (Walt.) Shinners
         SY=Gonolobus g. (Walt.) Perry
         SY=G. suberosus (L.) R. Br.
         SY=M. s. (L.) Shinners
         SY=Vincetoxicum s. (L.) Britt.
         SY=V. g. Walt.
   maritima (Jacq.) Woods.
         SY=Ibatia m. (Jacq.) Dcne.
   obliqua (Jacq.) Woods.
         SY=Gonolobus o. (Jacq.) Roemer &
           Schultes
         SY=G. shortii Gray
         SY=M. s. (Gray) Woods.
         SY=Odontostephana o. (Jacq.)
           Alexander
         SY=O. s. (Gray) Alexander

         SY=Vincetoxicum o. (Jacq.) Britt.
         SY=V. s. (Gray) Britt.
   parviflora (Torr.) Woods.
   parvifolia (Torr.) Woods.
         SY=Gonolobus californicus Jepson
         SY=G. p. Torr.
   pringlei (Gray) Woods.
         SY=Himantostemma p. Gray
   producta (Torr.) Woods.
         SY=Gonolobus p. Torr.
   pubiflora (Dcne.) Woods.
         SY=Edisonia p. (Dcne.) Small
   radiata Correll
   reticulata (Engelm. ex Gray) Woods.
         SY=Gonolobus r. Engelm. ex Gray
   sagittifolia (Gray) Woods.
         SY=Gonolobus s. Gray
   sintenisii (Schlechter) Woods.
         SY=Gonolobus s. Schlechter
         SY=Vincetoxicum s. (Schlechter)
           Britt.
   texensis Correll
   variifolia (Schlechter) Woods.
         SY=Gonolobus v. Schlechter
         SY=Vincetoxicum v. (Schlechter)
           Britt.
   woodsonii Shinners

Metaplexis R. Br.

   japonica (Thunb.) Makino

Morrenia Lindl.

   odorata (Hook. & Arn.) Lindl.

Oxypetalum R. Br.

   cordifolium (Vent.) Schlechter

Periploca L.

   graeca L.

Sarcostemma R. Br.

   clausum (Jacq.) Roemer & Schultes
         SY=Funastrum c. (Jacq.) Schlechter
   crispum Benth.
         SY=Funastrum c. (Benth.)
           Schlechter
         SY=S. lobata Waterfall
   cynanchoides Dcne.
     ssp. cynanchoides
         SY=Funastrum c. (Dcne.) Schlechter
         SY=F. c. var. subtruncatum
           (Robins. & Fern.) J.F.
           Macbr.
     ssp. hartwegii (Vail) R. Holm
         SY=Funastrum heterophyllum
           (Engelm.) Standl.
         SY=F. lineare (Dcne.) J.F. Macbr.
         SY=Philbertia heterophylla
           (Engelm. sensu Torr.) Jepson
         SY=P. hirtella (Gray) Parish
         SY=S. c. var. hartwegii (Vail)
           Shinners
   hirtellum (Gray) R. Holm
         SY=Funastrum h. (Gray) Schlechter
         SY=Philbertia h. (Gray) Parish
   torreyi (Gray) Woods.

ASTERACEAE

Acamptopappus Gray

    shockleyi Gray
    sphaerocephalus (Harvey & Gray) Gray
        var. hirtellus Blake
        var. sphaerocephalus

Acanthospermum Schrank

    australe (Loefl.) Kuntze
        SY=Melampodium a. Loefl.
    hispidum DC.
    humile (Sw.) DC.

Achillea L.

    filipendula Lam.
    ligustica All.
    millefolium L.
        var. alpicola (Rydb.) Garrett
          SY=A. a. (Rydb.) Rydb.
          SY=A. fusca Rydb.
          SY=A. lanulosa Nutt. ssp. a.
              (Rydb.) Keck
          SY=A. l. var. a. Rydb.
          SY=A. m. L. ssp. l. (Nutt.) Piper
              var. a. (Rydb.) Garrett
          SY=A. m. var. f. (Rydb.) G.N.
              Jones
        var. arenicola (Heller) Nobs
          SY=A. a. Heller
          SY=A. borealis Bong. ssp. a.
              (Heller) Keck
          SY=A. b. var. a. (Heller) J.T.
              Howell
          SY=A. m. var. maritima Jepson
          SY=A. subalpina Greene
         var. borealis (Bong.) Farw.
          SY=A. b. Bong.
           SY=A. b. ssp. typica Keck
          SY=A. m. ssp. atrotegula Boivin
          SY=A. m. ssp. a. var. fulva Boivin
          SY=A. m. ssp. a. var. parviligula
              Boivin
          SY=A. m. ssp. a. var. parvula
              Boivin
          SY=A. m. ssp. b. (Bong.) Breitung
        var. californica (Pollard) Jepson
          SY=A. borealis Bong. ssp. c.
              (Pollard) Keck
          SY=A. b. var. c. (Pollard) J.T.
              Howell
          SY=A. c. Pollard
        var. gigantea (Pollard) Nobs
          SY=A. g. Pollard
        var. lanulosa (Nutt.) Piper
          SY=A. angustissima Rydb.
          SY=A. asplenifolia sensu auctt.
              non Vent.
          SY=A. eradiata Piper
          SY=A. gracilis Raf.
          SY=A. lanulosa Nutt.
          SY=A. l. ssp. typica Keck
          SY=A. l. var. arachnoidea Lunell
          SY=A. l. var. eradiata (Piper)
              M.E. Peck
          SY=A. laxiflora Pollard &
              Cockerell
          SY=A. megacephala Raup
          SY=A. millefolium ssp. lanulosa
              (Nutt.) Piper
          SY=A. millefolium ssp. lanulosa
              var. megacephala (Raup)
              Boivin
          SY=A. millefolium ssp. megacephala
              (Raup) Argus
          SY=A. millefolium ssp.

             occidentalis (DC.) Hyl.
          SY=A. millefolium ssp.
              pallidotegula Boivin
          SY=A. millefolium ssp. p. var.
              megacephala (Raup) Boivin
          SY=A. millefolium ssp. p. var.
              russeolata Boivin
          SY=A. millefolium var.
              asplenifolia (Vent.) Farw.
          SY=A. millefolium var. gracilis
              (Raf.) Raf. ex DC.
          SY=A. millefolium var. lanulosa
              (Nutt.) Piper
          SY=A. millefolium var.
              occidentalis DC.
          SY=A. millefolium var. pacifica
              (Rydb.) G.N. Jones
          SY=A. millefolium var. rosea
              (Desf.) Torr. & Gray
          SY=A. occidentalis Raf. ex Rydb.
          SY=A. rosea Desf.
          SY=A. tomentosa Pursh, non L.
        var. litoralis Ehrend. ex Nobs
        var. millefolium
        var. nigrescens E. Mey.
          SY=A. n. (E. Mey.) Rydb.
        var. pacifica (Rydb.) G.N. Jones
          SY=A. p. Rydb.
        var. puberula (Rydb.) Nobs
          SY=A. p. Rydb.
    nobilis L.
    ptarmica L.
    sibirica Ledeb.
        SY=A. multiflora Hook.

Achyrachaena Schauer

    mollis Schauer

Acourtia D. Don

    microcephala DC.
        SY=Perezia m. (DC.) Gray
    nana (Gray) Reveal & King
        SY=Perezia n. Gray
    runcinata (D. Don) B.L. Turner
        SY=Perezia r. (D. Don) Gray
    thurberi (Gray) Reveal & King
        SY=Perezia t. Gray
    wrightii (Gray) Reveal & King
        SY=Perezia w. Gray

Adenocaulon Hook.

    bicolor Hook.

Adenostemma J.R. & G. Forst.

    lavenia (L.) Kuntze
    verbesina (L.) Schultz-Bip.

Ageratum L.

    conyzoides L.
    corymbosum Zuccagni
        SY=A. guatemalense M.F. Johnson
    houstonianum P. Mill.
    littorale Gray

Agoseris Raf.

    apargioides (Less.) Greene
        var. apargioides
          SY=A. hirsuta (Hook.) Greene
        var. eastwoodae (Fedde) Q. Jones
    arizonica Greene
    aurantiaca (Hook.) Greene

Agoseris (CONT.)
        var. aurantiaca
            SY=A. graminifolia Greene
            SY=A. rostrata Rydb.
        var. purpurea (Gray) Cronq.
    ?barbellulata Greene
    elata (Nutt.) Greene
    gaspensis Fern.
    glauca (Pursh) Raf.
        var. agrestis (Osterhout) Q. Jones ex
            Cronq.
            SY=A. a. Osterhout
        var. dasycephala (Torr. & Gray)
            Jepson
            SY=A. g. var. asper (Rydb.) Cronq.
            SY=A. g. var. pumila (Nutt.)
                Garrett
            SY=A. scorzonerifolia (Schrad.)
                Greene
        var. glauca
            SY=A. turbinata Rydb.
        var. laciniata (D.C. Eat.) Smiley
            SY=A. g. var. parviflora (Nutt.)
                Rydb.
            SY=A. p. (Nutt.) Greene
        var. monticola (Greene) Q. Jones ex
            Cronq.
    grandiflora (Nutt.) Greene
        SY=A. laciniata (Nutt.) Greene
        SY=A. plebeja Greene
    heterophylla (Nutt.) Greene
        var. crenulata Jepson
        var. heterophylla
            SY=A. h. ssp. californica (Nutt.)
                Piper
            SY=A. h. ssp. normalis Piper
            SY=A. h. var. cryptopleura Jepson
        var. kymapleura Greene
        var. turgida (Hall) Jepson
    maritima Sheldon
        SY=A. apargioides (Less.) Greene
                ssp. m. (Sheldon) Q. Jones
                ex Cronq.
    retrorsa (Benth.) Greene

Amblyolepis DC.

    setigera DC.
        SY=Helenium s. (DC.) Britt. &
            Rusby

Amblyopappus Hook. & Arn.

    pusillus Hook. & Arn.

Ambrosia L.

    acanthicarpa Hook.
        SY=Franseria a. (Hook.) Coville
        SY=Gaertneria a. (Hook.) Britt.
    ambrosioides (Cav.) Payne
        SY=Franseria a. Cav.
    artemisiifolia L.
        var. artemisiifolia
            SY=A. glandulosa Scheele
            SY=A. monophylla (Walt.) Rydb.
        var. elatior (L.) Descourtils
            SY=A. e. L.
        var. paniculata (Michx.) Blank.
    bidentata Michx.
    canescens Gray
    chamissonis (Less.) Greene
        SY=A. c. var. bipinnatisecta
            (Less.) J.T. Howell
        SY=Franseria c. Less.
        SY=F. c. ssp. b. (Less.) Wiggins &
            Stockwell

            SY=F. c. var. b. Less.
    cheiranthifolia Gray
    chenopodiifolia (Benth.) Payne
        SY=Franseria c. Benth.
    confertiflora DC.
        SY=Franseria c. (DC.) Rydb.
        SY=F. strigulosa Rydb.
        SY=Gaertneria tenuifolia Harvey &
            Gray
    cordifolia (Gray) Payne
        SY=Franseria c. Gray
    coronopifolia Torr. & Gray
        SY=A. psilostachya DC. var. c.
            (Torr. & Gray) Farw.
    deltoidea (Torr.) Payne
        SY=Franseria d. Torr.
    dumosa (Gray) Payne
        SY=Franseria d. Gray
    eriocentra (Gray) Payne
        SY=Franseria e. Gray
    grayi (A. Nels.) Shinners
    X helenae Rouleau [artemisiifolia var.
        elatior X trifida]
    hispida Pursh
    ilicifolia (Gray) Payne
        SY=Franseria i. Gray
    X intergradiens Wagner & Beals
        [artemisiifolia X psilostachya]
    linearis (Rydb.) Payne
        SY=Franseria l. Rydb.
    peruviana Willd.
    psilostachya DC.
        var. californica (Rydb.) Blake
        var. lindheimeriana (Scheele) Blank.
        var. psilostachya
            SY=A. rugelii Rydb.
    pumila (Nutt.) Gray
    tenuifolia Spreng.
    tomentosa Nutt.
        SY=Franseria discolor Nutt.
        SY=F. t. (Nutt.) A. Nels.
        SY=Gaertneria d. (Nutt.) Kuntze
        SY=G. t. (Nutt.) Heller
    trifida L.
        var. texana Scheele
            SY=A. aptera DC.
            SY=A. trifida var. a. (DC.) Kuntze
        var. trifida

Amphiachyris Nutt.

    dracunculoides (DC.) Nutt.
        SY=A. amoenum (Shinners) Solbrig
        SY=Gutierrezia d. (DC.) Blake
        SY=Xanthocephalum a. Shinners
        SY=X. d. (DC.) Shinners

Amphipappus Torr. & Gray

    fremontii Torr. & Gray
        ssp. fremontii
        ssp. spinosus (A. Nels.) Keck
            SY=A. f. var. s. (A. Nels.) C.L.
                Porter

Anaphalis DC.

    margaritaceae (L.) Benth. & Hook. f. ex
        C.B. Clarke
            SY=A. m. var. angustior (Miq.)
                Nakai
            SY=A. m. var. intercedens Hara
            SY=A. m. var. occidentalis Greene
            SY=A. m. var. revoluta Suksdorf
            SY=A. m. var. subalpina Gray
            SY=A. o. (Greene) Heller

Anisocoma Torr. & Gray

  acaulis Torr. & Gray

Antennaria Gaertn.

  acuta Rydb.
  affinis Fern.
  alpina (L.) Gaertn.
        var. media (Greene) Jepson
            SY=A. alaskana Malte
            SY=A. alpina var. canescens Lange
            SY=A. alpina var. compacta (Malte)
                Welsh
            SY=A. alpina var. frieseana
                Trautv.
            SY=A. alpina var. megacephala
                (Fern.) Welsh
            SY=A. alpina var. scabra (Greene)
                Jepson
            SY=A. alpina var. stolonifera
                (Porsild) Welsh
            SY=A. alpina var. ungavensis Fern.
            SY=A. angustifolia Ekman
            SY=A. arenicola Malte
            SY=A. atriceps Fern.
            SY=A. austromontana E. Nels.
            SY=A. bayardii Fern.
            SY=A. brevistyla Fern.
            SY=A. brunnescens Fern.
            SY=A. bocheriana Porsild
            SY=A. borealis Greene
            SY=A. cana (Fern. & Wieg.) Fern.
            SY=A. candida Greene
            SY=A. canescens (Lange) Malte
            SY=A. canescens var.
                pseudoporsildii Bocher
            SY=A. chlorantha Greene
            SY=A. compacta Malte
            SY=A. confusa Fern.
            SY=A. crymophila Porsild
            SY=A. densa Greene
            SY=A. densifolia Porsild
            SY=A. ekmaniana Porsild
            SY=A. frieseana (Trautv.) Ekman
            SY=A. f. ssp. alaskana (Malte)
                Hulten
            SY=A. f. ssp. alaskana var.
                beringensis Hulten
            SY=A. f. ssp. compacta (Malte)
                Hulten
            SY=A. f. var. megacephala (Fern.)
                Hulten
            SY=A. foggii Fern.
            SY=A. macounii Greene
            SY=A. media Greene
            SY=A. media ssp. ciliata E. Nels.
            SY=A. megacephala Fern.
            SY=A. modesta Greene
            SY=A. neoalaskana Porsild
            SY=A. neodioica Greene var.
                chlorantha (Greene) Boivin
            SY=A. pedunculata Porsild
            SY=A. pulchella Greene
            SY=A. pulvinata Greene
            SY=A. reflexa E. Nels.
            SY=A. rousseaui Porsild
            SY=A. scabra Greene
            SY=A. sornborgeri Fern.
            SY=A. stolonifera Porsild
            SY=A. subcanescens Ostenf. ex
                Malte
            SY=A. tomentella E. Nels.
            SY=A. ungavensis (Fern.) Malte
            SY=A. vexillifera Fern.
  anaphaloides Rydb.
            SY=A. pulcherrima (Hook.) Greene

        ssp. anaphaloides (Rydb.)
            W.A. Weber
  arcuata Cronq.
  argentea Benth.
            SY=A. luzuloides Torr. & Gray var.
                a. (Benth.) Gray
  corymbosa E. Nels.
            SY=A. dioica (L.) Gaertn. var. c.
                (E. Nels.) Jepson
            SY=A. hygrophila Greene
            SY=A. nardina Greene
  dimorpha (Nutt.) Torr. & Gray
            SY=A. d. var. integra Henderson
            SY=A. d. var. latisquama (Piper)
                M.E. Peck
            SY=A. d. var. macrocephala D.C.
                Eat.
            SY=A. l. Piper
            SY=A. m. (D.C. Eat.) Rydb.
            SY=Gnaphalium d. Nutt.
  dioica (L.) Gaertn.
            SY=A. dimorpha (Nutt.) Torr. &
                Gray var. flagellaris Gray
            SY=A. f. (Gray) Gray
            SY=A. insularis Greene
            SY=Gnaphalium dioicum L.
  geyeri Gray
            SY=Gnaphalium alienum Hook. & Arn.
  lanata (Hook.) Greene
            SY=A. carpathica (Wahlenb.) R. Br.
                var. l. Hook.
  luzuloides Torr. & Gray
            SY=A. argentea Benth. ssp.
                aberrans E. Nels.
            SY=A. l. var. oblanceolata (Rydb.)
                M.E. Peck
            SY=A. microcephala Gray
            SY=A. o. Rydb.
            SY=A. pyrimidata Greene
  manicouagana Landry
  microphylla Rydb.
            SY=A. acuminata Greene
            SY=A. alborosea A.E. & M.P.
                Porsild
            SY=A. angustifolia Rydb.
            SY=A. arida E. Nels.
            SY=A. arida var. humilis E. Nels.
            SY=A. bracteosa Rydb.
            SY=A. breitungii Porsild
            SY=A. concinna E. Nels.
            SY=A. dioica (L.) Gaertn. var.
                hyperborea Lange
            SY=A. d. var. rosea (Greene) D.C.
                Eat.
            SY=A. elegans Porsild
            SY=A. erigeroides Greene
            SY=A. foliacea Greene
            SY=A. f. var. humilis Rydb.
            SY=A. formosa Greene
            SY=A. groenlandica M.P. Porsild
            SY=A. hansii Kern.
            SY=A. hendersonii Piper
            SY=A. imbricata E. Nels.
            SY=A. incarnata Porsild
            SY=A. laingii Porsild
            SY=A. leontopodioides Cody
            SY=A. leuchippii M.P. Porsild
            SY=A. nitida Greene
            SY=A. oxyphylla Greene
            SY=A. parvifolia Greene
            SY=A. p. Nutt. var. bracteosa
                (Rydb.) Boivin
            SY=A. rosea Greene
            SY=A. r. ssp. divaricata E. Nels.
            SY=A. r. var. angustifolia (Rydb.)
                E. Nels.
            SY=A. r. var. imbricata (E. Nels.)

Antennaria   (CONT.)
                    E. Nels.
          SY=A. r. var. nitida (Greene)
                    Breitung
          SY=A. scariosa E. Nels.
          SY=A. speciosa E. Nels.
          SY=A. solstitialis Greene
          SY=A. sorida Greene
          SY=A. subviscosa Fern.
     monocephala D.C. Eat.
          SY=A. alpina (L.) Gaertn. var.
                    glabrata J. Vahl
          SY=A. angustata Greene
          SY=A. burwellensis Malte
          SY=A. congesta Malte
          SY=A. exilis Greene
          SY=A. fernaldiana Polunin
          SY=A. glabrata (J. Vahl) Greene
          SY=A. hudsonica Malte
          SY=A. m. ssp. angustata (Greene)
                    Hulten
          SY=A. m. ssp. m. var. exilis
                    (Greene) Hulten
          SY=A. m. ssp. philonipha (Porsild)
                    Hulten
          SY=A. m. var. latisquamea Hulten
          SY=A. nitens Greene
          SY=A. philonipha Porsild
          SY=A. pygmaea Fern.
          SY=A. shumaginensis Porsild
          SY=A. tansleyi Polunin
          SY=A. tweedsmurii Polunin
     neglecta Greene
          var. argillicola (Stebbins) Cronq.
          SY=A. virginica Stebbins
          SY=A. v. var. a. Stebbins
          var. athabascensis (Greene) Taylor &
                    MacBryde
          SY=A. a. Greene
          SY=A. campestris Rydb. var. a.
                    (Greene) Boivin
          SY=A. howellii Greene var. a.
                    (Greene) Boivin
          var. attenuata (Fern.) Cronq.
          SY=A. arcuata Cronq.
          SY=A. brainerdii Fern.
          SY=A. concolor Piper
          SY=A. neodioica Greene
          SY=A. neodioica var. attenuata
                    Fern.
          SY=A. neodioica var. chlorophylla
                    Fern.
          SY=A. neodioica var. grandis Fern.
          SY=A. neodioica var. interjecta
                    Fern.
          SY=A. pedicellata Greene
          SY=A. rhodantha Suksdorf
          SY=A. rupicola Fern.
          SY=A. russellii Boivin
          SY=A. stenolepis Greene
          var. gaspensis (Fern.) Cronq.
          SY=A. g. Fern.
          var. howellii (Greene) Cronq.
          SY=A. callilepis Greene
          SY=A. h. Greene
          SY=A. n. ssp. h. (Greene) Hulten
          var. neglecta
          SY=A. angustriarum Lunell
          SY=A. campestris Rydb.
          SY=A. chelonica Lunell
          SY=A. longifolia Greene
          SY=A. howellii Greene var.
                    campestris (Rydb.) Boivin
          SY=A. nebraskensis Greene
          SY=A. neglecta var. campestris
                    (Rydb.) Steyermark
          SY=A. petaloidea Fern.

          SY=A. p. var. scariosa Fern.
          SY=A. p. var. subcorymbosa Fern.
          var. randii (Fern.) Cronq.
          SY=A. canadensis Greene
          SY=A. neodioica Greene var. r.
                    (Fern.) Boivin
          SY=A. r. Fern.
     parvifolia Nutt.
          SY=A. appendiculata Fern.
          SY=A. aprica Greene
          SY=A. aprica var. minuscula
                    (Boivin) Boivin
          SY=A. aureola Lunell
          SY=A. dioica (L.) Gaertn. var.
                    marginata (Greene) Jepson
          SY=A. holmii Greene
          SY=A. marginata Greene
          SY=A. minuscula Boivin
          SY=A. obtusa Greene
          SY=A. pumila Greene
          SY=A. spathulata Fern.
     plantaginifolia (L.) Richards.
          var. ambigens (Greene) Cronq.
          SY=A. arnoglossa Greene var.
                    ambigens Greene
          SY=A. calophylla Greene
          SY=A. fallax Greene
          SY=A. f. var. c. (Greene) Fern.
          SY=A. farwellii Greene
          SY=A. munda Fern.
          SY=A. obovata E. Nels.
          SY=A. occidentalis Greene
          SY=A. parlinii Fern. var.
                    farwellii (Greene) Boivin
          var. arnoglossa (Greene) Cronq.
          SY=A. ampla Bush
          SY=A. arnoglossa Greene
          SY=A. denikeana Boivin
          SY=A. parlinii Fern.
          SY=A. parlinii var. arnoglossa
                    (Greene) Fern.
          var. plantaginifolia
          SY=A. caroliniana Rydb.
          SY=A. petiolata Fern.
          SY=A. plantaginifolia var.
                    petiolata (Fern.) Heller
          SY=Gnaphalium plantaginifolium L.
     porsildii Ekman
     pulcherrima (Hook.) Greene
          SY=A. carpathica (Wahlenb.) R. Br.
                    var. p. Hook.
          SY=A. eucosma Fern. & Wieg.
          SY=A. p. var. angustisquama
                    Porsild
          SY=A. p. var. sordida Boivin
     racemosa Hook.
          SY=A. eximia Greene
          SY=A. oblancifolia E. Nels.
          SY=A. pedicellata Greene
          SY=A. piperi Rydb.
     rosulata Rydb.
          SY=A. bakeri Greene
          SY=A. sierrae-blancae Rydb.
     soliceps Blake
     solitaria Rydb.
     stenophylla Gray
          SY=A. alpina (L.) Gaertn. var. s.
                    (Gray) Gray
          SY=A. leucophaea Piper
     suffrutescens Greene
     umbrinella Rydb.
          SY=A. aizoides Greene
          SY=A. albescens (E. Nels.) Rydb.
          SY=A. albicans Fern.
          SY=A. alpina (L.) Gaertn. var.
                    intermedia Rosenv.
          SY=A. arida E. Nels. ssp.

Antennaria (CONT.)
        viscidula E. Nels.
    SY=A. columnaris Fern.
    SY=A. confinis Greene
    SY=A. dioica (L.) Gaertn. var.
        kernensis Jepson
    SY=A. ellyae Porsild
    SY=A. flavescens Rydb.
    SY=A. fusca E. Nels.
    SY=A. gormanii St. John
    SY=A. intermedia (Rosenv.) M.P.
        Porsild
    SY=A. isolepis Greene
    SY=A. lanulosa Greene
    SY=A. maculata Greene
    SY=A. mucronata E. Nels.
    SY=A. pallida E. Nels.
    SY=A. peasei Fern.
    SY=A. pulvinata Greene pro parte
    SY=A. pulvinata ssp. albescens E.
        Nels.
    SY=A. sansonii Greene
    SY=A. sedoides Greene
    SY=A. straminea Fern.
    SY=A. wiegandii Fern.

Anthemis L.

  altissima L.
  arvensis L.
    var. agrestis (Wallr.) DC.
    var. arvensis
  cotula L.
    SY=Maruta c. (L.) DC.
  secundiramea Biv.
  tinctoria L.
    SY=Cota t. (L.) J. Gay

Antheropeas Rydb.

  lanosum (Gray) Rydb.
    SY=Eriophyllum l. (Gray) Gray
  wallacei (Gray) Rydb.
    SY=Eriophyllum w. (Gray) Gray
    SY=E. w. ssp. australe (Rydb.)
        Wiggins
    SY=E. w. var. calvescens Blake
    SY=E. w. var. rubellum (Gray) Gray

Aphanostephus DC.

  arizonicus Gray
  humilis (Benth.) Gray
  kidderi Blake
  pilosus Buckl.
  ramosissimus DC.
  riddellii Torr. & Gray
  skirrobasis (DC.) Trel.
    var. skirrobasis
    var. thalassius Shinners

Arctium L.

  lappa L.
  minus Bernh.
  X mixtum Nyman [minus X tomentosum]
  nemorosum Lej. & Court.
  X nothum (Ruhm.) Weiss [lappa X minus]
  tomentosum P. Mill.

Arctotis L.

  stoechadifolia Berg.
    var. grandis (Thunb.) Less.

X Argyrautia Sherff

X degeneri Sherff [Arygroxiphium grayanum X
  Dubautia sp.]

Argyroxiphium DC.

  caliginis Forbes emend. St. John
  forbesii St. John
  grayanum (Hbd.) Deg.
  X kai (Forbes) Deg. & Deg.
  kauense (Rock & Neal) Deg. & Deg.
  macrocephalum Gray
  sandwicense DC.
  virescens Hbd.
    var. paludosa St. John
    var. virescens

Arnica L.

  acaulis (Walt.) B.S.P.
  alpina (L.) Olin & Ladau
    var. alpina
    var. angustifolia (Vahl) Fern.
      SY=A. alpina ssp. angustifolia
        (Vahl) Maguire
      SY=A. alpina var. ungavensis
        Boivin
    var. attenuata (Greene) Ediger &
      Barkl.
      SY=A. alpina ssp. attenuata
        (Greene) Maguire
      SY=A. alpina var. linearis Hulten
      SY=A. alpina var. vestita Hulten
    var. plantaginea (Pursh) Ediger &
      Barkl.
      SY=A. alpina ssp. sornborgeri
        (Fern.) Maguire
      SY=A. p. Pursh
      SY=A. terrae-novae Fern.
    var. tomentosa (Macoun) Cronq.
      SY=A. a. ssp. t. (Macoun) Maguire
      SY=A. t. Macoun
  amplexicaulis Nutt.
      SY=A. a. ssp. genuina Maguire
      SY=A. a. ssp. prima (Maguire)
        Maguire
      SY=A. a. var. piperi St. John &
        Warren
      SY=A. a. var. prima (Maguire)
        Boivin
      SY=A. mollis Hook. var. aspera
        (Greene) Boivin
  cernua T.J. Howell
  chamissonis Less.
    ssp. chamissonis
      var. chamissonis
        SY=A. c. ssp. genuina Maguire
      var. interior Maguire
    ssp. foliosa (Nutt.) Maguire
      var. andina (Nutt.) Ediger & Barkl.
        SY=?A. c. ssp. f. var. jepsoniana
          Maguire
      var. bernardina (Greene) Maguire
      var. foliosa (Nutt.) Maguire
        SY=A. f. Nutt.
      var. incana (Gray) Hulten
        SY=A. c. ssp. i. (Gray) Maguire
      var. maguirei (A. Nels.) Maguire
  cordifolia Hook.
    var. cordifolia
      SY=A. c. ssp. genuina Maguire
      SY=A. c. ssp. whitneyi (Fern.)
        Maguire
      SY=A. w. Fern.
    var. pumila (Rydb.) Maguire
  discoidea Benth.
    var. alata (Rydb.) Cronq.
      SY=A. parviflora Gray ssp. a.

Arnica (CONT.)
        (Rydb.) Maguire
    var. discoidea
    var. eradiata (Gray) Cronq.
      SY=A. grayi Heller
      SY=A. parviflora Gray
diversifolia Greene
fulgens Pursh
lanceolata Nutt.
      SY=A. mollis Hook. var. petiolaris
        Fern.
latifolia Bong.
    var. gracilis (Rydb.) Cronq.
      SY=A. g. Rydb.
    var. latifolia
      SY=A. l. var. angustifolia Herder
lessingii (Torr. & Gray) Greene
      SY=A. l. ssp. norbergii Hulten &
        Maguire
lonchophylla Greene
      SY=A. alpina (L.) Olin & Ladau
        ssp. l. (Greene) Taylor &
        MacBryde
      SY=A. a. var. l. (Greene) Welsh
      SY=A. arnoglossa Greene
      SY=A. chionopappa Fern.
      SY=A. gaspensis Fern.
      SY=A. l. ssp. arnoglossa (Greene)
        Maguire
      SY=A. l. ssp. c. (Fern.) Maguire
      SY=A. l. var. arnoglossa (Maguire)
        Boivin
      SY=A. l. var. arnoglossa (Rydb.)
        Boivin
longifolia D.C. Eat.
      SY=A. l. ssp. genuina Maguire
      SY=A. l. ssp. myriadenia (Piper)
        Maguire
louiseana Farr
    var. griscomii (Fern.) Boivin
      SY=A. l. ssp. g. (Fern.) Maguire
    var. louiseana
      SY=A. l. var. genuina Maguire
    var. mendenhallii (Rydb.) Maguire
      SY=A. frigida C.A. Mey. ex Iljin
      SY=A. l. ssp. f. (C.A. Mey. ex
        Iljin) Maguire
      SY=A. l. var. f. (C.A. Mey. ex
        Iljin) Welsh
      SY=A. l. var. pilosa Maguire
mollis Hook.
      SY=A. m. var. sylvatica (Greene)
        Maguire
nevadensis Gray
X paniculata A. Nels. [cordifolia X parryi]
parryi Gray
    var. parryi
      SY=A. p. ssp. genuina Maguire
    var. sonnei (Greene) Cronq.
      SY=A. p. ssp. s. (Greene) Maguire
rydbergii Greene
sororia Greene
      SY=A. fulgens Pursh var. s.
        (Greene) Dougl. &
        Ruyle-Dougl.
spathulata Greene
    var. eastwoodiae (Rydb.) Ediger &
      Barkl.
    var. spathulata
tomentella Greene
unalaschcensis Less.
venosa Hall
viscosa Gray

Arnoseris Gaertn.

minima (L.) Schweig. & Koerte

Artemisia L.

abrotanum L.
      SY=A. procera Willd.
absinthium L.
    var. absinthium
    var. insipida Stechmann
alaskana Rydb.
aleutica Hulten
annua L.
arbuscula Nutt.
    ssp. arbuscula
      SY=A. tridentata Nutt. ssp. a.
        (Nutt.) Hall & Clements
    ssp. thermopola Beetle
arctica Less.
    ssp. arctica
      SY=A. norvegica Fries ssp.
        saxatilis (Bess. ex Hook.)
        Hall & Clements
      SY=A. n. var. s. (Bess. ex Hook.)
        Jepson
    ssp. beringensis (Hulten) Hulten
    ssp. comata (Rydb.) Hulten
      SY=A. norvegica Fries var. c.
        (Rydb.) Welsh
    ssp. saxicola (Rydb.) Hulten
argilosa Beetle
australis Less.
biennis Willd.
bigelovii Gray
californica Less.
campestris L.
    ssp. borealis (Pallas) Hall & Clements
      var. borealis (Pallas) M.E. Peck
        SY=A. b. Pallas
        SY=A. b. var. latisecta Fern.
        SY=A. c. ssp. canadensis (Michx.)
          Scoggan
        SY=A. campestre var. canadensis
          (Michx.) Welsh
        SY=A. canadensis Michx.
        SY=A. richardsoniana Bess.
        SY=A. spithamea Pursh
      var. purshii (Bess. ex Hook.) Cronq.
        SY=A. borealis Pallas ssp. p.
          (Bess. ex Hook.) Hulten
        SY=A. b. var. p. Bess. ex Hook.
      var. scouleriana (Bess.) Cronq.
        SY=A. ripicola Rydb.
      var. wormskioldii (Bess.) Cronq.
    ssp. campestris
    ssp. caudata (Michx.) Hall & Clements
      SY=A. caudata Michx.
      SY=A. caudata var. calvens Lunell
      SY=A. forwoodii S. Wats.
    ssp. pacifica (Nutt.) Hall & Clements
      SY=A. campestris var. douglasiana
        (Bess.) Boivin
      SY=A. c. var. pacifica (Nutt.)
        M.E. Peck
      SY=A. c. var. strutzia Welsh
      SY=A. camporum Rydb.
      SY=A. p. Nutt.
cana Pursh
    ssp. bolanderi (Gray) G.H. Ward
      SY=A. b. Gray
      SY=A. tridentata Nutt. ssp. b.
        (Gray) Hall & Clements
      SY=A. t. var. b. (Gray) McMinn
    ssp. cana
    ssp. viscidula (Osterhout) Beetle
      SY=A. c. var. v. Osterhout
carruthii Wood ex Carruthers
      SY=A. c. var. wrightii (Gray)
        Blake
      SY=A. kansana Britt.

Artemisia (CONT.)
  douglasiana Bess.
          SY=A. vulgaris L. var. d. (Bess.)
            St. John
  dracunculus L.
    ssp. dracunculus
          SY=A. dracunculoides Pursh
          SY=A. dracunculoides var.
            dracunculina (S. Wats.)
            Blake
          SY=A. glauca Pallas var.
            dracunculina (S. Wats.)
            Fern.
    ssp. glauca (Pallas) Hall & Clements
          SY=A. g. Pallas
  filifolia Torr.
  franserioides Greene
  frigida Willd.
  furcata Bieb.
      var. furcata
          SY=A. hyperborea Rydb.
          SY=A. trifurcata Steph. ex Spreng.
      var. heterophylla (Bess.) Hulten
  globularia Cham. ex Bess.
          SY=A. flava Jurtsev
          SY=A. g. var. lutea Hulten
  glomerata Ledeb.
  hillebrandii Skottsberg
  hookeriana Bess.
  kauaiensis (Skottsberg) Skottsberg
  krushiana Bess.
  laciniata Willd.
          SY=A. laciniatiformis Komarov
  lindleyana Bess.
          SY=A. prescottiana Bess.
  longifolia Nutt.
  longiloba (Osterhout) Beetle
          SY=A. spiciformis Osterhout var.
            l. Osterhout
  ludoviciana Nutt.
    ssp. albula (Woot.) Keck
          SY=A. l. var. a. (Woot.) Shinners
    ssp. candicans (Rydb.) Keck
          SY=A. diversifolia Rydb.
          SY=A. l. var. latiloba Nutt.
    ssp. incompta (Nutt.) Keck
          SY=A. l. var. i. (Nutt.) Cronq.
    ssp. ludoviciana
          SY=A. gnaphalodes Nutt.
          SY=A. herriotii Rydb.
          SY=A. l. var. latifolia (Bess.)
            Torr. & Gray
          SY=A. ludoviciana ssp. typica Keck
          SY=A. ludoviciana var. americana
            (Bess.) Fern.
          SY=A. ludoviciana var. brittonii
            (Rydb.) Fern.
          SY=A. ludoviciana var. gnaphalodes
            (Nutt.) Torr. & Gray
          SY=A. ludoviciana var. pabularis
            (Rydb.) Fern.
          SY=A. p. (A. Nels.) Rydb.
          SY=A. purshiana Bess.
    ssp. mexicana (Willd.) Keck
          SY=A. l. var. m. (Willd.) Fern.
          SY=A. m. Willd.
          SY=A. vulgaris L. var. m. (Willd.)
            Torr. & Gray
    ssp. redolens (Gray) Keck
          SY=A. l. var. r. (Gray) Shinners
    ssp. sulcata (Rydb.) Keck
  macrobotrys Ledeb.
  mauiensis (Gray) Skottsberg
      var. diffusa Skottsberg
      var. mauiensis
  michauxiana Bess.
          SY=A. vulgaris L. ssp. m. (Bess.)

            St. John
  nesiotica Raven
          SY=A. californica Less. var.
            insularis (Rydb.) Munz
  norvegica Fries
  nova A. Nels.
          SY=A. arbuscula Nutt. ssp. n. (A.
            Nels.) G.H. Ward
          SY=A. a. var. n. (A. Nels.) Cronq.
          SY=A. tridentata Nutt. ssp. n. (A.
            Nels.) Hall & Clements
  packardiae Grimes & Ertter
  palmeri Gray
  papposa Blake & Cronq.
  parryi Gray
  pattersonii Gray
  pedatifida Nutt.
  pontica L.
  porteri Cronq.
  pycnocephala (Less.) DC.
  pygmaea Gray
  rigida (Nutt.) Gray
  rothrockii Gray
          SY=A. tridentata Nutt. ssp. r.
            (Gray) Hall & Clements
          SY=A. t. var. r. (Gray) McMinn
  rupestris L.
    ssp. rupestris
    ssp. woodii Neilson
          SY=A. frigida Willd. var.
            williamsiae Welsh
  scopulorum Gray
  senjavinensis Bess.
  serrata Nutt.
  spinescens D.C. Eat.
  stelleriana Bess.
  suksdorfii Piper
  tilesii Ledeb.
    ssp. elatior (Torr. & Gray) Hulten
          SY=A. e. (Torr. & Gray) Rydb.
          SY=A. t. var. e. Torr. & Gray
    ssp. gormanii (Rydb.) Hulten
    ssp. tilesii
    ssp. unalaschcensis (Bess.) Hulten
          SY=A. t. var. aleutica (Hulten)
            Welsh
          SY=A. t. var. u. Bess.
          SY=A. u. Rydb. var. a. Hulten
  tridentata Nutt.
    ssp. tridentata
          SY=A. angustifolia (Gray) Rydb.
          SY=A. t. ssp. parishii (Gray) Hall
            & Clements
          SY=A. t. var. a. Gray
          SY=A. t. var. p. (Gray) Jepson
    ssp. vaseyana (Rydb.) Beetle
          SY=A. t. var. v. (Rydb.) Boivin
          SY=A. v. Rydb.
    ssp. wyomingensis Beetle & Young
  tripartita Rydb.
    ssp. rupicola Beetle
    ssp. tripartita
          SY=A. tridentata Nutt. ssp.
            trifida (Nutt.) Hall &
            Clements
          SY=A. trifida Nutt.
  tyrellii Rydb.
  unalaschcensis Rydb.
  vulgaris L.
      var. glabra Ledeb.
      var. latiloba Ledeb.
      var. vulgaris

Aster L.

  acuminatus Michx.
      var. acuminatus

Aster   (CONT.)
    var. magdalenensis Fern.
  adnatus Nutt.
  alpigenus (Torr. & Gray) Gray
    ssp. alpigenus
      SY=Haplopappus a. Torr. & Gray
    ssp. andersonii (Gray) Onno
      SY=A. alpigenus var. andersonii
        (Gray) M.E. Peck
      SY=A. andersonii (Gray) Gray
      SY=A. elatus (Greene) Cronq.
      SY=Erigeron andersonii Gray
      SY=Oreastrum e. Greene
    ssp. haydenii (Porter) Cronq.
      SY=A. h. Porter
  alpinus L.
    ssp. vierhapperi Onno
  X amethystinus Nutt. [ericoides X
    novae-angliae]
  anomalus Engelm.
  ascendens Lindl.
    SY=A. adscendens Lindl.
    SY=A. chilensis Nees ssp.
      adscendens (Lindl.) Cronq.
    SY=A. macounii Rydb.
    SY=A. subgriseus Rydb.
  avitus Alexander
  azureus Lindl.
    var. azureus
    var. poaceus (Burgess) Fern.
      SY=A. p. Burgess
      SY=A. vernalis Burgess
  behringensis Gandog.
  bernardinus Hall
    SY=A. defoliatus Parish
    SY=A. deserticola J.F. Macbr.
  bifoliatus (Walt.) Ahles
    SY=A. tortifolius Michx.
    SY=Conyza b. Walt.
    SY=Sericocarpus acutisquamosus
      (Nash) Small
    SY=S. b. (Walt.) Porter
  X blakei (Porter) House [acuminatus X
    nemoralis]
  borealis (Torr. & Gray) Prov.
    SY=A. franklinianus Rydb.
    SY=A. junceus sensu auctt. non
      Ait.
    SY=A. junciformis Rydb.
    SY=A. laxifolius Nees var. b.
      Torr. & Gray
  brachyactis Blake
    SY=Brachyactis angusta (Lindl.)
      Britt.
    SY=Tripolium a. Lindl.
  brickellioides Greene
    var. brickellioides
      SY=A. tomentellus sensu Frye &
       Rigg, non Hook.
      SY=Eucephalus t. (Frye & Rigg)
       Greene
    var. glabratus Greene
      SY=A. siskiyouensis A. Nels. &
       J.F. Macbr.
  campestris Nutt.
    var. bloomeri (Gray) Gray
      SY=A. b. Gray
    var. campestris
  carolinianus Walt.
  chapmanii Torr. & Gray
  chilensis Nees
    var. chilensis
    var. invenustus (Greene) Jepson
      SY=A. i. Greene
    var. lentus (Greene) Jepson
      SY=A. l. Greene
    var. medius Jepson

    var. sonomensis (Greene) Jepson
      SY=A. s. Greene
  ciliolatus Lindl.
    SY=A. lindleyanus Torr. & Gray
    SY=A. saundersii Burgess
  coerulescens DC.
    SY=A. praealtus Poir. var.
      texicola Wieg.
  X columbianus Piper [campestris X ericoides
    ssp. pansus]
  X commixtus (Nees) Kuntze [macrophyllus X
    spectabilis]
    SY=A. X mirabilis Torr. & Gray
    SY=Eurybia c. Nees
  concolor L.
    var. concolor
      SY=A. plumosus Small
    var. simulatus (Small) R.W. Long
      SY=A. s. Small
  conspicuus Lindl.
  cordifolius L.
    var. cordifolius
    var. furbishiae Fern.
    var. moratus Shinners
    var. polycephalus Porter
    var. racemiflorus Fern.
  curtisii Torr. & Gray
  curtus Cronq.
    SY=Sericocarpus rigidus Lindl.
  cusickii Gray
    SY=A. foliaceus Lindl. var. c.
      (Gray) Cronq.
  depauperatus (Porter) Fern.
    SY=A. ericoides sensu Ait. non L.
      var. d. Porter
  divaricatus L.
    var. chlorolepis (Burgess) Ahles
      SY=A. c. Burgess
    var. divaricatus
      SY=A. boykinii Burgess
      SY=A. carmesinus Burgess
      SY=A. castaneus Burgess
      SY=A. claytonii Burgess
      SY=A. excavatus Burgess
      SY=A. flexilis Burgess
      SY=A. stilettiformis Burgess
      SY=A. tenebrosus Burgess
  drummondii Lindl.
    SY=A. finkii Rydb.
    SY=A. sagittifolius Wedemeyer ex
      Willd. var. d. (Lindl.)
      Shinners
  dumosus L.
    var. dodgei Fern.
    var. dumosus
      SY=A. coridifolius Michx.
      SY=A. d. var. c. (Michx.) Torr. &
       Gray
      SY=A. d. var. gracilentus Torr. &
       Gray
      SY=A. pinifolius Alexander
    var. gracilipes Wieg.
      SY=A. g. (Wieg.) Alexander
    var. pergracilis Wieg.
    var. subulifolius Torr. & Gray
  eatonii (Gray) T.J. Howell
    SY=A. foliaceus Lindl. var. e.
      Gray
    SY=A. mearnsii Rydb.
    SY=A. oregonus sensu auctt. non
      (Nutt.) Torr. & Gray
  elliottii Torr. & Gray
  engelmannii (D.C. Eat.) Gray
    SY=A. elegans (Nutt.) Torr. & Gray
      var. engelmannii D.C. Eat.
  ericoides L.
    ssp. ericoides

Aster (CONT.)
  var. ericoides
   SY=A. multiflorus Ait.
   SY=A. m. var. exiguus Fern.
   SY=A. polycephalus Rydb.
  var. prostratus (Kuntze) Blake
   SY=A. exiguus (Fern.) Rydb.
    (misapplied)
   SY=A. multiflorus Ait. var. p.
    Kuntze
  ssp. pansus (Blake) A.G. Jones
   var. pansus (Blake) A.G. Jones
    SY=A. multiflorus Ait. var. p.
     Blake
    SY=A. p. (Blake) Cronq.
   var. stricticaulis (Torr. & Gray)
    F.C. Gates
    SY=A. multiflorus Ait. var. s.
     Torr. & Gray
    SY=A. s. (Torr. & Gray) Rydb.
eryngiifolius Torr. & Gray
  SY=A. spinulosus Chapman
falcatus Lindl.
 ssp. commutatus (Torr. & Gray) A.G.
  Jones
  var. commutatus (Torr. & Gray) A.G.
   Jones
   SY=A. c. (Torr. & Gray) Gray
   SY=A. c. var. polycephalus Blake
    non A. polycephalus Rydb.
   SY=A. cordineri A. Nels.
   SY=A. ericoides L. var. commutatus
    (Torr. & Gray) Boivin, pro
    parte
   SY=A. multiflorus Ait. var.
    commutatus Torr. & Gray
  var. crassulus (Rydb.) Cronq.
   SY=A. adsurgens sensu auctt. non
    Greene
   SY=A. crassulus Rydb.
 ssp. falcatus
   SY=A. elegantulus Porsild
   SY=A. ericoides L. var. commutatus
    (Torr. & Gray) Boivin, pro
    parte
   SY=A. ramulosus Lindl.
fendleri Gray
firmus Nees
  SY=A. lucidulus (Gray) Wieg.
  SY=A. puniceus L. var. f. (Nees)
   Torr. & Gray
  SY=A. p. var. l. Gray
foliaceus Lindl.
  var. apricus Gray
   SY=A. a. (Gray) Rydb.
  var. canbyi Gray
   SY=A. f. var. burkei Gray
   SY=A. phyllodes Rydb.
   SY=A. tweedyi Rydb.
  var. foliaceus
  var. lyallii (Gray) Cronq.
   SY=A. cusickii Gray var. l. Gray
  var. parryi (D.C. Eat.) Gray
   SY=A. ascendens Lindl. var. p.
    D.C. Eat.
   SY=A. diabolicus Piper
   SY=A. foliaceus var. frondeus Gray
   SY=A. frondeus (Gray) Greene
frondosus (Nutt.) Torr. & Gray
  SY=Brachyactis f. (Nutt.) Gray
  SY=Tripolium f. Nutt.
furcatus Burgess
georgianus Alexander
  SY=A. patens Ait. var. g.
   (Alexander) Cronq.
glaucescens (Gray) Blake
  SY=A. engelmannii (D.C. Eat.) Gray

  SY=Eucethalus g. (Gray) Greene
glaucodes Blake
 ssp. glaucodes
  var. formosus (Greene) Kittell
   SY=Eucephalus f. Greene
  var. glaucodes
   SY=A. glaucus sensu Torr. & Gray,
    non Nees
   SY=Eucephalus glaucus Nutt.
 ssp. pulcher Blake
   SY=A. glaucus sensu Torr. & Gray,
    non Nees var. wasatchensis
    M.E. Jones
   SY=A. w. (M.E. Jones) Blake
gormanii (Piper) Blake
  SY=Eucephalus g. Piper
gracilis Nutt.
grandiflorus L.
?gravesii Burgess
greatai Parish
hallii Gray
  SY=A. chilensis Nees ssp. h.
   (Gray) Cronq.
hesperius Gray
  var. hesperius
  var. laetevirens (Greene) Cronq.
   SY=A. l. Greene
   SY=A. osterhoutii Rydb.
  var. wootonii Greene
   SY=A. w. (Greene) Greene
infirmus Michx.
  SY=Doellingeria i. (Michx.) Greene
integrifolius Nutt.
X interior Wieg. [simplex X tradescantii]
  SY=A. simplex Willd. var. i.
   (Wieg.) Cronq.
intricatus (Gray) Blake
  SY=Bigelovia i. Gray
jessicae Piper
laevis L.
  var. concinnus (Willd.) House
   SY=A. c. Willd.
  var. geyeri Gray
   SY=A. g. (Gray) T.J. Howell
  var. laevis
   SY=A. falcidens Burgess
   SY=A. l. var. falcatus Farw.
   SY=A. steeleorum Shinners
X lanceolatus Willd. [borealis X simplex]
  SY=A. lamarckianus Nees
  SY=A. paniculatus sensu Lam. non
   P. Mill.
  SY=A. simplex Willd. var.
   ramosissimus (Torr. & Gray)
   Cronq.
  SY=A. tenuifolius sensu Torr. &
   Gray non L. var. r. Torr. &
   Gray
lateriflorus (L.) Britt.
  var. angustifolius Wieg.
   SY=A. agrostifolius Burgess
  var. flagellaris Shinners
  var. hirsuticaulis (Lindl.) Porter
   SY=A. h. Lindl.
  var. horizontalis (Desf.) Farw.
   SY=A. h. Desf.
   SY=A. l. var. pendulus (Ait.)
    Burgess
   SY=A. p. Ait.
  var. indutus Shinners
  var. lateriflorus
   SY=Solidago l. L.
  var. tenuipes Wieg.
   SY=A. acadiensis Shinners
laurentianus Fern.
  SY=A. l. var. contiguus Fern.
  SY=A. l. var. magdalenensis Fern.

Aster   (CONT.)
   ledophyllus (Gray) Gray
        var. covillei (Greene) Cronq.
           SY=A. c. Greene
        var. ledophyllus
           SY=A. engelmannii (D.C. Eat.) Gray
              var. l. Gray
   lemmonii Gray
   linariifolius L.
        var. linariifolius
           SY=Ionactis l. (L.) Greene
        var. victorinii Fern.
   longifolius Lam.
        SY=A. junceus Ait.
   X longulus Sheldon [borealis X puniceus]
   lowrieanus Porter
        var. incisus (Britt.) Porter
           SY=A. cordifolius L. var. i.
              Britt.
        var. lanceolatus Porter
        var. lowrieanus
           SY=A. cordifolius L. var.
              laevigatus Porter
           SY=A. plumarius Burgess
           SY=A. schistosus Steele
   X maccallae Rydb. [ciliolatus X
     subspicatus]
   macrophyllus L.
        var. apricensis Burgess
           SY=A. nobilis Burgess
        var. excelsior Burgess
        var. ianthinus (Burgess) Fern.
           SY=A. i. Burgess
           SY=A. multiformis Burgess
        var. macrophyllus
           SY=A. riciniatus Burgess
           SY=A. roscidus Burgess
           SY=A. violaris Burgess
        var. pinguifolius Burgess
        var. sejunctus Burgess
        var. velutinus Burgess
   modestus Lindl.
        SY=A. major (Hook.) Porter
        SY=A. unalaschkensis sensu Hook.
           non Less. var. major Hook.
   nebraskensis Britt.
        SY=A. praealtus Poir. var. n.
           (Britt.) Wieg.
        SY=A. woldenii Rydb.
   nemoralis Ait.
   novae-angliae L.
   novi-belgii L.
        var. elodes (Torr. & Gray) Gray
           SY=A. e. Torr. & Gray
        var. litoreus Gray
        var. novi-belgii
           SY=A. n-b. var. rosaceus Rouss.
   oblongifolius Nutt.
        SY=A. kumleinii Fries
        SY=A. o. var. angustatus Shinners
        SY=A. o. var. rigidulus Gray
   occidentalis (Nutt.) Torr. & Gray
        var. delectabilis (Hall) Ferris
           SY=A. d. Hall
        var. intermedius Gray
        var. occidentalis
           SY=A. ascendens Lindl. var.
              fremontii Torr. & Gray
           SY=A. f. (Torr. & Gray) Gray
           SY=A. spathulatus Lindl.
           SY=Tripolium o. Nutt.
        var. parishii (Gray) Ferris
           SY=A. fremontii (Torr. & Gray)
              Gray var. p. Gray
        var. yosemitanus (Gray) Cronq.
           SY=A. ascendens Lindl. var. y.
              Gray

        SY=A. paludicola Piper
        SY=A. y. (Gray) Greene
   ontarionis Wieg.
        SY=A. diffusus Ait. var.
           thyrsoides Gray
        SY=A. lateriflorus (L.) Britt.
           var. t. (Gray) Sheldon
        SY=A. missouriensis sensu Britt.
           non Kuntze
        SY=A. m. var. t. (Gray) Wieg.
        SY=A. pantotrichus Blake
        SY=A. p. var. t. (Gray) Blake
        SY=A. tradescantii L. var.
           thyrsoides (Gray) Boivin
   oregonensis (Nutt.) Cronq.
     ssp. californicus (Dur.) Keck
        SY=Sericocarpus c. Dur.
        SY=S. o. Nutt. ssp. c. (Dur.)
           Ferris
     ssp. oregonensis
        SY=Sericocarpus o. Nutt.
   paludosus Ait.
     ssp. hemisphericus (Alexander) Cronq.
        SY=A. gattingeri sensu Alexander
           non Kuntze
        SY=A. h. Alexander
        SY=A. pedionomus Alexander
        SY=A. verutifolius Alexander
        SY=Heleastrum h. (Alexander)
           Shinners
     ssp. paludosus
   parviceps (Burgess) Mackenzie & Bush
        SY=A. depauperatus (Porter) Fern.
           var. p. (Burgess) Fern.
        SY=A. ericoides sensu Ait. non L.
           var. p. Burgess
        SY=A. e. var. pusillus Gray
   patens Ait.
        var. floridanus R.W. Long
           SY=A. fontinalis Alexander
         var. gracilis Hook.
           SY=A. tenuicaulis (C. Mohr)
              Burgess
        var. patens
        var. patentissimus (Lindl.) Torr. &
           Gray
           SY=A. continuus Small
           SY=A. patentissimus Lindl.
        var. phlogifolius (Muhl.) Nees
           SY=A. phlogifolius Muhl.
   paternus Cronq.
        SY=Conyza asteroides L.
        SY=Sericocarpus a. (L.) B.S.P.
   paucicapitatus (B.L. Robins.) B.L. Robins.
        SY=A. engelmannii (D.C. Eat.) Gray
           var. p. B.L. Robins.
   pauciflorus Nutt.
        SY=A. hydrophilus Greene
   peirsonii C.W. Sharsmith
   perelegans A. Nels. & J.F. Macbr.
        SY=A. elegans sensu Torr. & Gray
           non Willd.
        SY=Eucephalus e. Nutt.
   pilosus Willd.
        var. demotus Blake
           SY=A. juniperinus Burgess
           SY=A. ramosissimus P. Mill.
        var. pilosus
           SY=A. ericoides sensu Ait. non L.
              var. platyphyllus Torr. &
              Gray
           SY=A. e. sensu Ait. non L. var.
              villosus (Michx.) Torr. &
              Gray
           SY=A. pilosus var. platyphyllus
              (Torr. & Gray) Blake
           SY=A. v. Michx.

Aster (CONT.)
   var. priceae (Britt.) Cronq.
    SY=A. kentuckiensis Britt.
    SY=A. priceae Britt.
 porteri Gray
 praealtus Poir.
   var. angustior Wieg.
   var. imbricatior Wieg.
   var. praealtus
   var. subasper (Lindl.) Wieg.
    SY=A. s. Lindl.
 pratensis Raf.
   SY=A. phyllolepis Torr. & Gray
 prenanthoides Muhl.
 pringlei (Gray) Britt.
   SY=A. ericoides sensu Ait. non L.
    var. p. Gray
   SY=A. faxonii Porter
   SY=A. pilosus Willd. var. pringlei
    (Gray) Blake
 puniceus L.
   var. perlongus Fern.
   var. puniceus
    SY=A. conduplicatus Burgess
    SY=A. forwoodii S. Wats.
    SY=A. p. var. oligocephalus Fern.
    SY=A. p. var. compactus Fern.
    SY=A. p. var. demissus (Lindl.)
     Fern.
 purpuratus Nees
   SY=A. attenuatus Lindl.
   SY=A. virgatus sensu Ell. non
    Moench
 racemosus Ell.
 radula Ait.
   var. radula
   var. strictus (Pursh) Gray
    SY=A. s. sensu Pursh non Poir.
 radulinus Gray
   SY=A. eliasii A. Nels.
 reticulatus Pursh
   SY=Doellingeria r. (Pursh) Greene
 riparius H.B.K.
 X sagittifolius Wedemeyer ex Willd.
  [?ciliolatus X cordifolius]
 sandwicensis (Gray) Hieron.
 scabricaulis Shinners
 schreberi Nees
   SY=A. chasei G.N. Jones
   SY=A. curvescens Burgess
   SY=A. glomeratus Bernh. ex Nees
 scopulorum Gray
 sericeus Vent.
 shortii Lindl.
   SY=A. camptosorus Small
 sibiricus L.
   var. meritus (A. Nels.) Raup
    SY=A. m. A. Nels.
   var. pygmaeus (Lindl.) Cody
    SY=A. p. Lindl.
   var. sibiricus
    SY=A. richardsonii Spreng.
 simmondsii Small
   SY=A. sulznerae Small
 simplex Willd.
   var. estuarinus Boivin
   var. simplex
    SY=A. eulae Shinners
    SY=A. paniculatus sensu Lam. non
     P. Mill. var. s. (Willd.)
     Burgess
 solidagineus Michx.
   SY=Conyza linifolia L.
   SY=Sericocarpus l. (L.) B.S.P.
 spatelliformis Burgess
 spectabilis Ait.
   var. cinerascens Blake

   var. spectabilis
    SY=A. smallii Alexander
   var. suffultus Fern.
 spinosus Benth.
 stenomeres Gray
 X subgeninatus (Fern.) Boivin [ciliolatus X
  tardiflorus]
   SY=A. foliaceus var. s. Fern.
 subspicatus Nees
   var. grayi (Suksdorf) Cronq.
    SY=A. g. Suksdorf
   var. subspicatus
    SY=A. douglasii Lindl.
    SY=A. oregonum (Nutt.) Torr. &
     Gray
    SY=Tripolium o. Nutt.
 subulatus Michx.
   var. australis (Gray) Shinners
    SY=A. exilis Ell. var. a. Gray
   var. euroauster Fern. & Grisc.
   var. ligulatus Shinners
    SY=A. exilis Ell.
    SY=A. inconspicuus Less.
    SY=A. neomexicanus Woot. & Standl.
   var. obtusifolius Fern.
   var. subulatus
 surculosus Michx.
 tardiflorus L.
   var. tardiflorus
    SY=A. crenifolius (Fern.) Cronq.
    SY=A. c. var. arcuans (Fern.)
     Cronq.
    SY=A. foliaceus sensu auctt. non
     Lindl.
    SY=?A. gaspensis Victorin
    SY=A. johannensis Fern.
    SY=A. novi-belgi sensu auctt. non
     L.
    SY=A. robynsianus Rouss.
    SY=A. rolandii Shinners
   var. vestitus Fern.
    SY=A. anticostensis Fern.
    SY=A. johannensis Fern. var.
     villicaulis (Gray) Fern.
    SY=A. longifolius Lam. var.
     villicaulis Gray
    SY=A. novi-belgi sensu Boivin non
     L. var. villicaulis (Gray)
     Boivin
 tataricus L.f.
 tenuifolius L.
   var. aphyllus R.W. Long
    SY=A. bracei Britt.
   var. tenuifolius
 texanus Burgess
   var. parviceps Shinners
   var. texanus
 tradescantii L.
   SY=A. dumosus L. var. strictior
    Torr. & Gray
   SY=A. saxatilis (Fern.) Blanch.
   SY=A. tradescantii var. saxatilis
    (Fern.) House
   SY=A. vimineus Lam. var. saxatilis
    Fern.
 turbinellus Lindl.
 umbellatus P. Mill.
   var. brevisquamis Fern.
    SY=A. humilis Willd.
    SY=Doellingeria h. (Willd.) Britt.
   var. latifolius Gray
    SY=A. sericocarpoides (Small) K.
     Schum.
    SY=Doellingeria s. Small
    SY=D. u. (P. Mill.) Nees var. l.
     (Gray) House
   var. pubens Gray

Aster   (CONT.)
          SY=A. pubentior Cronq.
        var. umbellatus
          SY=Doellingeria u. (P. Mill.) Nees
  undulatus L.
        var. asperulus (Torr. & Gray) Wood
          SY=A. asperifolius Burgess
          SY=A. diversifolius Michx.
          SY=A. linguiformis Burgess
          SY=A. mohrii Burgess
          SY=A. u. var. d. (Michx.) Gray
        var. loriformis Burgess
          SY=A. l. (Burgess) Burgess
        var. undulatus
          SY=A. claviger Burgess
          SY=A. corrigiatus Burgess
          SY=A. gracilescens Burgess
          SY=A. proteus Burgess
          SY=A. sylvestris Burgess
          SY=A. triangularis (Burgess)
              Burgess
          SY=A. truellius Burgess
          SY=A. u. var. triangularis Burgess
  urophyllus Lindl.
          SY=A. hirtellus Lindl.
          SY=A. sagittifolius sensu auctt.
              non Wedemeyer ex Willd.
          SY=A. s. var. u. (Lindl.) Burgess
  vialis (Bradshaw) Blake
          SY=Eucephalus v. Bradshaw
  vimineus Lam.
        var. subdumosus Wieg.
        var. vimineus
          SY=A. brachypholis Small
  walteri Alexander
          SY=A. squarrosus sensu Walt. non
              All.
  yukonensis Cronq.

Astranthium Nutt.

  integrifolium (Michx.) Nutt.
    ssp. ciliatum (Raf.) DeJong
          SY=A. i. var. c. (Raf.) Larsen
          SY=A. i. var. trifolium (Raf.)
              Shinners
    ssp. integrifolium
  robustum (Shinners) DeJong
          SY=A. integrifolium (Michx.) Nutt.
              var. r. Shinners

Atrichoseris Gray

  platyphylla Gray

Baccharis L.

  angustifolia Michx.
  bigelovii Gray
  brachyphylla Gray
  dioica Vahl
  douglasii DC.
  emoryi Gray
  glomeruliflora Pers.
  glutinosa (Ruiz & Pavon) Pers.
  halimifolia L.
        var. angustior DC.
        var. halimifolia
  havardii Gray
  neglecta Britt.
  pilularis DC.
        var. consanguinea (DC.) Kuntze
          SY=B. p. ssp. c. (DC.) C.B. Wolf
        var. pilularis
  plummerae Gray
  pteronioides DC.
  salicifolia (Ruiz & Pavon) Pers.

  salicina Torr. & Gray
  sarothroides Gray
  sergiloides Gray
  texana (Torr. & Gray) Gray
  thesioides H.B.K.
  viminea DC.
  wrightii Gray

Bahia Lag.

  absinthifolia Benth.
        var. absinthifolia
        var. dealbata (Gray) Gray
  bigelovii Gray
  biternata Gray
  dissecta (Gray) Britt.
  pedata Gray

Baileya Harvey & Gray ex Torr.

  multiradiata Harvey & Gray ex Torr.
          SY=B. m. var. pleniradiata Harvey
              & Gray
          SY=B. perennis (A. Nels.) Rydb.
  pauciradiata Harvey & Gray
  pleniradiata Harvey & Gray

Balduina Nutt.

  angustifolia (Pursh) B.L. Robins.
          SY=Actinospermum a. (Pursh) Torr.
              & Gray
  atropurpurea Harper
          SY=Endorima a. (Harper) Small
  uniflora Nutt.
          SY=Endorima u. (Nutt.) Barnh.

Balsamita P. Mill.

  major Desf.
          SY=Chrysanthemum balsamita (L.)
              Baill.
          SY=C. b. var. tanacetoides Boiss.
          SY=Tanacetum b. L.

Balsamorhiza Nutt.

  careyana Gray
        var. careyana
        var. intermedia Cronq.
  deltoidea Nutt.
  hirsuta Nutt.
  hookeri (Hook.) Nutt.
        var. hispidula (Sharp) Cronq.
          SY=B. h. Sharp
        var. hookeri
        var. idahoensis (Sharp) Cronq.
          SY=B. macrophylla Nutt. var. i.
              Sharp
        var. lagocephala (Sharp) Cronq.
          SY=B. h. var. l. Sharp
          SY=B. h. var. lanata Sharp
        var. neglecta (Sharp) Cronq.
          SY=B. hirsuta Nutt. var. n. Sharp
  incana Nutt.
  macrolepis Sharp
        var. macrolepis
        var. platylepis (Sharp) Ferris
          SY=B. p. Sharp
  macrophylla Nutt.
  rosea A. Nels. & J.F. Macbr.
  sagittata (Pursh) Nutt.
          SY=B. bonseri St. John
  serrata A. Nels. & J.F. Macbr.
  terebinthacea (Hook.) Nutt.
  X tomentosa Rydb. [incana X sagittata]

Bartlettia Gray

  scaposa Gray

Bebbia Greene

  juncea (Benth.) Greene
    var. aspera Greene

Bellis L.

  perennis L.

Benitoa Keck

  occidentalis (Hall) Keck

Berkheya Ehrh.

  heterophylla O. Hoffmann

Berlandiera DC.

  X betonicifolia (Hook.) Small [pumila X
   texana]
      SY=B. texana DC. var. b. (Hook.)
        Torr. & Gray
  X humilis Small [pumila X subacaulis]
  lyrata Benth.
    var. lyrata
    var. macrophylla Gray
  pumila (Michx.) Nutt.
      SY=B. dealbata (Torr. & Gray)
        Small
      SY=B. tomentosa (Pursh) Nutt.
      SY=B. t. var. d. Torr. & Gray
  subacaulis (Nutt.) Nutt.
  texana DC.

Bidens L.

  amplectens Sherff
  amplissima Greene
  aristosa (Michx.) Britt.
    var. aristosa
      SY=B. a. var. fritcheyi Fern.
      SY=B. a. var. mutica (Gray)
        Gattinger
    var. retrorsa (Sherff) Wunderlin
      SY=B. involucrata (Nutt.) Britt.
      SY=B. polylepis Blake
      SY=B. p. var. r. Sherff
  asplenioides Sherff
  asymetrica (Levl.) Sherff
    var. asymetrica
    var. subocculata Deg. & Sherff ex
      Sherff
  aurea (Ait.) Sherff
      SY=B. a. var. wrightii (Gray)
        Sherff
  X awaluana Deg., Deg. & Sherff [mauiensis
   var. cuneatoides X menziesii]
  bidentoides (Nutt.) Britt.
      SY=B. b. var. mariana (Blake)
        Sherff
      SY=B. m. Blake
  bigelovii Gray
  bipinnata L.
      SY=B. b. var. biternatoides Sherff
  campylotheca Schultz-Bip.
  cernua L.
      SY=B. c. var. dentata (Nutt.)
        Boivin
      SY=B. c. var. elliptica Wieg.
      SY=B. c. var. integra Wieg.
      SY=B. c. var. minima (Huds.) Pursh
      SY=B. c. var. oligodonta Fern. &

      St. John
      SY=B. c. var. radiatus DC.
      SY=B. glaucescens Greene
  cervicata Sherff
  coartata Sherff
  X conjunctata Sherff [mauiensis var.
   cuneatoides X wiebkei]
  connata Muhl. ex Willd.
      SY=B. c. var. ambiversa Fassett
      SY=B. c. var. anomala Farw.
      SY=B. c. var. fallax (Warnst.)
        Sherff
      SY=B. c. var. gracilipes Fern.
      SY=B. c. var. inundata Fern.
      SY=B. c. var. petiolata (Nutt.)
        Farw.
      SY=B. c. var. pinnata S. Wats.
      SY=B. c. var. submutica Fassett
  coronata (L.) Britt.
      SY=B. c. var. brachyodonta Fern.
      SY=B. c. var. tenuiloba (Gray)
        Sherff
      SY=B. c. var. trichosperma
        (Michx.) Fern.
      SY=B. trichosperma (Michx.) Britt.
  cosmoides (Gray) Sherff
    var. cosmoides
    var. refracta Hochr.
  ctenophylla Sherff
  X cuneata Sherff [mauiensis var.
   cuneatoides X molokaiensis]
  cynapiifolia H.B.K.
    var. cynapiifolia
    var. portoricensis (Spreng.) O.E.
      Schulz
    var. tenuis O.E. Schulz
  degeneri Sherff
    var. apioides Sherff
    var. degeneri
  X dimidiata Deg. & Sherff ex Sherff
   [cosmoides X sp.]
  discoidea (Torr. & Gray) Britt.
  X distans Sherff [coartata X skottsbergii]
  eatonii Fern.
    var. eatonii
    var. fallax Fern.
    var. illicita Blake
    var. interstes Fassett
    var. kennebecensis Fern.
    var. major Fassett
    var. mutabilis Fassett
    var. simulans Fassett
  fecunda Deg. & Sherff ex Sherff
  ferulifolia (Jacq.) DC.
  forbesii Sherff
  frondosa L.
      SY=B. f. var. anomala Porter ex
        Fern.
      SY=B. f. var. pallida Wieg.
      SY=B. f. var. puberula Wieg.
      SY=B. f. var. stenodonta Fern. &
        St. John
  fulvescens Sherff
  X graciloides Sherff [hillebrandiana X
   wiebkei]
  hawaiensis (Gray) Pilger
  heterodoxa (Fern.) Fern. & St. John
    var. agnostica Fern.
    var. atheistica Fern.
      SY=B. infirma Fern.
    var. heterodoxa
      SY=B. tripartita L. var. h. Fern.
    var. monardifolia Fern.
    var. orthodoxa Fern. & St. John
  heterosperma Gray
  hillebrandiana (Drake) Deg.
  hyperborea Greene

Bidens (CONT.)
    var. arcuans Fern.
    var. gaspensis Fern.
    var. hyperborea
        SY=B. h. var. colpophila (Fern. &
           St. John) Fern.
    var. svensonii Fassett
        SY=B. h. var. cathancensis Fern.
  laevis (L.) B.S.P.
        SY=B. elegans Greene
        SY=B. nashii Small
  lemmonii Gray
  leptocephala Sherff
  macrocarpa (Gray) Sherff
    var. macrocarpa
    var. ovatifolia (Gray) Sherff
X magnidisca Deg. & Sherff [fulvescens X
  macrocarpa]
  mauiensis (Gray) Sherff
    var. ciliata St. John
    var. cuneatoides Sherff
    var. forbesiana Sherff
    var. lanaiensis Sherff
    var. mauiensis
    var. media Sherff
  menziesii (Gray) Sherff
    var. filiformis Sherff
    var. lepida Deg. & Sherff
    var. leptodonta Sherff
    var. menziesii
  micrantha Gaud.
    var. caduca (Sherff) Sherff
    var. kaalana Sherff
    var. laciniata (Hbd.) Sherff
    var. micrantha
    var. rudimentifera Sherff
X micranthoides Sherff [forbesii X
  menziesii]
  mitis (Michx.) Sherff
      SY=B. m. var. leptophylla Torr. &
        Gray
  molokaiensis (Hbd.) Sherff
  napaliensis Sherff
  nematocera (Sherff) Sherff
  obtusiloba Sherff
  pentamera (Sherff) Deg. & Sherff
  perversa Deg. & Sherff
    var. aquilonalis Deg. & Sherff
    var. perversa
  pilosa L.
    var. minor (Blume) Sherff
    var. pilosa
    var. radiata Schultz-Bip.
X populifolia Sherff [coartata X
  molokaiensis]
  pulchella (Less.) Schultz-Bip.
  reptans (L.) G. Don
    var. reptans
    var. urbanii (Greenm.) O.E. Schulz
        SY=B. u. Greenm.
X salicoides Sherff [menziesii X wiebkei]
  sandvicensis Less.
    var. antrorsa Deg. & Sherff
    var. caduca Sherff
    var. imminuta Deg. & Sherff
    var. leiocarpa Deg. & Sherff
    var. sandvicensis
    var. setosa (Sherff) Sherff
  skottsbergii Sherff
    var. conglutinata (Deg. & Sherff)
      Sherff
    var. pololuensis Deg. & Sherff ex
      Sherff
    var. skottsbergii
  stokesii Sherff
  tenuisecta Gray
  torta Sherff

  tripartita L.
        SY=B. acuta (Wieg.) Britt.
        SY=B. comosa (Gray) Wieg.
  valida Sherff
  vulgata Greene
        SY=B. puberula Wieg.
        SY=B. v. var. p. (Wieg.) Greene
        SY=B. v. var. schizantha Lunell
X waianensis Sherff [fulvescens X
  menziesii]
  waimeana Sherff
  wiebkei Sherff

Bigelowia DC.

  nudata (Michx.) DC.
    ssp. australis L.C. Anders.
        SY=B. n. var. a. (L.C. Anders.)
          Shinners
    ssp. nudata
        SY=B. virgata (Nutt.) DC.
        SY=Chondrophora n. (Michx.) Britt.
        SY=C. v. (Nutt.) Greene
  nuttallii L.C. Anders.

Blennosperma Less.

  bakeri Heiser
  nanum (Hook.) Blake
    var. nanum
    var. robustum J.T. Howell

Blepharipappus Hook.

  scaber Hook.
    ssp. laevis (Gray) Keck
    ssp. scaber

Blepharizonia Greene

  plumosa (Kellogg) Greene
    ssp. plumosa
    ssp. viscida Keck

Blumea DC.

  laciniata (Roxb.) DC.

Boltonia L'Her.

  asteroides (L.) L'Her.
    var. asteroides
        SY=B. a. var. glastifolia (Hill)
          Fern.
    var. decurrens (Torr. & Gray) Engelm.
        SY=B. d. (Torr. & Gray) Wood
        SY=B. latisquama Gray var. d.
          (Torr. & Gray) Fern. &
          Grisc.
    var. latisquama (Gray) Cronq.
        SY=B. l. Gray
    var. recognita (Fern. & Grisc.)
      Cronq.
        SY=B. latisquama Gray var.
          microcephala Fern. & Grisc.
        SY=B. l. var. occidentalis Gray
        SY=B. l. var. r. Fern. & Grisc.
        SY=B. r. (Fern. & Grisc.) G.N.
          Jones
  caroliniana (Walt.) Fern.
        SY=B. ravenelii Fern. & Grisc.
  diffusa Ell.
    var. diffusa
    var. interior Fern. & Grisc.
        SY=B. i. (Fern. & Grisc.) G.N.
          Jones

Borrichia Adans.

    arborescens (L.) DC.
    X cubana Britt. & Blake [arboreseus X
      frutescens]
    frutescens (L.) DC.

Bradburia Torr. & Gray

    hirtella Torr. & Gray

Brickellia Ell.

    amplexicaulis B.L. Robins.
      var. amplexicaulis
      var. lanceolata (Gray) B.L. Robins.
    arguta B.L. Robins.
      var. arguta
      var. odontolepis B.L. Robins.
    atractyloides Gray
    baccharidea Gray
    betonicifolia Gray
    brachyphylla (Gray) Gray
      var. brachyphylla
      var. hinckleyi (Standl.) Flyr
        SY=B. h. Standl.
      var. terlinguensis Flyr
    californica (Torr. & Gray) Gray
      var. californica
      var. jepsonii B.L. Robins.
    chenopodina (Greene) B.L. Robins.
    chlorolepis (Woot. & Standl.) Shinners
        SY=Kuhnia c. Woot. & Standl.
        SY=K. rosmarinifolia Vent.
        SY=K. r. var. c. (Woot. & Standl.)
          Blake
    conduplicata (B.L. Robins.) B.L. Robins.
    cordifolia Ell.
        SY=Coleosanthus c. (Ell.) Kuntze
    coulteri Gray
    cylindracea Gray & Engelm.
    dentata (DC.) Schultz-Bip.
    desertorum Coville
    eupatorioides (L.) Shinners
      var. corymbulosa (Torr. & Gray)
        Shinners
        SY=Kuhnia e. L. var. c. Torr. &
          Gray
        SY=K. hitchcockii A. Nels.
      var. eupatorioides
        SY=Kuhnia e. L.
        SY=K. e. var. angustifolia Raf.
        SY=K. e. var. pyramidalis Raf.
        SY=K. glutinosa Ell.
      var. ozarkana (Shinners) Shinners
        SY=Kuhnia e. L. var. o. Shinners
      var. texana (Shinners) Shinners
        SY=Kuhnia e. L. var. t. Shinners
    fendleri Gray
        SY=Eupatorium f. (Gray) Gray
    floribunda Gray
    frutescens Gray
    grandiflora (Hook.) Nutt.
      var. grandiflora
        SY=Coleosanthus g. (Hook.) Kuntze
      var. petiolaris Gray
    greenei Gray
    incana Gray
    knappiana E. Drew
    laciniata Gray
    lemmonii Gray
      var. lemmonii
      var. wootonii (Greene) B.L. Robins.
    leptophylla (Scheele) Shinners
      var. leptophylla
        SY=Kuhnia l. Scheele
      var. mexicana (Shinners) Shinners

        SY=Kuhnia l. var. m. Shinners
    longifolia S. Wats.
    microphylla (Nutt.) Gray
    mosieri (Small) Shinners
        SY=Kuhnia eupatorioides L. var.
          floridana R.W. Long
        SY=K. e. var. gracilis Torr. &
          Gray
        SY=K. m. Small
    multiflora Kellogg
    nelsonii B.L. Robins.
    nevinii Gray
    oblongifolia Nutt.
      var. linifolia (D.C. Eat.) B.L.
        Robins.
      var. oblongifolia
        SY=B. o. var. typica B.L. Robins.
    parvula Gray
    pringlei Gray
    rusbyi Gray
    scabra (Gray) A. Nels. ex B.L. Robins.
        SY=B. microphylla (Nutt.) Gray
          var. s. Gray
    shineri M.E. Jones ex Flyr
        SY=Eupatorium parryi Gray
    simplex Gray
    squamulosa Gray
    venosa (Woot. & Standl.) B.L. Robins.
    veronicifolia (H.B.K.) Gray
      var. petrophila (B.L. Robins.) B.L.
        Robins.
      var. veronicifolia
    viejensis Flyr
    watsonii B.L. Robins.

Brintonia Greene

    discoidea (Ell.) Greene
        SY=Solidago d. Ell.

Cacalia L.

    atriplicifolia L.
        SY=C. rotundifolia (Raf.) House
        SY=Mesadenia a. (L.) Raf.
    auriculata DC.
      ssp. kamtschatica (Maxim.) Hulten
    decomposita Gray
        SY=Odontotrichum d. (Gray) Rydb.
    diversifolia Torr. & Gray
        SY=Mesadenia d. (Torr. & Gray)
          Greene
    floridana Gray
        SY=Mesadenia f. (Gray) Greene
    mühlenbergii (Schultz-Bip.) Fern.
        SY=C. reniformis Muhl.
        SY=Mesadenia r. (Muhl.) Raf.
    ovata Walt.
        SY=C. lanceolata Nutt.
        SY=C. l. var. elliottii (Harper)
          Kral & R.K. Godfrey
        SY=C. l. var. virescens (Harper)
          Shinners
        SY=Mesadenia e. Harper
        SY=M. l. (Nutt.) Raf.
        SY=M. maxima Harper
    plantaginea (Raf.) Shinners
        SY=C. paniculata Raf.
        SY=C. pteranthes Raf.
        SY=C. tuberosa Nutt.
        SY=Mesadenia t. (Nutt.) Britt.
    rugelia (Shuttlw. ex Chapman) Barkl. &
      Crong.
        SY=Senecio r. (Shuttlw. ex
          Chapman) Gray
    suaveolens L.
        SY=Synosma s. (L.) Britt.

Cacalia (CONT.)
  sulcata Fern.
           SY=Mesadenia s. (Fern.) Harper

Cacaliopsis Gray

   nardosmia (Gray) Gray
          SY=Cacalia n. Gray
          SY=Cacalia n. var. glabrata
             (Piper) Boivin
          SY=Cacaliopsis n. var. g. Piper
          SY=Luina n. (Gray) Cronq.
          SY=L. n. var. g. (Piper) Cronq.

Calendula L.

   arvensis L.
   officinalis L.

Callistephus Cass.

   chinensis (L.) Nees
          SY=Aster c. L.

Calycadenia DC.

   ciliosa Greene
   fremontii Gray
          SY=Hemizonia f. (Gray) Gray
          SY=H. multiglandulosa (DC.) Gray
             var. sparsa Gray
   hispida (Greene) Greene
    ssp. hispida
          SY=C. campestris Greene
          SY=Hemizonia h. Greene
    ssp. reducta Keck
   hooveri G.D. Carr
   mollis Gray
          SY=Hemizonia m. (Gray) Gray
   multiglandulosa DC.
    ssp. bicolor (Greene) Keck
    ssp. cephalotes (DC.) Keck
    ssp. robusta Keck
   oppositifolia (Greene) Greene
          SY=Hemizonia o. Greene
   pauciflora Gray
          SY=C. elegans Greene
          SY=C. p. var. e. (Greene) Jepson
          SY=C. ramulosa Greene
          SY=Hemizonia p. (Gray) Gray
   spicata (Greene) Greene
          SY=Hemizonia s. Greene
   truncata DC.
    ssp. microcephala Hall ex Keck
    ssp. scabrella (E. Drew) Keck
          SY=C. s. (E. Drew) Greene
          SY=C. t. var. s. (E. Drew) Jepson
          SY=Hemizonia s. E. Drew
          SY=Lagophylla s. (E. Drew) M.E.
             Jones
    ssp. truncata
          SY=Hemizonia t. (DC.) Gray
   villosa DC.
          SY=Hemizonia douglasii Gray

Calycoseris Gray

   parryi Gray
   wrightii Gray

Calyptocarpus Less.

   blepharolepis B.L. Robins.
   vialis Less.
          SY=Synedrella v. (Less.) Gray

Carduus L.

   acanthoides L.
   crispus L.
   lanceolatus L.
   nutans L.
    ssp. leiophyllus (Petrovic) Stojanov &
     Stef.
          SY=C. n. var. l. (Petrovic)
             Stojanov & Stojanov
          SY=C. n. var. vestitus (Hallier)
             Boivin
    ssp. macrolepis (Peterm.) Kazmi
          SY=C. m. Peterm.
    ssp. nutans
          SY=C. n. var. macrocephalus
             (Desf.) Boivin
   X orthocephalus Wallr. [acanthoides X
    nutans]
   pycnocephalus L.
   tenuiflorus W. Curtis

Carlina L.

   vulgaris L.

Carminatia Moc. ex DC.

   tenuiflora DC.

Carphephorus Cass.

   bellidifolius (Michx.) Torr. & Gray
   carnosus (Small) C.W. James
          SY=Litrisa c. Small
   corymbosus (Nutt.) Torr. & Gray
   odoratissimus (J.F. Gmel.) Herbert
          SY=Trilisa o. (J.F. Gmel.) Cass.
   paniculatus (J.F. Gmel.) Herbert
          SY=Trilisa p. (J.F. Gmel.) Cass.
   pseudoliatris Cass.
   tomentosus (Michx.) Torr. & Gray
          SY=C. t. var. walteri (Ell.) Fern.

Carphochaete Gray

   bigelovii Gray

Carthamus L.

   baeticus (Boiss. & Reut.) Nyman
   lanatus L.
   tinctorius L.

Centaurea L.

   americana Nutt.
   calcitrapa L.
   calcitrapoides L.
   cineraria L.
   cyanus L.
   diffusa Lam.
   diluta Ait.
   eriophora L.
   iberica Trev. ex Spreng.
   jacea L.
   macrocephala Puschk. ex Willd.
   maculosa Lam.
   melitensis L.
   montana L.
   moschata L.
          SY=Amberboa m. (L.) DC.
   muricata L.
   nervosa Willd.
   nigra L.
   nigrescens Willd.
          SY=C. dubia Suter ssp. vochinensis
             (Bernh. ex Reichenb.) Hayek
          SY=C. v. Bernh. ex Reichenb.

Centaurea (CONT.)
  paniculata L.
  phrygia L.
        SY=C. austriaca Willd.
  X pratensis Thuill. [jacea X nigra]
  repens L.
        SY=Acroptilon r. (L.) DC.
        SY=C. picris Pallas
  rothrockii Greenm.
  salmantica L.
        SY=Mantisalca s. (L.) Briq. &
             Cavillier
  scabiosa L.
  solstitialis L.
  sulphurea Willd.
  transalpina Schleich. ex DC.
        SY=C. dubia Suter
  triumfettii All.
        SY=C. variegata Lam.
        SY=C. v. var. squarrosa (Willd.)
             Boiss.

Centipeda Lour.

  minima (L.) A. Braun & Aschers.

Chaenactis DC.

  alpigena C.W. Sharsmith
  alpina (Gray) M.E. Jones
      var. alpina
        SY=C. douglasii (Hook.) Hook. &
          Arn. var. a. Gray
        SY=C. minuscula Greene
      var. leucopsis (Greene) Cockerell
      var. rubella (Greene) Stockwell
  artemisiifolia (Harvey & Gray ex Gray) Gray
  carphoclinia Gray
      var. attenuata (Gray) M.E. Jones
      var. carphoclinia
      var. peirsonii (Jepson) Munz
  cusickii Gray
  douglasii (Hook.) Hook. & Arn.
      var. achilleifolia (Hook. & Arn.) A.
        Nels.
        SY=C. brachiata Greene
        SY=C. b. var. stansburiana
             Stockwell
      var. douglasii
        SY=C. d. var. typicus Cronq.
        SY=C. suksdorfii Stockwell
      var. glandulosa Cronq.
      var. montana M.E. Jones
        SY=C. angustifolia Greene
        SY=C. cineria Stockwell
        SY=C. d. var. nana Stockwell
        SY=C. humilis Rydb.
        SY=C. panamintensis Stockwell
      var. rubricaulis (Rydb.) Ferris
        SY=C. r. Rydb.
  evermannii Greene
        SY=C. nevadensis (Kellogg) Gray
             var. mainsiana (A. Nels. &
             J.F. Macbr.) Stockwell
  fremontii Gray
        SY=C. furcata Stockwell
  glabriuscula DC.
      var. curta (Gray) Jepson
        SY=C. g. var. aurea (Greene)
             Stockwell
      var. denudata (Nutt.) Munz
      var. glabriuscula
      var. gracilenta (Greene) Keck
        SY=C. tanacetifolia Gray
        SY=C. t. var. gracilenta (Greene)
             Stockwell
      var. lanosa (DC.) Hall

        var. megacephala Gray
        SY=C. g. var. heterocarpha (Torr.
             & Gray ex Gray) Hall
        SY=C. h. Torr. & Gray ex Gray
      var. orcuttiana (Greene) Hall
        SY=C. o. (Greene) Parish
        SY=C. tenuifolia Nutt. var. o.
             Greene
      var. tenuifolia (Nutt.) Hall
        SY=C. t. Nutt.
  macrantha D.C. Eat.
  X mexicana Stockwell [?carphoclinia X
       stevioides]
  nevadensis (Kellogg) Gray
  nevii Gray
  parishii Gray
  ramosa Stockwell
  santolinoides Greene
        SY=C. s. var. indurata Stockwell
  stevioides Hook. & Arn.
      var. brachypappa (Gray) Hall
      var. stevioides
        SY=C. gillespiei Stockwell
        SY=C. latifolia Stockwell
      var. thornberi Stockwell
  suffrutescens Gray
        SY=C. s. var. incana Stockwell
  thompsonii Cronq.
  xantiana Gray

Chaetadelpha Gray

  wheeleri Gray ex S. Wats.

Chaetopappa DC.

  asteroides (Nutt.) DC.
        SY=C. a. var. grandis Shinners
        SY=C. a. var. imberbis Gray
  bellioides (Gray) Shinners
  effusa (Gray) Shinners
  hersheyi Blake
  parryi Gray

Chamaechaenactis Rydb.

  scaposa (Eastw.) Rydb.
      var. parva Preece & Turner
      var. scaposa

Chamaemelum P. Mill.

  fuscatum (Brot.) Vasc.
        SY=Anthemis f. Brot.
  mixta (L.) All.
        SY=Anthemis m. L.
        SY=Ormenis m. (L.) Dumort.
  nobilis (L.) All.
        SY=Anthemis n. L.

Chaptalia Vent.

  dentata (L.) Cass.
  nutans (L.) Polak
        SY=C. n. var. texana (Greene)
             Burkart
  seemanii (Schultz-Bip. ex Seem.) Hemsl.
        SY=C. alsophila Greene
        SY=C. leucocephala Greene
  tomentosa Vent.

Chondrilla L.

  juncea L.

Chrysactinia Gray

Chrysactinia (CONT.)
  mexicana Gray

Chrysanthemum L.

  anethifolium Brouss. ex Willd.
  carinatum Schousboe
  coronarium L.
  segetum L.

Chrysogonum L.

  virginianum L.
      var. australe (Alexander ex Small)
        Ahles
        SY=C. a. Alexander ex Small
      var. virginianum

Chrysoma Nutt.

  pauciflosculosa (Michx.) Greene
      SY=C. solidaginoides Nutt.
      SY=Solidago p. Michx.

Chrysopsis (Nutt.) Ell.

  cruiseana Dress
      SY=Heterotheca c. (Dress) Harms
  floridana Small
      SY=C. mariana (L.) Ell. var. f.
        (Small) Fern.
      SY=Heterotheca f. (Small) R.W.
        Long
      SY=H. m. (L.) Shinners ssp. f.
        (Small) Harms
  godfreyi Semple
  gossypina (Michx.) Ell.
      SY=C. arenicola Alexander
      SY=C. decumbens Chapman
      SY=C. pilosa (Walt.) Britt. non
        Nutt.
      SY=C. trichophylla Nutt.
      SY=Heterotheca g. (Michx.)
        Shinners
      SY=H. t. (Nutt.) Shinners
  hyssopifolia Nutt.
      SY=C. gigantea Small
      SY=C. mixta Dress
      SY=Heterotheca h. (Nutt.) R.W.
        Long
      SY=H. h. var. g. (Small) Harms
  lanuginosa Small
  latisquamea Pollard
      SY=Heterotheca l. (Pollard) Harms
  linearifolia Semple
    ssp. dressii Semple
    ssp. linearifolia
  mariana (L.) Ell.
      SY=Heterotheca m. (L.) Shinners
  pilosa Nutt.
      SY=C. nuttallii Britt.
      SY=Heterotheca p. (Nutt.) Shinners
  scabrella Torr. & Gray
      SY=Heterotheca s. (Torr. & Gray)
        Harms
  subulata Small
      SY=Heterotheca hyssopifolia Nutt.
        var. s. (Small) R.W. Long
      SY=H. s. (Small) Harms

Chrysothamnus Nutt.

  albidus (M.E. Jones ex Gray) Greene
  depressus Nutt.
  greenei (Gray) Greene
      SY=C. g. ssp. filifolius (Rydb.)
        Hall & Clements

      SY=C. g. var. f. (Rydb.) Blake
  humilis Greene
      SY=C. viscidiflorus (Hook.) Nutt.
        ssp. h. (Greene) Hall &
        Clements
      SY=C. v. var. h. (Greene) Jepson
  linifolius Greene
      SY=C. viscidiflorus (Hook.) Nutt.
        ssp. l. (Greene) Hall &
        Clements
      SY=C. v. var. l. (Greene) Kittell
  molestus (Blake) L.C. Anders.
      SY=C. viscidiflorus (Hook.) Nutt.
        var. m. Blake
  nauseosus (Pallas) Britt.
    ssp. albicaulis (Nutt.) Hall & Clements
      SY=C. n. ssp. speciosus (Nutt.)
        Hall & Clements
      SY=C. n. var. a. (Nutt.) Rydb.
      SY=C. n. var. s. (Nutt.) Hall
    ssp. arenarius L.C. Anders.
    ssp. bernardinus (Hall) Hall & Clements
    ssp. bigelovii (Gray) Hall & Clements
      SY=C. n. var. b. (Gray) Hall
      SY=C. n. var. glareosus (M.E.
        Jones) Hall
    ssp. ceruminosus (Dur. & Hilg.) Hall &
     Clements
    ssp. consimilis (Greene) Hall & Clements
      SY=C. n. ssp. pinifolius (Greene)
        Hall & Clements
      SY=C. n. var. artus (A. Nels.)
        Cronq.
      SY=C. n. var. c. (Greene) Hall
    ssp. graveolens (Nutt.) Piper
      SY=C. g. (Nutt.) Greene
      SY=C. n. var. glabratus (Gray)
        Cronq.
      SY=C. n. var. graveolens (Nutt.)
        Hall
      SY=C. n. var. petrophilus Cronq.
    ssp. hololeucus (Gray) Hall & Clements
      SY=C. n. ssp. gnaphalodes (Greene)
        Hall & Clements
      SY=C. n. var. g. (Greene) Hall
      SY=C. n. var. h. (Gray) Hall
    ssp. junceus (Greene) Hall & Clements
      SY=C. n. var. j. (Greene) Hall
    ssp. latisquameus (Gray) Hall & Clements
      SY=C. n. var. l. (Gray) Hall
    ssp. leiospermus (Gray) Hall & Clements
      SY=C. n. var. abbreviatus (M.E.
        Jones) Blake
      SY=C. n. var. l. (Gray) Hall
    ssp. mohavensis (Greene) Hall & Clements
    ssp. nanus (Cronq.) Keck
      SY=C. nauseosus var. nanus Cronq.
    ssp. nauseosus
      SY=C. frigidus Greene
      SY=C. n. var. typicus (Hall)
        Cronq.
    ssp. nitidus L.C. Anders.
    ssp. psilocarpus (Blake) L.C. Anders.
      SY=C. n. var. p. Blake
    ssp. salicifolius (Rydb.) Hall &
     Clements
      SY=C. n. var. s. (Rydb.) Hall
    ssp. texensis L.C. Anders. ined.
    ssp. turbinatus (M.E. Jones) Hall &
     Clements
      SY=C. n. var. t. (M.E. Jones)
        Blake
    ssp. washoensis L.C. Anders.
  paniculatus (Gray) Hall
  parryi (Gray) Greene
    ssp. affinis (A. Nels.) L.C. Anders.
     ined.

Chrysothamnus (CONT.)
    ssp. asper (Greene) Hall & Clements
        SY=C. p. var. a. (Greene) Munz
    ssp. attenuatus (M.E. Jones) Hall &
        Clements
        SY=C. p. var. a. (M.E. Jones)
            Kittell
    ssp. howardii (Parry) Hall & Clements
        SY=C. h. (Parry) Greene
        SY=C. p. var. h. (Parry) Kittell
    ssp. imulus Hall & Clements
    ssp. latior Hall & Clements
    ssp. monocephalus (A. Nels. & Kennedy)
        Hall & Clements
        SY=C. p. var. m. (A. Nels. &
            Kennedy) Jepson
    ssp. montanus L.C. Anders.
    ssp. nevadensis (Gray) Hall & Clements
        SY=C. p. var. n. (Gray) Jepson
    ssp. parryi
        SY=C. p. ssp. typicus Hall &
            Clements
    ssp. salmoensis L.C. Anders.
    ssp. vulcanicus (Greene) Hall & Clements
        SY=C. p. var. v. (Greene) Jepson
pulchellus (Gray) Greene
    ssp. baileyi (Woot. & Standl.) Hall &
        Clements
        SY=C. p. var. b. (Woot. & Standl.)
            Blake
    ssp. pulchellus
        SY=C. p. ssp. elatior (Standl.)
            Hall & Clements
spathulatus L.C. Anders.
        SY=C. viscidiflorus (Hook.) Nutt.
            var. ludens Shinners
teretifolius (Dur. & Hilg.) Hall
vaseyi (Gray) Greene
viscidiflorus (Hook.) Nutt.
    ssp. lanceolatus (Nutt.) Hall & Clements
        SY=C. v. ssp. elegans (Greene)
            Hall & Clements
        SY=C. v. var. e. (Greene) Blake
        SY=C. v. var. l. (Nutt.) Greene
    ssp. planifolius L.C. Anders.
    ssp. puberulus (D.C. Eat.) Hall &
        Clements
        SY=C. v. var. p. (D.C. Eat.)
            Jepson
    ssp. stenophyllus (Gray) Hall & Clements
        SY=C. axillaris Keck
        SY=C. v. var. s. (Gray) Hall
    ssp. viscidiflorus
        SY=C. v. ssp. latifolius (D.C.
            Eat.) Hall & Clements
        SY=C. v. ssp. pumilus (Nutt.) Hall
            & Clements
        SY=C. v. ssp. typicus Hall &
            Clements
        SY=C. v. var. l. (D.C. Eat.)
            Greene
        SY=C. v. var. p. (Nutt.) Jepson
        SY=C. v. var. t. (Hall & Clements)
            Cronq.

Cichorium L.

    endiva L.
    intybus L.

Cirsium P. Mill.

    acanthodontum Blake
        SY=C. oreganum Piper
    altissimum (L.) Spreng.
        SY=Carduus a. L.
        SY=Cirsium a. var. biltmoreanum

                Petrak
        SY=Cirsium iowense (Pammel) Fern.
    amblylepis Petrak
    andersonii (Gray) Petrak
    andrewsii (Gray) Jepson
        SY=Cnicus amplifolius Greene
    araneans Rydb.
        SY=Carduus araneosus Osterhout
    arcuum A. Nels.
    arizonicum (Gray) Petrak
    arvense (L.) Scop.
        var. argenteum (Vest) Fiori
        var. arvense
            SY=Carduus a. (L.) Robson
            SY=Cirsium incanum (Gmel.) Fisch.
            SY=Cirsium setosum (Willd.) Bieb.
        var. horridum Wimmer & Grab.
        var. integrifolium Wimmer & Grab.
        var. mite Wimmer & Grab.
        var. vestitum Wimmer & Grab.
    brevifolium Nutt.
        SY=Cirsium palousense Piper
    brevistylum Cronq.
    calcareum (M.E. Jones) Woot. & Standl.
        SY=Carduus truncatus Greene
        SY=Cirsium bipinnatum (Eastw.)
            Rydb.
        SY=Cirsium pulchellum (Greene)
            Woot. & Standl.
        SY=Cirsium p. ssp. b. (Eastw.)
            Petrak
        SY=Cirsium p. var. glabrescens
            Petrak
    californicum Gray
        var. bernardinum (Greene) Petrak
            SY=Cirsium c. ssp. pseudoreglense
                Petrak
        var. californicum
    callilepis (Greene) Jepson
        var. callilepis
        var. oregonense (Petrak) J.T. Howell
            SY=Cirsium remotifolium (Hook.)
                DC. ssp. o. Petrak
        var. pseudocarlinoides (Petrak) J.T.
            Howell
            SY=Cirsium remotifolium (Hook.)
                DC. ssp. p. Petrak
    campylon H.K. Sharsmith
    X canalense (Osterhout) Petrak
    canescens Nutt.
        SY=Cirsium nebraskense (Britt.)
            Lunell
        SY=C. nelsonii (Pammel) Petrak
        SY=C. plattense (Rydb.) Cockerell
    canovirens (Rydb.) Petrak
    canum (L.) All.
    carolinianum (Walt.) Fern. & Schub.
        SY=Carduus c. Walt.
        SY=Cirsium flaccidum Small
    centaureae (Rydb.) K. Schum.
    chellyense Moore & Frankton
    chuskaense Moore & Frankton
    ciliolatum (Henderson) J.T. Howell
        SY=Cirsium botrys Petrak
        SY=C. howellii Petrak
    clavatum (M.E. Jones) Petrak
    clokeyi Blake
    coloradense (Rydb.) Cockerell
    congdonii Moore & Frankton
    crassicaule (Greene) Jepson
    X crassum (Osterhout) Petrak
    cymosum (Greene) J.T. Howell
        SY=Cirsium triacanthum Petrak
    davisii Cronq.
    discolor (Muhl. ex Willd.) Spreng.
        SY=Carduus d. (Muhl. ex Willd.)
            Nutt.

Cirsium (CONT.)
   douglasii DC.
       var. canescens (Petrak) J.T. Howell
         SY=Cirsium breweri (Gray) Jepson
         SY=C. b. var. c. Petrak
         SY=C. b. var. lanosissimum Petrak
           pro parte
       var. douglasii
         SY=Cirsium breweri (Gray) Jepson
           var. wrangelii Petrak
   drummondii Torr. & Gray
         SY=Cirsium coccinatum Osterhout
   eatonii (Gray) B.L. Robins.
         SY=Cirsium eriocephalum Gray var.
           leiocephalum D.C. Eat.
   edule Nutt.
         SY=Carduus macounii Greene
         SY=Cirsium m. (Greene) Rydb.
   engelmannii Rydb.
         SY=Cirsium altissimum (L.) Spreng.
           var. filipendulum (Engelm.)
            Gray
         SY=C. f. Engelm.
         SY=C. terraenigrae Shinners
   erosum (Rydb.) K. Schum.
   flodmanii (Rydb.) Arthur
         SY=Cirsium nebraskense (Britt.)
           Lunell var. discissum Lunell
         SY=C. oblanceolatum (Rydb.) K.
           Schum.
   foliosum (Hook.) DC.
         SY=Cirsium f. var. minganense
           (Victorin) Boivin
   fontinale (Greene) Jepson
      var. fontinale
      var. obispoense J.T. Howell
   gilense Woot. & Standl.
   grahamii Gray
   griseum (Rydb.) K. Schum.
   hallii (Gray) M.E. Jones
   helenioides (L.) Hill
         SY=Cirsium heterophyllum (L.) Hill
   hillii (Canby) Fern.
         SY=Cirsium pumilum (Nutt.) Spreng.
           ssp. h. (Canby) Moore &
           Frankton
         SY=C. p. var. h. (Canby) Boivin
   hookerianum Nutt.
   horridulum Michx.
         SY=Carduus spinosissimus Walt.
   humboldtense Rydb.
         SY=Carduus nevadensis Greene
   hydrophilum (Greene) Jepson
      var. hydrophilum
         SY=Cirsium vaseyi (Gray) Jepson
           var. h. (Greene) Jepson
      var. vaseyi (Gray) J.T. Howell
         SY=Cirsium montigenum Petrak
         SY=Cirsium v. (Gray) Jepson
         SY=Cnicus breweri Gray var. v.
           Gray
   inornatum Woot. & Standl.
   kamtschaticum Ledeb.
   lacerum (Rydb.) Petrak
   laterifolium (Osterhout) Petrak
   lecontei Torr. & Gray
         SY=Carduus l. (Torr. & Gray)
           Pollard
   loncholepis Petrak
   longistylum Moore & Frankton
   mendocinum Petrak
   mexicanum DC.
   modestum (Osterhout) Cockerell
   mohavense (Greene) Petrak
   muticum Michx.
         SY=Carduus m. (Michx.) Pers.
         SY=Cirsium bigelowii DC.

         SY=Cirsium m. var. monticola Fern.
         SY=Cirsium m. var. subpinnatifidum
           (Britt.) Fern.
   navajoense Moore & Frankton
   neomexicanum Gray
   nidulum (M.E. Jones) Petrak
   nuttallii DC.
         SY=Carduus n. (DC.) Pollard
   occidentale (Nutt.) Jepson
      var. compacta Hoover
      var. occidentale
         SY=Cirsium coulteri Harvey & Gray
   ochrocentrum (Gray) Gray
   olivescens (Rydb.) Petrak
   oreophilum (Rydb.) K. Schum.
   osterhoutii (Rydb.) Petrak
   pallidum Woot. & Standl.
   palustre (L.) Scop.
   parryi (Gray) Petrak
   pastoris J.T. Howell
         SY=Cirsium candidissimum (Greene)
           A. Davids. & Moxley
   peckii Henderson
   perplexans (Rydb.) Petrak
   pitcheri (Torr. ex Eat.) Torr. & Gray
   praeteriens J.F. Macbr.
   proteanum J.T. Howell
         SY=Cirsium occidentale (Nutt.)
           Jepson var. venustum
           (Greene) Jepson
   pulcherrimum (Rydb.) K. Schum.
   pumilum (Nutt.) Spreng.
         SY=Carduus p. Nutt.
         SY=Cirsium odoratum (Muhl.) Petrak
   quercetorum (Gray) Jepson
      var. quercetorum
         SY=Cirsium q. var. mendocinum
           Petrak
      var. walkerianum (Petrak) Jepson
         SY=Cirsium w. Petrak
      var. xerolepis Petrak
   remotifolium (Hook.) DC.
      var. odontolepis Petrak
      var. remotifolium
         SY=Cirsium stenolepidum Nutt.
   repandum Michx.
         SY=Carduus r. (Michx.) Pers.
   rhothophilum Blake
         SY=Cirsium maritimum (Elmer)
           Petrak
   rothrockii (Gray) Petrak
         SY=Cirsium diffusum (Eastw.) Rydb.
         SY=C. pulchellum (Greene) Woot. &
           Standl. ssp. d. (Eastw.)
           Petrak
   rusbyi (Greene) Petrak
   rydbergii Petrak
         SY=Cirsium lactucinum Rydb.
   scabrum (Poir.) Bonnet & Barratte
   scapanolepis Petrak
         SY=Carduus spathulatus Osterhout
         SY=Cirsium spathulifolium Rydb.
   scariosum Nutt.
         SY=Cirsium butleri (Rydb.) Petrak
         SY=C. kelseyi (Rydb.) Petrak
         SY=C. magnificum (A. Nels.) Petrak
         SY=C. minganense Victorin
   scopulorum (Greene) Cockerell
         SY=Cirsium eriocephalum Gray
         SY=C. hesperium (Eastw.) Petrak
   subniveum Rydb.
   texanum Buckl.
         SY=Carduus austrinus Small
         SY=Cirsium t. var. stenolepis
           Shinners
   tioganum (Congd.) Petrak
         SY=Cirsium acaule All. var.

Cirsium (CONT.)
         americanum Gray
                  SY=Cirsium acaulescens (Gray) K.
                       Schum.
                  SY=Cirsium americanum (Gray) K.
                       Schum.
                  SY=Cirsium coloradense (Rydb.)
                       Cockerell ssp. acaulescens
                       (Gray) Petrak
                  SY=Cirsium c. ssp. longissimum
                       (Heller) Petrak
                  SY=Cirsium drummondii Torr. & Gray
                       ssp. lanatum Petrak
                  SY=Cirsium d. ssp. lanatum var.
                       oregonense Petrak
                  SY=Cirsium d. ssp. latisquamum
                       Petrak
                  SY=Cirsium d. ssp. vexans Petrak
                  SY=Cirsium d. var. acaulescens
                       (Gray) J.F. Macbr.
                  SY=Cirsium quercetorum (Gray)
                       Jepson var. citrinum Petrak
                  SY=Cnicus drummondii (Torr. &
                       Gray) Gray var. acaulescens
                       Gray
                  SY=Cnicus tioganus Congd.
     tracyi (Rydb.) Petrak
                  SY=Cirsium acuatum (Osterhout)
                       Petrak
                  SY=C. floccosum (Rydb.) Petrak
     turneri Warnock
     tweedyi (Rydb.) Petrak
                  SY=Cirsium hookerianum Nutt. var.
                       eriocephalum (Gray) A. Nels.
                  SY=C. polyphyllum (Rydb.) Petrak
     undulatum (Nutt.) Spreng.
                  SY=Carduus helleri Small
                  SY=Cirsium megacephalum (Gray)
                       Cockerell
                  SY=Cirsium ochrocentrum (Gray)
                       Gray var. h. (Small) Petrak
                  SY=Cirsium u. var. m. (Gray) Fern.
     utahense Petrak
                  SY=Carduus inamoenus Greene
                  SY=Cirsium wallowense M.E. Peck
     validus Greene
     X vancouverense Moore & Frankton
          [brevistylum X edule]
     vernale (Osterhout) Cockerell
     vinaceum Woot. & Standl.
     virginianum (L.) Michx.
                  SY=Carduus v. L.
                  SY=Cirsium revolutum (Small)
                       Petrak
     vittatum Small
                  SY=Carduus pinetorum Small
                  SY=Carduus smallii (Britt.) Ahles
                  SY=Cirsium horridulum Michx. var.
                       v. (Small) R.W. Long
                  SY=Cirsium s. Britt.
     vulgare (Savi) Tenore
                  SY=Cirsium lanceolatum (L.) Scop.
                  SY=?C. l. var. hypoleucum DC.
     wheeleri (Gray) Petrak
                  SY=Cirsium blumeri Petrak
                  SY=C. perennans (Greene) Woot. &
                       Standl.
     wrightii Gray

Clappia Gray

     suaedifolia Gray

Clibadium L.

     erosum (Sw.) DC.

Cnicus L.

     benedictus L.
                  SY=Centaurea b. (L.) L.

Conyza Less.

     apurensis H.B.K.
                  SY=C. subspathulatus Cronq.
                  SY=Erigeron a. (H.B.K.) Griseb.
                  SY=E. chinensis Jacq.
                  SY=E. spathulatus Vahl
                  SY=L. c. (Jacq.) Britt.
     bilbaoana Remy
                  SY=Erigeron b. (Remy) Cabrera
     bonariensis (L.) Cronq.
                  SY=Erigeron b. L.
                  SY=E. linifolius Willd.
                  SY=Leptilon b. (L.) Small
                  SY=L. l. (Willd.) Small
     canadensis (L.) Cronq.
          var. canadensis
                  SY=Erigeron c. L.
                  SY=Leptilon c. (L.) Britt.
          var. glabrata (Gray) Cronq.
                  SY=Erigeron c. L. var. g. Gray
          var. pusilla (Nutt.) Cronq.
                  SY=C. parva Cronq.
                  SY=Erigeron c. L. var. pusillus
                       (Nutt.) Boivin, non Ahles
                  SY=E. pusillus Nutt.
                  SY=Leptilon pusillum (Nutt.)
                       Britt.
     chilensis Spreng.
     coulteri Gray
                  SY=Eschenbachia c. (Gray) Rydb.
                  SY=?Leptilon integrifolium Woot. &
                       Standl.
     eriophylla (Gray) Cronq.
                  SY=Erigeron e. Gray
     filaginoides (DC.) Hieron.
     floribunda H.B.K.
                  SY=Erigeron f. (H.B.K.)
                       Schultz-Bip.
                  SY=Leptilon f. (H.B.K.)
                       Schultz-Bip.
     lyrata H.B.K.
                  SY=Eschenbachia l. (H.B.K.) Britt.
                       & Millsp.
     ?odorata Sesse & Moc.
     ramosissima Cronq.
                  SY=Erigeron divaricatus Michx.
                  SY=Leptilon d. (Michx.) Raf.
     schiedeana (Less.) Cronq.
                  SY=Erigeron s. Less.
     sophiaefolia H.B.K.
                  SY=C. coulteri Gray var.
                       tenuisecta Gray
                  SY=Eschenbachia t. (Gray) Woot. &
                       Standl.

Coreocarpus Benth.

     arizonicus (Gray) Blake

Coreopsis L.

     auriculata L.
     basalis (Otto & A. Dietr.) Blake
                  SY=Calliopsis b. Otto & A. Dietr.
                  SY=Coreopsis drummondii (D. Don)
                       Torr. & Gray
     bigelovii (Gray) Hall
                  SY=Leptosyne b. Gray
     californica (Nutt.) H.K. Sharsmith
     calliopsidea (DC.) Gray
     X delphinifolia Lam. [tripteris X

Coreopsis (CONT.)
    verticillata and perhaps major]
        SY=Coreopsis d. var. chlooidea
            Sherff
        SY=C. major Walt. var. linearis
            Small
    douglasii (DC.) Hall
        SY=Coreopsis stillmanii (Gray)
            Blake var. jonesii Sherff
        SY=Leptosyne d. DC.
    falcata Boynt.
    floridana E. B. Sm.
    gigantea (Kellogg) Hall
        SY=Leptosyne g. Kellogg
    gladiata Walt.
        SY=Coreopsis angustifolia Soland.
        SY=C. helianthoides Beadle
        SY=C. longifolia Small
        SY=C. l. var. godfreyi Sherff
    grandiflora Hogg ex Sweet
      var. grandiflora
        SY=Coreopsis g. var. pilosa Sherff
      var. harveyana (Gray) Sherff
        SY=Coreopsis heterolepis Sherff
      var. longipes (Hook.) Torr. & Gray
      var. saxicola (Alexander) E. B. Sm.
        SY=Coreopsis s. Alexander
        SY=C. s. var. duncanii Sherff
    hamiltonii (Elmer) H.K. Sharsmith
    integrifolia Poir.
    intermedia Sherff
    lanceolata L.
        SY=Coreopsis crassifolia Ait.
        SY=C. heterogyna Fern.
        SY=C. l. var. villosa Michx.
    latifolia Michx.
        SY=Leiodon l. (Michx.) Shuttlw.
    leavenworthii Torr. & Gray
        SY=Coreopsis angustata Greene
        SY=C. l. var. curtissii Sherff
        SY=C. l. var. garberi Gray
        SY=C. l. var. lewtonii (Small)
            Sherff
        SY=C. lewtonii Small
    linifolia Nutt.
        SY=Coreopsis gladiata Walt. var.
            l. (Nutt.) Cronq.
        SY=C. oniscicarpa Fern.
        SY=C. o. var. simulans Fern.
    major Walt.
        SY=Coreopsis m. var. oemleri
            (Ell.) Britt. ex Small &
            Vail
        SY=C. m. var. rigida (Nutt.)
            Boynt.
        SY=C. m. var. stellata (Nutt.)
            B.L. Robins.
    maritima (Nutt.) Hook. f.
    nudata Nutt.
    nuecensis Heller
    nuecensoides E. B. Sm.
    palmata Nutt.
    pubescens Ell.
      var. debilis (Sherff) E. B. Sm.
        SY=Coreopsis corninsularis Sherff
        SY=C. d. Sherff
      var. pubescens
      var. robusta Gray ex Eames
    pulchra Boynt.
    rosea Nutt.
    stillmanii (Gray) Blake
        SY=Leptosyne s. Gray
    tinctoria Nutt.
      var. atkinsoniana (Dougl. ex Lindl.)
        H.M. Parker
        SY=Coreopsis a. Dougl. ex Lindl.
      var. similis (Boynt.) H.M. Parker

        SY=Coreopsis s. Boynt.
      var. tinctoria
        SY=Coreopsis cardaminefolia
            (DCoreopsis) Nutt.
        SY=C. stenophylla Boynt.
    tripteris L.
        SY=Coreopsis t. var. deamii
            Standl.
        SY=C. t. var. intercedens Standl.
        SY=C. t. var. smithii Sherff
        SY=C. t. var. subrhomboidea Sherff
    verticillata L.
    wrightii (Gray) H.M. Parker
        SY=Coreopsis basalis (Otto & A.
            Dietr.) Blake var. w. (Gray)
            Blake
        SY=C. drummondii (D. Don) Torr. &
            Gray var. w. Gray

Corethrogyne DC.

    californica DC.
      var. californica
      var. lyonii Blake
      var. obovata (Benth.) Kuntze
        SY=C. o. Benth.
    filaginifolia (Hook. & Arn.) Nutt.
      var. bernardina (Abrams) Hall
      var. filaginifolia
        SY=C. tomentella (Hook. & Arn.)
            Torr. & Gray
      var. glomerata Hall
        SY=C. f. var. brevicula (Greene)
            Canby
      var. hamiltonensis Keck
      var. incana (Nutt.) Canby
        SY=C. i. Nutt.
      var. latifolia Hall
        SY=C. flagellaris Greene
      var. peirsonii Canby
      var. pinetorum I.M. Johnston
      var. robusta Greene
      var. sessilis (Greene) Canby
        SY=C. s. Greene
      var. virgata (Benth.) Gray
        SY=C. f. var. rigida Gray
        SY=C. floccosa Greene
      var. viscidula (Greene) Keck
        SY=C. scabra Greene
        SY=C. v. Greene
    leucophylla (Lindl.) Jepson
    linifolia (Hall) Ferris
        SY=C. filaginifolia (Hook. & Arn.)
            Nutt. var. l. Hall

Cosmos Cav.

    bipinnatus Cav.
    caudatus H.B.K.
        SY=Bidens c. (H.B.K.) Schultz-Bip.
    parviflorus (Jacq.) H.B.K.
    sulphureus Cav.

Cotula L.

    australis (Sieber ex Spreng.) Hook. f.
    coronopifolia L.

Crassocephalum Moench

    crepidioides (Benth.) S. Moore

Crepis L.

    acuminata Nutt.
      ssp. acuminata
        SY=C. a. ssp. typica Babcock &

Crepis (CONT.)
                Stebbins
    ssp. pluriflora Babcock & Stebbins
  atribarba Heller
    ssp. atribarba
        SY=C. a. ssp. typicus Babcock &
            Stebbins
    ssp. originalis Babcock & Stebbins
        SY=C. barbigera Leib. ex Coville
  bakeri Greene
    ssp. bakeri
    ssp. cusickii (Eastw.) Babcock &
        Stebbins
    ssp. idahoensis Babcock & Stebbins
  biennis L.
  bursifolia L.
  capillaris (L.) Wallr.
  elegans Hook.
  foetida L.
  intermedia Gray
  modocensis Greene
    ssp. glabraeosa (Piper) Babcock &
        Stebbins
    ssp. modocensis
        SY=C. m. ssp. typica Babcock &
            Stebbins
    ssp. rostrata (Coville) Babcock &
        Stebbins
        SY=C. m. var. r. (Coville) Boivin
    ssp. subacaulis (Kellogg) Babcock &
        Stebbins
  monticola Coville
  nana Richards.
    ssp. clivicola Leggett
    ssp. nana
        SY=C. n. ssp. typica Babcock
        SY=C. n. var. lyratifolia (Turcz.)
            Hulten
    ssp. ramosa Babcock & Stebbins
  nicaeensis Balbis ex Pers.
  occidentalis Nutt.
    ssp. conjuncta (Jepson) Babcock &
        Stebbins
    ssp. costata (Gray) Babcock & Stebbins
        SY=C. o. var. c. Gray
    ssp. occidentalis
        SY=C. o. ssp. typica Babcock &
            Stebbins
    ssp. pumila (Rydb.) Babcock & Stebbins
  pannonica (Jacq.) K. Koch
  pleurocarpa Gray
  pulchra L.
  rubra L.
  runcinata (James) Torr. & Gray
    ssp. andersonii (Gray) Babcock &
        Stebbins
    ssp. barberi (Greenm.) Babcock &
        Stebbins
    ssp. glauca (Nutt.) Babcock & Stebbins
        SY=C. g. (Nutt.) Torr. & Gray
    ssp. hallii Babcock & Stebbins
    ssp. hispidulosa (T.J. Howell) Babcock &
        Stebbins
        SY=C. platyphylla Greene
        SY=C. riparia A. Nels.
        SY=C. runcinata var. h. T.J.
            Howell
    ssp. imbricata Babcock & Stebbins
    ssp. runcinata
        SY=C. glaucella Rydb.
        SY=C. perplexans Rydb.
        SY=C. r. ssp. typica Babcock &
            Stebbins
  setosa Haller f.
  tectorum L.
  vesicaria L.
    ssp. haenseleri (Boiss. ex DC.) P.D.

        Sell
        SY=C. v. ssp. taraxacifolia
            (Thuill.) Thellung
        SY=C. v. var. t. (Thuill.) Boivin

Crocidum Hook.

  multicaule Hook.

Crupina (Pers.) DC.

  vulgaris Cass.

Cynara L.

  cardunculus L.
  scolymus L.

Dendranthema (DC.) Des Moulins

  arcticum (L.) Tzvelev
    ssp. polare (Hulten) Tzvelev
        SY=Chrysanthemum a. L. pro parte
        SY=C. a. var. p. (Hulten) Boivin

Dichaetophora Gray

  campestris Gray

Dicoria Torr. & Gray ex Torr.

  brandegei Gray
        SY=D. paniculata Eastw.
  canescens Torr. & Gray
    ssp. canescens
        SY=D. oblongifolia Rydb.
    ssp. clarkae (Kennedy) Keck
        SY=D. clarkae Kennedy
    ssp. hispidula (Rydb.) Keck
  wetherillii Eastw.

Dicranocarpus Gray

  parviflorus Gray

Dimeresia Gray

  howellii Gray

Dimorphotheca Vaill. ex Moench

  ecklonis DC.
  sinuata DC.

Doronicum L.

  orientale Hoffmann
        SY=D. caucasicum Bieb.
  pardalianches L.

Dracopis Cass.

  amplexicaulis (Vahl) Cass.
        SY=Rudbeckia a. Vahl

Dubautia Gaud.

  arborea (Gray) Keck
        SY=Railliardia a. Gray
  ciliolata (DC.) Keck
    var. ciliolata
        SY=Railliardia c. DC.
    var. juniperoides (Gray) Keck
        SY=R. c. var. j. Gray
    var. trinervia (Hbd.) Keck
        SY=Railliardia c. var. t. Hbd.
  demissifolia (Sherff) Keck

Dubautia (CONT.)
    var. demissifolia
        SY=Railliardia d. Sherff
        SY=?R. d. var. dolichophylla St.
           John
    var. verticillata Sherff
        SY=Railliardia d. Sherff var. v.
           Sherff
  X fallax Sherff [plantaginea X scabra]
        SY=Railliardia kohalae Skottsberg
        SY=Railliautia X f. (Sherff)
           Sherff
  X fucosa Sherff [linearis X scabra]
        SY=D. molokaiensis (Hbd.) Keck
        SY=D. m. var. oppositifolia
           (Sherff) Keck
        SY=Railliardia m. Hbd.
        SY=Railliardia m. var. o. Sherff
        SY=Railliautia X f. (Sherff)
           Sherff
knudsenii Hbd.
laevigata Gray
    var. laevigata
    var. parvifolia Sherff
latifolia (Gray) Keck
    SY=Railliardia l. Gray
    SY=R. l. var. h. Sherff
laxa Hook. & Arn.
    var. blakei Deg. & Sherff
    var. bryanii Deg. & Sherff
    var. greenwelliae Deg. & Sherff
    var. hirsuta Hbd.
    var. hispida Sherff
    var. intercedens Deg., Deg. & Sherff
    var. laxa
        SY=D. knudsenii Hbd. var. degeneri
           Sherff
    var. obovata Sherff
    var. pedicellata Rock
    var. pseudoplantaginea Skottsberg
    var. skottsbergii Sherff
    var. waianensis Deg. & Sherff
linearis (Gaud.) Keck
    var. linearis
        SY=Railliardia l. Gaud.
    var. opposita (Sherff) Keck
        SY=Railliardia l. Gaud. var. o.
           Sherff
magnifolia Sherff
X media Sherff [laevigata X laxa]
    SY=D. X mendacoides Deg. & Sherff
    SY=D. X mendax Sherff
menziesii (Gray) Keck
    var. angustifolia (Sherff) Keck
        SY=Railliardia m. Gray var. a.
           Sherff
    var. menziesii
        SY=D. rockii (Sherff) Keck
        SY=D. struthioloides (Gray) Keck
        SY=Railliardia m. Gray
        SY=R. r. Sherff
        SY=R. s. Gray
microcephala Skottsberg
    var. forbesii Sherff
    var. microcephala
montana (Mann) Keck
    var. longifolia (Sherff) Keck
        SY=Railliardia m. Mann var. l.
           Sherff
    var. montana
        SY=D. hillebrandii (Mann) Keck
        SY=Railliardia ciliolata DC. var.
           leptocephala Deg. & Sherff
        SY=R. h. Mann
        SY=R. m. Mann
        SY=R. X vafra Deg. & Sherff
nagatae St. John

paleata Gray
X paludicola Sherff [paleata X waialealae]
    SY=D. waialealae Rock var.
        megaphylla Sherff
plantaginea Gaud.
    var. acridentata Sherff
    var. glandulosa St. John
    var. pauoaensis St. John
    var. plantaginea
    var. platyphylla Hbd.
    var. strigosa Skottsberg
platyphylla (Gray) Keck
    var. leptophylla (Sherff) Keck
        SY=Railliardia p. Gray var. l.
           Sherff
    var. platyphylla
        SY=Railliardia p. Gray
        SY=R. p. var. trillioidea Deg. &
           Sherff
raillardioides Hbd.
reticulata (Sherff) Keck
    SY=Railliardia r. Sherff
scabra (DC.) Keck
    var. leiophylla (Gray) Keck
        SY=Railliardia s. DC. var. l. Gray
    var. munroi (Sherff) Keck
        SY=D. molokaiensis (Hbd.) Keck
           var. stipitata (Sherff) Keck
        SY=Railliardia lonchophylla Sherff
           var. stipitata (Sherff) Keck
        SY=R. molokaiensis Hbd. var.
           stipitata Sherff
        SY=R. scabra DC. var. munroi
           Sherff
sherffiana Fosberg
    SY=Railliardia s. (Fosberg) Sherff
    SY=R. s. var. paucinervia Deg. &
        Sherff
ternifolia (Sherff) Keck
    SY=Railliardia t. Sherff
thyrsiflora (Sherff) Keck
    var. cernua Sherff
        SY=Railliardia t. Sherff var. c.
           Sherff
    var. thyrsiflora
        SY=Railliardia t. Sherff
waialealae Rock

Dugaldia Cass.

hoopesii (Gray) Rydb.
    SY=Helenium h. Gray

Dyssodia Cav.

acerosa DC.
aurea (Gray) A. Nels.
    var. aurea
        SY=Thymophylla a. (Gray) Greene
    var. polychaeta (Gray) M.C. Johnston
        SY=D. p. (Gray) B.L. Robins.
        SY=Thymophylla p. Gray
concinna (Gray) B.L. Robins.
cooperi Gray
micropoides (DC.) Loes.
neomexicana (Gray) B.L. Robins.
papposa (Vent.) A.S. Hitchc.
    SY=Boebera p. (Vent.) Rydb.
pentachaeta (DC.) B.L. Robins.
  ssp. hartwegii (Gray) Strother
        SY=D. h. (Gray) B.L. Robins.
        SY=D. p. var. h. (Gray) Strother
  ssp. pentachaeta
    var. belenidium (DC.) Strother
        SY=D. b. (DC.) Macloskie
        SY=D. thurberi (Gray) Woot. &
           Standl.

Dyssodia (CONT.)
            var. pentachaeta
            var. puberula (Rydb.) Strother
    porophylloides Gray
    setifolia (Lag.) B.L. Robins.
            var. radiata (Gray) Strother
            var. setifolia
    tagetoides Torr. & Gray
            SY=Hymenatherum t. (Torr. & Gray)
                Gray
    tenuiloba (DC.) B.L. Robins.
        ssp. tenuiloba
            var. tenuiloba
                SY=Thymophylla t. (DC.) Small
            var. texana (Cory) Strother
                SY=D. texana Cory
            var. treculii (Gray) Strother
                SY=D. treculii (Gray) B.L. Robins.
        ssp. wrightii (Gray) Strother
                SY=D. t. var. w. (Gray) Strother
                SY=D. w. (Gray) B.L. Robins.
    tephroleuca Blake

Eastwoodia Brandeg.

    elegans Brandeg.

Eatonella Gray

    congdonii Gray
    nivea (D.C. Eat.) Gray

Echinacea Moench

    angustifolia DC.
            var. angustifolia
                SY=Brauneria a. (DC.) Heller
                SY=E. pallida Nutt. var. a. (DC.)
                    Cronq.
            var. strigosa R.L. McGregor
    atrorubens Nutt.
    laevigata (C.L. Boynt. & Beadle) Blake
                SY=E. purpurea (L.) Moench var. l.
                    (C.L. Boynt. & Beadle)
                    Cronq.
    pallida Nutt.
                SY=Brauneria p. (Nutt.) Britt.
    paradoxa (J.B.S. Norton) Britt.
            var. neglecta R.L. McGregor
            var. paradoxa
                SY=E. atrorubens Nutt. var. p.
                    (J.B.S. Norton) Cronq.
    purpurea (L.) Moench
                SY=Brauneria p. (L.) Britt.
                SY=E. p. var. arkansana Steyermark
    sanguinea Nutt.
    simulata R.L. McGregor
    tennesseensis (Beadle) Small
                SY=E. angustifolia DC. var. t.
                    (Beadle) Blake

Echinops L.

    exaltatus Schrad.
                SY=E. commutatus Juratzka
    rito L.
        ssp. ruthenicus (Bieb.) Nyman
                SY=E. ruthenicus Bieb.
    sphaerocephalus L.

Eclipta L.

    prostrata (L.) L.
                SY=E. alba (L.) Hassk.
                SY=Verbesina a. L.

Egletes Cass.

    prostrata (Sw.) Kuntze
    viscosa (L.) Less.

Elephantopus L.

    angustifolius Sw.
    carolinianus Raeusch.
            SY=E. flexuosus Raf.
            SY=E. violaceus Schulz-Bip.
    elatus Bertol.
            SY=E. e. var. intermedius Gleason
    mollis H.B.K.
            SY=E. hypomalacus Blake
            SY=E. martii Graham
            SY=E. pilosus Philipson
    nudatus Gray
    scaber L.
    spicatus Juss. ex Aubl.
            SY=Pseudoelephantopus s. (Juss. ex
                Aubl.) Rohr
    tomentosus L.
            SY=E. carolinianus Raeusch. var.
                simplex Nutt.
            SY=E. nudicaulis Poir.

Eleutheranthera Poit. ex Bosc

    ruderalis (Sw.) Schultz-Bip.

Emilia Cass.

    coccinea (Sims) G. Don
    fosbergerii D.H. Nicols.
            SY=E. javanica (Burm. f.) C.B.
                Robins.
    sonchifolia (L.) DC. ex Wight

Encelia Adans.

    californica Nutt.
    farinosa Gray ex Torr.
            var. farinosa
            var. phenicodonta (Blake) I.M.
                Johnston
            var. radians Brandeg. ex Blake
    frutescens (Gray) Gray
            var. frutescens
            var. resinosa M.E. Jones
    scaposa (Gray) Gray
            SY=E. s. var. stenophylla Shinners
    virginiensis A. Nels.
        ssp. actonii (Elmer) Keck
            SY=E. a. Elmer
            SY=E. frutescens (Gray) Gray var.
                a. (Elmer) Blake
        ssp. virginiensis
            SY=E. frutescens (Gray) Gray var.
                v. (A. Nels.) Blake

Enceliopsis (Gray) A. Nels.

    argophylla (D.C. Eat.) A. Nels.
            var. argophylla
            var. grandiflora (M.E. Jones) Jepson
    covillei (A. Nels.) Blake
    nudicaulis (Gray) A. Nels.
            var. corrugata Cronq.
            var. nudicaulis
    nutans (Eastw.) A. Nels.

Engelmannia Gray ex Nutt.

    pinnatifida Gray ex Nutt.

Enydra Lour.

    sessilis (Sw.) DC.

Erechtites Raf.

    arguta (A. Rich.) DC.
           SY=Senecio a. A. Rich.
    hieraciifolia (L.) Raf. ex DC.
        var. cacalioides (Fisch. ex Spreng.)
           Griseb.
        var. hieraciifolia
           SY=E. h. var. intermedia Fern.
           SY=E. h. var. praealta (Raf.)
              Fern.
        var. megalocarpa (Fern.) Cronq.
           SY=E. m. Fern.
    minima (Poir.) DC.
           SY=E. prenanthoides (A. Rich.) DC.
           SY=Senecio m. Poir.
    valerianifolia (Wolf) DC.

Ericameria Nutt.

    arborescens (Gray) Greene
           SY=Haplopappus a. (Gray) Hall
    austrotexana M.C. Johnston
    brachylepis (Gray) Hall
           SY=Haplopappus propinquus Blake
    cervina (S. Wats.) Rydb.
           SY=Haplopappus c. S. Wats.
    cooperi (Gray) Hall
           SY=Haplopappus c. (Gray) Hall
    cuneata (Gray) McClatchie
        var. cuneata
           SY=Haplopappus c. Gray
        var. macrocephala Urbatsch
        var. spathulata (Gray) McClatchie
           SY=Haplopappus c. Gray var. s.
              (Gray) Blake
    ericoides (Less.) Jepson
           SY=Diplopappus e. Less.
           SY=Haplopappus e. (Less.) Hook. &
              Arn.
           SY=?H. e. ssp. blakei C.B. Wolf
    fasciculata (Eastw.) J.F. Macbr.
           SY=Haplopappus eastwoodae Hall
    laricifolia (Gray) Shinners
           SY=Haplopappus l. Gray
    linearifolia (DC.) Urbatsch & Wussow
           SY=Haplopappus l. DC.
           SY=H. l. ssp. interior (Coville)
              Hall
           SY=H. l. var. i. (Coville) M.E.
              Jones
           SY=Stenotopsis i. (Coville) Rydb.
           SY=S. l. (DC.) Rydb.
    nana Nutt.
           SY=Haplopappus n. (Nutt.) D.C.
              Eat.
    palmeri (Gray) Hall
      ssp. palmeri
           SY=Haplopappus p. Gray
      ssp. pachylepis (Hall) Urbatsch ined.
           SY=H. pachylepis Hall
    parishii (Greene) Hall
           SY=Haplopappus p. (Greene) Blake
    pinifolia (Gray) Hall
           SY=Haplopappus p. Gray
    resinosa Nutt.
           SY=Haplopappus r. (Nutt.) Gray
    rydbergii (Blake) Urbatsch ined.
           SY=Haplopappus r. Blake
    triantha (Blake) Shinners
           SY=Haplopappus t. Blake

Erigeron L.

    abajoensis Cronq.
    acris L.
        var. debilis Gray

           SY=E. a. ssp. d. (Gray) Piper
           SY=E. angulosus Gaudin ssp. d.
              (Gray) Piper
           SY=E. d. (Gray) Rydb.
           SY=E. jucundus Greene
           SY=E. nivalis Nutt.
        var. elatus (Hook.) Cronq.
           SY=E. alpinus L. var. e. Hook.
           SY=E. e. (Hook.) Greene
           SY=E. e. (Hook.) Greene var.
              oligocephalus (Fern. &
              Wieg.) Fern.
        var. kamtschaticus (DC.) Herder
           SY=E. a. ssp. k. (DC.) Hara
           SY=E. a. ssp. politus (Fries)
              Schinz & Keller
           SY=E. a. var. asteroides (Andrz.
              ex Bess.) DC.
           SY=E. angulosus Gaudin var. k.
              (DC.) Hara
           SY=E. droebachianus O.F. Muell. ex
              Retz.
           SY=E. elongatus Ledeb.
            SY=E. k. DC.
            SY=E. p. Fries
    aequifolius Hall
    aliceae T.J. Howell
    allocotus Blake
    alpiniformis Cronq.
    annuus (L.) Pers.
           SY=E. a. var. discoideus (Victorin
             & Rouss.) Cronq.
    aphanactis (Gray) Greene
        var. aphanactis
           SY=E. concinnus (Hook. & Arn.)
             Torr. & Gray var. a. Gray
        var. congestus (Greene) Cronq.
    arenarioides (D.C. Eat. ex Gray) Rydb.
    argentatus Gray
    arizonicus Gray
    arthurii Boivin
    asperugineus (D.C. Eat.) Gray
    asperum Nutt.
           SY=E. glabellus Nutt. var.
              pubescens Hook. pro parte
    aureus Greene
        var. acutifolius Raup
        var. aureus
           SY=Haplopappus brandegei Gray
    barbellulatus Greene
    basalticus Hoover
    bellidiastrum Nutt.
        var. bellidiastrum
        var. robustus Cronq.
    bellioides DC.
    bigelovii Gray
    bloomeri Gray
        var. bloomeri
           SY=E. b. var. typicus Cronq.
        var. nudatus (Gray) Cronq.
        var. pubens Keck
    brandegei Gray
    breweri Gray
        var. breweri
        var. elmeri (Greene) Jepson
        var. jacinteus (Hall) Cronq.
        var. porphyreticus (M.F. Jones)
           Cronq.
    caespitosus Nutt.
    calvus Coville
    canus Gray
    cascadensis Heller
    cervinus Greene
    chrysopsidis Gray
      ssp. austinae (Greene) Cronq.
           SY=E. a. Greene
      ssp. chrysopsidis

Erigeron (CONT.)
      SY=E. c. var. brevifolius Piper
  clokeyi Cronq.
  compactus Blake
    var. compactus
    var. consimilis (Cronq.) Blake
      SY=E. consimilis Cronq.
  compositus Pursh
    var. compositus
      SY=E. c. var. typicus Hook.
    var. discoideus Gray
      SY=E. trifidus Hook.
    var. glabratus Macoun
      SY=E. c. var. multifidus (Rydb.)
        J.F. Macbr. & Payson
  corymbosus Nutt.
  coulteri Porter
  cronquistii Maguire
  cuneifolius DC.
  decumbens Nutt.
    ssp. decumbens
    ssp. robustior Cronq.
  delicatus Cronq.
  disparipilus Cronq.
  divergens Torr. & Gray
    var. cinereus (Gray) Gray
      SY=E. nudiflorus Buckl.
    var. divergens
      SY=E. d. var. typicus Cronq.
  eatonii Gray
    ssp. eatonii
      SY=E. e. ssp. typicus Cronq.
    ssp. plantagineus (Greene) Cronq.
      SY=E. e. var. p. (Greene) Cronq.
    ssp. villosus Cronq.
      SY=E. e. var. v. (Cronq.) Cronq.
  elatior (Gray) Greene
  elegantulus Greene
  engelmannii A. Nels.
    var. davisii (Cronq.) Cronq.
      SY=E. e. ssp. d. Cronq.
    var. engelmannii
      SY=E. e. ssp. typicus Cronq.
  evermannii Rydb.
  eximius Greene
      SY=E. superbus Greene
  filifolius (Hook.) Nutt.
    var. filifolius
      SY=E. f. var. typicus Cronq.
    var. robustior M.E. Peck
  flabellifolius Rydb.
  flagellaris Gray
      SY=E. f. var. typicus Cronq.
  flettii G.N. Jones
  flexuosus Cronq.
  foliosus Nutt.
    var. blochmanae (Greene) Hall
    var. confinis (T.J. Howell) Jepson
    var. covillei (Greene) Compton
    var. foliosus
    var. hartwegii (Greene) Jepson
    var. stenophyllus (Nutt.) Gray
 formosissimus Greene
    var. formosissimus
      SY=E. f. var. typicus Cronq.
    var. viscidus (Rydb.) Cronq.
      SY=E. glabellus Nutt. var. v.
        (Rydb.) Boivin
  garrettii A. Nels.
      SY=E. controversus Greene
  geiseri Shinners
  gilensis Woot. & Standl.
      SY=Achaetogeron chihuahuensis
        Larsen
  glabellus Nutt.
    var. glabellus
      SY=E. asperum Nutt. pro parte

      SY=E. g. ssp. typicus Cronq.
      SY=E. oblanceolatus Rydb.
    var. pubescens Hook.
      SY=E. anodonta Lunell
      SY=E. asperum Nutt. var. p.
        (Hook.) Breitung
      SY=E. g. ssp. p. (Hook.) Cronq.
      SY=E. oligodontus Lunell
  glaucus Ker-Gawl.
  gracilis Rydb.
  grandiflorus Hook.
  hessii Nesom
  howellii Gray
  hultenii Spongberg
  humilis Graham
      SY=E. unalaschkensis (DC.) Vierh.
      SY=E. uniflorus L. var.
        unalaschkensis (DC.) Ostenf.
  hyperboreus Greene
      SY=E. alaskanus Cronq.
  hyssopifolius Michx.
    var. hyssopifolius
    var. villicaulis Fern.
  inornatus (Gray) Gray
    var. angustatus Gray
    var. biolettii (Greene) Jepson
    var. inornatus
    var. reductus Cronq.
    var. viscidulus Gray
  jamaicensis L.
  jonesii Cronq.
  kachinensis Welsh & Moore
  karvinskianus DC.
  kuschei Eastw.
  lanatus Hook.
  lassenianus Greene
    var. deficiens Cronq.
    var. lassenianus
  latus (A. Nels. & J.F. Macbr.) Cronq.
  leibergii Piper
  leiomeris Gray
  lemmonii Gray
  linearis (Hook.) Piper
      SY=E. peucephyllus Gray
  lobatus A. Nels.
  lonchophyllus Hook.
    var. laurentianus Victorin
    var. lonchophyllus
      SY=E. minor (Hook.) Rydb.
  maguirei Cronq.
  mancus Rydb.
  melanocephalus (A. Nels.) A. Nels.
  mexiae K. Becker
  mimegletes Shinners
      SY=E. geiseri Shinners var.
        calcicola Shinners
  miser Gray
  modestus Gray
  muirii Gray
      SY=E. grandiflorus Hook. ssp. m.
        (Gray) Hulten
  multiceps Greene
  myrionactis Small
  nanus Nutt.
  nauseosus (M.E. Jones) A. Nels.
  nematophyllus Rydb.
  neomexicanus Gray
      SY=E. delphinifolius Willd. ssp.
        n. (Gray) Cronq. var.
        euneomexicanus Cronq.
  nevadincola Blake
  ochroleucus Nutt.
    var. ochroleucus
    var. scribneri (Canby ex Rydb.)
      Cronq.
  oreganus Gray
  oreophilus Greenm.

Erigeron (CONT.)
        SY=E. delphinifolius Willd. ssp.
           neomexicanus (Gray) Cronq.
           var. o. (Greenm.) Cronq.
ovinus Cronq.
oxyphyllus Greene
pallens Cronq.
parishii Gray
parryi Coult. & Rose
peregrinus (Pursh) Greene
    ssp. callianthemus (Greene) Cronq.
      var. angustifolius (Gray) Cronq.
      var. callianthemus (Greene) Cronq.
        SY=E. c. Greene
        SY=E. p. var. eucallianthemus
          Cronq.
      var. hirsutus Cronq.
      var. scaposus (Torr. & Gray) Cronq.
    ssp. peregrinus
      var. dawsonii Greene
      var. peregrinus
        SY=E. unalaschkensis Less.
      var. thompsonii (Blake ex J.W.
        Thompson) Cronq.
petiolaris Greene
petrophilus Greene
philadelphicus L.
    SY=E. p. var. glaber Henry
    SY=E. p. var. provancheri
      (Victorin & Rouss.) Boivin
    SY=E. philadelphicus var.
      scaturicola (Fern.) Fern.
    SY=E. provancheri Victorin &
      Rouss.
    SY=E. purpureus Ait.
pinnatisectus (Gray) A. Nels.
piperianus Cronq.
plateauensis Cronq.
    SY=E. lobatus A. Nels. var.
      warnockii Shinners
platyphyllus Greene
poliospermus Gray
    var. cereus Cronq.
    var. poliospermus
      SY=E. p. var. typicus Cronq.
pringlei Gray
proselyticus Nesom
    SY=E. flagellaris Gray var.
      trilobatus Maguire ex Cronq.
pulchellus Michx.
    var. brauniae Fern.
    var. pulchellus
    var. tolsteadii Cronq.
pulcherrimus Heller
    var. pulcherrimus
    var. wyomingia (Rydb.) Cronq.
pumilus Nutt.
    ssp. concinnoides Cronq.
      var. concinnoides Cronq.
        SY=E. concinnus (Hook. & Arn.)
          Torr. & Gray
        SY=E. p. ssp. concinnoides var.
          euconcinnoides Cronq.
      var. condensatus (D.C. Eat.) Cronq.
        SY=E. concinnus (Hook. & Arn.)
          Torr. & Gray var.
          condensatus D.C. Eat.
        SY=E. condensatus (D.C. Eat.)
          Greene
      var. subglaber Cronq.
        SY=?E. perglaber Blake
    ssp. intermedius Cronq.
      var. gracilior Cronq.
      var. intermedius Cronq.
        SY=E. p. ssp. i. var.
          euintermedius Cronq.
    ssp. pumilus

        SY=E. p. ssp. typicus Cronq.
purpuratus Greene
    var. dilatatus Boivin
    var. purpuratus
pygmaeus (Gray) Greene
quercifolius Lam.
radicatus Hook.
religiosus Cronq.
rhizomatus Cronq.
rusbyi Gray
rydbergii Cronq.
sanctarum S. Wats.
scotteri Boivin
simplex Greene
sionis Cronq.
speciosus (Lindl.) DC.
    var. ?conspicuus (Rydb.) Boivin
    var. macranthus (Nutt.) Cronq.
      SY=E. m. Nutt.
    var. speciosus
strigosus Muhl. ex Willd.
    var. beyrichii (Fisch. & Mey.) Torr.
      & Gray
      SY=E. ramosus (Walt.) B.S.P. var.
        b. (Fisch. & Mey.) Gray
    var. septentrionalis (Fern. & Wieg.)
      Fern.
    var. strigosus
      SY=E. ramosus (Walt.) B.S.P.
      SY=E. s. var. discoideus Robbins
        ex Gray
      SY=E. s. var. eligulatus Cronq.
      SY=E. s. var. typicus Cronq.
      SY=E. traversii Shinners
subglaber Cronq.
subtrinervis Rydb.
    var. conspicuus (Rydb.) Cronq.
      SY=E. speciosus (Lindl.) DC. var.
        c. (Rydb.) Breitung sensu
        auctt. non (Rydb.) Boivin
      SY=E. subtrinervis ssp. c. (Rydb.)
        Cronq.
    var. subtrinervis
      SY=E. s. ssp. typicus Cronq.
supplex Gray
tenellus DC.
tener (Gray) Gray
tenuis Torr. & Gray
tweedyi Canby
uintahensis Cronq.
uncialis Blake
    ssp. conjugans (Blake) Cronq.
      SY=E. u. var. c. Blake
    ssp. uncialis
uniflorus L.
    ssp. eriocephalus (J. Vahl) Cronq.
      SY=E. e. J. Vahl
      SY=E. u. ssp. u. var. e. (J. Vahl)
        Boivin
ursinus D.C. Eat.
utahensis Gray
    var. sparsifolius (Eastw.) Cronq.
    var. utahensis
      SY=E. u. var. tetrapleurus (Gray)
        Cronq.
vagus Payson
vernus (L.) Torr. & Gray
vetensis Rydb.
watsonii (Gray) Cronq.
yukonensis Rydb.
    SY=E. glabellus Nutt. var. y.
      (Rydb.) Hulten

Eriophyllum Lag.

ambiguum (Gray) Gray
    var. ambiguum

Eriophyllum (CONT.)
        var. paleaceum (Brandeg.) Ferris
    confertiflorum (DC.) Gray
            SY=E. c. var. latum Hall
            SY=E. c. var. laxiflorum Gray
    congdonii Brandeg.
            SY=E. nubigenum Greene var. c.
                    (Brandeg.) Constance
    jepsonii Greene
    lanatum (Pursh) Forbes
            var. achilleoides (DC.) Jepson
            var. aphanactis J.T. Howell
            var. arachnoideum (Fisch. & Lall.)
                Jepson
            var. croceum (Greene) Jepson
            var. cuneatum (Kellogg) Jepson
            var. grandiflorum (Gray) Jepson
            var. hallii Constance
            var. integrifolium (Hook.) Smiley
                SY=E. i. (Hook.) Greene
                SY=E. l. var. monoense (Rydb.)
                    Jepson
            var. lanatum
                SY=E. l. var. typicum Constance
            var. lanceolatum (T.J. Howell) Jepson
            var. leucophyllum (DC.) W.R. Carter
            var. obovatum (Greene) Hall
    latilobum Rydb.
    mohavense (I.M. Johnston) Jepson
    multicaule (DC.) Gray
    nevinii Gray
    nubigenum Greene
    pringlei Gray
    staechadifolium Lag.
            SY=E. s. var. artemisiifolium
                    (Less.) J.F. Macbr.
            SY=E. s. var. depressum Greene
    tanacetiflorum Greene
            SY=E. confertiflorum (DC.) Gray
                    var. t. (Greene) Jepson

Eupatorium L.

    adenophorum Spreng.
            SY=Ageratina a. (Spreng.) King &
                    H.E. Robins.
            SY=E. glandulosum H.B.K.
    album L.
        var. album
            SY=E. a. var. glandulosum (Michx.)
                    DC.
            SY=E. petaloideum Britt.
        var. subvenosum Gray
        var. vaseyi (Porter) Cronq.
            SY=E. a. var. monardifolium Fern.
    altissimum L.
            SY=Ageratina a. (L.) King & H.E.
                    Robins.
            SY=E. rugosum var. chlorolepis
                    Fern.
            SY=E. r. var. tomentellum (B.L.
                    Robins.) Blake
            SY=E. r. var. villicaule (Fern.)
                    Blake
            SY=E. saltuense Fern.
            SY=E. urticifolium Reichard var.
                    t. B.L. Robins.
    anomalum Nash
    aromaticum L.
        var. aromaticum
            SY=Ageratina a. (L.) King & H.E.
                    Robins.
            SY=E. latidens Small
        var. incisum Gray
    azureum DC.
    betonicifolium P. Mill.
            SY=Conoclinium b. (P. Mill.) King

                    & H.E. Robins.
    bigelovii Gray
    borinquense (Britt.) B.L. Robins.
            SY=Osmia b. Britt.
    cannabinum L.
    capillifolium (Lam.) Small
    chapmanii Small
    coelestinum L.
            SY=Conoclinium c. (L.) DC.
    compositifolium Walt.
    corymbosum Aubl.
            SY=Osmia c. (Aubl.) Britt. &
                    Wilson
    cuneifolium Willd.
            SY=?E. hyssopifolium L. var.
                    linearifolium (Walt.) Fern.
            SY=?E. l. Walt.
            SY=E. tortifolium Chapman
    dolicholepis (Urban) Britt.
    droserolepis B.L. Robins.
    dubium Willd. ex Poir.
            SY=Eupatoriadelphus d. (Willd. ex
                    Poir.) King & H.E. Robins.
    fernaldii Godfrey
    fistulosum Barratt
            SY=Eupatoriadelphus f. (Barratt)
                    King & H.E. Robins.
    frustratum B.L. Robins.
            SY=Osmia f. (B.L. Robins.) Small
    geraniifolium Urban
            SY=Osmia g. (Urban) Britt. &
                    Wilson
    glaucescens Ell.
    greggii Gray
    havanense H.B.K.
            SY=Ageratina h. (H.B.K.) King &
                    H.E. Robins.
    herbaceum (Gray) Greene
            SY=Ageratina h. (Gray) King & H.E.
                    Robins.
            SY=Eupatorium texense (Torr. &
                    Gray) Rydb.
    hyssopifolium L.
        var. hyssopifolium
            SY=Eupatorium h. var. calcaratum
                    Fern. & Schub.
            SY=E. lecheifolium Greene
        var. laciniatum Gray
            SY=Eupatorium torreyanum Short
    incarnatum Walt.
            SY=Fleischmannia i. (Walt.) King &
                    H.E. Robins.
    iresinoides Kunth
    ivifolium L.
            SY=Osmia i. (L.) Schultz-Bip.
    jucundum Greene
            SY=Ageratina j. (Greene) Clewell &
                    Wooton
    lemmonii B.L. Robins.
            SY=Ageratina l. (B.L. Robins.)
                    King & H.E. Robins.
    leptophyllum DC.
            SY=Eupatorium capillifolium (Lam.)
                    Small var. l. (DC.) Ahles
    leucolepis (DC.) Torr. & Gray
        var. leucolepis
        var. noviangliae Fern.
    luciae-brauniae Fern.
    macrophyllum L.
            SY=Hebeclinium m. (L.) DC.
    maculatum L.
        var. bruneri (Gray) Breitung
            SY=Eupatorium b. Gray
        var. foliosum (Fern.) Wieg.
        var. maculatum
            SY=Eupatoriadelphus m. (L.) King &
                    H.E. Robins.

Eupatorium (CONT.)
            SY=Eupatorium purpureum L. var. m.
                (L.) Darl.
    microstemon Cass.
            SY=Fleischmannia m. (Cass.) King &
                H.E. Robins.
    mikanioides Chapman
    mohrii Greene
    occidentale Hook.
            SY=Ageratina o. (Hook.) King &
                H.E. Robins.
    odoratum L.
            SY=Osmia o. (L.) Schultz-Bip.
    oteroi Monachino
    pauperculum Gray
            SY=Ageratina p. (Gray) King & H.E.
                Robins.
    perfoliatum L.
            var. colpophilum Fern. & Grisc.
            var. perfoliatum
    pinnatifidum Ell.
            SY=Eupatorium eugenei Small
            SY=E. pectinatum Small
    X polyneuron (F.J. Herm.) Wunderlin
        [chapmanii X ?serotinum]
            SY=Eupatorium cuneatum Engelm.
            SY=E. serotinum Michx. var. p.
                F.J. Herm.
    polyodon Urban
    portoricense Urban
            SY=Critonia p. (Urban) Britt. &
                Wilson
    purpureum L.
            var. amoenum (Pursh) Gray
            var. purpureum
            SY=Eupatoriadelphus p. (L.) King &
                H.E. Robins.
            SY=Eupatorium falcatum Michx.
            SY=Eupatorium trifoliatum L.
    pycnocephalum Less.
            SY=Fleischmannia p. (Less.) King &
                H.E. Robins.
    recurvans Small
    resinifluum Urban
            SY=Ageratina r. (Urban) King &
                H.E. Robins.
    resinosum Torr.
            var. kentuckiense Fern.
            var. resinosum
    riparium Regel
            SY=Ageratina r. (Regel) King &
                H.E. Robins.
    rothrockii Gray
            SY=Ageratina r. (Gray) King & H.E.
                Robins.
    rotundifolium L.
            var. cordigerum Fern.
            SY=Eupatorium c. (Fern.) Fern.
            var. ovatum (Bigelow) Torr.
            SY=Eupatorium pubescens Muhl.
            SY=E. r. ssp. o. (Bigelow)
                Montgomery & Fairbrothers
                pro parte
            var. rotundifolium
            var. scabridum (Ell.) Gray
            SY=Eupatorium s. Ell.
            var. saundersii (Porter) Cronq.
            SY=Eupatorium pilosum Walt.
            SY=E. verbenifolium Reichard
    rugosum Houtt.
            var. angustata (Gray) Blake
            SY=Ageratina altissima L. var.
                angustata (Gray) Clewell &
                Wooton
            var. roanense (Small) Fern.
            SY=Ageratina altissima (L.) King &
                H.E. Robins. var. roanense

                (Small) Clewell & Wooton
            SY=Eupatorium roanense Small
            SY=E. urticifolium Reichard
            var. rugosum
    semiserratum DC.
            SY=Eupatorium cuneifolium Willd.
                var. s. (DC.) Fern. & Grisc.
    serotinum Michx.
    sessilifolium L.
            var. brittonianum Porter
            var. sessilifolium
    shastense Taylor & Stebbins
    sinuatum Lam.
            SY=Osmia s. (Lam.) Britt. & Wilson
    solidaginifolium Gray
    sordidum Less.
    triplinerve Vahl
    vaseyi Porter
            SY=Eupatorium sessilifolium L.
                var. v. (Porter) Fern. &
                Grisc.
    villosum Sw.
    wrightii Gray
            SY=Ageratina w. (Gray) King & H.E.
                Robins.

Euryops Cass.

    multifidus (Thunb.) DC.

Euthamia Ell.

    galetorum Greene
            SY=E. graminifolia (L.) Nutt. ex
                Cass. var. galetorum
                (Greene) Friesner
            SY=Solidago galetorum (Greene)
                Friesner
            SY=S. graminifolia (L.) Salisb.
                var. galetorum (Greene)
                House
            SY=S. tenuifolia Pursh var.
                pycnocephala Fern.
    graminifolia (L.) Nutt. ex Cass.
            var. graminifolia
            SY=Solidago g. (L.) Salisb.
            var. major (Michx.) Moldenke
            SY=Solidago g. var. m. (Michx.)
                Fern.
            var. nuttallii (Greene) W. Stone
            SY=E. floribunda Greene
            SY=E. n. Greene
            SY=Solidago hirtella (Greene) Bush
            SY=S. g. var. n. (Greene) Fern.
            SY=S. g. var. polycephala Fern.
            SY=S. n. (Greene) Bush
    gymnospermoides Greene
            SY=E. camporum Greene
            SY=E. chrysothamnoides Greene
            SY=E. glutinosa Rydb.
            SY=E. media Greene
            SY=E. pulverulenta Greene
            SY=Solidago camporum (Greene) A.
                Nels.
            SY=S. chrysothamnoides (Greene)
                Bush
            SY=S. graminifolia (L.) Salisb.
                var. media (Greene) S.K.
                Harris
            SY=S. graminifolia var.
                gymnospermoides (Greene)
                Croat
            SY=S. gymnospermoides (Greene)
                Fern.
            SY=S. media (Greene) Bush
            SY=S. moseleyi Fern.
            SY=S. perglabra Friesner

Euthamia (CONT.)
          SY=S. texensis Friesner
X hirtipes Fern. [graminifolia var.
     nuttallii X tenuifolia]
leptocephala (Torr. & Gray) Greene
          SY=Solidago l. Torr. & Gray
occidentalis Nutt.
          SY=E. californica Gandog.
          SY=E. linarifolia Gandog.
          SY=Solidago o. (Nutt.) Torr. &
               Gray
tenuifolia (Pursh) Greene
          SY=E. microcephala Greene
          SY=E. microphylla Greene
          SY=E. minor (Michx.) Greene
          SY=E. remota Greene
          SY=Solidago caroliniana (L.)
               B.S.P.
          SY=S. microcephala (Greene) Bush
        , SY=S. microphylla (Greene) Bush
          SY=S. minor (Michx.) Fern.
          SY=S. r. (Greene) Friesner
          SY=S. t. Pursh

Evax Gaertn.

     acaulis (Kellogg) Greene
     candida (Torr. & Gray) Gray
     caulescens (Benth.) Gray
          var. caulescens
          var. humilis (Greene) Jepson
     multicaulis DC.
          SY=E. m. var. drummondii (Torr. &
               Gray) Gray
          SY=E. verna Raf.
          SY=Filaginopsis nivea Small
     prolifera Nutt. ex DC.
          SY=Filago p. (Nutt. ex DC.) Britt.
     sparsiflora (Gray) Jepson
          var. brevifolia (Gray) Jepson
          var. sparsiflora

Facelis Cass.

     retusa (Lam.) Schultz-Bip.
          SY=F. apiculata Cass.

Filago L.

     arizonica Gray
     californica Nutt.
     depressa Gray
     germanica (L.) Huds.
          SY=Gifola g. (L.) Dumort.
     minima Fries
     vulgaris Lam.
          SY=F. germanica L. non Huds.

Flaveria Juss.

     bidentis (L.) Kuntze
     brownii Powell
     campestris J.R. Johnston
     chlorifolia Gray
     floridana J.R. Johnston
     linearis Lag.
          SY=F. X latifolia (J.R. Johnston)
               Long & Rhamstine
          SY=F. latifolia (J.R. Johnston)
               Rydb.
     macdougallii Theroux, Pinkava & Keil
     trinervia (Spreng.) C. Mohr

Florestina Cass.

     tripteris DC.
          SY=Palafoxia t. (DC.) Shinners

          SY=P. t. var. brevis Shinners

Flourensia DC.

     cernua DC.

Gaillardia Foug.

     aestivalis (Walt.) H. Rock
          var. aestivalis
               SY=G. chrysantha Small
               SY=G. fastigiata Greene
               SY=G. lanceolata Michx.
               SY=G. l. var. f. (Greene)
                    Waterfall
               SY=G. rigida Small ex Rydb.
               SY=G. serotina (Walt.) H. Rock
          var. flavovirens (C. Mohr) Cronq.
               SY=G. lanceolata Michx. var. f. C.
                    Mohr
               SY=G. lutea Greene
     amblyodon J. Gay
     aristata Pursh
     arizonica Gray
          var. arizonica
          var. pringlei (Rydb.) Blake
               SY=G. p. Rydb.
     coahuilensis B.L. Turner
     flava Rydb.
     mexicana Gray
     multiceps Greene
          var. microcephala B.L. Turner
          var. multiceps
     parryi Greene
     pinnatifida Torr.
          var. linearis (Rydb.) Biddulph
          var. pinnatifida
               SY=G. gracilis A. Nels.
               SY=G. mearnsii Rydb.
     pulchella Foug.
          var. australis Turner & Whalen
          var. picta (Sweet) Gray
               SY=G. picta Sweet
          var. pulchella
               SY=G. drummondii (Hook.) DC.
               SY=G. neomexicana A. Nels.
               SY=G. villosa Rydb.
     spathulata Gray
     suavis (Gray & Engelm.) Britt. & Rusby
          SY=Agassizia s. Gray & Engelm.
          SY=G. trinervata Small

Galinsoga Ruiz & Pavon

     parviflora Cav.
          SY=G. p. var. semicalva Gray
          SY=G. s. (Gray) St. John & White
          SY=G. s. var. percalva Blake
     quadriradiata Ruiz & Pavon
          SY=G. aristulata Bickn.
          SY=G. bicolorata St. John & White
          SY=G. caracasana (DC.)
               Schultz-Bip.
          SY=G. ciliata (Raf.) Blake

Garberia Gray

     fruticosa (Nutt.) Gray

Gazania Gaertn.

     linearis (Thunb.) Druce
          SY=G. longiscapa DC.

Geraea Torr. & Gray

     canescens Torr. & Gray

Geraea (CONT.)
        var. canescens
        var. paniculata (Gray) Blake
    viscida (Gray) Blake

Glyptopleura D.C. Eat.

    marginata D.C. Eat.
    setulosa Gray

Gnaphalium L.

    arizonicum Gray
    beneolens A. Davids.
    bicolor Bioletti
    californicum DC.
    chilense Spreng.
        var. chilense
        var. confertifolium Greene
    collinum Labill.
    exilifolium A. Nels.
        SY=G. grayi A. Nels. & J.F. Macbr.
    hawaiiense Deg. & Sherff
    helleri Britt.
        var. helleri
        SY=G. obtusifolium L. var. h.
            (Britt.) Blake
        var. micradenium (Weatherby) Mahler
        SY=G. obtusifolium L. var. m.
            Weatherby
    indicum L.
    japonicum Thunb.
    leucocephalum Gray
    luteo-album L.
    microcephalum Nutt.
        var. microcephalum
        var. thermale (E. Nels.) Cronq.
        SY=G. t. E. Nels.
    norvegicum Gunn.
        SY=Omalotheca n. (Gunn.)
            Schultz-Bip. & F.W. Schultz
    obtusifolium L.
        var. obtusifolium
        var. praecox Fern.
        var. saxicola (Fassett) Cronq.
        SY=G. s. Fassett
    palustre Nutt.
    pensilvanicum Willd.
        SY=G. peregrinum Fern.
    portoricense Urban
    pringlei Gray
    purpureum L.
        var. falcatum (Lam.) Torr. & Gray
        SY=G. f. Lam.
        var. purpureum
        SY=Gamochaeta p. (L.) Cabrera
        SY=Gnaphalium americanum P. Mill.
        SY=Gnaphalium calviceps Fern.
        SY=Gnaphalium p. var. a. (P.
            Mill.) Klatt
        SY=Gnaphalium p. var. spathulatum
            (Lam.) Ahles
        SY=Gnaphalium spathulatum Lam.
        SY=Gnaphalium spicatum Lam.
        var. ustulatum (Nutt.) Boivin
    ramosissimum Nutt.
    sandwicensium Gaud.
        var. flagellare Sherff
        var. kilaueanum Deg. & Sherff
        var. lineatum Sherff
        var. molokaiense Deg. & Sherff
        var. sandwicensium
    supinum L.
        SY=Omalotheca s. (L.) DC.
    sylvaticum L.
        SY=Omalotheca s. (L.) Schultz-Bip.
            & F.W. Schultz

    uliginosum L.
    viscosum H.B.K.
        SY=Gnaphalium decurrens Ives
        SY=G. macounii Greene
    wrightii Gray

Gochnatia H.B.K.

    hypoleuca (DC.) Gray

Greenella Gray

    arizonica Gray
    discoidea Gray

Grindelia Willd.

    acutifolia Steyermark
    aphanactis Rydb.
    arizonica Gray
        var. arizonica
        var. dentata Steyermark
        var. microphylla Steyermark
        var. stenophylla Steyermark
    camporum Greene
        var. camporum
        var. parviflora Steyermark
    columbiana (Piper) Rydb.
        SY=G. nana Nutt. ssp. c. Piper
    decumbens Greene
        var. decumbens
        var. subincisa (Greene) Steyermark
    fastigiata Greene
    fraxino-pratensis Reveal & Barneby
    grandiflora Hook.
    gymnospermoides (Gray) Ruffin
        SY=Xanthocephalum g. (Gray) Benth.
            & Hook. ex Rothrock
    hallii Steyermark ex Rothrock
    havardii Steyermark
    hirsutula Hook. & Arn.
        ssp. hirsutula
        ssp. rubricaulis (DC.) Keck
    howellii Steyermark
    humilis Hook. & Arn.
    inornata Greene
        var. angusta Steyermark
        var. inornata
    integrifolia DC.
        var. integrifolia
        var. macrophylla (Greene) Cronq.
        SY=G. arenicola Steyermark
        SY=G. stricta DC.
        SY=G. s. ssp. blakei (Steyermark)
            Keck
        SY=G. s. ssp. venulosa (Jepson)
            Keck
    laciniata Rydb.
    lanceolata Nutt.
        SY=G. littoralis Steyermark
        SY=G. texana Scheele var.
            lanceolata (Nutt.) Shinners
    latifolia Kellogg
        var. latifolia
        var. platyphylla Greene
        SY=G. l. ssp. p. (Greene) Keck
    maritima (Greene) Steyermark
    microcephala DC.
        var. adenodonta Steyermark
        var. microcephala
        var. pusilla Steyermark
    nana Nutt.
        var. integerrima (Rydb.) Steyermark
        var. nana
        SY=G. n. var. integrifolia Nutt.
        SY=G. squarrosa (Pursh) Dunal var.
            integrifolia (Nutt.) Boivin

Grindelia (CONT.)
    oolepis Blake
    paludosa Greene
    procera Greene
    revoluta Steyermark
    robusta Nutt.
            var. bracteosa (J.T. Howell) Keck
            var. robusta
    scabra Greene
            SY=G. s. var. neomexicana (Woot. &
                Standl.) Steyermark
    squarrosa (Pursh) Dunal
            var. nuda (Wood) Gray
            var. quasiperennis Lunell
                SY=G. perennis A. Nels.
            var. serrulata (Rydb.) Steyermark
                SY=G. s. Rydb.
            var. squarrosa
    subalpina Greene
            var. erecta (A. Nels.) Steyermark
            var. subalpina

Guardiola Humb. & Bonpl.

    platyphylla Gray

Guizotia Cass.

    abyssinica (L.f.) Cass.

Gundlachia Gray

    corymbosa (Urban) Britt.

Gutierrezia Lag.

    bracteata Abrams
    californica (DC.) Torr. & Gray
    microcephala (DC.) Gray
            SY=G. linoides Greene
            SY=G. longipappa Blake
            SY=G. lucida Greene
            SY=G. sarothrae (Pursh) Britt. &
                Rusby var. m. (DC.) L.
                Benson
            SY=Xanthocephalum m. (DC.)
                Shinners
    sarothrae (Pursh) Britt. & Rusby
            var. pomariensis Welsh
                SY=X. s. var. p. (Welsh) Welsh
            var. sarothrae
                SY=G. diversifolia Greene
                SY=G. lepidota Greene
                SY=G. linearis Rydb.
                SY=G. tenuis Greene
                SY=Xanthocephalum s. (Pursh)
                    Shinners
    serotina Greene
    texana (DC.) Torr. & Gray
            SY=Xanthocephalum t. (DC.)
                Shinners

Gymnosperma Less.

    glutinosum (Spreng.) Less.
            SY=Gutierrezia g. (Spreng.)
                Schultz-Bip.
            SY=Selloa g. Spreng.
            SY=Xanthocephalum g. (Spreng.)
                Shinners
            SY=X. sphaerocephalum (Gray)
                Shinners

Gynura Cass.

    aurantiaca (Blume) DC.

Haploesthes Gray

    diaresis author ?
    greggii Gray
            var. texana (Coult.) I.M. Johnston

Haplopappus Cass.

    aberrans (A. Nels.) Hall
    acaulis (Nutt.) Gray
            var. acaulis
                SY=H. a. ssp. typicus Hall
                SY=Stenotus a. (Nutt.) Nutt.
            var. glabratus D.C. Eat.
    acradenius (Greene) Blake
        ssp. acradenius
        ssp. bracteosus (Greene) Hall
        ssp. eremophilus (Greene) Hall
    apargioides Gray
    armerioides (Nutt.) Gray
            SY=Stenotus a. Nutt.
    bloomeri (Hook.) Gray
            var. angustatus Gray
            var. bloomeri
                SY=Ericameria b. (Hook.) J.F.
                    Macbr.
            var. sonnii (Gray) Hall
    carthamoides (Hook.) Gray
            var. carthamoides
                SY=H. c. var. erythropappus
                    (Rydb.) St. John
                SY=H. c. var. typicus (Hall)
                    Cronq.
            var. cusickii Gray
                SY=H. carthamoides ssp. cusickii
                    (Gray) Hall
            var. rigidus (Rydb.) M.E. Peck
                SY=H. carthamoides ssp. r. (Rydb.)
                    Hall
    clementis (Rydb.) Blake
            SY=Pyrrocoma c. Rydb.
    contractus Hall
    croceus Gray
            var. croceus
                SY=H. c. ssp. typicus Hall
                SY=Pyrrocoma c. (Gray) Greene
            var. genuflexus (Greene) Blake
    divaricatus (Nutt.) Gray
            SY=Croptilon d. (Nutt.) Raf.
            SY=Isopappus d. (Nutt.) Torr. &
                Gray
    engelmannii (Gray) Hall
            SY=Oonopsis e. (Gray) Greene
    eximius Hall
    fremontii Gray
        ssp. fremontii
            SY=H. f. ssp. typicus Hall
            SY=Oonopsis foliosa (Gray) Greene
        ssp. monocephalus (A. Nels.) Hall
    gilmanii Blake
    greenei Gray
            var. greenei
            var. mollis Gray
    hallii Gray
    hirtus Gray
            var. hirtus
                SY=H. h. ssp. typicus Hall
            var. lanulosus (Greene) M.E. Peck
                SY=H. h. ssp. l. (Greene) Hall
            var. sonchifolius (Greene) M.E. Peck
                SY=H. h. ssp. s. (Greene) Hall
    insecticruris Henderson
            SY=H. integrifolius Porter ex Gray
                ssp. insecticruris
                (Henderson) Hall
    integrifolius Porter ex Gray
        ssp. integrifolius

Haplopappus (CONT.)
    SY=H. i. ssp. typicus Hall
    SY=Pyrrocoma i. (Porter ex Gray)
        Greene
  ssp. scaberulus (Greene) Hall
junceus Greene
lanceolatus (Hook.) Torr. & Gray
    var. lanceolatus
      SY=H. l. var. typicus (Hall)
        Cronq.
      SY=Pyrrocoma l. (Hook.) Greene
    var. sublanatus Cody
    var. tenuicaulis (D.C. Eat.) Gray
      SY=H. l. ssp. t. (D.C. Eat.) Hall
      SY=H. l. ssp. vaseyi (Parry) Hall
      SY=H. l. var. v. Parry
lanuginosus Gray
  ssp. andersonii (Rydb.) Hall
    SY=H. l. var. a. (Rydb.) Cronq.
  ssp. lanuginosus
    SY=H. l. ssp. typicus Hall
liatriformis (Greene) St. John
    SY=H. integrifolius Porter ex Gray
        ssp. l. (Greene) Hall
    SY=Pyrrocoma l. Greene
lucidus (Keck) Keck
lyallii Gray
    SY=Tonestus l. (Gray) A. Nels.
macleanii Brandeg.
macronema Gray
    var. canescens (A. Nels.) Cronq.
    var. macronema
      SY=H. m. ssp. typicus Hall
      SY=Macronema discoideum Nutt.
microcephalus Cronq.
multicaulis (Nutt.) Gray
ophitidis (J.T. Howell) Keck
parryi Gray
    SY=Oreochrysum p. (Gray) Rydb.
    SY=Solidago p. (Gray) Greene
peirsonii (Keck) J.T. Howell
pygmaeus (Torr. & Gray) Gray
    SY=Tonestus p. (Torr. & Gray) A.
      Nels.
racemosus (Nutt.) Torr.
  ssp. brachycephalus (A. Nels.) Hall
  ssp. congestus (Greene) Hall
  ssp. glomeratus (Nutt.) Hall
    SY=H. r. ssp. duriusculus (Greene)
      Hall
    SY=H. r. var. glomerellus Gray
  ssp. halophilus (Greene) Hall
  ssp. pinetorum Keck
  ssp. racemosus
  ssp. sessiliflorus (Greene) Hall
radiatus (Nutt.) Cronq.
    SY=H. carthamoides (Hook.) Gray
        var. maximus Gray
    SY=Pyrrocoma r. Nutt.
ravenii R.C. Jackson
rigidifolius E.B. Sm.
    SY=Croptilon divaricatus (Nutt.)
      Raf. var. hirtellum
      (Shinners) Shinners
salicinus Blake
scopulorum (M.E. Jones) Blake
    var. hirtellus Blake
    var. scopulorum
stenophyllus Gray
?subviscosus (Greene) Blake
suffruticosus (Nutt.) Gray
tenuisectus (Greene) Blake
uniflorus (Hook.) Torr. & Gray
  ssp. gossypinus (Greene) Hall
    SY=Pyrrocoma u. (Hook.) Greene
        ssp. g. Greene
  ssp. howellii (Gray) Hall

    SY=H. u. var. h. (Gray) M.E. Peck
  ssp. uniflorus
    SY=H. u. ssp. typicus Hall
    SY=Pyrrocoma u. (Hook.) Greene
validus (Rydb.) Cory
  ssp. graniticus E.B. Sm.
    SY=Croptilon divaricatum (Nutt.)
      Raf. var. g. (E.B. Sm.)
      Shinners
  ssp. torreyi E.B. Sm.
    SY=Croptilon divaricatum (Nutt.)
      Raf. var. hookerianum
      (Torr. & Gray) Shinners
    SY=H. d. (Nutt.) Gray var. h.
      (Torr. & Gray) Waterfall
  ssp. validus
    SY=Isopappus v. Rydb.
venetus (H.B.K.) Blake
  ssp. furfuraceus (Greene) Hall
    SY=H. v. var. argutus (Greene)
      Keck
  ssp. oxyphyllus (Greene) Hall
  ssp. sedoides Greene
    SY=H. v. var. s. (Greene) Munz
  ssp. venetus
  ssp. vernonioides (Nutt.) Hall
wardii (Gray) Dorn
?watsonii Gray

Hartwrightia Gray

  floridana Gray ex S. Wats.

Hazardia Greene

  brickellioides (Blake) W.D. Clark
    SY=Haplopappus b. Blake
  cana (Gray) Greene
    SY=Haplopappus c. (Gray) Blake
  detonsa (Greene) Greene
    SY=Haplopappus d. (Greene) Raven
  squarrosa (Hook. & Arn.) Greene
    var. grindelioides (DC.) W.D. Clark
      SY=Haplopappus s. Hook. & Arn.
        ssp. g. (DC.) Keck
    var. obtusa (Greene) Jepson
      SY=Haplopappus s. ssp. o. (Greene)
        Hall
    var. squarrosa
      SY=Haplopappus s. Hook. & Arn.
  stenolepis (Hall) Hoover
    SY=Haplopappus squarrosus Hook. &
      Arn. ssp. stenolepis Hall
  whitneyi (Gray) Greene
    var. discoidea (J.T. Howell) W.D.
    Clark
      SY=Haplopappus w. ssp. d. (J.T.
      Howell) Keck
    var. whitneyi
      SY=Haplopappus w. Gray

Hecastocleis Gray

  shockleyi Gray ex S. Wats.

Hedypnois Scop.

  cretica (L.) Dum.-Cours.

Helenium L.

  amarum (Raf.) H. Rock
    SY=H. tenuifolium Nutt.
  arizonicum Blake
  autumnale L.
    var. autumnale
      SY=H. a. var. canaliculatum (Lam.)

Helenium (CONT.)
                    Torr. & Gray
          SY=H. c. Lam.
          SY=H. latifolium P. Mill.
      var. fylesii Boivin
      var. grandiflorum (Nutt.) Torr. &
          Gray
          SY=H. macranthum Rydb.
      var. montanum (Nutt.) Fern.
          SY=H. m. Nutt.
      var. parviflorum (Nutt.) Fern.
          SY=H. p. Nutt.
  badium (Gray) Geene
          SY=H. amarum (Raf.) H. Rock var.
              b. (Gray) Waterfall
  bigelovii Gray
          SY=H. rivulare (Greene) Rydb.
  bolanderi Gray
          SY=Dugaldia grandiflora Rydb.
          SY=H. bigelovii Gray var. festivum
              Jepson
  brevifolium (Nutt.) Wood
          SY=H. curtisii Gray
  campestre Small
  drummondii H. Rock
          SY=H. fimbriatum sensu auctt. non
              (Michx.) Gray
          SY=Leptopoda f. Torr. & Gray
  elegans DC.
      var. amphilobum (Gray) Bierner
          SY=H. a. Gray
      var. elegans
  flexuosum Raf.
          SY=H. floridanum Fern.
          SY=H. godfreyi Fern.
          SY=H. nudiflorum Nutt.
  linifolium Rydb.
  microcephalum DC.
      var. microcephalum
      var. ooclinium (Gray) Bierner
          SY=H. o. Gray
  pinnatifidum (Nutt.) Rydb.
          SY=H. incisum (Torr. & Gray) Wood
          SY=H. vernale sensu auctt. non
              Walt.
  X polyphyllum Small [autumnale X flexuosum]
  puberulum DC.
          SY=H. decurrens (Less.) Vatke non
              Moench
  quadridentatum Labill.
  thurberi Gray
  vernale Walt.
          SY=H. helenium (Nutt.) Small
          SY=H. nuttallii Gray
  virginicum Blake

Helianthella Torr. & Gray

  californica Gray
      var. californica
      var. nevadensis (Greene) Jepson
      var. shastensis W.A. Weber
  castanea Greene
  microcephala (Gray) Gray
  parryi Gray
  quinquenervis (Hook.) Gray
  uniflora (Nutt.) Torr. & Gray
      var. douglasii (Torr. & Gray) W.A.
          Weber
      var. uniflora
          SY=Helianthus u. Nutt.

Helianthus L.

  agrestis Pollard
  X alexidis Boivin [maximilianii X
      nuttallii]

  X ambiguus (Gray) Britt. [divaricatus X
      giganteus]
  angustifolius L.
          SY=H. a. var. planifolius Fern.
  annuus L.
          SY=H. a. ssp. jaegeri (Heiser)
              Heiser
          SY=H. a. ssp. lenticularis
              (Dougl.) Cockerell
          SY=H. a. ssp. texanus Heiser
          SY=H. a. var. l. (Dougl.)
              Steyermark
          SY=H. a. var. macrocarpus (DC.)
              Cockerell
          SY=H. a. var. t. (Heiser) Shinners
          SY=H. aridus Rydb.
          SY=H. l. Dougl.
  anomalus Blake
  argophyllus Torr. & Gray
  arizonensis R.C. Jackson
  atrorubens L.
          SY=H. a. var. alsodes Fern.
  bolanderi Gray
          SY=H. exilis Gray
  X brevifolius E.E. Wats. [grosseserratus X
      mollis]
  californicus DC.
  carnosus Small
  ciliaris DC.
  X cinereus Torr. & Gray [mollis X
      occidentalis ssp. plantagineus]
  couplandii Boivin
  cusickii Gray
  debilis Nutt.
      ssp. cucumerifolius (Torr. & Gray)
          Heiser
          SY=H. c. Torr. & Gray
          SY=H. d. var. c. (Torr. & Gray)
              Gray
      ssp. debilis
      ssp. silvestris Heiser
          SY=H. d. var. s. (Heiser) Cronq.
      ssp. tardiflorus Heiser
          SY=H. d. var. t. (Heiser) Cronq.
      ssp. vestitus (E.E. Wats.) Heiser
          SY=H. d. var. v. (E.E. Wats.)
              Cronq.
          SY=H. v. E.E. Wats.
  decapetalus L.
          SY=H. scrophulariifolius Britt.
          SY=H. trachelifolius P. Mill.
  deserticola Heiser
  divaricatus L.
          SY=H. d. var. angustifolius Kuntze
  X divariserratus R.W. Long [divaricatus X
      grosseserratus]
  X doronicoides Lam. [giganteus X mollis]
          SY=H. pilosus Tausch
  eggertii Small
  floridanus Gray ex Chapman
  giganteus L.
          SY=H. alienus E.E. Wats.
          SY=H. borealis E.E. Wats.
          SY=H. g. ssp. a. (E.E. Wats.) R.W.
              Long
          SY=H. g. var. subtuberosus Britt.
          SY=H. nuttallii Torr. & Gray var.
              s. (Britt.) Boivin
          SY=H. s. (Britt.) Britt.
          SY=H. validus E.E. Wats.
  glaucophyllus D.M. Sm.
  X glaucus Small [divaricatus X
      microcephalus]
  gracilentus Gray
  grosseserratus Martens
          SY=H. g. ssp. maximus R.W. Long
          SY=H. g. var. hypoleucus Gray

Helianthus (CONT.)
        SY=H. instabilis E.E. Wats.
    heterophyllus Nutt.
    hirsutus Raf.
        SY=H. h. var. stenophyllus Torr. &
            Gray
        SY=H. h. var. trachyphyllus Torr.
            & Gray
        SY=H. s. (Torr. & Gray) E.E. Wats.
    X intermedius R.W. Long [grosseserratum X
        maximilianii]
    X kellermanii Britt. [grosseserratus X
        salicifolius]
    laciniatus Gray
        SY=H. crenatus R.C. Jackson
        SY=H. heiseri R.C. Jackson
    X laetiflorus Pers. [rigidus X tuberosus]
        SY=H. scaberrimus Ell.
    laevigatus Torr. & Gray
        SY=H. reindutus (Steele) E.E.
            Wats.
    longifolius Pursh
    X luxurians E.E. Wats. [giganteus X
        grosseserratus]
    maximilianii Schrad.
        SY=H. dalyi Britt.
    microcephalus Torr. & Gray
    mollis Lam.
        SY=H. m. var. cordatus S. Wats.
    neglectus Heiser
    niveus (Benth.) Brandeg.
        ssp. canescens (Gray) Heiser
            SY=H. c. (Gray) S. Wats.
            SY=H. canus (Britt.) Woot. &
                Standl.
            SY=H. petiolaris Nutt. var.
                canescens Gray
        ssp. tephrodes (Gray) Heiser
            SY=H. t. Gray
    nuttallii Torr. & Gray
        ssp. nuttallii
            SY=H. fascicularis Greene
            SY=H. n. ssp. canadensis R.W. Long
            SY=H. n. ssp. coloradensis
                (Cockerell) R.W. Long
        ssp. parishii (Gray) Heiser
            SY=H. p. Gray
        ssp. rydbergii (Britt.) R.W. Long
            SY=H. n. var. r. (Britt.) Boivin
            SY=H. r. Britt.
    occidentalis Riddell
        ssp. occidentalis
            SY=H. dowellianus M.A. Curtis
            SY=H. o. var. d. (M.A. Curtis)
                Torr. & Gray
        ssp. plantagineus (Torr. & Gray)
            Shinners
            SY=H. o. var. p. Torr. & Gray
    orgyaloides Cockerell
    paradoxus Heiser
    petiolaris Nutt.
        ssp. fallax Heiser
        ssp. petiolaris
    praecox Torr. & Gray
        ssp. hirtus (Heiser) Heiser
            SY=H. debilis Nutt. ssp. h. Heiser
        ssp. praecox
            SY=H. debilis Nutt. ssp. p. (Torr.
                & Gray) Heiser
        ssp. runyonii (Heiser) Heiser
            SY=H. debilis Nutt. ssp. r. Heiser
    ?praetermissus E.E. Wats.
    pumilus Nutt.
    radula (Pursh) Torr. & Gray
    resinosus Small
        SY=H. tomentosus sensu auctt. non
            Michx.

    rigidus (Cass.) Desf.
        ssp. rigidus
            SY=H. X laetiflorus Pers. var. r.
                (Cass.) Fern.
        ssp. subrhomboideus (Rydb.) Heiser
            SY=H. X laetiflorus Pers. var. s.
                (Rydb.) Fern.
            SY=H. r. var. s. (Rydb.) Cronq.
            SY=H. s. Rydb.
    salicifolius A. Dietr.
        SY=H. filiformis Small
        SY=H. orgyalis DC.
    schweinitzii Torr. & Gray
    silphioides Nutt.
        SY=H. atrorubens L. var. pubescens
            Kuntze
    simulans E.E. Wats.
    smithii Heiser
    strumosus L.
        SY=H. montanus E.E. Wats.
        SY=H. saxicola Small
    tuberosus L.
        SY=H. tomentosus Michx.
        SY=H. tuberosus var. subcanescens
            Gray
    X verticillatus E.E. Wats. [angustifolius X
        grosseserratus]

Helichrysum P. Mill.

    bracteatum (Vent.) Andr.
    foetidum (L.) Cass.

Heliopsis Pers.

    gracilis Nutt.
    helianthoides (L.) Sweet
        ssp. helianthoides
            SY=H. h. var. solidaginoides (L.)
                Fern.
        ssp. occidentalis T.R. Fisher
            SY=H. h. var. o. (T.R. Fisher)
                Steyermark
        ssp. scabra (Dunal) T.R. Fisher
            SY=H. h. var. s. (Dunal) Fern.
            SY=H. minor (Hook.) C. Mohr
            SY=H. s. Dunal
    parvifolia Gray

Hemizonia DC.

    arida Keck
    australis (Keck) Keck
    calyculata (Babcock & Hall) Keck
    clementina Brandeg.
    clevelandii Greene
    congesta DC.
    conjugens Keck
    corymbosa (DC.) Torr. & Gray
        ssp. corymbosa
        ssp. macrocephala (Nutt.) Keck
    fasciculata (DC.) Torr. & Gray
    fitchii Gray
    floribunda Gray
    halliana Keck
    kelloggii Greene
    laevis (Keck) Keck
    lobbii Greene
    lutescens (Greene) Keck
    luzulifolia DC.
        ssp. luzulifolia
        ssp. rudis (Benth.) Keck
    minthornii Jepson
    mohavensis Keck
    multicaulis Hook. & Arn.
        ssp. multicaulis
        ssp. vernalis Keck

Hemizonia (CONT.)
  pallida Keck
  paniculata Gray
      ssp. increscens Hall ex Keck
      ssp. paniculata
  parryi Greene
      ssp. congdonii (Robins. & Greenm.) Keck
      ssp. parryi
      ssp. rudis (Greene) Keck
  pentactis (Keck) Keck
  pungens (Hook. & Arn.) Torr. & Gray
      ssp. maritima (Greene) Keck
      ssp. pungens
      ssp. septentrionalis Keck
            SY=H. p. var. s. (Keck) Cronq.
  ramosissima Benth.
  tracyi (Babcock & Hall) Keck

Herrickia Woot. & Standl.

  horrida Woot. & Standl.
            SY=Aster h. (Woot. & Standl.)
                Blake

Hesperomannia Gray

  arborescens Gray
      ssp. arborescens
      ssp. bushiana (Deg.) Carlq.
      ssp. swezeyi (Deg.) Carlq.
  arbuscula Hbd.
      ssp. arbuscula
      ssp. oahuensis (Hbd.) Carlq.
          var. oahuensis
          var. pearsallii St. John
  lydgatei Forbes

Heterosperma Cav.

  pinnatum Cav.

Heterotheca Cass.

  bolanderi (Gray) Harms
            SY=Chrysopsis b. Gray
            SY=C. sessiliflora (Nutt.)
                Shinners var. b. (Gray) Gray
            SY=C. villosa (Pursh) Nutt. ex DC.
                var. b. (Gray) Gray ex
                Jepson
  breweri (Gray) Shinners
            SY=Chrysopsis b. Gray
            SY=C. b. var. multibracteata
                Jepson
            SY=C. gracilis Eastw.
            SY=C. wrightii Gray
  camphorata (Eastw.) Harms
            SY=Chrysopsis c. Eastw.
            SY=C. villosa (Pursh) Nutt. ex DC.
                var. c. (Eastw.) Jepson
  camporum (Greene) Shinners
            SY=Chrysopsis c. Greene
            SY=C. villosa (Pursh) Nutt. ex DC.
                var. c. (Greene) Cronq.
            SY=H. v. (Pursh) Shinners var. c.
                (Greene) Wunderlin
  canescens (DC.) Shinners
            SY=Chrysopsis berlandieri Greene
            SY=C. c. (DC.) Torr. & Gray non
                DC.
            SY=C. villosa (Pursh) Nutt. ex DC.
                var. c. (DC.) Gray
  chrysopsides DC.
  echioides (Benth.) Shinners
            SY=Chrysopsis e. Benth.
            SY=C. sessiliflora (Nutt.)
                Shinners var. e. (Benth.)

                Gray
            SY=C. villosa (Pursh) Nutt. ex DC.
                var. e. (Benth.) Gray
  fastigiata (Greene) Harms
            SY=Chrysopsis f. Greene
            SY=C. villosa (Pursh) Nutt. ex DC.
                var. f. (Greene) Hall
  fulcrata (Greene) Shinners
            SY=Chrysopsis cryptocephala Woot.
            SY=C. f. Greene
            SY=C. resinolens A. Nels.
            SY=C. senilis Woot.
  grandiflora Nutt.
            SY=H. floribunda Benth.
  horrida (Rydb.) Harms
            SY=Chrysopsis arida A. Nels.
            SY=C. canescens (DC.) Torr. & Gray
                var. nana Gray
            SY=C. h. Rydb.
  inuloides Cass.
      var. inuloides
      var. rosei Wagenkn.
  jonesii (Blake) Welsh & Atwood
  latifolia Buckl.
      var. arkansana Wagenkn.
      var. latifolia
      var. macgregoris Wagenkn.
  leptoglossa DC.
  oregona (Nutt.) Shinners
            SY=Ammodia o. Nutt.
            SY=Chrysopsis o. (Nutt.) Gray
            SY=C. o. var. compacta Keck
            SY=C. o. var. radis (Greene)
                Jepson
            SY=C. o. var. scaberrima Gray
  psammophila Wagenkn.
  rutteri (Rothrock) Shinners
  sessiliflora (Nutt.) Shinners
            SY=Chrysopsis californica Elmer
            SY=C. s. Nutt.
            SY=C. villosa (Pursh) Nutt. ex DC.
                var. s. (Nutt.) Gray
  stenophylla (Gray) Shinners
            SY=Chrysopsis hispida (Hook.) DC.
                var. s. Gray
            SY=C. scabrifolia A. Nels.
            SY=C. stenophylla (Gray) Greene
            SY=C. villosa (Pursh) Nutt. ex DC.
                var. stenophylla Gray
  subaxillaris (Lam.) Britt. & Rusby
      var. petiolaris Benke
            SY=H. s. var. procumbens Wagenkn.
      var. subaxillaris
            SY=Chrysopsis scabra (Pursh) Ell.
            SY=H. s. (Pursh) DC.
            SY=H. lamarckii Cass.
  villosa (Pursh) Shinners
      var. angustifolia (Rydb.) Harms
            SY=Chrysopsis a. Rydb.
            SY=C. v. (Pursh) Nutt. ex DC. var.
                a. (Rydb.) Cronq.
      var. depressa (Rydb.) Harms
            SY=Chrysopsis d. Rydb.
      var. foliosa (Nutt.) Harms
            SY=Chrysopsis f. (Nutt.) Shinners
            SY=C. hirsutissima Greene
            SY=C. mollis Nutt.
            SY=C. v. (Pursh) Nutt. ex DC. var.
                f. (Nutt.) D.C. Eat.
      var. glomerata (A. Nels.) Harms
            SY=Chrysopsis cooperi A. Nels.
      var. hispida (Hook.) Harms
            SY=Chrysopsis barbata Rydb.
            SY=C. butleri Rydb.
            SY=C. h. (Hook.) DC.
            SY=C. v. (Pursh) Nutt. ex DC. var.
                h. (Hook.) Gray

Heterotheca   (CONT.)
        SY=C. wisconsinensis Shinners
        SY=H. w. (Shinners) Shinners
      var. villosa
        SY=Chrysopsis bakeri Greene
        SY=C. ballardii Rydb.
        SY=C. caudata Rydb.
        SY=C. columbiana Greene
        SY=C. imbricata A. Nels.
        SY=C. v. (Pursh) Nutt. ex DC.
    viscida (Gray) Harms
        SY=Chrysopsis villosa (Pursh)
           Nutt. ex DC. var. viscida
           Gray
        SY=C. viscida (Gray) Greene

Hieracium L.

    acranthophorum Om.
      var. acranthophorum
      var. isortoquense Bocher
    albertinum Farr
    albiflorum Hook.
    X alleghaniense Britt. [gronovii X
      paniculatum]
    alpinum L.
    amitsokense (Almquist) Dahlst.
    angmagssalikense Om.
    argutum Nutt.
      var. argutum
      var. parishii (Gray) Jepson
    atratum Fries
    aurantiacum L.
    bolanderi Gray
        SY=H. siskiyouense M.E. Peck
    caespitosum Dumort.
        SY=H. pratense Tausch
    canadense Michx.
      var. canadense
        SY=H. columbianum Rydb.
        SY=H. scabriusculum Schwein. var.
           columbianum (Rydb.) Lepage
        SY=H. umbellatum L. var.
           scabriusculum Farw.
      var. divaricatum Lepage
      var. fasciculatum (Pursh) Fern.
      var. subintegrum Lepage
        SY=H. kalmii L. var. s. (Lepage)
           Lepage
    carneum Greene
    cynoglossoides Arv.-Touv. ex Gray
        SY=H. cusickii Gandog.
    devoldii Om.
    X dutillyanum Lepage [canadense X
      umbellatum]
    eugenii Om.
    X fassettii Lepage [kalmii X scabrum]
    fendleri Schultz-Bip.
      var. discolor Gray
      var. fendleri
      var. mogollense Gray
    X fernaldii Lepage [canadense X scabrum]
        SY=H. canadense Michx. var.
           hirtirameum Fern.
    X flagellare Willd. [caespitosum X
      pilosella]
      nm. amauracron (Missbach & Zahn) Lepage
        SY=H. f. ssp. a. Missbach & Zahn
      nm. cernuiforme (Naegeli & Peter) Lepage
        SY=H. f. ssp. c. Naegeli & Peter
      nm. glatzense (Naegeli & Peter) Lepage
        SY=H. f. ssp. g. Naegeli & Peter
      nm. pilosius Lepage
    X floribundum Wimmer & Grab. [caespitosum X
      lactucella]
        SY=H. auricula L.
        SY=H. X dorei Lapage

X fuscatrum Naegeli & Peter [aurantiacum X
    caespitosum]
gracile Hook.
    var. alaskanum Zahn
    var. detonsum (Gray) Gray
    var. gracile
        SY=H. g. var. densifloccosum
           (Zahn) Cronq.
        SY=H. triste Willd. ex Spreng.
           var. g. (Hook.) Gray
    var. yukonensis Porsild
greenii Gray
groenlandicum (Arv.-Touv.) Almquist
X grohii Lepage [canadense X lachenalii]
gronovii L.
        SY=H. g. var. foliosum Michx.
horridum Fries
hyparcticum Almquist
inuloides Tausch
ivigtutense (Almquist) Om.
kalmii L.
    var. fasciculatum (Pursh) Lepage
    var. kalmii
        SY=H. canadense Michx. var. k.
           (L.) Scoggan
lachenalii K.C. Gmel.
        SY=H. vulgatum Fries
lemmonii Gray
lividorubens (Almquist) Elfstr.
longiberbe T.J. Howell
longipilum Torr.
X marianum Willd. [gronovii X venosum]
        SY=H. pennsylvanicum Fries
megacephalon Nash
        SY=H. argyraeum Small
murorum L.
musartutense Om.
nepiocratum Om.
paniculatum L.
pilosella L.
    var. niveum Muell.-Arg.
    var. pilosella
piloselloides Vill.
        SY=H. florentinum All.
plicatum Lindb.
praealtum Vill. ex Gochnat
    var. decipiens W.D.J. Koch
    var. praealtum
pringlei Gray
rigorosum (Laestad.) Almquist
    var. nanusekense Om.
    var. rigorosum
    var. sermilikense Om.
    var. unanakense Om.
robinsonii (Zahn) Fern.
rusbyi Greene
sabaudum L.
scabrum Michx.
    var. intonsum Fern. & St. John
    var. leucocaule Fern. & St. John
    var. scabrum
    var. tonsum Fern. & St. John
scholanderi Om.
scouleri Hook.
    var. nudicaule (Gray) Cronq.
        SY=H. cynoglossoides Arv.-Touv. ex
           Gray var. n. Gray
    var. scouleri
        SY=H. chapacanum Zahn
        SY=H. parryi Zahn
X scribneri Small [paniculatum X venosum]
stelechodes Om.
stiptocaule Om.
X stoloniflorum Waldst. & Kit. [aurantiacum
    X pilosella]
    nm. cayouetteanum Lepage
    nm. laurentianum Lepage

Hieracium (CONT.)
  sylowii Om.
        var. norwagorum Om.
        var. sylowii
  traillii Greene
          SY=H. greenii Porter & Britt.
  tridentatum Fries
  trigonophorum Oskarss.
  triste Willd. ex Spreng.
        var. fulvum Hulten
        var. triste
          SY=H. t. var. tritiforme Zahn
  umbellatum L.
          SY=H. scabriusculum Schwein.
          SY=H. s. var. perhirsutum Lepage
          SY=H. s. var. saximontanum Lepage
          SY=H. s. var. scabrum (Schwein.)
               Lepage
  venosum L.
        var. nudicaule (Michx.) Farw.
        var. venosum
  wrightii (Gray) Robins. & Greenm.

Holocarpha (DC.) Greene

  heermannii (Greene) Keck
  macradenia (DC.) Greene
  obconica (Clausen & Keck) Keck
    ssp. autumnalis Keck
    ssp. obconica
  virgata (Gray) Keck
    ssp. elongata Keck
    ssp. virgata

Holozonia Greene

  filipes (Hook. & Arn.) Greene

Hulsea Torr. & Gray

  algida Gray
          SY=H. caespitosa A. Nels. &
               Kennedy
          SY=H. carnosa Rydb.
          SY=H. nevadensis Gandog.
  brevifolia Gray
  californica Torr. & Gray
  heterochroma Gray
  mexicana Rydb.
  nana Gray
          SY=H. n. var. larsenii Gray
          SY=H. volcanica Gandog.
  vestita Gray
    ssp. callicarpha (Hall) Wilken
          SY=H. c. (Hall) S. Wats.
          SY=H. v. var. c. Hall
    ssp. gabrielensis Wilken
    ssp. inyoensis (Keck) Wilken
          SY=H. californica Torr. & Gray
             ssp. i. Keck
          SY=H. i. (Keck) Munz
    ssp. parryi (Gray) Wilken
          SY=H. p. Gray
    ssp. pygmaea (Gray) Wilken
          SY=H. v. var. p. Gray
    ssp. vestita

Hymenoclea Torr. & Gray ex Gray

  monogyra Torr. & Gray ex Gray
  salsola Torr. & Gray ex Gray
        var. patula (A. Nels.) Peterson &
          Payne
          SY=H. fasciculata A. Nels.
        var. pentalepis (Rydb.) L. Benson
          SY=H. p. Rydb.
        var. salsola

Hymenopappus L'Her.

  artemisiifolius DC.
        var. artemisiifolius
        var. riograndensis B.L. Turner
  biennis B.L. Turner
  filifolius Hook.
        var. cinereus (Rydb.) I.M. Johnston
          SY=H. arenosus Heller
          SY=H. c. Rydb.
        var. eriopodus (A. Nels.) B.L. Turner
        var. filifolius
        var. idahoensis B.L. Turner
        var. lugens (Greene) Jepson
          SY=H. gloriosus Heller
          SY=H. l. Greene
        var. luteus (Nutt.) B.L. Turner
          SY=H. l. Nutt.
        var. megacephalus B.L. Turner
        var. nanus (Rydb.) B.L. Turner
        var. nudipes (Maguire) B.L. Turner
          SY=H. f. var. alpestris (Maguire)
               Shinners
        var. parvulus (Greene) B.L. Turner
        var. pauciflorus (I.M. Johnston) B.L.
          Turner
          SY=H. p. I.M. Johnston
        var. polycephalus (Osterhout) B.L.
          Turner
          SY=H. p. Osterhout
        var. tomentosus (Rydb.) B.L. Turner
  flavescens Gray
        var. cano-tomentosus Gray
          SY=H. robustus Greene
        var. flavescens
  mexicanus Gray
  newberryi (Gray) I.M. Johnston
          SY=Leucampyx n. Gray
  radiatus Rose
  scabiosaeus L'Her.
        var. corymbosus (Torr. & Gray) B.L.
          Turner
          SY=H. c. Torr. & Gray
        var. scabiosaeus
          SY=H. carolinensis (Lam.) Porter
  tenuifolius Pursh

Hymenothrix Gray

  loomisii Blake
  wislizenii Gray
  wrightii Gray

Hymenoxys Cass.

  acaulis (Pursh) Parker
        var. acaulis
          SY=Actinea a. (Pursh) Spreng.
          SY=Actinella a. (Pursh) Nutt.
          SY=Tetraneuris a. (Pursh) Greene
          SY=T. simplex A. Nels.
        var. arizonica (Greene) Parker
          SY=Actinea acaulis (Pursh) Spreng.
             var. arizonica (Greene)
             Blake
          SY=Tetraneuris a. Greene
        var. caespitosa (A. Nels.) Parker
          SY=Tetraneuris acaulis (Pursh)
             Greene var. c. A. Nels.
        var. glabra (Gray) Parker
          SY=Actinea herbacea (Greene) B.L.
             Robins.
          SY=Tetraneuris h. Greene
  anthemoides (Juss.) Cass.
  argentea (Gray) Parker
          SY=Tetraneuris a. (Gray) Greene
  bigelovii (Gray) Parker

Hymenoxys   (CONT.)
    brachyactis Woot. & Standl.
    brandegei (Porter ex Gray) Parker
    cooperi (Gray) Cockerell
            var. canescens (D.C. Eat.) Parker
                SY=Actinea canescens (D.C. Eat.)
                    Blake
            var. cooperi
                SY=Actinea c. (Gray) Kuntze
    depressa (Torr. & Gray ex Gray) Welsh &
        Reveal
                SY=Tetraneuris d. (Torr. & Gray ex
                    Gray) Greene
    grandiflora (Torr. & Gray ex Gray) Parker
                SY=Actinea g. (Torr. & Gray ex
                    Gray) Kuntze
    helenioides (Rydb.) Cockerell
    ivesiana (Greene) Parker
                SY=H. acaulis (Pursh) Parker var.
                    i. (Greene) Parker
                SY=Tetraneuris i. Greene
                SY=T. mancosensis A. Nels.
                SY=T. pilosa Greene
    lemmonii (Greene) Cockerell
    linearifolia Hook.
                SY=Tetraneuris l. (Hook.) Greene
    odorata DC.
    quinquesquamata Rydb.
    richardsonii (Hook.) Cockerell
            var. floribunda (Gray) Parker
                SY=H. olivacea Cockerell
            var. richardsonii
    rusbyi (Gray) Cockerell
    scaposa (DC.) Parker
            var. argyrocaulon Parker
            var. glabra (Nutt.) Parker
                SY=H. g. Nutt.
                SY=Tetraneuris fastigiata Greene
                SY=T. g. (Nutt.) Greene
                SY=T. stenophylla Rydb.
            var. scaposa
                SY=H. s. var. linearis (Nutt.)
                    Parker
                SY=Tetraneuris s. (DC.) Greene
            var. villosa Shinners
                SY=Tetraneuris s. (DC.) Greene
                    var. v. (Shinners) Shinners
    subintegra Cockerell
    texana (Coult. & Rose) Cockerell
    torreyana (Nutt.) Parker
                SY=Tetraneuris t. (Nutt.) Greene
    turneri Parker
    vaseyi (Gray) Cockerell

Hypochoeris L.

    brasiliensis (Less.) Hook. & Arn.
                SY=H. elata (Weddell) Griseb.
    glabra L.
    microcephala (Schultz-Bip.) Cabrera
            var. albiflora (Kuntze) Cabrera
            var. microcephala
    radicata L.
    tweedii Hook. & Arn.

Inula L.

    brittanica L.
    helenium L.
    salicina L.
    viscosa (L.) Ait.
                SY=Cupularia v. (L.) Gren. & Godr.
                SY=Dittrichia v. (L.) W. Greuter

Isocarpha R. Br.

    oppositifolia (L.) Cass.

Isocoma Nutt.

    coronopifolia (Gray) Greene
    drummondii (Torr. & Gray) Greene
                SY=Haplopappus d. (Torr. & Gray)
                    Blake
                SY=I. megalantha Shinners
    pluriflora (Torr. & Gray) Greene
                SY=Haplopappus p. (Torr. & Gray)
                    Hall
    wrightii (Gray) Rydb.
                SY=Haplopappus heterophyllus
                    (Gray) Blake

Iva L.

    acerosa (Nutt.) R.C. Jackson
                SY=Oxytenia a. Nutt.
    ambrosiaefolia (Gray) Gray
    angustifolia DC.
            var. angustifolia
            var. latior Shinners
                SY=I. texensis R.C. Jackson
    annua L.
            var. annua
                SY=I. ciliata Willd.
            var. caudata (Small) R.C. Jackson
                SY=I. c. Small
            var. macrocarpa (Blake) R.C. Jackson
                SY=I. ciliata Willd. var. m. Blake
    asperifolia Less.
    axillaris Pursh
                SY=I. a. ssp. robustior (Hook.)
                    Bassett
                SY=I. a. var. r. Hook.
    cheiranthifolia H.B.K.
    dealbata Gray
    frutescens L.
        ssp. frutescens
        ssp. oraria (Bartlett) R.C. Jackson
                SY=I. f. var. o. (Bartlett) Fern.
                    & Grisc.
    hayesiana Gray
    imbricata Walt.
    microcephala Nutt.
    nevadensis M.E. Jones
    xanthifolia Nutt.

Ixeris Cass.

    stolonifera Gray
                SY=Lactuca s. (Gray) Maxim.

Jamesianthus Blake & Sherff

    alabamensis Blake & Sherff

Jaumea Pers.

    carnosa (Less.) Gray

Keysseria Lauterb.

    ericii (Forbes) Cabrera
    helenae (Forbes & Lydgate) Cabrera
    lavandula St. John
    maviensis (Mann) Cabrera

Krigia Schreb.

    biflora (Walt.) Blake
                SY=Cynthia virginica (L.) D. Don
    caespitosa (Raf.) Chambers
                SY=K. oppositifolia Raf.
                SY=Serinia o. (Raf.) Kuntze
    dandelion (L.) Nutt.
                SY=Cynthia d. (L.) DC.

Krigia (CONT.)
  gracilis (DC.) Shinners
  montana (Michx.) Nutt.
          SY=Cynthia m. (Michx.) Standl.
  occidentalis Nutt.
          SY=Cymbia o. (Nutt.) Standl.
  virginica (L.) Willd.

Lacamothrix W.S. Davis ined

  carterae W.S. Davis ined.
  xantii (Gray) W.S. Davis ined.
          SY=Malacothrix x. Gray

Lactuca L.

  biennis (Moench) Fern.
          SY=L. spicata (Lam.) A.S. Hitchc.
             var. integrifolia (Gray)
             Britt.
  canadensis L.
      var. canadensis
          SY=L. c. var. integrifolia
             (Bigelow) Torr. & Gray
          SY=L. c. var. typica Wieg.
          SY=L. sagittifolia Ell.
      var. latifolia Kuntze
      var. longifolia (Michx.) Farw.
      var. obovata Wieg.
          SY=L. steelei Britt.
  floridana (L.) Gaertn.
      var. floridana
          SY=Mulgedium f. (L.) DC.
      var. villosa (Jacq.) Cronq.
          SY=Mulgedium v. (Jacq.) Small
          SY=L. v. Jacq.
  graminifolia Michx.
  hirsuta Muhl.
      var. albiflora (Torr. & Gray)
             Shinners
      var. hirsuta
      var. sanguinea (Bigelow) Fern.
  ludoviciana (Nutt.) Riddell
          SY=L. campestris Greene
          SY=L. c. var. typica Wieg.
  X morssii B.L. Robins. [biennis X
      canadensis]
  saligna L.
  sativa L.
          SY=L. scariola L. var. integrata
             Gren. & Godr.
          SY=L. scariola var. integrifolia
             (Bogenh.) G. Beck
  serriola L.
          SY=L. scariola L.
  spicata (Lam.) A.S. Hitchc.
          SY=Mulgedium s. (Lam.) Small
  tatarica (L.) C.A. Mey.
      ssp. pulchella (Pursh) Stebbins
          SY=L. oblongifolia Nutt.
          SY=L. p. (Pursh) DC.
          SY=L. t. var. heterophylla (Nutt.)
             Boivin
          SY=L. t. var. p. (Pursh) Breitung
  terrae-novae Fern.
  virosa L.

Lagascea Cav.

  decipiens Hemsl.
  mollis Cav.
          SY=Nocca m. (Cav.) Jacq.

Lagenophora Cass.

  viridis St. John

Lagophylla Nutt.

  congesta Greene
  dichotoma Benth.
  glandulosa Gray
      ssp. glandulosa
      ssp. serrata (Greene) Keck
  minor (Keck) Keck
  ramosissima Nutt.

Lapsana L.

  apogonoides Maxim.
  communis L.

Lasthenia Cass.

  burkei (Greene) Greene
  californica DC. ex Lindl.
          SY=Baeria chrysostoma Fisch. &
             Mey.
          SY=B. chrysostoma ssp. gracilis
             (DC.) Ferris
          SY=B. chrysostoma ssp. hirsutula
             (Greene) Ferris
          SY=B. chrysostoma var. g. (DC.)
             Hall
          SY=L. chrysostoma (Fisch. & Mey.)
             Greene
  chrysantha (Greene ex Gray) Greene
          SY=Crockeria c. Greene ex Gray
  conjugens Greene
          SY=Baeria fremontii (Torr. ex
             Gray) Gray var. c. (Greene)
             Ferris
  coronaria (Nutt.) Ornduff
          SY=Baeria californica (Hook.)
             Chambers
  debilis (Greene ex Gray) Ornduff
          SY=Baeria d. Greene ex Gray
  ferrisiae Ornduff
  fremontii (Torr. ex Gray) Greene
          SY=Baeria f. (Torr. ex Gray) Gray
          SY=B. f. var. heterochaeta Hoover
  glaberrima DC.
  glabrata Lindl.
      ssp. coulteri (Gray) Ornduff
          SY=L. g. var. c. Gray
      ssp. glabrata
  leptalea (Gray) Ornduff
          SY=Baeria l. (Gray) Gray
  macrantha (Gray) Greene
      ssp. bakeri (J.T. Howell) Ornduff
          SY=Baeria b. J.T. Howell
          SY=B. m. var. b. (J.T. Howell)
             Keck
      ssp. macrantha
          SY=Baeria m. (Gray) Gray
          SY=B. m. var. pauciaristata Gray
          SY=B. m. var. thalassophila J.T.
             Howell
      ssp. prisca Ornduff
  microglossa (DC.) Greene
          SY=Baeria m. (DC.) Greene
  minor (DC.) Ornduff
      ssp. maritima (Gray) Ornduff
          SY=Baeria maritima Gray
          SY=B. minor (DC.) Ferris ssp.
             maritima (Gray) Ferris
          SY=L. minor var. maritima (Gray)
             Cronq.
      ssp. minor
          SY=Baeria m. (DC.) Ferris
  platycarpha (Gray) Greene
          SY=Baeria p. (Gray) Gray

Launaea Cass.

Launaea (CONT.)
  intybacea (Jacq.) Beauv.
        SY=Brachyrhamphus i. (Jacq.) DC.
        SY=Lactuca i. Jacq.

Layia Hook. & Arn.

  carnosa (Nutt.) Torr. & Gray
  chrysanthemoides (DC.) Gray
    ssp. chrysanthemoides
    ssp. maritima Keck
  discoidea (Keck) Keck
  fremontii (Torr. & Gray) Gray
  gaillardioides (Hook. & Arn.) DC.
  glandulosa (Hook.) Hook. & Arn.
    ssp. glandulosa
    ssp. lutea Keck
  heterotricha (DC.) Hook. & Arn.
  hieracioides (DC.) Hook. & Arn.
  jonesii Gray
  leucopappa Keck
  munzii Keck
  paniculata Keck
  pentachaeta Gray
    ssp. albida Keck
    ssp. pentachaeta
  platyglossa (Fisch. & Mey.) Gray
    ssp. campestris Keck
    ssp. platyglossa
  septentrionalis Keck
  ziegleri Munz

Leontodon L.

  autumnalis L.
    var. autumnalis
      SY=Apargia a. (L.) Hoffmann
    var. pratensis (Link) W.D.J. Koch
  hispidus L.
    var. glabratus (W.D.J. Koch) Bisch.
      SY=L. hastilis L.
    var. hispidus
      SY=Apargia h. (L.) Willd.
      SY=L. hastilis L. var. vulgaris
        W.D.J. Koch
  nudicaulis (L.) Merat
      SY=Apargia n. (L.) Britt.
  taraxacoides (Vill.) Merat
      SY=L. leysseri (Wallr.) G. Beck
      SY=L. nudicaulis (L.) Merat ssp.
        t. (Vill.) Schinz & Thellung

Lepidospartum (Gray) Gray

  burgesii B.L. Turner
  latisquamum S. Wats.
  squamatum (Gray) Gray
      SY=L. s. var. palmeri (Gray) L.C.
        Wheeler

Leptocarpha DC.

  maculata (Michx.) Torr.

Lessingia Cham.

  germanorum Cham.
    var. germanorum
    var. parvula (Greene) J.T. Howell
    var. tenuis (Gray) J.T. Howell
      SY=L. t. (Gray) Coville
  glandulifera Gray
    var. glandulifera
      SY=L. germanorum Cham. var.
        glandulifera (Gray) J.T.
        Howell
    var. pectinata (Greene) Jepson

      SY=L. germanorum Cham. var. p.
        (Greene) J.T. Howell
    var. tomentosa (Greene) Ferris
      SY=L. germanorum Cham. var. t.
        (Greene) J.T. Howell
  hololeuca Greene
    var. arachnoidea (Greene) J.T. Howell
    var. hololeuca
  lemmonii Gray
    var. lemmonii
      SY=L. germanorum Cham. var. l.
        (Gray) J.T. Howell
    var. peirsonii (J.T. Howell) Ferris
      SY=L. germanorum Cham. var. p.
        J.T. Howell
    var. ramulosissima (A. Nels.) Ferris
      SY=L. germanorum Cham. var. r. (A.
        Nels.) J.T. Howell
  leptoclada Gray
  micradenia Greene
    var. arachnoidea (Greene) Ferris
    var. glabra (Keck) Ferris
      SY=L. ramulosa Gray var. glabrata
        Keck
    var. micradenia
      SY=L. ramulosa Gray var. m.
        (Greene) J.T. Howell
  nana Gray
  nemaclada Greene
    var. albiflora (Eastw.) J.T. Howell
    var. mendocina (Greene) J.T. Howell
    var. nemaclada
  ramulosa Gray
    var. adenophora (Greene) Gray
    var. ramulosa
  virgata Gray
    var. glomerata (Greene) J.T. Howell
    var. virgata

Leucanthemella Tzvelev

  serotina (L.) Tzvelev
      SY=Leucanthemum s. (L.) Stankov
      SY=Tanacetum s. (L.) Schultz-Bip.

Leucanthemopsis (Giroux) Heywood

  alpinum (L.) Heywood
      SY=Chrysanthemum a. L.

Leucanthemum P. Mill.

  integrifolium (Richards.) DC.
      SY=Chrysanthemum i. Richards.
  lacustre (Brot.) Samp.
      SY=Chrysanthemum l. Brot.
  maximum (Ramond) DC.
      SY=Chrysanthemum m. Ramond
  vulgare Lam.
      SY=Chrysanthemum leucanthemum L.
      SY=C. l. var. boecheri Boivin
      SY=C. l. var. pinnatifidum Lecoq &
        Lamotte
      SY=Leucanthemum l. (L.) Rydb.

Leucelene Greene

  ericoides (Torr.) Greene
      SY=Aster arenosus (Heller) Blake
      SY=A. leucelene Blake

Liatris Gaertn. ex Schreb.

  acidota Engelm. & Gray
      SY=Laciniaria a. (Engelm. & Gray)
        Kuntze
  angustifolia (Bush) Gaiser

Liatris (CONT.)
   aspera Michx.
       var. aspera
          SY=Laciniaria a. (Michx.) Greene
       var. intermedia (Lunell) Gaiser
       var. salutans (Lunell) Shinners
   X boykinii Torr. & Gray [elegans X
     tenuifolia]
          SY=Laciniaria b. (Torr. & Gray)
             Kuntze
   bracteata Gaiser
   chapmanii Torr. & Gray
          SY=Laciniaria c. (Torr. & Gray)
             Kuntze
   X creditonensis Gaiser [ligulistylis X
     squarrosa var. glabrata]
   cylindracea Michx.
          SY=Laciniaria c. (Michx.) Kuntze
   cymosa (H. Ness) K. Schum.
   densispicata (Bush) Gaiser
       var. densispicata
          SY=Liatris mucronata DC.
       var. interrupta Gaiser
   elegans (Walt.) Michx.
       var. carizzana Gaiser
       var. elegans
          SY=Laciniaria e. (Walt.) Kuntze
       var. flabellata (Small) Gaiser
          SY=Laciniaria f. Small
   X fallacior (Lunell) Rydb. [ligulistylis X
     punctata]
   X frostii Gaiser [aspera X pycnostachya]
   garberi Gray
          SY=Laciniaria chlorolepis Small
          SY=Laciniaria g. (Gray) Kuntze
          SY=Laciniaria nashii Small
   X gladewitzii (Farw.) Shinners [cylindracea
     X sphaeroidea]
   gracilis Pursh
          SY=Laciniaria g. (Pursh) Kuntze
          SY=Laciniaria laxa Small
   graminifolia (Walt.) Willd.
       var. dubia (Bart.) Gray
       var. elegantula (Greene) K. Schum.
       var. graminifolia
          SY=Laciniaria g. (Walt.) Kuntze
          SY=Laciniaria pilosa (Ait.) Heller
       var. lasia Fern. & Grisc.
       var. racemosa (DC.) Venard
       var. smallii (Britt.) Fern. & Grisc.
          SY=Laciniaria s. Britt.
       var. virgata (Nutt.) Fern.
   helleri (Porter) Porter
          SY=Laciniaria h. (Porter) Heller
   laevigata (Nutt.) Small
          SY=Liatris tenuifolia Nutt. var.
            l. (Nutt.) B.L. Robins.
   lancifolia (Greene) Kittell
          SY=Laciniaria kansana Britt.
   ligulistylis (A. Nels.) K. Schum.
   microcephala (Small) K. Schum.
          SY=Laciniaria m. Small
   ohlingerae (Blake) B.L. Robins.
          SY=Ammopursus o. (Blake) Small
   pauciflora Pursh
          SY=Laciniaria p. (Pursh) Kuntze
   provincialis Godfrey
   punctata Hook.
       var. mexicana Gaiser
       var. nebraskana Gaiser
       var. punctata
          SY=Laciniaria p. (Hook.) Kuntze
          SY=Liatris p. var. typica Gaiser
   pycnostachya Michx.
       var. lasiophylla Shinners
       var. pycnostachya
          SY=Laciniaria p. (Michx.) Kuntze

          SY=Liatris bebbiana Rydb.
   regiomontis (Small) K. Schum.
          SY=Laciniaria r. Small
   X ridgwayi Standl. [pycnostachya X
     squarrosa]
   scariosa (L.) Willd.
       var. nieuwlandii Lunell
          SY=Liatris borealis Nutt.
          SY=L. X n. (Lunell) Gaiser
          SY=L. novae-angliae Lunell var.
            nieuwlandii (Lunell)
            Shinners
       var. novae-angliae Lunell
          SY=L. n-a. (Lunell) Shinners
       var. scariosa
          SY=Laciniaria s. (L.) Hill
       var. virginiana (Lunell) Gaiser
   secunda Ell.
          SY=Laciniaria s. (Ell.) Small
   sphaeroidea Michx.
   spicata (L.) Willd.
          SY=Laciniaria s. (L.) Kuntze
          SY=Liatris s. var. resinosa
            (Nutt.) Gaiser
   squarrosa (L.) Michx.
       var. alabamensis (Alexander) Gaiser
          SY=Liatris glabrata Rydb. var. a.
            (Alexander) Shinners
       var. compacta Torr. & Gray
          SY=Liatris c. (Torr. & Gray) Rydb.
       var. glabrata (Rydb.) Gaiser
          SY=Liatris g. Rydb.
       var. hirsuta (Rydb.) Gaiser
          SY=Liatris h. Rydb.
       var. squarrosa
          SY=Laciniaria s. (L.) Hill
          SY=Liatris s. var. gracilenta
            Gaiser
   squarrulosa Michx.
          SY=Laciniaria ruthii Alexander
          SY=Laciniaria shortii Alexander
          SY=Laciniaria tracyi Alexander
          SY=Liatris earlei (Greene) K.
            Schum.
          SY=Liatris scabra (Greene) K.
            Schum.
          SY=Liatris scariosa (L.) Willd.
            var. squarrulosa (Michx.)
            Gray
   X steelei Gaiser [aspera X spicata]
   tenuifolia Nutt.
       var. quadrifolia Chapman
       var. tenuifolia
          SY=Laciniaria t. (Nutt.) Kuntze
   tenuis Shinners
   turgida Gaiser
          SY=Laciniaria pilosa sensu auctt.
            non (Ait.) Heller
   X weaveri Shinners [aspera X punctata]

Lindheimera Gray & Engelm.

   texana Gray & Engelm.

Lipochaeta DC.

   bryanii Sherff
   connata (Gaud.) DC.
       var. acris (Sherff) R.C. Gardner
          SY=L. a. Sherff
          SY=L. lobata (Gaud.) DC. var.
            incisior St. John
       var. connata
          SY=L. alata Sherff
          SY=L. a. var. acrior Sherff
          SY=L. alata var. pulchrior Sherff
          SY=L. profusa Sherff

Lipochaeta (CONT.)
        SY=L. profusa var. robustior Deg.
            & Sherff
degeneri Sherff
deltoidea St. John
dubia Deg. & Sherff
        SY=L. minuscula Deg. & Sherff
fauriei Levl.
heterophylla Gray
        SY=L. hastata Hbd.
        SY=L. heterophylla var.
            molokaiensis Sherff
integrifolia (Nutt.) Gray
        SY=L. i. var. argentea Sherff
        SY=L. i. var. gracilis Sherff
        SY=L. i. var. major Sherff
        SY=L. i. var. megacephala Deg. &
            Sherff
kamolensis Deg. & Sherff
lavarum (Gaud.) DC.
        SY=L. l. var. conferta Sherff
        SY=L. l. var. hillebrandiana
            Sherff
        SY=L. l. var. lanaiensis Sherff
        SY=L. lavarum var. longifolia
            Sherff
        SY=L. lavarum var. maneleana
            Sherff
        SY=L. lavarum var. ovata Sherff
        SY=L. lavarum var. salicifolia
            Sherff
        SY=L. lavarum var. skottsbergii
            Sherff
        SY=L. lavarum var. stearnsii Deg.
            & Sherff
lobata (Gaud.) DC.
      var. hastulatoides Deg. & Sherff
      var. leptophylla Deg. & Sherff
        SY=L. lobata var. grossedentata
            Deg. & Sherff
      var. lobata
        SY=L. calycosa Gray
        SY=L. l. var. albescens Sherff
        SY=L. l. var. aprevalliana (del
            Castillo) Sherff
        SY=L. l. var. denticulata (Wawra)
            Sherff
        SY=L. l. var. hastulata (Hook. &
            Arn.) Sherff
        SY=L. niihauensis St. John
        SY=L. trilobata St. John
micrantha (Nutt.) Gray
      var. exigua (Deg. & Sherff) R.C.
        Gardner
        SY=L. e. Deg. & Sherff
      var. micrantha
ovata R.C. Gardner
perdita Sherff
        SY=L. kawaihoaensis St. John
populifolia (Sherff) R.C. Gardner
        SY=L. subcordata Gray var. p.
            Sherff
X procumbens Deg. & Sherff [integrifolia X
   lobata]
remyi Gray
rockii Sherff
        SY=L. forbesii Sherff
        SY=L. f. var. sherffii Deg. & Clay
        SY=L. heterophylla Gray var.
            malvacea Deg. & Sherff
        SY=L. kahoolawensis Sherff
        SY=L. lobata (Gaud.) DC. var.
            makensis Deg. & Sherff
        SY=L. l. var. maunaloensis Sherff
        SY=L. r. var. dissecta Sherff
        SY=L. r. var. subovata Sherff
        SY=L. scabra St. John

        SY=L. succulenta (Hook. & Arn.)
            DC. var. trifida Sherff
subcordata Gray
      var. ?membranacea Sherff
      var. subcordata
        SY=L. flexuosa del Castillo
        SY=L. intermedia Deg. & Sherff
succulenta (Hook. & Arn.) DC.
        SY=L. connata (Gaud.) DC. var.
            decurrens Gray
        SY=L. c. var. littoralis Hbd.
        SY=L. lanceolata Nutt.
        SY=L. s. var. angustata Sherff
        SY=L. s. var. barclayi Sherff
        SY=L. s. var. decurrens (Gray)
            Sherff
        SY=L. variolosa Levl.
tenuifolia Gray
tenuis Deg. & Sherff
        SY=L. t. var. sellingii Deg. &
            Sherff
venosa Sherff
waimeaensis St. John

Logfia Cass.

   arvensis (L.) Holub
        SY=Filago a. L.
   gallica (L.) Coss. & Germ.
        SY=Filago g. L.

Luina Benth.

   hypoleuca Benth.
   serpentina Crong.

Luteidiscus St. John

   calcisabulorum St. John
   capillaris (Gaud.) St. John
        SY=Tetramolopium c. Gaud.
   rockii (Sherff) St. John
        SY=Tetramolopium r. Sherff

Lygodesmia D. Don

   aphylla (Nutt.) DC.
   arizonica S. Tomb
   dianthopsis (Eat.) S. Tomb
   grandiflora (Nutt.) Torr. & Gray
      var. grandiflora
      var. stricta Maguire
   juncea (Pursh) D. Don
   texana (Torr. & Gray) Greene
        SY=L. ramosissima Greene
        SY=L. wrightii (Gray) Shinners
        SY=Stephanomeria w. Gray

Machaeranthera Nees

   amplifolia Woot. & Standl.
   annua (Rydb.) Shinners
        SY=Haplopappus a. (Rydb.) Cory
        SY=H. phyllocephalus DC. ssp. a.
            (Rydb.) Hall
        SY=H. p. var. a. (Rydb.) Waterfall
        SY=Sideranthus a. Rydb.
   aquifolia Greene ex Woot. & Standl.
        SY=Aster a. (Greene ex Woot. &
            Standl.) Blake
   arida B.L. Turner & Horne
        SY=M. ammophila Reveal
        SY=M. arizonica Jackson & Johnston
        SY=Psilactis coulteri sensu auctt.
            non Gray
   aurea (Gray) Shinners
        SY=Haplopappus a. Gray

Machaeranthera (CONT.)
      SY=Sideranthus a. (Gray) Small
  australis (Greene) Shinners
      SY=Eriocarpum a. Greene
      SY=Haplopappus spinulosus (Pursh)
         DC. ssp. a. (Greene) Hall
      SY=M. pinnatifida (Hook.) Shinners
         ssp. p. var. chihuahuana
         B.L. Turner & Hartman
      SY=Sideranthus a. (Greene) Rydb.
  bigelovii (Gray) Greene
      SY=Aster b. Gray
      SY=A. pattersonii Gray
      SY=M. centaureoides Greene
      SY=M. p. (Gray) Greene
  blephariphylla (Gray) Shinners
      SY=Aster gymnocephalus sensu
         auctt. non Gray
      SY=Haplopappus b. Gray
      SY=M. correllii Shinners
  boltoniae (Greene) B.L. Turner & Horne
      SY=Aster b. Greene
      SY=Psilactis asteroides Gray
      SY=P. lepta Shinners
  brevilingulata (Schultz-Bip. ex Hemsl.)
    B.L. Turner & Horne
      SY=Psilactis b. Schultz-Bip. ex
         Hemsl.
  canescens (Pursh) Gray
    ssp. canescens
      SY=Aster attenuatus sensu Frye &
         Rigg, non Lindl.
      SY=A. c. Pursh
      SY=A. c. var. viscosus (Nutt.)
         Gray
      SY=A. inornatus Greene
      SY=A. leucanthemifolius Greene
      SY=Dieteria asteroides Torr.
      SY=D. c. (Pursh) Nutt.
      SY=D. pulverulenta Nutt.
      SY=D. sessiliflora (Nutt.) Greene
      SY=D. v. Nutt.
      SY=Diplopappus incanus Lindl.
      SY=M. asteroides (Torr.) Greene
      SY=M. attenuata (Frye & Rigg) T.J.
         Howell
      SY=M. incana (Lindl.) Greene
      SY=M. inornata Greene
      SY=M. leucanthemifolia (Greene)
         Greene
      SY=M. linearis sensu Rydb. non
         Greene
      SY=M. pinosa Elmer
      SY=M. pulverulenta (Nutt.) Greene
      SY=M. sessiliflora (Nutt.) Greene
      SY=M. spinulosa Greene non Amellus
         s. Pursh
      SY=M. viscosa (Nutt.) Greene
    ssp. ziegleri Munz
  coloradoensis (Gray) Osterhout
      SY=Aster c. Gray
      SY=Eriocarpum c. (Gray) Greene
      SY=Xylorhiza c. (Gray) Rydb.
  commixta Greene
      SY=M. superba A. Nels.
  gracilis (Nutt.) Shinners
      SY=Dieteria g. Nutt. non Aster g.
         Nutt.
      SY=Haplopappus g. (Nutt.) Gray
      SY=Sideranthus g. (Nutt.) A. Nels.
  grindelioides (Nutt.) Cronq.
    var. depressa (Maguire) Cronq. & Keck
      SY=Haplopappus nuttallii Torr. &
         Gray var. d. Maguire
    var. grindelioides
      SY=Eriocarpum g. Nutt.
      SY=Haplopappus nuttallii Torr. &

      Gray
      SY=Sideranthus g. (Nutt.) Britt.
  gymnocephala (DC.) Shinners
      SY=Haplopappus g. DC.
  havardii (Waterfall) Shinners
      SY=Haplopappus h. Waterfall
  kingii (D.C. Eat.) Cronq. & Keck
      SY=Aster k. D.C. Eat.
  laetevirens Greene
      SY=Aster leiodes Blake
      SY=M. canescens sensu D.C. Eat.
         non (Pursh) Gray
  lagunensis Keck
  linearis Greene
      SY=Aster cichoriaceus (Greene)
         Blake
      SY=A. l. (Greene) Cory
      SY=M. c. Greene
  mexicana B.L. Turner & Horne
  mucronata Greene
      SY=Aster adenolepis Blake
  parviflora Gray
      SY=Aster p. sensu Gray non Nees
      SY=A. parvulus Blake
  phyllocephala (DC.) Shinners
      SY=Eriocarpum megacephalum Nash
      SY=Haplopappus p. DC.
      SY=H. p. var. m. (Nash) Waterfall
      SY=M. p. var. m. (Nash) Shinners
      SY=Sideranthus m. (Nash) Small
  pinnatifida (Hook.) Shinners
    ssp. gooddingii (A. Nels.) B.L. Turner &
    Hartman
      var. gooddingii
        SY=Haplopappus g. (A. Nels.) Munz
         & Johnston
        SY=H. spinulosus (Pursh) DC. var.
         g. (A. Nels.) Blake
        SY=Sideranthus g. A. Nels.
      var. paradoxa B.L. Turner & Hartman
    ssp. pinnatifida
      var. glaberrima (Rydb.) B.L. Turner &
      Hartman
        SY=Haplopappus spinulosus (Pursh)
         DC. ssp. g. (Rydb.) Hall
        SY=H. s. var. g. (Rydb.) Blake
        SY=Sideranthus g. Rydb.
        SY=S. s. (Pursh) Sweet ex Rydb.
         var. g. (Rydb.) A. Nels.
      var. pinnatifida
        SY=Amellus spinulosus Pursh non M.
         spinulosus Greene
        SY=Aster p. (Hook.) Kuntze
        SY=Dieteria s. (Pursh) Nutt.
        SY=Diplopappus p. Hook.
        SY=Eriocarpum s. (Pursh) Greene
        SY=E. wootonii Greene
        SY=Haplopappus coulteri Harvey &
         Gray ex Gray
        SY=H. s. (Pursh) DC.
        SY=H. s. ssp. cotula (Small) Hall
        SY=H. s. ssp. laevis (Woot. &
         Standl.) Hall
        SY=H. s. var. canescens Gray
        SY=H. s. var. glaber Gray
        SY=H. s. var. turbinellus (Rydb.)
         Blake
        SY=H. texensis R.C. Jackson
        SY=M. l. (Woot. & Standl.)
         Shinners
        SY=M. pinnata (Nutt.) Shinners
        SY=M. texensis (R.C. Jackson)
         Shinners
        SY=Sideranthus cotula Small
        SY=S. l. Woot. & Standl.
        SY=S. machaeranthera Small
        SY=S. puberulus Rydb.

Machaeranthera (CONT.)
        SY=S. s. (Pursh) Sweet ex Rydb.
        SY=S. turbinellus Rydb.
        SY=S. w. (Greene) Standl.
        SY=Starkea pinnata Nutt.
    rubricaulis Rydb.
        SY=Aster rubrotinctus Blake
    scabrella (Greene) Shinners
        SY=Eriocarpum s. Greene
        SY=Haplopappus spinulosus (Pursh)
            DC. ssp. scabrellus (Greene)
            Hall
        SY=H. spinulosus var. scabrellus
            (Greene) Blake
        SY=M. pinnatifida (Hook.) Shinners
            ssp. gooddingii (A. Nels.)
            B.L. Turner & Hartman var.
            scabrella (Geene) B.L.
            Turner & Hartman
    shastensis Gray
      var. eradiata (Gray) Cronq. & Keck
        SY=Aster s. (Gray) Gray var. e.
            Gray
        SY=M. e. (Gray) T.J. Howell
      var. glossophylla (Piper) Cronq. &
        Keck
        SY=Aster g. Piper
        SY=A. s. (Gray) Gray var. g.
            (Piper) Cronq.
      var. latifolia (Cronq.) Cronq. & Keck
        SY=Aster s. (Gray) Gray var. l.
            Cronq.
      var. montana (Greene) Cronq. & Keck
        SY=M. m. Greene
      var. shastensis
        SY=Aster s. (Gray) Gray
    sonorae (Gray) Stucky
        SY=Aster s. Gray
    tagetina Greene
        SY=Aster t. (Greene) Blake
    tanacetifolia (H.B.K.) Nees
        SY=Aster t. H.B.K.
    tenuis (S. Wats.) B.L. Turner & Horne
        SY=Psilactis t. S. Wats.
    tephrodes (Gray) Greene
        SY=Aster canescens (Pursh) Gray
            var. t. Gray
        SY=A. t. (Gray) Blake

Madia Molina

    anomala Greene
    bolanderi (Gray) Gray
    capitata Nutt.
        SY=M. sativa Molina ssp. c.
            (Nutt.) Piper
        SY=M. s. var. congesta Torr. &
            Gray
    citrigracilis Keck
    citriodora Greene
    elegans D. Don ex Lindl.
      ssp. densifolia (Greene) Keck
        SY=M. e. var. d. (Greene) Jepson
      ssp. elegans
      ssp. vernalis Keck
      ssp. wheeleri (Gray) Keck
    exigua (Sm.) Gray
    glomerata Hook.
    gracilis (Sm.) Keck
      ssp. collina Keck
      ssp. gracilis
        SY=M. dissitiflora (Nutt.) Torr. &
            Gray
      ssp. pilosa Keck
    hallii Keck
    madioides (Nutt.) Greene
    minima (Gray) Keck

nutans (Greene) Keck
radiata Kellogg
rammii Greene
sativa Molina
subspicata Keck
yosemitana Parry ex Gray

Malacothrix DC.

    californica DC.
      var. californica
      var. glabrata D.C. Eat. ex Gray
        SY=M. g. (D.C. Eat. ex Gray) Gray
    clevelandii Gray
    coulteri Harvey & Gray
      var. cognata Jepson
      var. coulteri
    fendleri Gray
    floccifera (DC.) Blake
    foliosa Gray
      var. foliosa
      var. indecora (Greene) E. Williams
        SY=M. i. Greene
      var. squalida (Greene) E. Williams
        SY=M. insularis Greene var. s.
            (Greene) Ferris
        SY=M. s. Greene
    incana (Nutt.) Torr. & Gray
      var. incana
      var. succulenta (Elmer) E. Williams
        SY=M. s. Elmer
    inconspicua W.S. Davis ined.
    phaeocarpa W.S. Davis ined.
    philbrickii W.S. Davis ined.
    polycephala W.S. Davis ined.
    saxatilis (Nutt.) Torr. & Gray
      var. altissima (Greene) Ferris
      var. arachnoidea (McGregor) E.
        Williams
      var. commutata (Torr. & Gray) Ferris
      var. implicata (Eastw.) Hall
      var. saxatilis
      var. tenuifolia (Nutt.) Gray
    similis W.S. Davis & Raven
    sonchoides (Nutt.) Torr. & Gray
      var. sonchoides
      var. torreyi (Gray) E. Williams
        SY=M. t. Gray
    sonorae W.S. Davis & Raven
    stebbinsii W.S. Davis & Raven

Malperia S. Wats.

    tenuis S. Wats.

Marshallia Schreb.

    caespitosa Nutt. ex DC.
      var. caespitosa
      var. signata Beadle & F.E. Boynt.
    graminifolia (Walt.) Small
        SY=M. laciniarioides Small
        SY=M. williamsonii Small
    grandiflora Beadle & F.E. Boynt.
    mohrii Beadle & F.E. Boynt.
    obovata (Walt.) Beadle & F.E. Boynt.
      var. obovata
        SY=M. o. var. platyphylla (M.A.
            Curtis) Beadle & F.E. Boynt.
      var. scaposa Channell
    ramosa Beadle & F.E. Boynt.
    tenuifolia Raf.
    trinervia (Walt.) Trel. ex Banner & Coville

Matricaria L.

    chamomilla L.

Matricaria   (CONT.)
            SY=Chamomilla c. (L.) Rydb.
            SY=C. recutita (L.) Rauschert
            SY=M. c. var. coronata (J. Gay)
                Coss. & Germ.
            SY=M. r. L.
    courrantiana DC.
    maritima L.
        ssp. maritima
            SY=Chamomilla m. (L.) Rydb.
            SY=M. m. var. nana (Hook.) Boivin
        ssp. phaeocephala (Rupr.) Rauschert
            SY=M. ambigua (Ledeb.) Krylov
            SY=M. grandiflora (Hook.) Britt.
            SY=Tripleurospermum p. (Rupr.)
                Pobed.
            SY=M. m. var. nana (Hook.) Boivin
    matricarioides (Less.) Porter
            SY=Chamomilla suaveolens (Pursh)
                Rydb.
            SY=Lepidanthus s. (Pursh) Nutt.
            SY=M. s. (Pursh) Buch.
            SY=Santolina s. Pursh
            SY=Tanacetum s. (Pursh) Hook.
    occidentalis Greene
            SY=Chamomilla o. (Greene) Rydb.
    perforata Merat
            SY=Chamomilla inodora (L.) Gilib.
            SY=M. i. L.
            SY=M. maritima L. ssp. i. (L.)
                Clapham
            SY=M. m. var. agrestis (Knaf)
                Wilmott
            SY=Tripleurospermum i. (L.)
                Schultz-Bip.
    suffruticosa (L.) Druce

Megalodonta Greene

    beckii (Torr. ex Spreng.) Greene
        var. beckii
            SY=Bidens b. Torr. ex Spreng.
        var. hendersonii Sherff
        var. oregonensis Sherff

Melampodium L.

    cinereum DC.
            var. cinereum
            var. hirtellum Stuessy
            var. ramosissimum (DC.) Gray
    divaricatum (L.C. Rich.) DC.
    leucanthum Torr. & Gray
    longicorne Gray
    perfoliatum (Cav.) H.B.K.
    sericeum Lag.
            SY=M. hispidum H.B.K.

Melanthera Rohr

    angustifolia A. Rich.
    aspera (Jacq.) Small
        var. aspera
            SY=M. canescens (Kuntze) O.E.
                Schulz
            SY=M. deltoidea Michx.
            SY=M. montana O.E. Schulz
        var. glabriuscula (Kuntze) J.C. Parks
            SY=M. calcicola Britt.
            SY=M. confusa Britt.
    ligulata Small
    nivea (L.) Small
            SY=M. hastata Michx.
    parvifolia Small
            SY=M. radiata Small

Micropus L.

californicus Fisch. & Mey.
        var. californicus
        var. subvestitus Gray

Microseris D. Don

    acuminata Greene
    bigelovii (Gray) Schultz-Bip.
    borealis (Bong.) Schultz-Bip.
            SY=Apargidium b. (Bong.) Torr. &
                Gray
    campestris Greene
    decipiens Chambers
    douglasii (DC.) Schultz-Bip.
        ssp. douglasii
        ssp. platycarpha (Gray) Chambers
        ssp. tenella (Gray) Chambers
    elegans Greene ex Gray
    heterocarpa (Nutt.) Chambers
    howellii Gray
    laciniata (Hook.) Schultz-Bip.
        ssp. detlingii Chambers
        ssp. laciniata
            SY=Scorzonella l. (Hook.)
                Schultz-Bip.
            SY=S. l. var. pratensis (Greene)
                Jepson
            SY=S. procera Greene
        ssp. leptosepala (Nutt.) Chambers
            SY=Scorzonella laciniata (Hook.)
                Schultz-Bip. var. bolanderi
                (Gray) Jepson
            SY=S. leachiana M.F. Peck
            SY=S. leptosepala Nutt.
    lindleyi (DC.) Gray
            SY=M. linearifolia (DC.)
                Schultz-Bip.
            SY=Uropappus linearifolius (DC.)
                Nutt.
    nutans (Hook.) Schultz-Bip.
        ssp. nutans
            SY=Ptilocalais n. (Hook.) Greene
            SY=Scorzonella n. Hook.
            SY=S. n. var. major (Gray) M.E.
                Peck
        ssp. siskiyouensis Chambers
    paludosa (Greene) J.T. Howell
    sylvatica (Benth.) Schultz-Bip.

Mikania Willd.

    cordifolia (L. f.) Willd.
    fragilis Urban
    micrantha Kunth
            var. congesta (DC.) B.L. Robins.
            SY=M. c. DC.
            var. micrantha
    odoratissima Urban
    pachyphylla Urban
    porosa Urban
    scandens (L.) Willd.
            SY=M. batatifolia DC.
            SY=M. s. var. pubescens (Muhl.)
                Torr. & Gray
    stevensiana Britt.

Monolopia DC.

    gracilens Gray
    lanceolata Nutt.
    major DC.
    stricta Crum

Monoptilon Torr. & Gray

    bellidiforme Torr. & Gray ex Gray
    bellioides (Gray) Hall

Montanoa Ll. & Lex.

    hibiscifolia (Benth.) K. Koch

Munzothamnus Raven

    blairii (Munz & Johnston) Raven
        SY=Malacothrix b. (Munz &
          Johnston) Munz & Johnston

Mycelis Cass.

    muralis (L.) Dumort.
        SY=Lactuca m. (L.) Fresn.

Neurolaena R. Br.

    lobata (L.) Cass.

Nicolletia Gray

    edwardsii Gray
    occidentalis Gray

Nothocalais (Gray) Greene

    alpestris (Gray) Chambers
        SY=Agoseris a. (Gray) Greene
        SY=Microseris a. (Gray) Q. Jones
          ex Cronq.
    cuspidata (Pursh) Greene
        SY=Agoseris c. (Pursh) Raf.
        SY=Microseris c. (Pursh)
          Schultz-Bip.
    nigrescens (Henderson) Heller
        SY=Microseris n. Henderson
    troximoides (Gray) Greene
        SY=Microseris t. Gray
        SY=Scorzonella t. (Gray) Jepson

Onopordum L.

    acanthium L.
    tauricum Willd.

Orochaenactis Coville

    thysanocarpha (Gray) Coville

Osmadenia Nutt.

    tenella Nutt.
        SY=Calycadenia t. (Nutt.) Torr. &
          Gray
        SY=Hemizonia t. (Nutt.) Gray

Palafoxia Lag.

    arida B.L. Turner & M.I. Morris
        var. arida
        var. gigantea (M.E. Jones) B.L.
          Turner & M.I. Morris
          SY=P. linearis (Cav.) Lag. var. g.
            M.E. Jones
    callosa (Nutt.) Torr. & Gray
        SY=Othake c. (Nutt.) Bush
        SY=Palafoxia c. var. bella (Cory)
          Shinners
        SY=Polypteris c. (Nutt.) Gray
    feayi Gray
    hookeriana Torr. & Gray
        var. hookeriana
        var. minor Shinners
    integrifolia (Nutt.) Torr. & Gray
        SY=Polypteris i. Nutt.
    linearis (Cav.) Lag.
    reverchonii (Bush) Cory

riograndensis Cory
rosea (Bush) Cory
    var. macrolepis (Rydb.) B.L. Turner
        SY=Palafoxia m. (Rydb.) Cory
        SY=P. texana DC. var. m. (Rydb.)
          Shinners
    var. rosea
sphacelata (Nutt. ex Torr.) Cory
        SY=Othake s. (Nutt. ex Torr.)
          Rydb.
        SY=Polypteris s. (Nutt. ex Torr.)
          Trel.
texana DC.
    var. ambigua (Shinners) B.L. Turner
    var. texana
        SY=Palafoxia rosea (Bush) Cory
          var. papposa Shinners

Parthenice Gray

    mollis Gray

Parthenium L.

    alpinum (Nutt.) Torr. & Gray
        SY=Bolophyta a. Nutt.
    argentatum Gray
    confertum Gray
        var. confertum
        var. divaricatum Rollins
        var. intermedium Mears ined.
        var. lyratum (Gray) Rollins
          SY=P. hysterophorus L. var. l.
            Gray
          SY=P. l. (Gray) Gray
        var. microcephalum Rollins
    hysterophorus L.
        SY=P. lobatum Buckl.
    incanum H.B.K.
    integrifolium L.
        var. auriculatum (Britt.) Cornelius
          ex Cronq.
          SY=P. a. Britt.
          SY=P. hispidum Raf. var. a.
            (Britt.) Rollins
        var. henryanum Mears
        var. hispidum (Raf.) Mears
          SY=P. h. Raf.
          SY=P. repens Eggert
        var. integrifolium
        var. mabryanum Mears
    ligulatum (M.E. Jones) Barneby
        SY=P. alpinum (Nutt.) Torr. & Gray
          var. l. M.E. Jones
    radfordii Mears
    tetraneuris Barneby
        SY=P. alpinum (Nutt.) Torr. & Gray
          var. t. (Barneby) Rollins

Pectis L.

    angustifolia Torr.
        var. angustifolia
        var. fastigiata (Gray) Keil
        var. tenella (DC.) Keil
          SY=P. t. DC.
    carthusianorum Less.
    ciliaris L.
    coulteri Harvey & Gray
    cylindrica (Fern.) Rydb.
    febrifuga van Hall
    filipes Harvey & Gray
        var. filipes
        var. subnuda Fern.
    floribunda A. Rich.
    humifusa Sw.
    imberbis Gray

Pectis (CONT.)
  leptocephala (Cass.) Urban
  linearifolia Urban
  linifolia L.
  longipes Gray
  papposa Harvey & Gray
      var. grandis Keil
      var. papposa
  portoricensis Urban
  prostrata Cav.
      var. prostrata
      var. urceolata Fern.
  rusbyi Greene ex Gray
      SY=P. palmeri S. Wats.
  tenuicaulis Urban

Pentachaeta Nutt.

  alsinoides Greene
      SY=Chaetopappa a. (Greene) Keck
  aurea Nutt.
      SY=Chaetopappa a. (Nutt.) Keck
      SY=P. orcuttii Gray
      SY=P. paleacea Greene
  bellidiflora Greene
      SY=Chaetopappa b. (Greene) Keck
      SY=P. exilis Gray var. grayi
         Jepson
  exilis Gray
    ssp. aeolica Van Horn & Ornduff
    ssp. exilis
      SY=Aphantochaeta e. Gray
      SY=Chaetopappa e. (Gray) Keck
      SY=P. aphantochaeta Greene
      SY=P. e. var. a. Gray
      SY=P. e. var. discoidea Gray
  fragilis Brandeg.
      SY=Chaetopappa f. (Brandeg.) Keck
  lyonii Gray
      SY=Chaetopappa l. (Gray) Keck

Pentzia Thunb.

  incana (Thunb.) Kuntze

Pericome Gray

  caudata Gray
      SY=P. c. var. glandulosa (Goodman)
         Harrington

Perityle Benth.

  aglossa Gray
  angustifolia (Gray) Shinners
      SY=Laphamia a. Gray
      SY=L. a. ssp. laciniata (Gray)
         Niles
  bisetosa (Torr. ex Gray) Shinners
      var. appressa Powell
      var. bisetosa
        SY=Laphamia b. Torr. ex Gray
      var. scalaris Powell
  californica Benth.
      SY=P. plumigera Harvey & Gray
  cernua (Greene) Shinners
      SY=Laphamia c. Greene
  ciliata (L.H. Dewey) Rydb.
  cinerea (Gray) Powell
      SY=Laphamia c. Gray
  cochisensis (Niles) Powell
      SY=Laphamia c. Niles
  congesta (M.E. Jones) Shinners
      SY=Laphamia c. M.E. Jones
  coronopifolia Gray
  dissecta (Torr.) Gray
      SY=Laphamia d. Torr.

  emoryi Torr.
      SY=P. e. var. nuda (Torr.) Gray
  gilensis (M.E. Jones) J.F. Macbr.
      var. gilensis
        SY=Laphamia g. M.E. Jones
        SY=L. g. ssp. longilobus Niles
      var. salensis Powell
  gracilis (M.E. Jones) Rydb.
      SY=Laphamia fastigiata Brandeg.
      SY=L. g. M.E. Jones
      SY=P. f. (Brandeg.) Shinners
  inyoensis (Ferris) Powell
      SY=Laphamia i. Ferris
  lemmonii (Gray) J.F. Macbr.
  lindheimeri (Gray) Shinners
      var. halimifolia (Gray) Powell
        SY=Laphamia h. Gray
        SY=P. h. (Gray) Shinners
      var. lindheimeri
        SY=Laphamia halamifolia Gray ssp.
          l. (Gray) Niles
        SY=L. rotundata Rydb.
        SY=P. r. (Rydb.) Shinners
  megalocephala (S. Wats.) J.F. Macbr.
      var. intricata (Brandeg.) Powell
        SY=Laphamia i. Brandeg.
        SY=Laphamia m. ssp. i. (Brandeg.)
          Keck
        SY=P. i. (Brandeg.) Shinners
      var. megalocephala
        SY=Laphamia m. S. Wats.
      var. oligophylla Powell
  microglossa Benth.
      var. microglossa
        SY=P. spilanthoides (Schultz-Bip.)
          Rydb.
      var. saxosa (Brandeg.) Powell
  parryi Gray
  quinqueflora (Steyermark) Shinners
      SY=Laphamia q. Steyermark
  rupestris (Gray) Shinners
      var. albiflora Powell
      var. rupestris
        SY=Laphamia r. Gray
  saxicola (Eastw.) Shinners
      SY=Laphamia s. Eastw.
  stansburii (Gray) J.F. Macbr.
      SY=Laphamia s. Gray
  staurophylla (Barneby) Shinners
      SY=Laphamia s. Barneby
  tenella (M.E. Jones) J.F. Macbr.
      SY=Laphamia palmeri Gray
      SY=L. p. var. t. M.E. Jones
  turneri Powell
  vaseyi Coult.
  villosa (Blake) Shinners
      SY=Laphamia v. Blake
  vitreomontana Warnock
  warnockii Powell

Petasites P. Mill.

  frigidus (L.) Fries
      var. frigidus
        SY=P. alaskanus Rydb.
        SY=P. corymbosus (R. Br.) Rydb.
        SY=P. f. var. c. (R. Br.) Cronq.
        SY=P. f. var. hyperboreoides
          Hulten
        SY=P. gracilis Britt.
      var. nivalis (Greene) Cronq.
        SY=P. hyperboreus Rydb.
        SY=P. n. Greene
        SY=P. palmatus (Ait.) Gray var.
          frigidus Macoun
        SY=P. trigonophylla Greene
        SY=P. vitifolius Greene

Petasites (CONT.)
            var. palmatus (Ait.) Cronq.
                SY=P. arcticus Porsild
                SY=P. hookerianus (Nutt.) Rydb.
                SY=P. p. (Ait.) Gray
                SY=P. speciosus (Nutt.) Piper
        hybridus (L.) Gaertn., Mey. & Scherb.
                SY=P. vulgaris Hill
        japonicus (Sieb. & Zucc.) Maxim.
        sagittatus (Banks ex Pursh) Gray
                SY=P. dentatus Blank.
        X warrenii St. John [frigidus var. palmatus
            X sagittatus]

Petradoria Greene

        discoidea L.C. Anders.
                SY=Chrysothamnus gramineus Hall
        pumila (Nutt.) Greene
            ssp. graminea (Woot. & Standl.) L.C.
            Anders.
                SY=Solidago g. (Woot. & Standl.)
                        Blake
            ssp. pumila
                SY=Solidago petradoria Blake

Peucephyllum Gray

        schottii Gray

Phalacroseris Gray

        bolanderi Gray
                var. bolanderi
                var. coronata Hall

Phoebanthus Blake

        grandiflora (Torr. & Gray) Blake
        tenuifolia (Torr. & Gray) Blake

Picradeniopsis Rydb. ex Britt.

        oppositifolia (Nutt.) Rydb.
                SY=Bahia o. (Nutt.) Gray
        woodhousei (Gray) Rydb.
                SY=Bahia w. (Gray) Gray

Picris L.

        echioides L.
        hieracioides L.
            ssp. hieracioides
                var. alpina koidzumi
                var. hieracioides
            ssp. kamtschatica (Ledeb.) Hulten
                SY=P. k. Ledeb.
        sprengeriana (L.) Poir.

Pinaropappus Less.

        parvus Blake
        roseus (Less.) Less.
                var. foliosus (Heller) Shinners
                var. roseus

Piptocarpha R. Br.

        tetrantha Urban

Piptocoma Cass.

        antillana Urban
                SY=P. rufescens Cass.

Pityopsis Nutt.

adenolepis (Fern.) Semple ined.
                SY=Chryopsis a. Fern.
                SY=Heterotheca a. (Fern.) Ahles
        aspera (Shuttlw.) Small
                SY=Chrysopsis a. Shuttlw.
                SY=Heterotheca graminifolia
                        (Michx.) Shinners var. a.
                        (Shuttlw.) Gray
                SY=H. a. (Shuttlw.) Shinners
        falcata (Pursh) Small
                SY=Chrysopsis f. (Pursh) Ell.
        flexuosa (Nash) Small
                SY=Chrysopsis f. Nash
                SY=Heterotheca f. (Nash) Harms
        graminifolia (Michx.) Nutt.
            var. graminifolia
                SY=Chrysopsis g. (Michx.) Ell.
                SY=C. g. var. latifolia Fern.
                SY=C. nervosa (Willd.) Fern.
                SY=C. tracyi Small
                SY=Heterotheca g. (Michx.)
                        Shinners
                SY=H. g. var. t. (Small) P.W. Long
                SY=H. n. (Willd.) Shinners
                SY=P. t. (Small) Small
            var. microcephala (Small) Semple
            ined.
                SY=Chrysopsis m. Small
                SY=Heterotheca g. var. m. (Small)
                        Cronq.
                SY=H. m. (Small) Shinners
                SY=H. nervosa (Willd.) Shinners
                        var. m. (Small) Shinners ex
                        Ahles
                SY=P. m. (Small) Small
        oligantha (Chapman ex Torr. & Gray) Small
                SY=Chrysopsis o. Chapman ex Torr.
                        & Gray
                SY=Heterotheca o. (Chapman ex
                        Torr. & Gray) Harms
        pinifolia (Ell.) Nutt.
                SY=Chrysopsis p. Ell.
                SY=Heterotheca p. (Ell.) Ahles
        ruthii (Small) Small
                SY=Chrysopsis r. Small
                SY=Heterotheca r. (Small) Harms

Platyschkuhria Rydb.

        integrifolia (Gray) Rydb.
                var. desertorum (M.E. Jones) Ellison
                var. integrifolia
                SY=Bahia nudicaulis Gray
                var. oblongifolia (Gray) Ellison
                SY=Bahia o. (Gray) Gray
                var. ourolepis (Blake) Ellison
                SY=Bahia o. Blake

Pleurocoronis King & H.E. Robins.

        pluriseta (Gray) King & H.E. Robins.
                SY=Hofmeisteria p. Gray

Pluchea Cass.

        adnata (Humb. & Bonpl.) C. Mohr
        camphorata (L.) DC.
                SY=P. petiolata Cass.
                SY=P. viscida (Raf.) House
        foetida (L.) DC.
                var. foetida
                SY=P. tenuifolia Small
                var. imbricata Kearney
                SY=P. i. (Kearney) Nash
        X fosbergii Cooperrider & Galang
            [caroliniana X indica]
        indica (L.) Less.

Pluchea (CONT.)
  longifolia Nash
  odorata (L.) Cass.
        var. odorata
            SY=P. purpurascens (Sw.) DC.
        var. succulenta (Fern.) Cronq.
            SY=P. purpurascens (Sw.) DC. var.
                s. Fern.
  rosea Godfrey
  suaveolens (Vell.) Kuntze
            SY=P. quitoc DC.
  symphytifolia (P. Mill.) Gillis
            SY=Conyza carolinensis Jacq.
            SY=C. cortesii H.B.K.
            SY=C. s. P. Mill.
            SY=P. carolinensis (Jacq.) Sweet
            SY=P. cortesii (H.B.K.) DC.
            SY=P. odorata sensu auctt. non
                Cass.

Plummera Gray

  ambigens Blake
  floribunda Gray

Polymnia L.

  canadensis L.
            SY=P. c. var. radiata Gray
            SY=P. r. (Gray) Small
  laevigata Beadle
  uvedalia L.
        var. densipilis Blake
        var. floridana Blake
        var. uvedalia
            SY=Smallanthus u. (L.) Mackenzie

Porophyllum Adans.

  gracile Benth.
            SY=P. junciforme Greene
  greggii Gray
  leiocarpum (Urban) Rydb.
  ruderale (Jacq.) Cass.
      ssp. macrocephalum (DC.) R.R. Johnson
            SY=P. m. DC.
            SY=P. r. var. m. (DC.) Cronq.
      ssp. ruderale
            SY=P. porophyllum (L.) Kuntze
  scoparium Gray

Prenanthella Rydb.

  exigua (Gray) Rydb.
            SY=Lygodesmia e. Gray

Prenanthes L.

  alata (Hook.) D. Dietr.
            SY=P. lessingii Hulten
  alba L.
            SY=Nabalus a. (L.) Hook.
  altissima L.
            SY=Nabalus a. (L.) Hook.
            SY=P. a. var. cinnamomea Fern.
  aspera Michx.
            SY=Nabalus a. (Michx.) Torr. &
                Gray
  autumnalis Walt.
            SY=Nabalus virgatus (Michx.) DC.
            SY=P. v. Michx.
  barbata (Torr. & Gray) Milstead
            SY=P. serpentaria Pursh var. b.
                Torr. & Gray
  boottii (DC.) Gray
            SY=Nabalus b. DC.
  crepidinea Michx.

            SY=Nabalus c. (Michx.) DC.
  X mainensis Gray [racemosa X trifoliolata]
  nana (Bigelow) Torr.
            SY=Nabalus n. (Bigelow) DC.
            SY=P. trifoliolata (Cass.) Fern.
                var. n. (Bigelow) Fern.
  racemosa Michx.
      ssp. multiflora Cronq.
      ssp. racemosa
            SY=Nabalus r. (Michx.) DC.
  roanensis (Chickering) Chickering
            SY=Nabalus cylindrica Small
            SY=N. r. Chickering
            SY=P. c. (Small) F.L. Braun
  sagittata (Gray) A. Nels.
  serpentaria Pursh
            SY=Nabalus integrifolius Cass.
            SY=N. s. (Pursh) Hook.
            SY=P. i. (Cass.) Small
  trifoliolata (Cass.) Fern.
            SY=Nabalus t. Cass.

Prionopsis Nutt.

  ciliata Nutt.
            SY=Haplopappus c. (Nutt.) DC.

Proustia Lag.

  krugiana Urban

Psathyrotes Gray

  annua (Nutt.) Gray
            SY=Bulbostylis a. Nutt.
  pilifera Gray
  ramosissima (Torr.) Gray
  scaposa Gray

Pseudobahia Rydb.

  bahiifolia (Benth.) Rydb.
  heermannii (Dur.) Rydb.
  peirsonii Munz

Pseudoclappia Rydb.

  arenaria Rydb.
  watsonii Powell & B.L. Turner

Psilocarphus Nutt.

  brevissimus Nutt.
        var. brevissimus
        var. multiflorus Cronq.
  elatior Gray
  oregonus Nutt.
  tenellus Nutt.
        var. tenellus
        var. tenuis (Eastw.) Cronq.

Psilostrophe DC.

  bakeri Greene
  cooperi (Gray) Greene
  gnaphalodes DC.
  sparsiflora (Gray) A. Nels.
  tagetina (Nutt.) Greene
            SY=P. t. var. grandiflora (Rydb.)
                Heiser
            SY=P. t. var. lanata A. Nels.
  villosa Rydb.

Pterocaulon Ell.

  alopecuroideum (Lam.) DC.
  pycnostachyum (Michx.) Ell.

Pterocaulon   (CONT.)
    SY=P. undulatum (Walt.) C. Mohr
  virgatum (L.) DC.

Pulicaria Gaertn.

  arabica (L.) Cass.
      SY=Vicoa auriculata sensu auctt.
        non Cass.
  dysenterica (L.) Bernh.
      SY=Inula d. L.
  paludosa Link
      SY=P. hispanica (Boiss.) Boiss.

Pyrrhopappus DC.

  carolinianus (Walt.) DC.
      SY=Sitilias c. (Walt.) Raf.
  georgianus Shinners
      SY=P. carolinianus (Walt.) DC.
        var. g. (Shinners) Ahles
  grandiflorus (Nutt.) Nutt.
      SY=P. scaposus DC.
  multicaulis DC.
    var. geiseri (Shinners) Northington
      SY=P. g. Shinners
    var. multicaulis
      SY=Sitilias m. (DC.) Greene
  rothrockii Gray

Rafinesquia Nutt.

  californica Nutt.
  neomexicana Gray

Raillardella (Gray) Benth.

  argentea (Gray) Gray
  muirii Gray
  pringlei Greene
  scabrida Eastw.
  scaposa (Gray) Gray

Rainiera Greene

  stricta (Greene) Greene
      SY=Luina s. (Greene) B.L. Robins.

Ratibida Raf.

  columnifera (Nutt.) Woot. & Standl.
      SY=Lepachys c. (Nutt.) Rydb.
      SY=R. columnaris (Sims) D. Don
      SY=R. columnaris var. pulcherrima
        (DC.) D. Don
      SY=Rudbeckia columnifera Nutt.
  peduncularis (Torr. & Gray) Barnh.
    var. peduncularis
    var. picta (Gray) Sharp
  pinnata (Vent.) Barnh.
      SY=Rudbeckia p. Vent.
  tagetes (James) Barnh.
      SY=Rudbeckia t. James

Reichardia Roth

  picroides (L.) Roth

Remya Hbd. ex Benth. & Hook.

  kauaiensis Hbd.
    var. kauaiensis
    var. magnifica Deg. & Sherff
    var. magnifolia Deg. & Sherff
  mauiensis Hbd.

Rhagadiolus Scop.

stellatus (L.) Gaertn.
      SY=R. edulis Willd.

Rigiopappus Gray

  leptocladus Gray

Rolandra Rottb.

  fruticosa (L.) Kuntze

Rudbeckia L.

  auriculata (Perdue) Kral
      SY=R. fulgida Ait. var. a. Perdue
  bicolor Nutt.
  californica Gray
    var. californica
    var. glauca Blake
    var. intermedia Perdue
  fulgida Ait.
    var. deamii (Blake) Perdue
      SY=R. d. Blake
    var. fulgida
      SY=R. acuminata C.L. Boynt. &
        Beadle
      SY=R. foliosa C.L. Boynt. & Beadle
      SY=R. tenax C.L. Boynt. & Beadle
      SY=R. truncata Small
    var. palustris (Eggert) Perdue
      SY=R. p. Eggert
    var. spathulata (Michx.) Perdue
      SY=R. s. Michx.
    var. speciosa (Wenderoth) Perdue
      SY=R. s. Wenderoth
    var. sullivantii (C.L. Boynt. &
      Beadle) Cronq.
      SY=R. speciosa Wenderoth var.
        sullivantii (C.L. Boynt. &
        Beadle) B.L. Robins.
      SY=R. sullivantii C.L. Boynt. &
        Beadle
    var. umbrosa (C.L. Boynt. & Beadle)
    Cronq.
      SY=R. chapmanii C.L. Boynt. &
        Beadle
      SY=R. u. C.L. Boynt. & Beadle
  graminifolia (Torr. & Gray) C.L. Boynt. &
  Beadle
  grandiflora (Sweet) DC.
    var. alismifolia (Torr. & Gray)
    Cronq.
      SY=R. a. Torr. & Gray
    var. grandiflora
  heliopsidis Torr. & Gray
  hirta L.
    var. angustifolia (T.V. Moore) Perdue
      SY=R. divergens T.V. Moore
      SY=R. floridana T.V. Moore var. a.
        T.V. Moore
    var. floridana (T.V. Moore) Perdue
      SY=R. f. T.V. Moore
    var. hirta
      SY=R. amplectens T.V. Moore
      SY=R. brittonii Small
      SY=R. h. var. b. (Small) Fern.
      SY=R. h. var. monticola (Small)
        Fern.
      SY=R. m. Small
    var. pulcherrima Farw.
      SY=R. h. var. corymbifera Fern.
      SY=R. h. var. lanceolata (Bisch.)
        Core
      SY=R. h. var. sericea (T.V. Moore)
        Fern.
      SY=R. h. var. serotina (Nutt.)
        Core

Rudbeckia (CONT.)
            SY=R. longipes T.V. Moore
            SY=R. sericea T.V. Moore
            SY=R. serotina Nutt.
            SY=R. serotina var. c. (Fern.)
                Fern. & Schub.
            SY=R. serotina var. lanceolata
                (Bisch.) Fern. & Schub.
            SY=R. serotina var. sericea (T.V.
                Moore) Fern. & Schub.
    laciniata L.
            var. ampla (A. Nels.) Cronq.
            var. bipinnata Perdue
            var. digitata (P. Mill.) Fiori
                SY=R. heterophylla Torr. & Gray
            var. gaspereauensis Fern.
            var. humilis Gray
            var. laciniata
                SY=R. l. var. hortensis Bailey
            var. montana (Gray) Perdue
                SY=R. m. Gray
                SY=R. occidentalis Nutt. var. m.
                    (Gray) Perdue
    maxima Nutt.
    missouriensis Engelm.
    mohrii Gray
    mollis Ell.
    nitida Nutt.
            var. nitida
                SY=R. glabra DC.
            var. texana Perdue
    occidentalis Nutt.
            var. alpicola (Piper) Cronq.
                SY=R. a. Piper
            var. occidentalis
    subtomentosa Pursh
    triloba L.
            var. pinnatiloba Torr. & Gray
                SY=R. p. (Torr. & Gray) Beadle
            var. rupestris (Chickering) Gray
                SY=R. beadlei Small pro parte
                SY=R. r. Chickering
                SY=R. t. var. b. (Small) Fern. pro
                    parte
            var. triloba

Sachsia Griseb.

    polycephala Griseb.
            SY=S. bahamensis Urban

Salmea DC.

    scandens (L.) DC.

Santolina L.

    chamaecyparissus L.

Sanvitalia Lam.

    abertii Gray
    ocymoides DC.
    procumbens Lam.

Sartwellia Gray

    flaveriae Gray
            SY=S. puberula Rydb.
    gypsophila Powell & B.L. Turner

Saussurea DC.

    alpina (L.) DC.
    americana D.C. Eat.
    angustifolia (Willd.) DC.
            var. angustifolia

            var. viscida (Hulten) Welsh
                SY=S. v. Hulten
            var. yukonensis Porsild
                SY=S. viscida Hulten var. y.
                    (Porsild) Hulten
    glomerata Poir.
    nuda Ledeb.
            var. densa (Hook.) Hulten
                SY=S. d. (Hook.) Rydb.
            var. nuda
    X tschuktschorum Lipsch. [angustifolia X
        nuda var. densa]
    weberi Hulten

Schkuhria Roth

    anthemoidea (DC.) Coult.
            var. anthemoidea
            var. wrightii (Gray) Heiser
                SY=S. wislizenii Gray var.
                    wrightii (Gray) Blake
    multiflora Hook. & Arn.
            SY=Bahia neomexicana (Gray) Gray
    wislizenii Gray
            var. frustrata Blake
            var. wislizenii

Sclerocarpus Jacq.

    africanus Jacq.
    uniserialis (Hook.) Hemsl.

Sclerolepis Cass.

    uniflora (Walt.) B.S.P.

Scolymus L.

    hispanicus L.
    maculatus L.

Scorzonera L.

    hispanica L.
    laciniata L.
            SY=Podospermum l. (L.) DC.

Senecio L.

    actinella Greene
            SY=S. a. var. mogollonicus
                (Greene) Greenm.
    amplectens Gray
            var. amplectens
                SY=Ligularia a. (Gray) W.A. Weber
            var. holmii (Greene) Harrington
                SY=Ligularia h. (Greene) W.A.
                    Weber
                SY=S. h. Greene
    ampullaceus Hook.
    anonymus Wood
            SY=S. smallii Britt.
    antennariifolius Britt.
    aphanactis Greene
    arizonicus Greene
    aronicoides DC.
    astephanus Greene
    atratus Greene
    atropurpureus (Ledeb.) Fedtsch.
            SY=S. a. ssp. frigidus (Richards.)
                Hulten
            SY=S. a. ssp. f. var. ulmeri
                (Steffen) Porsild
            SY=S. a. ssp. tomentosus (Kjellm.)
                Hulten
            SY=S. a. ssp. t. var. dentatus
                (Gray) Hulten

Senecio (CONT.)
            SY=S. f. (Richards.) Less.
            SY=S. kjellmanii Porsild
    aureus L.
            SY=Packera a. (L.) Love & Love
            SY=S. a. var. aquilonius Fern.
            SY=S. aureus var. gracilis (Pursh)
                Hook.
            SY=S. aureus var. intercursus
                Fern.
            SY=S. g. Pursh
    ?bartianus Heller
    bernardinus Greene
            SY=S. ionophyllus Greene var.
                sparsilobatus (Parish) Hall
    bigelovii Gray
        var. bigelovii
            SY=Ligularia b. (Gray) W.A. Weber
        var. hallii Gray
            SY=Ligularia b. (Gray) W.A. Weber
                var. h. (Gray) W.A. Weber
    blochmanae Greene
    bolanderi Gray
        var. bolanderi
        var. harfordii (Greenm.) T.M. Barkl.
            SY=S. h. Greenm.
    breweri Burtt-Davy
    californicus DC.
    cannabinaefolius Hook. & Arn.
    canus Hook.
            SY=S. convallium Greenm.
            SY=S. hallii Britt.
            SY=S. h. var. discoides W.A. Weber
            SY=S. harbourii Rydb.
            SY=S. howellii Greene
            SY=S. purshianus Nutt.
    cardamine Greene
    cineraria DC.
    clarkianus Gray
    clevelandii Greene
            SY=S. c. var. heterophyllus Hoover
    clivorum Maxim.
    confusus Britten
    congestus (R. Br.) DC.
            SY=S. c. var. palustris (L.) Fern.
            SY=S. c. var. tonsus Fern.
            SY=S. p. (L.) Hook.
    crassulus Gray
            SY=S. c. var. cusickii (Piper)
                Greenm. ex Peck
    crocatus Rydb.
            SY=S. c. var. wolfii Greenm.
    cruentus DC.
    cymbalaria Pursh
            SY=Packera fernaldii (Greenm.)
                Love & Love
            SY=P. resedifolius (Less.) Love &
                Love
            SY=S. r. Less.
    cymbalarioides Buek
            SY=S. subnudus DC.
            SY=S. moresbiensis (Calder &
                Taylor) Dougl. &
                Ruyle-Dougl.
    cynthioides Greene
            SY=S. wrightii Greenm.
    debilis Nutt.
            SY=S. fedifolius Rydb.
    dimorphophyllus Greene
        var. dimorphophyllus
            SY=S. heterodoxus Greene ex Rydb.
        var. intermedius T.M. Barkl.
        var. paysonii T.M. Barkl.
    douglasii DC.
        var. douglasii
            SY=S. d. var. tularensis Munz
        var. longilobus (Benth.) L. Benson

            SY=S. l. Benth.
            SY=S. warnockii Shinners
        var. monoensis (Greene) Jepson
            SY=S. m. Greene
    elegans L.
            SY=S. crepidineus Greene
    elmeri Piper
    eremophilus Richards.
        var. eremophilus
            SY=S. glauciifolius Rydb.
        var. kingii (Rydb.) Greenm.
            SY=S. ambrosioides Rydb.
            SY=S. k. Rydb.
        var. macdougalii (Heller) Cronq.
            SY=S. m. Heller
    ertterae T.M. Barkl.
    eurycephalus Torr. & Gray ex Gray
        var. eurycephalus
        var. lewisrosei (J.T. Howell) T.M.
            Barkl.
    fendleri Gray
            SY=S. f. var. lanatus Osterhout
    flettii Wieg.
    foetidus T.J. Howell
        var. foetidus
        var. hydrophiloides (Rydb.) T.M.
            Barkl. ex Cronq.
            SY=S. h. Rydb.
            SY=S. pereziifolius Rydb.
    franciscanus Greene
    fremontii Torr. & Gray
        var. blitoides (Greene) Cronq.
            SY=S. b. Greene
            SY=S. carthamoides Greene
        var. fremontii
        var. occidentalis Gray
    fuscatus Hayek
            SY=S. lindstroemii (Ostenf.)
                Porsild
    ganderi T.M. Barkl. & Beauchamp
    glabellus Poir.
    greenei Gray
    hartianus Heller
    hartwegii Benth.
            SY=S. seemanii Schultz-Bip. ex
                Seem.
    hesperius Greene
    huachucanus Gray
    hydrophilus Nutt.
    hyperborealis Greenm.
            SY=Packera h. Love & Love
            SY=P. ogotorukensis (Packer) Love
                & Love
            SY=S. conterminus Greenm.
            SY=S. o. Packer
    imparipinnatus Klatt
            SY=S. greggii Rydb.
    indecorus Greene
            SY=S. pauciflorus Pursh var.
                fallax (Greenm.) Greenm.
    integerrimus Nutt.
        var. exaltatus (Nutt.) Cronq.
            SY=S. columbianus Greene
            SY=S. e. Nutt.
            SY=S. hookeri Torr. & Gray
            SY=S. i. var. vaseyi (Greenm.)
                Cronq.
            SY=S. v. Greenm.
        var. integerrimus
        var. major (Gray) Cronq.
        var. ochroleucus (Gray) Cronq.
        var. scribneri (Rydb.) T.M. Barkl.
            SY=S. scribneri Rydb.
    ionophyllus Greene
    jacobaea L.
    layneae Greene
    lemmonii Gray

Senecio (CONT.)
  lugens Richards.
          SY=S. integerrimus Nutt. var. l.
                (Richards.) Boivin
  lyonii Gray
  macounii Greene
          SY=S. ligulifolius Greene
  megacephalus Nutt.
  mikanioides Otto ex Walp.
  millefolium Torr. & Gray
          SY=S. memmingeri Britt.
  millelobatus Rydb.
  mohavensis Gray
  multicapitatus Greenm.
  multilobatus Torr. & Gray ex Gray
          SY=S. lynceus Greene
          SY=S. l. var. leucoreus Blake
          SY=S. stygius Greene
          SY=S. thornberi Greenm.
          SY=S. uintahensis (A. Nels.)
                Greenm.
  neomexicanus Gray
      var. metcalfei (Greene) T.M. Barkl.
      var. mutabilis (Greene) T.M. Barkl.
          SY=S. m. Greene
      var. neomexicanus
          SY=S. eurypterus Greenm.
          SY=S. n. var. griffithsii Greenm.
          SY=S. oresbius Greenm.
          SY=S. thurberi Gray
      var. toumeyi (Greene) T.M. Barkl.
  neowebsteri Blake
          SY=S. websteri Greenm.
  newcombei Greene
  obovatus Muhl. ex Willd.
          SY=S. o. var. elliottii (Torr. &
                Gray) Fern.
          SY=S. o. var. rotundus Britt.
          SY=S. r. (Britt.) Small
  parryi Gray
  pattersonianus Hoover
  pauciflorus Pursh
          SY=S. discoideus (Hook.) Britt.
  pauperculus Michx.
          SY=S. gaspensis Greenm.
          SY=S. g. var. firmifolius
                (Greenm.) Fern.
          SY=S. p. var. balsamitae (Muhl. ex
                Willd.) Fern.
          SY=S. p. var. crawfordii (Britt.)
                T.M. Barkl.
          SY=S. p. var. f. (Greenm.) Greenm.
          SY=S. p. var. neoscoticus Fern.
          SY=S. p. var. praelongus (Greenm.)
                House
          SY=S. p. var. thompsoniensis
                (Greenm.) Boivin
          SY=S. tweedyi Rydb.
  plattensis Nutt.
          SY=S. pseudotomentosus Mackenzie &
                Bush
  porteri Greene
  pseudaureus Rydb.
      var. flavulus (Greene) Greenm.
          SY=S. f. Greene
      var. pseudoaureus
          SY=S. pauciflorus Pursh var.
                jucundulus Jepson
      var. semicordatus (Mackenzie & Bush)
                T.M. Barkl.
          SY=S. aureus L. var. s. (Mackenzie
                & Bush) Greenm.
  pseudoarnica Less.
  pudicus Greene
          SY=Ligularia p. (Greene) W.A.
                Weber
  quaerens Greene

  quercetorum Greene
  rapifolius Nutt.
  riddellii Torr. & Gray
          SY=S. spartioides Torr. & Gray
                var. fremontii (Torr. &
                Gray) Greenm. ex L.O.
                Williams
          SY=S. s. var. parksii (Cory)
                Shinners
  sacramentanus Woot. & Standl.
  salignus DC.
  sandvicensis Less.
  sanguisorboides Rydb.
  schweinitzianus Nutt.
          SY=S. robbinsii Oakes ex Rusby
  scorzonella Greene
  serra Hook.
      var. admirabilis (Greene) A. Nels.
      var. serra
          SY=S. s. var. altior Jepson
  sheldonensis Porsild
  soldanella Gray
          SY=Ligularia s. (Gray) W.A. Weber
  spartioides Torr. & Gray
          SY=S. s. var. granularis Maguire &
                Holmgren ex Cronq.
  sphaerocephalus Greene
  squalidus L.
  streptanthifolius Greene
          SY=S. cymbalarioides (Torr. &
                Gray) Nutt.
          SY=S. c. var. borealis Greenm.
          SY=S. c. var. suksdorfii (Greenm.)
                M.E. Peck
          SY=S. leonardii Rydb.
          SY=S. oodes Rydb.
          SY=S. platylobus Rydb.
  sylvaticus L.
  taraxacoides (Gray) Greene
          SY=Ligularia t. (Gray) W.A. Weber
  tomentosus Michx.
          SY=S. alabamensis Britt. ex Small
  triangularis Hook.
          SY=S. t. var. angustifolius G.N.
                Jones
  tridenticulatus Rydb.
          SY=S. acutidens Rydb.
          SY=S. densus Greene
          SY=S. manitobensis Greenm.
  viscosus L.
  vulgaris L.
  werneriifolius (Gray) Gray
          SY=S. molinarius Greenm.
          SY=S. muirii Greenm.
          SY=S. saxosus Klatt
          SY=S. s. var. toiyabensis Greenm.
          SY=S. w. var. incertus Greenm.
  wootonii Greene
  yukonensis Porsild

Shinnersoseris S. Tomb

  rostrata (Gray) S. Tomb
          SY=Lygodesmia r. Gray

Sigesbeckia L.

  orientalis L.

Silphium L.

  albiflorum Gray
  asperrimum Hook.
          SY=S. dentatum Ell. var. gatesii
                (C. Mohr) Ahles
          SY=S. g. C. Mohr
          SY=S. radula Nutt.

Silphium (CONT.)
    asteriscus L.
        ssp. angustatum (Gray) T.R. Fisher &
            Speer ined.
                SY=S. asteriscus var. angustatum
                    Gray
        ssp. asteriscus
                SY=S. a. var. scabrum Nutt.
        ssp. dentatum (Ell.) T.R. Fisher & Speer
            ined.
                SY=S. a. var. laevicaule DC.
                SY=S. d. Ell.
                SY=S. elliottii Small
                SY=S. incisum Greene
                SY=S. nodum Small
        ssp. latifolium (Gray) Weber & T.R.
            Fisher ined.
                SY=S. confertifolium Small
                SY=S. glabrum Eggert
        ssp. trifoliatum (Ell.) Weber & T.R.
            Fisher ined.
                SY=S. atropurpureum Retz.
                SY=S. laevigatum Ell. non Pursh
                SY=S. t. L.
                SY=S. t. var. latifolium Gray
    brachiatum Gattinger
    compositum Michx.
        ssp. compositum
                SY=S. orae Small
        ssp. ovatifolium (Torr. & Gray) Sweeny &
            T.R. Fisher
                SY=S. o. (Torr. & Gray) Small
        ssp. reniforme (Raf. ex Nutt.) Sweeny &
            T.R. Fisher
                SY=S. c. var. r. (Raf. ex Nutt.)
                    Torr. & Gray
                SY=S. r. Raf. ex Nutt.
        ssp. venosum (Small) Sweeny & T.R.
            Fisher
                SY=S. lapsuum Small
                SY=S. v. Small
    gracile Gray
                SY=S. s. var. wrightii Perry
    integrifolium Michx.
            var. deamii Perry
            var. gattingeri Perry
            var. integrifolium
                SY=S. laevigatum Pursh
            var. laeve Torr. & Gray
            var. neglectum Settle & T.R. Fisher
    laciniatum L.
            var. laciniatum
            var. robinsonii Perry
    mohrii Small
    perfoliatum L.
        ssp. connatum (L.) Cruden
                SY=S. c. L.
                SY=S. p. var. c. (L.) Cronq.
        ssp. perfoliatum
    pinnatifidum Ell.
                SY=S. terebinthinaceum Jacq. var.
                    p. (Ell.) Gray
    reverchonii Bush
    scaberrimum Ell.
    simpsonii Greene
    speciosum Nutt.
    terebinthinaceum Jacq.
            var. lucy-brauniae Steyermark
            var. terebinthinaceum
                SY=S. rumicifolium Small

Silybum Adans.

    marianum (L.) Gaertn.
                SY=Mariana m. (L.) Hill

Simsia Pers.

calva (Engelm. & Gray) Gray
lagasciformis DC.
        SY=S. exaristata Gray
        SY=S. e. var. perplexa Blake

Solidago L.

albopilosa E.L. Braun
?anticostensis Fern.
arguta Ait.
    var. arguta
        SY=S. tarda Mackenzie
    var. boottii (Hook.) Palmer &
        Steyermark
        SY=S. a. ssp. b. (Hook.) G. Morton
        SY=S. a. var. neurolepis (Fern.)
            Steyermark
        SY=S. a. var. strigosa (Small)
            Steyermark
        SY=S. austrina Small
        SY=S. b. Hook.
        SY=S. b. var. brachyphylla
            (Chapman) Gray
        SY=S. boottii var. ludoviciana
            Gray
        SY=S. brachyphylla Chapman
        SY=S. dispersa Small
        SY=S. flavovirens Chapman
        SY=S. ludoviciana (Gray) Small
        SY=S. n. Fern.
        SY=S. perlonga Fern.
        SY=S. s. Small
    var. caroliniana Gray
        SY=S. a. ssp. c. (Gray) G. Morton
        SY=S. yadkinensis (Porter) Small
    var. harrissii (Steele) Cronq.
        SY=S. h. Steele
    ssp. ?pseudoyadkinensis G. Morton
arizonica (Gray) Woot. & Standl.
        SY=S. canadensis L. var. a. Gray
        SY=S. howellii Woot. & Standl.
X asperula Desf. [rugosa X sempervirens]
auriculata Shuttlw.
        SY=S. amplexicaulis Torr. & Gray
        SY=S. notabilis Mackenzie
X beaudryi Boivin [rugosa X uliginosa]
X bernardii Boivin [ptarmicoides X
    riddellii]
bicolor L.
buckleyi Torr. & Gray
caesia L.
        SY=S. c. var. axillaris Gray
        SY=S. flaccidifolia Small
        SY=S. latissimifolia P. Mill.
calcicola Fern.
californica Nutt.
canadensis L.
    var. bartramiana (Fern.) Beaudry
        SY=S. b. Fern.
    var. canadensis
    var. gilvocanescens Rydb.
        SY=S. g. (Rydb.) Smyth
        SY=S. pruinosa Greene
    var. hargeri Fern.
    var. salebrosa (Piper) M.E. Jones
        SY=S. c. ssp. elongata (Nutt.)
            Keck
        SY=S. c. ssp. s. (Piper) Keck
        SY=S. c. var. e. (Nutt.) M.E. Peck
        SY=S. e. Nutt.
        SY=S. lepida DC. var. e. (Nutt.)
            Fern.
        SY=S. l. var. fallax Fern.
    var. scabra (Muhl.) Torr. & Gray
        SY=S. altissima L.
        SY=S. a. var. canescens (Gray)
            M.C. Johnston

Solidago   (CONT.)
      SY=S. a. var. pluricephala M.C.
        Johnston
      SY=S. a. var. procera (Ait.) Fern.
      SY=S. hirsutissima P. Mill.
      SY=S. lunellii Rydb.
    var. subserrata (DC.) Cronq.
      SY=S. lepida DC.
      SY=S. l. var. molina Fern.
chlorolepis Fern.
confinis Gray
curtisii Torr. & Gray
    var. curtisii
      SY=S. lancifolia Torr. & Gray
      SY=S. monticola Torr. & Gray
    var. pubens (M.A. Curtis) Gray
      SY=S. p. M.A. Curtis
cutleri Fern.
deamii Fern.
drummondii Torr. & Gray
dumetorum Lunell
elliottii Torr. & Gray
    var. ascendens Fern.
    var. edisoniana Fern.
    var. elliottii
      SY=S. mirabilis Small
    var. pedicellata Fern.
erecta Pursh
X erskinei Boivin [canadensis X
  sempervirens]
X farwellii Fern. [purshii X uliginosa]
fistulosa P. Mill.
flexicaulis L.
    SY=S. latifolia L.
gattingeri Chapman
gigantea Ait.
    var. gigantea
    var. pitcheri (Nutt.) Shinners
    var. serotina (Ait.) Cronq.
      SY=S. g. var. leiophylla Fern.
      SY=S. s. Ait.
glomerata Michx.
gracillima Torr. & Gray
guiradonis Gray
harperi Mackenzie
hispida Muhl.
    var. arnoglossa Fern.
    var. hispida
      SY=S. bicolor L. var. concolor
        Torr. & Gray
      SY=S. b. var. ovalis Farw.
    var. lanata (Hook.) Fern.
      SY=S. bicolor L. var. l. (Hook.)
        Seymour
    var. tonsa Fern.
      SY=S. bicolor L. var. t. (Fern.)
        Boivin
houghtonii Torr. & Gray
juncea Ait.
    var. juncea
    var. neobohemica Fern.
X krotkovii Boivin
leavenworthii Torr. & Gray
X leiophallax Friesner [canadensis var.
  salebrosa X gigantea var. serotina]
X lutescens (Lindl. ex DC.) Boivin
  [ptarmicoides X riddellii]
      SY=Aster l. (Lindl. ex DC.) Torr.
        & Gray
      SY=A. ptarmicoides (Nees) Torr. &
        Gray var. l. (Lindl. ex DC.)
        Gray
      SY=Diplopappus l. Lindl. ex DC.
      SY=Unamia l. (Lindl. ex DC.) Rydb.
macrophylla Pursh
      SY=S. m. var. thyrsoidea (E. Mey.)
        Fern.

      SY=S. mensalis Fern.
X maheuxii Boivin [riddellii X rigida var.
  humilis]
missouriensis Nutt.
    var. extrareia Gray
    var. fasciculata Holz.
      SY=S. glaberrima Martens
      SY=S. g. var. moritura (Steele)
        Palmer & Steyermark
      SY=S. missouriensis var. g.
        (Martens) Rydb.
    var. missouriensis
    var. montana Gray
    var. tolmieana (Gray) Cronq.
mollis Bartl.
    var. angustata Shinners
    var. mollis
multiradiata Ait.
    var. arctica (DC.) Fern.
    var. multiradiata
    var. parviceps Fern.
    var. scopulorum Gray
      SY=S. ciliosa Greene
      SY=S. s. (Gray) A. Nels.
nana Nutt.
nemoralis Ait.
    var. longipetiolata (Mackenzie &
      Bush) Palmer & Steyermark
      SY=S. decemflora DC.
      SY=S. n. var. d. (DC.) Fern.
      SY=S. pulcherrima A. Nels.
    var. nemoralis
      SY=S. n. var. haleana Fern.
neomexicana (Gray) Woot. & Standl.
nitida Torr. & Gray
      SY=Oligoneuron n. (Torr. & Gray)
        Small
odora Ait.
    var. chapmanii (Torr. & Gray) Cronq.
      SY=S. c. Torr. & Gray
    var. odora
ohioensis Riddell
X ovata Friesner [sphacelata X ulmifolia]
patula Muhl.
    var. patula
    var. strictula Torr. & Gray
      SY=S. salicina Ell.
?pendula Small
petiolaris Ait.
    var. petiolaris
      SY=S. lindheimeriana Scheele
      SY=S. milleriana Mackenzie
    var. wardii (Britt.) Fern.
      SY=S. angusta Torr. & Gray
      SY=S. p. var. a. (Torr. & Gray)
        Gray
pinetorum Small
plumosa Small
porteri Small
ptarmicoides (Nees) Boivin
      SY=Asper p. (Nees) Torr. & Gray
      SY=A. p. var. georgianus Gray
      SY=Doellingeria p. Nees
      SY=Inula alba Nutt.
      SY=Unamia a. (Nutt.) Rydb.
puberula Nutt.
    var. puberula
    var. pulverulenta (Nutt.) Chapman
      SY=S. pulverulenta Nutt.
pulchra Small
purshii Porter
      SY=S. chrysolepis Fern.
radula Nutt.
    var. laeta (Greene) Fern.
    var. radula
    var. stenolepis Fern.
riddellii Frank

Solidago (CONT.)
    rigida L.
        var. glabrata E.L. Braun
            SY=Oligoneuron jacksonii (Kuntze)
                Small
            SY=S. corymbosa Ell.
            SY=S. j. (Kuntze) Fern.
        var. humilis Porteri
            SY=Oligoneuron canescens Rydb.
        var. rigida
            SY=Oligoneuron grandiflorum (Raf.)
                Small
            SY=O. r. (L.) Small
    roanensis Porter
            SY=S. maxonii Pollard
            SY=S. r. var. monticola (Torr. &
                Gray) Fern.
    rugosa Ait.
        var. aspera (Ait.) Fern.
            SY=S. a. Ait.
            SY=S. celtidifolia Small
            SY=S. r. ssp. a. (Ait.) Cronq.
            SY=S. r. var. c. (Small) Fern.
        var. rugosa
        var. sphagnophila Graves
            SY=S. aestivalis Bickn.
        var. villosa (Pursh) Fern.
    rupestris Raf.
    sciaphila Steele
    sempervirens L.
        var. mexicana (L.) Fern.
            SY=S. angustifolia Ell.
            SY=S. m. L.
            SY=S. petiolata P. Mill.
            SY=S. stricta Ait.
        var. sempervirens
    shortii Torr. & Gray
    simulans Fern.
    sparsiflora Gray
            SY=S. trinervata Greene
    spathulata DC.
        ssp. glutinosa (Nutt.) Keck
            SY=S. g. Nutt.
        ssp. randii (Porter) Cronq.
            var. racemosa (Greene) Gleason
                SY=S. racemosa Greene
            var. randii
                SY=S. glutinosa Nutt. var. r.
                    (Porter) Cronq.
                SY=S. r. (Porter) Britt.
        ssp. spathulata
            var. gillmanii (Gray) Cronq.
                SY=S. g. (Gray) Steele
                SY=S. racemosa Greene var. g.
                    (Gray) Fern.
            var. nana (Gray) Cronq.
                SY=S. decumbens Greene
                SY=S. d. var. oreophila (Rydb.)
                    Fern.
                SY=S. glutinosa Nutt. var. n.
                    (Gray) Cronq.
                SY=S. o. Rydb.
            var. neomexicana (Gray) Cronq.
            var. spathulata
            var. subcinerea Gray
    speciosa Nutt.
        var. jejunifolia (Steele) Cronq.
            SY=S. j. Steele
            SY=S. uliginosa Nutt. var. j.
                (Steele) Boivin
        var. pallida Porter
            SY=S. p. (Porter) Rydb.
        var. rigidiuscula Torr. & Gray
            SY=S. r. (Torr. & Gray) Porter
            SY=S. s. var. angustata Torr. &
                Gray
        var. speciosa

            SY=S. conferta P. Mill.
    spectabilis (D.C. Eat.) Gray
    sphacelata Raf.
            SY=Brachychaeta s. (Raf.) Britt.
    spithamaea M.A. Curtis
    squarrosa Muhl.
    tortifolia Ell.
    uliginosa Nutt.
        var. levipes (Fern.) Fern
            SY=S. uniligulata (DC.) Porter
                var. l. Fern.
        var. linoides (Torr. & Gray) Fern.
        var. peracuta (Fern.) Friesner
        var. terrae-novae (Torr. & Gray)
            Fern.
        var. uliginosa
            SY=S. neglecta Torr. & Gray
            SY=S. uniligulata (DC.) Porter
            SY=S. u. var. n. (Torr. & Gray)
                Fern.
    X ulmicaesia Friesner [caesia X ulmifolia]
    ulmifolia Muhl.
        var. microphylla Gray
            SY=S. delicatula Small
            SY=S. helleri Small
        var. palmeri Cronq.
        var. ulmifolia
    verna M.A. Curtis
    victorinii Fern.
    wrightii Gray
        var. adenophora Blake
            SY=S. bigelovii Gray
        var. wrightii

Soliva Ruiz & Pavon

    anthemifolia (Juss.) Less.
            SY=Gymnostyles a. Juss.
    daucifolia Nutt.
    mutisii H.B.K.
    pterosperma (Juss.) Less.
    sessilis Ruiz & Pavon
    stolonifera (Brot.) Loud.
            SY=Gymnostyles nasturtiifolia
                Juss.
            SY=S. n. (Juss.) DC.

Sonchus L.

    arvensis L.
        ssp. arvensis
            SY=S. a. var. glabrescens Guenth.,
                Grab. & Wimmer
            SY=S. a. var. shumovichii Bovin
        ssp. uliginosus (Bieb.) Nyman
            SY=S. u. Bieb.
    asper (L.) Hill
        ssp. asper
        ssp. glaucescens (Jord.) J. Ball
    oleraceus L.
    tenerrimus L.

Sphaeromeria Nutt.

    argentea Nutt.
            SY=Tanacetum nuttallii Torr. &
                Gray
    cana (D.C. Eat.) Heller
            SY=Tanacetum c. D.C. Eat.
    capitata Nutt.
            SY=Tanacetum c. (Nutt.) Torr. &
                Gray
    compacta (Hall) Holmgren, Schulz & Lowrey
            SY=Tanacetum c. Hall
    diversifolia (D.C. Eat.) Rydb.
            SY=Tanacetum d. D.C. Eat.
    potentilloides (Gray) Heller

Sphaeromeria (CONT.)
        var. nitrophila (Cronq.) Holmgren,
            Schulz & Lowry
            SY=Tanacetum p. (Gray) Gray var.
                n. Cronq.
        var. potentilloides
            SY=Tanacetum p. (Gray) Gray
    ruthiae Holmgren, Schulz & Lowrey
    simplex (A. Nels.) Heller
            SY=Tanacetum s. A. Nels.

Spilanthes Jacq.

    americana (Mutis) Hieron.
        var. americana
        var. repens (Walt.) A.H. Moore
        var. stolonifera (DC.) A.H. Moore
    iodiscaea A.H. Moore
    urens Jacq.

Stephanomeria Nutt.

    cichoriacea Gray
    cinerea (Blake) Blake
    diegensis Gottlieb
    elata Nutt.
    exigua Nutt.
        ssp. carotifera (Hoover) Gottlieb
            SY=S. c. Hoover
        ssp. coronaria (Greene) Gottlieb
            SY=S. e. var. c. (Greene) Jepson
        ssp. deanei (J.F. Macbr.) Gottlieb
            SY=S. e. var. d. J.F. Macbr.
        ssp. exigua
            SY=Lygodesmia bigelovii (Gray)
                Shinners
            SY=S. e. ssp. pentachaeta D.C.
                Eat.
            SY=S. e. var. p. (D.C. Eat.) Hall
        ssp. macrocarpa Gottlieb
    lactucina Gray
    lygodesmoides M.E. Jones
    malheurensis Gottlieb
    paniculata Nutt.
    parryi Gray
    pauciflora (Torr.) Nutt.
        var. parishii (Jepson) Munz
        var. pauciflora
            SY=Lygodesmia p. (Torr.) Shinners
            SY=Ptiloria p. (Torr.) Raf.
            SY=S. neomexicana (Greene) Cory
    runcinata Nutt.
            SY=Ptiloria ramosa Rydb.
    schottii Gray
    spinosa (Nutt.) S. Tomb
            SY=Lygodesmia s. Nutt.
    tenuifolia (Torr.) Hall
        var. myrioclada (D.C. Eat.) Cronq.
            SY=S. m. D.C. Eat.
        var. tenuifolia
            SY=Lygodesmia t. (Torr.) Shinners
    thurberi Gray
    virgata Benth.
        ssp. pleurocarpa (Greene) Gottlieb
        ssp. virgata
            SY=S. v. var. tomentosa (Greene)
                Munz

Stevia Cav.

    lemmonii Gray
    micrantha Lag.
    plummerae Gray
        var. alba Gray
        var. plummerae
    rhombifolia H.B.K.
    salicifolia Cav.

    serrata Cav.
        SY=S. s. var. ivifolia (Willd.)
            B.L. Robins.
    viscida H.B.K.

Stokesia L'Her.

    laevis (Hill) Greene

Struchium P. Br.

    sparganophorum (L.) Kuntze

Stylocline Nutt.

    amphibola (Gray) J.T. Howell
    filaginea (Gray) Gray
        var. depressa Jepson
        var. filaginea
    gnaphalioides Nutt.
    micropoides Gray
    psilocarphoides M.E. Peck

Synedrella Gaertn.

    nodiflora (L.) Gaertn.

Syntrichopappus Gray

    fremontii Gray
    lemmonii Gray

Tagetes L.

    erecta L.
    lemmonii Gray
    micrantha Cav.
    minuta L.
    patula L.

Tanacetum L.

    bipinnatum (L.) Schultz-Bip.
        ssp. bipinnatum
            SY=Chrysanthemum b. L.
        ssp. huronense (Nutt.) Breitung
            SY=Chrysanthemum b. L. ssp. h.
                (Nutt.) Hulten
            SY=T. h. Nutt.
            SY=T. h. var. bifarium Fern.
            SY=T. h. var. floccosum Raup
            SY=T. h. var. johannense Fern.
            SY=T. h. var. terrae-novae Fern.
    camphoratum Less.
    douglasii DC.
    parthenium (L.) Schultz-Bip.
            SY=Chrysanthemum p. (L.) Bernh.
            SY=Matricaria p. L.
    vulgare L.
            SY=Chrysanthemum v. (L.) Bernh.
            SY=T. v. var. crispum DC.

Taraxacum Wiggers

    amphiphron Bocher
    arcticum (Trautv.) Dahlst.
    arctogenum Dahlst.
    brachyceras Dahlst.
    californicum Munz & Johnston
    campylodes Hagl.
    carneocoloratum A. Nels.
    carthamopsis Porsild
    ceratophorum (Ledeb.) DC.
            SY=T. ambigens Fern.
            SY=T. a. var. flutius Fern.
            SY=T. dumetorum Greene
            SY=T. hyperboreum Dahlst.

Taraxacum   (CONT.)
        SY=T. longii Fern.
  croceum Dahlst.
        SY=T. lapponicum Kihlm. ex
              Hand.-Maz.
  curvidens M.P. Christens.
  cyclocentrum M.P. Christens.
  dahlstedtii Lindb. f.
  davidssonii M.P. Christens.
  devians Dahlst.
  eriophorum Rydb.
  eurylepium Dahlst.
  firmum Dahlst.
  hyparcticum Dahlst.
  integratum Hagl.
  islandiciforme Dahlst.
  kok-saghyz Rodin
  lacerum Greene
  laevigatum (Willd.) DC.
        SY=Leontodon erythrospermum
            (Andrz. ex Bess.) Britt.
        SY=T. e. Andrz. ex Bess.
  lateritium Dahlst.
  latilobum DC.
        SY=Leontodon l. (DC.) Britt.
  latispinulosum M.P. Christens.
  laurentianum Fern.
  lyratum (Ledeb.) DC.
        SY=T. alaskanum Rydb.
        SY=T. kamtschaticum Dahlst.
  malteanum Dahlst.
  naevosum Dahlst.
        SY=T. atroglaucum M.P. Christens.
        SY=T. dilutisquameum M.P.
            Christens.
  officinale Weber
        SY=Leontodon taraxacum L.
  palustre (Lyons) Symons
        SY=?T. officinale Weber var. p.
            (Sm.) Blytt
        SY=T. p. var. vulgare (Lam.) Fern.
  ?paucisquamosum M.E. Peck
  pellianum Porsild
  phymatocarpum J. Vahl
  pleniflorum M.P. Christens.
  pseudonorvegicum Dahlst.
  pumilum Dahlst.
  purpuridens Dahlst.
  retroflexum Lindb. f.
  rhodolepis Dahlst.
  scanicum Dahlst.
  scopulorum (Gray) Rydb.
  trigonolobum Dahlst.
  umbrinum Dahlst.
  undulatum Lindb. f. & Marklund
  vagans Hagl.

Tessaria Ruiz & Pavon

  sericea (Nutt.) Shinners
        SY=Pluchea s. (Nutt.) Cav.

Tetradymia DC.

  argyraea Munz & Roos
  axillaris A. Nels.
    var. axillaris
    var. longispina (M.E. Jones) Strother
        SY=T. spinosa Hook. & Arn. var. l.
            M.E. Jones
  canescens DC.
        SY=T. c. var. inermis (Rydb.)
            Payson
  comosa Gray
  filifolia Greene
  glabrata Torr. & Gray
  nuttallii Torr. & Gray

  spinosa Hook. & Arn.
  stenolepis Greene
  tetrameres (Blake) Strother
        SY=T. comosa Gray ssp. t. Blake

Tetragonotheca L.

  helianthoides L.
  ludoviciana (Torr. & Gray) Gray
  repanda (Buckl.) Small
  texana (Gray) Engelm. & Gray

Tetramolopium Nees

  arbusculum (Gray) Sherff
  arenarium (Gray) Hbd.
    var. arenarium
    var. confertum Sherff
    var. dentata Hbd.
  consanguineum (Gray) Hbd.
    var. consanguineum
    var. leptophyllum Sherff
  conyzoides (Gray) Hbd.
    var. conyzoides
    var. dentatum (Mann) Sherff
  filiforme Sherff
  humile (Gray) Hbd.
    var. humile
    var. skottsbergii Sherff
    var. sublaeve Sherff
  lepidotum (Less.) Sherff
    var. lepidotum
    var. luxurians (Hbd.) Sherff
  polyphyllum Sherff
  remyi (Gray) Hbd.
  tenerrimum (Less.) Nees

Thelesperma Less.

  ambiguum Gray
        SY=T. fraternum Shinners
        SY=T. megapotamicum var. a. (Gray)
            Shinners
  burridgeanum (Regel, Korn. & Rach.) Blake
  curvicarpum T.E. Melchert
  filifolium (Hook.) Gray
    var. filifolium
        SY=T. trifidum (Poir.) Britt.
    var. intermedium (Rydb.) Shinners
        SY=T. i. Rydb.
  flavodiscum (Shinners) B.L. Turner
  longipes Gray
  megapotamicum (Spreng.) Kuntze
        SY=T. gracile (Torr.) Gray
  nuecense B.L. Turner
  simplicifolium Gray
  subnudum Gray
    var. marginatum (Rydb.) T.E. Melchert
        SY=T. m. Rydb.
    var. subnudum

Thurovia Rose

  triflora Rose

Tithonia Desf. ex Juss.

  diversifolia (Hemsl.) Gray
  rotundifolia (P. Mill.) Blake
  thurberi Gray

Tolpis Adans.

  barbata (L.) Gaertn.
        SY=Crepis b. L.
  umbellata Bertol.

Townsendia Hook.

    alpigena Piper
        var. alpigina
        var. minima (Eastw.) Dorn
           SY=T. m. Eastw.
           SY=T. montana M.E. Jones var.
               minima (Eastw.) Beaman
    annua Beaman
    aprica Welsh & Reveal
    condensata Parry ex Gray
    eximia Gray
    exscapa (Richards.) Porter
           SY=T. intermedia Rydb.
           SY=T. sericea Hook.
    fendleri Gray
    florifer (Hook.) Gray
           SY=T. f. var. watsonii (Gray)
               Cronq.
    formosa Greene
    glabella Gray
    grandiflora Nutt.
    hookeri Beaman
    incana Nutt.
           SY=T. arizonica Gray
    jonesii (Beaman) Reveal
        var. jonesii
           SY=T. mensana M.E. Jones var. j.
               Beaman
        var. tumulosa Reveal
    leptotes (Gray) Osterhout
    mensana M.E. Jones
    montana M.E. Jones
    parryi D.C. Eat.
    rothrockii Gray ex Rothrock
    scapigera D.C. Eat.
    spathulata Nutt.
    strigosa Nutt.
    texensis Larsen

Tracyina Blake

    rostrata Blake

Tragopogon L.

    X crantzii Dichlt.
    dubius Scop.
      ssp. dubius
      ssp. major (Jacq.) Voll.
          SY=T. m. Jacq.
    mirabilis Rouy
    mirus Ownbey
    miscellus Ownbey
    porrifolius L.
    pratensis L.
      ssp. orientalis (L.) Celak.
      ssp. pratensis

Trichocoronis Gray

    riparia (Greene) Greene
    rivularis Gray
    wrightii (Torr. & Gray) Gray

Trichoptilium Gray

    incisum (Gray) Gray

Tridax L.

    procumbens L.

Trixis Sw.

    californica Kellogg
    inula Crantz

        SY=T. radialis (L.) Kuntze

Tussilago L.

    farfara L.

Urospermum Nutt.

    picroides (L.) Scop. ex F.W. Schmidt

Vanclevea Greene

    stylosa (Eastw.) Greene

Varilla Gray

    texana Gray

Venegasia DC.

    carpesioides DC.

Venidium Less.

    fastuosum (Jacq.) Stapf

Verbesina L.

    alata L.
        SY=Tepion a. (L.) Britt.
    alternifolia (L.) Britt.
        SY=Actinomeris a. (L.) DC.
        SY=Ridan a. (L.) Britt.
    aristata (Fll.) Heller
        SY=Actinomeris nudicaulis Nutt.
        SY=Helianthus a. Fll.
        SY=Pterophyton a. (Ell.) Alexander
        SY=V. n. (Nutt.) Gray
    chapmanii J.R. Coleman
    dissita Gray
    encelioides (Cav.) Benth. & Hook. f. ex
      Gray
      ssp. encelioides
          SY=Ximenesia e. Cav.
      ssp. exauriculata (Robins. & Greenm.)
        J.R. Coleman
          SY=V. encelioides var.
              exauriculata Robins. &
              Greenm.
          SY=V. exauriculata (Robins &
              Greenm.) Cockerell
          SY=Ximenesia exauriculata (Robins
              & Greenm.) Rydb.
    helianthoides Michx.
        SY=Actinomeris h. (Michx.) Nutt.
        SY=A. h. var. elliottii DC.
        SY=A. h. var. nuttallii DC.
        SY=A. oppositifolia DC.
        SY=Phaethusa h. (Michx.) Britt.
        SY=Pterophyton h. (Michx.)
             Alexander
    heterophylla (Chapman) Gray
        SY=Actinomeris pauciflora Nutt.
        SY=A. h. Chapman
        SY=Pterophyton h. (Chapman)
             Alexander
        SY=P. pauciflorum (Nutt.)
             Alexander
        SY=V. warei Gray
    lindheimeri Robins. & Greenm.
    longifolia Gray
    microptera DC.
    nana (Gray) Robins. & Greenm.
        SY=Wootonella n. (Gray) Standl.
        SY=Ximenesia encelioides Cav. var.
             n. Gray
        SY=X. n. (Gray) Shinners

Verbesina (CONT.)
    occidentalis (L.) Walt.
            SY=Phaethusa o. (L.) Britt.
    oreophila Woot. & Standl.
    rothrockii Robins. & Greenm.
    virginica L.
        var. laciniata (Poir.) Gray
            SY=Phaethusa l. (Poir.) Small
            SY=V. l. (Poir.) Nutt.
        var. virginica
            SY=Phaethusa v. (Walt.) Small
    walteri Shinners
            SY=Ridan paniculata (Walt.) Small

Vernonia Schreb.

    acaulis (Walt.) Gleason
    ?albicaulis Pers.
    angustifolia Michx.
        ssp. angustifolia
        ssp. mohrii (S.B. Jones) S.B. Jones &
            Faust
            SY=V. a. var. m. S.B. Jones
        ssp. scaberrima (Nutt.) S.B. Jones &
            Faust
            SY=V. a. var. s. (Nutt.) Gray
            SY=V. recurva Gleason
            SY=V. s. Nutt.
    arkansana DC.
            SY=V. crinita Raf.
    baldwinii Torr.
        ssp. baldwinii
        ssp. interior (Small) Faust
            SY=V. i. Small
    blodgettii Small
    borinquensis Urban
    cinerea (L.) Less.
            SY=Seneciodes c. (L.) Kuntze ex
                Post & Kuntze
            SY=V. c. var. parviflora (Reinw.)
                DC.
    X concinna Gleason [angustifolia X
        gigantea]
    X dissimilis Gleason [angustifolia X
        glauca]
    fasciculata Michx.
        ssp. corymbosa (Schwein. ex Keating)
            S.B. Jones
            SY=V. f. var. c. (Schwein. ex
                Keating) Schub.
        ssp. fasciculata
    flaccidifolia Small
    X georgiana Bartlett [acaulis X
        angustifolia]
    gigantea (Walt.) Trel. ex Branner & Coville
        ssp. gigantea
            SY=V. altissima Nutt.
            SY=V. a. var. taeniotrichia Blake
            SY=V. a. var. lilacina Clute
        ssp. ovalifolia (Torr. & Gray) Urbatsch
            SY=V. o. Torr. & Gray
    glauca (L.) Willd.
            SY=V. noveboracensis (L.) Michx.
                        var. tomentosa (Walt.)
                        Britt. & Brown
    X guadalupensis Heller [baldwinii ssp.
        interior X lindheimeri]
    larsenii B.L. King & S.B. Jones
            SY=V. leucophylla Larsen
    lettermannii Engelm. ex Gray
    lindheimeri Gray & Engelm. ex Gray
    marginata (Torr.) Raf.
            SY=V. m. var. tenuifolia (Small)
                        Shinners
    missurica Raf.
            SY=V. aborigina Gleason
    noveboracensis (L.) Michx.

            SY=V. harperi Gleason
    pulchella Small
    sericea L.C. Rich.
    texana (Gray) Small
    X vulturina Shinners [baldwinii ssp.
        interior X marginata]

Viguiera H.B.K.

    annua (M.E. Jones) Blake
    ciliata (Robins. & Greenm.) Blake
    cordifolia Gray
    deltoidea Gray
        var. deltoidea
            SY=V. d. var. genuina Blake
        var. parishii (Greene) Vasey & Rose
    dentata (Cav.) Spreng.
        var. dentata
        var. lancifolia Blake
    laciniata Gray
    longifolia (Robins. & Greenm.) Blake
    ludens (Shinners) M.C. Johnston
            SY=Helianthus l. Shinners
    multiflora (Nutt.) Blake
        var. multiflora
            SY=Heliomeris m. Nutt.
        var. nevadensis (A. Nels.) Blake
    ovalis Blake
    porteri (Gray) Blake
    reticulata S. Wats.
    soliceps Barneby
    stenoloba Blake

Wedelia Jacq.

    calycina L.C. Rich.
            SY=Stemmodontia c. (L.C. Rich.)
                        O.F. Schulz
    glauca (Ortega) Hoffmann
            SY=Pascalia g. Ortega
    gracilis L.C. Rich.
    hispida H.B.K.
            SY=Zexmenia h. (H.B.K.) Gray
    lanceolata DC.
    parviflora L.C. Rich.
    reticulata DC.
    trilobata (L.) A.S. Hitchc.

Whitneya Gray

    dealbata Gray

Wilkesia Gray

    gymnoxiphium Gray
    hobdyi St. John

Wyethia Nutt.

    amplexicaulis (Nutt.) Nutt.
    angustifolia (DC.) Nutt.
            SY=W. a. var. foliosa (Congd.)
                        Hall
    arizonica Gray
    bolanderi (Gray) W.A. Weber
    X cusickii Piper [amplexicaulis X
        helianthoides]
    elata Hall
    glabra Gray
    helenioides (DC.) Nutt.
    helianthoides Nutt.
    invenusta (Greene) W.A. Weber
    longicaulis Gray
    mollis Gray
    ovata Torr. & Gray
    reticulata Greene
    scabra Hook.

Wyethia (CONT.)
        var. attenuata W.A. Weber
        var. canescens W.A. Weber
        var. scabra

Xanthisma DC.

    texanum DC.
        ssp. drummondii (Torr. & Gray) Semple
            SY=X. t. var. d. (Torr. & Gray)
                Gray
        ssp. texanum
            var. orientalis Semple
            var. texanum

Xanthium L.

    spinosum L.
            SY=Acanthoxanthium s. (L.) Fourr.
            SY=X. s. var. inerme Bel.
    strumarium L.
        var. canadense (P. Mill.) Torr. &
            Gray
            SY=X. acerosum Greene
            SY=X. californicum Greene
            SY=X. californicum var.
                rotundifolium Widder
            SY=X. cavanillesii Schouw
            SY=X. commune Britt.
            SY=X. echinatum Murr.
            SY=X. italicum Moretti
            SY=X. oviforme Wallr.
            SY=X. pensylvanicum Wallr.
            SY=X. saccharatum Wallr.
            SY=X. speciosum Kearney
            SY=X. strumarium ssp. italicum
                (Moretti) D. Love
            SY=X. strumarium var. o. (Wallr.)
                M.E. Peck
            SY=X. strumarium var. p. (Wallr.)
                M.E. Peck
            SY=X. varians Greene
        var. glabratum (DC.) Cronq.
            SY=X. americanum Walt.
            SY=X. calvum Millsp. & Sherff
            SY=X. chasei Fern.
            SY=X. chinense P. Mill.
            SY=X. cylindraceum Millsp. &
                Sherff
            SY=X. echinellum Greene
            SY=X. globosum Shull
            SY=X. inflexum Mackenzie & Bush
            SY=X. orientale L.
        var. strumarium
            SY=X. ambrosioides Hook. & Arn.
            SY=X. cenchroides Millsp. & Sherff
            SY=X. curvescens Millsp. & Sherff
            SY=X. glanduliferum Greene
            SY=X. macounii Britt.
            SY=X. s. var. wootonii (Cockerell)
                M.E. Peck
            SY=X. w. Cockerell

Xanthocephalum Willd.

    wrightii (Gray) Gray

Xylorhiza Nutt.

    cognata (Hall) T.J. Wats.
            SY=Aster c. Hall
            SY=A. standleyi A. Davids.
            SY=Machaeranthera c. (Hall) Cronq.
                & Keck
            SY=X. s. (A. Davids.) A. Davids.
    confertifolia (Cronq.) T.J. Wats.
            SY=Machaeranthera glabriuscula

            (Nutt.) Cronq. & Keck var.
            c. Cronq.
    glabriuscula Nutt.
        var. glabriuscula
            SY=Aster g. (Nutt.) Torr. & Gray
            SY=A. g. var. parryi (Gray) Onno
            SY=A. p. Gray
            SY=A. xylorhiza (Nutt.) Torr. &
                Gray
            SY=Machaeranthera g. (Nutt.)
                Cronq. & Keck
            SY=M. g. var. villosa (Nutt.)
                Cronq. & Keck
            SY=X. g. var. v. (Nutt.) A. Nels.
            SY=X. v. Nutt.
        var. linearifolia T.J. Wats.
    orcuttii (Vasey & Rose) Greene
            SY=Aster o. Vasey & Rose
            SY=Machaeranthera o. (Vasey &
                Rose) Cronq. & Keck
    tortifolia (Torr. & Gray) Greene
        var. imberbis (Cronq.) T.J. Wats.
            SY=Machaeranthera t. (Torr. &
                Gray) Cronq. & Keck var. i.
                Cronq.
        var. tortifolia
            SY=Aster abatus Blake
            SY=A. goodingii Onno
            SY=A. mohavensis sensu Coville non
                Kuntze
            SY=A. t. sensu (Torr. & Gray) Gray
                non Michx.
            SY=Haplopappus t. Torr. & Gray
            SY=Machaeranthera t. (Torr. &
                Gray) Cronq. & Keck
            SY=X. lanceolata Rydb.
            SY=X. scopulorum A. Nels.
    venusta (M.E. Jones) Heller
            SY=Aster v. M.E. Jones
            SY=Machaeranthera v. (M.E. Jones)
                Cronq. & Keck
    wrightii (Gray) Greene
            SY=Aster w. Gray
            SY=Machaeranthera w. (Gray) Cronq.
                & Keck
            SY=Townsendia w. (Gray) Gray

Youngia Cass.

    japonica (L.) DC.
            SY=Crepis j. (L.) Benth.
    thunbergiana DC.

Zaluzania Pers.

    grayana Robins. & Greenm.

Zexmenia Ll. & Lex.

    brevifolia Gray
    podocephala Gray

Zinnia L.

    acerosa (DC.) Gray
            SY=Z. pumila Gray
    anomala Gray
    elegans Jacq.
            SY=Crassina e. (Jacq.) Kuntze
    grandiflora Nutt.
            SY=Crassina g. (Nutt.) Kuntze
    peruviana (L.) L.
            SY=Crassina multiflora (L.) Kuntze
            SY=Z. m. L.
            SY=Z. pauciflora L.

BALANOPHORACEAE

Scybalium Schott & Endl.

  jamaicense (Sw.) Schott & Endl.

BALSAMINACEAE

Impatiens L.

   aurella Rydb.
   balfourii Hook. f.
   balsamina L.
   capensis Meerb.
          SY=I. biflora Walt.
          SY=I. fulva Nutt.
          SY=I. nortonii Rydb.
   ecalcarata Blank.
   glandulifera Royle
          SY=I. roylei Walp.
   noli-tangere L.

          SY=I. occidentalis Rydb.
   oliveri C. Wright ex W. Wats.
   pallida Nutt.
   parviflora DC.

BASELLACEAE

Anredera Juss.

   baselloides (H.B.K.) Baill.
          SY=Boussingaultia b. H.B.K.
   cordifolia (Ten.) Steenis
          SY=Boussingaultia gracilis Miers
          SY=B. g. var. pseudobaselloides
              (Hauman) Bailey
   leptostachys (Moq.) Steenis
          SY=Boussingaultia l. Moq.
   vesicaria (Lam.) Gaertn. f.
          SY=A. scandens sensu auctt. non
              (L.) Moq.

Basella L.

   alba L.
          SY=B. rubra L.

BATACEAE

Batis P. Br.

   maritima L.

BEGONIACEAE

Begonia L.

   cucullata Willd.
          var. cucullata
          var. hookeri (A. DC.) L.B. Sm. &
             Schub.
          SY=B. semperflorens Link & Otto
   decandra Pavon ex A. DC.

Hillebrandia D. Oliver

   sandwicensis D. Oliver

BERBERIDACEAE

Achlys DC.

   triphylla (Sm.) DC.

Berberis L.

   canadensis P. Mill.
   fendleri Gray
   harrisoniana Kearney & Peebles
   X ottawensis Schneid. [thunbergii X
      vulgaris]
   thunbergii DC.
   vulgaris L.
   wilcoxii Kearney

Caulophyllum Michx.

   thalictroides (L.) Michx.
          var. giganteum Farw.
          var. thalictroides

Diphylleia Michx.

   cymosa Michx.

Epimedium L.

   X youngianum Fisch. & Mey. [diphyllum X
      grandiflorum]
          SY=E. y. var. niveum Stern.

Jeffersonia Bart.

   diphylla (L.) Pers.

Mahonia Nutt.

   amplectens Eastw.
          SY=Berberis a. (Eastw.) L.C.
              Wheeler
   aquifolium (Pursh) Nutt.
          SY=Berberis a. Pursh
   bealei (Fortune) Carr.
          SY=Berberis b. Fortune
   californica (Jepson) Ahrendt
          SY=Berberis c. Jepson
   dictyota (Jepson) Fedde
          SY=Berberis d. Jepson
   fremontii (Torr.) Fedde
          SY=Berberis f. Torr.
   haematocarpa (Woot.) Fedde
          SY=Berberis h. Woot.
   higginsae (Munz) Ahrendt
          SY=Berberis h. Munz
   nervosa (Pursh) Nutt.
          SY=Berberis n. Pursh
   nevinii (Gray) Fedde
          SY=Berberis n. Gray
   pinnata (Lag.) Fedde
          SY=Berberis p. Lag.
          SY=?B. p. ssp. insularis Munz
   piperiana Abrams
          SY=Berberis p. (Abrams) McMinn
   pumila (Greene) Fedde
          SY=Berberis p. Greene
   repens (Lindl.) G. Don
          SY=Berberis aquifolium (Pursh)
              Nutt. var. r. (Lindl.)

Mahonia (CONT.)
               Scoggan
           SY=B. r. Lindl.
  sonnei Abrams
           SY=Berberis s. (Abrams) McMinn
  swazeyi (Buckl.) Fedde
           SY=Berberis s. Buckl.
  trifoliata (Moric.) Fedde
      var. glauca I.M. Johnston
           SY=Berberis t. Moric. var. g.
              (I.M. Johnston) I.M.
              Johnston
      var. trifoliata
           SY=Berberis t. Moric.

Nandina Thunb.

  domestica Thunb.

Podophyllum L.

  peltatum L.

Vancouveria Morr. & Dec.

  chrysantha Greene
  hexandra (Hook.) Morr. & Dec.
           SY=V. brevicula Greene
           SY=V. parvifolia Greene
           SY=V. picta Greene
  planipetala Calloni

BETULACEAE

Alnus P. Mill.

  X fallacina Callier [incana ssp. rugosa X
    serrulata]
  glutinosa (L.) Gaertn.
           SY=A. alnus (L.) Britt.
  incana (L.) Moench
    ssp. rugosa (Du Roi) Clausen
           SY=A. i. var. americana Regel
           SY=A. r. (Du Roi) Spreng.
           SY=A. r. var. a. (Regel) Fern.
    ssp. tenuifolia (Nutt.) Breitung
           SY=A. i. ssp. rugosa Clausen var.
              occidentalis (Dippel) C.L.
              Hitchc.
           SY=A. i. ssp. t. var. virescens S.
              Wats.
           SY=A. X purpusii Callier
           SY=A. t. Nutt.
  maritima Muhl. ex Nutt.
           SY=A. metoporina Furlow
  oblongifolia Torr.
  rhombifolia Nutt.
           SY=A. r. var. bernardina Munz &
              Johnston
  rubra Bong.
           SY=A. oregona Nutt.
           SY=A. o. var. pinnatisecta Starker
  serrulata (Ait.) Willd.
           SY=A. incana (L.) Moench var. s.
              (Ait.) Boivin
           SY=A. noveboracensis Britt.
           SY=A. s. var. subelliptica Fern.
  viridis (Chaix) DC.
    ssp. crispa (Ait.) Turrill
           SY=A. c. (Ait.) Pursh
           SY=A. c. var. elongata Raup
           SY=A. c. var. mollis Fern.
           SY=A. v. ssp. fruticosa (Rupr.)
              Nyman

    ssp. sinuata (Regel) Love & Love
           SY=A. alnobetula (Ehrh.) K. Koch
              pro parte
           SY=A. crispa (Ait.) Pursh ssp.
              laciniata Hulten
           SY=A. c. ssp. s. (Regel) Hulten
           SY=A. s. (Regel) Rydb.
           SY=A. v. var. s. Regel

Betula L.

  alleghaniensis Britt.
      var. alleghaniensis
           SY=B. a. var. fallax (Fassett)
              Brayshaw
           SY=B. lutea Michx. f.
           SY=B. l. var. f. Fassett
      var. macrolepis (Fern.) Brayshaw
           SY=B. lutea Michx. f. var. m.
              Fern.
  X alpestris Fries [nana X pubescens ssp.
    tortuosa]
  borealis Spach
  X caerulea Blanch. [cordifolia X
    populifolia]
    nm. caerulea
    nm. grandis (Blanch.) Boivin
           SY=B. X c-g. Blanch.
    nm. cunninghamii Boivin
  cordifolia Regel
           SY=B. alba L. var. c. (Regel)
              Regel
           SY=B. papyrifera Marsh. var. c.
              (Regel) Fern.
  X dugleana Lepage [glandulosa X
    neoalaskana]
  X dutillyi Lepage [glandulosa X minor]
  X eastwoodiae (Sarg.) Dugle [glandulosa X
    occidentalis]
           SY=B. X arbuscula Dugle
  glandulosa Michx.
           SY=B. g. var. hallii (T.J. Howell)
              C.L. Hitchc.
  X hornei Butler [nana X papyrifera]
           SY=B. X beeniana A. Nels.
  X jackii Schneid. [lenta X pumila]
  lenta L.
  michauxii Spach
           SY=B. terrae-novae Fern.
  minor (Tuckerman) Fern.
           SY=B. papyrifera Marsh. var. m.
              (Tuckerman) Wats. & Coult.
           SY=B. saxophila Lepage
  nana L.
    ssp. exilis (Sukatschev) Hulten
           SY=B. glandulosa Michx. var.
              sibirica (Ledeb.) Schneid.
           SY=B. n. var. s. Ledeb.
    ssp. nana
  neoalaskana Sarg.
           SY=B. alaskana Sarg.
           SY=B. n. Sarg.
           SY=B. papyrifera Marsh. ssp.
              humilus (Regel) Hulten
           SY=B. p. var. h. (Regel) Fern. &
              Raup
           SY=B. p. var. n. (Sarg.) Raup
           SY=B. resinifera Britt.
  X neoborealis Lepage [borealis X
    glandulosa]
  nigra L.
  occidentalis Hook.
           SY=B. fontinalis Sarg.
           SY=B. o. var. inopina (Jepson)
              C.L. Hitchc.
           SY=B. papyrifera Marsh. ssp. o.
              (Hook.) Hulten

Betula   (CONT.)
        SY=B. p. var. o. (Hook.) Sarg.
  papyrifera Marsh.
      var. commutata (Regel) Fern.
        SY=B. alba L. var. c. Regel
      var. kenaica (W.H. Evans) A. Henry
        SY=B. k. W.H. Evans
        SY=B. neoalaskana Sarg. var. k.
          (W.H. Evans) Boivin
      var. papyrifera
        SY=B. p. var. elobata (Fern.)
          Sarg.
        SY=B. p. var. macrostachya Fern.
        SY=B. p. var. pensilis Fern.
  pendula Roth
      SY=B. verrucosa Ehrh.
  platyphylla Sukatschev
      var. japonica (Miq.) Hara
        SY=B. j. (Miq.) Sieb.
        SY=B. mandshurica (Regel) Nakai
          var. j. (Miq.) Rehd.
      var. kamtschatica (Regel) Hara
        SY=B. mandshurica (Regel) Nakai
          var. k. (Regel) Rehd.
      var. platyphylla
        SY=B. mandshurica (Regel) Nakai
      var. szechuanica Schneid.
        SY=B. mandshurica (Regel) Nakai
          var. s. (Schneid.) Rehd.
  populifolia Marsh.
  pubescens Ehrh.
    ssp. pubescens
      SY=B. alba L.
    ssp. tortuosa (Ledeb.) Schneid.
      SY=B. t. Ledeb.
  pumila L.
      var. glabra Regel
      var. glandulifera Regel
        SY=B. g. (Regel) Butler
        SY=B. glandulosa Michx. var.
          glandulifera (Regel) Gleason
        SY=B. nana L. var. glandulifera
          (Regel) Boivin
      var. pumila
      var. renifolia Fern.
        SY=B. nana L. var. r. (Fern.)
          Boivin
  X purpusii Schneid. [alleghaniensis X
    pumila var. glandulifera]
  X raymundii Lepage [glandulosa X pumila]
  X sandbergii Britt. [papyrifera X pumila
    var. glandulifera]
  X sargentii Dugle [glandulosa X pumila var.
    glandulifera]
  uber (Ashe) Fern.
      SY=B. lenta L. var. u. Ashe
  X uliginosa Dugle [neoalaskana X pumila
    var. glandulifera]
  X ungavensis Lepage [glandulosa X
    papyrifera]
  X utahensis (Britt.) Dugle [occidentalis X
    papyrifera]
        SY=B. andrewsii A. Nels.
        SY=B. X conmixta Sarg.
        SY=B. occidentalis Hook. var.
          fecunda Fern.
        SY=B. papyrifera Marsh. var.
          subcordata (Rydb.) Sarg.
        SY=B. X piperi (Britt.) C.L.
          Hitchc.
  X winteri Dugle [neoalaskana X papyrifera]

Carpinus L.

  caroliniana Walt.
      SY=C. c. var. virginiana (Marsh.)
        Fern.

Corylus L.

  americana Walt.
      SY=C. a. var. indehiscens Palmer &
        Steyermark
  avellana L.
  cornuta Marsh.
      var. californica (A. DC.) Sharp
        SY=C. californica (A. DC.) Rose
      var. cornuta
        SY=C. rostrata Ait.
      var. glandulosa Boivin

Ostrya Scop.

  chisosensis Correll
  knowltonii Coville
  virginiaia (P. Mill.) K. Koch
      SY=C. v. var. lasia Fern.

BIGNONIACEAE

Amphilophium Kunth

  paniculatum (L.) H.B.K.

Amphitecna Miers

  latifolia (P. Mill.) A.H. Gentry
      SY=Enallagma l. (P. Mill.) Small

Arrabidaea DC.

  chica Bureau

Bignonia L.

  capreolata L.
      SY=Anisostichus c. (L.) Bureau
      SY=A. crucigera (L.) Bureau

Campsis Lour.

  radicans (L.) Seem. ex Bureau
      SY=Bignonia r. L.
      SY=Tecoma r. (L.) Juss.

Catalpa Scop.

  bignonioides Walt.
      SY=C. catalpa (L.) Karst.
  longissima (Jacq.) Dum.-Cours.
  ovata G. Don
  speciosa (Warder ex Barney) Engelm.

Chilopsis D. Don

  linearis (Cav.) Sweet
      var. arcuata Fosberg
      var. glutinosa (Engelm.) Fosberg
      var. linearis
        SY=Bignonia l. Cav.

Crescentia L.

  cujete L.
  linearifolia Miers
  portoricensis Britt.

Cydista Miers

  aequinoctialis (L.) Miers

Distictis Mart. ex Meisn.

Distictis (CONT.)
     lactiflora (Vahl) DC.

Jacaranda Juss.

     mimosifolia D. Don
               SY=J. acutifolia sensu auctt. non
                  Humb. & Bonpl.

Macfadyena A. DC.

     unguis-cati (L.) A.H. Gentry
               SY=Batocydia u-c. (L.) Mart.
               SY=Bignonia u-c. L.
               SY=Doxantha u-c. (L.) Miers

Paulownia Sieb. & Zucc.

     tomentosa (Thunb.) Sieb. & Zucc. ex Steud.

Pyrostegia Presl

     venusta (Ker-Gawl.) Miers
               SY=Bignonia v. Ker-Gawl.
               SY=P. ignea (Vell.) Presl

Schlegelia Miq.

     brachyantha Griseb.
               SY=S. portoricensis (Urban) Britt.

Tabebuia Gomez ex DC.

     haemantha (Bertol. ex Spreng.) DC.
     heterophylla (DC.) Britt.
          ssp. heterophylla
               SY=T. lucida Britt.
               SY=T. pentaphylla (L.) Hemsl.
               SY=T. triphylla DC.
          ssp. pallida (Miers) Stehle
               SY=T. p. Miers
     rigida Urban
     rosea (Bertol.) DC.
     schumanniana Urban

Tecoma Juss.

     castanifolia (D. Don) Melchior
               SY=T. gaudichaudii DC.
     stans (L.) Juss. ex H.B.K.
          var. angustata Rehd.
          var. stans
               SY=Bignonia s. L.

Tynanthus Miers

     polyanthus (Bureau) Sandw.
               SY=Bignonia caryophyllus Bello
               SY=T. c. (Bello) Alain

BIXACEAE

Amoreuxia Moc. & Sesse ex DC.

     gonzalezii Sprague & Riley
     palmatifida Moc. & Sesse
     wrightii Gray

Bixa L.

     orellana L.

Cochlospermum Kunth ex DC.

vitifolium (Willd.) Spreng.

BOMBACACEAE

Ceiba P. Mill.

     pentandra (L.) Gaertn.

Ochroma Sw.

     pyramidale (Cav.) Urban

Pachira Aubl.

     insignis (Sw.) Sav.

Quararibea Aubl.

     turbinata (Sw.) Poir.

BORAGINACEAE

Amsinckia Lehm.

     douglasiana A. DC.
     eastwoodae J.F. Macbr.
               SY=A. intermedia Fisch. & Mey.
                  var. e. (J.F. Macbr.) Jepson
                  & Hoover
     furcata Suksdorf
               SY=A. vernicosa Hook. & Arn. var.
                  f. (Suksdorf) Hoover
     gloriosa Eastw. ex Suksdorf
     grandiflora Kleeb. ex Gray
     intermedia Fisch. & Mey.
               SY=A. arizonica Suksdorf
               SY=A. demissa Suksdorf
               SY=A. echinata Gray
               SY=A. intactilis J.F. Macbr.
               SY=A. intermedia var. e. (Gray)
                  Wiggins
               SY=A. microphylla Suksdorf
               SY=A. nana Suksdorf
               SY=A. rigida Suksdorf
     lunaris J.F. Macbr.
     lycopsoides (Lehm.) Lehm.
               SY=A. barbata Greene
               SY=A. hispida (Ruiz & Pavon) I.M.
                  Johnston
               SY=A. idahoensis M.E. Jones
               SY=A. parviflora Heller
     menziesii (Lehm.) A. Nels. & J.F. Macbr.
               SY=A. micrantha Suksdorf
     retrorsa Suksdorf
               SY=A. rugosa Rydb.
     spectabilis Fisch. & Mey.
               SY=A. scouleri I.M. Johnston
               SY=A. spectabilis var. bracteosa
                  (Gray) Boivin
               SY=A. spectabilis var. microcarpa
                  (Greene) Jepson & Hoover
               SY=A. spectabilis var. nicolai
                  (Jepson) I.M. Johnston ex
                  Munz
     tessellata Gray
     vernicosa Hook. & Arn.

Anchusa L.

     arvensis (L.) Bieb.
               SY=Lycopsis a. L.

Anchusa (CONT.)
   azurea P. Mill.
         SY=A. italica Retz.
   barrellieri (All.) Vitman
         SY=Buglossum b. All.
   capensis Thunb.
   officinalis L.
         SY=A. procera Bess.

Antiphytum DC.

   floribunda (Torr.) Gray
   heliotropioides A. DC.

Asperugo L.

   procumbens L.

Borago L.

   officinalis L.

Bothriospermum Bunge

   tenellum (Hornem.) Fisch. & Mey.

Bourreria P. Br.

   cassinifolia (A. Rich.) Griseb.
   ovata Miers
   radula (Poir.) G. Don
         SY=B. revoluta H.B.K.
         SY=B. succulenta Jacq. var.
              revoluta (H.B.K.) O.E.
              Schulz
   succulenta Jacq.
   virgata (Sw.) G. Don
         SY=B. domingensis (DC.) Griseb.

Buglossoides Moench

   arvense (L.) I.M. Johnston
         SY=Lithospermum a. L.

Cordia L.

   alba (Jacq.) Roemer & Schultes
         SY=Calyptracordia a. Jacq.
         SY=Cordia dentata Poir.
   alliodora (Ruiz & Pavon) Oken
         SY=Cerdana a. Ruiz & Pavon
   bahamensis Urban
         SY=Varronia b. (Urban) Millsp.
   bellonis Urban
         SY=Varronia b. (Urban) Britt.
   boissieri A. DC.
   borinquensis Urban
   collococca L.
         SY=Cordia glabra L.
   gerascanthus L.
   globosa (Jacq.) H.B.K.
         SY=Varronia g. Jacq.
   laevigata Lam.
         SY=Cordia nitida Vahl
   lima (Desv.) Roemer & Schultes
         SY=Varronia l. Desv.
   obliqua Willd.
   parvifolia A. DC.
   podocephala Torr.
   polycephala (Lam.) I.M. Johnston
         SY=Varronia corymbosa (L.) Desv.
   rickseckeri Millsp.
         SY=Sebesten r. (Millsp.) Britt.
   rupicola Urban
         SY=Varronia r. (Urban) Britt.
   sebestena L.
         SY=Sebesten s. (L.) Britt. ex

        Small
   stenophylla Alain
         SY=Varronia angustifolia West
   subcordata Lam.
   sulcata DC.
   wagnerorum Howard

Cryptantha Lehm. ex G. Don

   abata I.M. Johnston
   affinis (Gray) Greene
   albida (H.B.K.) I.M. Johnston
   ambigua (Gray) Greene
   angustifolia (Torr.) Greene
   aperta (Eastw.) Payson
         SY=Oreocarya a. Eastw.
   atwoodii Higgins
   bakeri (Greene) Payson
   barbigera (Gray) Greene
   barnebyi I.M. Johnston
   breviflora (Osterhout) Payson
   caespitosa (A. Nels.) Payson
   cana (A. Nels.) Payson
   capitata (Eastw.) I.M. Johnston
   celosioides (Eastw.) Payson
         SY=C. bradburiana Payson
         SY=C. macounii (Eastw.) Payson
         SY=C. nubigena (Greene) Payson
             var. m. (Eastw.) Boivin
         SY=C. sheldonii (Brand) Payson
         SY=Oreocarya c. Eastw.
         SY=O. glomerata (Pursh) Greene
         SY=O. m. Eastw.
         SY=O. s. Brand
   circumscissa (Hook. & Arn.) I.M. Johnston
      var. circumscissa
         SY=C. c. var. genuina I.M.
              Johnston
         SY=Greeneocharis c. (Hook. & Arn.)
              Rydb.
      var. hispida (J.F. Macbr.) I.M.
         Johnston
         SY=Greeneocharis c. (Hook. & Arn.)
              Rydb. var. h. J.F. Macbr.
      var. rosulata J.T. Howell
         SY=C. c. ssp. r. (J.T. Howell)
              Mathew & Raven
   clevelandii Greene
      var. clevelandii
      var. dissita (I.M. Johnston) Jepson &
         Hoover
      var. florosa I.M. Johnston
   compacta Higgins
   confertiflora (Greene) Payson
         SY=Oreocarya c. Greene
   corollata (I.M. Johnston) I.M. Johnston
   costata Brandeg.
   crassipes I.M. Johnston
   crassisepala (Torr. & Gray) Greene
      var. crassisepala
      var. elachantha I.M. Johnston
   crinita Greene
   crymophila I.M. Johnston
         SY=Oreocarya c. (I.M. Johnston)
              Jepson & Hoover
   decipiens (M.E. Jones) Heller
   dumetorum (Greene ex Gray) Greene
   echinella Greene
   elata (Eastw.) Payson
   excavata Brandeg.
   fendleri (Gray) Greene
   flaccida (Dougl.) Greene
   flava (A. Nels.) Payson
   flavoculata (A. Nels.) Payson
         SY=Oreocarya f. A. Nels.
   fulvocanescens (S. Wats.) Payson
      var. echinoides (M.E. Jones) Higgins

Cryptantha (CONT.)
          SY=C. e. (M.E. Jones) Payson
      var. fulvocanescens
          SY=Oreocarya f. (S. Wats.) Greene
ganderi I.M. Johnston
glomeriflora Greene
gracilis Osterhout
grahamii I.M. Johnston
hispidula Greene ex Brand
hoffmannii I.M. Johnston
          SY=Oreocarya h. (I.M. Johnston)
             Abrams
holoptera (Gray) J.F. Macbr.
hooveri I.M. Johnston
humilis (Gray) Payson
      var. commixta (J.F. Macbr.) Higgins
          SY=C. nana (Eastw.) Payson var. c.
             (J.F. Macbr.) Payson
      var. humilis
          SY=Oreocarya h. (Gray) Greene
      var. nana (Eastw.) Higgins
          SY=C. n. (Eastw.) Payson
          SY=C. n. var. typica Payson
      var. ovina (Payson) Higgins
          SY=C. nana (Eastw.) Payson var. o.
             Payson
      var. shantzii (Tidestrom) Higgins
          SY=C. nana (Eastw.) Payson var. s.
             (Tidestrom) Payson
hypsophila I.M. Johnston
inaequata I.M. Johnston
incana Greene
insolita (J.F. Macbr.) Payson
intermedia (Gray) Greene
      var. grandiflora (Rydb.) Cronq.
          SY=C. fragilis M.E. Peck
          SY=C. g. Rydb.
          SY=C. hendersonii (A. Nels.) Piper
      var. intermedia
          SY=C. barbigera (Gray) Greene var.
             fergusoniae J.F. Macbr.
interrupta (Greene) Payson
jamesii (Torr.) Payson
      var. abortiva (Greene) Payson
          SY=Oreocarya a. Greene
      var. disticha (Eastw.) Payson
      var. jamesii
          SY=Oreocarya j. var. typica Payson
      var. laxa (J.F. Macbr.) Payson
      var. multicaulis (Torr.) Payson
          SY=Oreocarya suffruticosa (Torr.)
             Greene
      var. pustulosa (Rydb.) Harrington
      var. setosa (M.E. Jones) I.M.
          Johnston
          SY=Oreocarya j. var. cinerea
             (Greene) Payson
johnstonii Higgins
jonesiana (Payson) Payson
kelseyana Greene
leiocarpa (Fisch. & Mey.) Greene
leucophaea (Dougl.) Payson
          SY=Oreocarya l. (Dougl.) Greene
longiflora (A. Nels.) Payson
mariposae I.M. Johnston
maritima (Greene) Greene
      var. maritima
      var. pilosa I.M. Johnston
mensana (M.E. Jones) Payson
mexicana (Brandeg.) I.M. Johnston
micrantha (Torr.) I.M. Johnston
      var. lepida (Gray) I.M. Johnston
          SY=C. m. ssp. l. (Gray) Mathew &
             Raven
          SY=Eremocarya m. (Torr.) Greene
             var. l. (Gray) J.F. Macbr.
      var. micrantha

          SY=Eremocarya m. (Torr.) Greene
micromeres (Gray) Greene
microstachys (Greene ex Gray) Greene
milobakeri I.M. Johnston
minima Rydb.
mirabunda Brand
mohavensis (Greene) Greene
muricata (Hook. & Arn.) A. Nels. & J.F.
    Macbr.
      var. clokeyi (I.M. Johnston) Jepson
          SY=C. c. I.M. Johnston
      var. denticulata (Greene) I.M.
          Johnston
      var. jonesii (Gray) I.M. Johnston
      var. muricata
nemaclada Greene
nevadensis A. Nels. & Kennedy
      var. nevadensis
      var. rigida I.M. Johnston
nubigena (Greene) Payson
          SY=Oreocarya n. Greene
oblata (M.E. Jones) Payson
ochroleuca Higgins
osterhoutii (Payson) Payson
oxygona (Gray) Greene
palmerii (Gray) Payson
          SY=C. coryi I.M. Johnston
paradoxa (A. Nels.) Payson
pattersonii (Gray) Greene
paysonii (J.F. Macbr.) I.M. Johnston
propria (A. Nels. & J.F. Macbr.) Payson
          SY=Oreocarya p. A. Nels. & J.F.
             Macbr.
pterocarya (Torr.) Greene
      var. cycloptera (Greene) J.F. Macbr.
      var. pterocarya
      var. purpusii Jepson
      var. stenoloba I.M. Johnston
pusilla (Torr. & Gray) Greene
racemosa (S. Wats.) Greene
rattanii Greene
          SY=C. corollata (I.M. Johnston)
             I.M. Johnston ssp. r.
             (Greene) Abrams
recurvata Coville
rollinsii I.M. Johnston
roosiorum Munz
rostellata (Greene) Greene
      var. rostellata
      var. spithamea (I.M. Johnston) Jepson
rugulosa (Payson) Payson
salmonensis (A. Nels. & J.F. Macbr.) Payson
scoparia A. Nels.
semiglabra Barneby
sericea (Gray) Payson
          SY=C. s. var. perennis (A. Nels.)
             Payson
          SY=Oreocarya s. (Gray) Greene
setosissima (Gray) Payson
shackletteana Higgins
similis Mathew & Raven
simulans Greene
sobolifera Payson
sparsiflora (Greene) Greene
spiculifera (Piper) Payson
          SY=Oreocarya s. Piper
stricta (Osterhout) Payson
subretusa I.M. Johnston
          SY=Oreocarya s. (I.M. Johnston)
             Abrams
tenuis (Eastw.) Payson
texana (A. DC.) Greene
thompsonii I.M. Johnston
          SY=Oreocarya t. (I.M. Johnston)
             Abrams
thyrsiflora (Greene) Payson
torreyana (Gray) Greene

Cryptantha (CONT.)
        var. pumila (Heller) I.M. Johnston
        var. torreyana
            SY=C. calycosa (Gray) Rydb.
            SY=C. t. var. c. (Gray) Greene
    traskiae I.M. Johnston
    tumulosa (Payson) Payson
        SY=Oreocarya t. Payson
    utahensis (Gray) Greene
    virgata (Porter) Payson
    virginensis (M.E. Jones) Payson
        SY=Oreocarya v. (M.E. Jones) J.F.
            Macbr.
    watsonii (Gray) Greene
    weberi I.M. Johnston
    wetherillii (Eastw.) Payson

Cynoglossum L.

    amabile Stapf & Drummond
    boreale Fern.
    furcatum Wallich
    grande Dougl. ex Lehm.
    microglochin Benth.
    occidentale Gray
    officinale L.
    virginianum L.
    wallichii G. Don
    zeylanicum (Vahl) Thunb.

Dasynotus I.M. Johnston

    daubenmirei I.M. Johnston

Echium L.

    coincyanum Lacaita
    creticum L.
        SY=E. australis Lam.
    fastuosum Ait.
    plantagineum L.
        SY=E. lycopsis L.
    pustulatum Sibthorp & Sm.
        SY=E. vulgare L. var. p. (Sibthorp
            & Sm.) Coincy
    vulgare L.

Ehretia P. Br.

    anacua (Teran & Berl.) I.M. Johnston

Eritrichium Schrad. ex Gaudin

    aretioides (Cham.) DC.
        SY=E. elongatum (Rydb.) Wight var.
            a. (Cham.) I.M. Johnston
        SY=E. nanum (Vill.) Schrad. ex
            Gaudin var. a. (Cham. Herder
    howardii (Gray) Rydb.
    nanum (L.) Schrad. ex Gaudin
        var. argenteum (Wight) I.M. Johnston
        var. chamissonis (DC.) Herder
            SY=E. c. DC.
        var. elongatum (Rydb.) Cronq.
            SY=E. e. (Rydb.) Wight
        var. nanum
    rupestre (Pallas) Bunge
    splendens Kearney

Hackelia Opiz

    amethystina J.T. Howell
    bella (J.F. Macbr.) I.M. Johnston
    besseyi (Rydb.) J.L. Gentry
        SY=H. grisea (Woot. & Standl.)
            I.M. Johnston
    brevicula (Jepson) J.L. Gentry

californica (Gray) I.M. Johnston
ciliata (Dougl. ex Lehm.) I.M. Johnston
cinerea (Piper) I.M. Johnston
cronquistii J.L. Gentry
        SY=H. patens (Nutt.) I.M. Johnston
            var. semiglabra Cronq.
cusickii (Piper) Brand
davisii Cronq.
deflexa (Wahlenb.) Opiz
    var. americana (Gray) Fern. & I.M.
        Johnston
        SY=H. a. (Gray) Fern.
        SY=H. d. ssp. a. (Gray) Hulten
        SY=Lappula a. (Gray) Rydb.
        SY=L. d. (Wahlenb.) Garcke pro
            parte
        SY=L. d. var. a. (Gray) Greene
diffusa (Dougl. ex Lehm.) I.M. Johnston
    var. arida (Piper) R.L. Carr
        SY=H. a. (Piper) I.M. Johnston
        SY=Lappula a Piper
    var. cottonii (Piper) R.L. Carr
    var. diffusa
        SY=H. saxatilis (Piper) Brand
        SY=Lappula d. (Dougl. ex Lehm.)
            Greene
floribunda (Lehm.) I.M. Johnston
        SY=H. leptophylla (Rydb.) I.M.
            Johnston
        SY=Lappula f. (Lehm.) Greene
gracilenta (Eastw.) I.M. Johnston
hirsuta (Woot. & Standl.) I.M. Johnston
hispida (Gray) I.M. Johnston
    var. disjuncta R.L. Carr
    var. hispida
        SY=Lappula h. (Gray) Greene
micrantha (Eastw.) J.L. Gentry
        SY=H. jessicae (McGregor) Brand
mundula (Jepson) Ferris
nervosa (Kellogg) I.M. Johnston
ophiobia R.L. Carr
patens (Nutt.) I.M. Johnston
    var. harrisonii J.L. Gentry
    var. patens
        SY=H. diffusa (Dougl. ex Lehm.)
            I.M. Johnston var.
            coerulescens (Rydb.) I.M.
            Johnston
pinetorum (Greene ex Gray) I.M. Johnston
    var. jonesii J.L. Gentry
    var. pinetorum
setosa (Piper) I.M. Johnston
sharsmithii I.M. Johnston
ursina (Greene ex Gray) I.M. Johnston
    var. diabolii J.L. Gentry
    var. pustulata (J.F. Macbr.) J.L.
        Gentry
    var. ursina
velutina (Piper) I.M. Johnston
        SY=H. longituba I.M. Johnston
venusta (Piper) St. John
virginiana (L.) I.M. Johnston
        SY=Lappula v. (L.) Greene

Harpagonella Gray

    palmeri Gray
        var. arizonica I.M. Johnston
        var. palmeri

Heliotropium L.

    amplexicaule Vahl
        SY=Cochranea anchusifolia (Poir.)
            Guerke
    angiospermum Murr.
        SY=Schobera a. (Murr.) Britt.

Heliotropium (CONT.)
  anomalum Hook. & Arn.
      var. anomalum
      var. argenteum Gray
  antillanum Urban
  arborescens L.
  confertifolium (Torr.) Gray
  convolvulaceum (Nutt.) Gray
      var. californicum (Greene) I.M.
        Johnston
        SY=Euploca convolvulacea Nutt.
            ssp. californica (Greene)
            Abrams
      var. convolvulaceum
        SY=Euploca c. Nutt.
  curassavicum L.
      var. curassavicum
      var. obovatum DC.
        SY=H. spathulatum Rydb.
      var. oculatum (Heller) I.M. Johnston
        SY=H. c. ssp. o. (Heller) Thorne
  europaeum L.
  fruticosum L.
        SY=H. phyllostachyum Torr.
  glabriusculum (Torr.) Gray
  greggii Torr.
  guanicense Urban
  indicum L.
        SY=Tiaridium i. (L.) Lehm.
  microphyllum Sw.
        SY=H. crispiflorum Urban
  molle (Torr.) I.M. Johnston
  polyphyllum Lehm.
      var. horizontale (Small) R.W. Long
        SY=H. h. Small
      var. polyphyllum
        SY=H. leavenworthii Torr.
  procumbens P. Mill.
        SY=H. inundatum Sw.
  racemosum Rose & Standl.
        SY=H. convolvulaceum (Nutt.) Gray
            var. r. (Rose & Standl.)
            I.M. Johnston
  tenellum (Nutt.) Torr.
        SY=Lithococca t. (Nutt.) Small
  ternatum Vahl
  texanum I.M. Johnston
  torreyi I.M. Johnston

Lappula Fabr.

  cenchrusoides A. Nels.
  diploloma (Fisch. & Mey.) Guerke
  echinata Gilib.
  redowskii (Hornem.) Greene
      var. redowskii
        SY=L. echinata Gilib. var.
            occidentalis (S. Wats.)
            Boivin
        SY=L. o. (S. Wats.) Greene
        SY=L. r. var. desertorum (Greene)
            I.M. Johnston
        SY=L. r. var. o. (S. Wats.) Rydb.
      var. texana (Scheele) Brand
        SY=L. occidentalis (S. Wats.)
            Greene var. cupulata (Gray)
            Higgins
        SY=L. r. var. c. (Gray) M.E. Jones
        SY=L. t. (Scheele) Britt.
        SY=L. t. var. coronata (Greene) A.
            Nels. & J.F. Macbr.
        SY=L. t. var. heterospermum
            (Greene) A. Nels. & J.F.
            Macbr.
        SY=L. t. var. homosperma (A.
            Nels.) A. Nels. & J.F.
            Macbr.

  squarrosa (Retz.) Dumort.
        SY=L. fremontii (Torr.) Greene
        SY=L. lappula (L.) Karst.
        SY=L. myosotis Moench

Lithospermum L.

  californicum Gray
  calycosum (J.F. Macbr.) I.M. Johnston
  canescens (Michx.) Lehm.
        SY=Batschia c. Michx.
  caroliniense (J.F. Gmel.) MacM.
        SY=Batschia c. J.F. Gmel.
        SY=L. croceum Fern.
  cobrense Greene
  confine I.M. Johnston
  incisum Lehm.
        SY=Batschia linearifolia (Goldie)
            Small
        SY=L. angustifolium Michx.
        SY=L. l. Goldie
        SY=L. mandanense Spreng.
  latifolium Michx.
  matamorense A. DC.
  mirabile Small
  multiflorum Torr. ex Gray
  officinale L.
  parksii I.M. Johnston
      var. parksii
      var. rugulosum I.M. Johnston
  ruderale Dougl. ex Lehm.
        SY=L. pilosum Nutt.
  tuberosum Rugel ex DC.
  viride Greene

Macromeria D. Don

  viridiflora DC.
      var. thurberi (Gray) I.M. Johnston
      var. viridiflora

Mertensia Roth

  alpina (Torr.) G. Don
  arizonica Greene
      var. arizonica
      var. grahamii L.O. Williams
      var. leonardii (Rydb.) I.M. Johnston
      var. subnuda (J.F. Macbr.) L.O.
        Williams
  bakeri Greene
      var. bakeri
      var. osterhoutii L.O. Williams
  bella Piper
  brevistyla S. Wats.
  campanulata A. Nels.
  ciliata (James ex Torr.) G. Don
      var. ciliata
        SY=M. c. var. latiloba L.O.
            Williams
        SY=M. c. var. subpubescens (Rydb.)
            J.F. Macbr. & Payson
      var. stomatechoides (Kellogg) Jepson
  drummondii (Lehm.) G. Don
  franciscana Heller
  fusiformis Greene
  humilis Rydb.
  lanceolata (Pursh) A. DC.
      var. brachyloba (Greene) A. Nels.
      var. fendleri Gray
      var. lanceolata
        SY=M. linearis Greene
      var. pubens (J.F. Macbr.) L.O.
        Williams
      var. secundorum Cockerell
  longiflora Greene
  macdougalii Heller

Mertensia (CONT.)
    maritima (L.) S.F. Gray
            var. asiatica (Takeda) Welsh
                SY=M. m. ssp. a. Takeda
            var. maritima
                SY=Pneumaria m. (L.) Hill
    oblongifolia (Nutt.) G. Don
            var. amoena (A. Nels.) L.O. Williams
            var. nevadensis (A. Nels.) L.O.
                Williams
            var. oblongifolia
                SY=M. oreophila L.O. Williams
    paniculata (Ait.) G. Don
            var. alaskana (Britt.) L.O. Williams
                SY=M. a. Britt.
            var. borealis (J.F. Macbr.) L.O.
                Williams
            var. eastwoodiae (J.F. Macbr.) Hulten
                SY=M. e. J.F. Macbr.
                SY=M. p. ssp. e. (J.F. Macbr.)
                    Welsh
            var. paniculata
                SY=M. palmeri A. Nels. & J.F.
                    Macbr.
    perplexa Rydb.
    pilosa (Cham.) DC.
    platyphylla Heller
            var. platyphylla
            var. subcordata (Greene) L.O.
                Williams
    toiyabensis J.F. Macbr.
    umbratilis Greenm.
    virginica (L.) Pers. ex Link
    viridis (A. Nels.) A. Nels.
            var. caelestina (A. Nels. &
                Cockerell) L.O. Williams
            var. cana (Rydb.) L.O. Williams
            var. dilatata (A. Nels.) L.O.
                Williams
            var. parvifolia L.O. Williams
            var. viridis
                SY=M. v. var. cynoglossoides
                    (Greene) J.F. Macbr.

Myosotis L.

    alpestris F.W. Schmidt
            SY=M. sylvatica Hoffmann var. a.
                (F.W. Schmidt) Koch
    arvensis (L.) Hill
            SY=M. scorpioides L. var. a. L.
    asiatica (Vesterg.) Schischkin &
        Sergievskaja
            SY=M. alpestris F.W. Schmidt ssp.
                asiatica Vesterg.
    azorica H.C. Wats. ex Hook.
    canescens (Michx.) Lehm.
    discolor Pers.
            SY=M. versicolor (Pers.) Sm.
    latifolia Poir.
    laxa Lehm.
    macrosperma Engelm.
            SY=M. virginica (L.) B.S.P. var.
                m. (Engelm.) Fern.
    scorpioides L.
            SY=M. palustris (L.) Hill
    stricta Link ex Roemer & Schultes
            SY=M. micrantha sensu auctt. non
                Pallas
    sylvatica Hoffmann
    verna Nutt.
            SY=M. virginica sensu auctt. non
                (L.) B.S.P.

Nonea Medic.

    lutea (Desr.) DC.

    rosea (Bieb.) Link
    vesicaria (L.) Reichenb.

Omphalodes P. Mill.

    aliena Gray
    linifolia (L.) Moench
            SY=Cynoglossum l. L.
    verna Moench

Onosmodium Michx.

    helleri Small
    hispidissimum Mackenzie
            SY=O. h. var. macrospermum
                Mackenzie & Bush
            SY=O. molle Michx. ssp. h.
                (Mackenzie) Boivin
            SY=O. molle var. h. (Mackenzie)
                Cronq.
    molle Michx.
        ssp. bejariense (A. DC.) Cochrane
            SY=O. b. A. DC.
            SY=O. m. var. b. (A. DC.) Cronq.
        ssp. molle
        ssp. occidentale (Mackenzie) Cochrane
            SY=O. m. var. o. (Mackenzie) I.M.
                Johnston
            SY=O. o. Mackenzie
            SY=O. o. var. sylvestre Mackenzie
        ssp. subsetosum (Mackenzie & Bush)
            Cochrane
            SY=O. m. var. s. (Mackenzie &
                Bush) Cronq.
            SY=O. s. Mackenzie & Bush
    virginianum (L.) A. DC.
            SY=O. v. var. hirsutum Mackenzie

Pectocarya DC. ex Meisn.

    heterocarpa (I.M. Johnston) I.M. Johnston
    linearis (Ruiz & Pavon) DC.
            var. ferocula I.M. Johnston
                SY=P. l. ssp. f. (I.M. Johnston)
                    Thorne
            var. linearis
    penicillata (Hook. & Arn.) A. DC.
            SY=P. linearis (Ruiz & Pavon) DC.
                var. p. (Hook. & Arn.) M.E.
                Jones
    platycarpa (Munz & Johnston) Munz &
        Johnston
            SY=P. linearis (Ruiz & Pavon) DC.
                var. p. (Munz & Johnston)
                Cronq.
    pusilla (A. DC.) Gray
    recurvata I.M. Johnston
    setosa Gray

Pentaglottis Tausch

    sempervirens (L.) Tausch ex L.H. Bailey
            SY=Anchusa s. L.
            SY=Caryolophora s. (L.) Fisch. &
                Trautv.

Plagiobothrys Fisch. & Mey.

    acanthocarpus (Piper) I.M. Johnston
            SY=Allocarya a. Piper
    arizonicus (Gray) Greene ex Gray
    austinae (Greene) I.M. Johnston
            SY=Allocarya a. Greene
    bracteatus (T.J. Howell) I.M. Johnston
            var. aculeolatus (Piper) I.M.
                Johnston
            var. bracteatus

Plagiobothrys (CONT.)
  SY=Allocarya b. T.J. Howell
 canescens Benth.
  var. canescens
  var. catalinensis (Gray) Jepson
   SY=P. arizonicus (Gray) Greene ex
    Gray var. c. Gray
 chorisianus (Cham.) I.M. Johnston
  var. chorisianus
   SY=Allocarya c. (Cham.) Greene
  var. hickmanii (Greene) I.M. Johnston
   SY=Allocarya c. (Cham.) Greene
    var. h. (Greene) Jepson
  var. undulatus (Piper) Higgins ined.
   SY=Allocarya u. Piper
   SY=P. u. (Piper) I.M. Johnston
 collinus (Phil.) I.M. Johnston
  var. californicus (Gray) Higgins
   SY=Allocarya californica (Fisch. &
    Mey.) Greene var. minuta
    (Piper) Jepson & Hoover
   SY=A. cooperi Greene
   SY=Echidiocarya californica Gray
   SY=P. californicus (Gray) Greene
  var. fulvescens (I.M. Johnston)
   Higgins
   SY=Echidiocarya californica Gray
    ssp. fulvescens (I.M.
    Johnston) Abrams
   SY=P. californicus var. f. I.M.
    Johnston
  var. gracilis (I.M. Johnston) Higgins
   SY=Echidiocarya californica Gray
    ssp. g. (I.M. Johnston)
    Abrams
   SY=P. californicus (Gray) Greene
    var. g. I.M. Johnston
  var. ursinus (Gray) Higgins
   SY=Echidiocarya californica Gray
    var. u. (Gray) Jepson
   SY=P. californicus (Gray) Greene
    var. u. (Gray) I.M. Johnston
 diffusus (Greene) I.M. Johnston
  SY=Allocarya d. Greene
  SY=P. torreyi (Gray) Gray var. d.
   (Greene) I.M. Johnston
 distantiflorus (Piper) I.M. Johnston ex
  M.E. Peck
 fulvus (Hook. & Arn.) I.M. Johnston
  var. campestris (Greene) I.M.
   Johnston
   SY=P. c. Greene
  var. fulvus
 glaber (Gray) I.M. Johnston
  SY=Allocarya g. (Gray) J.F. Macbr.
 glyptocarpus (Piper) I.M. Johnston
  var. glyptocarpus
   SY=Allocarya g. Piper
  var. modestus I.M. Johnston
   SY=Allocarya g. Piper ssp. m.
    (I.M. Johnston) Abrams
 greenei (Gray) I.M. Johnston
  SY=Allocarya g. (Gray) Greene
 hirtus (Greene) I.M. Johnston
  var. corallicarpa (Piper) I.M.
   Johnston
   SY=Allocarya h. Greene ssp. c.
    (Piper) Abrams
   SY=P. h. Piper
  var. figuratus (Piper) I.M. Johnston
   SY=Allocarya f. Piper
   SY=P. f. (Piper) I.M. Johnston ex
    M.E. Peck
  var. hirtus
   SY=Allocarya h. Greene
 hispidus Gray
 humistratus (Greene) I.M. Johnston

  SY=Allocarya h. Greene
 hystriculus (Piper) I.M. Johnston
  SY=Allocarya h. Piper
 infectivus I.M. Johnston
 jonesii Gray
 kingii (S. Wats.) Gray
  var. harknessii (Greene) Jepson
   SY=P. h. Greene
  var. kingii
 lamprocarpus (Piper) I.M. Johnston
  SY=Allocarya l. Piper
 leptocladus (Greene) I.M. Johnston
  SY=Allocarya l. Greene
  SY=P. orthocarpus (Greene) I.M.
   Johnston
 lithocaryus (Greene ex Gray) I.M. Johnston
  SY=Allocarya l. Greene ex Gray
 mollis (Gray) I.M. Johnston
  var. mollis
   SY=Allocarya m. (Gray) Greene
  var. vestitus (Greene) I.M. Johnston
   SY=Allocarya m. (Gray) Greene var.
    v. (Greene) Jepson
 myosotoides (Lehm.) Brand
 nothofulvus (Gray) Gray
 orientalis (L.) I.M. Johnston
 parishii I.M. Johnston
 pringlei Greene
 reticulatus (Piper) I.M. Johnston
 salsus (Brandeg.) I.M. Johnston
  SY=Allocarya s. Brandeg.
 scouleri (Hook. & Arn.) I.M. Johnston
  var. penicillatus (Greene) Cronq.
  var. scouleri
   SY=Allocarya californica (Fisch. &
    Mey.) Greene
   SY=Allocarya cognata Greene
   SY=A. cusickii Greene
   SY=A. granulata Piper
   SY=A. hispidula Greene
   SY=A. scopulorum Greene
   SY=A. scouleri (Hook. & Arn.)
    Greene
   SY=P. cognatus (Greene) I.M.
    Johnston
   SY=P. cusickii (Greene) I.M.
    Johnston
   SY=P. g. (Piper) I.M. Johnston
   SY=P. h. (Greene) I.M. Johnston
   SY=P. nelsonii (Greene) I.M.
    Johnston
   SY=P. reticulatus (Piper) I.M.
    Johnston var. rossianorum
    I.M. Johnston
   SY=P. scopulorum (Greene) I.M.
    Johnston
 scriptus (Greene) I.M. Johnston
  SY=Allocarya s. Greene
 shastensis Greene ex Gray
 stipitatus (Greene) I.M. Johnston
  var. micranthus (Piper) I.M. Johnston
   SY=Allocarya s. Greene ssp. m.
    Piper
  var. stipitatus
   SY=Allocarya s. Greene
 strictus (Greene) I.M. Johnston
  SY=Allocarya s. Greene
 tenellus (Nutt. ex Hook.) Gray
  SY=P. asper Greene
 tener (Greene) I.M. Johnston
  var. subglaber I.M. Johnston
  var. tener
   SY=Allocarya t. Greene
   SY=P. t. var. fallax I.M. Johnston
 torreyi (Gray) Gray
 trachycarpus (Gray) I.M. Johnston
  SY=Allocarya t. (Gray) Greene

Plagiobothrys (CONT.)
  uncinatus J.T. Howell

Rochefortia Sw.

  acanthophora (DC.) Griseb.
  cuneata Sw.

Symphytum L.

  asperum Lepechin
  officinale L.
    ssp. officinale
    ssp. uliginosum (Kern.) Nyman
      SY=S. u. Kern.
  tuberosum L.

Tiquilia Pers.

  canescens (A. DC.) A. Richards.
    var. canescens
      SY=Coldenia c. A. DC.
      SY=C. c. var. subnuda I.M.
        Johnston
    var. pulchella (I.M. Johnston) A.
      Richards.
      SY=Coldenia c. A. DC. var. p. I.M.
        Johnston
  gossypina (Woot. & Standl.) A. Richards.
      SY=Coldenia g. (Woot. & Standl.)
        I.M. Johnston
  greggii (Torr. & Gray) A. Richards.
      SY=Coldenia g. (Torr. & Gray) Gray
  hispidissima (Torr. & Gray) A. Richards.
      SY=Coldenia h. (Torr. & Gray) Gray
  latior (I.M. Johnston) A. Richards.
      SY=Coldenia hispidissima (Torr. &
        Gray) Gray var. l. I.M.
        Johnston
  mexicana (S. Wats.) A. Richards.
      SY=Coldenia m. S. Wats.
      SY=C. m. var. tomentosa (S. Wats.)
        I.M. Johnston
  nuttallii (Benth. ex Hook.) A. Richards.
      SY=Coldenia n. Benth. ex Hook.
  palmeri (Gray) A. Richards.
      SY=Coldenia p. Gray
  plicata (Torr.) A. Richards.
      SY=Coldenia p. (Torr.) Coville

Tournefortia L.

  bicolor Sw.
  filiflora Griseb.
  gnaphalodes (L.) R. Br.
      SY=Mallotonia g. (L.) Britt.
  hirsutissima L.
  maculata Jacq.
      SY=T. laurifolia sensu auctt. non
        Vent.
      SY=T. peruviana Poir. pro parte
  poliochros Spreng.
      SY=Myriopus p. (Spreng.) Small
  scabra Lam.
  sibirica L.
      SY=Argusia s. (L.) Dandy
  volubilis L.
      SY=Myriopus v. (L.) Small
      SY=T. microphylla Bertol.

Trigonotis Steven

  peduncularis (Trev.) Benth. ex Baker & S.
    Moore

BRASSICACEAE

Alliaria Scop.

  petiolata (Bieb.) Cavara & Grande
      SY=A. alliaria (L.) Britt.
      SY=A. officinalis Andrz. ex Bieb.
      SY=Sisymbrium a. (L.) Scop.

Alyssum L.

  alyssoides (L.) L.
      SY=A. calycinum L.
  americanum Greene
  desertorum Stapf
  minus (L.) Rothm.
      var. micranthum (A. Mey.) Dudley
      var. minus
  montanum L.
  murale Waldst. & Kit.
  strigosum Banks & Soland.
  szovitsianum Fisch. & Mey.

Anelsonia J.F. Macbr. & Payson

  eurycarpa (Gray) J.F. Macbr. & Payson
      SY=Phoenicaulis e. (Gray) Abrams
      SY=Parrya e. (Gray) Jepson

Aphragmus Andrz. ex DC.

  eschscholtzianus Andrz. ex DC.

Arabidopsis (DC.) Heynh.

  thaliana (L.) Heynh.
      SY=Arabis t. L.
      SY=Sisymbrium t. (L.) J. Gay &
        Monn.

Arabis L.

  aculeolata Greene
  alpina L.
      var. albida (Stev.) Paolette
      var. alpina
      var. glabrata Blytt
  arenicola (Richards.) Gelert
      var. arenicola
      var. pubescens (S. Wats.) Gelert
  blepharophylla Hook. & Arn.
  breweri S. Wats.
      var. austinae (Greene) Rollins
      var. breweri
      var. pecunaria Rollins
  canadensis L.
  caucasica Willd.
  cobrensis M.E. Jones
  constancei Rollins
      SY=A. suffrutescens S. Wats. var.
        perstylosa Rollins
  crandallii B.L. Robins.
  crucisetosa Constance & Rollins
  cusickii S. Wats.
  davidsonii Greene
  demissa Greene
      var. demissa
      var. languida Rollins
      var. russeola Rollins
  dispar M.E. Jones
  divaricarpa A. Nels.
      var. dacotica (Greene) Boivin
        SY=A. bourgovii Rydb.
      var. dechamplainii Boivin
      var. divaricarpa
        SY=A. brachycarpa (Torr. & Gray)

Arabis (CONT.)

        Britt.
        SY=A. confinis sensu auctt. non S.
          Wats.
        SY=A. d. var. typica Rollins
    var. interposita (Greene) Rollins
        SY=A. acutina Greene
        SY=A. confinis S. Wats. var. i.
          (Greene) Welsh & Reveal
    var. stenocarpa M. Hopkins
drummondii Gray
    var. connexa (Greene) Fern.
    var. drummondii
        SY=?A. confinis S. Wats.
fendleri (S. Wats.) Greene
    var. fendleri
    var. spatifolia (Rydb.) Rollins
fernaldiana Rollins
    var. fernaldiana
        SY=A. f. var. typica Rollins
    var. stylosa (S. Wats.) Rollins
fructicosa A. Nels.
furcata S. Wats.
    var. furcata
    var. olympica (Piper) Rollins
        SY=A. o. Piper
georgiana Harper
glabra (L.) Bernh.
    var. furcatipilis M. Hopkins
    var. glabra
        SY=Turritis g. L.
glaucovalvula M.E. Jones
gracilipes Greene
gunnisoniana Rollins
hastatula Greene
hirsuta (L.) Scop.
    var. adpressipilis (M. Hopkins)
        Rollins
        SY=A. pycnocarpa M. Hopkins var.
          a. M. Hopkins
    var. eschscholtziana (Andrz.) Rollins
        SY=A. h. ssp. e. (Andrz.) Hulten
    var. glabrata Torr. & Gray
    var. pycnocarpa (M. Hopkins) Rollins
        SY=A. h. ssp. p. (M. Hopkins)
          Hulten var. p. (M. Hopkins)
          Rollins
        SY=A. ovata Michx.
        SY=A. p. M. Hopkins
hoffmannii (Munz) Rollins
holboellii Hornem.
    var. collinsii (Fern.) Rollins
        SY=A. retrofracta Graham var. c.
          (Fern.) Boivin
    var. holboellii
    var. pendulocarpa (A. Nels.) Rollins
        SY=A. p. A. Nels.
    var. pinetorum (Tidestrom) Rollins
    var. retrofracta (Graham) Rydb.
        SY=A. r. Graham
        SY=A. r. var. multicaulis Boivin
        SY=A. secunda T.J. Howell
    var. tenuis Bocher
inyoensis Rollins
johnstonii Munz
koehleri T.J. Howell
    var. koehleri
    var. stipitata Rollins
laevigata (Muhl.) Poir.
    var. burkii Porter
        SY=A. b. (Porter) Small
        SY=A. serotina Steele
    var. laevigata
lemmonii S. Wats.
    var. depauperata (A. Nels. & Kennedy)
        Rollins
    var. drepanoloba (Greene) Rollins

    var. lemmonii
        SY=A. l. var. typica Rollins
    var. paddoensis Rollins
lignifera A. Nels.
lyallii S. Wats.
    var. lyallii
    var. nubigena (J.F. Macbr. & Payson)
        Rollins
lyrata L.
    var. glabra (DC.) M. Hopkins
    var. kamchatica Fisch. ex DC.
        SY=A. l. ssp. k. (Fisch. ex DC.)
          Hulten
    var. lyrata
mcdonaldiana Eastw.
microphylla Nutt.
    var. macounii (S. Wats.) Rollins
    var. microphylla
    var. saximontana Rollins
    var. thompsonii Rollins
missouriensis Greene
    var. deamii (M. Hopkins) M. Hopkins
        SY=A. viridis Harger var. d. M.
          Hopkins
    var. missouriensis
        SY=A. laevigata (Muhl.) Poir. var.
          m. (Greene) Ahles
        SY=A. viridis Harger
modesta Rollins
nuttallii B.L. Robins.
        SY=A. macella Piper
oregona Rollins
        SY=A. purpurascens T.J. Howell
oxylobula Greene
parishii S. Wats.
patens Sullivant
pendulina Greene
perennans S. Wats.
perstellata E.L. Braun
    var. ampla Rollins
    var. perstellata
petiolaris (Gray) Gray
platysperma Gray
    var. howellii (S. Wats.) Jepson
        SY=A. covillei Greene
        SY=A. h. S. Wats.
    var. platysperma
        SY=A. inamoena Greene
procurrens Waldst. & Kit.
puberula Nutt.
        SY=A. beckwithii S. Wats. pro
          parte
pulchra M.E. Jones ex S. Wats.
    var. gracilis M.E. Jones
    var. munciensis M.E. Jones
    var. pallens M.E. Jones
    var. pulchra
pygmaea Rollins
rectissima Greene
repanda S. Wats.
    var. greenei Jepson
    var. repanda
rigidissima Rollins
schistacea Rollins
selbyi Rydb.
serpenticola Rollins
shockleyi Munz
shortii (Fern.) Gleason
    var. phalacrocarpa (M. Hopkins)
        Steyermark
        SY=A. dentata (Torr.) Torr. & Gray
          var. p. M. Hopkins
        SY=A. perstellata E.L. Braun var.
          phalacrocarpa (M. Hopkins)
          Fern.
    var. shortii
        SY=A. dentata (Torr.) Torr. & Gray

Arabis   (CONT.)
         SY=A. perstellata E.L. Braun var.
            s. Fern.
   sparsiflora Nutt.
      var. arcuata (Nutt.) Rollins
         SY=A. maxima Greene
      var. atrorubens (Greene) Rollins
         SY=A. a. Greene
      var. californica Rollins
      var. columbiana (Macoun) Rollins
      var. sparsiflora
      var. subvillosa (S. Wats.) Rollins
         SY=A. campyloloba Greene
   subpinnatifida S. Wats.
   suffrutescens S. Wats.
      var. horizontalis (Greene) Rollins
      var. suffrutescens
   tricornuta Rollins

Armoracia Gaertn., Mey. & Scherb.

   aquatica (Eat.) Wieg.
         SY=Neobeckia a. (Eat.) Greene
         SY=Rorippa americana (Gray) Britt.
         SY=R. aquatica (Eat.) Palmer &
            Steyermark
   rusticana (Lam.) Gaertn., Mey. & Scherb.
         SY=A. armoracia (L.) Britt.
         SY=A. lapathifolia Gilib.
         SY=Rorippa a. (L.) A.S. Hitchc.

Athysanus Greene

   pusillus (Hook.) Greene
      var. glabrior S. Wats.
      var. pusillus

Aurinia Desv.

   petraea (Ard.) Schur
         SY=Alyssum p. Ard.
   saxatilis (L.) Desv.
         SY=Alyssum s. L.

Barbarea R. Br.

   orthoceras Ledeb.
      var. dolichocarpa Fern.
      var. orthoceras
         SY=B. americana Rydb.
   stricta Andrz.
         SY=Campe s. (Andrz.) W. Wight
   verna (P. Mill.) Aschers.
         SY=Campe v. (P. Mill.) Heller
   vulgaris R. Br.
      var. arcuata (Opiz ex J. & C. Presl)
        Fries
         SY=B. a. (Opiz ex J. & C. Presl)
            Reichemb.
         SY=Campe barbarea (L.) W. Wight
      var. brachycarpa Rouy & Foucaud
      var. longisiliquosa Carion
      var. sylvestris Fries
      var. vulgaris

Berteroa DC.

   incana (L.) DC.
         SY=Alyssum i. L.
   mutabilis (Vent.) DC.

Brassica L.

   elongata Ehrh.
   fruticulosa Cyrillo
   geniculata (Desf.) J. Ball
   integrifolia (West) O.E. Schulz

   juncea (L.) Czern.
      var. crispifolia Bailey
      var. japonica (Thunb.) Bailey
         SY=B. j. Thunb.
      var. juncea
         SY=Sinapis j. L.
   napus L.
      var. napobrassica (L.) Reichenb.
         SY=B. napobrassica (L.) P. Mill.
      var. napus
   nigra (L.) W.D.J. Koch
         SY=Sinapis n. L.
   oleracea L.
      var. capitata L.
      var. oleracea
   pekinensis (Lour.) Rupr.
   rapa L.
      ssp. olifera DC.
         SY=B. campestris L.
         SY=B. r. ssp. c. (L.) Clapham
         SY=Caulanthus sulfureus Payson
      ssp. rapa
         SY=B. campestris L. var. r. (L.)
            Hartman
      ssp. sylvestris (L.) Janchen
   tournefortii Gouan

Braya Sternb. & Hoppe

   americana (Hook.) Fern.
         SY=B. humilis (C.A. Mey.) B.L.
            Robins. var. a. (Hook.)
            Boivin
   glabella Richards.
         SY=B. bartlettiana Jordal
         SY=B. humilis (C.A. Mey.) B.L.
            Robins. var. g. (Richards.)
            Boivin
   henryae Raup
   humilis (C.A. Mey.) B.L. Robins.
      ssp. arctica (Bocher) Rollins
      ssp. humilis
         SY=B. h. var. laurentiana (Bocher)
            Boivin
      ssp. richardsonii (Rydb.) Hulten
         SY=Arabidopsis r. Rydb.
         SY=B. fernaldii Abbe
         SY=B. purpurascens (R. Br.) Bunge
            ex Ledeb. var. f. (Abbe)
            Boivin
         SY=B. r. (Rydb.) Fern.
      ssp. ventosa Rollins
   intermedia Sorensen
   linearis Rouy
   novae-angliae (Rydb.) Sorensen
      ssp. abbei Bocher
         SY=B. humilis (C.A. Mey.) B.L.
            Robins. var. a. (Bocher)
            Boivin
      ssp. novae-angliae
        var. interior Bocher
         SY=B. humilis (C.A. Mey.) B.L.
            Robins. var. i. (Bocher)
            Boivin
        var. novae-angliae
        SY=Arabidopsis n-a. (Rydb.) Britt.
        SY=B. humilis (C.A. Mey.) B.L.
          Robins. var. leiocarpa
          (Trautv.) Fern.
   pilosa Hook.
   purpurascens (R. Br.) Bunge ex Ledeb.
      var. longii (Fern.) Boivin
         SY=B. l. Fern.
      var. purpurascens
   thorild-wulffii Ostenf.
         SY=B. purpurascens (R. Br.) Bunge
           ex Ledeb. var. t-w.

Braya (CONT.)
                    (Ostenf.) Boivin

Bunias L.

     erucago L.
     orientalis L.

Cakile P. Mill.

     constricta Rodman
     edentula (Bigelow) Hook.
          ssp. edentula
               SY=C. americana Nutt.
               SY=C. e. ssp. californica (Heller)
                    Hulten
               SY=C. e. var. c. (Heller) Fern.
          ssp. harperi (Small) Rodman
               SY=C. h. Small
          ssp. lacustris (Fern.) Hulten
               SY=C. e. ssp. e. var. l. Fern.
     geniculata (B.L. Robins.) Millsp.
               SY=C. lanceolata (Willd.) O.E.
                    Schulz var. g. (B.L.
                    Robins.) Shinners
     lanceolata (Willd.) O.E. Schulz
          ssp. fusiformis (Greene) Rodman
               SY=C. chapmanii Millsp.
               SY=C. f. Greene
          ssp. lanceolata
          ssp. pseudoconstricta Rodman
     maritima Scop.
               SY=C. cakile (L.) Karst.

Calepina Adans.

     irregularis (Asso) Thellung

Camelina Crantz

     microcarpa Andrz. ex DC.
               SY=C. sativa (L.) Crantz ssp. m.
                    (Andrz. ex DC.) Schneid.
     sativa (L.) Crantz
               SY=C. dentata Pers.
               SY=C. parodii Ibarra & La Porte

Capsella Medic.

     bursa-pastoris (L.) Medic.
          var. bifida Crepin
          var. bursa-pastoris
               SY=Bursa b-p. (L.) Britt.
               SY=C. gracilis Gren.
     rubella Reut.

Cardamine L.

     angulata Hook.
     bellidifolia L.
          var. bellidifolia
          var. pachyphylla Coville & Leib.
          var. pinnatifida Hulten
          var. sinuata (J. Vahl) Lange
     breweri S. Wats.
          var. breweri
          var. leibergii (Holz.) C.L. Hitchc.
               SY=C. l. Holz.
          var. orbicularis (Greene) Detling
     bulbosa (Schreb.) B.S.P.
     californica (Nutt.) Greene
               SY=Dentaria c. Nutt.
               SY=D. integrifolia Nutt. var. c.
                    (Nutt.) Jepson
     cardiophylla Greene
               SY=Dentaria californica Nutt. var.
                    cardiophylla (Greene)

                    Detling
     clematitis Shuttlw. ex Gray
               SY=C. hugeri Small
     constancei Detling
     cordifolia Gray
          var. cordifolia
          var. incana Gray
     cuneata Greene
               SY=C. californica (Nutt.) Greene
                    ssp. cuneata (Greene) O.E.
                    Schulz
               SY=Dentaria californica Nutt. var.
                    cuneata (Greene) Detling
     debilis D. Don
     digitata Richards.
               SY=C. hyperborea O.E. Schulz
               SY=C. richardsonii Hulten
     douglassii (Torr.) Britt.
     flagellifera O.E. Schulz
     flexuosa With.
     hirsuta L.
     impatiens L.
     integrifolia (Nutt.) Greene
          var. integrifolia
               SY=Dentaria californica Nutt. var.
                    i (Nutt.) Detling
               SY=D. i. Nutt.
          var. sinuata (Greene) C.L. Hitchc.
               SY=D. californica Nutt. var. s.
                    (Greene) Detling
     konaensis St. John
     longii Fern.
     lyallii S. Wats.
               SY=C. cordifolia Gray var. l. (S.
                    Wats.) A. Nels. & J.F.
                    Macbr.
     macrocarpa Brandeg.
          var. macrocarpa
          var. texana Rollins
     micranthera Rollins
     microphylla J.E. Adams
               SY=C. minuta Willd.
     nuttallii Greene
          var. nuttallii
               SY=C. pulcherrima Greene var.
                    tenella (Pursh) C.L. Hitchc.
               SY=Dentaria t. Pursh
               SY=D. t. var. palmata Detling
          var. pulcherrima (Greene) Taylor &
               MacBryde
               SY=C. p. Greene
               SY=Dentaria tenella Pursh var. p.
                    (Greene) Detling
               SY=D. t. var. quercetorum (T.J.
                    Howell) Detling
     nymanii Gandog.
     occidentalis (S. Wats. ex B.L. Robins.)
          T.J. Howell
     oligosperma Nutt.
               SY=C. unijuga Rydb.
     parviflora L.
          ssp. parviflora
               var. arenicola (Britt.) O.E. Schulz
                    SY=C. a. Britt.
               var. parviflora
          ssp. virginica (L.) O.E. Schulz
     pattersonii Henderson
     penduliflora O.E. Schulz
     pensylvanica Muhl. ex Willd.
               SY=C. p. var. brittoniana Farw.
     pratensis L.
          var. angustifolia Hook.
               SY=C. p. ssp. a. (Hook.) O.E.
                    Schulz
          var. pratensis
     purpurea Cham. & Schlecht.
          var. albiflora Hulten

Cardamine (CONT.)
        var. purpurea
    regeliana Miq.
           SY=C. scutata Thunb.
    rotundifolia Michx.
    rupicola (Rydb.) C.L. Hitchc.
    umbellata Greene
           SY=C. oligosperma Nutt. var.
              kamtschatica (Regel) Detling
    vallicola Greene

Cardaria Desv.

    chalapensis (L.) Hand.-Maz.
           SY=C. draba (L.) Desv. var. repens
              (Schrenk) O.E. Schulz
           SY=Lepidium r. (Schrenk) Boiss.
    draba (L.) Desv.
           SY=Lepidium d. L.
    pubescens (C.A. Mey.) Jarmolenko
           SY=C. p. var. elongata Rollins
           SY=Hymenophysa p. C.A. Mey.

Caulanthus S. Wats.

    amplexicaulis S. Wats.
        var. amplexicaulis
        var. barbarae (J.T. Howell) Munz
    californicus (S. Wats.) Payson
    cooperi (S. Wats.) Payson
           SY=Thelypodium c. S. Wats.
    coulteri S. Wats.
    crassicaulis (Torr.) S. Wats.
           SY=Streptanthus c Torr.
    divaricatus Rollins
           SY=Thelypodiopsis d. (Rollins)
              Welsh & Reveal
    flavescens (Hook.) Payson
           SY=Thelypodium f. (Hook.) S. Wats.
    glaber (M.E. Jones) Rydb.
    glaucus S. Wats.
           SY=Streptanthus g. (S. Wats.)
              Jepson
    hallii Payson
    heterophyllus (Nutt.) Payson
    inflatus S. Wats.
    lasiophyllus (Hook. & Arn.) Payson
        var. inalienus (B.L. Robins.) Payson
           SY=Thelypodium l. (Hook. & Arn.)
              Greene var. i. B.L. Robins.
        var. lasiophyllus
           SY=Thelypodium l. (Hook. & Arn.)
              Greene
        var. rigidus (Greene) Payson
           SY=Thelypodium l. (Hook. & Arn.)
              Greene var. r. (Greene) B.L.
              Robins.
           SY=T. r. Greene
        var. utahensis (Rydb.) Payson
           SY=Thelypodium l. (Hook. & Arn.)
              Greene var. u. (Rydb.)
              Jepson
           SY=T. u. Rydb.
    lemmonii S. Wats.
           SY=C. coulteri S. Wats. var. l.
              (S. Wats.) Jepson
    major (M.E. Jones) Payson
    pilosus S. Wats.
           SY=Streptanthus p. (S. Wats.)
              Jepson
    simulans Payson
    stenocarpus Payson

Caulostramina Rollins

    jaegeri (Rollins) Rollins
           SY=Thelypodium j. Rollins

Chlorocrambe Rydb.

    hastata (S. Wats.) Rydb.

Chorispora R. Br.

    tenella (Pallas) DC.

Christolea Camb. ex Jacq.

    parryoides (Cham.) N. Busch

Cochlearia L.

    cyclocarpa Blake
    groenlandica L.
           SY=C. danica L.
           SY=C. officinalis L. ssp. arctica
              (Schlecht.) Hulten
           SY=C. o. ssp. oblongifolia (DC.)
              Hulten
           SY=C. officinalis var. a.
              (Schlecht.) Gelert ex
              Anders. & Hessel
           SY=Cochleariopsis g. (L.) Love &
              Love
           SY=Cochleariopsis g. ssp. a.
              (Schlecht.) Love & Love
           SY=Cochleariopsis g. ssp.
              oblongifolia (DC.) Love &
              Love
    officinalis L.
           SY=Cochlearia tridactylites Banks
    sessilifolia Rollins
           SY=Cochlearia officinalis L. ssp.
              arctica (Schlecht.) Hulten
              var. s. (Rollins) Hulten

Conringia Link

    orientalis (L.) Dumort.

Coronopus Zinn

    didymus (L.) Sm.
           SY=Carara d. (L.) Britt.
    squamatus (Forsk.) Aschers.
           SY=Carara coronopus (L.) Medic.
           SY=Coronopus procumbens Gilib.

Crambe L.

    maritima L.

Dentaria L.

    X anomala Eames [diphylla X heterophylla]
    diphylla Michx.
           SY=Cardamine d. (Michx.) Wood
           SY=D. incisa Small
    gemmata (Greene) T.J. Howell
    heterophylla Nutt.
           SY=Cardamine angustata O.E. Schulz
    X incisifolia Eames ex Britt. [laciniata X
        maxima]
    laciniata Muhl. ex Willd.
        var. coalescens Fern.
        var. laciniata
           SY=Cardamine concatenata (Michx.)
              Ahles
           SY=D. l. var. integra (O.E.
              Schulz) Fern.
    maxima Nutt.
    multifida Muhl.
           SY=Cardamine angustata O.E. Schulz
              var. m. (Muhl.) Ahles
           SY=D. furcata Small

Dentaria (CONT.)
  pachystigma S. Wats.
        var. dissectifolia Detling
        var. pachystigma
            SY=D. p. var. corymbosa (Jepson)
                Abrams

Descurainia Webb & Berth.

  californica (Gray) O.E. Schulz
  obtusa (Greene) O.E. Schulz
    ssp. adenophora (Woot. & Standl.)
        Detling
    ssp. brevisiliqua Detling
    ssp. obtusa
        SY=D. o. ssp. typica Detling
  pinnata (Walt.) Britt.
    ssp. brachycarpa (Richards.) Detling
        SY=D. b. (Richards.) O.E. Schulz
        SY=D. p. var. b. (Richards.) Fern.
        SY=Sophia b. (Richards.) Rydb.
    ssp. filipes (Gray) Detling
        SY=D. p. var. f. (Gray) M.E. Peck
        SY=Sophia f. (Gray) Heller
    ssp. glabra (Woot. & Standl.) Detling
        SY=D. p. var. g. (Woot. & Standl.)
            Shinners
    ssp. halictorum (Cockerell) Detling
        SY=D. p. var. h. (Cockerell) M.E.
            Peck
    ssp. intermedia (Rydb.) Detling
        SY=D. p. var. i. (Rydb.) C.L.
            Hitchc.
        SY=Sophia i. Rydb.
    ssp. menziesii (DC.) Detling
        SY=Sophia millefolia Rydb.
    ssp. nelsonii (Rydb.) Detling
        SY=D. p. var. n. (Rydb.) M.E. Peck
    ssp. ochroleuca (Woot.) Detling
        SY=D. p. var. o. (Woot.) Shinners
    ssp. paradisa (A. Nels. & Kennedy)
        Detling
        SY=D. p. (A. Nels. & Kennedy) O.E.
            Schulz
        SY=D. pinnata var. paradisa (A.
            Nels. & Kennedy) M.E. Peck
    ssp. paysonii Detling
        SY=D. p. var. p. (Detling) Welsh &
            Reveal
    ssp. pinnata
        var. osmiarum (Cockerell) Shinners
        var. pinnata
            SY=Sophia multifida Gilib.
            SY=S. p. (Walt.) T.J. Howell
  richardsonii (Sweet) O.E. Schulz
    ssp. incisa (Engelm.) Detling
        SY=D. r. var. sonnei (B.L.
            Robins.) C.L. Hitchc.
        SY=Sophia i. (Engelm.) Greene
    ssp. procera (Greene) Detling
        SY=D. r. var. brevipes (Nutt.)
            Welsh & Reveal
        SY=D. r. var. macrosperma O.E.
            Schulz
        SY=Sophia hartwegiana (Fourn.)
            Greene
    ssp. richardsonii
        SY=Sophia richardsoniana (Sweet)
            Rydb.
    ssp. viscosa (Rydb.) Detling
        SY=D. r. var. v. (Rydb.) M.E. Peck
  sophia (L.) Webb ex Prantl
        SY=Sophia s. (L.) Britt.
  sophioides (Fisch.) O.E. Schulz

Dimorphocarpa Rollins

  palmeri (Payson) Rollins
        SY=Dithyrea wislizenii Engelm.
            var. p. Payson
  pinnatifida Rollins
  wislizenii (Engelm.) Rollins
        SY=Dithyrea w. Engelm.
        SY=Dithyrea griffithsii Woot. &
            Standl.

Diplotaxis DC.

  erucoides (L.) DC.
  muralis (L.) DC.
  tenuifolia (L.) DC.

Dithyrea Harvey

  californica Harvey
  maritima A. Davids.
        SY=D. californica Harvey var. m.
            (A. Davids.) A. Davids.

Draba L.

  adamsii Ledeb.
  albertina Greene
        SY=D. stenoloba Ledeb. var. nana
            (O.E. Schulz) C.L. Hitchc.
  aleutica Ekman
  alpina L.
        var. alpina
            SY=D. eschscholtzii Pohle ex N.
                Busch
            SY=D. micropetala Hook.
            SY=D. pilosa M.F. Adams ex DC.
        var. nana Hook.
  apiculata C.L. Hitchc.
        var. apiculata
        var. daviesiae C.L. Hitchc.
            SY=D. densifolia Nutt. var.
                daviesiae (C.L. Hitchc.)
                Welsh & Reveal
  aprica Beadle
  arabisans Michx.
        var. arabisans
        var. canadensis (Burnet) Fern. &
            Knowlt.
  arctica J. Vahl
    ssp. arctica
    ssp. ostenfeldii (Ekman) Bocher
        SY=D. o. Ekman
  arctogena Ekman
  argyraea Rydb.
  arida C.L. Hitchc.
  asprella Greene
        var. asprella
            SY=D. a. var. typica C.L. Hitchc.
        var. kaibabensis C.L. Hitchc.
        var. stelligera O.E. Schulz
        var. zionensis (C.L. Hitchc.) Welsh &
            Reveal
            SY=D. z. C.L. Hitchc.
  asterophora Payson
        var. asterophora
        var. macrocarpa C.L. Hitchc.
  aurea Vahl
        var. aurea
            SY=D. minganensis (Victorin) Fern.
        var. leiocarpa (Payson & St. John)
            C.L. Hitchc.
  aureola S. Wats.
  barbata Pohle
  borealis DC.
        var. borealis
            SY=D. mccallae Rydb.
        var. maxima (Hulten) Welsh
            SY=D. m. Hulten

Draba (CONT.)
   brachycarpa Nutt. ex Torr. & Gray
   brachystylis Rydb.
   breweri S. Wats.
         SY=D. b. var. sublaxa Jepson
   cana Rydb.
         SY=D. stylaris J. Gay ex W.D.J.
           Koch
   chamissonis G. Don
   cinerea J.E. Adams
      var. cinerea
      var. ladogenesis Lindbl. f.
   corrugata S. Wats.
      var. corrugata
      var. saxosa (A. Davids.) Munz &
        Johnston
   corymbosa R. Br. ex DC.
         SY=D. bellii Holm
         SY=D. macrocarpa J.E. Adams
   crassa Rydb.
   crassifolia Graham
      var. crassifolia
         SY=D. c. var. typica C.L. Hitchc.
      var. nevadensis C.L. Hitchc.
   cruciata Payson
      var. cruciata
      var. integrifolia C.L. Hitchc. & C.W.
        Sharsmith
   cuneifolia Nutt. ex Torr. & Gray
      var. cuneifolia
         SY=D. c. var. helleri (Small) O.E.
           Schulz
         SY=D. c. var. leiocarpa O.E.
           Schulz
         SY=D. c. var. typica C.L. Hitchc.
      var. integrifolia S. Wats.
         SY=D. sonorae Greene var. i. (S.
           Wats.) O.E. Schulz
      var. sonorae (Greene) Parish
         SY=D. s. Greene
   densifolia Nutt.
         SY=D. caeruleomontana Payson & St.
           John
         SY=D. nelsonii J.F. Macbr. &
           Payson
         SY=D. sphaerula J.F. Macbr. &
           Payson
   douglasii Gray
      var. crockeri (Lemmon) C.L. Hitchc.
      var. douglasii
   exunguiculata (O.E. Schulz) C.L. Hitchc.
   fladnizensis Wulfen
   glabella Pursh
      var. glabella
         SY=D. hirta L.
         SY=D. h. var. laurentiana (Fern.)
           Boivin
         SY=D. l. Fern.
      var. megasperma (Fern. & Knowlt.)
        Fern.
         SY=D. m. Fern. & Knowlt.
      var. orthocarpa (Fern. & Knowlt.)
        Fern.
      var. pycnosperma (Fern. & Knowlt.)
        Mulligan
         SY=D. hirta L. var. p. (Fern. &
           Knowlt.) Boivin
         SY=D. p. Fern. & Knowlt.
   graminea Greene
   grayana (Rydb.) C.L. Hitchc.
   gredinii Ekman
   helleriana Greene
      var. bifurcata C.L. Hitchc.
      var. blumeri C.L. Hitchc.
      var. helleriana
      var. patens (Heller) O.E. Schulz
   howellii S. Wats.

      var. carnosula (O.E. Schulz) C.L.
        Hitchc.
      var. howellii
   hyperborea (L.) Desv.
   incana L.
      var. confusa (Ehrh.) Lilj.
      var. incana
   incerta Payson
         SY=D. peasei Fern.
   jaegeri Munz & Johnston
   juniperina Dorn
   kamtschatica (Ledeb.) N. Busch
         SY=D. nivalis Lilj. var. k.
           (Ledeb.) Pohle
   kananaskis Mulligan
   lactea M.F. Adams
         SY=D. allenii Fern.
         SY=D. fladnizensis Wulfen ex Jacq.
           var. heterotricha (Lindbl.)
           J. Ball
   lemmonii S. Wats.
      var. cyclomorpha (Payson) O.E. Schulz
      var. incrassata Rollins
      var. lemmonii
         SY=D. longisquamosa O.E. Schulz
   lonchocarpa Rydb.
      var. lonchocarpa
         SY=D. l. var. exigua O.E. Schulz
         SY=D. nivalis Lilj. var. elongata
           S. Wats.
         SY=D. n. var. exigua (O.E. Schulz)
           C.L. Hitchc.
      var. thompsonii (C.L. Hitchc.)
        Rollins
      var. vestita O.E. Schulz
   longipes Raup
   macounii O.E. Schulz
   maguirei C.L. Hitchc.
      var. burkei C.L. Hitchc.
      var. maguirei
   mogollonica Greene
   nemorosa L.
      var. leiocarpa Lindbl.
         SY=D. lutea Gilib.
      var. nemorosa
   nivalis Lilj.
      var. brevicula Rollins
      var. denudata (O.E. Schulz) C.L.
        Hitchc.
      var. nivalis
   norvegica Gunn.
      var. clivicola (Fern.) Boivin
         SY=D. c. Fern.
      var. hebecarpa (Lindbl.) O.E. Schulz
      var. norvegica
         SY=D. rupestris R. Br.
         SY=D. r. var. leiocarpa O.E.
           Schulz
      var. pleiophylla Fern.
      var. sornborgeri (Fern.) Boivin
   oblongata R. Br.
         SY=D. arctica J. Vahl ssp.
           groenlandica (Ekman) Bocher
         SY=D. g. Ekman
   ogilviensis Hulten
   oligosperma Hook.
         SY=D. o. var. subsessilis (S.
           Wats.) O.E. Schulz
         SY=D. s. S. Wats.
   oreibata J.F. Macbr. & Payson
   palanderiana Kjellm.
         SY=D. caesia J.E. Adams
   parryi Rydb.
   paucifructa Clokey & C.L. Hitchc.
   paysonii J.F. Macbr.
         SY=D. novolympica Payson & St.
           John

Draba (CONT.)
        SY=D. p. var. treleasii (O.E.
          Schulz) C.L. Hitchc.
  pectinipila Rollins
        SY=D. oligosperma Hook. var. p.
          (Rollins) C.L. Hitchc.
  petrophila Greene
    var. petrophila
    var. viridis (Heller) C.L. Hitchc.
  platycarpa Torr. & Gray
        SY=D. cuneifola Nutt. ex Torr. &
          Gray var. p. (Torr. & Gray)
          S. Wats.
  porsildii Mulligan
  praealta Greene
        SY=D. cascadensis Payson & St.
          John
  pseudopilosa Pohle
  pterosperma Payson
  quadricostata Rollins
  ramosissima Desv.
        SY=D. r. var. glabrifolia E.L.
          Braun
  rectifructa C.L. Hitchc.
  reptans (Lam.) Fern.
    ssp. reptans
        SY=D. caroliniana Walt.
        SY=D. r. var. typica C.L. Hitchc.
    ssp. stellifera (O.E. Schulz) Abrams
        SY=D. micrantha Nutt.
        SY=D. r. var. m. (Nutt.) Fern.
        SY=D. r. var. s. (O.E. Schulz)
          C.L. Hitchc.
  ruaxes Payson & St. John
        SY=D. exalata Ekman
        SY=D. ventosa Gray var. r. (Payson
          & St. John) C.L. Hitchc.
  sibirica (Pallas) Thellung
  sierrae C.W. Sharsmith
  smithii Gilg ex O.E. Schulz
  sobolifera Rydb.
  spectabilis Greene
    var. dasycarpa (O.E. Schulz) C.L.
      Hitchc.
    var. oxyloba (Greene) Gilg ex O.E.
      Schulz
    var. spectabilis
        SY=D. s. var. typica C.L. Hitchc.
  sphaerocarpa J.F. Macbr. & Payson
  sphaeroides Payson
    var. cusickii (B.L. Robins.) C.L.
      Hitchc.
        SY=D. c. B.L. Robins.
    var. sphaeroides
  standleyi J.F. Macbr. & Payson
  stenoloba Ledeb.
    var. ramosa C.L. Hitchc.
    var. stenoloba
  stenopetala Trautv.
    var. purpurea Hulten
    var. stenopetala
  streptocarpa Gray
    var. streptocarpa
    var. tonsa (Woot. & Standl.) O.E.
      Schulz
  subalpina Goodman & C.L. Hitchc.
  subcapitata Simm.
  ventosa Gray
  yukonensis Porsild

Dryopetalon Gray

  runcinatum Gray

Erophila DC.

  verna (L.) Chev.

    ssp. praecox (Steven) S.M. Walters
      SY=Draba v. L. var. aestivalis
        Lej.
      SY=D. v. var. boerhaavii van Hall
    ssp. spathulata (A.F. Lang) S.M. Walters
    ssp. verna
      SY=Draba v. L.

Eruca P. Mill.

  vesicaria (L.) Cav.
    ssp. sativa (P. Mill.) Thellung
      SY=Brassica eruca L.
      SY=E. e. (L.) Britt.
      SY=E. s. P. Mill.
    ssp. vesicaria

Erucastrum Presl

  gallicum (Willd.) O.E. Schulz
      SY=Brassica erucastrum L.
  nasturtiifolium (Poir.) O.E. Schulz

Erysimum L.

  ammophilum Heller
  arenicola S. Wats.
    var. arenicola
    var. torulosum (Piper) C.L. Hitchc.
      SY=E. t. Piper
  argillosum (Greene) Rydb.
      SY=E. capitatum (Dougl.) Greene
        var. a (Greene) R.J. Davis
  asperum (Nutt.) DC.
    var. angustifolia (Rydb.) Boivin
    var. asperum
      SY=Cheirinia a. (Nutt.) Britt.
      SY=E. asperrimum (Greene) Rydb.
  capitatum (Dougl.) Greene
    var. amoenum (Greene) R.J. Davis
    var. angustatum (Greene) G. Rossb.
      SY=E. a. Greene
    var. bealianum (Jepson) G. Rossb.
      SY=E. asperum (Nutt.) DC. var. b.
        Jepson
    var. capitatum
      SY=E. arkansanum Nutt.
      SY=E. asperum (Nutt.) DC. var. c.
        (Dougl.) Boivin
      SY=E. wheeleri Rothr.
    var. stellatum (J.T. Howell)
      Twisselmann
    var. washoensis G. Rossb.
  cheiranthoides L.
    ssp. altum Ahti
    ssp. cheiranthoides
      SY=Cheirinia c. (L.) Link
  cheirii (L.) Crantz
  concinnum Eastw.
  desertorum (Woot. & Standl.) G. Rossb.
  franciscanum G. Rossb.
    var. crassifolium G. Rossb.
    var. franciscanum
  hieraciifolium L.
  inconspicuum (S. Wats.) MacM.
    var. coarctatum (Fern.) G. Rossb.
      SY=E. c. Fern.
    var. inconspicuum
      SY=Cheirinia i. (S. Wats.) Britt.
      SY=E. parviflorum Nutt.
  insulare Greene
  menziesii (Hook.) Wettst.
  moniliforme Eastw.
  nivale (Greene) Rydb.
  occidentale (S. Wats.) B.L. Robins.
  pallasii (Pursh) Fern.
    var. bracteosum G. Rossb.

Erysimum (CONT.)
        var. ochroleucum Tolm.
        var. pallasii
    perenne (S. Wats. ex Coville) Abrams
            SY=E. asperum (Nutt.) DC. var. p.
                S. Wats. ex Coville
            SY=E. capitatum (Dougl.) Greene
                var. p. (S. Wats. ex
                Coville) R.J. Davis
    repandum L.
            SY=Cheirinia r. (L.) Link
    suffrutescens (Abrams) G. Rossb.
        var. grandifolium G. Rossb.
        var. lompocense G. Rossb.
        var. suffrutescens
            SY=E. concinnum Eastw. ssp. s.
                Abrams
    teretifolium Eastw.

Euclidium R. Br.

    syriacum (L.) R. Br.

Eutrema R. Br.

    edwardsii R. Br.
    penlandii Rollins

Glaucocarpum Rollins

    suffrutescens (Rollins) Rollins

Halimolobos Tausch

    diffusa (Gray) O.E. Schulz
        var. diffusa
        var. jaegeri (Munz) Rollins
    mollis (Hook.) Rollins
            SY=Arabis hookeri Lange
    perplexa (Henderson) Rollins
        var. lemhiensis C.L. Hitchc.
        var. perplexa
    virgata (Nutt.) O.E. Schulz
    whitedii (Piper) Rollins
            SY=Arabis w. Piper

Hesperis L.

    matronalis L.

Heterodraba Greene

    unilateralis (M.E. Jones) Greene
            SY=Athysanus u. (M.E. Jones)
                Jepson

Hirschfeldia Moench

    incana (L.) Lagreze-Fossat

Hymenolobus Nutt.

    procumbens (L.) Nutt. ex Torr. & Gray
            SY=Hutchinsia p. (L.) Desv.

Iberis L.

    amara L.
    umbellata L.

Idahoa A. Nels. & J.F. Macbr.

    scapigera (Hook.) A. Nels. & J.F. Macbr.

Iodanthus Torr. & Gray ex Steud.

    pinnatifidus (Michx.) Steud.

Ionopsidium (DC.) Reichenb.

    acaule (Desf.) Reichenb.

Isatis L.

    tinctoria L.

Leavenworthia Torr.

    alabamica Rollins
        var. alabamica
        var. brachystyla Rollins
    aurea Torr.
    crassa Rollins
        var. crassa
        var. elongata Rollins
    exigua Rollins
        var. exigua
        var. laciniata Rollins
        var. lutea Rollins
    stylosa Gray
    torulosa Gray
    uniflora (Michx.) Britt.
            SY=L. michauxii Torr.

Lepidium L.

    arbuscula Hbd.
    aucheri Boiss.
    austrinum Small
    banebyanum Reveal
    bidentatum Mont.
        var. o-waihiense (Cham. & Schlecht.)
            Fosberg
            SY=L. o-w. Cham. & Schlecht.
        var. remyi (Drake) Fosberg
            SY=L. r. Drake
    bourgeauanum Thellung
            SY=L. densiflorum Schrad. var. b.
                (Thellung) C.L. Hitchc.
    campestre (L.) R. Br.
    davisii Rollins
            SY=L. montana Nutt. ssp. d.
                (Rollins) C.L. Hitchc.
    densiflorum Schrad.
        var. densiflorum
            SY=L. d. var. typicum Thellung
            SY=L. neglectum Thellung
            SY=L. texanum Buckl.
        var. elongatum (Rydb.) Thellung
        var. macrocarpum Mulligan
        var. pubicarpum (A. Nels.) Thellung
        var. ramosum (A. Nels.) Thellung
    dictyotum Gray
        var. acutidens Gray
            SY=L. a. (Gray) T.J. Howell
        var. dictyotum
    flavum Torr.
        var. felipense C.L. Hitchc.
        var. flavum
    fremontii S. Wats.
    graminifolium L.
    heterophyllum (DC.) Benth.
    hirtum (L.) DC.
    jaredii Brandeg.
    lasiocarpum Nutt.
        var. georginum (Rydb.) C.L. Hitchc.
        var. lasiocarpum
            SY=L. l. var. typicum C.L. Hitchc.
        var. rotundum C.L. Hitchc.
        var. wrightii (Gray) C.L. Hitchc.
    latifolium L.
    latipes Hook.
    montanum Nutt.
        var. alpinum S. Wats.
        var. alyssoides (Gray) M.E. Jones

Lepidium   (CONT.)
        SY=L. m. ssp. a. (Gray) C.L.
           Hitchc.
     var. angustifolium C.L. Hitchc.
     var. canescens (Thellung) C.L.
       Hitchc.
        SY=L. m. ssp. c. (Thellung) C.L.
           Hitchc.
        SY=L. m. ssp. cinereum (C.L.
           Hitchc.) C.L. Hitchc.
     var. eastwoodiae (Woot.) C.L. Hitchc.
     var. glabrum C.L. Hitchc.
     var. heterophyllum (S. Wats.) C.L.
       Hitchc.
     var. integrifolium (Nutt.) C.L.
       Hitchc.
     var. jonesii (Rydb.) C.L. Hitchc.
        SY=L. j. Rydb.
     var. montanum
     var. neeseae Welsh & Reveal
     var. papilliferum (Henderson) C.L.
       Hitchc.
     var. spathulatum (B.L. Robins.) C.L.
       Hitchc.
     var. stellae Welsh & Reveal
   ssp. ?demissum C.L. Hitchc.
nanum S. Wats.
nitidum Nutt.
     var. howellii C.L. Hitchc.
     var. insigne Greene
     var. oreganum (T.J. Howell) C.L.
       Hitchc.
oblongum Small
        SY=L. bipinnatifidum sensu auctt.
           non Desv.
oxycarpum Torr. & Gray
perfoliatum L.
pinnatifidum Ledeb.
pinnatisectum (O.E. Schulz) C.L. Hitchc.
ramosissimum A. Nels.
        SY=L. divergens Osterhout
ruderale L.
sativum L.
serra Mann
sordidum Gray
strictum (S. Wats.) Rattan
        SY=L. pubescens sensu auctt. non
           Desv.
thurberi Woot.
virginicum L.
     var. medium (Greene) C.L. Hitchc.
        SY=L. idahoense Heller
        SY=L. m. Greene
     var. menziesii (DC.) C.L. Hitchc.
        SY=L. m. DC.
     var. pubescens (Greene) C.L. Hitchc.
        SY=L. medium Greene var. p.
          (Greene) B.L. Robins.
     var. robinsonii (Thellung) C.L.
       Hitchc.
     var. virginicum
        SY=L. v. var. typicum C.L. Hitchc.

Lesquerella S. Wats.

alpina (Nutt. ex Torr. & Gray) S. Wats.
   ssp. alpina
        SY=L. a. var. laevis (Payson) C.L.
           Hitchc.
        SY=L. a. var. spathulata (Rydb.)
           Payson
        SY=L. curvipes A. Nels.
        SY=L. spathulata Rydb.
        SY=L. subumbellata Rollins
   ssp. condensata (A. Nels.) Rollins &
     Shaw
        SY=L. a. var. c. (A. Nels.) C.L.

        Hitchc.
        SY=L. c. A. Nels.
   ssp. parvula (Greene) Rollins & Shaw
        SY=L. a. var. p. (Greene) Welsh &
           Reveal
angustifolia (Nutt.) S. Wats.
arctica (Wormskj. ex Hornem.) S. Wats.
   var. arctica
        SY=L. a. ssp. purshii (S. Wats.)
           Porsild
        SY=L. a. var. p. S. Wats.
        SY=L. p. (S. Wats.) Fern.
   var. scammanae Rollins
arenosa (Richards.) Rydb.
   var. arenosa
        SY=L. ludoviciana (Nutt.) S. Wats.
          var. a. (Richards.) S. Wats.
   var. argillosa Rollins & Shaw
argyraea (Gray) S. Wats.
arizonica S. Wats.
        SY=L. a. var. nudicaulis Payson
aurea Woot.
auriculata (Engelm. & Gray) S. Wats.
calcicola Rollins
calderi Mulligan & Porsild
        SY=L. arctica (Wormskj. ex
          Hornem.) S. Wats. ssp. c.
          (Mulligan & Porsild) Hulten
        SY=L. a. var. c. (Mulligan &
          Porsild) Welsh
carinata Rollins
cinerea S. Wats.
cordiformis (Rollins) Rollins & Shaw
        SY=L. kingii S. Wats. var. c.
          (Rollins) Maguire & Holmgren
        SY=L. k. var. nevadensis Maguire &
          Holmgren
densiflora (Gray) S. Wats.
densipila Rollins
douglasii S. Wats.
engelmannii (Gray) S. Wats.
fendleri (Gray) S. Wats.
filiformis Rollins
fremontii Rollins & Shaw
garrettii Payson
globosa (Desv.) S. Wats.
gooddingii Rollins & Shaw
gordonii (Gray) S. Wats.
gracilis (Hook.) S. Wats.
   ssp. gracilis
        SY=L. pallida (Torr. & Gray) S.
          Wats.
   ssp. nuttallii (Torr. & Gray) Rollins &
     Shaw
        SY=L. g. var. repanda (Nutt.)
          Payson
        SY=L. n. (Torr. & Gray) S. Wats.
        SY=L. r. (Nutt.) S. Wats.
grandiflora (Hook.) S. Wats.
hemiphysaria Maguire
   var. hemiphysaria
   var. lucens Welsh & Reveal
hitchcockii Munz
        SY=L. h. ssp. confluens Maguire &
          Holmgren
intermedia (S. Wats.) Heller
kingii S. Wats.
   ssp. bernardina (Munz) Munz
        SY=L. b. Munz
   ssp. diversifolia (Greene) Rollins &
     Shaw
        SY=L. d. Greene
        SY=L. k. var. sherwoodii (M.E.
          Peck) C.L. Hitchc.
        SY=L. occidentalis S. Wats. ssp.
          d. (Greene) Maguire &
          Holmgren

Lesquerella (CONT.)
            SY=L. o. var. d. (Greene) C.L.
                Hitchc.
            SY=L. s. M.E. Peck
    ssp. kingii
        var. cobrensis Rollins & Shaw
        var. kingii
    ssp. latifolia (A. Nels.) Rollins & Shaw
            SY=L. barnebyi Maguire
            SY=L. k. var. parviflora (Maguire
                & Holmgren) Welsh & Reveal
            SY=L. l. A. Nels.
            SY=L. occidentalis S. Wats. ssp.
                cusickii (M.E. Jones)
                Maguire var. p. Maguire &
                Holmgren
lasiocarpa (Hook. ex Gray) S. Wats.
    ssp. berlandieri (Gray) Rollins & Shaw
        var. berlandieri (Gray) Payson
        var. hispida (S. Wats.) Rollins &
            Shaw
    ssp. lasiocarpa
lata Woot. & Standl.
lescurii (Gray) S. Wats.
lindheimeri (Gray) S. Wats.
ludoviciana (Nutt.) S. Wats.
lyrata Rollins
macrocarpa A. Nels.
X maxima (Rollins) Rollins [densipila X
    stonensis]
            SY=L. densipila Rollins var. m.
                Rollins
mcvaughiana Rollins
montana (Gray) S. Wats.
multiceps Maguire
occidentalis S. Wats.
    ssp. cinerascens (Maguire & Holmgren)
        Rollins & Shaw
    ssp. occidentalis
            SY=L. cusickii M.E. Jones
            SY=L. o. ssp. cusickii (M.E.
                Jones) Maguire
            SY=L. o. var. c. (M.E. Jones) C.L.
                Hitchc.
ovalifolia Rydb.
    ssp. alba (Goodman) Rollins & Shaw
            SY=L. engelmannii ssp. a.
                (Goodman) Clarke
            SY=L. o. var. a. Goodman
    ssp. ovalifolia
            SY=L. engelmannii ssp. o. (Rydb.)
                Clarke
paysonii Rollins
perforata Rollins
pinetorum Woot. & Standl.
prostrata A. Nels.
pruinosa Greene
purpurea (Gray) S. Wats.
    ssp. foliosa (Rollins) Rollins & Shaw
    ssp. purpurea
rectipes Woot. & Standl.
recurvata (Engelm. ex Gray) S. Wats.
rubicundula Rollins
            SY=L. hitchcockii Munz ssp. r.
                (Rollins) Maguire & Holmgren
            SY=L. tumulosa (Barneby) Reveal
sessilis (S. Wats.) Small
stonensis Rollins
tenella A. Nels.
            SY=L. gordonii (Gray) S. Wats.
                var. sessilis S. Wats.
thamnophila Rollins & Shaw
utahensis Rydb.
valida Greene
            SY=L. lepodita Cory
wardii S. Wats.

Lobularia Desv.

    maritima (L.) Desv.
            SY=Koniga m. (L.) R. Br.

Lunaria L.

    annua L.
    rediviva L.

Lyrocarpa Harvey

    coulteri Hook. & Harvey
        var. coulteri
            SY=L. c. var. typica Rollins
        var. palmeri (S. Wats.) Rollins

Malcolmia (L.) R. Br.

    africana (L.) R. Br.
    maritima (L.) R. Br.

Mancoa Weddell

    pubens (Gray) Rollins

Matthiola R. Br.

    bicornis (Sibthorp & Sm.) DC.
            SY=M. longipetala (Vent.) DC. ssp.
                b. (Sibthorp & Sm.) P.W.
                Ball
    incana (L.) R. Br.
        var. annua (Sweet) Voss
        var. incana

Myagrum L.

    perfoliatum L.

Nasturtium R. Br.

    gambelii (S. Wats.) O.E. Schulz
            SY=Cardamine g. S. Wats.
    microphyllum (Boenn.) Reichenb.
            SY=Cardamine curvisiliqua Shuttlw.
                ex Chapman
            SY=N. officinale R. Br. var. m.
                (Boenn.) Thellung
            SY=Rorippa m. (Boenn.) Hyl.
    officinale R. Br.
            SY=N. o. var. siifolium
                (Reichenb.) W.D.J. Koch
            SY=Rorippa nasturtium-aquaticum
                (L.) Hayek
            SY=R. n-a. var. longisiliqua
                (Irmisch) Boivin
            SY=Sisymbrium n-a. L.
    X sterile (Airy-Shaw) Oefel.
            SY=Rorippa nasturtium-aquaticum
                (L.) Hayek var. s.
                (Airy-Shaw) Boivin
            SY=R. X s. Airy-Shaw

Nerisyrenia Greene

    camporum (Gray) Greene
    linearifolia (S. Wats.) Greene

Neslia Desv.

    paniculata (L.) Desv.

Parrya R. Br.

    arctica R. Br.
    nudicaulis (L.) Boiss.

Parrya (CONT.)
        ssp. interior Hulten
                SY=P. n. var. i (Hulten) Boivin
        ssp. nudicaulis
                var. grandiflora Hulten
                var. nudicaulis
                        SY=P. platycarpa Rydb.
                        SY=P. rydbergii Botsch.
        ssp. septentrionalis Hulten

Pennellia Nieuwl.

    longifolia (Benth.) Rollins
                SY=Thelypodium l. (Benth.) S.
                        Wats.
        micrantha (Gray) Nieuwl.
                SY=Thelypodium longifolium
                        (Benth.) S. Wats. var.
                        catalinense M.E. Jones
                SY=T. m. (Gray) S. Wats.

Phoenicaulis Nutt.

    cheiranthoides Nutt.
        ssp. cheiranthoides
        ssp. glabra (Jepson) Abrams
                SY=P. c. var. g. Jepson
        ssp. lanuginosa (S. Wats.) Abrams
                SY=P. c. var. l. (S. Wats.)
                        Rollins

Physaria Nutt. ex Torr. & Gray

    acutifolia Rydb.
                var. acutifolia
                        SY=P. australis (Payson) Rollins
                var. purpurea Welsh & Reveal
        alpestris Suksdorf
        bellii Mulligan
        brassicoides Rydb.
        chambersii Rollins
                SY=P. c. var. membranacea Rollins
        condensata Rollins
        didymocarpa (Hook.) Gray
                var. didymocarpa
                        SY=P. d. var. normalis Kuntze
                var. integrifolia Rollins
                var. lanata A. Nels.
                var. lyrata C.L. Hitchc.
        floribunda Rydb.
                SY=P. osterhoutii Payson
        geyeri (Hook.) Gray
                var. geyeri
                var. purpurea Rollins
        grahamii Morton
        newberryi Gray
                SY=P. didymocarpa (Hook.) Gray
                        var. n. (Gray) M.E. Jones
        oregona S. Wats.
        rollinsii Mulligan
        vitulifera Rydb.

Polyctenium Greene

    fremontii (S. Wats.) Greene
                var. bisulcatum (Greene) Rollins
                        SY=Smelowskia f. S. Wats. var. b.
                                (Greene) O.E. Schulz
                var. fremontii
                        SY=Smelowskia f. S. Wats.

Raphanus L.

    raphanistrum L.
    sativus L.
          .   SY=R. raphanistrum L. var. s. (L.)
                        G. Beck

Rapistrum Crantz

    perenne (L.) All.
    rugosum (L.) All.
                var. rugosum
                var. venosum (Pers.) DC.

Rhyncosinapis Hayek

    cheiranthos (Vill.) Dandy

Rorippa Scop.

    amphibia (L.) Bess.
    austriaca (Crantz) Bess.
    barbareifolia (DC.) Kitagawa
                SY=R. hispida (Desv.) Britt. var.
                        b. (DC.) Hulten
                SY=R. islandica (Oeder) Borbas
                        var. b. (DC.) Welsh
    calycina (Engelm.) Rydb.
    coloradensis R. Stuckey
    columbiae Suksdorf ex T.J. Howell
                SY=R. calycina (Engelm.) Rydb.
                        var. columbiae (Suksdorf ex
                        T.J. Howell) Rollins
    crystallina Rollins
    curvipes Greene
                var. alpina (S. Wats.) R. Stuckey
                        SY=R. a. (S. Wats.) Rydb.
                        SY=R. obtusa (Nutt.) Britt. var.
                                a. (S. Wats.) Britt.
                var. curvipes
                var. integra (Rydb.) R. Stuckey
                        SY=R. obtusa (Nutt.) Britt. var.
                                i. (Rydb.) Victorin
    curvisiliqua (Hook.) Bess. ex Britt.
                var. curvisiliqua
                        SY=Nasturtium c. (Hook.) Nutt.
                        SY=Radicula c. (Hook.) Greene
                        SY=Sisymbrium c. Hook.
                var. lyrata (Nutt.) R. Stuckey
                        SY=Rorippa l. (Nutt.) Greene
                var. nuttallii (S. Wats.) R. Stuckey
                var. occidentalis (Greene) R. Stuckey
                var. orientalis R. Stuckey
                var. procumbens R. Stuckey
                var. spatulata R. Stuckey
    indica (L.) Hiern
                var. apetala (DC.) Hochr.
                var. indica
                        SY=Radicula heterophylla (Blume)
                                Small
                        SY=Rorippa h. (Blume) R.O.
                                Williams
                        SY=Rorippa montana (Wallich) Small
    intermedia (Kuntze) R. Stuckey
    microsperma (DC.) Bailey
    microtitis (B.L. Robins.) Rollins
    palustris (L.) Bess.
        ssp. fernaldiana (Butters & Abbe)
                Jonsell
                        SY=Rorippa islandica (Oeder)
                                Borbas ssp. f. (Butters &
                                Abbe) Hulten
                        SY=R. i. var. f. Butters & Abbe
                        SY=R. p. ssp. glabra (O.E. Schulz)
                                R. Stuckey var. f. (Butters
                                & Abbe) R. Stuckey
        ssp. glabra (O.E. Schulz) R. Stuckey
                var. cernua (Nutt.) R. Stuckey
                var. dictyota (Greene) R. Stuckey
                var. glabra (O.E. Schulz) R. Stuckey
                        SY=Rorippa islandica (Oeder)
                                Borbas var. g. (O.F. Schulz)
                                Welsh & Reveal
                        SY=R. p. ssp. fernaldiana (Butters

Rorippa   (CONT.)
                            & Abbe) Jonsell var. g.
                            (O.E. Schulz) Taylor &
                            MacBryde
                var. glabrata (Lunell) R. Stuckey
                    SY=Rorippa hispida (Desv.) Britt.
                            var. glabrata Lunell
                    SY=R. islandica (Oeder) Borbas
                            var. glabrata (Lunell)
                            Butters & Abbe
            ssp. hispida (Desv.) Jonsell
                var. elongata R. Stuckey
                var. hispida (Desv.) Rydb.
                    SY=Radicula h. (Desv.) Britt.
                    SY=Rorippa h. (Desv.) Britt.
                    SY=Rorippa islandica (Oeder)
                            Borbas var. h. (Desv.)
                            Butters & Abbe
            ssp. occidentalis (S. Wats.) Abrams
                var. clavata (Rydb.) R. Stuckey
                var. occidentalis
                    SY=Rorippa islandica (Oeder)
                            Borbas var. o. (S. Wats.)
                            Butters & Abbe
            ssp. palustris
                var. palustris
                    SY=Radicula p. (L.) Moench
                    SY=Forippa islandica (Oeder)
                            Borbas var. microcarpa
                            (Regel) Fern.
                var. williamsii (Britt.) Hulten
        portoricensis (Spreng.) Stehle
                var. portoricensis
                    SY=Radicula p. (Spreng.) Britt.
                var. pumilum (O.E. Schulz) R. Stuckey
        prostrata (Bergeret) Schinz & Thellung
        ramosa Rollins
        sessiliflora (Nutt.) A.S. Hitchc.
                    SY=Radicula s. (Nutt.) Greene
        sinuata (Nutt.) A.S. Hitchc.
                    SY=Radicula s. (Nutt.) Greene
        sphaerocarpa (Gray) Britt.
                    SY=Radicula s. (Gray) Greene
        subumbellata Rollins
        sylvestris (L.) Bess.
                    SY=Radicula s. (L.) Druce
        tenerrima Greene
        teres (Michx.) R. Stuckey
                    SY=Radicula obtusa (Nutt.) Greene
                    SY=Radicula walteri (Ell.) Greene
                    SY=Rorippa o. (Nutt.) Britt.
                    SY=Rorippa w. (Ell.) C. Mohr
        truncata (Jepson) R. Stuckey

Schoenocrambe Greene

    linifolia (Nutt.) Greene
                    SY=Sisymbrium l. Nutt.

Selenia Nutt.

    aurea Nutt.
    dissecta Torr. & Gray
    grandis R.F. Martin
    jonesii Cory

Sibara Greene

    desertii (M.E. Jones) Rollins
                    SY=Arabis d. (M.E. Jones) Abrams
    filifolia (Greene) Greene
                    SY=Arabis f. Greene
    rosulata Rollins
    runcinata (S. Wats.) Rollins
                var. brachycarpa Rollins
                var. runcinata
                    SY=S. viereckii (O.E. Schulz)

                            Rollins
        virginica (L.) Rollins
                    SY=Arabis v. (L.) Poir.

Sinapis L.

    alba L.
                    SY=Brassica a. (L.) Rabenh.
                    SY=B. hirta Moench
    arvensis L.
                    SY=Brassica a. (L.) Rabenh.
                    SY=B. kaber (DC.) L.C. Wheeler
                    SY=B. k. var. pinnatifida (Stokes)
                            L.C. Wheeler
                    SY=B. k. var. schkuhriana
                            (Reichenb.) L.C. Wheeler

Sisymbrium L.

    altissimum L.
                    SY=Norta a. (L.) Britt.
    auriculatum Gray
    austriacum Jacq.
    irio L.
                    SY=Norta i. (L.) Britt.
    loeseli L.
    officinale (L.) Scop.
                var. leiocarpum DC.
                var. officinale
                    SY=Erysimum o. L.
    orientale L.
                    SY=Brassica kaber (DC.) L.C.
                            Wheeler var. o. (L.) Scoggan
    polyceratium L.

Smelowskia C.A. Mey.

    borealis (Greene) Drury & Rollins
                var. borealis
                var. jordalii Drury & Rollins
                var. koliana (Gombocz) Drury &
                            Rollins
                var. villosa Drury & Rollins
    calycina (Steph.) C.A. Mey. ex Ledeb.
                var. americana (Regel & Herder) Drury
                            & Rollins
                var. integrifolia (Seem.) Rollins
                    SY=S. c. ssp. i. (Seem.) Hulten
                var. media Drury & Rollins
                    SY=S. c. ssp. integrifolia (Seem.)
                            Hulten var. m. (Drury &
                            Rollins) Hulten
                var. porsildii Drury & Rollins
                    SY=S. c. ssp. integrifolia (Seem.)
                            Hulten var. p. (Drury &
                            Rollins) Hulten
    holmgrenii Rollins
    ovalis M.E. Jones
                var. congesta Rollins
                var. ovalis
    pyriformis Drury & Rollins

Stanleya Nutt.

    albescens M.E. Jones
    confertiflora (B.L. Robins.) T.J. Howell
    elata M.E. Jones
    pinnata (Pursh) Britt.
                var. bipinnata (Greene) Rollins
                var. gibberosa Rollins
                var. integrifolia (James) Rollins
                var. inyoensis (Munz & Roos) Reveal
                    SY=S. p. ssp. i. Munz & Roos
                var. pinnata
                    SY=S. p. var. typica Rollins
    tomentosa Parry
                var. runcinata (Rydb.) Rollins

Stanleya (CONT.)
        var. tomentosa
   viridiflora Nutt.

Streptanthella Rydb.

    longirostris (S. Wats.) Rydb.
         var. derelicta J.T. Howell
         var. longirostris

Streptanthus Nutt.

    albidus Greene
      ssp. albidus
           SY=S. glandulosus Hook. var. a.
               (Greene) Jepson
      ssp. peramoenus (Greene) Kruckeberg
    arizonicus S. Wats.
         var. arizonicus
         var. luteus Kearney & Peebles
    barbatus S. Wats.
    barbiger Greene
    batrachopus J.L. Morrison
    bernardinus (Greene) Parish
           SY=S. campestris S. Wats. var. b.
               (Greene) I.M. Johnston
    brachiatus F.W. Hoffman
    bracteatus Gray
    breweri Gray
         var. breweri
         var. hesperidis (Jepson) Jepson
           SY=S. h. Jepson
    callistus J.L. Morrison
    campestris S. Wats.
    carinatus C. Wright
    cordatus Nutt. ex Torr. & Gray
         var. cordatus
         var. piutensis J.T. Howell
    cutleri Cory
    diversifolius S. Wats.
    farnsworthianus J.T. Howell
    fenestratus (Greene) J.T. Howell
    glandulosus Hook.
      ssp. glandulosus
      ssp. pulchellus (Greene) Kruckeberg
           SY=S. g. var. p. (Greene) Jepson
      ssp. secundus (Greene) Kruckeberg
         var. hoffmanii Kruckeberg
         var. secundus (Greene) Kruckeberg
           SY=S. s. Greene
         var. sonomensis Kruckeberg
    gracilis Eastw.
    heterophyllus Nutt.
    hispidus Gray
    howellii S. Wats.
    hyacinthoides Hook.
    insignis Jepson
    maculatus Nutt.
    morrisonii F.W. Hoffman
      ssp. elatus F.W. Hoffman
      ssp. hirtiflorus F.W. Hoffman
      ssp. morrisonii
    niger Greene
           SY=S. glandulosus Hook. var. n.
               (Greene) Munz
    obtusifolius Hook.
    oliganthus Rollins
    platycarpus Gray
    polygaloides Gray
    sparsiflorus Rollins
    squamiformis Goodman
    tortuosus Kellogg
         var. flavescens Jepson
         var. orbiculatus (Greene) Hall
         var. pallidus Jepson
         var. suffrutescens (Greene) Jepson
         var. tortuosus

Subularia (L.) Forsk.

    aquatica L.
      ssp. americana Mulligan & Calder
           SY=S. aquatica var. americana
               (Mulligan & Calder) Boivin

Synthlipsis Gray

    greggii Gray

Teesdalia R. Br.

    nudicaulis (L.) R. Br.

Thellungiella O.E. Schulz

    salsuginea (Pallas) O.E. Schulz
           SY=Arabidopsis glauca (Nutt.)
               Rydb.
           SY=A. s. (Pallas) N. Busch
           SY=Sisymbrium s. Pallas

Thelypodiopsis Rydb.

    ambigua (S. Wats.) Al-Shehbaz
           SY=Sisymbrium a. (S. Wats.) Payson
    argillacea Welsh & Atwood
    aurea (Eastw.) Rydb.
           SY=Sisymbrium a. (Eastw.) Payson
    elegans (M.E. Jones) Rydb.
           SY=Sisymbrium e. (M.E. Jones)
               Payson
    juniperorum (Payson) Rydb.
           SY=Sisymbrium elegans (M.E. Jones)
               Payson var. j. (Payson)
               Harrington
    linearifolia (Gray) Al-Shehbaz
           SY=Hesperidanthus l. (Gray) Rydb.
           SY=Sisymbrium l. (Gray) Payson
    purpusii (Brandeg.) Rollins
           SY=Sisymbrium kearneyi Rollins
           SY=S. p. (Brandeg.) O.E. Schulz
    shinnersii (M.C. Johnston) Rollins
           SY=Sisymbrium s. M.C. Johnston
           SY=Thelypodium s. (M.C. Johnston)
               Rollins
    vaseyi (S. Wats.) Rollins
           SY=Sisymbrium v. S. Wats.

Thelypodium Endl.

    brachycarpum Torr.
    crispum Greene ex Payson
    eucosmum B.L. Robins.
    flexuosum B.L. Robins.
    howellii S. Wats.
      ssp. howellii
      ssp. spectabilis (M.E. Peck) Al-Shehbaz
           SY=T. h. var. s. M.E. Peck
    integrifolium (Nutt.) Endl.
      ssp. affine (Greene) Al-Shehbaz
           SY=T. i. var. a. (Greene) Welsh &
               Reveal
           SY=T. rhomboideum Greene
      ssp. complanatum Al-Shehbaz
           SY=T. i. var. c. (Al-Shehbaz)
               Welsh & Reveal
      ssp. gracilipes (B.L. Robins.)
         Al-Shehbaz
           SY=T. i. var. gracilipes (B.L.
               Robins.) Payson
      ssp. integrifolium
           SY=T. lilacinum Greene
           SY=T. l. var. subumbellatum Payson
      ssp. longicarpum Al-Shehbaz
    laciniatum (Hook.) Endl.

Thelypodium   (CONT.)
        SY=T. l. var. streptanthoides
           (Leib.) Payson
        SY=T. s. Lieb.
  laxiflorum Al-Shehbaz
        SY=Stanleyella wrightii (Gray)
           Rydb. var. tenella (M.E.
           Jones) Payson
        SY=T. w. Gray var. t. M.E. Jones
  lemmonii Greene
  milleflorum A. Nels.
        SY=T. laciniatum (Hook.) Endl.
           var. m. (A. Nels.) Payson
  paniculatum A. Nels.
        SY=T. sagittatum (Nutt.) Endl. ex
           Walp. var. crassicarpum
           Payson
  repandum Rollins
  rollinsii Al-Shehbaz
  sagittatum (Nutt.) Endl. ex Walp.
    ssp. ovalifolum (Rydb.) Al-Shehbaz
        SY=Sisymbrium s. var. o. (Rydb.)
           Welsh & Reveal
    ssp. sagittatum
        var. vermicularis Welsh & Reveal
  stenopetalum S. Wats.
  tenue Rollins
  texanum (Cory) Rollins
        SY=Stanleyella t. Cory
  wrightii Gray
    ssp. oklahomensis Al-Shehbaz
    ssp. wrightii
        SY=Stanleyella w. (Gray) Rydb.

Thlaspi L.

  alliaceum L.
  arcticum Porsild
  arvense L.
  montanum L.
    var. californicum (S. Wats.) P.
      Holmgren
        SY=T. c. S. Wats.
    var. fendleri (Gray) P. Holmgren
        SY=T. f. Gray
        SY=T. prolixum A. Nels.
        SY=T. stipitatum A. Nels.
    var. idahoense (Payson) P. Holmgren
        SY=T. fendleri Gray var. i.
           (Payson) C.L. Hitchc.
        SY=T. i. Payson
    var. montanum
        SY=T. alpestre L. var. glaucum A.
           Nels.
        SY=T. australe A. Nels.
        SY=T. cochleariforme DC.
        SY=T. f. var. coloradense (Rydb.)
           Maguire
        SY=T. f. var. g. (A. Nels.) C.L.
  •   Hitchc.
        SY=T. f. var. hesperium (Payson)
           C.L. Hitchc.
        SY=T. f. var. tenuipes Maguire
        SY=T. g. (A. Nels.) A. Nels.
        SY=T. g. var. h. Payson
        SY=T. g. var. pedunculatum Payson
    var. siskiyouense P. Holmgren
  parviflorum A. Nels.
  perfoliatum L.

Thysanocarpus Hook.

  amplectens Greene
  curvipes Hook.
    var. curvipes
    var. eradiatus Jepson
    var. longistylus Jepson

    elegans Fisch. & Mey.
        SY=T. curvipes Hook. var. e.
           (Fisch. & Mey.) B.L. Robins.
  laciniatus Nutt. ex Torr. & Gray
    var. conchuliferus (Greene) Jepson
        SY=T. c. Greene
    var. hitchcockii Munz
        SY=T. l. ssp. desertorum (Heller)
           Abrams
    var. laciniatus
        SY=T. l. var. crenatus (Nutt.)
           Brewer
        SY=T. l. var. emarginatus (Greene)
           Jepson
    var. ramosus (Greene) Munz
    var. rigidus Munz
  radians Benth.

Tropidocarpum Hook.

  capparideum Greene
  gracile Hook.
    var. dubium (A. Davids.) Jepson
    var. gracile

Warea Nutt.

  amplexifolia (Nutt.) Nutt.
        SY=W. auriculata Shinners
  carteri Small
  cuneifolia (Muhl. ex Nutt.) Nutt.
  sessilifolia Nash

BROMELIACEAE

Aechmea Ruiz & Pavon

  lingulata (L.) Baker
        SY=Wittmackia l. (L.) Mez
  nudicaulis (L.) Griseb.

Ananas P. Mill.

  comosus (L.) Merr.
        SY=A. ananas (L.) Cockerell

Bromelia L.

  pinguin L.

Catopsis Griseb.

  berteroniana (Schultes f.) Mez
  floribunda L.B. Sm.
        SY=C. nutans sensu Griseb.
  nitida (Hook.) Griseb.
  nutans (Sw.) Griseb.
  sessiliflora (Ruiz & Pavon) Mez

Guzmania Ruiz & Pavon

  berteroniana (Schultes f.) Mez
  erythrolepis Brongn. ex Planch.
  lingulata (L.) Mez
  monostachia (L.) Rusby ex Mez
    var. variegata Hortus ex Nash

Hechtia Klotzsch

  glomerata Zucc.
  scariosa L.B. Sm.
  texensis S. Wats.

Hohenbergia Schultes f.

Hohenbergia (CONT.)
    antillana Mez
    attenuata Britt.
    portoricensis Mez

Pitcairnia L'Her.

    angustifolia (Sw.) Redoute
            SY=P. latifolia Soland.

Tillandsia L.

    araujei Mez
    baileyi Rose ex Small
    balbisiana Schultes f.
    bartramii Ell.
            SY=T. juncea sensu auctt. non
                (Ruiz & Pavon) Poir.
            SY=T. myriophylla Small
            SY=T. simulata Small
    bulbosa Hook.
    circinata Schlecht.
    fasciculata Sw.
        var. clavispica Mez
        var. densispica Mez
            SY=T. hystricina Small
        var. fasciculata
        var. floridana L. B. Sm.
    festucoides Brongn. ex Baker
    flexuosa Sw.
            SY=T. aloifolia Hook.
    lindenii Regel
    lineatispica Mez
    polystachia (L.) L.
    pruinosa Sw.
    recurvata (L.) L.
            SY=Diaphoranthema r. (L.) Beer
    setacea Sw.
            SY=T. tenuifolia sensu Mez non L.
    tenuifolia L.
            SY=T. pulchella Hook.
    usneoides (L.) L.
            SY=Dendropogon u. (L.) Raf.
    utriculata L.
    valenzuelana A. Rich.
            SY=T. sublaxa Baker

Vriesea Lindl.

    macrostachya (Bello) Mez
            SY=Neovriesia m. (Bello) Britt.
    ringens (Griseb.) Harms
    sintenisii (Baker) L. B. Sm. & Pitt.
            SY=Thecophyllum s. (Baker) Mez

BRUNELLIACEAE

Brunellia Ruiz & Pavon

    comocladifolia Humb. & Bonpl.

BURMANNIACEAE

Apteria Nutt.

    aphylla (Nutt.) Barnh.
        var. aphylla
        var. hymenanthera (Miq.) Jonker
            SY=A. h. Miq.

Burmannia L.

    biflora L.
    capitata (Walt.) Mart.
    flava Mart.

Gymnosiphon Blume

    germainii Urban
            SY=Ptychomeria portoricensis
                (Urban) Schlechter
    sphaerocarpus Urban

Thismia Griffith

    americana N.E. Pfeiffer

BURSERACEAE

Bursera Jacq. ex L.

    fagaroides (H.B.K.) Engl.
        var. elongata McVaugh & Rzed.
            SY=B. odorata sensu auctt. non
                Brandeg.
    microphylla Gray
    simaruba (L.) Sarg.
            SY=Elaphrium s. (L.) Rose
            SY=Pistacia s. L.

Dacryodes Vahl

    excelsa Vahl

Tetragastris Gaertn.

    balsamifera (Sw.) Kuntze

BUTOMACEAE

Butomus L.

    umbellatus L.

Hydrocleys L.C. Rich.

    nymphoides (Humb. & Bonpl. ex Willd.) Buch.

BUXACEAE

Buxus L.

    citrifolia (Willd.) Spreng.
            SY=Tricera c. Willd.
    laevigata (Sw.) Spreng.
    portoricensis Alain
    sempervirens L.
    vahlii Baill.
            SY=Tricera v. (Baill.) Britt.

Pachysandra Michx.

    procumbens Michx.
    terminalis Sieb. & Zucc.

Simmondsia Nutt.

    chinensis (Link) Schneid.
            SY=Buxus c. Link
            SY=S. californica (Link) Nutt.

CACTACEAE

Ancistrocactus Britt. & Rose

    scheeri (Salm-Dyck) Britt. & Rose
        SY=A. brevihamatus (Engelm.)
           Britt. & Rose
    tobuschii (W.T. Marsh.) W.T. Marsh. ex
      Backeberg
    uncinatus (Galeotti) L. Benson
        var. uncinatus
          SY=Echinocactus u. Galeotti
          SY=Thelocactus u. (Galeotti) W.T.
           Marsh.
        var. wrightii (Engelm.) L. Benson

Ariocarpus Scheidw.

    fissuratus (Engelm.) K. Schum.

Cereus P. Mill.

    emoryi Engelm.
        SY=Bergerocereus e. (Engelm.)
          Britt. & Rose
    eriophorus N.E. Pfeiffer & Otto
        var. fragrans (Small) L. Benson
          SY=Harrisia f. Small
    giganteus Engelm.
        SY=Carnegiea g. (Engelm.) Britt. &
          Rose
    gracilis P. Mill.
        var. aboriginum (Small) L. Benson
          SY=Harrisia a. Small
        var. simpsonii (Small) L. Benson
          SY=Harrisia s. Small
    grandiflorus (L.) P. Mill.
        var. armatus (K. Schum.) L. Benson
          SY=Selenicereus coniflorus
            (Weingart) Britt. & Rose
        var. grandiflorus
          SY=S. g. (L.) Britt. & Rose
    grantianus (Britt.) Kelsey & Dayton
        SY=Leptocereus g. Britt.
    greggii Engelm.
        var. greggii
          SY=Peniocereus g. (Engelm.) Britt.
           & Rose
        var. transmontanus Engelm.
    hystrix (Haw.) Salm-Dyck
        SY=Lemaireocereus h. (Haw.) Britt.
          & Rose
    martinii Labouret
    pentagonus L.
        SY=Acanthocereus floridanus Small
        SY=A. p. (L.) Britt. & Rose
    peruvianus (L.) P. Mill.
    portoricensis (Britt.) Urban
        SY=Harrisia p. Britt.
    poselgeri (Lem.) Coult.
    pteranthus Link & Otto
        SY=Selenicereus p. (Link & Otto)
          Britt. & Rose
    quadricostatus Bello
        SY=Leptocereus q. (Bello) Britt. &
          Rose
    robinii (Lem.) L. Benson
        var. deeringii (Small) L. Benson
          SY=Cephalocereus d. Small
        var. robinii
          SY=Cephalocereus keyensis Britt. &
           Rose
          SY=Cereus r. var. k. (Britt. &
           Rose) L. Benson ex Long &
           Lakela
    royenii (L.) P. Mill.

        SY=Cephalocereus r. (L.) Britt. &
          Rose
    schottii Engelm.
        SY=Lophocereus s. (Engelm.) Britt.
          & Rose
    spinulosus DC.
    striatus Brandeg.
        SY=Cereus diguetii A. Weber
        SY=Wilcoxia d. (A. Weber) Peebles
    strictus DC.
        SY=Cephalocereus nobilis (Haw.)
          Britt. & Rose
        SY=Pilosocereus n. (Haw.) Byles &
          Rowley
    thurberi Engelm.
        SY=Lemaireocereus t. (Engelm.)
          Britt. & Rose
    trigonus Haw.
        SY=Hylocereus t. (Haw.) Safford
    undatus Haw.
        SY=Hylocereus u. (Haw.) Britt. &
          Rose

Coryphantha (Engelm.) Lem.

    cornifera (DC.) Orcutt
        var. echinus (Engelm.) L. Benson
          SY=C. e. (Engelm.) Britt. & Rose
          SY=C. pectinata (Engelm.) Britt. &
           Rose
    dasyacantha (Engelm.) Orcutt
        var. dasyacantha
          SY=Mammillaria d. Engelm.
        var. varicolor (Tiegel) L. Benson
          SY=Escobaria d. Britt. & Rose var.
           v. (Tiegel) D.R. Hunt
    duncanii (Hester) L. Benson
    hesteri Y. Wright
    macromeris (Engelm.) Orcutt
        var. macromeris
        var. runyonii (Britt. & Rose) L.
        Benson
          SY=C. r. Britt. & Rose
    minima Baird
    missouriensis (Sweet) Britt. & Rose
        var. caespitosa (Engelm.) L. Benson
          SY=C. similis (Engelm.) Britt. &
           Rose
          SY=Escobaria m. (Sweet) D.R. Hunt
           var. c. (Engelm.) D.R. Hunt
        var. marstonii (Clover) L. Benson
          SY=Escobaria missouriensis (Sweet)
           D.R. Hunt var. marstonii
           (Clover) D.R. Hunt
        var. missouriensis
          SY=Escobaria m. (Sweet) D.R. Hunt
          SY=Mammillaria m. Sweet
          SY=Neobesseya m. (Sweet) Britt. &
           Rose
        var. robustior (Engelm.) L. Benson
          SY=Escobaria m. (Sweet) D.R. Hunt
           var. r. (Engelm.) D.R. Hunt
          SY=Mammillaria similis Engelm.
           var. r. Engelm.
          SY=Neobesseya wissmannii (Hildmann
           ex K. Schum.) Britt. & Rose
    ramillosa Cutak
    recurvata (Engelm.) Britt. & Rose
        SY=Mammillaria r. Engelm.
    robbinsorum (W.H. Earle) Allan D. Zimmerman
    robertii Berger
        SY=Escobaria runyonii Britt. &
          Rose
    scheeri Lem.
        var. robustispina (Schott ex Engelm.)
          L. Benson
          SY=Cactus r. (Schott ex Engelm.)

Coryphantha (CONT.)
                 Kuntze
        SY=Coryphantha r. (Schott ex
           Engelm.) Britt. & Rose
        SY=Mammillaria r. Schott ex
           Engelm.
    var. scheeri
        SY=Coryphantha muehlenpfordtii
           (Poselg.) Britt. & Rose
    var. uncinata L. Benson
    var. valida (Engelm.) L. Benson
sneedii (Britt. & Rose) Berger
    var. leei (Rose) L. Benson
    var. sneedii
        SY=Escobaria s. Britt. & Rose
strobiliformis (Poselg.) Moran
    var. durispina (Quehl) L. Benson
    var. orcuttii (Boedeker) L. Benson
    var. strobiliformis
        SY=?Escobaria bella Britt. & Rose
        SY=E. tuberculosa (Engelm.) Britt.
           & Rose
sulcata (Engelm.) Britt. & Rose
    var. nickelsiae (K. Brandeg.) L.
      Benson
    var. sulcata
vivipara (Nutt.) Britt. & Rose
    var. alversonii (Coult.) L. Benson
        SY=Coryphantha a. (Coult.) Orcutt
        SY=Escobaria v. (Nutt.) Buxbaum
          var. a. (Coult.) D.R. Hunt
        SY=Mammillaria a. Coult. ex
          Zeissold
    var. arizonica (Engelm.) W.T. Marsh.
        SY=Coryphantha a. (Engelm.) Britt.
          & Rose
        SY=Escobaria v. (Nutt.) Buxbaum
          var. a. (Engelm.) D.R. Hunt
        SY=Mammillaria a. Engelm.
    var. bisbeeana (Orcutt) L. Benson
        SY=Escobaria v. (Nutt.) Buxbaum
          var. b. (Orcutt) D.R. Hunt
    var. desertii (Engelm.) W.T. Marsh.
        SY=Coryphantha d. (Engelm.) Britt.
          & Rose
        SY=Escobaria v. (Nutt.) Buxbaum
          var. d. (Engelm.) D.R. Hunt
    var. radiosa (Engelm.) Backeberg
        SY=Coryphantha neomexicana
          (Engelm.) Britt. & Rose
        SY=C. r. (Engelm.) Rydb.
        SY=Escobaria v. (Nutt.) Buxbaum
          var. r. (Engelm.) D.R. Hunt
        SY=Mammillaria n. Engelm.
    var. rosea (Clokey) L. Benson
        SY=Escobaria v. (Nutt.) Buxbaum
          var. r. (Clokey) D.R. Hunt
    var. vivipara
        SY=Mammillaria v. (Nutt.) Haw.

Echinocactus Link & Otto

asterias Zucc.
horizonthalonius Lem.
    var. horizonthalonius
    var. nicholii L. Benson
polycephalus Engelm. & Bigelow
    var. polycephalus
    var. xeranthemoides Coult.
        SY=E. x. (Coult.) Engelm. ex
          Coult.
texensis Hopffer
        SY=Homalocephala t. (Hopffer)
          Britt. & Rose

Echinocereus Engelm.

berlandieri (Engelm.) Engelm. ex Rumpl.
    var. angusticeps (Clover) L. Benson
        SY=E. a. Clover
        SY=E. blanckii (Poselg.) Poselg.
          ex Rumpl. var. a. (Clover)
          L. Benson
    var. berlandieri
        SY=E. blanckii sensu auctt. non
          (Poselg.) Poselg. ex Rumpl.
    var. papillosus (Linke ex Rumpl.) L.
      Benson
        SY=E. p. Linke ex Rumpl.
chloranthus Engelm.
    var. chloranthus
        SY=E. russanthus Weniger
    var. neocapillus Weniger
engelmannii Parry
    var. acicularis L. Benson
    var. armatus L. Benson
    var. chrysocentrus (Engelm. &
      Bigelow) Engelm. ex Rumpl.
    var. engelmannii
    var. howei L. Benson
    var. munzii (Parish) Pierce & Fosberg
        SY=E. m. (Parish) L. Benson
    var. nicholii L. Benson
    var. purpureus L. Benson
    var. variegatus (Engelm. & Bigelow)
      Engelm. ex Rumpl.
enneacanthus Engelm.
    var. brevispinus (W.O. Moore) L.
      Benson
    var. dubius (Engelm.) L. Benson
        SY=E. d. (Engelm.) Engelm. ex
          Rumpl.
    var. enneacanthus
    var. stramineus (Engelm.) L. Benson
        SY=E. s. (Engelm.) Engelm. ex
          Rumpl.
fasciculatus (Engelm.) L. Benson
    var. bonkerae (Thornb. & Bonker) L.
      Benson
    var. boyce-thompsonii (Orcutt) L.
      Benson
        SY=E. b-t. Orcutt
    var. fasciculatus
        SY=E. fendleri Engelm. var.
          robustus (Peebles) L. Benson
        SY=Mammillaria fasciculata Engelm.
fendleri (Engelm.) Engelm. ex Rumpl.
    var. fendleri
    var. kuenzleri (Castetter, Pierce &
      Schwerin) L. Benson
    var. rectispinus (Peebles) L. Benson
      ined.
hempelii F. Fobe
ledingii Peebles
lloydii Britt. & Rose
pectinatus (Scheidw.) Engelm.
    var. minor (Engelm.) L. Benson
    var. neomexicanus (Coult.) L. Benson
    var. pectinatus
        SY=E. p. var. ctenoides (Engelm.)
          Backeberg
    var. rigidissimus (Engelm.) Engelm.
      ex Rumpl.
    var. wenigeri L. Benson
pentalophus (DC.) Rumpl.
reichenbachii (Terscheck) Haage f. ex
    Britt. & Rose
    var. albertii L. Benson
        SY=?E. melanocentrus Lowry
    var. albispinus (Lahman) L. Benson
        SY=E. baileyi Rose
    var. chisosensis (W.T. Marsh.) L.
      Benson
    var. fitchii (Britt. & Rose) L.

Echinocereus (CONT.)
          Benson
          var. perbellus (Britt. & Rose) L.
          Benson
               SY=E. p. Britt. & Rose
          var. reichenbachii
               SY=E. caespitosus (Engelm.)
                    Engelm.
          var. robustus (Peebles) L. Benson
     triglochidiatus Engelm.
          var. arizonicus (Rose) L. Benson
          var. gonacanthus (Engelm. & Bigelow)
               L. Benson
          var. gurneyi L. Benson
          var. melanacanthus (Engelm.) L.
          Benson
               SY=Coryphantha vivipara (Nutt.)
                    Britt. & Rose var. aggregata
                    (Engelm.) W.T. Marsh.
               SY=E. coccineus (Engelm.) Engelm.
               SY=E. c. var. inermis (K. Schum.)
                    J.A. Purpus
               SY=?E. kunzei Guerke
               SY=E. t. var. i. (K. Schum.)
                    Rowley
               SY=Mammillaria a. Engelm.
          var. mojavensis (Engelm. & Bigelow)
               L. Benson
               SY=E. m. (Engelm. & Bigelow)
                    Rumpl.
          var. neomexicanus (Standl.) L. Benson
               SY=E. t. var. polyacanthus
                    (Engelm.) L. Benson
          var. paucispinus (Engelm.) Engelm. ex
               W.T. Marsh.
          var. triglochidiatus
     viridiflorus Engelm.
          var. correllii L. Benson
          var. cylindricus (Engelm.) Engelm. ex
               Rumpl.
          var. davisii (A.D. Houghton) W.T.
               Marsh.
          var. viridiflorus

Epithelantha A. Weber

     bokei L. Benson
               SY=E. micromeris (Engelm.) A.
                    Weber var. b. (L. Benson)
                    Glass & Foster
     micromeris (Engelm.) A. Weber

Ferocactus Britt. & Rose

     acanthodes (Lem.) Britt. & Rose
          var. acanthodes
               SY=Echinocactus a. Lem.
          var. lecontei (Engelm.) Lindsay
               SY=Echinocactus l. Engelm.
     covillei Britt. & Rose
     eastwoodiae (L. Benson) L. Benson ined.
               SY=F. acanthodes (Lem.) Britt. &
                    Rose var. e. L. Benson
     hamatacanthus (Muhlenpfordt) Britt. & Rose
          var. hamatacanthus
               SY=Echinocactus h. Muhlenpfordt
          var. sinuatus (A. Dietr.) L. Benson
     setispinus (Engelm.) L. Benson
               SY=Echinocactus s. Engelm.
     viridescens (Nutt.) Britt. & Rose
     wislizenii (Engelm.) Britt. & Rose
               SY=Echinocactus w. Engelm.

Lophophora Coult.

     williamsii (Lem.) Coult.

Mammillaria Haw.

     dioica K. Brandeg.
               SY=M. d. var. incerta (Parish)
                    Munz
     grahamii Engelm.
          var. grahamii
          var. oliviae (Orcutt) L. Benson
               SY=M. o. Orcutt
     heyderi Muhlenpfordt
          var. bullingtoniana Castetter, Pierce
               & Schwerin
          var. hemisphaerica (Engelm.) L.
          Benson
               SY=M. gummifera Engelm. var. h.
                    (Engelm.) L. Benson
          var. heyderi
               SY=M. gummifera Engelm. var.
                    applanata (Engelm.) L.
                    Benson
          var. macdougalii (Rose) L. Benson
               SY=M. gummifera Engelm. pro parte
               SY=M. g. var. m. (Rose) L. Benson
               SY=M. m. Rose
          var. meiacantha (Engelm.) L. Benson
               SY=M. gummifera Engelm. var. m.
                    (Engelm.) L. Benson
               SY=M. m. Engelm.
     lasiacantha Engelm.
     longimamma DC.
          var. sphaerica (A. Dietr.) K.
          Brandeg.
               SY=M. s. A. Dietr.
     mainiae K. Brandeg.
     microcarpa Engelm.
     nivosa Link
               SY=Neomammillaria n. (Link) Britt.
                    & Rose
     pottsii Scheer
     prolifera (P. Mill.) Haw.
          var. texana (Engelm.) Borg
     tetrancistra Engelm.
               SY=Phellosperma t. (Engelm.)
                    Britt. & Rose
     thornberi Orcutt
     viridiflora (Britt. & Rose) Boedeker
               SY=M. orestera L. Benson
     wrightii Engelm.
          var. wilcoxii (Toumey ex K. Schum.)
               W.T. Marsh.
               SY=M. meridiorosei Castetter,
                    Pierce & Schwerin
               SY=M. wilcoxii Toumey ex K. Schum.
          var. wrightii

Melocactus Link & Otto

     intortus (P. Mill.) Urban
               SY=Cactus i. P. Mill.

Neolloydia Britt. & Rose

     conoidea (DC.) Britt. & Rose
               SY=Mammillaria c. DC.
     erectocentra (Coult.) L. Benson
          var. acunensis (W.T. Marsh.) L.
          Benson
          var. erectocentra
               SY=Echinocactus e. Coult.
     gautii L. Benson
     intertexta (Engelm.) L. Benson
          var. dasyacantha (Engelm.) L. Benson
               SY=Echinomastus i. (Engelm.)
                    Britt. & Rose var. d.
                    (Engelm.) Backeberg
          var. intertexta
               SY=Echinocactus i. Engelm.

Neolloydia (CONT.)
  SY=Echinomastus i. (Engelm.)
    Britt. & Rose
johnsonii (Parry ex Engelm.) L. Benson
  SY=Echinocactus j. Parry ex
    Engelm.
  SY=Echinocactus j. var. lutescens
    Parish
  SY=Echinomastus j. (Parry ex
    Engelm.) E.M. Baxter
mariposensis (Hester) L. Benson
warnockii L. Benson

Opuntia P. Mill.

acanthocarpa Engelm. & Bigelow
  var. acanthocarpa
  var. coloradensis L. Benson
  var. ganderi (C.B. Wolf) L. Benson
    SY=O. a. ssp. g. C.B. Wolf
  var. major (Engelm. & Bigelow) L.
    Benson
  var. thornberi (Thornb. & Bonker) L.
    Benson
    SY=O. t. Thornb. & Bonker
arbuscula Engelm.
arenaria Engelm.
atrispina Griffiths
basilaris Engelm. & Bigelow
  var. aurea (E.M. Baxter) W.T. Marsh.
    SY=O. a. E.M. Baxter
  var. basilaris
    SY=O. b. var. ramosa Parish
    SY=O. b. var. whitneyana (E.M.
      Baxter) E.M. Baxter ex W.T.
      Marsh.
    SY=O. brachyclada Griffiths ssp.
      humistrata (Griffiths)
      Wiggins & Wolf
    SY=O. w. E.M. Baxter
    SY=O. w. var. albiflora E.M.
      Baxter
  var. brachyclada (Griffiths) Munz
    SY=O. brachyclada Griffiths
  var. longiareolata (Clover & Jotter)
    L. Benson
  var. treleasei (Coult.) Coult.
    SY=O. t. Coult.
bigelovii Engelm.
  var. bigelovii
  var. hoffmannii Fosberg
    SY=O. fosbergii C.B. Wolf
borinquensis Britt. & Rose
brasiliensis (Willd.) Berger
    SY=Brasiliopuntia b. (Willd.) Haw.
chlorotica Engelm. & Bigelow
clavata Engelm.
cochenillifera (L.) P. Mill.
    SY=Nopalea c. (L.) Salm-Dyck
cordobensis Speg.
cubensis Britt. & Rose
    SY=O. antillana Britt. & Rose
    SY=O. ochrocentra Small
echinocarpa Engelm. & Bigelow
  var. echinocarpa
  var. wolfii L. Benson
erinacea Engelm. & Bigelow
  var. columbiana (Griffiths) L. Benson
    SY=O. c. Griffiths
  var. erinacea
  var. hystricina (Engelm. & Bigelow)
    L. Benson
    SY=O. h. Engelm. & Bigelow
  var. ursina (A. Weber) Parish
    SY=O. rubrifolia Engelm. ex Coult.
    SY=O. u. A. Weber
  var. utahensis (Engelm.) L. Benson

    SY=O. erinacea Engelm. & Bigelow
      var. paucispina Dunkle
    SY=O. e. var. xanthostemma (K.
      Schum.) L. Benson
    SY=O. rhodantha K. Schum.
ficus-indica (L.) P. Mill.
    SY=O. demissa Griffiths
    SY=O. engelmannii Salm-Dyck
    SY=O. megacantha Salm-Dyck
    SY=O. occidentalis Engelm. &
      Bigelow
    SY=O. subarmata Griffiths
fragilis (Nutt.) Haw.
  var. brachyarthra (Engelm. & Bigelow)
    Coult.
  var. fragilis
    SY=O. f. var. denudata Wieg. &
      Backeberg
    SY=O. schweriniana K. Schum.
fulgida Engelm.
  var. fulgida
  var. mamillata (Schott) Coult.
humifusa (Raf.) Raf.
  var. ammophila (Small) L. Benson
    SY=O. a. Small
    SY=O. compressa (Salisb.) J.P.
      Macbr. var. a. (Small) L.
      Benson
    SY=O. impedata Small
    SY=O. lata Small
    SY=O. pisciformis Small
    SY=O. turgida Small
  var. austrina (Small) Dress
    SY=O. a. Small
    SY=O. compressa (Salisb.) J.F.
      Macbr. var. a. (Small) L.
      Benson
    SY=O. cumulicola Small
    SY=O. eburnispina Small
    SY=O. pollardii Britt. & Rose
  var. humifusa
    SY=O. calcicola Wherry
    SY=O. compressa (Salisb.) J.F.
      Macbr.
    SY=O. compressa var. microsperma
      (Engelm. & Bigelow) L.
      Benson
    SY=O. opuntia (L.) Karst.
    SY=O. rafinesquei Engelm.
imbricata Haw.
  var. argentea Anthony
  var. imbricata
    SY=O. arborescens Engelm.
kelvinensis V. & K. Grant
kleiniae DC.
  var. kleiniae
  var. tetracantha (Toumey) W.T. Marsh.
    SY=O. t. Toumey
leptocaulis DC.
leucotricha DC.
lindheimeri Engelm.
  var. cuija (Griffiths & Hare) L.
    Benson
    SY=O. cantabrigiensis Lynch
  var. lehmannii L. Benson
  var. lindheimeri
  var. linguiformis (Griffiths) L.
    Benson
    SY=O. linguiformis Griffiths
  var. tricolor (Griffiths) L. Benson
littoralis Engelm.
  var. austrocalifornica L. Benson &
    Walkington
  var. littoralis
  var. martiniana (L. Benson) L. Benson
  var. piercei (Fosberg) L. Benson &
    Walkington

Opuntia (CONT.)
            SY=O. covillei Britt. & Rose var.
                p. (Fosberg) Munz
        var. vaseyi (Coult.) L. Benson &
            Walkington
            SY=O. covillei Britt. & Rose
            SY=O. v. (Coult.) Britt. & Rose
    macrorhiza Engelm.
        var. macrorhiza
            SY=O. compressa (Salisb.) J.F.
                Macbr. var. m. (Engelm.) L.
                Benson
            SY=O. mackensenii Rose
            SY=O. plumbea Rose
            SY=O. tortispina Engelm. & Bigelow
        var. pottsii (Salm-Dyck) L. Benson
            SY=O. ballii Rose
            SY=O. p. Salm-Dyck
            SY=O. tenuispina Engelm. & Bigelow
    moniliformis (L.) Haw.
    munzii C.B. Wolf
    nicholii L. Benson
    oricola Philbrick
    parryi Engelm.
        var. parryi
            SY=O. echinocarpa Engelm. &
                Bigelow var. parkeri
                (Engelm.) Coult.
        var. serpentina (Engelm.) L. Benson
            SY=O. s. Engelm.
    phaeacantha Engelm.
        var. camanchica (Engelm. & Bigelow)
            L. Benson
            SY=O. c. Engelm. & Bigelow
        var. discata (Griffiths) L. Benson &
            Walkington
            SY=O. megacarpa Griffiths
        var. flavispina L. Benson
        var. laevis (Coult.) L. Benson
            SY=O. l. Coult.
        var. major Engelm.
            SY=O. canada Griffiths
            SY=O. gilvescens Griffiths
            SY=O. woodsii Backeberg
        var. mojavensis (Engelm.) Fosberg
            SY=O. m. Engelm.
        var. phaeacantha
            SY=O. dulcis Engelm.
        var. spinosibacca (Anthony) L. Benson
        var. superbospina (Griffiths) L.
            Benson
        var. wootonii (Griffiths) L. Benson
            SY=O. engelmannii Salm-Dyck var.
                w. (Griffiths) Fosberg
    polyacantha Haw.
        var. juniperina (Britt. & Rose) L.
            Benson
            SY=O. sphaerocarpa Engelm. &
                Bigelow
        var. polyacantha
        var. rufispina (Engelm. & Bigelow) L.
            Benson
            SY=O. rutila Nutt.
        var. trichophora (Engelm. & Bigelow)
            Coult.
            SY=O. t. (Engelm. & Bigelow)
                Britt. & Rose
    prolifera Engelm.
    pulchella Engelm.
            SY=O. barkleyana (Daston) Rowley
            SY=O. brachyrhopalica (Daston)
                Rowley
            SY=O. gracilicylindrica (Wieg. &
                Backeberg) Rowley
            SY=O. pygmaea (Wieg. & Backeberg)
                Rowley
            SY=O. spectatissima (Daston)

                Rowley
            SY=O. tuberculosirhopalica (Wieg.
                & Backeberg) Rowley
            SY=O. wiegandii (Backeberg) Rowley
    pusilla (Haw.) Haw.
            SY=O. drummondii Graham
            SY=O. tracyi Britt.
    ramosissima Engelm.
    repens Bello
    rubescens Salm-Dyck
    rufida Engelm.
    schottii Engelm.
        var. grahamii (Engelm.) L. Benson
            SY=O. g. Engelm.
        var. schottii
    spinosior (Engelm.) Toumey
    spinosissima (Martyn) P. Mill.
            SY=Consolea corallicola Small
    stanlyi Engelm.
        var. kunzei (Rose) L. Benson
            SY=O. k. Rose
            SY=O. wrightiana (E.M. Baxter)
                Peebles
        var. parishii (Orcutt) L. Benson
            SY=O. p. Orcutt
        var. peeblesiana L. Benson
        var. stanlyi
    stricta (Haw.) Haw.
        var. dillenii (Ker-Gawl.) L. Benson
            SY=O. atrocapensis Small
            SY=O. d. (Ker-Gawl.) Haw.
            SY=O. nitens Small
            SY=O. tunoidea Gibbes
            SY=O. zebrina Small
        var. stricta
            SY=O. bentonii Griffiths
            SY=O. macrarthra Gibbes
            SY=O. magnifica Small
    strigil Engelm.
        var. flexospina (Griffiths) L. Benson
        var. strigil
    ?tenuiflora Small
    tomentosa Salm-Dyck
    triacantha (Willd.) Sweet
            SY=O. abjecta Small
    tunicata (Lehm.) Link & Otto
        var. davisii (Engelm. & Bigelow) L.
            Benson
            SY=O. d. Engelm. & Bigelow
        var. tunicata
    ?turbinata Small
    versicolor Engelm.
    violacea Engelm.
        var. castetteri L. Benson
        var. gosseliniana (A. Weber) L.
            Benson
            SY=O. g. A. Weber
        var. macrocentra (Engelm.) L. Benson
            SY=O. m. Engelm.
        var. santa-rita (Griffiths & Hare) L.
            Benson
            SY=O. s-r. (Griffiths & Hare) Rose
        var. violacea
    vulgaris P. Mill.
    whipplei Engelm. & Bigelow
        var. multigeniculata (Clokey) L.
            Benson
        var. viridiflora (Britt. & Rose) L.
            Benson
            SY=O. v. Britt. & Rose
        var. whipplei
    wigginsii L. Benson

Pediocactus Britt. & Rose

    bradyi L. Benson
    knowltonii L. Benson

Pediocactus (CONT.)
    papyracanthus (Engelm.) L. Benson
            SY=Echinocactus p. Engelm.
    paradinei B.W. Benson
    peeblesianus (Croizat) L. Benson
            var. fickeiseniae L. Benson
            var. ?maianus L. Benson
            var. peeblesianus
    sileri (Engelm.) L. Benson
    simpsonii (Engelm.) Britt. & Rose
            var. minor (Engelm.) Cockerell
            var. robustior (Coult.) L. Benson
            var. simpsonii
            SY=Echinocactus s. Engelm.
    winkleri Heil

Pereskia (Plum. ex L.) P. Mill.

    aculeata P. Mill.
            SY=P. pereskia (L.) Karst.
    grandifolia Haw.

Rhipsalis Gaertn.

    baccifera (J. Mill.) Stearn
            SY=R. cassutha Gaertn.

Schlumbergera Lem.

    truncata (Haw.) Moran

Sclerocactus Britt. & Rose

    ?contortus Heil
    glaucus (J.A. Purpus) L. Benson
            SY=Pediocactus g. (J.A. Purpus)
                Arp
    mesae-verdae (Boiss. & Davids.) L. Benson
            SY=Echinocactus m-v. (Boiss. &
                Davids.) L. Benson
            SY=Pediocactus m-v. (Boiss. &
                Davids.) Arp
    parviflorus Clover & Jotter
            var. intermedius (Peebles) Woodruff &
                Benson
            SY=S. whipplei (Engelm. & Bigelow)
                Britt. & Rose var. i.
                (Peebles) L. Benson
            var. parviflorus
            SY=S. w. (Engelm.) Britt. & Rose
                var. roseus (Clover) L.
                Benson
    polyancistrus (Engelm. & Bigelow) Britt. &
        Rose
            SY=Echinocactus p. Engelm. &
                Bigelow
            SY=S. p. Britt. & Rose
    pubispinus (Engelm.) L. Benson
    spinosior (Engelm. ex Boiss.) Woodruff &
        Benson
            SY=Echinocactus whipplei Engelm. &
                Bigelow var. s. Engelm. ex
                Boiss.
            SY=S. pubispinus (Engelm.) L.
                Benson var. sileri L. Benson
            SY=S. whipplei Britt. & Rose var.
                spinosior Engelm. ex Boiss.
    ?terrae-canyonal Heil
    whipplei (Engelm. & Bigelow) Britt. & Rose
            var. ?hailii Castetter, Pierce &
                Schwerin
            var. ?reevsii Castetter, Pierce &
                Schwerin
            var. whipplei
            SY=Echinocactus w. Engelm. &
                Bigelow
            SY=Pediocactus w. (Engelm. &

                Bigelow) Arp
        wrightiae L. Benson

Thelocactus Britt. & Rose

    bicolor Galeotti
            var. flavidispinus Backeberg
            var. schottii (Engelm.) Krainz

CALLITRICHACEAE

Callitriche L.

    anceps Fern.
    fassettii Schotsman
    hermaphroditica L.
            SY=C. autumnalis L.
    heterophylla Pursh emend. Darby
            var. bolanderi (Hegelm.) Fassett
            SY=C. b. Hegelm.
            SY=C. h. ssp. b. (Hegelm.) Calder
                & Taylor
            var. heterophylla
    intermedia Hoffmann
        ssp. hamulata (Kuetz. ex W.D.J. Koch)
            Clapham
            SY=C. h. Kuetz. ex W.D.J. Koch
        ssp. intermedia
    longipedunculata Morong
    marginata Torr.
            SY=C. sepulta S. Wats.
    nuttallii Torr.
    peploides Nutt.
            SY=C. p. var. semialata Fassett
    stagnalis Scop.
    terrestris Raf. emend. Torr.
            SY=C. austinii Engelm.
            SY=C. deflexa A. Braun
            SY=C. d. var. a. (Engelm.) Hegelm.
            SY=C. d. var. subsessilis Fassett
    trochlearis Fassett
    verna L. emend. Kuetz.
            SY=C. palustris L.

CALYCANTHACEAE

Calycanthus L.

    floridus L.
            var. floridus
            SY=C. mohrii Small
            var. laevigatus (Willd.) Torr. & Gray
            SY=C. fertilis Walt.
            SY=C. nanus Loisel.
    occidentalis Hook. & Arn.

CALYCERACEAE

Acicarpha Juss.

    tribuloides Juss.

CAMPANULACEAE

Brighamia Gray

Brighamia (CONT.)
   citrina (Forbes & Lydgate) St. John
        var. citrina
        var. napaliensis St. John
   insignis Gray
   remyi St. John
   rockii St. John

Campanula L.

   americana L.
        SY=Campanula a. var. illinoensis
           (Fresn.) Farw.
        SY=Campanulastrum a. (L.) Small
   angustiflora Eastw.
      var. angustiflora
      var. exilis J.T. Howell
   aparinoides Pursh
        SY=Campanula a. var. grandiflora
           Holz.
        SY=C. a. var. uliginosa (Rydb.)
           Gleason
        SY=C. u. Rydb.
   aurita Greene
   bononiensis L.
   californica (Kellogg) Heller
   carpatica Jacq.
   chamissonis Fedorov
   divaricata Michx.
        SY=Campanula flexuosa Michx.
   exigua Rattan
   floridana S. Wats. ex Gray
        SY=Rotantha f. (S. Wats. ex Gray)
           Small
   gieseckiana Vest
     ssp. gieseckiana
      var. arctica (Lange) Bocher
        SY=C. rotundifolia L. var. a.
           Lange
      var. gieseckiana
     ssp. groenlandica (Berlin)·Bocher
   glomerata L.
   lactiflora Bieb.
   lasiocarpa Cham.
        SY=Campanula l. ssp. latisepala
           (Hulten) Hulten
        SY=C. latisepala Hulten
        SY=C. latisepala var. dubia Hulten
   latifolia L.
   medium L.
   parryi Gray
      var. idahoensis McVaugh
      var. parryi
   patula L.
   persicifolia L.
        SY=Campanula p. var. alba Hort.
   piperi T.J. Howell
   prenanthoides Dur.
        SY=Asyneuma p. (Dur.) McVaugh
   punctata Lam.
   rapunculoides L.
        SY=Campanula r. var. ucranica
           (Bess.) K. Koch
   reverchonii Gray
   robinsiae Small
        SY=Rotantha r. (Small) Small
   rotundifolia L.
        SY=Campanula alaskana (Gray) W.
           Wight ex J.P. Anders.
        SY=C. dubia A. DC.
        SY=C. heterodoxa Bong.
        SY=C. intercedens Witasek
        SY=C. petiolata A. DC.
        SY=C. r. var. alaskana Gray
        SY=C. r. var. i. (Witasek) Farw.
        SY=C. r. var. lancifolia Mert. &
           Koch

        SY=C. r. var. p. (A. DC.) J.K.
           Henry
        SY=C. sacajaweana M.E. Peck
        SY=C. velutina A. DC.
   scabrella Engelm.
   scouleri Hook. ex A. DC.
   shetleri Heckard
   trachelium L.
   uniflora L.
   wilkinsiana Greene

Clermontia Gaud.

   arborescens (Mann) Hbd.
   aspera E. Wimmer
   calophylla E. Wimmer
   carnifera Levl.
   clermontioides (Gaud.) Heller
      var. barbata (Rock) St. John
      var. clermontioides
      var. epiphytica Hochr.
      var. hirsutiflora Rock
   coerulea Hbd.
     ssp. brevidens (Skottsberg) St. John
      var. brevidens Skottsberg
      var. flavescens (E. Wimmer) St. John
     ssp. coerulea
      var. coerulea
      var. degeneri Skottsberg
      var. greenwelliana F. Wimmer
      var. parvifolia Rock
   convallis E. Wimmer
   drepanomorpha Rock
   forbesii St. John
   fulva Levl.
   furcata E. Wimmer
   grandiflora Gaud.
   haleakalensis Rock
   hanaensis St. John
   hawaiiensis (Hbd.) Rock
   hirsutinervis St. John
   kakeana Walp.
      var. kakeana
      var. orientalis St. John
   kohalae Rock
      var. hiloensis E. Wimmer
      var. kohalae
      var. robusta Rock
   konaensis St. John
   leptoclada Rock
      var. holopsila E. Wimmer
      var. leptoclada
      var. urceolata Rock
   lindseyana Rock
      var. lindseyana
        SY=C. hawaiiensis (Hbd.) Rock var.
           grandis Rock
      var. livida Rock
   loyana Rock
   micrantha (Hbd.) Rock
   molokaiensis St. John
   montis-loa Rock
      var. montis-loa
      var. tenuifolia Skottsberg
   multiflora Hbd.
   munroi St. John
   oblongifolia Gaud.
   pallida Hbd.
      var. pallida
      var. ramosissima Rock
   paradisia E. Wimmer
   parviflora Gaud. ex Gray
      var. calycina Rock
      var. grandis Rock
      var. intermedia Skottsberg
      var. parviflora
      var. umbraticola Skottsberg

Clermontia (CONT.)
    peleana Rock
    persicifolia Gaud.
    pyrularia Hbd.
    reticulata St. John
    rockiana E. Wimmer
    rosacea St. John
    samuelii Forbes
    siguliflora (Rock) Rock
    subpetiolata St. John
    tuberculata Forbes
        var. subtuberculata St. John
        var. tuberculata
    wailauensis St. John
    waimeae Rock
        var. longisepala Rock
        var. obovata Rock
        var. waimeae

Cyanea Gaud.

    aculeatiflora Rock
    acuminata (Gaud.) Hbd.
        var. acuminata
        var. calycina Hosaka
    angustifolia (Cham.) Hbd.
        var. angustifolia
        var. hillebrandii Rock
        var. isabella E. Wimmer
        var. lanaiensis Rock
        var. racemosa Hbd.
        var. tomentella Hbd.
    arborea (Mann) Hbd.
    asarifolia St. John
    asplenifolia (Mann) Hbd.
    atra Hbd.
        var. atra
        var. lobata Rock
    baldwinii Forbes & Munro
    bicolor St. John
    bishopii Rock
    bondiana (Rock) Rock
    bryanii Rock
    carlsonii Rock
    chockii Rock
    comata Hbd.
    copelandii Rock
    coriacea (Gray) Hbd.
        var. coriacea
        var. degeneriana E. Wimmer
        var. fauriei (Levl.) E. Wimmer
        var. gratiosa E. Wimmer
        var. hardyi (Rock) E. Wimmer
        var. serratifolia Rock
    coronata E. Wimmer
    crispihirta E. Wimmer
    degeneriana E. Wimmer
    densiflora (Rock) Rock
    dentata E. Wimmer
    dunbariae Rock
    fernaldii Rock
    ferox Hbd.
        var. ferox
        var. laevicalyx Skottsberg
    fissa (Mann) Hbd.
    floribunda E. Wimmer
    gayana Rock
        var. duvelii Rock
        var. gayana
        var. wainihaensis Rock
    gibsonii Hbd.
    giffardii Rock
    grimesiana Gaud.
        var. cylindrocalyx Rock
        var. grimesiana
        var. hirsutifolia Rock
        var. lydgatei Rock

        var. mauiensis Rock
        var. munroi Hosaka
        var. obovata St. John
    haleakalaensis St. John
    hamatiflora Rock
    hirtella (Mann) Hbd.
        var. hirtella
        var. striata E. Wimmer
        var. subglabra E. Wimmer
    holophylla Hbd.
        var. holophylla
        var. obovata Rock
    horrida (Rock) Deg. & Hosaka
    knudsenii Rock
        var. glabra E. Wimmer
        var. knudsenii
    kunthiana (Gaud.) Hbd.
    larrisonii Rock
    leptostegia Gray
        var. leptostegia
        var. velutina Skottsberg
    lindseyana Rock
        var. lindseyana
        var. livida Rock
    linearifolia Rock
    lobata Mann
        var. hamakuae Rock
        var. lobata
    longipedunculata Rock
    longissima (Rock) St. John
        SY=C. scabra Hbd. var. l. Rock
    macrostegia Hbd.
        var. macrostegia
        var. parvibracteata Rock
        var. viscosa Rock
    magnifica E. Wimmer
    mannii (Brigham) Hbd.
    mariana E. Wimmer
    marksii Rock
    mceldowneyi Rock
    megacarpa (Rock) Rock
    membranacea Rock
    multispicata Levl.
    nelsonii St. John
    noli-me-tangere Rock
    obtusa (Gray) Hbd.
    ovatisepala E. Wimmer
    pilosa Gray
    pinnatifida (Cham.) E. Wimmer
    platyphylla (Gray) Hbd.
    procera Hbd.
    profuga Forbes
    pulchra Rock
    pycnocarpa (Hbd.) E. Wimmer
        SY=C. arborea (Mann) Hbd. var. p.
          Hbd.
    quercifolia (Hbd.) E. Wimmer
        var. atropurpurea E. Wimmer
        var. quercifolia
    recta (Wawra) Hbd.
    regina (Wawra) Rock
    remyi Rock
    rivularis Rock
        SY=Delissea r. (Rock) E. Wimmer
    rockii E. Wimmer
    rollandioides Rock
    scabra Hbd.
        var. scabra
        var. sinuata Rock
        var. variabilis Rock
    shipmanii Rock
    solanacea Hbd.
    solenocalyx Hbd.
        var. latifolia E. Wimmer
        var. solenocalyx
    spathulata (Hbd.) Heller
    stictophylla Rock

Cyanea (CONT.)
        var. inermis Rock
        var. stictophylla
    submuricata E. Wimmer
    superba (Cham.) Gray
        var. superba
        var. velutina Rock
    sylvestris Heller
        var. eriantha (Skottsberg) E. Wimmer
        var. sylvestris
    tritomantha Gray
        var. lydgatei Rock
        var. tritomantha
    truncata (Rock) Rock
        var. juddii (Forbes) St. John
        var. truncata
    undulata Forbes
    wailauensis Rock

Delissea Gaud.

    argutidentata E. Wimmer
            SY=D. undulata Gaud. var. a. (E.
                Wimmer) E. Wimmer
    fallax Hbd.
    filigera Wawra
    laciniata Hbd.
        var. laciniata
        var. parvifolia Rock
    niihauensis St. John
    parviflora Hbd.
    rhytidosperma Mann
    sinuata Hbd.
        var. lanaiensis Rock
        var. sinuata
    subcordata Gaud.
        var. kauaiensis St. John
        var. obtusifolia Wawra
        var. subcordata
        var. waikaneensis St. John
        var. waralaeensis St. John
    undulata Gaud.

Downingia Torr.

    bacigalupii Weiler
    bella Hoover
    bicornuta Gray
        var. bicornuta
        var. picta Hoover
    concolor Greene
        var. brevior McVaugh
        var. concolor
        var. tricolor (Greene) Jepson
    cuspidata (Greene) Greene ex Jepson
    elegans (Dougl. ex Lindl.) Torr.
        var. brachypetala (Gandog.) McVaugh
        var. elegans
            SY=D. e. var. corymbosa (A. DC.)
                Gray
    insignis Greene
    laeta (Greene) Greene
    montana Greene
    ornatissima Greene
        var. eximia (Hoover) McVaugh
        var. ornatissima
    pulchella (Lindl.) Torr.
    pusilla (G. Don) Torr.
    yina Applegate
        var. major McVaugh
            SY=D. willamettensis M.E. Peck
        var. yina

Githopsis Nutt.

    calycina Benth.
    diffusa Gray

            SY=G. gilioides Ewan
    filicaulis Ewan
    latifolia Eastw.
    pulchella Vatke
    specularioides Nutt.
        ssp. candida Ewan
        ssp. specularioides

Heterocodon Nutt.

    rariflorum Nutt.
            SY=Specularia r. (Nutt.) McVaugh

Hippobroma G. Don

    longiflora (L.) G. Don
            SY=Isotoma l. (L.) Presl
            SY=Laurentia l. (L.) Engl.

Howellia Gray

    aquatilis Gray

Jasione L.

    montana L.

Legenere McVaugh

    limosa (Greene) McVaugh

Legousia Dur.

    speculum-veneris (L.) Chaix
            SY=Specularia s-v. (L.) Tanfani

Lobelia L.

    amoena Michx.
    anatina E. Wimmer
    appendiculata A. DC.
        var. appendiculata
        var. gattingeri (Gray) McVaugh
            SY=L. g. Gray
    berlandieri A. DC.
        var. berlandieri
        var. brachypoda (Gray) McVaugh
    boykinii Torr. & Gray
    brachypoda A. DC. ex Small
    brevifolia Nutt. ex A. DC.
    canbyi Gray
    cardinalis L.
        ssp. cardinalis
        ssp. graminea (Lam.) McVaugh
            var. graminea (Lam.) McVaugh
            var. multiflora (Paxton) McVaugh
            var. phyllostachya (Engelm.) McVaugh
                SY=L. splendens Willd.
            var. pseudosplendens McVaugh
    cliffortiana L.
    costata E. Wimmer
    dortmanna L.
    dunbariae Rock
    dunnii Greene
        var. serrata (Gray) McVaugh
    elongata Small
            SY=L. glandulifera (Gray) Small
    feayana Gray
    fenestralis Cav.
    flaccidifolia Small
    floridana Chapman
    gaudichaudii DC.
        var. gaudichaudii

        var. kauaensis Gray
        var. koolauensis Hosaka & Fosberg

Lobelia (CONT.)
        var. longibracteata Rock
georgiana McVaugh
glandulosa Walt.
gloria-montis Rock
        var. gloria-montis
        var. molokaiensis Deg.
grayana E. Wimmer
gruina Cav.
hillebrandii Rock
        var. hillebrandii
        var. monostachya Rock
        var. paniculata Rock
homophylla E. Wimmer
hypoleuca Hbd.
        var. heterocarpa E. Wimmer
        var. hypoleuca
        var. rockii St. John & Hosaka
inflata L.
kalmii L.
          SY=L. k. var. strictiflora Rydb.
          SY=L. s. (Rydb.) Lunell
laxiflora H.B.K.
        var. angustifolia A. DC.
niihauensis St. John
        var. forbesii St. John
        var. meridiana St. John
        var. niihauensis
nuttallii Roemer & Schultes
oahuensis Rock
paludosa Nutt.
portoricensis (Vatke) Urban
          SY=Tupa p. Vatke
puberula Michx.
        var. mineolana E. Wimmer
        var. puberula
        var. simulans Fern.
remyi Rock
reverchonii B.L. Turner
          SY=L. puberula Michx. var.
             pauciflora Bush
robusta Graham
        var. portoricensis (A. DC.) McVaugh
          SY=L. assurgens L. var. p. (A.
             DC.) Urban
        var. robusta
          SY=Tupa r. (Graham) A. DC.
rotundifolia Juss. ex A. DC.
siphilitica L.
          SY=L. s. var. ludoviciana A. DC.
X speciosa Sweet [cardinalis X siphilitica]
spicata Lam.
        var. campanulata McVaugh
        var. hirtella Gray
          SY=L. h. (Gray) Greene
        var. leptostachya (A. DC.) Mackenzie
          & Bush
          SY=L. l. A. DC.
        var. scaposa McVaugh
        var. spicata
          SY=L. bracteata Small
          SY=L. s. var. originalis McVaugh
          SY=L. s. var. parviflora Gray
tortuosa Heller
        var. intermedia St. John
        var. tortuosa
villosa (Rock) St. John & Hosaka
yuccoides Hbd.

Nemacladus Nutt.

   capillaris Greene
   glanduliferus Jepson
        var. glanduliferus
        var. orientalis McVaugh
   gracilis Eastw.
   interior (Munz) G.T. Robbins

          SY=N. rubescens Greene var. i.
            (Munz) McVaugh
   longiflorus Gray
        var. breviflorus McVaugh
        var. longiflorus
   montanus Greene
   pinnatifidus Greene
   ramosissimus Nutt.
   rigidus Curran
   rubescens Greene
        var. rubescens
        var. tenuis McVaugh
   secundiflorus G.T. Robbins
   sigmoideus G.T. Robbins
   twisselmannii J.T. Howell

Parishella Gray

   californica Gray

Platycodon A. DC.

   grandiflorum (Jacq.) A. DC.

Porterella Torr.

   carnosula (Hook. & Arn.) Torr.

Rollandia Gaud.

   angustifolia (Hbd.) Rock
        var. angustifolia
        var. ochreata E. Wimmer
   bidentata St. John
   calycina (Cham.) G. Don
        var. calycina
        var. kaalae (Wawra) E. Wimmer
   crispa Gaud.
        var. crispa
        var. muricata Rock
   degeneriana E. Wimmer
   humboldtiana Gaud.
   lanceolata Gaud.
        var. brevipes E. Wimmer
        var. glaberrima E. Wimmer
        var. kipapaensis Hosaka
        var. lanceolata
        var. rockii St. John & Hosaka
        var. tomentella (Wawra) E. Wimmer
        var. viridiflora Rock
   longiflora Wawra
   parvifolia Forbes
   pinnatifida (Cham.) G. Don
   purpurellifolia Rock
   sessilifolia Deg.
   st.-johnii Hosaka
   waianaeensis St. John

Sphenoclea Gaertn.

   zeylandica Gaertn.

Trematolobelia Zahlbr. ex Rock

   grandifolia (Rock) Deg.
   kauaiensis (Rock) Skottsberg
   macrostachys (Hook. & Arn.) Zahlbr.
        var. haleakalaensis St. John
        var. macrostachys
   wimmeri Deg. & Deg.

Triodanis Raf.

   coloradoensis (Buckl.) McVaugh
   holzingeri McVaugh
        SY=Specularia h. (McVaugh) Fern.
   lamprosperma McVaugh

Triodanis (CONT.)
          SY=Specularia l. (McVaugh) Fern.
   leptocarpa (Nutt.) Nieuwl.
          SY=Specularia l. (Nutt.) Gray
   perfoliata (L.) Nieuwl.
          var. biflora (Ruiz & Pavon) Bradley
             SY=Specularia b. (Ruiz & Pavon)
                   Fisch. & Mey.
             SY=Triodanis b. (Ruiz & Pavon)
                   Greene
          var. perfoliata
             SY=Specularia p. (L.) A. DC.
   texana McVaugh

Wahlenbergia Schrad. ex Roth

   gracilis (Forst. f.) A. DC.
   linarioides (Lam.) A. DC.
   marginata (Thunb.) A. DC.

CANELLACEAE

Canella P. Br.

   winterana (L.) Gaertn.
          SY=C. alba Murr.

Pleodendron v. Tiegh.

   macranthum (Baill.) v. Tiegh.

CANNACEAE

Canna L.

   coccinea P. Mill.
   edulis Ker-Gawl.
   flaccida Salisb.
   generalis Bailey
   glauca L.
   indica L.
   pertusa Urban
   sylvestris Roscoe

CAPPARIDACEAE

Atamisquea Miers ex Hook. & Arn.

   emarginata Miers ex Hook. & Arn.

Capparis L.

   amplissima Lam.
          SY=C. portoricensis Urban
   baducca L.
          SY=C. frondosa Jacq.
   cynophallophora L.
   flexuosa (L.) L.
   hastata Jacq.
          SY=C. coccolobifolia Mart. ex
                Eichl.
   incana H.B.K.
   indica (L.) Fawcett & Rendle
   sandwichiana DC.
          var. sandwichiana
          var. zohary Deg. & Deg.

Cleome L.

aculeata L.
diffusa Banks ex DC.
gynandra L.
          SY=Gynandropsis g. (L.) Brig.
hassleriana Chod.
          SY=C. houtteana sensu auctt. non
                Raf.
          SY=C. pungens Willd.
          SY=C. spinosa sensu auctt. non
                Jacq.
          SY=Neocleome s. (Jacq.) Small
isomeris Greene
          SY=Isomeris arborea Nutt.
          SY=I. a. var. angustata Parish
          SY=I. a. var. globosa Coville
          SY=I. a. var. insularis Jepson
jonesii (J.F. Macbr.) Tidestrom
lutea Hook.
          var. jonesii Macbr.
          var. lutea
multicaulis DC.
          SY=C. sonorae Sesse & Mocimo ex
                Gray
platycarpa Torr.
rutidosperma DC.
          SY=C. ciliata Schum. & Thonn.
serrata Jacq.
          SY=Neocleome s. (Jacq.) Small
serrulata Pursh
          SY=C. s. var. angusta (M.E. Jones)
                Tidestrom
sparsifolia S. Wats.
speciosa Raf.
          SY=C. s. H.B.K.
spinosa Jacq.
          ssp. sandwicensis (Gray) Iltis ined.
             SY=C. s. Gray
          ssp. spinosa
stenophylla Klotzsch ex Urban
viscosa L.
          SY=C. icosandra L.

Cleomella DC.

   angustifolia Torr.
   brevipes S. Wats.
   hillmanii A. Nels.
          SY=C. grandiflora (S. Wats.)
                Coville
          SY=C. macbrideana Payson
   longipes Torr. ex Hook.
   obtusifolia Torr. & Frem.
          SY=C. o. var. pubescens A. Nels.
   palmerana M.E. Jones
          SY=C. montrosae Payson
          SY=C. nana Eastw.
   parviflora Gray
   plocasperma S. Wats.
          SY=C. mojavensis Payson
          SY=C. oocarpa Gray
          SY=C. p. var. m. (Payson) Crum
          SY=C. p. var. stricta Crum
          SY=C. stenosperma Coville

Koeberlinia Zucc.

   spinosa Zucc.
          SY=K. s. var. tenuispina Kearney &
                Peebles

Morisonia L.

   americana L.

Oxystylis Torr. & Frem.

   lutea Torr. & Frem.

Polanisia Raf.

    dodecandra (L.) DC.
      ssp. dodecandra
          SY=P. graveolens Raf.
      ssp. riograndensis Iltis
      ssp. trachysperma (Torr. & Gray) Iltis
          SY=P. d. ssp. d. var. t. (Torr. &
              Gray) Iltis
          SY=P. t. Torr. & Gray
    erosa (Nutt.) Iltis
      ssp. breviglandulosa Iltis
      ssp. erosa
          SY=Cristatella e. Nutt.
    jamesii (Torr. & Gray) Iltis
          SY=Cristatella j. Torr. & Gray
    tenuifolia Le Conte ex Torr. & Gray
          SY=Aldenella t. (Le Conte ex Torr.
              & Gray) Greene
          SY=Cleome t. Le Conte ex Torr. &
              Gray
    uniglandulosa (Cav.) DC.
          SY=P. dodecandra (L.) DC. ssp. u.
              (Cav.) Iltis

Wislizenia Engelm.

    refracta Engelm.
      ssp. californica (Greene) Keller
          SY=W. c. Greene
      ssp. palmeri (Gray) Keller
          SY=W. divaricata Greene
      ssp. refracta
          SY=W. r. var. melilotoides
              (Greene) I.M. Johnston

CAPRIFOLIACEAE

Diervilla P. Mill.

    lonicera P. Mill.
          SY=D. diervilla (L.) MacM.
          SY=D. l. var. hypomalaca Fern.
    rivularis Gattinger
          SY=D. sessilifolia Buckl. var. r.
              (Gattinger) Ahles
    sessilifolia Buckl.

Linnaea L.

    borealis L.
      ssp. americana (Forbes) Hulten
          SY=L. a. Forbes
          SY=L. b. var. a. (Forbes) Rehd.
      ssp. borealis
      ssp. longiflora (Torr.) Hulten
          SY=L. b. var. l. Torr.

Lonicera L.

    albiflora Torr. & Gray
        var. albiflora
        var. dumosa (Gray) Rehd.
    arizonica Rehd.
    X bella Zabel [morrowii X tatarica]
    caerulea L.
        var. cauriana (Fern.) Boivin
          SY=L. cauriana Fern.
    canadensis Bartr.
          SY=Xylosteon ciliatum (Muhl.)
              Pursh
    caprifolium L.
    ciliosa (Pursh) DC.
    conjugialis Kellogg

dioica L.
    var. dasygyna (Rehd.) Gleason
    var. dioica
    var. glaucescens (Rydb.) Butters
      SY=L. g. Rydb.
    var. orientalis Gleason
etrusca Santi
flava Sims
    var. flava
    var. flavescens (Small) Gleason
      SY=L. flavida Cockerell
fragrantissima Lindl. & Paxton
      SY=Xylosteon f. (Lindl. & Paxton)
         Small
hirsuta Eat.
    var. hirsuta
    var. interior Gleason
    var. schindleri Boivin
hispidula (Lindl.) Dougl. ex Torr. & Gray
    var. hispidula
    var. vacillans (Benth.) Gray
interrupta Benth.
involucrata (Richards.) Banks ex Spreng.
    var. flavescens (Dippel) Rehd.
    var. involucrata
      SY=Distegia i. (Richards.)
           Cockerell
japonica Thunb.
    var. chinensis (P.W. Wats.) Baker
    var. japonica
      SY=Nintooa j. (Thunb.) Sweet
ledebourii Eschsch.
      SY=L. involucrata (Richards.)
         Banks ex Spreng. var. l.
         (Eschsch.) Zabel
maackii (Rupr.) Maxim.
morrowii Gray
X notha Zabel [ruprechtiana X tatarica]
oblongifolia (Goldie) Hook.
    var. altissima (Jennings) Rehd.
    var. oblongifolia
periclymenum L.
prolifera (Kirchn.) Rehd.
    var. glabra Gleason
      SY=L. sullivantii Gray
    var. prolifera
sempervirens L.
      SY=L. s. var. hirsutula Rehd.
      SY=L. s. var. minor Ait.
      SY=Phenianthus s. (L.) Raf.
standishii Jacques
subspicata Hook. & Arn.
    var. denudata Rehd.
    var. johnstonii Keck
    var. subspicata
tatarica L.
utahensis S. Wats.
villosa (Michx.) Roemer & Schultes
    var. calvescens (Fern. & Wieg.) Fern.
    var. fulleri Fern.
    var. solonis (Eat.) Fern.
    var. tonsa Fern.
    var. villosa
      SY=L. caerulea L. var. v. (Michx.)
         Torr. & Gray
xylosteum L.

Sambucus L.

    canadensis L.
      var. canadensis
      var. submollis Rehd.
    cerulea Raf.
      var. cerulea
        SY=S. glauca Nutt.
      var. neomexicana (Woot.) Rehd.
        SY=S. n. Woot.

Sambucus (CONT.)
        SY=S. n. var. vestita (Woot. &
          Standl.) Kearney & Peebles
      var. velutina (Dur. & Hilg.) Schwerin
        SY=S. v. Dur. & Hilg.
  ebulus L.
  mexicana Presl ex DC.
        SY=S. orbiculata Greene
        SY=S. rehderana Schwerin
  nigra L.
        SY=S. laciniata P. Mill.
  racemosa L.
    ssp. pubens (Michx.) House
      var. arborescens (Torr. & Gray) Gray
        SY=S. callicarpa Greene
        SY=S. p. Michx. var. a. Torr. &
          Gray
      var. leucocarpa (Torr. & Gray) Cronq.
      var. melanocarpa (Gray) McMinn
        SY=S. m. Gray
      var. microbotrys (Rydb.) Kearney &
        Peebles
        SY=S. m. Rydb.
      var. pubens (Michx.) Koehne
        SY=S. p. Michx.
    ssp. racemosa
  simpsonii Rehd.

Symphoricarpos Duham.

  acutus (Gray) Dieck
        SY=S. mollis Nutt. var. a. Gray
  albus (L.) Blake
      var. albus
        SY=S. racemosus Michx.
      var. laevigatus (Fern.) Blake
        SY=S. a. ssp. l. (Fern.) Hulten
        SY=S. rivularis Suksdorf
      var. pauciflorus (Britt.) Blake
        SY=S. p. Britt.
  guadalupensis Correll
  hesperius G.N. Jones
        SY=S. mollis Nutt. ssp. h. (G.N.
          Jones) Abrams ex Ferris
        SY=S. m. var. h. (G.N. Jones)
          Cronq.
  longiflorus Gray
  microphyllus H.B.K.
  mollis Nutt.
  occidentalis Hook.
  orbiculatus Moench
        SY=S. symphoricarpos (L.) MacM.
  oreophilus Gray
      var. oreophilus
        SY=S. rotundifolius Gray var. o.
          (Gray) M.E. Jones
        SY=S. vaccinoides Rydb.
      var. utahensis (Rydb.) A. Nels.
        SY=S. tetonensis A. Nels.
        SY=S. u. Rydb.
  palmeri G.N. Jones
  parishii Rydb.
  rotundifolius Gray

Triosteum L.

  angustifolium L.
      var. angustifolium
      var. eamesii Wieg.
  aurantiacum Bickn.
      var. aurantiacum
        SY=T. perfoliatum L. var. a.
          (Bickn.) Wieg.
      var. glaucescens Wieg.
        SY=T. perfoliatum L. var. g.
          (Wieg.) Wieg.
      var. illinoense (Wieg.) Palmer &

      Steyermark
        SY=T. i. (Wieg.) Rydb.
        SY=T. perfoliatum L. var. i. Wieg.
  perfoliatum L.

Viburnum L.

  acerifolium L.
      var. acerifolium
      var. densiflorum (Chapman) McAtee
        SY=V. d. Chapman
      var. glabrescens Rehd.
      var. ovatum McAtee
  australe Morton
        SY=V. affine Bush var. australe
          (Morton) McAtee
  bracteatum Rehd.
  cassinoides L.
      var. cassinoides
      var. harbisonii McAtee
  crenatum McAtee
  dentatum L.
      var. deamii (Rehd.) Fern.
        SY=V. carolinianum Ashe
        SY=V. c. var. cismontanum McAtee
        SY=V. carolinianum var. deamii
          (Rehd.) McAtee
        SY=V. dentatum var. indianense
          (Rehd.) Gleason
        SY=V. i. (Rehd.) McAtee
        SY=V. pubescens (Ait.) Pursh var.
          deamii Rehd.
        SY=V. p. var. i. Rehd.
      var. dentatum
        SY=V. d. var. semitomentosum
          Michx.
        SY=V. pubescens (Ait.) Pursh
        SY=V. s. (Michx.) Rehd.
      var. scabrellum Torr. & Gray
        SY=V. s. (Torr. & Gray) Chapman
        SY=V. s. var. ashei Bush
        SY=V. s. var. dilutum McAtee
      var. venosum (Britt.) Gleason
        SY=V. scabrellum (Torr. & Gray)
          Chapman var. v. (Britt.)
          McAtee
        SY=V. v. Britt.
  edule (Michx.) Raf.
        SY=V. pauciflorum (La Pylaie)
          Torr. & Gray
  ellipticum Hook.
  lantana L.
  lantanoides Michx.
        SY=V. alnifolium Marsh.
        SY=V. grandifolium Ait.
  lentago L.
  molle Michx.
        SY=V. ozarkense Ashe
  nitidum Ait.
        SY=V. cassinoides L. var. n.
          (Ait.) McAtee
  nudum L.
        SY=V. n. var. angustifolium Torr.
          & Gray
        SY=V. n. var. grandifolium Small
  obovatum Walt.
        SY=V. nashii Small
  opulus L.
  plicatum Thunb.
        SY=V. tomentosum Thunb.
  prunifolium L.
        SY=V. p. var. bushii (Ashe) Palmer
          & Steyermark
  rafinesquianum Schultes
      var. affine (Bush) House
        SY=V. a. Bush
      var. rafinesquianum

Viburnum (CONT.)
            SY=V. affine Bush var. hypomalacum
                Blake
    recognitum Fern.
            var. alabemense McAtee
            var. recognitum
                SY=V. dentatum L. var. lucidum
                    Ait.
    rufidulum Raf.
            SY=V. prunifolium L. var.
                ferrugineum Torr. & Gray
    sieboldii Miq.
    trilobum Marsh.
            SY=V. opulus L. var. americanum
                (P. Mill.) Ait.
            SY=V. o. var. t. (Marsh.) Clausen

Weigelia Thunb.

    floribunda (Sieb. & Zucc.) C.A. Mey.
            SY=W. florida (Sieb. & Zucc.) A.
                DC.

CARICACEAE

Carica L.

    papaya L.

CARYOPHYLLACEAE

Achyronychia Torr. & Gray

    cooperi Torr. & Gray

Agrostemma L.

    githago L.
    gracilis Boiss.

Alsinodendron Mann

    obovatum Sherff
            var. obovatum
            var. parvifolium Deg. & Sherff
    trinerve Mann

Arenaria L.

    aberrans M.E. Jones
    aculeata S. Wats.
            SY=A. kingii (S. Wats.) M.E. Jones
                var. glabrescens (S. Wats.)
                Maguire
            SY=A. salmonensis Henderson
    benthamii Fenzl ex Torr. & Gray
    capillaris Poir.
        ssp. americana Maguire
            SY=A. c. var. a. (Maguire) R.J.
                Davis
            SY=A. formosa (Fisch.) Regel
        ssp. capillaris
            SY=A. c. var. nardifolia (Ledeb.)
                Regel
    congesta Nutt. ex Torr. & Gray
            var. cephaloides (Rydb.) Maguire
            var. charlestonensis Maguire
            var. congesta
            var. crassula Maguire
            var. glandulifera Maguire
            var. lithophila (Rydb.) Maguire

            SY=A. c. var. expansa Maguire
            var. prolifera Maguire
            var. simulans Maguire
            var. subcongesta (S. Wats.) S. Wats.
                SY=A. burkei T.J. Howell
            var. suffrutescens (Gray) B.L.
                Robins.
            var. wheelerensis Maguire
    ?cumberlandensis B.E. Wofford & Kral
    eastwoodiae Rydb.
            var. adenophora Kearney & Peebles
            var. eastwoodiae
                SY=A. fendleri Gray var. e.
                    (Rydb.) Harrington
    fendleri Gray
            var. brevifolia (Maguire) Maguire
            var. fendleri
                SY=A. f. var. diffusa Porter
            var. porteri Rydb.
            var. tweedyi (Rydb.) Maguire
    ?fontinalis (Short & Peter) Shinners
            SY=Alsine f. (Short & Peter)
                Britt.
            SY=Stellaria f. (Short & Peter)
                B.L. Robins.
    franklinii Dougl. ex Hook.
            var. franklinii
            var. thompsonii M.E. Peck
    hookeri Nutt. ex Torr. & Gray
        ssp. desertorum (Maguire) W.A. Weber
            SY=Arenaria h. var. d. Maguire
        ssp. hookeri
        ssp. pinetorum (A. Nels.) W.A. Weber
            SY=Arenaria h. var. p. (A. Nels.)
                Maguire
    humifusa Wahlenb.
    kingii (S. Wats.) M.E. Jones
        ssp. compacta (Coville) Maguire
            SY=Arenaria c. Coville
        ssp. kingii
        ssp. plateauensis Maguire
        ssp. rosea Maguire
        ssp. uintahensis (A. Nels.) C.L. Hitchc.
            SY=Arenaria aculeata S. Wats. var.
                u. (A. Nels.) M.E. Peck
            SY=A. k. var. u. (A. Nels.)
                Maguire
            SY=A. u. A. Nels.
    lanuginosa (Michx.) Rohrb.
        ssp. lanuginosa
            var. lanuginosa
            var. longipedunculata Duncan
                SY=Arenaria confusa Rydb.
        ssp. saxosa (Gray) Maguire
            SY=Arenaria l. var. cinerascens
                (B.L. Robins.) Shinners
            SY=A. l. var. mearnsii (Woot. &
                Standl.) Kearney & Peebles
            SY=A. s. Gray
            SY=A. s. var. c. B.L. Robins.
    leptoclados (Reichenb.) Guss.
            SY=A. serpyllifolia L. var.
                tenuior Mert. & Koch
    livermorensis Correll
    longipedunculata Hulten
    ludens Shinners
    macradenia S. Wats.
        ssp. ferrisiae Abrams
        ssp. macradenia
            var. arcuifolia Maguire
            var. kuschei (Eastw.) Maguire
            var. macradenia
            var. parishiorum B.L. Robins.
    paludicola B.L. Robins.
    pseudofrigida Ostenf. & Dahl
            SY=A. ciliata L. var. p. (Ostenf.
                & Dahl) Boivin

Arenaria   (CONT.)
    pumicola Coville & Leib.
            var. californica Maguire
            var. pumicola
    serpyllifolia L.
    stenomeres Eastw.
    ursina B.L. Robins.
    verna L.

Cardionema DC.

    ramosissimum (Weinm.) A. Nels. & J.F.
        Macbr.

Cerastium L.

    adsurgens Greene
    aleuticum Hulten
            SY=C. beeringianum Cham. &
                Schlecht. var. a. (Hulten)
                Welsh
    alpinum L.
            var. alpinum
                SY=C. a. var. glanduliferum W.D.J.
                    Koch
                SY=C. a. var. lanatum (Lam.)
                    Hegetschw.
                SY=C. beeringianum Cham. &
                    Schlecht.
                SY=C. b. var. grandiflorum (Fenzl)
                    Hulten
                SY=C. pulchellum Rydb.
            var. capillare (Fern. & Wieg.) Boivin
                SY=C. beeringianum Cham. &
                    Schlecht. var. c. Fern. &
                    Wieg.
            var. strigosum Hulten
    arcticum Lange
            var. arcticum
            var. procerum (Lange) Hulten
            var. sordidum Hulten
            var. vestitum Hulten
    arvense L.
            var. arvense
                SY=C. a. var. villosum (Muhl.)
                    Hollick & Britt.
                SY=C. campestre Greene
                SY=C. velutinum Raf.
            var. villosissimum Pennell
            var. viscidulum Gremli
                SY=C. oreophilum Greene
    axillare Correll
    biebersteinii DC.
    brachypetalum Pers.
            SY=C. b. var. compactus B.L.
                Robins.
    cerastioides (L.) Britt.
            SY=Provancheria c. (L.) Boivin
            SY=Stellaria c. L.
    dichotomum L.
    diffusum Pers.
            SY=C. tetrandrum M.A. Curtis
    dubium (Bast.) O. Schwartz
    earlei Rydb.
            SY=C. beeringianum Cham. &
                Schlecht. ssp. e. (Rydb.)
                Hulten
    fisherianum Ser.
    fontanum Baumg.
        ssp. scandicum H. Gartner
                SY=C. vulgatum L. var. alpestre
                    Hartman
        ssp. triviale (Link) Jalas
                SY=C. holosteoides Fries
                SY=C. h. var. vulgare (Hartman)
                    Hyl.
                SY=C. vulgatum L. pro parte

                SY=C. vulgatum var. hirsutum Fries
                SY=C. vulgatum var. holosteoides
                    (Fries) Wahlenb.
    glomeratum Thuill.
            SY=C. g. var. apetalum (Dumort.)
                Fenzl
            SY=C. viscosum sensu auctt. non L.
    jenisejense Hulten
    maximum L.
    nutans Raf.
            var. brachypodum Engelm. ex Gray
                SY=C. b. (Engelm. ex Gray) B.L.
                    Robins.
            var. nutans
                SY=C. longepedunculatum Muhl.
            var. obtectum Kearney & Peebles
    pumilum W. Curtis
        ssp. pallens (F.W. Schultz) Schinz &
            Thellung
                SY=C. glutinosum Fries
        ssp. pumilum
    regelii Ostenf.
    semidecandrum L.
    siculum Guss.
    sordidum B.L. Robins.
    terrae-novae Fern. & Wieg.
    texanum Britt.
    tomentosum L.

Corrigiola L.

    littoralis L.

Dianthus L.

    armeria L.
    barbatus L.
    chinensis L.
    deltoides L.
    plumarius L
    repens Willd.
    sylvestris Wulfen

Drymaria Willd. ex Roemer & Schultes

    cordata (L.) Willd. ex Roemer & Schultes
    effusa Gray
            var. depressa (Greene) J. Duke
                SY=D. d. Greene
            var. effusa
    glandulosa Presl
            SY=D. fendleri S. Wats.
    gracilis Cham. & Schlecht.
    laxiflora Benth.
    leptophylla (Cham. & Schlecht.) Fenzl ex
        Rohrb.
            SY=D. tenella Gray
    molluginea (Lag.) Didr.
            SY=D. sperguloides Gray
    pachyphylla Woot. & Standl.

Gypsophila L.

    acutifolia Stevens ex Spreng.
    elegans Bieb.
    muralis L.
    paniculata L.
    perfoliata L.
            var. latifolia Maxim.
            var. perfoliata

Herniaria L.

    cinerea DC.
    glabra L.

Holosteum L.

Holosteum (CONT.)
  umbellatum L.

Honckenya Ehrh.

    peploides (L.) Ehrh.
      ssp. major (Hook.) Hulten
        SY=Arenaria peploides L. var. m.
            Hook.
        SY=A. p. var. maxima Fern.
        SY=H. p. var. major (Hook.) Abrams
      ssp. peploides
        SY=Arenaria p. L.
        SY=A. p. var. diffusa Hornem.
      ssp. robusta (Fern.) Hulten
        SY=Arenaria p. L. var. r. Fern.

Loeflingia L.

    squarrosa Nutt.
      ssp. artemisiarum Barneby & Twisselman
      ssp. cactorum Barneby & Twisselman
      ssp. squarrosa
        SY=L. pusilla Curran
        SY=L. texana Hook.

Lychnis L.

    alpina L.
        var. albiflora (Lange) Fern.
        var. alpina
          SY=Viscaria a. (L.) G. Don
        var. americana Fern.
          SY=Viscaria a. (Fern.) Buch.
    chalcedonica L.
    coronaria (L.) Desr.
    flos-cuculi L.
    viscaria L.

Minuartia L.

    alabamensis McCormick, Bozeman & Spongberg
        SY=Arenaria a. (McCormick, Bozeman
            & Spongberg) Wyatt
    arctica (Stev. ex Ser.) Aschers. & Graebn.
        SY=Arenaria a. Stev. ex Ser.
    austromontana S.J. Wolf
        SY=Arenaria rossii R. Br. ssp.
            columbiana sensu auctt. non
            (Raup) Maguire
        SY=A. r. var. c. sensu auctt. non
            Raup
    biflora (L.) Schinz & Thellung
        SY=Arenaria sajanensis Willd. ex
            Schlecht.
    californica (Gray) Mattf.
        SY=Arenaria c. (Gray) Brewer
    caroliniana (Walt.) Mattf.
        SY=Arenaria c. Walt.
        SY=Sabulina c. (Walt.) Small
    dawsonensis (Britt.) House
        SY=Arenaria d. Britt.
        SY=A. litorea Fern.
        SY=A. stricta Michx. ssp. d.
            (Britt.) Maguire
    douglasii (Fenzl ex Torr. & Gray) Mattf.
        var. douglasii
          SY=Arenaria d. Fenzl ex Torr. &
              Gray
        var. emarginata (H.K. Sharsmith)
            McNeill
          SY=A. d. var. emarginata H.K.
              Sharsmith
    drummondii (Shinners) McNeill
        SY=Arenaria d. Shinners
        SY=Stellaria nuttallii Torr. &
            Gray

    elegans (Cham. & Schlecht.) Shishkin
        SY=Arenaria r. R. Br. var. e.
            (Cham. & Schlecht.) Welsh
        SY=M. rossii (R. Br.) Graebn. var.
            e. (Cham. & Schlecht.)
            Hulten
    filiorum (Maguire) McNeill
        SY=Arenaria f. Maguire
    glabra (Michx.) Mattf.
        SY=Arenaria g. Michx.
        SY=A. groenlandica (Retz.) Spreng.
            var. glabra (Michx.) Fern.
        SY=Sabulina glabra (Michx.) Small
    godfreyi (Shinners) McNeill
        SY=Arenaria g. Shinners
        SY=M. uniflora sensu Mattf. non
            (Walt.) Mattf.
        SY=Sabulina u. sensu Small non
            (Walt.) Small
        SY=Stellaria paludicola Fern. &
            Schub.
    groenlandica (Retz.) Ostenf.
        SY=Arenaria g. (Retz.) Spreng.
        SY=Sabulina g. (Retz.) Small
    howellii (S. Wats.) Mattf.
        SY=Arenaria h. S. Wats.
    macrantha (Rydb.) House
        SY=Arenaria m. (Rydb.) A. Nels. ex
            Coult. & A. Nels.
    macrocarpa (Pursh) Ostenf.
        SY=Arenaria m. Pursh
    marcescens (Fern.) House
        SY=Arenaria laricifolia (L.) B.L.
            Robins. var. m. (Fern.)
            Boivin
        SY=A. m. Fern.
    michauxii (Fern.) Farw.
        var. michauxii
          SY=Arenaria stricta Michx.
          SY=A. s. ssp. macra (A. Nels. &
              J.F. Macbr.) Maguire
          SY=Sabulina s. (Michx.) Small
        var. texana (B.L. Robins.) Mattf.
          SY=Arenaria s. ssp. t. (B.L.
              Robins.) Maguire
          SY=A. s. var. t. B.L. Robins.
          SY=A. t. (B.L. Robins.) Britt.
    muriculata (Maguire) McNeill
        SY=Arenaria m. Maguire
    nuttallii (Pax) Briq.
      ssp. fragilis (Maguire & Holmgren)
          McNeill
        SY=Arenaria n. ssp. f. Maguire &
            Holmgren
        SY=A. n. var. f. (Maguire &
            Holmgren) C.L. Hitchc.
      ssp. gracilis (B.L. Robins.) McNeill
        SY=Arenaria n. ssp. g. (B.L.
            Robins.) Maguire
        SY=A. n. var. g. B.L. Robins.
      ssp. gregaria (Heller) McNeill
        SY=Arenaria n. ssp. g. (Heller)
            Maguire
        SY=A. n. var. g. (Heller) Jepson
      ssp. nuttallii
        SY=Arenaria n. Pax
        SY=M. pungens (Nutt.) Mattf.
    obtusiloba (Rydb.) House
        SY=Arenaria o. (Rydb.) Fern.
    patula (Michx.) Mattf.
        var. patula
          SY=Arenaria p. Michx.
          SY=Sabulina p. (Michx.) Small
        var. robusta (Steyermark) McNeill
          SY=Arenaria p. var. r.
              (Steyermark) Maguire
    pusilla (S. Wats.) Mattf.

Minuartia (CONT.)
        var. diffusa (Maguire) McNeill
           SY=A. p. var. d. Maguire
        var. pusilla
           SY=Arenaria p. S. Wats.
  rosei (Maguire & Barneby) McNeill
  rossii (R. Br.) Graebn.
        SY=Arenaria r. R. Br.
        SY=A. r. var. apetala Maguire
        SY=A. r. var. corollina Fenzl
        SY=A. r. var. daethiana Polunin
  rubella (Wahlenb.) Hiern
        SY=Arenaria propinqua Richards.
        SY=A. r. (Wahlenb.) Sm.
        SY=A. verna L. var. p. (Richards.)
           Fern.
        SY=Minuartia rossii (R. Br.)
           Graebn. var.
           orthotrichoides (Schischkin)
           Hulten pro parte
  stricta (Sw.) Hiern
        SY=Alsine uliginosa (Murr.) Britt.
        SY=Arenaria s. Michx. var. u.
           (Murr.) Boivin
        SY=Arenaria u. (Murr.) Schleich.
        SY=Stellaria u. Murr.
  tenella (Nutt.) Mattf.
        SY=Arenaria stricta Michx. ssp.
           macra (A. Nels. & J.F.
           Macbr.) Maguire
  uniflora (Walt.) Mattf.
        SY=Arenaria u. (Walt.) Muhl.
        SY=Sabulina brevifolia (Nutt.)
           Small
        SY=S. u. (Walt.) Small
  yukonensis Hulten
        SY=Arenaria laricifolia (L.) B.L.
           Robins. pro parte
        SY=A. l. var. hultenii Welsh pro
           parte
        SY=A. l. var. occulta (Ser.)
           Boivin pro parte

Moehringia L.

  lateriflora (L.) Fenzl
        SY=Arenaria l. L.
        SY=A. l. var. angustifolia (Regel)
           St. John
  macrophylla (Hook.) Fenzl
        SY=Arenaria m. Hook.

Moenchia Ehrh.

  erecta (L.) Gaertn., Mey., & Scherb.

Myosoton Moench

  aquaticum (L.) Moench
        SY=Alsine a. (L.) Britt.
        SY=Stellaria a. (L.) Scop.

Paronychia P. Mill.

  americana (Nutt.) Fenzl ex Walp.
    ssp. americana
        SY=Siphonychia a. (Nutt.) Torr. &
           Gray
    ssp. pauciflora (Small) Chaudhri
        SY=Siphonychia p. Small
  argyrocoma (Michx.) Nutt.
        SY=P. a. var. albimontana (Fern.)
           Maguire
  baldwinii (Torr. & Gray) Fenzl
    ssp. baldwinii
        SY=Anychiastrum b. (Torr. & Gray)
           Small

    ssp. riparia (Chapman) Chaudhri
        SY=Anychiastrum r. (Chapman) Small
        SY=P. r. Chapman
  canadensis (L.) Wood
        SY=Anychia c. (L.) B.S.P.
        SY=P. dichotoma (L.) A. Nels.
  chartacea Fern.
        SY=Nyachia pulvinata Small
        SY=P. p. (Small) Pax & Hoffmann
  chorizanthoides Small
  congesta Correll
  depressa Nutt. ex Torr. & Gray
        var. brevicuspis (A. Nels.) Chaudhri
        var. depressa
        var. diffusa (A. Nels.) Chaudhri
           SY=P. diffusa A. Nels.
  drummondii Torr. & Gray
    ssp. drummondii
    ssp. parviflora Chaudhri
  erecta (Chapman) Shinners
        var. corymbosa (Small) Chaudhri
           SY=Odontonychia c. Small
        var. erecta
           SY=Odontonychia e (Chapman) Small
  fastigiata (Raf.) Fern.
        var. fastigiata
           SY=Anychia polygonoides Raf.
           SY=P. f. var. typica Fern.
        var. nuttallii (Small) Fern.
        var. paleacea Fern.
  franciscana Eastw.
  herniarioides (Michx.) Nutt.
        SY=Gastronychia h. (Michx.) Small
  jamesii Torr. & Gray
        var. hirsuta Chaudhri
        var. jamesii
           SY=P. wardii Rydb.
        var. parviflora Chaudhri
        var. praelongifolia Correll
  jonesii M.C. Johnston
  lindheimeri Engelm. ex Gray
        var. lindheimeri
        var. longibracteata Chaudhri
  maccartii Correll
  montana (Small) Pax & Hoffmann
        SY=Anychiastrum m. Small
        SY=P. fastigiata (Raf.) Fern. var.
           pumila (Wood) Fern.
  monticola Cory
  nudata Correll
  patula Shinners
  pulvinata Gray
        var. longiaristata Chaudhri
        var. pulvinata
           SY=P. sessiliflora Nutt. ssp. p.
           (Gray) W.A. Weber
  rugelii Shuttlw. ex Chapman
        var. interior (Small) Chaudhri
           SY=Odontonychia i. Small
        var. rugelii
           SY=Gibbesia r. (Shuttlw. ex
           Chapman) Small
  sessiliflora Nutt.
  setacea Torr. & Gray
        var. longibracteata Chaudhri
        var. setacea
  virginica Spreng.
        var. parksii (Cory) Chaudhri
           SY=P. p. Cory
        var. virginica
           SY=P. v. var. scoparia (Small)
           Cory
  wilkinsonii S. Wats.

Petrorhagia (Ser. ex DC.) Link

  prolifer (L.) Ball & Heywood

Petrorhagia (CONT.)
    SY=Dianthus p. L.
    SY=Kohlrauschia velutina (Guss.)
      Reichenb.
    SY=Tunica p. (L.) Scop.
 saxifraga (L.) Link
    SY=Tunica s. (L.) Scop.

Polycarpon L.

 depressum Nutt.
 tetraphyllum (L.) L.

Sagina L.

 apetala Ard.
    SY=S. a. var. barbata Fenzl
 caespitosa (J. Vahl) Lange
    SY=S. nivalis (Lindbl.) Fries var.
      c. (J. Vahl) Boivin
 decumbens (Ell.) Torr. & Gray
  ssp. decumbens
    SY=S. d. var. smithii (Gray) S.
      Wats.
  ssp. occidentalis (S. Wats.) Crow
    SY=S. o. S. Wats.
 hawaiensis Pax
 japonica (Sw.) Ohwi
 maxima Gray
  ssp. crassicaulis (S. Wats.) Crow
    SY=S. c. S. Wats.
  ssp. maxima
    SY=S. crassicaulis S. Wats. var.
      litoralis (Hulten) Hulten
 nivalis (Lindbl.) Fries
    SY=S. intermedia Fenzl
 nodosa (L.) Fenzl
  ssp. borealis Crow
  ssp. nodosa
    SY=S. n. var. glandulosa (Bess.)
      Aschers.
    SY=S. n. var. pubescens L.
    SY=S. n. var. pubescens (Bess.)
      Mert. & Koch
 procumbens L.
    SY=S. p. var. compacta Lange
 saginoides (L.) Karst.
    SY=S. linnaei Presl
    SY=S. s. var. hesperia Fern.
 subulata (Sw.) Presl

Saponaria L.

 ocymoides L.
 officinalis L.
 pumilio (L.) Fenzl ex A. Braun
    SY=Silene p. (L.) Wulfen

Schiedea Cham. & Schlecht.

 adamantis St. John
 amplexicaulis Mann
 apokremnos St. John
 diffusa Gray
    var. angustifolia Wawra
    var. diffusa
    var. macraei Sherff
 globosa Mann
    var. foliosior Deg. & Sherff
    var. globosa
    var. graminifolia Deg. & Sherff
 haleakalensis Deg. & Sherff
 hawaiiensis Hbd.
 helleri Sherff
 hookeri Gray
    var. acrisepala Sherff
    var. hookeri

    var. intercedens Sherff
 implexa (Hbd.) Sherff
 kaalae Wawra
    var. acutifolia Sherff
    var. kaalae
 kealiae Caum & Hosaka
 ligustrina Cham. & Schlecht.
    var. ligustrina
    var. nematopoda Deg. & Sherff
 lychnoides Hbd.
 lydgatei Hbd.
    var. attenuata Deg. & Sherff
    var. lydgatei
 mannii St. John
 membranacea St. John
 menziesii Hook.
    var. forbesii Sherff
    var. menziesii
    var. molokaiensis Sherff
    var. spergulacea Hbd.
 nuttallii Hook.
    var. intermedia Hbd.
    var. lihuensis Sherff
    var. molokaiensis Sherff
    var. nuttallii
    var. pauciflora Deg. & Sherff
 pubescens Hbd.
    var. degeneri Sherff
    var. hillebrandii Sherff
    var. lanaiensis Sherff
    var. pubescens
    var. purpurascens Sherff
 remyi Mann
    var. foliosa Sherff
    var. multinervia Sherff
    var. remyi
 salicaria Hbd.
 sarmentosa Deg. & Sherff
 spergulina Gray
    var. degeneriana Sherff
    var. leiopoda Sherff
    var. major Sherff
    var. spergulina
 stellarioides Mann
    var. brevifolia Sherff
    var. hillebrandii Hochr.
    var. implexoides Sherff
    var. longifolia Sherff
    var. stellarioides
 verticillara F. Br.
 viscosa Mann
    var. laevis (Sherff) St. John
    var. viscosa

Scleranthus L.

 annuus L.
 perennis L.

Scopulophila M.E. Jones

 rixfordii (Brandeg.) Munz & Johnston

Silene L.

 acaulis (L.) Jacq.
    var. acaulis
    SY=S. a. var. exscapa (All.) DC.
    var. subacaulis (F.N. Williams) Fern.
     & St. John
    SY=S. a. ssp. s. (F.N. Williams)
      C.L. Hitchc. & Maguire
 alba (P. Mill.) Krause
    SY=Lychnis a. P. Mill.
    SY=L. loveae Boivin
    SY=Melandrium a. (P. Mill.) Garcke
    SY=M. dioicum (L.) Coss. & Germ.

Silene (CONT.)
  alexanderi Hbd.
  antirrhina L.
          SY=S. a. var. confinis Fern.
          SY=S. a. var. deaneana Fern.
          SY=S. a. var. depauperata Rydb.
          SY=S. a. var. divaricata B.L.
              Robins.
          SY=S. a. var. laevigata Engelm. &
              Gray
          SY=S. a. var. subglaber Engelm. &
              Gray
          SY=S. a. var. vaccarifolia Rydb.
  aperta Greene
  armeria L.
  bernardina S. Wats.
    ssp. bernardina
          SY=S. montana S. Wats. ssp. b. (S.
              Wats.) C.L. Hitchc. &
              Maguire
    ssp. maguirei Bocquet
      var. maguirei Bocquet
          SY=S. montana S. Wats.
      var. sierrae (C.L. Hitchc. &
          Macguire) Bocquet
          SY=S. montana S. Wats. ssp.
              montana var. s. C.L. Hitchc.
              & Maguire
  bridgesii Rohrb.
  californica Dur.
  campanulata S. Wats.
    ssp. campanulata
    ssp. glandulosa C.L. Hitchc. & Maguire
    ssp. greenei (S. Wats.) C.L. Hitchc. &
        Maguire
          SY=S. c. var. g. S. Wats.
  caroliniana Walt.
    ssp. caroliniana
    ssp. pensylvanica (Michx.) Clausen
          SY=S. c. var. p. (Michx.) Fern.
          SY=S. p. Michx.
    ssp. wherryi (Small) Clausen
          SY=S. c. var. w. (Small) Fern.
          SY=S. w. Small
  clokeyi C.L. Hitchc. & Maguire
  conica L.
  conoidea L.
  cryptopetala Hbd.
  cserei Baumg.
  degeneri Sherff
  dichotoma Ehrh.
  dioica (L.) Clairville
          SY=Lychnis d. L.
  douglasii Hook.
      var. douglasii
          SY=S. d. var. villosa C.L. Hitchc.
              & Maguire
          SY=S. lyallii S. Wats.
      var. monantha (S. Wats.) B.L. Robins.
      var. oraria (M.E. Peck) C.L. Hitchc.
          & Maguire
          SY=S. o. M.E. Peck
  drummondii Hook.
      var. drummondii
          SY=Lychnis d. (Hook.) S. Wats.
          SY=L. d. var. heterochroma Boivin
          SY=L. pudica Boivin
          SY=Melandrium d. (Hook.) Hulten
      var. kruckebergii Bocquet
      var. striata (Rydb.) Bocquet
          SY=Lychnis d. (Hook.) S. Wats.
              var. s. (Rydb.) Maguire
  gallica L.
          SY=S. anglica L.
  grayi S. Wats.
  hawaiiensis Sherff
      var. hawaiiensis

          var. kaupoana (Deg. & Sherff) Deg. &
              Sherff
  hookeri Nutt. ex Torr. & Gray
    ssp. bolanderi (Gray) Abrams
    ssp. hookeri
          SY=S. ingramii Tidestrom & Dayton
    ssp. pulverulenta (M.E. Peck) C.L.
        Hitchc. & Maguire
          SY=S. p. M.E. Peck
  invisa C.L. Hitchc. & Maguire
  involucrata (Cham. & Schlecht.) Bocquet
    ssp. elatior (Regel) Bocquet
          SY=L. furcata (Raf.) Fern. ssp. e.
              (Regel) Maguire
          SY=L. triflora R. Br. ex Sommerf.
              var. elatior (Regel) Boivin
          SY=Melandrium a. var. brachycalyx
              (Raup) Hulten
          SY=M. b. Raup
    ssp. involucrata
          SY=Lychnis affinis J. Vahl ex
              Fries
          SY=L. furcata (Raf.) Fern.
          SY=L. gillettii Boivin
          SY=Melandrium a. (J. Vahl ex
              Fries) J. Vahl
          SY=S. f. Raf.
  kingii (S. Wats.) Bocquet
          SY=Lychnis k. S. Wats.
          SY=Melandrium k. (S. Wats.) Tolm.
  laciniata Cav.
    ssp. brandegei C.L. Hitchc. & Maguire
    ssp. greggii (Gray) C.L. Hitchc. &
        Maguire
          SY=S. l. var. g. (Gray) S. Wats.
    ssp. major C.L. Hitchc. & Maguire
      var. angustifolia C.L. Hitchc. &
          Maguire
      var. latifolia C.L. Hitchc. & Maguire
      var. major C.L. Hitchc. & Maguire
  lanceolata Gray
      var. angustifolia Hbd.
      var. forbesii Sherff
      var. hillebrandii Sherff
      var. lanceolata
  lemmonii S. Wats.
  macrosperma (Porsild) Hulten
          SY=Lychnis apetala L. var. m.
              (Porsild) Boivin
          SY=Melandrium macrospermum Porsild
          SY=S. uralensis (Rupr.) Bocquet
              ssp. porsildii Bocquet
  marmorensis Kruckeberg
  menziesii Hook.
    ssp. dorrii (Kellogg) C.L. Hitchc. &
        Maguire
    ssp. menziesii
      var. menziesii
      var. viscosa (Greene) C.L. Hitchc. &
          Maguire
    ssp. williamsii (Britt.) Hulten
          SY=S. m. var. w. (Britt.) Boivin
          SY=S. w. Britt.
  multinervia S. Wats.
  nivea (Nutt.) Otth
  noctiflora L.
          SY=Melandrium n. (L.) Fries
  nuda (S. Wats.) C.L. Hitchc. & Maguire
    ssp. insectivora (Henderson) C.L.
        Hitchc. & Maguire
          SY=S. i. Henderson
    ssp. nuda
          SY=S. pectinata S. Wats.
  nutans L.
  occidentalis S. Wats.
    ssp. longistipitata C.L. Hitchc. &
        Maguire

Silene (CONT.)
      ssp. occidentalis
   oregana S. Wats.
   ovata Pursh
   parishii S. Wats.
            var. latifolia C.L. Hitchc. & Maguire
            var. parishii
            var. viscida C.L. Hitchc. & Maguire
   parryi (S. Wats.) C.L. Hitchc. & Maguire
              SY=S. douglasii Hook. var.
                    macounii (S. Wats.) B.L.
                    Robins.
              SY=S. m. S. Wats.
              SY=S. scouleri (Eastw.) C.L.
                    Hitchc. & Maguire ssp.
                    scouleri var. m. (S. Wats.)
                    Boivin
   pendula L.
   petersonii Maguire
            var. minor C.L. Hitchc. & Maguire
            var. petersonii
   plankii C.L. Hitchc. & Maguire
   polypetala (Walt.) Fern. & Schub.
              SY=S. baldwinii Nutt.
   rectiramea B.L. Robins.
   regia Sims
   repens Patrin ex Pers.
      ssp. australe C.L. Hitchc. & Maguire
              SY=S. r. var. a. (C.L. Hitchc. &
                    Maguire) C.L. Hitchc.
      ssp. purpurata (Greene) C.L. Hitchc. &
            Maguire
      ssp. repens
   rotundifolia Nutt.
   sargentii S. Wats.
              SY=S. watsonii B.L. Robins.
   scaposa B.L. Robins.
            var. lobata C.L. Hitchc. & Maguire
            var. scaposa
   scouleri (Eastw.) C.L. Hitchc. & Maguire
      ssp. hallii (S. Wats.) C.L. Hitchc. &
            Maguire
      ssp. pringlei (S. Wats.) C.L. Hitchc. &
            Maguire
            var. concolor (Greene) C.L. Hitchc. &
                  Maguire
            var. eglandulosa C.L. Hitchc. &
                  Maguire
            var. grisea C.L. Hitchc. & Maguire
            var. leptophylla C.L. Hitchc. &
                  Maguire
            var. pringlei
      ssp. scouleri
            var. pacifica (Eastw.) C.L. Hitchc.
              SY=S. grandis Eastw.
              SY=S. p. Eastw.
              SY=S. s. ssp. g. (Eastw.) C.L.
                    Hitchc. & Maguire
            var. scouleri
              SY=S. repens Patrin ex Pers. var.
                    costata (Williams) Boivin
   seeleyi Morton & Thompson
   sibirica (L.) Pers.
   soczavian (Schischkin) Bocquet
              SY=Melandrium s. Schischkin
   sorensenis (Boivin) Bocquet
              SY=Lychnis s. Boivin
              SY=L. triflora R. Br. ex Sommerf.
              SY=Melandrium t. (R. Br. ex
                    Sommerf.) J. Vahl
   spaldingii S. Wats.
   stellata (L.) Ait. f.
              SY=Lychnis s. var. scabrella
                    (Nieuwl.) Palmer &
                    Steyermark
   struthioloides Gray
   subciliata B.L. Robins.

   suksdorfii B.L. Robins.
   taimyrensis (Tolm.) Bocquet
              SY=Lychnis t. (Tolm.) Polunin
              SY=L. triflora R. Br. ex Sommerf.
                    ssp. dawsonii (B.L. Robins.)
                    Maguire
              SY=Melandrium ostenfeldii Porsild
              SY=M. taimyrense Tolm.
   taylorae (B.L. Robins.) Hulten
              SY=Lychnis t. B.L. Robins.
              SY=Melandrium t. (B.L. Robins.)
                    Tolm.
              SY=S. involucrata (Cham. &
                    Schlecht.) Bocquet ssp.
                    tenella (Tolm.) Bocquet
   thurberi S. Wats.
   uralensis (Rupr.) Bocquet
      ssp. attenuata (Farr) McNeill
              SY=Lychnis apetala L. ssp.
                    attenuata (Farr) Maguire
              SY=L. apetala var. attenuata
                    (Farr) C.L. Hitchc.
      ssp. montana (S. Wats.) McNeill
              SY=Lychnis apetala L. ssp. m. (S.
                    Wats.) Maguire
              SY=L. a. var. m. (S. Wats.) C.L.
                    Hitchc.
              SY=S. hitchguirei Bocquet
              SY=S. wahlbergella Chowdhuri ssp.
                    m. (S. Wats.) Hulten
      ssp. uralensis
              SY=Lychnis apetala L.
              SY=L. apetala var. arctica (Fries)
                    Cody
              SY=L. apetala var. glabra Regel
              SY=Melandrium apetalum (L.) Fenzl
              SY=M. apetalum ssp. arcticum
                    (Fries) Hulten
              SY=M. apetalum var. g. (Regel)
                    Hulten
              SY=S. wahlbergella Chowdhuri ssp.
                    attenuata (Farr) Hulten
              SY=S. w. ssp. arcticum (Fries)
                    Hulten
   verecunda S. Wats.
      ssp. andersonii (Clokey) C.L. Hitchc. &
            Maguire
              SY=S. a. Clokey
      ssp. platyota (S. Wats.) C.L. Hitchc. &
            Maguire
            var. platyota (S. Wats.) Jepson
              SY=S. occidentalis S. Wats. var.
                    nancta Jepson
              SY=S. p. S. Wats.
            var. eglandulosa C.L. Hitchc. &
                  Maguire
      ssp. verecunda
   virginica L.
            var. hallensis Pickens & Pickens
            var. robusta Strausbaugh & Core
            var. virginica
   vulgaris (Moench) Garcke
              SY=S. cucubalus Wibel
              SY=S. c. var. latifolia
                    (Reichenb.) G. Beck
              SY=S. l. (P. Mill.) Britten &
                    Rendle
   wrightii Gray

Spergula L.

   arvensis L.
            var. arvensis
            var. sativa (Boenn.) Mert. & Koch
              SY=S. s. Boenn.
   pentandra L.

Spergularia (Pers.) J. & C. Presl

    atrosperma R.P. Rossb.
    bocconii (Scheele) Aschers. & Graebn.
    canadensis (Pers.) G. Don
        var. canadensis
            SY=Tissa c. (Pers.) Britt.
        var. occidentalis R.P. Rossb.
    diandra (Guss.) Boiss.
            SY=S. salsuginea (Bunge) Fenzl
    echinosperma Celak.
            SY=S. salsuginea (Bunge) Fenzl
                var. bracteata B.L. Robins.
    macrotheca (Hornem.) Heynh.
        var. leucantha (Greene) B.L. Robins.
        var. longistyla R.P. Rossb.
        var. macrotheca
    marina (L.) Griseb.
        var. marina
            SY=S. m. var. leiosperma (Kindb.)
                Guerke
            SY=Tissa m. (L.) Britt.
        var. simonii Deg. & Deg.
        var. tenuis (Greene) R.P. Rossb.
    media (L.) Presl ex Griseb.
    platensis (St. Hil. & Juss.) Fenzl
    rubra (L.) J. & C. Presl
            SY=Tissa r. (L.) Britt.
    villosa (Pers.) Camb.

Stellaria L.

    alaskana Hulten
    alsine Grimm
    americana (Porter ex B.L. Robins.) Standl.
            SY=Alsine a. (Porter ex B.L.
                Robins.) Rydb.
            SY=Arenaria stephaniana (Willd.)
                Shinners var. a. (Porter ex
                B.L. Robins.) Shinners
    antillana Urban
            SY=Alsine a. (Urban) Britt. &
                Wilson
    arenicola Raup
            SY=S. longipes Goldie var. a.
                (Raup) Boivin
    calycantha (Ledeb.) Bong.
        var. bongardiana (Fern.) Fern.
            SY=S. borealis Bigelow var.
                bongardiana Fern.
            SY=S. sitchana Steud. var.
                bongardiana (Fern.) Hulten
        var. calycantha
            SY=Alsine borealis (Bigelow)
                Britt.
            SY=S. b. Bigelow
        var. floribunda (Fern.) Fern.
            SY=S. borealis Bigelow var. f.
                Fern.
        var. isophylla (Fern.) Fern.
        var. laurentiana Fern.
    ciliatosepala Trautv.
    corei Shinners
            SY=Alsine tennesseensis (C. Mohr)
                Small
            SY=S. pubera Michx. var. sylvatica
                (Beguinot) Weatherby
            SY=S. s. (Beguinot) Maguire
    crassifolia Ehrh.
        var. crassifolia
            SY=Alsine c. (Ehrh.) Britt.
        var. eriocalycina Schischkin
        var. linearis Fenzl
    crassipes Hulten
            SY=S. edwardsii R. Br. var. c.
                (Hulten) Boivin
    crispa Cham. & Schlecht.

    cuspidata Willd.
    dicranoides (Cham. & Schlecht.) Fenzl
            SY=Arenaria chamissonis Maguire
    fennica (Murb.) Perf.
            SY=Alsine glauca (With.) Britt.
            SY=S. g. With.
    graminea L.
        var. graminea
            SY=Alsine g. (L.) Britt.
        var. latifolia Peterm.
    holostea L.
            SY=Alsine h. (L.) Britt.
    humifusa Rottb.
        var. humifusa
            SY=Alsine h. (Rottb.) Britt.
        var. oblongifolia Fenzl
    irrigua Bunge
    jamesiana Torr.
            SY=Arenaria j. (Torr.) Shinners
    littoralis Torr.
    longifolia Muhl. ex Willd.
        var. atrata J.W. Moore
        var. eciliata (Boivin) Boivin
        var. longifolia
            SY=Alsine l. (Muhl. ex Willd.)
                Britt.
    longipes Goldie
        var. edwardsii (R. Br.) Gray
            SY=S. e. R. Br.
            SY=S. e. var. arctica (Schischkin)
                Hulten
            SY=S. laxmannii Fisch.
        var. laeta (Richards.) S. Wats.
            SY=S. laeta Richards.
            SY=S. longifolia Muhl. ex Willd.
                var. laeta (Richards.) S.
                Wats.
        var. longipes
            SY=Alsine l. (Goldie) Coville
            SY=A. s. (Richards.) Rydb.
            SY=S. stricta Richards.
        var. subvestita (Greene) Polunin
            SY=S. s. Greene
    media (L.) Vill.
            SY=Alsine m. L.
            SY=S. apetala sensu auctt. non
                Ucria
            SY=S. m. var. glaberrima G. Beck
    monantha Hulten
            SY=S. laeta Richards. var.
                altocaulis (Hulten) Boivin
            SY=S. longipes Goldie var. a.
                (Hulten) C.L. Hitchc.
            SY=S. m. ssp. atlantica Hulten
    muscorum Fassett
    neglecta Weihe
    nitens Nutt.
    obtusa Engelm.
            SY=S. viridula (Piper) St. John
    pallida (Dumort.) Pire
    palustris Retz.
    prostrata Baldw.
            SY=Alsine baldwinii Small
    pubera Michx.
            SY=Alsine p. (Michx.) Britt.
    ruscifolia Pallas
        ssp. aleutica Hulten
    simcoei (T.J. Howell) C.L. Hitchc.
            SY=S. borealis Bigelow var. s.
                (T.J. Howell) Fern.
            SY=S. calycantha (Ledeb.) Bong.
                ssp. interior Hulten
            SY=S. c. var. s. (T.J. Howell)
                Fern.
            SY=S. washingtoniana B.L. Robins.
    sitchana Steud.
            SY=S. borealis Bigelow var. s.

Stellaria (CONT.)
                    (Steud.) Fern.
            SY=S. calycantha (Ledeb.) Bong.
                var. s. (Steud.) Fern.
    umbellata Turcz. ex Kar. & Kir.
            SY=S. gonomischa Boivin

Stipulicida Michx.

    filiformis Nash
    setacea Michx.
            var. lacerata James
            var. setacea

Vaccaria Moench

    pyramidata Medic.
            SY=Saponaria vaccaria L.
            SY=V. segetalis (Neck.) Garcke ex
                Aschers.
            SY=V. v. (L.) Britt.
            SY=V. vulgaris Host

Velezia L.

    rigida L.

Wilhelmsia Reichenb.

    physodes (Fisch. ex Ser.) McNeill
            SY=Arenaria p. Fisch. ex Ser.

CASUARINACEAE

Casuarina L. ex Adans.

    cristata Miq.
    cunninghamiana Miq.
    equisetifolia L. ex J.R. & G. Forst.
            SY=C. litorea L.
    glauca Sieb.

CELASTRACEAE

Canotia Torr.

    holacantha Torr.

Cassine L.

    xylocarpa Vent.
            SY=Elaeodendrum x. (Vent.) DC.

Celastrus L.

    orbiculatus Thunb.
    scandens L.

Crossopetalum P. Br.

    ilicifolium (Poir.) Kuntze
            SY=C. floridanum J.R. Gardner
            SY=Rhacoma i. (Poir.) Trel.
    rhacoma Crantz
            SY=Rhacoma crossopetalum L.

Euonymus L.

    alatus (Thunb.) Regel
    americanus L.
    atropurpureus Jacq.

            var. atropurpureus
            var. cheatumii Lundell
    europaeus L.
    fortunei (Turcz.) Hand.-Maz.
    obovatus Nutt.
    occidentalis Nutt. ex Torr.
            var. occidentalis
            var. parishii (Trel.) Jepson
    phellomanus Loes. ex Diels

Gyminda (Griseb.) Sarg.

    latifolia (Sw.) Urban
            SY=G. l. var. glaucifolia Small

Hippocratea L.

    caribaea Urban
            SY=Pristimera c. (Urban) A.C. Sm.
    volubilis L.

Maytenus Molina

    cymosa Krug & Urban
    elliptica (Lam.) Krug & Urban
    elongata (Urban) Britt.
    phyllanthoides Benth.
    ponceana Britt.
    texana Lundell
            SY=M. phyllanthoides Benth. var.
                ovalifolia Loes.

Mortonia Gray

    greggii Gray
    sempervirens Gray
            SY=M. scabrella Gray
    utahensis (Coville ex Gray) A. Nels.
            SY=M. scabrella Gray var. u.
                Coville ex Gray

Pachistima Raf.

    canbyi Gray
    myrsinites (Pursh) Raf.

Perrottetia H.B.K.

    sandwicensis Gray
            var. sandwicensis
            var. tomentosa Deg. & Greenw.

Schaefferia Jacq.

    cuneifolia Gray
    frutescens Jacq.

Torralbasia Krug & Urban

    cuneifolia (C. Wright) Krug & Urban

CERATOPHYLLACEAE

Ceratophyllum L.

    demersum L.
    muricatum Cham.
            SY=C. demersum L. var. echinatum
                (Gray) Gray
            SY=C. e. Gray
            SY=C. floridanum Fassett
    submersum L.

CHENOPODIACEAE

Allenrolfea Kuntze

   occidentalis (S. Wats.) Kuntze

Aphanisma Nutt. ex Moq.

   blitoides Nutt.

Arthrocnemum Moq.

   subterminale (Parish) Standl.
         SY=Salicornia s. Parish

Atriplex L.

   acadiensis Taschereau
   acanthocarpa (Torr.) S. Wats.
   alaskensis S. Wats.
         SY=A. patula L. var. a. (S. Wats.)
           Welsh
   arenaria Nutt.
     ssp. aptera (A. Nels.) Hall & Clements
     ssp. arenaria
   argentea Nutt.
     ssp. argentea
       var. argentea
          SY=A. a. ssp. typica Hall &
           Clements
       var. caput-medusae (Eastw.) Fosberg
          SY=A. a. ssp. typica Hall &
           Clements var. c-m. (Eastw.)
           Fosberg
       var. hillmanii M.E. Jones
     ssp. expansa (S. Wats.) Hall & Clements
         SY=A. e. S. Wats.
         SY=A. e. var. mohavensis M.E.
           Jones
   barclayana (Benth.) D. Dietr.
   bonnevillensis C.A. Hanson
   californica Moq.
   canescens (Pursh) Nutt.
     ssp. aptera (A. Nels.) Hall & Clements
         SY=A. c. var. a. (A. Nels.) C.L.
           Hitchc.
     ssp. canescens
       var. canescens
          SY=A. c. var. macilenta Jepson
          SY=A. tetraptera (Benth.) Rydb.
       var. laciniata Parish
   confertifolia (Torr. & Frem.) S. Wats.
         SY=A. jonesii Standl.
   cordulata Jepson
   coronata S. Wats.
       var. coronata
       var. notatior Jepson
   corrugata S. Wats.
   coulteri (Moq.) D. Dietr.
   cuneata A. Nels.
         SY=A. nuttallii S. Wats. ssp. c.
           (A. Nels.) Hall & Clements
   drymarioides Standl.
   elegans (Moq.) D. Dietr.
       var. elegans
       var. fasciculata (S. Wats.) M.E.
         Jones
         SY=A. e. ssp. f. (S. Wats.) Hall &
           Clements
         SY=A. f. S. Wats.
       var. thornberi M.E. Jones
   falcata (M.E. Jones) Standl.
         SY=A. nuttallii S. Wats. ssp. f.
           (M.E. Jones) Hall & Clements
         SY=A. n. var. f. M.E. Jones
   franktonii Taschereau

   fruticulosa Jepson
   gardneri (Moq.) D. Dietr.
         SY=A. gordonii (Moq.) D. Dietr.
         SY=A. nuttallii S. Wats. ssp.
           gardneri (Moq.) Hall &
           Clements
         SY=A. n. var. gardneri (Moq.) Hall
           & Clements
   garrettii Rydb.
         SY=A. canescens (Pursh) Nutt. ssp.
           g. (Rydb.) Hall & Clements
         SY=A. c. var. g. (Rydb.) L. Benson
   glabriuscula Edmondston
   gmelinii C.A. Mey.
         SY=A. glabriuscula Edmondston var.
           oblanceolata Victorin &
           Rouss.
         SY=A. gmelinii var. zosterifolia
           (Hook.) Moq.
         SY=A. patula L. ssp. obtusa
           (Cham.) Hall & Clements
         SY=A. p. var. oblanceolata
           (Victorin & Rouss.) Boivin
         SY=A. p. var. obtusa (Cham.) M.E.
           Peck
         SY=A. p. var. z. (Hook.) C.L.
           Hitchc.
   graciliflora M.E. Jones
   griffithsii Standl.
         SY=A. torreyi (S. Wats.) S. Wats.
           var. g. (Standl.) G.D. Br.
   heterosperma Bunge
   holocarpa F. Muell.
   hortensis L.
         SY=A. h. var. atrosanguinea Hort.
   hymenelytra (Torr.) S. Wats.
   johnsonii C.B. Wolf
   klebergorum M.C. Johnston
   laciniata L.
         SY=A. sabulosa Rouy
   lampa (Moq.) Gillies
   lentiformis (Torr.) S. Wats.
     ssp. breweri (S. Wats.) Hall & Clements
         SY=A. b. S. Wats.
         SY=A. l. var. b. (S. Wats.) McMinn
     ssp. lentiformis
   leucophylla (Moq.) D. Dietr.
   lindleyi Moq.
   linearis S. Wats.
         SY=A. canescens (Pursh) Nutt. ssp.
           l. (S. Wats.) Hall &
           Clements
   littoralis L.
         SY=A. patula L. var. l. (L.) Gray
   matamorensis A. Nels.
   muelleri Benth.
   navajoensis C.A. Hanson
   X neomexicana Standl. [confertifolia X
     cuneata]
   nummularia Lindl.
   oblongifolia Waldst. & Kit.
   obovata Moq.
   pacifica A. Nels.
   parishii S. Wats.
   parryi S. Wats.
   patula L.
     ssp. ?hastata (L.) Hall & Clements
         SY=A. h. L.
         SY=A. p. var. h. (L.) Gray
     ssp. patula
         SY=A. p. var. bracteata Westlund
     ssp. spicata (S. Wats.) Hall & Clements
         SY=A. joaquiniana A. Nels.
   pentandra (Jacq.) Standl.
         SY=A. littoralis (Jacq.) Fawcett &
           Rendle
   phyllostegia (Torr.) S. Wats.

Atriplex (CONT.)
            SY=A. p. var. draconis (M.E.
                Jones) Fosberg
    pleiantha W.A. Weber
    polycarpa (Torr.) S. Wats.
    powellii S. Wats.
    praecox Hulphers
    prostrata Boucher ex DC.
            SY=A. latifolia Wahlenb.
    pusilla (Torr.) S. Wats.
    rosea L.
    saccaria S. Wats.
    semibaccata R. Br.
    serenana A. Nels.
        var. davidsonii (Standl.) Munz
            SY=A. d. Standl.
        var. serenana
    spinifera J.F. Macbr.
    subspicata (Nutt.) Rydb.
            SY=A. patula L. ssp. s. (Nutt.)
                Fosberg
            SY=A. p. var. s. (Nutt.) S. Wats.
    suckleyi Torr.
            SY=A. dioica (Nutt.) J.F. Macbr.
            SY=Endolepis d. (Nutt.) Standl.
    tatarica L.
    tenuissima A. Nels.
    texana S. Wats.
    torreyi (S. Wats.) S. Wats.
    triangularis Willd.
    tridentata Kuntze
            SY=A. nuttallii S. Wats. pro parte
            SY=A. n. ssp. t. (Kuntze) Hall &
                Clements
    truncata (Torr.) Gray
    tularensis Coville
    vallicola Hoover
    vesicaria Heward ex Benth.
    wardii Standl.
    watsonii A. Nels.
    welshii C.A. Hanson
            SY=?A. cuneata A. Nels. ssp.
                introgressa C.A. Hanson
    wolfii S. Wats.
    wrightii S. Wats.

Axyris L.

    amaranthoides L.

Bassia All.

    hirsuta (L.) Aschers.
            SY=Kochia h. (L.) Nolte
    hyssopifolia (Pallas) Kuntze
            SY=Echinopsilon h. (Pallas) Moq.

Beta L.

    vulgaris L.

Ceratoides (Tourn.) Gagnebin

    lanata (Pursh) J.T. Howell
        var. lanata
            SY=Eurotia l. (Pursh) Moq.
        var. subspinosa (Rydb.) J.T. Howell
            SY=Eurotia l. var. s. (Rydb.)
                Kearney & Peebles

Chenopodium L.

    acerifolium Andrz.
    albescens Small
    album L.
        var. album
            SY=C. a. var. polymorphum Aellen

        var. candicans Moq.
        var. lanceolatum (Muhl.) Coss. &
            Germ.
            SY=C. l. Muhl.
        var. microphyllum Boenn.
        var. stevensii Aellen
    ambrosioides L.
            SY=Ambrina a. (L.) Spach
            SY=C. a. ssp. euambrosioides
                Aellen
            SY=C. a. ssp. e. var.
                anthelminticum (L.) Aellen
            SY=C. ambrosioides ssp. e. var.
                typicum (Speg.) Aellen
            SY=C. ambrosioides var.
                anthelminticum (L.) Gray
            SY=C. ambrosioides var. chilense
                (Schrad.) Speg.
            SY=C. ambrosioides var. vagans
                (Standl.) J.T. Howell
    aristatum L.
    atrovirens Rydb.
            SY=C. aridum A. Nels.
            SY=C. fremontii S. Wats. var.
                atrovirens (Rydb.) Fosberg
    berlandieri Moq.
        var. berlandieri
            SY=C. album L. ssp. dacoticum
                Aellen
            SY=C. a. ssp. fallax Aellen
        var. boscianum (Moq.) H.A. Wahl
            SY=C. b. Moq.
        var. sinuatum (J. Murr) H.A. Wahl
            SY=C. b. ssp. pseudopetiolare
                Aellen
        var. zschackii (J. Murr) J. Murr
            SY=C. berlandieri var. farinosum
                (Ludwig) Aellen
            SY=C. b. ssp. platyphyllum
                (Issler) Ludwig
            SY=C. b. ssp. z. (J. Murr) Zobel
    bonus-henricus L.
    botryodes Sm.
            SY=C. chenopodioides (L.) Aellen
                var. degenianum (Aellen)
                Aellen
            SY=C. c. var. lengyelianum
                (Aellen) Aellen
            SY=C. humile sensu auctt. non Hook.
            SY=C. rubrum L. var. h. sensu
                auctt. non (Hook.) S. Wats.
    botrys L.
            SY=Botrydium b. (L.) Small
    bushianum Aellen
        var. acutidentatum Aellen
        var. bushianum
            SY=C. paganum Reichenb.
    californicum (S. Wats.) S. Wats.
    capitatum (L.) Aschers.
            SY=Blitum c. L.
    carnosulum Moq.
        var. carnosulum
        var. patagonicum (Phil.) H.A. Wahl
            SY=C. p. Phil.
    chenopodioides (L.) Aellen
    cycloides A. Nels.
    desiccatum A. Nels.
        var. desiccatum
            SY=C. leptophyllum (Moq.) Nutt. ex
                S. Wats. var. oblongifolium
                S. Wats.
            SY=C. pratericola Rydb. ssp. d.
                (A. Nels.) Aellen
            SY=C. p. var. o. (S. Wats.) H.A.
                Wahl
        var. leptophylloides (J. Murr) H.A.
            Wahl

Chenopodium (CONT.)
  SY=C. pratericola Rydb.
  SY=C. p. ssp. eupratericola Aellen
foggii H.A. Wahl
foliosum (Moench) Aschers.
  SY=C. virgatum L.
fremontii S. Wats.
  var. fremontii
  var. pringlei (Standl.) Aellen
giganteum D. Don
  SY=C. amaranticolor Coste & Reyn.
gigantospermum Aellen
  SY=C. hybridum L. ssp. g. (Aellen)
   Hulten
  SY=C. h. var. g. (Aellen) Rouleau
glaucum L.
  var. glaucum
   SY=C. g. ssp. euglaucum Aellen
  var. pulchrum Aellen
   SY=C. g. ssp. salinum (Standl.)
    Aellen
   SY=C. g. var. s. (Standl.) Boivin
   SY=C. s. Standl.
graveolens Willd.
  SY=A. incisum Poir.
hians Standl.
hybridum L.
incanum (S. Wats.) Heller
  var. elatum Crawford
  var. incanum
   SY=C. fremontii S. Wats. var. i.
    S. Wats.
  var. occidentale Crawford
incognitum H.A. Wahl
leptophyllum (Moq.) Nutt. ex S. Wats.
macrocalycium Aellen
macrospermum Hook. f.
  var. farinosum (S. Wats.) J.T. Howell
   SY=C. f. (S. Wats.) Standl.
  var. halophilum
missouriense Aellen
  SY=C. m. var. bushianum Aellen
  SY=C. paganum sensu auctt. non
   Reichenb.
multifidum L.
  SY=Roubieva m. (L.) Moq.
murale L.
neomexicanum Standl.
  SY=C. graveolens Willd. var. n.
   (Aellen) Aellen
  SY=C. incisum Poir. var. n. Aellen
nevadense Standl.
oahuense (Meyen) Aellen
  var. discosperma Fosberg
  var. oahuense
opulifolium Schrad. ex Koch & Ziz
  SY=?C. album L. var. viride (L.)
   Moq.
overi Aellen
pallescens Standl.
palmeri Standl.
  SY=C. arizonicum Standl.
pekeloi Deg., Deg. & Aellen
polyspermum L.
  var. acutifolium (Sm.) Gaud.
  var. obtusifolium Gaud.
  var. polyspermum
pumilio R. Br.
  SY=C. carinatum sensu auctt. non
   R. Br.
rubrum L.
schraderanum Schultes
  SY=C. foetidum Schrad.
serotinum L.
standleyanum Aellen
  SY=C. hybridum L. var. s. (Aellen)
   Fern.

strictum Roth
 ssp. glaucophyllum (Aellen) Aellen
  SY=C. s. var. g. (Aellen) H.A.
   Wahl
subglabrum (S. Wats.) A. Nels.
  SY=C. leptophyllum (Moq.) Nutt. ex
   S. Wats. var. s. S. Wats.
urbicum L.
  SY=C. u. var. intermedium (Mert. &
   Koch) W.D.J. Koch
X variabile Aellen [album X berlandieri
 var. zschackei]
  SY=C. X v. var. murrii Aellen
vulvaria L.
  SY=C. olidum S. Wats.
watsonii A. Nels.
  SY=C. glabrescens (Aellen) H.A.
   Wahl

Corispermum L.

 hyssopifolium L.
  SY=C. h. Lunell
  SY=C. h. L. var. rubricaule Hook.
  SY=C. simplicissimum Lunell
 nitidum Kit. ex Schultes
 orientale Lam.
  SY=C. emarginatum Rydb.
  SY=C. hyssopifolium L. var. e.
   (Rydb.) Boivin
  SY=C. o. var. e. (Rydb.) J.F.
   Macbr.
  SY=C. villosum Rydb.

Cycloloma Moq.

 atriplicifolium (Spreng.) Coult.
  SY=Kochia a. Spreng.

Grayia Hook. & Arn.

 brandegei Gray
 spinosa (Hook.) Moq.
  SY=Atriplex s. (Hook.) Collotzi

Halogeton C.A. Mey.

 glomeratus (Stephan ex Bieb.) C.A. Mey.

Kochia Roth

 americana S. Wats.
  SY=K. a. var. vestita S. Wats.
  SY=K. v. (S. Wats.) Rydb.
 californica S. Wats.
  SY=K. americana S. Wats. var. c.
   (S. Wats.) M.E. Jones
 iranica Bornm.
  SY=K. alata Bates
 scoparia (L.) Schrad.
  var. culta Farw.
  var. pubescens Fenzl
   SY=K. sieversiana (Pallas) C.A.
    Mey.
  var. scoparia
   SY=K. s. var. trichophila (Stapf)
    Bailey
   SY=K. t. Stapf

Monolepis Schrad.

 nuttalliana (Roemer & Schultes) Greene
 pusilla Torr. ex S. Wats.
 spathulata Gray

Nitrophila S. Wats.

Nitrophila (CONT.)
   mohavensis Munz & Roos
   occidentalis (Moq.) S. Wats.

Polycnemum L.

   arvense L.
   majus A. Braun
   verrucosum A.F. Lang

Salicornia L.

   europaea L.
        SY=?S. depressa Standl.
        SY=S. e. var. pachystachya (W.D.J.
           Koch) Fern.
        SY=S. e. var. prostrata sensu
           auctt. non (Pallas) Fern.
        SY=S. e. var. simplex (Pursh)
           Fern.
        SY=S. ramosissima Woods.
   rubra A. Nels.
        SY=S. europaea L. ssp. r. (A.
           Nels.) Breitung
        SY=S. e. var. prona (Lunell)
           Boivin
        SY=?S. e. var. prostrata (Pallas)
           Fern.
   virginica L.
        SY=S. bigelovii Torr.
        SY=S. mucronata Bigelow

Salsola L.

   collina Pallas
   iberica Sennen & Pau
        SY=S. kali L. ssp. ruthenica
           (Iljin) Soo
        SY=S. k. var. tenuifolia Tausch
   kali L.
        SY=S. k. var. caroliniana (Walt.)
           Nutt.
        SY=S. pestifer A. Nels.
   paulsenii Litv.
   soda L.

Sarcobatus Nees

   vermiculatus (Hook.) Torr.
      var. baileyi (Coville) Jepson
        SY=S. b. Coville
      var. vermiculatus

Sarcocornia A.J. Scott

   fruticosa (L.) A.J. Scott
        SY=Salicornia f. (L.) L.
   pacifica (Standl.) A.J. Scott
        SY=Salicornia p. Standl.
   perennis (P. Mill.) A.J. Scott
        SY=Arthrocnemum p. (P. Mill.) Moss
        SY=Salicornia ambigua Michx.
        SY=Salicornia p. P. Mill.
   utahensis (Tidestrom) A.J. Scott
        SY=Salicornia u. Tidestrom

Spinacea L.

   oleracea L.

Suaeda Forsk. ex Scop.

   americana (Pers.) Fern.
        SY=S. maritima (L.) Dumort. var.
           a. (Pers.) Boivin
   californica S. Wats.
      var. californica

      var. pubescens Jepson
        SY=S. taxifolia Standl. ssp.
           brevifolia (Standl.) Abrams
      var. taxifolia (Standl.) Munz
        SY=S. t. Standl.
   conferta (Small) I.M. Johnston
   depressa (Pursh) S. Wats.
      var. depressa
        SY=Dondia d. (Pursh) Britt.
        SY=S. calceoliformis (Hook.) Moq.
      var. erecta S. Wats.
        SY=S. minutiflora S. Wats.
   linearis (Ell.) Moq.
        SY=Dondia l. (Ell.) Heller
   maritima (L.) Dumort.
        SY=Dondia m. (L.) Druce
   mexicana (Standl.) Standl.
   occidentalis S. Wats.
   richii Fern.
   suffrutescens S. Wats.
      var. detonsa I.M. Johnston
        SY=S. duripes I.M. Johnston
        SY=S. nigrescens I.M. Johnston
        SY=S. n. var. glabra I.M. Johnston
      var. suffrutescens
   tampicensis (Standl.) Standl.
   torreyana S. Wats.
      var. ramosissima (Standl.) Munz
        SY=S. r. (Standl.) I.M. Johnston
      var. torreyana
        SY=Dondia fruticosa (S. Wats.)
           Standl.
        SY=S. f. (S. Wats.) Standl.
        SY=S. intermedia S. Wats.
        SY=S. nigra (Raf.) J.F. Macbr.

Suckleya Gray

   suckleyana (Torr.) Rydb.

Zuckia Standl.

   arizonica Standl.

CHLORANTHACEAE

Hedyosmum Sw.

   arborescens Sw.

CISTACEAE

Cistus L.

   villosus L.
      var. corsicus (Loisel.) Gross.
      var. tauricus Gross.
      var. undulatus Gross.

Helianthemum P. Mill.

   arenicola Chapman
        SY=Crocanthemum a. (Chapman)
           Barnh.
   bicknellii Fern.
        SY=Crocanthemum b. (Fern.) Barnh.
        SY=C. majus sensu Britt.
   canadense (L.) Michx.
        SY=Crocanthemum c. (L.) Britt.
        SY=H. c. var. sabulonum Fern.
   carolinianum (Walt.) Michx.

Helianthemum   (CONT.)
        SY=Crocanthemum c. (Walt.) Spach
    corymbosum Michx.
        SY=Crocanthemum c. (Michx.) Britt.
    dumosum (Bickn.) Fern.
        SY=Crocanthemum d. Bickn.
    georgianum Chapman
        SY=Crocanthemum g. (Chapman)
            Barnh.
    glomeratum (Lag.) Lag. ex Dunal
        SY=Crocanthemum g. (Lag.) Janchen
    greenei B.L. Robins.
    guttatum (L.) P. Mill.
    nashii Britt.
        SY=Crocanthemum n. (Britt.) Barnh.
        SY=C. thyrsoideum (Barnh.) Janchen
        SY=H. t. Barnh.
    propinquum Bickn.
        SY=Crocanthemum p. (Bickn.) Bickn.
    rosmarinifolium Pursh
        SY=Crocanthemum r. (Pursh) Janchen
        SY=?H. capitatum Nutt. ex Engelm.
            & Gray
    scoparium Nutt.
        SY=H. aldersonii Greene
        SY=H. s. var. a. (Greene) Munz
        SY=H. s. var. vulgare Jepson
        SY=H. suffrutescens Schreib.

Hudsonia L.

    ericoides L.
        SY=H. e. ssp. andersonii Nickerson
            & Skog
    montana Nutt.
        SY=H. e. ssp. m. (Nutt.) Nickerson
            & Skog
    tomentosa Nutt.
        var. intermedia Peck
        SY=H. ericoides L. ssp. i. (Peck)
            Nickerson & Skog
        SY=H. X intermedia sensu auctt.
            non (Peck) Erskine
        var. tomentosa
        SY=H. ericoides L. ssp. t. (Nutt.)
            Nickerson & Skog

Lechea L.

    cernua Small
    deckertii Small
        SY=L. myriophylla Small
    divaricata Shuttlw. ex Britt.
    intermedia Leggett
        var. depauperata Hogdon
        SY=L. minor L. var. d. (Hogdon)
            Boivin
        var. intermedia
        var. juniperina (Bickn.) B.L. Robins.
        SY=L. j. Bickn.
        var. laurentiana Hodgdon
    lakelae Wilbur
    maritima Leggett
        var. maritima
        SY=L. minor L. var. maritima
            (Leggett) Gray
        var. subcylindrica Hodgdon
        var. virginica Hodgdon
    mensalis Hodgdon
    minor L.
    pulchella Raf.
        var. moniliformis (Bickn.) Seymour
        SY=L. leggettii Britt. & Hollick
            var. m. (Bickn.) Hodgdon
        SY=L. m. Bickn.
        var. pulchella
        SY=L. leggettii Britt. & Hollick

        SY=L. l. var. ramosissima Hodgdon
    racemulosa Michx.
    san-sabeana (Buckl.) Hodgdon
    sessiliflora Raf.
        SY=L. exserta Small
        SY=L. patula Leggett
        SY=L. prismatica Small
    stricta Leggett
    tenuifolia Michx.
        SY=L. t. var. occidentalis Hodgdon
    torreyi Leggett ex Britt.
        var. congesta Hodgdon
        var. torreyi
    tripetala (Moc. & Sesse) Britt.
    villosa Ell.
        SY=L. minor L. var. v. (Ell.)
            Boivin
        SY=L. mucronata Raf.
        SY=L. v. var. macrotheca Hodgdon
        SY=L. v. var. schaffneri Hodgdon

CLETHRACEAE

Clethra L.

    acuminata Michx.
    alnifolia L.
        SY=C. a. var. tomentosa (Lam.)
            Michx.
        SY=C. t. Lam.

CLUSIACEAE

Calophyllum L.

    calaba L.
        SY=C. antillanum Britt.
    inophyllum L.

Clusia L.

    grisebachiana (Planch. & Triana) Alain
        SY=C. krugiana Urban
    grundlachii Stahl
    minor L.
    rosea Jacq.

Hypericum L.

    adpressum Bart.
    anagalloides Cham. & Schlecht.
    boreale (Britt.) Bickn.
    brachyphyllum (Spach) Steud.
        SY=H. aspalathoides Willd. pro
            parte
    buckleyi M.A. Curtis
    calycinum L.
    canadense L.
        var. canadense
        SY=H. c. var. galiiforme Fern.
        var. magninsulare Weatherby
    canariense L.
    chapmanii P. Adams
    cistifolium Lam.
        SY=H. opacum Torr. & Gray
    concinnum Benth.
    cumulicola (Small) P. Adams
        SY=Sanidophyllum c. Small
    degeneri Fosberg
    densiflorum Pursh
        SY=H. glomeratum Small

Hypericum (CONT.)
    denticulatum Walt.
            var. acutifolium (Ell.) Blake
                SY=H. a. Ell.
            var. denticulatum
                SY=H. d. var. ovalifolium (Britt.)
                    Blake
                SY=H. virgatum Lam.
            var. recognitum Fern. & Schub.
    diosmoides Griseb.
    dissimulatum Bickn.
    dolabriforme Vent.
            SY=H. bissellii B. L. Robins.
    drummondii (Grev. & Hook.) Torr. & Gray
            SY=Sarothra d. Grev. & Hook.
    edisonianum (Small) P. Adams & Robson
            SY=Ascyrum e. Small
    ellipticum Hook.
    exile P. Adams
    fasciculatum Lam.
    formosum H.B.K.
        ssp. formosum
        ssp. scouleri (Hook.) C.L. Hitchc.
            var. nortoniae (M.E. Jones) C.L.
                Hitchc.
            var. scouleri
                SY=H. s. Hook.
    frondosum Michx.
            SY=H. aureum Bartr.
            SY=H. splendens Small
    galioides Lam.
            SY=H. ambiguum Ell.
            SY=H. g. var. pallidum C. Mohr
    gentianoides (L.) B.S.P.
            SY=Sarothra g. L.
    gramineum Forst. f.
    graveolens Buckl.
    gymnanthum Engelm. & Gray
    hypericoides (L.) Crantz
            SY=Ascyrum h. L.
            SY=A. h. var. oblongifolium
                (Spach) Fern.
            SY=A. linifolium Spach
    kalmianum L.
    lissophloeus P. Adams
    lloydii (Svens.) P. Adams
    lobocarpum Gattinger
            SY=H. densiflorum Pursh var. l.
                (Gattinger) Svens.
            SY=H. oklahomense Palmer
    majus (Gray) Britt.
    microsepalum (Torr. & Gray) Gray ex S.
        Wats.
            SY=Crookea m. (Torr. & Gray) Small
    mitchellianum Rydb.
    X moserianum Andre [calycinum X patulum]
    mutilum L.
            SY=H. m. var. latisepalum Fern.
            SY=H. m. var. parviflorum (Willd.)
                Fern.
    myrtifolium Lam.
    nitidum Lam.
    nudiflorum Michx.
            SY=H. apocynifolium Small
    pauciflorum H.B.K.
    perforatum L.
    prolificum L.
            SY=H. spathulatum (Spach) Steud.
    pseudomaculatum Bush
            SY=H. punctatum Lam. var.
                pseudomaculatum (Bush) Fern.
    punctatum Lam.
            SY=H. subpetiolatum Bickn.
    pyramidatum Ait.
            SY=H. ascyron L.
    reductum P. Adams
            SY=H. aspalathoides Willd. pro

                    parte
    setosum L.
    sphaerocarpum Michx.
            SY=H. s. var. turgidum (Small)
                Svens.
            SY=H. t. Small
    stans (Michx.) P. Adams & Robson
            SY=Ascyrum cuneifolium Chapman
            SY=A. s. Michx.
    stragulum P. Adams & Robson
            SY=Ascyrum hypericoides L. var.
                multicaule (Michx.) Fern.
    suffruticosum P. Adams & Robson
            SY=Ascyrum pumilum Michx.
    tetrapetalum Lam.
            SY=Ascyrum t. (Lam.) Vail

Mammea L.

    americana L.

Rheedia L.

    hessii Britt.
    portoricensis Urban
            SY=R. acuminata (Spreng.) Triana &
                Planch.

Triandenum Raf.

    fraseri (Spach) Gleason
            SY=Hypericum virginicum L. var. f.
                (Spach) Fern.
    tubulosum (Walt.) Gleason
            SY=Hypericum t. Walt.
            SY=T. longifolium Small
    virginicum (L.) Raf.
            SY=Hypericum v. L.
    walteri (J.G. Gmel.) Gleason
            SY=Hypericum petiolatum Walt.
            SY=H. tubulosum Walt. var. w.
                (J.G. Gmel.) Lott
            SY=H. w. J.G. Gmel.
            SY=T. p. (Walt.) Britt.

COMBRETACEAE

Buchenavia Eichl.

    capitata (Vahl) Eichl.

Bucida L.

    buceras L.
    spinosa Jennings

Conocarpus L.

    erectus L.
            var. erectus
            var. sericeus Griseb.

Laguncularia Gaertn. f.

    racemosa (L.) Gaertn. f.

Terminalia L.

    catappa L.

COMMELINACEAE

Callisia Loefl.

    fragrans (Lindl.) Woods.
        SY=Spironema f. Lindl.
    monandra (Sw.) Schultes
        SY=Tradescantia m. Sw.
    repens L.

Campelia L.C. Rich.

    zanonia (L.) H.B.K.

Commelina L.

    benghalensis L.
    communis L.
        var. communis
        var. ludens (Miq.) C.B. Clarke
        SY=C. debilis Ledeb.
    dianthifolia Delile
        var. dianthifolia
        var. longispatha (Torr.) Brashier
    diffusa Burm. f.
        SY=C. caroliniana Walt.
        SY=C. longicaulis Jacq.
    erecta L.
        var. angustifolia (Michx.) Fern.
        SY=C. a. Michx.
        SY=C. crispa Woot.
        SY=C. e. var. c. (Woot.) Palmer &
           Steyermark
        var. deamiana Fern.
        var. erecta
    gigas Small
    virginica L.
        SY=C. elegans Kunth

Cuthbertia Small

    graminea Small
        SY=Tradescantia rosea Vent. var.
           g. (Small) Anderson & Woods.
    ornata Small
        SY=Tradescantia rosea Vent. var.
           o. (Small) Anderson & Woods.
    rosea (Vent.) Small
        SY=Tradescantia r. Vent.

Gibasis Raf.

    geniculata (Jacq.) Rohw.
        SY=Aneilema g. (Jacq.) Woods.
        SY=Tradescantia g. Jacq.

Leiandra Raf.

    cordifolia (Sw.) Raf.
        SY=Tradescantella floridana (S.
           Wats.) Small

Murdannia Royle

    keisak (Hassk.) Hand.-Maz.
        SY=Aneilema k. Hassk.
    nudiflora (L.) Brenan
        SY=Aneilema n. (L.) R. Br. ex
           Kunth
    spirata (L.) Brueckner

Phaeospherion Hassk.

    persicariifolius (DC.) C.B. Clarke
        SY=Athyrocarpus p. (DC.) Hemsl.

Rhoeo Hance ex Walp.

    spathacea (Sw.) Stearn

        SY=R. discolor (L'Her.) Hance ex
           Walp.
        SY=Tradescantia d. L'Her.

Tinantia Schweidw.

    anomala (Torr.) C.B. Clarke
        SY=Commelinantia a. (Torr.) Tharp

Tradescantia L.

    bracteata Small
    brevifolia (Torr.) Rose
        SY=Setcreasea b. (Torr.) Pilger
        SY=T. leiandra Torr. var. b. Torr.
    buckleyi (I.M. Johnston) D.R. Hunt
        SY=Setcreasea b. I.M. Johnston
    diffusa Bush
        SY=T. X pedicellata Celarier
    edwardsiana Tharp
    ernestiana Anderson & Woods.
    fluminensis Vell.
    gigantea Rose
    hirsuticaulis Small
    hirsutiflora Bush
    humilis Rose
    leiandra Torr.
        SY=Setcreasea l. (Torr.) Pilger
        SY=S. l. var. glandulosa Correll
    longipes Anderson & Woods.
    micrantha Torr.
    occidentalis (Britt.) Smyth
        var. melanthera MacRoberts
        var. occidentalis
        SY=T. o. var. typica Anderson &
           Woods.
        var. scopulorum (Rose) Anderson &
           Woods.
    ohiensis Raf.
        var. foliosa (Small) MacRoberts
        SY=T. f. Small
        var. ohiensis
        SY=T. canaliculata Raf.
        SY=T. incarnata Small
        SY=T. reflexa Raf.
        var. paludosa (Anderson & Woods.)
           MacRoberts
        SY=T. p. Anderson & Woods.
    ozarkana Anderson & Woods.
    pinetorum Greene
        SY=Aneilema p. (Greene) Matuda
    reverchonii Bush
    roseolens Small
        SY=T. longifolia Small
    subacaulis Bush
        SY=T. texana Bush
    subaspera Ker-Gawl.
        var. montana (Shuttlw. ex Small &
           Vail) Anderson & Woods.
        SY=T. m. Shuttlw. ex Small & Vail
        var. subaspera
        SY=T. pilosa Lehm.
        SY=T. s. var. typica Anderson &
           Woods.
    tharpii Anderson & Woods.
    virginiana L.
        SY=T. brevicaulis Raf.
    wrightii Rose & Bush

Tripogandra Raf.

    elongata (G.F.W. Mey.) Woods.
        SY=Tradescantia e. G.F.W. Mey.

Zebrina Schnizl.

    pendula Schnizl.

Zebrina (CONT.)
        var. quadrifolia Bailey

CONNARACEAE

Rourea Aubl.

    surinamensis Miq.

CONVOLVULACEAE

Aniseia Choisy

    martinicensis (Jacq.) Choisy

Argyreia Lour.

    nervosa (Burm. f.) Bojer
            SY=A. speciosa (L. f.) Sweet
            SY=Convolvulus n. Burm. f.
            SY=C. s. L. f.
            SY=Rivea n. (Burm. f.) Hallier. f.

Bonamia Thouars

    grandiflora (Gray) Hallier f.
    menziesii Gray
            var. menziesii
            var. rockii Myint
    ovalifolia (Torr.) Hallier f.
            SY=Breweria o. (Torr.) Gray

Calystegia R. Br.

    atriplicifolia Hallier f.
            SY=Convolvulus nyctagineus Greene
    catesbiana Pursh
        ssp. catesbiana
        ssp. sericata (House) Brummitt
            SY=Calystegia s. (House) Bell
            SY=Convolvulus s. House
    collina (Greene) Brummitt
        ssp. collina
            SY=Convolvulus malacophyllus
                Greene ssp. c. (Greene)
                Abrams
        ssp. tridactylosa (Eastw.) Brummitt
    fraterniflora (Mackenzie & Bush) Brummitt
            SY=Calystegia sepium (L.) R. Br.
                var. f. (Mackenzie & Bush)
                Shinners
            SY=Convolvulus f. (Mackenzie &
                Bush) Mackenzie & Bush
            SY=Convolvulus s. L. var. f.
                Mackenzie & Bush
    fulcrata (Gray) Brummitt
        ssp. fulcrata
            SY=Convolvulus fulcratus (Gray)
                Greene
            SY=Convolvulus luteolus Gray var.
                f. Gray
        ssp. malacophylla (Greene) Brummitt
            var. berryi (Eastw.) Brummitt
                SY=Convolvulus fulcratus (Gray)
                    Greene var. b. (Eastw.)
                    Jepson
            var. malacophylla
                SY=Calystegia m. (Greene) Munz
                SY=Convolvulus m. Greene
        ssp. pedicellata (Jepson) Brummitt
            SY=Calystegia malacophylla

                (Greene) Munz ssp. p.
                (Jepson) Munz
            SY=Convolvulus m. Greene ssp. p.
                (Jepson) Abrams
        ssp. tomentella (Greene) Brummitt
            var. deltoidea (Greene) Brummitt
                SY=Calystegia malacophylla
                    (Greene) Munz ssp. t.
                    (Greene) Munz var. d.
                    (Greene) Munz
                SY=Convolvulus fulcratus (Gray)
                    Greene var. d. (Greene)
                    Jepson
            var. tomentella (Greene) Brummitt
                SY=Calystegia malacophylla
                    (Greene) Munz ssp. t.
                    (Greene) Munz
                SY=Convolvulus t. Greene
    longipes (S. Wats.) Brummitt
            SY=Convolvulus l. S. Wats.
    macounii (Greene) Brummitt
            SY=Convolvulus interior House
    macrostegia (Greene) Brummitt
        ssp. arida (Greene) Brummitt
            SY=Convolvulus a. Greene
        ssp. cyclostegia (House) Brummitt
            SY=Convolvulus c. House
        ssp. intermedia (Abrams) Brummitt
            SY=Convolvulus aridus Greene ssp.
                i. Abrams
        ssp. longiloba (Abrams) Brummitt
            SY=Convolvulus aridus Greene ssp.
                l. Abrams
        ssp. macrostegia
            SY=Convolvulus m. Greene
        ssp. tenuifolia (Abrams) Brummitt
            SY=Convolvulus aridus Greene ssp.
                t. Abrams
    occidentalis (Gray) Brummitt
            SY=Convolvulus o. Gray
            SY=Convolvulus o. ssp.
                fruticetorum (Greene) Abrams
    peirsonii (Abrams) Brummitt
            SY=Convolvulus p. Abrams
    polymorpha (Greene) Munz
            SY=Convolvulus p. Greene
    pubescens Lindl.
            SY=Convolvulus japonicus Thunb.
            SY=Convolvulus pellitus Ledeb.
    purpurata (Greene) Brummitt
        ssp. purpurata
            SY=Convolvulus occidentalis Gray
                var. p. (Greene) J.T. Howell
        ssp. saxicola (Eastw.) Brummitt
            SY=Convolvulus occidentalis Gray
                ssp. s. (Eastw.) J.T. Howell
        ssp. solanensis (Jepson) Brummitt
            SY=Convolvulus occidentalis Gray
                var. solanensis (Jepson)
                J.T. Howell
    sepium (L.) R. Br.
        ssp. americana (Sims) Brummitt
            SY=Calystegia s. var. a. (Sims)
                Matsuda
            SY=Calystegia s. var. repens (L.)
                Gray
            SY=Convolvulus a. (Sims) Greene
            SY=Convolvulus r. L.
            SY=Convolvulus s. L. var. a. Sims
            SY=Convolvulus s. var. r. (L.)
                Gray
        ssp. binghamiae (Greene) Brummitt
            SY=Convolvulus b. Greene
            SY=Convolvulus s. L. var.
                dumetorum Pospichal
        ssp. limnophila (Greene) Brummitt
        ssp. sepium

Calystegia   (CONT.)
            SY=Convolvulus nashii House
            SY=Convolvulus s. L.
            SY=Convolvulus s. var. communis R.
                Tryon
    soldanella (L.) R. Br. ex Roemer & Schultes
            SY=Convolvulus s. L.
    spithamaea (L.) Pursh
        ssp. purshiana (Wherry) Brummitt
            SY=Calystegia s. var. pubescens
                Gray
            SY=Convolvulus purshianus Wherry
            SY=Convolvulus s. L. var.
                pubescens (Gray) Fern.
        ssp. spithamaea
            SY=Convolvulus s. L.
        ssp. stans (Michx.) Brummitt
    stebbinsii Brummitt
    subacaulis Hook. & Arn.
            SY=Convolvulus s. (Hook. & Arn.)
                Greene

Convolvulus L.

    althaeoides L.
    arvensis L.
            SY=C. ambigens House
            SY=C. incanus Vahl
            SY=Strophocaulos arvensis (L.)
                Small
    equitans Benth.
            SY=C. hermannioides Gray
    linearilobus Eastw.
    nodiflorus Desv.
            SY=Jacquemontia n. (Desv.) G. Don
    simulans Perry
    wallichianus Spreng.

Cressa L.

    depressa Goodding
            SY=C. truxillensis H.B.K. var.
                vallicola (Heller) Munz
    insularis House
    nudicaulis Griseb.
    truxillensis H.B.K.
            var. minima (Heller) Munz
            var. truxillensis

Cuscuta L.

    americana L.
    applanata Engelm.
    approximata Bab.
            var. approximata
            var. urceolata (Kunze) Yuncker
    attenuata Waterfall
    boldinghii Urban
    brachycalyx (Yuncker) Yuncker
            var. apodanthera (Yuncker) Yuncker
            var. brachycalyx
    californica Hook. & Arn.
            var. apiculata Engelm.
            var. californica
            var. papillosa Yuncker
    campestris Yuncker
            SY=C. pentagona Engelm. var.
                calycina Engelm.
            SY=Grammica campestris (Yuncker)
                Hadac & Chrtek
    cassytoides Nees ex Engelm.
    ceanothii Behr
            SY=C. subinclusa Dur. & Hilg.
    cephalanthii Engelm.
            SY=Grammica c. (Engelm.) Hadac &
                Chrtek
    compacta Juss.

            var. compacta
            var. efimbriata Yuncker
    corylii Engelm.
            SY=Grammica c. (Engelm.) Hadac &
                Chrtek
    cuspidata Engelm.
            SY=Grammica c. (Engelm.) Hadac &
                Chrtek
    decipiens Yuncker
    dentatasquamata Yuncker
    denticulata Engelm.
            SY=Grammica d. (Engelm.) W.A.
                Weber
    epilinum Weihe
    epithymum (L.) L.
    erosa Yuncker
    europaea L.
    exaltata Engelm.
    fasciculata Yuncker
    glabrior (Engelm.) Yuncker
            var. glabrior
            var. pubescens (Engelm.) Yuncker
    globulosa Benth.
    glomerata Choisy
    gronovii Willd.
            var. calyptrata Engelm.
            var. gronovii
            SY=Grammica g. (Willd. ex Roemer &
                Schultes) Hadac & Chrtek
            var. latiflora Engelm.
            SY=C. g. var. saururi (Engelm.)
                MacM.
    harperi Small
    howelliana Rubtzoff
    indecora Choisy
            var. bifida Yuncker
            var. indecora
            SY=Grammica i. (Choisy) W.A. Weber
            var. longisepala Yuncker
            var. neuropetala (Engelm.) Yuncker
            SY=Grammica indecora (Choisy) W.A.
                Weber ssp. n. (Engelm.) W.A.
                Weber
    jepsonii Yuncker
    leptantha Engelm.
    mitriformis Engelm.
    nevadensis I.M. Johnston
    obtusiflora H.B.K.
            var. glandulosa Engelm.
            SY=C. g. (Engelm.) Small
            var. obtusiflora
    occidentalis Millsp. ex Mill. & Nutt.
            SY=Grammica o. (Millsp. ex Mill. &
                Nutt.) Hadac & Chrtek
    odontolepis Engelm.
    pentagona Engelm.
            SY=Grammica p. (Engelm.) W.A.
                Weber
    planiflora Tenore
    plattensis A. Nels.
    polygonorum Engelm.
    rostrata Shuttlw. ex Engelm.
    runyonii Yuncker
    salina Engelm.
            var. major Yuncker
            var. papillata Yuncker
            var. salina
            SY=Grammica s. (Engelm.) Taylor &
                MacBryde
    sandwichiana Choisy
            var. kailuana Yuncker
            var. sandwichiana
    squamata Engelm.
    suaveolens Ser.
    suksdorfii Yuncker
            var. subpedicellata Yuncker
            var. suksdorfii

Cuscuta (CONT.)
  tuberculata Brandeg.
  umbellata H.B.K.
        SY=C. u. var. reflexa (Coult.)
            Yuncker
        SY=Grammica u. (H.B.K.) Hadac &
            Chrtek
  umbrosa Bey. ex Hook.
        SY=C. curta (Engelm.) Rydb.
        SY=C. gronovii Willd. var. c.
            Engelm.
        SY=C. megalocarpa Rydb.
        SY=Grammica u. (Bey. ex Hook.)
            Hadac & Chrtek
  vetchii Brandeg.
  warneri Yuncker

Dichondra J.R. & G. Forst.

  argentea Humb. & Bonpl. ex Willd.
  brachypoda Woot. & Standl.
  carolinensis Michx.
        SY=D. repens J.R. & G. Forst. var.
            c. (Michx.) Choisy
  donnelliana Tharp & M.C. Johnston
  micrantha Urban
  occidentalis House
  recurvata Tharp & M.C. Johnston
  repens J.R. & G. Forst.
  sericea Sw.
        SY=D. repens J.R. & G. Forst. var.
            s. (Sw.) Choisy

Evolvulus L.

  alsinoides (L.) L.
      var. angustifolia Torr.
        SY=E. alsinoides var. acapulcensis
            (Willd.) van Ooststr.
      var. grisebachianus Meisn.
      var. hirticaulis Torr.
      var. linifolius (L.) Baker
        SY=E. l. L.
  arizonicus Gray
      var. arizonicus
      var. laetus (Gray) van Ooststr.
  convolvuloides (Willd.) Stearn
        SY=E. glaber Spreng.
  grisebachii Peter
        SY=E. wrightii House
  nummularius L.
  pilosus Nutt.
        SY=E. nuttallianus Roemer &
            Schultes
  sericeus Sw.
      var. averyi Ward
      var. glaberrimus B.L. Robins.
        SY=E. macilentus Small
      var. sericeus
        SY=E. s. var. discolor (Benth.)
            Gray
  squamosus Britt.
  tenuis Mart.
      var. longifolius (Choisy) van
        Ooststr.
      var. tenuis

Ipomoea L.

  alba L.
        SY=Calonyction aculeatum (L.)
            House
  amnicola Morong
  aristolochiifolia (H.B.K.) G. Don
    ssp. aristolochiifolia
    ssp. fistulosa (Mart. ex Choisy) D.
      Austin

        SY=I. cardiophylla Gray
  barbatisepala Gray
  batatas (L.) Lam.
  cairica (L.) Sweet
      var. cairica
      var. hederacea Hallier f.
      var. lineariloba (Hbd.) Deg. & van
        Ooststr.
  calantha Griseb.
  capillacea (H.B.K.) G. Don
        SY=I. muricata Cav.
  carnea Jacq.
  coccinea L.
        SY=Quamoclit c. (L.) Moench
  coptica (L.) Roth ex Roemer & Schultes
  costellata Torr.
  cristulata Hallier f.
  egregia House
  fistulosa Mart. ex Choisy
        SY=I. crassicaulis (Benth.) B.L.
            Robins.
  gracilis R. Br.
  hederifolia L.
        SY=I. coccinea L. var. h. (L.)
            Gray
  horsfalliae Hook.
  indica (Burm. f.) Merr.
        SY=I. acuminata (Vahl) Roemer &
            Schultes
        SY=I. cathartica Poir.
        SY=I. congesta R. Br.
        SY=I. mutabilis Ker-Gawl.
        SY=Pharbitis cathartica (Poir.)
            Choisy
  krugii Urban
  lacunosa L.
  lemmonii Gray
  leptophylla Torr.
  leptotoma Torr.
      var. leptotoma
      var. wootonii E.H. Kelso
  lindheimeri Gray
  longifolia Benth.
  macrantha Roemer & Schultes
        SY=Calonyction tuba (Schlecht.)
            Colla
        SY=I. t. (Schlecht.) G. Don
  macrorhiza Michx.
  meyeri (Spreng.) G. Don
  microdactyla Griseb.
        SY=Exogonium m. (Griseb.) House
  X multifida (Raf.) Shinners [coccinea X
    quamoclit]
  nil (L.) Roth
        SY=I. barbigera Sims
        SY=I. hederacea (L.) Jacq.
        SY=I. h. var. integriuscula Gray
        SY=Pharbitis b. (Sims) G. Don
        SY=P. h. (L.) Choisy
        SY=P. n. (L.) Choisy
  nodiflorus Desv.
        SY=Jacquemontia n. (Desv.) G. Don
  obscura (L.) Ker-Gawl.
  ochroleuca Spanoghe
  palustris Urban
  pandurata (L.) G.F.W. Mey.
        SY=I. p. var. rubescens Choisy
  pes-caprae (L.) R. Br.
    ssp. brasiliensis (L.) van Ooststr.
        SY=I. b. (L.) Sweet
        SY=I. p-c. var. emarginata Hallier
          f.
    ssp. pes-caprae
  plummerae Gray
  pubescens Lam.
        SY=I. heterophylla Ortega
  purpurea (L.) Roth

Ipomoea (CONT.)
            SY=I. hirsutula Jacq. f.
            SY=I. p. var. diversifolia
                (Lindl.) O'Donell
            SY=Pharbitis p. (L.) Voigt
quamoclit L.
            SY=Quamoclit q. (L.) Britt.
            SY=Q. vulgaris Choisy
repanda Jacq.
            SY=Exogonium r. (Jacq.) Choisy
rupicola House
sagittata Poir.
?sectifolia Shinners
setifera Poir
            SY=I. rubra (Vahl) Millsp.
setosa Ker-Gawl.
            SY=I. melanotricha Brandeg.
shumardiana (Torr.) Shinners
steudelii Millsp.
            SY=Exogonium arenarium Choisy
stolonifera (Cyrillo) J.F. Gmel.
tenuiloba Torr.
tenuissima Choisy
thurberi Gray
tiliacea (Willd.) Choisy
trichocarpa Ell.
            SY=I. t. var. torreyana (Gray)
                Shinners
            SY=I. trifida (H.B.K.) G. Don
tricolor Cav.
triloba L.
tuboides Deg. & van Ooststr.
            var. pubescens Deg. & van Ooststr.
            var. tuboides
turbinata Lag.
            SY=I. muricata (L.) Jacq.
violacea L.
wrightii Gray
            SY=I. heptaphylla (Rottb. &
                Willd.) Voigt
            SY=I. spirale House

Jacquemontia Choisy

    cayensis Britt.
    cumanensis (H.B.K.) Kuntze
    curtissii Peter ex Hallier f.
    havanensis (Jacq.) Urban
            SY=J. jamaicensis (Jacq.) Hallier
                f.
    ovalifolia (Choisy) Hallier f.
        ssp. obcordata (Millsp.) Robertson
            SY=J. obcordata (Millsp.) House
            SY=J. subsalina Britt.
        ssp. ovalifolia
        ssp. sandwicensis (Gray) Robertson
            SY=J. o. var. tomentosa Choisy
            SY=J. s. Gray
            SY=J. s. var. t. (Choisy) Hbd.
    palmeri S. Wats.
    pentantha (Jacq.) G. Don
            SY=J. canescens (H.B.K.) Benth.
    pringlei Gray
    reclinata House
    solanifolia (L.) Hallier
            SY=Exogonium s. (L.) Britt.
            SY=Ipomoea s. L.
    tamnifolia (L.) Griseb.
            SY=Thyella t. (L.) Raf.
    verticillata (L.) Urban

Merremia Dennst. ex Endl.

    aegyptia (L.) Urban
            SY=Ipomoea a. L.
    dissecta (Jacq.) Hallier f.
            SY=Ipomoea d. (Jacq.) Pursh

            SY=I. sinuata Ortega
            SY=Operculina d. (Jacq.) House
hederacea (Burm. f.) Hallier f.
quinquefolia (L.) Hallier f.
            SY=Ipomoea q. L.
tridentata (L.) Hallier f.
    ssp. angustifolia (Jacq.) van Ooststr.
            SY=Ipomoea a. Jacq.
    ssp. tridentata
tuberosa (L.) Rendle
            SY=Ipomoea t. L.
            SY=Operculina t. (L.) Meisn.
umbellata (L.) Hallier f.
            SY=Ipomoea polyanthes Roemer &
                Schultes

Operculina Silva Manso

    pinnatifida (H.B.K.) O'Donell
            SY=Ipomoea p. (H.B.K.) G. Don
    triquetra (Vahl) Hallier f.
            SY=Ipomoea t. (Vahl) Roemer &
                Schultes

Petrogenia I.M. Johnston

    repens I.M. Johnston

Porana Burm. f.

    paniculata Roxb.

Stictocardia Hallier f.

    campanulata (L.) Merr.
            SY=Ipomoea c. L.
            SY=Rivea c. (L.) House
    tiliifolia (Desr.) Hallier f.
            SY=Convolvulus t. Desr.

Stylisma Raf.

    abdita Myint
            SY=Bonamia a. (Myint) R.W. Long
    aquatica (Walt.) Chapman
            SY=Bonamia a. (Walt.) Gray
            SY=Breweria a. (Walt.) Gray
            SY=Breweria michauxii Fern. &
                Schub.
    humistrata (Walt.) Chapman
            SY=Bonamia h. (Walt.) Gray
            SY=Breweria h. (Walt.) Gray
    patens (Desr.) Myint
        ssp. angustifolia (Nash) Myint
            SY=Bonamia p. (Desr.) Shinners
                var. a. (Nash) Shinners
            SY=S. a. (Nash) House
        ssp. patens
            SY=Bonamia p. (Desr.) Shinners
            SY=S. trichosanthes (Michx.) House
    pickeringii (Torr. ex M.A. Curtis) Gray
        var. pattersonii (Fern. & Schub.)
            Myint
            SY=Breweria pickeringii (Torr. ex
                M.A. Curtis) Gray var.
                pattersonii Fern. & Schub.
            SY=S. pattersonii (Fern. & Schub.)
                G.N. Jones
        var. pickeringii
            SY=Bonamia p. (Torr. ex M.A.
                Curtis) Gray
            SY=Breweria p. (Torr. ex M.A.
                Curtis) Gray
            SY=Breweria p. var. caesariense
                Fern. & Schub.
    villosa (Nash) House
            SY=Bonamia v. (Nash) K.A. Wilson

Stylisma (CONT.)
          SY=Breweria v. Nash

Turbina Raf.

    corymbosa (L.) Raf.
          SY=Ipomoea c. (L.) Roth

CORNACEAE

Cornus L.

    X acadiensis Fern. [alternifolia X sericea]
    alternifolia L. f.
          SY=Svida a. (L. f.) Small
    amomum P. Mill.
      ssp. amomum
          SY=Svida a. (P. Mill.) Small
      ssp. obliqua (Raf.) J.S. Wilson
          SY=C. a. var. schuetzeana (C.A.
             Mey.) Rickett
          SY=C. o. Raf.
    X arnoldiana Rehd. [amomum ssp. obliqua X
      foemina ssp. racemosa]
    asperifolia Michx.
          SY=Svida a. (Michx.) Small
    canadensis L.
        var. canadensis
          SY=Chamaepericlymenum c. (L.)
             Aschers. & Graebn.
        var. dutillyi (Lepage) Boivin
    drummondii C.A. Mey.
          SY=Cornus priceae Small
          SY=Svida p. (Small) Small
    florida L.
          SY=Cynoxylon f. (L.) Raf.
    foemina P. Mill.
      ssp. foemina
          SY=Cornus femina (P. Mill.) Small
          SY=C. stricta Lam.
          SY=Svida femina (P. Mill.) Small
          SY=S. s. (Lam.) Small
      ssp. microcarpa (Nash) J.S. Wilson
          SY=Svida m. (Nash) Small
      ssp. racemosa (Lam.) J.S. Wilson
          SY=Cornus r. Lam.
    glabrata Benth.
    nuttallii Audubon ex Torr. & Gray
    purpusii Koehne
    rugosa Lam.
    sericea L.
      ssp. occidentalis (Torr. & Gray) Fosberg
          SY=Cornus alba L. var. o. (Torr. &
             Gray) Boivin
          SY=C. o. (Torr. & Gray) Coville
          SY=C. stolonifera Michx. var. o.
             (Torr. & Gray) C.L. Hitchc.
      ssp. sericea
          SY=Cornus alba L. pro parte
          SY=C. a. var. baileyi (Coult. &
             Evans) Boivin
          SY=C. a. var. californica (C.A.
             Mey.) Boivin
          SY=C. a. var. interior (Rydb.)
             Boivin
          SY=C. b. Coult. & Evans
          SY=C. X c. C.A. Mey.
          SY=C. instolonea A. Nels.
          SY=C. interior (Rydb.) N. Petersen
          SY=C. sericea L. ssp. stolonifera
             (Michx.) Fosberg
          SY=C. stolonifera Michx.
          SY=C. stolonifera var. b. (Coult.
             & Evans) Drescher

          SY=Svida instolonea (A. Nels.)
             Rydb.
          SY=Svida stolonifera (Michx.)
             Rydb.
    sessilis Torr. ex Dur.
    X slavinii Rehd. [rugosa X sericea]
    suecica L.
          SY=Chamaepericlymenum s. (L.)
             Aschers. & Graebn.
          SY=Svida s. (L.) Holub
    X unalaschkensis Ledeb. [canadensis X
      suecica]

CORYNOCARPACEAE

Corynocarpus J.R. & G. Forst.

    laevigata J.R. & G. Forst.

CRASSULACEAE

Cotyledon L.

    orbiculata L.

Crassula L.

    aquatica (L.) Schoenl.
          SY=Tillaea a. L.
          SY=T. a. var. drummondii (Torr. &
             Gray) Jepson
          SY=T. d. Torr. & Gray
          SY=Tillaeastrum a. (L.) Britt.
          SY=Tillaeastrum vaillantii
             (Willd.) Britt.
    erecta (Hook. & Arn.) Berger
          SY=Tillaea e. Hook. & Arn.
          SY=T. e. var. eremica Jepson
    muscosa (L.) Roth
          SY=C. tillaea Lester-Garland
          SY=Tillaea m. L.

Diamorpha Nutt.

    smallii Britt.
          SY=D. cymosa (Nutt.) Britt. ex
             Small
          SY=Sedum c. (Nutt.) Frod.
          SY=S. s. (Britt.) Ahles

Dudleya Britt. & Rose

    abramsii Rose
      ssp. abramsii
      ssp. murina (Eastw.) Moran
    arizonica Rose
          SY=Echeveria a. (Rose) Kearney &
             Peebles
          SY=E. pulverulenta Nutt. ssp. a.
             (Rose) Clokey
    attenuata (S. Wats.) Moran
      ssp. orcuttii (Rose) Moran
          SY=Stylophyllum o. Rose
          SY=S. parishii Britt.
    bettinae Hoover
    blochmaniae (Eastw.) Moran
      ssp. blochmaniae
          SY=Hasseanthus b. (Eastw.) Rose
      ssp. insularis (Moran) Moran
    brevifolia (Moran) Moran
          SY=D. blochmaniae (Eastw.) Moran

Dudleya (CONT.)
                    ssp. brevifolia (Moran)
                        Moran
    caespitosa (Haw.) Britt. & Rose
            SY=D. cotyledon (Jacq.) Britt. &
                Rose
    candelabrum Rose
    collomiae Rose
            SY=Echeveria c. (Rose) Kearney &
                Peebles
    cymosa (Lem.) Britt. & Rose
        ssp. cymosa
            SY=D. angustiflora Rose
            SY=D. laxa (Lindl.) Britt. & Rose
            SY=D. nevadensis (S. Wats.) Britt.
                & Rose
        ssp. gigantea (Rose) Moran
            SY=D. g. Rose
        ssp. marcescens Moran
        ssp. minor (Rose) Moran
            SY=D. goldmanii Rose
            SY=D. nevadensis (S. Wats.) Britt.
                & Rose ssp. m. (Rose) Abrams
        ssp. ovatifolia (Britt.) Moran
        ssp. setchellii (Jepson) Moran
            SY=D. s. (Jepson) Britt. & Rose
    densiflora (Rose) Moran
            SY=Stylophyllum nudicaule Abrams
    edulis (Nutt.) Moran
            SY=Stylophyllum e. (Nutt.) Britt.
                & Rose
    farinosa (Lindl.) Britt. & Rose
            SY=D. compacta Rose
            SY=D. eastwoodiae Rose
            SY=D. septentrionalis Rose
    greenei Rose
    hassei (Rose) Moran
            SY=Stylophyllum h. Rose
    lanceolata (Nutt.) Britt. & Rose
            SY=D. brauntonii Rose
            SY=D. congesta Rose
            SY=D. lurida Rose
    multicaulis (Rose) Moran
            SY=Hasseanthus elongatus Rose
    nesiotica (Moran) Moran
    palmeri (S. Wats.) Britt. & Rose
    parva Rose & A. Davids.
    pulverulenta (Nutt.) Britt. & Rose
            SY=Echeveria p. Nutt.
    saxosa (M.E. Jones) Britt. & Rose
        ssp. aloides (Rose) Moran
            SY=D. delicata Rose
            SY=D. grandiflora Rose
        ssp. saxosa
    stolonifera Moran
    traskiae (Rose) Moran
            SY=Stylophyllum t. Rose
    variegata (S. Wats.) Moran
            SY=Hasseanthus v. (S. Wats.) Rose
    virens (Rose) Moran
            SY=Stylophyllum albidum Rose
            SY=S. insulare Rose
            SY=S. v. Rose
    viscida (S. Wats.) Moran
            SY=Stylophyllum v. (S. Wats.)
                Britt. & Rose

Echeveria DC.

    strictiflora Gray
            SY=Cotyledon s. (Gray) Baker

Graptopetalum Rose

    bartramii Rose
            SY=Echeveria b. (Rose) Kearney &
                Peebles

    rusbyi (Greene) Rose
            SY=Echeveria r. (Greene) A. Nels.
                & J.F. Macbr.
            SY=G. orpetii E. Walther

Kalanchoe Adans.

    brasiliensis Camb.
    crenata (Andr.) Haw.
            SY=Bryophyllum c. (Andr.) Baker
    daigremontiana Hamet & Perrier
            SY=Bryophyllum d. (Hamet &
                Perrier) Berger
    fedtschenkoi Hamet & Perrier
    gastonis-bonnieri Hamet & Perrier
    laxiflora Baker
    marmorta Baker
            SY=K. grandiflora A. Rich.
    pinnata (Lam.) Pers.
            SY=Bryophyllum p. (Lam.) S. Kurz
            SY=Cotyledon p. Lam.
    tubiflora (Harvey) Hamet
            SY=Bryophylum t. Harvey
            SY=K. verticillata Elliot

Lenophyllum Rose

    texanum (J.G. Sm.) Rose
            SY=Sedum t. J.G. Sm.
            SY=Villadia t. (J.G. Sm.) Rose

Parvisedum Clausen

    congdonii (Eastw.) Clausen
            SY=Sedella c. (Eastw.) Britt. &
                Rose
    leiocarpum (H.K. Sharsmith) Clausen
            SY=Sedella l. H.K. Sharsmith
    pentandrum (H.K. Sharsmith) Clausen
            SY=Sedella p. H.K. Sharsmith
    pumilum (Benth.) Clausen
            SY=Sedella p. (Benth.) Britt. &
                Rose

Sedum L.

    acre L.
    aizoon L.
    albomarginatum Clausen
    alboroseum Boreau
    album L.
    annuum L.
    borschii (Clausen) Clausen
    cockerellii Britt.
            SY=S. wootonii Britt.
    debile S. Wats.
            SY=Gormania d. (S. Wats.) Britt.
    diffusum S. Wats.
    divergens S. Wats.
    glaucophyllum Clausen
    ?griffithsii Rose
    havardii Rose
    hispanicum Juslen.
    hybridum L.
    integrifolium (Raf.) A. Nels. ex Coult. &
        A. Nels.
        ssp. integrifolium
            SY=Rhodiola i. Raf.
            SY=S. alaskanum (Rose) Rose ex
                Hutchinson
            SY=S. rosea (L.) Scop. ssp. i.
                (Raf.) Hulten
            SY=S. r. var. alaskanum (Rose)
                Berger
            SY=S. r. var. i. (Raf.) Berger
        ssp. leedyi (Rosendahl & Moore) Clausen
            SY=S. rosea (L.) Scop. var. l.

Sedum (CONT.)
                    Rosendahl & Moore
      ssp. neomexicanum (Britt.) Clausen
      ssp. procerum Clausen
  kamtschaticum Fish. & Mey.
      ssp. ellacombianum (Praeger) Clausen
  lanceolatum Torr.
      ssp. lanceolatum
      ssp. nesioticum (G.N. Jones) Clausen
            SY=S. l. var. n. (G.N. Jones) C.L.
                  Hitchc.
      ssp. subalpinum (Blank.) Clausen
            SY=S. s. Blank.
  laxum (Britt.) Berger
      ssp. eastwoodiae (Britt.) Clausen
      ssp. flavidum Denton
      ssp. heckneri (M.E. Peck) Clausen
            SY=S. h. M.E. Peck
      ssp. latifolium Clausen
      ssp. laxum
            SY=Gormania l. Britt.
            SY=S. l. ssp. perplexum Clausen
  leibergii Britt.
            SY=S. divaricatum S. Wats.
            SY=S. l. var. typicum Clausen
  moranense H.B.K.
  moranii Clausen
            SY=Cotyledon glanduliferum
                  Henderson
            SY=Gormania g. (Henderson) Abrams
            SY=S. g. (Henderson) M.E. Peck
  nevii Gray
  niveum A. Davids.
  nuttallianum Raf.
  oblanceolatum Clausen
  obtusatum Gray
      ssp. boreale Clausen
      ssp. obtusatum
            SY=Gormania o. (Gray) Britt.
      ssp. paradisum Denton
      ssp. retusum (Rose) Clausen
            SY=S. laxum (Britt.) Berger ssp.
                  r. (Rose) Clausen
  ochroleucum Chaix
            SY=S. anopetalum DC.
  oreganum Nutt.
      ssp. oreganum
            SY=Gormania o. (Nutt.) Britt.
      ssp. tenue Clausen
  oregonense (S. Wats.) M.E. Peck
            SY=Gormania watsonii Britt.
  parvum Hemley
      ssp. nanifolium (Frod.) Clausen
            SY=?S. robertsianum Alexander
  ?pinetorum Brandeg.
            SY=Congdonia p. (Brandeg.) Jepson
  praealtum DC.
  pulchellum Michx.
            SY=S. vigilimontis Small
  purpureum (L.) Schultes
            SY=S. telephium L. ssp. p. (L.)
                  Schinz & Keller
            SY=S. triphyllum (Haw.) S.F. Gray
  pusillum Michx.
  radiatum S. Wats.
      ssp. ciliosum (T.J. Howell) Clausen
            SY=S. douglasii Hook. ssp. c.
                  (T.J. Howell) Clausen
            SY=S. stenopetalum Pursh ssp. c.
                  (T.J. Howell) Clausen
      ssp. depauperatum Clausen
      ssp. radiatum
            SY=S. douglasii Hook. ssp. r. (S.
                  Wats.) Clausen
            SY=S. stenopetalum Pursh ssp. r.
                  (S. Wats.) Clausen
  reflexum L.

  rhodanthum Gray
            SY=Clementsia r. (Gray) Rose
  rosea (L.) Scop.
            SY=Rhodiola roanensis Britt.
            SY=R. rosea L.
            SY=S. rosea var. roanensis
                  (Britt.) Berger
  rupestre L.
  rupicolum G.N. Jones
            SY=S. lanceolatum Torr. var. r.
                  (G.N. Jones) C.L. Hitchc.
  sarmentosum Bunge
  sexangulare L.
  sieboldii Sweet
  spathulifolium Hook.
      ssp. pruinosum (Britt.) Clausen & Uhl
            SY=S. s. var. p. (Britt.) Boivin
      ssp. purdyi (Jepson) Clausen
            SY=S. p. Jepson
      ssp. spathulifolium
            SY=S. s. ssp. anomalum (Britt.)
                  Clausen & Uhl
      ssp. yosemitense (Britt.) Clausen
            SY=S. s. ssp. anomalum (Britt.)
                  Clausen & Uhl var. majus
                  Praeger
            SY=S. y. Britt.
  spectabile Boreau
  spurium Bieb.
  stelliforme S. Wats.
  stenopetalum Pursh
      ssp. monanthum (Suksdorf) Clausen
      ssp. stenopetalum
            SY=S. douglasii Hook.
  telephioides Michx.
            SY=Anacampseros t. (Michx.) Haw.
  telephium L.
      ssp. fabaria (W.D.J. Koch) Schinz &
            Keller
      ssp. telephium
  ternatum Michx.
  villosum L.
  wrightii Gray

Sempervivum L.

  heuffelii Schott
  tectorum L.

Villadia Rose

  squamulosa H.B.K.

CROSSOSOMATACEAE

Apacheria C. Timason

  chiricahuensis C.T. Mason

Crossosoma Nutt.

  bigelovii S. Wats.
      var. bigelovii
      var. glaucum (Small) Kearney &
            Peebles
  californicum Nutt.
  parviflorum Robins. & Fern.

Forsellesia (Gray) Greene

  clokeyi Ensign
  meionandra (Koehne) Heller
            SY=Glossopetalon m. Koehne
  nevadensis (Gray) Greene

Forsellesia (CONT.)
            SY=Glossopetalon n. Gray
    planitierum Ensign
            SY=Glossopetalon p. (Ensign) St.
                John
    pungens (Brandeg.) Heller
            var. glabra Ensign
            var. pungens
            SY=Glossopetalon p. Brandeg.
    spinescens (Gray) Greene
            SY=Glossopetalon s. Gray
    stipulifera (St. John) Ensign
            SY=Glossopetalon nevadense Gray
                var. s. (St. John) C.L.
                    Hitchc.
            SY=G. s. St. John
    texensis Ensign

CUCURBITACEAE

Anguria Jacq.

    cookiana Britt.
    ottoniana Schlecht.
    pedata (L.) Jacq.
    trifoliata L.
    trilobata (L.) Jacq.

Apodanthera Arn.

    undulata Gray

Brandegea Cogn.

    bigelovii (S. Wats.) Cogn.

Bryonia L.

    alba L.
    cretica L.
        ssp. dioica (Jacq.) Tutin
            SY=B. d. Jacq.

Cayaponia Silva Manso

    americana (Lam.) Cogn.
    grandifolia (Torr. & Gray) Small
            SY=C. boykinii (Torr. & Gray)
                Cogn.
    quinqueloba (Raf.) Shinners
    racemosa (P. Mill.) Cogn.

Citrullus Schrad.

    colocynthis (L.) Schrad.
            SY=C. vulgaris Schrad.
    lanatus (Thunb.) Matsumura & Nakai
            var. citroides (Bailey) Mansf.
            SY=C. citrullus (L.) Karst.
            SY=C. vulgaris Schrad. var.
                citroides Bailey
            var. lanatus

Coccinia Wight & Arn.

    grandis (L.) Voigt
            SY=C. cordifolia (L.) Cogn.

Corallocarpus Welw. ex Benth. & Hook.

    emetocatharticus (Gros.) Cogn.

Cucumis L.

anguria L.
dipsaceus C.G. Ehrenb. ex Spach
melo L.
            var. dudaim (L.) Naud.
            var. melo
myriocarpus E. Mey. ex Naud.
sativus L.

Cucurbita L.

digitata Gray
foetidissima H.B.K.
            SY=Pepo f. (H.B.K.) Britt.
maxima Duchesne
moschata (Duchesne) Duchesne ex Poir.
            SY=Pepo m. (Duchesne) Britt.
okeechobeensis Small
            SY=Pepo o. (Small) Bailey
palmata S. Wats.
            SY=C. californica Torr. ex S.
                Wats.
pepo L.
            var. ovifera (L.) Alef.
            var. pepo
texana Gray

Cyclanthera Schrad.

    dissecta (Torr. & Gray) Arn.

Echinocystis Torr. & Gray

    lobata (Michx.) Torr. & Gray
            SY=Micrampelis l. (Michx.) Greene

Echinopepon Naud.

    wrightii (Gray) S. Wats.

Fevillea L.

    cordifolia L.

Ibervillea Greene

    lindheimeri Greene ex Small
    tenuisecta (Gray) Small
    tripartita (Naud.) Small
            SY=I. tenella (Naud.) Small

Lagenaria Ser.

    siceraria (Molina) Standl.
            SY=Cucurbita lagenaria L.
            SY=L. leucantha (Duchesne) Rusby
            SY=L. vulgaris Ser.

Luffa (L.) P. Mill.

    acutangula (L.) Roxb.
    cylindrica (L.) Roemer

Marah Kellogg

    fabaceus (Naud.) Greene
            var. agrestis (Greene) Stocking
                SY=M. inermis (Cogn.) S.T. Dunn
            var. fabaceus
    gilensis Greene
    horridus (Congd.) S.T. Dunn
    macrocarpus (Greene) Greene
            var. macrocarpus
            var. major (S.T. Dunn) Stocking
                SY=M. guadalupensis (S. Wats.)
                    Greene
            var. micranthus Stocking
    oreganus (Torr. & Gray) T.J. Howell

Marah (CONT.)
                SY=Echinocystis o. (Torr. & Gray)
                    Cogn.
   watsonii (Cogn.) Greene

Melothria L.

   guadalupensis (Spreng.) Cogn.
   pendula L.
        var. aspera Cogn.
            SY=M. microcarpa Shuttlw.
            SY=M. nashii Small
        var. crassifolia (Small) Cogn.
            SY=M. c. Small
        var. pendula
            SY=M. p. var. chlorocarpa
                (Engelm.) Cogn.

Momordica L.

   balsamina L.
   charantia L.

Sechium P. Br.

   edule (Jacq.) Sw.

Sicyos L.

   ampelophyllus Woot. & Standl.
   angulatus L.
   atollensis St. John
   caumii St. John
   cucumerinus Gray
   glaber Woot.
   hillebrandii St. John
   hispidus Hbd.
   laciniatus L.
        SY=S. l. var. genuinus Cogn.
   lamoureuxii St. John
   laysanensis St. John
   macrophyllus Kunth
   maximowiczii Cogn.
   microcarpus Mann
   nihoaensis St. John
   niihauensis St. John
   pachycarpus Hook. & Arn.
   parviflorus Willd.
   remyanus Cogn.
   semitonsus St. John

Sicyosperma Gray

   gracile Gray

Skottsbergiliana St. John

   lasiocephala (Skottsberg) St. John
        SY=Sicyos l. Skottsberg
   partita St. John

Thladiantha Bunge

   dubia Bunge

Tumamoca Rose

   macdougalii Rose

CUNONIACEAE

Weinmannia L.

   pinnata L.

CYPERACEAE

Abildgaardia Vahl

   ovata (Burm. f.) Kral
        SY=A. monostachya (L.) Vahl
        SY=Fimbristylis m. (L.) Hassk.

Bulbostylis Kunth

   barbata (Rottb.) C.B. Clarke
        SY=Stenophyllus b. (Rottb.) Britt.
   capillaris (L.) C.B. Clarke
        SY=B. c. var. crebra Fern.
        SY=B. capillaris var. isopoda
            Fern.
        SY=Fimbristylis capillaris (L.)
            Gray
        SY=Stenophyllus capillaris (L.)
            Britt.
   ciliatifolia (Ell.) Fern.
        var. ciliatifolia
            SY=Stenophyllus c. (Ell.) C. Mohr
        var. coarctata (Ell.) Kral
            SY=B. coarctata (Ell.) Fern.
            SY=Stenophyllus coarctata (Ell.)
                Britt.
   curassavica (Britt.) Kukenth. & Ekman
        SY=Fimbristylis c. (Britt.) Alain
   funckii (Steud.) C.B. Clarke
   junciformis (H.B.K.) Lindm.
        SY=B. fendleri C.B. Clarke
        SY=B. j. var. ampliceps Kukenth.
   pauciflora (Liebm.) C.B. Clarke
        SY=Fimbristylis portoricensis
                (Britt.) Alain
        SY=Stenophyllus portoricensis
            Britt.
   schaffneri (Boeckl.) C.B. Clarke
   stenophylla (Ell.) C.B. Clarke
        SY=Stenophyllus s. (Ell.) Britt.
   vestita (Kunth) C.B. Clarke
        SY=Fimbristylis v. (Kunth) Hemsl.
   warei (Torr.) C.B. Clarke
        SY=Stenophyllus w. (Torr.) Britt.

Carex L.

   X abitibiana Lepage [aquatilis X stricta]
   aboriginum M.E. Jones
   abrupta Mackenzie
   abscondita Mackenzie
        SY=C. a. var. rostellata Fern.
   X absconditiformis Fern. [abscondita X
        laxiculmis]
   acutiformis Ehrh.
   acutinella Mackenzie
   adelostoma Krecz.
        SY=C. morrisseyi Porsild
   adusta Boott
   aenea Fern.
   X aestivaliformis Mackenzie [aestivalis X
        gracillima]
   aestivalis M.A. Curtis
   aggregata Mackenzie
        SY=C. sparganioides Willd. var. a.
            (Mackenzie) Gleason
   agrostoides Mackenzie
   alata Torr.
   albida Bailey
   albolutescens Schwein.
   albonigra Mackenzie
   albursina Sheldon
        SY=C. laxiflora Lam. var.
            latifolia Boott
   alligata Boott

Carex (CONT.)
      var. alligata
      var. degeneri R.W. Krauss
alma Bailey
alopecoidea Tuckerman
amphibola Steud.
      var. amphibola
      var. globosa (Bailey) Bailey
        SY=C. bulbostylis Mackenzie
      var. rigida (Bailey) Fern.
      var. turgida Fern.
        SY=C. grisea Wahlenb.
        SY=C. corrugata Fern.
amplectens Mackenzie
amplifolia Boott
amplisquama F.J. Herm.
anguillata Drej.
angustior Mackenzie
      SY=C. a. var. gracilenta Clausen &
        Wahlenb.
      SY=C. echinata Murr. var.
        angustata (Carey) Bailey
      SY=C. muricata L. var. angustata
        Carey
annectens (Bickn.) Bickn.
      SY=C. a. var. xanthocarpa (Bickn.)
        Wieg.
      SY=C. brachyglossa Mackenzie
anthoxanthea Presl
X anticostensis (Fern.) Lepage [rostrata X
    saxatilis var. miliaris]
    nm. anticostensis
    nm. inflatior Lepage
    nm. longidens Lepage
    nm. minor Lepage
aperta Boott
aquatilis Wahlenb.
      var. altior (Rydb.) Fern.
      var. aquatilis
        SY=C. a. var. substricta Kukenth.
        SY=C. s. (Kukenth.) Mackenzie
      var. stans (Drej.) Boott
        SY=C. a. ssp. s. (Drej.) Hulten
arapahoensis Clokey
arcta Boott
arctata Boott
arenaria L.
arenicola F. Schmidt
argyrantha Tuckerman
arkansana (Bailey) Bailey
artitecta Mackenzie
      var. artitecta
        SY=C. nigromarginata Schwein. var.
        muhlenbergii (Gray) Gleason
        SY=C. varia Muhl.
      var. subtilirostris F.J. Herm.
assiniboinensis W. Boott
athabascensis F.J. Herm.
atherodes Spreng.
athrostachya Olney
atlantica Bailey
      var. atlantica
      var. incomperta (Bickn.) F.J. Herm.
        SY=C. i. Bickn.
atratiformis Britt.
atrofusca Schkuhr
      var. atrofusca
      var. major (Boeckl.) Raymond
atrosquama Mackenzie
      SY=C. atrata L. ssp. atrosquama
        (Mackenzie) Hulten
      SY=C. atrata var. atrosquama
        (Mackenzie) Cronq.
aurea Nutt.
austrina (Small) Mackenzie
      SY=C. muhlenbergii Willd. var. a.
        Small

austrocaroliniana Bailey
backii Boott
      SY=C. durifolia Bailey
baileyi Britt.
baltzellii Chapman ex Dewey
barbarae Dewey
barrattii Schwein. & Torr.
bayardii Fern.
bebbii (Bailey) Fern.
bella Bailey
bicknellii Britt.
      var. bicknellii
      var. opaca F.J. Herm.
bicolor Bellardi ex All.
bigelowii Torr.
   ssp. bigelowii
      SY=C. consimilis Holm
   ssp. hyperborea (Drej.) Bocher
biltmoreana Mackenzie
bipartita Bellardi ex All.
      var. austromontana F.J. Herm.
        SY=C. lachenalii Schkuhr
      var. bipartita
blanda Dewey
      SY=C. laxiflora Lam. var. b.
        (Dewey) Boott
boecheriana Love, Love & Raymond
bolanderi Olney
      SY=C. deweyana Schwein. var. b.
        (Olney) W. Boott
bonanzensis Britt.
      SY=C. praeceptorum Mackenzie
bonplandii Kunth
brainerdii Mackenzie
brevicaulis Mackenzie
      SY=C. deflexa Hornem. var. b.
        (Mackenzie) Boivin
breviligulata Mackenzie
brevior (Dewey) Mackenzie ex Lunell
      SY=C. festucacea Willd. var. b.
        (Dewey) Fern.
brevipes W. Boott
breweri Boott
brittoniana Bailey
bromoides Willd.
brunnea Thunb.
brunnescens (Pers.) Poir.
   ssp. alaskana Kalela
   ssp. brunnescens
   ssp. pacifica Kalela
   ssp. sphaerostachya (Tuckerman) Kalela
      SY=C. b. var. s. (Tuckerman)
        Kukenth.
bullata Schkuhr
bushii Mackenzie
      SY=C. caroliniana Schwein. var.
        cuspidata (Dewey) Shinners
buxbaumii Wahlenb.
caesariensis Mackenzie
X calderi Boivin [canescens X heleonastes]
californica Bailey
campylocarpa Holm
      SY=C. c. ssp. affinis Maguire &
        Holmgren
canescens L.
   ssp. arctiformis (Mackenzie) Calder &
     Taylor
      var. arctiformis (Mackenzie) Calder &
        Taylor
        SY=C. a. Mackenzie
      var. disjuncta Fern.
   ssp. canescens
      var. canescens
        SY=C. curta Good.
      var. subloliacea (Laestad.) Hartman
        SY=C. lapponica O.F. Lang
capillaris L.

Carex (CONT.)
>   ssp. capillaris
>   SY=C. c. var. elongata Olney
>   SY=C. c. var. major Blytt
>   ssp. chlorostachys (Stev.) Love, Love &
>   Raymond
>   ssp. krausei (Boeckl.) Bocher
>   SY=C. c. var. k. (Boeckl.) Crantz
>   SY=C. k. Boeckl.
>   ssp. porsildiana (Polunin) Bocher
>   SY=C. krausei Boeckl. ssp. p.
>   (Polunin) Love, Love &
>   Raymond
> capitata L.
>   ssp. arctogena (H. Sm.) Hiitonen
>   SY=C. a. H. Sm.
>   SY=C. c. var. a. (H. Sm.) Hulten
>   ssp. capitata
> careyana Dewey
> caroliniana Schwein.
> caryophyllea Lat.
> castanea Wahlenb.
> cephalantha (Bailey) Bickn.
>   SY=C. laricina Mackenzie
>   SY=C. muricata L. var. c. Bailey
>   SY=C. m. var. l. (Mackenzie)
>   Gleason
> cephaloidea (Dewey) Dewey
>   SY=C. sparganioides Willd. var. c.
>   (Dewey) Carey
> cephalophora Willd.
> chapmanii Steud.
> cherokeensis Schwein.
> chihuahuensis Mackenzie
> chordorrhiza Ehrh. ex L. f.
> circinata C.A. Mey.
> collinsii Nutt.
> communis Bailey
> comosa Boott
> complanata Torr. & Hook.
> concinna R. Br.
> concinnoides Mackenzie
> congdonii Bailey
> conjuncta Boott
> X connectens Holmb. [limosa X paupercula]
> conoidea Willd.
>   SY=C. katahdinensis Fern.
> convoluta Mackenzie
>   SY=C. rosea Willd. var. pusilla
>   Peck
> crawei Dewey
> crawfordii Fern.
>   SY=C. c. var. vigens Fern.
> crebriflora Wieg.
> crinita Lam.
>   var. crinita
>   SY=C. c. var. brevicrinis Fern.
>   SY=C. c. var. minor Boott
>   SY=C. c. var. simulans Fern.
>   SY=C. nigromarginata Schwein. var.
>   m. (Boott) Gleason
>   var. porteri (Olney) Fern.
> X crinitoides Lepage [aquatilis X crinita]
> cristatella Britt.
> crus-corvi Shuttlw. ex Kunze
> X cryptochlaena Holm [lyngbyei X ramenskii]
> cryptolepis Mackenzie
>   SY=C. flava L. var. fertilis M.E.
>   Peck
>   SY=C. f. var. graminis Bailey
> cumulata (Bailey) Fern.
> cusickii Mackenzie
>   SY=C. obovoidea Cronq.
> dasycarpa Muhl.
> davisii Schwein. & Torr.
> davyi Mackenzie
> X deamii F.J. Herm. [shortiana X typhina]

debilis Michx.
>   var. debilis
>   var. intercursa Fern.
>   var. interjecta Bailey
>   var. pubera Gray
>   var. rudgei Bailey
>   SY=C. flexuosa Muhl.
>   var. strictior Bailey
> decomposita Muhl.
> deflexa Hornem.
> demareei Palmer
> demissa Hornem.
>   SY=C. oederi Retz.
>   SY=C. serotina Merat
> densa (Bailey) Bailey
> deweyana Schwein.
>   var. collectanea Fern.
>   var. deweyana
> diandra Schrank
> digitalis Willd.
>   SY=C. d. var. asymmetrica Fern.
>   SY=C. d. var. macropoda Fern.
> dioica L.
> disperma Dewey
> disticha Huds.
> divisa Huds.
> divulsa Stokes
> douglasii Boott
> dudleyi Mackenzie
> dutillyi O'Neill & Duman
> ebenea Rydb.
> eburnea Boott
> echinata Murr.
>   SY=C. josselynii (Fern.) Mackenzie
>   SY=C. leersii Willd.
> egglestonii Mackenzie
> eleocharis Bailey
>   SY=C. stenophylla Wahlenb. ssp. e.
>   (Bailey) Hulten
> eleusinoides Turcz. ex C.A. Mey.
>   SY=C. kokrinensis Porsild
> elliottii Schwein. & Torr.
> elyniformis Porsild
> elynoides Holm
> emmonsii Dewey
> emoryi Dewey
>   SY=C. stricta Lam. var. elongata
>   (Boeckl.) Gleason
> enanderi Hulten
>   SY=C. eurystachya F.J. Herm.
> engelmannii Bailey
> eurycarpa Holm
>   SY=C. oxycarpa Holm
> exilis Dewey
> X exsalina Lepage [bigelowii X paleacea]
> exserta Mackenzie
> exsiccata Bailey
>   SY=C. vesicaria L. var. major
>   Boott
> extensa Good.
> festivella Mackenzie
> festucacea Willd.
> feta Bailey
>   SY=C. straminea Willd. var. mixta
>   Bailey
> filifolia Nutt.
> X firmior (J.M. Norm.) Holmb. [limosa X
>   rariflora]
> fissa Mackenzie
>   var. aristata F.J. Herm.
>   var. fissa
> fissuricola Mackenzie
> flacca Schreb.
>   SY=C. glauca Scop.
> flaccosperma Dewey
>   var. flaccosperma
>   var. glaucodea (Tuckerman) Kukenth.

Carex (CONT.)
> SY=C. g. Tuckerman

flava L.
>> var. flava
>>> SY=C. flava var. rectirostra Gray
>>> SY=C. flava var. laxior (Kukenth.)
>>>> Gleason
>> var. gaspensis Fern.

X flavicans F. Nyl. [aquatilis X salina
> var. subspathacea]
>> SY=C. X halophila F. Nyl. nm. f.
>>> (F. Nyl.) Boivin

floridana Schwein.
> SY=C. nigromarginata Schwein. var.
>> f. (Schwein.) Kukenth.

foenea Willd.
>> var. foenea
>>> SY=C. siccata Dewey
>> var. tuberculata F.J. Herm.

foetida All.

folliculata L.
>> var. australis Bailey
>> var. folliculata

formosa Dewey

fracta Mackenzie

frankii Kunth

franklinii Boott
> SY=C. petricosa Dewey var. f.
>> (Boott) Boivin

garberi Fern.

X gardneri Lepage [paleacea X salina]

geophila Mackenzie

geyeri Boott

gigantea Rudge

gigas (Holm) Mackenzie

glacialis Mackenzie

glareosa Wahlenb.
> ssp. glareosa
>> var. amphigena Fern.
>>> SY=C. a. (Fern.) Mackenzie
>>> SY=C. bipartita Bellardi ex All.
>>>> var. a. (Fern.) Polunin
>> var. glareosa
>>> SY=C. bipartita Bellardi ex All.
>>>> var. g. (Wahlenb.) Polunin
> ssp. pribylovensis (Macoun) Chater &
>> Halliday
>>> SY=C. p. Macoun

glaucescens Ell.

globosa Boott

gmelinii Hook. & Arn.

gracilescens Steud.
> SY=C. laxiflora Lam. var.
>> gracillima (Boott) Robins. &
>> Fern.

gracilior Mackenzie

gracillima Schwein.
> SY=C. g. var. macerrima Fern. &
>> Wieg.

X grahamii Boott

X grantii Benn. [aquatilis X salina var.
> kattegatensis]

granularis Muhl. ex Willd.
>> var. granularis
>>> SY=C. rectior Mackenzie
>> var. haleana (Olney) Porter
>>> SY=C. h. Olney
>>> SY=C. shriveri Britt.

gravida Bailey
>> var. gravida
>> var. lunelliana (Mackenzie) F.J.
>>> Herm.
>>> SY=C. l. Mackenzie

grayi Carey
> SY=C. asa-grayi Bailey
> SY=C. g. var. hispidula Gray

X groenlandica Lange [bigelowii X nigra]

gynandra Schwein.
> SY=C. crinita Lam. var. g.
>> (Schwein.) Schwein. & Torr.

gynocrates Wormskj. ex Drej.
> SY=C. dioica L. ssp. g. (Wormskj.
>> ex Drej.) Hulten
> SY=C. d. var. g. (Wormskj. ex
>> Drej.) Ostenf.
> SY=C. parallela sensu auctt. non
>> (Laestad.) Sommerf.

gynodynama Olney

X haematolepis Drej. [bigelowii X lyngbyei]

halliana Bailey
> SY=C. oregonensis Olney ex Bailey

hallii Olney
> SY=C. parryana Dewey ssp. h.
>> (Olney) Murr.

X halophila F. Nyl.

harfordii Mackenzie

X hartii Dewey [lurida X retrorsa]

hassei Bailey
> SY=C. garberi Fern. ssp. bifaria
>> (Fern.) Hulten
> SY=C. g. var. b. Fern.

haydeniana Olney
> SY=C. macloviana d'Urv. ssp. h.
>> (Olney) Taylor & MacBryde
> SY=C. nubicola Mackenzie

haydenii Dewey
> SY=C. stricta Lam. var. decora
>> Bailey

heleonastes Ehrh.
> ssp. heleonastes
>> SY=C. curta Good. var. robustior
>>> (Kukenth.) Boivin
> ssp. neurochlaena (Holm) Bocher
>> SY=C. n. Holm

heliophila Mackenzie
> SY=C. pensylvanica Lam. var.
>> digyna Boeckl.

helleri Mackenzie

X helvola Blytt [bipartita var.
> austromontana X canescens]

hendersonii Bailey

heteroneura W. Boott
>> var. brevisquama F.J. Herm.
>>> SY=C. atrata L.
>>> SY=C. a. var. chalciolepis (Holm)
>>>> Kukenth.
>>> SY=C. a. var. erecta W. Boott
>> var. chalciolepis (Holm) F.J. Herm.
>>> SY=C. c. Holm
>> var. epapillosa (Mackenzie) F.J.
>>> Herm.
>>> SY=C. e. Mackenzie
>> var. heteroneura

hindsii C.B. Clarke

hirsutella Mackenzie
> SY=C. c. var. hirsuta (Bailey)
>> Gleason

hirta L.

hirtifolia Mackenzie

?hirtissima W. Boott

hitchcockiana Dewey

holostoma Drej.

hoodii Boott

hookerana Dewey

hormathodes Fern.
> SY=C. straminea Willd. var. invisa
>> W. Boott

hostiana DC.
>> var. hostiana
>> var. laurentiana Fern. & Wieg.

houghtonii Torr.
> SY=C. houghtoniana Dewey

howei Mackenzie
> SY=C. h. var. capillacea (Bailey)

Carex (CONT.)
                    Fern.
hyalina Boott
hyalinolepis Steud.
          SY=C. impressa Mackenzie
          SY=C. riparia M.A. Curtis
hystricina Muhl. ex Willd.
idahoa Bailey
          SY=C. parryana Dewey ssp. i.
              (Bailey) Murr.
illota Bailey
incurviformis Mackenzie
      var. danaensis (Stacey) F.J. Herm.
      var. incurviformis
          SY=C. maritima Gunn. var. i.
              (Mackenzie) Boivin
          SY=C. m. var. setina (Christ)
              Fern.
inops Bailey
          SY=C. pensylvanica Lam. var.
              vespertina Bailey
integra Mackenzie
interior Bailey
          SY=C. i. ssp. charlestonensis
              Clokey
interrupta Boeckl.
intumescens Rudge
          SY=C. i. var. fernaldii Bailey
jacobi-peteri Hulten
jamesii Schwein.
jepsonii J.T. Howell
jonesii Bailey
joorii Bailey
X josephi-schmittii Raymond [fernaldii X
   retrorsa]
kauaiensis R.W. Krauss
kelloggii W. Boott ex S. Wats.
          SY=C. lenticularis Michx. var.
              limnophila (Holm) Cronq.
X kenaica Lepage [ramenskii X salina var.
   subspathacea]
X knieskernii Dewey [arctata X castanea]
kobomugii Ohwi
lacustris Willd.
          SY=C. riparia M.A. Curtis var. l.
              (Willd.) Kukenth.
laeviconica Dewey
laeviculmis Meinsh.
laevivaginata (Kukenth.) Mackenzie
          SY=C. stipata Muhl. ex Willd. var.
              l. Kukenth.
langeana Fern.
lanuginosa Michx.
          SY=C. lasiocarpa Ehrh. var.
              latifolia (Boeckl.) Gleason
lasiocarpa Ehrh.
      var. americana Fern.
          SY=C. lanuginosa Michx. var. a.
              (Fern.) Boivin
          SY=C. lasiocarpa ssp. a. (Fern.)
              Love & Bernard
latebracteata Waterfall
laxa Wahlenb.
laxiculmis Schwein.
      var. copulata (Bailey) Fern.
          SY=C. X c. (Bailey) Mackenzie
      var. laxiculmis
laxiflora Lam.
      var. laxiflora
          SY=C. anceps Muhl.
          SY=C. l. var. patulifolia (Dewey)
              Carey
      var. serrulata F.J. Herm.
leavenworthii Dewey
          SY=C. cephalophora Willd. var.
              angustifolia Boott
          SY=C. c. var. l. (Dewey) Kukenth.

leiophylla Mackenzie
          SY=C. sabulosa Turcz. ssp. l.
              (Mackenzie) Porsild
lemmonii W. Boott
lenticularis Michx.
      var. albimontana Dewey
      var. blakei Dewey
      var. eucycla Fern.
      var. lenticularis
lepidocarpa Tausch
          SY=C. flava L. var. l. (Tausch)
              Godr.
          SY=C. f. var. nelmesiana (Raymond)
              Boivin
leporina L.
          SY=C. tracyi Mackenzie
leporinella Mackenzie
leptalea Wahlenb.
   ssp. harperi (Fern.) Calder & Taylor
          SY=C. h. Fern.
          SY=C. l. var. h. (Fern.) W. Stone
   ssp. leptalea
   ssp. pacifica Calder & Taylor
leptonervia Fern.
          SY=C. laxiflora Lam. var. varians
              Bailey
leptopoda Mackenzie
          SY=C. deweyana Schwein. ssp. l.
              (Mackenzie) Calder & Taylor
          SY=C. d. var. l. (Mackenzie)
              Boivin
leucodonta Holm
X leutzii Kneucker
      nm. leutzii
      nm. pseudofulva (Fern.) Boivin
          SY=C. X p. Fern.
      nm. xanthina (Fern.) Boivin
          SY=C. X x. Fern.
limnophila F.J. Herm.
limosa L.
X limula T. Fries [aquatilis X bigelowii]
livida (Wahlenb.) Willd.
      var. grayana (Dewey) Fern.
      var. livida
      var. radicaulis Paine
      var. rufiniformis Fern.
loliacea L.
lonchocarpa Willd.
longii Mackenzie
louisianica Bailey
lucorum Link
lugens Holm
lupuliformis Sartwell ex Dewey
lupulina Willd.
          SY=C. X macounii Dewey
          SY=C. l. var. pedunculata Gray
lurida Wahlenb.
luzulifolia W. Boott
luzulina Olney
      var. ablata (Bailey) F.J. Herm.
          SY=C. a. Bailey
      var. luzulina
lyngbyei Hornem.
          SY=C. cryptocarpa C.A. Mey.
          SY=C. l. var. robusta (Bailey)
              Cronq.
mackenziei Krecz.
macloviana d'Urv.
          SY=C. incondita F.J. Herm.
macrocephala Willd.
      var. bracteata Holm
      var. macrocephala
macrochaeta C.A. Mey.
magnifolia Mackenzie
X mainensis Porter [saxatilis var.
   militaris X vesicaria]
marina Dewey

Carex (CONT.)
    ssp. marina
            SY=C. amblyorhyncha Krecz.
    ssp. pseudolagopina (Sorensen) Bocher
        var. pseudolagopina (Sorensen) Bocher
            SY=C. amblyorhyncha Krecz. ssp. p.
                (Sorensen) Bocher
        var. solida (Sorensen) Bocher
mariposana Bailey ex Mackenzie
maritima Gunn.
            SY=C. incurva Lightf.
meadii Dewey
media R. Br.
?melanocarpa Cham.
melanostachya Willd.
membranacea Hook.
            SY=C. membranopacta Bailey
X mendica Lepage [salina var. kattegatensis
    X salina var. salina]
mendocinensis Olney
            SY=C. debiliformis Mackenzie
merritt-fernaldii Mackenzie
mertensii Prescott
mesochorea Mackenzie
meyenii Nees
michauxiana Boeckl.
            SY=C. abacta Bailey
microdonta Torr. & Hook.
microglochin Wahlenb.
microptera Mackenzie
        var. crassinervia F.J. Herm.
        var. microptera
            SY=C. macloviana d'Urv. var.
                microptera (Mackenzie)
                Boivin
microrhyncha Mackenzie
X minganinsularum Raymond [exilis X
    sterilis]
misandra R. Br.
misera Buckl.
miserabilis Mackenzie
mitchelliana M.A. Curtis
            SY=C. crinita Lam. var. m. (M.A.
                Curtis) Gleason
mohriana Mackenzie
molesta Mackenzie
montereyensis Mackenzie
montis-eeka Hbd.
muhlenbergii Willd.
        var. australis Olney
        var. enervis Boott
            SY=C. plana Mackenzie
        var. muhlenbergii
multicaulis Bailey
multicostata Mackenzie
            SY=C. pachycarpa Mackenzie
muricata L.
            SY=C. stellata sensu auctt. non
                MacKenzie
muriculata F.J. Herm.
muskingumensis Schwein.
nardina Fries
        var. atriceps Kukenth.
        var. hepburnii (Boott) Kukenth.
            SY=C. h. Boott
            SY=C. n. ssp. h. (Boott) Love,
                Love & Kapoor
        var. nardina
nealae R.W. Krauss
X nearctica Raymond [aquatilis var. stans X
    bigelowii]
nebraskensis Dewey
nelsonii Mackenzie
X neobigelowii Lepage [bigelowii X
    lenticularis]
X neofilipendula Lepage [aquatilis X
    paleacea]

X neomiliaris Lepage [aquatilis X saxatilis
    var. miliaris]
X neopaleacea Lepage [buxbaumii X paleacea]
X neorigida Lepage [bigelowii X salina]
nervina Bailey
nesophila Holm
            SY=C. microchaeta Holm
neurophora Mackenzie
nigra (L.) Reichard
            SY=C. acuta sensu auctt. non L.
            SY=C. goodenowii J. Gay
nigricans C.A. Mey.
nigromarginata Schwein.
normalis Mackenzie
norvegica Retz.
    ssp. inserrulata Kalela
            SY=C. n. var. inferalpina
                (Wahlenb.) Boivin
    ssp. norvegica
            SY=C. halleri Gunn.
            SY=C. vahlii Schkuhr
    ssp. stevenii (Holm) Murr.
            SY=C. media R. Br. var. s. (Holm)
                Fern.
nova Bailey
novae-angliae Schwein.
X nubens Lepage [salina var. kattegatensis
    X saxatilis var. miliaris]
nudata W. Boott
            SY=C. suborbiculata Mackenzie
nutans Host
obispoensis Stacey
obnupta Bailey
obtusata Lilj.
occidentalis Bailey
            SY=C. neomexicana Mackenzie
oklahomensis Mackenzie
            SY=C. stipata Muhl. ex Willd. var.
                o. (Mackenzie) Gleason
oligocarpa Willd.
oligosperma Michx.
        var. churchilliana Raymond
        var. oligosperma
X olneyi Boott [bullata X rostrata var.
    utriculata]
X oneillii Lepage [hindsii X rostrata]
onusta Mackenzie
oreocharis Holm
ormantha (Fern.) Mackenzie
ormostachya Wieg.
            SY=C. laxiflora Lam. var. o.
                (Wieg.) Gleason
oronensis Fern.
oxylepis Torr. & Hook.
        var. oxylepis
        var. pubescens J.K. Underwood
pachystachya Cham. ex Steud.
            SY=C. macloviana d'Urv. ssp. p.
                (Cham. ex Steud.) Hulten
            SY=C. p. var. gracilis (Olney)
                Mackenzie
            SY=C. p. var. monds-coulteri L.
                Kelso
pairaei F.W. Schultz
paleacea Wahlenb.
            SY=C. p. var. transatlantica Fern.
X paleacoides Lepage [glareosa ssp.
    glareosa var. amphigena X paleacea]
pallescens L.
            SY=C. p. var. neogaea Fern.
X paludivagans Drury [rostrata X rotundata]
panicea L.
            SY=C. p. var. microcarpa Sonder
X pannewitziana Figert [rostrata X
    vesicaria]
pansa Bailey
            SY=C. arenicola F. Schmidt ssp. p

Carex (CONT.)
                         (Bailey) T. Koyama & Calder
parryana Dewey
        var. brevisquama F.J. Herm.
        var. parryana
X patuensis Lepage [atratiformis X
    saxatilis var. miliaris]
paucicostata Mackenzie
pauciflora Lightf.
paucifructa Mackenzie
paupercula Michx.
        var. irrigua (Wahlenb.) Fern.
            SY=C. magellanica Lam. ssp. i.
                (Wahlenb.) Hulten
            SY=C. m. var. i. (Wahlenb.) B.S.P.
        var. pallens Fern.
        var. paupercula
            SY=C. magellanica sensu auctt. non
                Lam.
            SY=C. paupercula var. brevisquama
                Fern.
paysonis Clokey
            SY=C. tolmiei Boott
peckii Howe
            SY=C. albicans sensu auctt. non
                Willd.
            SY=C. nigromarginata Schwein. var.
                elliptica (Boott) Gleason
pedunculata Willd.
pelocarpa F.J. Herm.
pensylvanica Lam.
        var. distans Peck
        var. pensylvanica
perglobosa Mackenzie
X persalina Lepage [salina var. salina X
    salina var. subspathacea]
petasata Dewey
petricosa Dewey
        var. distichiflora (Boivin) Boivin
        var. misandroides (Fern.) Boivin
            SY=C. m. Fern.
        var. petricosa
phaeocephala Piper
            SY=C. eastwoodiana Stacey
phyllomanica W. Boott
X physocarpioides Lepage [rostrata X
    saxatilis var. major]
physorhyncha Liebm.
picta Steud.
X pieperiana Junge [flava X hostiana var.
    laurentiana]
piperi Mackenzie
pityophila Mackenzie
planostachys Kunze
plantaginea Lam.
platylepis Mackenzie
platyphylla Carey
plectocarpa F.J. Herm.
pluriflora Hulten
            SY=C. rariflora (Wahlenb.) Sm.
                var. p. (Hulten) Boivin
pluvia R.W. Krauss
        var. koolauensis R.W. Krauss
        var. pluvia
podocarpa R. Br.
            SY=C. montanensis Bailey
polymorpha Muhl.
polystachya Sw. ex Wahlenb.
praegracilis W. Boott
            SY=C. camporum Mackenzie
prairea Dewey
prarisa Dewey
prasina Wahlenb.
praticola Rydb.
            SY=C. p. var. subcoriacea F.J.
                Herm.
preslii Steud.

prionophylla Holm
projecta Mackenzie
proposita Mackenzie
pseudocyperus L.
X pseudohelvola Kihlm. [canescens X
    mackenziei]
pseudoscirpoidea Rydb.
            SY=C. scirpoidea Michx. var. p.
                (Rydb.) Cronq.
purpurifera Mackenzie
            SY=C. laxiflora Lam. var. p.
                (Mackenzie) Gleason
pyrenaica Wahlenb.
    ssp. micropoda (C.A. Mey.) Hulten
            SY=C. m. C.A. Mey.
    ssp. pyrenaica
X quebecensis Lepage [bigelowii X saxatilis
    var. rhomalea]
X quirponensis Fern. [atratiformis X
    norvegica]
radiata (Wahlenb.) Dewey
            SY=C. rosea Willd. var. radiata
                (Wahlenb.) Dewey
ramenskii Komarov
            SY=C. r. var. caudata Hulten
rariflora (Wahlenb.) Sm.
        var. androgyna Porsild
        var. rariflora
            SY=C. stygia Holm non Fries
raymondii Calder
            SY=C. atratiformis Britt. ssp. r.
                (Calder) Porsild
raynoldsii Dewey
reniformis (Bailey) Small
retroflexa Willd.
        var. retroflexa
        var. texensis (Torr.) Fern.
            SY=C. t. (Torr.) Bailey
retrorsa Schwein.
richardsonii R. Br.
roanensis F.J. Herm.
X rollandii Lepage [aquatilis X nigra]
            SY=Carex X aquanigra Boivin
rosea Willd.
rossii Boott ex Hook.
            SY=C. deflexa Hornem. var. r.
                (Boott ex Hook.) Bailey
rostrata Stokes ex With.
        var. ambigens Fern.
        var. anticostensis Fern.
        var. rostrata
        var. utriculata (Boott) Bailey
            SY=C. inflata Huds. var. u.
                (Boott) Druce
            SY=C. u. Boott
rotundata Wahlenb.
        var. compacta (R. Br.) Boivin
        var. rotundata
rousseaui Raymond
rufina Drej.
rugosperma Mackenzie
rupestris Bellardi ex All.
        var. drummondiana (Dewey) Bailey
            SY=C. d. Dewey
            SY=C. r. ssp. d. (Dewey) Holub
        var. rupestris
ruthii Mackenzie
            SY=C. muricata L. var. r.
                (Mackenzie) Gleason
sabulosa Turcz.
salina Wahlenb.
        var. kattegatensis (Fries) Almquist
            SY=C. recta Boott
        var. pseudofilipendula Kukenth.
        var. salina
        var. subspathacea (Wormskj.)
            Tuckerman

Carex (CONT.)
        SY=C. s. Wormskj.
  saliniformis Mackenzie
  X sardloaensis Dahl. [heteroneura var.
    brevisquama X stylosa]
  sartwelliana Olney
        SY=C. yosemitana Bailey
  sartwellii Dewey
      var. sartwellii
      var. stenorrhyncha F.J. Herm.
  saxatilis L.
      var. major Olney
          SY=C. ambusta Boott
          SY=C. physocarpa Presl
      var. miliaris (Michx.) Bailey
          SY=C. m. Michx.
      var. rhomalea Fern.
          SY=C. r. (Fern.) Mackenzie
      var. saxatilis
          SY=C. s. ssp. laxa (Trautv.)
            Kalela
  X saxenii Raymond [paleacea X salina var.
    kattegatensis]
    nm. dumanii (Lepage) Boivin
        SY=C. X d. Lepage
    nm. ferruginea Lepage
    nm. saxenii
  saximontana Mackenzie
        SY=C. backii Boott var. s.
          (Mackenzie) Boivin
  scabrata Schwein.
  scabriuscula Mackenzie
  schottii Dewey
  schweinitzii Dewey ex Schwein.
  scirpiformis Mackenzie
        SY=C. scirpoidea Michx. var.
          scirpiformis (Mackenzie)
          O'Neill & Duman
        SY=C. stenochlaena (Holm)
          Mackenzie
  scirpoidea Michx.
      var. convoluta Kukenth.
      var. curatorum (Stacey) Cronq.
          SY=C. c. Stacey
      var. scirpoidea
  scoparia Schkuhr ex Willd.
      var. scoparia
      var. tessellata Fern. & Wieg.
  scopulorum Holm
      var. bracteosa (Bailey) F.J. Herm.
          SY=C. gymnoclada Holm
      var. chimaphila (Holm) Kukenth.
          SY=C. c. Holm
      var. scopulorum
          SY=C. accedens Holm
  senta Boott
  seorsa Howe
  serratodens W. Boott
        SY=C. bifida W. Boott
  sheldonii Mackenzie
  shortiana Dewey
  silicea Olney
  simulata Mackenzie
  sitchensis Prescott
  socialis Mohlenbrock & Schwegm.
  X soerensenii Lepage [rariflora X salina
    var. subspathacea]
  soperi Raup
  sparganioides Willd.
  specifica Bailey
  spectabilis Dewey
  specuicola J.T. Howell
  spicata Huds.
  X spiculosa T. Fries [nigra X salina]
  spissa Bailey
  sprengelii Dewey ex Spreng.
  squarrosa L.

  X stansalina Lepage [aquatilis var. stans X
    salina]
  stenophylla Wahlenb.
      var. enervis (C.A. Mey.) Kukenth.
      var. stenophylla
  stenoptera Mackenzie
  stenoptila F.J. Herm.
  sterilis Willd.
        SY=C. elachycarpa Fern.
        SY=C. muricata L. var. s. (Willd.)
          Gleason
  stipata Muhl. ex Willd.
      var. maxima Chapman
          SY=C. s. var. uberior C. Mohr
          SY=C. u. (C. Mohr) Mackenzie
      var. stipata
  straminea Willd.
        SY=C. richii (Fern.) Mackenzie
  straminiformis Bailey
  striatula Michx.
        SY=C. laxiflora Lam. var.
          angustifolia Dewey
  stricta Lam.
      var. stricta
      var. strictior (Dewey) Carey
  styloflexa Buckl.
  stylosa C.A. Mey.
        SY=C. s. var. nigritella (Drej.)
          Fern.
  subbracteata Mackenzie
  suberecta (Olney) Britt.
  subfusca W. Boott
        SY=C. macloviana d'Urv. var. s.
          (W. Boott) Kukenth.
        SY=C. stenoptera Mackenzie
  X subimpressa Clokey [hyalinolepis X
    lanuginosa]
  X sublimosa Lepage [limosa X paleacea]
  X subnigra Lepage [nigra X paleacea]
  subnigricans Stacey
        SY=C. rachillis Maguire
  X subreducta Lepage [bigelowii X salina
    var. subspathacea]
  X subsalina Lepage [aquatilis X salina]
  X substans Lepage [aquatilis var. stans X
    salina var. subspathacea]
  X subviridula (Kukenth.) Fern. [flava X
    viridula]
  suksdorfii Kukenth.
  X sullivantii Boott [gracillima X
    hirtifolia]
  X supergoodenoughii (Kukenth.) Lepage
    [nigra X salina var. kattegatensis]
  supina Willd. ex Wahlenb.
      var. spaniocarpa (Steud.) Boivin
          SY=C. supina ssp. spaniocarpa
          (Steud.) Hulten
  svenonis Skottsberg
  swanii (Fern.) Mackenzie
        SY=C. virescens Willd. var. s.
          Fern.
  sychnocephala Carey
  sylvatica Huds.
  tahoensis Smiley
  tenax Chapman
  tenera Dewey
        SY=C. t. var. echinodes (Fern.)
          Wieg.
  teneriformis Mackenzie
  tenuiflora Wahlenb.
  terrae-novae Fern.
        SY=C. glacialis Mackenzie var.
          t-n. (Fern.) Boivin
  tetanica Schkuhr
  thurberi Dewey
  tincta Fern.
  tompkinsii J.T. Howell

Carex (CONT.)
    tonsa (Fern.) Bickn.
                SY=C. rugosperma Mackenzie var. t.
                    (Fern.) E.G. Voss
                SY=C. umbellata Schkuhr ex Willd.
                    var. t. Fern.
    torreyi Tuckerman
                SY=C. abbreviata Prescott
    torta Boott
    triangularis Boeckl.
    tribuloides Wahlenb.
            var. sangamonensis Clokey
            var. tribuloides
    triceps Michx.
    X trichina Fern. [tenuiflora X trisperma]
    trichocarpa Schkuhr
    triquetra Boott
    trisperma Dewey
            var. billingsii Knight
            var. trisperma
    tuckermanii Dewey
    tumulicola Mackenzie
    turgescens Torr.
    typhina Michx.
    ultra Bailey
    umbellata Schkuhr ex Willd.
                SY=C. abdita Bickn.
    X ungavensis Lepage [paleacea X stylosa]
    unilateralis Mackenzie
    ursina Dewey
    vaginata Tausch
                SY=C. saltuensis Bailey
    vallicola Dewey
            var. rusbyi (Mackenzie) F.J. Herm.
                SY=C. r. Mackenzie
            var. vallicola
    venusta Dewey
            var. minor Boeckl.
                SY=C. oblita Steud.
            var. venusta
    vernacula Bailey
                SY=C. foetida All. var. v.
                    (Bailey) Kukenth.
    verrucosa Muhl.
    vesicaria L.
            var. distenta Fries
            var. jejuna Fern.
            var. laurentiana Fern.
            var. monile (Tuckerman) Fern.
                SY=C. m. Tuckerman
            var. raeana (Boott) Fern.
                SY=C. r. Boott
            var. vesicaria
                SY=C. inflata Huds.
    vestita Willd.
    vexans F.J. Herm.
    vicaria Bailey
    virens Lam.
    virescens Willd.
                SY=C. costata Schw.
                SY=C. v. var. c. (Schw.) Dewey
    ?viridior Mackenzie
    viridula Michx.
                SY=C. oederi Retz. ssp. v.
                    (Michx.) Hulten
    vulpinoidea Michx.
            var. platycarpa Hall
            var. pycnocephala F.J. Herm.
            var. vulpinoidea
                SY=C. setacea Dewey
    wahuensis C.A. Mey.
            var. rubiginosa R.W. Krauss
            var. wahuensis
    walteriana Bailey
            var. brevis Bailey
            var. walteriana
    whitneyi Olney

                SY=C. flaccifolia Mackenzie
    wiegandii Mackenzie
    willdenowii Schkuhr
            var. megarrhyncha F.J. Herm.
            var. willdenowii
    williamsii Britt.
                SY=C. capillaris L. var. w.
                    (Britt.) Boivin
    woodii Dewey
                SY=C. colorata Mackenzie
                SY=C. tetanica Schkuhr var. w.
                    (Dewey) Wood
    wootonii Mackenzie
    xerantica Bailey

Cladium P. Br.

    californicum (S. Wats.) O'Neill
                SY=C. mariscus R. Br. var. c. S.
                    Wats.
                SY=Mariscus c. (S. Wats.) Britt.
    colocasia (L.) W. Wight
    jamaicensis Crantz
                SY=Mariscus j. (Crantz) Britt.
    leptostachyum Nees & Meyen
    mariscoides (Muhl.) Torr.
                SY=Mariscus m. (Muhl.) Kuntze
    mariscus R. Br.

Cymophyllus Mackenzie

    fraseri (Andr.) Mackenzie
                SY=Carex f. Andr.

Cyperus L.

    acuminatus Torr. & Hook.
                SY=C. virens Michx. var. arenicola
                    (Boeckl.) Shinners
    albomarginatus Mart. & Schrad. ex Nees
                SY=C. sabulosus sensu auctt. non
                    Mart. & Schrad.
    alternifolius L.
    amabilis Vahl
            var. amabilis
            var. macrostachyus Kukenth.
    amuricus Maxim.
                SY=C. microiria Steud.
    aristatus Rottb.
            var. aristatus
                SY=C. inflexus Muhl.
            var. runyonii O'Neill
    articulatus L.
    auriculatus Kunth
                SY=C. ferax L.C. Rich. var. a.
                    (Kunth) Kukenth.
    blodgettii Britt.
                SY=C. litoreus (C.B. Clarke)
                    Britt.
    brevifolioides Thieret & Delahoussaye
    brevifolius (Rottb.) Endl. ex Hassk.
                SY=Kyllinga b. Rottb.
    calcicola Britt.
    cephalanthus Torr. & Hook.
    compressus L.
    confertus Sw.
    congestus Vahl
    cuspidatus H.B.K.
    cylindricus (Ell.) Britt.
    cyperinus (Retz.) Sur.
    cyperoides (L.) Kuntze
    densicaespitosus Mattf. & Kukenth.
    dentatus Torr.
    diandrus Torr.
                SY=C. d. var. capitatus Britt.
    difformis L.
    digitatus Roxb.

Cyperus (CONT.)
   dipsaciformis Fern.
         SY=C. retrofractus (L.) Torr. var.
            d. (Fern.) Kukenth.
   distans L. f.
   distinctus Steud.
   eggersii Boeckl.
   elegans L.
   engelmannii Steud.
   eragrostis Lam.
         SY=C. vegetus Willd.
   erythrorhizos Muhl.
         SY=C. halei Torr.
   esculentus L.
      var. esculentus
      var. sativus Boeckl.
   fauriei Kukenth.
   fendlerianus Boeckl.
      var. debilis (Britt.) Kukenth.
         SY=C. rusbyi Britt.
      var. fendlerianus
         SY=Mariscus f. (Boeckl.) T. Koyama
   filicinus Vahl
   filiculmis Vahl
      var. filiculmis
         SY=C. bushii Britt.
         SY=C. f. var. oblitus Fern. &
            Grisc.
         SY=Mariscus f. (Vahl) T. Koyama
      var. macilentus Fern.
   filiformis Sw.
   flavescens L.
      var. flavescens
      var. poaeformis (Pursh) Fern.
   flavus (Vahl) Nees
         SY=C. cayennensis (Lam.) Britt.
   flexuosus Vahl
         SY=C. vahlii (Nees) Steud.
   floridanus Britt.
         SY=C. filiformis Sw. var.
            densiceps Kukenth.
   fuligineus Chapman
   fuscus L.
   giganteus Vahl
   globulosus Aubl.
         SY=C. multiflorus (Britt.) Small
   granitophilus McVaugh
   grayi Torr.
   grayioides Mohlenbrock
   haspan L.
         SY=C. h. var. americanus Boeckl.
   hermaphroditus (Jacq.) Standl.
         SY=C. dissitiflorus Torr.
         SY=C. thyrsiflorus Schlecht. &
            Cham.
   hillebrandii Boeckl.
      var. decipiens (Hbd.) Kukenth.
      var. helleri Kukenth.
      var. hillebrandii
      var. mauiensis (Hbd.) Kukenth.
   houghtonii Torr.
   huarmensis (H.B.K.) M.C. Johnston
   hypochlorus Hbd.
      var. brevior Kukenth.
      var. densispicatus Skottsberg
      var. hypochlorus
      var. kauaiensis Kukenth.
   imbricatus Retz.
         SY=C. radiatus Vahl
   iria L.
   javanicus Houtt.
   kyllinga Endl.
   laevigatus L.
         SY=C. careyi Britt.
   lancastriensis Porter
   lanceolatus Poir.
      var. compositus J. & C. Presl

      var. lanceolatus
         SY=C. densus Link
   lecontei Torr.
   ligularis L.
   louisianensis Thieret
   lupulinus (Spreng.) Marcks
   lutescens Torr. & Hook.
   luzulae (L.) Retz.
   macrocephalus Liebm.
   manimae H.B.K.
      var. asperrimus (Liebm.) Kukenth.
      var. manimae
   martindalei Britt.
   X mesochorus Geise [houghtonii X
     schweinitzii]
   metzii Mattf. & Kukenth.
   meyenianus Kunth
   mutisii (H.B.K.) Griseb.
      var. asper (Liebm.) Kukenth.
      var. mutisii
         SY=C. incompletus (Jacq.) Link
   nanus Willd.
      var. nanus
         SY=C. granularis (Desf.) Britt.
      var. subtenuis Kukenth.
         SY=C. lentiginosus Millsp. & Chase
         SY=C. tenuis Sw.
   neo-kunthianus Kukenth.
   X nieuwlandii Geise [flavescens X
     rivularis]
   niger Ruiz & Pavon
      var. capitatus (Britt.) O'Neill
         SY=C. melanostachys H.B.K.
   obtusatus (Presl) Mattf. & Kukenth.
         SY=Kyllinga pungens Link
   ochraceus Vahl
   odoratus L.
      var. acicularis (Schrad. ex Nees)
        O'Neill
      var. odoratus
         SY=C. ferax L.C. Rich.
         SY=C. ferruginescens Boeckl.
         SY=C. longispicatus J.B.S. Norton
         SY=C. speciosus Vahl
   onerosus M.C. Johnston
   ovularis (Michx.) Torr.
         SY=C. o. var. sphaericus Boeckl.
   oxylepis Nees ex Steud.
   papyrus L.
   parishii Britt.
   pennatiformis Kukenth.
      var. bryanii Kukenth.
      var. pennatiformis
   peruvianus (Lam.) F.N. Williams
         SY=Kyllinga p. Lam.
   phaeolepis Cherm.
   phleoides (Nees ex Steud.) H. Mann
      var. hawaiensis (Mann) Kukenth.
      var. phleoides
   pilosus Vahl
   planifolius L.C. Rich.
         SY=C. brunneus Sw.
   plukenetii Fern.
   polystachyos Rottb.
      var. filicinus Vahl
      var. laxiflorus Benth.
      var. miser Kukenth.
      var. pallidus Hbd.
      var. paniculatus (Rottb.) C.B. Clarke
         SY=C. p. Rottb.
      var. polystachyos
      var. texensis (Torr.) Fern.
         SY=C. microdontus Torr.
   pringlei Britt.
   prolixus Humb. & Kunth
   pseudovegetus Steud.
         SY=C. robustus Kunth

Cyperus (CONT.)
    pumilus L.
    reflexus Vahl
    refractus Engelm. ex Steud.
    regiomontanus Britt.
    retrofractus (L.) Torr.
        SY=C. r. var. hystricinus (Fern.)
            Kukenth.
        SY=C. h. Fern.
    retrorsus Chapman
        var. deeringianus (Britt. ex Small)
            Fern. ex Grisc.
        SY=C. d. Britt. ex Small
        SY=C. pollardii Britt. ex Small
        SY=C. winkleri Britt. ex Small
        var. retrorsus
        SY=C. nashii Britt.
        SY=C. ovularis (Michx.) Torr. var.
            cylindricus (Ell.) Torr.
        SY=C. r. var. n. (Britt.) Fern. &
            Grisc.
        SY=C. torreyi Britt.
        var. robustus (Boeckl.) Kukenth.
        SY=C. globulosus Aubl. var.
            robustus (Boeckl.) Shinners
    rivularis Kunth
        SY=C. niger Ruiz & Pavon var.
            castaneus (Pursh) Kukenth.
        SY=C. n. var. r. (Kunth) V. Grant
    rockii Kukenth.
    rotundus L.
    sandwicensis Kukenth.
        var. pseudo-prescottianus Kukenth.
        var. sandwicensis
    schweinitzii Torr.
        SY=C. s. var. uberior Kukenth.
        SY=Mariscus s. (Torr.) T. Koyama
    serotinus Rottb.
    seslerioides H.B.K.
    sesquiflorus (Torr.) Mattf. & Kukenth.
        SY=Kyllinga odorata Vahl
    setigerus Torr. & Hook.
        SY=C. hallii Britt.
        SY=C. praelongatus Steud.
    spectabilis Link
        SY=C. buckleyi Britt.
    sphacelatus Rottb.
    strigosus L.
        var. multiflorus Geise
        var. robustior Britt.
        var. stenolepis (Torr.) Kukenth.
        SY=C. hansenii Britt.
        SY=C. s. Torr.
        var. strigosus
    surinamensis Rottb.
    swartzii (Dietr.) Boeckl. & Kukenth.
    tenuifolius (Steud.) Dandy
        SY=Kyllinga pumila Michx.
    tetragonus Ell.
    trachysanthos Hook. & Arn.
    trinervis R. Br.
    uniflorus Boeckl.
        var. pseudothyrsiflorus Kukenth.
        var. uniflorus
    unioloides R. Br.
    urbanii Boeckl.
    virens Michx.
    wrightii Britt.

Dichromena Michx.

    colorata (L.) A.S. Hitchc.
        SY=Rhynchospora c. (L.) H.
            Pfeiffer
    floridensis Britt.
    latifolia Baldw. ex Ell.
    nivea (Boeckl.) Britt.

        SY=Rhynchospora n. Boeckl.

Dulichium L.C. Rich. ex Pers.

    arundinaceum (L.) Britt.

Eleocharis R. Br.

    acicularis (L.) Roemer & Schultes
        var. acicularis
            SY=E. a. var. occidentalis Svens.
            SY=E. a. var. typica Svens.
            SY=E. reverchonii Svens.
        var. gracilescens Svens.
        var. submersa (Hj. Nilss.) Svens.
    acutisquamata Buckl.
    albida Torr.
    atropurpurea (Retz.) Kunth
    austrotexana M.C. Johnston
    baldwiniana (Schultes) Torr.
    baldwinii (Torr.) Chapman
        SY=E. capillacea sensu auctt. non
            Kunth
        SY=E. prolifera Torr.
    bella (Piper) Svens.
        SY=E. acicularis (L.) Roemer &
            Schultes var. b. Piper
    bernardina Munz & Johnston
        SY=E. pauciflora (Lightf.) Link
            var. b. (Munz & Johnston)
            Svens.
    bolanderi Gray
        SY=E. montevidensis Kunth var. b.
            (Gray) V. Grant
    brachycarpa Svens.
    brittonii Svens. ex Small
        SY=E. microcarpa Torr. var. b.
            (Svens. ex Small) Svens.
    cancellata S. Wats.
    caribaea (Rottb.) Blake
    cellulosa Torr.
    cylindrica Buckl.
    decumbens C.B. Clarke
        SY=E. montevidensis Kunth var. d.
            (C.B. Clarke) V. Grant
    elegans (H.B.K.) Roemer & Schultes
    elliptica Kunth
        var. elliptica
            SY=E. acuminata (Muhl.) Nees
            SY=E. capitata (L.) R. Br. var.
                borealis Svens.
            SY=E. compressa Sullivant
            SY=E. compressa var. atrata Svens.
            SY=E. e. var. compressa
                (Sullivant) Drapalik &
                Mohlenbrock
            SY=E. tenuis (Willd.) Schultes
                var. atrata (Svens.) Boivin
            SY=E. t. var. b. (Svens.) Gleason
        var. pseudoptera (Weatherby) L. Harms
            SY=E. capitata (L.) R. Br. var. p.
                Weatherby
            SY=E. tenuis (Willd.) Schultes
                var. p. (Weatherby) Svens.
    elongata Chapman
    engelmannii Steud.
        SY=E. e. var. monticola (Fern.)
            Svens.
        SY=E. e. var. robusta Fern.
        SY=E. obtusa (Willd.) Schultes
            var. e. (Steud.) Gilly
    equisetoides (Ell.) Torr.
    erythropoda Steud.
        SY=E. calva Torr.
    fallax Weatherby
        SY=E. ambigens Fern.
    fistulosa (Poir.) Schultes

Eleocharis (CONT.)
    flavescens (Poir.) Urban
        var. flavescens
            SY=E. flaccida (Reichenb.) Urban
            SY=E. flaccida var. fuscescens
                Kukenth.
            SY=E. praticola Britt.
        var. thermalis (Rydb.) Cronq.
    geniculata (L.) Roemer & Schultes
    halophila Fern. & Brack.
    intermedia Schultes
    interstincta (Vahl) Roemer & Schultes
    kamtschatica (C.A. Mey.) Komarov
    lanceolata Fern.
            SY=E. obtusa (Willd.) Schultes
                var. l. (Fern.) Gilly
    macrostachya Britt.
            SY=E. calva Torr. var. australis
                (Nees) St. John
            SY=E. mamillata sensu auctt. non
                Lindb. f.
            SY=E. palustris (L.) Roemer &
                Schultes var. a. Nees
            SY=E. xyridiformis Fern. & Brack.
    melanocarpa Torr.
    microcarpa Torr.
            SY=E. m. var. filiculmis Torr.
            SY=E. torreyana Boeckl.
    minima Kunth
            SY=E. bicolor Chapman
            SY=E. m. var. ambigua (Steud.)
                Kukenth.
            SY=E. m. var. b. (Chapman) Svens.
            SY=E. uncialis Chapman ex Small
    montana (H.B.K.) Roemer & Schultes
            SY=E. m. var. nodulosa (Roth)
                Svens.
            SY=E. n. (Roth) Schultes
    montevidensis Kunth
            SY=E. arenicola Torr.
    mutata (L.) Roemer & Schultes
    nana Kunth
    nigrescens (Nees) Steud.
            SY=E. carolina Small
            SY=E. minutiflora Boeckl.
            SY=E. n. var. m. (Boeckl.) Svens.
    nitida Fern.
    obtusa (Willd.) Schultes
        var. detonsa (Gray) Drapalik &
            Mohlenbrock
        var. obtusa
            SY=E. diandra C. Wright
            SY=E. macounii Fern.
            SY=E. o. var. ellipsoidalis Fern.
            SY=E. o. var. gigantea (C.B.
                Clarke) Fern.
            SY=E. o. var. jejuna Fern.
            SY=E. o. var. peasei Svens.
            SY=E. ovata (Roth) Roemer &
                Schultes var. obtusa
                (Willd.) Kukenth. ex
                Skottsberg
        var. ovata (Roth) Drapalik &
            Mohlenbrock
            SY=E. ovata (Roth) Roemer &
                Schultes
            SY=E. ovata var. heuseri Uechtr.
    olivacea Torr.
            SY=E. flavescens (Poir.) Urban
                var. o. (Torr.) Gleason
    pachycarpa Desv.
    pachystyla (C. Wright) C.B. Clarke
    palmeri Svens.
    palustris (L.) Roemer & Schultes
    parishii Britt.
            SY=E. disciformis Parish
            SY=E. montevidensis Kunth var. d.

            (Parish) V. Grant
            SY=E. m. var. p. (Britt.) V. Grant
    parvula (Roemer & Schultes) Link ex Buff. &
        Fingerh.
            SY=E. coloradoensis (Britt.) Gilly
            SY=E. leptos (Steud.) Svens.
            SY=E. l. var. c. (Britt.) Svens.
            SY=E. l. var. johnstonii Svens.
            SY=E. p. var. anachaeta (Torr.)
                Svens.
            SY=E. p. var. c. (Britt.) Beetle
    pauciflora (Lightf.) Link
            SY=E. p. var. fernaldii Svens.
            SY=E. p. var. suksdorfiana
                (Beauv.) Svens.
            SY=E. quinqueflora (F.X. Hartmann)
                Schwarz ssp. f. (Svens.)
                Hulten
            SY=E. q. ssp. s. (Beauv.) Hulten
            SY=Scirpus nanus Spreng.
            SY=S. p. Lightf.
    quadrangulata (Michx.) Roemer & Schultes
            SY=E. q. var. crassior Fern.
    quinqueflora (F.X. Hartmann) Schwarz
            SY=Scirpus q. F.X. Hartmann
    radicans (Poir.) Kunth
            SY=E. acicularis (L.) Schultes &
                Roemer var. r. (Poir.)
                Britt.
    retroflexa (Poir.) Urban
    robbinsii Oakes
    rostellata (Torr.) Torr.
    sintenisii Boeckl.
            SY=E. yunquensis Britt.
    smallii Britt.
            SY=E. palustris (L.) Roemer &
                Schultes var. major Sonder
            SY=E. p. var. vigens Bailey
            SY=E. s. var. m. (Sonder) Seymour
    tenuis (Willd.) Schultes
            SY=E. capitata L.
    tortilis (Link) Roemer & Schultes
            SY=E. simplex (Ell.) A. Dietr.
    tricostata Torr.
    tuberculosa (Michx.) Roemer & Schultes
            SY=E. t. var. pubnicoensis Fern.
    uniglumis (Link) Schultes
    verrucosa (Svens.) L. Harms
            SY=E. capitata (L.) R. Br. var. v.
                Svens.
            SY=E. tenuis (Willd.) Schultes
                var. v. (Svens.) Svens.
    vivipara Link
            SY=E. curtisii Small
    wolfii Gray

Eriophorum L.

    alpinum L.
            SY=Leucocoma a. (L.) Rydb.
            SY=Scirpus hudsonianus (Michx.)
                Fern.
            SY=Trichophorum a. (L.) Pers.
    altaicum Meinsh.
        var. altaicum
        var. neogaeum Raymond
    angustifolium Honckeny
        ssp. scabriusculum Hulten
        ssp. subarcticum (Vassiljev) Hulten
            var. coloratum Hulten
            var. subarcticum (Vassiljev) Hulten
            SY=E. a. var. giganteus Hulten
            SY=E. a. var. majus Schultz
            SY=E. polystachyon L. pro parte
        ssp. triste (T. Fries) Hulten
            SY=E. t. (T. Fries) Hadac & A.
                Love

Eriophorum (CONT.)
    X beringianum Raymond [angustifolium ssp.
       subarctium X chamissonis]
    brachyantherum Trautv. & Meyer
            var. brachyantherum
                SY=E. opacum (Bjornstr.) Fern.
            var. pellucidum Lepage
    callitrix Cham.
            var. callitrix
            var. moravium Raymond
    chamissonis C.A. Mey.
                SY=E. russeolum Fries ssp.
                   rufescens (E. Anders.) Hyl.
    X churchillianum Lepage [angustifolium ssp.
       scabriusculum X vaginatum ssp. spissum]
    cringer (Gray) Beetle
                SY=Scirpus c. Gray
    gracile W.D.J. Koch
            var. caurianum Fern.
            var. gracile
    X medium E. Anders. [russeolum X
       scheuchzeri]
    X porsildii Raymond [chamissonis X
       vaginatum ssp. spissum]
    X pylaieanum Raymond [russeolum X vaginatum
       ssp. spissum]
    X rousseauianum Raymond [angustifolium X
       scheuchzeri]
    russeolum Fries
            var. albidum W. Nyl.
                SY=E. chamissonis C.A. Mey. var.
                   a. (W. Nyl.) Fern.
            var. majus Sommier
            var. russeolum
    scheuchzeri Hoppe
                SY=E. s. var. tenuifolia Ohwi
    X sorensenii Raymond [angustifolium ssp.
       triste X scheuchzeri]
    tenellum Nutt.
                SY=E. t. var. monticola Fern.
    vaginatum L.
        ssp. spissum (Fern.) Hulten
                SY=E. s. Fern.
                SY=E. v. var. s. (Fern.) Boivin
        ssp. vaginatum
    virginicum L.
    viridicarinatum (Engelm.) Fern.

Fimbristylis Vahl

    annua (All.) Roemer & Schultes
                SY=F. alamosana Fern.
                SY=F. baldwiniana (Schultes) Torr.
    atollensis St. John
    autumnalis (L.) Roemer & Schultes
                SY=F. a. var. mucronulata (Michx.)
                   Fern.
                SY=F. geminata (Nees) Kunth
    caroliniana (Lam.) Fern.
                SY=F. harperi Britt.
    castanea (Michx.) Vahl
    complanata (Retz.) Link
    cymosa R. Br.
    decipiens Kral
    dichotoma (L.) Vahl
                SY=F. diphylla (Retz.) Vahl
                SY=F. laxa Vahl
    ferruginea (L.) Vahl
    harrisii (Britt.) J.E. Adams
    hawaiiensis Hbd.
    inaguensis Britt.
    miliacea (L.) Vahl
    perpusilla Harper
    puberula (Michx.) Vahl
            var. interior (Britt.) Kral
                SY=F. i. Britt.
            var. puberula

                SY=F. anomala Boeckl.
                SY=F. castanea (Michx.) Vahl var.
                   p. (Michx.) Britt.
                SY=F. drummondii (Torr. & Hook.)
                   Boeckl.
    pycnocephala Hbd.
    quinquangularis (Vahl) Kunth
    schoenoides (Retz.) Vahl
    spadicea (L.) Vahl
    spathacea Roth
    thermalis S. Wats.
    tomentosa Vahl
                SY=E. pilosa Vahl
    vahlii (Lam.) Link

Fuirena Rottb.

    breviseta (Coville) Coville
    bushii Kral
    longa Chapman
    pumila (Torr.) Spreng.
    scirpoidea Michx.
    simplex Vahl
            var. aristulata (Torr.) Kral
            var. simplex
    squarrosa Michx.
                SY=F. hispida Ell.
    umbellata Rottb.
                SY=Scirpus fuirena T. Koyama

Gahnia J.R. & G. Forst.

    beecheyi H. Mann
    globosa H. Mann
    kauaiensis Benl
    lanaiensis Deg., Deg. & Kern

Hemicarpha Nees & Arn.

    drummondii Nees
                SY=H. aristulata (Coville) Smyth
                SY=H. micrantha (Vahl) Britt. var.
                   a. Coville
                SY=H. m. var. d. (Nees) Friedland
    micrantha (Vahl) Britt.
                SY=H. m. var. minor (Schrad.)
                   Friedland
                SY=Scirpus micranthus Vahl
    occidentalis Gray

Kobresia Willd.

    myosuroides (Vill.) Fiori & Paol.
                SY=Elyna bellardii (All.) Degl.
                SY=K. b. (All.) Degl. ex Loisel.
    sibirica Turcz.
                SY=K. bellardii (All.) Degl. ex
                   Loisel. var. macrocarpa
                   (Clokey ex Mackenzie)
                   Harrington
                SY=K. hyperborea Porsild
                SY=K. m. Clokey ex Mackenzie
                SY=K. schoenoides (C.A. Mey.)
                   Steud. pro parte
                SY=K. schoenoides var. lepagei
                   (Duman) Boivin
                SY=K. h. var. alaskana Duman
    simpliciuscula (Wahlenb.) Mackenzie
                SY=K. bipartita (All.) Dalla Torre
                SY=K. s. var. americana Duman

Lagenocarpus Nees

    guianensis Lindl. & Nees
                SY=L. portoricensis Britt.

Lipocarpha R. Br.

Lipocarpha (CONT.)
  maculata (Michx.) Torr.
  microcephala Kunth

Machaerina Vahl

    angustifolia (Gaud.) T. Koyama
    gahniiformis (Gaud.) Kern
          SY=Morelotia g. Gaud.
    mariscoides (Gaud.) Kern
      ssp. mariscoides
      ssp. meyenii (Kunth) T. Koyama
    restioides (Sw.) Vahl
          SY=Cladium r. (Sw.) Benth.
          SY=Mariscus r. (Sw.) Vahl

Oreobolus R. Br.

    furcatus Mann

Psilocarya Torr.

    nitens (Vahl) Wood
          SY=Rhynchospora n. (Vahl) Gray
          SY=P. portoricensis Britt.
    rufa Nees
          SY=Rhynchospora r. (Nees) Boeckl.
    schiedeana (Nees) Liebm.
          SY=Rhynchospora eximia (Nees)
             Boeckl.
    scirpoides Torr.
          SY=?P. corymbifera (C. Wright)
             Britt.
          SY=P. s. var. grimesii Fern. &
             Grisc.
          SY=Rhynchospora s. (Torr.) Gray

Remirea Aubl.

    maritima Aubl.
          SY=Cyperus pedunculatus (R. Br.)
             Kern

Rhynchospora Vahl

    alba (L.) Vahl
          SY=R. luquillensis Britt.
    baldwinii Gray
    brachychaeta C. Wright ex Sauvalle
          SY=R. blauneri Britt.
    breviseta (Gale) Channell
          SY=R. oligantha Gray var. b. Gale
    caduca Ell.
          SY=R. patula Gray
    californica Gale
    capillacea Torr.
          SY=R. c. var. laeviseta E.J. Hill
          SY=R. smallii Britt.
    capitellata (Michx.) Vahl
          SY=R. glomerata (L.) Vahl var. c.
             (Michx.) Kukenth.
          SY=R. g. var. leptocarpa Chapman
             ex Britt.
          SY=R. g. var. minor Britt.
          SY=R. l. (Chapman ex Britt.) Small
    careyana Fern.
    cephalantha Gray
        var. attenuata Gale
        var. cephalantha
          SY=R. axillaris (Lam.) Britt.
        var. microcephala (Britt.) Kukenth.
          SY=R. m. Britt.
        var. pleiocephala Fern. & Gale
    chalarocephala Fern. & Gale
        var. angusta Gale
        var. chalarocephala
    chapmanii M.A. Curtis

ciliaris (Michx.) C. Mohr
      SY=R. rappiana Small
compressa Carey ex Chapman
corniculata (Lam.) Gray
      var. corniculata
      var. interior Fern.
corymbosa (L.) Britt.
crinipes Gale
culixa Gale
curtissii Britt. ex Small
cyperoides (Sw.) Mart.
debilis Gale
      SY=R. trichodes C.B. Clarke
decurrens Chapman
divergens Chapman ex M.A. Curtis
elliottii A. Dietr.
      SY=R. schoenoides (Ell.) Wood
fascicularis (Michx.) Vahl
      var. distans (Michx.) Chapman
          SY=R. d. (Michx.) Vahl
      var. fascicularis
fernaldii Gale
filifolia Gray
      SY=R. fuscoides C.B. Clarke
fusca (L.) Ait. f.
gigantea Link
globularis (Chapman) Small
      var. globularis
          SY=R. cymosa Ell.
          SY=R. g. var. recognita Gale
          SY=R. obliterata Gale
      var. pinetorum (Small) Gale
          SY=R. p. Small
glomerata (L.) Vahl
      var. angusta Gale
      var. glomerata
          SY=R. g. var. paniculata Chapman
gracilenta Gray
grayi Kunth
harperi Small
harveyi W. Boott
      SY=R. earlei Britt.
      SY=R. planckii Britt.
hispidula (Vahl) Boeckl.
indianolensis Small
inexpansa (Michx.) Vahl
intermedia (Chapman) Britt.
intermixta C. Wright ex Sauvalle
      SY=R. bruneri Britt.
inundata (Oakes) Fern.
jamaicensis Britt.
knieskernii Carey
lavarum Gaud.
lindeniana Griseb.
      var. bahamensis (Britt.) Gale
          SY=R. b. Britt.
      var. lindeniana
macra (C.B. Clarke) Small
macrostachya Gray
      var. colpophila Fern. & Gale
      var. macrostachya
marisculus Lindl. & Nees
      SY=R. borinquensis Britt.
megalocarpa Gray
      SY=R. dodecandra Baldw. ex Gray
micrantha Vahl
microcarpa Baldw. ex Gray
      SY=R. edisoniana Small
miliacea (Lam.) Gray
mixta Britt. ex Small
      SY=R. prolifera Small
nervosa (Vahl) Boeckl.
      SY=Dichromena ciliata Vahl
odorata C. Wright ex Griseb.
      SY=R. stipitata Chapman
oligantha Gray
pallida M.A. Curtis

Rhynchospora (CONT.)
    perplexa Britt.
            var. perplexa
            var. virginiana Fern.
    pleiantha (Kukenth.) Gale
    plumosa Ell.
            SY=R. semiplumosa Gray
    podosperma C. Wright ex Sauvalle
            SY=R. filiformis Vahl
    polyphlla (Vahl) Vahl
    punctata Ell.
    pusilla Chapman ex M.A. Curtis
            SY=R. berteri (Spreng.) C.B.
                Clarke
    racemosa C. Wright ex Sauvalle
    radicans (Schlecht. & Cham.) N.E. Pfeiffer
            SY=Dichromena r. Schlecht. & Cham.
    rariflora (Michx.) Ell.
    rugosa (Vahl) Gale
            SY=R. glauca Vahl
    saxicola Small
            SY=R. globularis (Chapman) Small
                var. s. (Small) Kukenth.
    sclerioides Hook. & Arn.
    setacea (Berg.) Boeckl.
    sola Gale
    solitaria Harper
    spiciformis Hbd.
    stellata Griseb.
    stenophylla Chapman ex M.A. Curtis
    sulcata Gale
    tenuis Link
    thornei Kral
    torreyana Gray
    tracyi Britt.
    uniflora Boeckl.
            SY=R. elongata Boeckl.
    wrightiana Boeckl.
            SY=R. tenuis Baldw. ex Gray

Schoenus L.

    nigricans L.

Scirpus L.

    acutus Muhl. ex Bigelow
            SY=S. lacustris L. pro parte
            SY=S. occidentalis (S. Wats.)
                Chase
    americanus Pers.
            SY=S. a. ssp. monophyllus (Presl)
                T. Koyama var. m. (Presl) T.
                Koyama
            SY=S. chilensis Nees & Meyen ex
                Kunth
            SY=S. conglomeratus H.B.K.
            SY=S. m. Presl
            SY=S. olneyi Gray
            SY=S. pungens Vahl ssp. m. (Presl)
                Taylor & MacBryde
    ancistrochaetus Schuyler
    atrocinctus Fern.
            SY=S. cyperinus (L.) Kunth var.
                brachypodus (Fern.) Gilly
    atrovirens Willd.
    caespitosus L.
            SY=S. c. var. callosus Bigelow
            SY=S. caespitosus var. delicatulus
                Fern.
            SY=Trichophorum caespitosum (L.)
                Hartman
    californicus (C.A. Mey.) Steud.
            SY=S. c. var. tereticulmis
                (Steud.) Beetle
    cernuus Vahl
            var. californicus (Torr.) Beetle

            SY=S. cernuus ssp. californicus
                (Vahl) Thorne
    clementis M.E. Jones
    clintonii Gray
    confervoides Poir.
            SY=Websteria c. (Poir.) Hooper
            SY=W. submersa (Sauvalle) Britt.
    congdonii Britt.
    X contortus (Eames) T. Koyama [americanus X
        pungens]
    cubensis Poepp. & Kunth
    cylindricus (Torr.) Britt.
            SY=S. novae-angliae Britt.
            SY=S. robustus Pursh var. n-a.
                (Britt.) Beetle
            SY=S. subterminalis Torr. var. c.
                (Torr.) T. Koyama
    cyperinus (L.) Kunth
            SY=S. c. var. condensatus Fern.
            SY=S. c. var. laxus (Gray) Beetle
            SY=S. cyperinus var. pelius Fern.
            SY=S. cyperinus var. eriophorum
                (Michx.) Kuntze
            SY=S. cyperinus var. rubricosus
                (Fern.) Gilly
            SY=S. e. Michx.
            SY=S. r. Fern.
    deltarum Schuyler
    diffusus Schuyler
    divaricatus Ell.
    erectus Poir.
            SY=S. erismanae Schuyler
            SY=S. wilkensii Schuyler
    etuberculatus (Steud.) Kuntze
    expansus Fern.
            SY=S. sylvaticus L. pro parte
    flaccidifolius (Fern.) Schuyler
            SY=S. atrovirens Willd. var. f.
                Fern.
    fluviatilis (Torr.) Gray
    georgianus Harper
            SY=S. atrovirens Willd. var. g.
                (Harper) Fern.
    hallii Gray
            SY=S. supinus sensu auctt. non L.
            SY=S. s. var. h. (Gray) Gray
    hattorianus Makino
    heterochaetus Chase
            SY=S. lacustris L. pro parte
            SY=S. tenuiculmis (Sheldon) Sojak
    holoschoenus L.
    juncoides Roxb.
            SY=S. rockii Kukenth.
    koilolepis (Steud.) Gleason
            SY=S. carinatus (Hook. & Arn.)
                Gray
    lineatus Michx.
            SY=S. fontinalis Harper
    longii Fern.
    maritimus L.
            SY=Bolboschoenus paludosus Soo
            SY=S. fernaldii Bickn.
            SY=S. m. var. f. (Bickn.) Beetle
            SY=S. m. var. p. (A. Nels.)
                Kukenth.
            SY=S. pacificus Britt.
            SY=S. paludosus A. Nels.
            SY=S. paludosus var. atlanticus
                Fern.
    microcarpus Presl
            SY=S. m. var. longispicatus M.E.
                Peck
            SY=S. m. var. rubrotinctus (Fern.)
                M.E. Jones
            SY=S. r. Fern.
    molestus M.C. Johnston
    mucronatus L.

Scirpus (CONT.)
  nevadensis S. Wats.
         SY=Amphiscirpus n. (S. Wats.)
            Oteng. Yeboah
  pallidus (Britt.) Fern.
         SY=S. atrovirens Willd. var. p.
            Britt.
  pedicellatus Fern.
         SY=S. cyperinus (L.) Kunth var. p.
            (Fern.) Schuyler
         SY=S. p. var. pullus Fern.
  pendulus Muhl.
         SY=S. lineatus sensu auctt. non
            Michx.
  polyphyllus Vahl
  pungens Vahl
         SY=S. americanus Pers. var.
            longispicatus Britt.
         SY=S. a. var. polyphyllus
            (Boeckl.) Beetle
         SY=S. p. ssp. monophyllus (Presl)
            Taylor & MacBryde var.
            longisetus Benth. & F.
            Muell.
         SY=S. p. ssp. p. var.
            longispicatus (Britt.)
            Taylor & MacBryde
  purshianus Fern.
         SY=S. debilis Pursh non Lam.
         SY=S. juncoides Roxb. var.
            williamsii (Fern.) T. Koyama
         SY=S. smithii Gray var. setosus
            Fern.
         SY=S. smithii var. w. (Fern.)
            Beetle
  robustus Pursh
         SY=S. maritimus L. var. agonus
            (Fern.) Beetle
         SY=S. m. L. var. macrostachyus
            Michx.
  rollandii Fern.
         SY=S. pumilus Vahl
         SY=S. p. ssp. r. (Fern.) Raymond
         SY=S. p. var. r. (Fern.) Beetle
         SY=Trichophorum p. (Vahl) Schinz &
            Thellung
         SY=T. p. ssp. r. (Fern.) Taylor &
            MacBryde
         SY=T. p. var. r. (Fern.) Hulten
  X rubiginosus Beetle [acutus X
    californicus]
  rufus (Huds.) Schrad.
      var. neogaeus Fern.
         SY=Blysmus r. sensu auctt. non
            (Huds.) Link
  saximontanus Fern.
         SY=S. bergsonii Schuyler
         SY=S. supinus L. var. saximontanus
            (Fern.) T. Koyama
  setaceus L.
  smithii Gray
  subterminalis Torr.
  tabernaemontanii K.C. Gmel.
         SY=S. lacustris L. pro parte
         SY=?S. l. ssp. glaucus (Reichenb.)
            Hartman
         SY=S. l. ssp. creber (Vahl) T.
            Koyama
         SY=S. steinmetzii Fern.
         SY=S. validus Vahl
         SY=S. v. var. creber Fern.
  torreyi Olney
  tuberosus Desf.
         SY=S. maritimus L. var. t. (Desf.)
            Roemer & Schultes
  verecundus Fern.
         SY=S. planifolius Muhl.

Scleria Berg.

  baldwinii (Torr.) Steud.
         SY=S. costata (Britt.) Small
  bourgeaui Boeckl.
  bracteata Cav.
  ciliata Michx.
      var. ciliata
         SY=S. c. var. elliottii (Chapman)
            Fern.
         SY=S. e. Chapman
      var. glabra (Chapman) Fairey
         SY=S. brittonii Core
         SY=S. pauciflora Muhl. ex Willd.
            var. g. Chapman
  cubensis Boeckl.
  eggersiana Boeckl.
         SY=S. grisebachii C.B. Clarke
  georgiana Core
         SY=S. gracilis Ell.
  havanensis Britt.
  hirtella Sw.
         SY=S. distans Poir.
         SY=S. doradoensis Britt.
         SY=S. interrupta L.C. Rich.
  lithosperma (L.) Sw.
  microcarpa Nees ex Kunth
  minor (Britt.) W. Stone
  mitis Berg.
  nutans Willd. ex Kunth
  oligantha Michx.
  pauciflora Muhl. ex Willd.
      var. caroliniana (Willd.) Wood
      var. curtissii (Britt.) Fairey
         SY=S. c. Britt.
      var. pauciflora
         SY=S. p. var. kansana Fern.
  pterota Presl ex C.B. Clarke
      var. melaleuca (Reichenb. ex
         Schlecht. & Cham.) Standl.
         SY=S. m. Reichenb. ex Schlecht. &
            Cham.
      var. pterota
  purdiei C.B. Clarke
  reticularis Michx.
      var. pubescens Britt.
      var. pumila Britt.
      var. reticularis
         SY=S. muhlenbergii Steud.
         SY=S. r. var. anceps (Liebm.)
            Fairey
         SY=S. setacea Poir.
         SY=S. stevensiana Britt.
  scabriuscula Schlecht.
  scindens Nees
  secans (L.) Urban
  tenella Kunth
  testacea Nees
  triglomerata Michx.
         SY=S. flaccida Steud.
         SY=S. nitida Willd.
  verticillata Muhl. ex Willd.

Uncinia Pers.

  hamata (Sw.) Urban
  uncinata (L.f.) Kukenth.
      var. uliginosa Skottsberg
      var. uncinata

CYRILLACEAE

Cliftonia Banks ex Gaertn. f.

Cliftonia (CONT.)
     monophylla (Lam.) Britt. ex Sarg.

Cyrilla Garden ex L.

     racemiflora L.
          var. parvifolia (Shuttlw. ex Nash)
               Sarg.
               SY=C. arida Small
               SY=C. p. Shuttlw. ex Nash
               SY=C. r. var. subglobosa Fern.
          var. racemiflora

DATISCACEAE

Datisca L.

     glomerata (Presl) Baill.

DIAPENSIACEAE

Diapensia L.

     lapponica L.
          ssp. lapponica
          ssp. obovata (F. Schmidt) Hulten
               var. obovata F. Schmidt
                    SY=D. o. (F. Schmidt) Nakai
               var. rosea Hulten
                    SY=D. l. var. r. Hulten

Galax Raf.

     urceolata (Poir.) Brummitt
          SY=G. aphylla sensu auctt. non L.

Pyxidanthera Michx.

     barbulata Michx.
          SY=P. b. var. brevifolia (Wells)
               Ahles
          SY=P. brevifolia Wells

Shortia Torr. & Gray

     galacifolia Torr. & Gray
          var. brevistyla Davies
          var. galacifolia
               SY=Sherwoodia g. (Torr. & Gray)
                    House

DILLENIACEAE

Doliocarpus Roland.

     calinoides (Eichl.) Gilg
     major J.F. Gmel.
          SY=D. brevipedicellatus Garcke

DIOSCOREACEAE

Dioscorea L.

     alata L.
     altissima Lam.

batatas Dcne.
bulbifera L.
floridana Bartlett
          SY=D. villosa L. var. f.
               (Bartlett) Ahles
hirticaulis Bartlett
pentaphylla L.
pilosiuscula Bertero ex Spreng.
polygonoides Humb. & Bonpl. ex Willd.
quaternata (Walt.) J.F. Gmel.
          SY=D. glauca Muhl.
          SY=D. q. var. g. (Muhl.) Fern.
trifida L. f.
villosa L.
     var. glabrifolia (Bartlett) Fern.
     var. villosa
          SY=D. cayennensis Lam.

Rajania L.

     cordata L.

DIPSACACEAE

Cephalaria Schrad. ex Roemer & Schultes

     alpina (L.) Roemer & Schultes
     syriaca (L.) Roemer & Schultes

Dipsacus L.

     fullonum L.
          SY=D. sylvestris Huds.
     laciniatus L.
     sativus (L.) Honckeny

Knautia L.

     arvensis (L.) T. Coult.
          SY=Scabiosa a. L.

Scabiosa L.

     atropurpurea L.
     columbaria L.
     ochroleuca L.
     stellata L.

Succisa Haller.

     pratensis Moench
          SY=Scabiosa succisa L.

Succisella G. Beck

     inflexa (Kluk) G. Beck
          SY=Scabiosa australis Wulfen
          SY=Succisa a. (Wulfen) Reichenb.

DROSERACEAE

Dionaea Ell.

     muscipula Ell.

Drosera L.

     anglica Huds.
          SY=D. longifolia L.
     brevifolia Pursh
          SY=D. annua E.L. Reed

Drosera (CONT.)
          SY=D. leucantha Shinners
     capillaris Poir.
     filiformis Raf.
          var. filiformis
          var. tracyi (Macfarlane) Diels
               SY=D. t. Macfarlane
     X hybrida Macfarlane [filiformis X
          intermedia]
     intermedia Hayne
     linearis Goldie
     X obovata Mert. & Koch [anglica X
          rotundifolia]
     rotundifolia L.
          var. comosa Fern.
          var. gracilis Laestad.
          var. rotundifolia

EBENACEAE

Diospyros L.

     blancoi A. DC.
          SY=D. discolor Willd.
          SY=D. philippensis (Desr.) Guerke
     ferrea (Willd.) Bakh.
          SY=D. f. ssp. sandwicensis (A.
               DC.) Fosberg
          SY=D. f. ssp. s. var. s. (A. DC.)
               Bakh.
          SY=D. f. var. degeneri Fosberg
          SY=D. f. var. kauaiensis Fosberg
          SY=D. f. var. pubescens
               (Skottsberg) Fosberg
          SY=D. f. var. toppingii Fosberg
     hillebrandii (Seem.) Fosberg
     revoluta Poir.
          SY=D. ebenaster sensu auctt. non
               Retz.
     sintenisii (Krug & Urban) Standl.
          SY=Maba s. Krug & Urban
     texana Scheele
     virginiana L.
          SY=D. mosieri Small
          SY=D. v. var. m. (Small) Sarg.
          SY=D. v. var. platycarpa Sarg.
          SY=D. v. var. pubescens (Pursh)
               Dippel

ELAEAGNACEAE

Elaeagnus L.

     angustifolia L.
     commutata Bernh. ex Rydb.
          SY=E. argentea Pursh non Moench
     multiflora Thunb.
     orientalis L.
     pungens Thunb.
     umbellata Thunb.
          var. parvifolia (Royle) Schneid.
               SY=E. p. Royle

Hippophae L.

     rhamnoides L.

Shepherdia Nutt.

     argentea (Pursh) Nutt.
          SY=Elaeagnus utilis A. Nels.

          SY=Lepargyraea a. (Pursh) Greene
     canadensis (L.) Nutt.
          SY=Elaeagnus c. (L.) A. Nels.
          SY=Lepargyraea c. (L.) Greene
     rotundifolia Parry

ELAEOCARPACEAE

Elaeocarpus L.

     bifidus Hook. & Arn.

Muntingia L.

     calabura L.

Sloanea L.

     amygdalina Griseb.
     berteriana Choisy

ELATINACEAE

Bergia L.

     texana (Hook.) Walp.

Elatine L.

     ambigua Wight
     americana (Pursh) Arn.
          SY=E. triandra Schkuhr var. a.
               (Pursh) Fassett
     brachysperma Gray
          SY=E. triandra Schkuhr var. b.
               (Gray) Fassett
     californica Gray
     chilensis Gray
     minima (Nutt.) Fisch. & Mey.
     obovata (Fassett) Mason
     rubella Rydb.
     triandra Schkuhr
          SY=E. gracilis Mason

EMPETRACEAE

Ceratiola Michx.

     ericoides Michx.

Corema D. Don

     conradii (Torr.) Torr. ex Loud.

Empetrum L.

     eamesii Fern. & Wieg.
          ssp. atropurpureum (Fern. & Wieg.) D.
          Love
               SY=E. a. Fern. & Wieg.
               SY=E. nigrum L. var. a. (Fern. &
                    Wieg.) Boivin
               SY=E. n. var. purpureum (Raf.) DC.
               SY=E. rubrum Vahl var. a. (Fern. &
                    Wieg.) R. Good
          ssp. eamesii
               SY=E. nigrum L. var. e. (Fern. &
                    Wieg.) Boivin

Empetrum   (CONT.)
          SY=E. rubrum Vahl
          SY=E. r. ssp. e. (Fern. & Wieg.)
              R. Good
   nigrum L.
      ssp. hermaphroditum (Lange) Bocher
          SY=E. eamesii Fern. & Wieg. ssp.
              h. (Lange) D. Love
          SY=E. h. (Lange) Hagerup
      ssp. nigrum

EPACRIDACEAE

Styphelia Sm.

   douglasii (Gray) F. Muell. ex Skottsberg
   tameiameiae (Cham.) F. Muell.
       var. brownii (Gray) St. John
       var. hexamera Fosberg & Hosaka
       var. tameiameiae

ERICACEAE

Allotropa Torr. & Gray ex Gray

   virgata Torr. & Gray ex Gray

Andromeda L.

   polifolia L.
       var. glaucophylla (Link) DC.
          SY=A. g. Link
          SY=A. p. ssp. g. (Link) Hulten
       var. jamesiana (Lepage) Boivin
       var. polifolia
          SY=A. p. var. concolor Boivin

Arbutus L.

   arizonica (Gray) Sarg.
   menziesii Pursh
   texana Buckl.

Arctostaphylos Adans.

   ?acutifolia Eastw.
   alpina (L.) Spreng.
          SY=Mairania a. (L.) Desv.
   andersonii Gray
   auriculata Eastw.
   australis Eastw.
   canescens Eastw.
          SY=A. candidissima Eastw.
          SY=A. canescens var. candidissima
              (Eastw.) Munz
          SY=A. canescens var. sonomensis
              (Eastw.) J.E. Adams ex
              McMinn
   catalinae P.V. Wells
   cinerea T.J. Howell
   X coloradensis Rollins
          SY=A. nevadensis Gray var. c.
              (Rollins) Harrington
   columbiana Piper
          SY=A. c. var. tracyi (Eastw.) J.E.
              Adams
          SY=A. t. Eastw.
   confertiflora Eastw.
          SY=A. subcordata Eastw. var. c.
              (Eastw.) Munz
   cruzensis J.B. Roof

densiflora M.S. Baker
edmundsii J.T. Howell
          SY=A. e. var. parvifolia J.B. Roof
glandulosa Eastw.
   ssp. adamsii (Munz) Munz
          SY=A. g. var. a. Munz
   ssp. crassifolia (Jepson) P.V. Wells
          SY=A. g. var. c. Jepson
   ssp. glandulosa
          SY=A. g. var. campbellae (Eastw.)
              J.E. Adams ex McMinn
          SY=A. g. var. cushingiana (Eastw.)
              J.E. Adams ex McMinn
          SY=A. intricata T.J. Howell
          SY=A. i. var. oblongifolia (T.J.
              Howell) Munz
          SY=A. o. T.J. Howell
   ssp. howellii (Eastw.) P.V. Wells
          SY=A. g. var. h. (Eastw.) J.E.
              Adams ex McMinn
   ssp. mollis (J.E. Adams) P.V. Wells
          SY=A. g. var. m. J.E. Adams
   ssp. zacaensis (Eastw.) P.V. Wells
          SY=A. g. var. z. (Eastw.) J.E.
              Adams ex McMinn
glauca Lindl.
          SY=A. g. var. puberula J.T. Howell
glutinosa Schreib.
hispidula T.J. Howell
          SY=A. h. var. viscosissima M.E.
              Peck
          SY=A. stanfordiana Parry ssp. h.
              (T.J. Howell) J.E. Adams
hookeri D. Don
   ssp. franciscana (Eastw.) Munz
          SY=A. f. Eastw.
   ssp. hearstiorum (Hoover & J.B. Roof)
       P.V. Wells
          SY=A. hearstiorum Hoover & J.B.
              Roof
   ssp. hookeri
   ssp. montana (Eastw.) P.V. Wells
          SY=A. m. Eastw.
          SY=A. pungens H.B.K. var. m.
              (Eastw.) Munz
   ssp. ravenii P.V. Wells
hooveri P.V. Wells
imbricata Eastw.
          SY=A. andersonii Gray var. i.
              (Eastw.) J.E. Adams ex
              McMinn
insularis Greene
          SY=A. i. var. pubescens Eastw.
luciana P.V. Wells
manzanita Parry
   ssp. bakeri (Eastw.) P.V. Wells
          SY=A. b. Eastw.
          SY=A. stanfordiana Parry ssp. b.
              (Eastw.) J.E. Adams
   ssp. elegans (Eastw.) P.V. Wells
          SY=A. e. Eastw.
   ssp. laevigata (Eastw.) Munz
          SY=A. l. Eastw.
   ssp. manzanita
   ssp. roofii (Gankin) P.V. Wells
          SY=A. r. Gankin
   ssp. wieslanderi P.V. Wells
X media Greene [columbiana X uva-ursi]
mewukka Merriam
montereyensis Hoover
morroensis Wies. & Schreib.
myrtifolia Parry
nevadensis Gray
          SY=A. parvifolia T.J. Howell
nissenana Merriam
nummularia Gray
   ssp. nummularia

Arctostaphylos (CONT.)
      ssp. sensitiva (Jepson) P.V. Wells
            SY=A. n. var. s. (Jepson) McMinn
  obispoensis Eastw.
  otayensis Wies. & Schreib.
  ?pacifica J.B. Roof
  pajaroensis J.E. Adams
  pallida Eastw.
            SY=A. andersonii Gray var. p.
                (Eastw.) J.E. Adams ex
                McMinn
  parryana Lemmon
  patula Greene
      ssp. patula
      ssp. platyphylla (Gray) P.V. Wells
            SY=A. parryana Lemmon var.
                pinetorum (Rollins) Wies. &
                Schreib.
  pechoensis Dudley ex Abrams
  peninsularis P.V. Wells
  pilosula Jepson & Wies.
      ssp. pilosula
      ssp. pismoensis P.V. Wells
  pringlei Parry
      ssp. drupacea (Parry) P.V. Wells
            SY=A. d. (Parry) J.F. Macbr.
            SY=A. p. var. d. Parry
      ssp. pringlei
  pumila Nutt.
  pungens H.B.K.
  purissima P.V. Wells
  refugioensis Gankin
  regismontana Eastw.
  X repens (J.T. Howell) P.V. Wells
  rubra (Rehd. & Wilson) Fern.
            SY=A. alpina (L.) Spreng. var. r.
                (Rehd. & Wilson) Bean
            SY=Arctous erythrocarpa Small
  rudis Jepson & Wies.
  silvicola Jepson & Wies.
  stanfordiana Parry
  tomentosa (Pursh) Lindl.
      ssp. bracteosa (DC.) J.E. Adams
            SY=Arctostaphylos b. (DC.) Abrams
            SY=A. crustacea Eastw. var.
                tomentosiformis J.E. Adams
            SY=A. tomentosa var.
                tomentosiformis (J.E. Adams)
                Munz
      ssp. crinita (J.E. Adams ex McMinn)
          Gankin
      ssp. crustacea (Eastw.) P.V. Wells
            SY=Arctostaphylos c. Eastw.
            SY=A. c. var. rosei (Eastw.)
                McMinn
            SY=A. r. Eastw.
      ssp. eastwoodiana P.V. Wells
      ssp. insulicola P.V. Wells
      ssp. rosei (Eastw.) P.V. Wells
            SY=Arctostaphylos crustacea Eastw.
                var. r. (Eastw.) McMinn
            SY=A. r. Eastw.
      ssp. subcordata (Eastw.) P.V. Wells
            SY=Arctostaphylos s. Eastw.
      ssp. tomentosa
            SY=Arctostaphylos bracteosa (DC.)
                Abrams var. hebeclada (DC.)
                Eastw.
            SY=A. t. ssp. t. var. h. (DC.)
                Munz
            SY=A. t. var. trichoclada (DC.)
                Munz
  uva-ursi (L.) Spreng.
      ssp. adenotricha (Fern. & J.F. Macbr.)
          Calder & Taylor
            SY=Arctostaphylos u-u. var. a.
                Fern. & J.F. Macbr.

      ssp. coactilis (Fern. & J.F. Macbr.)
          Love, Love & Kapoor
            SY=Arctostaphylos u-u. var. c.
                Fern. & J.F. Macbr.
            SY=A. u-u. var. pacifica Hulten
      ssp. stipitata Packer & Denford
      ssp. uva-ursi
            SY=Uva-ursi uva-ursi (L.) Britt.
  virgata Eastw.
  viridissima (Eastw.) McMinn
            SY=Arctostaphylos pechoensis
                Dudley ex Abrams var. v.
                Eastw.
  viscida Parry
      ssp. mariposa (Dudley) P.V. Wells
            SY=Arctostaphylos m. Dudley
            SY=A. m. var. bivisa Jepson
      ssp. pulchella (T.J. Howell) P.V. Wells
            SY=Arctostaphylos p. T.J. Howell
      ssp. viscida

Befaria Mutis ex L.

  racemosa Vent.

Calluna Salisb.

  vulgaris (L.) Hull

Cassiope D. Don

  hypnoides (L.) D. Don
            SY=Harrimanella h. (L.) Coville
  lycopodioides (Pallas) D. Don
      ssp. cristipilosa Calder & Taylor
            SY=C. l. var. c. (Calder & Taylor)
                Boivin
      ssp. lycopodioides
  mertensiana (Bong.) D. Don
      var. gracilis (Piper) C.L. Hitchc.
      var. mertensiana
  stelleriana (Pallas) DC.
            SY=Harrimanella s. Pallas
  tetragona (L.) D. Don
      ssp. saximontana (Small) Porsild
            SY=C. t. var. s. (Small) C.L.
                Hitchc.
      ssp. tetragona

Chamaedaphne Moench

  calyculata (L.) Moench
      var. angustifolia (Ait.) Rehd.
            SY=Cassandra c. (L.) D. Don var.
                a. (Ait.) Seymour
      var. latifolia (Ait.) Fern.
            SY=Cassandra c. (L.) D. Don var.
                l. (Ait.) Seymour
      var. nana (Lodd.) E. Busch

Chimaphila Pursh

  maculata (L.) Pursh
      var. dasystemma (Torr.) Kearney &
          Peebles
      var. maculata
  menziesii (R. Br. ex D. Don) Spreng.
  umbellata (L.) Bart.
      ssp. acuta (Rydb.) Hulten
            SY=C. u. var. a. (Rydb.) Blake
      ssp. cisatlantica (Blake) Hulten
            SY=C. corymbosa Pursh
            SY=C. u. var. cisatlanticum Blake
      ssp. occidentalis (Rydb.) Hulten
            SY=C. o. Rydb.
            SY=C. u. var. o. (Rydb.) Blake

Cladothamnus Bong.

    pyroliflorus Bong.

Comarostaphylis Zucc.

    diversifolia (Parry) Greene
       ssp. diversifolia
               SY=Arctostaphylos d. (Parry) Parry
       ssp. planifolia (Jepson) G. Wallace ex
          Thorne
               SY=C. d. var. p. Jepson

Elliottia Muhl. ex Ell.

    racemosa Muhl. ex Ell.

Epigaea L.

    repens L.
               SY=E. r. var. glabrifolia Fern.

Erica L.

    cinerea L.
    lusitanica K. Rudolphi
    tetralix L.
    vagans L.

Gaultheria L.

    hispidula (L.) Muhl. ex Bigelow
               SY=Chiogenes h. (L.) Torr. & Gray
    humifusa (Graham) Rydb.
    miqueliana Takeda
    ovatifolia Gray
    procumbens L.
    shallon Pursh

Gaylussacia H.B.K.

    baccata (Wang.) K. Koch
               SY=Decachaena b. (Wang.) Small
    brachycera (Michx.) Gray
               SY=Buxella b. (Michx.) Small
    dumosa (Andr.) Torr. & Gray
          var. bigeloviana Fern.
               SY=Lasiococcus d. (Andr.) Small
                      var. b. (Fern.) Fern.
          var. dumosa
               SY=Lasiococcus d. (Andr.) Small
    frondosa (L.) Torr. & Gray ex Torr.
               SY=Decachaena f. (L.) Torr. & Gray
    mosieri Small
               SY=Lasiococcus m. (Small) Small
    nana (Gray) Small
               SY=Decachaena n. (Gray) Small
               SY=G. frondosa (L.) Torr. & Gray
                      var. n. Gray
    orocola Small
               SY=Lasiococcus o. (Small) Small
    tomentosa (Gray) Small
               SY=G. frondosa (L.) Torr. & Gray
                      var. t. Gray
               SY=Decachaena t. (Gray) Small
    ursina (M.A. Curtis) Torr. & Gray ex Gray
               SY=Decachaena u. (M.A. Curtis)
                      Small

Gonocalyx Planch. & Linden ex Lindl.

    concolor Nevl.
    portoricensis (Urban) A.C. Sm.
               SY=Ceratostema p. (Urban) Hoerold

Hemitomes Gray

congestum Gray

Kalmia L.

    angustifolia L.
          var. angustifolia
          var. carolina (Small) Fern.
               SY=K. c. Small
    cuneata Michx.
    hirsuta Walt.
               SY=Kalmiella h. (Walt.) Small
    latifolia L.
               SY=Kalmia l. var. laevipes Fern.
    microphylla (Hook.) Heller
          var. microphylla
               SY=Kalmia polifolia Wang. ssp. m.
                  (Hook.) Calder & Taylor
               SY=K. p. var. m. (Hook.) Rehd.
          var. occidentalis (Small) Ebinger
               SY=Kalmia m. ssp. o. (Small)
                  Taylor & MacBryde
               SY=K. o. Small
    polifolia Wang.
               SY=Kalmia p. var. rosmarinifolia
                  (Dum.-Cours.) Rehd.

Kalmiopsis Rehd.

    leachiana (Henderson) Rehd.

Ledum L.

    X columbianum Piper [glandulosum X
       groenlandicum]
               SY=L. glandulosum Nutt. ssp.
                  australe C.L. Hitchc.
               SY=L. g. ssp. c. (Piper) C.L.
                  Hitchc.
               SY=L. g. ssp. c. var. a. C.L.
                  Hitchc.
               SY=L. g. ssp. olivaceum C.L.
                  Hitchc.
               SY=L. g. var. c. (Piper) C.L.
                  Hitchc.
    glandulosum Nutt.
          var. californicum (Kellogg) C.L.
             Hitchc.
               SY=L. g. ssp. g. var. c. (Kellogg)
                  C.L. Hitchc.
          var. glandulosum
    groenlandicum Oeder
               SY=L. palustre L. ssp. g. (Oeder)
                  Hulten
               SY=L. p. var. latifolium (Jacq.)
                  Michx.
    palustre L.
       ssp. decumbens (Ait.) Hulten
               SY=L. d. (Ait.) Lodd. ex Steud.
               SY=L. p. L. var. d. Ait.

Leiophyllum (Pers.) Hedw. f.

    buxifolium (Berg.) Ell.
               SY=Dendrium b. (Berg.) Desv.
               SY=L. b. var. hugeri (Small)
                  Schneid.
               SY=L. b. var. prostratum (Loud.)
                  Gray
               SY=L. h. (Small) K. Schum.
               SY=L. lyonii Sweet

Leucothoe D. Don

    axillaris (Lam.) D. Don
               SY=L. a. var. ambigens Fern.
               SY=L. axillaris var. catesbaei
                  (Walt.) Gray

Leucothoe (CONT.)
          SY=L. c. Walt.
     davisiae Torr. ex Gray
     fontanesiana (Steud.) Sleumer
          SY=L. axillaris (Lam.) D. Don var.
               editorum (Fern. & Schub.)
               Ahles
          SY=L. e. Fern. & Schub.
     populifolia (Lam.) Dippel
          SY=L. acuminata (Ait.) D. Don
          SY=L. walteri (Willd.) Melvin
     racemosa (L.) Gray
          SY=Eubotrys elongata (Small) Small
          SY=E. r. var. e. (Small) Fern.
          SY=E. r. (L.) Nutt.
          SY=L. e. Small
          SY=L. r. var. projecta Fern.
     recurva (Buckl.) Gray
          SY=Eubotrys r. (Buckl.) Britt.

Loiseleuria Desv.

     procumbens (L.) Desv.
          SY=Azalea p. L.
          SY=Chamaecistus p. (L.) Kuntze

Lyonia Nutt.

     ferruginea (Walt.) Nutt.
          SY=Xolisma f. (Walt.) Heller
     fruticosa (Michx.) G.S. Torr. ex B.L.
       Robins.
          SY=Xolisma f. (Michx.) Nash
     ligustrina (L.) DC.
          var. foliosiflora (Michx.) Fern.
          SY=Arsenococcus frondosus (Pursh)
               Small
          SY=Xolisma foliosiflora (Michx.)
               Small
          SY=Xolisma l. (L.) Britt. var.
               foliosiflora (Michx.) C.
               Mohr
          var. ligustrina
          SY=Arsenococcus l. (L.) Small
          SY=L. l. var. capreifolia (S.
               Wats.) DC.
          SY=L. l. var. salicifolia (S.
               Wats.) DC.
          SY=Xolisma l. (L.) Britt.
     lucida (Lam.) K. Koch
          SY=Desmothamnus l. (Lam.) Small
          SY=Neopieris nitida (Bartr.)
               Britt.
     mariana (L.) D. Don
          SY=Neopieris m. (L.) Britt.
          SY=Xolisma m. (L.) Rehd.
     rubiginosa (Pers.) G. Don
          SY=Xolisma r. (Pers.) Small
          SY=X. stahlii (Urban) Small

Menziesia Sm.

     ferruginea Sm.
          SY=M. f. ssp. glabella (Gray)
               Calder & Taylor
          SY=M. f. var. g. (Gray) M.E. Peck
          SY=M. g. Gray
     pilosa (Michx.) Juss.

Moneses Salisb. ex S.F. Gray

     uniflora (L.) Gray
          ssp. reticulata (Nutt.) Calder & Taylor
               SY=M. u. var. r. (Nutt.) Blake
          ssp. uniflora
               SY=Pyrola u. L.

Monotropa L.

     hypopithys L.
          SY=Hypopithys americana (DC.)
               Small
          SY=H. fimbriata (Gray) T.J. Howell
          SY=H. lanuginosa (Michx.) Nutt.
          SY=H. latisquama Rydb.
          SY=H. monotropa Crantz
          SY=M. h. ssp. lanuginosa (Michx.)
               Hara
          SY=M. h. var. americana (DC.)
               Domin
          SY=M. h. var. latisquama (Rydb.)
               Kearney & Peebles
          SY=M. h. var. rubra (Torr.) Farw.
          SY=M. lanuginosa Michx.
          SY=M. latisquama (Rydb.) Hulten
     uniflora L.
          SY=M. brittonii Small

Monotropsis Schwein. ex Ell.

     odorata Schwein.
          SY=M. o. var. lehmaniae (Burnham)
               Ahles
          SY=M. l. Burnham
          SY=M. reynoldsiae (Gray) Heller

Ornithostaphylos Small

     oppositifolia (Parry) Small

Orthilia Raf.

     secunda (L.) House
          ssp. obtusata (Turcz.) Bocher
               SY=O. s. var. o. (Turcz.) House
               SY=Pyrola s. L. ssp. o. (Trucz.)
                    Hulten
               SY=P. s. var. o. Turcz.
          ssp. secunda
               SY=Pyrola s. L.
               SY=Ramischia s. (L.) Garcke

Oxydendrum DC.

     arboreum (L.) DC.

Phyllodoce Salisb.

     aleutica (Spreng.) Heller
     breweri (Gray) Heller
     caerulea (L.) Bab.
     empetriformis (Sm.) D. Don
     glanduliflora (Hook.) Coville
          SY=P. aleutica (Spreng.) Heller
               ssp. g. (Hook.) Hulten
     X intermedia (Hook.) Camp [empetriformis X
       glanduliflora]

Pieris D. Don

     floribunda (Pursh ex Sims) Benth. & Hook.
     phillyreifolia (Hook.) DC.
          SY=Ampelothamnus p. (Hook.) Small

Pityopus Small

     californica (Eastw.) Copeland f.

Pleuricospora Gray

     fimbriolata Gray
          SY=P. longipetala T.J. Howell

Pterospora Nutt.

Pterospora (CONT.)
  andromedea Nutt.

Pyrola L.

  americana Sweet
          SY=P. asarifolia Michx. ssp.
              americana (Sweet) Krisa
          SY=P. rotundifolia L. pro parte
          SY=P. r. var. americana (Sweet)
              Fern.
  asarifolia Michx.
        var. asarifolia
        var. purpurea (Bunge) Fern.
          SY=P. asarifolia Michx. var.
              incarnata (Fisch.) Fern.
          SY=P. uliginosa Torr. & Gray ex
              Torr.
  bracteata Hook.
          SY=P. asarifolia Michx. var. b.
              (Hook.) Jepson
  californica Krisa
  chlorantha Sw.
        var. chlorantha
          SY=P. virens Schweig. ex Schweig.
              & Koerte
          SY=P. v. var. saximontana Fern.
        var. convoluta (Bart.) Fern.
          SY=P. virens Schweig. ex Schweig.
              & Koerte var. c. (Bart.)
              Fern.
  dentata Sm.
        var. apophylla Copeland
        var. dentata
          SY=P. picta Sm. ssp. d. (Sm.)
              Piper
        var. integra Gray
          SY=P. picta Sm. ssp. i. (Gray)
              Piper
  elliptica Nutt.
  grandiflora Radius
        var. canadensis (Andres) Porsild
        var. grandiflora
  X meadia Sw. [grandiflora X minor]
  minor L.
          SY=Braxilia m. (L.) House
  oxypetala Austin
  picta Sm.
          SY=P. aphylla Sm.

Rhododendron L.

  alabamense Rehd.
          SY=Azalea a. (Rehd.) Small
  albiflorum Hook.
  arborescens (Pursh) Torr.
          SY=Azalea a. Pursh
          SY=R. a. var. richardsonii Rehd.
  atlanticum (Ashe) Rehd.
          SY=Azalea a. Ashe
  austrinum (Small) Rehd.
          SY=Azalea a. Small
  bakeri (Lemmon & McKay) Hume
          SY=R. cumberlandense E.L. Braun
  calendulaceum (Michx.) Torr.
          SY=Azalea c. Michx.
          SY=A. lutea sensu auctt. non L.
  camtschaticum Pallas
    ssp. camtschaticum
    ssp. glandulosum (Standl.) Hulten
  canadense (L.) Torr.
          SY=Azalea c. (L.) Kuntze
          SY=Rhodora c. L.
  canescens (Michx.) Sweet
        var. candidum (Small) Rehd.
          SY=Azalea candida Small
          SY=Rhododendron candidum (Small)

          Rehd.
        var. canescens
          SY=Azalea c. Michx.
  carolinianum Rehd.
  catawbiense Michx.
  chapmanii Gray
          SY=Rhododendron minus Michx. var.
              c. (Gray) Duncan & Pullen
  coryi Shinners
  flammeum (Michx.) Sarg.
          SY=Rhododendron speciosum (Willd.)
              Sweet
          SY=Azalea s. Willd.
  lapponicum (L.) Wahlenb.
        var. lapponicum
        var. parvifolium (J.F. Adams) Herder
  macrophyllum D. Don ex G. Don
          SY=Rhododendron californicum Hook.
  maximum L.
          SY=Rhododendron ashleyi Coker
  minus Michx.
  oblongifolium (Small) Millais
  occidentale (Torr. & Gray) Gray
        var. occidentale
        var. paludosum Jepson
        var. sonomense (Greene) Rehd.
  X pennsylvanicum (Gagle) Rehd. [atlanticum
      X periclymenoides]
  periclymenoides (Michx.) Shinners
          SY=Azalea nudiflora L.
          SY=Rhododendron n. (L.) Torr.
          SY=R. n. var. glandiferum (Porter)
              Rehd.
  prinophyllum (Small) Millais
          SY=Azalea p. Small
          SY=Rhododendron nudiflorum (L.)
              Torr. var. roseum (Loisel.)
              Wieg.
          SY=R. r. (Loisel.) Rehd.
  prunifolium (Small) Millais
          SY=Azalea p. Small
  serrulatum (Small) Millais
          SY=Azalea s. Small
          SY=Rhododendron viscosum (L.)
              Torr. var. s. (Small) Ahles
  vanhoeffenii Abrom.
  vaseyi Gray
          SY=Biltia v. (Gray) Small
  viscosum (L.) Torr.
          SY=Azalea v. L.
          SY=Rhododendron v. var. glaucum
              (Michx.) Gray
          SY=R. v. var. montanum Rehd.
          SY=R. v. var. nitidum (Pursh) Gray
          SY=R. v. var. tomentosum Rehd.
  X welleslyanum Waterer ex Rehd.
      [catawbiense X maximum]

Sarcodes Torr.

  sanguinea Torr.

Symphysia Presl

  racemosa (Vahl) Stearn
          SY=Hornemannia r. Vahl
          SY=Thibaudia krugii Urban &
              Hoerold
          SY=Vaccinium r. (Vahl) Wilbur &
              Luteyn

Vaccinium L.

  alaskense T.J. Howell
  angustifolium Ait.
          SY=V. a. var. hypolasium Fern.
          SY=V. a. var. laevifolium House

Vaccinium (CONT.)
            SY=V. a. var. nigrum (Wood) Dole
            SY=V. brittonii Porter ex Bickn.
            SY=V. lamarckii Camp
            SY=V. n. (Wood) Britt.
    arboreum Marsh.
        var. arboreum
            SY=Batodendron a. (Marsh.) Nutt.
        var. glaucescens (Greene) Sarg.
            SY=Batodendron andrachniforme
                Small
    berberifolium (Gray) Skottsberg
    boreale Hall & Aalders
    caespitosum Michx.
        var. caespitosum
            SY=V. arbuscula (Gray) Merriam
            SY=V. c. var. a. Gray
        var. paludicola (Camp) Hulten
            SY=V. nivictum Camp
    calycinum Sm.
        var. calycinum
        var. montanum (Wawra) Skottsberg
    coccinium Piper
    corymbosum L.
            SY=Cyanococcus amoenus (Ait.)
                Small
            SY=C. atrococcus (Gray) Small
            SY=C. corymbosus (L.) Rydb.
            SY=C. cuthbertii Small
            SY=C. elliottii (Chapman) Small
            SY=C. fuscatus (Ait.) Small
            SY=C. margarettae (Ashe) Small
            SY=C. simulatus (Small) Small
            SY=C. virgatus (Ait.) Small
            SY=V. amoenum Ait.
            SY=V. arkansanum Ashe
            SY=V. ashei Rehd.
            SY=V. atrococcum (Gray) Heller
            SY=V. australe Small
            SY=V. caesariense Mackenzie
            SY=V. constablaei Gray
            SY=V. corymbosum var. albiflorum
                (Hook.) Fern.
            SY=V. corymbosum var. amoenum
                (Ait.) Gray
            SY=V. corymbosum var. atrococcum
                Gray
            SY=V. corymbosum var. glabrum Gray
            SY=V. elliottii Chapman
            SY=V. formosum Andr.
            SY=V. marianum Watson
            SY=V. simulatum Small
            SY=V. virgatum Ait.
            SY=V. v. var. ozarkense Ashe
            SY=V. v. var. speciosum Palmer
    crassifolium Andr.
            SY=Herpothamnus c. (Andr.) Small
    darrowii Camp
            SY=V. myrsinites Lam. var. glaucum
                Gray
    deliciosum Piper
    dentatum Sm.
        var. argutidens Skottsberg
        var. dentatum
        var. lanceolatum (Gray) Skottsberg
        var. minutifolium Skottsberg
    erythrocarpum Michx.
            SY=Hugeria e. (Michx.) Small
            SY=Oxycoccus e. (Michx.) Pers.
    geminiflorum H.B.K.
    globulare Rydb.
    hirsutum Buckl.
            SY=Cyanococcus h. (Buckl.) Small
    macrocarpon Ait.
            SY=Oxycoccus m. (Ait.) Pursh
    membranaceum Dougl. ex Hook.
    myrsinites Lam.

            SY=Cyanococcus m. (Lam.) Small
            SY=V. nitidum Andr.
    myrtilloides Michx.
            SY=Cyanococcus canadensis (Kalm ex
                A. Rich.) Rydb.
            SY=V. angustifolium Ait. var. m.
                (Michx.) House
            SY=V. canadense Kalm ex A. Rich.
    myrtillus L.
        ssp. myrtillus
        ssp. oreophilum (Rydb.) Love, Love &
            Kapoor
            SY=V. o. Rydb.
    X nubigenum Fern. [caespitosum X
        ovalifolium]
    ovalifolium Sm.
    ovatum Pursh
        var. ovatum
        var. saporosum Jepson
    oxycoccos L.
        var. intermedium Gray
            SY=Oxycoccus i. (Gray) Rydb.
            SY=O. palustris Pers.
            SY=O. p. var. i. (Gray) T.J.
                Howell
            SY=O. p. var. ovaliifolius
                (Michx.) Seymour
            SY=O. quadripetalus Gilib.
            SY=V. oxycoccus var. ovaliifolium
                Michx.
        var. microphyllum (Lange) Rouss. &
            Raymond
            SY=Oxycoccus palustris Pers. ssp.
                m. (Lange) Love & Love
        var. oxycoccos
            SY=Oxycoccus o. (L.) MacM.
            SY=O. microcarpus Turcz. ex Rupr.
            SY=V. m. (Turcz. ex Rupr.)
                Schmalhausen ex Busch
    pahalae Skottsberg
    pallidum Ait.
            SY=Cyanococcus p. (Ait.) Small
            SY=C. liparus Small
            SY=C. vacillans (Kalm ex Torr.)
                Rydb.
            SY=C. subcordatus Small
            SY=C. tallapusae Coville
            SY=V. alto-montanum Ashe
            SY=V. corymbosum L. var. p. (Ait.)
                Gray
            SY=V. vacillans Kalm ex Torr.
            SY=V. v. var. crinitum Fern.
            SY=V. v. var. missouriense Ashe
            SY=V. viride Ashe
    parviflorum Gray
    parvifolium Sm.
    peleanum Skottsberg
    reticulatum Sm.
    scoparium Leib.
    stamineum L.
        var. melanocarpum C. Mohr
            SY=Polycodium depressum Small
            SY=P. m. (C. Mohr) Small
            SY=V. m. (C. Mohr) C. Mohr
        var. stamineum
            SY=Polycodium ashei Harbison
            SY=P. candicans Small
            SY=P. floridanum (Nutt.) Greene
            SY=P. leptosepalum Small
            SY=P. macilentum Small
            SY=P. neglectum Small
            SY=P. s. (L.) Greene
            SY=V. caesium Greene
            SY=V. n. (Small) Fern.
            SY=V. s. var. candicans (Small) C.
                Mohr
            SY=V. s. var. interius (Ashe)

Vaccinium (CONT.)
                    Palmer & Steyermark
          SY=V. s. var. n. (Small) Deam
   tenellum Ait.
          SY=Cyanococcus t. (Ait.) Small
   uliginosum L.
      ssp. microphyllum Lange
          SY=V. u. ssp. gaultherioides
                (Bigelow) S. Young
      ssp. occidentale (Gray) Hulten
          SY=V. o. Gray
          SY=V. u. var. o. (Gray) Hara
          SY=V. u. var. salicinum (Cham.)
                Hulten
      ssp. pedris (Harshberger) S. Young
      ssp. pubescens (Wormsk. ex Hornem.) S.
         Young
          SY=V. u. ssp. alpinum (Bigelow)
                Hulten
          SY=V. u. var. a. Bigelow
   vitis-idaea L.
      ssp. minus (Lodd.) Hulten
          SY=V. v-i. var. m. Lodd.
          SY=V. v-i. var. punctata Moench

Xylococcus Nutt.

   bicolor Nutt.

Zenobia D. Don

   pulverulenta (Bartr. ex Willd.) Pollard
          SY=Z. cassinefolia (Vent.) Pollard

ERIOCAULACEAE

Eriocaulon L.

   cinereum R. Br.
   compressum Lam.
          var. compressum
          var. harperi Moldenke
   decangulare L.
          var. decangulare
          var. latifolium Chapman
          var. minor Moldenke
   kornickianum van Huerck & Muell.-Arg.
   lineare Small
   microcephalum H.B.K.
   parkeri B.L. Robins.
          SY=E. septangulare With. var. p.
             (B.L. Robins.) Boivin &
             Cayouette
   ravenelii Chapman
   septangulare With.
          SY=E. pellucidum Michx.
   texense Koern.

Lachnocaulon Kunth

   anceps (Walt.) Morong
          SY=L. floridanum Small
          SY=L. glabrum Koern.
   beyrichianum Sporleder ex Koern.
   digynum Koern.
   engleri Ruhl.
          var. caulescens Moldenke
          var. engleri
   minus (Chapman) Small
          SY=L. eciliatum Small

Syngonanthus Ruhl.

   flavidulus (Michx.) Ruhl.

ERYTHROXYLACEAE

Erythroxylum P. Br.

   areolatum L.
   rotundifolium Lunan
          SY=E. brevipes DC.
   rufum Cav.
   urbanii O.E. Schulz

EUPHORBIACEAE

Acalypha L.

   alopecuroidea Jacq.
   berteroana Muell.-Arg.
   bisetosa Bertol.
   californica Benth.
   chamaedrifolia (Lam.) Muell.-Arg.
   deamii (Weatherby) Ahles
          SY=A. rhomboidea Raf. var. d.
                Weatherby
   gracilens Gray
      ssp. gracilens
          SY=A. g. var. delzii L. Mill.
          SY=A. g. var. fraseri
                (Muell.-Arg.) Weatherby
      ssp. monococca (Engelm. & Gray) G.L.
         Webster
          SY=A. g. var. m. Engelm. & Gray
          SY=A. m. (Engelm. & Gray) L. Mill.
   hederacea Torr.
   indica L.
   lindheimeri Muell.-Arg.
          SY=A. l. var. major Pax & Hoffman
   monostachya Cav.
   neomexicana Muell.-Arg.
   ostryifolia Riddell
          SY=A. caroliniana Ell.
   poiretii Spreng.
   portoricensis Muell.-Arg.
   pringlei S. Wats.
   radians Torr.
   rhomboidea Raf.
   setosa A. Rich.
   virginica L.

Adelia L.

   ricinella L.
          SY=Ricinella r. (L.) Britt.
   vaseyi (Coult.) Pax & Hoffmann

Alchornea Sw.

   latifolia Sw.

Alchorneopsis Muell.-Arg.

   portoricensis Urban

Aleurites J.R. & G. Forst.

   fordii Hemsl.
   moluccana (L.) Willd.

Andrachne L.

   arida (Warnock & M.C. Johnston) G.L.
      Webster
          SY=Savia a. Warnock & M.C.
                Johnston
   phyllanthoides (Nutt.) Coult.

Andrachne (CONT.)
          SY=Savia p. (Nutt.) Pax & Hoffmann

Antidesma L.

    bunias (L.) Spreng.
    crenatum St. John
    X kapuae Rock [platyphyllum X pulvinatum]
    platyphyllum Mann
          var. hamakuaense Fosberg
          var. hillebrandii Pax & Hoffmann
          var. platyphyllum
          var. subamplexicaule Sherff
    pulvinatum Hbd.
          var. contractum Deg. & Sherff
          var. leiogonum Sherff
          var. pulvinatum

Argythamnia P. Br.

    aphoroides Muell.-Arg.
    argyraea Cory
    blodgettii (Torr.) Chapman
          SY=Ditaxis b. (Torr.) Pax
    brandegei Millsp.
          var. brandegei
              SY=Ditaxis b. (Millsp.) Rose &
                  Standl.
          var. intonsa (I.M. Johnston) Ingram
              SY=Ditaxis b. (Millsp.) Rose &
                  Standl. var. i. I.M.
                  Johnston
    californica Brandeg.
          SY=Ditaxis c. (Brandeg.) Pax &
              Hoffmann
    candicans Sw.
    clariana Jepson
          SY=Ditaxis adenophora (Gray) Pax &
              Hoffmann
    cyanophylla (Woot. & Standl.) Ingram
          SY=Ditaxis c. Woot. & Standl.
    fasciculata (Vahl) Muell.-Arg.
          SY=Ditaxis f. Vahl
    humilis (Engelm. & Gray) Muell.-Arg.
          var. humilis
              SY=Ditaxis h. (Engelm. & Gray) Pax
          var. laevis (Torr.) Shinners
              SY=A. l. Torr.
              SY=Ditaxis l. (Torr.) Heller
    lanceolata (Benth.) Muell.-Arg.
          SY=Ditaxis l. (Benth.) Pax &
              Hoffmann
    mercurialina (Nutt.) Muell.-Arg.
          var. mercurialina
              SY=Ditaxis m. (Nutt.) Coult.
          var. pilosissima (Benth.) Shinners
    neomexicana Muell.-Arg.
          SY=Ditaxis n. (Muell.-Arg.) Heller
    serrata (Torr.) Muell.-Arg.
          SY=Ditaxis s. (Torr.) Heller
    simulans Ingram
    stahlii Urban

Bernardia Houst. ex P. Br.

    dichotoma (Willd.) Muell.-Arg.
          SY=Adelia bernardia L.
    incana Morton
    myricifolia (Scheele) S. Wats.
    obovata I.M. Johnston

Bischofia Blume

    javanica Blume

Breynia J.R. & G. Forst.

disticha J.R. & G. Forst.
          SY=B. nivosa (W.J. Sm.) Small

Caperonia St. Hil.

    castaniifolia (L.) St. Hil.
    palustris (L.) St. Hil.
          SY=Croton p. L.

Chamaesyce S.F. Gray

    abramsiana (L.C. Wheeler) Burch ined.
          SY=Euphorbia a. L.C. Wheeler
    acuta (Engelm.) Millsp.
          SY=Euphorbia a. Engelm.
    adenoptera (Bertol.) Small
      ssp. pergamena (Small) Burch
    albomarginata (Torr. & Gray) Small
          SY=Euphorbia a. Torr. & Gray
    angusta (Engelm.) Small
          SY=Euphorbia a. Engelm.
    arizonica (Engelm.) Arthur
          SY=C. versicolor (Greene) J.B.S.
              Norton
          SY=Euphorbia a. Engelm.
          SY=E. v. Greene
    arnottiana (Endl.) Deg. & Deg.
          var. arnottiana
              SY=Euphorbia a. Endl.
              SY=E. hookeri (Steud.) Arthur
          var. integrifolia (Hbd.) Deg. & Deg.
              SY=Euphorbia a. Endl. var. i.
                  (Hbd.) St. John
    articulata (Aubl.) Britt.
          SY=C. linearis (Retz.) Millsp.
          SY=C. vahlii (Willd.) P. Wilson
          SY=Euphorbia a. Aubl.
          SY=E. l. Retz.
    astyla (Engelm. ex Boiss.) Millsp.
          SY=Euphorbia a. Engelm. ex Boiss.
    atrococca (Heller) Croizat & Deg.
          var. atrococca
              SY=Euphorbia a. Heller
          var. kilaueana (Sherff) Deg. & Deg.
              SY=Euphorbia a Heller var. k.
                  Sherff
          var. kokeeana (Sherff) Deg. & Deg.
              SY=Euphorbia a. Heller var. k.
                  Sherff
    berteriana (Balbis) Millsp.
          SY=Euphorbia b. Balbis
          SY=E. stipitata Millsp.
    blodgettii (Engelm. ex A.S. Hitchc.) Small
          SY=C. nashii Small
    bombensis (Jacq.) Dugand
          SY=C. ammannioides (H.B.K.) Small
          SY=C. ingallsii Small
          SY=Euphorbia a. H.B.K.
          SY=E. b. Jacq.
    capitellata (Engelm.) Millsp.
          SY=C. pycnanthema (Engelm.)
              Millsp.
          SY=Euphorbia c. Engelm.
          SY=E. p. Engelm.
    carunculata (Waterfall) Shinners
          SY=Euphorbia c. Waterfall
    celastroides (Boiss.) Croizat & Deg.
          var. amplectens (Sherff) Deg. & Deg.
              SY=Euphorbia c. Boiss. var. a.
                  Sherff
          var. celastroides
              SY=Euphorbia c. Boiss.
              SY=?E. c. var. hathewayi Sherff
              SY=?E. c. var. laehiensis Deg.,
                  Deg. & Sherff
              SY=?E. c. var. pseudoniuensis Deg.
                  & Sherff

Chamaesyce (CONT.)
  var. halawana (Sherff) Deg. & Deg.
   SY=Euphorbia c. Boiss. var. h.
    Sherff
  var. hanapepensis (Sherff) Deg. &
   Deg.
   SY=Euphorbia c. Boiss. var. h.
    Sherff
  var. haupuana (Sherff) Deg. & Deg.
   SY=Euphorbia c. Boiss. var. h.
    Sherff
  var. humbertii (Sherff) Deg. & Deg.
   SY=Euphorbia c. Boiss. var. h.
    Sherff
  var. ingrata (Deg. & Sherff) Deg. &
   Deg.
   SY=Euphorbia c. Boiss. var. i.
    Deg. & Sherff
  var. kaenana (Sherff) Deg. & Deg.
   SY=Euphorbia c. Boiss. var. k.
    Sherff
  var. kealiana (Sherff) Deg. & Deg.
   SY=Euphorbia c. Boiss. var. k.
    Sherff
  var. kohalana (Sherff) Deg. & Deg.
   SY=Euphorbia c. Boiss. var. k.
    Sherff
  var. lorifolia (Gray ex Mann) Deg. &
   Deg.
   SY=Euphorbia c. Boiss. var. l.
    (Gray ex Mann) Sherff
  var. mauiensis (Sherff) Deg. & Deg.
   SY=Euphorbia c. Boiss. var. m.
    Sherff
  var. moomomiana (Sherff) Deg. & Deg.
   SY=Euphorbia c. Boiss. var. m.
    Sherff
  var. nelsonii author ?
  var. nematopoda (Sherff) Deg. & Deg.
   SY=Euphorbia c. Boiss. var. n.
    Sherff
  var. niuensis (Sherff) Deg. & Deg.
   SY=Euphorbia c. Boiss. var. n.
    Sherff
  var. odonatoides (Deg. & Sherff) Deg.
   & Deg.
   SY=Euphorbia c. Boiss. var. o.
    Deg. & Sherff
  var. saxicola (Deg. & Sherff) Deg. &
   Deg.
   SY=Euphorbia c. Boiss. var. s.
    Deg. & Sherff
  var. stokesii (Forbes) Deg. & Deg.
   SY=Euphorbia c. Boiss. var. s.
    (Forbes) Sherff
  var. waikoluensis (Sherff) Deg. &
   Deg.
   SY=Euphorbia c. Boiss. var. w.
    Sherff
 chaetocalyx (Boiss.) Woot. & Standl.
   SY=C. fendleri (Torr. & Gray)
    Small var. c. (Boiss.)
    Shinners
   SY=Euphorbia f. Torr. & Gray var.
    c. Boiss.
   SY=E. f. var. triliqulata L.C.
    Wheeler
 ?chiogenes Small
 cinerascens (Engelm.) Small
   SY=Euphorbia c. Engelm.
 clusiifolia (Hook. & Arn.) Arthur
   SY=Euphorbia c. Hook. & Arn.
 conferta Small
 cordifolia (Ell.) Small
   SY=Euphorbia c. Ell.
 cowellii Millsp.
 cumulicola Small

 degeneri (Sherff) Croizat & Deg.
  var. degeneri
   SY=Euphorbia d. Sherff
  var. molokaiensis (Sherff) Deg. &
   Croizat
   SY=E. d. Sherff var. m. Sherff
 deltoidea (Engelm. ex Chapman) Small
  ssp. deltoidea
   SY=C. adhaerens Small
  ssp. serpyllum (Small) Burch
   SY=C. s. Small
 ?deppeana (Boiss.) Millsp.
   SY=Euphorbia d. Boiss.
 fendleri (Torr. & Gray) Small
   SY=Euphorbia f. Torr. & Gray
   SY=E. f. var. typica L.C. Wheeler
 florida (Engelm.) Millsp.
   SY=Euphorbia f. Engelm.
 forbesii (Sherff) Croizat & Deg.
   SY=Euphorbia f. Sherff
 garberi (Engelm. ex Chapman) Small
   SY=C. adicioides Small
   SY=C. brachypoda Small
   SY=C. mosieri Small
   SY=Euphorbia g. Engelm. ex Chapman
 geyeri (Engelm.) Small
   SY=Euphorbia g. Engelm.
   SY=E. g. var. wheeleriana Warnock
    & M.C. Johnston
 glyptosperma (Engelm.) Small
   SY=Euphorbia g. Engelm.
 golondrina (L.C. Wheeler) Shinners
   SY=Euphorbia g. L.C. Wheeler
 gracillima (S. Wats.) Millsp.
   SY=Euphorbia g. S. Wats.
 halemanui (Sherff) Croizat & Deg.
   SY=Euphorbia h. Sherff
 hillebrandii (Levl.) Croizat & Deg.
  var. hillebrandii
   SY=Euphorbia h. Levl.
  var. palikeana (Deg. & Sherff) Deg. &
   Deg.
   SY=Euphorbia h. Levl. var. p. Deg.
    & Sherff
  var. waimanoana (Sherff) Deg. &
   Croizat
   SY=Euphorbia h. Levl. var. w.
    Sherff
 hirta (L.) Millsp.
   SY=Euphorbia h. L.
 hooveri (L.C. Wheeler) Burch ined.
   SY=Euphorbia h. L.C. Wheeler
 humistrata (Engelm. ex Gray) Small
   SY=Euphorbia h. Engelm. ex Gray
 hypericifolia (L.) Millsp.
   SY=C. glomerifera Millsp.
   SY=Euphorbia g. (Millsp.) L.C.
    Wheeler
   SY=E. h. L.
 hyssopifolia (L.) Small
   SY=C. brasilensis (Lam.) Small
   SY=Euphorbia b. Lam.
   SY=E. h. L.
 indivisa (Engelm.) Millsp.
   SY=Euphorbia i. (Engelm.)
    Tidestrom
 jejuna (Warnock & M.C. Johnston) Shinners
   SY=Euphorbia j. Warnock & M.C.
    Johnston
 laredana (Millsp.) Small
   SY=Euphorbia l. Millsp.
 lasiocarpa (Klotzsch) Arthur
   SY=Euphorbia l. Klotzsch
 lata (Engelm.) Small
   SY=Euphorbia l. Engelm.
 maculata (L.) Small
   SY=C. mathewsii Small

Chamaesyce (CONT.)
    SY=C. supina (Raf.) Moldenke
    SY=C. tracyi Small
    SY=Euphorbia maculata L.
    SY=E. s. Raf.
  melanadenia (Torr.) Millsp.
    SY=Euphorbia m. Torr.
  mendezii (Boiss.) Millsp.
    SY=Euphorbia m. Boiss.
  mesembryanthemifolia (Jacq.) Dugand
    SY=C. buxifolia (Lam.) Small
    SY=Euphorbia b. Lam.
  micromera (Boiss. ex Engelm.) Woot. &
    Standl.
    SY=Euphorbia m. Boiss. ex Engelm.
  missurica (Raf.) Shinners
    SY=C. m. var. calciola Shinners
    SY=C. zygophylloides (Boiss.)
      Small
    SY=Euphorbia m. Raf.
  monensis Millsp.
  multiformis (Hook. & Arn.) Croizat & Deg.
    var. haleakalana (Sherff) Deg. & Deg.
      SY=Euphorbia m. Hook. & Arn. var.
        h. Sherff
    var. kaalana (Sherff) Deg. & Deg.
      SY=Euphorbia m. Hook. & Arn. var.
        k. Sherff
    var. kapuleiensis (Deg. & Sherff)
      Deg. & Deg.
      SY=Euphorbia m. Hook. & Arn. var.
        k. Deg. & Sherff
    var. manoana (Sherff) Deg. & Deg.
      SY=Euphorbia m. Hook. & Arn. var.
        m. Sherff
    var. microphylla (Boiss.) Deg. & Deg.
      SY=Euphorbia multiformis Hook. &
        Arn. var. microphylla Boiss.
    var. multiformis
      SY=Euphorbia m. Hook. & Arn.
      SY=E. m. var. mohihiensis Sherff
    var. perdita (Sherff) Deg. & Deg.
      SY=Euphorbia m. Hook. & Arn. var.
        p. Sherff
    var. sparsiflora (Heller) Deg. & Deg.
      SY=Euphorbia m. Hook. & Arn. var.
        s. (Heller) Sherff
    var. tomentella (Boiss.) Deg. & Deg.
      SY=Euphorbia m. Hook. & Arn. var.
        t. Boiss.
  nutans (Lag.) Small
    SY=C. preslii (Guss.) Arthur
    SY=Euphorbia n. Lag.
    SY=E. p. Guss.
  ocellata (Dur. & Hilg.) Millsp.
    SY=Euphorbia o. Dur. & Hilg.
    SY=E. o. var. arenicola (Parish)
      Jepson
    SY=E. o. var. rattanii (S. Wats.)
      L.C. Wheeler
  olowaluana (Sherff) Croizat & Deg.
    var. gracilis (Rock) Deg. & Deg.
      SY=Euphorbia o. Sherff var. g.
        (Rock) Sherff
    var. lepidifolia (Deg. & Sherff) Deg.
      & Deg.
      SY=Euphorbia o. Sherff var. l.
        Deg. & Sherff
    var. olowaluana
      SY=Euphorbia o. Sherff
  ophthalmica (Pers.) Burch
    SY=C. gemella (Lag.) Small
    SY=Euphorbia o. Lag.
    SY=E. hirta L. var. procumbens
      (DC.) N.E. Br.
    SY=E. o. Pers.
  parishii (Greene) Millsp. ex Parish

    SY=Euphorbia p. Greene
  parryi (Engelm.) Rydb.
    SY=Euphorbia p. Engelm.
  pediculifera (Engelm.) Rose & Standl.
    SY=Euphorbia p. Engelm.
  perennans Shinners
    SY=Euphorbia p. (Shinners) Warnock
      & M.C. Johnston
  pinetorum Small
  platysperma (Engelm.) Shinners
    SY=Euphorbia p. Engelm.
  polycarpa (Benth.) Millsp.
    var. hirtella (Boiss.) Millsp. ex
      MacDougal
      SY=Euphorbia p. Benth. var. h.
        Boiss.
    var. polycarpa
      SY=Euphorbia p. Benth.
    var. simulans (L.C. Wheeler) Shinners
      SY=Euphorbia p. Benth. var. s.
        L.C. Wheeler
      SY=E. s. (L.C. Wheeler) Warnock &
        M.C. Johnston
  polygonifolia (L.) Small
    SY=Euphorbia p. L.
  porteriana Small
    var. keyensis (Small) Burch
      SY=C. k. Small
    var. porteriana
    var. scoparia (Small) Burch
      SY=C. s. Small
  prostrata (Ait.) Small
    SY=Euphorbia chamaesyce L.
    SY=E. p. Ait.
  remyi (Gray ex Boiss.) Croizat & Deg.
    var. hanaleiensis (Sherff) Deg. &
      Deg.
      SY=Euphorbia r. Gray ex Boiss.
        var. h. Sherff
    var. kahiliana (Sherff) Deg. & Deg.
      SY=Euphorbia r. Gray ex Boiss.
        var. k. Sherff
    var. kauaiensis (Deg. & Sherff) Deg.
      & Deg.
      SY=Euphorbia r. Gray ex Boiss.
        var. k. Deg. & Sherff
    var. leptopoda (Sherff) Deg. & Deg.
      SY=Euphorbia r. Gray ex Boiss.
        var. l. Sherff
    var. lydgatei (Sherff) Deg. & Deg.
      SY=Euphorbia r. Gray ex Boiss.
        var. l. Sherff
    var. molesta (Sherff) Deg. & Deg.
      SY=Euphorbia r. Gray ex Boiss.
        var. m. Sherff
    var. olokelensis (Skottsberg &
      Sherff) Deg. & Deg.
      SY=Euphorbia r. Gray ex Boiss.
        var. o. Skottsberg & Sherff
    var. pteropoda (Sherff) Deg. & Deg.
      SY=Euphorbia r. Gray ex Boiss.
        var. p. Sherff
    var. remyi
      SY=Euphorbia r. Gray ex Boiss.
    var. wahiawana (Sherff) Deg. & Deg.
      SY=Euphorbia r. Gray ex Boiss.
        var. w. Sherff
    var. waimeana (Sherff) Deg. & Deg.
      SY=Euphorbia r. Gray ex Boiss.
        var. w. Sherff
    var. wilkesii (Sherff) Deg. & Deg.
      SY=Euphorbia r. Gray ex Boiss.
        var. w. Sherff
  revoluta (Engelm.) Small
    SY=Euphorbia r. Engelm.
  rockii (Forbes) Croizat & Deg.
    SY=Euphorbia r. Forbes

Chamaesyce (CONT.)
  serpens (H.B.K.) Small
        SY=Euphorbia s. H.B.K.
  serpyllifolia (Pers.) Small
        SY=C. albicaulis (Rydb.) Rydb.
        SY=C. neomexicana (Greene) Standl.
        SY=Euphorbia n. Greene
        SY=E. s. Pers.
        SY=E. s. var. hirtula (Engelm.)
            L.C. Wheeler
  serrula (Engelm.) Woot. & Standl.
        SY=Euphorbia s. Engelm.
  setiloba (Engelm.) Millsp. ex Parish
        SY=Euphorbia s. Engelm.
  skottsbergii (Sherff) Croizat & Deg.
    var. audens (Sherff) Deg. & Deg.
        SY=Euphorbia s. Sherff var. a.
            Sherff
    var. skottsbergii
        SY=C. s. var. kalaeloana (Sherff)
            Deg. & Deg.
        SY=Euphorbia s. Sherff
        SY=E. s. var. k. Sherff
        SY=?E. s. var. vaccinioides Sherff
  stictospora (Engelm.) Small
        SY=Euphorbia s. Engelm.
  theriaca (L.C. Wheeler) Shinners
        SY=Euphorbia t. L.C. Wheeler
  thymifolia (L.) Millsp.
        SY=Euphorbia t. L.
  torralbasii (Urban) Millsp.
        SY=Euphorbia t. Urban
  trachysperma (Engelm.) Millsp.
        SY=Euphorbia t. Engelm.
  turpinii (Boiss.) Millsp.
        SY=C. albescens (Urban) Millsp.
        SY=C. anegadensis Millsp.
        SY=C. portoricensis (Urban)
            Millsp.
        SY=Euphorbia p. Urban
        SY=E. p. var. albescens Urban
  vallis-mortae Millsp.
        SY=Euphorbia v-m. (Millsp.) J.T.
            Howell
  vermiculata (Raf.) House
        SY=C. rafinesquii (Greene) Arthur
        SY=Euphorbia v. Raf.
  villifera (Scheele) Small
        SY=Euphorbia v. Scheele

Claoxylon A. Juss.

  helleri Sherff
  sandwicense Muell.-Arg.
    var. degeneri Sherff
    var. hillebrandii Sherff
    var. magnifolium Sherff
    var. sandwicense Muell.-Arg.
        SY=C. s. var. glabrescens (Sherff)
            Sherff
    var. tomentosum Hbd.

Cnidoscolus Pohl

  aconitifolius (P. Mill) I.M. Johnston
  angustidens Torr.
  stimulosus (Michx.) Engelm. & Gray
        SY=Bivonea s. (Michx.) Raf.
  texanus (Muell.-Arg.) Small
        SY=Jatropha t. Muell.-Arg.

Codiaeum Juss.

  variegatum (L.) Blume

Croton L.

  alabamensis E.A. Sm. ex Chapman
  argyranthemus Michx.
  astroites Dry.
  betulinus Vahl
  californicus Muell.-Arg.
    var. californicus
    var. longipes (M.E. Jones) Ferguson
        SY=C. l. M.E. Jones
    var. mohavensis Ferguson
    var. tenuis (S. Wats.) Ferguson
  capitatus Michx.
    var. capitatus
    var. lindheimeri (Engelm. & Gray)
        Muell.-Arg.
        SY=C. c. var. albinoides
            (Ferguson) Shinners
        SY=C. engelmannii Ferguson
        SY=C. e. var. a. Ferguson
        SY=C. l. (Engelm. & Gray) Wood
  ciliatoglandulifer Ortega
  cortesianus H.B.K.
  coryi Croizat
  dioicus Cav.
  discolor Willd.
  elliottii Chapman
  fishlockii Britt.
  flavens Jacq.
    var. rigidus Muell.-Arg.
        SY=C. r. (Muell.-Arg.) Britt.
  fruticulosus Torr.
  glandulosus L.
    var. floridanus (Ferguson) R.W. Long
        SY=C. fergusonii Small
        SY=C. floridanus Ferguson
    var. glandulosus
        SY=C. arenicola Small
    var. lindheimeri Muell.-Arg.
    var. pubentissimus Croizat
    var. septentrionalis Muell.-Arg.
    var. simpsonii Ferguson
  humilis L.
        SY=C. berlandieri Torr.
  impressus Urban
  leucophyllus Muell.-Arg.
  lindheimerianus Scheele
    var. lindheimerianus
    var. tharpii M.C. Johnston
  linearis Jacq.
  lobatus L.
  lucidus L.
  monanthogynus Michx.
  nummulariifolius A. Rich.
  ovalifolius Vahl
  parksii Croizat
  poecilanthus Urban
  pottsii (Klotzsch) Muell.-Arg.
    var. pottsii
        SY=C. corymbulosus Engelm.
    var. thermophilus (M.C. Johnston)
        M.C. Johnston
  punctatus Jacq.
  sancti-lazari Croizat
  soliman Cham. & Schlecht.
  sonorae Torr.
  suaveolens Torr.
  texensis (Klotzsch) Muell.-Arg. ex DC.
  torreyanus Muell.-Arg.
  trinitatus Millsp.
        SY=C. miquelensis Ferguson
  wigginsii L.C. Wheeler

Crotonopsis Michx.

  elliptica Willd.
  linearis Michx.

Dalechampia L.

Dalechampia   (CONT.)
  scandens L.

Ditaxis Vahl ex A. Juss.

  ?diversiflora Clokey

Ditta Griseb.

  myricoides Griseb.

Drypetes Vahl

  alba Poit.
  diversifolia Krug & Urban
  glauca Vahl
  ilicifolia Krug & Urban
  lateriflora (Sw.) Krug & Urban

Eremocarpus Benth.

  setigerus (Hook.) Benth.

Euphorbia L.

  agraria Bieb.
          SY=E. podperae Croizat
  alta J.B.S. Norton
  antisyphilitica Zucc.
  arundelana Bartlett
          SY=Tithymalopsis a. (Bartlett)
            Small
  bicolor Engelm. & Gray
  bifida Hook. & Arn.
  bifurcata Engelm.
  bilobata Engelm.
  brachycera Engelm.
  chamaesula Boiss.
  commutata Engelm.
          SY=E. c. var. erecta J.B.S. Norton
          SY=Galarhoeus austrina (Small)
            Small
          SY=G. c. (Engelm.) Small
          SY=Tithymalus a. Small
          SY=T. c. (Engelm.) Klotzsch &
            Garcke
  corollata L.
      var. corollata
          SY=E. apocynifolia Small
          SY=E. c. var. angustifolia Ell.
          SY=E. discoidalis Chapman
          SY=Tithymalopsis apocyniflora
            (Small) Small
          SY=T. c. (L.) Klotzsch
          SY=T. d. (Chapman) Small
          SY=T. olivacea Small
      var. mollis Millsp.
      var. paniculata (Ell.) Boiss.
          SY=Tithymalopsis p. (Ell.) Small
      var. zinniiflora (Small) Ahles
          SY=E. z. Small
          SY=Tithymalopsis z. (Small) Small
  cotinifolia L.
  crenulata Engelm.
  curtisii Engelm.
          SY=Tithymalopsis c. (Engelm.)
            Small
          SY=T. eriogonoides Small
          SY=T. exserta Small
  cyparissias L.
          SY=Galarhoeus c. (L.) Small
          SY=Tithymalus c. (L.) Hill
  delicatula (Boiss.) Millsp.
  dictyosperma Fisch. & Mey.
          SY=E. arkansana Engelm. & Gray
          SY=Galarhoeus a. (Engelm. & Gray)
            Small

          SY=Tithymalus a. (Engelm. & Gray)
            Klotzsch & Garcke
  eriantha Benth.
  esula L.
          SY=Galarhoeus e. (L.) Rydb.
          SY=Tithymalus e. (L.) Hill
  exigua L.
  exserta (Small) Burch ined.
          SY=Tithymalopsis e. Small
  exstipulata Engelm.
      var. exstipulata
      var. lata Warnock & M.C. Johnston
  falcata L.
          SY=Tithymalus f. (L.) Klotzsch &
            Garcke
  floridana Chapman
          SY=Galarhoeus f. (Chapman) Small
  gracilior Cronq.
  haeeleeleana Herbst
  helioscopia L.
          SY=Galarhoeus h. (L.) Haw.
          SY=Tithymalus h. (L.) Hill
  helleri Millsp.
  hexagona Nutt. ex Spreng.
          SY=Zygophyllidium h. (Nutt. ex
            Spreng.) Small
  incisa Engelm.
      var. incisa
      var. mollis (J.B.S. Norton) L.C.
        Wheeler
  innocua L.C. Wheeler
  inundata Torr.
          SY=Galarhoeus i. (Torr.) Small
  ipecacuanhae L.
          SY=E. gracilis Ell.
          SY=Tithymalopsis g. (Ell.) Small
          SY=T. i. (L.) Small
  kuwaleana Deg. & Sherff
  lactea Haw.
  lathyris L.
          SY=Galarhoeus l. (L.) Haw.
          SY=Tithymalus l. (L.) Hill
  longicruris Scheele
  lucida Waldst. & Kit.
          SY=Galarhoeus l. (Waldst. & Kit.)
            Rydb.
          SY=Tithymalus l. (Waldst. & Kit.)
            Klotzsch & Garcke
  lurida Engelm.
  marginata Pursh
          SY=Agaloma m. (Pursh) Love & Love
          SY=Dichrophyllum m. (Pursh)
            Klotzsch & Garcke
          SY=Lepadena m. (Pursh) Nieuwl.
  marilandica Greene
          SY=T. m. (Greene) Small
  mercurialina Michx.
          SY=Tithymalopsis m. (Michx.) Small
  milii Des Moulins
  misera Benth.
  missurica Raf.
      var. intermedia (Engelm.) L.C.
        Wheeler
          SY=Chamaesyce petaloidea (Engelm.)
            Small
          SY=E. p. Engelm.
      var. missurica
  neowawraea Rock
          SY=Drypetes phyllanthoides (Rock)
            Sherff
          SY=Neowawrea p. Rock
  nephradenia Barneby
  neriifolia L.
  oblongata Griseb.
  obtusata Pursh
          SY=Galarhoeus o. (Pursh) Small
          SY=Tithymalus o. (Pursh) Klotzsch

Euphorbia (CONT.)
         & Garcke
  odontadenia Boiss.
  oerstedianum (Klotzsch & Garcke) Boiss.
        SY=Dichylium o. (Klotzsch &
          Garcke) Britt.
  palmeri Engelm.
    var. palmeri
    var. subpubens (Engelm.) L.C. Wheeler
  peplidion Engelm.
  peplus L.
        SY=Galarhoeus p. (L.) Haw.
        SY=Tithymalus p. (L.) Hill
  petiolaris Sims
        SY=Aklema p. (Sims) Millsp.
  platyphylla L.
        SY=Galarhoeus p. (L.) Small
        SY=Tithymalus p. (L.) Hill
  plummerae S. Wats.
  polyphylla Engelm.
        SY=Tithymalopsis p. (Engelm.)
          Small
  purpurea (Raf.) Fern.
        SY=Galarhoeus darlingtonii (Gray)
          Small
        SY=Tithymalus d. (Gray) Small
  robusta (Engelm.) Small
        SY=Tithymalus r. (Engelm.) Small
  roemeriana Scheele
  segetalis L.
  spathulata Lam.
        SY=Tithymalus missouriensis
          (J.B.S. Norton) Small
  strictior Holz.
  telephioides Chapman
        SY=Galarhoeus t. (Chapman) Small
  tetrapora Engelm.
  tirucalli L.
  trichotoma H.B.K.
        SY=Galarhoeus t. (H.B.K.) Small
  wrightii Torr. & Gray

Flueggea Willd.

  virosa (Willd.) Baill.
        SY=Conami portoricensis (Kuntze)
          Britt.

Garcia Rohr

  nutans Vahl
        SY=G. mayana Britt.

Gymnanthes Sw.

  lucida Sw.
        SY=Ateramnus l. (Sw.) Rothm.

Hippomane L.

  mancinella L.

Hura L.

  crepitans L.

Hyeronima Allem.

  clusioides (Tul.) Muell.-Arg.

Jatropha L.

  cardiophylla (Torr.) Muell.-Arg.
  cathartica Teran & Berl.
  cinerea (Ortega) Muell.-Arg.
  cuneata Wiggins & Rollins
  curcas L.

        SY=Curcas c. (L.) Britt. & Millsp.
  dioica Sesse ex Cerv.
    var. dioica
    var. graminea McVaugh
  gossypiifolia L.
    var. elegans (Klotzsch) Muell.-Arg.
    var. gossypiifolia
        SY=Adenoropium g. (L.) Pohl
  hernandiifolia Vent.
        SY=Curcas h. (Vent.) Britt.
  integerrima Jacq.
    var. hastata (Jacq.) Fosberg
        SY=Adenoropium h. (Jacq.) Britt. &
          Wilson
    var. integerrima
  macrorhiza Benth.
    var. macrorhiza
    var. septemfida Engelm.
  multifida L.
        SY=Adenoropium m. (L.) Pohl
  podagrica Hook.

Julocroton Mart.

  argenteus (L.) Didr.

Manihot P. Mill.

  davisiae Croizat
  esculenta Crantz
        SY=Jatropha manihot L.
        SY=M. m. (L.) Cockerell
        SY=M. utilissima Pohl
  glaziovii Muell.-Arg.
  grahamii Hook.
  walkerae Croizat

Margaritaria L. f.

  nobilis L. f.
        SY=Phyllanthus n. (L. f.)
          Muell-Arg.

Mercurialis L.

  annua L.

Pedilanthus Neck.

  tithymaloides (L.) Poit.
    ssp. angustifolius (Poit.) Dressler
        SY=P. a. Poit.
        SY=P. t. var. a. (Poit.) Griseb.
    ssp. padifolius (L.) Dressler
        SY=P. p. (L.) Poit.
    ssp. parasiticus (Klotzsch & Garcke)
      Dressler
        SY=P. latifolius Millsp. & Britt.
    ssp. smallii (Millsp.) Dressler
        SY=P. s. Millsp.
        SY=Tithymalus s. (Millsp.) Small
    ssp. tithymaloides

Phyllanthus L.

  abnormis Baill.
    var. abnormis
        SY=P. drummondii Small
        SY=P. garberi Small
    var. riograndensis G.L. Webster
  acidus (L.) Skeels
        SY=Acidus distichus (L.)
          Muell.-Arg.
        SY=Cicca a. (L.) Merr.
        SY=C. d. L.
  amarus Schum. & Thonn.
  caroliniensis Walt.

Phyllanthus (CONT.)
    ssp. caroliniensis
    ssp. saxicola (Small) G.L. Webster
            SY=P. pruinosus sensu auctt. non
                Poepp. ex Rich.
            SY=P. s. Small
    cuneifolius (Britt.) Croizat
            SY=Andrachne c. Britt.
    debilis Klein ex Willd.
    distichus Hook. & Arn.
            SY=P. sandwicensis Muell.-Arg.
            SY=P. s. var. degeneri Sherff
            SY=P. s. var. ellipticus
                Muell.-Arg.
    ericoides Torr.
    fraternus G.L. Webster
    juglandifolius Willd.
            SY=Asterandra grandifolia (L.)
                Britt.
    leibmannianus Muell.-Arg.
        ssp. platylepis (Small) G.L. Webster
            SY=P. p. Small
    nirurii L.
        ssp. lathyroides (H.B.K.) G.L. Webster
            SY=P. l. H.B.K.
    pentaphyllus C. Wright ex Griseb.
        ssp. pentaphyllus
            var. floridanus G.L. Webster
        ssp. ?polycladus (Urban) G.L. Webster
            SY=P. polycladus Urban
    polygonoides Nutt. ex Spreng.
    pudens L.C. Wheeler
    pulcher Muell.-Arg.
    stipulatus (Raf.) G.L. Webster
            SY=P. diffusus Klotzsch
    tenellus Roxb.
    urinaria L.

Poinsettia Graham

    cyathophora (Murr.) Klotzsch & Garcke
            SY=Euphorbia c. Murr.
            SY=E. heterophylla L. var. c.
                (Murr.) Griseb.
            SY=E. h. var. graminifolia
                (Michx.) Engelm.
    dentata (Michx.) Klotzsch & Garcke
            SY=Euphorbia d. Michx.
            SY=E. d. var. cuphosperma
                (Engelm.) Fern.
            SY=E. d. var. gracillima Millsp.
            SY=P. c. (Engelm.) Small
    heterophylla (L.) Klotzsch & Garcke
            SY=Euphorbia geniculata Ortega
            SY=E. h. L.
            SY=P. g. (Ortega) Klotzsch &
                Garcke
    pinetorum Small
            SY=Euphorbia p. (Small) G.L.
                Webster
    pulcherrima (Willd. ex Klotzsch) Graham
            SY=Euphorbia p. (Willd. ex
                Klotzsch) Graham
    radians (Benth.) Klotzsch & Garcke
            SY=Euphorbia r. Benth.

Reverchonia Gray

    arenaria Gray

Ricinus L.

    communis L.

Sapium P. Br.

    biloculare (S. Wats.) Pax

    caribaeum Urban
    glandulosum (L.) Morong
    jamaicense Sw.
    laurocerasus Desf.
    sebiferum (L.) Roxb.
            SY=Croton s. L.
            SY=Triadica s. (L.) Small

Savia Willd.

    bahamensis Britt.
    sessiliflora (Sw.) Willd.

Sebastiania Spreng.

    corniculata (Vahl) Muell.-Arg.
    fruticosa (Bartr.) Fern.
            SY=S. ligustrina (Michx.)
                Muell.-Arg.

Securinega Comm. ex Juss.

    acidoton (L.) Fawcett & Rendle

Stillingia Garden ex L.

    aquatica Chapman
    linearifolia S. Wats.
    paucidentata S. Wats.
    spinulosa Torr.
    sylvatica Garden ex L.
        ssp. sylvatica
            SY=S. angustifolia (Torr.) Engelm.
                ex S. Wats.
            SY=S. spathulata (Muell.-Arg.)
                Small
            SY=S. sylvatica var. salicifolia
                Torr.
        ssp. tenuis (Small) D.J. Rogers
            SY=S. t. Small
    texana I.M. Johnston
            SY=S. sylvatica Garden ex L. ssp.
                tenuis (Small) D.J. Rogers
                var. linearifolia
                Muell.-Arg.
    treculiana (Muell.-Arg.) I.M. Johnston

Tetracoccus Engelm. ex Parry

    dioicus Parry
    fasciculatus (S. Wats.) Croizat
        var. hallii (Brandeg.) Dressler
            SY=Halliophytum h. (Brandeg.) I.M.
                Johnston
            SY=H. f. var. h. (Brandeg.) McMinn
            SY=T. h. Brandeg.
    ilicifolius Coville & Gilman

Tragia L.

    amblyodonta (Muell.-Arg.) Pax & Hoffmann
    betonicifolia Nutt.
            SY=T. urticifolia Michx. var.
                texana Shinners
    brevispica Engelm. & Gray
    cordata Michx.
            SY=T. macrocarpa Willd.
    glanduligera Pax & Hoffmann
    laciniata (Torr.) Muell.-Arg.
    nepetifolia Cav.
    nigricans Bush
    ramosa Torr.
            SY=T. angustifolia Nutt.
            SY=T. stylaris Muell.-Arg.
    saxicola Small
    smallii Shinners
    urens L.

Tragia (CONT.)
        SY=T. linearifolia Ell.
    urticifolia Michx.
    volubilis L.

FABACEAE

Abrus Adans.

    praecatorius L.
            SY=A. abrus (L.) W. Wight

Acacia P. Mill.

    anegadensis Britt.
    angustissima (P. Mill.) Kuntze
            var. angustissima
            var. chisosiana Isely
            var. hirta (Nutt.) B.L. Robins.
                SY=A. h. (Nutt.) Torr. & Gray
                SY=Acaciella h. (Nutt.) Britt. &
                    Rose
            var. shrevei (Britt. & Rose) Isely
                SY=Acacia lemmonii Rose
                SY=Acaciella s. Britt. & Rose
            var. suffrutescens (Rose) Isely
                SY=Acacia a. var. cuspidata
                    (Schlecht.) L. Benson pro
                    parte
                SY=Acacia c. Schlecht. pro parte
            var. texensis (Torr. & Gray) Isely
                SY=Acacia a. var. cuspidata
                    (Schlecht.) L. Benson pro
                    parte
                SY=A. c. Schlecht. pro parte
                SY=A. t. Torr. & Gray
    auriculiformis A. Cunningham ex Benth.
    baileyana F. Muell.
    berlandieri Benth.
            SY=Acacia emoryana Benth.
    choriophylla Benth.
    constricta Benth.
            var. constricta
            var. paucispina Woot. & Standl.
    cornigera (L.) Willd.
    cyanophylla Lindl.
            SY=Acacia saligna Hort.
    dealbata Link
            SY=Acacia decurrens Willd. var.
                    dealbata (Link) F. Muell.
    decurrens Willd.
    elata A. Cunningham ex Benth.
    farnesiana (L.) Willd.
            SY=Vachellia f. (L.) Wight & Arn.
    greggii Gray
            var. arizonica Isely
            var. greggii
    kauaiensis Hbd.
    koa Gray
            var. hawaiiensis Rock
            var. koa
            var. lanaiensis Rock
            var. weimeae Hochr.
    koaia Hbd.
    longifolia (Andr.) Willd.
            SY=Acacia latifolia Hort.
    macracantha Humb. & Bonpl. ex Willd.
    mearnsii de Wildeman
            SY=Acacia decurrens Willd. var.
                    mollis Lindl.
    melanoxylon R. Br.
    millefolia S. Wats.
    muricata (L.) Willd.
    neovernicosa Isely

            SY=Acacia vernicosa Standl.
    nilotica (L.) Delile
    pinetorum F.J. Herm.
            SY=Vachellia insularis Small
            SY=V. peninsularis Small
    retinodes Schlecht.
    richei Gray
            SY=Acacia confusa Merr.
    rigidula Benth.
            SY=Acacia amentacea DC.
    riparia H.B.K.
    roemeriana Scheele
            SY=Acacia malacophylla (Benth.)
                    Britt. & Rose
    schaffneri (S. Wats.) F.J. Herm.
    schottii Torr.
    scleroxyla Tuss.
    smallii Isely
            SY=Vachellia densiflora (Alexander
                    ex Small) Cory
    suma (Roxb.) Kurz
            SY=Acacia polycantha Willd.
    tortuosa (L.) Willd.
    wrightii Benth.

Adenanthera L.

    pavonina L.

Aeschynomene L.

    americana L.
            var. americana
            var. glandulosa (Poir.) Rudd
    hystrix Poir.
            SY=Secula h. (Poir.) Small
    indica L.
    portoricensis Urban
            SY=A. gracilis sensu auctt. non
                    Vogel
    pratensis Small
    sensitiva Sw.
    villosa Poir.
    virginica (L.) B.S.P.
    viscidula Michx.
            SY=Secula v. (Michx.) Small

Albizia Durz.

    julibrissin Durz.
    lebbeck (L.) Benth.
    lophantha (Willd.) Benth.
            SY=A. distachya (Vent.) J.F.
                    Macbr.

Alhagi Gagnebin

    pseudalhagii (Bieb.) Desv.
            SY=A. camelorum Fisch.

Alysicarpus Neck. ex Desv.

    rugosus (Willd.) DC.
    vaginalis (L.) DC.

Amorpha L.

    californica Nutt.
            var. californica
            SY=A. c. var. hispidula (Greene)
                    Palmer
            var. napensis Jepson
    canescens Pursh
            SY=A. brachycarpa Palmer
    crenulata Rydb.
    fruticosa L.
            SY=A. angustifolia (Pursh) Boynt.

Amorpha (CONT.)
        SY=A. bushii Rydb.
        SY=A. croceolanata P.W. Wats.
        SY=A. curtissii Rydb.
        SY=A. dewinkeleri Small
        SY=A. f. var. a. Pursh
        SY=A. f. var. croceolanata (P.W.
            Wats.) Schneid.
        SY=A. f. var. emarginata Pursh
        SY=A. f. var. oblongifolia Palmer
        SY=A. f. var. occidentalis
            (Abrams) Kearney & Peebles
        SY=A. f. var. tennesseensis
            (Shuttlw. ex Kunze) Palmer
        SY=A. occidentalis Abrams
        SY=A. o. var. arizonica (Rydb.)
            Palmer
        SY=A. o. var. e. (Pursh) Palmer
        SY=A. t. Shuttlw. ex Kunze
        SY=A. virgata Small
   georgiana Wilbur
      var. confusa Wilbur
        SY=A. cyanostachya M.A. Curtis
      var. georgiana
   glabra Desf. ex Poir.
   herbacea Walt.
      var. floridana (Rydb.) Wilbur
        SY=A. f. Rydb.
      var. herbacea
   laevigata Nutt. ex Torr. & Gray
   nana Nutt.
   nitens Boynt.
   X notha Palmer [canescens X fruticosa]
   ouchitensis Wilbur
   paniculata Torr. & Gray
   roemeriana Scheele
        SY=A. texana Buckl.
   schwerinii Schneid.

Amphicarpaea Ell. ex Nutt.

   bracteata (L.) Fern.
        SY=A. b. var. comosa (L.) Fern.
        SY=A. c. (L.) G. Don
        SY=Falcata c. (L.) Kuntze
        SY=F. pitcheri (Torr. & Gray)
            Kuntze

Anadenanthera Speg.

   peregrina (L.) Speg.
        SY=Piptadenia p. (L.) Benth.

Andira Juss.

   inermis (W. Wright) H.B.K. ex DC.
        SY=A. jamaicensis (W. Wright)
            Urban
        SY=Geoffrea i. W. Wright

Anthyllis L.

   vulneraria L.

Apios Fabr.

   americana Medic.
      var. americana
        SY=Glycine apios L.
      var. turrigera Fern.
   priceana B.L. Robins.
        SY=Glycine p. (B.L. Robins.)
            Britt.

Arachis L.

   hypogaea L.

Astragalus L.

   aboriginum Richards.
        SY=A. a. var. glabriusculus
            (Hook.) Rydb.
        SY=A. a. var. lepagei (Hulten)
            Boivin
        SY=A. a. var. richardsonii
            (Sheldon) Boivin
        SY=A. forwoodii S. Wats.
        SY=A. f. var. wallowensis (Rydb.)
            M.E. Peck
        SY=A. g. (Hook.) Gray var. major
            Gray
        SY=A. linearis (Rydb.) Porsild
        SY=A. scrupulicola Fern. & Weath.
        SY=Atelophragma a. (Richards.)
            Rydb.
   accidens S. Wats.
      var. accidens
      var. hendersonii (S. Wats.) M.E.
        Jones
        SY=Astragalus watsonii Sheldon
   accumbens Sheldon
   acutirostris S. Wats.
   adanus A. Nels.
   adsurgens Pallas
      var. robustior Hook.
        SY=Astragalus a. ssp. r. (Hook.)
            Welsh
        SY=A. striatus Nutt. ex Torr. &
            Gray
        SY=A. sulphurescens Rydb.
      var. tananaicus (Hulten) Barneby
        SY=Astragalus a. ssp. viciifolius
            (Hulten) Welsh
   aequalis Clokey
   agnicidus Barneby
   agrestis Dougl. ex D. Don
        SY=Astragalus danicus Retz. var.
            dasyglottis (Fisch. ex DC.)
            Boivin
        SY=A. dasyglottis Fisch. ex DC.
        SY=A. goniatus Nutt. ex Torr. &
            Gray
        SY=A. hypoglottis Hook.
   albens Greene
   albulus Woot. & Standl.
   allochrous Gray
   alpinus L.
      var. alpinus
        SY=Astragalus a. ssp. alaskanus
            Hulten
        SY=Astragalus alpinus ssp.
            arcticus (Bunge) Hulten
        SY=Atelophragma alpinum (L.) Rydb.
      var. brunetianus Fern.
        SY=Astragalus alpinus var.
            labradoricus (DC.) Fern.
   altus Woot. & Standl.
   alvordensis M.E. Jones
   amblytropis Barneby
   americanus (Hook.) M.E. Jones
        SY=Astragalus frigidus (L.) Gray
            pro parte
        SY=A. f. (L.) Gray var. a. (Hook.)
            S. Wats.
        SY=A. f. var. gaspensis (Rouss.)
            Fern.
        SY=Phaca americana (Hook.) Rydb.
   amnis-amissi Barneby
   amphioxys Gray
      var. amphioxys
        SY=Astragalus a. var. melanocalyx
            (Rydb.) Tidestrom
      var. modestus Barneby
      var. vespertinus (Sheldon) M.E. Jones

Astragalus (CONT.)
  ampullarius S. Wats.
  andersonii Gray
  anisus M.E. Jones
  applegatii M.E. Peck
  aquilonius (Barneby) Barneby
  aretioides (M.E. Jones) Barneby
  argophyllus Nutt. ex Torr. & Gray
        var. argophyllus
            SY=Xylophacos a. (Nutt. ex Torr. &
                Gray) Rydb.
        var. martinii M.E. Jones
            SY=Astragalus a. var.
                pephragmenoides Barneby
        var. panguicensis (M.E. Jones) M.E.
            Jones
            SY=Astragalus p. (M.E. Jones) M.E.
                Jones
  aridus Gray
  arizonicus Gray
  arrectus Gray
            SY=Astragalus palousensis Piper
  arthurii M.E. Jones
  asclepiadoides M.E. Jones
  asymmetricus Sheldon
  atratus S. Wats.
        var. atratus
        var. inseptus Barneby
        var. mensanus M.E. Jones
            SY=Astragalus m. (M.E. Jones)
                Abrams
        var. owyheensis (A. Nels. & J.F.
            Macbr.) M.E. Jones
            SY=Astragalus o. A. Nels. & J.F.
                Macbr.
  atropubescens Coult. & Fish.
            SY=Astragalus arrectus Gray var.
                kelseyi (Rydb.) M.E. Jones
  atwoodii Welsh & Thorne
  austinae Gray ex Brewer & S. Wats.
  ?australis (L.) Lam.
  barnebyi Welsh & Atwood
  barrii Barneby
  beathii C.L. Porter
  beatleyae Barneby
  beckwithii Torr. & Gray
        var. beckwithii
        var. purpureus M.E. Jones
        var. sulcatus Barneby
        var. weiserensis M.E. Jones
            SY=Astragalus w. (M.E. Jones)
                Abrams
  bernardinus M.E. Jones
  bicristatus Gray
  bisulcatus (Hook.) Gray
        var. bisulcatus
            SY=Astragalus diholcos Tidestrom
            SY=Diholcos b. (Hook.) Rydb.
        var. haydenianus (Gray) Barneby
            SY=Astragalus h. Gray
        var. nevadensis (M.E. Jones) Barneby
  bodinii Sheldon
            SY=Astragalus b. var. yukonis
                (M.E. Jones) Boivin
            SY=A. stragulus Fern.
            SY=A. y. M.E. Jones
            SY=Phaca b. (Sheldon) Rydb.
  bolanderi Gray
  bourgovii Gray
  brandegei Porter
  brauntonii Parish
  brazoensis Buckl.
  breweri Gray
  bryantii Barneby
  californicus (Gray) Greene
  callithrix Barneby
  calycosus Torr. ex S. Wats.

        var. calycosus
        var. mancus (Rydb.) Barneby
        var. monophyllidius (Rydb.) Barneby
        var. scaposus (Gray) M.E. Jones
  camptopus Barneby
  canadensis L.
        var. brevidens (Gandog.) Barneby
            SY=Astragalus b. (Gandog.) Rydb.
        var. canadensis
            SY=Astragalus c. var. carolinianus
                (L.) M.E. Jones
            SY=A. canadensis var. longilobus
                Fassett
            SY=A. carolinianus L.
            SY=A. halei Rydb.
        var. mortonii (Nutt.) S. Wats.
            SY=Astragalus m. Nutt.
  caricinus (M.E. Jones) Barneby
            SY=Astragalus lyallii Gray var. c.
                M.E. Jones
  casei Gray
  castaneiformis S. Wats.
        var. castaneiformis
            SY=Astragalus c. var. typicus
                Barneby
        var. consobrinus Barneby
  castetteri Barneby
  ceramicus Sheldon
        var. apus Barneby
        var. ceramicus
        var. filifolius (Gray) F.J. Herm.
            SY=Astragalus longifolius (Pursh)
                Rydb.
            SY=Phaca l. (Pursh) Nutt.
  cerussatus Sheldon
  chamaeleuce Gray
  chamaemeniscus Barneby
  chinensis L. f.
  chloodes Barneby
  cibarius Sheldon
  cicer L.
  cimae M.E. Jones
        var. cimae
        var. sufflatus Barneby
  clarianus Jepson
  clevelandii Greene
  cobrensis Gray
        var. cobrensis
        var. maguirei Kearney
  coccineus Brandeg.
  collinus (Hook.) Don
        var. collinus
        var. laurentii (Rydb.) Barneby
            SY=Astragalus l. (Rydb.) M.E. Peck
  coltonii M.E. Jones
        var. coltonii
        var. moabensis M.E. Jones
            SY=Astragalus canovirens (Rydb.)
                Barneby
  columbianus Barneby
  congdonii S. Wats.
  conjunctus S. Wats.
            SY=Astragalus reventus Gray var.
                c. (S. Wats.) M.E. Jones
  convallarius Greene
        var. convallarius
        var. finitimus Barneby
        var. scopulorum Barneby
  cottamii Welsh
  cottonii M.E. Jones
  crassicarpus Nutt.
        var. berlandieri Barneby
            SY=Astragalus mexicanus A. DC.
            SY=Geoprumnon m. (A. DC.) Rydb.
        var. cavus Barneby
        var. crassicarpus
            SY=Astragalus coryocarpus

Astragalus (CONT.)
    Ker-Gawl.
     SY=A. succulentus Richards.
     SY=Geoprumnon crassicarpum (Nutt.)
      Rydb.
     SY=G. s. (Richards.) Rydb.
    var. paysonii (E.H. Kelso) Barneby
    var. trichocalyx (Nutt.) Barneby ex
     Gleason
     SY=Astragalus mexicanus A. DC.
      var. t. (Nutt.) Fern.
     SY=A. t. Nutt.
cremnophylax Barneby
    var. cremnophylax
    var. myriorrhaphis Barneby
cronquistii Barneby
crotalariae (Benth.) Gray
curtipes Gray
curvicarpus (Heller) J.F. Macbr.
    var. brachycodon (Barneby) Barneby
     SY=Astragalus whitedii Piper var.
      b. Barneby
    var. curvicarpus
    var. subglaber (Rydb.) Barneby
     SY=Astragalus s. (Rydb.) M.E. Peck
cusickii Gray
    var. cusickii
    var. flexilipes Barneby
cyaneus Gray
cymboides M.E. Jones
    SY=Astragalus amphioxys Gray var.
     cymbellus (M.E. Jones) M.E.
     Jones
deanei (Rydb.) Barneby
desereticus Barneby
desperatus M.E. Jones
    var. conspectus Barneby
    var. desperatus
     SY=Astragalus d. var. petrophilus
      M.E. Jones
deterior (Barneby) Barneby
detritalis M.E. Jones
diaphanus Dougl. ex Hook.
    SY=Astragalus d. var. diurnus (S.
     Wats.) Barneby ex M.E. Peck
    SY=A. diurnus S. Wats.
    SY=A. drepanolobus Gray
didymocarpus Hook. & Arn.
    var. didymocarpus
     SY=Astragalus catalinensis Nutt.
     SY=A. d. var. daleoides Barneby
    var. dispermus (Gray) Jepson
     SY=Astragalus dispermus Gray
    var. milesianus (Rydb.) Jepson
    var. obispoensis (Rydb.) Jepson
distortus Torr. & Gray
    var. distortus
     SY=Holophacos d. (Torr. & Gray)
      Rydb.
    var. engelmannii (Sheldon) M.E. Jones
     SY=Astragalus e. Sheldon
diversifolius Gray
douglasii (Torr. & Gray) Gray
    var. douglasii
     SY=Astragalus d. var. megalophysus
      (Rydb.) Munz & McBurney
    var. parishii (Gray) M.E. Jones
     SY=Astragalus p. Gray
    var. perstrictus (Rydb.) Munz &
     McBurney ex Munz
     SY=Astragalus parishii Gray ssp.
      perstrictus (Rydb.) Abrams
drabelliformis Barneby
drummondii Dougl. ex Hook.
    SY=Tium d. (Dougl. ex Hook.) Rydb.
duchesnensis M.E. Jones
eastwoodae M.E. Jones

egglestonii (Rydb.) Kearney & Peebles
emoryanus (Rydb.) Cory
    var. emoryanus
    var. terlinguensis (Cory) Barneby
endopterus (Barneby) Barneby
    SY=Astragalus wootonii Sheldon
      var. e. Barneby
ensiformis M.E. Jones
episcopus S. Wats.
    SY=Astragalus kaibensis M.E. Jones
eremiticus Sheldon
    SY=Astragalus e. var. malheurensis
     (Heller) Barneby
    SY=A. e. var. spencianus M.E.
     Jones
    SY=A. m. Heller
eucosmus B.L. Robins.
    SY=Astragalus e. ssp. sealei
     (Lepage) Hulten
    SY=Astragalus e. var. facinorum
     Fern.
    SY=Astragalus parviflorus (Pursh)
     Nutt.
    SY=Astragalus s. Lepage
    SY=Atelophragma elegans (Hook.)
     Rydb.
eurekensis M.E. Jones
falcatus Lam.
feensis M.E. Jones
    SY=Astragalus sanctae-fidei
     Tidestrom
filipes Torr. ex Gray
    SY=Astragalus f. var. residuus
     Jepson
    SY=A. macgregorii (Rydb.)
     Tidestrom
    SY=A. stenophyllus Torr. & Gray
    SY=A. s. var. f. (Torr. ex Gray)
     Tidestrom
flavus Nutt. ex Torr. & Gray
    var. argillosus (M.E. Jones) Barneby
    var. candicans Gray
     SY=Astragalus confertiflorus Gray
    var. flavus
     SY=Astragalus confertiflorus Gray
      var. flaviflorus (Kuntze)
      M.E. Jones
flexuosus (Hook.) G. Don
    var. diehlii (M.E. Jones) Barneby
    var. flexuosus
     SY=Astragalus f. var. elongatus
      (Hook.) M.E. Jones
     SY=A. f. var. sierrae-blancae
      (Rydb.) Barneby
     SY=Pisophaca f. (Hook.) Rydb.
    var. greenei (Gray) Barneby
     SY=A. stictocarpus (Rydb.)
      Tidestrom
fucatus Barneby
funereus M.E. Jones
gambelianus Sheldon
    SY=Astragalus g. ssp. elmeri
     (Greene) Abrams
geyeri Gray
    var. geyeri
    var. triquetrus (Gray) M.E. Jones
     SY=Astragalus t. Gray
gibbsii Kellogg
giganteus S. Wats.
gilensis Greene
gilmanii Tidestrom
gilviflorus Sheldon
    SY=Astragalus triphyllus Pursh
glycyphyllos L.
gracilis Nutt.
grayi Parry ex S. Wats.
gypsodes Barneby

Astragalus (CONT.)
  hallii Gray
          var. fallax (S. Wats.) Barneby
              SY=Astragalus famelicus Sheldon
          var. hallii
  hamiltonii C.L. Porter
  harrisonii Barneby
  hartwegii Benth.
  hoodianus T.J. Howell
              SY=Astragalus conjunctus S. Wats.
                  var. oxytropidoides M.E.
                  Jones
              SY=A. reventus Gray var. o. (M.E.
                  Jones) C.L. Hitchc.
              SY=Cnemidophacos knowlesianus
                  Rydb.
  hornii Gray
  howellii Gray
  humillimus Gray ex Brandeg.
  humistratus Gray
          var. crispulus Barneby
          var. hosackiae (Greene) M.E. Jones
          var. humistratus
              SY=Astragalus datilensis (Rydb.)
                  Tidestrom
          var. humivagans (Rydb.) Barneby
          var. sonorae (Gray) M.E. Jones
          var. tenerrimus M.E. Jones
  hyalinus M.E. Jones
  hypoxylus S. Wats.
  inflexus Dougl. ex Hook.
  insularis Kellogg
          var. harwoodii Munz & McBurney ex
              Munz
              SY=Astragalus h. (Munz & McBurney
                  ex Munz) Abrams
          var. insularis
  inversus M.E. Jones
  inyoensis Sheldon
  iodanthus S. Wats.
          var. iodanthus
              SY=Astragalus i. var. diaphanoides
                  Barneby
          var. vipereus Barneby
  iodopetalus (Rydb.) Barneby
  iselyi Welsh
  jaegerianus Munz
  jejunus S. Wats.
  johannis-howellii Barneby
  kentrophyta Gray
          var. coloradoensis M.E. Jones
          var. danaus (Barneby) Barneby
          var. douglasii Barneby
          var. elatus (S. Wats.) Barneby
              SY=Astragalus impensis (Sheldon)
                  Woot. & Standl.
          var. implexus (Canby) Barneby
              SY=Astragalus tegetarius S. Wats.
          var. jessiae (M.E. Peck) Barneby
              SY=Astragalus j. M.E. Peck
          var. kentrophyta
              SY=Astragalus viridis (Nutt.)
                  Sheldon
              SY=Kentrophyta montana Nutt. ex
                  Torr. & Gray
          var. neomexicanus (Barneby) Barneby
          var. ungulatus M.E. Jones
  lancearius Gray
  layneae Greene
  leibergii M.E. Jones
              SY=Astragalus arrectus Gray var.
                  l. (M.E. Jones) M.E. Jones
  lemmonii Gray
  lentiformis Gray ex Brewer & S. Wats.
  lentiginosus Dougl. ex Hook.
          var. albiflorus (Gray) Schoener
              SY=Astragalus l. var. diphysus

                  (Gray) M.E. Jones
          var. albifolius M.E. Jones
              SY=Astragalus a. (M.E. Jones)
                  Abrams
          var. ambiguus Barneby
          var. antonius Barneby
          var. araneosus (Sheldon) Barneby
              SY=Astragalus a. Sheldon
          var. australis Barneby
          var. borreganus M.E. Jones
              SY=Astragalus agninus Jepson
              SY=A. arthu-schottii Gray
              SY=A. coulteri Benth.
              SY=A. l. var. c. (Benth.) M.E.
                  Jones
          var. chartaceus M.E. Jones
          var. coachellae Barneby ex Shreve &
              Wiggins
          var. floribundus Gray
          var. fremontii (Gray) S. Wats.
              SY=Astragalus f. Gray
              SY=A. f. ssp. eremicus (Sheldon)
                  Abrams
          var. idriensis M.E. Jones
              SY=Astragalus i. (M.E. Jones)
                  Abrams
              SY=A. tehatchapiensis (Rydb.)
                  Tidestrom
          var. ineptus (Gray) M.E. Jones
              SY=Astragalus i. Gray
          var. kennedyi (Rydb.) Barneby
          var. kernensis (Jepson) Barneby
              SY=Astragalus l. var.
                  charlestonensis (Clokey)
                  Barneby
          var. latus (M.E. Jones) M.E. Jones
          var. lentiginosus
              SY=Astragalus l. var. carinatus
                  M.E. Jones
          var. macrolobus (Rydb.) Barneby
          var. maricopae Barneby
          var. micans Barneby
          var. nigricalycis M.E. Jones
              SY=Astragalus n. (M.E. Jones)
                  Abrams
          var. oropedii Barneby
          var. palans (M.E. Jones) M.E. Jones
          var. piscinensis Barneby
          var. platyphyllidius (Rydb.) M.E.
              Peck
          var. salinus (T.J. Howell) Barneby
              SY=Astragalus s. T.J. Howell
          var. scorpionis M.E. Jones
              SY=Astragalus l. var. tremuletorum
                  Barneby
          var. semotus Jepson
          var. sesquimetralis (Rydb.) Barneby
          var. sierrae M.E. Jones
          var. stramineus (Rydb.) Barneby
          var. toyabensis Barneby
          var. ursinus (Gray) Barneby
          var. variabilis Barneby
          var. vitreus Barneby
          var. wilsonii (Greene) Barneby
          var. yuccanus M.E. Jones
  leptaleus Gray
  leptocarpus Torr. & Gray
  leucolobus S. Wats. ex M.E. Jones
  limnocharis Barneby
  lindheimeri Engelm. ex Gray
  linifolius Osterhout
  loanus Barneby
              SY=Astragalus newberryi Gray var.
                  wardianus Barneby
  lonchocarpus Torr.
  lotiflorus Hook.
              SY=Astragalus l. var. nebraskensis

Astragalus (CONT.)
                    Bates
                    SY=A. l. var. reverchonii (Gray)
                        M.E. Jones
                    SY=Batidophaca l. (Hook.) Rydb.
    lutosus M.E. Jones
    lyallii Gray
    macrodon (Hook. & Arn.) Gray
    magdalenae Greene
            var. niveus (Rydb.) Barneby
                    SY=Astragalus n. (Rydb.) Barneby
            var. peirsonii (Munz & McBurney)
                Barneby
                    SY=Astragalus p. Munz & McBurney
    malacoides Barneby
    malacus Gray
    marianus (Rydb.) Barneby
    megacarpus (Nutt.) Gray
    michauxii (Kuntze) F.J. Herm.
                    SY=Tium m. (Kuntze) Rydb.
    microcymbus Barneby
    microcystis Gray
    micromerius Barneby
    miguelensis Greene
    minthorniae (Rydb.) Jepson
            var. gracilior (Barneby) Barneby
            var. minthorniae
            var. villosus Barneby
    misellus S. Wats.
            var. misellus
                    SY=Astragalus howellii Gray var.
                        aberrans (M.E. Jones) C.L.
                        Hitchc.
            var. pauper Barneby
    miser Dougl. ex Hook.
            var. crispatus (M.E. Jones) Cronq.
                    SY=Astragalus decumbens (Nutt. ex
                        Torr. & Gray) Gray var. c.
                        (M.E. Jones) Cronq. &
                        Barneby
            var. decumbens (Nutt.) Cronq.
                    SY=Astragalus d. (Nutt.) Gray
            var. hylophilus (Rydb.) Barneby
            var. miser
                    SY=Astragalus strigosus Coult. &
                        Fish.
            var. oblongifolius (Rydb.) Cronq.
                    SY=Astragalus decumbens (Nutt. ex
                        Torr. & Gray) Gray var. o.
                        (Rydb.) Cronq.
            var. praeteritus Barneby
            var. serotinus (Gray) Barneby
                    SY=Astragalus decumbens (Nutt. ex
                        Torr. & Gray) Gray var. s.
                        (Gray) M.E. Jones
                    SY=A. s. Gray
            var. tenuifolius (Nutt.) Barneby
    missouriensis Nutt.
            var. amphibolus Barneby
            var. mimetes Barneby
            var. missouriensis
                    SY=Xylophacos m. (Nutt.) Rydb.
    moencoppensis M.E. Jones
    mohavensis S. Wats.
            var. hemigyrus (Clokey) Barneby
            var. mohavensis
    mokiacensis Gray
                    SY=Astragalus lentiginosus Dougl.
                        ex Hook. var. m. (Gray) M.E.
                        Jones
    mollissimus Torr.
            var. bigelovii (Gray) Barneby ex B.L.
                Turner
                    SY=Astragalus b. Gray
                    SY=A. b. var. typicus Barneby
            var. coryi Tidestrom
            var. earlei (Greene ex Rydb.)

                    Tidestrom
            var. marcidus (Greene ex Rydb.)
                Barneby ex B.L. Turner
            var. matthewsii (S. Wats.) Barneby
                    SY=Astragalus m. S. Wats.
            var. mogollonicus (Greene) Barneby
                    SY=Astragalus bigelovii Gray var.
                        m. (Greene) Barneby
            var. mollissimus
            var. thompsonae (S. Wats.) Barneby
                    SY=Astragalus bigelovii Gray var.
                        t. (S. Wats.) M.E. Jones
                    SY=A. t. S. Wats.
    molybdenus Barneby
                    SY=Astragalus plumbeus Barneby
    monoensis Barneby
    monumentalis Barneby
    mulfordae M.E. Jones
    musimonum Barneby
    musiniensis M.E. Jones
    naturitensis Payson
    neglectus (Torr. & Gray) Sheldon
                    SY=Phaca n. Torr. & Gray
    nelsonianus Barneby
    neomexicanus Woot. & Standl.
    nevinii Gray
    newberryi Gray
            var. blyae (Rydb.) Barneby
            var. newberryi
    nidularius Barneby
    nothoxys Gray
    nudisiliquus A. Nels.
    nutans M.E. Jones
                    SY=Astragalus chuckwallae Abrams
    nuttallianus DC.
            var. austrinus (Small) Barneby ex
                Shreve & Wiggins
                    SY=Astragalus a. (Small) O.E.
                        Schulz
            var. cedrosensis M.E. Jones
            var. imperfectus (Rydb.) Barneby
            var. macilentus (Small) Barneby
            var. micranthiformis Barneby
            var. nuttallianus
            var. pleianthus (Shinners) Barneby
            var. trichocarpus Torr. & Gray
            var. zapatanus Barneby
    nuttallii (Torr. & Gray) J.T. Howell
            var. nuttallii
                    SY=Astragalus menziesii Gray
            var. virgatus (Gray) Barneby
                    SY=Astragalus menziesii Gray ssp.
                        v. (Gray) Abrams
    nutzotinensis Rouss.
    nyensis Barneby
    obcordatus Ell.
                    SY=Phaca o. (Ell.) Rydb.
    obscurus S. Wats.
    oniciformis Barneby
    oocalycis M.E. Jones
    oocarpus Gray
    oophorus S. Wats.
            var. caulescens (M.E. Jones) M.E.
                Jones
            var. clokeyanus Barneby
            var. lonchocalyx Barneby
            var. oophorus
    oreganus Nutt. ex Torr. & Gray
    osterhoutii M.E. Jones
    oxyphysus Gray
    pachypus Greene
            var. jaegeri Munz & McBurney ex Munz
            var. pachypus
    palmeri Gray
                    SY=Astragalus vaseyi S. Wats.
                    SY=A. v. var. johnstonii Munz &
                        McBurney ex Munz

Astragalus (CONT.)
            SY=A. v. var. metanus (M.E. Jones)
                Munz & McBurney
    panamintensis Sheldon
    pardalinus (Rydb.) Barneby
    parryi Gray
    pattersonii Gray ex Brand
    pauperculus Greene
            SY=Astragalus bruceae (M.E. Jones)
                Abrams
    paysonii (Rydb.) Barneby
    peckii Piper
    pectinatus Dougl. ex G. Don
            SY=Cnemidophacos p. (Dougl. ex G.
                Don) Rydb.
    perianus Barneby
    phoenix Barneby
    pictiformis Barneby
    pinonis M.E. Jones
    plattensis Nutt. ex Torr. & Gray
            SY=Astragalus pachycarpus Torr. &
                Gray
            SY=Geoprumnon plattense (Nutt. ex
                Torr. & Gray) Rydb.
    platytropis Gray
    polaris Benth. ex Hook.
    pomonensis M.E. Jones
    porrectus S. Wats.
    praelongus Sheldon
        var. ellisiae (Rydb.) Barneby ex B.L.
            Turner
        var. lonchopus Barneby
        var. praelongus
            SY=Astragalus pattersonii Gray ex
                Brand var. praelongus
                (Sheldon) M.E. Jones
            SY=A. recedens (Greene ex Rydb.)
                C.L. Porter
    preussii Gray
        var. laxiflorus Gray
            SY=Astragalus crotalariae (Benth.)
                Gray var. davidsonii (Rydb.)
                Munz & McBurney
        var. preussii
            SY=Astragalus p. var. latus M.E.
                Jones
    proimanthus Barneby
    proximus (Rydb.) Woot. & Standl.
    pseudiodanthus Barneby
    pterocarpus S. Wats.
    pubentissimus Torr. & Gray
    pulsiferae Gray
        var. pulsiferae
        var. suksdorfii (T.J. Howell) Barneby
    puniceus Osterhout
        var. gertrudis (Greene) Barneby
        var. puniceus
    purshii Dougl. ex Hook.
        var. concinnus Barneby
        var. glareosus (Dougl.) Barneby
            SY=Astragalus g. Dougl.
            SY=A. ventosus Suksdorf ex Rydb.
        var. lagopinus (Rydb.) Barneby
            SY=Astragalus l. (Rydb.) M.E. Peck
        var. lectulus (S. Wats.)M.E. Jones
            SY=Astragalus jonesii Abrams
            SY=A. l. S. Wats.
        var. ophiogenes (Barneby) Barneby
            SY=Astragalus o. Barneby
        var. pumilio Barneby
        var. purshii
            SY=Astragalus incurvus (Rydb.)
                Abrams
            SY=A. p. var. interior M.E. Jones
        var. tinctus M.E. Jones
            SY=Astragalus candelarius Sheldon
            SY=A. leucolobus S. Wats. ex M.E.

            Jones ssp. consectus
                (Sheldon) Abrams
            SY=A. p. var. longilobus M.E.
                Jones
    pycnostachyus Gray
        var. lanosissimus (Rydb.) Munz &
            McBurney ex Munz
        var. pycnostachyus
    racemosus Pursh
        var. longisetus M.E. Jones
        var. racemosus
        var. treleasei C.L. Porter
    rafaelensis M.E. Jones
    rattanii Gray
        var. jepsonianus Barneby
        var. rattanii
    ravenii Barneby
    recurvus Greene
    reflexus Torr. & Gray
    remotus (M.E. Jones) Barneby
    reventiformis (Rydb.) Barneby
            SY=Astragalus reventus Gray var.
                canbyi M.E. Jones
    reventus Gray
    riparius Barneby
    ripleyi Barneby
    robbinsii (Oakes) Gray
        var. alpiniformis (Rydb.) Barneby
            SY=Astragalus a. Rydb.
        var. fernaldii (Rydb.) Barneby
            SY=Astragalus eucosmus B.L.
                Robins. var. f. (Rydb.)
                Boivin
            SY=A. f. (Rydb.) H.F. Lewis
        var. harringtonii (Rydb.) Barneby
            SY=Astragalus h. (Rydb.) Hulten
            SY=Atelophragma h. Rydb.
        var. jesupii Egglest. & Sheldon
            SY=Astragalus j. (Egglest. &
                Sheldon) Britt.
        var. minor (Hook.) Barneby
            SY=Astragalus blakei Egglest.
            SY=A. collieri (Rydb.) Porsild
            SY=A. macounii Rydb.
            SY=A. r. var. b. (Egglest.)
                Barneby
        var. occidentalis S. Wats.
            SY=Astragalus o. (S. Wats.) M.E.
                Jones
        var. robbinsii
    rusbyi Greene
    sabulonum Gray
    sabulosus M.E. Jones
    salmonis M.E. Jones
    saurinus Barneby
    scaphoides (M.E. Jones) Rydb.
    schmollae C.L. Porter
    sclerocarpus Gray
    scopulorum Porter
    sepultipes (Barneby) Barneby
    serenoi (Kuntze) Sheldon
        var. serenoi
        var. sordescens Barneby
    sericoleucus Gray
            SY=Astragalus sericea sensu auctt.
                non DC.
            SY=Orophaca sericea sensu aucct.
                non (Nutt.) Britt.
    serpens M.E. Jones
    sesquiflorus S. Wats.
            SY=Astragalus s. var. brevipes
                Barneby
    sheldonii (Rydb.) Barneby
            SY=Astragalus conjunctus S. Wats.
                var. s. (Rydb.) M.E. Peck
            SY=A. reventus Gray var. s.
                (Rydb.) C.L. Hitchc.

Astragalus (CONT.)
  shevockii Barneby
  shortianus Nutt. ex Torr. & Gray
  siliceus Barneby
  simplicifolius (Nutt.) Gray
  sinuatus Piper
          SY=Astragalus whitedii Piper
  solitarius M.E. Peck
  sophoroides M.E. Jones
  soxmaniorum Lundell
  spaldingii Gray
  sparsiflorus Gray
          var. majusculus Gray
          var. sparsiflorus
  spatulatus Sheldon
          SY=Astragalus caespitosus (Nutt.)
                  Gray
          SY=A. s. var. uniflorus (Rydb.)
                  Barneby
          SY=Homalobus c. Nutt.
          SY=Orophaca c. (Nutt.) Britt.
  speirocarpus Gray
  sterilis Barneby
  straturensis M.E. Jones
  striatiflorus M.E. Jones
  subcinereus Gray
          SY=Astragalus sileranus M.E. Jones
          SY=A. sileranus var. cariacus M.E.
                  Jones
  subvestitus (Jepson) Barneby
  succumbens Dougl. ex Hook.
  tegetarioides M.E. Jones
  tenellus Pursh
          SY=Astragalus t. var. strigulosus
                  (Rydb.) F.J. Herm.
          SY=Homalobus t. (Pursh) Britt.
  tener Gray
          var. tener
          var. titi (Eastw.) Barneby
  tennesseensis Gray ex Chapman
          SY=Geoprumnon t. (Gray ex Chapman)
                  Rydb.
  tephrodes Gray
          var. brachylobus (Gray) Barneby
          SY=Astragalus arrectus Gray var.
                  pephragmenus M.E. Jones
          var. chloridae (M.E. Jones) Barneby
          var. eurylobus Barneby
          var. remulcus (Gray) Barneby
          var. tephrodes
          SY=Astragalus t. var. typicus
                  Barneby
  terminalis S. Wats.
  tetrapterus Gray
          SY=Astragalus cinerascens (Rydb.)
                  Tidestrom
          SY=A. t. var. capricornus M.E.
                  Jones
          SY=A. t. var. cinerascens (Rydb.)
                  Barneby
  thurberi Gray
  tidestromii (Rydb.) Clokey
  titanophilus Barneby
          SY=Astragalus convallarius Greene
                  var. foliolatus Barneby
  toanus M.E. Jones
  toquimanus Barneby
  traskiae Eastw.
  tricarinatus Gray
  trichopodus (Nutt.) Gray
          var. lonchus (M.E. Jones) Barneby
          SY=Astragalus leucopsis (Torr. &
                  Gray) Torr.
          SY=A. t. ssp. leucopsis (Torr. &
                  Gray) Thorne
          var. phoxus (M.E. Jones) Barneby
          SY=Astragalus antisellii Gray

          SY=A. gaviotus Elmer
          SY=A. t. ssp. a. (Gray) Thorne
          var. trichopodus
          SY=A. t. var. capillipes (M.E.
                  Jones) M.E. Jones
  tridactylicus Gray
          SY=Astragalus sericoleucus Gray
                  var. t. (Gray) M.E. Jones
  troglodytus S. Wats.
  tweedyi Canby
  tyghensis M.E. Peck
          SY=Astragalus spaldingii Gray var.
                  t. (M.E. Peck) C.L. Hitchc.
  umbellatus Bunge
          SY=Astragalus frigidus (L.) Gray
                  pro parte
  umbraticus Sheldon
  uncialis Barneby
  utahensis (Torr.) Torr. & Gray
  vaccarum Gray
  vallaris M.E. Jones
  vexilliflexus Sheldon
          var. nubilus Barneby
          var. vexilliflexus
          SY=Homalobus v. (Sheldon) Rydb.
  villosus Michx.
          SY=Phaca intonsa (Sheldon) Rydb.
  wardii Gray
  waterfallii Barneby
  webberi Gray
  wetherillii M.E. Jones
  whitneyi Gray
          var. confusus Barneby
          var. lenophyllus (Rydb.) Barneby
          var. siskiyouensis (Rydb.) Barneby
          SY=Astragalus w. ssp. s. (Rydb.)
                  Abrams
          var. sonneanus (Greene) Jepson
          SY=Astragalus w. ssp. hookerianus
                  (Torr. & Gray) Abrams
          var. whitneyi
          SY=Astragalus w. ssp. pinosus
                  (Elmer) Abrams
  williamsii Rydb.
  wingatanus S. Wats.
  wittmannii Barneby
  woodruffii M.E. Jones
  wootonii Sheldon
          SY=Astragalus w. var. typicus
                  Barneby
  wrightii Gray
  xiphoides (Barneby) Barneby
          SY=Astragalus convallarius Greene
                  var. x. Barneby
  zionis M.E. Jones

Baptisia Vent.

  alba (L.) R. Br.
          SY=B. albescens Small
  arachnifera Duncan
  australis (L.) R. Br.
          var. australis
          var. minor (Lehm.) Fern.
          SY=B. m. Lehm.
          SY=B. m. var. aberrans Larisey
  X bicolor Greenm. & Larisey [australis var.
                  minor X leucophaea]
  bracteata Muhl. ex Ell.
          SY=B. leucophaea Nutt. var. b.
                  (Muhl. ex Ell.) Isely
  bushii Small
  calycosa Canby
          var. calycosa
          var. villosa Canby
          SY=B. hirsuta Small
  cinerea (Raf.) Fern. & Schub.

Baptisia (CONT.)
        SY=B. villosa (Walt.) Nutt.
cuneata Small
X deamii Larisey [lactea X tinctoria var.
  creba]
X fragilis Larisey [lactea X sphaerocarpa]
X fulva Larisey [alba X perfoliata]
intercalata Larisey
X intermedia Larisey [leucophaea var.
  glabrescens X sphaerocarpa]
lactea (Raf.) Thieret
        SY=B. leucantha Torr. & Gray
        SY=B. leucantha var. divaricata
           Larisey
        SY=B. leucantha var. pauciflora
           Larisey
        SY=B. pendula Larisey
        SY=B. pendula var. macrophylla
           Larisey
        SY=B. pendula var. obovata Larisey
        SY=B. psammophila Larisey
lanceolata (Walt.) Ell.
        SY=B. elliptica Small
        SY=B. e. var. tomentosa Larisey
lecontei Torr. & Gray
leucophaea Nutt.
    var. glabrescens Larisey
    var. laevicaulis Canby
        SY=B. laevicaulis (Canby) Small
    var. leucophaea
macilenta Small ex Larisey
megacarpa Chapman ex Torr. & Gray
        SY=B. riparia Larisey
        SY=B. r. var. minima Larisey
microphylla Nutt.
nuttalliana Small
perfoliata (L.) R. Br.
pinetorum Larisey
serenae M.A. Curtis
simplicifolia Croom
sphaerocarpa Nutt.
        SY=B. viridis Larisey
X stricta Larisey [bracteata X
  sphaerocarpa]
X sulphurea Engelm. [lactea X sphaerocarpa]
tinctoria (L.) R. Br.
    var. crebra Fern.
    var. projecta Fern.
    var. tinctoria
        SY=B. gibbesii Small

Barbieria DC.

  pinnata (Pers.) Baill.

Bauhinia L.

  lunarioides Gray ex S. Wats.
        SY=B. congesta (Britt. & Rose)
           Lundell
        SY=B. macrantha Benth. ex Hemsl.
        SY=B. m. var. grayana Wunderlin
  monandra Kurz
  pauletia Pers.
  tomentosa L.
  ungula Jacq.

Brongniartia H.B.K.

  minutifolia S. Wats.

Caesalpinia L.

  bonduc (L.) Roxb.
        SY=C. crista sensu auctt. non L.
        SY=Guilandina b. L.
        SY=G. c. (L.) Small

        SY=Ticanto nuga (L.) Medic.
  brachycarpa (Gray) Fisher
        SY=Hoffmanseggia b. Gray
  caudata (Gray) Fisher
        SY=Hoffmanseggia c. Gray
  coriaria (Jacq.) Willd.
        SY=Libidibia c. (Jacq.) Schlecht.
  culebrae (Britt. & Wilson) Alain
  decapetala (Roth) Alston
        SY=Biancaea sepiaria (Roxb.)
           Todaro
        SY=C. s. Roxb.
  divergens Urban
        SY=Guilandina d. (Urban) Britt.
  drummondii (Torr. & Gray) Fisher
        SY=C. texensis (Fisher) Fisher
        SY=Hoffmanseggia d. Torr. & Gray
  gilliesii (Wallich ex Hook.) A. Dietr.
        SY=Poinciana g. Wallich ex Hook.
  jamesii (Torr. & Gray) Fisher
        SY=Hoffmanseggia j. Torr. & Gray
  major (Medic.) Dandy & Exell
        SY=Guilandina ovalifolia (Urban)
           Britt.
  melanosperma (Eggers) Urban
        SY=Guilandina m. Eggers
  mexicana Gray
        SY=Poinciana m. (Gray) Britt. &
           Rose
  monensis Britt.
  parryi (Fisher) Eifert
        SY=Hoffmanseggia p. (Fisher) B.L.
           Turner
  pauciflora (Griseb.) C. Wright ex Sauvalle
  peninsularis (Britt.) Eifert
  phyllanthoides Standl.
  portoricensis (Britt. & Wilson) Alain
        SY=Guilandina p. Britt. & Wilson
  pulcherrima (L.) Sw.
        SY=Poinciana p. L.
  virgata Fisher
        SY=Hoffmanseggia microphylla Torr.
  wootonii (Britt.) Eifert
        SY=C. atropunctata Eifert
        SY=Hoffmanseggia melanosticta
           (Schauer) Gray

Cajanus DC.

  cajan (L.) Druce
        SY=Cajan c. (L.) Millsp.

Calliandra Benth.

  biflora Tharp
  caracasana (Jacq.) Benth.
  conferta Benth.
  eriophylla Benth.
    var. chamaedrys Isely
    var. eriophylla
  haematostoma (Bertero ex DC.) Benth.
        SY=Anneslia h. (Bertero ex DC.)
           Britt.
  humilis Benth.
    var. humilis
        SY=C. herbacea Engelm. ex Gray
    var. reticulata (Gray) L. Benson
        SY=C. r. Gray
  portoricensis (Jacq.) Benth.
        SY=Anneslia p. (Jacq.) Britt.
  purpurea (L.) Benth.
        SY=Anneslia p. (L.) Britt.
  schottii Torr. ex S. Wats.

Calopogonium Desv.

  caeruleum (Benth.) Hemsl.

Calopogonium   (CONT.)
          SY=Stenolobium c. Benth.
     mucunoides Desv.
          SY=C. orthocarpum Urban

Canavalia DC.

     brasiliensis Mart. ex Benth.
     cathartica Thouars
     centralis St. John
     dictyota Piper
     ensiformis (L.) DC.
     forbesii St. John
     galeata (Gaud.) Vogel
     gladiata (Jacq.) DC.
     haleakalaensis St. John
     hawaiiensis Deg., Deg. & Sauer
     iaoensis St. John
     kauaiensis Sauer
     kauensis St. John
     lineata (Thunb.) DC.
     makahaensis St. John
     molokaiensis Deg., Deg. & Sauer
     munroi (Deg. & Deg.) St. John
     napaliensis St. John
     nitida (Cav.) Piper
          SY=C. rusiosperma Urban
     nualoloensis St. John
     peninsularis St. John
     pubescens Hook. & Arn.
          SY=C. lanaiensis (Rock) Deg. &
               Deg.
     rockii St. John
     rosea (Sw.) DC.
          SY=C. maritima (Aubl.) Thouars
     sanguinea St. John
     sericea Gray
     stenophylla St. John

Caragana Fabr.

     arborescens Lam.
     aurantica Koehne
     frutex (L.) K. Koch

Cassia L.

     absus L.
     aeschynomene DC.
          SY=C. leschenaultiana DC.
          SY=C. l. var. a. (DC.) Benth.
          SY=Chamaecrista a. (DC.) Greene
     alata L.
          SY=Herpetica a. (L.) Raf.
     angustifolia Vahl
          SY=Senna a. (Vahl) Batka
     antillana (Britt. & Rose) Alain
          SY=Chamaefistula a. Britt. & Rose
     armata S. Wats.
     aspera Muhl. ex Ell.
          SY=Chamaecrista a. (Muhl. ex Ell.)
               Greene
     bacillaris L. f.
          SY=Cassia gaudichaudii Hook. &
               Arn.
          SY=Chamaefistula b. (L. f.) G. Don
     bauhinioides Gray
          SY=Cassia b. var. arizonica B.L.
               Robins. ex J.F. Macbr.
     bicapsularis L.
          SY=Adipera b. (L.) Britt. & Rose
     calycioides DC. ex Colladon
          SY=Cassia aristellata (Pennell)
               Cory & Parks
     chapmanii Isely
          SY=Cassia bahamensis P. Mill.
          SY=Peiranisia b. (P. Mill.) Britt.

               & Rose
     coluteoides Colladon
     corymbosa Lam.
          SY=Adipera c. (Lam.) Britt. & Rose
     covesii Gray
     deeringiana (Small & Pennell) J.F. Macbr.
          SY=Chamaecrista d. Small & Pennell
     didymobotrya Fresn.
     diffusa DC.
          SY=Chamaecrista chamaecrista (L.)
               Britt.
     diphylla L.
     durangensis Rose
          var. iselyi Irwin & Barneby
     ekmaniana Urban
     emarginata L.
          SY=Isandrina e. (L.) Britt. & Rose
     exunguis Urban
     fasciculata Michx.
          var. brachiata (Pollard) Pullen ex
               Isely
          SY=Cassia b. (Pollard) J.F. Macbr.
          SY=Chamaecrista b. Pollard
          var. fasciculata
          SY=Cassia f. var. depressa
               (Pollard) J.F. Macbr.
          SY=Cassia f. var. macrosperma
               Fern.
          SY=Cassia f. var. robusta
               (Pollard) J.F. Macbr.
          SY=Cassia r. Pollard
          SY=Chamaecrista d. (Pollard)
               Greene
          SY=Chamaecrista f. (Michx.) Greene
          SY=Chamaecrista r. (Pollard)
               Pollard ex Heller
          var. puberula (Greene) J.F. Macbr.
          SY=Cassia f. var. depressa
               (Pollard) J.F. Macbr.
          SY=Cassia mississippiensis Pollard
          SY=Chamaecrista littoralis Pollard
          SY=Chamaecrista m. (Pollard)
               Pollard ex Heller
          var. rostrata (Woot. & Standl.) B.L.
               Turner
          SY=Cassia r. (Woot. & Standl.)
               Tidestrom
     fistula L.
     floribunda Cav.
          SY=Adipera laevigata (Willd.)
               Britt. & Rose
          SY=Cassia l. Willd.
     fruticosa P. Mill.
     glandulosa L.
          var. glandulosa
          var. swartzii (Wikstr.) J.F. Macbr.
          SY=Chamaecrista s. (Wikstr.)
               Britt.
     grammica Spreng.
          SY=Chamaecrista g. (Spreng.)
               Pollard
     grandis L. f.
     granulata (Urban) J.F. Macbr.
          SY=Chamaecrista g. (Urban) Britt.
     greggii Gray
     hebecarpa Fern.
          SY=Cassia h. var. longipila E.L.
               Braun
     hirsuta L.
          SY=Ditremexa h. (L.) Britt. & Rose
     italica L.
          SY=Cassia obovata Colladon
          SY=Senna o. (Colladon) Batka
     javanica L.
     keyensis (Pennell) J.F. Macbr.
          SY=Chamaecrista k. Pennell
     leptadenia Greenm.

Cassia   (CONT.)
    leptocarpa Benth.
        var. glaberrima M.E. Jones
    ligustrina L.
        SY=Ditremexa l. (L.) Britt. & Rose
    lindheimerana Scheele
    marilandica L.
        SY=Cassia medsgeri Shafer
        SY=Ditremexa marilandica (L.)
            Britt. & Rose
        SY=D. medsgeri (Shafer) Britt. &
            Rose
    mirabilis (Pollard) Urban
        SY=Chamaecrista m. Pollard
    nictitans L.
        SY=Cassia mohrii Pollard
        SY=Cassia n. var. hebecarpa Fern.
        SY=Cassia n. var. leiocarpa Fern.
        SY=Chamaecrista m. (Pollard) Small
            ex Britt. & Rose
        SY=Chamaecrista n. (L.) Moench
        SY=Chamaecrista procumbens (L.)
            Greene
    obtusifolia L.
        SY=Cassia tora L.
        SY=Emelista t. (L.) Britt. & Rose
    occidentalis L.
        SY=Ditremexa o. (L.) Britt. & Rose
    orcuttii (Britt. & Rose) B.L. Turner
    patellaria DC.
    pilosa L.
    pilosior (J.F. Macbr.) Irwin & Barneby
    polyphylla Jacq.
        SY=Peiranisia p. (Jacq.) Britt. &
            Rose
    portoricensis Urban
        SY=Chamaecrista p. (Urban) Cook &
            Collins
    pumilio Gray
    ripleyi Irwin & Barneby
    roemerana Scheele
    rotundifolia Pers.
        SY=Chamaecrista r. (Pers.) Greene
    serpens L.
    siamea Lam.
    stahlii Urban
        SY=Adipera s. (Urban) Britt. &
            Rose
    surattensis Burm. f.
        var. suffruticosa (Koenig ex Roth)
            J.R. Sealy ex Isely
            SY=Cassia suffruticosa Koenig ex
                Roth
        var. surattensis
            SY=Cassia glauca Lam.
    texana Buckl.
    tomentosa L. f.
    wislizenusii Gray
    wrightii Gray

Centrosema (DC.) Benth.

    arenicola (Small) F.J. Herm.
        SY=Bradburya a. Small
    floridanum (Britt.) Lakela
        SY=Bradburya f. Britt.
    plumieri (Turp. ex Pers.) Benth.
        SY=Bradburya p. (Turp. ex Pers.)
            Kuntze
    pubescens Benth.
        SY=Bradburya p. (Benth.) Kuntze
    virginianum (L.) Benth.
        SY=Bradburya v. (L.) Kuntze
        SY=C. v. var. ellipticum (DC.)
            Fern.

Cercis L.

canadensis L.
    var. canadensis
    var. mexicana (Rose) M. Hopkins
    var. texensis (S. Wats.) M. Hopkins
occidentalis Torr. ex Gray

Chapmannia Torr. & Gray

    floridana Torr. & Gray

Christia Moench

    verspertilionis (L. f.) Bakh. f.
        SY=Lourea v. (L. f.) Desv.

Cicer L.

    arietinum L.

Cladrastis Raf.

    kentukea (Dum.-Cours.) Rudd
        SY=C. lutea (Michx. f.) K. Koch

Clitoria L.

    fragrans Small
        SY=Martiusia f. (Small) Small
    laurifolia Poir.
        SY=Martiusia l. (Poir.) Britt.
    mariana L.
        SY=Martiusia m. (L.) Small
    rubiginosa Juss.
        SY=Martiusia r. (Juss.) Britt.
    ternatea L.

Cologania Kunth

    angustifolia H.B.K.
        SY=C. longifolia Gray
    lemmonii Gray
    pallida Rose
    pulchella H.B.K.

Colutea L.

    arborescens L.

Coronilla L.

    scorpioides (L.) W.D.J. Koch
    varia L.

Corynella DC.

    paucifolia DC.

Coursetia DC.

    axillaris Coult. & Rose
    microphylla Gray

Cracca Benth.

    caribaea (Jacq.) Benth.
        SY=Benthamantha c. (Jacq.) Kunth
        SY=C. edwardsii Gray pro parte
        SY=Galega c. Jacq.
        SY=Tephrosia c. (Jacq.) DC.

Crotalaria L.

    assamica Benth.
    brevidens Benth.
        SY=C. intermedia Kotschy
    incana L.
    juncea L.

Crotalaria (CONT.)
  lanceolata E. Mey.
  longirostrata Hook. & Arn.
  lotifolia L.
          SY=C. l. var. eggersii Senn &
                  Vieques
  pallida Ait.
          SY=C. mucronata Desv.
          SY=C. striata DC.
  pilosa P. Mill.
  pumila Ortega
  purshii DC.
          SY=C. p. var. bracteolifera Fern.
  retusa L.
  rotundifolia (Walt.) Poir.
        var. rotundifolia
          SY=C. linaria Small
          SY=C. maritima Chapman
          SY=C. m. var. l. (Small) Senn
        var. vulgaris Windler
          SY=C. angulata P. Mill. sensu Senn
  sagittalis L.
          SY=C. s. var. blumeriana Senn
          SY=C. s. var. fruticosa (P. Mill.)
                  Fawcett & Rendle
          SY=C. s. var. oblonga Michx.
  spectabilis Roth
          SY=C. retzii A.S. Hitchc.
  stipularia Desv.
          SY=C. s. var. serpyllifolia DC.
  verrucosa L.
  zanzibarica Benth.
          SY=C. usaramoensis Baker f.

Cynometra L.

  portoricensis Krug & Urban

Cytisus L.

  canariensis (L.) Kuntze
  X dallimorei Rolfe [multiflorus X scoparius
      var. andreanus]
  linifolia (L.) Webb & Berth.
          SY=C. l. (L.) Lam.
  monspessulanus L.
          SY=Teline m. (L.) K. Koch
  multiflorus (L'Her. ex Ait.) Sweet
  proliferus L.f.
  X racemosus Nichols.
  scoparius (L.) Link
        var. andreanus (Puiss.) Dippel
        var. scoparius
  stenopetalus (Webb) Christ
          SY=C. maderensis Masf.

Dalbergia L.f.

  amerimnon Benth.
          SY=Amerimnon brownii Jacq.
  ecastophyllum (L.) Taubert
          SY=Ecastophyllum e. (L.) Britt.
  monetaria L.f.
          SY=Securidaca volubilis L.
  sissoo Roxb. ex DC.

Dalea L. ex Juss.

  albiflora Gray
          SY=D. ordiae Gray
          SY=Petalostemon pilosulus Rydb.
  aurea Nutt.
          SY=Parosela a. (Nutt.) Britt.
  ?ballii Gray
  bartonii Barneby
  bicolor Humb. & Bonpl. ex Willd.
        var. argyraea (Gray) Barneby

          SY=D. a. Gray
  brachystachya Gray
          SY=D. lemmonii Parry
  candida (Michx.) Willd.
        var. candida
          SY=Petalostemon c. Michx.
        var. oligophylla (Torr.) Shinners
          SY=Petalostemon c. (Willd.) Michx.
                  var. o. (Torr.) F.J. Herm.
          SY=Petalostemon occidentale
                  (Heller) Fern.
          SY=P. oligophyllum Torr.
  carnea (Michx.) Poir.
        var. albida (Torr. & Gray) Barneby
          SY=Petalostemon a. (Torr. & Gray)
                  Small
        var. carnea
          SY=Petalostemon c. Michx.
        var. gracilis (Nutt.) Barneby
          SY=Petalostemon g. Nutt.
  carthagenensis (Jacq.) J.F. Macbr.
        var. carthagenensis
          SY=D. c. ssp. domingensis (DC.)
                  Clausen
          SY=Parosela d. (DC.) Millsp.
        var. floridana (Rydb.) Barneby
          SY=Parosela f. Rydb.
        var. portoricana Barneby
  compacta Spreng.
        var. compacta
          SY=Petalostemon c. (Spreng.)
                  Swezey
          SY=P. decumbens Nutt.
        var. pubescens (Gray) Barneby
          SY=Petalostemon pulcherrimum
                  (Heller) Heller
  cylindriceps Barneby
  emarginata (Torr. & Gray) Shinners
          SY=Petalostemon e. Torr. & Gray
  enneandra Nutt.
          SY=D. laxifolia Pursh
          SY=Parosela e. (Nutt.) Britt.
  exigua Barneby
          SY=Petalostemon exile Gray
  exserta (Rydb.) Gentry
  feayi (Chapman) Barneby
          SY=Petalostemon f. Chapman
  filiformis Gray
  flavescens (S. Wats.) Welsh
          SY=D. epica Welsh
          SY=Petalostemon f. S. Wats.
  foliosa (Gray) Barneby
          SY=Petalostemon f. Gray
  formosa Torr.
  frutescens Gray
          SY=D. f. var. laxa B.L. Turner
  gattingeri (Heller) Barneby
          SY=Petalostemon g. Heller
  grayi (Vail) L.O. Williams
  greggii Gray
  hallii Gray
  jamesii (Torr.) Torr. & Gray
  lachnostachys Gray
  lanata Spreng.
        var. lanata
          SY=Parosela l. (Spreng.) Britt.
        var. terminalis (M.E. Jones) Barneby
          SY=D. glaberrima S. Wats.
          SY=D. t. M.E. Jones
  laniceps Barneby
  lasiathera Gray
  leporina (Ait.) Bullock
          SY=D. alopecuroides Willd.
          SY=D. lagopus (Cav.) Willd.
          SY=Parosela a. (Willd.) Rydb.
  lumholtzii Robbins & Fern.
  mollis Benth.

Dalea (CONT.)
  mollissima (Rydb.) Munz
          SY=D. mollis Benth. var.
              mollissima (Rydb.) Munz
          SY=D. neomexicana (Gray) Cory ssp.
              mollissima (Rydb.) Wiggins
  multiflora (Nutt.) Shinners
          SY=Petalostemon m. Nutt.
  nana Torr.
      var. carnescens (Rydb.) Kearney &
          Peebles
          SY=D. n. var. elatior Gray ex B.L.
              Turner
      var. nana
          SY=Parosela n. (Torr.) Heller
  neomexicana (Gray) Cory
      var. longipila (Rydb.) Barneby
          SY=D. l. (Rydb.) Cory
      var. neomexicana
  obovata (Torr. & Gray) Shinners
          SY=Petalostemon o. Torr. & Gray
  ornata (Dougl. ex Hook.) Eat. & Wright
          SY=Petalostemon o. Dougl. ex Hook.
  ?pedunculata Pursh
          SY=Orbexilum p. (Pursh) Rydb.
          SY=Psoralea p. (Pursh) Poir.
  phleoides (Torr. & Gray) Shinners
      var. microphylla (Torr. & Gray)
          Barneby
          SY=Petalostemon m. (Torr. & Gray)
              Heller
      var. phleoides
          SY=Petalostemon glandulosum Coult.
              & Fish.
          SY=P. p. Torr. & Gray
  pinnata (Walt. ex J.F. Gmel.) Barneby
      var. adenopoda (Rydb.) Barneby
          SY=Kuhnistera a. Rydb.
          SY=K. truncata Small
      var. pinnata
          SY=Kuhnistera p. (Walt. ex J.F.
              Gmel.) Kuntze
          SY=Petalostemon p. (Walt. ex J.F.
              Gmel.) Blake
      var. trifoliata (Chapman) Barneby
          SY=Petalostemon t. (Chapman)
              Wemple
  pogonathera Gray
      var. pogonathera
      var. walkerae (Tharp & Barkl.) B.L.
          Turner
  polygonoides Gray
          SY=D. p. var. anomala (M.E. Jones)
              Morton
  pringlei Gray
  pulchra H.C. Gentry
  purpurea Vent.
      var. arenicola (Wemple) Barneby
          SY=Petalostemon a. Wemple
      var. purpurea
          SY=Petalostemon molle Rydb.
          SY=P. p. (Vent.) Rydb.
          SY=P. p. var. m. (Rydb.) Boivin
  reverchonii (S. Wats.) Shinners
          SY=Petalostemon r. S. Wats.
  sabinalis (S. Wats.) Shinners
          SY=Petalostemon s. S. Wats.
  scandens (P. Mill.) Clausen
      var. paucifolia (Coult.) Barneby
          SY=D. thyrsiflora Gray
  scariosa S. Wats.
          SY=Petalostemon prostratum Woot. &
              Standl.
          SY=P. s. (S. Wats.) Wemple
  searlsiae (Gray) Barneby
          SY=Petalostemon s. Gray
  tentaculoides H.C. Gentry

  tenuifolia (Gray) Shinners
          SY=Petalostemon t. Gray
  tenuis (Coult.) Shinners
          SY=Petalostemon stanfieldii Small
          SY=P. t. (Coult.) Heller
  urceolata Greene
  versicolor Zucc.
    ssp. versicolor
      var. sessilis (Gray) Barneby
          SY=D. wislizenusii Gray ssp. s.
              (Gray) H.C. Gentry
          SY=D. w. var. sanctae-crucis
              (Rydb.) Kearney & Peebles
  villosa (Nutt.) Spreng.
      var. grisea (Torr. & Gray) Barneby
          SY=Petalostemon g. Torr. & Gray
      var. villosa
          SY=Petalostemon v. Nutt.
  wrightii Gray

Delonix Raf.

  regia (Bojer ex Hook.) Raf.
          SY=Poinciana r. Bojer ex Hook.

Desmanthus Willd.

  brevipes B.L. Turner
  cooleyi (Eat.) Trel.
  covillei (Britt. & Rose) Wiggins ex B.L.
      Turner
  illinoensis (Michx.) MacM. ex B.L. Robins.
      & Fern.
          SY=Acuan i. (Michx.) Kuntze
          SY=Mimosa i. Michx.
  leptolobus Torr. & Gray
          SY=Acuan l. (Torr. & Gray) Kuntze
  obtusus S. Wats.
  reticulatus Benth.
  velutinus Scheele
  virgatus (L.) Willd.
      var. acuminatus (Benth.) Isely
          SY=D. a. Benth.
      var. depressus (Humb. & Bonpl. ex
          Willd.) B.L. Turner
          SY=Acuan d. (Humb. & Bonpl. ex
              Willd.) Kuntze
      var. glandulosus B.L. Turner
      var. virgatus
          SY=Acuan v. (L.) Medic.

Desmodium Desv.

  adscendens (Sw.) DC.
          SY=Meibomia a. (Sw.) Kuntze
  affine Schlecht.
          SY=Meibomia a. (Schlecht.) Kuntze
  angustifolium (H.B.K.) DC.
          SY=Meibomia a. (H.B.K.) Kuntze
  arizonicum S. Wats.
          SY=Meibomia a. (S. Wats.) Vail
  axillare (Sw.) DC.
      var. acutifolium (Kuntze) Urban
          SY=Meibomia umbrosa Britt.
      var. axillare
          SY=D. a. var. obtusifolia (Kuntze)
              Urban
          SY=Meibomia a. (Sw.) Kuntze
          SY=M. a. var. o. Kuntze
      var. stoloniferum (Rich. ex Poir.)
          Schub.
          SY=D. a. var. sintensii Urban
          SY=Meibomia sintenisii (Urban)
              Britt.
  barbatum (L.) Benth.
          SY=Meibomia b. (L.) Kuntze
  batocaulon Gray

Desmodium (CONT.)
        SY=Meibomia b. (Gray) Kuntze
  canadense (L.) DC.
        SY=Meibomia c. (L.) Kuntze
  canescens (L.) DC.
        SY=D. c. var. hirsutum (Hook.)
            B.L. Robins.
        SY=D. c. var. villosissimum Torr.
            & Gray
        SY=Meibomia c. (L.) Kuntze
  ciliare (Muhl. ex Willd.) DC.
     var. ciliare
        SY=Meibomia c. (Muhl. ex Willd.)
            Blake
     var. lancifolium Fern. & Schub.
  cinerascens Gray
        SY=Meibomia c. (Gray) Kuntze
  cubense Griseb.
        SY=Meibomia c. (Griseb.) Schindl.
  cuspidatum (Muhl. ex Willd.) Loud.
     var. cuspidatum
        SY=D. bracteosum (Michx.) DC.
        SY=D. grandiflorum (Walt.) DC.
        SY=Meibomia b. (Michx.) Kuntze
     var. longifolium (Torr. & Gray)
        Schub.
        SY=D. bracteosum (Michx.) DC. var.
            l. (Torr. & Gray) B.L.
            Robins.
        SY=D. l. (Torr. & Gray) Smyth
        SY=Meibomia l. (Torr. & Gray) Vail
  fernaldii Schub.
        SY=D. rhombifolium sensu auctt.
            non Ell.
        SY=Meibomia r. sensu Vail
  floridanum Chapman
        SY=Meibomia f. (Chapman) Kuntze
        SY=M. rhombifolia Vail pro parte
  glabellum (Michx.) DC.
        SY=D. dillenii Darl. pro parte
        SY=Meibomia g. (Michx.) Kuntze
  glabrum (P. Mill.) DC.
        SY=D. mollis (Vahl) DC.
        SY=Hedysarum m. Vahl
        SY=Meibomia g. (P. Mill.) Kuntze
        SY=M. mollis (Vahl) Kuntze
  glutinosum (Muhl. ex Willd.) Wood
        SY=D. acuminatum (Michx.) DC.
        SY=Meibomia a. (Michx.) Blake
        SY=M. grandiflora sensu auctt. non
            (Walt.) Kuntze
  grahamii Gray
        SY=Meibomia g. (Gray) Kuntze
  gramineum Gray
        SY=D. angustifolium (H.B.K.) DC.
            var. g. (Gray) Schub.
        SY=Meibomia g. (Gray) Kuntze
  humifusum (Muhl.) Beck
        SY=Meibomia h. (Muhl.) Kuntze
  illinoense Gray
        SY=Meibomia i. (Gray) Kuntze
  incanum DC.
        SY=D. canum (J.F. Gmel.) Schinz &
            Thellung
        SY=D. frutescens (Jacq.) Schindl.
        SY=D. supinum DC.
        SY=Meibomia c. (J.F. Gmel.) Blake
        SY=M. i. (DC.) Vail
        SY=M. s. (Sw.) Britt.
  intortum (P. Mill.) Urban
        SY=Meibomia i. (P. Mill.) Blake
  laevigatum (Nutt.) DC.
        SY=Meibomia l. (Nutt.) Kuntze
  lindheimeri Vail
        SY=Meibomia l. (Vail) Vail
  lineatum DC.
        SY=D. arenicola (Vail) F.J. Herm.

        SY=D. l. var. polymorpha Gray
        SY=Meibomia a. Vail
        SY=Meibomia l. (DC.) Kuntze
        SY=M. l. var. p. Vail
        SY=M. p. (Gray) Small
  marilandicum (L.) DC.
        SY=Meibomia m. (L.) Kuntze
  metcalfei (Rose & Painter) Kearney &
    Peebles
        SY=Meibomia m. Rose & Painter
  neomexicanum Gray
        SY=Meibomia n. (Gray) Kuntze
  nudiflorum (L.) DC.
        SY=Meibomia n. (L.) Kuntze
  nuttallii (Schindl.) Schub.
        SY=Meibomia n. Schindl.
  obtusum (Muhl. ex Willd.) DC.
        SY=D. rigidum (Ell.) DC.
        SY=Meibomia o. (Muhl. ex Willd.)
            Vail
        SY=M. r. (Ell.) Kuntze
  ochroleucum M.A. Curtis
        SY=Meibomia o. (M.A. Curtis)
            Kuntze
  paniculatum (L.) DC.
     var. epetiolatum Schub.
     var. paniculatum
        SY=D. dichromum Shinners
        SY=D. dillenii Darl.
        SY=D. p. var. angustifolium Torr.
            & Gray
        SY=D. p. var. dillenii (Darl.)
            Isely
        SY=D. p. var. pubens Torr. & Gray
        SY=Meibomia chapmanii (Britt.)
            Small
        SY=M. dillenii (Darl.) Kuntze
        SY=M. paniculata (L.) Kuntze
        SY=M. pubens (Torr. & Gray) Rydb.
  pauciflorum (Nutt.) DC.
        SY=Meibomia p. (Nutt.) Kuntze
  perplexum Schub.
        SY=D. dillenii Darl. pro parte
  procumbens (P. Mill.) A.S. Hitchc.
     var. exiguum (Gray) Schub.
     var. procumbens
        SY=D. spirale (Sw.) DC.
        SY=D. sylvaticum Benth.
        SY=Meibomia p. (P. Mill.) Britt.
        SY=M. p. var. sylvatica (Benth.)
            Schindl.
        SY=M. spirale (Sw.) Kuntze
  psilocarpum Gray
  psilophyllum Schlecht.
        SY=D. wrightii Gray
        SY=Meibomia p. (Schlecht.) Kuntze
  retinens Schlecht.
        SY=D. wislizenii Engelm. ex Gray
        SY=Meibomia r. (Schlecht.) Kuntze
        SY=M. wrightii (Engelm. ex Gray)
            Kuntze
  rosei Schub.
  rotundifolium DC.
        SY=D. michauxii (Vail) Daniels
        SY=Meibomia m. Vail
        SY=M. r. (DC.) Kuntze
  scopulorum S. Wats.
        SY=D. wigginsii Schub.
  scorpiurus (Sw.) Desv.
        SY=Meibomia s. (Sw.) Kuntze
  sessilifolium (Torr.) Torr. & Gray
        SY=Meibomia s. (Torr.) Kuntze
  strictum (Pursh) DC.
        SY=Meibomia s. (Pursh) Kuntze
  tenuifolium Torr. & Gray
        SY=Meibomia t. (Torr. & Gray)
            Kuntze

Desmodium (CONT.)
  tortuosum (Sw.) DC.
              SY=D. annuum Gray pro parte
              SY=D. purpureum (P. Mill.) Fawcett
                  & Rendle
              SY=Meibomia p. (P. Mill.) Kuntze
              SY=M. t. (Sw.) Kuntze
  triflorum (L.) DC.
              SY=Meibomia t. (L.) Kuntze
              SY=Sagotia t. (L.) Duchass. &
                  Walp.
  tweedyi Britt.
              SY=Meibomia t. (Britt.) Vail
  uncinatum (Jacq.) DC.
              SY=Meibomia u. (Jacq.) Kuntze
  viridiflorum (L.) DC.
              SY=Meibomia v. (L.) Kuntze
  wydlerianum Urban
              SY=Meibomia w. (Urban) Britt.

Dichrostachys (DC.) Wight & Arn.

  cinerea (L.) Wight & Arn.

Dioclea H.B.K.

  multiflora (Torr. & Gray) C. Mohr
  reflexa Hook. f.

Diphysa Jacq.

  thurberi (Gray) Rydb.

Dolichos L.

  lablab L.
              SY=Lablab niger Medic.
              SY=L. purpureus (L.) Sweet

Entada Adans.

  polyphylla Benth.

Eriosema (DC.) Reichenb.

  crinatum (H.B.K.) G. Don

Errazurizia Phil.

  rotundata (Woot.) Barneby
              SY=Parryella r. Woot.

Erythrina L.

  berteroana Urban
  corallodendrum L.
          var. connata Krukoff
          var. corallodendrum
  crista-galli L.
              SY=Micropteryx c. (L.) Walp.
  eggersii Krukoff & Moldenke
              SY=E. horrida Eggers
  flabelliformis Kearney
  glauca Willd.
  herbacea L.
              SY=E. arborea (Chapman) Small
  poeppigiana (Walp.) O.F. Cook
  tahitensis Nadeaud
              SY=E. sandwicensis Deg.
              SY=E. s. var. luteosperma St. John

Eysenhardtia H.B.K.

  polystachya (Ortega) Sarg.
  spinosa Engelm. ex Gray
  texana Scheele
              SY=E. angustifolia Pennell

Galactia P. Br.

  brachypoda Torr. & Gray
  ?brevipes Small
  canescens Benth.
  dubia DC.
  eggersii Urban
  elliottii Nutt.
          var. elliottii
          var. leavenworthii Torr. & Gray
  erecta (Walt.) Vail
  fasciculata Vail
  floridana Torr. & Gray
          var. floridana
          var. longiracemosa Vail
          var. microphylla Vail
  glabella Michx.
  heterophylla Gray
              SY=G. grayi Vail
  longifolia Benth. ex Hoehne
  marginalis Benth.
  minor Duncan
  mollis Michx.
  parvifolia A. Rich.
  pinetorum Small
  prostrata Small
  regularis (L.) B.S.P.
              SY=G. mississippiensis (Vail)
                  Rydb.
              SY=G. volubilis (L.) Britt. var.
                  m. Vail
  spiciformis Torr. & Gray
  striata (Jacq.) Urban
  texana (Scheele) Gray
  volubilis (L.) Britt.
              SY=G. macreei M.A. Curtis sensu
                  auctt.
  wrightii Gray
          var. mollissima Kearney & Peebles
          var. wrightii

Galega L.

  officinalis L.

Genista L.

  tinctoria L.

Genistidium I.M. Johnston

  dumosum I.M. Johnston

Gleditsia L.

  aquatica Marsh.
  X texana Sarg. [aquatica X triacanthos]
  triacanthos L.
              SY=G. t. var. inermis Pursh

Gliricidia H.B.K. ex Endl.

  sepium (Jacq.) Kunth ex Griseb.

Glycine L.

  max (L.) Merr.
  soja (L.) Sieb. & Zucc.
              SY=G. ussuriensis Regel & Maack

Glycyrrhiza L.

  glabra L.
  lepidota (Nutt.) Pursh
          var. glutinosa (Nutt.) S. Wats.
          var. lepidota

Gymnocladus Lam.

   dioicus (L.) K. Koch

Haematoxylon L.

   campechianum L.

Hedysarum L.

   alpinum L.
      var. alpinum
        SY=H. a. ssp. americanum (Michx.
          ex Pursh) Fedtsch.
        SY=H. alpinum var. americanum
          Michx. ex Pursh
        SY=H. americanum (Michx. ex Pursh)
          Britt.
      var. grandiflorum Rollins
        SY=H. hedysaroides sensu auctt.
          non (L.) Schinz & Thellung
        SY=H. truncatum Eastw.
      var. philosocia (A. Nels.) Rollins
        SY=H. p. A. Nels.
   boreale Nutt.
     ssp. boreale
       var. boreale
         SY=H. b. var. cinerascens (Rydb.)
          Rollins
         SY=H. b. var. obovatum Rollins
         SY=H. b. var. rivulare (L.O.
          Williams) Northstrom
         SY=H. b. var. typicum Rollins
         SY=H. b. var. utahensis (Rydb.)
          Rollins
         SY=H. c. Rydb.
         SY=H. mackenzii Richards. var.
          fraseri Boivin
      var. gremiale (Rollins) Northstrom &
      Welsh
        SY=H. g. Rollins
     ssp. mackenzii (Richards.) Welsh
        SY=H. b. var. m. (Richards.) C.L.
          Hitchc.
        SY=H. m. Richards.
   occidentale Greene
      var. conone Welsh
      var. occidentale
        SY=H. uintahense A. Nels.
   sulphurescens Rydb.

Hoffmanseggia Cav.

   drepanocarpa Gray
   glauca (Ortega) Eifert
      SY=H. densiflora Benth. ex Gray
      SY=H. d. var. capitata Fisher
      SY=H. d. var. demissa (Gray)
        Fisher
      SY=H. d. var. pringlei Fisher
      SY=H. d. var. stricta (Benth.)
        Fisher
      SY=H. galcaria Cav.
   oxycarpa Benth. ex Gray
      SY=Caesalpinia o. (Benth. ex Gray)
        Fisher
   repens (Eastw.) Cockerell
   tenella Tharp & Williams

Hymenaea L.

   courbaril L.

Indigofera L.

   caroliniana P. Mill.
   endecaphylla Jacq.

guatemalensis Moc. & Sesse
hirsuta Harvey
keyensis Small
kirilowii Maxim. ex Palibine
lindheimeriana Scheele
miniata Ortega
      var. leptosepala (Nutt.) B.L. Turner
        SY=I. l. Nutt.
      var. miniata
      var. texana (Buckl.) B.L. Turner
parviflora K. Heyne
sphaerocarpa Gray
suffruticosa P. Mill.
sumatrana Gaertn.
tinctoria L.

Inga P. Mill.

   fagifolia (L.) Willd.
      SY=I. laurina (Sw.) Willd.
   vera Willd.
      SY=I. inga (L.) Britt.

Kummerowia Schindl.

   stipulacia (Maxim.) Makino
      SY=Lespedeza s. Maxim.
   striata (Thunb.) Schindl.
      SY=Lespedeza s. (Thunb.) Hook. &
        Arn.

Lathyrus L.

   angulatus L.
   aphaca L.
   arizonicus Britt.
   bijugatus White
      SY=L. b. var. sandbergii White
   brachycalyx Rydb.
     ssp. brachycalyx
        SY=L. ornatus Nutt.
     ssp. eucosmus (Butters & St. John) Welsh
        SY=L. e. Butters & St. John
     ssp. zionis (C.L. Hitchc.) Welsh
        SY=L. z. C.L. Hitchc.
   cicera L.
   delnorticus C.L. Hitchc.
   graminifolius (S. Wats.) White
   hirsutus L.
   hitchcockianus Barneby & Reveal
   holochlorus (Piper) C.L. Hitchc.
   japonicus Willd.
      var. glaber (Ser.) Fern.
        SY=L. maritimus (L.) Bigelow
        SY=L. m. var. g. (Ser.) Eames
      var. japonicus
        SY=L. j. var. aleuticus (Greene)
         Fern.
      var. ?parviflorus Fassett
      var. pellitus Fern.
        SY=L. maritimus (L.) Bigelow var.
         p. (Fern.) Gleason
      var. pubescens Hartman
        SY=L. maritimus (L.) Bigelow ssp.
         p. (Hartman) Regel
   jepsonii Greene
     ssp. californicus (S. Wats.) C.L.
      Hitchc.
        SY=L. watsonii White
     ssp. jepsonii
   laetiflorus Greene
     ssp. alefeldii (White) Bradshaw
        SY=L. a. White
     ssp. barbarae (White) C.L. Hitchc.
     ssp. glaber C.L. Hitchc.
     ssp. laetiflorus
   lanszwertii Kellogg

Lathyrus (CONT.)
    ssp. aridus (Piper) Bradshaw
        SY=L. l. var. a. (Piper) Jepson
    ssp. lanszwertii
latifolius L.
leucanthus Rydb.
      var. laetivirens (Greene ex Rydb.)
        C.L. Hitchc.
        SY=L. laetivirens Greene ex Rydb.
      var. leucanthus
littoralis (Nutt. ex Torr. & Gray) Endl.
longipes White
nevadensis S. Wats.
    ssp. cusickii (S. Wats.) C.L. Hitchc.
        SY=L. c. S. Wats.
    ssp. lanceolatus (T.J. Howell) C.L.
      Hitchc.
      var. nuttallii (S. Wats.) C.L.
        Hitchc.
        SY=L. n. S. Wats.
      var. parkeri (St. John) C.L. Hitchc.
      var. pilosellus (M.E. Peck) C.L.
        Hitchc.
      var. puniceus C.L. Hitchc.
    ssp. nevadensis
ochroleucus Hook.
odoratus L.
palustris L.
      var. myrtifolius (Muhl.) Gray
        SY=L. m. Muhl.
        SY=Orobus m. (Muhl.) A. Hall
      var. palustris
        SY=L. p. var. linearifolius Ser.
      var. pilosus (Cham.) Ledeb.
        SY=L. palustris ssp. pilosus
          (Cham.) Hulten
        SY=L. palustris var. macranthus
          (White) Fern.
        SY=L. palustris var. retusus Fern.
          & St. John
parvifolius S. Wats.
        SY=L. schaffneri Rydb.
pauciflorus Fern.
    ssp. brownii (Eastw.) Piper
    ssp. pauciflorus
      var. pauciflorus
        SY=L. p. var. tenuior (Piper) St.
          John
      var. utahensis (M.E. Jones) C.L.
        Hitchc.
        SY=L. p. ssp. u. (M.E. Jones)
          Piper
polymorphus Nutt.
    ssp. incanus (Sm. & Rydb.) C.L. Hitchc.
        SY=L. i. (Sm. & Rydb.) Rydb.
    ssp. polymorphus
      var. hapemanii (A. Nels.) C.L.
        Hitchc.
      var. polymorphus
        SY=L. decaphyllus Pursh
polyphyllus Nutt. ex Torr. & Gray
pratensis L.
pusillus Ell.
rigidus White
sativus L.
sphaericus Retz.
splendens Kellogg
sulphureus Brewer ex Gray
      var. argillaceus Jepson
      var. sulphureus
        SY=L. nevadensis S. Wats. ssp.
          stipulaceous (White)
          Bradshaw
sylvestris L.
tingitanus L.
torreyi Gray
tracyi Bradshaw

tuberosus L.
venosus Muhl. ex Willd.
    ssp. arkansanus (Fassett) C.L. Hitchc.
        SY=L. v. var. a. Fassett
    ssp. venosus
      var. intonsus (Butters & St. John)
        C.L. Hitchc.
        SY=L. oreophyllus Woot. & Standl.
        SY=L. v. var. meridionalis Butters
          & St. John
      var. venosus
vestitus Nutt. ex Torr. & Gray
    ssp. bolanderi (S. Wats.) C.L. Hitchc.
        SY=L. b. S. Wats.
        SY=L. peckii Piper
    ssp. ochropetalus (Piper) C.L. Hitchc.
    ssp. puberulus (White ex Greene) C.L.
      Hitchc.
        SY=L. v. var. violaceus (Greene)
          Abrams
    ssp. vestitus

Lens P. Mill.

    culinaris Medic.
        SY=Ervum lens L.

Lespedeza Michx.

    X acuticarpa Mackenzie & Bush [violacea X
      virginica]
    angustifolia (Pursh) Ell.
        SY=L. hirta (L.) Hornem. var.
          intercursa Fern.
    bicolor Turcz.
    X brittonii Bickn. [procumbens X virginica]
        SY=L. procumbens Michx. var.
          elliptica Blake
    capitata Michx.
        SY=L. c. var. stenophylla Bissell
          & Fern.
        SY=L. c. var. velutina (Bickn.)
          Fern.
        SY=L. c. var. vulgaris Torr. &
          Gray
    cuneata (Dum.-Cours.) G. Don
    cyrtobotrya Miq.
    daurica (Laxm.) Schindl.
    hirta (L.) Hornem.
      ssp. curtisii Clewell
      ssp. hirta
        SY=L. capitata Michx. var.
          calycina (Schindl.) Fern.
        SY=L. polystachya Michx.
    intermedia (S. Wats.) Britt.
    leptostachya Engelm.
    X longifolia DC. [capitata X hirta]
        SY=L. hirta (L.) Hornem. var.
          dissimulans Fern.
        SY=L. h. var. l. (DC.) Fern.
        SY=L. l. DC.
    X manniana Mackenzie & Bush [capitata X
      violacea]
        SY=L. nuttallii Darl. var. m.
          (Mackenzie & Bush) Gleason
    X neglecta (Britt.) Mackenzie & Bush
      [stuevei X virginica]
        SY=L. stuevei Nutt. var.
          angustifolia Britt.
    X nuttallii Darl. [hirta X intermedia]
    X oblongifolia (Britt.) W. Stone
        SY=L. hirta (L.) Hornem. var.
          appressipilis Blake
    procumbens Michx.
    repens (L.) Bart.
    X simulata Mackenzie & Bush [capitata X
      virginica]

Lespedeza (CONT.)
  stuevei Nutt.
  texana Britt.
  thunbergii (DC.) Nakai
  violacea (L.) Pers.
        SY=L. frutescens (L.) Britt.
        SY=L. prairea (Mackenzie & Bush)
           Britt.
  virginica (L.) Britt.

Leucaena Benth.

  leucocephala (Lam.) de Wit
        SY=L. glauca sensu auctt. non (L.)
           Benth.
  pulverulenta (Schlecht.) Benth.
  retusa Benth.

Lonchocarpus H.B.K.

  domingensis (Turp.) DC.
  glaucifolius Urban
  latifolius (Willd.) DC.
  pentaphyllus (Poir.) DC.
  violaceus Kunth

Lotus L.

  aboriginus Jepson
        SY=Hosackia rosea Eastw.
        SY=L. crassifolius (Benth.) Greene
           var. subglaber (Ottley) C.L.
           Hitchc.
        SY=L. stipularis (Benth.) Greene
           var. subglaber Ottley
  alamosanus (Rose) Gentry
  angustissimus L.
  argophyllus (Gray) Greene
    ssp. adsurgens (Dunkle) Raven
        SY=L. argophyllus var. adsurgens
           Dunkle
    ssp. argophyllus
        SY=Hosackia a. Gray
        SY=H. fremontii (Gray) Abrams
        SY=L. a. var. f. (Gray) Ottley
    ssp. decorus (I.M. Johnston) Munz
        SY=Hosackia argophylla Gray var.
           d. I.M. Johnston
        SY=L. a. var. d. (I.M. Johnston)
           Ottley
    ssp. niveus (Greene) Munz
        SY=Hosackia n. (Greene) S. Wats.
        SY=L. a. var. n. (Greene) Ottley
        SY=L. n. Greene
    ssp. ornithopus (Greene) Raven
        SY=Hosackia o. Greene
        SY=H. o. ssp. venusta (Eastw.)
           Abrams
        SY=L. a. var. o. (Greene) Ottley
  argyraeus (Greene) Greene
    ssp. argyraeus
        SY=Hosackia a. Greene
    ssp. multicaulis (Ottley) Munz
        SY=Hosackia wrightii Gray var. m.
           Ottley
  balsamiferus (Kellogg) Greene
        SY=Hosackia stipularis Benth. ssp.
           b. (Kellogg) Abrams
  benthamii Greene
        SY=Hosackia cystoides Benth.
  corniculatus L.
  crassifolius (Benth.) Greene
        SY=Hosackia c. Benth.
  cupreus Greene
        SY=Hosackia c. (Greene) Smiley
  davidsonii Greene
        SY=Hosackia sulphurea (Greene)

           Abrams
  denticulatus (E. Drew) Greene
        SY=Hosackia d. E. Drew
  formosissimus Greene
        SY=Hosackia gracilis Benth.
  grandiflorus (Benth.) Greene
    var. grandiflorus
        SY=Hosackia g. Benth.
    var. mutabilis Ottley
        SY=Hosackia leucophaea (Greene)
           Abrams
  greenei (Woot. & Standl.) Ottley ex Kearney
    & Peebles
  hamatus Greene
  haydonii (Orcutt) Greene
        SY=Hosackia h. Orcutt
  heermannii (Dur. & Hilg.) Greene
    var. eriophorus (Greene) Ottley
        SY=Hosackia tomentosa Hook. & Arn.
    var. heermannii
        SY=Hosackia tomentosa Hook. & Arn.
           ssp. glabriuscula (Hook. &
           Arn.) Abrams
  helleri Britt.
        SY=Acmispon h. (Britt.) Small
  humistratus Greene
        SY=Hosackia brachycarpa Benth.
  junceus (Benth.) Greene
    var. biolettii Ottley
    var. junceus
        SY=Hosackia j. Benth.
  krylovii Schischkin & Sergievskaja
  longebracteatus Rydb.
  mearnsii Britt. ex Greene
  micranthus Benth.
        SY=Hosackia parviflora Benth.
  neo-incanus Munz
        SY=Hosackia incana Torr.
  neomexicanus Greene
  nevadensis (S. Wats.) Greene
    var. douglasii (Greene) Ottley
        SY=Hosackia decumbens Benth.
        SY=L. douglasii Greene
    var. nevadensis
        SY=Hosackia n. (S. Wats.) Parish
        SY=L. douglasii (Benth.) Greene
           var. n. (S. Wats.) Ottley
  nuttallianus Greene
        SY=Hosackia prostrata Nutt.
  oblongifolius (Benth.) Greene
    var. nevadensis (Gray) Munz
        SY=L. o. var. torreyi (Gray)
           Ottley
        SY=L. t. (Gray) Greene
    var. oblongifolius
        SY=Hosackia lathyroides Dur. &
           Hilg.
        SY=H. o. Benth.
  oroboides (H.B.K.) Ottley ex Kearney &
    Peebles
  parviflorus Desf.
  pedunculatus Cav.
        SY=L. uliginosus sensu auctt. non
           Schkuhr
  pinnatus Hook.
        SY=Hosackia p. (Hook.) Abrams
  procumbens (Greene) Greene
    var. jepsonii (Ottley) Ottley
        SY=Hosackia sericea Benth. ssp. j.
           (Ottley) Abrams
    var. procumbens
        SY=Hosackia sericea Benth.
  purshianus (Benth.) Clem. & Clem.
    var. glaber (Nutt.) Munz
    var. purshianus
        SY=Acmispon americanum (Nutt.)
           Rydb.

Lotus (CONT.)
   SY=Hosackia a. (Nutt.) Piper
   SY=L. a. (Nutt.) Bisch.
   SY=L. unifoliolatus (Hook.) Benth.
 rigidus (Benth.) Greene
   SY=Hosackia r. Benth.
 rubellus (Nutt.) Greene
   SY=Hosackia r. Nutt.
 rubriflorus H.K. Sharsmith
 salsuginosus Greene
  ssp. brevivexillus (Ottley) Munz
    SY=Hosackia humilis (Greene)
     Abrams
    SY=L. s. var. b. Ottley
  ssp. salsuginosus
    SY=Hosackia maritima Nutt.
 scoparius (Nutt.) Ottley
  ssp. brevialatus (Ottley) Munz
    SY=Hosackia glabra (Vogel) Torr.
     var. b. (Ottley) Abrams
    SY=L. s. var. b. Ottley
  ssp. scoparius
    SY=Hosackia glabra (Vogel) Torr.
  ssp. traskiae (Eastw. ex Abrams) Raven
    SY=Hosackia dendroidea (Greene)
     Abrams
    SY=L. s. var. d. (Greene) Ottley
 stipularis (Benth.) Greene
   SY=Hosackia s. Benth.
 strigosus (Nutt.) Greene
  var. hirtellus (Greene) Ottley
   SY=Hosackia s. Nutt. var. h.
    (Greene) Hall
  var. strigosus
   SY=Hosackia s. Nutt.
 subpinnatus Lag.
   SY=Hosackia s. (Lag.) Torr. & Gray
 tenuis Waldst. & Kit. ex Willd.
 tomentellus Greene
   SY=Hosackia t. (Greene) Abrams
 utahensis Ottley
 wrightii (Gray) Greene
   SY=Hosackia w. Gray
 yollabolliensis Munz

Lupinus L.

 abramsii C.P. Sm.
 adsurgens E. Drew
  var. adsurgens
   SY=L. debilis Eastw.
   SY=L. d. Greene ex C.F. Baker
   SY=L. pendletonii Heller
  var. lilacinus (Heller) C.P. Sm.
   SY=L. alcis-montis C.P. Sm.
   SY=L. aliceae C.P. Sm.
   SY=L. brandegeei Eastw.
   SY=L. l. Heller
  var. undulatus C.P. Sm.
   SY=L. nemoralis Greene
 affinis J.G. Agardh
   SY=L. nanus Dougl. var. carnosulus
    (Greene) C.P. Sm.
 agardhianus Heller
   SY=L. concinnus J.G. Agardh var.
    a. (Heller) C.P. Sm.
   SY=L. gracilis J.G. Agardh non
    Nutt.
 albicaulis Dougl. ex Hook.
  var. albicaulis
   SY=L. falcifer Nutt. ex Torr. &
    Gray
   SY=L. quercetorum Heller
   SY=L. wolfianus C.P. Sm.
  var. shastensis (Heller) C.P. Sm.
   SY=L. s. Heller
 albifrons Benth. ex Lindl.

  var. albifrons
  var. collinus Greene
  var. douglasii (J.G. Agardh) C.P. Sm.
  var. eminens (Greene) C.P. Sm.
   SY=L. a. ssp. e. (Greene) D. Dunn
  var. fissicalyx (Heller) C.P. Sm.
  var. flumineus C.P. Sm.
 X alpestris A. Nels. [argenteus X caudatus]
   SY=L. acclivatatis C.P. Sm.
   SY=L. adscendens Rydb.
   SY=L. alexanderae C.P. Sm.
   SY=L. aliesicola C.P. Sm.
   SY=L. annieae C.P. Sm.
   SY=L. argenteus Pursh var.
    aristovatus C.P. Sm.
   SY=L. argenteus var. krauchianus
    C.P. Sm.
   SY=L. argenteus var. macounii
    (Rydb.) R.J. Davis
   SY=L. argenteus var. prati-harti
    C.P. Sm.
   SY=L. argenteus var. submanens
    C.P. Sm.
   SY=L. argenteus var. wallianus
    C.P. Sm.
   SY=L. calcicola C.P. Sm.
   SY=L. capitis-amniculi C.P. Sm.
   SY=L. clokeyanus C.P. Sm.
   SY=L. flavopinum C.P. Sm.
   SY=L. junipericola C.P.Sm.
   SY=L. lariversianus C.P. Sm.
   SY=L. laxus Rydb.
   SY=L. macounii Rydb.
   SY=L. patulipes C.P. Sm.
   SY=L. populorum C.P. Sm.
   SY=L. pulcherrimus Rydb.
   SY=L. sicco-silvae C.P. Sm.
   SY=L. trainianus C.P. Sm.
 alpicola Henderson
 ammophilus Greene
 amphibius Suksdorf
 andersonii S. Wats.
   SY=L. rinae Eastw.
 angustiflorus Eastw.
   SY=L. andersonii S. Wats. var.
    christinae (Heller) Munz
   SY=L. c. Heller
 antoninus Eastw.
 apertus Heller
   SY=L. andersonii S. Wats. var.
    apertus (Heller) C.P. Sm.
 arboreus Sims
  var. arboreus
  var. eximius (Burtt-Davy) C.P. Sm.
 arbustus Dougl. ex Lindl.
  ssp. arbustus
   var. arbustus
    SY=L. laxiflorus Dougl. ex Lindl.
     var. a. (Dougl. ex Lindl.)
     M.E. Jones
   var. montanus (T.J. Howell) D. Dunn
    SY=L. laxiflorus Dougl. ex Lindl.
     var. cognatus C.P. Sm.
  ssp. calcaratus (Kellogg) D. Dunn
   SY=L. c. Kellogg
   SY=L. laxiflorus Dougl. ex Lindl.
    var. c. (Kellogg) C.P. Sm.
  ssp. neolaxiflorus D. Dunn
   SY=L. amniculi-putori C.P. Sm.
   SY=L. augustii C.P. Sm.
   SY=L. caudatus Kellogg var.
    submanens C.P. Sm.
   SY=L. lyleianus C.P. Sm.
   SY=L. laxiflorus Dougl. ex Lindl.
    var. lyleianus C.P. Sm.
   SY=L. mackeyi C.P. Sm.
   SY=L. standingii C.P. Sm.

Lupinus (CONT.)
            SY=L. stipaphilus C.P. Sm.
            SY=L. stockii C.P. Sm.
            SY=L. wenachensis Eastw.
            SY=L. yakimensis C.P. Sm.
    ssp. pseudoparviflorus (Rydb.) D. Dunn
            SY=L. laxiflorus Dougl. ex Lindl.
                var. elmerianus C.P. Sm.
            SY=L. l. var. p. (Rydb.) C.P. Sm.
                & St. John
            SY=L. mucronulatus J.T. Howell
                var. umatillensis C.P. Sm.
            SY=L. laxispicatus Rydb.
            SY=L. scheuberae Rydb.
    ssp. silvicola (Heller) D. Dunn
            SY=L. lassenensis Eastw.
            SY=L. laxiflorus Dougl. ex Lindl.
                var. s. (Heller) C.P. Sm.
    arcticus S. Wats.
        ssp. arcticus
            SY=L. borealis Heller
            SY=L. donnellyensis C.P. Sm.
            SY=L. gakonensis C.P. Sm.
            SY=L. multicaulis C.P. Sm.
            SY=L. multifolius C.P. Sm.
            SY=L. nootkatensis Donn ex Sims
                var. kyellmannii Ostenf.
            SY=L. polyphyllus Lindl. ssp.
                arcticus (S. Wats.) Phillips
            SY=L. relictus Hulten
            SY=L. yukonensis Greene
        ssp. canadensis (C.P. Sm.) D. Dunn
            SY=L. latifolius J.G. Agardh var.
                c. C.P. Sm.
        ssp. subalpinus (Piper & B.L. Robins.)
            D. Dunn
            SY=L. arcticus S. Wats. var. s.
                (Piper & B.L. Robins.) C.P.
                Sm.
            SY=L. glacialis C.P. Sm.
            SY=L. latifolius J.G. Agardh var.
                s. (Piper & B.L. Robins.)
                C.P. Sm.
            SY=L. s. Piper & B.L. Robins.
            SY=L. volcanicus Greene
            SY=L. v. var. rupestricola C.P.
                Sm.
    argenteus Pursh
        ssp. argenteus
            var. argenteus
            SY=L. a. Pursh var. decumbens
                (Torr.) S. Wats.
            SY=L. corymbosus Heller
            SY=L. d. Torr.
            SY=L. foliosus (Nutt. ex Torr. &
                Gray) Nutt. ex Hook.
            SY=L. garrettianus C.P. Sm.
            SY=L. laxiflorus Dougl. ex Lindl.
                var. corymbosus Jepson
            SY=L. l. var. f. Nutt. ex Torr. &
                Gray
            var. tenellus (Dougl. ex G. Don) D.
                Dunn
            SY=L. argenteus Pursh var.
                stenophyllus (Nutt. ex
                Rydb.) R.J. Davis
            SY=L. cariciformes C.P. Sm.
            SY=L. charlestonensis C.P. Sm.
            SY=L. clarkensis C.P. Sm.
            SY=L. edward-palmeri C.P. Sm.
            SY=L. foliosus Nutt. ex Rydb. var.
                stenophyllus Nutt. ex Rydb.
            SY=L. fremontensis C.P. Sm.
            SY=L. hullianus C.P. Sm.
            SY=L. lanatocarinatus C.P. Sm.
            SY=L. laxiflorus Dougl. ex Lindl.
                non American authors

            SY=L. laxiflorus var. t. Dougl. ex
                Torr. & Gray
            SY=L. montis-cookii C.P. Sm.
            SY=L. populorum C.P. Sm.
            SY=L. stenophyllus Nutt. ex Rydb.
            SY=L. t. Dougl. ex G. Don
            SY=L. t. Dougl. ex J.G. Agardh
        ssp. ingratus (Greene) Harmon
            SY=L. i. Greene
        ssp. rubricaulis (Greene) Hess & D. Dunn
            SY=L. caudatus Kellogg var. r.
                (Greene) C.P. Sm.
            SY=L. r. Greene
        ssp. spathulatus (Rydb.) Hess & D. Dunn
            SY=L. alsophilus Greene
            SY=L. s. Rydb.
            SY=L. s. var. boreus C.P. Sm.
    aridus Dougl. ex Lindl.
        ssp. aridus
            SY=L. lepidus Dougl. ex Lindl.
                var. a. (Dougl. ex Lindl.)
                Jepson
            SY=L. l. ssp. a. (Dougl. ex
                Lindl.) Detling
        ssp. ashlandensis Cox
        ssp. lenorensis (C.P. Sm.) Cox
            SY=L. l. C.P. Sm.
        ssp. loloensis Cox
    arizonicus (S. Wats.) S. Wats.
        ssp. arizonicus
            var. arizonicus
            SY=L. concinnus J.G. Agardh var.
                a. S. Wats.
            SY=L. sparsiflorus Benth. var. a.
                (S. Wats.) C.P. Sm.
            SY=L. subhirsutus A. Davids.
            var. barbatulus (Thornb.) I.M.
                Johnston
            SY=L. sparsiflorus Benth. var. b.
                Thornb.
        ssp. lagunensis (M.E. Jones) Christian &
            D. Dunn
            SY=L. bartolomei M.E. Jones
            SY=L. l. M.E. Jones
            SY=L. sparsiflorus Benth. var.
                insignitus C.P. Sm.
        ssp. setosissimus (C.P. Sm.) Christian &
            D. Dunn
            SY=L. sparsiflorus Benth. var.
                setosissimus C.P. Sm.
        ssp. sonorensis Christian & D. Dunn
    bakeri Greene
        ssp. amplus (Greene) Fleak & D. Dunn
            SY=L. a. Greene
            SY=L. comatus Rydb.
            SY=L. habrocomus Greene
        ssp. bakeri
            SY=L. arcenthinus Greene
            SY=L. dichrous Greene
    barbigeri S. Wats.
    benthamii Heller
        var. benthamii
            SY=L. leptophyllus Benth. non
                Schlecht. & Cham.
        var. opimus C.P. Sm.
    bicolor Lindl.
        ssp. bicolor
            SY=L. hirsutulus Greene
            SY=L. micranthus Dougl. non Guss.
                var. b. (Lindl.) S. Wats.
            SY=L. strigulosus Gandog.
        ssp. marginatus D. Dunn
        ssp. microphyllus (S. Wats.) D. Dunn
            SY=L. bicolor var. m. (S. Wats.)
                C.P. Sm.
            SY=L. micranthus Dougl. var.
                microphyllus S. Wats.

Lupinus (CONT.)
    ssp. pipersmithii (Heller) D. Dunn
        SY=L. bicolor var. p. (Heller)
            C.P. Sm.
        SY=L. p. Heller
    ssp. tridentatus (Eastw. ex C.P. Sm.) D.
      Dunn
      var. rostratus (Eastw.) Jeps.
        SY=L. b. var. r. (Eastw.) Jeps.
        SY=L. r. Eastw.
      var. tridentatus (Eastw. ex C.P. Sm.)
        D. Dunn
        SY=L. b. var. tetraspermus C.P.
            Sm.
        SY=L. b. var. tridentatus Eastw.
            ex C.P. Sm.
    ssp. umbellatus (Greene) D. Dunn
      var. trifidus (Torr. ex S. Wats.)
        C.P. Sm.
        SY=L. b. var. t. (Torr. ex S.
            Wats.) C.P. Sm.
        SY=L. micranthus Dougl. non Guss.
            var. t. Torr. ex S. Wats.
        SY=L. t. Torr. ex S. Wats.
      var. umbellatus (Greene) C.P. Sm.
        SY=L. plebeius Greene ex C.F.
            Baker
        SY=L. sabulosus Heller
        SY=L. u. Greene
bingenensis Suksdorf
    var. bingenensis
        SY=L. b. var. albus Suksdorf
        SY=L. b. var. roseus Suksdorf
        SY=L. leucopsis J.G. Agardh var.
            b. (Suksdorf) C.P. Sm.
    var. dubius C.P. Sm.
        SY=L. l. var. hendersonianus C.P.
            Sm.
        SY=L. l. var. shermanensis C.P.
            Sm.
    var. subsaccatus Suksdorf
        SY=L. sulphureus Dougl. ex Hook.
            ssp. subsaccatus (Suksdorf)
            Phillips
        SY=L. sulphureus var. subsaccatus
            (Suksdorf) C.L. Hitchc.
brevicaulis S. Wats.
    SY=L. dispersus Heller
    SY=L. scaposus Rydb.
brevior (Jepson) Christian & D. Dunn
    SY=L. concinnus J.G. Agardh var.
        b. (Jepson) D. Dunn
    SY=L. sparsiflorus Benth. var. b.
        Jepson
breweri Gray
    var. breweri
    var. bryoides C.P. Sm.
        SY=L. tegiticulus Eastw.
    var. grandiflorus C.P. Sm.
        SY=L. b. var. clokeyanus C.P. Sm.
        SY=L. campbellae Eastw.
        SY=L. c. Eastw. var. bernardianus
            Eastw.
    var. parvulus C.P. Sm.
burkei S. Wats.
    ssp. burkei
        SY=L. polyphyllus Lindl. var. b.
            (S. Wats.) C.L. Hitchc.
    ssp. caeruleomontanus D. Dunn & Cox
caespitosus Nutt. ex Torr. & Gray
    var. caespitosus
        SY=L. lepidus Dougl. ex Lindl.
            ssp. c. (Nutt. ex Torr. &
            Gray) Detling
        SY=L. marleanus C.P. Sm.
    var. utahensis (S. Wats.) Cox
        SY=L. amniculi-cervi C.P. Sm.

        SY=L. aridus Dougl. ex Lindl. var.
            u. S. Wats.
        SY=L. lepidus Dougl. ex Lindl.
            var. u. (S. Wats.) C.L.
            Hitchc.
        SY=L. psoralioides Pollard
        SY=L. sinus-meyersi C.P. Sm.
        SY=L. watsonii Heller
caudatus Kellogg
    ssp. argophyllus (Gray) Phillips
        SY=L. a. (Gray) Cockerell
        SY=L. aduncus Greene
        SY=L. decumbens Nutt. var.
            argophyllus Gray
        SY=L. helleri Greene
        SY=L. laxiflorus Dougl. ex Lindl.
            var. argophyllus (Gray) M.E.
            Jones
    ssp. caudatus
        SY=L. argentinus Rydb.
        SY=L. gayophytophilus C.P. Sm.
        SY=L. holosericeus Nutt. ex Torr.
            & Gray var. utahensis S.
            Wats.
        SY=L. lupinus Rydb.
        SY=L. meionanthus Gray var.
            heteranthus S. Wats.
        SY=L. montis-lieratatis C.P. Sm.
        SY=L. rosei Eastw.
        SY=L. stinchfieldiae C.P. Sm.
        SY=L. u. (S. Wats.) Moldenke pro
            parte
    ssp. cutleri (Eastw.) Hess & D. Dunn
        SY=L. c. Eastw.
    ssp. montigenus (Heller) Hess & D. Dunn
        SY=L. m. Heller
cervinus Kellogg
chamissonis Eschsch.
chihuahuensis S. Wats.
citrinus Kellogg
    var. citrinus
    var. deflexus (Congd.) Jepson
        SY=L. d. Congd.
concinnus J.G. Agardh
    ssp. concinnus
    ssp. optatus (C.P. Sm.) D. Dunn
        SY=L. c. var. o. C.P. Sm.
    ssp. orcuttii (S. Wats.) D. Dunn
        SY=L. o. S. Wats.
        SY=L. c. var. o. (S. Wats.) C.P.
            Sm.
        SY=L. micensis M.E. Jones
confertus Kellogg
    SY=L. aridus Dougl. ex Lindl. var.
        c. (Kellogg) C.P. Sm.
    SY=L. c. var. ramosus (Jepson)
        Eastw.
    SY=L. lepidus Dougl. ex Lindl.
        ssp. c. (Kellogg) Detling
    SY=L. l. var. c. (Kellogg) C.P.
        Sm.
    SY=L. l. var. r. Jepson
congdonii (C.P. Sm.) D. Dunn
    SY=L. micranthus Dougl. var. c.
        C.P. Sm.
covillei Greene
    SY=L. dasyphyllus Greene
crassus Payson
croceus Eastw.
    var. croceus
    var. pilosellus (Eastw.) Munz
        SY=L. p. Eastw.
culbertsonii Greene
    ssp. culbertsonii
        SY=L. lepidus Dougl. ex Lindl.
            var. c. (Greene) C.P. Sm.
    ssp. hypolasius (Greene) Cox

Lupinus (CONT.)
          SY=L. brunneo-maculatus Eastw.
          SY=L. crassulus Greene ex C.P.
                Baker
          SY=L. h. Greene
  cumulicola Small
  cusickii S. Wats.
     ssp. abortivus (Greene) Cox
          SY=L. a. Greene
          SY=L. aridus Dougl. ex Lindl. var.
                abortivus (Greene) C.P. Sm.
     ssp. brachypodus (Piper) Cox
          SY=L. b. Piper
     ssp. cusickii
          SY=L. aridus Dougl. ex Lindl. var.
                c. (S. Wats.) C.P. Sm.
          SY=L. lepidus Dougl. ex Lindl.
                ssp. c. (S. Wats.) Detling
          SY=L. l. var. c. (S. Wats.) C.L.
                Hitchc.
          SY=L. longivallis C.P. Sm.
  dalesae Eastw.
          SY=L. formosus Greene var.
                clemensae C.P. Sm.
  densiflorus Benth.
     var. aureus (Kellogg) Munz
          SY=L. d. var. menziesii (J.G.
                Agardh) C.P. Sm.
          SY=L. d. var. perfistulosus C.P.
                Sm.
          SY=L. m. J.G. Agardh
     var. austrocollium C.P. Sm.
          SY=L. d. ssp. a. (C.P. Sm.) D.
                Dunn
     var. densiflorus
          SY=L. d. var. latilabris C.P. Sm.
          SY=L. d. var. scopulorum C.P. Sm.
          SY=L. d. var. stenopetalus C.P.
                Sm.
          SY=L. d. var. tracyi C.P. Sm.
     var. glareosus (Elmer) C.P. Sm.
          SY=L. g. Elmer
     var. lacteus (Kellogg) C.P. Sm.
          SY=L. arenicola Heller
          SY=L. d. var. altus C.P. Sm.
          SY=L. d. var. dudleyi C.P. Sm.
          SY=L. d. var. latidens C.P. Sm.
          SY=L. d. var. mcgregorii C.P. Sm.
          SY=L. d. var. persecundus C.P. Sm.
          SY=L. d. var. sublanatus C.P. Sm.
          SY=L. d. var. vastiticola C.P. Sm.
          SY=L. d. var. versatabilis C.P.
                Sm.
     var. palustris (Kellogg) C.P. Sm.
          SY=L. d. var. crinitus Eastw. ex
                C.P. Sm.
          SY=L. d. var. curvicarinus C.P.
                Sm.
          SY=L. d. var. stanfordianus C.P.
                Sm.
          SY=L. d. var. trichocalyx C.P. Sm.
  depressus Rydb.
          SY=L. argenteus Pursh var. d.
                (Rydb.) C.L. Hitchc.
  diffusus Nutt.
          SY=L. villosus Willd. ssp. d.
                (Nutt.) Phillips
          SY=L. v. var. d. (Nutt.) Torr. &
                Gray
  duranii Eastw.
  elatus I.M. Johnston
          SY=L. albicaulis Dougl. ex Hook.
                var. e. (I.M. Johnston)
                Jepson
          SY=L. formosus Greene var. e.
                (I.M. Johnston) C.P. Sm.
  elmeri Greene

          SY=L. albicaulis Dougl. ex Hook.
                var. sylvestris (E. Drew)
                Greene
          SY=L. s. E. Drew non Lam.
  evermannii Rydb.
          SY=L. sparhawkianus C.P. Sm.
  excubitus M.E. Jones
     var. austromontanus (Heller) C.P. Sm.
          SY=L. a. Heller
     var. excubitus
     var. hallii (Abrams) C.P. Sm.
          SY=L. e. ssp. h. (Abrams) D. Dunn
          SY=L. h. Abrams
          SY=L. paynei A. Davids.
     var. johnstonii C.P. Sm.
     var. medius (Jepson) Munz
          SY=L. albifrons Benth. ex Lindl.
                var. m. Jepson
          SY=L. grayii (S. Wats.) S. Wats.
                var. m. (Jepson) C.P. Sm.
  fissicalyx Heller
          SY=L. albifrons Benth. ex Lindl.
                var. f. (Heller) C.P. Sm.
  flavoculatus Heller
          SY=L. rubens Rydb. var. f.
                (Heller) C.P. Sm.
  formosus Greene
     var. formosus
          SY=L. lutosus Heller
     var. robustus
          SY=L. albipilosus Heller
          SY=L. bridgesii (S. Wats.) Heller
          SY=L. caeruleus Heller
          SY=L. greenei Heller non A. Nels.
          SY=L. navaliformis Heller
          SY=L. navicularis Heller
  fulcratus Greene
  garfieldensis C.P. Sm.
          SY=L. sericeus Pursh ssp. a.
                Phillips
          SY=L. s. var. asotinensis
                (Phillips) C.L. Hitchc.
  gormanii Piper
          SY=L. pumicola Heller
  gracilentus Greene
  grayi (S. Wats.) S. Wats.
          SY=L. andersonii S. Wats. var. g.
                S. Wats.
          SY=?L. falsoformis C.P. Sm.
          SY=?L. falsograyi C.P. Sm.
          SY=L. ione-grisetae C.P. Sm.
          SY=L. louise-bucariae C.P. Sm.
  guadalupensis Greene
          SY=L. aliclementinus C.P. Sm.
          SY=L. moranii Dunkle
  hartmannii C.P. Sm.
  havardii S. Wats.
  hillii Greene
     var. arizonicus (C.P. Sm.) Harmon
          SY=L. ingratus Greene var. a. C.P.
                Sm.
     var. hillii
          SY=L. marcusianus C.P. Sm.
     var. osterhoutianus (C.P. Sm.) Harmon
          SY=L. o. C.P. Sm.
  hirsutissimus Benth.
  holmgrenanus C.P. Sm.
  holosericeus Nutt. ex Torr. & Gray
          SY=L. lacuum-trinitatum C.P. Sm.
          SY=L. minearanus C.P. Sm.
          SY=L. multicincinnus C.P. Sm.
          SY=L. summae C.P. Sm.
  horizontalis Heller
     var. horizontalis
     var. platypetalus C.P. Sm.
  huachucanus M.E. Jones
          SY=L. platanophilus M.E. Jones

Lupinus (CONT.)
  hyacinthinus Greene
              SY=L. albicaulis Dougl. ex Hook.
                    var. h. (Greene) Jepson
              SY=L. formosus Greene var. h.
                    (Greene) C.P. Sm.
  X inyoensis Heller [caudatus X palmeri]
  johannis-howellii C.P. Sm.
  jonesii Rydb.
  kingii S. Wats.
        var. agillaceus (Woot. & Standl.)
              C.P. Sm.
        var. kingii
              SY=L. capitatus Greene
              SY=L. sileri S. Wats.
  klamathensis Eastw.
  kuschei Eastw.
              SY=L. sericeus Pursh var. k.
                    (Eastw.) Boivin
  lapidicola Heller
  latifolius J.G. Agardh
    ssp. dudleyi (C.P. Sm.) Kenney & Dunn
              SY=L. d. (C.P. Sm.) Eastw.
              SY=L. l. var. d. C.P. Sm.
    ssp. latifolius
              SY=L. caudiciferus Eastw.
              SY=L. columbianus Heller
              SY=L. confusus Heller non Rose
              SY=L. cytisoides J.G. Agardh
              SY=L. latifolius var. columbianus
                    (Heller) C.P. Sm.
              SY=L. l. var. ligulatus (Greene)
                    C.P. Sm.
              SY=L. ligulatus Greene
              SY=L. perennis L. ssp. latifolius
                    (J.G. Agardh) Phillips
              SY=L. rivularis Dougl. ex Lindl.
                    var. latifolius (J.G.
                    Agardh) Jepson
              SY=L. lasiotropis Greene ex Eastw.
    ssp. leucanthus (Rydb.) Kenney & D. Dunn
              SY=L. leucanthus Rydb.
              SY=L. latifolius var. parishii
                    sensu auctt. non C.P. Sm.
    ssp. longipes (Greene) Kenney & D. Dunn
              SY=L. columbianus Heller var.
                    simplex C.P. Sm.
              SY=L. edwin-livingstoni C.P. Sm.
              SY=L. l. Greene
              SY=L. pennellianus Heller ex
                    Eastw.
              SY=L. wyethii S. Wats. var.
                    hansenii C.P. Sm.
    ssp. parishii (C.P. Sm.) Kenney & D.
        Dunn
              SY=L. l. var. p. C.P. Sm.
              SY=L. p. Eastw.
    ssp. viridifolius (Heller) Kenney & D.
        Dunn
        var. barbatus (Henderson) Munz
              SY=L. b. (Henderson) Heller
              SY=L. latifolius var. b.
                    (Henderson) Kenney & Dunn
              SY=L. ligulatus Greene var. b.
                    Henderson
              SY=L. rivularis Dougl. ex Lindl.
                    var. b. (Henderson) C.P. Sm.
        var. viridifolius (Heller) Kenney &
              D. Dunn
              SY=L. latifolius var. v. (Heller)
                    C.P. Sm.
              SY=L. rivularis Dougl. ex Lindl.
                    var. v. (Heller) Jepson
              SY=L. v. Heller
    ssp. wigginsii (C.P. Sm.) Kenney & D.
        Dunn
              SY=L. latifolius var. w. C.P. Sm.

  lemmonii C.P. Sm.
  lepidus Dougl. ex Lindl.
  leucophyllus Dougl. ex Lindl.
    ssp. erectus (Henderson) Harmon
              SY=L. e. Henderson
              SY=L. l. var. tenuispicus (A.
                    Nels.) C.P. Sm.
              SY=L. t. A. Nels.
    ssp. leucophyllus
        var. belliae C.P. Sm.
        var. canescens (T.J. Howell) C.P. Sm.
              SY=L. c. T.J. Howell
        var. leucophyllus
              SY=L. cyaneus Rydb.
              SY=L. enodatus C.P. Sm.
              SY=L. forslingii C.P. Sm.
              SY=L. holosericeus Nutt. ex Torr.
                    & Gray var. amblyophyllus
                    B.L. Robins.
              SY=L. l. var. plumosus (Dougl. ex
                    Lindl.) B.L. Robins.
              SY=L. l. var. retrorsus
                    (Henderson) C.P. Sm.
              SY=L. macrostachys Rydb. non Rusby
              SY=L. p. Dougl. ex Lindl.
              SY=L. retrorsus Henderson
  leutescens C.P. Sm.
  littoralis Dougl.
  longifolius (S. Wats.) Abrams
              SY=L. chamissonis Eschsch. var. l.
                    S. Wats.
              SY=L. mollisifolius A. Davids.
  ludovicinaus Greene
  luteolus Kellogg
  lyallii Gray
    ssp. alcis-temporis (C.P. Sm.) Cox
              SY=L. a-t. C.P. Sm.
    ssp. lyallii
        var. danaus (Gray) S. Wats.
              SY=L. d. Gray
              SY=L. d. var. bicolor Eastw.
        var. fruticulosus (Greene) C.P. Sm.
              SY=L. f. Greene
              SY=L. perditorum Greene
        var. lyallii
        var. macroflorus Cox
        var. roguensis Cox
        var. villosus Jepson
    ssp. minutifolius (Eastw.) Cox
              SY=L. m. Eastw.
    ssp. subpandens C.P. Sm. ex D. Dunn
    ssp. washoensis (Heller) Cox
              SY=L. aridus Dougl. ex Lindl. var.
                    w. (Heller) C.P. Sm.
              SY=L. w. Heller
  magnificus M.E. Jones
        var. glarecola M.E. Jones
              SY=L. kerrii Eastw.
        var. hesperius (Heller) C.P. Sm.
              SY=L. h. Heller
        var. magnificus
  malacophyllus Greene
  meionanthus Gray
  minimus Dougl.
              SY=L. ovinus Greene
              SY=L. piperi B.L. Robins.
  mollis Heller
  nanus Dougl.
    ssp. latifolius (Benth. ex Torr.) D.
        Dunn
              SY=L. n. var. l. Benth. ex Torr.
    ssp. menkerae (C.P. Sm.) D. Dunn
              SY=L. n. var. m. C.P. Sm.
    ssp. nanus
  neomexicanus Greene
              SY=L. blumeri Greene
  nootkatensis Donn ex Sims

Lupinus (CONT.)
    var. fruticosus Sims
        SY=L. arboreus Sims var. f. (Sims)
          S. Wats.
        SY=L. n. var. glaber Hook.
        SY=L. n. var. unalaskensis S.
          Wats.
    var. nootkatensis
        SY=L. albertensis C.P. Sm.
        SY=L. kiskensis C.P. Sm.
        SY=L. n. var. ethel-looffii C.P.
          Sm.
        SY=L. n. var. henry-looffii C.P.
          Sm.
        SY=L. n. var. perlanatus C.P. Sm.
        SY=L. perennis L. ssp. n. (Donn ex
          Sims) Phillips
        SY=L. trifurcatus C.P. Sm.
obtusilobus Heller
        SY=L. ornatus Dougl. ex Lindl.
          var. obtusilobus (Heller)
          C.P. Sm.
odoratus Heller
    var. odoratus
    var. pilosellus C.P. Sm.
onustus S. Wats.
        SY=L. mucronulatus T.J. Howell
        SY=L. oreganus Heller var.
          pusillulus C.P. Sm.
        SY=L. pinetorum M.E. Jones non
          Heller
        SY=L. sulphureus Dougl. ex Hook.
          ssp. delnortensis (Eastw.)
          Phillips
        SY=L. thompsonianus C.P. Sm.
        SY=L. violaceus Heller
        SY=L. v. var. d. Eastw.
        SY=L. v. var. shastensis Eastw.
oreganus Heller
    var. kincaidii C.P. Sm.
        SY=L. sulphureus Dougl. ex Hook.
          ssp. k. (C.P. Sm. SY=L. s.
          var. k. (C.P. Sm.) C.L.
          Hitchc.
    var. oreganus
        SY=P. amabilis Heller
        SY=P. biddlei Henderson ex C.P.
          Sm.
ornatus Dougl. ex Lindl.
        SY=L. hellerae Heller
        SY=L. minimus Dougl. var. h.
          (Heller) C.P. Sm.
pachylobus Greene
        SY=L. micranthus (Dougl. non
          Guss.) var. p. (Greene)
          Jepson
padre-crowleyi C.P. Sm.
        SY=L. dedeckerae Munz & D. Dunn
pallidus Brandeg.
        SY=L. concinnus J.G. Agardh var.
          desertorum (Heller) C.P. Sm.
        SY=L. c. var. p. (Brandeg.) C.P.
          Sm.
        SY=L. d. Heller
palmeri S. Wats.
parviflorus Nutt. ex Hook. & Arn.
    ssp. floribundus (Greene) Harmon
        SY=L. f. Greene
    ssp. myrianthus (Greene) Harmon
        var. fulvomaculatus (Payson) Harmon
          SY=L. f. Payson
        var. myrianthus
          SY=L. leptostachys Greene
          SY=L. m. Greene
    ssp. parviflorus
        SY=L. allimicranthus C.P. Sm.
        SY=L. argenteus Pursh ssp. p.

        (Nutt. ex Hook. & Arn.)
        Phillips
        SY=L. argenteus var. p. (Nutt. ex
          Hook. & Arn.) C.L. Hitchc.
peirsonii Mason
perennis L.
    ssp. gracilis (Nutt.) D. Dunn
        SY=L. g. Nutt. non J.G. Agardh
        SY=L. nuttallii S. Wats.
        SY=L. p. var. g. (Nutt.) Chapman
    ssp. perennis
        var. occidentalis S. Wats.
        var. perennis
plattensis S. Wats.
        SY=L. glabratus (S. Wats.) Rydb.
        SY=L. ornatus Dougl. ex Lindl.
          var. g. S. Wats.
polycarpus Greene
        SY=L. micranthus Dougl. non Guss.
polyphyllus Lindl.
    ssp. bernardianus (Abrams ex Eastw.)
        Munz
        SY=L. b. Abrams ex Eastw.
        SY=L. superbus Heller var. b.
          Abrams ex C.P. Sm.
    ssp. polyphyllus
        var. grandifolius (Lindl. ex J.G.
          Agardh) Torr. & Gray
          SY=L. g. Lindl. ex J.G. Agardh
          SY=L. macrophyllus Benth.
          SY=L. magnus Greene
        var. pallidipes (Heller) C.P. Sm.
          SY=L. pallidipes Heller
        var. polyphyllus
          SY=L. matanuskensis C.P. Sm.
          SY=L. pseudopolyphyllus C.P. Sm.
          SY=L. stationis C.P. Sm.
    ssp. superbus (Heller) Munz
        SY=L. apodotropis Heller
        SY=L. alilatissimus C.P. Sm.
        SY=L. carolus-bucarii C.P. Sm.
        SY=L. elongatus Greene ex Heller
        SY=L. lacus-huntingtoni C.P. Sm.
        SY=L. meli-campestris C.P. Sm.
        SY=L. perglaber Eastw.
        SY=L. piperitus A. Davids.
        SY=L. piperitus var. sparsipilosus
          Eastw.
        SY=L. prato-lacunosum C.P. Sm.
        SY=L. procerus Heller
        SY=L. sabulii C.P. Sm.
        SY=L. sabulii var. subpersistens
          C.P. Sm.
        SY=L. superbus Heller var.
          elongatus (Greene ex Heller)
          C.P. Sm.
        SY=L. viridicalyx C.P. Sm.
pratensis Heller
    var. eriostachyus C.P. Sm.
    var. pratensis
        SY=L. sellulus Kellogg var. elatus
          Eastw.
prunophilus M.E. Jones
        SY=L. arcticus S. Wats. var. p.
          (M.E. Jones) C.P. Sm.
        SY=L. polyphyllus Lindl. var.
          prunophilus (M.E. Jones)
          Phillips
        SY=L. wyethii S. Wats. var.
          prunophilus (M.E. Jones)
          C.P. Sm.
punto-reyesensis C.P. Sm.
pusillus Pursh
    ssp. intermontanus (Heller) D. Dunn
        SY=L. i. Heller
        SY=L. p. var. i. (Heller) C.P. Sm.
    ssp. pusillus

Lupinus (CONT.)
    ssp. rubens (Rydb.) D. Dunn
        SY=L. odoratus Heller var. r.
          (Rydb.) Jepson
        SY=L. r. Rydb.
  rivularis Dougl. ex Lindl.
    SY=L. lignipes Heller
  roseolus Rydb.
  ruber Heller
    SY=L. microcarpus Sims var. r.
      (Heller) C.P. Sm.
  sabinianus Dougl. ex Lindl.
    SY=L. sabinii Dougl. ex Hook.
    SY=L. sericeus Pursh ssp. sabinii
      (Dougl. ex Hook.) Phillips
  saxosus T.J. Howell
    var. saxosus
    var. subsericeus (B.L. Robins. ex
      Piper) C.P. Sm.
      SY=L. subsericeus B.L. Robins. ex
        Piper
  sellulus Kellogg
    ssp. sellulus
      var. artulus (Jepson) Eastw.
        SY=L. lepidus Dougl. ex Lindl.
          var. a. Jepson
      var. lobbii (Gray ex S. Wats.) Cox
        SY=L. aridus Dougl. ex Lindl. var.
          lobbii Gray ex S. Wats.
        SY=L. chionophilus Greene ex C.F.
          Baker
        SY=L. lepidus Dougl. ex Lindl.
          var. lobbii (Gray ex S.
          Wats.) C.L. Hitchc.
        SY=L. lobbii Gray ex Greene
        SY=L. lyallii Gray var. lobbii
          (Gray ex S. Wats.) C.P. Sm.
      var. medius (Detling) Cox
        SY=L. lepidus Dougl. ex Lindl.
          ssp. m. Detling
        SY=L. l. var. m. (Detling) C.L.
          Hitchc.
      var. sellulus
        SY=L. aridus Dougl. ex Lindl. var.
          torreyi (Gray ex S. Wats.)
          C.P. Sm.
        SY=L. lepidus Dougl. ex Lindl.
          var. t. (Gray ex S. Wats.)
          Jepson
    ssp. ursinus (Eastw.) Munz
      SY=L. sellulus var. u. (Eastw.)
        Cox
      SY=L. u. Eastw.
  sericatus Kellogg
  sericeus Pursh
    ssp. huffmanii (C.P. Sm.) Fleak & D.
      Dunn
      SY=L. aegra-ovium C.P. Sm.
      SY=L. h. C.P. Sm.
      SY=L. larsonanus C.P. Sm.
      SY=L. puroviridus C.P. Sm.
      SY=L. quercus-jugi C.P. Sm.
      SY=L. rickeri C.P. Sm.
    ssp. marianus (Rydb.) Fleak & D. Dunn
      SY=L. m. Rydb.
    ssp. sericeus
      var. egglestonianus C.P. Sm.
        SY=L. buckinghamii C.P. Sm.
        SY=L. flavicaulis Rydb.
        SY=L. huilcoflorus C.P. Sm.
        SY=L. ramosus A. Nels.
        SY=L. spiraeaphilus C.P. Sm.
        SY=L. tuckeranus C.P. Sm.
      var. flexuosus (Lindl. ex J.G.
        Agardh) C.P. Sm.
        SY=L. fikeranus C.P. Sm.
        SY=L. flexuosus Lindl. ex J.G.

        Agardh
        SY=L. s. var. fikeranus (C.P. Sm.)
          C.L. Hitchc.
        SY=L. s. var. subflexuosus St.
          John & Warren
        SY=L. subulatus Rydb.
      var. maximus C.P. Sm. ex Fleak & D.
        Dunn
      var. sericeus
        SY=L. aliumbellatus C.P. Sm.
        SY=L. amniculi-salicis C.P. Sm.
        SY=L. blankinshipii Heller
        SY=L. falsocomatus C.P. Sm.
        SY=L. herman-workii C.P. Sm.
        SY=L. jonesii Blank. non Rydb.
        SY=L. leucopsis J.G. Agardh
  shockleyi S. Wats.
  sierrae-blancae Woot. & Standl.
    ssp. aquilinus (Woot. & Standl.) Fleak &
      D. Dunn
      SY=L. a. Woot. & Standl.
    ssp. sierrae-blancae
      SY=L. laetus Woot. & Standl.
  sparsiflorus Benth.
    ssp. inopinatus (C.P. Sm.) Dziekanowski
      & D. Dunn
      SY=L. s. var. i. C.P. Sm.
    ssp. mohavensis Dziekanowski & D. Dunn
    ssp. pondii (Greene) Dziekanowski & D.
      Dunn
      SY=L. p. Greene
      SY=L. s. var. p. (Greene) C.P. Sm.
    ssp. sparsiflorus
  spectabilis Hoover non C.P. Sm.
    SY=L. nanus Dougl. var. perlasius
      C.P. Sm.
  stiversii Kellogg
  subcarnosus Hook.
  sublanatus Eastw.
  subvexus C.P. Sm.
    var. albilanatus C.P. Sm.
    var. phoeniceus C.P. Sm.
    var. subvexus
      SY=L. s. var. insularis C.P. Sm.
      SY=L. s. var. nigrescens C.P. Sm.
    var. transmontanus C.P. Sm.
      SY=L. s. var. flutivalis C.P. Sm.
      SY=L. s. var. liebergii C.P. Sm.
  succulentus Dougl. ex W.D.J. Koch
    var. brandegeei C.P. Sm.
    var. succulentus
      SY=L. s. var. laynae C.P. Sm.
  suksdorfii B.L. Robins.
  sulphureus Dougl. ex Hook.
    SY=L. s. var. applegateianus C.P.
      Sm.
    SY=L. s. var. echlerianus C.P. Sm.
  texensis Hook.
  tidestromii Greene
    var. laynae (Eastw.) Munz
    var. tidestromii
  tracyi Eastw.
  truncatus Nutt. ex Hook. & Arn.
    SY=L. t. var. burlewii C.P. Sm.
  uncinalis S. Wats.
  vallicola Heller
    ssp. apricus (Greene) D. Dunn
      SY=L. a. Greene
      SY=L. nanus Dougl. var. a.
        (Greene) C.P. Sm.
      SY=L. v. var. a. (Greene) C.P. Sm.
    ssp. vallicola
      SY=L. blaisdellii Eastw.
      SY=L. nanus Dougl. var. v.
        (Heller) C.P. Sm.
      SY=L. persistens Heller non Rose
  variicolor Steud.

Lupinus (CONT.)
             SY=L. franciscanus Greene
             SY=L. michneri Greene
             SY=L. versicolor Lindl. non Sweet
    villosus Willd.
    volcanicus Greene
    westianus Small
    wyethii S. Wats.
        ssp. tetonensis (E. Nels.) Cox & D. Dunn
             SY=L. arcticus S. Wats. var. t.
                 (E. Nels.) C.P. Sm.
             SY=L. humicola A. Nels. var. t. E.
                 Nels.
        ssp. wyethii
             SY=L. arcticus S. Wats. var.
                 humicola A. Nels.
             SY=L. candicans Rydb.
             SY=L. diversalpicola C.P. Sm.
             SY=L. flavescens Rydb.
             SY=L. h. A. Nels.
             SY=L. rydbergii Blank.

Lysiloma Benth.

    latisiliqua (L.) Benth.
             SY=L. bahamensis Benth.
    microphylla Benth.
        var. thornberi (Britt. & Rose) Isely
             SY=L. t. Britt. & Rose

Machaerium Pers.

    lunatum (L.f.) Ducke
             SY=Drepanocarpus l. (L.f.) G.F.W.
                 Mey.

Macroptilium (Benth.) Urban

    lathyroides (L.) Urban
             SY=Phaseolus l. L.

Marina Liebm.

    calycosa (Gray) Barneby
             SY=Dalea c. Gray
    diffusa (Moric.) Barneby
             SY=Dalea d. Moric.
    orcuttii (S. Wats.) Barneby
    parryi (Torr. & Gray) Barneby
             SY=Dalea p. Torr. & Gray

Medicago L.

    arabica (L.) Huds.
    falcata L.
    laciniata (L.) P. Mill.
    littoralis Rhode ex Loisel.
    lupulina L.
        var. cupaniana (Guss.) Boiss.
        var. lupulina
             SY=M. l. var. glandulosa Neilr.
    minima (L.) Bartalini
             SY=M. m. var. compacta Neyraut
             SY=M. m. var. longiseta DC.
             SY=M. m. var. pubescens Webb
    orbicularis (L.) Bartalini
    polymorpha L.
             SY=M. hispida Gaertn.
             SY=M. h. var. apiculata (Willd.)
                 Urban
             SY=M. h. var. confinis (W.D.J.
                 Koch) Burnat
             SY=M. p. var. brevispina (Benth.)
                 Heyn
             SY=M. p. var. ciliaris (Ser.)
                 Shinners
             SY=M. p. var. polygyra (Urban)

             Shinners
             SY=M. polymorpha var. tricycla
                 (Gren. & Godr.) Shinners
             SY=M. polymorpha var. vulgaris
                 (Benth.) Shinners
    praecox DC.
    sativa L.
    secundiflora Dur.
    turbinata (L.) All.

Melilotus P. Mill.

    alba Medic.
        var. alba
        var. annua Coe
    altissima Thuill.
    elegans Salzm. ex Ser.
    indica (L.) All.
    neapolitana Ten.
    officinalis (L.) Pallas
    sulcata Desf.
    wolgica Poir.

Mezoneuron Desf.

    kavaiense (Mann) Hbd.

Mimosa L.

    biuncifera Benth.
        var. biuncifera
        var. glabrescens Gray
    borealis Gray
    ceratonia L.
    dysocarpa Benth.
        var. dysocarpa
        var. wrightii (Gray) Kearney &
            Peebles
    emoryana Benth.
    grahamii Gray
        var. grahamii
        var. lemmonii (Gray) Kearney &
            Peebles
    laxiflora Benth.
    malacophylla Gray
    pigra L.
        var. berlandieri (Gray) B.L. Turner
        var. pigra
    pudica L.
        var. pudica
        var. unijuga (Duchass. & Walp.)
            Griseb.
    strigillosa Torr. & Gray
    warnockii B.L. Turner
    wherryana (Britt.) Standl.
    zygophylla Benth.

Mucuna Adans.

    deeringiana (Bort) Merr.
             SY=Stizolobium d. Bort
    gigantea (Willd.) DC.
    pruriens (L.) DC.
             SY=Stizolobium pruritum (Wight)
                 Piper
    sloanei Fawcett & Rendle
    urens (L.) Medic.

Myrospermum Jacq.

    frutescens Jacq.

Neorudolphia Britt.

    volubilis (Willd.) Britt.

Neptunia Lour.

Neptunia (CONT.)
  lutea (Leavenworth) Benth.
          SY=N. l. var. multipinnatifida
              B.L. Turner
  plena (L.) Benth.
  pubescens Benth.
      var. microcarpa (Rose) Windler
          SY=N. palmeri Britt. & Rose
      var. pubescens
          SY=N. floridana Small
          SY=N. p. var. f. (Small) B.L.
              Turner
          SY=N. p. var. lindheimeri (B.L.
              Robins.) B.L. Turner

Nissolia Jacq.

  platycalyx S. Wats.
  schottii (Torr.) Gray
  wislizenii (Gray) Gray

Olneya Gray

  tesota Gray

Onobrychis P. Mill.

  viciifolia Scop.

Ononis L.

  repens L.
  spinosa L.

Ormosia G. Jackson

  krugii Urban

Ornithopus L.

  perpusillus L.
          SY=O. roseus Dufour
  pinnatus (P. Mill.) Druce
  sativus Brot.

Oxyrhynchus Brandeg.

  volubilis Brandeg.

Oxytropis DC.

  arctica R. Br.
      var. arctica
          SY=O. coronaminis Fern.
      var. barnebyana Welsh
      var. koyukukensis (Porsild) Welsh
          SY=O. k. Porsild
  bellii (Britt. ex Macoun) Palibine
          SY=O. arctica R. Br. var. b.
              (Britt. ex Macoun) Boivin
  besseyi (Rydb.) Blank.
      var. argophylla (Rydb.) Barneby
          SY=Aragallus a. Rydb.
      var. besseyi
          SY=Aragallus b. Rydb.
      var. fallax Barneby
      var. obnapiformis (C.L. Porter) Welsh
          SY= O. o. C.L. Porter
      var. salmonensis Barneby
      var. ventosa (Greene) Barneby
          SY=Aragallus v. Greene
  campestris (L.) DC.
      var. chartacea (Fassett) Barneby
          SY=O. chartacea Fassett
      var. columbiana (St. John) Barneby
          SY=O. c. St. John
      var. cusickii (Greenm.) Barneby

          SY=O. alpicola (Rydb.) M.E. Jones
          SY=O. campestris var. rydbergii
              (A. Nels.) P.J. Davis
          SY=O. cusickii Greenm.
          SY=O. rydbergii A. Nels.
      var. davisii Welsh
      var. dispar (A. Nels.) Barneby
          SY=O. dispar (A. Nels.) K. Schum.
      var. gracilis (A. Nels.) Barneby
          SY=O. c. ssp. g. (A. Nels.) Boivin
          SY=O. c. var. cervinus (Greene)
              Boivin
          SY=O. g. (A. Nels.) K. Schum.
          SY=O. luteola (Greene) Piper &
              Beattie
          SY=O. villosa (Rydb.) K. Schum.
      var. johannensis Fern.
          SY=O. j. (Fern.) Heller
      var. jordalii (Porsild) Welsh
          SY=O. campestris ssp. j. (Porsild)
              Hulten
          SY=O. j. Porsild
      var. terrae-novae (Fern.) Barneby
          SY=O. t-n. Fern.
      var. varians (Rydb.) Barneby
          SY=O. hyperborea Porsild
          SY=O. v. (Rydb.) K. Schum.
  deflexa (Pallas) DC.
      var. deflexa
      var. foliolosa (Hook.) Barneby
          SY=O. d. var. capitata Boivin
          SY=O. f. Hook.
      var. sericea Torr. & Gray
          SY=O. d. var. parviflora Boivin
  huddelsonii Porsild
  jonesii Barneby
  kobukensis Welsh
  kokrinensis Porsild
  lagopus Nutt.
      var. atropurpurea (Rydb.) Barneby
          SY=Aragallus a. Rydb.
      var. conjugens Barneby
      var. lagopus
          SY=Spiesia l. (Nutt.) Kuntze
  lambertii Pursh
      var. articulata (Greene) Barneby
          SY=Aragallus a. Greene
      var. bigelovii Gray
          SY=O. patens (Rydb.) A. Nels.
      var. lambertii
          SY=O. involuta (A. Nels.) K.
              Schum.
  maydelliana Trautv.
          SY=O. campestris (L.) DC. ssp.
              melanocephala Hook.
          SY=O. c. var. glabrata Hook.
          SY=O. g. (Hook.) A. Nels.
          SY=O. maydelliana ssp.
              melanocephala (Hook.)
              Porsild
  mertensiana Turcz.
  multiceps Torr. & Gray
  nana Nutt.
  nigrescens (Pallas) Fisch. ex DC.
      var. lonchopoda Barneby
      var. nigrescens
          SY=O. glaberrima Hulten
          SY=O. n. ssp. bryophila (Greene)
              Hulten
          SY=O. n. ssp. pygmaea (Pallas)
              Hulten
          SY=O. p. (Pallas) Fern.
      var. uniflora (Hook.) Barneby
          SY=O. arctobia Bunge
          SY=O. n. ssp. a. (Bunge) Hulten
          SY=O. n. var. a. (Bunge) Gray
  oreophila Gray

Oxytropis (CONT.)
       var. juniperina Welsh
       var. oreophila
           SY=Aragallus o. (Gray) A. Nels.
           SY=Spiesia o. Kuntze
  parryi Gray
  podocarpa Gray
       SY=O. p. var. inflata (Hook.)
           Boivin
  ripara Litv.
  scammaniana Hulten
  sericea Nutt. ex Torr. & Gray
       var. sericea
           SY=O. pinetorum (Heller) K. Schum.
       var. spicata (Hook.) Barneby
           SY=O. macounii (Greene) Rydb.
           SY=O. spicata (Hook.) Standl.
  splendens Dougl. ex Hook.
       SY=Aragallus s. (Dougl. ex Hook.)
           Greene
       SY=Astragalus s. (Dougl. ex Hook.)
           Tidestrom
       SY=O. richardsonii (Hook.) K.
           Schum
       SY=O. s. var. r. Hook.
       SY=O. s. var. vestita Hook.
  viscida Nutt. ex Torr. & Gray
       var. hudsonica (Greene) Barneby
           SY=Aragallus h. Greene
           SY=?O. borealis DC.
           SY=O. h. (Greene) Fern.
           SY=?O. leucantha (Pallas) Pers.
       var. subsucculenta (Hook.) Barneby
           SY=O. glutinosa Porsild
           SY=O. uralensis C.A. Mey. var. s.
           Hook.
       var. visida
           SY=Aragallus viscidulus Rydb.
           SY=A. v. var. depressus Rydb.
           SY=O. gaspensis Fern. & Kelsey
           SY=O. ixodes Butters & Abbe
           SY=O. leucantha (Pallas) Pursh
              var. d. (Rydb.) Boivin
           SY=O. l. var. g. (Fern. & Kelsey)
           Boivin
           SY=O. l. var. i. (Butters & Abbe)
           Boivin
           SY=O. l. var. leuchippiana Boivin
           SY=O. leucantha var. magnifica
           Boivin
           SY=O. leucantha var. v. (Nutt.)
           Boivin
           SY=O. sheldonensis Porsild
           SY=O. verruculosa Porsild
           SY=O. viscidula Tidestrom ssp.
           sulphurea Porsild

Pachyrhizus L.C. Rich. ex DC.

  erosus (L.) Urban
       SY=Cacara e. (L.) Kuntze

Parkinsonia L.

  aculeata L.
  florida (Benth. ex Gray) S. Wats.
       SY=Cercidium f. Benth. ex Gray
  microphylla Torr.
       SY=Cercidium m. (Torr.) Rose &
           I.M. Johnston
  texana (Gray) S. Wats.
       var. macrum (I.M. Johnston) Isely
           SY=Cercidium m. I.M. Johnston
       var. texanum
           SY=Cercidium t. Gray

Parryella Torr. & Gray ex Gray

filifolia Torr. & Gray

Peltophorum (T. Vogel) Walp.

  pterocarpum (DC.) Backer ex K. Heyne
       SY=P. inerme (Roxb.) Naves ex
           Vill.

Peteria P. Br.

  scoparia Gray

Phaseolus L.

  acutifolius Gray
       var. acutifolius
       var. latifolius Freeman
       var. tenuifolius Gray
  adenanthus G.W.F. Mey.
  angustissimus Gray
       var. angustissimus
       var. latus M.E. Jones
  atropurpureus DC.
  coccineus L.
  grayanus Woot. & Standl.
  heterophyllus Willd.
       var. heterophyllus
       var. rotundifolius (Gray) Piper
  limensis Macfad.
  lunatus L.
       var. lunatus
       var. lunonanus Bailey
  metcalfei Woot. & Standl.
  parvulus Greene
  peduncularis H.B.K.
       SY=Vigna p. (H.B.K.) Fawcett &
           Rendle
  polystachyus (L.) B.S.P.
       var. aquilonius Fern.
       var. polystachyus
  ritensis M.E. Jones
  schottii Benth.
       SY=P. trichocarpus C. Wright
  sinuatus (Nutt.) Torr. & Gray
  ?smilacifolius Pollard
  vulgaris L.
       var. humilis Alef.
       var. vulgaris
  wrightii Gray

Pickeringia Nutt. ex Torr. & Gray

  montana Nutt. ex Torr. & Gray
    ssp. montana
    ssp. tomentosa (Abrams) Abrams

Pictetia DC.

  aculeata (Vahl) Urban

Piscidia L.

  carthagenensis Jacq.
  piscipula (L.) Sarg.
       SY=Ichthyomethia p. (L.) A.S.
           Hitchc.

Pisum L.

  sativum L.
       var. arvense (L.) Poir.
       var. humile Poir.
       var. macrocarpon Ser.
       var. sativum

Pithecellobium Mart.

Pithecellobium   (CONT.)
   arboreum (L.) Urban
   dulce (Roxb.) Benth.
   flexicaule (Benth.) Coult.
   graciliflorum Blake
   keyense Britt. ex Britt. & Rose
          SY=P. guadalupense (Pers.) Chapman
   pallens (Benth.) Standl.
   saman (Jacq.) Benth.
          SY=Samanea s. (Jacq.) Merr.
   unguis-cati (L.) Benth.

Pongamia Vent.

   pinnata (L.) Merr.

Prosopis L.

   glandulosa Torr.
       var. glandulosa
          SY=P. chilensis (Molina) Stuntz
                  var. g. (Torr.) Standl.
          SY=P. juliflora (Sw.) DC. var. g.
                  (Torr.) Cockerell
          SY=P. odorata Torr. & Frem.
       var. prostrata Burkart
       var. torreyana (L. Benson) M.C.
          Johnston
          SY=P. juliflora (Sw.) DC. var. t.
                  L. Benson
   laevigata (Humb. & Bonpl. ex Willd.) M.C.
      Johnston
   limensis Benth.
   pallida (Humb. & Bonpl. ex Willd.) H.B.K.
   pubescens Benth.
   reptans Benth.
       var. cinerascens (Gray) Burkart
          SY=P. c. (Gray) Benth.
       var. reptans
   strombulifera (Lam.) Benth.
   velutina Woot.
          SY=Neltuma v. (Woot.) Britt. &
                  Rose
          SY=P. chilensis (Molina) Stuntz
                  var. v. (Woot.) Standl.
          SY=P. juliflora (Sw.) DC. var. v.
                  (Woot.) Sarg.

Psoralea L.

   americana L.
          SY=?Cullen a. (L.) Rydb.
   argophylla Pursh
          SY=Psoralidium a. (Pursh) Rydb.
   aromatica Payson
   bituminosa L.
          SY=Asphalthium b. (L.) Kuntze
   californica S. Wats.
   canescens Michx.
          SY=Pediomelum c. (Michx.) Rydb.
   castorea S. Wats.
   collina Rydb.
   corylifolia L.
   cuspidata Pursh
   cyphocalyx Gray
   digitata Nutt. ex Torr. & Gray
       var. digitata
       var. parvifolia Shinners
   epipsila Barneby
   esculenta Pursh
          SY=Pediomelum e. (Pursh) Rydb.
   hypogaea Nutt. ex Torr. & Gray
       var. hypogaea
       var. scaposa Gray
          SY=Psoralea s. (Gray) J.F. Macbr.
          SY=P. s. var. breviscapa Shinners
   juncea Eastw.

lanceolata Pursh
       var. lanceolata
          SY=Psoralidium l. (Pursh) Rydb.
       var. purshii (Vail) Piper
          SY=Psoralea l. ssp. scabra (Nutt.)
                  Piper
          SY=P. s. Nutt.
       var. stenophylla (Rydb.) Toft & Welsh
          SY=Psoralidium s. (Rydb.) Rydb.
   latestipulata Shinners
       var. appressa Ockendon
       var. latestipulata
   linearifolia Torr. & Gray
   lupinellus Michx.
          SY=Rhytidomene l. (Michx.) Rydb.
   macrophylla (Rowlee) Small
          SY=Orbexilum m. (Rowlee) Rydb.
   macrostachya DC.
       var. longiloba (Rydb.) J.F. Macbr.
          SY=Hoita l. Rydb.
       var. macrostachya
          SY=Hoita hallii Rydb.
          SY=H. villosa Greene
          SY=Psoralea douglasii Greene
       var. rhombifolia Torr.
          SY=Hoita r. (Torr.) Rydb.
   megalantha Woot. & Standl.
   mephitica S. Wats.
          SY=Psoralea m. var. retrorsa
                  (Rydb.) Kearney & Peebles
   onobrychis Nutt.
          SY=Orbexilum o. (Nutt.) Rydb.
   orbicularis Lindl.
   pariensis Welsh & Atwood
   physodes Dougl. ex Hook.
   psoralioides (Walt.) Cory
       var. eglandulosa (Ell.) Freeman
       var. gracile (Chapman) Freeman
          SY=Orbexilum g. (Chapman) Rydb.
       var. psoralioides
   reverchonii S. Wats.
   rhombifolia Torr. & Gray
   rigida Parish
   rydbergii Cory
   simplex Nutt. ex Torr. & Gray
          SY=Orbexilum s. (Nutt. ex Torr. &
                  Gray) Rydb.
   stipulata Torr. & Gray
   strobilina Hook. & Arn.
   subacaulis Torr. & Gray
          SY=Pediomelum s. (Torr. & Gray)
                  Rydb.
   subulata Bush
   tenuiflora Pursh
       var. bigelovii (Rydb.) J.F. Macbr.
       var. floribunda (Nutt.) Rydb.
          SY=Psoralea f. Nutt.
          SY=P. obtusifolia Torr. & Gray
       var. tenuiflora
   trinervata (Rydb.) Standl.
   virgata Nutt.
          SY=Orbexilum v. (Nutt.) Rydb.

Psorothamnus Rydb.

   arborescens (Torr. ex Gray) Barneby
       var. arborescens
          SY=Dalea a. Torr. ex Gray
          SY=D. fremontii Torr. ex Gray var.
                  saundersii (Parish) Munz
       var. minutifolius (Parish) Barneby
          SY=Dalea fremontii Torr. ex Gray
                  var. m. (Parish) L. Benson
       var. pubescens (Parish) Barneby
          SY=Dalea amoena S. Wats.
          SY=D. a. var. p. (Parish) Peebles
          SY=D. fremontii Torr. ex Gray var.

Psorothamnus (CONT.)
                    p. (Parish) L. Benson
          var. simplifolius (Parish) Barneby
               SY=Dalea californica S. Wats.
     emoryi (Gray) Rydb.
          SY=Dalea e. Gray
     fremontii (Torr. ex Gray) Barneby
          var. attenuatus Barneby
          var. fremontii
               SY=Dalea f. Torr. ex Gray
               SY=D. f. var. johnsonii (S. Wats.)
                    Munz
     kingii (S. Wats.) Barneby
          SY=Dalea k. S. Wats.
     polydenius (Torr. ex S. Wats.) Rydb.
          var. jonesii Barneby
          var. polydenius
               SY=Dalea p. Torr. ex S. Wats.
     schottii (Torr.) Barneby
          SY=Dalea s. Torr.
               SY=D. s. var. puberula (Parish)
                    Munz
     scoparius (Gray) Rydb.
          SY=Dalea s. Gray
     spinosus (Gray) Barneby
          SY=Dalea s. Gray
     thompsonae (Vail) Welsh & Atwood
          var. thompsonae
               SY=Dalea t. (Vail) L.O. Williams
          var. whitingii (Kearney & Peebles)
               Barneby
               SY=Dalea w. Kearney & Peebles

Pterocarpus Jacq.

     officinalis Jacq.

Pueraria DC.

     lobata (Willd.) Ohwi
          SY=P. thunbergiana (Sieb. & Zucc.)
               Benth.
     phaseoloides (Roxb.) Benth.

Rhynchosia Lour.

     americana (P. Mill.) M.C. Metz
     caribaea (Jacq.) DC.
     cinerea Nash
     cytisoides (Bertol.) Wilbur
          SY=Pitcheria galactoides Nutt.
          SY=R. g. (Nutt.) Endl. ex Walp.
     difformis (Ell.) DC.
          SY=R. lewtonii (Vail) Small
          SY=R. tomentosa (L.) Hook. & Arn.
               var. volubilis (Michx.)
               Torr. & Gray
     edulis Griseb.
          SY=R. rariflora Standl.
     latifolia Nutt. ex Torr. & Gray
          SY=Dolicholus l. (Nutt. ex Torr. &
               Gray) Vail
     michauxii Vail
     minima (L.) DC.
          SY=Dolicholus m. (L.) Medic.
          SY=R. m. var. diminifolia Walraven
     parvifolia DC.
          SY=Leucopterum p. (DC.) Small
     phaseoloides (Sw.) DC.
          SY=Dolicholus pyramidalis sensu
               auctt. non (Lam.) Britt. &
               Wilson
     reniformis (Pursh) DC.
          SY=Dolicholus simplicifolius
               (Walt.) Vail
          SY=R. intermedia (Torr. & Gray)
               Small

               SY=R. s. (Walt.) Wood
     reticulata (Sw.) DC.
          SY=Dolicholus r. (Sw.) Millsp.
     senna Gillies ex Hook.
          var. angustifolia (Gray) Grear
               SY=R. texana Torr. & Gray
               SY=R. t. var. a. Gray
     swartzii (Vail) Urban
          SY=Dolicholus s. Vail
     tomentosa (L.) Hook. & Arn.
          var. mollissima (Ell.) Torr. & Gray
               SY=R. m. (Ell.) S. Wats.
          var. monophylla (Ell.) Torr. & Gray
          var. tomentosa
               SY=Dolicholus erectus (Walt.) Vail
               SY=D. t. (L.) Vail
               SY=R. e. (Walt.) DC.
               SY=R. t. var. e. (Walt.) Torr. &
                    Gray

Robinia L.

     X albicans Ashe [boyntonii X pseudoacacia]
     X ambigua Poir. [pseudoacacia X viscosa]
               SY=R. a. var. bella-rosea
                    (Nichols.) Rehd.
     boyntonii Ashe
     elliottii (Chapman) Ashe
     hartwegii Koehne
     hispida L.
          var. fertilis (Ashe) Clausen
               SY=R. f. Ashe
          var. hispida
               SY=R. grandiflora Ashe
               SY=R. longiloba Ashe
               SY=R. pallida Ashe
               SY=R. pedunculata Ashe
               SY=R. speciosa Ashe
     X holdtii Beissn. [neomexicana X
          pseudoacacia]
     kelseyi Hort. ex Hutchinson
     X margaretta Ashe [hispida X pseudoacacia]
     nana Ell.
     neomexicana Gray
          SY=R. luxurians (Dieck) Schneid.
          SY=R. n. var. l. Dieck
          SY=R. n. var. subvelutina (Rydb.)
               Kearney & Peebles
     pseudoacacia L.
          var. pseudoacacia
          var. rectissima (L.) Raber
     rusbyi Woot. & Standl.
     salavinii Rehd. [kelseyi X pseudoacacia]
     viscosa Vent. ex Vauq.
          var. hartwegii (Koehne) Ashe
          var. viscosa

Sabinea DC.

     florida (Vahl) DC.
     punicea Urban

Schrankia Willd.

     hystricina (Small ex Britt. & Rose) Standl.
     latidens (Small) K. Schum.
     microphylla (Dry.) J.F. Macbr.
          SY=Leptoglottis angustisiliqua
               Britt. & Rose
          SY=L. chapmanii Small ex Britt. &
               Rose
          SY=L. m. (Dry.) Britt. & Rose
          SY=Morongia angustata (Torr. &
               Gray) Britt.
          SY=M. m. (Dry.) Britt.
          SY=S. angustata Torr. & Gray
     nuttallii (DC. ex Britt. & Rose) Standl.

Schrankia   (CONT.)
          SY=Leptoglottis n. DC. ex Britt. &
              Rose
occidentalis (Woot. & Standl.) Standl.
portoricensis Urban
          SY=Morongia p. (Urban) Britt.
roemeriana (Scheele) Blank.
uncinata Willd.
          SY=Morongia u. (Willd.) Britt.

Scorpiurus L.

   muricatus L.
          SY=S. subvillosus L.
          SY=S. sulcatus L.

Securigera DC.

   securidaca (L.) Deg. & Dorf.
          SY=Bonaveria s. (L.) Reichenb.

Sesbania Scop.

   bispinosa (Jacq.) Steud. ex Fawcett &
      Rendle
          SY=Sesban b. (Jacq.) Rydb.
   drummondii (Rydb.) Cory
          SY=Daubentonia d. Rydb.
   emerus (Aubl.) Urban
          SY=Sesban e. (Aubl.) Britt. &
              Wilson
   grandiflora (L.) Pers.
          SY=Agati g. (L.) Desv.
   macrocarpa Muhl. ex Raf.
          SY=Sesban exalta (Raf.) Rydb.
          SY=Sesbania e. (Raf.) Cory
   punicea (Cav.) Benth.
          SY=Daubentonia p. (Cav.) DC.
   sericea (Willd.) Link
   sesban (L.) Merr.
   tomentosa Hook. & Arn.
      var. molokaiensis Deg. & Sherff
      var. tomentosa
   vesicaria (Jacq.) Ell.
          SY=Glottidium v. (Jacq.) Harper

Sophora L.

   affinis Torr. & Gray
   arizonica S. Wats.
          SY=S. formosa Kearney & Peebles
   chrysophylla (Salisb.) Seem.
      ssp. chrysophylla
         var. chrysophylla
         var. makuaensis Chock
      ssp. circularis Chock
         var. circularis Chock
         var. kauensis Chock
      ssp. glabrata (Gray) Chock
         var. glabrata (Gray) Chock
         var. grisea (Deg. & Sherff) Chock
         var. lanaiensis Chock
         var. ovata Chock
      ssp. unifoliata (Rock) Chock
         var. elliptica Chock
         var. kanaioensis Chock
         var. unifoliata (Rock) Chock
   gypsophila B.L. Turner & Powell
         var. guadalupensis B.L. Turner &
              Powell
         var. gypsophila
   leachiana M.E. Peck
   nuttalliana B.L. Turner
          SY=S. sericea Nutt.
   secundiflora (Ortega) Lag. ex DC.
   stenophylla Gray ex Ives
   tomentosa L.

Spartium L.

   junceum L.

Sphaerophysa DC.

   salsula (Pallas) DC.
          SY=Swainsona s. (Pallas) Taubert

Sphinctospermum Rose

   constrictum (S. Wats.) Rose

Stahlia Bello

   monosperma (Tul.) Urban

Strongylodon Vogel

   ruber Vogel

Strophostyles Ell.

   helvola (L.) Ell.
      var. helvola
      var. missouriensis (S. Wats.) Britt.
   leiosperma (Torr. & Gray) Piper
          SY=Phaseolus l. Torr. & Gray
          SY=S. pauciflora (Benth.) S. Wats.
   umbellata (Muhl. ex Willd.) Britt.
      var. paludigena Fern.
      var. umbellata

Stylosanthes Sw.

   biflora (L.) B.S.P.
          SY=S. b. var. hispidissima
              (Michx.) Pollard & Ball
          SY=S. riparia Kearney
          SY=S. r. var. setifera Fern.
   calcicola Small
   hamata (L.) Taubert
   viscosa Sw.

Tamarindus L.

   indica L.

Tephrosia Pers.

   angustissima Shuttlw. ex Chapman
          SY=Cracca a. (Shuttlw. ex Chapman)
              Kuntze
          SY=T. purpurea (Shuttlw.) B.L. Robins.
   candida DC.
          SY=Xiphocarpus c. (DC.) Endl.
   chrysophylla Pursh
          SY=Cracca carpenteri Rydb.
          SY=C. chapmanii (Vail) Small
          SY=C. chrysophylla (Pursh) Kuntze
          SY=T. carpenteri (Rydb.) Killip
   cinerea (L.) Pers.
          SY=Cracca c. (L.) Morong
   corallicola (Small) Leon
          SY=Cracca c. Small
   curtissii (Small) Shinners
          SY=Cracca c. Small
   florida (Dietr.) C.E. Wood
      var. florida
          SY=Cracca ambigua (M.A. Curtis)
              Kuntze
          SY=Galega f. Dietr.
          SY=T. a. (M.A. Curtis) Chapman
      var. gracillima (B.L. Robins.)
          Shinners
          SY=Cracca g. (B.L. Robins.) Heller
          SY=T. ambigua (M.A. Curtis) Kuntze

Tephrosia (CONT.)
          var. g. B.L. Robins.
    hispidula (Michx.) Pers.
          SY=Cracca h. (Michx.) Kuntze
          SY=T. elegans Nutt.
    leiocarpa Gray
          SY=Cracca l. (Gray) Kuntze
          SY=T. affinis S. Wats.
          SY=T. viridis M.E. Jones
    lindheimeri Gray
    mohrii (Rydb.) Godfrey
          SY=Cracca m. Rydb.
    onobrychoides Nutt.
          SY=Cracca o. (Nutt.) Kuntze
          SY=C. texana Rydb.
          SY=T. angustifolia Featherman
          SY=T. multiflora Featherman
          SY=T. t. (Rydb.) Cory
    potosina Brandeg.
          SY=Cracca p. (Brandeg.) Standl.
    purpurea (L.) Pers.
          SY=Cracca p. L.
    rugelii Shuttlw. ex B.L. Robins.
          SY=Cracca r. (Shuttlw. ex B.L.
              Robins.) Heller
    seminole Shinners
    senna H.B.K.
          SY=Cracca cathartica (Sesse &
              Moc.) Britt. & Millsp.
    sessiflora (Poir.) Urban
    spicata (Walt.) Torr. & Gray
          SY=Cracca s. (Walt.) Kuntze
          SY=C. flexuosa (Vail) Heller
          SY=Galega s. Walt.
          SY=T. f. Chapman
          SY=T. s. var. semitonsa Fern.
    tenella Gray
    thurberi (Rydb.) C.E. Wood
          SY=Cracca t. Rydb.
    virginiana (L.) Pers.
          SY=Cracca latidens Small
          SY=C. v. L.
          SY=T. l. (Small) Standl.
          SY=T. v. var. glabra Nutt. ex
              Torr. & Gray
          SY=T. v. var. holosericea (Nutt.)
              Torr. & Gray

Teramnus P. Br.

    labialis (L. f.) Spreng.
    uncinatus (L.) Sw.

Thermopsis R. Br. ex Ait. & Ait. f.

    caroliniana M.A. Curtis
    divaricarpa A. Nels.
    fraxinifolia (Nutt.) M.A. Curtis
    gracilis T.J. Howell
          var. argentata (Greene) Jepson
              SY=T. a. Greene
          var. gracilis
          var. venosa (Eastw.) Jepson
              SY=T. subglabra Henderson
    macrophylla Hook. & Arn.
          var. agnina J.T. Howell
          var. macrophylla
          var. semota Jepson
          var. velutina (Greene) Larisey
    mollis (Michx.) M.A. Curtis ex Gray
          SY=T. hugeri Small
    montana Nutt. ex Torr. & Gray
          var. montana
          var. ovata (B.L. Robins.) St. John
              SY=T. o. (B.L. Robins.) Rydb.
    pinetorum Greene
    rhombifolia (Nutt. ex Pursh) Nutt. ex

          Richards.
          var. annulocarpa (A. Nels.) L.O.
              Williams
          var. arenosa (A. Nels.) Larisey
          var. rhombifolia
    villosa (Walt.) Fern. & Schub.

Trifolium L.

    albopurpureum Torr. & Gray
          SY=T. a. var. neolagopus (Loja.)
              McDermott
          SY=T. macraei Hook. & Arn. var. a.
              (Torr. & Gray) Greene
    alexandrinum L.
    amabile H.B.K.
    amoenum Greene
    andersonii Gray
        ssp. andersonii
        ssp. beatleyae J. Gillett
        ssp. monoense (Greene) J. Gillett
              SY=T. m. Greene
    andinum Nutt.
    angustifolium L.
    appendiculatum Loja.
    arvense L.
    attenuatum Greene
    aureum Pollich
          SY=T. agrarium L.
    barbigerum Torr.
          var. andrewsii Gray
              SY=T. grayi Loja.
    beckwithii Brewer ex S. Wats.
    bejariense Moric.
    bifidum Gray
          SY=T. b. var. decipiens Greene
    bolanderi Gray
    brandegei S. Wats.
    breweri S. Wats.
    campestre Schreb.
          SY=T. procumbens sensu auctt. non
              L.
    carolinianum Michx.
    ciliolatum Benth.
    cyathiferum Lindl.
    dasyphyllum Torr. & Gray
        ssp. anemophilum (Greene) J. Gillett
        ssp. dasyphyllum
        ssp. uintense (Rydb.) J. Gillett
    dedeckerae J. Gillett
    depauperatum Desv.
          SY=T. amplectens Torr. & Gray
          SY=T. d. var. laciniatum (Greene)
              Jepson
          SY=T. stenophyllum Nutt.
    dichotomum Hook. & Arn.
          SY=T. macraei Hook. & Arn. var. d.
              (Hook. & Arn.) Brewer ex S.
              Wats.
          SY=T. petrophilum Heller
    douglasii House
    dubium Sibthorp
    eriocephalum Nutt.
        ssp. arcuatum (Piper) J. Gillett
              SY=T. e. var. piperi J.S. Martin
        ssp. cascadense J. Gillett
        ssp. cusickii (Piper) J. Gillett
              SY=T. e. var. c. (Piper) J.S.
                  Martin
              SY=T. e. var. harneyense (T.J.
                  Howell) McDermott
        ssp. eriocephalum
        ssp. martinii J. Gillett
        ssp. villiferum (House) J. Gillett
              SY=T. e. var. v. (House) J.S.
                  Martin
    fistulosum Vaughan

Trifolium (CONT.)
  fragiferum L.
          SY=T. f. ssp. bonannii (Presl)
                Sojak
  fucatum Lindl.
          SY=T. flavulum Greene
          SY=T. fucatum var. gambelii
                (Nutt.) Jepson
          SY=T. fucatum var. virescens
                (Greene) Jepson
          SY=T. g. Nutt.
  glomeratum L.
  gracilentum Torr. & Gray
          SY=T. g. var. inconspicuum Fern.
  gymnocarpon Nutt.
    ssp. gymnocarpon
          SY=T. g. var. subcaulescens (Gray)
                A. Nels.
          SY=T. s. Gray
    ssp. plummerae (S. Wats.) J. Gillett
          SY=T. g. var. p. (S. Wats.) J.S.
                Martin
  haydenii Porter
  hirtum All.
  howellii S. Wats.
  hybridum L.
          SY=T. h. ssp. elegans (Savi)
                Aschers. & Graebn.
          SY=T. h. var. e. (Savi) Boiss.
          SY=T. h. var. pratense Rabenh.
  incarnatum L.
          SY=T. i. var. elatius Gibelli &
                Belli
  kingii S. Wats.
    ssp. kingii
    ssp. macilentum (Greene) J. Gillett
          SY=T. m. Greene
  lacerum Greene
          SY=T. arizonicum Greene
  lappaceum L.
  latifolium (Hook.) Greene
  leibergii A. Nels. & J.F. Macbr.
  lemmonii S. Wats.
  longipes Nutt.
    ssp. atrorubens (Greene) J. Gillett
          SY=T. l. var. a. (Greene) Jepson
    ssp. caurinum (Piper) J. Gillett
          SY=T. l. var. multiovulatum
                (Henderson) C.L. Hitchc.
    ssp. elmeri (Greene) J. Gillett
          SY=T. l. var. e. (Greene)
                McDermott
    ssp. hansenii (Greene) J. Gillett
          SY=T. l. var. h. (Greene) Jepson
    ssp. longipes
    ssp. multipedunculatum (Kennedy) J.
        Gillett
          SY=T. m. Kennedy
    ssp. oreganum (T.J. Howell) J. Gillett
          SY=T. o. T.J. Howell
    ssp. pedunculatum (Rydb.) J. Gillett
          SY=T. l. var. p. (Rydb.) C.L.
                Hitchc.
    ssp. pygmaeum (Gray) J. Gillett
          SY=T. l. var. p. Gray
          SY=T. l. var. rusbyi (Greene)
                Harrington
          SY=T. r. Greene
    ssp. reflexum (A. Nels.) J. Gillett
          SY=T. l. var. rydbergii A. Nels.
          SY=T. oreganum T.J. Howell var.
                rydbergii Greene
    ssp. shastense (House) J. Gillett
          SY=T. l. var. s. (House) Jepson
  lupinaster L.
  macraei Hook. & Arn.
  macrocephalum (Pursh) Poir.

  medium L.
  microcephalum Pursh
  microdon Hook. & Arn.
          SY=T. m. var. pilosum Eastw.
  monanthum Gray
      var. eastwoodianum J.S. Martin
          SY=T. m. var. tenerum (Eastw.)
                Parish
      var. grantianum (Heller) Parish
      var. monanthum
      var. parvum (Kellogg) McDermott
  nanum Torr.
  neurophyllum Greene
  obtusiflorum Hook.
          SY=T. polyodon Greene
          SY=T. tridentatum Lindl. var. o.
                (Hook.) S. Wats.
  oliganthum Steud.
          SY=T. variegatum Nutt. var.
                pauciflorum (Nutt.)
                McDermott
  olivaceum Greene
          SY=T. columbianum Greene
          SY=T. o. var. c. (Greene) Jepson
          SY=T. o. var. griseum Jepson
  ornithopioides L.
          SY=Trigonella o. (L.) DC.
  owyheense Gilkey
  palmeri S. Wats.
  parryi Gray
    ssp. montanense (Rydb.) J. Gillett
    ssp. parryi
    ssp. salictorum (Greene ex Rydb.) J.
        Gillett
  pinetorum Greene
  plumosum Dougl. ex Hook.
    ssp. amplifolium (J.S. Martin) J.
        Gillett
          SY=T. p. var. a. J.S. Martin
    ssp. plumosum
  polymorphum Poir.
          SY=T. amphianthum Torr. & Gray
  pratense L.
          SY=T. p. var. frigidum Gaudin
          SY=T. p. var. sativum (P. Mill.)
                Schreb.
  productum Greene
          SY=T. kingii S. Wats. var. p.
                (Greene) Jepson
  reflexum L.
          SY=T. r. var. glabrum Loja.
  repens L.
          SY=T. saxicola Small
  resupinatum L.
  rollinsii J. Gillett
  stoloniferum Muhl.
  striatum L.
  subterraneum L.
  thompsonii Morton
  X trichocalyx Heller
  tridentatum Lindl.
          SY=T. t. var. aciculare (Nutt.)
                McDermott
  truncatum (Greene) Greene
          SY=T. amplectens Torr. & Gray var.
                hydrophilum (Green) Jepson
          SY=T. a. var. t. (Greene) Jepson
  variegatum Nutt.
      var. melanthum (Hook. & Arn.) Greene
          SY=T. m. Hook. & Arn.
          SY=T. v. var. major Loja.
      var. rostratum (Greene) C.L. Hitchc.
      var. variegatum
          SY=T. geminiflorum Greene
          SY=T. trilobatum Jepson
          SY=T. v. var. t. (Jepson)
                McDermott

Trifolium (CONT.)
 vesiculosum Savi
 virginicum Small
 wormskjoldii Lehm.
            SY=T. fendleri Greene
            SY=T. fimbriatum Lindl.
            SY=T. spinulosum Dougl. ex Hook.
            SY=T. willdenovii Spreng.

Trigonella L.

   caerulea (L.) Ser.
   corniculata (L.) L.
   foenum-graecum L.
   monspeliaca L.

Ulex L.

   europaeus L.
   minor Roth
            SY=U. nanus T.R. Forst. ex Symons

Vicia L.

   acutifolia Ell.
   americana Muhl. ex Willd.
      ssp. americana
            SY=V. a. ssp. oregana (Nutt.)
                  Abrams
            SY=V. a. var. o. (Nutt.) A. Nels.
            SY=V. a. var. truncata (Nutt.)
                  Brewer
            SY=V. a. var. villosa (Kellogg)
                  F.J. Herm.
            SY=V. californica Greene
            SY=V. c. var. madrensis Jepson
            SY=V. o. Nutt.
      ssp. mexicana C.R. Gunn
      ssp. minor (Hook.) C.R. Gunn
            SY=Lathyrus linearis Nutt.
            SY=V. a. var. angustifolia Nees
            SY=V. a. var. l. (Nutt.) S. Wats.
            SY=V. a. var. m. Hook.
            SY=V. l. (Nutt.) Greene
            SY=V. sparsifolia Nutt.
            SY=V. s. var. truncata (Nutt.) S.
                  Wats.
            SY=V. trifida Rydb.
            SY=V. trifida Dietr.
   articulata Hornem.
   benghalensis L.
   caroliniana Walt.
            SY=V. hugeri Small
   cracca L.
            SY=V. semicincta Greene
   disperma DC.
   ervilia (L.) Willd.
   faba L.
   floridana S. Wats.
   grandiflora Scop.
      var. grandiflora
      var. kitaibeliana W.D.J. Koch
   hassei S. Wats.
            SY=V. exigua Nutt. var. h. (S.
                  Wats.) Jepson
            SY=V. e. var. californica Torr.
   hirsuta (L.) S.F. Gray
   hybrida L.
   lathyroides L.
   leucophaea Greene
   ludoviciana Nutt.
      ssp. leavenworthii (Torr. & Gray)
            Lassetter & Gunn
            SY=V. leavenworthii Torr. & Gray
      ssp. ludoviciana
            SY=V. exigua Nutt.
            SY=V. l. var. laxiflora Shinners

            SY=V. ludoviciana var. texana
                  (Torr. & Gray) Shinners
            SY=V. producta Rydb.
            SY=V. t. (Torr. & Gray) Small
   lutea L.
   menziesii Spreng.
   minutiflora Dietr.
            SY=V. micrantha Nutt. ex Torr. &
                  Gray
            SY=V. reverchonii S. Wats.
   monantha Retz.
   narbonensis L.
   nigricans Hook. & Arn.
      ssp. gigantea (Hook.) Lassetter & Gunn
            SY=V. g. Hook.
      ssp. nigricans
   ocalensis Godfrey
   pannonica Crantz
   pulchella H.B.K.
   sativa L.
      ssp. nigra (L.) Ehrh.
            SY=V. angustifolia L.
      ssp. sativa
            SY=V. angustifolia L. var.
                  segetalis (Thuill.) W.D.J.
                  Koch
            SY=V. a. var. uncinata (Desv.)
                  Rouy
            SY=V. sativa var. a. (L.) Wahlenb.
            SY=V. sativa var. linearis Lange
            SY=V. sativa var. segetalis
                  (Thuill.) Ser.
   sepium L.
      var. montana W.D.J. Koch
      var. sepium
   tenuifolia Roth
            SY=V. cracca L. ssp. t. (Roth)
                  Gaudin
            SY=V. c. var. t. (Roth) G. Beck
   tetrasperma (L.) Moench
      var. gracilis (Loisel.) Aschers. &
            Graebn.
      var. tenuissima Druce
      var. tetrasperma
   villosa Roth
      ssp. varia (Host) Corb.
            SY=V. dasycarpa Ten.
      ssp. villosa
            SY=V. v. var. glabrescens W.D.J.
                  Koch

Vigna Savi

   angularis (Willd.) Ohwi & Ohashi
            SY=Phaseolus a. (Willd.) W. Wight
   antillana (Urban) Fawcett & Rendle
   luteola (Jacq.) Benth.
            SY=V. repens (L.) Kuntze
   marina (Burm.) Merr.
   o-wahuensis Vogel
   sandwicensis Gray
      var. heterophylla Rock
      var. sandwicensis
   unguiculata (L.) Walp.
            SY=V. sinensis (L.) Savi
   vexillata (L.) A. Rich.

Wisteria Nutt.

   floribunda (Willd.) DC.
            SY=Kraunhia f. (Willd.) Taubert
   frutescens (L.) Poir.
            SY=Kraunhia f. (L.) Greene
   macrostachya Nutt.
            SY=Kraunhia m. (Nutt.) Small
            SY=W. frutescens (L.) Poir. var.
                  m. (Nutt.) Torr. & Gray

Wisteria (CONT.)
      sinensis (Sims) Sweet

Zornia J.F. Gmel.

      bracteata (Walt.) J.F. Gmel.
      diphylla (L.) Pers.
      gemella (Willd.) Vogel
      leptophylla (Benth.) Pittier
            SY=Z. diphylla (L.) Pers. var. l.
                  Benth.
      reticulata Sm.

FAGACEAE

Castanea P. Mill.

      alnifolia Nutt.
            var. alnifolia
            var. floridana Sarg.
            SY=C. f. (Sarg.) Ashe
            SY=C. pumila (L.) P. Mill. var.
                  margaretta Ashe
      dentata (Marsh.) Borkh.
      mollissima Blume
      X neglecta Dode [dentata X pumila]
      pumila (L.) P. Mill.
            var. ashei Sudworth
            SY=C. a. (Sudworth) Sudworth
            var. ozarkensis (Ashe) Tucker
            SY=C. X alabamensis Ashe
            SY=C. ozarkensis Ashe
            SY=C. o. var. arkansana Ashe
            var. pumila

Castanopsis (D. Don) Spach

      chrysophylla (Dougl. ex Hook.) A. DC.
            SY=C. c. (Dougl. ex Hook.) A. DC.
                  var. minor (Benth.) A. DC.
            SY=Chrysolepis c. (Dougl. ex
                  Hook.) Hjelmquist
            SY=Chrysolepis c. var. m. (Benth.)
                  Munz
      sempervirens (Kellogg) Dudley
            SY=Chrysolepis s. (Kellogg)
                  Hjelmquist

Fagus L.

      grandifolia Ehrh.
            SY=F. g. var. caroliniana (Loud.)
                  Fern. & Rehd.
            SY=F. g. var. heterophylla Camp
      sylvatica L.
            SY=F. s. var. atropunicea Weston

Lithocarpus Blume

      densiflora (Hook. & Arn.) Rehd.
            var. densiflora
            var. echinoides (R. Br.) Abrams

Quercus L.

      agrifolia Nee
            var. agrifolia
            SY=Q. pricei Sudworth
            var. frutescens Engelm.
            var. oxyadenia (Torr.) J.T. Howell
      alba L.
            var. alba
            var. subcaerulea Pickens & Pickens
            var. subflavea Pickens & Pickens

      X alvordiana Eastw. [douglasii X turbinella
            ssp. californica]
                  SY=Q. dumosa Nutt. var. a.
                        (Eastw.) Jepson
      arizonica Sarg.
      arkansana Sarg.
                  SY=Q. caput-rivuli Ashe
      X asheana Little [incana X laevis]
                  SY=Q. ashei Trel.
      X atlantica Ashe [incana X laurifolia]
                  SY=Q. X sublaurifolia Trel. ex
                        Palmer
      austrina Small
      X beadlei Trel. [alba X michauxii]
      X beaumontiana Sarg. [falcata X laurifolia]
      X bebbiana Schneid. [alba X macrocarpa]
      X benderi Baenitz [coccinea X rubra]
      X bernardiensis W. Wolf [montana X
            stellata]
      bicolor Willd.
      X bimundorum Palmer [alba X robur]
      X blufftonensis Trel. [falcata X laevis]
      breviloba (Torr.) Sarg.
                  SY=Q. durandii Buckl. var. b.
                        (Torr.) Palmer
                  SY=Q. sinuata Walt. var. b.
                        (Torr.) C.H. Muller
      X brittonii W.T. Davis [ilicifolia X
            marilandica]
      X burnetensis Little [macrocarpa X
            virginiana]
      X bushii Sarg. [marilandica X velutina]
      X byarsii Sudworth [macrocarpa X michauxii]
      X caduca Trel. [incana X nigra]
      X caesariensis Moldenke [falcata X
            ilicifolia]
      X capesii W. Wolf [nigra X phellos]
      chapmanii Sarg.
      X chasei McMinn, Babcock, & Righter
            [agrifolia X kelloggii]
      chrysolepis Liebm.
            var. chrysolepis
                  SY=Q. wilcoxii Rydb.
            var. nana (Jepson) Jepson
      coccinea Muenchh.
            var. coccinea
                  SY=Q. richteri Baenitz
            var. tuberculata Sarg.
      X cocksii Sarg. [laurifolia X velutina]
      X columnaris Laughlin [palustris X rubra]
      X comptonae Sarg. [lyrata X virginiana]
      X cravenensis Little [incana X marilandica]
                  SY=Q. X carolinensis Trel.
      X deamii Trel. [macrocarpa X muhlenbergii]
                  SY=Q. X fallax Palmer
      X demarei Ashe [nigra X velutina]
      depressipes Trel.
      X discreta Laughlin [shumardii X velutina]
      X diversiloba Tharp ex A. Camus [laurifolia
            X marilandica]
      douglasii Hook. & Arn.
      dumosa Nutt.
            var. dumosa
            var. elegantula Jepson
            var. kinselae C.H. Muller
      dunnii Kellogg
                  SY=Q. chrysolepis Liebm. var.
                        palmeri (Engelm.) Sarg.
                  SY=Q. p. Engelm.
      durandii Buckl.
      durata Jepson
      X egglestonii Trel. [imbricaria X
            shumardii]
                  SY=Q. X shirlingii Bush
      ellipsoidalis E.J. Hill
      emoryi Torr.
      engelmannii Greene

Quercus   (CONT.)
    X eplingii C.H. Muller [douglasii X ?
        garryana]
    X exacta Trel. [imbricaria X palustris]
    falcata Michx.
              SY=Q. f. var. triloba (Michx.)
                   Nutt.
              SY=Q. t. Michx.
    X faxonii Trel. [alba X prinoides]
    X fernaldii Trel. [ilicifolia X rubra]
              SY=Q. X lowellii Sarg.
    X fernowii Trel. [alba X stellata]
    X filialis Little [phellos X velutina]
              SY=Q. X dubia Ashe
              SY=Q. X inaequalis Palmer &
                   Steyermark
    X fontana Laughlin [coccinea X velutina]
    gambelii Nutt.
              SY=Q. eastwoodiae Rydb.
              SY=Q. gunnisonii (Torr.) Rydb.
              SY=Q. leptophylla Rydb.
              SY=Q. nitescens Rydb.
              SY=Q. novo-mexicana (A. DC.) Rydb.
              SY=Q. submollis Rydb.
              SY=Q. utahensis (A. DC.) Rydb.
              SY=Q. vreelandii Rydb.
    X ganderi C.B. Wolf [agrifolia var.
        oxyadenia X kelloggii]
    X garlandensis Palmer [falcata X nigra]
    garryana Dougl. ex Hook.
        var. breweri (Engelm.) Jepson
              SY=Q. b. Engelm.
              SY=Q. oerstediana R. Br.
        var. garryana
        var. semota Jepson
    geminata Small
              SY=Q. virginiana P. Mill. var. g.
                   (Small) Sarg.
    georgiana M.A. Curtis
    X giffordii Trel. [ilicifolia X phellos]
    glaucoides Mart. & Gal.
              SY=Q. laceyi Small
    graciliformis C.H. Muller
    X grandidentata Ewan [dumosa X engelmannii]
    gravesii Sudworth
    grisea Liebm.
    X guadalupensis Sarg. [macrocarpa X
        stellata]
    X harbisonii Sarg. [stellata X virginiana]
    X hastingsii Sarg. [marilandica X
        shumardii]
    havardii Rydb.
    X hawkinsiae Sudworth [rubra X velutina]
              SY=Q. X porteri Trel.
    hemisphaerica Bartr.
    X heterophylla Michx. f. [phellos X rubra]
    hinckleyi C.H. Muller
    X howellii Tucker [dumosa X garryana]
    X humidicola Palmer [bicolor X lyrata]
    hypoleucoides A. Camus
    ilicifolia Wang.
    imbricaria Michx.
    incana Bartr.
              SY=Q. cinerea Michx.
    X incomita Palmer [falcata X marilandica]
    X inconstans Palmer [gravesii X
        hypoleucoides]
              SY=Q. X livermorensis C.H. Muller
    intricata Trel.
    X introgressa P.M. Thompson [bicolor X
        (muhlenbergii X prinoides)]
    X jackiana Schneid. [alba X bicolor]
    X jolonensis Sarg. [douglasii X lobata]
    X jorrii Trel. [falcata X shumardii]
    kelloggii Newberry
              SY=Q. californica (Torr.) Cooper
    laevis Walt.

    laurifolia Michx.
              SY=Q. obtusa (Willd.) Ashe
              SY=Q. phellos L. var. l. (Michx.)
                   Chapman
              SY=Q. succulenta Small
              SY=Q. virginiana P. Mill. var.
                   maritima (Michx.) Sarg.
    X leana Nutt. [imbricaria X velutina]
    lobata Nee
              SY=Q. hindsii Benth.
              SY=Q. l. var. argillara Jepson
              SY=Q. l. var. insperata Jepson
              SY=Q. l. var. turbinata Jepson
              SY=Q. l. var. walteri Jepson
    X ludoviciana Sarg. [pagoda X phellos]
              SY=Q. X subfalcata Trel.
    lyrata Walt.
    macdonaldii Greene
              SY=Q. dumosa Nutt. var. m.
                   (Greene) Jepson
    X macnabiana Sudworth [durandii X stellata]
              SY=Q. X mahlonii Palmer
    macrocarpa Michx.
              SY=Q. m. var. depressa (Nutt.)
                   Engelm.
              SY=Q. m. var. oliviformis (Michx.
                   f.) Gray
    margaretta Ashe
              SY=Q. drummondii Liebm.
              SY=Q. stellata Wang. var. araniosa
                   Sarg.
              SY=Q. s. var. m. (Ashe) Sarg.
    marilandica Muenchh.
              SY=Q. neoashei Bush
    X megaleia Laughlin [lyrata X macrocarpa]
    X mellichampii Trel. ex Sarg. [laevis X
        laurifolia]
    michauxii Nutt.
              SY=Q. houstoniana C.H. Muller
    minima (Sarg.) Small
              SY=Q. virginiana P. Mill. var. m.
                   Sarg.
    mohriana Buckl. ex Rydb.
    montana Willd.
              SY=Q. prinus L.
    X moreha Kellogg [kelloggii X wislizenusii]
    X moultonensis Ashe [phellos X shumardii]
    muhlenbergii Engelm.
              SY=Q. alexanderi Britt.
              SY=Q. prinoides Willd. var.
                   acuminata (Michx.) Gleason
    X munzii Tucker [lobata X turbinella ssp.
        californica]
    X mutabilis Palmer & Steyermark [palustris
        X shumardii var. schneckii]
    myrtifolia Willd.
    X neopalmeri Sudworth ex Palmer [nigra X
        shumardii]
    X neo-tharpii A. Camus [minima X stellata]
    X nessiana Palmer [bicolor X virginiana]
    nigra L.
              SY=Q. microcarya Small
              SY=Q. n. var. heterophylla (Ait.)
                   Ashe
    nuttallii Palmer
              SY=Q. n. var. cachensis Palmer
    oblongifolia Torr.
    oglethorpensis Duncan
    X organensis Trel. [arizonica X grisea]
    X oviedoensis Sarg. [incana X myrtifolia]
    pagoda Raf.
              SY=Q. falcata Michx. var.
                   leucophylla (Ashe) Palmer &
                   Steyermark
              SY=Q. f. var. pagodifolia Ell.
              SY=Q. pagodifolia (Ell.) Ashe
    X palaeolithicola Trel. [ellipsoidalis X

Quercus (CONT.)
    velutina]
  X palmeriana A. Camus [falcata X
    imbricaria]
        SY=Q. X anceps Palmer
  palustris Muenchh.
  parvula Greene
  X pauciloba Rydb. [gambelii X turbinella]
        SY=Q. fendleri Leibm.
        SY=Q. undulata Torr.
        SY=Q. venustula Greene
  phellos L.
  X podophylla Trel. [incana X velutina]
  prinoides Willd.
        SY=Q. p. var. rufescens Rehd.
  pumila Walt.
  pungens Liebm.
        var. pungens
        var. vaseyana (Buckl.) C.H. Muller
  X rehderi Trel. [ilicifolia X velutina]
  X riparia Laughlin [rubra X shumardii]
  X robbinsii Trel. [coccinea X ilicifolia]
  robur L.
  X robusta C.H. Muller [emoryi X gravesii]
  rolfsii Small
  rubra L.
        var. borealis (Michx. f.) Farw.
        SY=Q. b. Michx. f.
        var. rubra
        SY=Q. borealis Michx. f. var.
            maxima (Marsh.) Sarg.
        SY=Q. m. (Marsh.) Ashe
  X rudkinii Britt. [marilandica X phellos]
  rugosa Nee
        SY=Q. diversicolor Trel.
        SY=Q. reticulata Humb. & Bonpl.
  X runcinata (A. DC.) Engelm. [imbricaria X
    rubra]
  sadleriana R. Br. Campst.
  X sargentii Rehd. [montana X robur]
  X saulii Schneid. [alba X montana]
  X schochiana Dieck ex Palmer [palustris X
    phellos]
  X schuettei Trel. [bicolor X macrocarpa]
        SY=Q. X hillii Trel.
  shumardii Buckl.
        var. acerifolia Palmer
        var. schneckii (Britt.) Sarg.
        SY=Q. schneckii Britt.
        var. shumardii
        var. stenocarpa Laughlin
        var. texana (Buckl.) Ashe
        SY=Q. shumardii var. microcarpa
            (Torr.) Shinners
        SY=Q. t. Buckl.
  similis Ashe
        SY=Q. mississippiensis Ashe
        SY=Q. stellata Wang. var. m.
            (Ashe) Little
  sinuata Walt.
  X smallii Trel. [georgiana X marilandica]
  stellata Wang.
        var. boyntonii (Beadle) Sarg.
        SY=Q. b. Beadle
        var. paludosa Sarg.
        SY=Q. ashei Sterrett
        var. stellata
        SY=Q. s. var. attenuata Sarg.
        SY=Q. s. var. parviloba Sarg.
  X stelloides Palmer [prinoides X stellata]
  X sterilis Trel. ex Palmer [marilandica X
    nigra]
  X sterretii Trel. [lyrata X stellata]
  X subconvexa Tucker [durata X garryana]
  X subintegra Trel. [falcata X incana]
  X substellata Trel. [bicolor X stellata]
  tardifolia C.H. Muller

  X tharpii C.H. Muller [emoryi X
    graciliformis]
  tomentella Engelm.
  X tottenii Melvin [lyrata X michauxii]
  toumeyi Sarg.
  X townei Palmer [dumosa X lobata]
  X tridentata (A. DC.) Engelm. [imbricaria X
    marilandica]
  turbinella Greene
        ssp. ajoensis (C.H. Muller) Felger &
            Lowe
        SY=Q. a. C.H. Muller
        SY=Q. t. var. a. (C.H. Muller)
            Little
        ssp. californica Tucker
        ssp. turbinella
        SY=Q. dumosa Nutt. var. t.
            (Greene) Jepson
  vaccinifolia Kellogg
  X vaga Palmer & Steyermark [palustris X
    velutina]
  velutina Lam.
        SY=Q. v. var. missouriensis Sarg.
  X venulosa Ashe [arkansana X cinerea]
  virginiana P. Mill.
        var. fusiformis (Small) Sarg.
        SY=Q. f. Small
        var. virginiana
  X walteriana Ashe [laevis X nigra]
  X willdenowiana (Dippel) Zabel [falcata X
    velutina]
        SY=Q. X pinetorum Moldenke
  wislizenusii A. DC.
        var. frutescens Engelm.
        var. wislizenusii
        SY=Q. shrevei C.H. Muller

FLACOURTIACEAE

Banara Aubl.

    portoricensis Krug & Urban
    vanderbiltii Urban

Casearia Jacq.

    aculeata Jacq.
    arborea (L.C. Rich.) Urban
    decandra Jacq.
    guianensis (Aubl.) Urban
    sylvestris Sw.

Flacourtia L'Her.

    indica (Burm. f.) Merr.
    inermis Roxb.

Homalium Jacq.

    leiogynum Blake
    racemosum Jacq.
        SY=H. pleiandrum Blake

Laetia Loefl. ex L.

    procera (Poepp. & Endl.) Fichl.
        SY=Casearia bicolor (Poepp. &
            Endl.) Urban
        SY=L. b. (Poepp. & Endl.) Fichl.

Lunania Hook.

    buchii Urban

Prockia P. Br. ex L.

   crucis L.

Samyda L.

   dodecandra Jacq.
   spinulosa Vent.

Xylosma Forst. f.

   buxifolium Gray
        SY=Myroxylon b. (Gray) Krug &
           Urban
   crenatum (St. John) St. John
   flexuosa (H.B.K.) Kuntze
   hawaiiense Seem.
      var. hawaiiense
      var. hillebrandii (Wawra) Sleumer
   pachyphyllum (Krug & Urban) Urban
        SY=Myroxylon p. Krug & Urban
   schaefferioides Gray
   schwaneckeanum (Krug & Urban) Urban
        SY=Myroxylon s. Krug & Urban

FLAGELLARIACEAE

Joinvillea Gaud. ex Brongn. & Grisc.

   ascendens Gaud. ex Brongn. & Grisc.

FOUQUIERIACEAE

Fouquieria H.B.K.

   splendens Engelm.

FRANKENIACEAE

Frankenia L.

   grandifolia Cham. & Schlecht.
      var. campestris Gray
      var. grandifolia
   jamesii Torr. ex Gray
   johnstonii Correll
   palmeri S. Wats.
   pulverulenta L.

GARRYACEAE

Garrya Dougl. ex Lindl.

   buxifolia Gray
   elliptica Dougl. ex Lindl.
   fadyenia Hook.
   flavescens S. Wats.
     ssp. congdonii (Eastw.) Dahling
        SY=G. c. Eastw.
     ssp. flavescens
     ssp. pallida (Eastw.) Dahling
        SY=G. f. var. p. (Eastw.) Bacig.
          ex Ewan
   fremontii Torr.
        SY=G. f. var. laxa Eastw.

ovata Benth.
     ssp. goldmanii (Woot. & Standl.) Dahling
        SY=G. g. Woot. & Standl.
     ssp. lindheimeri (Torr.) Dahling
        SY=G. l. Torr.
     ssp. ovata
   veatchii Kellogg
   wrightii Torr.

GENTIANACEAE

Bartonia Muhl. ex Willd.

   paniculata (Michx.) Muhl.
     ssp. iodandra (B.L. Robins.) J. Gillett
        SY=B. i. B.L. Robins.
        SY=B. p. var. intermedia Fern.
        SY=B. p. var. iodandra (B.L.
          Robins.) Fern.
        SY=B. p. var. sabulonensis Fern.
        SY=B. virginica (L.) B.S.P. var.
          s. (Fern.) Boivin
     ssp. paniculata
        SY=B. lanceolata Small
        SY=B. virginica (L.) B.S.P. var.
          p. (Michx.) Boivin
   texana Correll
   verna (Michx.) Muhl.
   virginica (L.) B.S.P.

Centaurium Hill

   beyrichii (Torr. & Gray) B.L. Robins.
        SY=C. b. var. glanduliferum
          Correll
   brittonii Millsp. & Greenm.
   calycosum (Buckl.) Fern.
      var. arizonicum (Gray) Tidestrom
      var. calycosum
        SY=C. c. var. breviflorum Shinners
      var. nanum (Gray) B.L. Robins.
   davyi (Jepson) Abrams
   erythraea Rafn
        SY=C. minus Moench
        SY=C. umbellatum Gilib.
   exaltatum (Griseb.) W. Wight ex Piper
        SY=C. nuttallii (S. Wats.) Heller
   floribundum (Benth.) B.L. Robins.
   littorale (D. Turner) Gilmour
   muhlenbergii (Griseb.) W. Wight ex Piper
        SY=C. curvistamineum (Wittr.)
          Abrams
   namophilum Reveal, Broome & Beatley
   nudicale (Engelm.) B.L. Robins.
   pulchellum (Sw.) Druce
   sebaeoides (Griseb.) Druce
   spicatum (L.) Fritsch
   texense (Griseb.) Fern.
   trichanthum (Griseb.) B.L. Robins.
   venustum (Gray) B.L. Robins.
     ssp. abramsii Munz
     ssp. venustum

Enicostema Blume

   littorale Blume
        SY=E. verticillatum (L.) Gilg

Eustoma Salisb.

   exaltatum (L.) Salisb. ex G. Don
        SY=E. barkleyi Standl. ex Shinners
   grandiflorum (Raf.) Shinners
        SY=E. russellianum (Hook.) G. Don

Eustoma   (CONT.)
                ex Sweet

Frasera Walt.

     albicaulis Dougl. ex Griseb.
         var. albicaulis
            SY=Swertia a. (Dougl. ex Griseb.)
                Kuntze
            SY=S. bethelii St. John
            SY=S. californica St. John
            SY=S. modocensis St. John
            SY=S. sierrae St. John
         var. columbiana (St. John) C.L.
            Hitchc.
            SY=Swertia c. St. John
         var. cusickii (Gray) C.L. Hitchc.
            SY=Swertia c. (Gray) St. John
            SY=S. nitida (Benth.) Jepson ssp.
                c. (Gray) Abrams
         var. idahoensis (St. John) C.L.
            Hitchc.
            SY=Swertia i. St. John
         var. nitida (Benth.) C.L. Hitchc.
            SY=F. a. ssp. n. (Benth.) D.M.
                Post
            SY=F. n. Benth.
            SY=Swertia eastwoodae St. John
            SY=S. n. (Benth.) Jepson
     albomarginata S. Wats.
         var. albomarginata
            SY=Swertia a. (S. Wats.) Kuntze
         var. induta (Tidestrom) Card
     caroliniensis Walt.
            SY=Swertia c. (Walt.) Kuntze
     coloradensis (Rogers) D.M. Post
            SY=Swertia c. Rogers
     fastigiata (Pursh) Heller
            SY=Swertia f. Pursh
     gypsicola (Barneby) D.M. Post
     montana Mulford
     neglecta Hall
            SY=Swertia n. (Hall) Jepson
     pahutensis Reveal
     paniculata Torr.
            SY=Swertia utahensis (M.E. Jones)
                St. John
     parryi Torr.
            SY=Swertia p. (Torr.) Kuntze
     puberulenta A. Davids.
            SY=Swertia p. (A. Davids.) Jepson
     speciosa Dougl. ex Griseb.
            SY=Swertia radiata (Kellogg)
                Kuntze
            SY=S. r. var. macrophylla (Greene)
                St. John
     tubulosa Coville
            SY=Swertia t. (Coville) Jepson
     umpquaensis Peck & Applegate
            SY=Swertia u. (Peck & Applegate)
                St. John

Gentiana L.

     affinis Griseb.
            SY=Dasystephana a. (Griseb.) Rydb.
            SY=D. interrupta (Greene) Rydb.
            SY=G. a. var. forwoodii (Gray)
                Kusnez.
            SY=G. a. var. major A. Nels. &
                Kennedy
            SY=G. a. var. ovata Gray
            SY=G. a. var. parvidentata Kusnez.
            SY=G. bigelovii Gray
            SY=G. f. Gray
            SY=G. interrupta Greene
            SY=G. oregana Engelm. ex Gray

            SY=G. rusbyi Greene
            SY=Pneumonanthe a. (Griseb.) W.A.
                Weber
     alba Muhl.
            SY=G. flavida Gray
     algida Pallas
            SY=Gentianodes a. (Pallas) Love &
                Love
            SY=Gentianodes romanzovii (Bunge)
                Ledeb.
     andrewsii Griseb.
         var. andrewsii
            SY=Dasystephana a. (Griseb.) Small
            SY=Pneumonanthe a. (Griseb.) W.A.
                Weber
         var. dakotica A. Nels.
     aquatica L.
            SY=Chondrophylla fremontii (Torr.)
                A. Nels.
            SY=Ciminalis f. (Torr.) W.A. Weber
            SY=Gentiana f. Torr.
     austromontana Pringle & Sharp
     autumnalis L.
            SY=Dasystephana porphyrio (J.F.
                Gmel.) Small
            SY=Gentiana p. J.F. Gmel.
     X billingtonii Farw. [andrewsii X
        puberulenta]
     bisetaea T.J. Howell
     calycosa Griseb.
            SY=Gentiana c. var. obtusiloba
                (Rydb.) C.L. Hitchc.
            SY=G. c. var. xantha A. Nels.
            SY=Pneumonanthe c. (Griseb.)
                Greene
     catesbaei Walt.
            SY=Dasystephana latifolia
                (Chapman) Small
            SY=D. parvifolia (Chapman) Small
            SY=Gentiana c. var.
                nummulariifolia Fern.
            SY=G. elliottii Chapman
            SY=G. e. var. l. Chapman
     clausa Raf.
     X curtisii J. Pringle [alba X puberulenta]
     decora Pollard
            SY=Dasystephana d. (Pollard) Small
     douglasiana Bong.
     glauca Pallas
     X grandilacustris J. Pringle [andrewsii X
        rubricaulis]
     linearis Froel.
            SY=Dasystephana l. (Froel.) Britt.
            SY=Gentiana saponaria L. var. l.
                (Froel.) Griseb.
     newberryi Gray
     nivalis L.
     X pallidocyanea J. Pringle [alba X
        andrewsii]
     parryi Engelm.
            SY=Gentiana calycosa Griseb. var.
                asepala (Maguire) C.L.
                Hitchc.
     pennelliana Fern.
            SY=Dasystephana tenuifolia (Raf.)
                Pennell
     platypetala Griseb.
     prostrata Haenke ex Jacq.
            SY=Ciminalis p. (Haenke ex Jacq.)
                Love & Love
            SY=Gentiana p. var. americana
                Engelm.
     puberulenta J. Pringle
            SY=Gentiana puberula sensu auctt.
                non Michx.
     rubricaulis Schwein.
            SY=Dasystephana grayi (Kusnez.)

Gentiana (CONT.)
                    Britt.
          SY=Gentiana linearis Froel. var.
              latifolia Gray
          SY=G. linearis var. lanceolata
              Gray
    saponaria L.
          SY=Dasystephana puberula (Michx.)
              Small as to type, non sensu
              Small
          SY=Dasystephana s. (L.) Small
          SY=Gentiana cherokeensis (W.P.
              Lemmon) Fern.
          SY=G. puberula Michx. as to type,
              non sensu Small
    sceptrum Griseb.
          SY=Gentiana menziesii Griseb.
          SY=G. s. var. cascadensis M.E.
              Peck
          SY=G. s. var. humilis Engelm. ex
              Gray
    setigera Gray
    tiogana Heller
    villosa L.
          SY=Dasystephana v. (L.) Small
          SY=Gentiana deloachii (W.P.
              Lemmon) Shinners

Gentianella Moench

    amarella (L.) Borner
      ssp. acuta (Michx.) J. Gillett
          SY=Amarella acuta (Michx.) Raf.
          SY=A. plebia (Cham.) Greene
          SY=A. p. var. holmii (Wettst.)
              Rydb.
          SY=A. strictiflora (Rydb.) Greene
          SY=Gentiana acuta Michx.
          SY=Gentiana amarella L. pro parte
          SY=Gentiana amarella ssp. acuta
              (Michx.) Hulten
          SY=Gentiana amarella ssp. acuta
              var. acuta (Michx.) Herder
          SY=Gentiana amarella ssp.
              heterosepala (Engelm.) J.
              Gillett
          SY=Gentiana amarella var. stricta
              (Griseb.) S. Wats.
          SY=Gentiana plebeja Cham. ex Bunge
          SY=Gentiana p. var. holmii Wettst.
          SY=Gentiana strictiflora (Rydb.)
              A. Nels.
          SY=Gentianella amarella ssp. acuta
              var. plebeja (Cham.) Hulten
          SY=Gentianella amarella var. acuta
              (Michx.) Herder
      ssp. heterosepala (Engelm.) J. Gillett
          SY=Amarella scopulorum Greene
          SY=Gentiana h. Engelm.
          SY=Gentiana s. (Greene) Tidestrom
      ssp. wrightii (Gray) J. Gillett
          SY=Gentiana w. Gray
    aurea (L.) H. Sm. ex Hyl.
          SY=Gentiana a. L.
    auriculata (Pallas) J. Gillett
    microcalyx (Lemmon) J. Gillett
          SY=Gentiana m. Lemmon
    propinqua (Richards.) J. Gillett
      ssp. aleutica (Cham. & Schlecht.) J.
          Gillett
          SY=Gentiana a. Cham. & Schlecht.
          SY=Gentianella p. var. a. (Cham. &
              Schlecht.) Welsh
      ssp. propinqua
          SY=Gentiana arctophila Griseb.
          SY=Gentiana p. Richards.
          SY=Gentiana p. ssp. arctophila

                    (Griseb.) Hulten
    quinquefolia (L.) Small
      ssp. occidentalis (Gray) J. Gillett
          SY=Gentiana q. L. var. o. Gray
          SY=Gentianella o. (Gray) Small
      ssp. quinquefolia
          SY=Gentiana q. L.
    tenella (Rottb.) Borner
      ssp. pribilofii J. Gillett
      ssp. tenella
          SY=Comastoma t. (Rottb.) Toyokuni
          SY=Gentiana t. Rottb.
    tortuosa (M.E. Jones) J. Gillett
          SY=Gentiana t. M.F. Jones
    wislizenii (Engelm.) J. Gillett
          SY=Gentiana w. Engelm.

Gentianopsis Ma

    barbellata (Engelm.) Iltis
          SY=Gentiana b. Engelm.
          SY=Gentianella b. (Engelm.) J.
              Gillett
    crinita (Froel.) Ma
          SY=Anthopogon crinitum (Froel.)
              Raf.
          SY=Gentiana c. Froel.
          SY=Gentiana ventricosa Griseb.
          SY=Gentianella c. (Froel.) G. Don
          SY=Gentianella c. ssp. nevadensis
              (Gilg) Weaver & Rudenberg
    detonsa (Rottb.) Ma
      ssp. detonsa
          SY=Gentiana barbata Froel.
          SY=Gentiana d. Rottb.
          SY=Gentiana d. var. groenlandicum
              Victorin
          SY=Gentiana richardsonii Porsild
          SY=Gentianella d. (Rottb.) G. Don
      ssp. yukonensis (J. Gillett) J. Gillett
          SY=Gentianella d. (Rottb.) G. Don
              ssp. y. J. Gillett
    holopetala (Gray) Iltis
          SY=Gentiana h. (Gray) Holm
          SY=Gentianella detonsa (Rottb.) G.
              Don ssp. h. (Gray) J.
              Gillett
    macountii (Holm) Iltis
          SY=Anthopogon tonsum (Lunell)
              Rydb.
          SY=Gentiana crinita Froel. var. t.
              (Lunell) Boivin
          SY=Gentiana gaspensis Victorin
          SY=Gentiana macounii Holm
          SY=Gentianella c. (Froel.) G. Don
              ssp. m. (Holm) J. Gillett
          SY=Gentianopsis procera (Holm) Ma
              ssp. m. (Holm) Iltis
    macrantha (D. Don) Iltis
          SY=Gentiana grandis (Gray) Holm
          SY=Gentianella detonsa (Rottb.)
              G. Don ssp. superba (Greene)
              J. Gillett
    nesophila (Holm) Iltis
          SY=Gentiana detonsa Rottb. var. n.
              (Holm) Boivin
          SY=Gentiana n. Holm
          SY=Gentianella d. (Rottb.) G. Don
              ssp. n. (Holm) J. Gillett
    procera (Holm) Ma
          SY=Gentiana crinita Froel. var.
              browniana (Hook.) Boivin
          SY=Gentiana p. Holm
          SY=Gentianella c. (Froel.) G. Don
              ssp. p. (Holm) J. Gillett
    raupii (Porsild) Iltis

Gentianopsis (CONT.)
            SY=Gentiana detonsa Rottb. var. r.
                (Porsild) Boivin
            SY=Gentiana r. Porsild
            SY=Gentianella d. (Rottb.) G. Don
                ssp. r. (Porsild) J. Gillett
    simplex (Gray) Iltis
            SY=Gentiana s. Gray
            SY=Gentianella s. (Gray) J.
                Gillett
    thermalis (Kuntze) Iltis
            SY=Gentiana detonsa Rottb. var.
                unicaulis (A. Nels.) C.L.
                Hitchc.
            SY=Gentiana elegans A. Nels.
            SY=Gentiana t. Kuntze
            SY=Gentianella d. (Rottb.) G. Don
                ssp. e. (A. Nels.) J.
                Gillett
    victorinii (Fern.) Iltis
            SY=Gentiana v. Fern.
            SY=Gentianella crinita (Froel.) G.
                Don ssp. v. (Fern.) J.
                Gillett

Halenia Borkh.

    deflexa (Sm.) Griseb.
      ssp. brentoniana (Griseb.) J. Gillett
            SY=H. d. var. b. (Griseb.) Gray
      ssp. deflexa
    recurva (Sm.) Allen

Leiphaimos Schlecht. & Cham.

    aphylla (Jacq.) Gilg
    parasitica Schlecht. & Cham.
    portoricensis Britt.

Lisianthus P. Br.

    laxiflorus Urban

Lomatogonium A. Braun

    rotatum (L.) Fries ex Nyman
            SY=L. r. ssp. tenuifolium
                (Griseb.) Porsild
            SY=Pleurogyne r. (L.) Griseb.

Microcala Hoffmgg. & Link

    quadrangularis (Lam.) Griseb.
            SY=Cicendia q. (Lam.) Griseb.
            SY=Exacum chilense Bert.
            SY=E. inflatum Hook. & Arn.
            SY=Gentiana q. Lam.

Obolaria L.

    virginica L.

Sabatia Adans.

    angularis (L.) Pursh
    arenicola Greenm.
            SY=S. carnosa Small
    bartramii Wilbur
            SY=S. decandra (Walt.) Harper
            SY=S. dodecandra (L.) B.S.P. var.
                coriacea (Ell.) Ahles
    brachiata Ell.
    brevifolia Raf.
            SY=S. elliottii Steud.
            SY=S. paniculata (Michx.) Pursh
    calycina (Lam.) Heller
            SY=S. cubensis (Griseb.) Urban

            SY=S. gracilis (Michx.) Salisb.
                var. cubensis Griseb.
    campanulata (L.) Torr.
            SY=S. c. var. gracilis (Michx.)
                Fern.
            SY=S. g. (Michx.) Salisb.
    campestris Nutt.
    capitata (Raf.) Blake
            SY=Lapithea c. (Raf.) Small
            SY=L. boykinii (Gray) Small
            SY=S. b. Gray
    difformis (L.) Druce
            SY=S. lanceolata (Walt.) Torr. &
                Gray
    dodecandra (L.) B.S.P.
      var. dodecandra
      var. foliosa (Fern.) Wilbur
            SY=S. f. Fern.
            SY=S. harperi Small
            SY=S. obtusata Blake
    gentianoides Ell.
            SY=Lapithea g. (Ell.) Griseb.
    grandiflora (Gray) Small
            SY=S. alainii Victorin
            SY=S. campanulata (L.) Torr. var.
                g. (Gray) Blake
            SY=S. gracilis (Michx.) Salisb.
                var. grandiflora Gray
    kennedyana Fern.
            SY=S. dodecandra (L.) B.S.P. var.
                k. (Fern.) Ahles
    macrophylla Hook.
      var. macrophylla
      var. recurvans (Small) Wilbur
            SY=S. r. Small
    quadrangula Wilbur
            SY=S. paniculata (Michx.) Pursh
    stellaris Pursh
            SY=S. maculata (Benth.) Benth. &
                Hook.
            SY=S. palmeri Gray
            SY=S. purpusii Brandeg.
            SY=S. simulata Britt.

Schultesia Mart.

    brachyptera Cham.
    heterophylla Miq.

Swertia L.

    perennis L.
            SY=S. p. var. obtusa (Ledeb.)
                Griseb.

GERANIACEAE

Erodium L'Her. ex Ait.

    botrys (Cav.) Bertol.
    brachycarpum (Godr.) Thellung
            SY=E. obtusiplicatum (Maire,
                Weiller & Wilczek) J.T.
                Howell
    ciconium (L.) L'Her.
    cicutarium (L.) L'Her.
      ssp. jacquinianum (Fisher, C.A. Mey. &
          Ave-Lall.) Briq.
            SY=E. aethiopicum (Lam.) Brumh. &
                Thellung
    cygnorum Nees
    laciniatum (Cav.) Willd.
    macrophyllum Hook. & Arn.
      var. californicum (Greene) Jepson

Erodium (CONT.)
        var. macrophyllum
    malacoides (L.) Willd.
    moschatum (L.) L'Her.
            var. moschatum
            var. praecox Lange
    stephenianum Willd.
    texanum Gray

Geranium L.

    arboreum Gray
            SY=Neurophyllodes a. (Gray) Deg.
    atropurpureum Heller
    attenuilobum G.N. Jones & F.F. Jones
    bicknellii Britt.
            var. bicknellii
                SY=G. nemorale Suksdorf
                SY=G. n. var. b. (Britt.) Fern.
            var. longipes (S. Wats.) Fern.
                SY=G. carolinianum L. var. l. S.
                    Wats.
    caespitosum James
    californicum G.N. Jones & F.F. Jones
            SY=G. concinnum G.N. Jones & F.F.
                Jones
    carolinianum L.
            var. australe (Benth.) Fosberg
            var. carolinianum
            var. confertiflorum Fern.
            var. sphaerospermum (Fern.) Breitung
                SY=G. s. Fern.
    columbinum L.
    cowenii Rydb.
            SY=G. fremontii Torr. ex Gray var.
                c. (Rydb.) Harrington
    cuneatum Hook.
        ssp. cuneatum
                SY=G. c. var. menziesii Gray
                SY=G. c. var. rockii Skottsberg
        ssp. hololeucum (Gray) Carlq. & Bissing
                SY=G. c. var. h. Gray
        ssp. hypoleucum (Gray) Carlq. & Bissing
                SY=G. c. var. h. Gray
        ssp. tridens (Hbd.) Carlq. & Bissing
                SY=G. c. var. t. (Hbd.) Fosberg
                SY=G. t. Hbd.
    dissectum L.
            SY=G. laxum Hanks
    eremophilum Woot. & Standl.
    erianthum DC.
            SY=G. pratense L. var. e. (DC.)
                Boivin
    fremontii Torr. ex Gray
    homeanum Turcz.
            SY=G. glabratum (Hook.) Small
    humile Hbd.
        ssp. humile
        ssp. kauaiense (Rock) Carlq. & Bissing
                SY=G. h. var. k. Rock
    ibericum Cav.
            SY=G. nepalense Sweet var.
                thunbergii (Sieb. & Zucc.)
                Kudo
    lentum Woot. & Standl.
    lucidum L.
    maculatum L.
    marginale Rydb. ex Hanks & Small
    molle L.
    multiflorum Gray
        ssp. multiflorum
                SY=Neurophyllodes m. (Gray) Deg. &
                    Greenw.
        ssp. ovatifolium (Gray) Carlq. & Bissing
                SY=G. m. var. canum Hbd.
                SY=G. m. var. forbesii (Deg. &
                    Deg.) St. John

            SY=G. m. var. o. (Gray) Fosberg
            SY=G. m. var. superbum (Deg., Deg.
                & Greenw.) St. John
            SY=G. o. Gray
            SY=Neurophyllodes o. (Gray) Deg. &
                Greenw.
    nepalense Sweet
    oreganum T.J. Howell
    parryi (Engelm.) Heller
    potentilloides L'Her. ex DC.
            SY=G. microphyllum Hook. f.
    pratense L.
    pusillum L.
    pyrenaicum Burm. f.
    retrorsum L'Her. ex DC.
    richardsonii Fisch. & Trautv.
    robertianum L.
            SY=Robertiella r. (L.) Hanks
    rotundifolium L.
    sanguineum L.
    sibiricum L.
    solanderi Carolin
    sylvaticum L.
    texanum (Trel.) Heller
    toquimense A. Holmgren & N. Holmgren
    viscosissimum Fisch. & Mey.
            var. nervosum (Rydb.) C.L. Hitchc.
                SY=G. canum Rydb.
                SY=G. n. Rydb.
                SY=G. strigosus St. John
            var. viscosissimum
    wislizenii S. Wats.

Pelargonium L'Her.

    capitatum (L.) L'Her.
    X domesticum Bailey [angulosum var.
        grandiflorum X cucullatum]
    graveolens L'Her.
    grossularioides (L.) L'Her.
    hortorum Bailey
    inodorum Willd.
    inquinans (L.) L'Her.
    peltatum (L.) L'Her.
    vitifolium (L.) L'Her.
    zonale (L.) L'Her.

GESNERIACEAE

Alloplectus Mart.

    ambiguus Urban
            SY=Crantzia a. (Urban) Britt.

Columnea L.

    tulae Urban

Cyrtandra J.R. & G. Forst.

    adpressipilosa St. John
    alata St. John & Storey
    alnea St. John
    ambigua (Hbd.) St. John & Storey
    arcuata St. John
    arguta (Gray) C.B. Clarke
    atomigyna St. John & Storey
    axilliflora St. John & Storey
    baccifera C.B. Clarke
    basipartita St. John
    begoniifolia Hbd.
    biserrata Hbd.
    brevicalyx (Hbd.) St. John
    brevicornuta St. John

Cyrtandra (CONT.)
    bryanii St. John & Storey
    calpidicarpa (Rock) St. John & Storey
    calycoschiza (St. John & Storey) St. John
    campaniformis St. John
    carinata St. John & Storey
    caudatisepala St. John
    caulescens Rock
    charadraia St. John
    chartacea St. John & Storey
    christophersenii St. John & Storey
    cladantha Skottsberg
    collarifera St. John & Storey
    confertiflora (Wawra) C.B. Clarke
    conradtii Rock
    cordifolia Gaud.
        var. brevipilita St. John
        var. cordifolia
        var. gynoglabra Rock
    cornuta St. John
    crassior St. John & Storey
    crenata St. John & Storey
    cupuliformis St. John & Storey
    cyaneoides Rock
    degenerans (Wawra) Heller
    dentata St. John & Storey
    ellipticifolia St. John
    elliptisepala St. John
    elstonii Hochr.
    ferricolorata St. John
    ferruginosa St. John & Storey
    filipes Hbd.
    forbesii St. John & Storey
    fosbergii St. John & Storey
    frederickii St. John & Storey
    fusiformis St. John & Storey
    garberi St. John
    garnotiana Gaud.
    gayana Heller
        var. gayana
        var. macrocarpa Skottsberg
    georgiana Forbes
    giffardii Rock
    glauca Drake
    gracilis Hbd.
    grandiflora Gaud.
    grayana Hbd.
        var. grayana
        var. lanaiensis Rock
        var. latifolia Hbd.
        var. linearifolia Rock
        var. nervosa Rock
    grossecrenata St. John & Storey
    halawensis Rock
    hashimotoi Rock
    hawaiensis C.B. Clarke
    hii Forbes
    hirsutula St. John & Storey
    hobdyi St. John
    honolulensis Wawra
    hosakae St. John & Storey
    hyperdasa St. John
    infrapallida St. John
    infundibuliformis St. John
    intonsa St. John & Storey
    intrapilosa St. John
    intravillosa St. John & Storey
    kaalae St. John & Storey
    kahanaensis St. John & Storey
    kahukuensis St. John & Storey
    kailuaensis St. John
    kalichii Wawra
    kaluanuiensis St. John
    kaneoheensis St. John
    kauaiensis Wawra
    kaulantha St. John & Storey
    kealiae Wawra

    kipahuluensis St. John
    kipapaensis St. John & Storey
    knudsenii Rock
    kohalae Rock
    koolauensis St. John & Storey
    laevis St. John
    latebrosa Hbd.
        var. latebrosa
        var. subglabra Hbd.
    laxiflora Mann
    lessoniana Gaud.
        var. angustifolia Hbd.
        var. intrapubens St. John
        var. koolauloaensis St. John
        var. lessoniana
    leucocalyx St. John
    limosiflora Rock
    linearis St. John
    longicalyx St. John
    longifolia (Wawra) Hbd.
        var. arborescens (Wawra) C.B. Clarke
        var. degenerans (Wawra) C.B. Clarke
        var. longifolia
        var. parallela C.B. Clarke
        var. wahiawae Rock
    longiloba St. John
    lydgatei Hbd.
    lysiosepala (Gray) C.B. Clarke
        var. fauriei (Levl.) Rock
        var. grayi (C.B. Clarke) Rock
        var. haleakalensis Rock
        var. hawaiiensis Skottsberg
        var. latifolia Rock
        var. lysiosepala
        var. pilosa Hbd.
    macraei Gray
    macrantha C.B. Clarke
    macrocalyx Hbd.
    malacophylla C.B. Clarke
        var. erosa Rock
        var. malacophylla
    mannii St. John & Storey
    mauiensis Rock
        var. mauiensis
        var. truncata Rock
    megastigmata St. John
    menziesii Hook. & Arn.
    montis-loa Rock
    munroi Forbes
    niuensis St. John
    nubincolens St. John
    nutans St. John
    oahuensis Levl.
    oblanceolata St. John & Storey
    oenobarba Mann
        var. herbacea (Wawra) Heller
        var. oenobarba
        var. petiolaris Wawra
    olivacea St. John
    olona Forbes
    opaeulea St. John & Storey
    oulophylla St. John & Storey
    paloloensis St. John & Storey
    paludosa Gaud.
        var. haupuensis Rock
        var. integrifolia Hbd.
        var. irrostrata St. John
        var. microcarpa Wawra
        var. paludosa
        var. subherbacea Wawra
    partita St. John
    pearsallii St. John
    perstaminodica St. John
    pickeringii Gray
        var. pickeringii
        var. waiheae Rock
    piligyna St. John & Storey

Cyrtandra (CONT.)
    platyphylla Gray
            var. brevipes Skottsberg
            var. hiloensis Rock
            var. membranacea Rock
            var. parviflora Rock
            var. platyphylla
            var. robusta Rock
            var. stylopubens Rock
    plurifolia St. John & Storey
    poamohoensis St. John & Storey
    polyantha C.B. Clarke
    procera Hbd.
    propinqua Forbes
    pruinosa St. John & Storey
    pubens St. John
    pupukeaensis St. John & Storey
    ramosissima Rock
    reflexa St. John & Storey
    rivularis St. John & Storey
    rockii St. John & Storey
    rotata St. John
    sandwicensis (Levl.) St. John & Storey
    scabrella C.B. Clarke
    sessilis St. John & Storey
    skottsbergii St. John & Storey
    spathulata St. John
    stupantha St. John & Storey
    subcordata St. John
    subintegra St. John
    subrecta St. John
    subumbellata (Hbd.) St. John & Storey
            var. intonsa St. John
            var. subumbellata
    ternata St. John
    tintinnabula Rock
    triflora Gaud.
    tristis Hbd.
    turbiniformis St. John & Storey
    umbracculiflora Rock
    vaniotii Levl.
    villicalyx St. John & Storey
            var. pubentigyna St. John & Storey
            var. villicalyx
    villosa St. John & Storey
    villosiflora St. John
    viridiflora St. John & Storey
    waianaeensis St. John & Storey
    waianuensis Rock
    wainihaensis Levl.
    waiolanii Wawra
            var. capitata Hbd.
            var. waiolanii
    waiomaoensis St. John
    wawrae C.B. Clarke
    wilderi St. John & Storey

Gesneria L.

    citrina Urban
    cuneifolia (DC.) Fritsch
    pauciflora Urban
    pedunculosa (DC.) Fritsch
            SY=Pentarhaphia albiflora Dcne.
    viridiflora (Dcne.) Kuntze
        ssp. sintenisii (Urban) L. Skog
            SY=Duchartrea s. (Urban) Britt.
            SY=G. s. Urban

Rhytidophyllum Mart.

    auriculatum Hook.

GOODENIACEAE

Scaevola L.

    X cerasifolia Skottsberg [gaudichaudiana X
        mollis]
    chamissoniana Gaud.
            var. bracteosa Hbd.
            var. caerulescens Levl.
            var. chamissoniana
            var. cylindrocarpa (Hbd.) Krause
            var. hitchcockii Skottsberg
            var. piccoi Deg. & Deg.
    coriacea Nutt.
    gaudichaudiana Cham.
            var. gaudichaudiana
            var. pilosa Krause
            var. stenolithos Skottsberg
    gaudichaudii Hook. & Arn.
            var. dentata Krause
            var. gaudichaudii
    glabra Hook. & Arn.
    kauaiensis (Deg.) St. John
    kilaueae Deg.
            var. kilaueae
            var. powersii Deg. & Deg.
    mollis Hook. & Arn.
    plumieri (L.) Vahl
    procera Hbd.
            var. procera
            var. pseudomollis Skottsberg
    skottsbergii St. John
    taccada (Gaertn.) Roxb.
            var. bryanii St. John
            var. fauriei (Levl.) St. John
            var. sericea (Vahl) St. John
            var. taccada

HAEMODORACEAE

Lachnanthes Ell.

    caroliana (Lam.) Dandy
            SY=Gyrotheca tinctoria (Walt.)
                    Salisb.
            SY=L. t. (Walt.) Ell.

Lophiola Ker-Gawl.

    aurea Ker-Gawl.
            SY=L. americana (Pursh) Wood
            SY=L. septentrionalis Fern.

Odontostomum Torr.

    hartwegii Torr.

Xiphidium Aubl.

    ceruleum Aubl.

HALORAGIDACEAE

Gonocarpus Thunb.

    chinensis (Lour.) Orchard

Gunnera L.

    eastwoodae St. John
    kaalensis (Krajina) St. John
    kauaiensis Rock
    makahaensis St. John

Gunnera (CONT.)
    mauiensis (Krajina) St. John
    molokaiensis St. John
    petaloidea Gaud.

Haloragis J.P. & G. Forst.

    erecta (Murr.) Schindl.

Myriophyllum L.

    alterniflorum DC.
            SY=M. a. var. americanum Pugsley
    aquaticum (Vell.) Verdc.
            SY=M. brasiliense Camb.
            SY=M. proserpinacoides Gillies ex
                Hook. & Arn.
    elatinoides Gaud.
    exalbescens Fern.
            SY=M. e. var. magdalenense (Fern.)
                A. Love
            SY=M. m. Fern.
            SY=M. spicatum L. ssp. e. (Fern.)
                Hulten
            SY=M. s. var. capillaceum Lange
            SY=M. s. var. e. (Fern.) Jepson
    farwellii Morong
    heterophyllum Michx.
    hippuroides Nutt. ex Torr. & Gray
    humile (Raf.) Morong
    laxum Shuttlw. ex Chapman
    pinnatum (Walt.) B.S.P.
            SY=M. scabratum Michx.
    spicatum L.
    tenellum Bigelow
    verticillatum L.
            SY=M. v. var. cheneyi Fassett
            SY=M. v. var. intermedium W.D.J.
                Koch
            SY=M. v. var. pectinatum Wallr.
            SY=M. v. var. pinnatifidum Wallr.

Proserpinaca L.

    palustris L.
        var. amblyogona Fern.
            SY=P. a. (Fern.) Small
        var. crebra Fern. & Grisc.
        var. palustris
            SY=P. intermedia Mackenzie
            SY=P. p. var. latifolia Schindl.
            SY=P. platycarpa Small
    pectinata Lam.

HAMAMELIDACEAE

Fothergilla Murr.

    gardenii Murr.
            SY=F. parvifolia Kearney
    major (Sims) Lodd.

Hamamelis L.

    vernalis Sarg.
            SY=H. v. var. tomentella (Rehd.)
                Palmer
    virginiana L.
            SY=H. macrophylla Pursh
            SY=H. v. var. henryi Jenne
            SY=H. v. var. m. (Pursh) Nutt.
            SY=H. v. var. parvifolia Nutt.

Liquidambar L.

    styraciflua L.

HERNANDIACEAE

Hernandia L.

    sonora L.

HIPPOCASTANACEAE

Aesculus L.

    X arnoldiana Sarg. [glabra X (flava X
        pavia)]
    X bushii Schneid. [glabra X pavia]
            SY=A. X mississippiensis Sarg.
    californica (Spach) Nutt.
    flava Soland.
            SY=A. octandra Marsh.
            SY=A. o. var. vestita Sarg.
            SY=A. o. var. virginica Sarg.
    glabra Willd.
        var. arguta (Buckl.) B.L. Robins.
            SY=A. a. Buckl.
            SY=A. g. var. buckleyi Sarg.
        var. glabra
            SY=A. g. var. leucodermis Sarg.
            SY=A. g. var. micrantha Sarg.
            SY=A. g. var. monticola Sarg.
            SY=A. g. var. pallida (Willd.)
                Kirchn.
            SY=A. g. var. sargentii Rehd.
    hippocastanum L.
    X marylandica Booth ex Dippel [flava X
        glabra]
    X mutabilis (Spach) Scheele [pavia X
        sylvatica]
            SY=A. X harbisonii Sarg.
    X neglecta Lindl. [flava X sylvatica]
            SY=A. X glaucescens Sarg.
    parviflora Walt.
    pavia L.
            SY=A. austrina Small
            SY=A. discolor Pursh
            SY=A. d. var. mollis (Raf.) Sarg.
            SY=A. p. var. d. (Pursh) Gray
            SY=A. p. var. flavescens (Sarg.)
                Correll
    sylvatica Bartr.
            SY=A. georgiana Sarg.
            SY=A. s. var. lanceolata (Sarg.)
                Bartr.
    X worlitzensis Koehne [flava X (pavia X
        sylvatica)]
            SY=A. X dupontii Sarg.

HIPPURIDACEAE

Hippuris L.

    montana Ledeb.
    tetraphylla L. f.
    vulgaris L.

HYDROCHARITACEAE

Blyxa Noronha ex Thouars

    aubertii L.C. Rich.

Egeria Planch.

    densa Planch.
           SY=Anacharis d. (Planch.) Victorin
           SY=Elodea d. (Planch.) Caspary
           SY=Philotria d. (Planch.) Small

Elodea L.C. Rich.

    bifoliata St. John
    brandegeae St. John
    canadensis L.C. Rich.
           SY=Anacharis c. (L.C. Rich.)
               Planch.
           SY=A. c. var. planchonii (Caspary)
               Victorin
           SY=E. p. Caspary
           SY=Philotria c. (L.C. Rich.)
               Britt.
    columbiana St. John
    linearis (Rydb.) St. John
           SY=Philotria l. Rydb.
    longivaginata St. John
    nevadensis (Planch.) St. John
    nuttallii (Planch.) St. John
           SY=Anacharis n. Planch.
           SY=A. occidentalis (Pursh)
               Victorin
           SY=E. o. (Pursh) St. John
           SY=Philotria angustifolia (Muhl.)
               Britt.
           SY=P. n. (Planch.) Rydb.
           SY=P. minor (Engelm.) Small
    schweinitzii (Planch.) Caspary

Halophila Thouars

    aschersonii Ostenf.
    decipiens Ostenf.
           SY=H. baillonis Aschers.
    engelmannii Aschers.
    hawaiiana Doty & Stone
           SY=H. ovalis (R. Br.) Hook. f.
               ssp. h. (Doty & Stone) den
               Hartog

Hydrilla L.C. Rich.

    verticillata (L.f.) Royle

Hydrocharis L.

    morsus-ranae L.

Limnobium L.C. Rich.

    spongia (Bosc) Steud.
    stoloniferum (G.F.W. Mey.) Griseb.
           SY=Hydromystria s. G.F.W. Mey.

Ottelia Pers.

    alismoides (L.) Pers.

Thalassia Banks & Soland. ex Koenig

    testudina Koenig

Vallisneria L.

    americana Michx.
    neotropicalis Victorin
    spiralis L.

HYDROPHYLLACEAE

Draperia Torr.

    systyla (Gray) Torr.

Ellisia L.

    nyctelea L.
           SY=E. n. var. coloradensis Brand

Emmenanthe Benth.

    penduliflora Benth.
    rosea (Brand) Constance
           SY=E. penduliflora Benth. var. r.
               Brand

Eriodictyon Benth.

    altissimum P.V. Wells
    angustifolium Nutt.
           SY=E. a. var. amplifolium Brand
    californicum (Hook. & Arn.) Torr.
    capitatum Eastw.
    crassifolium Benth.
        var. crassifolium
        var. denudatum Abrams
        var. nigrescens Brand
    lanatum (Brand) Abrams
           SY=E. trichocalyx Heller ssp. l.
               (Brand) Munz
           SY=E. t. var. l. (Brand) Jepson
    tomentosum Benth.
    traskiae Eastw.
           SY=E. tomentosum Benth. ssp.
               smithii Munz
           SY=E. tomentosum ssp. traskiae
               (Eastw.) Munz
    trichocalyx Heller

Eucrypta Nutt.

    chrysanthemifolia (Benth.) Greene
        var. bipinnatifida (Torr.) Constance
        var. chrysanthemifolia
    micrantha (Torr.) Heller

Hesperochiron S. Wats.

    californicus (Benth.) S. Wats.
           SY=H. c. var. incanus (Greene)
               Brand
           SY=H. c. var. watsonianus (Greene)
               Brand
    pumilus (Griseb.) Porter

Hydrolea L.

    corymbosa J.F. Macbr. ex Ell.
           SY=Nama c. (J.F. Macbr. ex Ell.)
               Kuntze
    ovata Nutt.
           SY=Nama o. (Nutt.) Britt.
    quadrivalvis Walt.
           SY=Nama q. (Walt.) Kuntze
           SY=Nyctelea nyctelea (L.) Britt.
    spinosa L.
    uniflora Raf.
           SY=H. affinis Gray
           SY=Nama a. (Gray) Kuntze

Hydrophyllum L.

    appendiculatum Michx.
           SY=Decemium a. (Michx.) Small

Hydrophyllum (CONT.)
    canadense L.
    capitatum Dougl. ex Benth.
            var. alpinum S. Wats.
            var. capitatum
            var. thompsonii (M.E. Peck) Constance
    fendleri (Gray) Heller
            var. albifrons (Heller) J.F. Macbr.
            var. fendleri
    macrophyllum Nutt.
    occidentale (S. Wats.) Gray
            SY=H. o. var. watsonii Gray
    tenuipes Heller
    virginianum L.
            SY=H. v. var. atranthum
                (Alexander) Constance

Lemmonia Gray

    californica Gray

Nama L.

    aretioides (Hook. & Arn.) Brand
            SY=N. a. var. multiflorum (Heller)
                Jepson
    carnosum (Woot.) C.L. Hitchc.
            SY=N. stenophyllum Gray ex Hemsl.
                var. egenum J.F. Macbr.
    demissum Gray
            SY=N. d. var. covillei Brand
            SY=N. d. var. desertii Brand
    densum Lemmon
            var. densum
            var. parviflorum (Greenm.) C.L.
                Hitchc.
    depressum Lemmon ex Gray
    dichotomum (Ruiz & Pavon) Choisy
    havardii Gray
    hispidum Gray
            var. hispidum
                SY=N. tenue (Woot. & Standl.)
                    Tidestrom
            var. mentzelii Brand
                SY=N. foliosum (Woot. & Standl.)
                    Tidestrom
            var. revolutum Jepson
            var. spathulatum (Torr.) C.L. Hitchc.
    jamaicense L.
            SY=Marilaunidium j. (L.) Kuntze
    lobbii Gray
    parvifolium (Torr.) Greenm.
    pusillum Lemmon ex Gray
    retrorsum J.T. Howell
    rothrockii Gray
    sandwicense Gray
            SY=N. s. var. laysanicum Brand
    serpylloides Hemsl.
            SY=N. s. var. velutinum C.L.
                Hitchc.
    stenocarpum Gray
    stenophyllum Gray ex Hemsl.
    stevensii C.L. Hitchc.
    torynophyllum Greene
    undulatum H.B.K.
    xylopodum (Woot. & Standl.) C.L. Hitchc.

Nemophila Nutt. ex W. Bart.

    aphylla (L.) Brummitt
            SY=N. microcalyx (Nutt.) Fisch. &
                Mey.
            SY=N. triloba (Raf.) Thieret
    breviflora Gray
    heterophylla Fisch. & Mey.
            SY=N. h. var. tenera (Eastw.) A.
                Nels. & J.F. Macbr.

    kirtleyi Henderson
    maculata Benth. ex Lindl.
    menziesii Hook. & Arn.
            var. atomaria (Fisch. & Mey.)
                Chandler
                SY=N. m. ssp. a. (Fisch. & Mey.)
                    Brand
            var. integrifolia Parish
                SY=N. m. ssp. i. (Parish) Munz
                SY=N. m. ssp. i. var. annulata
                    Chandler
                SY=N. m. ssp. i. var. incana Brand
                SY=N. m. ssp. i. var. intermedia
                    (Bioletti) Brand
                SY=N. m. var. rotata (Eastw.)
                    Chandler
            var. menziesii
                SY=N. m. var. venosa (Jepson)
                    Brand
    parviflora Dougl. ex Benth.
            var. austiniae (Eastw.) Brand
            var. parviflora
            var. quercifolia (Eastw.) Chandler
    pedunculata Dougl. ex Benth.
    phacelioides Nutt. ex Bart.
    pulchella Eastw.
            var. fremontii (Elmer) Constance
            var. gracilis (Eastw.) Constance
            var. pulchella
    spatulata Coville

Phacelia Juss.

    adenophora J.T. Howell
            SY=Miltitzia glandulifera (Torr.)
                Heller
    affinis Gray
            SY=P. a. var. patens J.T. Howell
    alba Rydb.
            SY=P. neomexicana Thurb. ex Torr.
                var. a. (Rydb.) Brand
    amabilis Constance
    ambigua M.E. Jones
            var. ambigua
                SY=P. crenulata Torr. var. a.
                    (M.E. Jones) J.F. Macbr.
            var. minutiflora (J. Voss) Atwood
                SY=P. m. J. Voss
    anelsonii J.F. Macbr.
    argentea A. Nels. & J.F. Macbr.
    argillacea Atwood
    arizonica Gray
            SY=P. popei Torr. & Gray var. a.
                (Gray) J. Voss
    austromontana J.T. Howell
            SY=P. lobata (A. Davids.) Jepson
    bakeri (Brand) J.F. Macbr.
    barnebyana J.T. Howell
    beatleyae Reveal & Constance
    bicolor Torr. ex S. Wats.
    bipinnatifida Michx.
            SY=P. b. var. plummeri Wood
            SY=P. brevistyla Buckl.
    bolanderi Gray
    bombycina Woot. & Standl.
            SY=P. tenuipes Woot. & Standl.
    brachyloba (Benth.) Gray
    breweri Gray
    californica Cham.
    calthifolia Brand
    campanularia Gray
        ssp. campanularia
        ssp. vasiformis Gillett
    capitata Kruckeberg
    cephalotes Gray
    cicutaria Greene
            var. cicutaria

Phacelia   (CONT.)
          var. heliophila (J.F. Macbr.) J.T.
             Howell
                SY=P. vallis-mortae J. Voss var.
                   h. (J.F. Macbr.) J. Voss
          var. hispida (Gray) J.T. Howell
                SY=P. c. ssp. h. (Gray) Beauchamp
          var. hubbyi (J.F. Macbr.) J.T. Howell
    ciliata Benth.
          var. ciliata
          var. opaca J.T. Howell
    cinerea Eastw. ex J.F. Macbr.
    coerulea Greene
    congdonii Greene
                SY=P. divaricata (Benth.) Gray
                   var. c. (Greene) Munz
    congesta Hook.
                SY=P. c. var. dissecta Gray
    constancei Atwood
    cookei Constance & Heckard
    corrugata A. Nels.
                SY=P. crenulata Torr.
          var. corrugata (A. Nels.) Brand
    corymbosa Jepson
    crenulata Torr.
          var. angustifolia Atwood
          var. crenulata
                SY=P. c. var. funerea J. Voss ex
                   Munz
                SY=P. c. var. vulgaris Brand
    cryptantha Greene
                SY=P. c. var. derivata J. Voss
    curvipes Torr.
    dalesiana J.T. Howell
    davidsonii Gray
                SY=P. curvipes Torr. var.
                   macrantha (Parish) Munz
    demissa Gray
          var. demissa
                SY=P. d. var. typica J.T. Howell
          var. heterotricha J.T. Howell
    denticulata Osterhout
    distans Benth.
                SY=P. d. var. australis Brand
    divaricata (Benth.) Gray
    douglasii (Benth.) Torr.
                SY=P. d. var. petrophila Jepson
    dubia (L.) Trel.
          var. dubia
          var. georgiana McVaugh
    egena (Greene ex Brand) J.T. Howell
    eisenii Brandeg.
                SY=P. e. var. brandegeana J.T.
                   Howell
    filiformis Brand
    fimbriata Michx.
    floribunda Greene
    formosula Osterhout
    franklinii (R. Br.) Gray
    fremontii Torr.
    frigida Greene
       ssp. dasyphylla (J.F. Macbr.) Heckard
       ssp. frigida
    geraniifolia Brand
                SY=P. perityloides Coville var.
                   jaegeri Munz
    gilioides Brand
    glaberrima (Torr.) J.T. Howell
    glabra Nutt.
    glandulifera Piper
    glandulosa Nutt.
    glechomifolia Gray
    grandiflora (Benth.) Gray
    greenei J.T. Howell
    grisea Gray
    gymnoclada Torr.
                SY=P. crassifolia Torr. ex S.

Wats.
    hardhamiae Munz
    hastata Dougl. ex Lehm.
       ssp. compacta (Brand) Heckard
                SY=P. h. var. c. (Brand) Cronq.
       ssp. hastata
                SY=P. alpina Rydb.
                SY=P. h. var. a. (Rydb.) Cronq.
                SY=P. h. var. leucophylla (Torr.)
                   Cronq.
                SY=P. l. Torr.
                SY=P. l. var. a. (Rydb.) Dundas
                SY=P. l. var. suksdorfii J.F.
                   Macbr.
    heterophylla Pursh
       ssp. heterophylla
                SY=P. h. var. typica Dundas
       ssp. virgata (Greene) Heckard
    hirsuta Nutt.
    howelliana Atwood
    humilis Torr. & Gray
          var. dudleyi J.T. Howell
          var. humilis
    hydrophylloides Torr. ex Gray
    idahoensis Henderson
    imbricata Greene
       ssp. bernardina (Greene) Heckard
                SY=P. californica Cham. var. b.
                   (Greene) Jepson
       ssp. imbricata
       ssp. patula (Brand) Heckard
                SY=P. i. var. p. (Brand) Jepson
    incana Brand
    inconspicua Greene
    indecora J.T. Howell
    infundibuliformis Torr.
    insularis Munz
          var. continentis J.T. Howell
                SY=P. divaricata (Benth.) Gray
                   var. c. (J.T. Howell) Munz
          var. insularis
                SY=P. divaricata (Benth.) Gray
                   var. i. (Munz) Munz
    integrifolia Torr.
          var. integrifolia
                SY=P. i. var. typica J. Voss
          var. texana (J. Voss) Atwood
                SY=P. t. J. Voss
    intermedia Woot.
    inundata J.T. Howell
                SY=Miltitzia parviflora (Gray)
                   Brand
    inyoensis (J.F. Macbr.) J.T. Howell
                SY=Miltitzia i. J.F. Macbr.
    ivesiana Torr.
                SY=P. i. var. typica J.T. Howell
    laxa Small
    laxiflora J.T. Howell
    leibergii Brand
                SY=P. bicolor Torr. ex S. Wats.
                   var. l. (Brand) A. Nels. &
                   J.F. Macbr.
    lemmonii Gray
    lenta Piper
    leonis J.T. Howell
    leptosepala Rydb.
                SY=P. hastata Dougl. ex Lehm. var.
                   l. (Rydb.) Cronq.
    linearis (Pursh) Holz.
    longipes Torr. ex Gray
    lutea (Hook. & Arn.) J.T. Howell
          var. calva Cronq.
          var. lutea
                SY=Miltitzia l. (Hook. & Arn.) A.
                   DC.
          var. purpurascens J.T. Howell
    lyallii (Gray) Rydb.

Phacelia (CONT.)
  lyonii (Gray) Rydb.
  maculata Wood
          SY=P. dubia (L.) Trel. var. fallax
             (Fern.) Gleason
  malvifolia Cham.
          SY=P. m. var. loasifolia (Benth.)
             Brand
  mammalariensis Atwood
  marcescens Eastw. ex J.F. Macbr.
  minor (Harvey) Thellung ex F. Zimmerman
  minutissima Henderson
  mohavensis Gray
  mollis J.F. Macbr.
  mustelina Coville
  mutabilis Greene
          SY=P. californica Cham. var.
             jacintensis Dundas
          SY=P. heterophylla Pursh var.
             pseudohispida (Brand) Cronq.
  nashiana Jepson
  neglecta M.E. Jones
  nemoralis Greene
    ssp. nemoralis
    ssp. oregonensis Heckard
  neomexicana Thurb. ex Torr.
          SY=P. n. var. pseudo-arizonica
             (Brand) J. Voss
  nevadensis J.T. Howell
  novenmillensis Munz
  oreopola Heckard
    ssp. oreopola
    ssp. simulans Heckard
  orogenes Brand
  pachyphylla Gray
  pallida I.M. Johnston
  palmeri Torr. ex S. Wats.
          SY=P. p. var. foetida (Goodding)
             Brand
  parishii Gray
  parryi Torr.
  patuliflora (Engelm. & Gray) Gray
        var. patuliflora
        var. teucriifolia (I.M. Johnston)
             Constance
  peckii J.T. Howell
  pedicellata Gray
  pediculoides (J.T. Howell) Constance
          SY=P. ivesiana Torr. var. p. J.T.
             Howell
  peirsoniana J.T. Howell
  perityloides Coville
  phacelioides (Benth.) Brand
  platyloba Gray
  popei Torr. & Gray
          SY=P. depauperata Woot. & Standl.
          SY=P. p. var. similis (Woot. &
             Standl.) J. Voss
  pringlei Gray
  procera Gray
  pulchella Gray
        var. gooddingii (Brand) J.T. Howell
        var. pulchella
          SY=P. p. var. typica J.T. Howell
  purpusii Brandeg.
  purshii Buckl.
          SY=P. bicknellii Small
          SY=P. boykinii (Gray) Small
  quickii J.T. Howell
  racemosa (Kellogg) Brandeg.
  rafaelensis Atwood
  ramosissima Dougl. ex Lehm.
        var. ramosissima
          SY=P. eremophila Greene
          SY=P. r. var. e. (Greene) J.F.
             Macbr.
          SY=P. r. var. montereyensis Munz

          SY=P. r. var. valida M.E. Peck
        var. suffrutescens Parry
          SY=P. r. var. austrolitoralis Munz
          SY=P. s. Parry
  ranunculacea (Nutt.) Constance
          SY=P. covillei S. Wats.
  rattanii Gray
  robusta (J.F. Macbr.) I.M. Johnston
  rotundifolia Torr. ex S. Wats.
  rupestris Greene
          SY=P. congesta Hook. var. r.
             (Greene) J.F. Macbr.
  salina (A. Nels.) J.T. Howell
  saxicola Gray
  scopulina (A. Nels.) J.T. Howell
          SY=Miltitzia s. (A. Nels.) Rydb.
          SY=P. lutea (Hook. & Arn.) J.T.
             Howell var. s. (A. Nels.)
             Cronq.
  sericea (Graham) Gray
    ssp. ciliosa (Rydb.) Gillett
          SY=P. s. var. c. Rydb.
    ssp. sericea
          SY=P. s. var. caespitosa Brand
  serrata J. Voss
  splendens Eastw.
  stebbinsii Constance & Heckard
  stellaris Brand
          SY=P. douglasii (Benth.) Torr.
             var. cryptantha Brand
  strictiflora (Engelm. & Gray) Gray
        var. connexa Constance
        var. lundelliana Constance
        var. robbinsii Constance
        var. strictiflora
  suaveolens Greene
          SY=P. keckii Munz & Johnston
          SY=P. s. ssp. k. (Munz & Johnston)
             Thorne
          SY=P. s. var. k. (Munz & Johnston)
             J.T. Howell
  submutica J.T. Howell
  tanacetifolia Benth.
  tetramera J.T. Howell
          SY=Miltitzia pusilla (Gray) Brand
  thermalis Greene
  umbrosa Greene
  utahensis J. Voss
  vallicola Congd. ex Brand
  vallis-mortae J. Voss
  verna J.T. Howell
  viscida (Benth. ex Lindl.) Torr.
  vossii Atwood
  welshii Atwood

Pholistoma Lilja ex Lindl.

  auritum (Lindl.) Lilja ex Lindl.
        var. arizonicum (M.E. Jones)
             Constance
        var. auritum
  membranaceum (Benth.) Constance
  racemosum (Nutt.) Constance

Romanzoffia Cham.

  californica Greene
          SY=R. suksdorfii Greene
  sitchensis Bong.
  tracyi Jepson
  unalaschensis Cham. ex Nees
        var. glabriuscula Hulten
        var. unalaschensis

Tricardia Torr. ex Gray

  watsonii Torr. ex S. Wats.

Turricula J.F. Macbr.

    parryi (Gray) J.F. Macbr.

Wigandia H.B.K.

    caracasana Kunth
        var. macrophylla Brand

ICACINACEAE

Mappia Jacq.

    racemosa Jacq.

Ottoschulzia Urban

    rhodoxylon (Urban) Urban

ILLICIACEAE

Illicium L.

    floridanum Ellis
    parviflorum Michx. ex Vent.

IRIDACEAE

Alophia Herbert

    drummondii (Graham) R.C. Foster
        SY=Eustylis purpurea (Herbert)
           Engelm. & Gray
        SY=Herbertia d. (Graham) Small
        SY=Nemastylis p. Herbert
        SY=Tigridia p. (Herbert) Shinners

Belamcanda Adans.

    chinensis (L.) DC.
        SY=Gemmingia c. (L.) Kuntze

Chasmanthe N.E. Br.

    aethiopica (L.) N.E. Br.

Crocosmia Planch.

    X crocosmiflora (V. Lemoine ex E. Morr.)
      N.E. Br. [aurea X pottsii]
        SY=Tritonia c. V. Lemoine ex E.
           Morr.

Crocus L.

    angustifolius Weston
        SY=C. susianus Ker-Gawl.
        SY=C. vernus P. Mill.
    flavus Weston
        SY=C. maesiacus Ker-Gawl.
    sativus L.
    vernus (L.) J. Hill
        var. neapolitanus Ker-Gawl.
        var. vernus

Eleutherine Herbert

    bulbosa (P. Mill.) Urban

        SY=Galatea b. (P. Mill.) Britt.
        SY=Sisyrinchium b. P. Mill.

Freesia Eckl. ex Klatt

    refracta (Jacq.) Klatt

Gladiolus L.

    byzantinus P. Mill.
    X colvillei Sweet [cardinalis X tristis]
    X grandovensis Van Houtte [cardinalis X
        natalensis]
    X hortulanus Bailey [natalensis X various
        spp.]
    papilio Hook. f.
    segetum Ker-Gawl.

Herbertia Sweet

    lahue (Molina) Goldblatt
      ssp. caerulea (Herbert) Goldblatt
        SY=Alophia drummondii sensu auctt.
          non (Graham) Small
        SY=H. c. (Herbert) Herbert
        SY=Trifurcia c. Herbert
        SY=T. l. (Molina) Goldblatt ssp.
          c. (Herbert) Goldblatt

Iris L.

    aphylla L.
        SY=I. duerinckii Buckl.
    bracteata S. Wats.
    brevicaulis Raf.
        SY=I. brevipes Small
        SY=I. foliosa Mackenzie & Bush
        SY=I. mississippiensis Alexander
    chamaeiris Bertol.
    chrysophylla T.J. Howell
    cristata Soland.
        var. alba Dykes
        var. cristata
        SY=Neubeckia c. (Soland.) Alef.
    douglasiana Herbert
        SY=I. d. var. oregonensis R.C.
          Foster
    ensata Thunb.
        SY=I. kaempferi Sieb.
    fernaldii R.C. Foster
    foetidissima L.
    fulva Ker-Gawl.
        SY=I. ecristata Alexander
    X fulvala Dykes [brevicaulis X fulva]
    X germanica L.
    giganticaerulea Small
        SY=I. alticristata Small
        SY=I. aurilinea Alexander
        SY=I. citricristata Small
        SY=I. elephantina Small
        SY=I. fluviatilis Small
        SY=I. miraculosa Small
        SY=I. paludicola Alexander
        SY=I. parvicaerulea Alexander
        SY=I. venulosa Alexander
        SY=I. wherryana Small
    hartwegii Baker
      ssp. australis (Parish) Lenz
        SY=I. h. var. a. Parish
      ssp. columbiana Lenz
      ssp. hartwegii
      ssp. pinetorum (Eastw.) Lenz
    hexagona Walt.
        var. flexicaulis (Small) R.C. Foster
        var. hexagona
        SY=I. alabamensis Small
        SY=?I. rivularis Small

Iris (CONT.)
  var. savannarum (Small) R.C. Foster
    SY=I. albispiritus Small
    SY=I. s. Small
innominata Henderson
kimballiae Small
lacustris Nutt.
longipetala Herbert
macrosiphon Torr.
    SY=I. amabilis Eastw.
    SY=I. californica Leichtl.
missouriensis Nutt.
    SY=I. m. var. arizonicus (Dykes)
      R.C. Foster
    SY=I. m. var. pelogonus (Goodding)
      R.C. Foster
munzii R.C. Foster
nelsonii Randolph
ochroleuca L.
orientalis P. Mill.
pallida Lam.
prismatica Pursh
  var. austrina Fern.
  var. prismatica
pseudacorus L.
pumila L.
  var. flaviflora Fuss.
  var. pumila
purdyi Eastw.
X robusta E. Anders. [versicolor X
 virginica var. shrevei]
X sancti-cyrii Rouss. [setosa var.
 canadensis X versicolor]
sanguinea Donn
setosa Pallas ex Link
  var. canadensis M. Foster
    SY=I. hookeri Penny
  var. interior E. Anders.
    SY=I. s. ssp. i. (E. Anders.)
      Hulten
    SY=I. s. var. platyrhyncha Hulten
  var. setosa
siberica L.
tenax Dougl. ex Lindl.
 ssp. klamathensis Lenz
 ssp. tenax
    SY=I. t. ssp. gormanii Piper
    SY=I. t. var. g. (Piper) R.C.
      Foster
tenuis S. Wats.
tenuissima Dykes
 ssp. purdyiformis (R.C. Foster) Lenz
 ssp. tenuissima
thompsonii R.C. Foster
    SY=I. tenax Dougl. ex Lindl. ssp.
     thompsonii (R.C. Foster) Q.
     Clarkson
tridentata Pursh
    SY=I. tripetala Walt.
variegata L.
verna L.
  var. smalliana Fern. ex M.E. Edwards
  var. verna
    SY=Neubeckia v. (L.) Alef.
versicolor L.
  var. blandescens Nieuwl.
  var. versicolor
X vinicolor Small [fulva X giganticaerulea]
    SY=I. chrysophoenicia Small
    SY=I. lilacinaurea Alexander
virginica L.
  var. shrevei (Small) E. Anders.
    SY=I. s. Small
    SY=I. versicolor L. var. s.
     (Small) Boivin
  var. virginica
    SY=I. acleantha Small

    SY=I. albilinea Alexander
    SY=I. amnicola Alexander
    SY=I. atrocyanea Small
    SY=I. atroenantha Small
    SY=I. auralata Small
    SY=I. bifurcata Small
    SY=I. callilopha Alexander
    SY=I. callirhodea Alexander
    SY=I. cerasioides Alexander
    SY=I. chlorolopha Small
    SY=I. chrysaeola Small
    SY=I. chrysolopha Small
    SY=I. citriviola Small
    SY=I. crocinubia Alexander
    SY=I. cyanantha Alexander
    SY=I. cyanochrysea Small
    SY=I. dewinkeleri Small
    SY=I. fourchiana Small
    SY=I. fulvaurea Small
    SY=I. fumiflora Alexander
    SY=I. fumifulva Small
    SY=I. fuscaurea Small
    SY=I. fuscirosea Small
    SY=I. fuscisanguinea Alexander
    SY=I. fuscivenosa Small
    SY=I. gentilliana Alexander
    SY=I. georgiana Britt.
    SY=I. ianthina Alexander
    SY=I. iochroma Small
    SY=I. iocyanea Small
    SY=I. iodantha Alexander
    SY=I. ioleuca Alexander
    SY=I. iophaea Alexander
    SY=I. lancipetala Alexander
    SY=I. ludoviciana Small
    SY=I. marplei Alexander
    SY=I. moricolor Small
    SY=I. oenantha Small
    SY=I. oolopha Alexander
    SY=I. pallidirosea Alexander
    SY=I. parvirosea Small
    SY=I. phoenicis Small
    SY=I. pseudocristata Small
    SY=I. purpurissata Small
    SY=I. pyrrholopha Alexander
    SY=I. regalis Small
    SY=I. regifulva Alexander
    SY=I. rhodantha Alexander
    SY=I. rhodochrysea Small
    SY=I. rosiflora Small
    SY=I. rosilutea Alexander
    SY=I. rosipurpurea Alexander
    SY=I. rubea Alexander
    SY=I. rubicunda Small
    SY=I. rubrolilacina Alexander
    SY=I. salmonicolor Small
    SY=I. schizolopha Alexander
    SY=I. subfulva Small
    SY=I. thomasii Small
    SY=I. tyriana Small
    SY=I. violipurpurea Small
    SY=I. violilutea Alexander
    SY=I. violivenosa Small
    SY=I. viridis Alexander
    SY=I. viridivinea Small

Libertia Spreng.

 formosa Graham

Nemastylis Nutt.

 floridana Small
 geminiflora Nutt.
    SY=N. acuta (Bartr.) Herbert
 nuttallii Pickering ex R.C. Foster
 tenuis (Herbert) Benth.

Nemastylis (CONT.)
        ssp. pringlei (S. Wats.) Goldblatt
                SY=N. t. var. p. (S. Wats.) R.C.
                        Foster

Romulea Maratti

    rosea (L.) Ecklon

Sisyrinchium L.

    acre Mann
    albidum Raf.
                SY=S. capillare Bickn.
                SY=S. scabrellum Bickn.
    angustifolium P. Mill.
                SY=S. bermudiana sensu auctt. non
                        L.
                SY=S. graminoides Bickn.
    arenicola Bickn.
                SY=S. floridanum Bickn.
                SY=S. incrustatum Bickn.
                SY=S. rufipes Bickn.
    arizonicum Rothrock
                SY=Oreolirion a. (Rothrock) Bickn.
    atlanticum Bickn.
                SY=S. mucronatum Michx. var. a.
                        (Bickn.) Ahles
    bellum S. Wats.
    biforme Bickn.
    californicum (Ker-Gawl.) Dry.
                SY=Hydastylus borealis Bickn.
                SY=H. brachypus Bickn.
                SY=H. c. (Ker-Gawl.) Salisb.
                SY=S. boreale (Bickn.) Henry
    campestre Bickn.
            var. campestre
                SY=S. flaviflorum Bickn.
            var. kansanum Bickn.
                SY=S. k. (Bickn.) Alexander
    cernuum (Bickn.) Kearney & Peebles
    demissum Greene
            var. amethystinum (Bickn.) Kearney &
                    Peebles
            var. demissum
                SY=S. longipedunculatum Bickn.
                SY=S. radicatum Bickn.
    dimorphum R. Oliver
    douglasii A. Dietr.
            var. douglasii
                SY=Olsynium d. (A. Dietr.) Bickn.
            var. inflatum (Suksdorf) P. Holmgren
                SY=S. i. (Suksdorf) St. John
    elmeri Greene
                SY=Hydastylus e. (Greene) Bickn.
    ensigerum Bickn.
    exile Bickn.
                SY=S. brownei Small
                SY=S. micranthum sensu auctt. non
                        Cav.
    farwellii Bickn.
    fibrosum Bickn.
    funereum Bickn.
    fuscatum Bickn.
    groenlandicum Bocher
    halophilum Greene
    hastile Bickn.
    heterocarpum Bickn.
    hitchcockii Henderson
    idahoense Bickn.
            var. idahoense
                SY=S. birameum Piper
            var. macounii (Bickn.) Henderson
                SY=S. m. Bickn.
            var. occidentale (Bickn.) Henderson
                SY=S. o. Bickn.
            var. segetum (Bickn.) Henderson

                SY=S. s. Bickn.
    intermedium Bickn.
    langloisii Greene
                SY=S. canbyi Bickn.
    laxum Otto ex Sims
    littorale Greene
    longipes (Bickn.) Kearney & Peebles
    macrocarpon Bickn.
    minus Engelm. & Gray
    montanum Greene
            var. crebrum Fern.
                SY=S. bermudiana L. var. c.
                        (Fern.) Boivin
            var. montanum
                SY=S. strictum Bickn.
    mucronatum Michx.
    nashii Bickn.
    pruinosum Bickn.
                SY=S. brayi Bickn.
                SY=S. colubriferum Bickn.
                SY=S. helleri Bickn.
                SY=S. varians Bickn.
    rosulatum Bickn.
    sagittiferum Bickn.
    sarmentosum Suksdorf ex Greene
    septentrionale Bickn.
    solstitiale Bickn.
    xerophyllum Greene

Sphenostigma Baker

    coelestinum (Bartr. ex Willd.) R.C. Foster
                SY=Nemastylis c. (Bartr. ex
                        Willd.) Nutt.
                SY=Salpingostylis c. (Bartr. ex
                        Willd.) Small

Trimezia Salisb. ex Herbert

    martinicensis (Jacq.) Herbert
                SY=Iris m. Jacq.

Watsonia P. Mill.

    bulbillifera Matthews & L. Bolus

Xiphion P. Mill.

    tingitana Boiss. & Reut.
    xiphium L.

JUGLANDACEAE

Carya Nutt.

    aquatica (Michx. f.) Nutt.
                SY=C. a. var. australis Sarg.
                SY=Hicoria aquatica (Michx. f.)
                        Britt.
    X brownii Sarg. [cordiformis X illinoensis]
                SY=C. X b. var. varians Sarg.
    X collina Laughlin [texana X tomentosa]
    cordiformis (Wang.) K. Koch
                SY=C. c. var. latifolia Sarg.
                SY=Hicoria c. (Wang.) Britt.
    X demareei Palmer [cordiformis X glabra]
    X dunbarii Sarg. [laciniosa X ovata]
    floridana Sarg.
                SY=Hicoria f. (Sarg.) Sudworth
    glabra (P. Mill.) Sweet
            var. glabra
                SY=C. microcarpa (Nutt.) Britt.
                SY=Hicoria g. (P. Mill.) Britt.
            var. hirsuta (Ashe) Ashe

Carya (CONT.)
        SY=C. ovalis (Wang.) Sarg. var. h.
          (Ashe) Sarg.
        SY=C. X o. (Wang.) Sarg. nm. h.
          (Ashe) Boivin
        SY=Hicoria g. (P. Mill.) Britt.
          var. h. Ashe
      var. megacarpa (Sarg.) Sarg.
        SY=C. magnifloridana Murrill
        SY=Hicoria austrina Small
illinoensis (Wang.) K. Koch
        SY=C. oliviformis (Michx. f.)
          Nutt.
        SY=C. pecan (Marsh.) Engl. &
          Graebn.
        SY=Hicoria p. (Marsh.) Britt.
laciniosa (Michx. f.) Loud.
        SY=Hicoria l. (Michx. f.) Sarg.
X laneyi Sarg. [cordiformis X ovata]
        SY=C. X l. var. chateaugayensis
          Sarg.
        SY=C. X l. nm. c. (Sarg.) Boivin
X lecontei Little [aquatica X illinoensis]
        SY=Hicoria texana Le Conte
leiodermis Sarg.
X ludoviciana (Ashe) Little [aquatica X
   texana]
myristiciformis (Michx. f.) Nutt.
        SY=Hicoria m. (Michx. f.) Britt.
X nussbaumeri Sarg. [illinoensis X
   laciniosa]
ovalis (Wang.) Sarg.
        SY=C. glabra (P. Mill.) Sweet var.
          odorata (Marsh.) Little
        SY=C. ovalis var. mollis (Ashe)
          Sudworth
        SY=C. ovalis var. obcordata (Muhl.
          & Willd.) Sarg.
        SY=C. ovalis var. obovalis Sarg.
        SY=C. ovalis var. odorata (Marsh.)
          Sarg.
        SY=C. X ovalis nm. odorata
          (Marsh.) Boivin
        SY=Hicoria ovalis (Wang.) Ashe
ovata (P. Mill.) K. Koch
      var. australis (Ashe) Little
        SY=C. carolinae-septentrionalis
          (Ashe) Engl. & Graebn.
        SY=Hicoria c-s. Ashe
      var. ovata
        SY=C. alba (L.) Nutt. pro parte
        SY=C. o. var. fraxinifolia Sarg.
        SY=C. o. var. nuttallii Sarg.
        SY=C. o. var. pubescens Sarg.
        SY=Hicoria alba (L.) Britt. pro
          parte
        SY=H. borealis Ashe
        SY=H. o. (P. Mill.) Britt.
pallida (Ashe) Engl. & Graebn.
        SY=Hicoria p. Ashe
X schneckii Sarg. [illinoensis X tomentosa]
texana Buckl.
        SY=C. buckleyi Dur.
        SY=C. b. var. arkansana Sarg.
        SY=C. glabra (P. Mill.) Sweet var.
          villosa (Sarg.) B.L. Robins.
        SY=C. t. var. a. (Sarg.) Little
        SY=C. t. var. villosa (Sarg.)
          Little
        SY=Hicoria a. (Sarg.) Ashe
        SY=H. texana sensu auctt. non Le
          Conte
        SY=H. villosa (Sarg.) Ashe
tomentosa (Poir.) Nutt.
        SY=C. alba (P. Mill.) K. Koch
        SY=C. t. var. subcoriacea (Sarg.)
          Palmer & Steyermark

        SY=Hicoria t. (Poir.) Raf.

Juglans L.

   ailanthifolia Carr.
   californica S. Wats.
   cinerea L.
        SY=Wallia c. (L.) Alef.
   hindsii (Jepson) Jepson ex R. E. Sm.
   jamaicensis C. DC.
   major (Torr.) Heller
        SY=J. elaeopyren Dode
        SY=J. rupestris Engelm. var. m.
          Torr.
   microcarpa Berl.
      var. microcarpa
        SY=J. rupestris Engelm. ex Torr.
      var. stewartii (I.M. Johnston) W.
        Manning
   nigra L.
        SY=Wallia n. (L.) Alef.

JUNCACEAE

Juncus L.

   abjectus F.J. Herm.
   abortivus Chapman
        SY=J. pelocarpus E. Mey. var.
          crassicaudex Engelm.
   acuminatus Michx.
   acutiflorus Ehrh.
   acutus L.
      var. sphaerocarpus Engelm.
   albescens (Lange) Fern.
        SY=J. triglumis L. ssp. a. (Lange)
          Hulten
        SY=J. t. var. a. Lange
   X alpiniformis Fern. [alpinus X
      articulatus]
   alpinus Vill.
      ssp. alpinus
        SY=J. alpinoarticulatus Chaix
        SY=J. alpinus var. fuscescens
          Fern.
      ssp. nodulosus (Wahlenb.) Lindm.
        var. alpestris (Hartman) Hyl.
        var. nodulosus
        var. rariflorus Hartman
        SY=J. richardsonianus Schultes
        var. uniceps Hartman
   arcticus Willd.
      ssp. alaskanus Hulten
        SY=J. arcticus var. alaskanus
          (Hulten) Welsh
        SY=J. balticus Willd. var.
          alaskanus (Hulten) Porsild
      ssp. arcticus Willd.
   articulatus L.
        SY=J. a. var. obtusatus Engelm.
   balticus Willd.
      var. balticus
        SY=J. arcticus Willd. var. b.
          (Willd.) Trautv.
      var. littoralis Engelm.
        SY=J. arcticus Willd. ssp. l.
          (Engelm.) Hulten
      var. montanus Engelm.
        SY=J. ater Rydb.
        SY=J. arcticus Willd. ssp. ater
          (Rydb.) Hulten
        SY=J. a. var. m. (Engelm.) Welsh
      var. stenocarpus Buch. & Fern.
      var. vallicola Rydb.

Juncus (CONT.)
   biglumis L.
   bolanderi Engelm.
   brachycarpus Engelm.
   brachycephalus (Engelm.) Buch.
   brachyphyllus Wieg.
           SY=J. kansanus F.J. Herm.
   brevicaudatus (Engelm.) Fern.
   bryoides F.J. Herm.
   bufonius L.
           var. bufonius
           var. congdonii (S. Wats.) J.T. Howell
           var. congestus Wahlenb.
           var. halophilus Buch. & Fern.
           var. occidentalis F.J. Herm.
               SY=J. sphaerocarpus sensu auctt.
                   non Nees
           var. ranarius (Perr. & Song.) Hayek
               SY=J. r. Perr. & Song.
   caesariensis Coville
   canadensis J. Gay ex Laharpe
           var. canadensis
           var. euroauster Fern.
           var. sparsiflorus Fern.
   capillaris F.J. Herm.
   capitatus Weigel
   castaneus Sm.
       ssp. castaneus
           var. castaneus
           var. pallidus Hook.
       ssp. leucochlamys (Zinz. ex Krecz.)
           Hulten
           SY=J. l. Zinz. ex Krecz.
   chlorocephalus Engelm.
   compressus Jacq.
           SY=J. bulbosus L.
           SY=J. supinus Moench
   confusus Coville
   cooperi Engelm.
   coriaceus Mackenzie
           SY=J. setaceus Rostk.
   covillei Piper
           var. covillei
           var. obtusatus (Engelm.) C.L. Hitchc.
               SY=J. o. Engelm.
   debilis Gray
   dichotomus Ell.
   diffusissimus Buckl.
   drummondii E. Mey.
           var. drummondii
           var. subtriflorus (E. Mey.) C.L.
               Hitchc.
   dubius Engelm.
   effusus L.
           var. brunneus Engelm.
           var. compactus Hoppe
               SY=J. e. var. caeruleomontanus St.
                   John
           var. conglomeratus (L.) Engelm.
               SY=J. c. L.
           var. costulatus Fern.
           var. decipiens Buch.
           var. effusus
           var. exiguus Fern. & Wieg.
           var. gracilis Hook.
           var. pacificus Fern. & Wieg.
           var. pylaei (Laharpe) Fern. & Wieg.
           var. solutus Fern. & Wieg.
           var. subglomeratus Lam. & DC.
   elliottii Chapman
           var. elliottii
           var. polyanthemus C. Mohr
   ensifolius Wikstr.
           var. ensifolius
               SY=J. xiphioides E. Mey. var.
                   triandrus Engelm.
           var. major Hook.

falcatus E. Mey.
       var. falcatus
       var. sitchensis Buch.
           SY=J. f. ssp. s. (Buch.) Hulten
filiformis L.
filipendulus Buckl.
X fulvescens Fern. [articulatus X
   brevicaudatus]
georgianus Coville
gerardii Loisel.
       var. gerardii
       var. pedicellatus Fern.
glomeratus Batson
greenei Oakes & Tuckerman
griscomii Fern.
gymnocarpus Coville
haenkei E. Mey.
           SY=J. arcticus Willd. var.
               sitchensis Engelm.
           SY=J. balticus Willd. var. h. (E.
               Mey.) Buch.
hallii Engelm.
hemiendytus F.J. Herm.
howellii F.J. Herm.
inflexus L.
interior Wieg.
       var. arizonicus (Wieg.) F.J. Herm.
           SY=J. a. Wieg.
       var. interior
       var. neomexicanus (Wieg.) F.J. Herm.
           SY=J. n. Wieg.
kelloggii Engelm.
           SY=J. brachystylus (Engelm.) Piper
leiospermus F.J. Herm.
X lemieuxii Boivin [articulatus X
   canadensis]
lesueurii Boland.
longii Fern.
longistylis Torr.
       var. longistylis
       var. scabratus F.J. Herm.
macrandrus Coville
macrophyllus Coville
marginatus Rostk.
       var. marginatus
           SY=J. aristulatus Michx.
           SY=J. biflorus Ell.
           SY=J. m. var. b. (Ell.) Wood
       var. setosus Coville
           SY=J. s. (Coville) Small
maritimus Lam.
megacephalus M. A. Curtis
megaspermus F.J. Herm.
mertensianus Bong.
mexicanus Willd.
           SY=J. balticus Willd. var. m.
               (Willd.) Kuntze
militaris Bigelow
nevadensis S. Wats.
       var. badius (Suksdorf) C.L. Hitchc.
           SY=J. b. Suksdorf
           SY=J. mertensianus Bong. ssp.
               gracilis (Engelm.) F.J.
               Herm. var. b. (Suksdorf)
               F.J. Herm.
       var. columbianus (Coville) St. John
           SY=J. c. Coville
           SY=J. mertensianus Bong. ssp.
               gracilis (Engelm.) F.J.
               Herm. var. c. (Coville) St.
               John
       var. inventus (Henderson) C.L.
           Hitchc.
       var. nevadensis
           SY=J. duranii Ewan
           SY=J. mertensianus Bong. ssp.
               gracilis (Engelm.) F.J.

Juncus (CONT.)
            Herm.
            SY=J. m. ssp. g. var. d. (Ewan)
                F.J. Herm.
nodatus Coville
            SY=J. acuminatus Michx. var.
                robustus Engelm.
            SY=J. r. (Engelm.) Coville
X nodosiformis Fern. [apinus X nodosus]
nodosus L.
        var. meridianus F.J. Herm.
oronensis Fern.
orthophyllus Coville
oxymeris Engelm.
parryi Engelm.
patens E. Mey.
pelocarpus E. Mey.
        var. pelocarpus
        var. sabulonensis St. John
pervetus Fern.
phaeocephalus Engelm.
        var. paniculatus Engelm.
        var. phaeocephalus
            SY=J. p. var. glomeratus Engelm.
planifolius R. Br.
platyphyllus (Wieg.) Fern.
            SY=J. dichotomus Ell. var. p.
                Wieg.
            SY=J. tenuis Willd. var. p.
                (Wieg.) F.J. Herm.
polycephalus Michx.
regelii Buch.
            SY=J. jonesii Rydb.
repens Michx.
roemerianus Scheele
rugulosus Engelm.
saximontanus A. Nels.
            SY=J. ensifolius Wikstr. var.
                brunnescens (Rydb.) Crong.
            SY=J. e. var. montanus (Engelm.)
                C.L. Hitchc.
            SY=J. s. var. robustior M.E. Peck
scirpoides Lam.
            SY=J. s. var. compositus Harper
            SY=J. s. var. meridionalis Buch.
secundus Beauv. ex Poir.
slwookoorum S.B. Young
squarrosus L.
stygius L.
    ssp. americanus (Buch.) Hulten
            SY=J. s. var. a. Buch.
subcaudatus (Engelm.) Coville & Blake
        var. planisepalus Fern.
        var. subcaudatus
subtilis E. Mey.
supiniformis Engelm.
            SY=J. oreganus S. Wats.
tenuis Willd.
        var. congestus Engelm.
            SY=J. occidentalis (Coville) Wieg.
        var. tenuis
            SY=J. macer S.F. Gray
            SY=J. t. var. anthelatus Wieg.
            SY=J. t. var. multicornis E. Mey.
        var. uniflorus (Farw.) Farw.
            SY=J. dudleyi Wieg.
            SY=J. t. var. d. (Wieg.) F.J.
                Herm.
        var. williamsii Fern.
texanus (Engelm.) Coville
textilis Buch.
torreyi Coville
tracyi Rydb.
trifidus L.
        var. monanthos (Jacq.) Bluff &
            Fingerhuth
        var. trifidus

triformis Engelm.
triglumis L.
trigonocarpus Steud.
tweedyi Rydb.
uncialis Greene
validus Coville
        var. fascinatus M.C. Johnston
        var. validus
            SY=J. crassifolius Buch.
            SY=J. platycephalus Michx.
vaseyi Engelm.
            SY=J. greenei Oakes & Tuckerman
                var. v. (Engelm.) Boivin
xiphioides E. Mey.

Luzula DC.

acuminata Raf.
        var. acuminata
            SY=Juncoides saltuense (Fern.)
                Heller
            SY=L. carolinae S. Wats. var. s.
                (Fern.) Fern.
            SY=L. pilosa (L.) Willd.
            SY=L. p. var. s. (Fern.) Boivin
            SY=L. s. Fern.
        var. carolinae (S. Wats.) Fern.
            SY=Juncoides c. (S. Wats.) Kuntze
            SY=L. c. S. Wats.
arctica Blytt
    ssp. arctica
            SY=Juncoides a. (Blytt) Coville
            SY=L. nivalis (Laestad.) Beurling
    ssp. latifolia (Kjellm.) Porsild
            SY=L. hyperborea R. Br. var. l.
                (Kjellm.) Boivin
            SY=L. nivalis (Laestad.) Beurling
                var. l. (Kjellm.) Samuelsson
            SY=L. tundricola Gorodk.
arcuata (Wahlenb.) Sw.
    ssp. arcuata
    ssp. unalaschcensis (Buch.) Hulten
            SY=L. a. var. u Buch.
bulbosa (Wood) Rydb.
            SY=Juncoides b. (Wood) Small
            SY=L. campestris (L.) DC. var. b.
                Wood
            SY=L. multiflora (Retz.) Lej. var.
                b. (Wood) F.J. Herm.
campestris (L.) DC.
        var. campestris
            SY=Juncoides c. (L.) Kuntze
        var. pallescens Wahlenb.
            SY=L. p. (Wahlenb.) Bess.
confusa Lindeberg
divaricata S. Wats.
            SY=L. parviflora (Ehrh.) Desv.
                ssp. d. (S. Wats.) Hulten
            SY=L. p. var. d. (S. Wats.) Boivin
echinata (Small) F.J. Herm.
        var. echinata
            SY=Juncoides e. Small
            SY=L. campestris (L.) DC. var. e.
                (Small) Fern. & Wieg.
        var. mesochorea F.J. Herm.
forsteri (Sm.) DC.
glabrata (Hoppe) Desv.
            SY=L. hitchcockii Hamet-Ahti
groenlandica Bocher
        var. fuscoatra Bocher
        var. groenlandica
hawaiiensis Buch.
        var. glabrata (Hbd.) Deg. & Deg.
        var. hawaiiensis
        var. oahuensis (Deg. & Fosberg) Deg.
            & Deg.
hyperborea R. Br.

Luzula (CONT.)
                SY=Juncoides h. (R. Br.) Sheldon
    luzuloides (Lam.) Dandy & Wilmott
                SY=Juncoides nemorosum (Pollard)
                    Kuntze
    multiflora (Retz.) Lej.
        ssp. comosa (E. Mey.) Hulten
            var. acadiensis Fern.
            var. comosa (E. Mey.) Hulten
                SY=L. campestris (L.) DC. var.
                    congesta (Thuill.) E. Mey.
                SY=L. comosa E. Mey.
                SY=L. comosa var. congesta
                    (Thuill.) S. Wats.
                SY=L. intermedia (Thuill.) A.
                    Nels.
                SY=L. m. var. comosa (E. Mey.)
                    Fern. & Wieg.
                SY=L. m. var. congesta (Thuill.)
                    Koch
                SY=L. orestera C.W. Sharsmith
        ssp. frigida (Buch.) Krecz.
                SY=L. campestris (L.) DC. var. f.
                    Buch.
                SY=L. m. var. f. (Buch.)
                    Samuelsson
                SY=L. m. var. fusconigra Celak.
                SY=L. sudetica (Willd.) DC.
                SY=L. s. var. frigida (Buch.)
                    Fern.
        ssp. multiflora
            var. contracta Samuelsson
            var. kjellmaniana Taylor & MacBryde
                SY=L. k. sensu auctt. non Miyabe &
                    Kudo
                SY=L. m. var. k. sensu Samuelsson
            var. minor (Satake) Taylor & MacBryde
                SY=L. campestris (L.) DC. var.
                    minor (Satake) Welsh
                SY=L. kobayasii Satake
                SY=L. k. var. minor Satake
                SY=L. multiflora var. k. (Satake)
                    Samuelsson
            var. multiflora
                SY=L. campestris (L.) DC. var. m.
                    (Retz.) Celak.
    parviflora (Ehrh.) Desv.
        ssp. fastigiata (E. Mey.) Hamet-Ahti
        ssp. melanocarpa (Michx.) Hamet-Ahti
                SY=L. p. var. m. (Michx.) Buch.
        ssp. parviflora
                SY=Juncoides p. (Ehrh.) Coville
    rufescens Fisch. ex E. Mey.
                SY=L. pilosa (L.) Willd. var. r.
                    (Fisch. ex E. Mey.) Boivin
    spadicea (All.) DC.
                SY=L. piperi (Coville) M.E. Jones
                SY=L. wahlenbergii Rupr.
                SY=L. w. ssp. p. (Coville) Hulten
    spicata (L.) DC.
                SY=Juncoides s. (L.) Kuntze
    subcapitata (Rydb.) Harrington
    subcongesta (S. Wats.) Jepson
    subsessilis (S. Wats.) Buch.
    sudetica (Willd.) DC.

KRAMERIACEAE

Krameria L.

    glandulosa Rose & Painter
                SY=K. parvifolia Benth. var. g.
                    (Rose & Painter) J.F. Macbr.
    grayi Rose & Painter

    ixina L.
    lanceolata Torr.
                SY=K. secundiflora sensu auctt.
                    non DC.
                SY=K. spathulata Small
    parvifolia Benth.
            var. imparata J.F. Macbr.
                SY=K. i. (J.F. Macbr.) Britt.
            var. parvifolia
    ramosissima (Gray) S. Wats.

LAMIACEAE

Acanthomintha (Gray) Gray

    ilicifolia (Gray) Gray
    lanceolata Curran
    obovata Jepson
        ssp. duttonii Abrams
        ssp. obovata

Acinos P. Mill.

    arvensis (Lam.) Dandy
                SY=Calamintha acinos (L.)
                    Clairville ex Gaud.
                SY=Clinopodium acinos (L.) Kuntze
                SY=Satureja acinos (L.) Scheele

Agastache Clayton ex Gronov.

    breviflora (Gray) Epling
    cana (Hook.) Woot. & Standl.
    cusickii (Greenm.) Heller
            var. cusickii
            var. parva Cronq.
    foeniculum (Pursh) Kuntze
                SY=A. anethiodora (Nutt.) Britt.
    mearnsii Woot. & Standl.
                SY=A. pallidiflora (Heller) Rydb.
                    ssp. m. (Woot. & Standl.)
                    Lint & Epling
    micrantha (Gray) Woot. & Standl.
    nepetoides (L.) Kuntze
    occidentalis (Piper) Heller
    pallida (Lindl.) Cory
                SY=A. barberi (B.L. Robins.)
                    Epling
    pallidiflora (Heller) Rydb.
        ssp. neomexicana (Brig.) Lint & Epling
            var. havardii (Gray) R.W. Sanders
                ined.
                SY=A. breviflora (Gray) Epling
                    var. h. (Gray) Shinners
                SY=A. p. ssp. h. (Gray) Lint &
                    Epling
            var. neomexicana (Brig.) Lint &
                Epling
                SY=A. n. (Brig.) Standl.
        ssp. pallidiflora
            var. gilensis R.W. Sanders ined.
            var. greenei (Brig.) R.W. Sanders
                ined.
                SY=A. g. (Brig.) Woot. & Standl.
            var. pallidiflora
                SY=A. p. ssp. typica Lint & Epling
    parvifolia Eastw.
    pringlei (Brig.) Lint & Epling
            var. verticillata (Woot. & Standl.)
                R.W. Sanders ined.
                SY=A. v. Woot. & Standl.
    rupestris (Greene) Standl.
    scrophulariifolia (Willd.) Kuntze
                SY=A. s. var. mollis (Fern.)

Agastache (CONT.)
              Heller
    urticifolia (Benth.) Kuntze
          var. glaucifolia (Heller) Cronq.
          var. urticifolia
    wrightii (Greenm.) Woot. & Standl.

Ajuga L.

    chamipitys (L.) Schreb.
    genevensis L.
    reptans L.

Ballota L.

    nigra L.
          var. alba (L.) Sm.
          var. foetida (Hayek) Vis.
             SY=B. n. ssp. f. Hayek
          var. nigra

Blephilia Raf.

    ciliata (L.) Benth.
    hirsuta (Pursh) Benth.
          var. glabrata Fern.
          var. hirsuta

Brazoria Engelm. & Gray

    arenaria Lundell
    pulcherrima Lundell
    scutellarioides Engelm. & Gray
    truncata (Benth.) Engelm. & Gray

Calamintha P. Mill.

    arkansana (Nutt.) Shinners
          SY=Clinopodium a. (Nutt.) House
          SY=Clinopodium glabrum (Nutt.)
             Kuntze
          SY=Satureja a. (Nutt.) Briq.
          SY=S. glabella (Michx.) Briq. var.
             angustifolia (Torr.) Svens.
          SY=S. glabra (Nutt.) Fern.
    ashei (Weatherby) Shinners
          SY=Clinopodium a. (Weatherby)
             Shinners
          SY=Satureja a. Weatherby
    chandleri Brandeg.
          SY=Satureja c. (Brandeg.) Druce
    coccinea (Nutt.) Benth.
          SY=Clinopodium c. (Nutt.) Kuntze
          SY=Clinopodium macrocalyx Small
          SY=Satureja c. (Nutt.) Bertol.
    dentatum Chapman
          SY=Clinopodium d. (Chapman) Kuntze
          SY=Satureja d. (Chapman) Briq.
    georgiana (Harper) Shinners
          SY=Clinopodium g. Harper
          SY=Satureja caroliniana (Michx.)
             Briq.
          SY=S. g. (Harper) Ahles
    glabella (Michx.) Benth.
          SY=Clinopodium g. (Michx.) Kuntze
          SY=Satureja g. (Michx.) Briq.
    nepeta (L.) Savi
       ssp. glandulosa (Riquien) P.W. Ball
          SY=Satureja calamintha (L.)
             Scheele
          SY=S. c. var. g. (Riquien) Briq.
       ssp. nepeta
          SY=Clinopodium n. (L.) Kuntze
          SY=Satureja calamintha (L.)
             Scheele var. n. (L.) Briq.
          SY=S. c. var. nepetoides (Jord.)
             Briq.

          SY=S. nepeta (L.) Scheele
    sylvatica Bromf.
       ssp. ascendens (Jord.) P.W. Ball
          SY=Calamintha a. Jord.
          SY=C. officinalis Moench
          SY=Satureja s. (Bromf.) K. Maly
             ssp. a. (Jord.) Taylor &
             MacBryde
       ssp. sylvatica
          SY=Saureja s. (Bromf.) K. Maly

Cedronella Moench

    canariensis (L.) Willd. ex Webb & Berth.

Clinopodium L.

    gracile Kuntze
    vulgare L.
          SY=Satureja v. (L.) Fritsch
          SY=S. v. var. diminuta (Simon)
             Fern. & Wieg.
          SY=S. v. var. neogaea Fern.

Coleus Lour.

    amboinicus Lour.
    blumei Benth.
    pumilus Blanco

Collinsonia L.

    canadensis L.
    serotina Walt.
          SY=C. canadensis L. var. punctata
             (Ell.) Gray
          SY=C. p. Ell.
    tuberosa Michx.
    verticillata Baldw. ex Ell.
          SY=Micheliella v. (Baldw. ex Ell.)
             Briq.

Conradina Gray

    brevifolia Shinners
    canescens (Torr. & Gray) Gray
          SY=C. puberula Small
    glabra Shinners
    grandiflora Small
    verticillata Jennison
          SY=C. montana Small

Cunila L.

    origanoides (L.) Britt.
          SY=Mappia o. (L.) House

Dicerandra Benth.

    densiflora Benth.
    frutescens Shinners
    immaculata Lakela
    linearifolia (Ell.) Benth.
    odoratissima Harper

Dracocephalum L.

    moldavica L.
          SY=Moldavica m. (L.) Britt.
    parviflorum Nutt.
          SY=Moldavica p. (Nutt.) Britt.
    thymiflorum L.
          SY=Moldavica t. (L.) Rydb.

Elsholtzia Willd.

    ciliata (Thunb.) Hyl.

Elsholtzia (CONT.)
            SY=E. cristata Willd.

Galeopsis L.

    angustifolia Ehrh. ex Hoffmann
            SY=G. ladanum L. var. a. (Ehrh. ex
                Hoffmann) Wallr.
    bifida Boenn.
            SY=G. tetrahit L. var. b. (Boenn.)
                Lej. & Court.
    ladanum L.
            var. ladanum
            var. latifolia (Hoffmann) Wallr.
    pubescens Bess.
    speciosa P. Mill.
    tetrahit L.
            var. arvensis Schlecht.
            var. tetrahit

Glechoma L.

    hederacea L.
            var. hederacea
                SY=Nepeta h. (L.) Trevisan
            var. micrantha Moric.
                SY=G. h. var. parviflora (Benth.)
                    House

Haplostachys (Gray) Hbd.

    bryanii Sherff
            var. bryanii
            var. microdonta Sherff
            var. robusta Sherff
    haplostachya (Gray) St. John
            var. angustifolia (Sherff) St. John
            var. haplostachya
            var. leptostachya (Hbd.) St. John
    linearifolia (Drake) Sherff
            var. linearifolia
            var. rosmarinifolia (Hbd.) Sherff
    munroi Forbes
    truncata (Gray) Hbd.

Hedeoma Pers.

    acinoides Scheele
    apiculatum W.S. Stewart
    costatum Gray
            var. costatum
            var. pulchellum (Greene) Irving
                SY=H. p. Greene
    dentatum Torr.
    diffusum Greene
    drummondii Benth.
            SY=H. longiflora Rydb.
    hispidum Pursh
    hyssopifolium Gray
    molle Torr.
            SY=Poliomintha m. (Torr.) Gray
    nanum (Torr.) Briq.
        ssp. californicum W.S. Stewart
        ssp. macrocalyx W.S. Stewart
        ssp. nanum
            SY=H. n. ssp. typicum W.S. Stewart
    oblongifolium (Gray) Heller
    pilosum Irving
    plicatum Torr.
    pulcherrimum Woot. & Standl.
    pulegioides (L.) Pers.
    reverchonii (Gray) Gray
            var. reverchonii
                SY=H. drummondii Benth. var. r.
                    Gray
                SY=H. latum Small
            var. serpyllifolium (Small) Irving

            SY=H. s. Small
    todsenii Irving

Hyptis Jacq.

    alata (Raf.) Shinners
            var. alata
                SY=H. radiata Willd.
            var. stenophylla Shinners
    americana Briq.
            SY=H. scoparia Poit.
    atrorubens Poit.
    capitata Jacq.
    emoryi Torr.
    escobilla Urban
    lantanifolia Poit.
    mutabilis (L.C. Rich.) Briq.
            SY=H. m. var. spicata (Poit.)
                Briq.
    pectinata (L.) Poit.
    spicigera Lam.
    suaveolens (L.) Poit.
    verticillata Jacq.

Hyssopus L.

    officinalis L.

Lamiastrum Heister ex Fabr.

    galeobdolon (L.) Ehrend. & Polatschek
            SY=Lamium g. (L.) L.

Lamium L.

    album L.
    amplexicaule L.
            var. album Pickens & Pickens
            var. amplexicaule
    hybridum Vill.
    maculatum L.
    mollucellifolium Fries
    purpureum L.

Lavandula L.

    angustifolia P. Mill.
            SY=L. spica L.

Leonotis (Pers.) R. Br.

    leonurus (L.) Ait. f.
    nepetifolia (L.) Ait. f.

Leonurus L.

    cardiaca L.
        ssp. cardiaca
        ssp. villosus (Desf. ex Spreng.) Hyl.
            SY=L. c. var. v. (Desf. ex
                Spreng.) Benth.
    marrubiastrum L.
    sibiricus L.

Lepechinia Willd.

    calycina (Benth.) Epling
    cardiophylla Epling
    fragrans (Greene) Epling
    ganderi Epling
    hastata (Gray) Epling

Leucas R. Br.

    martinicensis (Jacq.) R. Br.

Lycopus L.

Lycopus (CONT.)
   americanus Muhl. ex Bart.
         SY=L. sinuatus Ell.
         SY=L. a. var. longii Benner
         SY=L. a. var. scabrifolius Fern.
   amplectens Raf.
         SY=L. a. var. pubens (Britt.)
           Fern.
         SY=L. p. Britt.
         SY=L. sessilifolius Gray
   angustifolius Ell.
         SY=L. rubellus Moench var. a.
           (Ell.) Ahles
   asper Greene
         SY=L. lucidus Turcz. ex Benth.
           ssp. americanus (Gray)
           Hulten
         SY=L. l. var. americanus Gray
   cokeri Ahles
   europaeus L.
         SY=L. e. ssp. mollis (Kern.)
           Rothm. ex Skalicky
         SY=L. e. var. m. (Kern.) Briq.
   laurentianus Rolland-Germain
         SY=L. americanus Muhl. ex
           Bart. var. l. (Rolland-
           Germain) Boivin

   rubellus Moench
         SY=L. r. var. arkansanus (Fresn.)
           Benner
         SY=L. r. var. lanceolatus Benner
         SY=L. velutinus Rydb.
   X sherardii Steele [uniflorus X virginicus]
   uniflorus Michx.
         SY=L. virginicus L. var.
           pauciflorus Benth.
   virginicus L.

Macbridea Ell. ex Nutt.

   alba Chapman
   caroliniana (Walt.) Blake
         SY=M. pulchra Ell.

Marrubium L.

   vulgare L.

Marsypianthes Mart. ex Benth.

   chamaedrys (Vahl) Kuntze

Meehania Britt.

   cordata (Nutt.) Britt.

Melissa L.

   officinalis L.

Mentha L.

   aquatica L.
   arvensis L.
     ssp. haplocalyx Briq.
         SY=M. a. ssp. borealis (Michx.)
           Taylor & MacBryde
         SY=M. a. var. glabrata (Benth.)
           Fern.
         SY=M. a. var. villosa (Benth.)
           S.R. Stewart
         SY=M. glabrior (Hook.) Rydb.
         SY=M. penardii (Briq.) Rydb.
   canadensis L.
         SY=M. arvensis sensu auctt. non L.
         SY=M. a. var. c. (L.) Fern.

         SY=M. a. var. lanata Piper
         SY=M. a. var. sativa Benth.
   X gentilis L. [arvensis X spicata]
         SY=M. cardiaca Gerarde ex Baker
         SY=M. g. nm. c. (Gerarde ex Baker)
           Boivin
   X ?muelleriana F.W. Schultz [arvensis X
     suaveolens]
   nemorosa Willd.
   X piperita L. [aquatica X spicata]
     nm. citrata (Ehrh.) Briq.
         SY=M. X c. Ehrh.
     nm. piperita
   pulegium L.
   X rotundifolia (L.) Huds. [longifolia X
     suaveolens]
   spicata L.
         SY=M. cordifolia sensu auctt.
         SY=M. crispa L.
         SY=M. longifolia sensu auctt. non
           (L.) Huds.
         SY=M. l. var. mollissima sensu
           auctt. non (Borkh.) Rouy
         SY=M. l. var. undulata sensu
           auctt. non (Willd.) Fiori &
           Paol.
         SY=M. viridis L.
   suaveolens Ehrh.
         SY=M. rotundifolia sensu auctt.
   X verticillata L. [aquatica X arvensis]
         SY=M. sativa L.
         SY=M. X v. var. peduncularis
           (Boreau) Rouy
   X villosa Huds. [spicata X suaveolens]
         SY=M. X v. nm. alopecuroides
           (Hull) Briq.

Micheliella Briq.

   ?anisata (Sims) Briq.

Micromeria Benth.

   brownei (Sw.) Benth.
     var. brownei
         SY=Satureja b. (Sw.) Briq.
     var. pilosiuscula Gray
         SY=M. p. (Gray) Small

Molucella L.

   laevis L.

Monarda L.

   bradburiana Beck
         SY=M. fistulosa Sims
         SY=M. rigida Raf.
         SY=M. villosa Martens
   citriodora Cerv. ex Lag.
     ssp. austromontana (Epling) Scora
         SY=M. a. Epling
     ssp. citriodora
       var. attenuata Scora
       var. citriodora
         SY=M. aristada Nutt.
         SY=M. dispersa Small
         SY=M. tenuiaristata (Gray) Small
       var. parva Scora
   clinopodia L.
         SY=M. allophylla Michx.
   clinopodioides Gray
   didyma L.
   dressleri Scora
   fistulosa L.
       var. brevis Fosberg & Artz
       var. fistulosa

Monarda (CONT.)
      var. longepetiolata Boivin
      var. maheuxii Boivin
      var. menthifolia (Graham) Fern.
        SY=M. m. Graham
      var. mollis (L.) Benth.
        SY=M. m. L.
        SY=M. scabra Beck
      var. rubra Gray
  lindheimeri Engelm. & Gray
  media Willd.
  X medioides Duncan [fistulosa var. mollis X
    media]
  pectinata Nutt.
  punctata L.
      var. arkansana (McClintock & Epling)
        Shinners
        SY=M. p. ssp. a. McClintock &
          Epling
      var. coryi (McClintock & Epling) Cory
        SY=M. p. ssp. c. McClintock &
          Epling
      var. fruticulosa (Epling) Scora
        SY=M. f. Epling
      var. immaculata (Pennell) Scora
        SY=M. p. ssp. i. Pennell
      var. intermedia (McClintock & Epling)
        Waterfall
        SY=M. p. ssp. i. McClintock &
          Epling
      var. lasiodonta Gray
        SY=M. l. (Gray) Small
      var. maritima Cory
        SY=M. m. (Cory) Correll
      var. occidentalis (Epling) Palmer &
        Steyermark
        SY=M. p. ssp. o. Epling
      var. punctata
        SY=M. p. var. leucantha Nash
      var. stanfieldii (Small) Cory
      var. villicaulis (Pennell) Shinners
        SY=M. p. ssp. v. Pennell
  russeliana Nutt. ex Sims
      SY=M. virgata Raf.
  stipitatoglandulosa Waterfall
  viridissima Correll

Monardella Benth.

  antonina Hardham
  arizonica Epling
  benitensis Hardham
  breweri Gray
  candicans Benth.
  cinerea Abrams
  crispa Elmer
  douglasii Benth.
      var. douglasii
      var. venosa (Torr.) Jepson
  exilis (Gray) Greene
  hypoleuca Gray
    ssp. hypoleuca
    ssp. lanata (Abrams) Munz
      SY=M. l. Abrams
  lanceolata Gray
      var. glandulifera I.M. Johnston
      var. lanceolata
      var. microcephala Gray
  leucocephala Gray
  linoides Gray
    ssp. linoides
    ssp. oblonga (Greene) Abrams
    ssp. stricta (Parish) Epling
    ssp. viminea (Greene) Abrams
  macrantha Gray
    ssp. hallii (Abrams) Abrams
      SY=M. m. var. h. Abrams

    ssp. macrantha
  nana Gray
    ssp. arida (Hall) Abrams
    ssp. leptosiphon (Torr.) Abrams
    ssp. nana
    ssp. tenuiflora (S. Wats.) Abrams
  odoratissima Benth.
    ssp. australis (Abrams) Epling
    ssp. discolor (Greene) Epling
      SY=M. o. var. d. (Greene) St. John
    ssp. glauca (Greene) Epling
      SY=M. o. var. g. (Greene) St. John
      SY=M. purpurea T.J. Howell
    ssp. odoratissima
      SY=M. o. ssp. euodoratissima
        Epling
    ssp. pallida (Heller) Epling
      SY=M. o. var. ovata (Greene)
        Jepson pro parte
    ssp. parvifolia (Greene) Epling
      SY=M. o. var. p. (Greene) Jepson
    ssp. pinetorum (Heller) Epling
  palmeri Gray
  pringlei Gray
  robisonii Epling
  stebbinsii Hardham
  subglabra (Hoover) Hardham
      SY=M. villosa Benth. var. s.
        Hoover
  undulata Benth.
      var. frutescens Hoover
      var. undulata
  villosa Benth.
    ssp. neglecta (Greene) Epling
    ssp. sheltonii (Torr.) Epling
    ssp. subserrata (Greene) Epling
    ssp. villosa
      var. franciscana (Elmer) Epling
      var. obispoensis Hoover
      var. villosa
  viridis Jepson
    ssp. saxicola (I.M. Johnston) Ewan
      SY=M. s. I.M. Johnston
    ssp. viridis

Mosla (Benth.) Maxim.

  dianthera (Buch.) Maxon

Nepeta L.

  cataria L.
  grandiflora Bieb.
  mussinii Spreng. ex Henckel

Ocimum L.

  basilicum L.
  micranthum Willd.
  sanctum L.

Origanum L.

  vulgare L.

Perilla L.

  frutescens (L.) Britt.
      var. crispa (Benth.) Deane
      var. frutescens

Phlomis L.

  fruticosa L.
  tuberosa L.

Phyllostegia Benth.

Phyllostegia (CONT.)
    bracteata Sherff
    brevidens Gray
            var. ambigua Gray
            var. brevidens
            var. degeneri Sherff
            var. expansa Sherff
            var. heterodoxa Sherff
            var. hirsutula (Hbd.) Sherff
            var. longipes (Hbd.) Sherff
            var. pauciflora Sherff
            var. pubescens Sherff
    degeneri Sherff
    electra Forbes
    floribunda Benth.
    forbesii (Sherff) St. John
            SY=P. floribunda Benth. var.
                forbesii Sherff
    glabra (Gaud.) Benth.
            var. glabra
            var. lanaiensis Sherff
            var. macraei (Benth.) Sherff
    grandiflora (Gaud.) Benth.
    helleri Sherff
            var. helleri
            var. imminuta Sherff
    hillebrandii Mann ex Hbd.
    hirsuta Benth.
            var. hirsuta
            var. laxior Deg. & Sherff
    hispida Hbd.
    knudsenii Hbd.
    lantanoides Sherff
    ledyardii St. John
    longimontis St. John
    macrophylla (Gaud.) Benth.
            var. macrophylla
            var. phytolaccoides Sherff
            var. remyi Sherff
            var. velutina Sherff
    mannii Sherff
    mollis Benth.
            var. fagerlindii Sherff
            var. glabrescens Deg. & Sherff
            var. hochreutineri Sherff
            var. lydgatei Sherff
            var. micrantha Sherff
            var. mollis
            var. skottsbergii Sherff
                SY=P. m. var. resinosa Fosberg
    parviflora (Gaud.) Benth.
            var. canescens Sherff
            var. glabriuscula Gray
            var. honolulensis (Wawra) Sherff
            var. major Sherff
            var. parviflora
    racemosa Benth.
            var. bryanii Sherff
            var. racemosa
    rockii Sherff
    stachyoides Gray
            var. hitchcockii Sherff
            var. stachyoides
    variabilis Bitter
    vestita Benth.
    villosa St. John
    waimeae Wawra
    wawrana Sherff
    yamaguchii Hosaka & Deg.

Physostegia Benth.

    angustifolia Fern.
            SY=P. edwardsiana Shinners
    correllii (Lundell) Shinners
            SY=Dracocephalum c. Lundell
    digitalis Small

godfreyi Cantino
intermedia (Nutt.) Engelm. & Gray
            SY=Dracocephalum i. Nutt.
            SY=P. micrantha Lundell
leptophylla Small
            SY=Dracocephalum l. (Small) Small
            SY=D. veroniciforme (Small) Small
            SY=P. aboriginorum Fern.
            SY=P. v. Small
parviflora Nutt. ex Gray
            SY=Dracocephalum nuttallii Britt.
            SY=P. n. (Britt.) Fassett
            SY=P. virginiana (L.) Benth. var.
                p. (Nutt. ex Gray) Boivin
pulchella Lundell
purpurea (Walt.) Blake
            SY=P. obovata (Ell.) Godfrey ex
                Weatherby
virginiana (L.) Benth.
            SY=Dracocephalum denticulatum Ait.
            SY=D. formosius (Lunell) Rydb.
            SY=D. purpureum (Walt.) McClintock
            SY=D. v. L.
            SY=P. d. (Ait.) Britt.
            SY=P. f. Lunell
            SY=P. praemorsa Shinners
            SY=P. serotina Shinners
            SY=P. speciosa (Sweet) Sweet
            SY=P. v. var. elongata Boivin
            SY=P. v. var. f. (Lunell) Boivin
            SY=P. v. var. granulosa (Fassett)
                Fern.
            SY=P. v. var. ledinghamii Boivin
            SY=P. v. var. speciosa (Sweet)
                Gray

Piloblephis Raf.

    rigida (Bart. ex Benth.) Raf.
            SY=Pycnothymus r. (Bart. ex
                Benth.) Small
            SY=Satureja r. Bart. ex Benth.

Plectranthus L'Her.

    parviflorus Willd.

Pogogyne Benth.

    abramsii J.T. Howell
    clareana J.T. Howell
    douglasii Benth.
        ssp. douglasii
        ssp. parviflora (Benth.) J.T. Howell
    nudiuscula Gray
    serpylloides (Torr.) Gray
    zizyphoroides Benth.

Poliomintha Gray

    glabrescens Gray
    incana (Torr.) Gray

Prunella L.

    laciniata L.
    vulgaris L.
            var. aleutica Fern.
                SY=P. v. ssp. a. (Fern.) Hulten
            var. atropurpurea Fern.
            var. elongata Benth.
                SY=P. v. ssp. lanceolata (Bart.)
                    Hulten
                SY=P. v. var. l. (Bart.) Fern.
            var. hispida Benth.
            var. parviflora (Poir.) DC.
            var. vulgaris

Prunella (CONT.)
              SY=P. v. var. calvescens Fern.
              SY=P. v. var. minor Sm.
              SY=P. v. var. nana Clute
              SY=P. v. var. rouleauiana Victorin

Pycnanthemum Michx.

   albescens Torr. & Gray
              SY=Koellia a. (Torr. & Gray)
                    Kuntze
              SY=K. pauciflora Small
   californicum Torr.
   clinopodioides Torr. & Gray
              SY=Koellia c. (Torr. & Gray)
                    Kuntze
   curvipes (Greene) Grant & Epling
              SY=Keollia c. Greene
              SY=K. multiflora Small
   flexuosum (Walt.) B.S.P.
              SY=Koellia hugeri Small
              SY=K. hyssopifolium (Benth.)
                    Britt.
              SY=P. hyssopifolium Benth.
   floridanum Grant & Epling
   incanum (L.) Michx.
          var. incanum
              SY=Koellia beadlei Small
              SY=K. dubia (Gray) Small
              SY=K. incana (L.) Kuntze
              SY=P. b. (Small) Fern.
          var. loomisii (Nutt.) Fern.
              SY=P. l. Nutt.
          var. puberulum (Grant & Epling) Fern.
              SY=P. p. Grant & Epling
   monotrichum Fern.
   montanum Michx.
              SY=Koellia m. (Michx.) Kuntze
   muticum (Michx.) Pers.
              SY=Koellia m. (Michx.) Kuntze
   nudum Nutt.
              SY=Koellia n. (Nutt.) Kuntze
   pilosum Nutt.
              SY=Koellia p. (Nutt.) Britt.
   pycnanthemoides (Leavenworth) Fern.
          var. pycnanthemoides
              SY=Koellia p. (Leavenworth) Kuntze
          var. viridifolium Fern.
              SY=P. tullia Benth.
              SY=P. v. (Fern.) Grant & Epling
   setosum Nutt.
              SY=Koellia aristata (Michx.)
                    Kuntze
              SY=P. a. Michx.
              SY=P. umbratile Fern.
   tenuifolium Schrad.
              SY=Koellia flexuosa sensu auctt.
                    non (Walt.) MacM.
              SY=P. f. sensu auctt. non (Walt.)
                    B.S.P.
   torrei Benth.
   verticillatum (Michx.) Pers.
              SY=Koellia leptodon (Gray) Small
              SY=K. v. (Michx.) Kuntze
              SY=P. l. Gray
              SY=P. torrei Benth. var. l. (Gray)
                    Boomhour
              SY=P. virginianum (L.) Durand &
                    Jackson var. verticillatum
                    (Michx.) Boivin
   virginianum (L.) Durand & Jackson
              SY=Koellia v. (L.) MacM.

Rhododon Epling

   ciliatus (Benth.) Epling
              SY=Hedeoma angulata Tharp

              SY=H. texana Cory

Rosmarinus L.

   officinalis L.

Salazaria Torr.

   mexicana Torr.

Salvia L.

   aethiopis L.
   amissa Epling
   apiana Jepson
          var. apiana
          var. compacta Munz
   arizonica Gray
   azurea Michx. & Lam.
          var. azurea
          var. grandiflora Benth.
              SY=S. a. ssp. intermedia Epling
              SY=S. a. ssp. pitcheri (Torr.)
                    Epling
              SY=S. p. Torr.
   ballotiflora Benth.
   X bernardina Parish ex Greene [columbariae
       X mellifera]
   blodgettii Chapman
   brandegei Munz
              SY=S. mellifera Greene ssp.
                    revoluta (Brandeg.) Abrams
   carduacea Benth.
   ?chapmanii Gray
   clevelandii (Gray) Greene
   coccinea Juss. ex J. Murr
   columbariae Benth.
          var. columbariae
          var. ziegleri Munz
   davidsonii Greenm.
   dolichantha (Cory) Whitehouse
   dorrii (Kellogg) Abrams
          ssp. argentea (Rydb.) Munz
              SY=S. carnosa Dougl. ex Benth.
                    ssp. a. (Rydb.) Epling
              SY=S. c. var. a. (Rydb.) McMinn
          ssp. carnosa (Dougl. ex Benth.) Abrams
              SY=S. c. Dougl. ex Benth.
              SY=S. d. var. c. (Dougl. ex
                    Benth.) Cronq.
          ssp. dorrii
              SY=S. carnosa Dougl. ex Hall ssp.
                    pilosa (Gray) Epling
              SY=?S. d. var. gracilior M.E. Peck
          ssp. gilmanii (Epling) Abrams
          ssp. mearnsii (Britt.) McClintock
              SY=S. carnosa Dougl. ex Hall ssp.
                    m. (Britt.) Epling
   earlei Woot. & Standl.
   engelmannii Gray
   eremostachya Jepson
   farinacea Benth.
          var. farinacea
          var. latifolia Shinners
   funerea M.E. Jones
   glutinosa L.
   greatae Brandeg.
   greggii Gray
   henryi Gray
   hispanica L.
   lemmonii Gray
   leptophylla Benth.
   leucophylla Greene
   longistyla Benth.
   lycioides Gray
   lyrata L.
   mellifera Greene

Salvia (CONT.)
  micrantha Vahl
  microphylla Benth.
          SY=S. grahamii Benth.
  misella Kunth
  mohavensis Greene
  munzii Epling
          SY=S. mellifera Greene ssp.
              jonesii (Munz) Abrams
  nemorosa L.
          SY=S. sylvestris L.
  occidentalis Sw.
  officinalis L.
  pachyphylla Epling ex Munz
  X palmeri (Gray) Greene [apiana X
      clevelandii]
  parryi Gray
  penstemonoides Kunth & Bouche
  pinguifolia (Fern.) Woot. & Standl.
  pratensis L.
  privoides Benth.
  ramosissima Fern.
  reflexa Hornem.
          SY=S. lancifolia Poir.
  regla Cav.
  roemeriana Scheele
  sclarea L.
  serotina L.
  sonomensis Greene
  spathacea Greene
  splendens Ker-Gawl.
  subincisa Benth.
  summa A. Nels.
  texana (Scheele) Torr.
  thomasiana Urban
  tiliifolia Vahl
  urticifolia L.
  vaseyi (Porter) Parish
  verbenacea L.
  verticillata L.
  vinacea Woot. & Standl.

Satureja L.

  chamissonis (Benth.) Briq.
  douglasii (Benth.) Briq.
  hortensis L.
  mimuloides (Benth.) Briq.

Scutellaria L.

  alabamensis Alexander
  altamaha Small
  altissima L.
  angustifolia Pursh
  antirrhinoides Benth.
  arenicola Small
  arguta Buckl.
          SY=S. saxatilis Riddell var.
              pilosior Benth.
  australis (Fassett) Epling
          SY=S. parvula Michx. var. a.
              Fassett
  austinae Eastw.
  bolanderi Gray
    ssp. austromontana Epling
    ssp. bolanderi
  brittonii Porter
  bushii Britt.
  californica Gray
  cardiophylla Engelm. & Gray
  X churchilliana Fern. [galericulata X
      lateriflora]
  drummondii Benth.
  elliptica Muhl.
      var. elliptica
          SY=S. ovalifolia Pers.

          SY=S. o. ssp. mollis Epling
          SY=S. pilosa Michx.
      var. hirsuta (Short & Peter) Fern.
          SY=S. ovalifolia Pers. ssp. h.
              (Short & Peter) Epling
  floridana Chapman
  galericulata L.
          SY=S. epilobiifolia A. Hamilton
          SY=S. g. var. e. (A. Hamilton)
              Jordal
          SY=S. g. var. pubescens Benth.
  glabriuscula Fern.
  havanensis Jacq.
  incana Biehler
      var. australis (Epling) Collins ined.
          SY=S. altamaha Small ssp. a.
              Epling
      var. incana
      var. punctata (Chapman) C. Mohr
          SY=S. p. (Chapman) Leonard
  integrifolia L.
          SY=S. incana Biehler. ssp. hispida
              (Benth.) Epling
          SY=S. integrifolia var. h. Benth.
  laevis Shinners
  lateriflora L.
      var. grohii Boivin
      var. lateriflora
  leonardii Epling
          SY=S. parvula Michx. var. l.
              (Epling) Fern.
  mellichampii Small
  microphylla Moc. & Sesse ex Benth.
  montana Chapman
          SY=S. serrata Andr. var. m.
              (Chapman) Penl.
  multiglandulosa (Kearney) Small ex Harper
          SY=S. integrifolia L. var. m.
              Kearney
  muriculata Epling
  nana Gray
      var. nana
      var. sappharina Barneby
  nervosa Pursh
          SY=S. n. var. calvifolia Fern.
  ocmulgee Small
  ovata Hill
    ssp. bracteata (Benth.) Epling
          SY=S. o. var. b. Benth.
    ssp. cuthbertii (Alexander) Epling
          SY=S. c. Alexander
    ssp. mexicana Epling
    ssp. ovata
          SY=S. cordifolia Muhl.
          SY=S. o. ssp. mississippiensis
              (Mart.) Epling
          SY=S. o. var. calcarea (Epling)
              Gleason
          SY=S. o. var. versicolor (Nutt.)
              Fern.
    ssp. pseudoarguta Epling
          SY=S. o. var. p. (Epling) Core
    ssp. rugosa (Wood) Epling
          SY=S. o. var. r. (Wood) Fern.
    ssp. rupestris Epling
    ssp. venosa Epling
    ssp. virginiana Epling
          SY=S. o. var. v. (Epling) Core
  parvula Michx.
          SY=S. ambigua Nutt.
          SY=S. nervosa Pursh var. ambigua
              (Nutt.) Fern.
  potosina Brandeg.
    ssp. parviflora Epling
    ssp. platyphylla Epling
  pseudoserrata Epling
  resinosa Torr.

Scutellaria (CONT.)
   saxatilis Riddell
   serrata Andr.
   siphocamyploides Vatke
          SY=S. angustifolia Pursh var.
            canescens Gray
   tesselata Epling ex Kearney & Peebles
   thieretii Shinners
   tuberosa Benth.
      ssp. similis (Jepson) Epling
          SY=S. t. var. s. Jepson
      ssp. tuberosa
          SY=S. t. ssp. australis (Jepson)
            Epling
   wrightii Gray
          SY=S. brevifolia Gray
          SY=S. resinosa Torr. var. b.
            (Gray) Penl.

Sideritis L.

   lanata L.
   montana L.
   romana L.

Stachydeoma Small

   graveolens (Chapman) Small
          SY=Hedeoma g. Chapman

Stachys L.

   ajugoides Benth.
   albens Gray
   annua (L.) L.
   arvensis (L.) L.
   aspera Michx.
          SY=S. ambigua (Gray) Britt.
          SY=S. hyssopifolia Michx. var.
            ambigua Gray
   bigelovii Gray
   bullata Benth.
   byzantina K. Koch
          SY=S. olympica Poir.
   chamissonis Benth.
   clingmanii Small
   coccinea Jacq.
   cooleyae Heller
   crenata Raf.
          SY=S. agraria sensu auctt. non
            Cham. & Schlecht.
   cretica L.
          SY=S. germanica L. var. italica
            (P. Mill.) Briq.
          SY=S. i. P. Mill.
   drummondii Benth.
   eplingii J. Nelson
   floridana Shuttlw.
          SY=S. sieboldii Miq.
   germanica L.
   hyssopifolia Michx.
   latidens Small
   lythroides Small
   mexicana Benth.
          SY=S. ciliata Dougl.
          SY=S. emersonii Piper
   nuttallii Shuttlw. ex Benth.
          SY=S. cordata Riddell
          SY=S. riddellii House
   officinalis (L.) Trev.
          SY=Betonica o. L.
   palustris L.
      var. elliptica Clos
      var. homotricha Fern.
          SY=S. h. (Fern.) Rydb.
          SY=S. p. var. nipigonensis
            Jennings

      var. macrocalyx Jennings
      var. palustris
          SY=S. arenicola Britt.
      var. petiolata Clos
      var. phaneropoda Weatherby
      var. pilosa (Nutt.) Fern.
          SY=S. palustris ssp. pilosa
            (Nutt.) Epling
          SY=S. pilosa Nutt.
          SY=S. scopulorum Greene
      var. segetum (Mutel) Grogn.
   pycnantha Benth.
   recta L.
   rigida Nutt. ex Benth.
      ssp. lanata Epling
      ssp. quercetorum (Heller) Epling
      ssp. rigida
      ssp. rivularis (Heller) Epling
   rothrockii Gray
   stricta Greene
   subcordata Rydb.
   sylvatica L.
   tenuifolia Willd.
      var. hispida (Pursh) Fern.
          SY=S. h. Pursh
          SY=S. palustris L. var. h. (Pursh)
            Boivin
          SY=S. salvioides Small
          SY=S. t. var. platyphylla Fern.
      var. perlonga Fern.
      var. tenuifolia
   teucriformis Rydb.

Stenogyne Benth.

   affinis Forbes
      var. affinis
      var. degeneri Sherff
      var. retrorsa Deg. & Sherff
   angustifolia Gray
      var. angustifolia
      var. hillebrandii Sherff
      var. mauiensis Sherff
      var. meeboldii Sherff
      var. spathulata Sherff
   bifida Hbd.
   biflora (Sherff) St. John
          SY=S. scrophularioides Benth. var.
            b. Sherff
   calaminthoides Gray
      var. calaminthoides
      var. oxyodonata Sherff
      var. subrotunda Sherff
      var. waimeana Sherff
   calycosa Sherff
   cinerea Hbd.
   cranwelliae Sherff
   crenata Gray
      var. crenata
      var. muricata Deg. & Sherff
   diffusa Gray
      var. diffusa
      var. glabra Wawra
   glabrata (Hbd.) Sherff
   haliakalae Wawra
   hirsutula St. John
   kaalae Wawra
      var. kaalae
      var. latisepala Sherff
   kamehamehae Wawra
      var. albiflora Sherff
      var. kamehamehae
   kanehoana Deg. & Sherff
   kealiae Wawra
      var. angustata Deg. & Sherff
      var. kealiae
   macrantha Benth.

Stenogyne (CONT.)
         var. amicarum Deg. & Sherff
         var. gracilis Sherff
         var. grayi Mann
         var. latifolia Hbd.
         var. macrantha
microphylla Benth.
mollis (Sherff) St. John
oxygona Deg. & Sherff
purpurea Mann
         var. brevipedunculata Wawra
         var. forbesii Sherff
         var. leptophylla Sherff
         var. purpurea
rotundifolia Gray
         var. oblonga Sherff
         var. rotundifolia
rugosa Benth.
         var. rugosa
         var. subulata Sherff
            SY=S. kaalae Wawra var. coriacea
                Deg. & Sherff
?salicifolia author ?
            SY=S. angustifolia Gray var. s.
                Sherff
scandens Sherff
scrophularioides Benth.
         var. nelsonii (Benth.) Sherff
         var. remyi Sherff
         var. scrophularioides
         var. skottsbergii Sherff
serpens Hbd.
sessilis Benth.
         var. hexantha Sherff
         var. hexanthoides Deg. & Sherff
         var. lanaiensis Sherff
         var. sessilis
         var. wilkesii Sherff
sherffii Deg.
sororia Sherff
vagans Hbd.
viridis Hbd.

Synandra Nutt.

   hispidula (Michx.) Baill.

Teucrium L.

   botrys L.
   canadense L.
         var. boreale (Bickn.) Shinners
            SY=T. c. ssp. viscidum (Piper)
                Taylor & MacBryde pro parte
            SY=T. c. var. occidentale (Gray)
                McClintock & Epling
            SY=T. o. Gray
            SY=T. o. var. b. (Bickn.) Fern.
         var. canadense
            SY=T. c. var. angustatum Gray
            SY=T. c. var. littorale (Bickn.)
                Fern.
            SY=T. l. Bickn.
         var. hypoleucum Griseb.
            SY=T. c. var. nashii (Kearney)
                Shinners
            SY=T. n. Kearney
         var. virginicum (L.) Eat.
   cubense Jacq.
      ssp. cubense
      ssp. depressum (Small) McClintock &
         Epling
            SY=T. c. var. densum Jepson
            SY=T. depressum Small
      ssp. laevigatum (Vahl) McClintock &
         Epling
            SY=T. c. var. l. (Vahl) Shinners

glandulosum Kellogg
laciniatum Torr.
scorodonia L.

Thymus L.

   praecox Opiz
      ssp. arcticus (Dur.) Jalas
            SY=T. serpyllum sensu auctt. non
                L.
   pulegioides L.
            SY=T. serpyllum L. ssp. chamaedrys
                (Fries) Voll.
            SY=T. s. var. albus Hort.
   vulgaris L.

Trichostema L.

   arizonicum Gray
   austromontanum F.H. Lewis
      ssp. austromontanum
      ssp. compactum F.H. Lewis
   brachiatum L.
            SY=Isanthus b. (L.) B.S.P.
            SY=I. b. var. linearis Fassett
   dichotomum L.
            SY=T. d. var. puberulum Fern. &
                Grisc.
   lanatum Benth.
   lanceolatum Benth.
   laxum Gray
   micranthum Gray
   oblongum Benth.
   ovatum Curran
   parishii Vasey
   rubisepalum Elmer
   setaceum Houtt.
            SY=T. dichotomum L. var. lineare
                (Walt.) Gleason
            SY=T. l. Walt.
   simulatum Jepson
   suffrutescens Kearney

LARDIZABALACEAE

Akebia Dcne.

   quinata (Houtt.) Dcne.

LAURACEAE

Aniba Aubl.

   bracteata (Nees) Mez

Beilschmiedea Nees

   pendula (Sw.) Hemsl.
            SY=Hufelandia p. (Sw.) Nees

Cassytha L.

   filiformis L.

Cinnamomum Schaeffer

   camphora (L.) J. Presl
            SY=Camphora c. (L.) Karst.
   elongatum (Vahl) Kosterm.
            SY=Phoebe e. (Vahl) Nees
   mexicanum (Meisn.) Kosterm.

Cinnamomum   (CONT.)
                SY=Phoebe m. Meisn.
    montanum (Sw.) Bercht. & Presl
                SY=Phoebe m. (Sw.) Griseb.

Cryptocarya R. Br.

    mannii Hbd.
    oahuensis (Deg.) Fosberg

Licaria Aubl.

    brittoniana Allen & Gregory
    salicifolia (Sw.) Kosterm.
                SY=Acrodiclidium s. (Sw.) Griseb.
    triandra (Sw.) Kosterm.
                SY=Misanteca t. (Sw.) Mez

Lindera (Adans.) Thunb.

    benzoin (L.) Blume
        var. benzoin
                SY=Benzoin aestivale (L.) Nees
        var. pubescens (Palmer & Steyermark)
                Rehd.
    melissifolium (Walt.) Blume
                SY=Benzoin m. (Walt.) Nees

Litsea Lam.

    aestivalis (L.) Fern.
                SY=Glabrabia geniculata (Walt.)
                Britt.

Nectandra Roland. ex Rottb.

    antillana Meisn.
    coriacea (Sw.) Griseb.
    krugii Mez
    membranacea (Sw.) Griseb.
    patens (Sw.) Griseb.
    sintenisii Mez

Ocotea Aubl.

    cuneata (Griseb.) Urban
    floribunda (Sw.) Mez
    foeniculacea Mez
    leucoxylon (Sw.) Gomez
    moschata (Meisn.) Mez
    portoricensis Mez
    spathulalta Mez
    wrightii (Meisn.) Mez

Persea P. Mill.

    americana P. Mill.
                SY=P. persea (L.) Cockerell
    borbonia (L.) Spreng.
        var. borbonia
                SY=P. littoralis Small
                SY=Tamala b. (L.) Raf.
                SY=T. l. (Small) Small
        var. humilis (Nash) Kopp
                SY=P. h. Nash
                SY=Tamala h. (Nash) Small
        var. pubescens (Pursh) Little
                SY=P. palustris (Raf.) Sarg.
                SY=P. pubescens (Pursh) Sarg.
                SY=Tamala pubescens (Pursh) Small
    krugii Mez
    urbaniana Mez
                SY=P. portoricensis Britt. &
                Wilson

Sassafras Trew

albidum (Nutt.) Nees
                SY=S. a. var. molle (Raf.) Fern.
                SY=S. sassafras (L.) Karst.

Umbellularia (Nees) Nutt.

    californica (Hook. & Arn.) Nutt.
        var. californica
        var. fresnensis Eastw.

LEITNERIACEAE

Leitneria Chapman

    floridana Chapman

LEMNACEAE

Lemna L.

    gibba L.
    minor L.
                SY=L. cyclostasa (Ell.) Chev.
    minuta H.B.K.
                SY=L. aequinoctialis Welw.
                SY=L. minima Phil.
                SY=L. minuscula Herter
                SY=L. valdiviana Phil. var. minima
                (Phil.) Hegelm.
    obscura (Austin) Daubs
                SY=L. minor L. var. o. Austin
    perpusilla Torr.
    trinervis (Austin) Small
    trisulca L.
    valdiviana Phil.
                SY=L. v. var. abbreviata Hegelm.

Spirodela Schleid.

    polyrhiza (L.) Schleid.
    punctata (Mey.) C.H. Thompson
                SY=S. oligorrhiza (Kurz) Hegelm.

Wolffia Horkel ex Schleid.

    arrhiza (L.) Horkel ex Wimmer
                SY=W. cylindracea Hegelm.
    braziliensis Weddell
                SY=W. papulifera C.H. Thompson
    columbiana Karst.
                SY=Bruniera c. (Karst.) Nieuwl.
    punctata Griseb.
                SY=Bruniera p. (Griseb.) Nieuwl.

Wolffiella Hegelm.

    gladiata (Hegelm.) Hegelm.
                SY=Wolffia floridana J.D. Sm. ex
                Hegelm.
                SY=Wolffiella f. (J.D. Sm. ex
                Hegelm.) C.H. Thompson
    lingulata (Hegelm.) Hegelm.
    oblonga (Phil.) Hegelm.

LENNOACEAE

Ammobroma Torr. ex Gray

Ammobroma (CONT.)
  sonorae Torr. ex Gray

Pholisma Nutt. ex Hook.

  arenarium Nutt. ex Hook.

LENTIBULARIACEAE

Pinguicula L.

  caerulea Walt.
  ionantha Godfrey
  lutea Walt.
  macroceras Link
        var. macroceras
            SY=P. vulgaris L. ssp. m. (Link)
                Calder & Taylor
            SY=P. v. var. m. (Link) Herder
        var. microceras (Cham.) Caspar
  planifolia Chapman
  primuliflora Wood & Godfrey
  pumila Michx.
  villosa L.
  vulgaris L.

Utricularia L.

  amethystina Salzm. ex St. Hil. & Girard
            SY=Calpidisca standleyae Barnh.
  biflora Lam.
            SY=U. pumila Walt.
  cornuta Michx.
            SY=Stomoisia c. (Michx.) Raf.
  fibrosa Walt.
  floridana Nash
  foliosa L.
  geminiscapa Benj.
  gibba L.
            SY=U. obtusa Sw.
  inflata Walt.
  intermedia Hayne
  juncea Vahl
            SY=Stomoisia j. (Vahl) Barnh.
            SY=S. virgatula (Barnh.) Barnh.
            SY=U. v. Barnh.
  macrorhiza Le Conte
            SY=V. vulgaris L. ssp. m. (Le
                Conte) Clausen
  minor L.
            SY=U. occidentalis Gray
  ochroleuca R. Hartman
  olivacea Wright ex Griseb.
            SY=Biovularia o. (Wright ex
                Griseb.) Kam.
  purpurea Walt.
            SY=Vesiculina p. (Walt.) Raf.
  pusilla Vahl
            SY=Setiscapella p. (Vahl) Barnh.
  radiata Small
            SY=U. inflata Walt. var. minor
                Chapman
  resupinata B.D. Greene
            SY=Lenticula r. (B.D. Greene)
                Barnh.
  simulata Pilger
            SY=Aranella fimbriata (H.B.K.)
                Barnh.
            SY=U. f. H.B.K.
  subulata L.
            SY=Setiscapella cleistogama (Gray)
                Barnh.
            SY=S. s. (L.) Barnh.
            SY=U. c. (Gray) Britt.

  vulgaris L.
            SY=U. v. var. americana Gray

LILIACEAE

Aletris L.

  aurea Walt.
  bracteata Northrop
  farinosa L.
  lutea Small
  obovata Nash
  X tottenii F.T. Br. [lutea X obovata]

Allium L.

  aaseae Ownbey
  acuminatum Hook.
            SY=A. a. var. cuspidatum Fern.
  ampeloprasum L.
        var. ampeloprasum
        var. atroviolaceum (Boiss.) Regel
  amplectens Torr.
  anceps Kellogg
  atrorubens S. Wats.
        var. atrorubens
        var. inyonis (M.E. Jones) Ownbey &
            Aase
            SY=A. a. ssp. i. (M.E. Jones)
                Traub
            SY=A. i. M.E. Jones
  bigelovii S. Wats.
  bisceptrum S. Wats.
        var. bisceptrum
        var. palmeri (S. Wats.) Cronq.
            SY=A. p. S. Wats.
  bolanderi S. Wats.
        var. bolanderi
        var. stenanthum (E. Drew) Jepson
  brandegei S. Wats.
            SY=A. diehlii M.E. Jones
  brevistylum S. Wats.
  burdickii (Hanes) A.G. Jones
            SY=A. tricoccum Ait. var. b. Hanes
  burlewii A. Davids.
  campanulatum S. Wats.
            SY=A. austinae M.E. Jones
  canadense L.
        var. canadense
            SY=A. acetabulum Raf.
        var. ecristatum (M.E. Jones) Ownbey
            SY=A. c. ssp. e. (M.E. Jones)
                Traub & Ownbey
            SY=A. reticulatum G. Don var. e.
                M.E. Jones
        var. fraseri Ownbey
            SY=A. acetabulum Raf. var. f.
                (Ownbey) Shinners
            SY=A. c. ssp. f. (Ownbey) Traub &
                Ownbey
        var. hyacinthoides (Bush) Ownbey
            SY=A. c. ssp. h. (Bush) Traub &
                Ownbey
        var. lavandulare (Bates) Ownbey
            SY=A. c. ssp. l. (Bates) Traub &
                Ownbey
        var. mobilense (Regel) Ownbey
            SY=A. arenicola Small
            SY=A. c. ssp. m. (Regel) Traub &
                Ownbey
            SY=A. microscordion Small
            SY=A. mobilense Regel
            SY=A. mutabile Michx.
  carinatum L.

Allium (CONT.)
  cepa L.
        var. cepa
        var. solaninum Alef.
        var. viviparum M.C. Metz
            SY=A. c. var. bulbifera Bailey
  cernuum Roth
        var. cernuum
            SY=A. allegheniense Small
            SY=A. oxyphilum Wherry
        var. neomexicanum (Rydb.) J.F. Macbr.
        var. obtusum Cockerell ex J.F. Macbr.
  coryi M.E. Jones
  cratericola Eastw.
  crenulatum Wieg.
            SY=A. cascadense M.E. Peck
            SY=A. watsonii T.J. Howell
  crispum Greene
            SY=A. peninsulare Lemmon ex Greene
                var. c. (Greene) Jepson
  cuthbertii Small
  davisiae M.E. Jones
  dichlamydeum Greene
  dictuon St. John
  douglasii Hook.
        var. columbianum Ownbey & Mingrone
        var. constrictum Mingrone & Ownbey
        var. douglasii
        var. nevii (S. Wats.) Ownbey &
            Mingrone
            SY=A. n. S. Wats.
  drummondii Regel
            SY=A. nuttallii S. Wats.
  elmendorfii Ownbey
  falcifolium Hook. & Arn.
            SY=A. breweri S. Wats.
  fibrillum M.E. Jones
  fimbriatum S. Wats.
        var. abramsii Ownbey & Aase
        var. denticulatum Ownbey & Aase
        var. diabolense Ownbey & Aase
        var. fimbriatum
        var. mohavense (Tidestrom) Jepson
            SY=A. f. ssp. m. (Tidestrom) Traub
                & Ownbey
            SY=A. m. Tidestrom
        var. munzii Ownbey & Aase
        var. parryi (S. Wats.) Ownbey & Aase
            SY=A. f. ssp. p. (S. Wats.) Traub
                & Ownbey
            SY=A. p. S. Wats.
        var. purdyi (Eastw.) Ownbey & Aase
            SY=A. f. ssp. p. (Eastw.) Traub &
                Ownbey
        var. sharsmithae Ownbey & Aase
  fistulosum L.
  geyeri S. Wats.
        var. geyeri
        var. tenerum M.E. Jones
            SY=A. g. ssp. t. (M.E. Jones)
                Traub & Ownbey
            SY=A. rubrum Osterhout
  glandulosum Link & Otto
            SY=A. rhizomatum Woot. & Standl.
  gooddingii Ownbey
  haematochiton S. Wats.
  hickmanii Eastw.
  hoffmanii Ownbey
  howellii Eastw.
        var. clokeyi Ownbey & Aase
        var. howellii
        var. sanbenitense (Traub) Ownbey &
            Aase
            SY=A. h. ssp. s. (Traub) Traub &
                Ownbey
  hyalinum Curran
  jacintense (Munz) Mingrone & Ownbey

  kunthii G. Don
  lacunosum S. Wats.
        var. lacunosum
        var. micranthum Eastw.
  lemmonii S. Wats.
  macropetalum Rydb.
  macrum S. Wats.
  madidum S. Wats.
  membranaceum Ownbey
  monticola A. Davids.
        var. keckii (Munz) Ownbey & Aase
        var. monticola
  neapolitanum Cyrillo
  nevadense S. Wats.
        var. cristatum (S. Wats.) Ownbey
            SY=A. c. S. Wats.
        var. nevadense
  nigrum L.
  obtusum Lemmon
  oleraceum L.
  parishii S. Wats.
  parvum Kellogg
  passeyi N. Holmgren & A. Holmgren
  peninsulare Lemmon ex Greene
  perdulce S.V. Fraser
        var. perdulce
        var. sperryi Ownbey
  platycaule S. Wats.
  pleianthum S. Wats.
  plummerae S. Wats.
  porrum L.
  praecox Brandeg.
  punctum Henderson
  robinsonii Henderson
  runyonii Ownbey
  sanbornii Wood
        var. congdonii Jepson
            SY=A. intactum Jepson
            SY=A. s. ssp. i. (Jepson) Traub
        var. jepsonii Ownbey & Aase
        var. sanbornii
        var. tuolumnense Ownbey & Aase
  sativum L.
  schoenoprasum L.
            SY=A. s. var. laurentianum Fern.
            SY=A. s. var. sibiricum (L.)
                Hartman
            SY=A. sibiricum L.
  scillioides Dougl. ex S. Wats.
  serra McNeal & Ownbey
  serratum S. Wats.
  simillimum Henderson
  siskiyouense Ownbey
  speculae Ownbey & Aase
  stellatum Ker-Gawl.
  textile A. Nels. & J.F. Macbr.
            SY=A. geyeri S. Wats. var. t. (A.
                Nels. & J.F. Macbr.) Boivin
            SY=A. reticulatum G. Don
  tolmiei Baker ex S. Wats.
        var. persimile Ownbey
        var. platyphyllum (Tidestrom) Ownbey
        var. tolmiei
            SY=A. cusickii S. Wats.
  tribracteatum Torr.
  tricoccum Ait.
            SY=Validallium t. (Ait.) Small
  triquetrum L.
  unifolium Kellogg
  validum S. Wats.
  victorialis L.
            SY=A. v. ssp. platyphyllum Hulten
  vineale L.
    ssp. kochii (Lange) Richter
    ssp. vineale
  yosemitense Eastw.

Aloe L.

    barbadensis P. Mill.
          SY=A. perfoliata L. var. vera L.
          SY=A. v. (L.) Webb
          SY=A. vulgaris Lam.

Alstroemeria L.

    pulchella L. f.

Amianthium Gray

    muscaetoxicum (Walt.) Gray
          SY=Chrosperma m. (Walt.) Kuntze
          SY=Zigadenus m. (Walt.) Regel

Androstephium Torr.

    breviflorum S. Wats.
    coeruleum (Scheele) Torr.

Anthericum L.

    chandleri Greenm. & Thompson
    torreyi Baker

Asparagus L.

    asparagoides (L.) W. Wight
    densiflorus (Kunth) Jessop
          SY=A. sprengeri Regel
    officinalis L.
    setaceus (Kunth) Jessop
          SY=A. plumosus Baker

Asphodelus L.

    fistulosus L.

Astelia Banks & Soland. ex R. Br.

    argyrocoma Heller ex Skottsberg
    degeneri Skottsberg
    forbesii Skottsberg
      ssp. fallax Skottsberg
      ssp. forbesii
      ssp. nivea Skottsberg
      ssp. pachysperma Skottsberg
    menziesiana Sm.
         var. depauperata Skottsberg
         var. menziesiana
    veratroides Gaud.
      ssp. macrosperma Skottsberg
      ssp. veratroides
         var. gracilis Skottsberg
         var. veratroides
    waialealae Wawra

Bloomeria Kellogg

    crocea (Torr.) Coville
         var. aurea (Kellogg) Ingram
          SY=B. a. Kellogg
         var. crocea
         var. montana (Greene) Ingram
    humilis Hoover

Brodiaea Sm.

    appendiculata Hoover
    californica Lindl.
    coronaria (Salisb.) Engl.
      ssp. coronaria
      ssp. rosea (Greene) Niehaus
         SY=B. c. var. r. (Greene) Hoover
         SY=B. r. (Greene) Baker

    elegans Hoover
      ssp. elegans
         SY=B. e. var. mundula (Jepson)
            Hoover
      ssp. hooveri Niehaus
    filifolia S. Wats.
    insignis (Jepson) Niehaus
    jolonensis Eastw.
    kinkiensis Niehaus
    leptandra (Greene) Baker
         SY=B. californica Lindl. var. l.
           (Greene) Hoover
    minor (Benth.) S. Wats.
         SY=B. m. var. nana (Hoover) Hoover
    orcuttii (Greene) Baker
    pallida Hoover
    purdyi Eastw.
    stellaris S. Wats.
    terrestris Kellogg
      ssp. kernensis (Hoover) Niehaus
         SY=B. coronaria (Salisb.) Engl.
           var. k. Hoover
         SY=B. elegans Hoover var.
           australis Hoover
      ssp. terrestris
         SY=B. coronaria (Salisb.) Engl.
           var. macropoda (Torr.)
           Hoover

Calochortus Pursh

    albus Dougl. ex Benth.
         SY=C. a. var. rubellus Greene
    amabilis Purdy
    ambiguus (M.E. Jones) Ownbey
    amoenus Greene
    apiculatus Baker
    aureus S. Wats.
         SY=C. nuttallii Torr. & Gray var.
          a. (S. Wats.) Ownbey
    bruneaunis A. Nels. & J.F. Macbr.
         SY=C. nuttallii Torr. & Gray var.
          b. (A. Nels. & J.F. Macbr.)
          Ownbey
    catalinae S. Wats.
    clavatus S. Wats.
      ssp. clavatus
         SY=C. c. var. avius Jepson
      ssp. gracilis Ownbey
      ssp. pallidus (Hoover) Munz
      ssp. recurvifolius (Hoover) Munz
    coeruleus (Kellogg) S. Wats.
         var. coeruleus
         var. fimbriatus Ownbey
         var. nanus (Wood) Ownbey
         SY=C. elegans Pursh var. n. Wood
         SY=C. n. (Wood) Piper
         var. westonii (Eastw.) Ownbey
    concolor (Baker) Purdy
    dunnii Purdy
    elegans Pursh
         var. elegans
         var. oreophilus Ownbey
         var. selwayensis (St. John) Ownbey
    eurycarpus S. Wats.
    excavatus Greene
    flexuosus S. Wats.
    greenei S. Wats.
    gunnisonii S. Wats.
         var. gunnisonii
         var. perpulcheri Cockerell
    howellii S. Wats.
    indecorus Ownbey & M.E. Peck
    invenustus Greene
    kennedyi Porter
         SY=C. k. var. munzii Jepson
    leichtlinii Hook. f.

Calochortus (CONT.)
  lobbii (Baker) Purdy
  longebarbatus S. Wats.
          var. longebarbatus
          var. peckii Ownbey
  luteus Dougl. ex Lindl.
  lyallii Baker
  macrocarpus Dougl.
          var. macrocarpus
              SY=C. douglasianus Schultes f.
          var. maculosus (A. Nels. & J.F.
              Macbr.) A. Nels. & J.F. Macbr. ex
              J.F. Macbr.
  minimus Ownbey
  monanthus Ownbey
  monophyllus (Lindl.) Lem.
  nitidus Dougl.
  nudus S. Wats.
              SY=C. shastensis Purdy
  nuttallii Torr. & Gray
              SY=C. rhodothecus Clokey
  obispoensis Lemmon
  palmeri S. Wats.
          var. munzii Ownbey
          var. palmeri
              SY=C. paludicola A. Davids.
  panamintensis (Ownbey) Reveal
              SY=C. nuttallii Torr. & Gray var.
                  p. Ownbey
  persistens Ownbey
  plummerae Greene
  pulchellus Dougl. ex Benth.
  simulans (Hoover) Munz
  splendens Dougl. ex Benth.
              SY=C. davidsonianus Abrams
  striatus Parish
  subalpinus Piper
  superbus Purdy ex J.T. Howell
              SY=C. luteus Dougl. ex Lindl. var.
                  citrinus S. Wats.
  tiburonensis A.J. Hill
  tolmiei Hook. & Arn.
              SY=C. maweanus Leichtl.
              SY=C. purdyi Eastw.
  umbellatus Wood
  uniflorus Hook. & Arn.
  venustus Dougl. ex Benth.
              SY=C. v. var. carolii Cockerell
              SY=C. v. var. purpurascens S.
                  Wats.
              SY=C. v. var. sulphureus Purdy
  vestae Purdy
              SY=C. luteus Dougl. ex Lindl. var.
                  oculatus S. Wats.
  weedii Wood
          var. intermedius Ownbey
          var. vestus Purdy
          var. weedii

Camassia Lindl.

  cusickii S. Wats.
  howellii S. Wats.
  leichtlinii (Baker) S. Wats.
    ssp. leichtlinii
    ssp. suksdorfii (Greenm.) Gould
              SY=C. l. var. s. (Greenm.) C.L.
                  Hitchc.
  quamash (Pursh) Greene
    ssp. azurea (Heller) Gould
              SY=C. q. var. a. (Heller) C.L.
                  Hitchc.
    ssp. breviflora Gould
              SY=C. q. var. b. (Gould) C.L.
                  Hitchc.
    ssp. intermedia Gould
              SY=C. q. var. i. (Gould) C.L.

        Hitchc.
    ssp. linearis Gould
    ssp. maxima Gould
              SY=C. q. var. m. (Gould) Boivin
    ssp. quamash
    ssp. utahensis Gould
              SY=C. q. var. u. (Gould) C.L.
                  Hitchc.
    ssp. walpolei (Piper) Gould
  scilloides (Raf.) Cory
              SY=C. angusta (Engelm. & Gray)
                  Blank.
              SY=Quamasia hyacinthina (Raf.)
                  Britt.

Chamaelirium Willd.

  luteum (L.) Gray
              SY=C. obovale Small

Chlorogalum (Lindl.) Kunth

  angustifolium Kellogg
  grandiflorum Hoover
  parviflorum S. Wats.
  pomeridianum (DC.) Kunth
          var. divaricatum (Lindl.) Hoover
          var. minus Hoover
          var. pomeridianum
  purpureum Brandeg.
          var. purpureum
          var. reductum Hoover

Clintonia Raf.

  andrewsiana Torr.
  borealis (Ait.) Raf.
  umbellulata (Michx.) Morong
              SY=C. alleghaniensis Harned
              SY=Xeniatrum u. (Michx.) Small
  uniflora (Schultes) Kunth

Colchicum L.

  autumnale L.

Convallaria L.

  majalis L.
  montana Raf.
              SY=C. majalis L. var. montana
                  (Raf.) Ahles

Cooperia Herbert

  drummondii Herbert
              SY=Zephyranthes brazosensis
                  (Herbert) Traub
  jonesii Cory
              SY=Zephyranthes j. (Cory) Traub
  pedunculata Herbert
              SY=Zephyranthes drummondii D. Don
  smallii Alexander
              SY=Zephyranthes s. (Alexander)
                  Traub
  traubii Hayward

Crinum L.

  americanum L.
  bulbispermum (Burm. f.) Milne-Redhead &
    Schweickerdt
              SY=C. longifolium (L.) Thunb.
  strictum Herbert
          var. strictum
          var. traubii Moldenke
  zeylanicum (L.) L.

Crinum (CONT.)
        SY=C. latifolia L. var. z. (L.)
           Hook. f. ex Timer

Curculigo Gaertn.

   capitulata (Lour.) Kuntze
   scorzonerifolia (Lam.) Baker

Dianella Lam.

   sandwicensis Hook. & Arn.

Dichelostemma Kunth

   ida-maia (Wood) Greene
        SY=Brevoortia i-m. Wood
        SY=Brodiaea i-m. (Wood) Greene
   lacuna-vernalis Lenz
   multiflorum (Benth.) Heller
        SY=Brodiaea m. Benth.
   pulchellum (Salisb.) Heller
      var. pauciflora (Torr.) Hoover
        SY=Brodiaea pulchella (Salisb.)
          Greene var. pauciflora
          (Torr.) Morton
      var. pulchellum
        SY=Brodiaea capitata Benth.
        SY=B. congesta Sm.
        SY=B. p. (Salisb.) Greene
        SY=D. capitatum (Benth.) Wood
   venustum (Greene) Hoover
        SY=Brevoortia v. Greene
        SY=Brodiaea v. (Greene) Greene
   volubilis (Kellogg) Heller
        SY=Brodiaea v. (Kellogg) Baker
        SY=D. californicum (Torr.) Wood

Disporum Salisb. ex D. Don

   hookeri (Torr.) Nichols.
      var. hookeri
      var. oreganum (S. Wats.) Q. Jones
        SY=D. o. (S. Wats.) W. Mill.
      var. trachyandrum (Torr.) Q. Jones
        SY=D. t. (Torr.) Britt.
   lanuginosum (Michx.) Nichols.
   maculatum (Buckl.) Britt.
   parvifolium (S. Wats.) Britt.
   smithii (Hook.) Piper
   trachycarpum (S. Wats.) Benth. & Hook. f.
        SY=D. t. var. subglabrum E.H.
          Kelso

Endymion Dumort.

   nonscripta (L.) Garcke
        SY=Scilla n. (L.) Hoffmgg. & Link

Eremocrinum M.E. Jones

   albomarginatum (M.E. Jones) M.E. Jones

Erythronium L.

   albidum Nutt.
      var. albidum
      var. mesochoreum (Knerr) Rickett
        SY=E. a. var. coloratum Sterns
        SY=E. m. Knerr
   americanum Ker-Gawl.
        SY=E. a. ssp. harperi (W. Wolf)
          Parks & Hardin
        SY=E. a. var. rubrum Farw.
   californicum Purdy
   citrinum S. Wats.
   grandiflorum Pursh

      var. candidum (Piper) Abrams
        SY=E. g. ssp. c. Piper
        SY=E. g. var. idahoense (St. John)
          G. N. Jones
      var. chrysandrum (Applegate) Scoggan
        SY=E. g. ssp. c. Applegate
        SY=E. parviflorum (S. Wats.)
          Goodding
      var. grandiflorum
      var. nudipetalum (Applegate) C.L.
        Hitchc.
        SY=E. g. ssp. n. Applegate
      var. pallidum St. John
    ssp. ?pusateri Munz & Howell
   helenae Applegate
   hendersonii S. Wats.
   howellii S. Wats.
   klamathense Applegate
   montanum S. Wats.
   multiscapoideum (Kellogg) A. Nels. &
     Kennedy
   oregonum Applegate
     ssp. leucandrum (Applegate) Applegate
        SY=E. giganteum sensu auctt. non
          Lindl. ssp. l. Applegate
     ssp. oregonum
        SY=E. giganteum sensu auctt. non
          Lindl.
   propullans Gray
   purpurascens S. Wats.
   revolutum Sm.
   rostratum W. Wolf
   tuolumnense Applegate
   umbilicatum Parks & Hardin
     ssp. monostolum Parks & Hardin
     ssp. umbilicatum

Fritillaria L.

   agrestis Greene
   atropurpurea Nutt.
        SY=F. adamantina M.F. Peck
   biflora Lindl.
   brandegei Eastw.
   camschatcensis (L.) Ker-Gawl.
   eastwoodiae Macfarlane
        SY=F. phaeanthera Eastw.
   falcata (Jepson) D.E. Beetle
   gentneri Gilkey
   glauca Greene
   grayana Reichenb. f. & Baker
        SY=F. roderickii Knight
   lanceolata Pursh
        SY=F. camschatcensis (L.)
          Ker-Gawl. var. floribunda
          (Benth.) Boivin
        SY=F. mutica Lindl.
        SY=F. m. var. gracilis (S. Wats.)
          Jepson
   liliacea Lindl.
   micrantha Heller
        SY=F. multiflora Kellogg
        SY=F. parviflora Torr.
   pinetorum A. Davids.
        SY=F. atropurpurea Nutt. var. p.
         (A. Davids.) I.M. Johnston
   pluriflora Torr. ex Benth.
   pudica (Pursh) Spreng.
   purdyi Eastw.
   recurva Benth.
        SY=F. r. var. coccinea Greene
   striata Eastw.
   viridea Kellogg

Galanthus L.

   nivalis L.

Habranthus Herbert

    texanus (Herbert) Herbert ex Steud.
          SY=Zephyranthes t. Herbert

Harperocallis McDaniel

    flava McDaniel

Hastingsia S. Wats.

    alba (Dur.) S. Wats.
          SY=Schoenolirion a. Dur.
    bracteosa S. Wats.
          SY=Schoenolirion b. (S. Wats.)
             Jepson

Helonias L.

    bullata L.

Hemerocallis L.

    fulva (L.) L.
        var. fulva
        var. kwanso Regel
    lilio-asphodelus L. emend. Hyl.
          SY=H. flava (L.) L.
          SY=H. l. var. f. L.

Hesperocallis Gray

    undulata Gray

Hippeastrum Herbert

    puniceum (Lam.) Kuntze

Hosta Tratt.

    lancifolia (Thunb. ex Houtt.) Engl.
          SY=H. japonica (Thunb. ex Houtt.)
             Voss
          SY=Niobe j. (Thunb. ex Houtt.)
             Nash
    plantaginea (Lam.) Aschers.
    ventricosa (Salisb.) Stearn
          SY=Niobe coerula (Andr.) Nash

Hymenocallis Salisb.

    caribaea (L.) Herbert
    caroliniana (L.) Herbert
          SY=H. occidentalis (Le Conte)
             Kunth
    choctawensis Traub
    coronaria (Le Conte) Kunth
    crassifolia Herbert
    duvalensis Traub
    eulae Shinners
    floridana (Raf.) Morton
          SY=H. tridentata Small
    galvestonensis (Herbert) Baker
    henryae Traub
    humilis S. Wats.
    kimballiae Small
    ?laciniata Small
    latifolia (P. Mill.) M. Roemer
          SY=H. caymanensis Herbert
          SY=H. collieri Small
          SY=H. keyensis Small
    liriosome (Raf.) Shinners
    moldenkiana Traub
    palmeri S. Wats.
    palusvirensis Traub
    puntagordensis Traub
    pymaea Traub

    rotata (Ker-Gawl.) Herbert
          SY=H. bidentata Small
    traubii Moldenke

Hypoxis L.

    hirsuta (L.) Coville
          SY=H. decumbens L.
          SY=H. h. var. leptocarpa (Engelm.
             & Gray) Fern.
          SY=H. l. (Engelm. & Gray) Small
          SY=H. rigida Chapman
    juncea Sm.
          SY=H. wrightii (Baker) Brackett
    mexicana Schultes
    micrantha Pollard
    sessilis L.
          SY=H. longii Fern.

Ipheion Raf.

    uniflorum (Lindl.) Raf.
          SY=Brodiaea u. (Lindl.) Engl.

Leucocrinum Nutt. ex Gray

    montanum Nutt. ex Gray

Leucojum L.

    aestivum L.
        var. aestivum
        var. pulchellum author?

Lilium L.

    bolanderi S. Wats.
    bulbiferum L.
    canadense L.
        var. canadense
          SY=L. c. var. rubrum Britt.
        var. editorum Fern.
    catesbaei Walt.
      ssp. asperellum Wherry
      ssp. catesbaei
        var. catesbaei
        var. longii Fern.
    columbianum Hanson ex Baker
          SY=L. canadense L. var.
              parviflorum Hook.
    grayi S. Wats.
    humboldtii Roezl & Leichtl. ex Duchartre
        var. bloomerianum (Kellogg) Jepson
        var. humboldtii
        var. ocellatum (Kellogg) Elwes
          SY=L. h. ssp. o. (Kellogg) Thorne
    iridollae Henry
    kelleyanum Lemmon
    kelloggii Purdy
    lancifolium Thunb.
          SY=L. tigrinum Ker-Gawl.
    maritimum Kellogg
    martagon L.
    michauxii Poir.
          SY=L. carolinianum Michx.
          SY=L. superbum L. var. c. (Michx.)
             Chapman
    michiganense Farw.
          SY=L. canadense L. ssp. m. (Farw.)
             Boivin & Cody
          SY=L. c. var. umbelliferum (Farw.)
             Boivin
          SY=L. m. var. uniflorum Farw.
    occidentale Purdy
    pardalinum Kellogg
    parryi S. Wats.
        var. kessleri A. Davids.

Lilium (CONT.)
        var. parryi
    parvum Kellogg
    philadelphicum L.
        var. andinum (Nutt.) Ker-Gawl.
            SY=L. montanum A. Nels.
            SY=L. p. var. m. (A. Nels.) Wherry
            SY=L. umbellatum Pursh
        var. philadelphicum
    pitkinense Beane & Vollmer
    rubescens S. Wats.
    superbum L.
    vollmeri Eastw.
    washingtonianum Kellogg
        var. minus Purdy
        var. purpurascens Stearn
        var. washingtonianum
    wigginsii Beane & Vollmer

Lloydia Salisb. ex Reichenb.

    serotina (L.) Salisb. ex Reichenb.
        ssp. flava Calder & Taylor
            SY=L. s. var. f. (Calder & Taylor)
                Boivin
        ssp. serotina

Lycoris Herbert

    radiata (L'Her.) Herbert

Maianthemum Wiggers

    canadense Desf.
        var. canadense
            SY=Unifolium c. (Desf.) Greene
        var. interius Fern.
    dilatatum (Wood) A. Nels. & J.F. Macbr.
            SY=M. bifolium DC. var.
                kamtschaticum (J.G. Gmel.)
                Jepson
            SY=M. k. (J.G. Gmel) Nakai

Medeola L.

    virginiana L.

Melanthium L.

    hybridum Walt.
            SY=M. latifolium Desr.
    virginicum L.
            SY=M. dispersum Small
            SY=Veratrum v. (L.) Ait.

Milla Cav.

    biflora Cav.

Muilla S. Wats.

    clevelandii (S. Wats.) Hoover
            SY=Bloomeria c. S. Wats.
    coronata Greene
    maritima (Torr.) S. Wats.
            SY=M. serotina Greene
    transmontana Greene

Muscari P. Mill.

    armeniacum Leichtl. ex Baker
    atlanticum Boiss. & Reut.
            SY=M. racemosum sensu auctt. non
                (L.) P. Mill.
    botryoides (L.) P. Mill.
    comosum (L.) P. Mill.

Narcissus L.

    X incomparabilis P. Mill. [poeticus X
        pseudonarcissus]
    jonquilla L.
    X medioluteus P. Mill. [poeticus X tazetta]
            SY=N. biflorus W. Curtis
            SY=N. poetaz Hort. ex Bailey
    poeticus L.
    pseudonarcissus L.
    tazetta L.

Narthecium Huds.

    americanum Ker-Gawl.
            SY=Abama a. (Ker-Gawl.) Morong
            SY=A. montana Small
            SY=N. ossifragum Huds. var. a.
                (Ker-Gawl.) Gray
    californicum Baker

Nothoscordum Kunth

    bivalve (L.) Britt.
            SY=Allium b. (L.) Kuntze
    inodorum (Ait.) Nichols.
            SY=Allium i. Ait.
            SY=N. fragrans (Vent.) Kunth
    texanum M.E. Jones

Ornithogalum L.

    nutans L.
    pyrenaicum L.
    umbellatum L.

Pancratium L.

    declinatum Jacq.
            SY=Hymenocallis d. (Jacq.) M.
                Roemer

Polygonatum P. Mill.

    biflorum (Walt.) Ell.
        var. biflorum
            SY=Convallaria b. Walt.
            SY=?P. canaliculatum sensu auctt.
                non (Muhl.) Pursh
        var. commutatum (Schultes f.) Morong
            SY=P. c. (Schultes f.) A. Dietr.
        var. hebetifolium R.R. Gates
        var. melleum (Farw.) R. Ownbey
            SY=P. m. Farw.
        var. necopinum R. Ownbey
    cobrense (Woot. & Standl.) R.R. Gates
    latifolium (Jacq.) Desf.
    multiflorum (L.) All.
    pubescens (Willd.) Pursh

Schoenocaulon Gray

    drummondii Gray
    dubium (Michx.) Small
    texanum Scheele

Schoenolirion Torr. ex Dur.

    albiflorum (Raf.) R.R. Gates
            SY=Oxytria a. (Raf.) Pollard
            SY=S. elliottii Gray
    croceum (Michx.) Wood
            SY=Oxytria c. (Michx.) Raf.
    wrightii Sherman
            SY=Oxytria texana (Scheele)
                Pollard
            SY=S. t. (Scheele) Gray

Scilla L.

    sibirica Andr.

Scoliopus Torr.

    bigelovii Torr.
    hallii S. Wats.

Smilacina Desf.

    racemosa (L.) Desf.
        SY=S. amplexicaulis Nutt. ex Baker
        SY=S. a. var. glabra J.F. Macbr.

        SY=S. r. var. a. (Nutt. ex Baker)
            S. Wats.
        SY=S. r. var. cylindrata Fern.
        SY=S. r. var. glabra (J.F. Macbr.)
            St. John
        SY=S. r. var. jenkinsii Boivin
        SY=S. r. var. typica Fern.
        SY=Vagnera a. (Nutt. ex Baker)
            Greene
        SY=V. australis Rydb.
        SY=V. r. (L.) Morong
    stellata (L.) Desf.
        var. sessilifolia (Baker) Henderson
        SY=S. sessilifolia (Baker) Nutt.
        var. stellata
        SY=S. s. var. crassa Victorin
        SY=Vagnera s. (L.) Morong
    trifolia (L.) Desf.
        SY=Vagnera t. (L.) Morong

Stenanthium (Gray) Kunth

    gramineum (Ker-Gawl.) Morong
        var. gramineum
        var. micranthum Fern.
        var. robustum (S. Wats.) Fern.
        SY=S. r. S. Wats.
    occidentale Gray
        SY=Stenanthella o. (Gray) Rydb.

Streptopus Michx.

    amplexifolius (L.) DC.
        var. americanus Schultes
        SY=S. amplexifolius var.
            denticulatus Fassett
        SY=S. amplexifolius var.
            grandiflorus Fassett
        SY=Tortipes amplexifolius (L.)
            Small pro parte
        SY=Uvularia amplexifolius L. pro
            parte
        var. chalazatus Fassett
        var. papillatus Ohwi
    X oreopolus Fern. [amplexifolius X roseus]
        SY=S. amplexifolius (L.) DC. var.
            o. (Fern.) Fassett
    roseus Michx.
        var. curvipes (Vail) Fassett
        SY=S. c. Vail
        SY=S. r. ssp. c. (Vail) Hulten
        var. longipes (Fern.) Fassett
        var. perspectus Fassett
        var. roseus
    streptopoides (Ledeb.) Frye & Rigg
        ssp. streptopoides
        var. brevipes (Baker) Fassett
        SY=Krushea s. (Ledeb.) Kearney pro
            parte
        SY=S. s. ssp. b. (Baker) Calder &
            Taylor

Tofieldia Huds.

    coccinea Richards.
    glabra Nutt.
    glutinosa (Michx.) Pers.
        ssp. absona C.L. Hitchc.
        SY=T. g. var. a. (C.L. Hitchc.)
            Davis
        ssp. brevistyla C.L. Hitchc.
        SY=T. g. var. b. (C.L. Hithc.)
            C.L. Hitchc.
        SY=T. g. var. intermedia (Rydb.)
            Boivin
        ssp. glutinosa
        SY=T. racemosa (Walt.) B.S.P. var.
            g. (Michx.) Ahles
        SY=Triantha g. (Michx.) Baker
        ssp. montana C.L. Hitchc.
        SY=Tofieldia g. var. m. (C.L.
            Hitchc.) Davis
        ssp. occidentalis (S. Wats.) C.L.
        Hitchc.
        SY=Tofieldia g. var. o. (S. Wats.)
            C.L. Hitchc.
        SY=T. o. S. Wats.
    pusilla (Michx.) Pers.
        SY=Tofieldia palustris Huds.
    racemosa (Walt.) B.S.P.
        SY=Triantha r. (Walt.) Small
    tenuifolia (Michx.) Utech
        SY=Pleea t. Michx.

Tricyrtis Wallich

    hirta (Thunb.) Hook.

Trillium L.

    albidum J.D. Freeman
    angustipetalum (Torr.) J.D. Freeman
        SY=T. chloropetalum (Torr.) T.J.
            Howell var. a. (Torr.) Munz
    catesbaei Ell.
    cernuum L.
        var. cernuum
        var. macranthum Eames & Wieg.
        var. tangerae Wherry
    chloropetalum (Torr.) T.J. Howell
        var. chloropetalum
        var. giganteum (Hook. & Arn.) Munz
        SY=T. g. (Hook. & Arn.) Heller
    cuneatum Raf.
        SY=T. hugeri Small
    decipiens J.D. Freeman
    decumbens Harbison
    discolor Wray ex Hook.
    erectum L.
        var. album (Michx.) Pursh
        var. erectum
        var. flavum Eason
        var. sulcatum Barksdale
    flexipes Raf.
        SY=T. cernuum L. var. declinatum
            Gray
        SY=T. d. (Gray) Gleason
        SY=T. erectum L. var. blandum
            Jennison
        SY=T. gleasonii Fern.
    foetidissimum J.D. Freeman
    gracile J.D. Freeman
    grandiflorum (Michx.) Salisb.
    kurabayashii J.D. Freeman
    lancifolium Raf.
        SY=T. lanceolatum Boykin ex S.
            Wats.
        SY=T. recurvatum Beck var.
            lanceolatum S. Wats.

Trillium (CONT.)
  ludovicianum Harbison
  luteum (Muhl.) Harbison
          SY=T. cuneatum Raf. var. l.
              (Muhl.) Ahles
          SY=T. viride Beck var. l. (Muhl.)
              Gleason
  maculatum Raf.
  nivale Riddell
  ovatum Pursh
    ssp. oettingeri Munz & Thorne
    ssp. ovatum
  persistens Duncan
  petiolatum Pursh
  pusillum Michx.
          var. ozarkanum (Palmer & Steyermark)
              Steyermark
          SY=T. o. Palmer & Steyermark
          var. pusillum
          var. virginianum Fern.
  recurvatum Beck
  reliquum J.D. Freeman
  rivale S. Wats.
  sessile L.
  simile Gleason
  stamineum Harbison
  texanum Buckl.
  underwoodii Small
  undulatum Willd.
          var. pubescens author ?
          var. undulatum
  vaseyi Harbison
          SY=T. erectum L. var. v.
              (Harbison) Ahles
  viride Beck
  viridescens Nutt.

Triteleia Dougl. ex Lindl.

  bridgesii (S. Wats.) Greene
          SY=Brodiaea b. S. Wats.
  clementina Hoover
          SY=Brodiaea c. (Hoover) Munz
  crocea (Wood) Greene
          var. crocea
          SY=Brodiaea c. (Wood) S. Wats.
          var. modesta (Hall) Hoover
          SY=Brodiaea c. var. m. (Hall) Munz
          SY=T. m. (Hall) Abrams
  dudleyi Hoover
          SY=Brodiaea d. (Hoover) Munz
  gracilis (S. Wats.) Greene
          SY=Brodiaea g. S. Wats.
          SY=T. montana Hoover
  grandiflora Lindl.
          SY=Brodiaea douglasii (Lindl.) S.
              Wats.
          SY=B. g. (Lindl.) J.F. Macbr.
  hendersonii Greene
          var. hendersonii
          SY=Brodiaea h. (Greene) S. Wats.
          SY=B. h. var. galeae M.E. Peck
          var. leachiae (M.E. Peck) Hoover
          SY=Brodiaea l. M.E. Peck
  howellii (S. Wats.) Greene
          SY=Brodiaea douglasii (Lindl.) S.
              Wats. var. h. (S. Wats.)
              M.E. Peck
          SY=B. h. S. Wats.
          SY=T. bicolor (Suksdorf) Abrams
          SY=T. grandiflora Lindl. var. h.
              (S. Wats.) Hoover
  hyacinthina (Lindl.) Greene
          var. greenei Hoover
          SY=Brodiaea h. (Lindl.) Baker var.
              g. (Hoover) Munz
          SY=Hesperoscordum lilacinum

              (Greene) Heller ex Abrams
          var. hyacinthina
          SY=Brodiaea dissimulata M.E. Peck
          SY=B. h. (Lindl.) Baker
          SY=B. h. (Lindl.) Baker var.
              lactea Baker
          SY=Hesperoscordum h. Lindl.
  ixioides (Ait. f.) Greene
    ssp. analina (Greene) Lenz
          SY=Brodiaea lutea (Lindl.) Morton
              var. a. (Greene) Munz
          SY=B. scabra (Greene) Baker var.
              a. (Greene) M.E. Peck
          SY=Calliprora s. Greene var. a.
              Greene
          SY=T. i. var. a. (Greene) Hoover
    ssp. cookii (Hoover) Lenz
          SY=Brodiaea lutea (Lindl.) Morton
              var. c. (Hoover) Munz
          SY=T. i. var. c. Hoover
    ssp. ixioides
          SY=Brodiaea lutea (Lindl.) Morton
          SY=Calliprora i. (Ait. f.) Greene
    ssp. scabra (Greene) Lenz
          SY=Brodiaea lutea (Lindl.) Morton
              var. s. (Greene) Munz
          SY=Brodiaea s. (Greene) Baker
          SY=Calliprora s. Greene
          SY=T. s. (Greene) Hoover
  laxa Benth.
          SY=Brodiaea l. (Benth.) S. Wats.
          SY=B. l. var. candida (Greene)
              Jepson
          SY=B. l. var. nimia Jepson
          SY=B. l. var. tracyi Jepson
          SY=T. angustiflora Heller
  lemmonae (S. Wats.) Greene
          SY=Brodiaea l. S. Wats.
  lugens Greene
          SY=Brodiaea l. (Greene) Baker
          SY=Calliprora ixioidea (Ait. f.)
              Greene var. l. (Greene)
              Abrams
  peduncularis Lindl.
          SY=Brodiaea p. (Lindl.) S. Wats.
  unifolia (Lenz) Lenz
          SY=T. ixioides (Ait. f.) Greene
              ssp. u. Lenz
  versicolor Hoover
          SY=Brodiaea v. (Hoover) Munz

Triteleiopsis Hoover

  palmeri (S. Wats.) Hoover

Tulipa L.

  gesneriana L.
  sylvestris L.

Uvularia L.

  floridana Chapman
          SY=Oakesiella f. (Chapman) Small
  grandiflora Sm.
  perfoliata L.
  puberula Michx.
          SY=Oakesiella p. (Michx.) Small
          SY=U. carolina (J.F. Gmel.) Wilbur
          SY=U. nitida (Britt.) Mackenzie
          SY=U. pudica sensu Fern.
          SY=U. pudica sensu Fern. var. n.
              (Britt.) Fern.
  sessilifolia L.
          SY=Oakesiella s. (L.) S. Wats.

Veratrum L.

Veratrum (CONT.)
    album L.
        ssp. album
        ssp. oxysepalum (Turcz.) Hulten
    californicum Dur.
            var. californicum
            var. caudatum (Heller) C.L. Hitchc.
                SY=V. c. Heller
    fimbriatum Gray
    insolitum Jepson
    parviflorum Michx.
                SY=Melanthium p. (Michx.) S. Wats.
    tenuipetalum Heller
    viride Ait.
        ssp. eschscholtzii (Gray) Love & Love
                SY=V. eschscholtzianum (Roemer &
                    Schultes) Rydb.
                SY=V. eschscholtzii Gray
                SY=V. eschscholtzii var.
                    incriminatum Boivin
                SY=V. v. var. eschscholtzii (Gray)
                    Breitung
        ssp. viride
    woodii J.W. Robbins
                SY=V. intermedium Chapman

Xerophyllum Michx.

    asphodeloides (L.) Nutt.
    tenax (Pursh) Nutt.

Zephyranthes Herbert

    atamasco (L.) Herbert
                SY=Atamosco a. (L.) Greene
    bifolia (Aubl.) M. Roemer
    candida (Lindl.) Herbert
                SY=Atamosco c. (Lindl.) Small
    citrina Baker
    grandiflora Herbert
                SY=Atamosco carinata (Herbert) P.
                    Wilson
    longifolia Hemsl.
                SY=Atamosco l. (Hemsl.) Cockerell
    puertoricensis Traub
                SY=Atamosco tubispatha sensu
                    auctt. non (L'Her.) Maza
                SY=Z. t. (L'Her.) Herbert
    pulchella J.G. Sm.
    refugiensis F.B. Jones
    rosea Lindl.
                SY=Atamosco r. (Lindl.) Greene
    simpsonii Chapman
                SY=Atamosco s. (Chapman) Greene
    treatiae S. Wats.
                SY=Atamosco t. (S. Wats.) Greene
    tubispatha Herbert
                SY=?Z. insularum Hume

Zigadenus Michx.

    brevibracteatus (M.E. Jones) Hall
    densus (Desr.) Fern.
                SY=Tracyanthus angustifolius
                    (Michx.) Small
    elegans Pursh
        ssp. elegans
                SY=Anticlea e. (Pursh) Rydb.
                SY=Z. alpina Blank.
        ssp. glaucus (Nutt.) Hulten
                SY=Anticlea chlorantha (Richards.)
                    Rydb.
                SY=Z. g. Nutt.
    exaltatus Eastw.
    fontanus Eastw.
                SY=Z. venenosus S. Wats. var. f.
                    (Eastw.) Preece

fremontii (Torr.) Torr. ex S. Wats.
        var. fremontii
                SY=Z. f. var. inezianus Jepson
                SY=Z. f. var. salsus Jepson
        var. minor (Hook. & Arn.) Jepson
    glaberrimus Michx.
    leimanthoides Gray
                SY=Oceanorus l. (Gray) Small
    micranthus Eastw.
                SY=Z. venenosus S. Wats. var. m.
                    (Eastw.) Jepson
    nuttallii Gray ex S. Wats.
                SY=Toxicoscordion n. (Gray ex S.
                    Wats.) Rydb.
    paniculatus (Nutt.) S. Wats.
                SY=Toxicoscordion p. (Nutt.) Rydb.
    vaginatus (Rydb.) J.F. Macbr.
    venenosus S. Wats.
        var. gramineus (Rydb.) Walsh ex M.F.
            Peck
                SY=Toxicoscordion g. (Rydb.) Rydb.
                SY=Z. g. Rydb.
                SY=Z. intermedius Rydb.
        var. venenosus
                SY=Toxicoscordion v. (S. Wats.)
                    Rydb.
    virescens (H.B.K.) J.F. Macbr.
    volcanicus Benth.

LIMNANTHACEAE

Floerkea Willd.

    proserpinacoides Willd.
                SY=F. occidentalis Rydb.

Limnanthes R. Br.

    alba Hartw.
        var. alba
        var. versicolor (Greene) C.T. Mason
                SY=L. v. (Greene) Rydb.
    bakeri J.T. Howell
    douglasii R. Br.
        var. douglasii
                SY=L. howelliana Abrams
        var. nivea C.T. Mason
        var. rosea (Hartw.) C.T. Mason
                SY=L. r. Hartw.
        var. sulphurea C.T. Mason
    floccosa T.J. Howell
        ssp. bellingeriana (M.F. Peck) Arroyo
                SY=L. b. M.E. Peck
        ssp. californica Arroyo
        ssp. floccosa
        ssp. grandiflora Arroyo
        ssp. pumila (T.J. Howell) Arroyo
                SY=L. p. T.J. Howell
    gracilis T.J. Howell
        var. gracilis
        var. parishii (Jepson) C.T. Mason
                SY=L. versicolor (Greene) Rydb.
                    var. p. Jepson
    macounii Trel.
    montana Jepson
    striata Jepson
    vinculans Ornduff

LINACEAE

Hesperolinon (Gray) Small

Hesperolinon (CONT.)
    adenophyllum (Gray) Small
            SY=Linum a. Gray
    bicarpellatum (H.K. Sharsmith) H.K.
        Sharsmith
            SY=Linum b. H.K. Sharsmith
    breweri (Gray) Small
            SY=Linum b. Gray
    californicum (Benth.) Small
            SY=Linum c. Benth.
            SY=L. c. var. confertum Gray ex
                    Trel.
    clevelandii (Greene) Small
            SY=Linum c. Greene
            SY=L. c. var. petrophilum Jepson
    congestum (Gray) Small
            SY=Linum californicum Benth. var.
                    congestum (Gray) Jepson
            SY=L. congestum Gray
    didymocarpum H.K. Sharsmith
    disjunctum H.K. Sharsmith
    drymarioides (Curran) Small
            SY=Linum d. Curran
    micranthum (Gray) Small
            SY=Linum m. Gray
    spergulinum (Gray) Small
            SY=Linum s. Gray
    tehamense H.K. Sharsmith

Linum L.

    alatum (Small) Winkl.
            SY=Cathartolinum a. Small
    arenicola (Small) Winkl.
            SY=Cathartolinum a. Small
    aristatum Engelm.
            SY=Cathartolinum a. (Engelm.)
                    Small
    australe Heller
        var. australe
            SY=Cathartolinum a. (Heller) Small
            SY=L. aristatum Engelm. var.
                    australe (Heller) Kearney &
                    Peebles
        var. glandulosum Rogers
    austriacum L.
            SY=L. perenne L. var. a. (L.)
                    Schiede
    bienne P. Mill.
            SY=L. angustifolium Huds.
    carteri Small
        var. carteri
            SY=Cathartolinum c. (Small) Small
            SY=L. rigidum Pursh var. c.
                    (Small) Rogers
        var. smallii Rogers
    catharticum L.
            SY=Cathartolinum c. (L.) Small
    elongatum (Small) Winkl.
            SY=Cathartolinum e. Small
    floridanum (Planch.) Trel.
        var. chrysocarpum Rogers
        var. floridanum
            SY=Cathartolinum f. (Planch.)
                    Small
            SY=C. macrosepalum Small
            SY=L. virginianum L. var. f.
                    Planch.
    grandiflorum Desf.
    hudsonioides Planch.
            SY=Cerastium clawsonii Correll
    imbricatum (Raf.) Shinners
            SY=Cathartolinum multicaule
                    (Hook.) Small
    intercursum Bickn.
            SY=Cathartolinum i. (Bickn.) Small
    kingii S. Wats.

            SY=Cathartolinum sedoides (Porter)
                    Small
            SY=L. k. var. s. Porter
    lewisii Pursh
        var. alpicola Jepson
            SY=L. l. var. saxosum Maguire &
                    Holmgren
        var. lepagei (Boivin) Rogers
            SY=L. lepagei Boivin
            SY=L. lewisii ssp. lepagei
                    (Boivin) Mosquin
            SY=L. perenne L. var. lepagei
                    (Boivin) Boivin
        var. lewisii
            SY=L. perenne L. ssp. l. (Pursh)
                    Hulten
            SY=L. p. var. l. (Pursh) Fat. &
                    Wright
    lundellii Rogers
    macrocarpum Rogers
    medium (Planch.) Britt.
        var. medium
            SY=Cathartolinum m. (Planch.)
                    Small
            SY=L. striatum Walt. var. m.
                    (Planch.) Boivin
            SY=L. virginianum L. var. m.
                    Planch.
        var. texanum (Planch.) Fern.
            SY=Cathartolinum curtissii (Small)
                    Small
            SY=L. striatum Walt. var. t.
                    (Planch.) Boivin
    neomexicanum Greene
            SY=Cathartolinum n. (Greene) Small
    perenne L.
    pratense (J.B.S. Norton) Small
            SY=L. lewisii Pursh var. p. J.B.S.
                    Norton
    puberulum (Engelm.) Heller
            SY=Cathartolinum p. (Engelm.)
                    Small
            SY=L. rigidum Pursh var. p.
                    Engelm.
    rigidum Pursh
        var. berlandieri (Hook.) Torr. & Gray
            SY=Cathartolinum b. (Hook.) Small
            SY=C. sanctum (Small) Small
            SY=L. b. Hook.
            SY=L. s. Small
        var. compactum (A. Nels.) Rogers
            SY=Cathartolinum c. (A. Nels.)
                    Small
            SY=L. c. A. Nels.
        var. filifolium Shinners
        var. rigidum
            SY=Cathartolinum earlei Small
            SY=C. r. (Pursh) Small
    rupestre (Gray) Engelm. ex Gray
    schiedeanum Schlect. & Cham.
            SY=Linum greggii Engelm.
    striatum Walt.
            SY=Cathartolinum s. (Walt.) Small
            SY=L. s. var. multijugum Fern.
    subteres (Trel.) Winkl.
            SY=Linum aristatum Engelm. var. s.
                    Trel.
            SY=L. leptopoda A. Nels.
    sulcatum Riddell
        var. harperi (Small) Rogers
            SY=Cathartolinum h. (Small) Small
        var. sulcatum
            SY=Cathartolinum s. (Riddell)
                    Small
    trigynum L.
            SY=L. gallicum L.
    usitatissimum L.

Linum (CONT.)
          var. humile (P. Mill.) Pers.
          var. usitatissimum
     vernale Woot.
     virginianum L.
          SY=Cathartolinum v. (L.) Reichenb.
     westii Rogers

Radiola Hill

     linoides Roth
          SY=Millegrana radiola (L.) Druce

Sclerolinon Rogers

     digynum (Gray) Rogers
          SY=Linum d. Gray

LOASACEAE

Cevallia Lag.

     sinuata Lag.

Eucnide Zucc.

     bartonioides Zucc.
     rupestris (Baill.) Thompson & Ernst
          SY=Sympetaleia r. (Baill.) Gray ex
               S. Wats.
     urens (Parry ex Gray) Parry

Mentzelia L.

     affinis Greene
     albescens (Gillies) Griseb.
     albicaulis (Dougl. ex Hook.) Torr. & Gray
          var. albicaulis
          var. ctenophora (Rydb.) J. Darl.
     argillosa J. Darl.
     aspera L.
     asperula Woot. & Standl.
     californica Thompson & Roberts
     chrysantha Engelm. ex Brandeg.
          SY=M. lutea Greene
     congesta (Nutt.) Torr. & Gray
          var. congesta
          var. davidsoniana (Abrams) J.F.
               Macbr.
     crocea Kellogg
          SY=M. lindleyi Torr. & Gray ssp.
               c. (Kellogg) C.B. Wolf
     cronquistii Thompson & Zavortink
     decapetala (Pursh) Urban & Gilg
          SY=Nuttallia d. (Pursh) Greene
     densa Greene
     desertorum (A. Davids.) Thompson & Roberts
          SY=Acrolasia d. A. Davids.
     dispersa S. Wats.
          var. compacta (A. Nels.) J.F. Macbr.
          var. dispersa
          var. latifolia (Rydb.) J.F. Macbr.
          var. obtusa Jepson
     eremophila (Jepson) Thompson & Roberts
     floridana Nutt. ex Torr. & Gray
     gracilenta (Nutt.) Torr. & Gray
     hirsutissima S. Wats.
          var. stenophylla (Urban & Gilg) I.M.
               Johnston
     humilis (Gray) J. Darl.
     incisa Urban & Gilg
     integra (M.E. Jones) Tidestrom
          SY=M. multiflora (Nutt.) Gray var.
               i. M.E. Jones

involucrata S. Wats.
     var. involucrata
     var. megalantha I.M. Johnston
jonesii (Urban & Gilg) Thompson & Roberts
          SY=M. nitens Greene var. j. (Urban
               & Gilg) J. Darl.
laciniata (Rydb.) J. Darl.
laevicaulis (Dougl. ex Hook.) Torr. & Gray
     var. laevicaulis
          SY=M. l. var. acuminata (Rydb.) A.
               Nels. & J.F. Macbr.
     var. parviflora (Dougl. ex Hook.)
          C.L. Hitchc.
          SY=M. brandegei S. Wats.
leucophylla Brandeg.
lindheimeri Urban & Gilg
lindleyi Torr. & Gray
longiloba J. Darl.
mexicana Thompson & Zavortink
micrantha (Hook. & Arn.) Torr. & Gray
mojavensis Thompson & Roberts
mollis M.E. Peck
montana (A. Davids.) A. Davids.
          SY=Acrolasia m. A. Davids.
multicaulis (Osterhout) A. Nels.
multiflora (Nutt.) Gray
          SY=M. nuda (Pursh) Torr. & Gray
               var. rusbyi (Woot.)
               Harrington
          SY=M. pumila (Nutt.) Torr. & Gray
               var. m. (Nutt.) Urban & Gilg
          SY=M. r. Woot.
          SY=M. speciosa Osterhout
nitens Greene
     var. leptocaulis J. Darl.
     var. nitens
nuda (Pursh) Torr. & Gray
     var. nuda
          SY=Nuttallia n. (Pursh) Greene
     var. stricta (Osterhout) Harrington
          SY=M. s. (Osterhout) Stevens ex
               Jeffs & Little
          SY=Nuttallia s. (Osterhout) Greene
obscura Thompson & Roberts
oligosperma Nutt.
oreophila J. Darl.
packardiae Glad
pectinata Kellogg
          SY=M. gracilenta (Nutt.) Torr. &
               Gray var. p. (Kellogg)
               Jepson
perennis Woot.
pinetorum Heller
polita A. Nels.
pterosperma Eastw.
puberula J. Darl.
pumila (Nutt.) Torr. & Gray
          SY=M. p. var. procera (Woot. &
               Standl.) J. Darl.
ravenii Thompson & Roberts
reflexa Coville
reverchonii (Urban & Gilg) Thompson &
     Zavortink
saxicola Thompson & Zavortink
sinuata (Rydb.) R.J. Hill
springeri (Standl.) Tidestrom
strictissima (Woot. & Standl.) J. Darl.
texana Urban & Gilg
thompsonii Glad
torreyi Gray
     var. acerosa (M.E. Jones) Barneby
     var. torreyi
tricuspis Gray
tridentata (A. Davids.) Thompson & Roberts
          SY=M. tricuspis Gray var.
               brevicornuta I.M. Johnston
veatchiana Kellogg

Mentzelia   (CONT.)
          SY=M. albicaulis (Dougl. ex Hook.)
             Torr. & Gray var. gracilis
             (Rydb.) J. Darl.
          SY=M. a. var. v. (Kellogg) Urban &
             Gilg
          SY=M. gracilenta (Nutt.) Torr. &
             Gray var. v. (Kellogg)
             Jepson

Petalonyx Gray

   linearis Greene
   nitidus S. Wats.
   parryi Gray
   thurberi Gray
      ssp. gilmanii (Munz) Davis & Thompson
         SY=P. g. Munz
      ssp. thurberi

LOGANIACEAE

Buddleja L.

   asiatica Lour.
   davidii Franch.
   lindleyana Fortune ex Lindl.
         SY=Adenoplea l. (Fortune ex
            Lindl.) Small
   marrubiifolia Benth.
   racemosa Torr.
      ssp. incana (Torr.) Norm.
         SY=B. r. var. i. Torr.
      ssp. racemosa
   scordioides H.B.K.
   sessiliflora H.B.K.
   utahensis Coville

Chilianthus Burchell

   oleaceus Burchell

Cynoctonum J.F. Gmel.

   mitreola (L.) Britt.
   sessilifolium (Walt.) Jaume St. Hil.
      var. angustifolia Torr. & Gray
        SY=C. a. (Torr. & Gray) Small
      var. microphyllum R.W. Long
      var. sessilifolium
   succulentum R.W. Long

Emorya Torr.

   suaveolens Torr.

Gelsemium Juss.

   rankinii Small
   sempervirens (L.) St. Hil.

Labordia Gaud.

   baillonii St. John
   cyrtandrae (Baill.) St. John
      var. cyrtandrae
      var. nahikuana Sherff
   decurrens Sherff
      var. decurrens
      var. pocillata Sherff
   degeneri Sherff
      var. degeneri
      var. subcarinata Sherff
   fagraeoidea Gaud.

      var. conferta Sherff
      var. fagraeoidea
      var. hillebrandii Sherff
      var. humei Sherff
      var. jugorum Sherff
      var. longisepala Sherff
      var. multinervia Sherff
      var. saint-johniana Sherff
      var. septentrionalis Sherff
      var. sessilis (Gray) Sherff
      var. simulans Deg. & Sherff
      var. waianaeana Sherff
   glabra Hbd.
      var. glabra
      var. latisepala Sherff
      var. orientalis Sherff
   hedyosmifolia Baill.
      var. centralis (Skottsberg) St. John
      var. grayana (Hbd.) Sherff
      var. hedyosmifolia
      var. hosakana Sherff
      var. kilaueana Sherff
      var. magnifolia Deg. & Sherff
      var. robusta Sherff
      var. rockii author ?
      var. skottsbergii Sherff
   helleri Sherff
      var. helleri
      var. macrocarpa Sherff
   hirtella Mann
      var. haleakalana Sherff
      var. hirtella
      var. hispidior Sherff
      var. imbricata Deg. & Sherff
      var. laevis Sherff
      var. laevisepala Sherff
      var. microcalyx Hbd.
      var. microphylla Hbd.
      var. sororia Sherff
   hymenopoda Deg. & Sherff
   kaalae Forbes
      var. brachypoda Sherff
      var. fosbergii Sherff
      var. kaalae
      var. kauaiensis Sherff
      var. mendax Sherff
   lydgatei Forbes
   mauiensis Sherff
   membranacea Mann
      var. exigua Sherff
      var. membranacea
   molokaiana Baill.
      var. bryanii Sherff
      var. congesta Deg. & Sherff
      var. lophocarpa (Hbd.) Sherff
      var. molokaiana
      var. munroi Sherff
      var. phyllocalyx (Hbd.) Sherff
      var. setosa Deg. & Sherff
   nelsonii St. John
   olympiana Sherff
   pallida Mann
      var. hispidula Sherff
      var. pallida
   pedunculata Sherff
   pumila (Hbd.) Skottsberg
   tinifolia Gray
      var. euphorbioidea Sherff
      var. forbesii Sherff
      var. haupuensis Sherff
      var. honoluluensis Sherff
      var. lanaiensis Sherff
      var. leptantha Sherff
      var. microgyna Deg. & Sherff
      var. parvifolia Sherff
      var. tenuifolia Deg. & Sherff
      var. tinifolia

Labordia (CONT.)
        var. waialuana Sherff
   triflora Hbd.
   venosa Sherff
   waialealae Wawra
   wawrana Sherff

Polypremum L.

   procumbens L.

Spigelia L.

   anthelmia L.
   gentianoides Chapman
   lindheimeri Gray
   loganioides (Torr. & Gray) A. DC.
           SY=Coelostylis l. Torr. & Gray
   marilandica L.
   texana (Torr. & Gray) A. DC.

LORANTHACEAE

Arceuthobium Bieb.

   abietinum Engelm. ex Munz
   americanum Nutt. ex Engelm.
   apachecum Hawksworth & Wiens
   blumeri A. Nels.
   californicum Hawksworth & Wiens
   campylopodum Engelm.
   cyanocarpum Coult. & A. Nels.
   divaricatum Engelm.
   douglasii Engelm. ex L.C. Wheeler
   gillii Hawksworth & Wiens
   laricis (Piper) St. John
   microcarpum (Engelm.) Hawksworth & Wiens
   occidentale Engelm.
   pusillum M.E. Peck
           SY=Razoumofskya p. (M.E. Peck)
               Kuntze
   tsugense (Rosendahl) G.N. Jones
   vaginatum (Willd.) J. Presl
      ssp. cryptopodum (Engelm.) Hawksworth &
         Wiens
           SY=A. c. Engelm.

Dendropemon (Blume) Reichenb.

   bicolor Krug & Urban
           SY=Phthirusa b. (Krug & Urban)
               Engl.
   caribaeum Kurg & Urban
           SY=Phthirusa c. (Krug & Urban)
               Engl.
   purpureum (L.) Krug & Urban
           SY=Phthirusa p. (L.) Engl.
   sintenisii Krug & Urban
           SY=Phthirusa s. (Krug & Urban)
               Engl.

Dendrophthora Eichl.

   domingensis (Spreng.) Eichl.
   flagelliformis (Lam.) Krug & Urban
           SY=D. wrightii Eichl.

Eremolepis Griseb.

   wrightii Griseb.
           SY=Ixidium w. (Griseb.) Eichl.

Eubrachion Hook. f.

ambiguum (Hook. & Arn.) Engl.

Korthalsella v. Tiegh.

   complanata (v. Tiegh.) Engl.
   cylindrica (v. Tiegh.) Engl.
           var. cylindrica
           var. teres (Wawra) St. John
   degeneri Danser
   latissima (v. Tiegh.) Danser
           var. crassa (v. Tiegh.) Danser
           var. latissima
   platycaula (Bertero ex v. Tiegh.) Engl.
   remyana v. Tiegh.
           var. remyana
           var. wawrae (v. Tiegh.) Danser
   remyi (v. Tiegh.) Skottsberg

Phoradendron Nutt.

   berterianum (DC.) Griseb.
           SY=P. dichotomum (Bertero) Krug &
               Urban
   bolleanum (Seem.) Eichl.
      ssp. bolleanum
      ssp. densum (Torr.) Wiens
           SY=P. b. var. d. (Torr.) Fosberg
           SY=P. d. Torr.
      ssp. pauciflorum (Torr.) Wiens
           SY=P. b. var. p. (Torr.) Fosberg
           SY=P. p. Torr.
   californicum Nutt.
           SY=P. c. var. distans Trel.
           SY=P. c. var. leucocarpum (Trel.)
               Jepson
   capitellatum Torr. ex Trel.
           SY=P. bolleanum (Seem.) Eichl.
               var. c. (Torr. ex Trel.)
               Kearney & Peebles
   chrysocarpum Krug & Urban
           SY=Viscum anceps DC.
   helleri Trel.
   hexastichum (DC.) Griseb.
   juniperinum Engelm. ex Gray
      ssp. juniperinum
           SY=P. j. var. ligatum (Trel.)
               Fosberg
           SY=P. l. Trel.
      ssp. libocedrii (Engelm.) Wiens
           SY=P. j. var. l. Engelm.
           SY=P. l. (Engelm.) T.J. Howell
   piperoides (H.B.K.) Trel.
   racemosum (Aubl.) Krug & Urban
   randiae (Bello) Britt.
   rubrum (L.) Griseb.
   serotinum (Raf.) M.C. Johnston
           SY=P. eatonii Trel.
           SY=P. flavescens (Pursh) Nutt.
           SY=P. f. var. orbiculatum
               (Engelm.) Engelm.
           SY=P. macrotomum Trel.
           SY=P. s. var. m. (Trel.) M.C.
               Johnston
   tetrapterum Krug & Urban
   tomentosum (DC.) Engelm. ex Gray
      ssp. macrophyllum (Engelm.) Wiens
           SY=P. cockerellii Trel.
           SY=P. coloradense Trel.
           SY=P. flavescens (Pursh) Nutt.
               var. m. Engelm.
           SY=P. longispicum Trel.
           SY=P. m. (Engelm.) Cockerell
      ssp. tomentosum
           SY=P. flavescens (Pursh) Nutt.
               var. pubescens Engelm. ex
               Gray
           SY=P. serotinum (Raf.) M.C.

Phoradendron (CONT.)
           Johnston var. p. (Engelm.
           ex Gray) M.C. Johnston
  trinervium (Lam.) Griseb.
  villosum (Nutt.) Nutt.
    ssp. coryae (Trel.) Wiens
        SY=P. c. Trel.
        SY=P. havardianum Trel.
    ssp. villosum
        SY=P. flavescens (Pursh) Nutt.
        var. v. (Nutt.) Engelm.

Viscum L.

  album L.

LYTHRACEAE

Ammannia L.

  auriculata Willd.
      var. arenaria (H.B.K.) Koehne
      var. auriculata
  coccinea Rottb.
      SY=A. c. var. purpurea (Lam.)
        Koehne
  latifolia L.
      SY=A. koehnei Britt.
      SY=A. teres Raf.
      SY=A. t. var. exauriculata Fern.

Cuphea P. Br.

  aspera Chapman
      SY=Parsonsia lythroides Small
  carthagenensis (Jacq.) J.F. Macbr.
      SY=Parsonsia balsamona (Cham. &
        Schlecht.) Standl.
  glutinosa Cham. & Schlecht.
  hyssopifolia H.B.K.
  micrantha H.B.K.
      SY=Parsonsia m. (H.B.K.) Jennings
  parsonia (L.) R. Br. ex Steud.
      SY=Parsonsia p. (L.) Britt.
  procumbens Cav.
      SY=Parsonsia p. (Cav.) Small
  viscosissima Jacq.
      SY=C. petiolata (L.) Koehne
      SY=Parsonsia p. (L.) Rusby
  wrightii Gray
      var. nematopetala Bacig.
      var. wrightii

Decodon J.F. Gmel.

  verticillatus (L.) Ell.
      var. laevigatus Torr. & Gray
      var. verticillatus

Didiplis Raf.

  diandra (DC.) Wood
      SY=Peplis d. (DC.) Nutt.

Ginoria Jacq.

  rohrii (Vahl) Koehne

Heimia Link

  salicifolia (H.B.K.) Link

Lagerstroemia L.

indica L.
speciosa (L.) Pers.

Lawsonia L.

  inermis L.

Lythrum L.

  alatum Pursh
      var. alatum
      var. lanceolatum (Ell.) Torr. & Gray
        ex Roth
        SY=L. l. Ell.
  californicum Torr. & Gray
  curtisii Fern.
  dacotanum Nieuwl.
  flagellare Shuttlw. ex Chapman
      SY=L. vulneraria Ait.
  hyssopifolia L.
      SY=L. adsurgens Greene
  lineare L.
  maritimum H.B.K.
  ovalifolium Koehne
  portula (L.) D.A. Webb
      SY=Peplis p. L.
  salicaria L.
      SY=L. s. var. gracilior Turcz.
      SY=L. s. var. tomentosum (P.
        Mill.) DC.
      SY=L. s. var. vulgare DC.
  tribracteatum Salzm. ex Spreng.
  virgatum L.

Nesaea Comm. ex H.B.K.

  longipes Gray
      SY=Heimia l. (Gray) Cory

Rotala L.

  catholica (Cham. & Schlecht.) van Leeuwen
  indica (Willd.) Koehne
  ramosior (L.) Koehne
      SY=R. dentifera (Gray) Koehne
      SY=R. r. var. interior Fern. &
        Grisc.
      SY=R. r. var. typica Fern. &
        Grisc.

MAGNOLIACEAE

Liriodendron L.

  tulipifera L.

Magnolia L.

  acuminata (L.) L.
      var. acuminata
        SY=M. a. var. ozarkensis Ashe
        SY=Tulipastrum a. (L.) Small
      var. subcordata (Spach) Dandy
        SY=M. a. var. cordata (Michx.)
          Sarg.
        SY=M. c. Michx.
        SY=Tulipastrum c. (Michx.) Small
  ashei Weatherby
      SY=M. macrophylla Michx. ssp. a.
        (Weatherby) Spongberg
  fraseri Walt.
  grandiflora L.
  macrophylla Michx.
  portoricensis Bello

Magnolia (CONT.)
  pyramidata Bartr.
  splendens Urban
  tripetala L.
  virginiana L.
          SY=M. v. var. australis Sarg.
          SY=M. v. var. parva Ashe

MALPIGHIACEAE

Aspicarpa L.C. Rich.

  hirtella L.C. Rich.
  humilis (Benth.) Juss.
  hyssopifolia Gray
  longipes Gray

Banisteriopsis C.B. Robins. & Small

  lucida (L.C. Rich.) Small
          SY=Banisteria l. L.C. Rich.

Brachypterys A. Juss.

  ovata (Cav.) Small

Bunchosia L.C. Rich. ex A. Juss.

  glandulifera (Jacq.) H.B.K.
  glandulosa (Cav.) L.C. Rich.
  nitida (Jacq.) DC.

Byrsonima L.C. Rich. ex A. Juss.

  coriacea (Sw.) DC.
          SY=B. spicata (Cav.) DC.
  crassifolia (L.) H.B.K.
  horneana Britt. & Small
  lucida DC.
          SY=B. cuneata (Turcz.) P. Wilson
  ophiticola Small
  wadsworthii Little

Galphimia Cav.

  angustifolia Benth.
          SY=Thryallis a. (Benth.) Kuntze
  glauca Cav.
          SY=Thryallis g. (Cav.) Kuntze
  gracilis Bartl.
          SY=Thryallis g. (Bartl.) Kuntze

Heteropteris Kunth

  laurifolia (L.) A. Juss.
          SY=Banisteria l. L.
  purpurea (L.) Kunth
          SY=Banisteria p. L.
  wydleriana A. Juss.
          SY=Banisteria w. (A. Juss.) C.B.
              Robins.

Janusia A. Juss.

  gracilis Gray

Malpighia L.

  coccigera L.
  fucata Ker-Gawl.
  glabra L.
  infestissima (A. Juss.) L.C. Rich.
  linearis Jacq.
  pallens Small

  punicifolia L.
  shaferi Britt. & Wilson
  thompsonii Britt. & Small

Stigmatophyllum A. Juss.

  cordifolium Ndz.
  diversifolium A. Juss.
          SY=S. ledifolium (H.B.K.) Small
  periplocifolium (Desf.) A. Juss.
          SY=S. lingulatum (Poir.) Small
  puberum (L.C. Rich.) A. Juss.
  tomentosum (Desf.) Ndz.

Tetrapteris Cav.

  buxifolia Cav.
  citrifolia (Sw.) Pers.
          SY=T. inaequalis Cav.

MALVACEAE

Abelmoschus Medic.

  esculentus (L.) Moench
          SY=Hibiscus e. L.
  manihot (L.) Medic.
          SY=Hibiscus m. L.
  moschatus Medic.
          SY=Hibiscus abelmoschus L.

Abutilon P. Mill.

  abutiloides (Jacq.) Strong ined.
          SY=A. americanum (L.) Sweet
          SY=A. jacquinii G. Don
          SY=A. lignosum (Cav.) G. Don
  berlandieri Gray ex S. Wats.
          SY=A. californicum sensu auctt.
              non Benth.
  eremitopetalum Caum
  glabriflorum Hochr.
  grandifolium (Willd.) Sweet
  hirtum (Lam.) Sweet
  hulseanum (Torr. & Gray) Torr. ex Chapman
          SY=A. commutatum K. Schum.
          SY=?A. pauciflorum Sweet
  hypoleucum Gray
  incanum (Link) Sweet
          SY=A. pringlei Hochr.
  indicum (L.) Sweet
  malacum S. Wats.
  menziesii Seem.
  palmeri Gray
  parishii S. Wats.
  parvulum Gray
  permolle (Willd.) Sweet
  pictum (Gilies ex Hook.) Walp.
          SY=A. striatum Dickson ex Lindl.
  reventum S. Wats.
  sandwicense (Deg.) Christoph.
          var. sandwicense
          var. welchii Christoph.
  sonorae Gray
  texensis Torr. & Gray
  theophrasti Medic.
          SY=A. abutilon (L.) Rusby
          SY=A. avicennae Gaertn.
  thurberi Gray
  trisulcatum (Jacq.) Urban
          SY=A. triquetrum (L.) Sweet
  umbellatum (L.) Sweet
  virginianum Krapov.
          SY=Sida eggersii F.G. Baker

Abutilon (CONT.)
  wrightii Gray

Alcea L.

   rosea L.
            SY=Althaea r. (L.) Cav.

Allosidastrum (Hochr.) Krapov. & Fryxell

   pyramidatum (Desportes ex Cav.) Krapov. &
      Fryxell ined.
            SY=Sida p. Desportes ex Cav.

Allowissadula Bates

   holosericea (Scheele) Bates
            SY=Abutilon marshii Standl.
            SY=Wissadula h. (Scheele) Gracke
   lozanii (Rose) Bates
            SY=Pseudabutilon l. (Rose) R.E.
                Fries

Althaea L.

   cannabina L.
   hirsuta L.
   officinalis L.

Anoda Cav.

   abutiloides Gray
   acerifolia Cav.
            SY=A. reticulata S. Wats.
   crenatiflora Ortega
   cristata (L.) Schlecht.
            SY=A. c. var. brachyanthera
                (Reichenb.) Hochr.
            SY=A. c. var. digitata (Gray)
                Hochr.
   pentaschista Gray
            SY=A. p. var. obtusior B.L.
                Robins.
   thurberi Gray
   wrightii Gray

Bastardia H.B.K.

   bivalvis (Cav.) H.B.K.
   viscosa (L.) H.B.K.

Batesimalva Fryxell

   violacea (Rose) Fryxell
            SY=Gaya v. Rose

Callirhoe Nutt.

   alcaeoides (Michx.) Gray
   bushii Fern.
            SY=C. papaver (Cav.) Gray var. b.
                (Fern.) Waterfall
            SY=C. involucrata (Nutt. ex Torr.
                & Gray) Gray var. b. (Fern.)
                R.F. Martin
   digitata Nutt.
            var. digitata
            var. stipulata Waterfall
   involucrata (Nutt. ex Torr. & Gray) Gray
            var. involucrata
            var. lineariloba (Torr. & Gray) Gray
   leiocarpa R.F. Martin
            SY=C. pedata Gray
   papaver (Cav.) Gray
   scabriuscula B.L. Robins.
   triangulata (Leavenworth) Gray

Cienfuegosia Cav.

   drummondii (Gray) Lewton
   heterophylla (Vent.) Garcke
   yucatanensis Millsp.

Eremalche Greene

   exilis (Gray) Greene
            SY=Malvastrum e. Gray
   kernensis C.B. Wolf
            SY=Malvastrum k. (C.B. Wolf) Munz
   parryi (Greene) Greene
            SY=Malvastrum p. Greene
   rotundifolia (Gray) Greene
            SY=Malvastrum r. Gray

Fryxellia Bates

   pygmaea (Correll) Bates
            SY=Anoda p. Correll

Gossypium L.

   barbadense L.
            SY=G. brasiliense Macfad.
            SY=G. peruvianum Cav.
   hirsutum L.
            SY=G. h. var. punctatum (K.
                Schum.) Hutchinson
            SY=G. purpurascens Poir.
   thurberi Todaro
            SY=Thurberia thespesioides Gray
   tomentosum Nutt. ex Seem.
            SY=G. sandvicense Parl.

Herissantia Medic.

   crispa (L.) Briz.
            SY=Abutilon c. (L.) Sweet
            SY=Bogenhardia c. (L.) Kearney
            SY=Gayoides c. (L.) Small

Hibiscadelphus Rock

   bombycinus Forbes
   distans Bishop & Herbst
   giffardianus Rock
   hualalaiensis Rock
   wilderianus Rock

Hibiscus L.

   acetosella Welw. ex Hiern
   aculeatus Walt.
            SY=H. scaber Michx.
   arnottianus Gray
            var. arnottianus
            var. punaluuensis (Skottsberg) Deg. &
                Deg.
   bifurcatus Cav.
   biseptus S. Wats.
   brackenridgei Gray
            var. brackenridgei
            var. mokuleiana Roe
            var. molokaianus Rock
   brasiliensis L.
            SY=H. phoeniceus Jacq.
   cannabinus L.
   cardiophyllus Gray
   clayi Deg. & Deg.
   clypeatus L.
   coccineus (Medic.) Walt.
            SY=H. semilobatus Chapman
   coulteri Harvey ex Gray
   dasycalyx Blake & Shiller
   denudatus Benth.

Hibiscus (CONT.)
            SY=H. d. var. involucellatus Gray
    elatus Sw.
            SY=Paritium e. (Sw.) G. Don
    furcellatus Desr.
            SY=H. youngianus Gaud. ex Hook. &
                Arn.
    grandiflorus Michx.
    immaculatus Roe
    kahilii Forbes
    kokio Hbd. ex Wawra
            var. kokio
            var. pekeloi Deg. & Deg.
            var. pukoonis Caum
    laevis All.
            SY=H. militaris Cav.
    lambertianus H.B.K.
            SY=H. cubensis A. Rich.
    lasiocarpus Cav.
            SY=H. californicus Kellogg
            SY=H. leucophyllus Shiller
            SY=H. platanoides Greene
    moscheutos L.
            SY=H. incanus Wendl. f.
            SY=H. m. ssp. incanus (Wendl. f.)
                Ahles
            SY=H. m. ssp. palustris (L.)
                Clausen
            SY=?H. m. var. purpurascens Sweet
            SY=H. oculiroseus Britt.
            SY=H. opulifolius Greene
            SY=H. palustris L.
            SY=H. pinetorum Greene
    mutabilis L.
    newhousei Roe
    oahuensis Deg. & Deg.
    poeppigii (Spreng.) Garcke
            SY=H. pilosus sensu auctt. non
                (Sw.) Fawcett & Rendle
    rockii Deg. & Deg.
    roeatae St. John
    rosa-sinensis L.
    sabdariffa L.
    saintjohnianus Roe
    schizopetalus (Masters) Hook. f.
    syriacus L.
    tiliaceus L.
            SY=Pariti grande Britt. ex Small
            SY=Paritium t. (L.) St. Hil.
    trilobus Aubl.
    trionum L.
            SY=Trionum t. (L.) Woot. & Standl.
    ula Deg. & Deg.
    vitifolius L.
    waimeae Heller
            var. hannerae Deg. & Deg.
            var. waimeae

Horsfordia Gray

    alata (S. Wats.) Gray
    newberryi (S. Wats.) Gray

Iliamna Greene

    bakeri (Jepson) Wiggins
    crandallii (Rydb.) Wiggins
    grandiflora (Rydb.) Wiggins
            SY=I. angulata Greene
    latibracteata Wiggins
    longisepala (Torr.) Wiggins
    remota Greene
            SY=I. corei Sherff
            SY=Phymosia r. (Greene) Britt.
    rivularis (Dougl. ex Hook.) Greene
            var. diversa (A. Nels.) Wiggins
            var. rivularis

            SY=I. acerifolia Greene

Kokia Lewt.

    cookei Deg.
    drynarioides (Seem.) Lewt.
    kauaiensis (Rock) Deg. & Duvel
    lanceolata Lewt.

Kosteletzkya Presl

    depressa (L.) Blanch, Fryxell & Bates ined.
            SY=K. pentasperma (Bertero ex DC.)
                Griseb.
    smilacifolia Gray
    virginica (L.) Presl ex Gray
            var. althiifolia Chapman
                SY=K. a. (Chapman) Rusby
            var. aquilonia Fern.
            var. virginica

Lavatera L.

    arborea L.
    assurgentiflora Kellogg
    cretica L.
    thuringiaca L.
    trimestris L.

Malachra L.

    alceifolia Jacq.
    capitata L.
    fasciata Jacq.
            SY=M. alceifolia Jacq. var. f.
                (Jacq.) Robyns
    radiata L.
    urens Poit.

Malacothamnus Greene

    abbottii (Eastw.) Kearney
            SY=Malvastrum a. Eastw.
    aboriginum (B.L. Robins.) Greene
            SY=Malvastrum a. B.L. Robins.
    clementinus (Munz & Johnston) Kearney
            SY=Malvastrum c. Munz & Johnston
    davidsonii (B.L. Robins.) Greene
            SY=Malvastrum d. B.L. Robins.
    densiflorus (S. Wats.) Greene
            var. densiflorus
                SY=Malvastrum d. S. Wats.
            var. viscidus (Abrams) Kearney
                SY=Malvastrum d. var. v. (Abrams)
                    Estes
    fasciculatus (Nutt. ex Torr. & Gray) Greene
            SY=Malacothamnus arcuatus (Greene)
                Greene
            SY=Malacothamnus f. ssp.
                catalinensis (Eastw.) Thorne
            SY=Malacothamnus f. ssp.
                laxiflorus (Gray) Thorne
            SY=Malacothamnus f. var. c.
                (Eastw.) Kearney
            SY=Malacothamnus f. var. l. (Gray)
                Kearney
            SY=Malacothamnus f. var.
                nesioticus (B.L. Robins.)
                Kearney
            SY=Malacothamnus f. var. nuttallii
                (Abrams) Kearney
            SY=Malacothamnus hallii (Eastw.)
                Kearney
            SY=Malacothamnus mendocinensis
                (Eastw.) Kearney
            SY=Malva f. Nutt. ex Torr. & Gray
            SY=Malvastrum a. (Greene) B.L.

Malacothamnus (CONT.)
                  Robins.
         SY=Malvastrum f. (Nutt. ex Torr. &
              Gray) Greene
         SY=Malvastrum f. var. l. (Gray)
              Munz & Johnston
         SY=Malvastrum h. Eastw.
         SY=Malvastrum m. Eastw.
         SY=Malvastrum nesioticus B.L.
              Robins.
   foliosus (S. Wats.) Kearney
         SY=Malacothamnus paniculatus
              (Gray) Kearney
   fremontii (Torr. ex Gray) Greene
         SY=Malacothamnus f. ssp.
              cercophorus (B.L. Robins.)
              Munz
         SY=Malacothamnus helleri (Eastw.)
              Kearney
         SY=Malacothamnus niveus (Eastw.)
              Kearney
         SY=Malacothamnus orbiculatus
              (Greene) Greene
         SY=Malvastrum f. Torr. ex Gray
         SY=Malvastrum f. ssp. exfibulosum
              (Jepson) Wiggins
         SY=Malvastrum f. var. c. B.L.
              Robins.
         SY=Malvastrum f. var. n. (Eastw.)
              McMinn
         SY=Malvastrum o. Greene
   jonesii (Munz) Kearney
         SY=Malacothamnus gracilis (Eastw.)
              Kearney
         SY=Malvastrum g. Eastw.
         SY=Malvastrum j. Munz
   marrubioides (Dur. & Hilg.) Greene
         SY=Malvastrum m. Dur. & Hilg.
   palmeri (S. Wats.) Greene
         SY=Malacothamnus p. var.
              involucratus (B.L. Robins.)
              Kearney
         SY=Malacothamnus p. var. lucianus
              Kearney
         SY=Malvastrum p. S. Wats.
         SY=Malvastrum p. var. i. (B.L.
              Robins.) McMinn
   parishii (Eastw.) Kearney

Malva L.

   alcea L.
   crispa (L.) L.
         SY=M. verticillata L. var. c. L.
   moschata L.
   neglecta Wallr.
         SY=M. rotundifolia sensu auctt.
              non L.
   nicaeensis All.
   parviflora L.
   rotundifolia L.
         SY=M. pusilla Sm.
   sylvestris L.
         SY=M. mauritiana L.
         SY=M. s. ssp. m. (L.) Boiss.
         SY=M. s. var. m. (L.) Boiss.
   verticillata L.

Malvastrum Gray

   americanum (L.) Torr.
         SY=M. spicatum (L.) Gray
   aurantiacum (Scheele) Walp.
         SY=M. wrightii Gray
   bicuspidatum (S. Wats.) Rose
   corchorifolium (Desv.) Britt. ex Small
   coromandelianum (L.) Garcke

Malvaviscus Guettard

   arboreus Cav.
         var. arboreus
         var. drummondii (Torr. & Gray) Schery
              SY=M. d. Torr. & Gray
         var. mexicanus Schlecht.
   penduliflorus DC.

Malvella Jaubert & Spach

   lepidota (Gray) Fryxell
         SY=Sida l. Gray
   leprosa (Ortega) Krapov.
         SY=Sida hederacea (Dougl.) Torr. &
              Gray
         SY=S. l. (Ortega) K. Schum.
         SY=?S. l. var. depauperata (Gray)
              I. Clem.
         SY=S. l. var. h. (Dougl.) K.
              Schum.
   sagittifolia (Gray) Fryxell
         SY=Sida lepidota Gray var. s. Gray
         SY=S. leprosa (Ortega) Krapov.
              var. s. (Gray) I. Clem.
         SY=S. s. (Gray) Cory

Meximalva Fryxell

   filipes (Gray) Fryxell
         SY=Sida f. Gray

Modiola Moench

   caroliniana (L.) G. Don

Napaea L.

   dioica L.

Pavonia Cav.

   fruticosa (P. Mill.) Fawcett & Rendle
         SY=Typhalaea f. (P. Mill.) Britt.
   hastata Cav.
   lasiopetala Scheele
   paniculata Cav.
         SY=Lebretonia p. (Cav.) Britt.
         SY=P. corymbosa (Sw.) Willd.
         SY=P. wrightii Gray
   scabra (B. Vogel) Ciferri
         SY=Malache s. B. Vogel
   spicata Cav.
         SY=Pavonia racemosa Sw.
   spinifex (L.) Cav.

Rynchosida Fryxell ined.

   physocalyx (Gray) Fryxell ined.
         SY=Sida p. Gray

Sida L.

   acuta Burm. f.
         SY=S. carpinifolia L. f.
   aggregata Presl
         SY=S. setifera Presl
   ciliaris L.
         var. ciliaris
         var. mexicana (Moric.) Shinners
   cordata (Burm. f.) Borss.
         SY=S. humilis Cav.
   cordifolia L.
   elliottii Torr. & Gray
         SY=S. leptophylla Small
         SY=S. rubromarginata Nash
   fallax Walp.

Sida (CONT.)
       var. fallax
       var. kauaiensis Hochr.
  glabra P. Mill.
  glomerata Comm. ex Cav.
  glutinosa Comm. ex Cav.
  grayana I. Clem.
       SY=S. cuneifolia Gray non Roxb.
       SY=S. helleri Rose
  hermaphrodita (L.) Rusby
  inflexa Fern.
  jamaicensis L.
  ledyardii St. John
  lindheimeri Engelm. & Gray
  longipes Gray
  meyeniana Walp.
  nelsonii St. John
  neomexicana Gray
  procumbens Sw.
       SY=S. filicaulis Torr. & Gray
       SY=S. filiformis Moric.
       SY=S. supina L'Her.
  rhombifolia L.
  salviifolia Presl
       SY=S. erecta Macfad.
       SY=S. holwayi Baker & Rose
  spinosa L.
       var. angustifolia (Lam.) Griseb.
       var. spinosa
          SY=S. alba L.
  tragiifolia Gray
  urens L.

Sidalcea Gray

  calycosa M.E. Jones
    ssp. calycosa
    ssp. rhizomata (Jepson) Munz
       SY=S. r. Jepson
  campestris Greene
  candida Gray
       var. candida
       var. glabrata C.L. Hitchc.
  covillei Greene
       SY=S. neomexicana Gray var. c.
          (Greene) Roush
  cusickii Piper
       SY=S. c. ssp. purpurea C.L.
          Hitchc.
       SY=S. oregana (Nutt. ex Torr. &
          Gray) Gray var. c. (Piper)
          Roush
  diploscypha (Torr. & Gray) Gray
  glaucescens Greene
  hartwegii Gray ex Benth.
  hendersonii S. Wats.
  hickmanii Greene
    ssp. anomala C.L. Hitchc.
    ssp. hickmanii
    ssp. parishii (B.L. Robins.) C.L.
      Hitchc.
       SY=S. h. var. p. B.L. Robins.
    ssp. viridis C.L. Hitchc.
  hirsuta Gray
  hirtipes C.L. Hitchc.
  keckii Wiggins
  malachroides (Hook. & Arn.) Gray
  malviflora (DC.) Gray ex Benth.
    ssp. asprella (Greene) C.L. Hitchc.
       SY=S. a. Greene
    ssp. californica (Nutt. ex Torr. & Gray)
      C.L. Hitchc.
       SY=S. m. var. c. (Nutt. ex Torr. &
          Gray) Jepson
    ssp. celata (Jepson) C.L. Hitchc.
       SY=S. m. var. c. Jepson
    ssp. dolosa C.L. Hitchc.

    ssp. elegans (Greene) C.L. Hitchc.
    ssp. laciniata C.L. Hitchc.
       var. laciniata C.L. Hitchc.
       var. sancta C.L. Hitchc.
    ssp. malviflora
    ssp. nana (Jepson) C.L. Hitchc.
       SY=S. reptans Greene var. n.
          Jepson
    ssp. patula C.L. Hitchc.
    ssp. purpurea C.L. Hitchc.
    ssp. rostrata (Eastw.) Wiggins
    ssp. sparsifolia C.L. Hitchc.
       var. hirsuta C.L. Hitchc.
       var. sparsifolia C.L. Hitchc.
       var. stellata C.L. Hitchc.
       var. uliginosa C.L. Hitchc.
    ssp. virgata (T.J. Howell) C.L. Hitchc.
       SY=S. v. T.J. Howell
  multifida Greene
  nelsoniana Piper
  neomexicana Gray
    ssp. crenulata (A. Nels.) C.L. Hitchc.
       SY=S. n. var. c. (A. Nels.) C.L.
          Hitchc.
    ssp. neomexicana
    ssp. thurberi (B.L. Robins. ex Gray)
      C.L. Hitchc.
       SY=S. n. var. parviflora Greene
  oregana (Nutt. ex Torr. & Gray) Gray
    ssp. eximia (Greene) C.L. Hitchc.
       SY=S. e. Greene
    ssp. hydrophila (Heller) C.L. Hitchc.
    ssp. oregana
       var. calva C.L. Hitchc.
       var. maxima (M.E. Peck) C.L. Hitchc.
       var. nevadensis C.L. Hitchc.
       var. oregana
       var. procera C.L. Hitchc.
    ssp. spicata (Regel) C.L. Hitchc.
       SY=S. o. var. s. (Regel) Jepson
       SY=S. s. (Regel) Greene
       SY=S. s. var. tonsa M.E. Peck
    ssp. valida (Greene) C.L. Hitchc.
       SY=S. spicata (Regel) Greene var.
          v. (Greene) Wiggins
       SY=S. v. Greene
  pedata Gray
  ranunculacea Greene
  reptans Greene
  robusta Heller ex Roush
  setosa C.L. Hitchc.
    ssp. querceta C.L. Hitchc.
    ssp. setosa
  stipularis D. Dietr.

Sidastrum E.G. Baker

  multiflorum (Jacq.) Fryxell
       SY=Sida acuminata DC.
       SY=Sidastrum a. (DC.) Fryxell
  paniculatum (L.) Fryxell ined.
       SY=Sida p. L.

Sidopsis Rydb.

  hispida (Pursh) Rydb.
       SY=Malvastum angustum Gray
       SY=Sphaeralcea a. (Gray) Fern.

Sphaeralcea St. Hil.

  ambigua Gray
    ssp. ambigua
    ssp. monticola Kearney
       SY=S. a. var. aculeata Jepson
    ssp. rosacea (Munz & Johnston) Kearney
    ssp. rugosa Kearney

Sphaeralcea (CONT.)
   angustifolia (Cav.) G. Don
      ssp. angustifolia
         var. angustifolia
         var. oblongifolia (Gray) Shinners
             SY=S. a. var. lobata (Woot.)
                 Kearney
      ssp. cuspidata (Gray) Kearney
         SY=Phymosia c. (Gray) Britt.
         SY=S. a. var. c. Gray
         SY=S. c. (Gray) Britt.
   caespitosa M.E. Jones
   coccinea (Nutt.) Rydb.
      ssp. coccinea
         SY=Malvastrum c. (Nutt.) Gray
      ssp. dissecta (Nutt.) Kearney
         SY=S. c. var. d. (Nutt.) Garrett
      ssp. elata (E.G. Baker) Kearney
         SY=S. c. var. e. (E.G. Baker)
               Kearney
   coulteri (S. Wats.) Gray
   digitata (Greene) Rydb.
      ssp. digitata
      ssp. tenuipes (Woot. & Standl.) Kearney
         SY=S. d. var. angustiloba (Gray)
             Shinners
         SY=S. d. var. t. (Woot. & Standl.)
             Kearney
   emoryi Torr.
      ssp. arida (Rose) Kearney
         SY=S. e. var. a. (Rose) Kearney
      ssp. emoryi
      ssp. nevadensis Kearney
         SY=S. e. var. n. (Kearney) Kearney
      ssp. variabilis (Cockerell) Kearney
         SY=S. e. var. californica (Parish)
             Shinners
         SY=S. e. var. v. (Cockerell)
             Kearney
   fendleri Gray
      ssp. albescens Kearney
         SY=S. f. var. a. (Kearney) Kearney
      ssp. elongata Kearney
         SY=S. f. var. e. (Kearney) Kearney
      ssp. fendleri
      ssp. tripartita (Woot. & Standl.)
         Kearney
         SY=S. f. var. t. (Woot. & Standl.)
             Kearney
      ssp. venusta Kearney
         SY=S. f. var. v. (Kearney) Kearney
   grossulariifolia (Hook. & Arn.) Rydb.
      ssp. grossulariifolia
      ssp. pedata (Torr. ex Gray) Kearney
         SY=S. g. var. p. (Torr. ex Gray)
             Kearney
   hastulata Gray
   incana Torr.
      ssp. cuneata Kearney
         SY=S. i. var. c. (Kearney) Kearney
      ssp. incana
   laxa Woot. & Standl.
   leptophylla (Gray) Rydb.
   lindheimeri Gray
   munroana (Dougl. ex Lindl.) Spach ex Gray
      ssp. munroana
      ssp. subrhomboidea (Rydb.) Kearney
         SY=S. m. var. s. (Rydb.) Kearney
   orcuttii Rose
   parvifolia A. Nels.
   pedatifida Gray
   procera Porter
   rusbyi Gray
      ssp. eremicola (Jepson) Kearney
      ssp. gilensis Kearney
         SY=S. r. var. g. (Kearney) Kearney
      ssp. rusbyi

   subhastata Coult.
      ssp. connata Kearney
         SY=S. s. var. c. (Kearney) Kearney
      ssp. latifolia Kearney
         SY=S. s. var. l. (Kearney) Kearney
      ssp. martii (Cockerell) Kearney
         SY=S. s. var. m. (Cockerell)
             Kearney
      ssp. pumila (Woot. & Standl.) Kearney
         SY=S. s. var. p. (Woot. & Standl.)
             Kearney
      ssp. subhastata
      ssp. thyrsoidea Kearney
         SY=S. s. var. t. (Kearney) Kearney
   wrightii Gray

Thespesia Soland. ex Correa

   grandiflora DC.
         SY=Montezuma speciosissima Moc. &
             Sesse ex DC.
   populnea (L.) Soland. ex Correa

Urena L.

   lobata L.
         SY=U. trilobata Vell.
   sinuata L.
         SY=U. lobata L. var. s. (L.)
             Hochr.

Wissadula Medic.

   amplissima (L.) R.E. Fries
   contracta (Link) R.E. Fries
   periplocifolia (L.) Presl ex Thwaites

MARANTACEAE

Calathea G.F.W. Mey.

   allouia (Aubl.) Lindl.
   lutea (Aubl.) G.F.W. Mey.

Maranta L.

   arundinacea L.

Thalia L.

   dealbata Roscoe
         SY=T. barbata Small
   geniculata L.

MARCGRAVIACEAE

Marcgravia L.

   rectiflora Triana & Planch.
   sintenisii Urban

MARTYNIACEAE

Craniolaria L.

   annua L.

Ibicella Van Eselt.

Ibicella   (CONT.)
   lutea (Lindl.) Van Eselt.
            SY=Martynia l. Lindl.

Martynia L.

   annua L.

Proboscidea Schmidel

   altheifolia (Benth.) Dcne.
            SY=Martynia a. Benth.
            SY=M. arenaria Engelm.
            SY=P. arenaria (Engelm.) Dcne.
   fragrans (Lindl.) Dcne.
            SY=Martynia f. Lindl.
   louisianica (P. Mill.) Thellung
            SY=Martynia l. P. Mill.
   parviflora (Woot.) Woot. & Standl.
            SY=Martynia p. Woot.
            SY=P. crassibracteata Correll
   sabulosa Correll
   spicata Correll

MAYACACEAE

Mayaca Aubl.

   aubletii Michx.
   fluviatilis Aubl.

MELASTOMATACEAE

Acisanthera P. Br.

   quadrata Pers.
            SY=A. acisanthera (L.) Britt.

Arthrostema Pavon ex D. Don

   ciliatum Ruiz & Pavon

Calycogonium DC.

   krugii Cogn.
   squamulosum Cogn.

Clidemia D. Don

   domingensis (DC.) Cogn.
   hirta (L.) D. Don
   polystachya (Naud.) Cogn.
   strigillosa (Sw.) DC.
   umbrosa (Sw.) Cogn.

Conostegia D. Don

   puberula Naud.
            SY=C. hotteana Urban & Ekman

Henriettella Naud.

   fascicularis (Sw.) C. Wright ex Sauvalle
            SY=Henriettea f. (Sw.) Gomez
   macfadyenii Triana
            SY=Henriettea m. (Triana) Alain
   membranifolia Cogn.
            SY=Henriettea m. (Cogn.) Alain
   triflora (Vahl) Triana
            SY=Henriettea t. (Vahl) Alain

Heterocentron Hook. & Arn.

   macrostachyum Naud.
   subtriplinervium (Link & Otto) A. Braun &
      Bouche

Heterotrichum DC.

   angustifolium DC.
   cymosum (Wendl.) Urban

Mecranium Hook. f.

   amygdalinum (Desr.) C. Wright ex Sauvalle

Melastoma L.

   malabathricum L.

Miconia Ruiz & Pavon

   affinis DC.
            SY=M. microcarpa DC.
   foveolata Cogn.
   impetiolaris (Sw.) D. Don ex DC.
   laevigata (L.) DC.
   mirabilis (Aubl.) L.O. Williams
            SY=Tamonea guianensis Aubl.
   ottoschulzii Urban & Ekman
   pachyphylla Cogn.
   prasina (Sw.) DC.
   punctata (Desr.) D. Don ex DC.
   pycnoneura Urban
   racemosa (Aubl.) DC.
   rubiginosa (Bonpl.) DC.
   serrulata (DC.) Naud.
            SY=Tamonea macrophylla (D. Don)
                  Krasser
   sintenisii Cogn.
   subcorymbosa Britt.
   tetrandra (Sw.) D. Don ex Loud.
   tetrasoma Naud.
   thomasiana DC.

Mouriri Aubl.

   domingensis (Tussac) Spach
   helleri Britt.

Nepsera Naud.

   aquatica (Aubl.) Naud.

Ossaea DC.

   krugiana Cogn.
   scabrosa (L.) DC.
   scalpta (Vent.) DC.
            SY=O. domingensis Cogn.

Oxyspora DC.

   paniculata (D. Don) DC.

Pterolepis (DC.) Miq.

   glomerata (Rottb.) Miq.

Rhexia L.

   alifanus Walt.
   aristosa Britt.
   cubensis Griseb.
   lutea Walt.
   mariana L.
         var. interior (Pennell) Kral &
            Bostick

Rhexia (CONT.)
          SY=R. i. Pennell
      var. mariana
          SY=R. delicatula Small
          SY=R. filiformis Small
          SY=R. lanceolata Walt.
          SY=R. m. var. exalbida Michx.
          SY=R. m. var. leiosperma Fern. &
              Grisc.
      var. ventricosa (Fern. & Grisc.) Kral
          & Bostick
          SY=R. v. Fern. & Grisc.
    nashii Small
          SY=R. mariana L. var. purpurea
              Michx.
    nuttallii C.W. James
          SY=R. serrulata Nutt.
    parviflora Chapman
    petiolata Walt.
          SY=R. ciliosa Michx.
    salicifolia Kral & Bostick
    virginica L.
          SY=R. stricta Pursh
          SY=R. v. var. purshii (Spreng.)
              C.W. James
          SY=R. v. var. septemnervia (Walt.)
              Pursh

Tetrazygia L.C. Rich. ex DC.

    angustifolia (Sw.) DC.
    bicolor (P. Mill.) Cogn.
    biflora (Cogn.) Urban
          SY=Menendezia b. (Cogn.) Britt.
    crotonifolia (Desv.) DC.
    elaeagnoides (Sw.) DC.
    stahlii Cogn.
          SY=Menendezia s. (Cogn.) Britt.
    urbanii Cogn.
          SY=Menendezia u. (Cogn.) Britt.

Tibouchina Aubl.

    urvilleana (DC.) Cogn.

MELIACEAE

Cedrela P. Br.

    odorata L.

Guarea Allem. ex L.

    guidonia (L.) Sleumer
          SY=G. guara (Jacq.) P. Wilson
          SY=G. trichilioides L.
    ramiflora Vent.

Melia L.

    azedarach L.
          SY=M. a. var. umbraculifera Knox

Swietenia Jacq.

    macrophylla G. King
    mahogani (L.) Jacq.

Trichilia P. Br.

    hirta L.
    pallida Sw.
    triacantha Urban

MENISPERMACEAE

Calycocarpum (Nutt.) Spach

    lyonii (Pursh) Gray

Cissampelos L.

    pareira L.

Cocculus DC.

    carolinus (L.) DC.
          SY=Epibaterium c. (L.) Britt.
    diversifolius DC.
    ferrandianus Gaud.
    integer Hbd.
    lonchophyllus (Miers) Hbd.
    virgatus Hbd.

Hyperbaena Miers ex Benth.

    domingensis (DC.) Benth.
    laurifolia (Poir.) Urban

Menispermum L.

    canadense L.

MENYANTHACEAE

Fauria Franch.

    crista-galli (Menzies ex Hook.) Makino
          SY=Nephrophyllidium c-g. (Menzies
              ex Hook.) Gilg

Menyanthes L.

    trifoliata L.
          var. minor Raf.
          var. trifoliata

Nymphoides Seguier

    aquatica (Walt.) Kuntze
    cordata (Ell.) Fern.
          SY=N. lacunosa (Vent.) Kuntze
    indica (L.) Kuntze
          SY=N. humboldtiana (H.B.K.) Kuntze
    peltata (J.G. Gmel.) Kuntze
          SY=Limnanthemum p. J.G. Gmel.
          SY=N. nymphaeoides (L.) Britt.

MORACEAE

Artocarpus J. R. & G. Forst.

    altilis (Parkinson) Fosberg
          SY=A. communis J. R. & G. Forst.
    heterophyllus Lam.

Brosimum Sw.

    alicastrum Sw.

Broussonetia L'Her. ex Vent.

    papyrifera (L.) Vent.
          SY=Papyrius p. (L.) Kuntze

Cannabis L.

    sativa L.
      ssp. sativa
        var. sativa
        var. spontanea Vavilov

Cecropia Loefl.

    peltata L.

Chlorophora Gaud.

    tinctoria (L.) Gaud. ex Benth. & Hook. f.

Cudrania Trecul

    tricuspidata (Carr.) Bureau ex Lavallee

Dorstenia L.

    contrajerva L.

Fatoua Gaud.

    villosa (Thunb.) Nakai

Ficus L.

    aurea Nutt.
    carica L.
    citrifolia P. Mill.
        SY=F. brevifolia Nutt.
        SY=F. laevigata Vahl
    elastica Roxb. ex Hornem.
    obtusifolia H.B.K.
        SY=F. urbaniana Warb.
    palmata Forsk.
        SY=F. pseudocarica Miq.
    perforata L.
        SY=F. sintenisii Warb.
    pumila L.
    stahlii Warb.
    trigonata L.
        SY=F. crassinervia Desf.

Humulus L.

    japonicus Sieb. & Zucc.
        SY=H. scandens (Lour.) Merr.
    lupulus L.
      var. lupuloides E. Small
        SY=H. americanus Nutt.
      var. lupulus
      var. neomexicanus A. Nels. &
        Cockerell
      var. pubescens E. Small

Maclura Nutt.

    pomifera (Raf.) Schneid.
        SY=Toxylon p. Raf. ex Sarg.

Morus L.

    alba L.
        SY=M. a. var. tatarica (L.) Ser.
        SY=M. t. L.
    microphylla Buckl.
        SY=M. confinis Greene
        SY=M. crataegifolia Greene
        SY=M. grisea Greene
        SY=M. radulina Greene
    nigra L.
    rubra L.
      var. rubra
      var. tomentosa (Raf.) Bureau

Pseudolmedia Trecul

    spuria (Sw.) Griseb.

Pseudomorus Bureau

    brunoniana (Endl.) Bureau

Streblus Lour.

    sandwicensis (Deg.) St. John
        SY=Pseudomorus brunoniana (Endl.)
          Bureau var. s. (Deg.)
          Skottsburg
        SY=P. s. Deg.

Trophis P. Br.

    racemosa (L.) Urban

MORINGACEAE

Moringa Adans.

    oleifera Lam.
        SY=M. moringa (L.) Millsp.

MUSACEAE

Heliconia L.

    caribaea Lam.
        SY=Bihai bihai (L.) Griggs

Musa L.

    acuminata Colla
        SY=M. cavendishii Lamb. ex Paxton
    X paradisiaca L. [acuminata X balbisiana]
        SY=M. p. var. normalis Kuntze
        SY=M. X sapientum L.

MYOPORACEAE

Bontia L.

    daphnoides L.

Myoporum Banks & Soland. ex Forst. f.

    laetum Forst. f.
    sandwicense (A. DC.) Gray
      ssp. sandwicense
        var. degeneri Webster
        var. fauriei (Levl.) Kraenzlin
        var. lanaiense Webster
        var. sandwicense
        var. stellatum Webster
      ssp. st.-johnii Webster

MYRICACEAE

Comptonia L'Her.

    peregrina (L.) Coult.

Comptonia   (CONT.)
            SY=C. p. var. asplenifolia (L.)
                 Fern.
            SY=Myrica a. L.
            SY=M. a. var. tomentosa (Chev.)
                 Gleason
            SY=M. p. (L.) Kuntze

Myrica L.

    californica Cham.
    cerifera L.
            SY=Cerothamnus c. (L.) Small
            SY=C. pumilus (Michx.) Small
            SY=M. c. var. p. Michx.
            SY=M. pusilla Raf.
    faya Ait.
    gale L.
        var. gale
            SY=Gale palustris (Lam.) Chev.
        var. subglabra (Chev.) Fern.
        var. tomentosa C. DC.
    hartwegii S. Wats.
    heterophylla Raf.
            SY=Cerothamnus caroliniensis (P.
                 Mill.) Tidestrom
            SY=M. h. var. curtisii (Chev.)
                 Fern.
    holdridgeana Lundell
    holdridgii Lundell
    inodora Bartr.
            SY=Cerothamnus i. (Bartr.) Small
    X macfarlanei Youngken [cerifera X
        pensylvanica]
    pensylvanica Loisel.

MYRSINACEAE

Ardisia Sw.

    escallonoides Schiede & Deppe ex Schlecht.
        & Cham.
            SY=Icacorea paniculata (Nutt.)
                 Sudworth
    glauciflora Urban
            SY=Icacorea g. (Urban) Britt.
    luquillensis (Britt.) Alain
            SY=Icacorea l. Britt.
    obovata Desv.
            SY=Icacorea guadalupensis Britt.
    solanacea Roxb.
            SY=A. polycephala Wight non Wall.

Embelia Burm. f.

    hillebrandii Mez
    pacifica Hbd.

Grammadenia Benth.

    sintenisii (Urban) Mez

Myrsine L.

    alyxifolia Hosaka
    degeneri Hosaka
    denticulata (Wawra) Hosaka
    emarginata (Rock) Hosaka
    fernseei (Mez) Hosaka
    helleri (Deg. & Deg.) St. John
            SY=M. hosakae Wilbur
    juddii Hosaka
    kauaiensis Hbd.
    knudsenii (Rock) Hosaka

    kokeeana Hosaka
    lanaiensis Hbd.
        var. lanaiensis
        var. oahuensis Hosaka
    lessertiana A. DC.
    linearifolia Hosaka
        var. linearifolia
        var. nittae Hosaka
    meziana (Levl.) Wilbur
        var. fosbergii (Hosaka) Wilbur
        var. meziana
    mezii Hosaka
    petiolata Hosaka
    pukooensis (Levl.) Hosaka
    punctata (Levl.) Wilbur
    st.-johnii Hosaka
    sandwicensis A. DC.
        var. mauiensis Levl.
        var. sandwicensis
    wawraea (Mez) Hosaka

Parathesis (A. DC.) Hook. f.

    crenulata (Vent.) Hook. f.
            SY=Ardisia c. Vent.
            SY=P. serrulata (Sw.) Mez

Rapanea Aubl.

    coriacea (Sw.) Mez
            SY=R. ferruginea (Ruiz & Pavon)
                 Mez
    punctata (Lam.) Lundell
            SY=Myrsine guianensis (Aubl.)
                 Kuntze
            SY=R. g. Aubl.

Stylogyne A. DC.

    lateriflora (Sw.) Mez

Wallenia Sw.

    pendula (Urban) Mez
            SY=Petesioides p. (Urban) Britt.
    yunquense (Urban) Mez
            SY=Petesioides y. (Urban) Britt.

MYRTACEAE

Calyptranthes Sw.

    dumetorum Alain
    kiaerskovii Krug & Urban
    krugii Kiaersk.
    luquillensis Alain
    pallens (Poir.) Griseb.
    peduncularis Alain
    potoricensis Britt.
    sintenisii Kiaersk.
    thomasiana Berg
    triflora Alain
    zuzygium (L.) Sw.

Eucalyptus L'Her.

    camaldulensis Dehnhardt
    globulus Labill.
    polyanthemos Schauer
    tereticornis Sm.

Eugenia L.

    aeruginea DC.

Eugenia   (CONT.)
   anthera Small
   apiculata DC.
            SY=Luma a. (DC.) Burret
   axillaris (Sw.) Willd.
   bellonis Krug & Urban
            SY=Myrtus b. (Krug & Urban) Burret
   biflora (L.) DC.
            SY=E. lancea Poir.
   boqueronensis Britt.
   borinquensis Britt.
   confusa DC.
   cordata (Sw.) DC.
         var. cordata
         var. sintenisii (Kiaersk.) Krug &
            Urban
            SY=E. s. Kiaersk.
   corozalensis Britt.
   domingensis Berg
   eggersii Kiaersk.
   fajardensis Krug & Urban
   foetida Pers.
            SY=E. buxifolia (Sw.) Willd.
            SY=E. myrtoides Poir.
   glabrata (Sw.) DC.
   haematocarpa Alain
   koolauensis Deg.
   ligustrina (Sw.) Willd.
   margarettae Alain
   molokaiana Wilson & Rock
   monticola (Sw.) DC.
         var. latifolia Krug & Urban
         var. monticola
   procera (Sw.) Poir.
   pseudopsidium Jacq.
   reinwardtiana (Blume) DC.
   rhombea (Berg) Krug & Urban
   sandwicensis Gray
   serrasuela Krug & Urban
   sessiliflora Vahl
   stahlii (Kiaersk.) Krug & Urban
   stewardsonii Britt.
   underwoodii Britt.
   uniflora L.
   xerophytica Britt.

Gomidesia Berg

   lindeniana Berg

Leptospermum J.R. & G. Forst.

   scoparium J.R. & G. Forst.

Marlierea Camb.

   sintenisii Kiaersk.
            SY=Plinia s. (Kiaersk.) Britt.

Melaleuca L.

   quinquenervia (Cav.) Blake
            SY=M. leucadendra sensu auctt. non
               (L.) L.

Metrosideros Banks ex Gaertn.

   macropus Hook. & Arn.
   polymorpha Gaud.
      ssp. glaberrima (Levl.) Skottsberg
         var. glaberrima (Levl.) Skottsberg
            SY=M. collina (J.R. & G. Forst.)
               Gray ssp. c. var. g. (Levl.)
               Rock
         var. sericea (Rock) Skottsberg
      ssp. glabrifolia (Heller) Skottsberg
         var. glabrifolia (Heller) Skottsberg

            SY=M. collina (J.R. & G. Forst.)
               Gray ssp. c. var. g.
               (Heller) Rock
         var. parviflora Skottsberg
      ssp. imbricata (Rock) Skottsberg
            SY=M. collina (J.R. & G. Forst.)
               Gray ssp. c. var. i. Rock
      ssp. incana (Levl.) Skottsberg
         var. fauriei (Levl.) Skottsberg
            SY=M. collina (J.R. & G. Forst.)
               Gray ssp. polymorpha (Gaud.)
               Rock. var. f. (Levl.) Rock
         var. incana (Levl.) Skottsberg
            SY=M. collina (J.R. & G. Forst.)
               Gray ssp. c. var. i. (Levl.)
               Rock
         var. pumila (Heller) Skottsberg
            SY=M. collina (J.R. & G. Forst.)
               Gray ssp. c. var. pumila
               (Heller) Rock
            SY=M. pumila (Heller) Hochr.
            SY=M. pumila var. makanoiensis
               Hochr.
      ssp. micrantha Skottsberg
      ssp. polymorpha
         var. macrostemon Skottsberg
         var. nuda Skottsberg
         var. polymorpha
            SY=M. collina (J.R. & G. Forst.)
               Gray ssp. c. var. newellii
               Rock
            SY=M. c. ssp. c. var. prostrata
               Rock
            SY=M. c. ssp. polymorpha (Gaud.)
               Rock
            SY=M. c. ssp. polymorpha var. nuda
               Skottsberg
            SY=M. c. var. haleakalensis Rock
            SY=M. c. var. hemilanata Hochr.
            SY=M. c. var. macrophylla Rock
         var. pseudorugosa Skottsberg
         var. subimbricata Skottsberg
   rugosa Gray
   tremuloides (Heller) P. Knuth
   waialealae (Rock) Rock

Myrcia DC. ex Guill.

   citrifolia (Aubl.) Urban
            SY=Aulomyrcia c. (Aubl.) Amsh.
   deflexa (Poir.) DC.
   fallax (A. Rich.) DC.
            SY=M. berberis DC.
   leptoclada DC.
   paganii Krug & Urban
   splendens (Sw.) DC.

Myrcianthes Berg

   fragrans (Sw.) McVaugh
         var. fragrans
            SY=Anamomis dicrana (Berg) Britt.
            SY=A. f. (Sw.) Griseb.
            SY=Eugenia d. Berg
         var. simpsonii (Small) R.W. Long
            SY=Anamomis s. Small
            SY=Eugenia s. (Small) Sarg.

Myrciaria Berg

   floribunda (West ex Willd.) Berg
            SY=Eugenia f. West ex Willd.

Pimenta Lindl.

   dioica (L.) Merr.
            SY=Pimenta pimenta (L.) Cockerell

Pimenta (CONT.)
  racemosa (P. Mill.) J.W. Moore
        var. grisea (Kiaersk.) Fosberg
           SY=Amomis g. (Kiaersk.) Britt.
        var. racemosa
           SY=Amomis caryophyllata (Jacq.)
              Krug & Urban

Pseudanamomis Kausel

  umbellulifera (H.B.K.) Kausel
        SY=Anamomis u. (H.B.K.) Britt.

Psidium L.

  amplexicaule Pers.
  cattleianum Sabine
        var. cattleianum
           SY=P. littorale Raddi var.
              longipes (Berg) Fosberg
        var. littorale (Raddi) Fosberg
           SY=P. l. Raddi
  guajava L.
  longipes (Berg) McVaugh
        SY=Eugenia bahamensis Kiaersk.
        SY=E. l. Berg
        SY=Mosiera b. (Kiaersk.) Small
        SY=Mosiera l. (Berg) Small
        SY=Myrtus b. (Kiaersk.) Urban
        SY=Myrtus verrucosa Berg
  sintenisii (Kiaersk.) Alain
        SY=Calyptropsidium s. Kiaersk.

Rhodomyrtus (DC.) Hassk.

  tomentosus (Ait.) Hassk.

Siphoneugenia Berg

  densiflora Berg
        SY=Plinia dussii (Krug & Urban)
           Urban

Syzygium P. Br.

  cumini (L.) Skeels
        SY=Eugenia c. (L.) Druce
  grande (Wight) Walp.
        SY=Eugenia g. Wight
  jambos (L.) Alston
        SY=Eugenia j. L.
        SY=Jambosa j. (L.) Millsp.
  malaccensis (L.) Merr. & Perry
        SY=Eugenia m. L.
        SY=Jambosa m. (L.) DC.

NAJADACEAE

Najas L.

  ancistrocarpa Magnus
        SY=N. conferta A. Braun
  flexilis (Willd.) Rostk. & Schmidt
        SY=N. caespitosus (Maguire) Reveal
        SY=N. f. ssp. c. Maguire
        SY=N. f. var. congesta Farw.
        SY=N. f. var. robusta Morong
  gracillima (A. Braun) Magnus
  graminea Delile
  guadalupensis (Spreng.) Magnus
        var. floridana Haynes & Wentz
        var. guadalupensis
        var. muenscheri (Clausen) Haynes
           SY=N. m. Clausen

        var. olivacea (Rosendahl & Butters)
           Haynes
           SY=N. o. Rosendahl & Butters
  marina L.
        SY=N. gracilis (Morong) Small
        SY=N. major All.
        SY=N. major var. angustifolia A.
           Braun
        SY=N. m. var. recurvata Dudley
  minor All.
  wrightiana A. Braun

NYCTAGINACEAE

Abronia Juss.

  alba Eastw.
        SY=A. umbellata Lam. ssp. a.
           (Eastw.) Munz
  alpina Brandeg.
  ameliae Lundell
  ammophila Greene
        SY=A. fendleri Standl.
  angustifolia Greene
        SY=A. a. var. arizonica (Standl.)
           Kearney & Peebles
        SY=A. torreyi Standl.
  bigelovii Heimerl
  carletonii Coult. & Fish.
  elliptica A. Nels.
        SY=A. fragrans Nutt. ex Hook. var.
           e. (A. Nels.) M.E. Jones
        SY=A. pumila Rydb.
        SY=A. salsa Rydb.
        SY=A. ramosa Standl.
  fragrans Nutt. ex Hook.
        SY=A. f. var. glaucescens A. Nels.
  gracilis Benth.
  insularis Standl.
  latifolia Eschsch.
  macrocarpa L.A. Gal.
  maritima Nutt. ex S. Wats.
  mellifera Dougl. ex Hook.
  minor Standl.
  nana S. Wats.
    ssp. covillei (Heimerl) Munz
        SY=A. n. var. c. (Heimerl) Munz
    ssp. nana
        SY=A. n. var. lanciformis M.E.
           Jones
  neurophylla Standl.
  platyphylla Standl.
        SY=A. umbellata Lam. ssp. p.
           (Standl.) Munz
  pogonantha Heimerl
  turbinata Torr. ex S. Wats.
        SY=A. exalata Standl.
        SY=A. orbiculata Standl.
  umbellata Lam.
    ssp. acutalata (Standl.) Tillett
        SY=A. a. Standl.
        SY=A. u. var. a. (Standl.) C.L.
           Hitchc.
    ssp. breviflora (Standl.) Munz
        SY=A. b. Standl.
    ssp. umbellata
    ssp. variabilis (Standl.) Munz
        SY=A. v. Standl.
  villosa S. Wats.
        var. aurita (Abrams) Jepson
           SY=A. pinetorum Abrams
        var. villosa

Acleisanthes Gray

Acleisanthes (CONT.)
  acutifolia Standl.
  anisophylla Gray
  crassifolia Gray
  longiflora Gray
  obtusa (Choisy) Standl.
  wrightii (Gray) Benth. & Hook.

Allionia L.

  choisyi Standl.
  cristata (Standl.) Standl.
  incarnata L.

Ammocodon Standl.

  chenopodioides (Gray) Standl.
        SY=Selinocarpus c. Gray

Anulocaulis Standl.

  annulata (Coville) Standl.
        SY=Boerhavia a. Coville
  eriosolenus (Gray) Standl.
        SY=Boerhavia e. Gray
  gypsogenus Waterfall
  leiosolenus (Torr.) Standl.
        var. lasianthus I.M. Johnston
        var. leiosolenus
        SY=Boerhavia l. Torr.
  reflexus I.M. Johnston

Boerhavia L.

  anisophylla Torr.
  coulteri (Hook. f.) S. Wats.
  diffusa L.
        var. diffusa
        SY=B. coccinea P. Mill.
        var. gymnocarpa Heimerl
        var. pseudotetrandra Heimerl
        var. sandwicensis Heimerl
        var. tetrandra (Forst. f.) Heimerl
  erecta L.
  gracillima Heimerl
  intermedia M.E. Jones
        SY=B. erecta L. var. i. (M.E.
            Jones) Kearney & Peebles
  linearifolia Gray
        SY=B. lindheimeri Standl.
        SY=B. tenuifolia Gray
  mathisana author ?
  megaptera Standl.
  pterocarpa S. Wats.
  purpurascens Gray
  scandens L.
        SY=Commicarpus s. (L.) Standl.
  spicata Choisy
        SY=B. torreyana (S. Wats.) Standl.
  triquetra S. Wats.
  wrightii Gray

Bouganvillea Comm. ex Juss.

  glabra Choisy

Cyphomeris Standl.

  crassifolia (Standl.) Standl.
  gypsophiloides (Mart. & Gal.) Standl.
        SY=Boerhavia g. (Mart. & Gal.)
            Coult.

Guapira Aubl.

  discolor (Spreng.) Little
        SY=G. bracei (Britt.) Little

        SY=G. longifolia (Heimerl) Little
        SY=Pisonia d. Spreng.
        SY=P. discolor Spreng. var. l.
            Heimerl
        SY=Torrubia b. Britt.
        SY=T. d. (Spreng.) Britt.
        SY=T. l. (Heimerl) Britt.
  fragrans (Dum.-Cours.) Little
        SY=Torrubia f. (Dum.-Cours.)
            Standl.
  globosa (Small) Little
        SY=Torrubia g. Small
  obtusata (Jacq.) Little
        SY=Pisonia o. Jacq.
        SY=Torrubia o. (Jacq.) Britt.

Mirabilis L.

  aggregata (Ortega) Cav.
        SY=Allionia a. (Ortega) Spreng.
  albida (Walt.) Heimerl
        SY=Allionia a. Walt.
        SY=M. a. var. lata Shinners
        SY=M. a. var. uniflora Heimerl
        SY=Oxybaphus a. (Walt.) Sweet
  alipes (S. Wats.) Pilz
        SY=Hermidium a. S. Wats.
        SY=H. a. var. pallidum C.L. Porter
  bigelovii Gray
        var. aspera (Greene) Munz
        var. bigelovii
        var. retrorsa (Heller) Munz
  californica Gray
        var. californica
        SY=M. laevis (Benth.) Curran
        SY=Oxybaphus l. Benth.
        var. cedrosensis (Standl.) J.F.
            Macbr.
        SY=M. laevis (Benth.) Curran var.
            c. (Standl.) Munz
        var. cordifolia Dunkle
  carletonii (Standl.) Standl.
        SY=Allionia c. Standl.
        SY=Oxybaphus c. (Standl.)
            Weatherby
  ciliata (Standl.) Standl.
        SY=Allionia c. Standl.
  coahuilensis (Standl.) Standl.
        SY=Allionia c. Standl.
  coccineus (Torr.) Benth. & Hook.
        SY=Oxybaphus c. Torr.
  collina Shinners
  decipiens (Standl.) Standl.
        SY=Allionia d. Standl.
        SY=Oxybaphus linearis (Pursh) B.L.
            Robins. var. d. (Standl.)
            Kearney & Peebles
  decumbens (Nutt.) Daniels
        SY=Allionia d. (Nutt.) Spreng.
  diffusa (Heller) C.F. Reed
  dumetorum Shinners
  eutricha Shinners
  exaltata (Standl.) Standl.
        SY=Allionia e. Standl.
  gausapoides (Standl.) Standl.
        SY=M. linearis (Pursh) Heimerl
            var. subhispida Heimerl
  gigantea (Standl.) Shinners
        SY=Allionia g. Standl.
  glabra (S. Wats.) Standl.
        SY=Allionia g. (S. Wats.) Kuntze
        SY=Oxybaphus g. S. Wats.
  glabrifolia (Ortega) I.M. Johnston
        SY=Allionia texensis (Coult.)
            Small
        SY=M. corymbosa Cav.
  grayana (Standl.) Standl.

Mirabilis   (CONT.)
            SY=Allionia g. Standl.
  greenei S. Wats.
  hirsuta (Pursh) MacM.
            SY=Allionia h. Pursh
            SY=Oxybaphus h. (Pursh) Sweet
  jalapa L.
  lindheimeri (Standl.) Shinners
  linearis (Pursh) Heimerl
            SY=Allionia l. Pursh
            SY=M. hirsuta (Pursh) MacM. var.
                l. (Pursh) Boivin
            SY=M. lanceolata (Rydb.) Standl.
            SY=Oxybaphus angustifolium Sweet
            SY=O. lanceolatus (Rydb.) Standl.
            SY=O. linearis (Pursh) B.L.
                Robins.
  longiflora L.
            SY=M. l. var. wrightiana (Gray)
                Kearney & Peebles
            SY=M. w. Gray
  macfarlanei Constance & Rollins
  multiflora (Torr.) Gray
      var. glandulosa (Standl.) J.F. Macbr.
      var. multiflora
            SY=Quamoclidion cordifolium
                Osterhout
      var. pubescens S. Wats.
            SY=M. froebelii (Behr) Greene
            SY=M. f. var. glabrata (Standl.)
                Jepson
  nyctaginea (Michx.) MacM.
            SY=Allionia n. Michx.
            SY=Oxybaphus n. (Michx.) Sweet
  oblongifolia (Gray) Heimerl
            SY=Allionia comata Small
            SY=M. c. (Small) Standl.
            SY=Oxybaphus c. (Small) Weatherby
  oxybaphoides (Gray) Gray ex Torr.
            SY=Allioniella o. (Gray) Rydb.
  pauciflora (Buckl.) Standl.
            SY=Oxybaphus p. Buckl.
  pseudaggregata Heimerl
  pudica Barneby
  pumila (Standl.) Standl.
            SY=Allionia p. Standl.
            SY=Oxybaphus p. (Standl.) Standl.
  rotata (Standl.) I.M. Johnston
            SY=Allionia r. Standl.
  rotundifolia (Greene) Standl.
            SY=Allionia r. Greene
            SY=Oxybaphus r. (Greene) Standl.
  tenuiloba S. Wats.
  violacea (L.) Heimerl
            SY=Allionia v. L.

Neea Ruiz & Pavon

  buxifolia (Hook. f.) Heimerl

Nyctaginia Choisy

  capitata Choisy

Okenia Schlecht. & Cham.

  hypogaea Schlecht. & Cham.

Pisonia L.

  aculeata L.
      var. aculeata
            SY=P. helleri Standl.
      var. macranthocarpa J.D. Sm.
  albida (Heimerl) Britt.
  brunoniana Endl.
  floridana Britt.

            SY=Torrubia f. (Britt.) Britt.
  rotundata Griseb.
  sandwicensis Hbd.
  subcordata Sw.
  umbellifera (J.R. & G. Forst.) Seem.

Selinocarpus Gray

  angustifolius Torr.
  diffusus Gray
  lanceolatus Woot.
  nevadensis (Standl.) Fowler & Turner
            SY=S. diffusus Gray ssp. n.
                Standl.
  parvifolius (Torr.) Standl.

Tripterocalyx (Torr.) Hook.

  carnea (Greene) L.A. Gal.
      var. carnea
            SY=Abronia c. Greene
      var. wootonii (Standl.) L.A. Gal.
            SY=T. w. Standl.
  crux-maltae (Kellogg) Standl.
            SY=Abronia c-m. Kellogg
  micranthus (Torr.) Hook.
            SY=Abronia cycloptera Gray
            SY=A. m. var. pedunculata M.E.
                Jones
            SY=T. c. (Gray) Standl.
            SY=T. p. (M.E. Jones) Standl.

NYMPHAEACEAE

Brasenia Schreb.

  schreberi J.F. Gmel.

Cabomba Aubl.

  caroliniana Gray
  piauhyensis Gardn.
  pulcherrima (Harper) Fassett
            SY=C. caroliniana Gray var. p.
                Harper

Nelumbo Adans.

  lutea (Willd.) Pers.
            SY=N. pentapetala (Walt.) Fern.
  nucifera Gaertn.

Nuphar Sm.

  luteum (L.) Sibthorp & Sm.
      ssp. macrophyllum (Small) E.O. Beal
            SY=N. advena (Soland.) R. Br. var.
                tomentosum Nutt. ex Torr. &
                Gray
            SY=N. fluviatile (Harper) Standl.
            SY=N. X interfluitans Fern.
            SY=N. microcarpum (Mill. &
                Standl.) Standl.
            SY=N. ovatum (Mill. & Standl.)
                Standl.
            SY=N. puteorum Fern.
            SY=Nymphaea chartacea Mill. &
                Standl.
            SY=Nymphaea f. Harper
            SY=Nymphaea macrophylla Small
      ssp. orbiculatum (Small) F.C. Beal
            SY=Nuphar o. (Small) Standl.
            SY=Nymphaea bombycina (Mill. &
                Standl.) Standl.

Nuphar   (CONT.)
          SY=Nymphaea o. Small
     ssp. ozarkanum (Mill. & Standl.) E.O.
          Beal
               SY=Nuphar o. (Mill. & Standl.)
                    Standl.
     ssp. polysepalum (Engelm.) E.O. Beal
               SY=Nuphar p. Engelm.
               SY=Nymphaea p. (Engelm.) Greene
     ssp. pumilum (Timm) E.O. Beal
               SY=Nuphar microphyllum (Pers.)
                    Fern.
               SY=Nuphar X rubrodiscum (Morong)
                    Fern.
               SY=Nymphaea r. (Morong) Greene
     ssp. sagittifolium (Walt.) E.O. Beal
               SY=Nuphar s. (Walt.) Pursh
               SY=Nymphaea s. Walt.
     ssp. ulvaceum (Mill. & Standl.) E.O.
          Beal
               SY=Nymphaea u. Mill. & Standl.
     ssp. variegatum (Dur.) E.O. Beal
               SY=Nuphar advena (Soland.) R. Br.
               SY=Nuphar a. var. fraternum (Mill.
                    & Standl.) Standl.
               SY=Nuphar v. Engelm. ex Dur.
               SY=Nymphaea f. Mill. & Standl.
               SY=Nymphozanthus a. (Soland.) Fern.

Nymphaea L.

     alba L.
     amazonum Mart. & Zucc.
               SY=Castalia a. (Mart. & Zucc.)
                    Britt. & Wilson
     ampla (Salisb.) DC.
          var. ampla
               SY=Castalia a. Salisb.
          var. pulchella (DC.) Caspary
               SY=Castalia p. (DC.) Britt.
               SY=N. p. DC.
     blanda G.F.W. Mey.
          var. fenzliana (Lehm.) Caspary
     capensis Thunb.
     elegans Hook.
               SY=Castalia e. (Hook.) Greene
     jamesoniana Planch.
               SY=Castalia j. (Planch.) Britt. &
                    Wilson
     mexicana Zucc.
               SY=Castalia flava (Leitner) Greene
     odorata Ait.
          var. godfreyi Ward
          var. odorata
               SY=Castalia lekophylla Small
               SY=C. minor (Sims) DC.
               SY=C. o. (Ait.) Woodv. & Wood
               SY=C. reniformis DC.
               SY=C. tuberosa (Paine) Greene
               SY=N. X daubeniana O. Thomas
               SY=N. o. var. gigantea Tricker
               SY=N. o. var. m. Sims
               SY=N. o. var. maxima (Conrad)
                    Boivin
               SY=N. o. var. rosea Pursh
               SY=N. o. var. stenopetala Fern.
               SY=N. X thiona Ward
               SY=N. tuberosa Paine
     rudgeana G.F.W. Mey.
               SY=Castalia r. (G.F.W. Mey.)
                    Britt. & Wilson
     tetragona Georgi
               SY=Castalia t. (Georgi) Lawson
               SY=N. t. ssp. leibergii (Morong)
                    Porsild
               SY=N. t. var. l. (Morong) Boivin

NYSSACEAE

Nyssa L.

     aquatica L.
               SY=N. uniflora Wang.
     ogeche Bartr. ex Marsh.
               SY=N. acuminata Small
     sylvatica Marsh.
          var. biflora (Walt.) Sarg.
               SY=N. b. Walt.
               SY=N. ursina Small
          var. sylvatica
               SY=N. s. var. caroliniana (Poir.)
                    Fern.
               SY=N. s. var. dilatata Fern.
               SY=N. s. var. typica Fern.

OCHNACEAE

Ouratea Aubl.

     ilcifolia (DC.) Baill.
     littoralis Urban
     striata (v. Tiegh.) Urban

Sauvagesia L.

     erecta L.

OLACACEAE

Schoepfia Schreb.

     arenaria Britt.
     chrysophylloides (A. Rich.) Planch.
     obovata C. Wright
     schreberi J.F. Gmel.

Ximenia L.

     americana L.
               SY=X. X inermis L.

OLEACEAE

Chionanthus L.

     pygmaeus Small
     virginicus L.
          var. maritima Pursh
          var. virginicus

Forestiera Poir.

     acuminata (Michx.) Poir.
               SY=F. a. var. vestita Palmer
     angustifolia Torr.
               SY=F. texana Cory
     eggersiana Krug & Urban
     ligustrina (Michx.) Poir.
               SY=F. autumnalis (Michx.) Poir.
     neomexicana Gray
          var. arizonica Gray
          var. neomexicana
     phillyreoides (Benth.) Torr.
               SY=F. shrevei Standl.

Forestiera (CONT.)
   puberula Eastw.
   pubescens Nutt.
      var. glabrifolia Shinners
      var. pubescens
         SY=F. sphaerocarpa Torr.
   reticulata Torr.
   rhamnifolia Griseb.
   segregata (Jacq.) Krug & Urban
      var. pinetorum (Small) M.C. Johnston
         SY=F. p. Small
      var. segregata
         SY=F. globularis Small
         SY=F. porulosa (Michx.) Poir.

Forsythia Vahl

   ovata Nakai
   suspensa (Thunb.) Vahl
   viridissima Lindl.

Fraxinus L.

   americana L.
      var. americana
         SY=F. a. var. crassifolia Sarg.
         SY=F. a. var. curtissii (Vasey)
            Small
         SY=F. a. var. microcarpa Gray
      var. biltmoreana (Beadle) J. Wright
         ex Fern.
         SY=F. b. Beadle
   anomala Torr. ex S. Wats.
      var. anomala
      var. lowellii (Sarg.) Little
         SY=F. l. Sarg.
   berlandieriana A. DC.
   caroliniana P. Mill.
         SY=F. c. var. cubensis (Griseb.)
            Lingelsh.
         SY=F. caroliniana var.
            oblanceolata (M.A. Curtis)
            Fern. & Schub.
         SY=F. pauciflora Nutt.
   cuspidata Torr.
         SY=F. c. var. macropetala (Eastw.)
            Rehd.
   dipetala Hook. & Arn.
         SY=F. d. var. trifoliolata Torr.
         SY=F. t. (Torr.) Lewis & Epling
   excelsior L.
   gooddingii Little
   greggii Gray
   latifolia Benth.
   nigra Marsh.
   papillosa Lingelsh.
   pennsylvanica Marsh.
         SY=F. campestris Britt.
         SY=F. darlingtonii Britt.
         SY=F. lanceolata Borkh.
         SY=F. p. var. austinii Fern.
         SY=F. p. var. integerrima (Vahl)
            Fern.
         SY=F. p. var. lanceolata (Borkh.)
            Sarg.
         SY=F. p. var. subintegerrima
            (Vahl) Fern.
         SY=F. smallii Britt.
   profunda (Bush) Bush
         SY=F. michauxii Britt.
         SY=F. tomentosa Michx. f.
   quadrangulata Michx.
   texensis (Gray) Sarg.
         SY=F. americana L. var. t. Gray
   velutina Torr.
         SY=F. pennsylvanica Marsh. ssp. v.
            (Torr.) G.N. Mill.

         SY=F. v. var. coriacea (S. Wats.)
            Rehd.
         SY=F. v. var. glabra Rehd.
         SY=F. v. var. toumeyi (Britt.)
            Rehd.

Haenianthus Griseb.

   salicifolius Griseb.
      var. obovatus (Krug & Urban) Knobl.
         SY=H. o. Krug & Urban
      var. salicifolius

Jasminum L.

   amplexicaule Wallich ex Don
         SY=J. undulatum Ker-Gawl.
   dichotomum Vahl
   fluminense Vell.
         SY=J. azoricum L.
   mesnyi Hance
   multiflorum (Burm. f.) Andr.
         SY=J. pubescens (Retz.) Willd.
   nitidum Skan
   officinalis L.
      var. grandiflorum (L.) Bailey
         SY=J. g. L.
      var. officinalis
   sambac (L.) Soland.

Ligustrum L.

   amurense Carr.
   japonicum Thunb.
   lucidum Ait. f.
   obtusifolium Sieb. & Zucc.
   ovalifolium Hassk.
   quihoui Carr.
   sinense Lour.
   vulgare L.

Linociera Sw.

   axilliflora Griseb.
         SY=Mayepea a. (Griseb.) Krug &
            Urban
   caribaea (Jacq.) Knobl.
         SY=Mayepea c. (Jacq.) Kuntze
   domingensis (Lam.) Knobl.
         SY=Mayepea d. (Lam.) Krug & Urban
   holdridgii Camp & Monachino
   ligustrina Sw.

Menodora Humb. & Bonpl.

   decemfida (Gill) Gray
      var. longifolia Steyermark
   heterophylla Moric.
   longiflora (Engelm.) Gray
   scabra (Engelm.) Gray
      var. laevis (Woot. & Standl.)
         Steyermark
      var. longituba Steyermark
      var. ramosissima Steyermark
      var. scabra
   scoparia Engelm. ex Gray
   spinescens Gray
      var. mohavensis Steyermark
      var. spinescens

Olea L.

   europaea L.

Osmanthus Lour.

   americanus (L.) Benth. & Hook. f. ex Gray

Osmanthus (CONT.)
        var. americanus
            SY=Amarolea a. (L.) Small
            SY=O. floridana Chapman
        var. megacarpus (Small) P.S. Greene
            SY=Amarolea m. Small
            SY=O. m. (Small) Small ex Little
    sandwicensis (Gray) Knobl.

Syringa L.

    X persica L. [afghanica X laciniata]
    vulgaris L.

ONAGRACEAE

Boisduvalia Spach

    cleistogama Curran
    densiflora (Lindl.) S. Wats.
        SY=B. d. var. pallescens Suksdorf
        SY=B. d. var. salicina (Nutt. ex
            Torr. & Gray) Munz
    glabella (Nutt.) Walp.
        SY=B. g. var. campestris (Jepson)
            Jepson
    macrantha Heller
        SY=B. pallida Eastw.
    stricta (Gray) Greene

Calylophus Spach

    berlandieri Spach
        ssp. berlandieri
            SY=C. drummondianus Spach ssp. b.
                (Spach) Towner & Raven
            SY=Oenothera serrulata Nutt. ssp.
                drummondii (Torr. & Gray)
                Munz
        ssp. pinifolius (Engelm. ex Gray) Towner
            SY=C. drummondianus Spach
            SY=C. serrulatus (Nutt.) Raven
                var. spinulosus (Nutt. ex
                Torr. & Gray) Shinners
            SY=Oenothera serrulata Nutt. ssp.
                p. (Engelm. ex Gray) Munz
    hartwegii (Benth.) Raven
        ssp. fendleri (Gray) Towner & Raven
            SY=Oenothera f. Gray
            SY=O. h. Benth. var. f. (Gray)
                Gray
        ssp. filifolius (Eastw.) Towner & Raven
            SY=C. h. var. f. (Eastw.) Shinners
            SY=Oenothera f. (Eastw.) Tidestrom
            SY=O. h. Benth. var. f. (Eastw.)
                Munz
        ssp. hartwegii
            SY=Oenothera h. Benth.
            SY=O. h. var. typica Munz
        ssp. maccartii (Shinners) Towner & Raven
            SY=C. h. var. m. Shinners
            SY=Oenothera pringlei (Munz) Munz
            SY=O. greggii Gray var. pringlei
                Munz
        ssp. pubescens (Gray) Towner & Raven
            SY=C. h. var. p. (Gray) Shinners
            SY=Galpinsia interior Small
            SY=Oenothera greggii Gray
            SY=O. g. var. lampasana (Buckl.)
                Munz
    lavandulifolius (Torr. & Gray) Raven
        SY=C. hartwegii (Benth.) Raven
            ssp. l. (Torr. & Gray)
            Towner & Raven

            SY=C. h. var. l. (Torr. & Gray)
                Shinners
            SY=Galpinsia l. (Torr. & Gray)
                Small
            SY=Oenothera l. Torr. & Gray
            SY=O. l. var. glandulosa Munz
            SY=O. l. var. typica Munz
    serrulatus (Nutt.) Raven
        SY=C. australis Towner & Raven
        SY=Meriolix s. (Nutt.) Walp.
        SY=Oenothera s. Nutt.
        SY=O. s. var. typica Munz
    toumeyi (Small) Towner
        SY=Oenothera hartwegii Benth. var.
            t. (Small) Munz
        SY=O. t. (Small) Tidestrom
    tubicula (Gray) Raven
        SY=Oenothera t. Gray

Camissonia Link

    andina (Nutt.) Raven
        SY=Oenothera a. Nutt.
    arenaria (A. Nels.) Raven
        SY=Oenothera a. (A. Nels.) Raven
        SY=O. cardiophylla Torr. var.
            splendens Munz & Johnston
    benitensis Raven
    bistorta (Nutt. ex Torr. & Gray) Raven
        SY=Oenothera b. Nutt. ex Torr. &
            Gray
        SY=O. b. var. veitchiana Hook.
    boothii (Dougl. ex Lehm.) Raven
        ssp. alyssoides (Hook. & Arn.) Raven
            SY=Oenothera a. Hook. & Arn.
            SY=O. a. var. villosa S. Wats
            SY=O. boothii Dougl. ex Lehm. ssp.
                a. (Hook. & Arn.) Munz
        ssp. boothii
            SY=Oenothera b. Dougl. ex Lehm.
        ssp. condensata (Munz) Raven
            SY=Oenothera b. Dougl. ex Lehm.
                ssp. c. (Munz) Munz
            SY=O. decorticans (Hook. & Arn.)
                Greene var. c. Munz
        ssp. decorticans (Hook & Arn.) Raven
            SY=Oenothera b. Dougl. ex Lehm.
                ssp. d. (Hook. & Arn.) Munz
            SY=O. d. (Hook. & Arn.) Greene
        ssp. desertorum (Munz) Raven
            SY=Oenothera b. Dougl. ex Lehm.
                ssp. d. (Munz) Munz
            SY=O. decorticans (Hook. & Arn.)
                Greene var. desertorum Munz
        ssp. intermedia (Munz) Raven
            SY=Oenothera b. Dougl. ex Lehm.
                ssp. i. Munz
        ssp. inyoensis (Munz) Munz
            SY=Oenothera b. Dougl. ex Lehm.
                ssp. i. Munz
        ssp. rutila (A. Davids.) Munz
            SY=Oenothera b. Dougl. ex Lehm.
                ssp. r. (A. Davids.) Munz
            SY=O. decorticans (Hook. & Arn.)
                Greene var. r. (A. Davids.)
                Munz
    breviflora (Torr. & Gray) Raven
        SY=Oenothera b. Torr. & Gray
        SY=Taraxia b. (Torr. & Gray) Nutt.
            ex Small
    brevipes (Gray) Raven
        ssp. arizonica (Raven) Raven
            SY=Oenothera b. Gray ssp. a. Raven
        ssp. brevipes
            SY=Oenothera b. Gray
        ssp. pallidula (Munz) Raven
            SY=Oenothera b. Gray ssp. p.

Camissonia (CONT.)
        (Munz) Raven
     SY=O. b. var. p. Munz
     SY=O. p. (Munz) Munz
  californica (Nutt. ex Torr. & Gray) Raven
     SY=Oenothera c. (Nutt. ex Torr. &
        Gray) Greene
     SY=O. leptocarpa Greene
  campestris (Greene) Raven
    ssp. campestris
     SY=Oenothera c. Greene
     SY=O. c. Greene ssp. parishii
        (Abrams) Munz
     SY=O. dentata Cav. var. c.
        (Greene) Jepson
     SY=O. d. var. johnstonii Munz
     SY=O. d. var. p. (Abrams) Munz
    ssp. obispoensis Raven
  cardiophylla (Torr.) Raven
    ssp. cardiophylla
     SY=Oenothera c. Torr.
     SY=O. c. var. typica Munz
    ssp. robusta (Raven) Raven
     SY=Oenothera c. Torr. ssp. r.
        Raven
  chamaenerioides (Gray) Raven
     SY=Oenothera c. Gray
  cheiranthifolia (Hornem. ex Spreng.)
    Raimann
    ssp. cheiranthifolia
     SY=Oenothera c. Hornem. ex Spreng.
     SY=O. c. var. nitida (Greene) Munz
    ssp. suffruticosa (S. Wats.) Raven
     SY=Oenothera c. Hornem. ex Spreng.
        var. s. (S. Wats.) Munz
  claviformis (Torr. & Frem.) Raven
    ssp. aurantiaca (S. Wats.) Raven
     SY=Oenothera c. Torr. & Frem. ssp.
        a. (S. Wats.) Raven
     SY=O. c. var. a. (S. Wats.) Munz
    ssp. claviformis
     SY=Oenothera c. Torr. & Frem.
    ssp. cruciformis (Kellogg) Raven
     SY=Oenothera claviformis Torr. &
        Frem. ssp. cruciformis
        (Kellogg) Raven
     SY=O. claviformis var. citrina
        Raven
     SY=O. claviformis var. cruciformis
        (Kellogg) Munz
    ssp. funerea (Raven) Raven
     SY=Oenothera claviformis Torr. &
        Frem. ssp. f. Raven
    ssp. integrior (Raven) Raven
     SY=Oenothera claviformis Torr. &
        Frem. ssp. i. Raven
     SY=O. c. var. purpurascens (S.
        Wats.) Munz
     SY=O. scapoidea Nutt. ex Torr. &
        Gray var. p. S. Wats.
    ssp. lancifolia (Heller) Raven
     SY=Oenothera claviformis Torr. &
        Frem. ssp. l. (Heller) Raven
    ssp. peeblesii (Munz) Raven
     SY=Oenothera claviformis Torr. &
        Frem. ssp. p. (Munz) Raven
     SY=O. c. var. p. Munz
    ssp. peirsonii (Munz) Raven
     SY=Oenothera claviformis Torr. &
        Frem. ssp. p. (Munz) Raven
     SY=O. c. var. p. Munz
    ssp. rubescens (Raven) Raven
     SY=Oenothera claviformis Torr. &
        Frem. ssp. r. Raven
    ssp. yumae (Raven) Raven
     SY=Oenothera claviformis Torr. &
        Frem. ssp. y. Raven

confertiflora (Raven) Raven
     SY=Oenothera c. Raven
confusa Raven
contorta (Dougl.) Kearney
     SY=Oenothera c. Dougl.
     SY=O. cruciata (S. Wats.) Munz
eastwoodiae (Munz) Raven
     SY=Oenothera e. (Munz) Raven
exilis (Raven) Raven
     SY=Oenothera e. Raven
gouldii Raven
     SY=Oenothera g. (Raven) Welsh &
        Atwood
graciliflora (Hook. & Arn.) Raven
     SY=Oenothera g. Hook. & Arn.
guadalupensis (S. Wats.) Raven
    ssp. clementina (Raven) Raven
     SY=Oenothera g. S. Wats. pro parte
     SY=O. g. S. Wats. ssp. c. Raven
hardhamiae Raven
heterochroma (S. Wats.) Raven
     SY=C. h. ssp. monoensis (Munz)
        Raven
     SY=Oenothera h. S. Wats.
     SY=O. h. S. Wats. ssp. m. (Munz)
        Raven
     SY=O. h. var. m. Munz
hilgardii (Greene) Raven
     SY=Oenothera andina Nutt. var. h.
        (Greene) Munz
     SY=O. contorta Dougl. ex Hook.
        var. h. (Greene) Munz
     SY=O. h. Greene
hirtella (Greene) Raven
     SY=Oenothera h. Greene
     SY=O. micrantha Hornem. ex Spreng.
        var. jonesii (Levl.) Munz
ignota (Jepson) Raven
     SY=Oenothera i. (Jepson) Munz
     SY=O. micrantha Hornem. ex Spreng.
        var. i. Jepson
integrifolia Raven
intermedia Raven
kernensis (Munz) Raven
    ssp. gilmanii (Munz) Raven
     SY=Oenothera dentata Cav. var. g.
        Munz
     SY=O. k. Munz ssp. g. (Munz) Munz
     SY=O. k. ssp. mojavensis Munz
    ssp. kernensis
     SY=Oenothera k. Munz
lacustris Raven
lewisii Raven
luciae Raven
megalantha (Munz) Raven
     SY=Oenothera m. (Munz) Raven
micrantha (Hornem. ex Spreng.) Raven
     SY=Oenothera m. Hornem. ex Spreng.
minor (A. Nels.) Raven
     SY=Oenothera m. (A. Nels.) Munz
     SY=O. m. var. cusickii Munz
multijuga (S. Wats.) Raven
     SY=Oenothera m. S. Wats.
     SY=O. m. var. parviflora (S.
        Wats.) Munz
     SY=O. m. var. typica Munz
munzii (Raven) Raven
     SY=Oenothera m. Raven
nevadensis (Kellogg) Raven
     SY=Oenothera n. Kellogg
ovata (Nutt. ex Torr. & Gray) Raven
     SY=Oenothera o. Nutt. ex Torr. &
        Gray
pallida (Abrams) Raven
    ssp. hallii (A. Davids.) Raven
     SY=Oenothera h. (A. Davids.) Munz
     SY=O. bistorta Nutt. ex Torr. &

Camissonia (CONT.)
        Gray var. h. (A. Davids.)
        Jepson
   ssp. pallida
      SY=Oenothera abramsii J.F. Macbr.
      SY=O. micrantha Hornem. ex Spreng.
        var. exfoliata (A. Nels.)
        Munz
  palmeri (S. Wats.) Raven
      SY=Oenothera p. S. Wats.
  parryi (S. Wats.) Raven
      SY=Oenothera p. S. Wats.
  parvula (Nutt. ex Torr. & Gray) Raven
  pterosperma (S. Wats.) Raven
      SY=Oenothera p. S. Wats.
  pubens (S. Wats.) Raven
      SY=Oenothera contorta Dougl. ex
        Hook. var. p. (S. Wats.)
        Coville
      SY=O. p. (S. Wats.) Munz
  pusilla Raven
      SY=Oenothera contorta Dougl. ex
        Hook. var. flexuosa (A.
        Nels.) Munz
  pygmaea (Dougl. ex Lehm.) Raven
      SY=Oenothera boothii Dougl. ex
        Lehm. var. p. (Dougl. ex
        Lehm.) Torr. & Gray
      SY=O. p. Dougl. ex Lehm.
  refracta (S. Wats.) Raven
      SY=Oenothera r. S. Wats.
  robusta Raven
  scapoidea (Nutt. ex Torr. & Gray) Raven
   ssp. brachycarpa (Raven) Raven
      SY=Oenothera s. Nutt. ex Torr. &
        Gray ssp. b. Raven
   ssp. macrocarpa (Raven) Raven
      SY=Oenothera s. Nutt. ex Torr. &
        Gray ssp. m. Raven
   ssp. scapoidea
      SY=Oenothera s. Nutt. ex Torr. &
        Gray
      SY=O. s. var. seorsa (A. Nels.)
        Munz
   ssp. utahensis (Raven) Raven
      SY=Oenothera s. Nutt. ex Torr. &
        Gray ssp. u. Raven
  sierrae Raven
   ssp. alticola Raven
   ssp. sierrae
  specuicola (Raven) Raven
   ssp. hesperia (Raven) Raven
      SY=Oenothera s. Raven ssp. h.
        Raven
   ssp. specuicola
      SY=Oenothera s. Raven
  strigulosa (Fisch. & Mey.) Raven
      SY=Oenothera contorta Dougl. ex
        Hook. var. s. (Fisch. &
        Mey.) Munz
  subacaulis (Pursh) Raven
      SY=Oenothera heterantha Nutt. ex
        Torr. & Gray
      SY=O. s. (Pursh) Garrett
  tanacetifolia (Torr. & Gray) Raven
   ssp. quadriperforata Raven
   ssp. tanacetifolia
      SY=Oenothera t. Torr. & Gray
  walkeri (A. Nels.) Raven
   ssp. tortilis (Jepson) Raven
      SY=Oenothera w. (A. Nels.) Raven
        ssp. t. (Jepson) Raven
   ssp. walkeri
      SY=Oenothera multijuga S. Wats.
        var. orientalis Munz
      SY=O. w. (A. Nels.) Raven

Circaea L.
  alpina L.
   ssp. alpina
   ssp. pacifica (Aschers. & Magnus) Raven
      SY=C. a. var. p. (Aschers. &
        Magnus) M.E. Jones
      SY=C. p. Aschers. & Magnus
  X intermedia Ehrh. [alpina X lutetiana]
      SY=C. canadensis (L.) Hill sensu
        Fern.
  lutetiana (L.) Aschers. & Magnus
   ssp. canadensis (L.) Aschers. & Magnus
      SY=C. X c. var. virginiana Fern.
      SY=C. latifolia Hill
      SY=C. quadrisulcata (Maxim.)
        Franch. & Sav. var. c. (L.)
        Hara
      SY=C. q. var. lutetiana L. pro
        parte

Clarkia Pursh
  affinis Lewis & Lewis
  amoena (Lehm.) A. Nels. & J.F. Macbr.
   ssp. amoena
      SY=Godetia a. (Lehm.) G. Don
   ssp. caurina (Abrams) Lewis & Lewis
      SY=C. a. var. c. (Abrams) C.L.
        Hitchc.
      SY=C. a. var. pacifica (M.E. Peck)
        C.L. Hitchc.
   ssp. huntiana (Jepson) Lewis & Lewis
   ssp. lindleyi (Dougl.) Lewis & Lewis
      SY=C. a. var. l (Dougl.) C.L.
        Hitchc.
   ssp. whitneyi (Gray) Lewis & Lewis
      SY=Godetia w. (Gray) T. Moore
  arcuata (Kellogg) A. Nels. & J.F. Macbr.
      SY=Godetia hispidula S. Wats.
  australis E. Small
  biloba (Dur.) A. Nels. & J.F. Macbr.
   ssp. australis Lewis & Lewis
   ssp. biloba
      SY=Godetia b. (Dur.) S. Wats.
   ssp. brandegeae (Jepson) Lewis & Lewis
  borealis E. Small
   ssp. arida E. Small
   ssp. borealis
  bottae (Spach) Lewis & Lewis
      SY=C. deflexa (Jepson) Lewis &
        Lewis
      SY=Godetia b. Spach
      SY=G. b. var. d. (Jepson) C.L.
        Hitchc.
  breweri (Gray) Greene
  calientensis Vasek
  concinna (Fisch. & Mey.) Greene
  cylindrica (Jepson) Lewis & Lewis
   ssp. clavicarpa W.S. Davis
   ssp. cylindrica
      SY=Godetia c. (Jepson) C.L.
        Hitchc.
  davyi (Jepson) Lewis & Lewis
      SY=Godetia quadrivulnera (Dougl.)
        Spach var. d. Jepson
  delicata (Abrams) A. Nels. & J.F. Macbr.
  dudleyana (Abrams) J.F. Macbr.
      SY=Godetia d. Abrams
  epilobioides (Nutt.) A. Nels. & J.F. Macbr.
      SY=Godetia e. (Nutt.) S. Wats.
  exilis Lewis & Vasek
  franciscana Lewis & Raven
  gracilis (Piper) A. Nels. & J.F. Macbr.
   ssp. albicaulis (Jepson) Lewis & Lewis
      SY=Godetia amoena (Lehm.) G. Don
        var. albicaulis Jepson

Clarkia (CONT.)
    ssp. gracilis
        SY=Godetia amoena (Lehm.) G. Don
           var. concolor Jepson
        SY=G. a. var. g. (Piper) C.L.
           Hitchc.
    ssp. sonomensis (C.L. Hitchc.) Lewis &
      Lewis
        SY=Godetia amoena (Lehm.) G. Don
           var. s. C.L. Hitchc.
    ssp. tracyi (Jeps.) Abdel-Hameed & Snow
        SY=G. cylindrica (Jepson) C.L.
           Hitchc. var. tracyi Jepson
  imbricata Lewis & Lewis
  jolonensis Parnell
  lassenensis (Eastw.) Lewis & Lewis
        SY=Godetia l. Eastw.
  lewisii Raven & Parnell
  lingulata Lewis & Lewis
  mildrediae (Heller) Lewis & Lewis
  modesta Jepson
  mosquinii E. Small
    ssp. mosquinii
    ssp. xerophila E. Small
  nitens Lewis & Lewis
  prostrata Lewis & Lewis
  pulchella Pursh
  purpurea (W. Curtis) A. Nels. & J.F. Macbr.
    ssp. purpurea
        SY=Godetia p. (W. Curtis) G. Don
    ssp. quadrivulnera (Dougl.) Lewis &
      Lewis
        SY=C. q. (Dougl.) A. Nels. & J.F.
           Macbr.
        SY=Godetia p. (W. Curtis) G. Don
           var. parviflora (S. Wats.)
           C.L. Hitchc.
        SY=G. q. (Dougl.) Spach
        SY=G. q. var. vacensis Jepson
    ssp. viminea (Dougl.) Lewis & Lewis
        SY=C. v. (Dougl.) A. Nels. & J.F.
           Macbr.
        SY=Godetia v. (Dougl.) Spach
  rhomboidea Dougl.
  rostrata W.S. Davis
  rubicunda (Lindl.) Lewis & Lewis
        SY=C. r. ssp. blasdalei (Jepson)
           Lewis & Lewis
  similis Lewis & Ernst
  speciosa Lewis & Lewis
    ssp. immaculata Lewis & Lewis
    ssp. polyantha Lewis & Lewis
    ssp. speciosa
        SY=Godetia parviflora (Hook. &
           Arn.) Jepson
        SY=G. p. var. luteola C.L. Hitchc.
        SY=G. p. var. margaritae (Jepson)
           C.L. Hitchc.
  springvillensis Vasek
  stellata Mosquin
  tembloriensis Vasek
  unguiculata Lindl.
        SY=C. elegans Dougl.
  virgata Greene
  williamsonii (Dur. & Hilg.) Lewis & Lewis
        SY=Godetia viminea (Dougl.) Spach
           var. congdonii Jepson
        SY=G. v. var. incerta Jepson
  xantiana Gray

Epilobium L.

  anagallidifolium Lam.
        SY=E. alpinum L. pro parte
        SY=E. anagallidifolium var.
           pseudo-scaposum (Hausskn.)
           Hulten

angustifolium L.
  ssp. angustifolium
        SY=Chamaenerion a. (L.) Scop.
        SY=Chamerion a. (L.) Holub
        SY=Chamerion spicatum (Lam.) S.F.
           Gray
        SY=E. a. ssp. a. var. intermedium
           (Wormskj.) Fern.
  ssp. circumvagum Mosquin
        SY=E. a. ssp. c. var. macrophyllum
           (Hausskn.) Fern.
        SY=E. a. ssp. m. (Hausskn.) Hulten
        SY=E. a. var. abbreviatum (Lunell)
           Munz
        SY=E. a. var. platyphyllum
           (Daniels) Fern.
canum (Greene) Raven
  ssp. angustifolium (Keck) Raven
        SY=Zauschneria californica Presl
        SY=Z. californica ssp. a. Keck
  ssp. canum
        SY=Zauschneria c. Greene
  ssp. garrettii (A. Nels.) Raven
        SY=Zauschneria g. A. Nels.
  ssp. latifolia (Hook.) Raven
        SY=Zauschneria californica Presl
           ssp. l. (Hook.) Keck
        SY=Z. l. (Hook.) Greene
        SY=Z. l. var. arizonica (A.
           Davids.) Hilend
  ssp. mexicana (Presl) Raven
        SY=Zauschneria californica Presl
           ssp. m. (Presl) Raven
  ssp. septentrionale (Keck) Raven
        SY=Zauschneria s. Keck
ciliatum Raf.
  ssp. ciliatum
        SY=E. adenocaulon Hausskn.
        SY=E. a. var. ecomosum (Fassett)
           Munz
        SY=E. a. var. holosericeum (Trel.)
           Munz
        SY=E. a. var. parishii (Trel.)
           Munz
        SY=E. a. var. perplexans Trel.
        SY=E. americanum Hausskn.
        SY=E. brevistylum Barbey
        SY=E. b. var. ursinum (Parish ex
           Trel.) Jepson
        SY=E. californicum Hausskn.
        SY=E. californicum Hausskn. var.
           h. (Trel.) Munz
        SY=E. ciliatum var. e. (Fassett)
           Boivin
        SY=E. delicatum Trel.
        SY=E. e. (Fassett) Fern.
        SY=F. glandulosum Lehm. var.
           adenocaulon (Hausskn.) Fern.
        SY=E. g. var. macounii (Trel.)
           C.L. Hitchc.
        SY=E. l. var. m. Trel.
        SY=F. watsonii Barbey var.
           parishii (Trel.) C.L.
           Hitchc.
  ssp. glandulosum (Lehm.) Hoch & Raven
        SY=F. adenocaulon Hausskn. var.
           cinerascens (Piper) M.E.
           Peck
        SY=F. a. var. occidentale Trel.
        SY=E. boreale Hausskn.
        SY=E. g. Lehm.
        SY=E. g. var. cardiophyllum Fern.
        SY=E. g. var. o. (Trel.) Fern.
        SY=E. watsonii Barbey var. o.
           (Trel.) C.L. Hitchc.
  ssp. watsonii (Barbey) Hoch & Raven
        SY=E. franciscanum Barbey

Epilobium (CONT.)
        SY=E. w. Barbey
        SY=E. w. var. f. (Barbey) Jepson
    clavatum Trel.
        SY=E. alpinum L. var. albiflorum
            (Suksdorf) C.L. Hitchc.
        SY=E. alpinum var. c. (Trel.) C.L.
            Hitchc.
        SY=E. c. var. glareosum (G.N.
            Jones) Munz
    coloratum Biehler
    davuricum Fisch.
        SY=E. arcticum Samuelsson
        SY=E. d. Fisch. var. a.
            (Samuelsson) Polunin
        SY=E. palustre L. var. d. (Fisch.)
            Welsh
    foliosum (Nutt. ex Torr. & Gray) Suksdorf
        SY=E. minutum Lindl. ex Hook. var.
            f. Nutt. ex Torr. & Gray
    glaberrimum Barbey
        var. fastigiatum (Nutt.) Trel.
            SY=E. platyphyllum Rydb.
        var. glaberrimum
    halleanum Hausskn.
    hirsutum L.
    hornemannii Reichenb.
        ssp. behringianum (Hausskn.) Hoch &
            Raven
            SY=E. alpinum L. var. sertulatum
                (Hausskn.) Welsh
            SY=E. b. Hausskn.
            SY=E. s. Hausskn.
        ssp. hornemannii
            SY=E. alpinum L. var. nutans
                (Hornem.) Hook.
    lactiflorum Hausskn.
        SY=E. alpinum L. var. l.
            (Hausskn.) C.L. Hitchc.
    latifolium L.
        SY=Chamaenerion l. (L.) Sweet
        SY=Chamerion l. (L.) Holub
    leptocarpum Hausskn.
    leptophyllum Raf.
        SY=E. nesophilum (Fern.) Fern.
        SY=E. n. var. sabulonense Fern.
        SY=E. palustre L. var. s. (Fern.)
            Boivin
        SY=E. rosmarinifolium Pursh
        SY=E. treleasianum Levl. pro parte
    luteum Pursh
        SY=E. treleasianum Levl. pro parte
    minutum Lindl. ex Hook.
    mirabile Trel. ex Piper
    nevadense Munz
    nivium Brandeg.
    obcordatum Gray
        ssp. obcordatum
        ssp. siskiyouense Munz
            SY=E. o. var. laxum (Hausskn.)
                Dempster ex Jepson
    oreganum Greene
        SY=E. exaltatum E. Drew
    oregonense Hausskn.
        SY=E. alpinum L. var. gracillimum
            (Trel.) C.L. Hitchc.
    palustre L.
        SY=E. lineare Muhl.
        SY=E. oliganthum Michx.
        SY=E. p. var. grammadophyllum
            Hausskn.
        SY=E. p. var. labradoricum
            Hausskn.
        SY=E. p. var. lapponicum Wahlenb.
        SY=E. p. var. longirameum Fern. &
            Wieg.
        SY=E. p. var. oliganthum (Michx.)

            Fern.
        SY=E. pylaieanum Fern.
        SY=E. wyomingense A. Nels.
    paniculatum Nutt. ex Torr. & Gray
        SY=E. p. var. hammondii (T.J.
            Howell) M.E. Peck
        SY=E. p. var. jucundum (Gray)
            Trel.
        SY=E. p. var. laevicaule (Rydb.)
            Munz
        SY=E. p. var. subulatum (Hausskn.)
            Fern.
        SY=E. p. var. tracyi (Rydb.) Munz
    parviflorum Schreb.
    pringleanum Hausskn.
        SY=E. brevistylum Barbey var.
            subfalcatum (Trel.) Munz
        SY=E. b. var. tenue (Trel.) Jepson
        SY=E. glandulosum Lehm. var. t.
            (Trel.) C.L. Hitchc.
        SY=E. p. var. t. (Trel.) Munz
    rigidum Hausskn.
        SY=E. r. var. canescens Trel.
    saximontanum Hausskn.
        SY=E. drummondii Hausskn.
        SY=E. glandulosum Lehm. var.
            brionense Fern.
        SY=E. scalare Fern.
        SY=E. steckerianum Fern.
    strictum Muhl.
        SY=E. densum Raf.
    suffruticosum Nutt.
    X wisconsinense Ugent [ciliatum X
        coloratum]

Fuchsia L.

    magellanica Lam.

Gaura L.

    angustifolia Michx.
        SY=G. eatonii Small
        SY=G. simulans Small
    biennis L.
    boquillensis Raven & Gregory
    brachycarpa Small
    calcicola Raven & Gregory
    coccinea Pursh
        SY=G. c. var. arizonica Munz
        SY=G. c. var. epilobioides
            (H.B.K.) Munz
        SY=G. c. var. glabra (Lehm.) Torr.
            & Gray
        SY=G. c. var. parvifolia (Torr.)
            Rickett
        SY=G. c. var. typica Munz
        SY=G. g. Lehm.
        SY=G. odorata Sesse ex Lag.
    demareei Raven & Gregory
    drummondii (Spach) Torr. & Gray
    filipes Spach
        SY=G. f. var. major Torr. & Gray
        SY=G. michauxii Spach
    hexandra Ortega
        ssp. gracilis (Woot. & Standl.) Raven &
            Gregory
            SY=G. g. Woot. & Standl.
            SY=G. g. var. typica Munz
    lindheimeri Engelm. & Gray
    longiflora Spach
        SY=G. biennis L. var. pitcheri
            Torr. & Gray
        SY=G. filiformis Small
        SY=G. p. (Torr. & Gray) Small
    macrocarpa Rothrock
    mckelveyae (Munz) Raven & Gregory

Gaura (CONT.)
            SY=G. villosa Torr. var. m. Munz
    neomexicana Woot.
        ssp. coloradensis (Rydb.) Raven &
            Gregory
            SY=G. n. var. c. (Rydb.) Munz
        ssp. neomexicana
            SY=G. n. var. typica Munz
    parviflora Dougl.
            SY=G. p. var. lachnocarpa
                Weatherby
            SY=G. p. var. typica Munz
    sinuata Nutt. ex Ser.
    suffulta Engelm. ex Gray
        ssp. nealleyi (Coult.) Raven & Gregory
            SY=G. n. Coult.
        ssp. suffulta
    triangulata Buckl.
            SY=G. tripetala var. triangulata
                (Buckl.) Munz
    villosa Torr.
        ssp. parksii (Munz) Raven & Gregory
            SY=G. v. var. p. Munz
        ssp. villosa
            SY=G. v. var. arenicola Munz
            SY=G. v. var. typica Munz

Gayophytum A. Juss.

    decipiens Lewis & Szweykowski
    diffusum Torr. & Gray
        ssp. diffusum
        ssp. parviflorum lewis & Szweykowski
            SY=G. lasiospermum Greene
            SY=G. l. var. hoffmannii Munz
            SY=G. nuttallii Torr. & Gray var.
                abramsii Munz
            SY=G. n. var. intermedium (Rydb.)
                Munz
    eriospermum Coville
    heterozygum Lewis & Szweykowski
            SY=G. diffusum Torr. & Gray var.
                villosum Munz
    humile Juss.
            SY=G. nuttallii Torr. & Gray
    oligospermum Lewis & Szweykowski
    racemosum Nutt. ex Torr. & Gray
            SY=G. helleri Rydb.
            SY=G. h. var. glabrum Munz
            SY=G. humile Juss. var. hirtellum
                Munz
            SY=G. r. var. caesium (Torr. &
                Gray) Munz
    ramosissimum Torr. & Gray

Heterogaura Rothrock

    heterandra (Torr.) Coville
            SY=Gaura h. Torr.

Ludwigia L.

    alata Ell.
            SY=L. simulata Small
    alternifolia L.
            SY=L. a. var. linearifolia Britt.
            SY=L. a. var. pubescens Palmer &
                Steyermark
            SY=L. a. var. typica Munz
    arcuata Walt.
            SY=Ludwigiantha a. (Walt.) Small
    bonariensis (M. Micheli) Hara
            SY=Jussiaea neglecta Small
    brevipes (B.H. Long) Eames
    curtissii Chapman
    decurrens Walt.
            SY=Jussiaea d. (Walt.) DC.

            SY=J. erecta L.
            SY=Ludwigia e. (L.) Hara
    glandulosa Walt.
            SY=Ludwigia g. var. torreyi Munz
            SY=L. g. var. typica Munz
    hirtella Raf.
    X lacustris Eames [brevipes X palustris]
    lanceolata Ell.
    leptocarpa (Nutt.) Hara
            SY=Jussiaea l. Nutt.
            SY=Ludwigia l. var. meyeriana
                (Kuntze) Alain
    linearis Walt.
            SY=Ludwigia l. var. puberula
                Engelm. & Gray
    linifolia Poir.
    maritima Harper
    microcarpa Michx.
    octovalvis (Jacq.) Raven
        ssp. octovalvis
            SY=Epilobium suffruticosum Nutt.
                var. o. (Jacq.) Munz
            SY=Jussiaea angustifolia Lam.
            SY=Ludwigia o. var. ligustrifolia
                (H.B.K.) Griseb.
            SY=L. o. var. macropoda (Presl)
                Shinners
            SY=L. o. var. octofila (DC.) Alain
        ssp. sessiliflora (M. Micheli) Raven
            SY=Jussiaea scabra Willd.
            SY=Ludwigia o. var. sessiliflora
                (M. Micheli) Shinners
    palustris (L.) Ell.
            SY=Isnardia p. L.
            SY=Ludwigia p. var. americana
                (DC.) Fern. & Grisc.
            SY=L. p. var. nana Fern. & Grisc.
            SY=L. p. var. pacifica Fern. &
                Grisc.
    peploides (H.B.K.) Raven
        ssp. glabrescens (Kuntze) Raven
            SY=Jussiaea diffusa sensu auctt.
                non Forsk.
            SY=J. repens L.
            SY=J. r. var. g. Kuntze
            SY=Ludwigia p. var. g. (Kuntze)
                Shinners
        ssp. montevidensis (Spreng.) Raven
            SY=Jussiaea repens L. var. m.
                (Spreng.) Munz
            SY=Ludwigia p. var. m. (Spreng.)
                Shinners
        ssp. peploides
            SY=Jussiaea p. H.B.K.
            SY=J. repens L. var. p. (H.B.K.)
                Griseb.
    peruviana (L.) Hara
            SY=Jussiaea p. L.
    pilosa Walt.
    polycarpa Short & Peter
    repens Forst.
            SY=Isnardia intermedia Small &
                Alexander
            SY=I. repens (Forst.) DC.
            SY=Ludwigia nutans Ell.
            SY=L. n. var. stipitata Fern. &
                Grisc.
            SY=L. r. var. rotundata (Griseb.)
                Gomez
            SY=L. repens var. s. (Fern. &
                Grisc.) Munz
    simpsonii Chapman
    spathulata Torr. & Gray
            SY=Isnardia s. (Torr. & Gray)
                Small
    spathulifolia Small
    sphaerocarpa Ell.

Ludwigia (CONT.)
            SY=Ludwigia s. var. deamii Fern. &
                Grisc.
            SY=L. s. var. jungens Fern. &
                Grisc.
            SY=L. s. var. macrocarpa Fern. &
                Grisc.
    suffruticosa Walt.
    uruguayensis (Camb.) Hara
            SY=Jussiaea grandiflora Michx.
            SY=J. michauxiana Fern.
            SY=J. u. Camb.
    virgata Michx.

Oenothera L.

    albicaulis Pursh
            SY=Anogra a. (Pursh) Britt.
            SY=O. ctenophylla (Woot. &
                Standl.) Tidestrom
    argillicola Mackenzie
            SY=O. a. var. pubescens Core &
                Davis
    avita (W. Klein) W. Klein
        ssp. arizonica (Munz) W. Klein
            SY=O. deltoides Torr. & Frem. var.
                a. Munz
        ssp. avita
            SY=O. californica (S. Wats.) S.
                Wats. ssp. a. W. Klein
        ssp. eurekensis (Munz & Roos) W. Klein
            SY=O. deltoides Torr. & Frem. ssp.
                e. Munz & Roos
    biennis L.
        ssp. austromontana Munz
        ssp. biennis
            SY=O. b. ssp. caeciarum Munz
            SY=O. b. var. nutans (Atkinson &
                Bartlett) Wieg.
            SY=O. b. var. pycnocarpa (Atkinson
                & Bartlett) Wieg.
            SY=O. muricata L.
            SY=O. n. Atkinson & Bartlett
            SY=O. p. Atkinson & Bartlett
        ssp. centralis Munz
            SY=O. rubricapitata R.R. Gates
    brachycarpa Gray
            SY=Lavauxia b. (Gray) Britt.
            SY=Megapterium b. (Gray) Levl.
            SY=O. b. var. wrightii (Gray)
                Levl.
    caespitosa Nutt.
        ssp. australis Woot. & Standl.
            SY=O. c. var. a. (Woot. & Standl.)
                Munz
        ssp. caespitosa
            SY=Pachylophus c. (Nutt.) Raimann
        ssp. crinita (Rydb.) Munz
            SY=O. caespitosa var. crinita
                (Rydb.) Munz
        ssp. eximia (Gray) Munz
            SY=O. caespitosa var. e. (Gray)
                Munz
        ssp. jonesii (Munz) Munz
            SY=O. caespitosa var. j. Munz
        ssp. marginata (Nutt.) Munz
            SY=O. caespitosa var. longiflora
                (Heller) Munz
            SY=O. c. var. m. (Nutt.) Munz
        ssp. montana (Nutt.) Munz
            SY=O. caespitosa var. m. (Nutt.)
                Dur.
            SY=O. c. var. psammophila (A.
                Nels. & J.F. Macbr.) Munz
            SY=Pachylophus m. (Nutt.) A. Nels.
        ssp. purpurea (S. Wats.) Munz
            SY=O. caespitosa var. p. (S.

                Wats.) Munz
    californica (S. Wats.) S. Wats.
            SY=O. c. var. glabrata Munz
    canescens Torr. & Frem.
            SY=Gaurella c. (Torr. & Frem.)
                Small
    cavernae Munz
    coronopifolia Torr. & Gray
            SY=Anogra c. (Torr. & Gray) Britt.
    deltoides Torr. & Frem.
        ssp. ambigua (S. Wats.) W. Klein
            SY=O. d. var. decumbens (S. Wats.)
                Munz
        ssp. cognata (Jepson) W. Klein
            SY=C. d. var. c. (Jepson) Munz
        ssp. deltoides
            SY=O. d. var. cineracea (Jepson)
                Munz
            SY=O. d. var. typica Munz
        ssp. howellii (Munz) W. Klein
            SY=O. d. var. h. Munz
        ssp. piperi (Munz) W. Klein
            SY=O. d. var. p. Munz
    engelmannii (Small) Munz
    erythrosepala Borbas
    flava (A. Nels.) Garrett
            SY=Lavauxia f. A. Nels.
            SY=O. hamata (Woot. & Standl.)
                Tidestrom
    fremontii S. Wats.
            SY=Megapterium f. (S. Wats.)
                Britt.
    fruticosa L.
        ssp. fruticosa
            SY=Kneiffia allenii (Britt.) Small
            SY=K. arenicola Small
            SY=K. brevistipata (Pennell) Munz
            SY=K. f. (L.) Raimann
            SY=K. linearis (Michx.) Spach
            SY=K. longipedicellata Small
            SY=K. riparia (Nutt.) Small
            SY=K. semiglandulosa Pennell
            SY=K. subglobosa Small
            SY=O. arenicola (Small) Coker
            SY=O. f. var. eamesii (B.L.
                Robins.) Blake
            SY=O. f. var. goodmanii Munz
            SY=O. f. var. humifusa Allen
            SY=O. f. var. linearis (Michx.) S.
                Wats.
            SY=O. f. var. microcarpa Fern.
            SY=O. f. var. subglobosa (Small)
                Munz
            SY=O. f. var. unguiculata Fern.
            SY=O. f. var. vera Hook.
            SY=O. tetragona Roth ssp. glauca
                (Michx.) Munz var. riparia
                (Nutt.) Munz
            SY=O. t. ssp. t. var. brevistipata
                (Pennell) Munz
            SY=O. t. ssp. t. var. longistipata
                (Pennell) Munz
            SY=O. t. ssp. t. var. sharpii Munz
            SY=O. t. ssp. t. var. velutina
                (Pennell) Munz
        ssp. glauca (Michx.) Straley
            SY=Kneiffia g. (Michx.) Spach
            SY=K. hybrida (Michx.) Small
            SY=K. latifolia Rydb.
            SY=K. tetragona (Roth) Pennell
            SY=O. t. Roth
            SY=O. t. ssp. g. (Michx.) Munz
            SY=O. t. var. fraseri (Pursh) Munz
            SY=O. t. var. hybrida (Michx.)
                Fern.
            SY=O. t. var. latifolia (Rydb.)
                Fern.

Oenothera (CONT.)
  grandiflora L'Her.
  grandis (Britt.) Smyth
            SY=O. laciniata Hill var.
                  grandiflora (S. Wats.) B.L.
                  Robins.
            SY=Raimannia grandis (Britt.) Rose
  havardii S. Wats.
  hookeri Torr. & Gray
    ssp. angustifolia (R.R. Gates) Munz
            SY=O. h. var. a. R.R. Gates
    ssp. grisea (Bartlett) Munz
    ssp. hewettii Cockerell
            SY=O. hookeri var. hewettii
                  (Cockerell) Cockerell
    ssp. hirsutissima (Gray) Munz
            SY=O. biennis L. var. hirsutissima
                  Gray
            SY=O. hookeri var. hirsutissima
                  (Gray) Munz
    ssp. hookeri
            SY=O. biennis L. var. h. (Torr. &
                  Gray) Boivin
    ssp. montereyensis Munz
    ssp. ornata (A. Nels.) Munz
            SY=O. h. var. o. (A. Nels.) Munz
            SY=O. o. (A. Nels.) Rydb.
    ssp. venusta (Bartlett) Munz
    ssp. wolfii Munz
  humifusa Nutt.
            SY=Raimannia h. (Nutt.) Rose
            SY=R. mollissima (L.) Sprague &
                  Riley sensu auctt.
  jamesii Torr. & Gray
  kunthiana (Spach) Munz
  laciniata Hill
            SY=O. amplexicaulis (Woot. &
                  Standl.) Tidestrom
            SY=O. l. ssp. pubescens (Willd.)
                  Munz
            SY=O. l. var. p. (Willd.) Munz
            SY=O. mexicana Spach
            SY=Raimannia l. (Hill) Rose
  latifolia (Rydb.) Munz
            SY=O. pallida Lindl. ssp. l.
                  (Rydb.) Munz
  linifolia Nutt.
            SY=Kneiffia l. (Nutt.) Spach
            SY=O. l. var. glandulosa Munz
            SY=Peniophyllum l. (Nutt.) Pennell
  longissima Rydb.
    ssp. clutei (A. Nels.) Munz
            SY=O. l. var. c. (A. Nels.) Munz
    ssp. longissima
            SY=O. l. ssp. typica Munz
  missouriensis Sims
    var. incana Gray
            SY=O. macrocarpa Nutt. var. i.
                  (Gray) Reveal
    var. missouriensis
            SY=Megapterium m. (Sims) Spach
            SY=O. macrocarpa Nutt.
    var. oklahomensis (J.B.S. Norton)
          Munz
            SY=Megapterium o. J.B.S. Norton
            SY=O. macrocarpa Nutt. var. o.
                  (J.B.S. Norton) Reveal
  neomexicana (Small) Munz
  nuttallii Sweet
            SY=Anogra n. (Sweet) Spach
  organensis Munz
            SY=O. macrosiphon Woot. & Standl.
  pallida Lindl.
    ssp. gypsophila (Eastw.) Munz & W. Klein
            SY=O. runcinata (Engelm.) Munz
                  var. g. (Eastw.) Munz
    ssp. pallida

            SY=O. p. ssp. p. var. idahoensis
                  Munz
            SY=O. p. var. typica Munz
    ssp. runcinata (Engelm.) Munz & W. Klein
            SY=O. r. (Engelm.) Munz
            SY=O. r. var. brevifolia (Engelm.)
                  Munz
            SY=O. r. var. leucotricha (Woot. &
                  Standl.) Munz
            SY=O. r. var. typica Munz
    ssp. trichocalyx (Nutt. ex Torr. & Gray)
          Munz & W. Klein
            SY=O. t. Nutt. ex Torr. & Gray
  parviflora L.
    ssp. angustissima (R.R. Gates) Munz
            SY=O. p. var. a. (R.R. Gates)
                  Wieg.
    ssp. parviflora
      var. oakesiana (J.W. Robbins) Fern.
            SY=O. cruciata (S. Wats.) Munz
                  var. stenopetala (Bickn.)
                  Fern.
            SY=O. o. J.W. Robbins
      var. parviflora
            SY=O. cruciata (S. Wats.) Munz
                  var. sabulonensis Fern.
            SY=O. cymatilis Bartlett
  perennis L.
            SY=Kneiffia p. (L.) Pennell
            SY=K. pumila (L.) Spach
            SY=O. perennis var. rectipilis
                  Blake
            SY=O. perennis var. typica Munz
  pilosella Raf.
    ssp. pilosella
            SY=Kneiffia pratensis Small
            SY=K. sumstinei Jennings
    ssp. sessilis (Pennell) Straley
            SY=O. s. (Pennell) Munz
  platanorum Raven & Parnell
  primiveris Gray
            SY=O. p. ssp. bufonis (M.E. Jones)
                  Munz
            SY=O. p. ssp. caulescens (Munz)
                  Munz
            SY=O. p. var. c. Munz
  psammophila A. Nels. & J.F. Macbr.
  rhombipetala Nutt. ex Torr. & Gray
            SY=O. heterophylla Spach var. r.
                  (Nutt. ex Torr. & Gray)
                  Fosberg
            SY=Raimannia curtissii Rose
            SY=R. r. (Nutt. ex Torr. & Gray)
                  Rose
  rosea L'Her. ex Ait.
  spachiana Torr. & Gray
            SY=O. drummondii Hook.
            SY=Raimannia d. (Hook.) Rose
  speciosa Nutt.
            SY=Hartmannia s. (Nutt.) Small
            SY=O. delessertiana Steud.
            SY=O. s. Nutt. var. childsii
                  (Bailey) Munz
  stricta Ledeb. ex Link
  taraxacoides (Woot. & Standl.) Munz
  tetraptera Cav.
  texensis Raven & Parnell
  triloba Nutt.
            SY=Lavauxia t. (Nutt.) Spach
            SY=L. watsonii (Britt.) Small
  villosa Thunb.
    ssp. canovirens (Steele) D. Dietr. &
          Raven
            SY=O. albinervis R.R. Gates
            SY=O. biennis L. var. canescens
                  Torr. & Gray
            SY=O. canovirens Steele

Oenothera (CONT.)
          SY=O. strigosa (Rydb.) Mackenzie &
                Bush ssp. canovirens
                (Steele) Munz
     ssp. cheradophila (Bartlett) D. Dietr. &
          Raven
          SY=O. c. Bartlett
          SY=O. strigosa (Rydb.) Mackenzie &
                Bush ssp. c. (Bartlett)
                Munz
     ssp. strigosa (Rydb.) D. Dietr. & Raven
          SY=O. biennis L. var. s. (Rydb.)
                Piper
          SY=O. depressa Greene ssp. s.
                (Rydb.) Taylor & MacBryde
          SY=O. procera Woot. & Standl.
          SY=O. rydbergii House
          SY=O. s. (Rydb.) Mackenzie & Bush
     ssp. villosa
          SY=O. depressa Greene
     xylocarpa Coville

Stenosiphon Spach

     virgatus Spach
          SY=S. linifolius (Nutt.) Britt.

ORCHIDACEAE

Amerorchis Hulten

     rotundifolia (Banks ex Pursh) Hulten
          SY=Orchis r. Banks ex Pursh
          SY=O. r. var. lineata Mousley

Anoectochilus Blume

     sandvicensis Lindl.
          SY=?A. apiculatus L.O. Williams &
                Fosberg

Aplectrum (Nutt.) Torr.

     hyemale (Muhl. ex Willd.) Nutt.

Arethusa L.

     bulbosa L.

Arundina Blume

     bambusifolia (Roxb.) Lindl.

Basiphyllaea Schlechter

     corallicola (Small) Ames

Bletia Ruiz & Pavon

     patula Hook.
          SY=B. p. var. alba A.D. Hawkes
     purpurea (Lam.) DC.

Brachionidium Lindl.

     ciliatum Garay
     parvum Cogn.

Brassavola R. Br.

     cucullata (L.) R. Br.
          SY=Epidendrum c. L.
     nodosa (L.) Lindl.

Brassia R. Br.

     caudata (L.) Lindl.

Bulbophyllum Thouars

     pachyrrachis (A. Rich.) Griseb.
          SY=Bolbophyllaria p. (A. Rich.)
                Reichenb. f.

Calopogon R. Br.

     barbatus (Walt.) Ames
          SY=Limodorum parviflorum (Lindl.)
                Nash
     multiflorus Lindl.
          SY=Limodorum m. (Lindl.) C. Mohr
          SY=L. pinetorum Small
     pallidus Chapman
          SY=Limodorum p. (Chapman) C. Mohr
     tuberosus (L.) B.S.P.
          SY=C. pulchellus (Salisb.) R. Br.
          SY=C. p. var. latifolius (St.
                John) Fern.
          SY=C. p. var. simpsonii (Chapman)
                Ames
          SY=C. t. var. l. (St. John) Boivin
          SY=Limodorum s. (Chapman) Small
          SY=L. t. L.

Calypso Salisb.

     bulbosa (L.) Oakes
          var. americana (R. Br.) Luer
          SY=Cytherea b. (L.) House pro
                parte
          var. occidentalis (Holz.) Boivin
          SY=Calypso b. ssp. o. (Holz.)
                Calder & Taylor

Campylocentrum Benth.

     fasciolum (Lindl.) Cogn.
          SY=C. sullivanii Fawcett & Rendle
     jamaicense (Reichenb. f.) Benth.
     monteverdii Rolfe
     pachyrrhizum (Reichenb. f.) Rolfe
     pygmaeum Cogn.

Cephalanthera L.C. Rich.

     austiniae (Gray) Heller
          SY=Eburophyton a. (Gray) Heller

Cleistes L.C. Rich. ex Lindl.

     divaricata (L.) Ames
          SY=C. d. var. bifaria Fern.
          SY=Pogonia d. (L.) R. Br.

Cochleanthes Raf.

     flabelliformis (Sw.) P.E. Schultes & Garay
          SY=Warszewiczella f. (Sw.) Cogn.

Coelia Lindl.

     triptera Brongn.
          SY=Hormidium t. (Brongn.) Cogn.

Coeloglossum Hartman

     viride (L.) Hartman
          var. virescens (Muhl. ex Willd.) Luer
          SY=C. bracteatum (Muhl. ex Willd.)
                Parl.
          SY=C. viride ssp. b. (Muhl. ex

Coeloglossum (CONT.)
                    Willd.) Hulten
            SY=Habenaria b. (Muhl. ex Willd.)
                    R. Br.
            SY=H. viridis (L.) R. Br. var. b.
                    (Muhl. ex Willd.) Gray
        var. viride
            SY=C. v. var. islandicum (Lindl.)
                    Schulze
            SY=Habenaria v. (L.) R. Br.

Comparettia Poepp. & Endl.

    falcata Poepp. & Endl.

Corallorhiza Gagnebin

    maculata (Raf.) Raf.
            SY=C. m. var. flavida (M.E. Peck)
                    Cockerell
            SY=C. m. var. immaculata M.E. Peck
            SY=C. m. var. intermedia Farw.
            SY=C. m. var. occidentalis
                    (Lindl.) Cockerell
            SY=C. m. var. punicea (Bartlett)
                    Weatherby & Adams
    mertensiana Bong.
            SY=C. maculata (Raf.) Raf. ssp.
                    mertensiana (Bong.) Calder &
                    Taylor
    odontorhiza (Willd.) Nutt.
            SY=C. micrantha Chapman
    striata Lindl.
        var. striata
        var. vreelandii (Rydb.) L.O. Williams
            SY=C. bigelovii S. Wats.
    trifida (L.) Chatelain
            SY=C. corallorrhiza (L.) Karst.
            SY=C. t. var. verna (Nutt.) Fern.
    wisteriana Conrad

Corymborchis Thouars

    flava (Sw.) Kuntze

Cranichis Sw.

    diphylla Sw.
    muscosa Sw.

Cypripedium L.

    acaule Ait.
            SY=Fissipes a. (Ait.) Small
    X andrewsii A.M. Fuller [candidum X
        parviflorum]
            SY=C. X a. nm. landonii (Garay)
                    Boivin
    arietinum R. Br.
            SY=Criosanthes a. (R. Br.) House
    calceolus L.
    californicum Gray
    candidum Muhl. ex Willd.
    fasciculatum Kellogg ex S. Wats.
    X favillianum J.T. Curtis [candidum X
        pubescens]
            SY=Cypripedium X andrewsii A.M.
                    Fuller nm. f. (J.T. Curtis)
                    Boivin
    guttatum Sw.
        var. guttatum
        var. yatabeanum (Makino) Pfitzer
            SY=Cypripedium g. ssp. y. (Makino)
                    Hulten
    montanum Dougl. ex Lindl.
    parviflorum Salisb.
            SY=Cypripedium calceolus L. ssp.

                    p. (Salisb.) Hulten
            SY=C. c. var. p. (Salisb.) Fern.
    passerinum Richards.
            SY=Cypripedium p. var. minganense
                    Victorin
    planipetalum (Fern.) Morris & Eames
            SY=Cypripedium calceolus L. var.
                    p. (Fern.) Victorin & Rouss.
    pubescens Willd.
            SY=Cypripedium calceolus L. var.
                    p. (Willd.) Correll
            SY=C. parviflorum Salisb. var.
                    pubescens (Willd.) Knight
    reginae Walt.

Cyrtopodium R. Br.

    andersonii (Lamb. ex Andr.) R. Br.
    punctatum (L.) Lindl.

Dactylorhiza Neck. ex Nevski

    aristata (Fisch. ex Lindl.) Soo
        var. aristata
            SY=Orchis a. Fisch. ex Lindl.
        var. kodiakensis Luer & Luer f.
    maculata (L.) Soo

Dendrobium Sw.

    scopa Lindl.

Dichaea Lindl.

    echinocarpa (Sw.) Lindl.
    hystricina Reichenb. f.
    muricata (Sw.) Lindl.

Dilomilis Raf.

    montana (Sw.) Summerhayes
            SY=Octadesmia m. (Sw.) Benth.

Domingoa Schlechter

    haematochila (Reichenb. f.) Carabia
            SY=D. hymenodes (Reichenb. f.)
                    Schlechter

Elleanthus Presl

    linifolius Presl

Eltroplectris Raf.

    calcarata (Hook. f.) Raf.
            SY=Centrogenium setaceum (Lindl.)
                    Schlechter
            SY=E. acuminata Raf.
            SY=Pelexia s. Lindl.

Encyclia Hook.

    bifida (Aubl.) Britt. & Wilson
            SY=Encylcia papilionacea (Vahl)
                    Schlechter
            SY=Epidendrum b. Aubl.
            SY=Epidendrum brittonianum A.D.
                    Hawkes
    boothiana (Lindl.) Dressler
        var. boothiana
            SY=Epicladium b. (Lindl.) Small
        var. erythronioides (Small) Luer
    cochleata (L.) Dressler
        var. cochleata
            SY=Anacheilium c. (L.) Small
            SY=Epidendrum c. L.

Encyclia   (CONT.)
      var. triandra (Ames) Dressler
         SY=Epidendrum c. L. var. t. Ames
  fucata (Lindl.) Britt. & Millsp.
         SY=Epidendrum f. Lindl.
  gravida (Lindl.) Schlechter
         SY=Epidendrum oncidioides Lindl.
            var. g. Lindl.
  krugii (Bello) Britt. & Wilson
         SY=Epidendrum k. Bello
  pygmaea (Hook.) Dressler
         SY=Hormidium p. (Hook.) Benth. &
            Hook.
  sintenisii (Reichenb. f.) Britt.
         SY=Epidendrum s. Reichenb. f.
  tampensis (Lindl.) Small
         SY=Epidendrum t. Lindl.

Epidendrum L.

  acunae Dressler
  anceps Jacq.
         SY=Amphiglottis a. (Jacq.) Britt.
  belvederense Fawcett & Rendle
  ciliare L.
         SY=Auliza c. (L.) Salisb.
  conopseum R. Br.
         SY=Amphiglottis c. (R. Br.) Small
  difforme Jacq.
         SY=Amphiglottis corymbosa (Lindl.)
            Britt.
         SY=A. d. (Jacq.) Britt.
         SY=Epidendrum moirianum A.D.
            Hawkes
  lacerum Lindl.
         SY=Amphiglottis l. (Lindl.) Britt.
  miserrimum Reichenb. f.
         SY=Jacquiniella m. (Reichenb. f.)
            Stehle
  nocturnum Jacq.
         SY=Amphiglottis n. (Jacq.) Britt.
  nodosum L.
  X obrienanum Rolfe [ibaguense X evectum]
  pallidiflorum Hook.
         SY=Amphiglottis p. (Hook.) Britt.
  patens Sw.
  ramosum Jacq.
         SY=Spathiger r. (Jacq.) Britt.
  rigidum Jacq.
         SY=Spathiger r. (Jacq.) Britt.
  secundum Jacq.
         SY=Amphiglottis s. (Jacq.) Britt.
  strobiliferum Reichenb. f.
         SY=Spathiger s. (Reichenb. f.)
            Small
  vicentinum Lindl.
         SY=Epidendropsis v. (Lindl.)
            Dunsterville & Garay

Epipactis Sw.

  gigantea Dougl. ex Hook.
  helleborine (L.) Crantz
         SY=E. latifolia (L.) All.
         SY=Serapias h. L.

Erythrodes Blume

  hirtella (Sw.) Fawcett
         SY=Physurus h. (Sw.) Lindl.
  plantaginea (L.) Fawcett
         SY=Physurus p. (L.) Lindl.

Eulophia R. Br. ex Lindl.

  alta (L.) Fawcett & Rendle
         SY=Platypus a. (L.) Small

Eurystyles Wawra

  ananassocomos (Reichenb. f.) Schlechter

Galeandra Lindl.

  beyrichii Reichenb. f.

Galearis Raf.

  spectabilis (L.) Raf.
         SY=Galeorchis s. (L.) Rydb.
         SY=Orchis s. L.

Goodyera R. Br.

  oblongifolia Raf.
         SY=G. decipiens (Hook.) F.T.
            Hubbard
         SY=G. o. var. reticulata Boivin
         SY=Peramium d. (Hook.) Piper
  pubescens (Willd.) R. Br.
         SY=Peramium p. (Willd.) MacM.
  repens (L.) R. Br.
         SY=G. ophioides (Fern.) Rydb.
         SY=G. r. var. o. Fern.
         SY=Peramium o. (Fern.) Rydb.
  tesselata Lodd.
         SY=Peramium t. (Lodd.) Heller

Govenia Lindl.

  utriculata (Sw.) Lindl.

Gymnadenia R. Br.

  conopsea (L.) R. Br.
         SY=Habenaria c. (L.) Benth.

Habenaria Willd.

  alata Hook.
  distans Griseb.
  dussii Cogn.
  eustachya Reichenb. f.
  monorrhiza (Sw.) Reichenb. f.
  odontopetala Reichenb. f.
         SY=H. strictissima Reichenb. f.
            var. o. (Reichenb. f.) L.O.
            Williams
         SY=Habenella o. (Reichenb. f.)
            Small
  quinqueseta (Michx.) Eat.
      var. macroceratitis (Willd.) Luer
         SY=Habenaria habenaria (L.) Small
      var. quinqueseta
  repens Nutt.

Harrisella Fawcett & Rendle

  filiformis (Sw.) Cogn.
  porrecta (Reichenb. f.) Fawcett & Rendle

Hexalectris Raf.

  grandiflora (A. Rich. & Gal.) L.O. Williams
  nitida L.O. Williams
  revoluta Correll
  spicata (Walt.) Barnh.
  warnockii Ames & Correll

Ionopsis H.B.K.

  satyrioides (Sw.) Reichenb. f.
  utricularioides (Sw.) Lindl.

Isochilus R. Br.

Isochilus (CONT.)
   linearis (Jacq.) R. Br.

Isotria Raf.

   medeoloides (Pursh) Raf.
   verticillata (Muhl. ex Willd.) Raf.

Jacquiniella Schlechter

   globosa (Jacq.) Schlechter
   teretifolia (Sw.) Britt. & Wilson

Koellensteinia Reichenb. f.

   graminea (Lindl.) Reichenb. f.

Leochilus Kn. & Westc.

   labiatus (Sw.) Kuntze

Lepanthes Sw.

   dodiana Stimson
   eltoroensis Stimson
   rubripetala Stimson
   rupestris Stimson
   sanguinea Hook.
   selenipetala Reichenb. f.
   veleziana Stimson
         var. retusicolumna Stimson
         var. veleziana
   woodburyana Stimson

Lepanthopsis Ames

   melanantha (Reichenb. f.) Ames
         SY=Lepanthes harrisii Fawcett &
            Rendle

Liparis L.C. Rich.

   eggersii Reichenb. f.
   hawaiensis Mann
   lilifolia (L.) L.C. Rich. ex Lindl.
   loeselii (L.) L.C. Rich.
   nervosa (Thunb.) Lindl.
         SY=L. elata Lindl.
   vexillifera (Ll. & Lex.) Cogn.

Listera R. Br.

   auriculata Wieg.
         SY=Ophrys a. (Wieg.) House
   australis Lindl.
         SY=Ophrys a. (Lindl.) House
   borealis Morong
   caurina Piper
         SY=Ophrys c. (Piper) Rydb.
   convallarioides (Sw.) Nutt.
         SY=Ophrys c. (Sw.) W. Wight
   cordata (L.) R. Br.
         var. cordata
            SY=Ophrys c. L.
         var. nephrophylla (Rydb.) Hulten
            SY=L. c. ssp. n. (Rydb.) Love &
               Love
   ovata (L.) R. Br.
   smallii Wieg.
         SY=Ophrys s. (Wieg.) House
   X veltmanii Case [auriculata X
      convallarioides]

Lycaste Lindl.

   barringtoniae (Sm.) Lindl.

Macradenia R. Br.

   lutescens R. Br.

Malaxis Soland. ex Sw.

   brachypoda (Gray) Fern.
         SY=M. monophyllos (L.) Sw. ssp. b.
            (Gray) Love & Love
         SY=M. m. var. b. (Gray) F. Morris
            & Eames
   corymbosa (S. Wats.) Kuntze
   diphyllos Cham.
         SY=M. monophyllos (L.) Sw. var. d.
            (Cham.) Luer
   ehrenbergii (Reichenb. f.) Kuntze
   macrostachya (Lex.) Kuntze
         SY=M. soulei L.O. Williams
   monophyllos (L.) Sw.
   paludosa (L.) Sw.
         SY=Hammarbya p. (L.) Kuntze
   spicata Sw.
         SY=M. floridana (Chapman) Kuntze
   tenuis (S. Wats.) Ames
   unifolia Michx.
         SY=M. bayardii Fern.
         SY=Microstylis u. (Michx.) B.S.P.

Maxillaria Ruiz & Pavon

   coccinea (Jacq.) L.O. Williams
         SY=Ornithidium c. (Jacq.) Salisb.
   crassifolia (Lindl.) Reichenb. f.
         SY=M. sessilis (Sw.) Fawcett &
            Rendle
   purpurea (Spreng.) Ames & Correll
         SY=M. conferta (Griseb.) C.
            Schweinf.
         SY=Ornithidium c. Griseb.

Nidema Britt. & Millsp.

   ottonis (Reichenb. f.) Britt. & Millsp.
         SY=Epidendrum o. Reichenb. f.

Oeceoclades Lindl.

   maculata (Lindl.) Lindl.
         SY=Eulophidium m. (Lindl.) Pfitzer

Oncidium Sw.

   alatum Cogn.
   altissimum (Jacq.) Sw.
   bahamense Nash ex Britt. & Millsp.
   carthaginense (Jacq.) Sw.
   ensatum Lindl.
   floridanum Ames
   leiboldii Reichenb. f.
   lemonianum Lindl.
         SY=O. intermedium Bertero
   luridum Lindl.
         SY=O. undulatum (Sw.) Salisb.
   prionichilum Kraenzlin
   variegatum (Sw.) Willd.

Phaius Lour.

   tancarvilleae (Banks ex L'Her.) Blume

Piperia Rydb.

   elegans (Lindl.) Rydb.
         var. elata (Jepson) Luer
            SY=Habenaria unalascensis
               (Spreng.) S. Wats. var.
               elata (Jepson) Correll

Piperia (CONT.)
          SY=P. lancifolia Rydb.
          SY=P. leptopetala Rydb.
          SY=P. longispica Dur.
          SY=Platanthera u. (Spreng.) Kurtz
              ssp. e. (Jepson) Taylor &
              MacBryde
     var. elegans
          SY=Habenaria e. (Lindl.) Boland.
          SY=Piperia michaelii (Greene)
              Rydb.
          SY=P. multiflora Rydb.
  maritima Rydb.
          SY=Habenaria elegans (Lindl.)
              Boland. var. m. (Rydb.) Ames
          SY=H. greenei Jepson
          SY=H. unalascensis (Spreng.) S.
              Wats. var. m. (Greene)
              Correll
          SY=Platanthera u. (Spreng.) Kurtz
              ssp. m. (Rydb.) de Filipps
          SY=Platanthera u. var. m. (Rydb.)
              Correll
  unalaschensis (Spreng.) Rydb.
          SY=Habenaria u. (Spreng.) S. Wats.
          SY=Platanthera cooperi (S. Wats.)
              Rydb.
          SY=Platanthera u. (Spreng.) Kurtz

Platanthera L.C. Rich.

  albida (L.) Lindl.
     var. straminea (Fern.) Luer
          SY=Habenaria s. Fern.
          SY=Leucorchis a. (L.) E. Mey. ssp.
              s. (Fern.) A. Love
  X andrewsii (M. White) Luer [lacera X
     psycodes]
          SY=Habenaria X a. M. White
  X bicolor (Raf.) Luer [blephariglottis X
     ciliaris]
  blephariglottis (Willd.) Lindl.
     var. blephariglottis
          SY=Blephariglottis b. (Willd.)
              Rydb.
          SY=Habenaria b. (Willd.) Hook.
     var. conspicua (Nash) Luer
          SY=Blephariglottis c. (Nash) Small
          SY=Habenaria b. (Willd.) Hook.
              var. c. (Nash) Ames
  X canbyi (Ames) Luer [blephariglottis X
     cristata]
          SY=Habenaria X c. Ames
  X chapmanii (Small) Luer [ciliaris X
     cristata]
          SY=Blephariglottis c. Small
          SY=Habenaria X c. (Small) Ames
  chorisiana (Cham.) Reichenb.
          SY=Habenaria c. Cham.
  ciliaris (L.) Lindl.
          SY=Blephariglottis c. (L.) Rydb.
          SY=Habenaria c. (L.) R. Br.
  clavellata (Michx.) Luer
          SY=Gymnadeniopsis c. (Michx.)
              Rydb.
          SY=Habenaria c. (Michx.) Spreng.
          SY=H. c. var. ophioglossoides
              Fern.
          SY=H. c. var. wrightii Olive
  cristata (Michx.) Lindl.
          SY=Blephariglottis c. (Michx.)
              Raf.
          SY=Habenaria c. (Michx.) R. Br.
  dilatata (Pursh) Lindl. ex Beck
     var. albiflora (Cham.) Ledeb.
          SY=Habenaria d. (Pursh) Hook. var.
              a. (Cham.) Correll

          SY=Limnorchis d. (Pursh) Rydb.
              ssp. a. (Cham.) Love & Simon
     var. dilatata
          SY=Habenaria d. (Pursh) Hook.
          SY=Limnorchis d. (Pursh) Rydb.
          SY=P. d. var. angustifolia Hook.
     var. leucostachys (Lindl.) Luer
          SY=Habenaria d. (Pursh) Hook. var.
              l. (Lindl.) Ames
          SY=H. l. (Lindl.) S. Wats.
          SY=Limnorchis graminifolia Rydb.
  flava (L.) Lindl.
     var. flava
          SY=Habenaria f. (L.) R. Br.
          SY=H. scutellata (Nutt.) F. Morris
          SY=Perularia bidentata (Ell.)
              Small
          SY=Perularia f. (L.) Farw.
          SY=Perularia s. (Nutt.) Small
     var. herbiola (R. Br.) Luer
          SY=Habenaria f. (L.) R. Br. var.
              h. (R. Br.) Ames & Correll
  grandifolia (Bigelow) Lindl.
          SY=Blephariglottis g. (Bigelow)
              Rydb.
          SY=Habenaria fimbriata (Dry.) R.
              Br.
          SY=H. g. (Bigelow) Torr.
          SY=H. psycodes (L.) Spreng. var.
              g. (Bigelow) Gray
  hookeri (Torr. ex Gray) Lindl.
          SY=Habenaria h. Torr. ex Gray
          SY=H. h. var. abbreviata Fern.
          SY=Lysias hookeriana (Torr. ex
              Gray) Rydb.
  hyperborea (L.) Lindl.
     var. gracilis (Lindl.) Luer
          SY=Habenaria sparsiflora S. Wats.
              var. laxiflora (Rydb.)
              Correll
          SY=P. dilatata (Pursh) Lindl. ex
              Beck var. g. Ledeb.
          SY=Platanthera g. Lindl.
     var. huronensis (Nutt.) Luer
          SY=Habenaria h. (Nutt.) Spreng.
          SY=H. hyperborea (L.) R. Br. ex
              Ait. var. huronensis (Nutt.)
              Farw.
     var. hyperborea
          SY=Habenaria h. (L.) R. Br.
          SY=Limnorchis h. (L.) Rydb.
          SY=L. viridiflora (Cham.) Rydb.
     var. purpurascens (Rydb.) Luer
     var. viridiflora (Cham.) Luer
          SY=Habenaria holochila Hbd.
          SY=H. hyperborea (L.) R. Br. ex
              Ait. var. v. (Cham.) Welsh
          SY=Platanthera holochila (Hbd.)
              Kraenzlin
  integra (Nutt.) Gray ex Beck
          SY=Gymnadeniopsis i. (Nutt.) Rydb.
          SY=Habenaria i. (Nutt.) Spreng.
  integrilabia (Correll) Luer
          SY=Habenaria blephariglottis
              (Willd.) Hook. var. i.
              Correll
  lacera (Michx.) G. Don
     var. lacera
          SY=Blephariglottis l. (Michx.)
              Farw.
          SY=Habenaria l. (Michx.) R. Br.
     var. terrae-novae (Fern.) Luer
          SY=Habenaria l. (Michx.) Lodd.
              var. t-n. Fern.
  leucophaea (Nutt.) Lindl.
          SY=Blephariglottis l. (Nutt.)
              Farw.

Platanthera (CONT.)
        SY=Habenaria l. (Nutt.) Gray
  limosa Lindl.
        SY=Habenaria l. (Lindl.) Hemsl.
        SY=Limnorchis thurberi (Gray)
            Rydb.
  X media (Rydb.) Luer [dilatata X
   hyperborea]
        SY=Habenaria dilatata (Pursh)
            Hook. var. m. (Rydb.) Hulten
        SY=H. X m. (Rydb.) Niles
        SY=Limnorchis m. Rydb.
        SY=Platanthera convallariifolia
            (Fisch.) Lindl. var.
            dilatatoides Hulten
        SY=P. dilatata (Pursh) Lindl. ex
            Beck var. chlorantha Hulten
  nivea (Nutt.) Luer
        SY=Gymnadeniopsis n. (Nutt.) Rydb.
        SY=Habenaria n. (Nutt.) Spreng.
  obtusata (Banks ex Pursh) Lindl.
        SY=Habenaria o. (Banks ex Pursh)
            Richards.
        SY=H. o. var. collectanea Fern.
        SY=Lysias o. (Banks ex Pursh)
            Rydb.
        SY=Lysiella o. (Banks ex Pursh)
            Britt. & Rydb.
        SY=Platanthera o. ssp. oligantha
            (Turcz.) Hulten
  orbiculata (Pursh) Lindl.
      var. macrophylla (Goldie) Luer
        SY=Habenaria m. Goldie
        SY=H. o. (Pursh) Torr. var. m.
            (Goldie) Boivin
        SY=Platanthera m. (Goldie) Lindl.
      var. orbiculata
        SY=Habenaria o. (Pursh) Torr.
        SY=H. o. var. lehorsii Fern.
        SY=Lysias o. (Pursh) Rydb.
  peramoena (Gray) Gray
        SY=Blephariglottis p. (Gray) Rydb.
        SY=Habenaria p. Gray
  psycodes (L.) Lindl.
        SY=Blephariglottis p. (L.) Rydb.
        SY=Habenaria p. (L.) Spreng.
  sparsiflora (S. Wats.) Schlechter
      var. brevifolia (Greene) Luer
        SY=Habenaria s. S. Wats. var. b.
            (Greene) Correll
      var. ensifolia (Rydb.) Luer
      var. sparsiflora
        SY=Habenaria s. S. Wats.
        SY=Limnorchis s. (S. Wats.) Rydb.
  stricta Lindl.
        SY=Habenaria saccata Greene
        SY=Limnorchis saccata (Greene)
            Love & Simon
        SY=L. stricta (Lindl.) Rydb.
        SY=Platanthera saccata (Greene)
            Hulten
  tipuloides (L.) Lindl.
      var. behringiana (Rydb.) Hulten
        SY=Habenaria b. (Rydb.) Ames

Platythelys Garay

  querceticola (Lindl.) Garay
        SY=Erythrodes q. (Lindl.) Ames
        SY=Physurus q. Lindl.
        SY=Physurus sagraeanus A. Rich.

Pleurothallis R. Br.

  appendiculata Cogn.
  coriacea Bello
  crassipes Lindl.

  foliata Griseb.
  gelida Lindl.
  pruinosa Lindl.
  racemiflora Hook. ex Lindl.
        SY=P. longissima Lindl.
  ruscifolia (Jacq.) R. Br.
  urbaniana Reichenb. f.
  wilsonii Lindl.

Pogonia Juss.

  ophioglossoides (L.) Juss.
        SY=P. o. var. brachypogon Fern.

Polyradicion Garay

  lindenii (Lindl.) Garay
        SY=Polyrrhiza l. (Lindl.) Cogn.

Polystachya Hook.

  cerea Lindl.
        SY=P. minor Fawcett & Rendle
  concreta (Jacq.) Garay & Sweet
        SY=P. extinctoria Reichenb. f.
        SY=P. flavescens (Lindl.) J.J. Sm.
        SY=P. luteola (Sw.) Hook.
        SY=P. minuta (Aubl.) Frappier ex
            Cordemoy

Ponthieva R. Br.

  racemosa (Walt.) C. Mohr
      var. brittonae (Ames) Luer
        SY=P. b. Ames
      var. racemosa
  ventricosa (Griseb.) Fawcett & Rendle

Prescottia Lindl.

  oligantha (Sw.) Lindl.
  stachyoides (Sw.) Lindl.

Psilochilus Barb. Rodr.

  macrophyllus (Lindl.) Ames
        SY=Pogonia m. Lindl.

Pteroglossaspis Rehb. f.

  ecristata (Fern.) Rolfe
        SY=Eulophia e. (Fern.) Ames
        SY=Triorchos e. (Fern.) Small

Restrepiella Garay & Dunsterville

  ophiocephala (Lindl.) Garay & Dunsterville

Scaphyglottis Peopp. & Endl.

  modesta (Reichenb. f.) Schlechter
        SY=Tetragamestus m. Reichenb. f.

Spathoglottis Blume

  elmeri Ames
  plicata Blume

Spiranthes L.C. Rich.

  adnata (Sw.) Fawcett
        SY=Pelexia a. (Sw.) Spreng.
  amabilis Ames
  brevilabris Lindl.
      var. brevilabris
        SY=S. gracilis (Bigelow) Beck var.
            b. (Lindl.) Correll

Spiranthes (CONT.)
        var. floridana (Wherry) Luer
                SY=Ibidium f. Wherry
                SY=S. gracilis (Bigelow) Beck var.
                        f. (Wherry) Correll
casei Catling & Cruise
        SY=S. X intermedia sensu auctt.
                non Ames
cernua (L.) L.C. Rich.
        SY=Ibidium c. (L.) House
cinnabarina (Ll. & Lex.) Hemsl.
costaricensis Reichenb. f.
cranichoides (Griseb.) Cogn.
        SY=Beadlea c. (Griseb.) Small
        SY=Cyclopogon c. (Griseb.)
                Schlechter
durangensis Ames & C. Schwienf.
elata (Sw.) L.C. Rich.
        SY=Beadlea e. (Sw.) Small
        SY=Cyclopogon e. (Sw.) Schlechter
fawcettii Cogn.
        SY=Hapalorchis tenuis (Lindl.)
                Schlechter
graminea Lindl.
X intermedia Ames [lacera var. gracilis X
        vernalis]
lacera (Raf.) Raf.
        var. gracilis (Bigelow) Luer
                SY=Ibidium beckii (Lindl.) House
                SY=I. g. (Bigelow) House
                SY=S. b. Lindl.
                SY=S. g. (Bigelow) Beck
        var. lacera
laciniata (Small) Ames
        SY=Ibidium l. (Small) House
lanceolata (Aubl.) Leon
        var. lanceolata
                SY=S. orchioides (Sw.) A. Rich.
                SY=Stenorrhynchus l. (Aubl.) L.C.
                        Rich.
                SY=Stenorrhynchus o. (Sw.) L.C.
                        Rich.
        var. luteoalba (Reichenb. f.) Luer
        var. paludicola Luer
longilabris Lindl.
        SY=Ibidium l. (Lindl.) House
lucayana (Britt.) Cogn.
        SY=Mesadenus l. (Britt.)
                Schlechter
lucida (H.H. Eat.) Ames
        SY=Ibidium plantagineum (Raf.)
                House
magnicamporum Sheviak
michuacana (Ll. & Lex.) Hemsl.
ochroleuca (Rydb.) Rydb.
        SY=Spiranthes cernua (L.) L.C.
                Rich. var. o. (Rydb.) Ames
odorata (Nutt.) Lindl.
        SY=Spiranthes cernua (L.) L.C.
                Rich. var. o. (Nutt.)
                Correll
ovalis Lindl.
        SY=Ibidium o. (Lindl.) House
parasitica A. Rich. & Gal.
parksii Correll
polyantha Reichenb. f.
porrifolia Lindl.
        SY=Spiranthes romanzoffiana Cham.
                var. p. (Lindl.) Ames &
                Correll
praecox (Walt.) S. Wats.
        SY=Ibidium p. (Walt.) House
romanzoffiana Cham.
        SY=Ibidium strictum (Rydb.) House
        SY=Spiranthes s. Rydb.
speciosa (J.F. Gmel.) A. Rich.
        SY=Stenorrhynchus s. (J.F. Gmel.)

                L.C. Rich.
X steigeri Correll [cernua X romanzoffiana]
torta (Thunb.) Garay & Sweet
        SY=Ibidium tortile (Sw.) House
        SY=Spiranthes tortilis (Sw.) L.C.
                Rich.
tuberosa Raf.
        SY=Spiranthes grayi Ames
        SY=S. t. var. g. (Ames) Fern.
vernalis Engelm. & Gray
        SY=Ibidium v. (Engelm. & Gray)
                House

Stelis Sw.

    perpusilliflora Cogn.

Tetramicra Lindl.

    canaliculata (Aubl.) Urban
    elegans (Hamilton) Cogn.

Tipularia Nutt.

    discolor (Pursh) Nutt.
            SY=T. unifolia (Muhl.) B.S.P.

Triphora Nutt.

    craigheadii Luer
    gentianoides (Sw.) Ames & Schlechter
            SY=T. cubensis (Reichenb. f.) Ames
    latifolia Luer f.
    surinamensis (Lindl.) Britt.
    trianthophora (Sw.) Rydb.
            SY=T. t. var. schaffneri Camp
    yucatanensis Ames
            SY=T. rickettii Luer

Tropidia Lindl.

    polystachya (Sw.) Ames

Vanilla P. Mill.

    barbellata Reichenb. f.
            SY=V. articulata Northrop
    dilloniana Correll
    eggersii Rolfe
    mexicana P. Mill.
            SY=V. inodora Schiede
    phaeantha Reichenb. f.
    planifolia Andr.
            SY=V. vanilla (L.) Britt.

Wullschlaegelia Reichenb. f.

    aphylla (Sw.) Reichenb. f.

Xylobium Lindl.

    palmifolium (Sw.) Benth.

Zeuxine Lindl.

    strateumatica (L.) Schlechter

OROBANCHACEAE

Boschniakia C.A. Mey. ex Bong.

    hookeri Walp.
    rossica (Cham. & Schlect.) Fedtsch.
    strobilacea Gray

Conopholis Wallr.

    alpina Liebm.
        var. mexicana (Gray ex S. Wats.)
           Haynes
           SY=C. m. Gray ex S. Wats.
    americana (L.) Wallr.
           SY=Orobanche multiflora Nutt. var.
               xanthochroa (A. Nels. &
               Cockerell) Munz
           SY=O. x. A. Nels. & Cockerell

Epifagus Nutt.

    virginiana (L.) Bart.
           SY=Leptamnium v. (L.) Raf.

Orobanche L.

    bulbosa (Gray) G. Beck
    californica Cham. & Schlecht.
      ssp. californica
           SY=O. grayana G. Beck var.
               nelsonii Munz
           SY=O. g. var. violacea (Eastw.)
               Munz
      ssp. condensa Heckard
      ssp. feudgei (Munz) Heckard
           SY=O. grayana G. Beck var. f. Munz
      ssp. grandis Heckard
      ssp. grayana (G. Beck) Heckard
           SY=O. c. var. g. (G. Beck) Cronq.
           SY=O. g. G. Beck
      ssp. jepsonii (Munz) Heckard
           SY=O. grayana G. Beck var. j. Munz
    cooperi (Gray) Heller
      ssp. cooperi
           SY=Aphyllon c. Gray
           SY=Myzorrhiza c. (Gray) Rydb.
           SY=O. ludoviciana Nutt. var. c.
               (Gray) G. Beck
      ssp. latiloba (Munz) Collins ined.
           SY=O. ludoviciana var. latiloba
               Munz
    corymbosa (Rydb.) Ferris
      ssp. corymbosa
           SY=O. californica Cham. &
               Schlecht. var. corymbosa
               (Rydb.) Munz
      ssp. mutabilis Heckard
    dugesii (S. Wats.) Munz
           SY=Aphyllon d. S. Wats.
    fasciculata Nutt.
      var. fasciculata
           SY=Anoplanthus f. (Nutt.) Walp.
           SY=O. f. var. typica Achey
           SY=Thalesia f. (Nutt.) Britt.
      var. franciscana Achey
      var. lutea (Parry) Achey
           SY=Thalesia l. (Parry) Rydb.
      var. subulata Goodman
    ludoviciana Nutt.
      ssp. ludoviciana
           SY=Aphyllon arenosum Suksdorf
           SY=Conopholis l. (Nutt.) Wood
           SY=Myzorrhiza l. (Nutt.) Rydb.
           SY=O. l. var. a. (Suksdorf) Cronq.
           SY=O. l. var. genuina G. Beck
           SY=O. multiflora Nutt. var. a.
               (Suksdorf) Munz
      ssp. multiflora (Nutt.) Collins ined.
           SY=Myzorrhiza m. (Nutt.) Rydb.
           SY=O. l. var. m. (Nutt.) G. Beck
           SY=O. m. Nutt.
           SY=O. m. var. pringlei Munz
           SY=O. m. var. typica Munz
           SY=Phelipaea erianthera Engelm.

minor Sutton
multicaulis Brandeg.
    ssp. multicaulis
           SY=O. m. var. genuina Munz
    ssp. palmeri (Munz) Collins ined.
           SY=O. m. var. p. Munz
parishii (Jepson) Heckard
    ssp. brachyloba Heckard
    ssp. parishii
           SY=O. californica Cham. &
               Schlecht. var. p. Jepson
pinorum Geyer ex Hook.
purpurea Jacq.
ramosa L.
riparia Collins ined.
uniflora L.
    var. minuta (Suksdorf) G. Beck
           SY=O. u. ssp. occidentalis
               (Greene) Abrams ex Ferris
           SY=O. u. var. o. (Greene) Taylor &
               MacBryde
    var. purpurea (Heller) Achey
    var. sedii (Suksdorf) Achey
    var. terrae-novae (Fern.) Munz
           SY=O. t. Fern.
    var. uniflora
           SY=O. u. var. typica Achey
           SY=Thalesia u. (L.) Raf.
valida Jepson
           SY=O. ludoviciana Nutt. var. v.
               (Jepson) Munz
           SY=O. v. (Jepson) Jepson
vallicola (Jepson) Heckard
           SY=O. californica Cham. &
               Schlecht. var.
               claremontensis Munz

OXALIDACEAE

Oxalis L.

    acetosella L.
           SY=O. a. ssp. montana (Raf.)
               Hulten ex D. Love
           SY=O. a. var. rhodantha (Fern.) R.
               Knuth
           SY=O. m. Raf.
    albicans H.B.K.
      ssp. albicans
      ssp. californica (Abrams) Eiten
           SY=O. c. (Abrams) R. Knuth
      ssp. pilosa (Nutt.) Eiten
           SY=O. p. Nutt.
    alpina (Rose) R. Knuth
           SY=O. metcalfei (Small) R. Knuth
    amplifolia (Trel.) R. Knuth
    barrelieri L.
    berlandieri Torr.
    caerulea (Small) R. Knuth
    corniculata L.
      var. atropurpurea Planch.
      var. corniculata
           SY=O. c. var. macrantha Trel.
           SY=O. langloisii (Small) Fedde
           SY=O. recurva Ell. var. m. (Trel.)
               Wieg.
           SY=O. repens Thunb.
           SY=Xanthoxalis c. (L.) Small
           SY=X. l. Small
           SY=X. m. (Trel.) Small
      var. lupulina (R. Knuth) Zucc.
      var. viscidula Wieg.
    corymbosa DC.
           SY=Ionoxalis martiana (Zucc.)

Oxalis (CONT.)
                        Small
                SY=O. m. Zucc.
        decaphylla H.B.K.
                SY=O. grayii (Rose) R. Knuth
        dichondrifolia Gray
        dillenii Jacq.
            ssp. dillenii
                SY=O. d. var. radicans Shinners
                SY=O. florida Salisb.
            ssp. filipes (Small) Eiten
                SY=O. brittoniae Small
                SY=O. f. Small
                SY=O. florida Salisb. var. filipes
                        (Small) Ahles
                SY=Xanthoxalis b. Small
                SY=X. filipes (Small) Small
        drummondii Gray
        europaea Jord.
        grandis Small
                SY=Xanthoxalis g. (Small) Small
        hirta L.
        incarnata L.
        intermedia A. Rich.
                SY=Ionoxalis i. (A. Rich.) Small
        latifolia H.B.K.
        laxa Hook. & Arn.
        oregana Nutt. ex Torr. & Gray
                SY=O. acetosella L. ssp. o. (Nutt.
                        ex Torr. & Gray) D. Love
                SY=O. o. var. smallii (R. Knuth)
                        M.E. Peck
        pes-caprae L.
                SY=Bulboxalis cernua (Thunb.)
                        Small
                SY=O. c. Thunb.
        priceae Small
            ssp. colorea (Small) Eiten
                SY=Xanthoxalis c. Small
            ssp. priceae
                SY=O. florida Salisb. var. recurva
                        (Ell.) Ahles
                SY=O. r. Ell.
                SY=Xanthoxalis hirsuticaulis Small
                SY=X. p. (Small) Small
                SY=X. r. (Ell.) Small
            ssp. texana (Small) Eiten
                SY=O. t. (Small) Fedde
        purpurea L.
        rubra St. Hil.
        rugeliana Urban
        stricta L.
            var. piletocarpa Wieg.
            var. stricta
                SY=O. europaea Jord. var. bushii
                        (Small) Wieg.
                SY=Xanthoxalis b. Small
                SY=X. cymosa Small
                SY=X. interior Small
                SY=X. rufa Small
                SY=X. s. (L.) Small
        suksdorfii Trel.
        trilliifolia Hook.
                SY=Hesperoxalis t. (Hook.) Small
        violacea L.
            var. trichophora Fassett
            var. violacea
                SY=Ionoxalis v. (L.) Small

PAEONIACEAE

Paeonia L.

    brownii Dougl. ex Hook.

    californica Nutt. ex Torr. & Gray
            SY=P. brownii Dougl. ex Hook. ssp.
                    c. (Nutt. ex Torr. & Gray)
                    Abrams
    lactiflora Pallas
    officinalis L.

PANDANACEAE

Freycinetia Gaud.

    arborea Gaud.

Pandanus L. f.

    tectorius Parkinson ex Zucc.
            var. chamissonis (Gaud.) Stone
                SY=P. c. Gaud.
            var. douglasii (Gaud.) Stone
                SY=P. d. Gaud.
            var. laevigatus (Martelli) Stone
                SY=P. odoratissimus L. f. var. l.
                        Martelli
            var. menziesii (Gaud.) Stone
                SY=P. m. Gaud.
            var. oahuensis (Martelli) Stone
                SY=P. odoratissimus L. f. var. o.
                        Martelli
            var. sandvicensis Warb.

PAPAVERACEAE

Adlumia Raf. ex DC.

    fungosa (Ait.) Greene ex B.S.P.

Arctomecon Torr. & Frem.

    californica Torr. & Frem.
    humilis Coville
    merriamii Coville

Argemone L.

    aenea G.B. Ownbey
    albiflora Hornem.
        ssp. albiflora
                SY=A. alba Lestib. f.
        ssp. texana G.B. Ownbey
                SY=A. a. var. t. (G.B. Ownbey)
                        Shinners
    arizonica G.B. Ownbey
    aurantiaca G.B. Ownbey
    chisosensis G.B. Ownbey
    corymbosa Greene
        ssp. arenicola G.B. Ownbey
                SY=A. c. var. a. (G.B. Ownbey)
                        Shinners
        ssp. corymbosa
    glauca L. ex Pope
            var. decipiens G.B. Ownbey
            var. glauca
            var. inermis Deg. & Deg.
    gracilenta Greene
    hispida Gray
                SY=A. bipinnatifida Greene
                SY=A. platyceras Link & Otto var.
                        h. (Gray) Prain
    mexicana L.
                SY=A. leiocarpa Greene
                SY=A. m. var. ochroleuca (Sweet)

Argemone (CONT.)
        Lindl.
        SY=A. o. Sweet
        SY=A. o. var. stenophylla (Prain)
           Shinners
  munita Dur. & Hilg.
    ssp. argentea G.B. Ownbey
        SY=A. m. var. a. (G.B. Ownbey)
           Shinners
    ssp. munita
    ssp. robusta G.B. Ownbey
        SY=A. m. var. r. (G.B. Ownbey)
           Shinners
    ssp. rotundata (Rydb.) G.B. Ownbey
        SY=A. m. var. r. (Rydb.) Shinners
        SY=A. r. Rydb.
  pleiacantha Greene
    ssp. ambigua G.B. Ownbey
        SY=A. p. var. a. (G.B. Ownbey)
           Shinners
    ssp. pinnathisecta G.B. Ownbey
        SY=A. pleiacantha var.
           pinnathisecta (G.B. Ownbey)
           Shinners
    ssp. pleiacantha
  polyanthemos (Fedde) G.B. Ownbey
        SY=A. intermedia sensu auctt. non
           Sweet
        SY=A. i. var. p. Fedde
        SY=A. platyceras sensu auctt. non
           Link & Otto
  sanguinea Greene
        SY=A. platyceras Link & Otto var.
           rosea Coult.
  squarrosa Greene
    ssp. glabrata G.B. Ownbey
        SY=A. s. var. g. (G.B. Ownbey)
           Shinners
    ssp. squarrosa

Bocconia L.

  frutescens L.

Canbya Parry ex Gray

  aurea S. Wats.
  candida Parry

Chelidonium L.

  majus L.
    var. laciniatum (P. Mill.) Syme
    var. majus
    var. plenum Wehrhahn

Corydalis Vent.

  aquae-gelidae M.E. Peck & Wilson
  aurea Willd.
    ssp. aurea
        SY=Capnoides a. (Willd.) Kuntze
    ssp. occidentalis (Engelm.) G.B. Ownbey
        SY=Capnoides montanum (Engelm.)
           Britt.
        SY=Corydalis a. var. o. Engelm.
        SY=Corydalis m. Engelm.
  caseana Gray
    ssp. brachycarpa (Rydb.) G.B. Ownbey
        SY=Capnoides b. Rydb.
    ssp. brandegei (S. Wats.) G.B. Ownbey
        SY=Corydalis b. S. Wats.
    ssp. caseana
        SY=Corydalis bidwelliae S. Wats.
    ssp. cusickii (S. Wats.) G.B. Ownbey
        SY=Corydalis caseana var. cusickii
           (S. Wats.) C.L. Hitchc.

        SY=C. cusickii S. Wats.
    ssp. hastata (Rydb.) G.B. Ownbey
        SY=Capnoides h. Rydb.
        SY=Corydalis c. var. h. (Rydb.)
           C.L. Hitchc.
  crystallina Engelm.
        SY=Capnoides c. (Engelm.) Kuntze
  curvisiliqua Engelm.
    ssp. curvisiliqua
    ssp. grandibracteata (Fedde) G.B. Ownbey
        SY=Corydalis c. var. g. Fedde
  flavula (Raf.) DC.
        SY=Capnoides f. (Raf.) Kuntze
  lutea (L.) DC.
  micrantha (Engelm.) Gray
    ssp. australis (Chapman) G.B. Ownbey
        SY=Capnoides campestre Britt.
        SY=Capnoides halei Small
        SY=Corydalis aurea Willd. var.
           australis Chapman
        SY=Corydalis c. (Britt.) Buchh. &
           Palmer
        SY=Corydalis h. (Small) Fern. &
           Schub.
        SY=Corydalis m. var. a. (Chapman)
           Shinners
    ssp. micrantha
        SY=Capnoides m. (Engelm.) Britt.
        SY=Corydalis aurea Willd. var. m.
           Engelm.
    ssp. texensis G.B. Ownbey
        SY=Corydalis m. var. t. (G.B.
           Ownbey) Shinners
  pauciflora (Steph.) Pers.
        SY=Corydalis p. var. albiflora
           Porsild
        SY=C. p. var. chamissonis Fedde
  scouleri Hook.
  sempervirens (L.) Pers.
        SY=Capnoides s. (L.) Borckh.

Dendromecon Benth.

  rigida Benth.
    ssp. harfordii (Kellogg) Raven
        SY=D. h. Kellogg
    ssp. rhamnoides (Greene) Thorne
        SY=D. rigida var. rhamnoides
           (Greene) Munz
    ssp. rigida

Dicentra Bernh.

  canadensis (Goldie) Walp.
        SY=Bicuculla c. (Goldie) Millsp.
  chrysantha (Hook. & Arn.) Walp.
  cucullaria (L.) Bernh.
        SY=Bicuculla c. (L.) Millsp.
        SY=D. c. var. occidentalis (Rydb.)
           M.E. Peck
        SY=D. o. (Rydb.) Fedde
  eximia (Ker-Gawl.) Torr.
        SY=Bicuculla e. (Ker-Gawl.)
           Millsp.
  formosa (Haw.) Walp.
    ssp. formosa
        SY=D. f. ssp. f. var. brevifolia
           Henderson
        SY=D. f. ssp. f. var. brevipes
           Henderson
    ssp. oregona (Eastw.) Munz
        SY=D. o. Eastw.
  nevadensis Eastw.
        SY=D. formosa (Haw.) Walp. ssp. n.
           (Eastw.) Munz
  ochroleuca Engelm.
  pauciflora S. Wats.

Dicentra (CONT.)
    uniflora Kellogg

Eschscholtzia Cham.

    caespitosa Benth.
        ssp. caespitosa
            SY=E. rhombipetala Greene
        ssp. kernensis Munz
    californica Cham.
        ssp. californica
            SY=E. c. var. crocea (Benth.)
                Jepson
            SY=E. c. var. douglasii (Benth.)
                Gray
            SY=E. c. var. maritima (Greene)
                Jepson
            SY=E. c. var. peninsularis
                (Greene) Munz
        ssp. mexicana (Greene) C. Clark
            SY=E. m. Greene
    covillei Greene
            SY=E. minutiflora S. Wats. var.
                darwinensis M.E. Jones
    elegans Greene
    glyptosperma Greene
    hypecoides Benth.
            SY=E. caespitosa Benth. var. h.
                (Benth.) Gray
    lemmonii Greene
        var. asprella (Greene) Jepson
        var. laxa Greene
        var. lemmonii
    lobbii Greene
    minutiflora S. Wats.
    parishii Greene
    procera Greene
    ramosa Greene

Fumaria L.

    capreolata L.
    martinii Clavaud
    officinalis L.
    parviflora Lam.

Glaucium P. Mill.

    corniculatum (L.) J.H. Rudolph
    flavum Crantz

Hesperomecon Greene

    linearis (Benth.) Greene
            SY=Meconella l. (Benth.) A. Nels.
                & J.F. Macbr.
            SY=M. l. var. pulchella (Greene)
                Jepson

Hunnemannia Sweet

    fumariifolia Sweet

Macleaya R. Br.

    cordata (Willd.) R. Br.
            SY=Bocconia c. Willd.
            SY=B. japonica Andre

Meconella Nutt.

    californica Torr. & Frem.
            SY=M. collina Greene
            SY=M. octandra Greene
    denticulata Greene
            SY=M. oregana Nutt. var. d.
                (Greene) Jepson

        oregana Nutt.

Papaver L.

    alboroseum Hulten
    alpinum L.
            SY=P. nudicaule L. ssp. radicatum
                (Rottb.) Fedde var.
                pseudocorylalifolium Fedde
            SY=P. pygmaeum Rydb.
    argemone L.
    californicum Gray
            SY=P. lemmonii Greene
    croceum Ledeb.
    dubium L.
    hybridum L.
            SY=P. apulum Ten. var. micranthum
                sensu auctt. non (Boreau)
                Fedde
    lapponicum (Tolm.) Nordh.
        ssp. occidentale (Lundstr.) Knaben
            SY=P. alaskanum Hulten
            SY=P. cornwallisensis D. Love
            SY=P. denalii Gjaerevoll
            SY=P. freedmanianum D. Love
            SY=P. kluanensis D. Love
            SY=P. l. ssp. porsildii Knaben
            SY=P. nigroflavum D. Love
            SY=P. nudicaule L. ssp. radicatum
                (Rottb.) Fedde var.
                coloradense Fedde
            SY=P. nudicaule ssp. r. var.
                columbianum Fedde
            SY=P. nudicaule var. r. sensu
                auctt. non (Rottb.) DC.
            SY=P. r. sensu auctt. non Rottb.
            SY=P. r. Rottb. ssp. l. Tolm.
            SY=P. r. ssp. o. Lundstr.
            SY=P. r. ssp. porsildii (Knaben)
                D. Love
    macounii Greene
            SY=P. alaskanum Hulten var.
                macranthum Hulten
            SY=P. hultenii Knaben
            SY=P. h. var. salmonicolor Hulten
            SY=P. keelei Porsild
            SY=P. m. var. discolor Hulten
            SY=P. microcarpum sensu auctt. non
                DC.
            SY=P. scammianum D. Love
    mcconnellii Hulten
    nudicaule L.
    orientale L.
    rhoeas L.
    somniferum L.
    walpolei Porsild
            SY=P. w. var. sulphureo-maculata
                Hulten

Platystemon Benth.

    arizonicus Greene
    californicus Benth.
        var. californicus
        var. ciliatus Dunkle
        var. crinitus Greene
        var. horridulus (Greene) Jepson
        var. nutans Brandeg.
        var. ornithopus (Greene) Munz
    confinis Greene
    mohavensis Greene

Roemeria Medic.

    refracta (Stev.) DC.

Romneya Harvey

Romneya (CONT.)
  coulteri Harvey
  trichocalyx Eastw.
        SY=R. coulteri Harvey var. t.
          (Eastw.) Jepson

Sanguinaria L.

  canadensis L.
        SY=S. c. var. rotundifolia
          (Greene) Fedde

Stylomecon G. Taylor

  heterophylla (Benth.) G. Taylor

Stylophorum Nutt.

  diphyllum (Michx.) Nutt.

PASSIFLORACEAE

Passiflora L.

  affinis Engelm.
  anadenia Urban
  bilobata Juss.
  bryonioides H.B.K.
  edulis Sims
  filipes Benth.
  foetida L.
      var. arizonica Killip
      var. foetida
      var. gossypifolia (Desv.) Masters
      var. hispida (DC. ex Triana &
        Planch.) Killip ex Gleason
      var. riparia (C. Wright) Killip
  incarnata L.
  laurifolia L.
  lutea L.
      var. glabriflora Fern.
      var. lutea
  maliformis L.
  mexicana Juss.
  mollissima (H.B.K.) Bailey
  morifolia Masters
      SY=P. warmingii Masters
  multiflora L.
  murucuja L.
  pallens Poepp. ex Masters
  quadrangularis L.
  rubra L.
  serrato-digitata Ruiz & Pavon ex DC.
  sexflora Juss.
  suberosa L.
      SY=P. pallida L.
  subpeltata Ortega
  tenuiloba Engelm.
  tuberosa Jacq.
  tulae Urban

PEDALIACEAE

Ceratotheca Endl.

  triloba E. Mey.

Sesamum L.

  indicum L.
      SY=S. orientale L.

PHYTOLACCACEAE

Agdestis Moc. & Sesse

  clematidea Moc. & Sesse ex DC.

Microtea Sw.

  debilis Sw.
  portoricensis Urban

Petiveria L.

  alliacea L.

Phaulothamnus Gray

  spinescens Gray

Phytolacca L.

  americana L.
  heteropetala H. Walt.
  octandra L.
      SY=P. acinosa sensu Pope
      SY=P. icosandra L.
  rigida Small
  rivinoides Kunth & Bouche
  sandwicensis Endl.
      var. puberulenta (Deg.) St. John
      var. sandwicensis
        SY=P. brachystachys Moq.

Rivina L.

  humilis L.

Stegnosperma Benth.

  cubense A. Rich.

Trichostigma A. Rich.

  octandrum (L.) H. Walt.

PIPERACEAE

Peperomia Ruiz & Pavon

  alata Ruiz & Pavon
  alternifolia Yuncker
  cogniauxii Urban
  cookiana C. DC.
      var. cookiana
      var. flavinerva (C. DC.) Yuncker
      var. minutilimba Yuncker
      var. ovatilimba (C. DC.) Yuncker
      var. pukooana (C. DC.) Yuncker
  cornifolia (Sw.) A. Dietr.
  cranwelliae Yuncker
  degeneri Yuncker
  dextrilaeva St. John
  distachya (L.) A. Dietr.
  eekana C. DC.
  ellipticibacca C. DC.
  emarginella (Sw. ex Wikstr.) C. DC.
  erythroclada C. DC.
      var. erythroclada
      var. picta (Hbd.) Yuncker
  expallescens C. DC.
      var. brevipilosa Yuncker
      var. expallescens
  fauriei Levl.

Peperomia   (CONT.)
   floridana Small
         SY=P. magnoliifolia (Jacq.) A.
           Dietr.
         SY=Rhynchophorum f. (Small) Small
   forbesii Yuncker
   glabella (Sw.) A. Dietr.
   globulanthera C. DC.
   haupuensis St. John
   hawaiensis C. DC.
   helleri C. DC.
       var. grossa Yuncker
       var. helleri
       var. knudsenii (C. DC.) Yuncker
       var. subovata Yuncker
   hernandiifolia (Vahl) A. Dietr.
   hesperomannii Wawra
       var. brevifolia Yuncker
       var. hesperomannii
   hirtipetiola C. DC.
       var. hirtipetiola
       var. longilimba (C. DC.) Yuncker
   humilis (Vahl) A. Dietr.
       SY=Micropiper h. (Vahl) Small
   hypoleuca Miq.
       var. hypoleuca
       var. pluvigaudens (C. DC.) Yuncker
   kalihiana Yuncker
   kipahuluensis St. John & C. Lamoureux
   kokeana Yuncker
   koolauana C. DC.
   kulensis Yuncker
   latifolia Miq.
   leptostachya Hook. & Arn.
   ligustrina Hbd.
       var. ligustrina
       var. oopuolana Yuncker
   lilifolia C. DC.
       var. honokahauana Yuncker
       var. lilifolia
       var. nudilimba (C. DC.) Yuncker
       var. obtusata Yuncker
       var. psilostigma (C. DC.) Yuncker
   macraeana C. DC.
   maculosa (L.) Hook.
   mapulehuana Yuncker
   mauiensis Wawra
   maunakeana C. DC.
   megapoda Trel.
   membranacea Hook. & Arn.
       var. brevifolia Yuncker
       var. membranacea
       var. puukukuiana Yuncker
       var. waimeana Yuncker
   myrtifolia (Vahl) A. Dietr.
   oahuensis C. DC.
       var. oahuensis
       var. st.-johnii Yuncker
   obtusifolia (L.) A. Dietr.
       SY=Rhynchophorum o. (L.) Small
   parvula Hbd.
   pellucida (L.) H.B.K.
   plinervata St. John
   pololuana Yuncker
   quadrifolia (L.) H.B.K.
   remyi C. DC.
       var. remyi
       var. waipioana Yuncker
   rhombea Ruiz & Pavon
       SY=P. myrtillus Miq.
   rigidilimba C. DC.
   robustior Urban
   rockii C. DC.
   rotundifolia (L.) H.B.K.
   sandwicensis Miq.
       var. robusta Wawra
       var. sandwicensis

   serpens (Sw.) Loud.
   simplex Buch.-Ham.
   sintenisii C. DC.
       SY=P. dendrophila Schlecht.
   spathulifolia Small
       SY=Rhynchophorum s. (Small) Small
   subpetiolata Yuncker
   tenella (Sw.) A. Dietr.
   tetraphylla (Forst. f.) Hook. & Arn.
       var. elongata (Hbd.) Deg. & Deg.
       var. parvifolia (C. DC.) Deg. & Deg.
       var. tetraphylla
   treleasei Yuncker
   trichostigma C. DC.
   urocarpa Fisch. & Mey.
   waihoiana St. John
   waikamoiana Yuncker
   waipioana Yuncker
   wheeleri Britt.
   yabucoana Urban & C. DC.

Piper L.

   aduncum L.
   amalago L.
   auritum H.B.K.
   blattarum Spreng.
   dilatatum L.C. Rich.
   glabrescens (Miq.) C. DC.
       SY=P. treleaseanum Britt. & Wilson
   hispidum Sw.
       SY=P. scabrum Sw.
   leptostachyon Nutt.
       SY=Micropiper l. (Nutt.) Small
   marginatum Jacq.
   methysticum Forst. f.
   swartzianum (Miq.) C. DC.
   tuberculatum Jacq.
   wydlerianum (Miq.) C. DC.
       SY=P. citrifolium Lam.

Pothomorphe Miq.

   peltata (L.) Miq.
       SY=Piper p. L.
   umbellata (L.) Miq.
       SY=Piper u. L.

PITTOSPORACEAE

Pittosporum Banks ex Gaertn.

   acuminatum Mann
       var. acuminatum
       var. degeneri Sherff
       var. leptopodum Sherff
       var. magnifolium Sherff
       var. waimeanum Sherff
   acutisepalum (Hbd.) Sherff
   amplectens Sherff
   argentifolium Sherff
       var. argentifolium
         SY=P. insigne Hbd. var. fosbergii
          Sherff
       var. rockii Sherff
       var. sessile Sherff
   cladanthum Sherff
       var. cladanthum
         SY=P. cauliflorum Mann var. fulvum
          Hbd.
       var. gracilipes Sherff
       var. skottsbergii St. John
         SY=P. cladanthum var. reticulatum
          (Skottsberg) Sherff

Pittosporum (CONT.)
   confertiflorum Gray
            SY=P. cauliflorum Mann
            SY=P. cauliflorum var.
                  cladanthoides Sherff
            SY=P. cauliflorum var.
                  pedicellatum Sherff
            SY=P. confertiflorum var. longipes
                  Sherff
            SY=P. confertiflorum var. mannii
                  Sherff
            SY=P. confertiflorum var.
                  microphyllum Sherff
            SY=P. halophiloides Sherff
   dolosum Sherff
         var. aquilonium Sherff
         var. dolosum
   flocculosum (Hbd.) Sherff
   forbesii Sherff
   gayanum Rock
         var. gayanum
         var. skottsbergii Sherff
         var. waialealae Rock
   glabrum Hook. & Arn.
         var. glabrum
         var. intermedium Sherff
         var. spathulatum (Mann) Sherff
            SY=P. g. var. glomeratum (Hbd.)
                  Sherff
            SY=P. insigne Hbd. var. micranthum
                  Sherff
         var. tinifolium
   halophilum Rock, emend Sherff
   hawaiiense Hbd.
   helleri Sherff
   hosmeri Rock
         var. hosmeri
         var. longifolium Rock
         var. saint-johnii Sherff
   insigne Hbd.
         var. hillebrandii (Levl.) Sherff
         var. insigne
            SY=P. i. var. lydgatei Sherff
         var. pelekunuanum Sherff
   kahananum Sherff
   kauaiense Hbd.
            SY=P. k. var. phaeocarpum Sherff
            SY=P. k. var. repens Sherff
   monae St. John
   napaliense Sherff
   sulcatum Sherff
         var. remyi Sherff
         var. sulcatum
            SY=P. s. var. rumicifolium Sherff
   terminalioides Planch. ex Gray
            SY=P. t. var. lanaiense Sherff
            SY=P. t. var. macrocarpum Sherff
            SY=P. t. var. macropus Skottsberg
            SY=P. t. var. mauiense Sherff
   tobira (Thunb.) Ait.
   undulatum Vent.

Sollya Lindl.

   heterophylla Lindl.
            SY=S. fusiformis (Labill.) Briq.

PLANTAGINACEAE

Littorella Berg.

   americana Fern.
            SY=L. uniflora sensu auctt. non
                  (L.) Aschers.

            SY=L. u. var. a. (Fern.) Gleason

Plantago L.

   altissima L.
   arenaria Waldst. & Kit.
            SY=P. hybrida W. Bart.
            SY=P. indica L.
            SY=P. psyllium L.
   argyraea Morris
   aristata Michx.
            SY=P. a. var. nuttallii (Rapin)
                  Morris
   asiatica L.
            SY=P. major L. var. a. (L.) Dcne.
   australis Lam.
      ssp. hirtella (H.B.K.) Rahn
            SY=P. h. H.B.K.
            SY=P. h. ssp. galeottiana (Dcne.)
                  Thorne
            SY=P. h. var. g. (Dcne.) Pilger
            SY=P. h. var. mollior Pilger
   bigelovii Gray
      ssp. bigelovii
      ssp. californica (Greene) Bassett
            SY=P. c. Greene
            SY=P. hookeriana Fisch. & Mey.
                  var. c. (Greene) Poe
   canescens J.E. Adams
            SY=P. c. var. cylindrica (Macoun)
                  Boivin
            SY=P. septata Morris ex Rydb.
   cordata Lam.
   coronopus L.
            SY=P. c. ssp. commutata (Guss.)
                  Pilger
   debilis R. Br.
   elongata Pursh
      ssp. elongata
      ssp. pentasperma Bassett
   erecta Morris
      ssp. erecta
      ssp. rigidior Pilger
   eriopoda Torr.
            SY=P. shastensis Greene
   firma Kunze ex Walp.
            SY=P. truncata Cham. ssp. f.
                  (Kunze ex Walp.) Pilger
   glabrifolia (Rock) Pilger
   grayana Pilger
         var. abrotanelloides Skottsberg
         var. grayana
   hawaiensis (Gray) Pilger
         var. hawaiensis
         var. laxa Pilger
   helleri Small
   heterophylla Nutt.
   hillebrandii Pilger
   hookeriana Fisch. & Mey.
         var. hookeriana
         var. nuda (Gray) Poe
   insularis Eastw.
         var. fastigiata (Morris) Jepson
         var. insularis
            SY=P. scariosa Morris
   krajinai Pilger
   lanceolata L.
         var. lanceolata
         var. sphaerostachya Mert. & Koch
   macrocarpa Cham. & Schlecht.
   major L.
         var. intermedia (Gilib.) Pilger
            SY=P. m. ssp. i. (Gilib.) Arcang.
         var. major
         var. pachyphylla Pilger
         var. pilgeri Domin
         var. scopulorum Fries & Broberg

Plantago (CONT.)
         SY=P. halophila Bickn.
   maritima L.
         var. californica (Fern.) Pilger
            SY=P. juncoides Lam. var. c. Fern.
         var. juncoides (Lam.) Gray
            SY=P. juncoides Lam.
            SY=P. j. var. decipiens (Barneoud)
               Fern.
            SY=P. j. var. glauca (Hornem.)
               Fern.
            SY=P. j. var. laurentiana Fern.
            SY=P. m. ssp. borealis (Lange)
               Blytt & Dahl
            SY=P. m. ssp. juncoides (Lam.)
               Hulten
            SY=P. oliganthos Roemer & Schultes
            SY=P. o. var. fallax Fern.
   media L.
            SY=P. m. var. monnieri (Giraud.)
               Roug.
   melanochrous Pilger
   muscicola (Rock) Pilger
   pachyphylla Gray
         var. maviensis Gray
         var. pachyphylla
         var. rotundifolia Wawra
   patagonica Jacq.
         var. breviscapa (Shinners) Shinners
            SY=P. purshii Roemer & Schultes
               var. b. Shinners
         var. patagonica
            SY=P. p. var. gnaphaloides (Nutt.)
               Gray
            SY=P. purshii Roemer & Schultes
            SY=P. purshii var. oblonga
               (Morris) Shinners
            SY=P. purshii var. picta (Morris)
               Pilger
         var. spinulosa (Dcne.) Gray
            SY=P. purshii Roemer & Schultes
               var. s. (Dcne.) Shinners
            SY=P. s. Dcne.
   princeps Cham. & Schlecht.
         var. acaulis Wawra
         var. anomala Rock
         var. denticulata Hbd.
         var. elata Wawra
         var. hirtella Gray
         var. laxifolia Gray
         var. longibracteata Mann
         var. princeps
         var. queleniana (Gaud.) Rock
   pusilla Nutt.
         var. major Engelm.
         var. pusilla
   rhodosperma Dcne.
   rugelii Dcne.
         var. asperula Farw.
         var. rugelii
   sparsiflora Michx.
   subnuda Pilger
   tweedyi Gray
            SY=P. eriopoda Torr. var. t.
               (Gray) Boivin
   virginica L.
            SY=P. v. var. viridescens Fern.
   wrightiana Dcne.

PLATANACEAE

Platanus L.

   hybrida Brot.

            SY=P. acerifolia (Ait.) Willd.
   occidentalis L.
         SY=P. o. var. glabrata (Fern.)
            Sarg.
   racemosa Nutt.
   wrightii P. Wats.

PLUMBAGINACEAE

Armeria (DC.) Willd.

   maritima (P. Mill.) Willd.
      ssp. californica (Boiss.) Porsild
            SY=A. arctica (Cham.) Wallr. ssp.
               c. (Boiss.) Abrams
            SY=A. m. var. c. (Boiss.) G.H.M.
               Lawrence
      ssp. interior (Raup) Porsild
            SY=A. m. var. i. (Raup) G.H.M.
               Lawrence
            SY=Statice i. Raup
      ssp. labradorica (Wallr.) Hulten
            SY=A. l. Wallr.
            SY=A. l. var. submutica (Blake)
               H.F. Lewis
            SY=A. m. var. l. (Wallr.) G.H.M.
               Lawrence
            SY=A. vulgaris Willd. pro parte
      ssp. maritima
            SY=A. m. var. pubescens (Sowerby)
               Bab.
      ssp. purpurea (W.D.J. Koch) Love & Love
            SY=A. arctica (Cham.) Wallr.
            SY=A. m. ssp. a. (Cham.) Hulten
            SY=A. m. var. p. (W.D.J. Koch)
               G.H.M. Lawrence
      ssp. sibirica (Turcz. ex Boiss.) Nyman
            SY=A. m. var. s. (Turcz. ex
               Boiss.) G.H.M. Lawrence
            SY=A. scabra Pallas ssp. sibirica
               (Turcz. ex Boiss.) Hyl.

Limonium P. Mill.

   californicum (Boiss.) Heller
            SY=L. c. var. mexicanum (Blake)
               Munz
            SY=L. m. Blake
   carolinianum (Walt.) Britt.
            SY=L. angustatum (Gray) Small
            SY=L. c. var. a. (Gray) Blake
            SY=L. c. var. angustifolium Blake
            SY=L. c. var. compactum Shinners
            SY=L. carolinianum var. nashii
               (Small) Boivin
            SY=L. carolinianum var.
               obtusilobum (Blake) Ahles
            SY=L. carolinianum var.
               trichogonum (Blake) Boivin
            SY=L. n. Small
            SY=L. n. var. albiflorum (Raf.)
               House
            SY=L. n. var. angustatum (Gray)
               Ahles
            SY=L. n. var. t. Blake
            SY=L. o. Blake
   limbatum Small
            SY=L. l. var. glabrescens Correll
   perezii (Stapf) F.T. Hubbard
            SY=Statice p. Stapf
   reniforme (Girard) Linchevskii
            SY=L. perfoliatum (Karelin ex
               Boiss.) Kuntze
   sinuatum (L.) P. Mill.

Limonium   (CONT.)
                    SY=Statice s. L.
   vulgare P. Mill.

Plumbago L.

   auriculata Lam.
                    SY=P. capensis Thunb.
   scandens L.
   zeylanica L.

POACEAE

Aegilops L.

   crassa Boiss.
   cylindrica Host
                    SY=Triticum c. (Host) Ces.
                    SY=A. c. var. rubiginosa Popova
   ovata L.
   triuncialis L.

Aegopogon Humb. & Bonpl. ex Willd.

   tenellus (DC.) Trin.
                    SY=A. t. var. abortivus (Fourn.)
                       Beetle

X Agroelymus G. Camus ex A. Camus

   X bergrothii Rouss. [Agropyron repens X
      Elymus arenarius]
   X bowdenii Boivin [Agropyron smithii X
      Elymus innovatus]
   X covillensis Lepage [Agropyron violaceum X
      Elymus innovatus]
   X hirtiflorus (A.S. Hitchc.) Bowden
      [Agryopyron trachycaulum X Elymus
      innovatus]
                    SY=X A. ontariensis Lepage
                    SY=Elymus h. A.S. Hitchc.
   X hultenii Melderis [Agropyron violaceum X
      Elymus mollis]
   X jamesensis Lepage [Agropyron trachycaulum
      X Elymus mollis]
                    SY=X A. adamsii Rouss.
                    SY=X A. a. nm. jamesensis (Lepage)
                       Lepage
                    SY=X A. a. nm. longispica Lepage
                    SY=X A. a. nm. semiaelvus Lepage
                    SY=X A. j. var. anticostensis
                       Lepage
                    SY=X A. j. var. stoloniferus
                       Lepage
   X mossii Lepage [Agropyron trachycaulum X
      Elymus canadensis]
                    SY=X A. cayouetteorum Boivin
   X palmerensis Lepage [Agropyron sericeum X
      Elymus sibiricus]
                    SY=X A. hodgsonii Lepage
   X turneri Lepage [Agropyron dasystachyum X
      Elymus innovatus]
   X ungavensis (Louis-Marie) Lepage
      [Agropyron violaceum X Elymus mollis]

X Agrohordeum G. Camus ex A. Camus

   X jordalii Melderis [Agropyron sericeum X
      Hordeum brachyantherum]
   X langei author ? [Agropyron repens X
      Hordeum secalinum]
                    SY=Hordeopyrum rouxii (Gren. &
                       Duval) Simonet
   X macounii (Vasey) Lepage [Agropyron

   trachycaulum X Hordeum jubatum]
      nm. macounii
                    SY=Elymus m. Vasey
      nm. valencianum Bowden
   X pilosilemma Mitchell & Hodgson [Agropyron
      sericeum X Hordeum jubatum]

Agropyron Gaertn.

   arizonicum Scribn. & Sm.
                    SY=A. spicatum (Pursh) Scribn. &
                       Sm. var. a. (Scribn. & Sm.)
                       M.E. Jones
   bakeri E. Nels.
                    SY=A. trachycaulum (Link) Malte ex
                       H.F. Lewis var. b. (E.
                       Nels.) Boivin
   X brevifolium Scribn. [scribneri X
      violaceum or trachycaulum]
   dasystachyum (Hook.) Scribn. & Sm.
      ssp. albicans (Scribn. & Sm.) Dewey
         var. albicans
                    SY=A. a. Scribn. & Sm.
         var. griffithsii Scribn. & Sm. ex
            Piper
                    SY=A. a. Scribn. & Sm. var. g.
                       (Scribn. & Sm. ex Piper)
                       Beetle
                    SY=A. g. Scribn. & Sm. ex Piper
      ssp. dasystachyum
         var. dasystachyum
                    SY=A. elmeri Scribn.
                    SY=A. lanceolatum Scribn. & Sm.
         var. riparium (Scribn. & Sm.) Bowden
                    SY=A. r. Scribn. & Sm.
      ssp. psammophilum (Gillett & Senn) Dewey
                    SY=A. d. var. p. (Gillett & Senn)
                       E.G. Voss
                    SY=A. p. Gillett & Senn
      ssp. yukonense (Scribn. & Merr.) Dewey
                    SY=A. y. Scribn. & Merr.
   desertorum (Fisch. ex Link) Schultes
   elongatum (Host) Beauv.
                    SY=A. varnense sensu auctt. non
                       (Vel.) Hayek
   intermedium (Host) Beauv.
      var. intermedium
                    SY=A. glaucum (Desf. ex DC.)
                       Roemer & Schultes
      var. trichophorum (Link) Halac.
                    SY=A. t. (Link) Richter
   littorale Dum.
                    SY=A. pungens sensu auctt. non
                       (Pers.) Roemer & Schultes
                    SY=A. p. var. acadiense (F.T.
                       Hubbard) Fern.
                    SY=A. pycnanthemum Godr.
                    SY=Elymus pycnanthemum (Godr.)
                       Melderis
   parishii Scribn. & Sm.
                    SY=A. laeve (Scribn. & Sm.) A.S.
                       Hitchc.
                    SY=A. p. var. l. Scribn. & Sm.
   pectiniforme Roemer & Schultes
                    SY=A. cristatiforme Sarkar
                    SY=A. cristatum (L.) Gaertn. ssp.
                       pectinatum (Bieb.) Tzuel.
   pringlei (Scribn. & Sm.) A.S. Hitchc.
   X pseudorepens Scribn. & Sm. [dasystachyum
      ssp. psammophilum X trachycaulum]
                    SY=A. X p. var. magnum Scribn. &
                       Sm.
                    SY=A. X p. nm. sennii Boivin
                    SY=A. X p. nm. vulpinum (Rydb.)
                       Bowden
                    SY=A. v. (Rydb.) A.S. Hitchc.
                    SY=Elymus v. Rydb.

Agropyron (CONT.)
   repens (L.) Beauv.
        SY=A. r. var. subulatum (Schreb.)
           Roemer & Schultes
        SY=Elymus r. (L.) Gould
        SY=Elytrigia r. (L.) Desv. ex
           Nevski
   scribneri Vasey
   sericeum A.S. Hitchc.
        SY=A. dasystachyum (Hook.) Scribn.
           & Sm. var. s. sensu auctt.
           non (A.S. Hitchc.) Boivin
        SY=A. macrourum sensu auctt. non
           (Turcz.) Drobov
   sibiricum (Willd.) Beauv.
   smithii Rydb.
        SY=A. molle (Scribn. & Sm.) Rydb.
        SY=A. s. var. m. (Scribn. & Sm.)
           M.E. Jones
        SY=A. s. var. palmeri (Scribn. &
           Sm.) Heller
   spicatum (Pursh) Scribn. & Sm.
        SY=A. inerme (Scribn. & Sm.) Rydb.
        SY=A. s. var. i. (Scribn. & Sm.)
           Heller
        SY=A. s. var. pubescens Elmer
   trachycaulum (Link) Malte ex H.F. Lewis
     var. glaucum (Pease & Moore) Malte
        SY=A. pauciflorum (Schwein.) A.S.
           Hitchc. ex Silveus var. g.
           (Pease & Moore) Taylor &
           MacBryde
     var. latiglume (Scribn. & Sm.) Beetle
        SY=A. caninum (L.) Beauv. ssp.
           majus (Vasey) C.L. Hitchc.
           var. l. (Scribn. & Sm.) C.L.
           Hitchc.
        SY=A. l. (Scribn. & Sm.) Rydb.
     var. trachycaulum
        SY=A. caninum (L.) Beauv. ssp.
           majus (Vasey) C.L. Hitchc.
        SY=A. c. ssp. m. var. andinum
           (Scribn. & Sm.) C.L. Hitchc.
        SY=A. c. ssp. m. var. hornemannii
           (Koch) Pease & Moore
        SY=A. c. ssp. m. var. mitchellii
           Welsh
        SY=A. pauciflorum (Schwein.) A.S.
           Hitchc. ex Silveus
        SY=A. p. ssp. majus (Vasey)
           Melderis
        SY=A. p. ssp. novae-angliae
           (Scribn.) Melderis
        SY=A. p. var. n-a. (Scribn.)
           Taylor & MacBryde
        SY=A. p. ssp. teslinense (Porsild
           & Senn) Melderis
        SY=A. subsecundum (Link) A.S.
           Hitchc. var. a. (Scribn. &
           Sm.) A.S. Hitchc.
        SY=A. tenerum Vasey
        SY=A. teslinense Porsild & Senn
        SY=A. trachycaulum var. majus
           (Vasey) Fern.
        SY=A. trachycaulum var. n-a.
           (Scribn.) Fern.
        SY=A. violaceum (Hornem.) Lange
           ssp. a. (Scribn. & Sm.)
           Melderis
        SY=A. v. var. a. Scribn. & Sm.
        SY=Roegneria p. (Schwein.) Hyl.
     var. unilaterale (Vasey) Malte
        SY=A. caninum (L.) Beauv. ssp.
           majus (Vasey) C.L. Hitchc.
           var. u. (Vasey) C.L. Hitchc.
        SY=A. subsecundum (Link) A.S.
           Hitchc.

   violaceum (Hornem.) Lange
     var. alboviride (Hulten) Melderis
     var. violaceum
        SY=A. alaskanum Scribn. & Merr.
        SY=A. boreale (Turcz.) Drobov ex
           Polunin ssp. a. (Scribn. &
           Merr.) Melderis
        SY=A. b. ssp. hyperarcticum
           (Polunin) Melderis
        SY=A. b. var. a. (Scribn. & Merr.)
           Welsh
        SY=A. b. var. h. (Polunin) Welsh
        SY=Roegneria b. (Turcz.) Nevski
           var. h. (Polunin) Melderis
        SY=R. v. (Hornem.) Melderis
     var. virescens Lange
        SY=Roegneria v. (Lange) Bocher

X Agrositanion Bowden

   X saundersii (Vasey) Bowden [Agropyron
     trachycaulum X Sitanion hystrix]
        SY=Agropyron s. (Vasey) A.S.
           Hitchc.
   X saxicola (Scribn. & Sm.) Bowden
     [Agropyron spicatum X Sitanion hystrix]
        SY=Agropyron saxicola (Scribn. &
           Sm.) Piper

Agrostis L.

   aequivalvis (Trin.) Trin.
        SY=Podagrostis a. (Trin.) Scribn.
           & Merr.
   alaskana Hulten
        SY=A. melaleuca (Trin.) A.S.
           Hitchc.
   altissima (Walt.) Tuckerman
        SY=A. elata (Pursh) Trin.
        SY=A. perennans (Walt.) Tuckerman
           var. e. (Pursh) A.S. Hitchc.
   aristiglumis Swallen
   avenacea J.G. Gmel.
        SY=A. retrofracta Willd.
   blasdalei A.S. Hitchc.
     var. blasdalei
        SY=A. breviculmis sensu auctt. non
           A.S. Hitchc.
     var. marinensis Crampton
   californica Trin.
        SY=A. densiflora Vasey
        SY=A. glomerata sensu auctt. non
           (Presl) Kunth
   canina L.
     var. canina
        SY=A. c. var. varians (Thuill.)
           Ducomm.
     var. montana Hartman
        SY=A. stricta J.F. Gmel.
   capillaris L.
        SY=A. sylvatica Huds.
        SY=A. tenuis Sibthorp
        SY=A. t. var. aristata (Parnell)
           Druce
        SY=A. t. var. hispida (Willd.)
           Philipson
        SY=A. t. var. pumila (L.) Druce
   clivicola Crampton
     var. clivicola
     var. punta-reyesensis Crampton
   diegoensis Vasey
   elliottiana Schultes
   exarata Trin.
     var. exarata
        SY=A. asperifolia Trin.
     var. minor Hook.
        SY=A. e. ssp. e. var. m. (Hook.)

Agrostis (CONT.)
                    C.L. Hitchc.
            SY=A. e. ssp. m. (Hook.) C.L.
                Hitchc.
            SY=A. e. var. purpurascens Hulten
        var. monolepis (Torr.) A.S. Hitchc.
            SY=A. e. ssp. e. var. m. (Torr.)
                A.S. Hitchc.
        var. pacifica Vasey
    exigua Thurb.
    fallax Hbd.
    filicumis M.E. Jones
            SY=A. clavata sensu auctt. non
                Trin.
            SY=A. idahoensis Nash
    gigantea Roth
            SY=A. g. var. dispar (Michx.)
                Philipson
    hallii Vasey
        var. hallii
        var. pringlei (Scribn.) A.S. Hitchc.
    hendersonii A.S. Hitchc.
            SY=A. microphylla Steud. var. h.
                (A.S. Hitchc.) Beetle
    hiemalis (Walt.) B.S.P.
            SY=A. antecedens Bickn.
    hooveri Swallen
    howellii Scribn.
    humilis Vasey
            SY=A. clavata Trin.
            SY=Podagrostis h. (Vasey)
                Bjoerkman
    inflata Scribn.
    lepida A.S. Hitchc.
    longiligula A.S. Hitchc.
        var. australis J.T. Howell
        var. longiligula
    mertensii Trin.
            SY=A. borealis Hartman
            SY=A. b. var. americana (Scribn.)
                Fern.
            SY=A. b. var. paludosa (Scribn.)
                Fern.
            SY=A. rupestris sensu auctt. non
                All.
    microphylla Steud.
        var. major Vasey
        var. microphylla
            SY=A. m. var. intermedia Beetle
    nebulosa Boiss. & Reut.
    nigra With.
    oregonensis Vasey
    pallens Trin.
    perennans (Walt.) Tuckerman
            SY=A. oreophila Trin.
            SY=A. p. var. aestivalis Vasey
            SY=A. schweinitzii Trin.
    rossae Vasey
    sandwicensis Hbd.
    scabra Willd.
        var. geminata (Trin.) Swallen
            SY=A. g. Trin.
            SY=A. hiemalis (Walt.) B.S.P. var.
                g. (Trin.) A.S. Hitchc.
        var. scabra
            SY=A. h. var. tenuis (Tuckerman)
                Gleason
        var. septentrionalis Fern.
    semiverticillata (Forsk.) C. Christens.
            SY=A. verticillata Vill.
            SY=Polypogon s. (Forsk.) Hyl.
    stolonifera L.
        var. major (Gaudin) Farw.
            SY=A. alba sensu auctt. non L.
        var. palustris (Huds.) Farw.
            SY=A. alba L. var. p. (Huds.)
                Pers.

            SY=A. maritima Lam.
            SY=A. p. Huds.
            SY=A. s. var. compacta Hartman
        var. stolonifera
            SY=A. alba L. var. s. (L.) Sm.
    tandilensis (Kuntze) Parodi
            SY=A. kennedyana Beetle
    thurberiana A.S. Hitchc.
            SY=Podagrostis t. (A.S. Hitchc.)
                Hulten
    trinii Turcz.
    variabilis Rydb.

Aira L.

    caryophyllea L.
            SY=Aspris c. (L.) Nash
    elegans Willd. ex Gaudin
            SY=Aira capillaris Host
            SY=Aspris c. (Host) A.S. Hitchc.
    praecox L.
            SY=Aspris p. (L.) Nash

Allolepis Soderstrom & Decker

    texana (Vasey) Soderstrom & Decker
            SY=Distichlis t. (Vasey) Scribn.

Alopecurus L.

    aequalis Sobol.
        var. aequalis
            SY=A. a. var. natans (Wahlenb.)
                Fern.
            SY=A. aristulatus Michx.
        var. sonomensis Rubtzoff
    alpinus Sm.
        var. alpinus
            SY=A. occidentalis Scribn. &
                Tweedy
        var. glaucus (Less.) Krylov
            SY=A. a. ssp. g. (Less.) Hulten
            SY=A. g. Less.
        var. stejnegeri (Vasey) Hulten
    arundinaceus Poir.
            SY=A. ventricosus Pers.
    carolinianus Walt.
            SY=A. ramosus Poir.
    creticus Trin.
    geniculatus L.
        var. geniculatus
            SY=A. pallescens Piper
        var. microstachyus Uechtr.
    howellii Vasey
    myosuroides Huds.
    pratensis L.
    rendlei Eig, Britt. & For.
    saccatus Vasey

Ammophila Host

    arenaria (L.) Link
    breviligulata Fern.
            SY=A. champlainensis Seymour

Ampelodesmos Link

    mauritanicus (Poir.) Dur. & Schinz

Amphibromus Nees

    scabrivalvis (Trin.) Swallen

Amphicarpum Kunth

    muhlenbergianum (Schultes) A.S. Hitchc.
            SY=Amphicarpon floridanum Chapman

Amphicarpum (CONT.)
  purshii Kunth
          SY=Amphicarpon amphicarpon (Pursh)
              Nash

Andropogon L.

   arctatus Chapman
   bicornis L.
   brachystachyus Chapman
   cabanisii Hack.
   campyloracheus Nash
   capillipes Nash
          SY=A. virginicus L. var. glaucus
              Hack.
   elliottii Chapman
          SY=A. e. var. gracilior Hack.
          SY=A. e. var. projectus Fern. &
              Grisc.
   floridanus Scribn.
   gerardii Vitman
       var. chrysocomus (Nash) Fern.
          SY=A. c. Nash
       var. gerardii
          SY=A. furcatus Muhl.
          SY=A. provincialis Lam.
       var. paucipilus (Nash) Fern.
          SY=A. g. var. incanescens (Hack.)
              Boivin
          SY=A. hallii Hack.
          SY=A. h. var. i. Hack.
          SY=A. p. Nash
   glomeratus (Walt.) B.S.P.
          SY=A. virginicus L. var.
              abbreviatus (Hack.) Fern. &
              Grisc.
          SY=A. v. var. corymbosus (Chapman)
              Fern. & Grisc.
   leucostachys H.B.K.
   longiberbis Hack.
   mohrii (Hack.) Hack. ex Vasey
   perangustatus Nash
   subtenuis Nash
   ternarius Michx.
          SY=A. t. var. glaucescens
              (Scribn.) Fern. & Grisc.
   tracyi Nash
   virgatus Desv. ex Hamilton
   virginicus L.
       var. glaucopsis (Ell.) A.S. Hitchc.
       var. hirsutior (Hack.) A.S. Hitchc.
       var. virginicus
          SY=A. v. var. tetrastachyus (Ell.)
              Hack.

Anthaenantia Beauv.

   rufa (Ell.) Schultes
   villosa (Michx.) Beauv.

Anthephora Schreb.

   hermaphrodita (L.) Kuntze

Anthoxanthum L.

   aristatum Boiss.
          SY=A. odoratum L. var. puelii
              (Lecoq & Lamotte) Coss. &
              Durieu
          SY=A. p. Lecoq & Lamotte
   odoratum L.
     ssp. alpinum (Love & Love) Hulten
     ssp. odoratum

Apera Adans.

   interrupta (L.) Beauv.
          SY=Agrostis i. L.
   spica-venti (L.) Beauv.
          SY=Agrostis s-v. L.

Arctagrostis Griseb.

   latifolia (R. Br.) Griseb.
       var. angustifolia (Nash) Hulten
          SY=A. a. Nash
       var. arundinacea (Trin.) Griseb.
          SY=A. a. Nash var. crassispica
              Bowden
          SY=A. l. ssp. nahanniensis Porsild
       var. latifolia
          SY=A. poaeoides Nash
       var. longiglumis Polunin

Arctophila Rupr. ex Anderss.

   fulva (Trin.) Rupr. ex Anderss.
          SY=Colpodium f. (Trin.) Griseb.

Aristida L.

   adscensionis L.
       var. abortiva Beetle
       var. adscensionis
          SY=A. fasciculata Torr.
       var. modesta Hack.
   affinis (Schultes) Kunth
          SY=A. palustris (Chapman) Vasey
   arizonica Vasey
   barbata Fourn.
   basiramea Engelm. ex Vasey
   californica Thurb.
   chaseae A.S. Hitchc.
   cognata Trin. & Rupr.
   condensata Chapman
          SY=A. c. var. combsii (Scrib. &
              Ball) Henr.
   curtissii (Gray) Nash
          SY=A. dichotoma Michx. var. c.
              Gray
   desmantha Trin. & Rupr.
   dichotoma Michx.
   divaricata Humb. & Bonpl. ex Willd.
   fendleriana Steud.
   floridana (Chapman) Vasey
   glabrata (Vasey) A.S. Hitchc.
   glauca (Nees) Walp.
          SY=A. necopina Shinners
          SY=A. purpurea Nutt. var. g.
              (Nees) A. Holmgren & N.
              Holmgren
          SY=A. reverchonii Vasey
   gyrans Chapman
   hamulosa Henr.
   lanosa Muhl. ex Ell.
          SY=A. l. var. macera Fern. &
              Grisc.
   longespica Poir.
       var. geniculata (Raf.) Fern.
          SY=A. intermedia Scribn. & Ball
       var. longespica
   longiseta Steud.
       var. longiseta
          SY=A. purpurea Nutt. var.
              rariflora A.S. Hitchc.
          SY=A. p. var. l. (Steud.) Vasey
       var. robusta Merr.
          SY=A. purpurea Nutt. var. r.
              (Merr.) A. Holmgren & N.
              Holmgren
   mohrii Nash
   oligantha Michx.
   orcuttiana Vasey

Aristida (CONT.)
        SY=A. schiedeana sensu auctt. non
                Trin. & Rupr.
pansa Woot. & Standl.
        var. dissita (I.M. Johnston) Beetle
                SY=A. d. I.M. Johnston
        var. pansa
patula Chapman ex Nash
portoricensis Pilger
purpurascens Poir.
        SY=A. p. var. minor Vasey
purpurea Nutt.
        SY=A. p. var. laxiflora Merr.
ramosissima Engelm. ex Gray
refracta Griseb.
rhizomophora Swallen
roemeriana Scheele
simpliciflora Chapman
spiciformis Ell.
stricta Michx.
tenuispica A.S. Hitchc.
ternipes Cav.
        var. minor (Vasey) A.S. Hitchc.
        var. ternipes
tuberculosa Nutt.
virgata Trin.
wrightii Nash
        var. parishii (A.S. Hitchc.) Gould
                SY=A. p. A.S. Hitchc.
        var. wrightii

Arrhenatherum Beauv.

    elatius (L.) Beauv. ex J. & C. Presl
        var. biaristatum (Peterm.) Peterm.
        var. bulbosum (Willd.) Spenner
        var. elatius

Arthraxon Beauv.

    hispidus (Thunb.) Makino
        SY=A. ciliaris Beauv. ssp.
                langsdorfii (Trin.) Hack.
                var. cryptatherus Hack.
        SY=A. h. var. cryptatherus (Hack.)
                Honda

Arthrostylidium Rupr.

    capillifolium Griseb.
    multispicatum Pilger
    sarmentosum Pilger

Arundinaria Michx.

    gigantea (Walt.) Muhl.
        ssp. gigantea
        ssp. tecta (Walt.) McClure
                SY=A. t. (Walt.) Muhl.
                SY=Arundo t. Walt.

Arundinella Raddi

    confinis (Schultes) A.S. Hitchc. & Chase

Arundo L.

    donax L.
        var. donax
        var. versicolor (P. Mill.) Stokes

Avena L.

    barbata Pott ex Link
    brevis Roth
    fatua L.
        SY=A. f. var. glabrata Peterm.
        SY=A. f. var. vilis (Wallr.)
                Hausskn.
    hybrida Peterm.
    sativa L.
        SY=A. byzantina K. Koch
        SY=A. fatua L. var. s. (L.)
                Hausskn.
        SY=A. sativa L. var. orientalis
                (Schreb.) Alef.
    sterilis L.
    strigosa Schreb.

Avenochloa Holub

    hookeri (Scribn.) Holub
        SY=Avena h. Scribn.
        SY=Helictotrichon h. (Scribn.)
                Henr.

Axonopus Beauv.

    affinis Chase
    aureus Beauv.
    compressus (Sw.) Beauv.
        SY=Anastrophus c. (Sw.) Schlecht.
                ex Doell
    furcatus (Flugge) A.S. Hitchc.
        SY=Anastrophus f. (Flugge) Nash

Bambusa Schreb.

    bambos (L.) Voss
    multiplex (Lour.) Raeusch.
    vulgaris Schrad. ex J.C. Wendl.

Beckmannia Host

    syzigachne (Steud.) Fern.
        ssp. baicalensis (Kusnez.) Koyama &
                Kawano
                SY=B. eruciformis (L.) Host ssp.
                        b. (Kusnez.) Hulten
        ssp. syzigachne
                SY=B. eruciformis (L.) Host
                SY=B. s. var. uniflora (Scribn. ex
                        Gray) Boivin

Blepharidachne Hack.

    bigelovii (S. Wats.) Hack.
    kingii (S. Wats.) Hack.

Blepharoneuron Nash

    tricholepis (Torr.) Nash

Bothriochloa Kuntze

    alta (A.S. Hitchc.) Henr.
        SY=Andropogon a. A.S. Hitchc.
    barbinodis (Lag.) Herter
        var. barbinodis
                SY=Andropogon b. Lag.
        var. palmeri (Hack.) de Wet
                SY=B. p. (Hack.) Gould
        var. perforatus (Trin. ex Fourn.)
                Gould
                SY=Andropogon p. Trin. ex Fourn.
    bladhii (Retz.) S.T. Blake
        SY=Andropogon intermedia R. Br.
        SY=B. i. (R. Br.) A. Camus
    caucasica (Trin.) C.E. Hubbard
        SY=Andropogon c. Trin.
    decipens (Hack.) C.E. Hubbard
    edwardsiana (Gould) Parodi
        SY=Andropogon e. Gould
    erianthoides (F. Muell.) C.E. Hubbard

Bothriochloa (CONT.)
   exaristata (Nash) Henr.
         SY=Andropogon e. (Nash) A.S.
            Hitchc.
   hybrida (Gould) Gould
         SY=Andropogon h. Gould
   ischaemum (L.) Keng
      var. ischaemum
         SY=Amphilophis i. (L.) Nash
         SY=Andropogon i. L.
      var. songarica (Rupr. ex Fisch. &
        Mey.) Celarier & Harlan
         SY=Andropogon i. L. var. s. Rupr.
            ex Fisch. & Mey.
   pertusa (L.) A. Camus
         SY=Andropogon p. (L.) Willd.
   saccharoides (Sw.) Rydb.
      var. longipaniculata (Gould) Gould
         SY=Andropogon s. var. l. Gould
      var. saccharoides
         SY=Amphilophis s. (Sw.) Nash
         SY=Andropogon s. Sw.
      var. torreyana (Steud.) Gould
         SY=Andropogon s. var. t. (Steud.)
            Hack.
   springfieldii (Gould) Parodi
         SY=Andropogon s. Gould
   wrightii (Hack.) Henr.
         SY=Andropogon w. Hack.

Bouteloua Lag.

   americana (L.) Scribn.
   aristidoides (H.B.K.) Griseb.
      var. aristidoides
      var. arizonica M.E. Jones
   barbata Lag.
         SY=B. arenosa Vasey
   breviseta Vasey
         SY=B. ramosa Scribn. ex Vasey
   chondrosioides (H.B.K.) Benth. ex S. Wats.
   curtipendula (Michx.) Torr.
      var. caespitosa Gould & Kapadia
      var. curtipendula
         SY=Atheropogon c. (Michx.) Fourn.
   eludens Griffiths
   eriopoda (Torr.) Torr.
   gracilis (Willd. ex H.B.K.) Lag. ex Steud.
      var. gracilis
         SY=B. oligostachya (Nutt.) Torr.
            ex Gray
      var. stricta (Vasey) A.S. Hitchc.
   hirsuta Lag.
      var. glandulosa (Cerv.) Gould
         SY=B. g. (Cerv.) Swallen
      var. hirsuta
   juncea (Desv. ex Beauv.) A.S. Hitchc.
   kayi Warnock
   parryi (Fourn.) Griffiths
   pectinata Featherly
         SY=B. hirsuta Lag. var. p.
           (Featherly) Cory
   radicosa (Fourn.) Griffiths
   repens (H.B.K.) Scribn. & Merr.
         SY=B. filiformis (Fourn.)
           Griffiths
         SY=B. heterostega (Trin.)
           Griffiths
   rigidiseta (Steud.) A.S. Hitchc.
   rothrockii Vasey
   simplex Lag.
   trifida Thurb.
   uniflora Vasey
   warnockii Gould & Kapadia

Brachiaria (Trin.) Griseb.

adspersa (Trin.) Parodi
         SY=Panicum a. Trin.
arizonica (Scribn. & Merr.) S.T. Blake
         SY=Panicum a. Scribn. & Merr.
ciliatissima (Buckl.) Chase
eruciformis (Sm.) Griseb.
fasiculata (Sw.) S.T. Blake
         SY=Panicum f. Sw.
         SY=P. f. var. reticulatum (Torr.)
           Beal
mollis (Sw.) Parodi
         SY=Panicum m. Sw.
mutica (Forsk.) Stapf
         SY=Panicum barbinode Trin.
         SY=P. m. Forsk.
         SY=P. purpurascens Raddi
plantaginea (Link) A.S. Hitchc.
platyphylla (Griseb.) Nash
         SY=B. extensa Chase
reptans (L.) Gard. & C.E. Hubbard
         SY=Panicum prostratum Lam.
         SY=P. r. L.
subquadripara (Trin.) A.S. Hitchc.
texana (Buckl.) S.T. Blake
         SY=Panicum t. Buckl.

Brachyelytrum Beauv.

   erectum (Schreb.) Beauv.
      var. erectum
         SY=B. aristosum (Michx.) Trel.
         SY=B. e. var. septentrionale Babel
      var. glabratum (Vasey) Koyama &
      Kawano
         SY=B. aristosum (Michx.) Trel.
           var. g. Vasey

Brachypodium Beauv.

   distachyon (L.) Beauv.
   ?pinnatum (L.) Beauv.
   sylvaticum (Huds.) Beauv.

Briza L.

   maxima L.
   media L.
   minor L.

Bromus L.

   alopecuros Poir.
   altissimus Pursh
         SY=B. latiglumis (Shear) A.S.
           Hitchc.
         SY=B. purgans sensu auctt. non L.
   anomalus Rupr. ex Fourn.
         SY=Bromopsis a. (Rupr. ex Fourn.)
           Holub
         SY=Bromopsis porteri (Coult.)
           Holub
          SY=Bromus p. (Coult.) Nash
   arenarius Labill.
   arizonicus (Shear) Stebbins
   arvensis L.
   ?biebersteinii Roemer & Schultes
         SY=Bromopsis b. (Roemer &
           Schultes) Holub
   breviaristatus Buckl.
         SY=Bromus carinatus Hook. & Arn.
           var. linearis Shear
         SY=B. subvelutinus Shear
   briziformis Fisch. & Mey.
   carinatus Hook. & Arn.
         SY=Bromus c. var. californicus
           (Nutt.) Shear
         SY=B. carinatus var. hookerianus

Bromus (CONT.)
>
>   (Thurb.) Shear
>   SY=Ceratochloa carinata (Hook. &
>   Arn.) Tutin
catharticus Vahl
>   SY=Bromus haenkeanus (Presl) Kunth
>   SY=B. unioloides (Willd.) Raspail
>   SY=B. willdenowii Kunth
>   SY=Ceratochloa c. (Vahl) Herter
>   SY=C. u. (Willd.) Beauv.
ciliatus L.
>   SY=Bromopsis c. (L.) Holub
>   SY=Bromus canadensis Michx.
>   SY=Bromus ciliatus var. genuinus
>   Fern.
>   SY=Bromus ciliatus var. intonsus
>   Fern.
>   SY=Bromus dudleyi Fern.
>   SY=Bromus richardsonii Link var.
>   pallidus (Hook.) Shear
commutatus Schrad.
>   SY=Bromus c. var. apricorum
>   Simonkai
danthoniae Trin.
diandrus Roth
>   SY=Anisantha d. (Roth) Tutin
>   SY=Bromus d. Roth var. gussonii
>   (Parl.) Coss. & Durieu
>   SY=B. g. Parl.
>   SY=B. rigidus Roth var. g. (Parl.)
>   Coss. & Durieu
erectus Huds.
>   SY=Bromopsis e. (Huds.) Fourr.
frondosus (Shear) Woot. & Standl.
>   SY=Bromopsis f. (Shear) Holub
grandis (Shear) A.S. Hitchc.
>   SY=Bromopsis g. (Shear) Holub
hordeaceus L.
>   ssp. hordeaceus
>   SY=Bromus mollis L.
>   SY=?B. thominii Ard. ex Nyman
>   ssp. pseudothominii P. Sm. emend H.
>   Scholz
>   SY=Bromus mollis L. var.
>   leiostachys Hartman
inermis Leyss.
>   ssp. inermis
>   var. divaricatus Pohl
>   var. inermis
>   SY=Bromopsis i. (Leyss.) Holub
>   ssp. pumpellianus (Scribn.) Wagnon
>   var. arcticus (Shear ex Scribn. &
>   Merr.) Wagnon
>   SY=Bromus i. var. aristatus Schur
>   SY=B. p. Scribn. var. a. (Shear ex
>   Scribn. & Merr.) Porsild
>   var. pumpellianus (Scribn.) Wagnon
>   SY=Bromopsis p. (Scribn.) Holub
>   SY=Bromus p. Scribn.
>   SY=Bromus p. ssp. dicksonii
>   Mitchell & Wilton
>   var. purpurascens (Hook.) Wagnon
>   SY=Bromus i. ssp. pumpellianus
>   var. tweedyi (Scribn.) C.L.
>   Hitchc.
>   SY=B. pumpellianus Scribn. var.
>   tweedyi Scribn.
>   var. villosissimus Hulten
>   SY=Bromus pumpellianus Scribn.
>   var. v. Hulten
japonicus Thunb. ex Murr.
>   SY=Bromus j. var. porrectus Hack.
>   SY=B. patulus Mert. & Koch
kalmii Gray
>   SY=Bromopsis k. (Gray) Holub
>   SY=Bromus ciliatus L. var.
>   laeviglumis Scribn. ex Shear

>   SY=Bromus purgans L.
>   SY=Bromus p. var. l. (Scribn. ex
>   Shear) Swallen
laciniatus Beal
laevipes Shear
lanatipes (Shear) Rydb.
>   SY=Bromopsis l. (Shear) Holub
>   SY=Bromus anomalus Rupr. ex Fourn.
>   var. l. (Shear) A.S. Hitchc.
lanceolatus Roth
>   SY=Bromus macrostachys Desf.
lepidus Holmb.
madritensis L.
>   SY=Anisantha m. (L.) Nevski
marginatus Nees ex Steud.
>   SY=Bromus m. var. latior Shear
>   SY=B. m. var. seminudus Shear
>   SY=B. sitchensis Trin. var. m.
>   (Nees ex Steud.) Boivin
>   SY=Ceratochloa m. (Nees ex Steud.)
>   Jackson
maritimus (Piper) A.S. Hitchc.
>   SY=Bromus carinatus Hook. & Arn.
>   var. m. (Piper) C.L. Hitchc.
molliformis Lloyd
mucroglumis Wagnon
>   SY=Bromopsis m. (Wagnon) Holub
nottowayanus Fern.
>   SY=Brompsis n. (Fern.) Holub
orcuttianus Vasey
>   var. hallii A.S. Hitchc.
>   var. orcuttianus
>   SY=Bromopsis o. (Vasey) Holub
pacificus Shear
>   SY=Bromopsis p. (Shear) Holub
polyanthus Scribn.
>   SY=Bromus paniculatus (Shear)
>   Rydb.
>   SY=Ceratochloa polyantha (Scribn.)
>   Tzveleo
pseudolaevipes Wagnon
>   SY=Bromopsis p. (Wagnon) Holub
pubescens Muhl. ex Willd.
>   SY=Bromopsis p. (Muhl. ex Willd.)
>   Holub
>   SY=Bromus purgans sensu auctt. non
>   L.
racemosus L.
ramosus Huds.
>   SY=Bromopsis r. (Huds.) Holub
>   SY=Bromus asper Murr.
richardsonii Link
>   SY=Bromopsis r. (Link) Holub
rigidus Roth
>   SY=Anisantha r. (Roth) Nevski
>   SY=Bromus maximus Desf.
rubens L.
>   SY=Anisantha r. (L.) Nevski
scoparius L.
secalinus L.
>   ssp. grossus (Desf. ex Lam. & DC.)
>   Richter
>   SY=Bromus g. Desf. ex Lam. & DC.
>   SY=B. secalinus L. var. velutinus
>   (Schrad.) Koch
>   ssp. secalinus
>   SY=Bromus s. var. hirsutus Kindb.
sitchensis Trin.
>   var. aleutensis (Trin. ex Griseb.)
>   Hulten
>   SY=Bromus a. Trin. ex Griseb.
>   var. sitchensis
squarrosus L.
stamineus Desv.
sterilis L.
>   SY=Anisantha s. (L.) Nevski
suksdorfii Vasey

Bromus (CONT.)
        SY=Bromopsis s. (Vasey) Holub
  tectorum L.
      var. hirsutus Regel
      var. tectorum
         SY=Anisantha t. (L.) Nevski
         SY=Bromus t. var. glabratus
            Spenner
         SY=B. t. var. nudus Klett &
            Richter
  texensis (Shear) A.S. Hitchc.
         SY=Bromopsis t. (Shear) Holub
  trinii Desv.
         SY=Bromus t. var. excelsus Shear
         SY=Trisetobromus hirtus (Trin.)
            Nevski
  vulgaris (Hook.) Shear
         SY=Bromopsis v. (Hook.) Holub
         SY=Bromus v. var. eximius Shear
         SY=Bromus v. var. robustus Shear

Buchloe Engelm.

  dactyloides (Nutt.) Engelm.
         SY=Bulbilis d. (Nutt.) Raf. ex
            Kuntze

Calamagrostis Adans.

  bolanderi Thurb.
  breweri Thurb.
  cainii A.S. Hitchc.
  californica Kearney
  canadensis (Michx.) Beauv.
      var. arcta Stebbins
      var. canadensis
         SY=C. c. ssp. c. var. pallida
            (Vasey & Scribn.) Stebbins
         SY=C. c. var. acuminata Vasey ex
            Shear & Rydb.
         SY=C. c. var. pallida (Vasey &
            Scribn.) Stebbins
         SY=C. c. var. robusta Vasey
         SY=C. c. var. typica Stebbins
         SY=C. inexpansa (Munro) A.S.
            Hitchc. var. robusta
            (Vasey) Stebbins
      var. imberbis (Stebbins) C.L. Hitchc.
      var. macouniana (Vasey) Stebbins
         SY=C. m. Vasey
      var. scabra (Presl) A.S. Hitchc.
         SY=C. c. ssp. langsdorfii (Link)
            Hulten
         SY=C. c. var. l. (Link) Inman
         SY=C. l. (Link) Trin.
  cinnoides (Muhl.) Bart.
  crassiglumis Thurb.
  densa Vasey
  deschampsioides Trin.
  epigeios (L.) Roth
         SY=C. e. var. georgica (K. Koch)
            Ledeb.
  fernaldii Louis-Marie
  foliosa Kearney
  hillebrandii (Munro) A.S. Hitchc.
  holmii Lange
         SY=C. chordorrhiza Porsild
         SY=C. neglecta (Ehrh.) Gaertn.,
            Mey., & Scherb. var.
            borealis Kearney
  howellii Vasey
  hyperborea Lange
  insperata Swallen
  koelerioides Vasey
  labradorica Kearney
  lactea Beal
         SY=C. canadensis (Michx.) Beauv.

         var. l. (Beal) C.L. Hitchc.
  lacustris (Kearney) Nash
         SY=C. pickeringii Gray var. l.
            (Kearney) A.S. Hitchc.
  lapponica (Wahlenb.) Hartman
      var. groenlandica Lange
      var. lapponica
      var. nearctica Porsild
  laricina (Louis-Marie) Lalonde
         SY=C. purpurascens R. Br. var. l.
            Louis-Marie
  lepageana Louis-Marie
  montanensis (Scribn.) Scribn.
  neglecta (Ehrh.) Gaertn., Mey., & Scherb.
      var. gracilis (Scribn.) Scribn.
      var. neglecta
         SY=C. expansa (Munro) A.S. Hitchc.
         SY=C. inexpansa Gray
         SY=C. i. var. barbulata Kearney
         SY=C. i. var. brevior (Vasey)
            Stebbins
         SY=C. i. var. novae-angliae
            Stebbins
         SY=C. n. var. micrantha (Kearney)
            Stebbins
         SY=C. stricta (Timm) Koel.
  nubila Louis-Marie
  nutkaensis (Presl) Steud.
  ophitidis (J.T. Howell) Nygren
         SY=C. purpurascens R. Br. var. o.
            J.T. Howell
  perplexa Scribn.
  pickeringii Gray
         SY=C. p. var. debilis (Kearney)
            Fern. & Wieg.
  poluninii Sorensen
  porteri Gray
  purpurascens R. Br.
      var. maltei Polunin
      var. purpurascens
         SY=C. p. ssp. arctica (Vasey)
            Hulten
         SY=C. p. ssp. tasuensis Calder &
            Taylor
         SY=C. p. var. t. (Calder & Taylor)
            Boivin
         SY=C. vaseyi Beal
  rubescens Buckl.
  scopulorum M.E. Jones
  scribneri Beal
  sesquiflora (Trin.) Tzveleo
  tweedyi (Scribn.) Scribn.

Calamovilfa (Gray) Hack. ex Scribn. &
Southworth

  arcuata K.E. Rogers
  brevipilis (Torr.) Scribn.
         SY=C. b. var. calvipes Fern.
         SY=C. b. var. heterolepis Fern.
  curtissii (Vasey) Scribn.
  gigantea (Nutt.) Scribn. & Merr.
         SY=Calamagrostis g. (Nutt.)
            Scribn. & Merr.
  longifolia (Hook.) Scribn.
      var. longifolia
      var. magna Scribn. & Merr.

Catabrosa Beauv.

  aquatica (L.) Beauv.
      var. aquatica
         SY=C. a. var. uniflora S.F. Gray
      var. laurentiana Fern.

Catapodium Link

Catapodium (CONT.)
    rigidum (L.) C.E. Hubbard ex Dony
              SY=Scleropoa r. (L.) Griseb.

Cathestecum Presl

    erectum Vasey & Hack.

Cenchrus L.

    agrimonioides Trin.
              SY=C. a. var. laysanensis F. Br.
              SY=C. pedunculata Deg. & Whitney
    biflorus Roxb.
              SY=C. barbatus Schum.
              SY=C. catharticus Delile
    brownii Roemer & Schultes
              SY=C. viridis Spreng.
    ciliaris L.
              SY=Pennisetum c. (L.) Link
    echinatus L.
              SY=C. e. var. hillebrandianus
                  (A.S. Hitchc.) F. Br.
    gracillimus Nash
    incertus M.A. Curtis
              SY=?C. carolinianus Walt.
              SY=C. parviceps Shinners
              SY=C. pauciflorus Benth.
    longispinus (Hack.) Fern.
    myosuroides H.B.K.
              SY=Cenchropsis m. (H.B.K.) Nash
    setigerus Vahl
    tribuloides L.

Chasmanthium Link

    latifolium (Michx.) Yates
              SY=Uniola l. Michx.
    laxum (L.) Yates
              SY=Uniola l. (L.) B.S.P.
    nitidum (Baldw. ex Ell.) Yates
              SY=Uniola n. Baldw. ex Ell.
    ornithorhynchum (Steud.) Yates
              SY=Uniola o. Steud.
    sessiliflorum (Poir.) Yates
              SY=Uniola longifolia Scribn.
              SY=U. s. Poir.

Chloris Sw.

    andropogonoides Fourn.
    canterai Arech.
    capensis (Houtt.) Thellung
    chloridea (Presl) A.S. Hitchc.
    ciliata Sw.
    crinita Lag.
              SY=Trichloris c. (Lag.) Parodi
              SY=T. mendocina (Phil.) Kurtz
    cucullata Bisch.
    dandyana C.D. Adams
              SY=C. polydactyla (L.) Sw.
              SY=C. barbata (L.) Sw.
    divaricata R. Br.
              SY=C. cynodontoides Balansa
    gayana Kunth
    inflata Link
              SY=C. paraguaiensis Steud.
    pluriflora (Fourn.) Clayton
              SY=Trichloris p. Fourn.
    prieurii Kunth
    radiata (L.) Sw.
    sagraeana A. Rich.
    subdolichostachya C. Muell.
              SY=C. latisquamea Nash
    submutica H.B.K.
    texensis Nash
              SY=C. truncata R. Br.

    ventricosa R. Br.
    verticillata Nutt.
    virgata Sw.

Chrysopogon Trin.

    aciculatus (Retz.) Trin.
    fulvus (Spreng.) Chiov.
    pauciflorus (Chapman) Benth. ex Vasey
              SY=Rhaphis p. (Chapman) Nash

Chusquea Kunth

    abietifolia Griseb.
    coronalis Soderstrom & Calderon

Cinna L.

    arundinacea L.
          var. arundinacea
          var. inexpansa Fern. & Grisc.
    latifolia (Trev. ex Goepp.) Griseb.

Coelorachis Brongn.

    cylindrica (Michx.) Nash
              SY=Manisuris campestris (Nutt.)
                  A.S. Hitchc.
              SY=M. cylindrica (Michx.) Kuntze
    rugosa (Nutt.) Nash
              SY=Manisuris r. (Nutt.) Kuntze
    tessellata (Steud.) Nash
              SY=Manisuris t. (Steud.) Scribn.
    tuberculosa (Nash) Nash
              SY=Manisuris t. Nash

Coix L.

    lacryma-jobi L.

Coleanthus Seidel

    subtilis (Tratt.) Seidel

Colpodium Trin.

    vahlianum (Liebm.) Nevski
    wrightii Scribn. & Merr.

Coridochloa Nees

    cimicina (L.) Nees ex Jackson

Cortaderia Stapf

    atacamensis (Phil.) Pilger
    dioica (Spreng.) Speg.
              SY=C. selloana (Schultes) Aschers.
                  & Graebn.

Corynephorus Beauv.

    canescens (L.) Beauv.

Cottea Kunth

    pappophoroides Kunth

Crypsis Ait.

    aculeata (L.) Ait.
    alopecuroides (Piller & Mitterp.) Schrad.
              SY=Heleochloa a. (Piller &
                  Mitterp.) Host
    niliacea Fig. & de Not.
    schoenoides (L.) Lam.
              SY=Heleochloa s. (L.) Host

Ctenium Panzer

    aromaticum (Walt.) Wood
        SY=Campulosus a. (Walt.) Scribn.
    floridanum (A.S. Hitchc.) A.S. Hitchc.
        SY=Campulosus f. A.S. Hitchc.

Cutandia Willk.

    memphitica (Spreng.) Richter

Cymbopogon Spreng.

    citratus (DC.) Stapf
    nardus (L.) Rendle
    refractus (R. Br.) A. Camus

Cynodon L.C. Rich.

    aethiopicus Clayton & Harlan
    dactylon (L.) Pers.
        var. aridus Harlan & de Wet
        var. dactylon
        SY=Capriola d. (L.) Kuntze
    nlemfuensis Vanderyst
        var. robustus Clayton & Harlan
    plectostachyus (K. Schum.) Pilger
    transvaalensis Burtt-Davy

Cynosurus L.

    cristatus L.
    echinatus L.

Dactylis L.

    aschersoniana Graebn.
    glomerata L.
        SY=D. g. var. ciliata Peterm.
        SY=D. g. var. detonsa Fries
        SY=D. g. var. vivipara Parl.

Dactyloctenium Willd.

    aegyptium (L.) Beauv.

Danthonia Lam. & DC.

    X allenii Austin [compressa X spicata]
    californica Boland.
        SY=D. americana Scribn.
        SY=D. c. var. a. (Scribn.) A.S.
            Hitchc.
    canadensis Baum & Findlay
    compressa Austin
    intermedia Vasey
        SY=D. i. var. cusickii T.A.
            Williams
    parryi Scribn.
    pilosa R. Br.
    purpurea (Thunb.) Beauv. ex Roemer &
      Schultes
    semiannularis (Labill.) R. Br.
    sericea Nutt.
        SY=D. epilis Scribn.
        SY=D. s. var. e. (Scribn.) Gleason
    spicata (L.) Beauv. ex Roemer & Schultes
        SY=D. s. var. longipila Scribn. &
            Merr.
        SY=D. s. var. pinetorum Piper
        SY=D. thermalis Scribn.
    unispicata (Thurb.) Munro ex Macoun

Dendrocalamus Nees

    gigantea Munro

Deschampsia Beauv.

    alpina (L.) Roemer & Schultes
    atropurpurea (Wahlenb.) Scheele
        var. atropurpurea
        SY=Aira a. Wahlenb.
        SY=Vahlodea a. (Wahlenb.) Fries
        var. latifolia (Hook.) Scribn. ex
          Macoun
        SY=Vahlodea a. (Wahlenb.) Fries
          ssp. l. (Hook.) Porsild
        var. paramushirensis Kudo
        SY=Vahlodea a. (Wahlenb.) Fries
          ssp. p. (Kudo) Hulten
    australis Nees ex Steud.
      ssp. australis
      ssp. nubigena (Hbd.) Skottsberg
        var. gracilis Skottsberg
        var. nubigena
        var. tenuissima (Skottsberg)
          Skottsberg
    brevifolia R.Br.
    cespitosa (L.) Beauv.
      ssp. beringensis (Hulten) W.E. Lawrence
        SY=D. b. Hulten
        SY=D. c. var. arctica Vasey
      ssp. cespitosa
        var. cespitosa
        SY=Aira c. L.
        SY=D. c. ssp. genuina (Reichenb.)
          W.E. Lawrence
        SY=D. c. var. abbei Boivin
        SY=D. c. var. alpicola (Rydb.)
          Love, Love & Kapoor
        SY=D. c. var. intercotidalis
          Boivin
        SY=D. c. var. littoralis (Reut.)
          Richter
        SY=D. c. var. longiflora Beal
        SY=D. c. var. maritima Vasey
        SY=D. c. var. parviflora (Thuill.)
          Coss. & Germ.
        var. glauca (Hartman) Lindm. f.
      ssp. holciformis (Presl) W.E. Lawrence
        SY=Aira h. (Presl) Steud.
        SY=D. h. Presl
      ssp. orientalis Hulten
    congestiformis Booth
    danthonioides (Trin.) Munro ex Benth.
        SY=Aira d. Trin.
        SY=D. calycina Presl
        SY=D. d. var. gracilis (Vasey)
          Munz
    elongata (Hook.) Munro ex Benth.
        SY=Aira e. Hook.
    flexuosa (L.) Trin.
        var. flexuosa
        SY=Aira f. L.
        var. montana (L.) Ducomm.
        var. pallida Berl.
    mackenzieana Raup
        SY=D. cespitosa (L.) Beauv. var.
          m. (Raup) Boivin
    pallens Hbd.
    pumila (Ledeb.) Ostenf.

Diarrhena Beauv.

    americana Beauv.
        var. americana
        SY=Diarina festucoides Raf.
        SY=Korycarpus arundinaceus Zea ex
          Lag.
        var. obovata Gleason

Dichanthelium (A.S. Hitchc. & Chase) Gould

Dichanthelium (CONT.)
   aciculare (Desv. ex Poir.) Gould & Clark
          SY=D. angustifolium (Ell.) Gould
          SY=Panicum aciculare Desv. ex
              Poir.
          SY=P. angustifolium Ell.
          SY=P. arenicoloides Ashe
          SY=P. bennettense W.V. Brown
          SY=P. fusiforme A.S. Hitchc.
          SY=P. nemopanthum Ashe
          SY=P. neuranthum Griseb.
          SY=P. orthophyllum Ashe
          SY=P. ovinum Scribn. & Sm.
          SY=P. pinetorum Swallen
   acuminatum (Sw.) Gould & Clark
     var. acuminatum
          SY=D. lanuginosum (Ell.) Gould
          SY=D. l. var. fasciculatum (Torr.)
              Spellenberg
          SY=D. l. var. sericeum (Schmoll)
              Spellenberg
          SY=D. l. var. thermale (Boland.)
              Spellenberg
          SY=Panicum a. Sw.
          SY=P. benneri Fern.
          SY=P. brodiei St. John
          SY=P. chrysopsidifolium Nash
          SY=P. ferventicola Schmoll
          SY=P. ferventicola var. papillosum
              Schmoll
          SY=P. ferventicola var. sericeum
              Schmoll
          SY=P. glutinoscabrum Fern.
          SY=P. huachucae Ashe
          SY=P. h. var. fasciculatum (Torr.)
              F.T. Hubbard
          SY=P. languidum A.S. Hitchc. &
              Chase
          SY=P. lanuginosum Ell.
          SY=P. lanuginosum var.
              fasciculatum (Torr.) Fern.
          SY=P. lanuginosum var. h. (Ashe)
              A.S. Hitchc.
          SY=P. lanuginosum var.
              septentrionale (Fern.) Fern.
          SY=P. lanuginosum var.
              tennesseense (Ashe) Gleason
          SY=P. lassenianum Schmoll
          SY=P. occidentale Scribn.
          SY=P. pacificum A.S. Hitchc. &
              Chase
          SY=P. scoparioides Ashe
          SY=P. shastense Scribn. & Merr.
          SY=P. subvillosum Ashe
          SY=P. tennesseense Ashe
          SY=P. thermale Boland.
          SY=P. villosissimum Nash var.
              scoparioides (Ashe) Fern.
     var. densiflorum (Rand & Redf.) Gould
       & Clark
          SY=Panicum eatonii Nash
          SY=P. nitidum Lam. var.
              densiflorum Rand & Redf.
          SY=P. n. var. octonodum (Sm.)
              Scribn. & Merr.
          SY=P. o. Sm.
          SY=P. paucipilum Nash
          SY=P. spretum Schultes
     var. implicatum (Scribn.) Gould &
       Clark
          SY=D. meridionale (Ashe) Freckmann
          SY=Panicum albemarlense Ashe
          SY=P. auburne Ashe
          SY=P. curtifolium Nash
          SY=P. i. Scribn.
          SY=P. lanuginosum Ell. var. i.
              (Scribn.) Fern.

          SY=P. leucothrix Nash
          SY=P. m. Ashe
          SY=P. m. var. albemarlense (Ashe)
              Fern.
     var. lindheimeri (Nash) Gould & Clark
          SY=D. l. (Nash) Gould
          SY=D. lanuginosum (Ell.) Gould
              var. lindheimeri (Nash)
              Fern.
          SY=Panicum lanuginosum Ell. var.
              lindheimeri (Nash) Fern.
          SY=P. lindheimeri Nash
     var. longiligulatum (Nash) Gould &
       Clark
          SY=Panicum l. Nash
     var. thurowii (Scribn. & Sm.) Gould &
       Clark
          SY=Panicum t. Scribn. & Sm.
     var. villosum (Gray) Gould & Clark
          SY=D. lanuginosum (Ell.) Gould
              var. villosissimum (Nash)
              Gould
          SY=D. villosissimum (Nash)
              Freckmann
          SY=D. villosissimum (Nash)
              Freckmann var. praecocius
              (A.S. Hitchc. & Chase)
              Freckmann
          SY=Panicum euchlamydeum Shinners
          SY=P. l. Ell. var. praecocius
              (A.S. Hitchc. & Chase)
              McNeill & Dore
          SY=P. nitidum Lam. var. villosum
              Gray
          SY=P. p. A.S. Hitchc. & Chase
          SY=P. pseudopubescens Nash
          SY=P. villosissimum Nash var.
              pseudopubescens Fern.
     var. wrightianum (Scribn.) Gould &
       Clark
          SY=Panicum deminutivum Peck
          SY=P. minutulum Desv.
          SY=P. strictum Bosc ex Roemer &
              Schultes
          SY=P. w. Scribn.
   boreale (Nash) Freckmann
          SY=Panicum bicknellii Nash
          SY=P. bicknellii var. bushii
              (Nash) Farw.
          SY=P. boreale Nash
          SY=P. boreale var. michiganense
              Farw.
          SY=P. bushii Nash
          SY=P. calliphyllum Ashe
   boscii (Poir.) Gould & Clark
          SY=Panicum b. Poir.
          SY=P. b. var. molle (Vasey) A.S.
              Hitchc. & Chase
   clandestinum (L.) Gould
          SY=Panicum c. L.
   commutatum (Schultes) Gould
          SY=Panicum ashei Pearson ex Ashe
          SY=P. c. Schultes
          SY=P. c. var. a. (Pearson ex Ashe)
              Fern.
          SY=P. c. var. joorii (Vasey) Fern.
          SY=P. equilaterale Scribn.
          SY=P. j. Vasey
          SY=P. mutabile Scribn. & Sm.
   conjugens (Skottsberg) Clark & Gould
          SY=Panicum c. Skottsberg
   consanguineum (Kunth) Gould & Clark
          SY=Panicum c. Kunth
          SY=P. deamii A.S. Hitchc. & Chase
          SY=P. georgianum Ashe
   cynodon (Reichardt) Clark & Gould
          SY=Panicum imbricatum Hbd.

Dichanthelium (CONT.)
        SY=P. i. var. molokaiense
           Skottsberg
        SY=P. lustriale St. John
  depauperatum (Muhl.) Gould
        SY=Panicum d. Muhl.
        SY=P. d. var. involutum (Torr.)
           Wood
        SY=P. d. var. psilophyllum Fern.
        SY=P. strictum Pursh var.
           psilophyllum (Fern.) Farw.
  dichotomum (L.) Gould
      var. breve (A.S. Hitchc. & Chase)
        Gould & Clark
        SY=Panicum b. A.S. Hitchc. & Chase
      var. dichotomum
        SY=Panicum annulum Ashe
        SY=P. a. var. glabrescens Gleason
        SY=P. barbulatum Michx.
        SY=P. caerulescens Hack. ex A.S.
           Hitchc.
        SY=P. clutei Nash
        SY=P. d. L.
        SY=P. d. var. b. (Michx.) Wood
        SY=P. lucidum Ashe
        SY=P. l. var. opacum Fern.
        SY=P. mattamuskeetense Ashe
        SY=P. m. var. clutei (Nash) Fern.
        SY=P. microcarpon Muhl. ex Ell.
        SY=P. nitidum Lam.
        SY=P. nudicaule Vasey
        SY=P. roanokense Ashe
        SY=P. sphagnicola Nash
        SY=P. yadkinense Ashe
      var. ensifolium (Baldw. ex Ell.)
        Gould & Clark
        SY=D. e. (Baldw. ex Ell.) Gould
        SY=Panicum chamaelonche Trin.
        SY=P. e. Baldw. ex Ell.
        SY=P. flavovirens Nash
        SY=P. vernale A.S. Hitchc. & Chase
      var. glabrifolium (Nash) Gould &
        Clark
        SY=Panicum g. Nash
      var. tenue (Muhl.) Gould & Clark
        SY=Panicum concinnius A.S. Hitchc.
           & Chase
        SY=P. tenue Muhl.
        SY=P. trifolium Nash
  erectifolium (Nash) Gould & Clark
        SY=Panicum e. Nash
  forbesii (A.S. Hitchc.) Clark & Gould
        SY=Panicum f. A.S. Hitchc.
  hillebrandianum (A.S. Hitchc.) Clark &
    Gould
        SY=Panicum h. A.S. Hitchc.
        SY=P. h. var. gracilis Skottsberg
  issachnoides (Munro ex Hbd.) Clark & Gould
        SY=Panicum i. Munro ex Hbd.
        SY=P. i. var. kilohanae Skottsberg
  koolauense (St. John & Hosaka) Clark &
    Gould
        SY=Panicum k. St. John & Hosaka
        SY=P. oreoboloides (Whitney)
           Skottsberg
        SY=P. o. var. subimbricatum
           Skottsberg
  latifolium (L.) Gould & Clark
        SY=Panicum l. L.
  laxiflorum (Lam.) Gould
        SY=Panicum l. Lam.
        SY=P. l. var. strictirameum (A.S.
           Hitchc. & Chase) Fern.
        SY=P. xalapense H.B.K.
  leibergii (Vasey) Freckman
        SY=Panicum l. (Vasey) Scribn.
        SY=P. l. var. baldwinii Lepage

        SY=P. scoparium Lam. var. l. Vasey
        SY=P. scribnerianum Nash var. l.
           (Vasey) Scribn.
  leucoblepharis (Trin.) Gould & Clark
      var. glabrescens (Griseb.) Gould &
        Clark
        SY=Panicum dichotomum L. var. g.
           Griseb.
        SY=P. polycaulon Nash
      var. leucoblepharis
        SY=Panicum ciliatum Ell.
        SY=P. l. Trin.
      var. pubescens (Vasey) Gould & Clark
        SY=Panicum laxiflorum Lam. var. p.
           Vasey
        SY=P. strigosum Muhl.
  linearifolium (Scribn.) Gould
        SY=D. perlongum (Nash) Freckmann
        SY=Panicum l. Scribn.
        SY=P. l. var. werneri (Scribn.)
           Fern.
        SY=P. perlongum Nash
        SY=P. strictum var. l. Farw.
        SY=P. s. var. perlongum (Nash)
           Farw.
        SY=P. w. Scribn.
  malacophyllum (Nash) Gould
        SY=Panicum m. Nash
  nodatum (A.S. Hitchc. & Chase) Gould
        SY=Panicum n. A.S. Hitchc. & Chase
  oligosanthes (Schultes) Gould
      var. oligosanthes
        SY=Panicum o. Schultes
      var. scribnerianum (Nash) Gould
        SY=Panicum helleri Nash
        SY=P. o. Schultes var. h. (Nash)
           Fern.
        SY=P. o. var. scribnerianum (Nash)
           Fern.
        SY=P. scoparium S. Wats. ex Nash
           non Lam.
        SY=P. scribnerianum Nash
      var. wilcoxianum (Vasey) Gould &
        Clark
        SY=D. w. (Vasey) Freckmann
        SY=Panicum w. Vasey
        SY=P. w. var. bretungii Boivin
  ovale (Ell.) Gould & Clark
      var. addisonii (Nash) Gould & Clark
        SY=D. commonsianum (Ashe)
           Freckmann
        SY=Panicum addisonii Nash
        SY=P. alabamense Ashe
        SY=P. columbianus Scribn. var.
           commonsianum (Ashe) McNeill
           & Dore
        SY=P. commonsianum Ashe
        SY=P. commonsianum var. addisonii
           (Nash) Fern.
        SY=P. commonsianum var. addisonii
           (Nash) Pohl
        SY=P. mundum Fern.
        SY=P. owenae Bickn.
        SY=P. wilmingtonense Ashe
      var. ovale (Ell.) Gould & Clark
        SY=Panicum o. Ell.
        SY=P. malacon Nash
  pedicellatum (Vasey) Gould
        SY=Panicum p. Vasey
  ravenelii (Scribn. & Merr.) Gould
        SY=Panicum r. Scribn. & Merr.
  sabulorum (Lam.) Gould & Clark
      var. patulum (Scribn. & Merr.) Gould
        & Clark
        SY=Panicum columbianum Scribn.
           var. siccanum (A.S. Hitchc.
           & Chase) Boivin

Dichanthelium (CONT.)
    SY=P. commonsianum Ashe var.
       euchlamydeum (Shinners) Pohl
    SY=P. e. Shinners
    SY=P. lancearium Trin.
    SY=P. l. var. p. (Scribn. & Merr.)
       Fern.
    SY=P. lanuginosum Ell. var.
       siccanum A.S. Hitchc. &
       Chase
    SY=P. nashianum Scribn.
    SY=P. n. var. p. Scribn. & Merr.
    SY=P. onslowense Ashe
    SY=P. patentifolium Nash
    SY=P. patulum (Scribn. & Merr.)
       A.S. Hitchc.
    SY=P. tsugetorum Nash
    SY=P. webberianum Nash
  var. thinium (A.S. Hitchc. & Chase)
    Gould & Clark
    SY=D. columbianum (Scribn.)
       Freckmann
    SY=Panicum columbianum Scribn.
    SY=P. c. var. oricola (A.S.
       Hitchc. & Chase) Fern.
    SY=P. c. var. t. A.S. Hitchc. &
       Chase
    SY=P. heterophyllum Bosc ex Nees
    SY=P. h. var. t. (A.S. Hitchc. &
       Chase) F.T. Hubbard
    SY=P. oricola A.S. Hitchc. & Chase
    SY=P. portoricense Desv. ex
       Hamilton
    SY=P. unciphyllum Trin. var. t.
       A.S. Hitchc. & Chase
scabriusculum (Ell.) Gould & Clark
    SY=Panicum cryptanthum Ashe
    SY=P. lanuginosum Bosc ex Spreng.
    SY=P. recognitum Fern.
    SY=P. s. Ell.
scoparium (Lam.) Gould
    SY=Panicum s. Lam.
sphaerocarpon (Ell.) Gould
  var. isophyllum (Scribn.) Gould &
    Clark
    SY=D. s. var. polyanthes
       (Schultes) Gould
    SY=Panicum microcarpon Muhl. non
       Muhl. ex Ell.
    SY=P. m. Muhl. var. i. Scribn.
    SY=P. polyanthes Schultes
  var. sphaerocarpon
    SY=Panicum s. Ell.
    SY=P. s. var. inflatum (Scribn. &
       Sm.) A.S. Hitchc. & Chase
xanthophysum (Gray) Freckmann
    SY=Panicum x. Gray

Dichanthium Willem.

  annulatum (Forsk.) Stapf
    SY=Andropogon a. Forsk.
  aristatum (Poir.) C.E. Hubbard
    SY=Andropogon nodosus (Willem.)
       Nash
  caricosum (L.) A. Camus
  sericeum (R. Br.) A. Camus
    SY=Andropogon s. R. Br.

Diectomis Kunth

  fastigiata (Sw.) H.B.K.

Digitaria Heister ex Fabr.

  albicoma Swallen
  arenicola (Swallen) Beetle

    SY=Leptoloma a. Swallen
    SY=L. cognatum (Schultes) Chase
       var. a. (Swallen) Gould
  argillacea (A.S. Hitchc. & Chase) Fern.
    SY=Syntherisma a. A.S. Hitchc. &
       Chase
  bicornis (Lam.) Roemer & Schultes
  californica (Benth.) Henr.
    SY=Trichachne c. (Benth.) Chase
  ciliaris (Retz.) Koel.
    SY=D. adscendens (H.B.K.) Henr.
    SY=D. sanguinalis (L.) Scop. var.
       c. (Retz.) Parl.
    SY=D. s. var. marginata (Link)
       Fern.
    SY=Syntherisma m. (Link) Nash
  cognatum (Schultes) Pilger
    SY=Leptoloma c. (Schultes) Chase
  decumbens Stent
  diversiflora Swallen
  dolichophylla Henr.
  eggersii (Hack.) Henr.
    SY=Trichachne e. (Hack.) Henr.
    SY=Valota e. (Hack.) A.S. Hitchc.
       & Chase
  filiformis (L.) Koel.
    SY=D. laeviglumis Fern.
    SY=Syntherisma f. (L.) Nash
  floridana A.S. Hitchc.
    SY=Syntherisma f. (A.S. Hitchc.)
       A.S. Hitchc.
  gracillima (Scribn.) Fern.
    SY=Syntherisma g. (Scribn.) Nash
  hitchcockii (Chase) Stuckert
    SY=Trichachne h. (Chase) Chase
  horizontalis Willd.
    SY=Syntherisma digitatum (Sw.)
       A.S. Hitchc.
  insularis (L.) Mez ex Ekman
    SY=Trichachne i. (L.) Nees
    SY=Valota i. (L.) Chase
  ischaemum (Schreb. ex Schweig.) Schreb. ex
    Muhl.
  var. ischaemum
    SY=Syntherisma i. (Schreb. ex
       Schweig.) Nash
  var. mississippiensis (Gattinger)
    Fern.
  longiflora (Retz.) Pers.
  microbachne (Presl) Henr.
    SY=D. pruriens (Fisch. ex Trin.)
       Buse var. m. (Presl) Fosberg
  panicea (Sw.) Urban
    SY=Syntherisma p. (Sw.) Nash
  patens (Swallen) Henr.
    SY=Trichachne p. Swallen
  pauciflora A.S. Hitchc.
    SY=Syntherisma p. (A.S. Hitchc.)
       A.S. Hitchc.
  pruriens (Fisch. ex Trin.) Buse
  runyonii A.S. Hitchc.
  sanguinalis (L.) Scop.
    SY=Syntherisma s. (L.) Dulac
  serotina (Walt.) Michx.
    SY=Syntherisma s. Walt.
  similis Beetle ex Gould
    SY=Trichachne affinis Swallen
  simpsonii (Vasey) Fern.
    SY=Syntherisma s. (Vasey) Nash
  subcalva A.S. Hitchc.
  swalleniana Henr.
  texana A.S. Hitchc.
  tomentosa Koen. ex Rottl.
    SY=Leptoloma t. (Koen. ex Rottl.)
       Gould
  villosa (Walt.) Pers.
    SY=D. filiformis (L.) Koel. var.

Digitaria (CONT.)
                        v. (Walt.) Fern.
                SY=Syntherisma v. Walt.
        violascens Link
                SY=D. ischaemum (Schreb. ex
                        Schweig.) Schreb. ex Muhl.
                        var. v. (Link) Radford
                SY=Syntherisma chinensis (Nees)
                        A.S. Hitchc.

Diplachne Beauv.

        acuminata Nash
                SY=Leptochloa a. (Nash)
                        Mohlenbrock
                SY=L. fascicularis (Lam.) Beauv.
                        var. a. (Nash) Gleason
        dubia (Kunth) Scribn.
                SY=Leptochloa d. (Kunth) Nees
        fascicularis (Lam.) Beauv.
                SY=Leptochloa f. (Lam.) Gray
        maritima Bickn.
                SY=Leptochloa fascicularis (Lam.)
                        Beauv. var. m. (Bickn.)
                        Gleason
        panicoides (Presl) McNeill
                SY=D. halei Nash
                SY=Leptochloa floribunda Doell
                SY=L. p. (Presl) A.S. Hitchc.
                SY=Megastachya p. Presl
        uninervia (Presl) Parodi
                SY=Leptochloa u. (Presl) A.S.
                        Hitchc. & Chase
        viscida Scribn.
                SY=Leptochloa v. (Scribn.) Beal

Dissanthelium Trin.

        californicum (Nutt.) Benth.

Dissochondrus (Hbd.) Kuntze ex Hack.

        biflorus (Hbd.) Kuntze ex Hack.

Distichlis Raf.

        spicata (L.) Greene
                var. nana Beetle
                var. spicata
                        SY=D. s. var. borealis (Presl)
                                Beetle
                        SY=D. s. var. divaricata Beetle
                        SY=D. s. var. stolonifera Beetle
                var. stricta (Torr.) Beetle
                        SY=D. spicata ssp. stricta (Torr.)
                                Thorne
                        SY=D. stricta (Torr.) Rydb.
                        SY=D. stricta var. dentata (Rydb.)
                                C.L. Hitchc.

Dupontia R. Br.

        fisheri R. Br.
                var. aristata Malte ex Polunin
                var. fisheri
        psilosantha Rupr.
                SY=D. fisheri R. Br. ssp. p.
                        (Rupr.) Hulten

Echinochloa Beauv.

        colona (L.) Link
        crusgalli (L.) Beauv.
                var. crusgalli
                        SY=E. pungens (Poir.) Rydb. var.
                                coarctata Fern. & Grisc.
                var. frumentacea (Roxb.) W. Wight

                SY=E. f. (Roxb.) Link
                var. oryzicola (Vasing) Ohwi
                        SY=E. o. (Vasing) Vasing
        crus-pavonis (H.B.K.) Schultes
                var. crus-pavonis
                var. macera (Wieg.) Gould
                        SY=E. crusgalli (L.) Beauv. var.
                                zelayensis (H.B.K.) A.S.
                                Hitchc.
                        SY=E. z. (H.B.K.) Schultes
        muricata (Beauv.) Fern.
                var. microstachya Wieg.
                        SY=E. crusgalli (L.) Beauv. var.
                                mitis (Pursh) Peterm.
                        SY=E. microstachya (Wieg.) Rydb.
                        SY=E. muricata var. occidentalis
                                Wieg.
                        SY=E. o. (Wieg.) Rydb.
                        SY=E. pungens (Poir.) Rydb. var.
                                microstachya (Wieg.) Fern. &
                                Grisc.
                        SY=E. p. var. multiflora (Wieg.)
                                Fern. & Grisc.
                var. muricata
                        SY=E. m. var. ludoviciana Wieg.
                        SY=E. pungens (Poir.) Rydb.
                        SY=E. p. var. l. (Wieg.) Fern. &
                                Grisc.
                var. wiegandii Fassett
                        SY=E. w. (Fassett) McNeill & Dore
        paludigena Wieg.
        polystachya (H.B.K.) A.S. Hitchc.
        spectabilis (Nees) Link
        walteri (Pursh) Heller

Ehrharta Thunb.

        calycina Sm.
        capensis Thunb.
        erecta Lam.

Eleusine Gaertn.

        coracana (L.) Gaertn.
                var. africana (Kennedy & O'Byrne)
                        Hilu & de Wet
                        SY=E. a. Kennedy & O'Byrne
                var. coracana
        indica (L.) Gaertn.
        tristachya (Lam.) Lam.

Elionurus Humb. & Bonpl. ex Willd.

        barbiculmis Hack.
                SY=E. b. var. parviflorus Scribn.
        tripsacoides Humb. & Bonpl. ex Willd.

X Elyhordeum Mansf.

        X arcuatum Mitchell & Hodgson [Elymus
                sibiricus X Hordeum jubatum]
        X berkeleyanum (Bowden) Bowden [Elymus
                condensatus X Hordeum jubatum ssp.
                breviaristatum]
                SY=X Elymordeum b. Bowden
        X dakotense (Bowden) Bowden [Elymus
                canadensis X Hordeum jubatum]
                SY=X Elymordeum d. Bowden
        X dutillyanum (Lepage) Bowden [Elymus
                mollis X Hordeum jubatum]
                nm. dutillyanum
                        SY=X Elymordeum d. Lepage
                nm. littorale (Hodgson & Mitchell)
                        Bowden
                        SY=X Elymordeum l. Hodgson &
                                Mitchell
        X iowense Pohl [Elymus villosus X Hordeum

Elyhordeum X   (CONT.)
    jubatum]
          SY=X Elymordeum i. nm. pubescens
             Bowden
  X montanense (Scribn.) Bowden [Elymus
    virginicus X Hordeum jubatum]
          SY=X Elymordeum m. Scribn.
          SY=Hordeum m. Scribn.
          SY=H. pammelii Scribn. & Ball
  X piperi (Bowden) Bowden [Elymus
    triticoides X Hordeum jubatum]
          SY=X Elymordeum p. Bowden
  X schaackianum (Bowden) Bowden [Elymus
    hirsutus X Hordeum jubatum ssp.
    breviaristatum]
          SY=X Elymordeum s. Bowden
  X stebbinsianum (Bowden) Bowden [Elymus
    glaucus X Hordeum jubatum ssp.
    breviaristatum]
          SY=X Elymordeum s. Bowden
  X triploideum (Bowden) Bowden [Elymus
    racemosus X Hordeum vulgare]

Elymus L.

  X aleuticus (Hulten) Bowden [hirsutus X
    mollis]
  ambiguus Vasey & Scribn.
      var. ambiguus
      var. salmonis C.L. Hitchc.
      var. strigosus (Rydb.) A.S. Hitchc.
  angustus Trin.
  arenarius L.
      var. villosissimus (Scribn.) Polunin
          SY=E. a. var. brevispicus (Scribn.
            & Sm.) Boivin
  canadensis L.
      var. canadensis
          SY=E. brachystachys Scribn. & Ball
          SY=E. c. var. b. (Scribn. & Ball)
            Farw.
          SY=E. c. var. robustus (Scribn. &
            Sm.) Mackenzie & Bush
          SY=E. crescendus L.C. Wheeler
          SY=E. philadelphicus L.
          SY=E. r. Scribn. & Sm.
      var. interruptus (Buckl.) Church
          SY=E. diversiglumis Scribn. & Ball
          SY=E. i. Buckl.
      var. wiegandii (Fern.) Bowden
          SY=E. w. Fern.
  cinereus Scribn. & Merr.
      var. cinereus
      var. pubens (Piper) C.L. Hitchc.
          SY=E. condensatus Presl var. p.
            Piper
          SY=E. piperi Bowden
  condensatus Presl
  flavescens Scribn. & Sm.
          SY=E. arenicola Scribn. & Sm.
  glaucus Buckl.
      var. glaucus
          SY=E. g. var. breviaristatus
            Burtt-Davy
      var. jepsonii Burtt-Davy
          SY=E. g. ssp. j. (Burtt-Davy)
            Gould
          SY=E. g. var. tenuis Vasey
      var. virescens (Piper) Bowden
          SY=E. g. ssp. v. (Piper) Gould
          SY=E. v. Piper
  hirsutus Presl
  innovatus Beal
      var. innovatus
      var. velutinus (Bowden) Hulten ex
          Taylor & MacBryde
  junceus Fisch.

          SY=Psathyrostachys j. (Fisch.)
            Nevski
  X maltei Bowden [canadensis X virginicus]
          SY=E. arenarius L. var. simulans
            (Bowden) Boivin
          SY=E. X m. nm. brownii Bowden
          SY=E. X m. nm. churchii Bowden
          SY=E. X m. nm. s. Bowden
  mollis Trin.
    ssp. mollis
          SY=E. arenarius L. ssp. m. (Trin.)
            Hulten
          SY=E. a. var. m. (Trin.) Koidzumi
          SY=E. a. var. scabrinervis
            (Bowden) Boivin
    ssp. villosissimus (Scribn.) A. Love
  pacificus Gould
  riparius Wieg.
          SY=Elymus canadensis L. var. r.
            (Wieg.) Boivin
  salina M.E. Jones
          SY=Elymus ambiguus Vasey & Scribn.
            var. s. (M.E. Jones) C.L.
            Hitchc.
  sibiricus L.
  simplex Scribn. & Williams
          SY=Elymus triticoides Buckl. var.
            s. (Scribn. & Williams) A.S.
            Hitchc.
  svensonii Church
  triticoides Buckl.
      var. pubescens A.S. Hitchc.
      var. triticoides
          SY=Elymus t. ssp. multiflorus
            Gould
  X uclueletensis Bowden [glaucus X mollis]
  X vancouverensis (Vasey) Bowden [mollis X
    triticoides]
          SY=Elymus X v. nm. californicus
            Bowden
          SY=E. X v. nm. crescentianus
            Bowden
  villosus Muhl. ex Willd.
          SY=Elymus arkansanus Scribn. &
            Ball
  virginicus L.
      var. halophilus (Bickn.) Wieg.
          SY=Elymus h. Bickn.
      var. jenkinsii Bowden
      var. submuticus Hook.
          SY=Elymus curvatus Piper
          SY=E. s. (Hook.) Smyth & Smyth
      var. virginicus
          SY=Elymus australis Scribn. & Ball
          SY=E. glabriflorus (Vasey) Scribn.
            & Ball
          SY=E. hirsutiglumis Scribn.
          SY=E. jejunus (Ramaley) Rydb.
          SY=E. striatus Willd.
          SY=E. v. var. a. (Scribn. & Ball)
            A.S. Hitchc.
          SY=E. v. var. g. (Vasey) Bush
          SY=E. v. var. intermedius (Vasey)
            Bush
          SY=E. v. var. j. (Ramaley) Bush

X Elysitanion Bowden

  X aristatum (Merr.) Bowden [Elymus cinereus
    X Sitanion hystrix]
          SY=Elymus X a. Merr.
          SY=Elymus glaucus Buckl. var. a.
            (Merr.) A.S. Hitchc.
  X hansenii (Scribn.) Bowden [Elymus glaucus
    X Sitanion hystrix or S. jubatum]
          SY=Sitanion anomalum J.G. Sm.
          SY=S. h. (Scribn.) J.G. Sm.

Enneapogon Desv. ex Beauv.

    desvauxii Beauv.

Eragrostis von Wolf

    atropioides Hbd.
    atrovirens (Desv.) Trin.
          SY=E. chariis sensu auctt. non
            (Schultes) A.S. Hitchc.
          SY=E. nutans sensu auctt. non
            (Retz.) Trin. ex Steud.
    bahiensis Schrad. ex Schultes
    barrelieri Daveau
    brownei (Kunth) Nees
    capillaris (L.) Nees
    cilianensis (All.) E. Mosher
          SY=E. major Host
          SY=E. megastachya (Koel.) Link
    ciliaris (L.) R. Br.
        var. brachystachya Boiss.
        var. ciliaris
        var. laxa Kuntze
    cumingii Steud.
          SY=E. simplex Scribn.
    curtipedicellata Buckl.
    curvula (Schrad.) Nees
          SY=E. chloromelas Steud.
    cyperoides (Thunb.) Beauv.
    deflexa A.S. Hitchc.
    echinochloidea Stapf
    elliottii S. Wats.
          SY=E. acuta A.S. Hitchc.
    elongata (Willd.) Jacq.
    erosa Scribn.
    fosbergii Whitney
    frankii C.A. Mey. ex Steud.
          SY=E. f. var. brevipes Fassett
    gangetica (Roxb.) Steud.
          SY=E. stenophylla sensu auctt. non
            Hochst. ex Miq.
    glomerata (Walt.) L.H. Dewey
    glutinosa (Sw.) Trin.
    grandis Hbd.
        var. grandis
        var. oligantha Hbd.
        var. polyantha Hbd.
    hirsuta (Michx.) Nees
        var. laevivaginata Fern.
    hosakai Deg.
    hypnoides (Lam.) B.S.P.
    intermedia A.S. Hitchc.
    lehmanniana Nees
    leptophylla A.S. Hitchc.
    leptostachya (R. Br.) Steud.
    lugens Nees
    lutescens Scribn.
    mauiensis A.S. Hitchc.
    mexicana (Hornem.) Link
    minor Host
          SY=E. eragrostis (L.) Beauv.
          SY=E. poaeoides Beauv. ex Roemer &
            Schultes
    monticola (Gaud.) Hbd.
    neomexicana Vasey
    niihauensis Whitney
    obtusiflora (Fourn.) Scribn.
    orcuttiana Vasey
    palmeri S. Wats.
    paupera Jedw.
    pectinacea (Michx.) Nees
          SY=E. caroliniana (Spreng.)
            Scribn.
          SY=E. diffusa Buckl.
          SY=E. purshii Hort. ex Schrad.
    pilosa (L.) Beauv.
        var. perplexa (L.H. Harvey) S.D. Koch

          SY=E. perplexa L.H. Harvey
        var. pilosa
          SY=E. multicaulis Steud.
    plana Nees
    prolifera (Sw.) Steud.
          SY=E. domingensis (Pers.) Steud.
    refracta (Muhl.) Scribn.
    reptans (Michx.) Nees
          SY=E. weigeltiana (Reichenb. ex
            Trin.) Bush
          SY=Neeragrostis r. (Michx.) Nicora
    scaligera Salzm. ex Steud.
    secundiflora Presl
        ssp. oxylepis (Torr.) S.D. Koch
          SY=E. o. (Torr.) Torr.
        ssp. secundiflora
          SY=E. beyrichii J.G. Sm.
          SY=E. oxylepis (Torr.) Torr. var.
            b. (J.G. Sm.) Shinners
    sessilispica Buckl.
          SY=Acamptoclados s. (Buckl.) Nash
    silveana Swallen
    spectabilis (Pursh) Steud.
          SY=E. s. var. sparsihirsuta Farw.
    spicata Vasey
    suaveolens Becker ex Claus
    superba Peyr.
    swallenii A.S. Hitchc.
    teff (Zuccagni) Trotter
          SY=E. abyssinica (Jacq.) Link
    tenella (L.) Beauv. ex Roemer & Schultes
          SY=E. amabilis (L.) Wight & Arn.
    tephrosanthos Schultes
          SY=E. arida A.S. Hitchc.
    tracyi A.S. Hitchc.
    trichocolea Hack. & Arech.
        var. floridana (A.S. Hitchc.)
          Witherspoon
          SY=E. f. A.S. Hitchc.
        var. trichocolea
    trichodes (Nutt.) Wood
        var. pilifera (Scheele) Fern.
          SY=E. p. Scheele
        var. trichodes
    unioloides (Retz.) Nees
    urbaniana A.S. Hitchc.
    variabilis (Gaud.) Hbd.
    virescens Presl
    whitneyi Fosberg
        var. caumii Fosberg
        var. whitneyi

Eremochloa Buse

    ophiuroides (Munro) Hack.

Eremopoa Rosh.

    persica (Trin.) Rosh.

Eremopyrum (Ledeb.) Jaubert & Spach

    orientale (L.) Jaubert & Spach
          SY=Agropyron o. (L.) Roemer &
            Schultes
    squarrosum (Link) Jaubert & Spach
          SY=Agropyron s. (Link) Roth
    triticeum (Gaertn.) Nevski
          SY=Agropyron t. Gaertn.

Erianthus Michx.

    alopecuroides (L.) Ell.
          SY=E. a. var. hirsutis Nash
          SY=E. divaricatus (L.) A.S.
            Hitchc.
    brevibarbis Michx.

Erianthus (CONT.)
  coarctatus Fern.
        var. coarctatus
        var. elliottianus Fern.
  contortus Baldw.
  giganteus (Walt.) Muhl.
        SY=E. saccharoides Michx.
        SY=E. s. var. compactus (Nash)
                Fern.
  laxus Nash
  ravennae (L.) Beauv.
        var. purpurascens (Anderss.) Hack.
        var. ravennae
  strictus Baldw.
  tracyi Nash

Eriochloa H.B.K.

  aristata Vasey
  contracta A.S. Hitchc.
  lemmonii Vasey & Scribn.
        var. gracilis (Fourn.) Gould
           SY=E. g. (Fourn.) A.S. Hitchc.
        var. lemmonii
           SY=E. g. var. minor (Vasey) A.S.
                Hitchc.
  michauxii (Poir.) A.S. Hitchc.
        var. michauxii
        var. simpsonii A.S. Hitchc.
  polystachya H.B.K.
           SY=E. subglabra (Nash) A.S.
                Hitchc.
  procera (Retz.) C.E. Hubbard
  punctata (L.) Desv. ex Hamilton
  sericea (Scheele) Munro
  villosa (Thunb.) Kunth

Eriochrysis Beauv.

  cayennensis Beauv.

Erioneuron Nash

  grandiflorum (Vasey) Tateoka
           SY=E. avenaceum (H.B.K.) Tateoka
                var. g. (Vasey) Gould
           SY=E. a. var. nealleyi (Vasey)
                Gould
           SY=E. n. (Vasey) Tateoka
           SY=Tridens n. (Vasey) Woot. &
                Standl.
  pilosum (Buckl.) Nash
           SY=Tridens p. (Buckl.) A.S.
                Hitchc.
  pulchellum (H.B.K.) Tateoka
           SY=Tridens p. (H.B.K.) A.S.
                Hitchc.
           SY=Triodia p. H.B.K.

Euclasta Franch.

  condylotricha (Hochst. ex Steud.) Stapf
           SY=Andropogon c. Hochst. ex Steud.

Eustachys Desv.

  distichophylla (Lag.) Nees
           SY=Chloris d. Lag.
  floridana Chapman
           SY=Chloris f. (Chapman) Wood
  glauca Chapman
           SY=Chloris g. (Chapman) Wood
  neglecta (Nash) Nash
           SY=Chloris n. Nash
  petraea (Sw.) Desv.
           SY=Chloris p. Sw.
  retusa (Lag.) Kunth

           SY=Chloris argentina (Hack.) Lillo
                & Parodi

Festuca L.

  alaskana (Holmen) Krajina ined.
           SY=F. ovina L. ssp. a. Holmen
           SY=F. o. var. a. (Holmen) Welsh
  altaica Trin.
     ssp. altaica
     ssp. scabrella (Torr. & Hook.) Hulten
           SY=F. a. var. major (Vasey)
                Gleason
           SY=F. hallii (Vasey) Piper
           SY=F. s. Torr. & Hook.
           SY=F. s. var. m. Vasey
  amethystina L.
  arizonica Vasey
  arundinacea Schreb.
           SY=F. elatior L. var. a. (Schreb.)
                Wimmer
  baffinensis Polunin
  brachyphylla Schultes
        var. brachyphylla
           SY=F. brevifolia R. Br.
           SY=F. ovina L. var. b. (Schultes)
                Piper
           SY=F. o. var. brevifolia (R. Br.)
                S. Wats.
        var. groenlandica Schol.
  californica Vasey
        var. californica
        var. parishii (Piper) A.S. Hitchc.
  dasyclada Hack. ex Beal
  drymeia Mert. & Koch
  duriuscula L.
        var. cinerea (Vill.) Krajina
           SY=F. c. Vill.
           SY=F. glauca Lam.
           SY=F. ovina L. var. g. (Lam.)
                W.D.J. Koch
        var. duriuscula
           SY=F. ovina L. var. d. (L.) W.D.J.
                Koch
  elatior L.
           SY=F. pratensis Huds.
  elmeri Scribn. & Merr.
        var. conferta (Hack. ex Beal) A.S.
                Hitchc.
           SY=F. e. ssp. luxurians Piper
        var. elmeri
  gigantea (L.) Vill.
  hawaiiensis A.S. Hitchc.
  hyperborea Holmen
  idahoensis Elmer
           SY=F. i. var. oregona (Hack. ex
                Beal) C.L. Hitchc.
           SY=F. ingrata (Hack. ex Beal)
                Rydb.
           SY=F. occidentalis Hook. var.
                ingrata (Hack. ex Beal)
                Boivin
           SY=F. ovina L. var. columbiana
                Beal
           SY=F. ovina var. ingrata Hack. ex
                Beal
           SY=F. ovina var. oregona Hack. ex
                Beal
  kashmiriana Stapf
  ligulata Swallen
  obtusa Biehler
  occidentalis Hook.
           SY=F. ovina L. var. polyphylla
                Vasey ex Beal
  ovina L.
  paradoxa Desv.
           SY=F. nutans Biehler

Festuca (CONT.)
          SY=F. shortii Kunth ex Wood
     rubra L.
          var. arenaria (Osbeck) Fries
               SY=F. richardsonii Hook ssp.
                    cryophila (Krecz. & Bobr.)
                    Love & Love
               SY=F. rubra ssp. richardsonii
                    (Hook.) Hulten
               SY=F. rubra ssp. lanuginosa Mert.
                    & Koch
               SY=F. rubra var. littoralis Vasey
                    ex Beal
          var. commutata Gaudin
               SY=F. fallax Thuill.
          var. heterophylla (Lam.) Mutel
               SY=F. h. Lam.
          var. juncea (Hack.) Richter
          var. multiflora (Hoffmann) Asch. &
               Graebn.
               SY=F. aucta Krecz. & Bobr.
               SY=F. m. Hoffmann
               SY=F. r. ssp. a. (Krecz. & Bobr.)
                    Hulten
          var. mutica Hartman
          var. prolifera Piper
               SY=F. p. (Piper) Fern.
               SY=F. p. var. lasiolepis Fern.
          var. rubra
               SY=F. rubra ssp. pruinosa (Hack.)
                    Piper
     saximontana Rydb.
               SY=F. ovina L. ssp. montana
                    (Rydb.) St.-Yves var.
                    rydbergii St.-Yves
               SY=F. o. var. s. (Rydb.) Gleason
     sororia Piper
     subulata Trin.
     subuliflora Scribn.
     tenuifolia Sibthorp
               SY=F. capillata Lam.
               SY=F. ovina L. var. c. (Lam.)
                    Alef.
     thurberi Vasey
     valesiaca Schleich. ex Gaudin
     versuta Beal
     viridula Vasey
               SY=F. howellii Hack. ex Beal
     X viviparoidea Krajina ined. [baffinensis X
          brachyphylla]

Garnotia Brongn.

     acutigluma (Steud.) Ohwi
               SY=G. sandwicensis Hbd.

Gastridium Beauv.

     ventricosum (Gouan) Schinz & Thellung

Glyceria R. Br.

     acutiflora Torr.
               SY=Panicularia a. (Torr.) Kuntze
     arkansana Fern.
               SY=G. septentrionalis A.S. Hitchc.
                    var. a. (Fern.) Steyermark &
                    Kucera
     borealis (Nash) Batchelder
               SY=Panicularia b. Nash
     canadensis (Michx.) Trin.
          var. canadensis
               SY=Panicularia c. (Michx.) Kuntze
          var. laxa (Scribn.) A.S. Hitchc.
               SY=G. X l. (Scribn.) Scribn.
               SY=Panicularia l. Scribn.
     declinata Brebiss.

     elata (Nash) M.E. Jones
     fluitans (L.) R. Br.
               SY=Panicularia f. (L.) Kuntze
     X gatineauensis Bowden [melicaria X
          striata]
     grandis S. Wats.
          var. grandis
               SY=G. maxima (Hartman) Holmb. ssp.
                    g. (S. Wats.) Hulten
               SY=G. m. var. americana (Torr.)
                    Boivin
               SY=Panicularia g. (S. Wats.) Nash
          var. komarovii L. Kelso
     leptostachya Buckl.
     melicaria (Michx.) F. T. Hubbard
               SY=Panicularia m. (Michx.) A.S.
                    Hitchc.
     nubigena W.A. Anderson
     obtusa (Muhl.) Trin.
               SY=Panicularia o. (Muhl.) Kuntze
     occidentalis (Piper) J.C. Nels.
     otisii A.S. Hitchc.
     plicata T. Fries
     pulchella (Nash) K. Schum.
     septentrionalis A.S. Hitchc.
               SY=Panicularia s. (A.S. Hitchc.)
                    Bickn.
     spectabilis Mert. & Koch
     striata (Lam.) A.S. Hitchc.
               SY=G. nervata (Willd.) Trin.
               SY=G. s. ssp. stricta (Scribn.)
                    Hulten
               SY=G. stricta (Scribn.) Fern.
               SY=Panicularia n. (Willd.) Kuntze
               SY=P. striata (Lam.) A.S. Hitchc.

Gymnopogon Beauv.

     ambiguus (Michx.) B.S.P.
     brevifolius Trin.
     chapmanianus A.S. Hitchc.
     floridanus Swallen
     foliosus (Willd.) Nees

Gynerium Willd. ex Beauv.

     sagittatum (Aubl.) Beauv.

Hackelochloa Kuntze

     granularis (L.) Kuntze
               SY=Rytilix g. (L.) Skeels

Helictotrichon Bess. ex Schultes

     dahuricum (Komarov) Kitagawa
     mortonianum (Scribn.) Henr.
     pubescens (Huds.) Pilger
               SY=Avena p. Huds.

Hemarthria R. Br.

     altissima (Poir.) Stapf & C.E. Hubbard
               SY=Manisuris a. (Poir.) A.S.
                    Hitchc.

Heteropogon Pers.

     contortus (L.) Beauv. ex Roemer & Schultes
     melanocarpus (Ell.) Benth.

Hierochloe R. Br.

     alpina (Willd.) Roemer & Schultes
          ssp. alpina
               SY=Savastana a. (Willd.) Scribn.
          ssp. orthantha (Sorensen) G. Weim.

Hierochloe   (CONT.)
                SY=H. a. var. o. (Sorensen) Hulten
                SY=H. monticola (Bigelow) Love &
                     Love
                SY=H. o. Sorensen
    hirta (Schrank) Borbas
      ssp. arctica (Presl) G. Weim.
      ssp. hirta
                SY=H. odorata (L.) Beauv. ssp. h.
                     Schrank
    occidentalis Buckl.
                SY=H. macrophylla Thurb. ex
                     Boland.
    odorata (L.) Beauv.
                SY=H. nashii (Bickn.) Kaczmarek
                SY=H. o. var. fragrans (Willd.)
                     Richter
                SY=Savastana n. Bickn.
                SY=S. o. (L.) Scribn.
                SY=Torresia o. (L.) A.S. Hitchc.
    pauciflora R. Br.
                SY=Savastana p. (R. Br.) Scribn.

Hilaria H.B.K.

    belangeri (Steud.) Nash
          var. belangeri
          var. longifolia (Vasey) A.S. Hitchc.
    jamesii (Torr.) Benth.
    mutica (Buckl.) Benth.
    rigida (Thurb.) Benth. ex Scribn.
    swallenii Cory

Holcus L.

    lanatus L.
                SY=Notholcus l. (L.) Nash
    mollis L.

Hordeum L.

    arizonicum Covas
    brachyantherum Nevski
                SY=H. boreale Scribn. & Sm.
                SY=H. jubatum L. var. boreale
                     (Scribn. & Sm.) Boivin
                SY=H. nodosum L. var. boreale
                     (Scribn. & Sm.) A.S. Hitchc.
    californicum Covas & Stebbins
    depressum (Scribn. & Sm.) Rydb.
    geniculatum All.
                SY=H. gussonianum Parl.
                SY=H. hystrix Roth
    glaucum Steud.
                SY=H. stebbinsii Covas
    jubatum L.
      ssp. breviaristatum Bowden
      ssp. X intermedium Bowden
      ssp. jubatum
                SY=H. caespitosum Scribn.
                SY=H. j. var. c. (Scribn.) A.S.
                     Hitchc.
    leporinum Link
                SY=H. murinum L. ssp. l. (Link)
                     Aschers. & Graebn.
    marinum Huds.
    murinum L.
    pusillum Nutt.
          var. pubens A.S. Hitchc.
          var. pusillum
    vulgare L.
          var. trifurcatum (Schlecht.) Alef.
          var. vulgare
                SY=H. distichon L.

Hymenachne Beauv.

    amplexicaulis (Rudge) Nees

Hyparrhenia Anderss. ex Stapf

    hirta (L.) Stapf
    rufa (Nees) Stapf

Hystrix Moench

    californica (Boland. ex Thurb.) Kuntze
    patula Moench
          var. bigeloviana (Fern.) Deam
                SY=Elymus hystrix L. var. b.
                     (Fern.) Bowden
          var. patula
                SY=Elymus hystrix L.
                SY=H. h. (L.) Millsp.

Ichnanthus Beauv.

    axillaris (Nees) A.S. Hitchc. & Chase
    nemorosus (Sw.) Doell
    pallens (Sw.) Munro

Imperata Cyrillo

    brasiliensis Trin.
    brevifolia Vasey
                SY=I. hookeri (Anderss.) Rupr. ex
                     Hack.
    contracta (H.B.K.) A.S. Hitchc.
    cylindrica (L.) Beauv.

Isachne R. Br.

    angustifolia Nash
    distichophylla Munro ex Hbd.
    pallens Hbd.

Ischaemum L.

    byrone (Trin.) A.S. Hitchc.

Koeleria Pers.

    asiatica Domin
    cristata (L.) Pers.
                SY=K. c. var. longifolia Vasey ex
                     Burtt-Davy
                SY=K. c. var. pinetorum Abrams
                SY=K. gracilis Pers.
                SY=K. macrantha (Ledeb.) Schultes
                SY=K. nitida Nutt.
                SY=K. pyramidata (Lam.) Beauv.
    gerardii (Vill.) Shinners
                SY=K. phleoides (Vill.) Pers.

Lagurus L.

    ovatus L.

Lamarckia Moench

    aurea (L.) Moench

Lasiacis (Griseb.) A.S. Hitchc.

    divaricata (L.) A.S. Hitchc.
    harrisii Nash
    ligulata A.S. Hitchc. & Chase
    sloanei (Griseb.) A.S. Hitchc.
    sorghoidea (Desv.) A.S. Hitchc. & Chase
          var. patentiflora (A.S. Hitchc. &
               Chase) Davidse
                SY=L. p. A.S. Hitchc. & Chase
          var. sorghoidea
                SY=L. maculata (Aubl.) Urban

Leersia Sw.

   hexandra Sw.
         SY=Homalocenchrus h. (Sw.) Kuntze
   lenticularis Michx.
         SY=Homalocenchrus l. (Michx.)
             Kuntze
   monandra Sw.
         SY=Homalocenchrus m. (Sw.) Kuntze
   oryzoides (L.) Sw.
         SY=Homalocenchrus o. (L.) Pollard
   virginica Willd.
         SY=Homalocenchrus v. (Willd.)
             Britt.
         SY=L. v. var. ovata (Poir.) Fern.

Leptochloa Beauv.

   chloridiformis (Hack.) Parodi
   domingensis (Jacq.) Trin.
   filiformis (Lam.) Beauv.
         SY=L. attenuata (Nutt.) Steud.
         SY=L. f. var. a. (Nutt.)
             Steyermark & Kucera
   nealleyi Vasey
   panicea (Retz.) Ohwi
   scabra Nees
   virgata (L.) Beauv.

Leptochloopsis Yates

   virgata (Poir.) Yates
         SY=Uniola v. (Poir.) Griseb.

Leptocoryphium Nees

   lanatum (H.B.K.) Nees

Lepturus R. Br.

   filiformis (Roth) Trin.
   repens (Forst. f.) R. Br.
         var. cinereus (Burcham) Fosberg
         SY=L. c. Burcham
         var. occidentalis Fosberg
         var. repens
         var. subulatus Fosberg

Leucopoa Griseb.

   kingii (S. Wats.) W.A. Weber
         SY=Festuca confinis Vasey
         SY=F. k. S. Wats.
         SY=Hesperochloa k. (S. Wats.)
             Rydb.

Limnodea L.H. Dewey

   arkansana (Nutt.) L.H. Dewey

Lithachne Beauv.

   pauciflora (Sw.) Poir.

Lolium L.

   multiflorum Lam.
         SY=L. m. var. diminutum Mutel
         SY=L. m. var. ramosum Guss. ex
             Arcang.
         SY=L. m. var. muticum DC.
         SY=L. perenne L. ssp. multiflorum
             (Lam.) Husnot
   perenne L.
         SY=L. p. var. aristatum Willd.
         SY=L. p. var. cristatum Pers. ex
             Jackson

         persicum Boiss. & Hohen. ex Boiss.
         SY=L. dorei Boivin
         SY=L. d. var. laeve Boivin
   rigidum Gaudin
         SY=L. strictum Presl
   subulatum Vis.
   temulentum L.
         SY=L. t. var. arvense (With.) Bab.
         SY=L. t. var. leptochaeton A.
             Braun
         SY=L. t. var. macrochaeton A.
             Braun

Luziola Juss.

   bahiensis (Steud.) A.S. Hitchc.
   fluitans (Michx.) Terrell & H. Robins.
         SY=Hydrochloa caroliniensis Beauv.
   peruviana J.F. Gmel.

Lycurus H.B.K.

   phleoides H.B.K.

Melica L.

   altissima L.
   aristata Thurb. ex Boland.
   bulbosa Geyer, Porter & Coult.
         SY=M. bella Piper
         SY=M. bella ssp. intonsa Piper
   californica Scribn.
      var. californica
      var. nevadensis Boyle
   frutescens Scribn.
   fugax Boland.
         SY=M. f. ssp. madophylla Piper
         SY=M. f. var. inexpansa Suksdorf
         SY=M. f. var. macbridei (Rowland
             ex A. Nels.) Beetle
   geyeri Munro
      var. aristulata J.T. Howell
      var. geyeri
   harfordii Boland.
         SY=M. h. var. minor Vasey
   imperfecta Trin.
      var. flexuosa Boland.
      var. imperfecta
      var. minor Scribn.
      var. refracta Thurb.
   inflata (Boland.) Vasey
         SY=M. bulbosa Geyer, Porter &
             Coult. var. i. (Boland.)
             Boyle
   montezumae Piper
   mutica Walt.
   nitens (Scribn.) Nutt. ex Piper
   porteri Scribn.
      var. laxa Boyle
      var. porteri
   smithii (Porter ex Gray) Vasey
         SY=Avena s. Porter ex Gray
   spectabilis Scribn.
   stricta Boland.
         SY=M. s. var. albicaulis Boyle
   subulata (Griseb.) Scribn.
      var. pammelii (Scribn.) C.L. Hitchc.
      var. subulata
   torreyana Scribn.

Melinis Beauv.

   minutiflora Beauv.

Mibora Adans.

   minima (L.) Desv.

Microchloa R. Br.

   kunthii Desv.

Microlaena R. Br.

   stipoides (Labill.) R. Br.

Microstegium Nees

   vimineum (Trin.) A. Camus
       SY=Eulalia v. (Trin.) Kuntze
       SY=E. v. var. variabilis Kuntze
       SY=M. vimineum var. imberbe (Nees)
          Honda

Milium L.

   effusum L.
      var. cisatlanticum Fern.
      var. effusum

Miscanthus Anderss.

   floridulus (Lab.) Warb. ex Schum. & Laut.
   nepalensis (Trin.) Hack.
   sacchariflorus (Maxim.) Hack.
   sinensis Anderss.
       SY=M. s. var. variegatus Beal
       SY=M. s. var. zebrinus Beal

Molinia Schrank

   caerulea (L.) Moench
       SY=Aira c. L.

Monanthochloe Engelm.

   littoralis Engelm.

Monerma Beauv.

   cylindrica (Willd.) Coss. & Durieu
       SY=Lepturus c. (Willd.) Trin.

Muhlenbergia Schreb.

   andina (Nutt.) A.S. Hitchc.
       SY=M. comata (Thurb.) Thurb. ex
          Benth.
   appressa C.O. Goodding
   arenacea (Buckl.) A.S. Hitchc.
   arenicola Buckl.
   arizonica Scribn.
   arsenei A.S. Hitchc.
   asperifolia (Nees & Meyen) Parodi
       SY=Sporobolus a. (Nees & Meyen)
          Nees
   brevis C.O. Goodding
   bushii Pohl
       SY=M. brachyphylla Bush
   californica Vasey
   capillaris (Lam.) Trin.
   curtifolia Scribn.
   X curtisetosa (Scribn.) Pohl [schreberi X
     frondosa]
       SY=M. schreberi J.F. Gmel. var. c.
         (Scribn.) Steyermark &
         Kucera
   cuspidata (Torr.) Rydb.
       SY=M. brevifolia (Nutt.) M.E.
         Jones
   depauperata Scribn.
   dubia Fourn.
   dubioides C.O. Goodding
   dumosa Scribn. ex Vasey
   eludens C.G. Reeder

   emersleyi Vasey
   expansa (Poir.) Trin.
   filiculmis Vasey
   filiformis (Thurb.) Rydb.
       SY=M. f. var. fortis E.H. Kelso
       SY=M. simplex (Scribn.) Rydb.
   filipes M.A. Curtis
       SY=M. capillaris (Lam.) Trin. var.
         f. (M.A. Curtis) Chapman ex
         Beal
   fragilis Swallen
   frondosa (Poir.) Fern.
       SY=M. commutata (Scribn.) Bush
   glabriflora Scribn.
   glauca (Nees) B.D. Jackson
       SY=M. lemmonii Scribn.
   glomerata (Willd.) Trin.
       SY=M. g. var. cinnoides (Link)
         F.J. Herm.
       SY=M. racemosa (Michx.) B.S.P.
         var. c. (Link) Boivin
   gooddingii Soderstrom
   involuta Swallen
   jonesii (Vasey) A.S. Hitchc.
   lindheimeri A.S. Hitchc.
   longiligula A.S. Hitchc.
   metcalfei M.E. Jones
   mexicana (L.) Trin.
       SY=M. ambigua Torr.
       SY=M. foliosa (Roemer & Schultes)
         Trin.
       SY=M. f. ssp. a. (Torr.) Scribn.
       SY=M. f. ssp. setiglumis (S.
         Wats.) Scribn.
   microsperma (DC.) Kunth
   minutissima (Steud.) Swallen
       SY=Sporobolus confusus sensu
         auctt. non (Fourn.) Vasey
   montana (Nutt.) A.S. Hitchc.
   monticola Buckl.
   parviglumis Vasey
   pauciflora Buckl.
   pectinata C.O. Goodding
   polycaulis Scribn.
   porteri Scribn. ex Beal
   pulcherrima Scribn.
   pungens Thurb.
   racemosa (Michx.) B.S.P.
   repens (Presl) A.S. Hitchc.
   reverchonii Vasey & Scribn.
   richardsonis (Trin.) Rydb.
       SY=M. squarrosa (Trin.) Rydb.
   rigens (Benth.) A.S. Hitchc.
       SY=Epicampes r. Benth.
       SY=M. marshii I.M. Johnston
       SY=M. mundula I.M. Johnston
   rigida (H.B.K.) Kunth
   schreberi J.F. Gmel.
       SY=M. palustris Scribn.
       SY=M. s. var. p. (Scribn.) Scribn.
   setifolia Vasey
   sinuosa Swallen
   sobolifera (Muhl.) Trin.
       SY=M. s. var. setigera Scribn.
   sylvatica (Torr.) Torr. ex Gray
      var. robusta Fern.
      var. sylvatica
       SY=M. umbrosa Scribn.
   tenuiflora (Willd.) B.S.P.
      var. tenuiflora
      var. variabilis (Scribn.) Pohl
   tenuifolia (H.B.K.) Kunth
   texana Buckl.
   thurberi Rydb.
   torreyana (Schultes) A.S. Hitchc.
       SY=Sporobolus t. (Schultes) Nash
   torreyi (Kunth) A.S. Hitchc. ex Bush

Muhlenbergia (CONT.)
        SY=M. gracillima Torr.
  uniflora (Muhl.) Fern.
        SY=M. u. var. terrae-novae Fern.
        SY=Sporobolus u. (Muhl.) Scribn. &
            Merr.
  utilis (Torr.) A.S. Hitchc.
  villosa Swallen
  virescens (H.B.K.) Kunth
  wolfii (Vasey) Rydb.
        SY=Sporobolus ramulosus sensu
            auctt. non (H.B.K.) Kunth
  wrightii Vasey
  xerophila C.O. Goodding

Munroa Torr.

  squarrosa (Nutt.) Torr.
        SY=M. s. var. floccuosa Vasey ex
            Beal

Nardus L.

  stricta L.

Nassella (Trin.) Desv.

  chilensis (Trin.) E. Desv.

Neostapfia Burtt-Davy

  colusana (Burtt-Davy) Burtt-Davy
        SY=Anthochloa c. (Burtt-Davy)
            Scribn.

Neyraudia Hook. f.

  reynaudiana (Kunth) Keng

Olyra L.

  latifolia L.

Oplismenus Beauv.

  hirtellus (L.) Beauv.
    ssp. hirtellus
        SY=O. burmannii sensu auctt. non
            (Retz.) Beauv.
    ssp. setarius (Lam.) Mez ex Ekman
        SY=O. s. (Lam.) Roemer & Schultes

Orcuttia Vasey

  californica Vasey
    var. californica
        SY=O. c. var. inaequalis (Hoover)
            Hoover
    var. viscida Hoover
  greenei Vasey
  mucronata Crampton
  pilosa Hoover
  tenuis A.S. Hitchc.

Oryza L.

  latifolia Desv.
  sativa L.

Oryzopsis Michx.

  asperifolia Michx.
  bloomeri (Boland.) Ricker
  canadensis (Poir.) Torr.
        SY=Stipa c. Poir.
  exigua Thurb.
  hymenoides (Roemer & Schultes) Ricker

    var. contracta B.L. Johnson
        SY=O. c. (B.L. Johnson) Schlechter
    var. hymenoides
        SY=Eriocoma cuspidata Nutt.
        SY=Stipa h. Roemer & Schultes
  kingii (Boland.) Beal
  micrantha (Trin. & Rupr.) Thurb.
  miliacea (L.) Benth. & Hook. ex Aschers. &
    Schweinf.
  pungens (Torr. ex Spreng.) A.S. Hitchc.
  racemosa (Sm.) Ricker
  swallenii C.L. Hitchc. & Spellenberg

Panicum L.

  abscissum Swallen
  alakaiense Skottsberg
  albomarginatum Nash
  amarum Ell.
    var. amarulum (A.S. Hitchc. & Chase)
      P.G. Palmer
        SY=P. amarulum A.S. Hitchc. &
            Chase
    var. amarum
  anceps Michx.
  antidotale Retz.
  bartowense Scribn. & Merr.
  beecheyi Hook. & Arn.
  bergii Arech.
  boliviense Hack.
  brachyanthum Steud.
  bulbosum H.B.K.
        SY=P. b. var. minus Vasey
  capillare L.
    var. capillare
        SY=P. c. var. agreste Gattinger
    var. occidentale Rydb.
        SY=P. barbipulvinatum Nash
  capillarioides Vasey
  carteri Hosaka
  chapmanii Vasey
  coerulescens Hack.
  colliei Endl.
  coloratum L.
  combsii Scribn. & Ball
        SY=P. longifolium Torr. var. c.
            (Scribn. & Ball) Fern.
  degeneri Potztal
  dichotomiflorum Michx.
    var. dichotomiflorum
        SY=P. d. var. geniculatum (Wood)
            Fern.
        SY=P. d. var. imperiorum Fern.
    var. puritanorum Svens.
  diffusum Sw.
  elephantipes Nees
  fauriei A.S. Hitchc.
  flexile (Gattinger) Scribn.
  gattingeri Nash
        SY=P. capillare L. var. campestre
            Gattinger
  ghiesbreghtii Fourn.
  glutinosum Sw.
  gouinii Fourn.
  gymnocarpon Ell.
  hallii Vasey
    var. filipes (Scribn.) Waller
        SY=P. f. Scribn.
    var. hallii
  havaiense Reichardt
  havardii Vasey
  hemitomon Schultes
  heupueo St. John
  hians Ell.
        SY=Steinchisma h. (Ell.) Nash
  hillmanii Chase
  hirstii Swallen

Panicum (CONT.)
  hirsutum Sw.
  hirticaule Presl
          SY=P. pampinosum A.S. Hitchc. &
              Chase
  konaense Whitney & Hosaka
  lacustre A.S. Hitchc. & Ekman
  lamiatile St. John
  laxum Sw.
  lepidulum A.S. Hitchc. & Chase
  lithophilum Swallen
  maximum Jacq.
  miliaceum L.
  molokaiense Deg. & Whitney
  moomomiense St. John
  nephelophilum Gaud.
          var. nephelophilum
          var. rhyacophilum Hbd.
  niihauense St. John
  nubigenum Kunth
  obtusum H.B.K.
  parvifolium Lam. ex Spreng.
  pellitoides F. Br. & St. John
  pellitum Trin.
  philadelphicum Bernh. ex Trin.
  pilcomayense Hack.
  plenum A.S. Hitchc. & Chase
  pseudagrostis Trin.
  ramosius A.S. Hitchc.
  ramosum L.
  repens L.
  rhizomatum A.S. Hitchc. & Chase
          SY=P. anceps Michx. var. r. (A.S.
              Hitchc. & Chase) Fern.
  rigidulum Bosc ex Nees
          SY=P. agrostoides Spreng.
          SY=P. a. var. condensum (Nash)
              Fern.
          SY=P. a. var. ramosius (C. Mohr)
              Fern.
          SY=P. c. Nash
          SY=P. longifolium Torr.
          SY=P. l. var. pubescens (Vasey)
              Fern.
          SY=P. l. var. tusketense Fern.
          SY=P. r. var. c. (Nash)
              Mohlenbrock
  schiffneri Hack.
  sonorum Beal
  stevensianum A.S. Hitchc. & Chase
  stipitatum Nash
          SY=P. agrostoides Spreng. var.
              elongatum Scribn.
  stramineum A.S. Hitchc. & Chase
  tenerum Bey. ex Trin.
          SY=P. stenodes sensu auctt. non
              Griseb.
  tenuifolium Hook. & Arn.
  torridum Gaud.
  trichanthum Nees
  trichoides Sw.
  tuckermanii Fern.
          SY=P. philadelphicum Bernh. ex
              Trin. var. t. (Fern.)
              Steyermark & Schmoll
  urvilleanum Kunth
  verrucosum Muhl.
  virgatum L.
          var. cubense Griseb.
              SY=P. v. var. obtusum Wood
          var. spissum Linder
          var. virgatum
  xerophilum (Hbd.) A.S. Hitchc.

Pappophorum Schreb.

  bicolor Fourn.

pappiferum (Lam.) Kuntze
        SY=P. alopecuroideum Vahl
vaginatum Buckl.
        SY=P. mucronulatum sensu auctt.
            non Nees

Parapholis C.E. Hubbard

  incurva (L.) C.E. Hubbard
          SY=Pholiurus i. (L.) Schinz &
              Thellung

Paspalidium Stapf

  geminatum (Forsk.) Stapf
          var. geminatum
              SY=Panicum g. Forsk.
          var. paludivagum (A.S. Hitchc. &
              Chase) Gould
              SY=Panicum p. A.S. Hitchc. & Chase
              SY=Paspalidium p. (A.S. Hitchc. &
                  Chase) Parodi
  radiatum Vickery
          SY=Panicum r. (Vickery) St. John

Paspalum L.

  acuminatum Raddi
  almum Chase
  bifidum (Bertol.) Nash
          SY=P. b. var. projectum Fern.
  blodgettii Chapman
          SY=P. simpsonii Nash
  boscianum Flugge
  caespitosum Flugge
          SY=P. poiretii Roemer & Schultes
  clavuliferum C. Wright
  conjugatum Berg.
  convexum Humb. & Bonpl. ex Flugge
  decumbens Sw.
  densum Poir.
  difforme Le Conte
  dilatatum Poir.
  dissectum (L.) L.
  distichum L.
          var. distichum
              SY=P. vaginatum Sw.
          var. indutum Shinners
  fimbriatum H.B.K.
  floridanum Michx.
          var. floridanum
          var. glabratum Engelm. ex Vasey
              SY=P. g. (Engelm. ex Vasey) C.
                  Mohr
  fluitans (Ell.) Kunth
          SY=P. mucronatum Muhl.
          SY=P. repens sensu auctt. non
              Berg.
  giganteum Baldw. ex Vasey
  hartwegianum Fourn.
  hydrophilum Henr.
  intermedium Munro ex Morong
  laeve Michx.
          var. circulare (Nash) Fern.
              SY=P. c. Nash
          var. laeve
          var. pilosum Scribn.
              SY=P. longipilum Nash
              SY=P. plenipilum Nash
  langei (Fourn.) Nash
  laxum Lam.
          SY=P. glabrum Poir.
  lividum Trin.
  malacophyllum Trin.
  melanospermum Desv.
  millegrana Schrad.
  minus Fourn.

Paspalum (CONT.)
  molle Poir.
        SY=P. portoricense Nash
  monostachyum Vasey
  notatum Flugge
      var. notatum
      var. saurae Parodi
  orbiculare Forst. f.
  orbiculatum Poir.
  paniculatum L.
  parviflorum Rohde ex Flugge
  paspaloides (Michx.) Scribn.
        SY=Digitaria p. Michx.
        SY=P. distichum sensu auctt. non
          L.
  paucispicatum Vasey
  pleostachyum Doell
  plicatulum Michx.
        SY=P. texanum Swallen
  praecox Walt.
        SY=P. lentiferum Lam.
        SY=P. p. var. curtisianum (Steud.)
          Vasey
  pubiflorum Rupr. ex Fourn.
      var. glabrum Vasey ex Scribn.
        SY=P. geminum Nash
        SY=P. laeviglume Scribn. ex Nash
      var. pubiflorum
  racemosum Lam.
  rupestre Trin.
        SY=P. leoninum Chase
  saugetii Chase
  scrobiculatum L.
  secans A.S. Hitchc. & Chase
  separatum Shinners
  setaceum Michx.
      var. ciliatifolium (Michx.) Vasey
        SY=P. c. Michx.
        SY=P. propinquum Nash
      var. longepedunculatum (Le Conte)
        Wood
        SY=P. l. Le Conte
      var. muhlenbergii (Nash) D. Banks
        SY=P. ciliatifolium Michx. var. m.
          (Nash) Fern.
        SY=P. m. Nash
        SY=P. pubescens Muhl.
        SY=P. s. var. calvescens Fern.
      var. psammophilum (Nash) D. Banks
        SY=P. p. Nash
      var. rigidifolium (Nash) D. Banks
        SY=P. r. Nash
      var. setaceum
        SY=P. debile Michx.
      var. stramineum (Nash) D. Banks
        SY=P. bushii Nash
        SY=P. ciliatifolium Michx. var.
          stramineum (Nash) Fern.
        SY=P. stramineum Nash
      var. supinum (Bosc ex Poir.) Trin.
        SY=P. s. Bosc ex Poir.
      var. villosissimum (Nash) D. Banks
  unispicatum (Scribn. & Merr.) Nash
  urvillei Steud.
        SY=P. larranagai Arech.
  virgatum L.
  virletii Fourn.

Pennisetum L.C. Rich. ex Pers.

  alopecuroides (L.) Spreng.
  americanum (L.) Leecke
        SY=P. glaucum (L.) R. Br. pro
          parte non L.
  clandestinum Hochst. ex Chiov.
  nervosum (Nees) Trin.
        SY=Cenchrus n. (Nees) Kuntze

  purpureum Schum.
  setaceum (Forsk.) Chiov.
  setosum (Sw.) L.C. Rich.
  villosum R. Br. ex Fresn.
        SY=Cenchrus longisetus M.C.
          Johnston

Phalaris L.

  angusta Nees ex Trin.
  aquatica L.
        SY=P. stenoptera Hack.
        SY=P. tuberosa L.
        SY=P. t. var. hirtiglumis Balt. &
          Trab
        SY=P. t. var. s. (Hack.) A.S.
          Hitchc.
  arundinacea L.
        SY=P. a. var. picta L.
  brachystachys Link
  californica Hook. & Arn.
  canariensis L.
  caroliniana Walt.
  lemmonii Vasey
  minor Retz.
  paradoxa L.
        SY=P. p. var. praemorsa (Lam.)
          Coss. & Durieu

Pharus P. Br.

  glaber H.B.K.
  latifolius L.
  parvifolius Nash

Phippsia (Trin.) R. Br.

  algida (Phipps) R. Br.
      var. algida
      var. algidiformis (H. Sm.) Boivin

Phleum L.

  alpinum L.
        SY=P. a. var. commutatum (Gaudin)
          Griseb.
        SY=P. c. Gaudin
        SY=P. c. var. americanum (Fourn.)
          Hulten
  arenarium L.
  graecum Boiss. ex Heldr.
  paniculatum Huds.
  pratense L.
      var. nodosum (L.) Huds.
      var. pratense
  subulatum (Savi) Aschers. & Graebn.
        SY=P. bellardii Willd.

Phragmites Adans.

  australis (Cav.) Trin. ex Steud.
        SY=P. communis (L.) Trin.
        SY=P. c. var. berlandieri (Fourn.)
          Fern.
        SY=P. phragmites (L.) Karst.

Phyllostachys Sieb. & Zucc.

  aurea A. & C. Riviere
  aureosulcata McClure
  bambusoides Sieb. & Zucc.
  nigra (Lodd.) Munro
      var. henonis (Mitf.) Stapf ex Rendle

Piptochaetium Presl

  fimbriatum (H.B.K.) A.S. Hitchc.

Piptochaetium (CONT.)
        SY=P. f. var. confine I.M.
           Johnston

Pleuropogon R. Br.

   californicus (Nees) Benth. ex Vasey
       SY=Lophochlaena c. Nees
   davyi L. Benson
       SY=Lophochlaena californicus Nees
          var. d. (L. Benson) Love &
          Love
   hooverianus (L. Benson) J.T. Howell
       SY=Lophochlaena refracta Gray var.
          h. (L. Benson) Love & Love
   oregonus Chase
   refractus (Gray) Benth. ex Vasey
       SY=Lophochlaena r. Gray
   sabinii R. Br.

Poa L.

   abbreviata R. Br.
     ssp. abbreviata
     ssp. jordalii (Porsild) Hulten
       SY=P. j. Porsild
   albescens A.S. Hitchc.
   alpina L.
      var. alpina
      var. vivipara L.
   alsodes Gray
   ampla Merr.
       SY=P. confusa Rydb.
   annua L.
      var. annua
      var. aquatica Aschers.
      var. reptans Hausskn.
   arachnifera Torr.
   arctica R. Br.
     ssp. arctica
       SY=P. aperta Scribn. & Merr.
       SY=P. arctica var. glabriflora
          Rosh.
       SY=P. arctica var. vivipara Hook.
       SY=P. cenisia All. var. arctica
          (R. Br.) Richter
       SY=P. longipila Nash
     ssp. caespitans Nannf.
       SY=P. arctica var. c. (Nannf.)
          Boivin
     ssp. grayana (Vasey) Love, Love & Kapoor
       SY=P. g. Vasey
     ssp. longiculmis Hulten
     ssp. williamsii (Nash) Hulten
       SY=P. w. Nash
   arida Vasey
       SY=P. pseudopratensis Scribn. &
          Rydb.
   atropurpurea Scribn.
   autumnalis Muhl. ex Ell.
   bigelovii Vasey & Scribn.
   bolanderi Vasey
   brachyanthera Hulten
   bulbosa L.
   canbyi (Scribn.) Piper
       SY=P. laevigata Scribn.
   chaixii Vill.
   chapmaniana Scribn.
   compressa L.
   confinis Vasey
   curta Rydb.
   curtifolia Scribn.
   cusickii Vasey
     ssp. cusickii
     ssp. pubens Keck
   cuspidata Nutt.
       SY=P. brachyphylla Schultes

douglasii Nees
   ssp. douglasii
   ssp. macrantha (Vasey) Keck
       SY=P. d. var. m. (Vasey) Boivin
       SY=P. m. Vasey
eminens J. Presl
epilis Scribn.
       SY=P. cusickii Vasey var. e.
          (Scribn.) C.L. Hitchc.
       SY=P. c. var. purpurascens (Vasey)
          C.L. Hitchc.
eyerdamii Hulten
fendleriana (Steud.) Vasey
   var. fendleriana
   var. longiligula (Scribn. & Williams)
     Gould
       SY=P. l. Scribn. & Williams
   var. wyomingensis T.A. Williams
fernaldiana Nannf.
       SY=P. laxa Haenke ssp. f. (Nannf.)
          Hyl.
fibrata Swallen
gaspensis Fern.
glauca Vahl
       SY=P. ammophila Porsild
       SY=P. g. ssp. conferta (Blytt)
          Lindm.
       SY=P. g. var. c. (Blytt) Nannf.
glaucantha Gaudin
       SY=P. glauca Vahl ssp. glaucantha
          (Gaudin) Lindm.
       SY=P. scopulorum Butters & Abbe
glaucifolia Scribn. & Williams
gracillima Vasey
   var. gracillima
       SY=P. secunda sensu auctt. non
          Presl var. stenophylla
          (Vasey) Beetle
       SY=P. tenerrima Scribn.
   var. multnomae (Piper) C.L. Hitchc.
hartzii Gandog.
hispidula Vasey
howellii Vasey & Scribn.
       SY=P. bolanderi Vasey var. h.
          (Vasey & Scribn.) M.E. Jones
incurva Scribn. & Williams
       SY=P. secunda sensu auctt. non
          Presl var. i. (Scribn. &
          Williams) Beetle
interior Rydb.
       SY=P. nemoralis L. var. i. (Rydb.)
          Butters & Abbe
involuta A.S. Hitchc.
kelloggii Vasey
labradorica Steud.
lanata Scribn. & Merr.
   var. lanata
       SY=P. arctica R. Br. var. l.
          (Scribn. & Merr.) Boivin
   var. vivipara Hulten
languida A.S. Hitchc.
       SY=P. debilis Torr.
laxa Haenke
   var. flexuosa Sm.
       SY=P. f. (Sm.) Muhl.
       SY=P. l. ssp. f. (Sm.) Hyl.
   var. laxa
laxiflora Buckl.
leibergii Scribn.
       SY=P. hansenii Scribn.
leptocoma Trin.
lettermanii Vasey
       SY=Puccinellia l. (Vasey) Ponert
macrocalyx Trautv. & C.A. Mey.
       SY=Poa hispidula Vasey var.
          aleutica Hulten
       SY=P. norbergii Hulten

Poa   (CONT.)
  macroclada Rydb.
  malacantha Komarov
  mannii Munro
  marcida A.S. Hitchc.
      SY=Poa saltuensis Fern. & Wieg.
        var. m. (A.S. Hitchc.)
         Boivin
  merrilliana A.S. Hitchc.
  montevansii L. Kelso
  napensis Beetle
  nemoralis L.
  nervosa (Hook.) Vasey
    var. nervosa
    var. wheeleri (Vasey) C.L. Hitchc.
      SY=Poa w. Vasey
  nevadensis Vasey ex Scribn.
    var. juncifolia (Scribn.) Beetle
      SY=Poa brachyglossa Piper
      SY=P. j. Scribn.
      SY=P. j. ssp. porteri Keck
    var. nevadensis
  occidentalis Vasey
  paludigena Fern. & Wieg.
  palustris L.
      SY=Poa crocata Michx.
      SY=P. triflora Gilib.
  pattersonii Vasey
      SY=Poa abbreviata R. Br. ssp. p.
        (Vasey) Love, Love & Kapoor
  paucispicula Scribn. & Merr.
      SY=Poa leptocoma Trin. var. p.
        (Scribn. & Merr.) C.L.
        Hitchc.
  piperi A.S. Hitchc.
  pratensis L.
    ssp. agassizensis (Boivin & D. Love)
      Taylor & MacBryde
      SY=Poa a. Boivin & D. Love
    ssp. alpigena (Fries) Hiitonen
      SY=Poa a. (Fries) Lindm. f.
      SY=P. a. var. colpodea (Fries)
        Schol.
      SY=P. p. ssp. p. var. c. (Fries)
        Schol.
      SY=P. p. var. a. Fries
    ssp. angustifolia (L.) Gaudin
      SY=Poa a. L.
    ssp. pratensis
      SY=Poa p. ssp. p. var. domestica
        Laestad.
      SY=P. p. ssp. p. var. gelida
        (Roemer & Schultes) Bocher
      SY=P. p. ssp. p. var. iantha
        Wahlenb.
      SY=P. p. ssp. p. var. rigens
        Wahlenb.
      SY=P. p. ssp. p. var. vivipara
        (Malmgr.) Boivin
  pringlei Scribn.
  pseudoabbreviata Rosh.
  reflexa Vasey & Scribn.
  rhizomata A.S. Hitchc.
  rupicola Nash ex Rydb.
      SY=Poa glauca Vahl var. r. (Nash
        ex Rydb.) Boivin
  saltuensis Fern. & Wieg.
      SY=Poa s. var. microlepis Fern. &
        Wieg.
  sandbergii Vasey
      SY=Poa secunda sensu auctt. non
        Presl
  sandvicensis (Reichenb.) A.S. Hitchc.
  scabrella (Thurb.) Benth. ex Vasey
      SY=Poa buckleyana Nash
  siphonoglossa Hack.
  stenantha Trin.

  subcaerulea Sm.
  suksdorfii (Beal) Vasey ex Piper
  sylvestris Gray
  tracyi Vasey
  trivialis L.
  turneri Scribn.
  unilateralis Scribn.
      SY=Poa pachypholis Piper
  vaseyochloa Scribn.
      SY=Poa porsildii Gjaerevoll
  wolfii Scribn.

Polypogon Desf.

  australis Brongn.
  elongatus H.B.K.
  interruptus H.B.K.
      SY=P. lutosus sensu auctt. non
        (Poir.) A.S. Hitchc.
  maritimus Willd.
  monspeliensis (L.) Desf.

Polytrias Hack.

  praemorsa (Nees) Hack.

Pseudosasa Makino ex Nakai

  japonica (Sieb. & Zucc. ex Steud.) Makino
      SY=Arundinaria j. Sieb. & Zucc. ex
        Steud.
      SY=Sasa j. (Sieb. & Zucc. ex
        Steud.) Makino ex Sieb. &
        Zucc.

Puccinellia Parl.

  agrostoides Sorensen
  ambigua Sorensen
  americana Sorensen
  andersonii Swallen
      SY=P. glabra Swallen
      SY=P. triflora Swallen
  arctica (Hook.) Fern. & Weath.
      SY=P. borealis Swallen
  bruggemannii Sorensen
  coarctata Fern. & Weath.
    var. coarctata
    var. pseudofasciculata Sorensen
  contracta (Lange) Sorensen
  deschampsioides Sorensen
  distans (Jacq.) Parl.
    var. angustifolia (L.) Holmb.
    var. distans
      SY=P. d. var. tenuis (Uechtr.)
        Fern. & Weath.
  fasciculata (Torr.) Bickn.
      SY=P. borreri (Bab.) A.S. Hitchc.
  fernaldii (A.S. Hitchc.) E.G. Voss
      SY=Glyceria f. (A.S. Hitchc.) St.
        John
      SY=G. pallida (Torr.) Trin. var.
        f. A.S. Hitchc.
      SY=Torreyochloa f. (A.S. Hitchc.)
        Church
      SY=T. p. (Torr.) Church var. f.
        (A.S. Hitchc.) Dore
  geniculata (Turcz.) Krecz.
  grandis Swallen
  groenlandica Sorensen
  hauptiana (Krecz.) Krecz.
  hultenii Swallen
  interior Sorensen
  kamtschatica Holmb.
    var. kamtschatica
    var. sublaevis Holmb.
  langeana (Berlin) Sorensen

Puccinellia (CONT.)
    ssp. alaskana (Scribn. & Merr.) Sorensen
        SY=P. paupercula (Holm) Fern. &
          Weath. var. a. (Scribn. &
          Merr.) Fern. & Weath.
    ssp. asiatica Sorensen
    ssp. langeana
  laurentiana Fern. & Weath.
  lemmonii (Vasey) Scribn.
  macra Fern. & Weath.
  maritima (Huds.) Parl.
  nutkaensis (Presl) Fern. & Weath.
  nuttalliana (Schultes) A.S. Hitchc.
        SY=P. airoides (Nutt.) Wats. &
          Coult.
        SY=P. cusickii Weatherby
        SY=P. lucida Fern. & Weath.
  parishii A.S. Hitchc.
  paupercula (Holm) Fern. & Weath.
        SY=P. distans (Jacq.) Parl. var.
          minor (S. Wats.) Boivin
        SY=P. p. var. longiglumis Fern. &
          Weath.
  phryganodes (Trin.) Scribn. & Merr.
  porsildii Sorensen
  pumila (Vasey) A.S. Hitchc.
        SY=P. angustata (R. Br.) Nash
  rosenkrantzii Sorensen
  rupestris (With.) Fern. & Weath.
  sibirica Holmb.
  simplex Scribn.
  vacillans (T. Fries) Schol.
        SY=Maltea X v. (T. Fries) Boivin
  vaginata (Lange) Fern. & Weath.
    var. paradoxa Sorensen
      var. vaginata
  vahliana (Liebm.) Scribn. & Merr.

Redfieldia Vasey

  flexuosa (Thurb.) Vasey

Reimarochloa A.S. Hitchc.

  oligostachya (Munro) A.S. Hitchc.

Rhynchelytrum Nees

  repens (Willd.) C.E. Hubbard
        SY=P. roseum (Nees) Stapf & C.E.
          Hubbard ex Bews
        SY=Tricholaena repens (Willd.)
          A.S. Hitchc.
        SY=T. rosea Nees

Rottboellia L. f.

  exaltata L. f.
        SY=Manisuris e. (L. f.) Kuntze

Saccharum L.

  barberi Jeswiet
  officinarum L.
  sinense Roxb.
  spontaneum L.

Sacciolepis Nash

  indica (L.) Chase
  striata (L.) Nash
        SY=Panicum aquaticum Bosc ex
          Spreng. non Poir.

Schedonnardus Steud.

  paniculatus (Nutt.) Trel.

Schismus Beauv.

  arabicus Nees
  barbatus (L.) Thellung

Schizachne Hack.

  purpurascens (Torr.) Swallen
    var. pubescens Dore
    var. purpurascens
        SY=Avena torreyi Nash

Schizachyrium Nees

  brevifolium (Sw.) Nees ex Buse
        SY=Andropogon b. Sw.
  cirratum (Hack.) Woot. & Standl.
        SY=Andropogon c. Hack.
  gracile (Spreng.) Nash
        SY=Andropogon g. Spreng.
  hirtiflorum Nees
        SY=Andropogon domingensis (Spreng.
          ex Schultes) F.T. Hubbard
          non Steud.
        SY=A. h. (Nees) Kunth
        SY=A. h. var. feensis (Fourn.)
          Hack.
        SY=S. d. (Spreng. ex Schultes)
          Nash
        SY=S. f. (Fourn.) A. Camus
  maritimum (Chapman) Nash
        SY=Andropogon m. Chapman
  niveum (Swallen) Gould
        SY=Andropogon n. Swallen
  rhizomatum (Swallen) Gould
        SY=Andropogon r. Swallen
  sanguineum (Retz.) Alston
    var. brevipedicellatum (Beal) Hatch
        SY=Andropogon hirtiflorus (Nees)
          Kunth var. b. Beal
    var. oligostachyum (Chapman) Hatch
        SY=Andropogon hirtiflorus (Nees)
          Kunth var. o. (Chapman)
          Hack.
        SY=A. o. Chapman
    var. sanguineum
        SY=Andropogon semiberbis (Nees)
          Kunth
        SY=S. semiberbe Nees
  scoparium (Michx.) Nash
    var. divergens (Hack.) Gould
        SY=Andropogon d. (Hack.) Anderss.
          ex A.S. Hitchc.
        SY=A. s. Michx. ssp. maritimum
          (Chapman) Nash var. d. Hack.
    var. frequens (F.T. Hubbard) Gould
        SY=Andropogon s. Michx. var. f.
          F.T. Hubbard
    var. littorale (Nash) Gould
        SY=Andropogon l. Nash
        SY=A. s. Michx. var. ducis Fern. &
          Grisc.
        SY=A. s. var. l. (Nash) A.S.
          Hitchc.
        SY=S. l. (Nash) Bickn.
    var. neomexicanum (Nash) Gould
        SY=Andropogon s. Michx. var. n.
          (Nash) A.S. Hitchc.
    var. scoparium
        SY=Andropogon praematurus Fern.
        SY=A. s. Michx.
        SY=A. s. var. polycladus Scribn. &
          Ball
        SY=A. s. var. septentrionalis
          Fern. & Grisc.
    var. virile (Shinners) Gould
        SY=Andropogon s. var. v. Shinners

Schizachyrium (CONT.)
    sericatum (Swallen) Gould
            SY=Andropogon s. Swallen
    stoloniferum Nash
        var. stoloniferum
            SY=Andropogon s. (Nash) A.S.
                Hitchc.
        var. wolfei DeSelm
    tenerum Nees
            SY=Andropogon t. (Nees) Kunth

Schizostachyum Nees

    glaucifolium (Rupr.) Munro

Sclerochloa Beauv.

    dura (L.) Beauv.

Scleropogon Phil.

    brevifolius Phil.

Scolochloa Link

    festucacea (Willd.) Link
            SY=Fluminea f. (Willd.) A.S.
                Hitchc.

Scribneria Hack.

    bolanderi (Thurb.) Hack.

Secale L.

    cereale L.
    montanum Guss.

Setaria Beauv.

    adhaerens (Forsk.) Chiov.
    arizonica Rominger
    barbata (Lam.) Kunth
            SY=Chaetochloa b. (Lam.) A.S.
                Hitchc. & Chase
    chapmanii (Vasey) Pilger
    corrugata (Ell.) Schultes
            SY=Chaetochloa c. (Ell.) Scribn.
            SY=C. hispida Scribn. & Merr.
    faberi Herrm.
    firmula (A.S. Hitchc. & Chase) Pilger
            SY=Panicum f. A.S. Hitchc. & Chase
    geniculata (Lam.) Beauv.
            SY=Chaetochloa g. (Lam.) Millsp. &
                Chase
            SY=C. imberbis (Poir.) Scribn.
    glauca (L.) Beauv.
            SY=Chaetochloa g. (L.) Scribn.
            SY=C. lutescens (Weigel) F.T.
                Hubbard
            SY=Pennisetum typhoides sensu
                auctt. non (Burm.) Stapf &
                C.E. Hubbard
            SY=S. l. (Weigel) Stuntz
    grisebachii Fourn.
    italica (L.) Beauv.
        var. italica
            SY=Chaetochloa i. (L.) Scribn.
        var. metzeri (Koern.) Jav.
        var. stramineofructa (F.T. Hubbard)
            Bailey
    leucopila (Scribn. & Merr.) K. Schum.
    liebmannii Fourn.
    macrosperma (Scribn. & Merr.) K. Schum.
            SY=Chaetochloa m. Scribn. & Merr.
    macrostachya H.B.K.
    magna Griseb.

            SY=Chaetochloa m. (Griseb.)
                Scribn.
    nigrirostris (Nees) Dur. & Schinz
            SY=Chaetochloa n. (Nees) Skeels
    palmifolia (Koenig) Stapf
    pradana (Leon) Leon
    ramiseta (Scribn.) Pilger
            SY=Panicum r. Scribn.
    rariflora Mikan ex Trin.
            SY=Chaetochloa r. (Mikan ex Trin.)
                A.S. Hitchc. & Chase
    reverchonii (Vasey) Pilger
            SY=Panicum r. Vasey
    scheelei (Steud.) A.S. Hitchc.
    setosa (Sw.) Beauv.
        var. leiophylla (Nees) Arech.
            SY=S. l. (Nees) Kunth
        var. setosa
            SY=Chaetochloa s. (Sw.) Scribn.
    sphacelata (Schum.) Stapf & C.E. Hubbard
    tenacissima Schrad.
            SY=Chaetochloa t. (Schrad.) A.S.
                Hitchc. & Chase
    texana W.H.P. Emery
    utowanaea (Scrib. ex Millsp.) Pilger
            SY=Panicum u. Scribn. ex Millsp.
    verticillata (L.) Beauv.
        var. ambigua (Guss.) Parl.
            SY=S. viridis (L.) Beauv. var. a.
                (Guss.) Coss. & Durieu
        var. verticillata
            SY=Chaetochloa v. (L.) Scribn.
            SY=S. carnei A.S. Hitchc.
    villosissima (Scribn. & Merr.) K. Schum.
    viridis (L.) Beauv.
        var. major (Gaudin) Pospichal
        var. viridis
            SY=Chaetochloa v. (L.) Scribn.
            SY=S. v. var. breviseta (Doell)
                A.S. Hitchc.
            SY=S. v. var. weimannii (Roemer &
                Schultes) Borbas
    vulpiseta (Lam.) Roemer & Schultes
            SY=Chaetochloa v. (Lam.) A.S.
                Hitchc. & Chase

Sieglingia Bernh.

    decumbens (L.) Bernh.

Sinocalamus McClure

    oldhamii (Munro) McClure

Sitanion Raf.

    hordeoides Suksdorf
            SY=S. hystrix (Nutt.) J.G. Sm.
                var. hordeoides (Suksdorf)
                C.L. Hitchc.
    hystrix (Nutt.) J.G. Sm.
        var. brevifolium (J.G. Sm.) C.L.
            Hitchc.
            SY=Elymus longifolium (J.G. Sm.)
                Gould
            SY=S. l. J.G. Sm.
        var. hystrix
            SY=S. elymoides Raf.
            SY=S. h. var. californicum (J.G.
                Sm.) F.D. Wilson
    jubatum J.G. Sm.

X Sitordeum Bowden

    X californicum Bowden [Hordeum jubatum ssp.
        breviaristatum X Sitanion jubatum or S.
        hystrix]

Sorghastrum Nash

    elliottii (C. Mohr) Nash
    nutans (L.) Nash
        SY=S. avenaceum (Michx.) Nash
    parviflorum (Desv.) A.S. Hitchc. & Chase
    secundum (Ell.) Nash
    setosum (Griseb.) A.S. Hitchc.

Sorghum Moench

    almum L.
    bicolor (L.) Moench
      ssp. arundinaceum (Desv.) de Wet &
        Harlan
      ssp. bicolor
            SY=Holcus sorghum L.
            SY=S. b. var. aethiopicum (Hack.)
                de Wet & Huckabay
            SY=S. b. var. caffrorum (Retz.)
                Mohlenbrock
            SY=S. b. var. virgatum (Hack.) de
                Wet & Huckabay
            SY=S. caffrorum (Retz.) Beauv.
            SY=S. cernuum (Ard.) Host
            SY=S. dochna (Forsk.) Snowden
            SY=S. d. var. technicum (Koern. &
                Wern.) Snowden
            SY=S. durra (Forsk.) Stapf
            SY=S. lanceolatum Stapf
            SY=S. saccharatum (L.) Moench
            SY=S. subglabrescens Schweinf. &
                Aschers.
            SY=S. virgatum (Hack.) Stapf
            SY=S. vulgare Pers.
            SY=S. vulgare var. caffrorum
                (Retz.) Hubbard & Rehd.
            SY=S. vulgare var. durra (Forsk.)
                Hubbard & Rehd.
            SY=S. vulgare var. roxburghii
                (Stapf) Haines
            SY=S. vulgare var. saccharatum
                (L.) Boerl.
            SY=S. vulgare var. technicum
                (Koern.) Jav.
      ssp. drummondii (Steud.) de Wet & Harlan
            SY=S. b. var. sudanense (Piper)
                A.S. Hitchc.
            SY=S. b. var. d. (Steud.)
                Mohlenbrock
            SY=S. d. Steud.
            SY=S. s. (Piper) Stapf
            SY=S. vulgare Pers. var. d.
                (Steud.) Hack. ex Chiov.
    halepense (L.) Pers.
        SY=Holcus h. L.
    longifolium author ?

Spartina Schreb.

    alterniflora Loisel.
        SY=S. a. var. glabra (Muhl.) Fern.
        SY=S. a. var. pilosa (Merr.) Fern.
    anglica C.E. Hubbard
    bakeri Merr.
    X caespitosa (A.A. Eat.) Fern. [patens X
      pectinata]
    cynosuroides (L.) Roth
        SY=S. c. var. polystachya (Michx.)
            Beal ex Fern.
    foliosa Trin.
    gracilis Trin.
    maritima (Curtis) Fern.
        SY=S. stricta (Ait.) Roth
    patens (Ait.) Muhl.
        SY=S. p. var. juncea (Michx.) A.S.
            Hitchc.

        SY=S. p. var. monogyna (M.A.
            Curtis) Fern.
    pectinata Link
        SY=S. michauxiana A.S. Hitchc.
        SY=S. p. var. suttiei (Farw.)
            Fern.
    spartinae (Trin.) Merr.
    townsendii H. & J. Groves

Sphenopholis Scribn.

    filiformis (Chapman) Scribn.
    nitida (Biehler) Scribn.
    obtusata (Michx.) Scribn.
      var. major (Torr.) K.S. Erdman
        SY=S. intermedia (Rydb.) Rydb.
        SY=S. i. var. pilosa Dore
        SY=S. longiflora (Vasey) A.S.
            Hitchc.
      var. obtusata
        SY=S. o. var. lobata (Trin.)
            Scribn.
        SY=S. o. var. pubescens (Scribn. &
            Merr.) Scribn.
    X pallens (Biehler) Scribn. [obtusata X
      pensylvanica]
    pensylvanica (L.) A.S. Hitchc.
        SY=Trisetum p. (L.) Beauv. ex
            Roemer & Schultes

Spodiopogon Trin.

    aureus Hook. & Arn.

Sporobolus R. Br.

    africanus (Poir.) Robyns & Tournay
    airoides (Torr.) Torr.
    asper (Michx.) Kunth
      var. asper
        SY=S. a. var. hookeri (Trin.)
            Vasey
        SY=S. a. var. pilosus (Vasey) A.S.
            Hitchc.
        SY=S. p. Vasey
      var. drummondii (Trin.) Vasey
        SY=S. attenuatus Nash
        SY=S. d. (Trin.) Vasey
      var. macer (Trin.) Shinners
        SY=S. m. (Trin.) A.S. Hitchc.
    buckleyi Vasey
    clandestinus (Biehler) A.S. Hitchc.
        SY=S. asper (Michx.) Kunth var.
            canovirens (Nash) Shinners
        SY=S. a. var. clandestinus
            (Biehler) Shinners
        SY=S. canovirens Nash
        SY=S. clandestinus var. canovirens
            (Nash) Steyermark & Kucera
        SY=S. longifolius (Torr.) Wood
    contractus A.S. Hitchc.
        SY=S. cryptandrus (Torr.) Gray
            var. strictus Scribn.
    cryptandrus (Torr.) Gray
        SY=S. c. ssp. fuscicolus (Hook.)
            Jones & Fassett
        SY=S. c. var. occidentalis Jones &
            Fassett
    cubensis A.S. Hitchc.
    curtissii (Vasey) Small ex Scribn.
    domingensis (Trin.) Kunth
    flexuosus (Thurb. ex Vasey) Rydb.
    floridanus Chapman
    giganteus Nash
    heterolepis (Gray) Gray
    indicus (L.) R. Br.
        SY=S. angustus Buckl.

Sporobolus   (CONT.)
        SY=S. berteroanus (Trin.) A.S.
             Hitchc. & Chase
        SY=S. elongatus sensu auctt. non
             R. Br.
        SY=S. jacquemontii Kunth
        SY=S. poiretii (Roemer & Schultes)
             A.S. Hitchc.
    interruptus Vasey
    junceus (Michx.) Kunth
        SY=S. gracilis (Trin.) Merr.
    microspermus (Lag.) A.S. Hitchc.
        SY=Muhlenbergia debilis (H.B.K.)
             Kunth
    nealleyi Vasey
    neglectus Nash
        SY=S. vaginiflorus (Torr. ex Gray)
             Wood var. n. (Nash) Scribn.
    ozarkanus Fern.
        SY=S. neglectus Nash var. o.
             (Fern.) Steyermark & Kucera
        SY=S. vaginiflorus (Torr. ex Gray)
             Wood var. o. (Fern.)
             Shinners
    patens Swallen
    pulvinatus Swallen
    purpurascens (Sw.) Hamilton
    pyramidatus (Lam.) A.S. Hitchc.
        SY=S. argutus (Nees) Kunth
    silveanus Swallen
    tenuissimus (Schrank) Kuntze
        SY=S. muralis (Raddi) A.S. Hitchc.
             & Chase
    teretifolius Harper
    texanus Vasey
    tharpii A.S. Hitchc.
    vaginiflorus (Torr. ex Gray) Wood
        SY=S. v. var. inaequalis Fern.
    virginicus (L.) Kunth
    wrightii Munro ex Scribn.
        SY=S. airoides (Torr.) Torr. var.
             w. (Munro ex Scribn.) Gould

Stenotaphrum Trin.

    secundatum (Walt.) Kuntze

Stipa L.

    arida M.E. Jones
    avenacea L.
    avenacioides Nash
    brachychaeta Godr.
    californica Merr. & Burtt-Davy
    cernua Stebbins & Love
    columbiana Macoun
        var. columbiana
            SY=S. minor (Vasey) Scribn.
            SY=S. occidentalis Thurb. ex S.
                 Wats. var. m. (Vasey) C.L.
                 Hitchc.
        var. nelsonii (Scribn.) A.S. Hitchc.
            SY=S. n. Scribn.
            SY=S. occidentalis Thurb. ex S.
                 Wats. var. n. (Scribn.) C.L.
                 Hitchc.
    comata Trin. & Rupr.
        var. comata
            SY=S. c. ssp. intonsa Piper
        var. intermedia Scribn. & Tweedy
    coronata Thurb.
        var. coronata
        var. depauperata (M.E. Jones) A.S.
             Hitchc.
            SY=S. parishii Vasey
    curvifolia Swallen
    diegoensis Swallen

    eminens Cav.
    hendersonii (Vasey) Muhl.
        SY=Oryzopsis h. Vasey
    latiglumis Swallen
    lemmonii (Vasey) Scribn.
        var. lemmonii
            SY=S. l. var. jonesii Scribn.
        var. pubescens Crampton
    lepida A.S. Hitchc.
        var. andersonii (Vasey) A.S. Hitchc.
        var. lepida
    lettermanii Vasey
    leucotricha Trin. & Rupr.
    litoralis Phil.
    lobata Swallen
    neesiana Trin. & Rupr.
    neomexicana (Thurb.) Scribn.
    nevadensis B.L. Johnson
    occidentalis Thurb. ex S. Wats.
        var. occidentalis
        var. pubescens (Vasey) Maze, Taylor &
             MacBryde
            SY=S. elmeri Piper & Brodie ex
                 Scribn.
    pinetorum M.E. Jones
    porteri Rydb.
        SY=Ptilagrostis p. (Rydb.) W.A.
             Weber
    pringlei Scribn.
    pulchra A.S. Hitchc.
    richardsonii Link
    robusta (Vasey) Scribn.
        SY=S. vaseyi Scribn.
    scribneri Vasey
    spartea Trin.
        var. curtiseta A.S. Hitchc.
        var. spartea
    speciosa Trin. & Rupr.
    stillmanii Boland.
    tenuissima Trin.
    thurberiana Piper
    viridula Trin.
    webberi (Thurb.) B.L. Johnson
        SY=Oryzopsis w. (Thurb.) Benth. ex
             Vasey
    williamsii Scribn.

X Stiporyzopsis B.L. Johnson & Fogler

    X bloomeri (Boland.) B.L. Johnson
        [Oryzopsis hymenoides X Stipa
        occidentalis]
    X caduca (Beal) B.L. Johnson & Fogler
        [Oryzopsis hymenoides X Stipa viridula]

Swallenia Soderstrom & Decker

    alexandrae (Swallen) Soderstrom & Decker
        SY=Ectosperma a. Swallen

Taeniatherum Nevski

    caput-medusae (L.) Nevski
        SY=Elymus c-m. L.

Themeda Forsk.

    quadrivalvis (L.) Kuntze

Torreyochloa Church

    californica (Beetle) Church
        SY=Glyceria c. Beetle
        SY=Puccinellia c. (Beetle) Munz
    erecta (A.S. Hitchc.) Church
        SY=Glyceria e. A.S. Hitchc.
        SY=Puccinellia e. (A.S. Hitchc.)

Torreyochloa (CONT.)
        Munz
    pallida (Torr.) Church
            SY=Glyceria p. (Torr.) Trin.
            SY=Panicularia p. (Torr.) Kuntze
            SY=Puccinellia p. (Torr.) Clausen
    pauciflora (Presl) Church
            var. holmii (Beal) Taylor & MacBryde
            SY=Puccinellia p. (Presl) Munz
                    var. holmii (Beal) C.L.
                    Hitchc.
            var. microtheca (Buckl.) Taylor &
            MacBryde
            SY=Puccinellia p. (Presl) Munz
                    var. m. (Buckl.) C.L.
                    Hitchc.
            var. pauciflora
            SY=Glyceria p. Presl
            SY=Puccinellia p. (Presl) Munz

Trachypogon Nees

    montufarii (H.B.K.) Nees
    secundus (Presl) Scribn.

Tragus Haller

    berteronianus Schultes
            SY=Nazia aliena sensu auctt. non
                    (Spreng.) Scribn.
    racemosus (L.) All.
            SY=Nazia r. (L.) Kuntze

Trichoneura Anderss.

    elegans Swallen

Tridens Roemer & Schultes

    albescens (Vasey) Woot. & Standl.
            SY=Rhombolytrum a. (Vasey) Nash
    ambiguus (Ell.) Schultes
            SY=Triodia elliottii Bush
    buckleyanus (L.H. Dewey) Nash
    carolinianus (Steud.) Henr.
            SY=Triodia drummondii Scribn. &
                    Kearney
    congestus (L.H. Dewey) Nash
    eragrostoides (Vasey & Scribn.) Nash
            SY=Triodia e. Vasey & Scribn.
    flavus (L.) A.S. Hitchc.
            var. chapmanii (Small) Shinners
            SY=Tridens c. (Small) Chase
            SY=Triodia c. (Small) Bush
            var. flavus
            SY=Triodia f. (L.) Smyth
    muticus (Torr.) Nash
            var. elongatus (Buckl.) Shinners
            SY=Tridens e. (Buckl.) Nash
            SY=Triodia e. (Buckl.) Scribn.
            var. muticus
            SY=Triodia m. (Torr.) Scribn.
    oklahomensis (Featherly) Featherly
    strictus (Nutt.) Nash
            SY=Triodia s. (Nutt.) Benth. ex
                    Vasey
    texanus (S. Wats.) Nash

Triplasis Beauv.

    americana Beauv.
    purpurea (Walt.) Chapman
            SY=T. intermedia Nash

Tripogon Roemer & Schultes

    spicatus (Nees) Ekman

Tripsacum L.

    dactyloides (L.) L.
            SY=T. d. var. occidentale Cutler &
                    Anders.
    floridanum Porter ex Vasey
    lanceolatum Rupr. ex Fourn.

Trisetum Pers.

    aureum (Ten.) Ten.
    canescens Buckl.
            SY=T. cernuum Trin. ssp. canescens
                    (Buckl.) Calder & Taylor
            SY=T. cernuum var. canescens
                    (Buckl.) Beal
            SY=T. cernuum var. projectum
                    (Louis-Marie) Beetle
    cernuum Trin.
    flavescens (L.) Beauv.
    glomeratum (Kunth) Trin.
    inaequale Whitney
    interruptum Buckl.
    melicoides (Michx.) Scribn.
            SY=Graphephorum m. (Michx.) Desv.
            SY=T. m. var. majus (Gray) A.S.
                    Hitchc.
    montanum Vasey
    orthochaetum A.S. Hitchc.
    sibiricum Rupr.
        ssp. litoralis (Rupr.) Rosh.
    spicatum (L.) Richter
            var. alaskanum (Nash) Malte ex
                    Louis-Marie
            SY=T. s. ssp. a. (Nash) Hulten
            var. congdonii (Scribn. & Merr.) A.S.
                    Hitchc.
            SY=T. s. ssp. c. (Scribn. & Merr.)
                    Hulten
            var. maidenii (Gandog.) Fern.
            var. spicatiforme Hulten
            var. spicatum
            SY=T. s. ssp. majus (Rydb.) Hulten
            SY=T. s. var. m. (Rydb.) Farw.
            var. villosissimum (Lange)
                    Louis-Marie
    triflorum (Bigelow) Love & Love
        ssp. molle (Michx.) Love & Love
            SY=T. spicatum (L.) Richter ssp.
                    m. (Michx.) Hulten
            SY=T. s. var. m. (Michx.) Beal
        ssp. triflorum
            SY=T. spicatum (L.) Richter var.
                    pilosiglume Fern.
    wolfii Vasey

Triticum L.

    aestivum L.
            SY=T. aestivum L. var. spelta (L.)
                    Bailey
            SY=T. compactum Host
            SY=T. macha Dekap. & Menah.
            SY=T. sativum Lam.
            SY=T. spelta L.
            SY=T. sphaerococcum Percival
            SY=T. vulgare Vill.
    dicoccoides Koern.
    monococcum L.
    turgidum L.
            SY=T. dicoccum Schrank
            SY=T. durum Desf.
            SY=T. polonicum L.

Uniola L.

    paniculata L.

Vaseyochloa A.S. Hitchc.

    multinervosa (Vasey) A.S. Hitchc.

Ventenata Koel.

    dubia (Leers) Coss. & Durieu

Vetiveria Bory

    zizanioides (L.) Nash
        SY=Anatherum z. (L.) A.S. Hitchc.
          & Chase

Vulpia K.C. Gmel.

    bromoides (L.) S.F. Gray
        SY=Bromus dertonensis All.
        SY=Festuca b. L.
        SY=F. d. (All.) Aschers. & Graebn.
        SY=V. d. (All.) Gola
    megalura (Nutt.) Rydb.
        SY=Festuca m. Nutt.
        SY=V. myuros (L.) K.C. Gmel. var.
          hirsuta Hack.
    microstachys (Nutt.) Munro ex Benth.
        var. ciliata (Beal) Lonard & Gould
          SY=Festuca eastwoodae Piper
          SY=F. grayi (Abrams) Piper
          SY=F. m. Nutt. var. ciliata Beal
          SY=V. e. (Piper) Henr.
          SY=V. g. (Abrams) Henr.
        var. confusa (Piper) Lonard & Gould
          SY=Festuca c. Piper
          SY=F. suksdorfii Piper ex Suksdorf
          SY=F. tracyi A.S. Hitchc.
          SY=V. c. (Piper) Henr.
          SY=V. t. (A.S. Hitchc.) Henr.
        var. microstachys
          SY=Festuca arida Elmer
          SY=F. m. Nutt.
          SY=V. a. (Elmer) Henr.
        var. pauciflora (Scribn. ex Beal)
          Lonard & Gould
          SY=Festuca microstachys Nutt. var.
            p. Beal
          SY=F. m. Nutt. var. simulans
            (Hoover) Hoover
          SY=F. pacifica Piper
          SY=F. pacifica var. s. Hoover
          SY=F. reflexa Buckl.
          SY=V. pacifica (Piper) Rydb.
          SY=V. r. (Buckl.) Rydb.
    myuros (L.) K.C. Gmel.
        var. hirsuta Hack.
          SY=Festuca megalura Nutt.
          SY=F. megalura var. h. (Hack.)
            Aschers. & Graebn.
          SY=V. megalura (Nutt.) Rydb.
        var. myuros
          SY=Festuca m. L.
    octoflora (Walt.) Rydb.
        var. glauca (Nutt.) Fern.
          SY=Festuca gracilenta Buckl.
          SY=F. o. Walt. var. g. (Nutt.)
            Fern.
          SY=F. o. var. tenella (Willd.)
            Fern.
          SY=F. t. Willd.
          SY=F. t. var. g. Nutt.
          SY=V. o. var. t. (Willd.) Fern.
        var. hirtella (Piper) Henr.
          SY=Festuca o. Walt. ssp. h. Piper
          SY=F. o. var. h. (Piper) A.S.
            Hitchc.
        var. octoflora
          SY=Festuca o. Walt.

          SY=F. o. var. aristulata Torr. ex
            L. H. Dewey
    sciurea (Nutt.) Henr.
        SY=Festuca s. Nutt.
        SY=V. elliotea (Raf.) Fern.

Willkommia Hack.

    texana A.S. Hitchc.

Zea L.

    mays L.
    mexicana (Schrad.) Kuntze
        SY=Euchlaena m. Schrad.
        SY=Z. mays L. ssp. mexicana
          (Schrad.) Iltis
    perennis (A.S. Hitchc.) Reeves &
        Manglesdorf
        SY=Euchlaena p. A.S. Hitchc.

Zizania L.

    aquatica L.
        var. angustifolia A.S. Hitchc.
          SY=Z. palustris L.
        var. aquatica
        var. brevis Fassett
        var. interior Fassett
          SY=Z. palustris L. var. i.
            (Fassett) Dore
        var. subbrevis Boivin
    texana A.S. Hitchc.

Zizaniopsis Doell & Aschers.

    miliacea (Michx.) Doell & Aschers.

Zoysia Willd.

    japonica Steud.
    matrella (L.) Merr.
    tenuifolia Willd. ex Trin.

PODOCARPACEAE

Podocarpus L'Her. ex Labill.

    coriaceus L.C. Rich.
        SY=Nageia c. (L.C. Rich.) Kuntze

PODOSTEMACEAE

Podostemon Michx.

    ceratophyllum Michx.
        SY=P. abrotanoides Nutt.

POLEMONIACEAE

Allophyllum (Nutt.) A. & V. Grant

    divaricatum (Nutt.) A. & V. Grant
        SY=Gilia gilioides (Benth.) Greene
          ssp. volcanica (Brand) Mason
          & A. Grant
    gilioides (Benth.) A. & V. Grant
        SY=Gilia g. (Benth.) Greene

Allophyllum (CONT.)
    glutinosum (Benth.) A. & V. Grant
            SY=Gilia gilioides (Benth.) Greene
                  ssp. glutinosa (Benth.)
                    Mason & A. Grant
    integrifolium (Brand) A. & V. Grant
    violaceum (Heller) A. & V. Grant
            SY=Gilia v. Heller

Collomia Nutt.

    debilis (S. Wats.) Greene
        var. camporum Payson
        var. debilis
            SY=C. d. var. integra Payson
            SY=C. d. var. trifida Payson
            SY=C. d. var. typica Payson
        var. ipomoea Payson
        var. larsenii (Gray) Brand
            SY=C. l. Gray
    diversifolia Greene
    grandiflora Dougl. ex Lindl.
    heterophylla Dougl. ex Hook.
    linearis Nutt.
    macrocalyx Leib. ex Brand
    mazama Coville
    rawsoniana Greene
    tenella Gray
    tinctoria Kellogg
    tracyi Mason

Eriastrum Woot. & Standl.

    abramsii (Elmer) Mason
    brandegeae Mason
    densifolium (Benth.) Mason
        ssp. austromontanum (Craig) Mason
        ssp. densifolium
        ssp. elongatum (Benth.) Mason
        ssp. mohavensis (Craig) Mason
        ssp. sanctorum (Milliken) Mason
    diffusum (Gray) Mason
            SY=E. d. ssp. jonesii Mason
    eremicum (Jepson) Mason
        ssp. eremicum
        ssp. yageri (M. E. Jones) Mason
    filifolium (Nutt.) Woot. & Standl.
            SY=Gilia f. Nutt.
    hooveri (Jepson) Mason
    luteum (Benth.) Mason
    pluriflorum (Heller) Mason
        ssp. pluriflorum
        ssp. sherman-hoytae (Craig) Mason
    sapphirinum (Eastw.) Mason
        ssp. dasyanthum (Brand) Mason
        ssp. sapphirinum
            SY=E. s. ssp. ambiguum (M.E.
                  Jones) Mason
            SY=E. s. ssp. gymnocephalum
                  (Brand) Mason
    sparsiflorum (Eastw.) Mason
        ssp. harwoodii (Craig) H.K. Harrison
            SY=E. diffusum (Gray) Mason ssp.
                  h. (Craig) Mason
        ssp. sparsiflorum
    tracyi Mason
    virgatum (Benth.) Mason
    wilcoxii (A. Nels.) Mason
            SY=E. sparsiflorum (Eastw.) Mason
                  var. w. (A. Nels.) Cronq.

Gilia Ruiz & Pavon

    achilleifolia Benth.
        ssp. achilleifolia
        ssp. multicaulis (Benth.) V. & A. Grant
            SY=G. m. Benth.
            SY=G. m. ssp. pedunculata (Eastw.)
                  Mason & A. Grant
            SY=G. m. var. alba Milliken
    aliquanta A. & V. Grant
        ssp. aliquanta
        ssp. breviloba A. & V. Grant
    angelensis V. Grant
    australis (Mason & A. Grant) V. Grant
            SY=G. splendens Dougl. ex Lindl.
                  ssp. a. Mason & A. Grant
    austrooccidentalis (A. & V. Grant) A. & V.
        Grant
    brecciarum M.E. Jones
        ssp. argusana A. & V. Grant
        ssp. brecciarum
        ssp. neglecta A. & V. Grant
    caespitosa Gray
    campanulata Gray
    cana (M.E. Jones) Heller
        ssp. bernardina A. & V. Grant
        ssp. cana
            SY=G. latiflora (Gray) Gray ssp.
                  c. (M.E. Jones) Mason & A.
                  Grant
        ssp. speciformis A. & V. Grant
        ssp. speciosa (Jepson) A. & V. Grant
            SY=G. latiflora (Gray) Gray ssp.
                  s. (Jepson) Mason & A. Grant
        ssp. triceps (Brand) A. & V. Grant
            SY=G. latiflora (Gray) Gray ssp.
                  t. (Brand) Mason & A. Grant
    capillaris Kellogg
    capitata Sims
        ssp. abrotanifolia (Nutt. ex Greene) V.
            Grant
        ssp. capitata
        ssp. chamissonis (Greene) V. Grant
            SY=G. achilleifolia Benth. ssp. c.
                  (Greene) Brand
            SY=G. a. var. c. (Greene) A. Nels.
                  & J.F. Macbr.
        ssp. mediomontana V. Grant
        ssp. pacifica V. Grant
        ssp. pedemontana V. Grant
        ssp. staminea (Greene) V. Grant
            SY=G. achilleifolia Benth. ssp. s.
                  (Greene) Mason & A. Grant
        ssp. tomentosa (Eastw. ex Brand) V.
            Grant
    caruifolia Abrams
    clivorum (Jepson) V. Grant
    clokeyi Mason
    diegensis (Munz) A. & V. Grant
    filiformis Parry ex Gray
    flavocincta A. Nels.
        ssp. australis (A. & V. Grant) Day & V.
            Grant
            SY=G. ophthalmoides Brand ssp. a.
                  A. & V. Grant
        ssp. flavocincta
            SY=G. opthalmoides Brand ssp. f.
                  (A. Nels.) A. & V. Grant
    formosa Greene ex Brand
    haydenii Gray
            SY=G. montezumae Tidestrom &
                  Dayton
            SY=G. subnuda Torr. ex Gray ssp.
                  h. (Gray) Brand
    hutchinsifolia Rydb.
            SY=G. leptomeria Gray ssp. rubella
                  (Brand) Mason & A. Grant
    incisa Benth.
    inconspicua (Sm.) Sweet
    insignis (Brand) Cory & Parks
            SY=G. rigidula Benth. ssp. i.
                  Brand
    interior (Mason & A. Grant) A. Grant

Gilia (CONT.)
            SY=G. tenuiflora Benth. ssp. i.
            Mason & A. Grant
  inyoensis I.M. Johnston
  jacens A. & V. Grant
  latiflora (Gray) Gray
    ssp. cosana A. & V. Grant
    ssp. cuyamensis A. & V. Grant
    ssp. davyi (Milliken) A. & V. Grant
    ssp. elongata A. & V. Grant
    ssp. excellens (Brand) A. & V. Grant
    ssp. latiflora
  latifolia S. Wats. ex Parry
  leptalea (Gray) Greene
    ssp. bicolor Mason & A. Grant
    ssp. leptalea
    ssp. pinnatisecta Mason & A. Grant
  leptantha Parish
    ssp. leptantha
          SY=G. latiflora (Gray) Gray ssp.
          leptantha (Parish) Mason &
          A. Grant
    ssp. pinetorum A. & V. Grant
    ssp. purpusii (Milliken) A. & V. Grant
          SY=G. latiflora (Gray) Gray ssp.
          p. (Milliken) Mason & A.
          Grant
    ssp. salticola (Eastw.) A. & V. Grant
    ssp. transversa A. & V. Grant
  leptomeria Gray
          SY=G. subacaulis Rydb.
  ludens Shinners
  malior Day & V. Grant
  mcvickerae M.E. Jones
          SY=G. calcarea M.E. Jones
          SY=G. pinnatifida Nutt. var. c.
          (M.E. Jones) Brand
  mexicana A. & V. Grant
  micromeria Gray
          SY=G. leptomeria Gray ssp. m.
          (Gray) Mason & A. Grant
          SY=G. l. var. m. (Gray) Cronq.
  millefoliata Fisch. & Mey.
          SY= G. multicaulis Benth. ssp.
          millefoliata (Fisch. & Mey.)
          Mason & A. Grant
  minor A. & V. Grant
  modocensis Eastw.
  nevinii Gray
          SY=G. multicaulis Benth. ssp. n.
          (Gray) Mason & A. Grant
  nyensis Reveal
  ochroleuca M.E. Jones
    ssp. bizonata A. & V. Grant
    ssp. exilis (Gray) A. & V. Grant
          SY=G. abramsii (Brand) Mason & A.
          Grant
          SY=G. a. ssp. integrifolia Mason &
          A. Grant
          SY=G. latiflora (Gray) Gray ssp.
          e. (Gray) Mason & A. Grant
    ssp. ochroleuca
    ssp. vivida (A. & V. Grant) A. & V.
      Grant
          SY=G. leptantha Parish ssp. v. A.
          & V. Grant
  ophthalmoides Brand
  penstemonoides M.E. Jones
  perennans Shinners
  pinnatifida Nutt.
          SY=G. sedifolia Brandeg.
  rigidula Benth.
    ssp. acerosa (Gray) Wherry
          SY=G. a. (Gray) Britt.
          SY=G. r. var. a. Gray
    ssp. rigidula
  ripleyi Barneby

  scopulorum M.E. Jones
  sinuata Dougl. ex Benth.
  splendens Dougl. ex Lindl.
    ssp. grantii (Brand) V. & A. Grant
    ssp. splendens
          SY=G. s. ssp. grinnellii (Brand)
          Mason & A. Grant
  stellata Heller
  stenothyrsa Gray
  stewartii I.M. Johnston
  subnuda Torr. ex Gray
          SY=G. s. ssp. superba (Eastw.)
          Brand
  tenerrima Gray
  tenuiflora Benth.
    ssp. amplifaucalis A. & V. Grant
    ssp. arenaria (Benth.) A. & V. Grant
    ssp. hoffmannii (Eastw.) A. & V. Grant
    ssp. tenuiflora
  tetrabeccia A. & V. Grant
  transmontana (Mason & A. Grant) A. & V.
    Grant
          SY=G. ochroleuca M.E. Jones ssp.
          t. Mason & A. Grant
  tricolor Benth.
    ssp. diffusa (Congd.) Mason & A. Grant
    ssp. tricolor
  triodon Eastw.
  tweedyi Rydb.
          SY=G. sinuata Dougl. ex Benth.
          var. t. (Rydb.) Cronq.

Gymnosteris Greene

  nudicaulis (Hook. & Arn.) Greene
          SY=G. n. var. pulchella (Greene)
          Brand
  parvula (Rydb.) Heller

Ipomopsis Michx.

  aggregata (Pursh) V. Grant
    ssp. aggregata
          SY=Gilia a. (Pursh) Spreng.
          SY=G. a. ssp. euaggregata Brand
          SY=G. a. ssp. formosissima
          (Greene) Wherry
          SY=G. arizonica (Greene) Rydb.
          ssp. texana (Greene) Wherry
          SY=G. t. (Greene) Woot. & Standl.
          SY=I. aggregata var. t. (Greene)
          Shinners
    ssp. arizonica (Greene) V. & A. Grant
          SY=Gilia aggregata (Pursh) Spreng.
          var. arizonica (Greene)
          Fosberg
          SY=G. arizonica (Greene) Rydb.
    ssp. attenuata (Gray) V. & A. Grant
          SY=Gilia aggregata (Pursh) Spreng.
          var. attenuata Gray
          SY=G. attenuata (Gray) A. Nels.
    ssp. bridgesii (Gray) V. & A. Grant
          SY=Gilia aggregata (Pursh) Spreng.
          var. b. Gray
          SY=I. b. (Gray) Wherry
  candida (Rydb.) W.A. Weber
          SY=Gilia c. Rydb.
          SY=G. c. ssp. collina (Greene)
          Wherry
          SY=G. candida ssp. vera Wherry
  congesta (Hook.) V. Grant
    ssp. congesta
          SY=Gilia c. Hook.
          SY=G. c. var. burleyana (A. Nels.)
          Constance & Rollins
    ssp. crebrifolia (Nutt.) Day
          SY=Gilia congesta Hook. var.

Ipomopsis (CONT.)
            crebrifolia (Nutt.) Gray
   ssp. frutescens (Rydb.) Day
      SY=Gilia congesta Hook. var. f.
         (Rydb.) Cronq.
      SY=I. f. (Rydb.) V. Grant
   ssp. montana (A. Nels. & Kennedy) V.
    Grant
      SY=Gilia congesta Hook. var. m.
        (A. Nels. & Kennedy)
        Constance & Rollins
   ssp. palmifrons (Brand) Day
      SY=Gilia congesta Hook. ssp. p.
        Brand
      SY=G. c. var. p. (Brand) Cronq.
   ssp. pseudotypica (Constance & Rollins)
    Day
      SY=Gilia congesta Hook. var. p.
        Constance & Rollins
   ssp. viridis (Cronq.) Day
      SY=Gilia congesta Hook. var. v.
        Cronq.
depressa (M.E. Jones ex Gray) V. Grant
      SY=Gilia d. M.E. Jones ex Gray
gunnisonii (Torr. & Gray) V. Grant
      SY=Gilia g. Torr. & Gray
havardii (Gray) V. Grant
      SY=Gilia h. Gray
laxiflora (Coult.) V. Grant
      SY=Gilia l. (Coult.) Osterhout
longiflora (Torr.) V. Grant
      SY=Gilia l. (Torr.) G. Don
macombii (Torr. ex Gray) V. Grant
      SY=Gilia m. Torr. ex Gray
      SY=G. pringlei Gray
minutiflora (Benth.) V. Grant
      SY=Gilia m. Benth.
multiflora (Nutt.) V. Grant
      SY=Gilia campylantha Woot. &
        Standl.
      SY=G. m. Nutt.
polyantha (Rydb.) V. Grant
      SY=Gilia brachyantha Woot. &
        Standl.
      SY=G. p. Rydb.
      SY=G. p. var. whitingii Kearney &
        Peebles
polycladon (Torr.) V. Grant
      SY=Gilia p. Torr.
pumila (Nutt.) V. Grant
      SY=Gilia p. Nutt.
roseata (Rydb.) V. Grant
      SY=Gilia r. Rydb.
rubra (L.) Wherry
      SY=Gilia r. (L.) Heller
spicata (Nutt.) V. Grant
   ssp. capitata (Gray) V. Grant
      SY=Gilia cephaloidea Rydb. pro
        parte
      SY=G. s. var. capitata Gray
      SY=G. s. var. cephaloidea (Rydb.)
        Constance & Rollins pro
        parte
      SY=G. s. var. orchidacea (Brand)
        Cronq.
      SY=I. globularis (Brand) W.A.
        Weber
   ssp. spicata
      SY=G. s. Nutt.
      SY=G. s. var. tridactyla (Rydb.)
        Constance & Rollins
tenuifolia (Gray) V. Grant
      SY=Loeselia t. Gray
tenuituba (Rydb.) V. Grant
      SY=Gilia aggregata (Pursh) Spreng.
        var. macrosiphon Kearney &
        Peebles

      SY=G. t. Rydb.
thurberi (Torr. ex Gray) V. Grant
      SY=Gilia t. Torr. ex Gray
wrightii (Gray) Gould
      SY=Gilia w. Gray

Langloisia Greene

  matthewsii (Gray) Greene
  punctata (Gray ex Coville) Goodding
      SY=L. lanata Brand
  schottii (Torr.) Greene
      SY=L. flaviflora A. Davids.
  setosissima (Torr. & Gray) Greene

Leptodactylon Hook. & Arn.

  caespitosum Nutt.
  californicum Hook. & Arn.
    ssp. californicum
    ssp. glandulosum (Eastw.) Mason
  jaegeri (Munz) Wherry
  pungens (Torr.) Nutt. ex Rydb.
    ssp. hallii (Parish) Mason
       SY=Gilia h. Parish
    ssp. hookeri (Dougl. ex Hook.) Wherry
       SY=L. p. ssp. brevifolium (Rydb.)
         Wherry
       SY=L. p. var. h. (Dougl. ex Hook.)
         Jepson
    ssp. pulchriflora (Brand) Mason
       SY=L. lilacinum Greene ex Baker
       SY=L. pungens ssp. squarrosum
         (Gray) Tidestrom
    ssp. pungens
       SY=L. hazelae M.E. Peck
       SY=L. p. ssp. eupungens (Brand)
         Wherry
  watsonii (Gray) Rydb.

Linanthus Benth.

  acicularis Greene
  ambiguus (Rattan) Greene
  androsaceus (Benth.) Greene
    ssp. androsaceus
    ssp. croceus (Milliken) Mason
    ssp. laetus (Benth.) Mason
    ssp. luteolus (Greene) Mason
       SY=L. a. ssp. plaskettii (Eastw.)
        Mason
    ssp. luteus (Benth.) Mason
    ssp. micranthus (Steud.) Mason
  arenicola (M.E. Jones) Jepson & V. Bailey
  aureus (Nutt.) Greene
    ssp. aureus
    ssp. decorus (Gray) Mason
       SY=L. a. var. d. (Gray) Jepson
  bakeri Mason
  bellus (Gray) Greene
  bicolor (Nutt.) Greene
    ssp. bicolor
    ssp. minimus Mason
       SY=L. b. var. m. (Mason) Cronq.
  bigelovii (Gray) Greene
      SY=Gilia b. Gray
  bolanderi (Gray) Greene
  breviculus (Gray) Greene
      SY=L. b. ssp. royalis (Brand)
        Mason
  ciliatus (Benth.) Greene
    var. ciliatus
       SY=Gilia c. Benth.
    var. neglectus (Greene) Jepson
       SY=L. n. Greene
  concinnus Milliken
  demissus (Gray) Greene

Linanthus (CONT.)
    dianthiflorus (Benth.) Greene
        ssp. dianthiflorus
        ssp. farinosus (Brand) Mason
    dichotomus Benth.
        ssp. dichotomus
        ssp. meridianus (Eastw.) Mason
    filipes (Benth.) Greene
    floribundus (Gray) Greene ex Milliken
        ssp. floribundus
            SY=Linanthastrum f. (Gray) Wherry
            SY=Linanthastrum nuttallii (Gray)
                Ewan ssp. f. (Gray) Ewan
            SY=Linanthus n. (Gray) Greene ex
                Milliken ssp. f. (Gray) Munz
        ssp. glabrus Patterson
        ssp. hallii (Jepson) Mason
    grandiflorus (Benth.) Greene
    harknessii (Curran) Greene
        ssp. condensatus Mason
        ssp. harknessii
    jonesii (Gray) Greene
            SY=Linanthus bigelovii (Gray)
                Greene var. j. (Gray) Jepson
                & Mason
    killipii Mason
    lemmonii (Gray) Greene
    liniflorus (Benth.) Greene
        ssp. liniflorus
        ssp. pharnaceoides (Benth.) Mason
            SY=Linanthus l. var. p. (Benth.)
                Gray
            SY=L. p. (Benth.) Greene
    maculatus (Parish) Milliken
    melingii (Wiggins) V. Grant
            SY=Linanthastrum m. (Wiggins)
                Wherry
    montanus (Greene) Greene
    nudatus Greene
    nuttallii (Gray) Greene ex Milliken
        ssp. nuttallii
            SY=Leptodactylon n. (Gray) Rydb.
            SY=Linanthastrum n. (Gray) Ewan
        ssp. pubescens Patterson
        ssp. tenuilobus Patterson
    oblanceolatus (Brand) Eastw. ex Jepson
            SY=Linanthus tularensis (Brand)
                Mason
            SY=L. t. ssp. culbertsonii (Brand)
                Mason
    orcuttii (Parry & Gray) Jepson
            SY=Linanthus o. ssp. pacificus
                (Milliken) Mason
    pachyphyllus Patterson
    parryae (Gray) Greene
    pygmaeus (Brand) J.T. Howell
        ssp. continentalis Raven
        ssp. pygmaeus
    rattanii (Gray) Greene
    septentrionalis Mason
            SY=Linanthus harknessii (Curran)
                Greene var. s. (Mason)
                Jepson & V. Bailey
    serrulatus Greene

Loeselia L.

    glandulosa (Cav.) G. Don
    mexicana (Lam.) Brand
    scariosa (Mart. & Gal.) Walp.

Microsteris Greene

    gracilis (Hook.) Greene
        ssp. gracilis
            SY=Phlox g. (Hook.) Greene
        ssp. humilis (Greene) V. Grant

            SY=M. g. var. humilior (Hook.)
                Cronq.
            SY=M. humilis (Greene) Greene
            SY=M. micrantha (Kellogg) Greene
            SY=Phlox g. (Hook.) Greene ssp.
                humilis (Greene) Mason
            SY=P. g. var. humilior (Hook.)
                Boivin

Navarretia Ruiz & Pavon

    atractyloides (Benth.) Hook. & Arn.
    bakeri Mason
    breweri (Gray) Greene
    cotulifolia (Benth.) Hook. & Arn.
    divaricata (Torr. ex Gray) Greene
        var. divaricata
        var. vividior (Jepson & V. Bailey)
            Mason
    eriocephala Mason
    filicaulis (Torr.) Greene
    fossalis Moran
    hamata Greene
        ssp. foliacea (Greene) Mason
            SY=N. h. var. f. (Greene) Thorne
        ssp. hamata
        ssp. leptantha (Greene) Mason
    heterandra Mason
    heterodoxa (Greene) Greene
        ssp. heterodoxa
        ssp. rosulata (Brand) Mason
    intertexta (Benth.) Hook.
            SY=N. minima Nutt. var. i.
                (Benth.) Boivin
    jepsonii V. Bailey ex Jepson
    leucocephala Benth.
    mellita Greene
    minima Nutt.
    mitracarpa Greene
        ssp. jaredii (Eastw.) Mason
            SY=N. j. Eastw.
        ssp. mitracarpa
    nigelliformis Greene
    pauciflora Mason
    peninsularis Greene
    plieantha Mason
    prolifera Greene
        ssp. lutea (Brand) Mason
            SY=N. p. var. l. Brand
        ssp. prolifera
    propinqua Suksdorf
            SY=N. intertexta (Benth.) Hook.
                var. p. (Suksdorf) Brand
    prostrata (Gray) Greene
    pubescens (Benth.) Hook. & Arn.
    setiloba Coville
    squarrosa (Eschsch.) Hook. & Arn.
    subuligera Greene
    tagetina Greene
    viscidula Benth.
        ssp. purpurea (Greene) Mason
        ssp. viscidula

Phlox L.

    aculeata A. Nels.
    adsurgens Torr. ex Gray
    albomarginata M.E. Jones
        ssp. albomarginata
            SY=P. a. ssp. minor (M.E. Jones)
                Wherry
            SY=P. a. ssp. vera Wherry
        ssp. diapensioides (Rydb.) Wherry
    alyssifolia Greene
        ssp. abdita (A. Nels.) Wherry
        ssp. alyssifolia
            SY=P. a. ssp. vera Wherry

Phlox (CONT.)
    ssp. collina (Rydb.) Wherry
  amabilis Brand
  amoena Sims
    ssp. amoena
    ssp. lighthipei (Small) Wherry
  amplifolia Britt.
  andicola Nutt. ex Gray
    ssp. andicola
      SY=P. a. ssp. planitiarum (A.
        Nels.) Wherry
      SY=P. a. ssp. typica Wherry
    ssp. parvula Wherry
  austromontana Coville
    ssp. austromontana
      SY=P. a. ssp. vera Wherry
    ssp. densa (Brand) Wherry
    ssp. prostrata (E. Nels.) Wherry
      SY=P. a. var. p. E. Nels.
  bifida Beck
    ssp. bifida
      SY=P. b. ssp. arkansana Marsh. pro
        parte
      SY=P. b. var. glandifera Wherry
    ssp. stellaria (Gray) Wherry
      SY=P. b. var. cedaria (Brand)
        Fern.
      SY=P. b. var. s. (Gray) Wherry
      SY=P. s. Gray
  bryoides Nutt.
  buckleyi Wherry
  caespitosa Nutt.
    ssp. caespitosa
      SY=P. c. ssp. eucaespitosa Brand
    ssp. condensata (Gray) Wherry
      SY=P. condensata (Gray) E. Nels.
    ssp. platyphylla Wherry
    ssp. pulvinata Wherry
      SY=P. p. (Wherry) Cronq.
      SY=P. sibirica L. ssp. p. (Wherry)
        W.A. Weber
  carolina L.
    ssp. alta Wherry
    ssp. angusta Wherry
      SY=P. c. var. a. (Wherry)
        Steyermark
    ssp. carolina
    ssp. turritella Wherry
  caryophylla Wherry
  cluteana A. Nels.
  colubrina Wherry & Constance
  covillei E. Nels.
  cuspidata Scheele
      SY=P. c. var. humilis Whitehouse
      SY=P. c. var. grandiflora
        Whitehouse
  diffusa Benth.
    ssp. diffusa
    ssp. longistylis Wherry
      SY=P. d. var. l. (Wherry) M.E.
        Peck
    ssp. scleranthifolia (Rydb.) Wherry
    ssp. subcarinata Wherry
      SY=P. cyanea Eastw.
  dispersa C.W. Sharsmith
  divaricata L.
    ssp. divaricata
    ssp. laphamii (Wood) Wherry
      SY=P. d. var. l. Wood
  dolichantha Gray
  douglasii Hook.
      SY=P. d. ssp. eudouglasii Brand
  drummondii Hook.
    ssp. drummondii
      SY=P. d. ssp. d. var. peregrina
        Shinners
      SY=P. goldsmithii Whitehouse

    ssp. johnstonii (Wherry) Wherry
      SY=P. j. Wherry
    ssp. mcallisterii (Whitehouse) Wherry
      SY=P. d. var. m. (Whitehouse)
        Shinners
    ssp. tharpii (Whitehouse) Wherry
      SY=P. glabriflora (Brand)
        Whitehouse ssp. t.
        (Whitehouse) Wherry
      SY=P. t. Whitehouse
    ssp. wilcoxiana (Bogusch) Wherry
      SY=P. d. var. w. (Bogusch)
        Whitehouse
  floridana Benth.
      SY=P. f. ssp. bella Wherry
  glaberrima L.
    ssp. glaberrima
      SY=P. g. var. melampyrifolia
        (Salisb.) Wherry
    ssp. interior (Wherry) Wherry
      SY=P. g. var. i. Wherry
    ssp. triflora (Michx.) Wherry
      SY=P. carolina L. var. t. (Michx.)
        Steyermark
  glabriflora (Brand) Whitehouse
    ssp. glabriflora
    ssp. littoralis (Cory) Wherry
      SY=P. l. (Cory) Whitehouse
  gladiformis (M.E. Jones) E. Nels.
  X glutinosa Buckl. [divaricata ssp.
    laphamii X pilosa]
  grahamii Wherry
  grayi Woot. & Standl.
      SY=P. longifolia Nutt. ssp.
        brevifolia (Gray) Mason
  griseola Wherry
    ssp. griseola
    ssp. tumulosa Wherry
  hendersonii (E. Nels.) Cronq.
      SY=P. douglasii Hook. ssp. h. (E.
        Nels.) Wherry
  hirsuta E. Nels.
      SY=P. stansburyi (Torr.) Heller
        var. h. (E. Nels.) Jepson
  hoodii Richards.
    ssp. canescens (Torr. & Gray) Wherry
      SY=P. c. Torr. & Gray
      SY=P. h. var. c. (Torr. & Gray)
        M.E. Peck
    ssp. glabrata (E. Nels.) Wherry
    ssp. hoodii
      SY=P. h. ssp. genuina Wherry
    ssp. lanata (Piper) Munz
      SY=P. l. Piper
    ssp. muscoides (Nutt.) Wherry
      SY=P. m. Nutt.
    ssp. viscidula (Wherry) Wherry
      SY=P. h. var. v. Wherry
  idahonis Wherry
  jonesii Wherry
  kelseyi Britt.
    ssp. glandulosa Wherry
    ssp. kelseyi
      SY=P. k. ssp. genuina Wherry
    ssp. salina (M.E. Jones) Wherry
  longifolia Nutt.
    ssp. calva Wherry
    ssp. cortezana (A. Nels.) Wherry
    ssp. humilis (Dougl. ex Hook.) Wherry
    ssp. longifolia
      SY=P. l. ssp. typica Wherry
      SY=P. l. var. puberula E. Nels.
  longipilosa Waterfall
  maculata L.
    ssp. maculata
      SY=P. m. var. odorata (Sweet)
        Wherry

Phlox (CONT.)
  ssp. pyramidalis (Sm.) Wherry
    SY=P. m. var. purpurea Fern.
    SY=P. m. ssp. pyramidalis (Sm.)
      Wherry
    SY=P. m. var. pyramidalis (Sm.)
      Wherry
    SY=P. pyramidalis Sm.
 mesoleuca Greene
 missoulensis Wherry
    SY=P. kelseyi Britt. var. m.
      (Wherry) Cronq.
 mollis Wherry
    SY=P. subulata L. var. ciliata
      (Brand) Wherry
 multiflora A. Nels.
  ssp. depressa (E. Nels.) Wherry
    SY=P. m. ssp. costata (Rydb.)
      Wherry
  ssp. multiflora
    SY=P. m. ssp. typica Wherry
  ssp. patula (A. Nels.) Wherry
    SY=P. p. A. Nels.
 nana Nutt.
 nivalis Lodd. ex Sweet
  ssp. hentzii (Nutt.) Wherry
    SY=P. h. Nutt.
  ssp. nivalis
  ssp. texensis Lundell
 oklahomensis Wherry
 ovata L.
 paniculata L.
 peckii Wherry
 pilosa L.
  ssp. deamii Levin
  ssp. detonsa (Gray) Wherry
    SY=P. p. var. d. Gray
  ssp. fulgida (Wherry) Wherry
    SY=P. p. var. f. Wherry
  ssp. latisepala Wherry
    SY=P. p. var. aspera (E. Nels.)
      Wherry ex Gould
  ssp. ozarkana (Wherry) Wherry
    SY=P. p. var. o. Wherry
  ssp. pilosa
    SY=P. argillacea Clute & Ferris
    SY=P. p. var. virens (Michx.)
      Wherry
    SY=P. villosissima (Gray) Small
      pro parte
  ssp. pulcherrima Lundell
    SY=P. pilosa var. amplexicaulis
      (Raf.) Wherry
  ssp. riparia Wherry
  ssp. sangamonensis Levin & Sm.
 pulchra Wherry
 richardsonii Hook.
  ssp. alaskensis (Jordal) Wherry
    SY=P. sibirica L. var. a. (Jordal)
      Boivin
  ssp. richardsonii
    SY=P. sibirica L. ssp. r. (Hook.)
      Hulten
    SY=P. s. var. r. (Hook.) Welsh
 rigida Benth. ex DC.
    SY=P. douglasii Hook. ssp. r.
      (Benth. ex DC.) Wherry
 roemeriana Scheele
 X rugelii Brand [amoena X divaricata]
 sibirica L.
  ssp. borealis (Wherry) Shetler
    SY=P. b. Wherry
    SY=P. s. var. b. (Wherry) Boivin
  ssp. sibirica
 speciosa Pursh
  ssp. lanceolata (E. Nels.) Wherry
  ssp. lignosa Brand

  ssp. nitida (Suksdorf) Wherry
    SY=P. s. var. n. Suksdorf
  ssp. occidentalis (Dur. ex Torr.) Wherry
    SY=P. s. var. o. (Dur. ex Torr.)
      M.E. Peck
  ssp. speciosa
  ssp. woodhousei (Gray) Wherry
    SY=P. s. ssp. euspeciosa Brand
    SY=P. s. var. w. Gray
    SY=P. w. (Gray) E. Nels.
 stansburyi (Torr.) Heller
    SY=P. s. ssp. eustansburyi Brand
    SY=P. s. var. brevifolia (Gray) E.
      Nels.
 stolonifera Sims
 subulata L.
  ssp. australis (Wherry) Wherry
    SY=P. s. var. a. Wherry
  ssp. brittonii (Small) Wherry
    SY=P. b. Small
    SY=P. s. var. b. (Small) Wherry
  ssp. subulata
 superba Brand
 tenuifolia E. Nels.
 triovulata Thurb. ex Torr.
    SY=P. nana Nutt. ssp. glabella
      (Gray) Brand
 variabilis Brand
  ssp. nudata Wherry
  ssp. variabilis
 viridis E. Nels.
  ssp. compacta (Brand) Wherry
    SY=P. longifolia Nutt. ssp. c.
      (Brand) Wherry
  ssp. longipes (M.E. Jones) Wherry
    SY=P. longifolia Nutt. ssp.
      longipes (M.E. Jones) Wherry
  ssp. viridis
    SY=P. longifolia Nutt. ssp. v. (E.
      Nels.) Wherry
    SY=P. l. var. v. (E. Nels.)
      Peabody
 viscida E. Nels.

Polemonium L.

 boreale J.E. Adams
  ssp. boreale
    SY=P. b. ssp. b. var.
      villosissimum Hulten
    SY=P. b. ssp. richardsonii
      (Graham) J.P. Anders.
  ssp. macranthum (Cham.) Hulten
 brandegei (Gray) Greene
 caeruleum L.
  ssp. amygdalinum (Wherry) Munz
    SY=P. helleri Brand
    SY=P. intermedium (Brand) Rydb.
  ssp. villosum (J.H. Rudolph ex Georgi)
    Brand
    SY=P. acutiflorum Willd. ex Roemer
      & Schultes
 californicum Eastw.
    SY=P. columbianum Rydb.
 carneum Gray
    SY=P. c. ssp. luteum (Gray) Brand
 chartaceum Mason
 cuspidatum Sieb. & Zucc.
 delicatum Rydb.
  ssp. delicatum
    SY=P. d. ssp. typicum Wherry
    SY=P. pulcherrimum Hook. ssp. d.
      (Rydb.) Brand
    SY=P. p. var. d. (Rydb.) Cronq.
  ssp. scopulinum (Greene) Wherry
 elegans Greene
 eximium Greene

Polemonium (CONT.)
  foliosissimum Gray
        var. alpinum (Brand) Wherry
            SY=P. f. ssp. albiflorum (Eastw.)
                Brand
            SY=P. f. ssp. decurrens (Brand)
                Wherry
        var. flavum (Greene) Anway
            SY=P. filicinum Greene
            SY=P. flavum Greene
        var. foliosissimum
            SY=P. f. ssp. robustum (Rydb.)
                Brand
            SY=P. f. ssp. verum Wherry
        var. molle (Greene) Anway
  longii Fern.
  micranthum Benth.
            SY=Polemoniella m. (Benth.) Heller
  nevadense Wherry
  occidentale Greene ex Brand
        var. lacustre (Wherry) Lakela
        var. occidentale
            SY=Polemonium caeruleum L. ssp. o.
                (Greene ex Brand) J.F.
                Davids.
            SY=P. o. ssp. typicum Wherry
  pauciflorum S. Wats.
      ssp. hinckleyi (Standl.) Wherry
            SY=Polemonium h. Standl.
      ssp. pauciflorum
            SY=Polemonium p. ssp. typicum
                Wherry
  pectinatum Greene
  pulcherrimum Hook.
        var. calycinum (Eastw.) Brand
            SY=Polemonium p. ssp. tricolor
                (Eastw.) Brand
        var. pilosum (Greenm.) J.F. Davids.
        var. pulcherrimum
            SY=Polemonium berryi Eastw.
            SY=P. haydenii A. Nels.
            SY=P. shastense Baker ex Eastw.
  reptans L.
            SY=Polemonium r. var. villosum
                E.L. Braun
  vanbruntiae Britt.
            SY=Polemonium caeruleum L. ssp. v.
                (Britt.) J.F. Davids.
  viscosum Nutt.
            SY=Polemonium v. ssp. lemmonii
                (Brand) Wherry
            SY=P. v. ssp. mellitum (Gray) J.F.
                Davids.
            SY=P. v. ssp. genuinum Wherry

POLYGALACEAE

Monnina Ruiz & Pavon

    wrightii Gray

Polygala L.

    aboriginum Small
    acanthoclada Gray
        var. acanthoclada
        var. intricata Eastw.
    alba Nutt.
        var. alba
        var. suspecta S. Wats.
    balduinii Nutt.
        var. balduinii
            SY=Pilostaxis b. (Nutt.) Small
        var. carteri (Small) R.R. Sm. & Ward

            SY=Pilostaxis c. (Small) Small
        var. chlorgena Torr. & Gray
    barbeyana Chod.
    boykinii Nutt.
        var. boykinii
        var. sparsifolia Wheelock
            SY=Polygala flagellaris Small
            SY=P. praetervisa Chod.
        var. suborbicularis R.W. Long
    brevifolia Nutt.
    brizoides St. Hil.
    californica Nutt.
    chapmanii Torr. & Gray
    cornuta Kellogg
        var. cornuta
        var. fishiae (Parry) Jepson
            SY=Polygala c. ssp. f. (Parry)
                Munz
            SY=P. f. Parry
        var. pollardii Munz
            SY=Polygala c. ssp. p. (Munz) Munz
    cowellii (Britt.) Blake
            SY=Phlebotaenia c. Britt.
    crenata C.W. James
    cruciata L.
            SY=Polygala c. var. aquilonia
                Fern. & Schub.
            SY=P. c. var. cuspidata (Hook. &
                Arn.) Wood
            SY=P. ramosior (Nash) Small
    curtissii Gray
    cymosa Walt.
            SY=Pilostaxis c. (Walt.) Small
    glandulosa H.B.K.
    glochidiata H.B.K.
    grandiflora Walt.
        var. angustifolia Torr. & Gray
        var. grandiflora
            SY=Asemeia cumulicola Small
            SY=A. g. (Walt.) Small
            SY=A. miamiensis Small
        var. leiodes Blake
            SY=Asemeia l. (Blake) Small
    hecatantha Urban
    hemipterocarpa Gray
    hookeri Torr. & Gray
    incarnata L.
            SY=Galypola i. (L.) Nieuwl.
    leptocaulis Torr. & Gray
    leptostachys Shuttlw.
    lewtonii Small
    lindheimeri Gray
    longa Blake
    longicaulis H.B.K.
    lutea L.
            SY=Pilostaxis l. (L.) Small
    macradenia Gray
    maravillasensis Correll
    mariana P. Mill.
            SY=Polygala harperi Small
    minutifolia Rose
    nana (Michx.) DC.
            SY=Pilostaxis n. (Michx.) Raf.
    nuttallii Torr. & Gray
    obscura Benth.
    orthotricha Blake
    ovatifolia Gray
    palmeri S. Wats.
    paniculata L.
    paucifolia Willd.
            SY=Triclisperma p. (Willd.)
                Nieuwl.
    penaea L.
            SY=Badiera p. (L.) DC.
            SY=B. portoricensis Britt.
            SY=Polygala portoricensis (Britt.)
                Blake

Polygala (CONT.)
  piliophora Blake
  polygama Walt.
          SY=Polygala p. var. obtusata Chod.
          SY=P. p. var. ramulosa Farw.
  racemosa Blake
  ramosa Ell.
          SY=Pilostaxis r. (Ell.) Small
  reducta Blake
  rimulicola Steyermark
  rugelii Shuttlw. ex Chapman
          SY=Pilostaxis r. (Shuttlw. ex
                Chapman) Small
  rusbyi Greene
  sanguinea L.
          SY=Polygala viridescens L.
  scoparioides Chod.
  senega L.
          var. latifolia Torr. & Gray
          var. senega
  serpyllifolia House
  setacea Michx.
  smallii R.R. Sm. & Ward
          SY=Pilostaxis arenicola (Small)
                Small
  subspinosa S. Wats.
          var. heterorhyncha Barneby
          var. subspinosa
  tweedyi Britt.
  verticillata L.
          var. ambigua (Nutt.) Wood
          SY=Polygala a. Nutt.
          var. dolichoptera Fern.
          var. isocycla Fern.
          var. sphenostachya Pennell
          var. verticillata
          SY=Polygala pretzii Pennell
  vulgaris L.

Securidaca L.

  diversifolia (L.) Blake
          SY=Elsota d. (L.) Blake
  virgata Sw.
          SY=Elsota v. (Sw.) Kuntze

POLYGONACEAE

Antigonon Endl.

  flavescens S. Wats.
  guatemalense Meisn.
          SY=A. macrocarpum Britt. & Small
  leptopus Hook. & Arn.
          SY=Coreulum l. (Hook. & Arn.)
                Stuntz

Brunnichia Banks ex Gaertn.

  ovata (Walt.) Shinners
          SY=B. cirrhosa Gaertn.

Centrostegia Gray ex Benth.

  insignis (Curran) Heller
          SY=Chorizanthe i. Curran
          SY=Oxytheca i. (Curran) Goodman
  leptoceras Gray
          SY=Chorizanthe l. (Gray) S. Wats.
          SY=Eriogonella l. (Gray) Goodman
  thurberi Gray ex Benth.
          SY=Chorizanthe t. (Gray ex Benth.)
                S. Wats.
  vortriedei (Brandeg.) Goodman

          SY=Chorizanthe v. Brandeg.

Chorizanthe R. Br. ex Benth.

  angustifolia Nutt.
          SY=C. a. var. eastwoodae Goodman
  biloba Goodman
          SY=C. palmeri S. Wats. var. b.
                (Goodman) Munz
  blakleyi Hardham
  brevicornu Torr.
          var. brevicornu
          var. spathulata (Small ex Rydb.) C.L.
                Hitchc.
          SY=C. b. ssp. s. (Small ex Rydb.)
                Munz
          SY=C. s. Small ex Rydb.
  breweri S. Wats.
  clevelandii Parry
  corrugata (Torr.) Torr. & Gray
          SY=Acanthogonum c. Torr.
  cuspidata S. Wats.
          var. cuspidata
          var. marginata Goodman
          var. villosa (Eastw.) Munz
          SY=C. v. Eastw.
  diffusa Benth.
          var. diffusa
          var. nivea (Curran) Hoover
          SY=C. n. (Curran) Heller
          SY=C. pungens Benth. var. n.
                Curran
  douglasii Benth.
  fimbriata Nutt.
          var. fimbriata
          var. laciniata (Torr.) Jepson
  howellii Goodman
  leptotheca Goodman
  membranacea Benth.
          SY=Eriogonella m. (Benth.) Goodman
  obovata Goodman
  orcuttiana Parry
  palmeri S. Wats.
  parryi S. Wats.
          var. fernandia (S. Wats.) Jepson
          var. parryi
  polygonoides Torr. & Gray
          var. longispina (Goodman) Munz
          SY=Acanthogonum p. (Torr. & Gray)
                Goodman var. l. Goodman
          SY=C. p. ssp. l. (Goodman) Munz
          var. polygonoides
          SY=Acanthogonum p. (Torr. & Gray)
                Goodman
  procumbens Nutt.
          var. albiflora Goodman
          var. procumbens
  pungens Benth.
          var. hartwegii (Benth.) Goodman
          var. pungens
  rectispina Goodman
  rigida (Torr.) Torr. & Gray
          SY=Acanthogonum r. Torr.
  robusta Parry
  spinosa S. Wats.
          SY=Eriogonella s. (S. Wats.)
                Goodman
  staticoides Benth.
          var. brevispina Goodman
          var. compacta (Goodman) Reveal
          SY=C. chrysacantha Goodman
          SY=C. chrysacantha var. compacta
                Goodman
          SY=C. s. ssp. chrysacantha
                (Goodman) Munz
          var. elata Goodman
          var. latiloba Goodman

Chorizanthe (CONT.)
      var. staticoides
         SY=C. discolor Nutt.
  stellulata Benth.
  uniaristata Torr. & Gray
  valida S. Wats.
  ventricosa Goodman
        SY=C. palmeri S. Wats. var. v.
           (Goodman) Munz
  watsonii Torr. & Gray
  wheeleri S. Wats.
  xantii S. Wats.
      var. leucotheca Goodman
        SY=C. x. ssp. l. (Goodman) Munz
      var. xantii

Coccoloba P. Br.

  costata C. Wright ex Sauvalle
        SY=C. rupicola Urban
  diversifolia Jacq.
        SY=C. laurifolia sensu auctt. non
           Jacq.
  krugii Lindau
  microstachya Willd.
        SY=C. obtusifolia sensu auctt. non
           Jacq.
  pallida C. Wright
  pubescens L.
        SY=C. grandifolia Jacq.
  pyrifolia Desf.
  rugosa Desf.
  sintenisii Urban ex Lindau
  swartzii Meisn.
      var. portoricensis Meisn.
      var. swartzii
        SY=C. borinquensis Britt.
  tenuifolia L.
  uvifera (L.) L.
  venosa L.

Dedeckera Reveal & J.T. Howell

  eurekensis Reveal & J.T. Howell

Emex Campd.

  australis Steinh.
  spinosa (L.) Campd.

Eriogonum Michx.

  abertianum Torr.
      var. abertianum
        SY=E. a. var. gillespiei Fosberg
        SY=E. a. var. neomexicanum Gandog.
        SY=E. a. var. villosum Fosberg
      var. cyclosepalum (Greene) Fosberg
  acaule Nutt.
        SY=E. caespitosum Nutt. var. a.
           (Nutt.) R.J. Davis
  alatum Torr.
      var. alatum
        SY=E. a. ssp. triste (S. Wats.) S.
           Stokes
        SY=E. t. S. Wats.
      var. glabriusculum Torr.
      var. mogollense S. Stokes ex M.E.
        Jones
  aliquantum Reveal
  allenii S. Wats.
  alpinum Engelm.
  ammophilum Reveal
  ampullaceum J.T. Howell
        SY=E. mohavense S. Wats. ssp. a.
           (J.T. Howell) S. Stokes
  androsaceum Benth.

        SY=E. flavum Nutt. var. a.
           (Benth.) M.E. Jones
  anemophilum Greene
        SY=E. ochrocephalum S. Wats. ssp.
           a. (Greene) S. Stokes
  angulosum Benth.
  annuum Nutt.
  apachense Reveal
  apiculatum S. Wats.
  apricum J.T. Howell
      var. apricum
      var. prostratum Myatt
  arborescens Greene
  aretioides Barneby
  argillosum J.T. Howell
  argophyllum Reveal
  arizonicum S. Stokes ex M.E. Jones
  baileyi S. Wats.
      var. baileyi
        SY=E. vimineum Dougl. ex Benth.
          var. b. (S. Wats.) R.J.
          Davis
      var. divaricatum (Gandog.) Reveal
        SY=E. commixtum Greene ex
          Tidestrom
        SY=E. vimineum Dougl. ex Benth.
          var. c. (Greene ex
          Tidestrom) S. Stokes
  batemanii M.E. Jones
  beatleyae Reveal
  bicolor M.E. Jones
  bifurcatum Reveal
  X blissianum Mason [arborescens X
  giganteum]
  brachyanthum Coville
        SY=E. baileyi S. Wats. var.
          brachyanthum (Coville)
          Jepson
  brachypodum Torr. & Gray
        SY=E. deflexum Torr. var. b.
          (Torr. & Gray) Munz
        SY=E. parryi Gray
  brandegei Rydb.
  breedlovei (J.T. Howell) Reveal
      var. breedlovei
      var. shevockii J.T. Howell
  brevicaule Nutt.
      var. brevicaule
        SY=E. b. ssp. grangerense (M.E.
          Jones) S. Stokes
        SY=E. campanulatum Nutt.
        SY=E. nudicaule (Torr.) Small ssp.
          garrettii S. Stokes
        SY=E. n. ssp. parleyense S. Stokes
      var. cottamii (S. Stokes) Reveal
        SY=E. tenellum Torr. ssp. c. S.
          Stokes
      var. laxifolium (Torr. & Gray) Reveal
        SY=E. chrysocephalum Gray
        SY=E. c. ssp. bannockense S.
          Stokes
        SY=E. nudicaule (Torr.) Small ssp.
          angustum (M.E. Jones) S.
          Stokes
        SY=E. n. ssp. pumilum (S. Stokes)
          S. Stokes
      var. micranthum (Nutt.) Reveal
        SY=E. b. ssp. orendense (A. Nels.)
          S. Stokes
      var. wasatchense (M.E. Jones) Reveal
        SY=E. w. M.E. Jones
  butterworthianum J.T. Howell
  caespitosum Nutt.
        SY=E. sphaerocephalum Dougl. ex
          Benth. var. sericoleucum
          (Greene) S. Stokes
  caninum (Greene) Munz

Eriogonum (CONT.)
          SY=E. vimineum Dougl. ex Benth.
                var. californicum Gandog.
          SY=E. v. var. caninum Greene
  capillare Small
  cernuum Nutt.
        var. cernuum
          SY=E. c. ssp. tenue (Torr. & Gray)
                S. Stokes
          SY=E. c. var. t. Torr. & Gray
        var. viminale (S. Stokes) Reveal
          SY=E. c. ssp. v. S. Stokes
  chrysops Rydb.
          SY=E. ovalifolium Nutt. ssp. c.
                (Rydb.) S. Stokes ex M.E.
                Peck
  cinereum Benth.
  cithariforme S. Wats.
          SY=E. c. var. agninum (Greene)
                Reveal
          SY=E. gracile Benth. var. c. (S.
                Wats.) Munz
          SY=E. g. var. polygonoides (S.
                Stokes) Munz
  clavellatum Small
  collinum S. Stokes ex M.E. Jones
  coloradense Small
  compositum Dougl. ex Benth.
        var. compositum
        var. lancifolium St. John & Warren
        var. leianthum Hook.
          SY=E. umbellatum Torr. var.
                monocephalum Torr. & Gray
          SY=E. u. var. tolmieanum (Hook.)
                M.E. Jones
  concinnum Reveal
  congdonii (S. Stokes) Reveal
          SY=E. ternatum T.J. Howell var. c.
                (S. Stokes) J.T. Howell
  contiguum (Reveal) Reveal
          SY=E. inflatum Torr. & Frem. var.
                c. Reveal
  contortum Small ex Rydb.
  correllii Reveal
  corymbosum Benth.
        var. corymbosum
          SY=E. effusum Nutt. ssp. durum S.
                Stokes
        var. davidsei Reveal
        var. divaricatum Torr. & Gray
          SY=E. divergens Small
        var. erectum Reveal & Brotherson
          SY=E. effusum Nutt. ssp. salinum
                (A. Nels.) S. Stokes
          SY=E. s. A. Nels.
        var. glutinosum (M.E. Jones) M.E.
              Jones
          SY=E. aureum M.E. Jones
          SY=E. a. var. glutinosum M.E.
                Jones
          SY=E. effusum Nutt. ssp. nelsonii
                (L.O. Williams) S. Stokes
          SY=E. microthecum Nutt. var.
                crispum (L.O. Williams) S.
                Stokes
        var. matthewsae Reveal
        var. orbiculatum (S. Stokes) Reveal &
              Brotherson
          SY=E. effusum Nutt. ssp. o. S.
                Stokes
        var. revealianum (Welsh) Reveal
          SY=E. r. Welsh
        var. velutinum Reveal
  covilleanum Eastw.
  crocatum A. Davids.
  cronquistii Reveal
  cusickii M.E. Jones

          SY=E. chrysocephalum Gray ssp.
                cusickii (M.E. Jones) S.
                Stokes
  darrovii Kearney
  dasyanthemum Torr. & Gray
  davidsonii Greene
          SY=E. molestum S. Wats. var. d.
                (Greene) Jepson
          SY=E. vimineum Dougl. ex Benth.
                ssp. juncinellum (Gandog.)
                S. Stokes
  deflexum Torr.
        var. baratum (Elmer) Reveal
          SY=E. d. ssp. b. (Elmer) Munz
        var. deflexum
          SY=E. clutei Rydb.
        var. nevadense Reveal
        var. turbinatum (Small) Reveal
  densum Greene
          SY=E. vimineum Dougl. ex Benth.
                var. d. (Greene) S. Stokes
  deserticola S. Wats.
  desertorum (Maguire) R.J. Davis
          SY=E. chrysocephalum Gray ssp. d.
                Maguire
  diclinum Reveal
  divaricatum Hook.
  douglasii Benth.
        var. douglasii
          SY=E. caespitosum Nutt. var. d.
                (Benth.) M.E. Jones
        var. sublineare (S. Stokes) Reveal
          SY=E. d. var. tenue (Small) C.L.
                Hitchc.
          SY=E. strictum Benth. var. t.
                (Small) S. Stokes
          SY=E. t Small
  X duchesnense Reveal [brevicaule var.
        laxifolium X corymbosum var. erectum]
          SY=E. corymbosum Benth. var.
                albogilvum Reveal
  eastwoodianum J.T. Howell
          SY=E. covilleanum Eastw. ssp.
                adsurgens (Jepson) Abrams
  effusum Nutt.
        var. effusum
          SY=E. microthecum Nutt. var. e.
                (Nutt.) Torr. & Gray
        var. rosmarinoides Benth.
  elatum Dougl. ex Benth.
        var. elatum
        var. villosum Jepson
  elegans Greene
          SY=E. baileyi S. Wats. ssp. e.
                (Greene) Munz
  elongatum Benth.
  ephedroides Reveal
  eremicola J.T. Howell & Reveal
  eremicum Reveal
  ericifolium Torr. & Gray
        var. ericifolium
          SY=E. mearnsii Parry
        var. pulchrum (Eastw.) Reveal
          SY=E. m. var. p. (Eastw.) Kearney
                & Peebles
        var. thornei Reveal & Henrickson
  esmeraldense S. Wats.
        var. esmeraldense
        var. toiyabense J.T. Howell
  exilifolium Reveal
  fasciculatum Benth.
        var. fasciculatum
        var. flavoviride Munz & Johnston
          SY=E. fasciculatum ssp.
                flavoviride (Munz &
                Johnston) S. Stokes
        var. foliolosum (Nutt.) S. Stokes ex

Eriogonum (CONT.)
        Abrams
        SY=E. fasciculatum ssp. foliolosum
           (Nutt.) S. Stokes
      var. polifolium (Benth.) Torr. & Gray
        SY=E. fasciculatum ssp. p.
           (Benth.) S. Stokes
        SY=E. f. var. revolutum (Goodding)
           S. Stokes
   fendlerianum (Benth.) Small
        SY=E. ainsliei Woot. & Standl.
   flavum Nutt.
      var. aquilinum Reveal
      var. flavum
        SY=E. f. ssp. crassifolium
           (Benth.) S. Stokes
        SY=E. f. var. c. (Benth.) Benth.
      var. piperi (Greene) M.E. Jones
        SY=E. f. ssp. p. (Greene) S.
           Stokes
        SY=E. f. var. linguifolium Gandog.
        SY=E. p. Greene
      var. polyphyllum (Small ex Rydb.)
           M.E. Jones
   foliosum S. Wats.
   giganteum S. Wats.
      var. compactum Dunkle
        SY=E. g. ssp. c. (Dunkle) Munz
      var. formosum K. Brandeg.
        SY=E. g. ssp. f. (K. Brandeg.)
           Raven
      var. giganteum
   gilmanii S. Stokes
   glandulosum (Nutt.) Nutt. ex Benth.
        SY=E. carneum (J.T. Howell) Reveal
        SY=E. g. var. c. J.T. Howell
        SY=Oxytheca g. Nutt.
   gordonii Benth.
   gossypinum Curran
   gracile Benth.
   gracilipes S. Wats.
        SY=E. kennedyi Porter ex S. Wats.
           ssp. g. (S. Wats.) S. Stokes
        SY=E. ochrocephalum S. Wats. var.
           g. (S. Wats.) J.T. Howell
   gracillimum S. Wats.
   grande Greene
      var. grande
        SY=E. latifolium Sm. ssp. g.
           (Greene) S. Stokes
      var. rubescens (Greene) Munz
        SY=E. g. ssp. r. (Greene) Munz
        SY=E. latifolium Sm. var. r.
           (Greene) S. Stokes
        SY=E. r. Greene
      var. timorum Reveal
        SY=E. g. ssp. t. (Reveal) Munz
   grayi Reveal
   greggii Torr. & Gray
   gypsophilum Woot. & Standl.
   havardii S. Wats.
        SY=E. leucophyllum Woot. & Standl.
   heermannii Dur. & Hilg.
      var. argense (M.E. Jones) Munz
        SY=E. h. ssp. a. (M.E. Jones) Munz
        SY=E. howellii S. Stokes
      var. clokeyi Reveal
      var. floccosum Munz
        SY=E. heermannii ssp. f. (Munz)
           Munz
      var. heermannii
      var. humilius (S. Stokes) Reveal
        SY=E. heermannii ssp. humilius S.
           Stokes
      var. occidentale S. Stokes
      var. subracemosum (S. Stokes) Reveal
        SY=E. howellii S. Stokes var. s.

        S. Stokes
      var. sulcatum (S. Wats.) Munz &
        Reveal
        SY=E. heermannii ssp. s. (S.
           Wats.) S. Stokes
        SY=E. s. S. Wats.
   hemipterum (Torr.) S. Stokes
   heracleoides Nutt.
      var. angustifolium (Nutt.) Torr. &
        Gray
      var. heracleoides
        SY=E. h. var. minus Benth.
      var. leucophaeum Reveal
   hieracifolium Benth.
        SY=E. pannosum Woot. & Standl.
   hirtellum J.T. Howell & Bacig.
   hirtiflorum Gray ex S. Wats.
        SY=Oxytheca h. (Gray ex S. Wats.)
           Greene
   hoffmannii S. Stokes
      var. hoffmannii
      var. robustius S. Stokes
        SY=E. h. ssp. r. (S. Stokes) Munz
   holmgrenii Reveal
   hookeri S. Wats.
   howellianum Reveal
   humivagans Reveal
   hylophilum Reveal & Brotherson
   incanum Torr. & Gray
        SY=E. marifolium Torr. & Gray var.
           i. (Torr. & Gray) M.E. Jones
   inerme (S. Wats.) Jepson
      var. hispidulum Goodman
        SY=E. i. ssp. h. (Goodman) Munz
      var. inerme
        SY=Oxytheca i. S. Wats.
   inflatum Torr. & Frem.
      var. deflatum I.M. Johnston
      var. fusiforme (Small) Reveal
        SY=E. f. Small
      var. inflatum
   insigne S. Wats.
        SY=E. deflexum Torr. ssp.
           exaltatum (M.E. Jones) S.
           Stokes
        SY=E. d. ssp. i. (S. Wats.) S.
           Stokes
   intermontanum Reveal
   intrafractum Coville & Morton
   jamesii Benth.
      var. flavescens S. Wats.
        SY=E. arcuatum Greene
        SY=E. bakeri Greene
      var. jamesii
      var. rupicola Reveal
      var. simplex Gandog.
      var. undulatum (Benth.) S. Stokes ex
        M.E. Jones
        SY=E. u. Benth.
      var. wootonii Reveal
      var. xanthum (Small) Reveal
        SY=E. flavum Nutt. var. x. (Small)
           S. Stokes
   jonesii S. Wats.
   kearneyi Tidestrom
      var. kearneyi
        SY=E. dudleyanum S. Stokes
        SY=E. nodosum Small var. k.
           (Tidestrom) S. Stokes
      var. monoense (S. Stokes) Reveal
        SY=E. k. ssp. m. (S. Stokes) Munz
           ex Reveal
        SY=E. nodosum Small ssp. m. S.
           Stokes
   kelloggii Gray
   kennedyi Porter ex S. Wats.
      var. alpigenum Munz & Johnston

Eriogonum   (CONT.)
        SY=E. k. ssp. a. (Munz & Johnston)
          Munz
    var. austromontanum Munz & Johnston
        SY=E. k. ssp. a. (Munz & Johnston)
          S. Stokes
    var. kennedyi
    var. pinicola Reveal
    var. purpusii (Brandeg.) Reveal
        SY=E. k. ssp. p. (Brandeg.) Munz
kingii Torr. & Gray
lachnogynum Torr. ex Benth.
        SY=E. tetraneuris Small
lagopus Rydb.
        SY=E. multiceps Nees ssp. canum S.
          Stokes
        SY=E. pauciflorum Pursh var. c.
          (S. Stokes) Reveal
lancifolium Reveal & Brotherson
latens Jepson
        SY=E. elatum Dougl. ex Benth. ssp.
          glabrescens S. Stokes
        SY=E. monticola S. Stokes
latifolium Sm.
lemmonii S. Wats.
leptocladon Torr. & Gray
    var. leptocladon
        SY=E. effusum Nutt. var. shandsii
          S. Stokes
    var. papiliunculum Reveal
    var. ramosissimum (Eastw.) Reveal
        SY=E. r. Eastw.
leptophyllum (Torr. & Gray) Woot. & Standl.
libertinus Reveal
lobbii Torr. & Gray
    var. lobbii
    var. robustum (Greene) M.E. Jones
        SY=E. r. Greene
loganum A. Nels.
lonchophyllum Torr. & Gray
    var. lonchophyllum
        SY=E. salicinum Greene
        SY=E. sarothriforme Gandog.
        SY=E. scoparium Small
        SY=E. tristichum Small
    var. nudicaule (Torr.) Reveal
        SY=E. n. (Torr.) Small
longifolium Nutt.
    var. gnaphalifolium Gandog.
        SY=E. floridanum Small
    var. harperi (Goodman) Reveal
        SY=E. h. Goodman
    var. longifolium
        SY=E. l. var. lindheimeri Gandog.
        SY=E. longifolium var.
          plantagineum Engelm. & Gray
        SY=E. vespinum Shinners
luteolum Greene
maculatum Heller
        SY=E. angulosum Benth. ssp. m.
          (Heller) S. Stokes
        SY=E. a. var. m. (Heller) Jepson
        SY=E. cernuum Nutt. ssp.
          acutangulum (Gandog.) S.
          Stokes
mancum Rydb.
marifolium Torr. & Gray
microthecum Nutt.
    var. alpinum Reveal
    var. ambiguum (M.E. Jones) Reveal
        SY=E. m. ssp. aureum (M.E. Jones)
          S. Stokes
    var. corymbosoides Reveal
    var. foliosum (Torr. & Gray) Reveal
        SY=E. m. ssp. intermedium S.
          Stokes
        SY=E. m. var. friscanum (M.E.

          Jones) S. Stokes
        SY=E. m. var. macdougalii
          (Gandog.) S. Stokes
        SY=E. simpsonii Benth.
    var. johnstonii Reveal
    var. lapidicola Reveal
    var. laxiflorum Hook.
        SY=E. m. ssp. confertiflorum
          (Benth.) S. Stokes
        SY=E. m. ssp. l. (Hook.) S. Stokes
        SY=E. m. var. c. (Benth.) Torr. &
          Gray
    var. microthecum
        SY=E. m. var. idahoense (Rydb.) S.
          Stokes
    var. panamintense S. Stokes
mohavense S. Wats.
molestum S. Wats.
mortonianum Reveal
multiflorum Benth.
nanum Reveal
natum Reveal
nealleyi Coult.
X nebraskense Rydb. [effusum X pauciflorum]
nervulosum (S. Stokes) Reveal
        SY=E. ursinum S. Wats. var. n. S.
          Stokes
nidularium Coville
        SY=E. vimineum Dougl. ex Benth.
          ssp. n. (Coville) S. Stokes
niveum Dougl. ex Benth.
        SY=E. n. ssp. decumbens (Benth.)
          S. Stokes
        SY=E. n. var. d. (Benth.) Torr. &
          Gray
        SY=E. n. var. dichotomum (Dougl.
          ex Benth.) S. Stokes ex M.E.
          Jones
        SY=E. strictum Benth. var.
          lachnostegium Benth.
nortonii Greene
novonudum M.E. Peck
nudum Dougl. ex Benth.
    var. auriculatum (Benth.) Tracy ex
        Jepson
        SY=E. latifolium Sm. ssp. a.
          (Benth.) S. Stokes
    var. decurrens (S. Stokes) M.L.
        Bowerman
    var. deductum (Greene) Jepson
        SY=E. latifolium Sm. ssp. d.
          (Greene) S. Stokes
    var. indictum (Jepson) Reveal
        SY=E. latifolium Sm. ssp. i.
          (Jepson) S. Stokes
    var. murinum Reveal
    var. nudum
        SY=E. latifolium Sm. ssp. n.
          (Dougl. ex Benth.) S. Stokes
    var. oblongifolium S. Wats.
        SY=E. harfordii Small
        SY=E. latifolium Sm. ssp.
          sulphureum (Greene) S.
          Stokes
    var. pauciflorum S. Wats.
        SY=E. n. ssp. p. (S. Wats.) Munz
    var. pubiflorum Benth.
    var. scapigerum (Eastw.) Jepson
        SY=E. latifolium Sm. var. s.
          (Eastw.) S. Stokes
        SY=E. s. Eastw.
    var. westonii (S. Stokes) J.T. Howell
        SY=E. gramineum S. Stokes
        SY=E. latifolium Sm. ssp. saxicola
          (Heller) S. Stokes
        SY=E. n. ssp. s. (Heller) Munz
nummulare M.E. Jones

Eriogonum (CONT.)
    nutans Torr. & Gray
            var. glabratum Reveal
            var. nutans
                SY=E. deflexum Torr. ssp. ultrum
                    S. Stokes
                SY=E. n. var. brevipedicellatum S.
                    Stokes
        ochrocephalum S. Wats.
            var. calcareum (S. Stokes) M.E. Peck
                SY=E. o. ssp. c. S. Stokes
            var. ochrocephalum
        ordii S. Wats.
        ostlundii M.E. Jones
                SY=E. spathuiforme Rydb.
                SY=E. spathulatum Gray ssp.
                    spathuiforme (Rydb.) S.
                    Stokes
        ovalifolium Nutt.
            var. calestrinum Reveal
            var. depressum Blank.
            var. eximium (Tidestrom) J.T. Howell
                SY=E. o. ssp. e. (Tidestrom) S.
                    Stokes
            var. macropodum (Gandog.) Reveal
                SY=E. ochroleucum Small ex Rydb.
                SY=E. ovalifolium ssp. ochroleucum
                    (Small ex Rydb.) S. Stokes
                SY=E. ovalifolium var. ochroleucum
                    (Small ex Rydb.) M.E. Peck
            var. multiscapum Gandog.
            var. nivale (Canby) M.E. Jones
                SY=E. n. Canby
                SY=E. rhodanthum A. Nels. &
                    Kennedy
            var. ovalifolium
                SY=E. davisianum S. Stokes
                SY=E. o. var. celsum A. Nels.
                SY=E. o. var. orthocaulon (Small)
                    C.L. Hitchc.
                SY=E. o. var. purpureum (Nutt.)
                    Dur.
            var. vineum (Small) Jepson
                SY=E. o. ssp. v. (Small) S. Stokes
        palmerianum Reveal
                SY=E. baileyi S. Wats. var.
                    tomentosum S. Wats.
        panamintense Morton
            var. mensicola (S. Stokes) Reveal
                SY=E. m. S. Stokes
                SY=E. p. ssp. m. (S. Stokes) Munz
            var. panamintense
                SY=E. racemosum Nutt. var.
                    desertorum S. Stokes
        panguicense (M.E. Jones) Reveal
            var. alpestre (S. Stokes) Reveal
                SY=E. chrysocephalum Gray ssp. a.
                    S. Stokes
            var. panguicense
                SY=E. spathulatum Gray var. p.
                    (M.E. Jones) S. Stokes
        parishii S. Wats.
        parvifolium Sm.
                SY=E. p. ssp. lucidum J.T. Howell
                    ex S. Stokes
                SY=E. p. ssp. paynei C.B. Wolf ex
                    Munz
                SY=E. parvifolium var. lucidum
                    (J.T. Howell ex S. Stokes)
                    Reveal
                SY=E. parvifolium var. paynei
                    (C.B. Wolf ex Munz) Reveal
        pauciflorum Pursh
            var. gnaphalodes (Benth.) Reveal
            var. pauciflorum
                SY=E. depauperatum Small
                SY=E. multiceps Nees

        pedunculatum S. Stokes
        pelinophilum Reveal
        pendulum S. Wats.
        pharnaceoides Torr.
            var. cervinum Reveal
            var. pharnaceoides
        plumatella Dur. & Hilg.
                SY=E. p. var. jaegeri (Munz &
                    Johnston) S. Stokes ex Munz
        polycladon Benth.
        polypodum Small
        prattenianum Dur.
        procidum Reveal
        puberulum S. Wats.
                SY=E. p. var. venosum S. Stokes
        pusillum Torr. & Gray
                SY=E. reniforme Torr. & Frem. ssp.
                    p. (Torr. & Gray) S. Stokes
        pyrolifolium Hook.
            var. coryphaeum Torr. & Gray
                SY=E. p. var. bellingerianum M.E.
                    Peck
            var. pyrolifolium
        racemosum Nutt.
        reniforme Torr. & Frem.
                SY=E. r. var. comosum M.E. Jones
        ripleyi J.T. Howell
        rixfordii S. Stokes
                SY=E. deflexum Torr. ssp. r. (S.
                    Stokes) Munz
        rosense A. Nels. & Kennedy
                SY=E. ochrocephalum S. Wats. ssp.
                    agnellum (Jepson) S. Stokes
        roseum Dur. & Hilg.
                SY=E. virgatum Benth.
        rotundifolium Benth.
        rubricaule Tidestrom
        rupinum Reveal
        salicornioides Gandog.
        saurinum Reveal
        saxatile S. Wats.
        scabrellum Reveal
        scopulorum Reveal
        shockleyi S. Wats.
            var. longilobum (M.E. Jones) Reveal
                SY=E. s. ssp. l. (M.E. Jones) S.
                    Stokes
            var. shockleyi
                SY=E. s. ssp. candidum (M.E.
                    Jones) S. Stokes
        siskiyouense Small
        smithii Reveal
        spathulatum Gray
                SY=E. nudicaule (Torr.) Small ssp.
                    ochroflorum S. Stokes
        spergulinum Gray
            var. pratense (S. Stokes) J.T. Howell
            var. reddingianum (M.E. Jones) J.T.
                Howell
                SY=E. s. ssp. r. (M.E. Jones) Munz
            var. spergulinum
                SY=Oxytheca s. (Gray) Greene
        sphaerocephalum Dougl. ex Benth.
            var. fasciculifolium (A. Nels.) S.
                Stokes
                SY=E. fruticulosum S. Stokes
            var. halimioides (Gandog.) S. Stokes
            var. sphaerocephalum
                SY=E. s. var. brevifolium S.
                    Stokes ex M.E. Jones
                SY=E. s. var. geniculatum (Nutt.)
                    S. Stokes
                SY=E. s. var. megacephalum (Nutt.)
                    Stokes ex M.E. Jones
        strictum Benth.
            ssp. proliferum (Torr. & Gray) S. Stokes
                var. anserinum (Greene) R.J. Davis

Eriogonum (CONT.)
    SY=E. flavissimum Gandog.
    SY=E. ovalifolium Nutt. ssp. f.
      (Gandog.) S. Stokes
    SY=E. s. var. f. (Gandog.) C.L.
      Hitchc.
  var. glabrum C.L. Hitchc.
  var. greenei (Gray) Reveal
  var. proliferum (Torr. & Gray) Reveal
    SY=E. bellum S. Stokes
    SY=E. p. Torr. & Gray
    SY=E. s. var. cusickii (Gandog.)
      S. Stokes
  ssp. strictum
subreniforme S. Wats.
    SY=E. filicaule S. Stokes
suffruticosum S. Wats.
temblorense J.T. Howell & Twisselmann
tenellum Torr.
  var. platyphyllum (Torr. ex Benth.)
    Torr.
  var. ramosissimum Benth.
  var. tenellum
ternatum T.J. Howell
thomasii Torr.
thompsonae S. Wats.
  var. albiflorum Reveal
  var. atwoodii Reveal
  var. thompsonae
thurberi Torr.
thymoides Benth.
    SY=E. sphaerocephalum Dougl. ex
      Benth. var. minimum (Small)
      S. Stokes
    SY=E. t. ssp. congestum S. Stokes
tomentosum Michx.
trichopes Torr.
    SY=E. t. ssp. minus (Benth.) S.
      Stokes
tripodum Greene
truncatum Torr. & Gray
tumulosum (Barneby) Reveal
    SY=E. villiflorum Gray var. t.
      Barneby
twisselmanii (J.T. Howell) Reveal
umbellatum Torr.
  var. aureum (Gandog.) Reveal
    SY=E. u. var. intectum A. Nels.
  var. bahiiforme (Torr. & Gray) Jepson
    SY=E. u. ssp. b. (Torr. & Gray)
      Munz
  var. chlorothamnus Reveal
  var. cognatum (Greene) Reveal
    SY=E. c. Greene
  var. covillei (Small) Munz & Reveal
    SY=E. c. Small
    SY=E. u. ssp. c. (Small) Munz
  var. deadereticum Reveal
  var. devestivum Reveal
    SY=E. u. ssp. aridum (Greene) S.
      Stokes
    SY=E. u. var. a. (Greene) R.J.
      Davis
  var. dircocephalum Gandog.
  var. glaberrimum (Gandog.) Reveal
  var. hausknechtii (Dammer) M.E. Jones
  var. hypoleium (Piper) C.L. Hitchc.
  var. majus Hook.
    SY=E. heracleoides Nutt. var.
      subalpinum (Greene) S.
      Stokes
    SY=E. s. Greene
    SY=E. u. ssp. m. (Hook.) Piper
    SY=E. u. var. s. (Greene) M.E.
      Jones
  var. minus I.M. Johnston
    SY=E. u. ssp. m. (I.M. Johnston)

    Munz
  var. munzii Reveal
    SY=E. u. ssp. m. (Reveal) Thorne
      ex Munz
  var. nevadense Gandog.
  var. polyanthum (Benth.) M.F. Jones
    SY=E. u. ssp. dumosum (Greene) S.
      Stokes
    SY=E. u. ssp. p. (Benth.) S.
      Stokes
    SY=E. u. var. modocense (Greene)
      S. Stokes
  var. porteri (Small) S. Stokes
  var. speciosum (E. Drew) S. Stokes
  var. stellatum (Benth.) M.F. Jones
    SY=E. u. ssp. s. (Benth.) S.
      Stokes
    SY=E. u. var. chrysanthum Gandog.
    SY=E. u. var. croceum (Small) S.
      Stokes ex R.J. Davis
  var. subaridum S. Stokes
    SY=E. u. ssp. ferrissii (A. Nels.)
      S. Stokes
    SY=E. u. ssp. s. (S. Stokes) Munz
  var. torreyanum (Gray) M.E. Jones
    SY=E. t. Gray
  var. umbellatum
  var. vernum Reveal
  var. versicolor S. Stokes
    SY=E. u. ssp. v. (S. Stokes) Munz
ursinum S. Wats.
vestitum J.T. Howell
villiflorum Gray
vimineum Dougl. ex Benth.
    SY=E. v. var. shoshonense (A.
      Nels.) S. Stokes
viridescens Heller
    SY=F. angulosum Benth. ssp. v.
      (Heller) S. Stokes
    SY=E. bidentatum Jepson
viridulum Reveal
viscidulum J.T. Howell
visheri A. Nels.
watsonii Torr. & Gray
    SY=E. deflexum Torr. ssp. w.
      (Torr. & Gray) S. Stokes
    SY=F. d. var. multipedunculatum
      (S. Stokes) C.L. Hitchc.
    SY=E. d. var. w. (Torr. & Gray)
      R.J. Davis
wetherillii Eastw.
    SY=F. sessile S. Stokes ex M.E.
      Jones
wrightii Torr. ex Benth.
  var. membranaceum S. Stokes ex Jepson
    SY=E. w. ssp. m. (S. Stokes ex
      Jepson) S. Stokes
  var. nodosum (Small) Reveal
    SY=E. n. Small
    SY=E. w. ssp. n. (Small) Munz
  var. olanchense (J.T. Howell) Reveal
  var. pringlei (Coult. & Fish.) Reveal
    SY=F. p. Coult. & Fish.
  var. subscaposum S. Wats.
    SY=E. w. ssp. s. (S. Wats.) S.
      Stokes
  var. trachygonum (Torr. ex Benth.)
    Jepson
    SY=F. w. ssp. t. (Torr. ex Benth.)
      S. Stokes
  var. wrightii
    SY=E. w. ssp. glomerulum S. Stokes
zionis J.T. Howell
  var. coccineum J.T. Howell
  var. zionis

Fagopyrum P. Mill.

Fagopyrum (CONT.)
  esculentum Moench
        SY=F. fagopyrum (L.) Karst.
        SY=F. sagittatum Gilib.
        SY=Polygonum f. L.
  tataricum (L.) Gaertn.
        SY=Polygonum t. L.

Gilmania Coville

  luteola (Coville) Coville

Goodmania Reveal & Ertter

  luteola (Parry) Reveal & Ertter
        SY=Oxytheca l. Parry

Hollisteria S. Wats.

  lanata S. Wats.

Koenigia L.

  islandica L.
        SY=Macounastrum i. (L.) Small

Lastarriaea Remy

  coriacea (Goodman) Hoover
        SY=Chorizanthe c. Goodman
        SY=C. lastarriaea Parry
        SY=L. chilensis sensu auctt. non
            Remy

Mucronea Benth.

  californica Benth.
    var. californica
        SY=Chorizanthe c. (Benth.) Gray
    var. suksdorfii (J.F. Macbr.) Goodman
        SY=Chorizanthe c. (Benth.) Gray
            var. s. J.F. Macbr.
  perfoliata (Gray) Heller
    var. opaca Hoover
        SY=Chorizanthe p. Gray var. o.
            (Hoover) Munz
    var. perfoliata
        SY=Chorizanthe p. Gray

Nemacaulis Nutt.

  denudata Nutt.
    var. denudata
    var. gracilis Goodman & L. Benson

Oxyria Hill

  digyna (L.) Hill

Oxytheca Nutt.

  caryophylloides Parry
  dendroidea Nutt.
        SY=Eriogonum d. (Nutt.) S. Stokes
        SY=E. d. var. foliosa (Nutt.) M.E.
            Jones
        SY=E. d. var. hillmanii S. Stokes
        SY=O. f. Nutt.
  emarginata Hall
  parishii Parry
    var. abramsii (McGregor) Munz
        SY=Eriogonum a. (McGregor) S.
            Stokes
        SY=O. a. McGregor
        SY=O. p. ssp. a. (McGregor) Munz
    var. cienegensis Ertter
    var. goodmaniana Ertter

    var. parishii
  perfoliata Torr. & Gray
  trilobata Gray
  watsonii Torr. & Gray

Polygonella Michx.

  americana (Fisch. & Mey.) Small
        SY=Gonopyrum a. Fisch. & Mey.
  articulata (L.) Meisn.
        SY=Delopyrum a. (L.) Small
        SY=Polygonum a. L.
  ciliata Meisn.
    var. basiramia (Small) Horton
        SY=Delopyrum b. Small
    var. ciliata
        SY=Delopyrum c. (Meisn.) Small
  fimbriata (Ell.) Horton
    var. fimbriata
        SY=Polygonum f. Ell.
        SY=Thysanella f. (Ell.) Gray
    var. robusta (Small) Horton
        SY=Thysanella r. Small
  gracilis (Nutt.) Meisn.
        SY=Delopyrum filiforme Small
        SY=D. gracile (Nutt.) Small
        SY=Polygonum g. Nutt.
  macrophylla Small
  myriophylla (Small) Horton
        SY=Dentoceras m. Small
  parksii Cory
  polygama (Vent.) Engelm. & Gray
        SY=Polygonella brachystachya
            Meisn.
        SY=Polygonella croomii Chapman
        SY=Polygonum p. Vent.

Polygonum L.

  achoreum Blake
  acuminatum H.B.K.
        SY=Persicaria a. (H.B.K.) Maza
  alpinum All.
        SY=Polygonum alaskanum (Small) W.
            Wight ex Harshberger
        SY=P. alaskanum var. glabrescens
            Hulten
  amphibium L.
    var. emersum Michx.
        SY=Persicaria amphibium (L.) S.F.
            Gray pro parte
        SY=Persicaria coccinea (Muhl. ex
            Willd.) Greene
        SY=Persicaria muhlenbergii (S.
            Wats.) Small
        SY=Polygonum c. Muhl. ex Willd.
        SY=Polygonum c. var. pratincola
            (Greene) Stanford
        SY=Polygonum c. var. terrestre
            Willd.
        SY=Polygonum m. S. Wats.
        SY=Polygonum m. var. t. (Willd.)
            Trel.
    var. stipulaceum (Coleman) Fern.
        SY=Polygonum coccineum Muhl. ex
            Willd. var. hartwrightii
            (Gray) Biss.
        SY=P. c. var. rigidulum (Sheldon)
            Stanford
        SY=P. fluitans Eat.
        SY=P. natans (Michx.) Eat.
  arenarium Waldst. & Kit.
  arenastrum Jord. ex Boreau
  argyrocoleon Steud. ex Kunze
  arifolium L.
    var. arifolium
        SY=Tracaulon a. (L.) Raf.

Polygonum (CONT.)
    var. pubescens (Keller) Fern.
      SY=Polygonum a. var. lentiforme
        Fern. & Grisc.
aubertii Henry
    SY=Bilderdykia a. (Henry) Moldenke
    SY=Fallopia a. (Henry) A. Love
aviculare L.
    var. aviculare
      SY=Polygonum a. var. erectum
        (Roth) W.D.J. Koch
      SY=P. heterophyllum Lindl.
    var. vegetum Ledeb.
      SY=Polygonum monspeliensis Thieb.
baldschuanicum Regel
    SY=Reynoutria b. (Regel) Shinners
bidwelliae S. Wats.
bistorta L.
    var. bistorta
    var. plumosum (Small) Boivin
      SY=Polygonum b. ssp. p. (Small)
        Hulten
bistortoides Pursh
      SY=Bistorta b. (Pursh) Small
      SY=Polygonum b. var. linearifolium
        (S. Wats.) Small
      SY=P. b. var. oblongifolium
        (Meisn.) St. John
      SY=P. cephalophorum Greene
      SY=P. glastifolium Greene
      SY=P. vulcanicum Greene
bolanderi Brewer ex Gray
boreale (Lange) Small
buxiforme Small
      SY=Polygonum aviculare L. var.
        littorale (Link) Mert.
      SY=P. l. (Link) W.D.J. Koch
caespitosum Blume
    var. caespitosum
    var. longisetum (de Bruyn) A.N.
      Stewart
      SY=Polygonum l. de Bruyn
californicum Meisn.
    SY=Polygonum greenii S. Wats.
campanulatum Hook. f.
capitatum Buch.-Ham. ex D. Don
careyi Olney
    SY=Persicaria c. (Olney) Greene
cascadense W.H. Baker
caurianum B.L. Robins.
    SY=Polygonum humifusum Pallas
cilinode Michx.
    SY=Bilderdykia c. (Michx.) Greene
    SY=Polygonum c. var. laevigatum
      Fern.
    SY=Reynoutria c. (Michx.) Shinners
    SY=Tiniaria c. (Michx.) Small
confertiflorum Nutt. ex Piper
    SY=Polygonum esotericum L.C.
      Wheeler
    SY=P. imbricatum Nutt.
    SY=P. watsonii Small
convolvulus L.
    var. convolvulus
      SY=Bilderdykia c. (L.) Dumort.
      SY=Fallopia c. (L.) A. Love
      SY=Tiniaria c. (L.) Webb & Moq.
    var. subalatum Lej. & Court.
cuspidatum Sieb. & Zucc.
    SY=Pleuropterus zuccarinii Small
    SY=Reynoutria japonica Houtt.
davisiae Brewer ex Gray
densiflorum Meisn.
    SY=Persicaria d. (Meisn.) Moldenke
    SY=Persicaria portoricensis
      (Bertero) Small
    SY=Polygonum glabrum Willd.

douglasii Greene
    var. austinae (Greene) M.E. Jones
      SY=Polygonum a. Greene
    var. douglasii
      SY=P. emaciatum A. Nels.
    var. johnstonii Munz
      SY=Polygonum sawatchense Small
    var. latifolium (Engelm.) Greene
      SY=Polygonum buxiforme Small var.
        montanum (Small) R.J. Davis
      SY=Polygonum m. (Small) Greene
engelmannii Greene
      SY=Polygonum microspermum Small
erectum L.
fowleri B.L. Robins.
      SY=Polygonum buxifolium Nutt. ex
        Bong.
fusiforme Greene
glaucum Nutt.
      SY=Polygonum maritimum sensu
        auctt. non L.
graminifolium Wierzb. ex Heuffel
heterosepalum M.E. Peck & Ownbey
hirsutum Walt.
      SY=Persicaria h. (Walt.) Small
hydropiper L.
      SY=Persicaria h. (L.) Opiz
      SY=Polygonum h. var. projectum
        Stanford
hydropiperoides Michx.
    var. digitatum Fern.
    var. hydropiperoides
      SY=Persicaria h. (Michx.) Small
      SY=Persicaria paludicola Small
      SY=Persicaria persicarioides
        (H.B.K.) Small
      SY=Polygonum h. var. asperifolium
        Stanford
      SY=Polygonum h. var. bushianum
        Stanford
      SY=Polygonum h. var. euronotorum
        Fern.
      SY=Polygonum h. var. strigosum
        (Small) Stanford
      SY=Polygonum persicarioides H.B.K.
    var. psilostachyum St. John
      SY=Polygonum h. var. breviciliatum
        Fern.
kelloggii Greene
      SY=Polygonum unifolium Small
lapathifolium L.
    var. lapathifolium
      SY=Persicaria l. (L.) S.F. Gray
      SY=Persicaria tomentosa (Schrank)
        Bickn.
      SY=Polygonum l. var. nodosum
        (Pers.) Small
      SY=Polygonum l. var. ovatum A.
        Braun
      SY=Polygonum l. var. prostratum
        Wimmer
      SY=Polygonum incarnatum Ell.
      SY=Polygonum nodosum Pers.
      SY=Polygonum pensylvanicum (L.)
        Small ssp. oneillii
        (Brenckle) Hulten
      SY=Polygonum scabrum Moench
      SY=Polygonum tomentosum Schrank
    var. salicifolium Sibthorp
      SY=Polygonum incanum Willd.
      SY=P. l. var. i. (Willd.) W.D.J.
        Koch
leptocarpum B.L. Robins.
majus (Meisn.) Piper
      SY=Polygonum punctatum Ell. var.
        m. (Meisn.) Fassett
marinense Martens & Raven

Polygonum (CONT.)
   meisnerianum Cham. & Schlecht.
       var. beyrichianum (Cham. & Schlecht.)
         Meisn.
       var. meisnerianum
   minimum S. Wats.
   montereyense Brenckle
   neglectum Bess.
         SY=Polygonum aviculare L. var.
           angustissimum Meisn.
         SY=P. provinciale K. Koch
         SY=P. rurivagum Jord. ex Boreau
   nepalense Meisn.
   newberryi Small
         SY=Polygonum n. var. glabrum G.N.
           Jones
   nuttallii Small
   opelousanum Riddell ex Small
       var. adenocalyx Stanford
         SY=Polygonum hydropiperoides
           Michx. var. a. (Stanford)
           Gleason
       var. opelousanum
         SY=Persicaria hydropiperoides
           (Michx.) Small var. o.
           (Riddell ex Small) J.S.
           Wilson
         SY=Persicaria o. (Riddell ex
           Small) Small
         SY=Polygonum h. Michx. var. o.
           (Riddell ex Small) Stone
   orientale L.
         SY=Persicaria o. (L.) Spach
   otophyllum Fedde
   oxyspermum C.A. Mey. & Bunge ex Ledeb.
     ssp. oxyspermum
     ssp. raii (Bab.) Webb & Chater
         SY=Polygonum r. Bab.
   paronychia Cham. & Schlecht.
   parryi Greene
   patulum Bieb.
         SY=Polygonum bellardii All. sensu
           auctt.
   pensylvanicum (L.) Small
         SY=Persicaria longistyla (Small)
           Small
         SY=Persicaria mississippiensis
           (Stanford) Small
         SY=Persicaria pensylvanica (L.)
           Small
         SY=Polygonum bicorne Raf.
         SY=Polygonum l. Small
         SY=Polygonum l. var. omissum
           (Greene) Stanford
         SY=Polygonum mexicanum sensu Small
         SY=Polygonum mississippiensis
           Stanford
         SY=Polygonum mississippiensis var.
           interius Stanford
         SY=Polygonum p. var. durum
           Stanford
         SY=Polygonum p. var. eglandulosum
           J.C. Myers
         SY=Polygonum p. var. genuinum
           Fern.
         SY=Polygonum p. var. laevigatum
           Fern.
         SY=Polygonum p. var. rosiflorum
           J.B.S. Norton
   perfoliatum L.
   persicaria L.
       var. angustifolium Beckh.
       var. persicaria
         SY=Persicaria maculata (Raf.) S.F.
           Gray
         SY=Persicaria p. (L.) Small
         SY=Persicaria vulgaris Webb & Moq.

         SY=Polygonum dubium Stein
         SY=Polygonum minus Huds.
         SY=Polygonum minus var.
           subcontinuum (Meisn.) Fern.
         SY=Polygonum puritanorum Fern.
       var. ruderale (Salisb.) Meisn.
   phytolaccifolium Meisn. ex Small
         SY=Aconogonum p. (Meisn. ex Small)
           Small ex Rydb.
   polycnemoides Jaubert & Spach
       var. oliveri Jaubert & Spach
       var. polycnemoides
   polygaloides Meisn.
   polystachyum Wallich ex Meisn.
         SY=Aconogonum p. (Wallich ex
           Meisn.) Kral
   punctatum Ell.
       var. confertiflorum (Meisn.) Fassett
         SY=Polygonum p. var. leptostachyum
           (Meisn.) Small
       var. ellipticum Fassett
       var. punctatum
         SY=Persicaria p. (Ell.) Small
         SY=Persicaria p. var. eciliata
           Small
         SY=Polygonum acre H.B.K.
         SY=Polygonum p. var. parviflorum
           Fassett
         SY=Polygonum punctatum var. parvum
           Victorin & Rouss.
   ramosissimum Michx.
       var. prolificum Small
         SY=Polygonum p. (Small) B.L.
           Robins.
       var. ramosissimum
         SY=Polygonum allocarpum Blake
         SY=P. atlanticum (B.L. Robins.)
           Bickn.
         SY=P. autumnale Brenckle
         SY=P. exsertum Small
         SY=P. interior Brenckle
         SY=P. triangulum Bickn.
   robustius (Small) Fern.
         SY=Persicaria r. (Small) Bickn.
         SY=Polygonum punctatum Ell. var.
           majus (Meisn.) Fassett
   sachalinense F. Schmidt ex Maxim.
         SY=Reynoutria s. (F. Schmidt ex
           Maxim.) Nakai
   sagittatum L.
       var. gracilentum Fern.
       var. sagittatum
         SY=Tracaulon s. (L.) Small
         SY=Truellum s. (L.) Sojak
   scandens L.
       var. cristatum (Engelm. & Gray)
         Gleason
         SY=Bilderdykia c. (Engelm. & Gray)
           Greene
         SY=Polygonum c. Engelm. & Gray
         SY=Reynoutria s. (L.) Shinners
           var. c. (Engelm. & Gray)
           Shinners
         SY=Tiniaria c. (Engelm. & Gray)
           Small
       var. dumetorum (L.) Gleason
         SY=Bilderdykia d. (L.) Dumort.
         SY=Polygonum d. L.
         SY=Reynoutria s. (L.) Shinners
           var. d. (L.) Shinners
         SY=Tiniaria d. (L.) Opiz
       var. scandens
         SY=Bilderdykia s. (L.) Greene
         SY=Fallopia s. (L.) Holub
         SY=Polygonum dumetorum L. var. s.
           (L.) Gray
         SY=Reynoutria s. (L.) Shinners

Polygonum   (CONT.)
            SY=Tiniaria s. (L.) Small
    segetum H.B.K.
            SY=Persicaria s. (H.B.K.) Small
    setaceum Baldw. ex Ell.
        var. interjectum Fern.
            SY=Polygonum s. var. tonsum Fern.
        var. setaceum
            SY=Persicaria s. (Baldw. ex Ell.)
                Small
            SY=Polygonum hydropiperoides
                Michx. var. s. (Baldw. ex
                Ell.) Gleason
    shastense Brewer ex Gray
    spergulariiforme Meisn. ex Small
    striatulum B.L. Robins.
            SY=Polygonum braziliense K. Koch
            SY=P. camporum Meisn.
    tenue Michx.
        var. protrusum Fern.
        var. tenue
    texense M.C. Johnston
    triandrum Coolidge
    utahense Brenckle & Cottam
    virginianum L.
            SY=Antenoron v. (L.) Roberty &
                Vautier
            SY=Polygonum v. var. glaberrimum
                (Fern.) Steyermark
            SY=Tovara v. (L.) Raf.
            SY=T. v. var. g. Fern.
    viviparum L.
        var. macounii (Small) Hulten
        var. viviparum
            SY=Bistorta v. (L.) S.F. Gray
            SY=Polygonum fugax Small
            SY=P. v. var. alpinum Wahlenb.

Pterostegia Fisch. & Mey.

    drymarioides Fisch. & Mey.

Rheum L.

    rhabarbarum L.
    rhaponticum L.

Rumex L.

    acetosa L.
        ssp. acetosa
        ssp. alpestris (Scop.) A. Love
            SY=R. a. Scop.
        ssp. ambiguus (Gren.) A. Love
        ssp. arifolius (All.) Blytt & Dahl
        ssp. lapponicus Hiitonen
        ssp. thyrsiflorus (Fingerhuth) Hayek
            SY=R. t. Fingerhuth
    acetosella L.
        ssp. acetosella
            SY=Acetosella a. (L.) Small
            SY=A. vulgaris (Koch) Fourn.
            SY=R. a. var. pyrenaeus (Pourret)
                Timbal-Lagrave
        ssp. angiocarpus (Murb.) Murb.
            SY=R. angiocarpus Murb.
    X acutus L. [crispus X obtusifolius]
            SY=R. X crispo-obtusifolius Meisn.
    albescens Hbd.
    X alexidis Boivin [maritimus X
        stenophyllus]
    alpinus L.
    altissimus Wood
            SY=R. ellipticus Greene
    angiocarpus Murb.
    arcticus Trautv.
        ssp. arcticus

        ssp. pseudoxyris Tolm.
    brownii Campd.
    californicus Rech. f.
            SY=R. salicifolius Weinm. var.
                denticulatus Torr.
    chrysocarpus Moris
            SY=R. berlandieri Meisn.
    confertus Willd.
    X confusus Simonkai [crispus X patientia]
    conglomeratus Murr.
    crassus Rech. f.
    crispus L.
    densiforus Osterhout
            SY=R. praecox Rydb.
            SY=R. pycanthus Rech. f.
    dentatus L.
        ssp. dentatus
        ssp. klotzschianus (Meisn.) Rech. f.
    fascicularis Small
    floridanus Meisn.
    X franktonis Boivin [pseudonatronatus X
        triangulivalvis]
    frutescens Thouars
            SY=R. cuneifolius Campd.
    giganteus Ait.
        var. giganteus
        var. nelsonii Deg. & Deg.
    graminifolius Rudolph ex Lamb.
    hastatulus Baldw. ex Ell.
    hymenosepalus Torr.
        var. hymenosepalus
            SY=R. h. var. eu-hymenosepalus
                Rech. f.
        var. salinus (A. Nels.) Rech. f.
    kerneri Borbas
    lacustris Greene
    longifolius DC.
            SY=R. domesticus Hartman
    maritimus L.
        var. athrix St. John
        var. fueginus (Phil.) Dusen
            SY=R. f. Phil.
            SY=R. m. ssp. f. (Phil.) Hulten
        var. maritimus
        var. persicarioides (L.) R.S.
            Mitchell
            SY=R. persicarioides L.
    nematopodus Rech. f.
    obovatus Danser
    obtusifolius L.
        ssp. agrestis (Fries) Danser
        ssp. obtusifolius
        ssp. sylvestris (Wallr.) Rech.
            SY=R. o. var. s. (Wallr.) Koch
    occidentalis S. Wats.
        var. fenestratus (Greene) Lepage
            SY=R. f. Greene
        var. labradoricus (Rech. f.) Lepage
            SY=R. o. var. procerus (Greene)
                J.T. Howell
        var. occidentalis
    orbiculatus Gray
        var. borealis Rech. f.
            SY=R. britannica sensu auctt. non
                L.
        var. orbiculatus
    orthoneurus Rech. f.
    pallidus Bigelow
    paraguayensis Parodi
    patientia L.
    paucifolius Nutt. ex S. Wats.
        ssp. gracilescens (Rech. f.) Rech. f.
            SY=Acetosa g. (Rech. f.) A. Love &
                Evenson
            SY=R. p. var. g. Rech. f.
        ssp. paucifolius
            SY=Acetosa p. (Nutt. ex S. Wats.)

Rumex (CONT.)
                        A. Love
  pratensis Mert. & Koch
  pseudonatronatus Borbas
              SY=R. fennicus (Murb.) Murb.
  pulcher L.
    ssp. divaricatus (L.) Murb.
    ssp. pulcher
  rugosus Campd.
  salicifolius Weinm.
  sanguineus L.
  sibiricus Hulten
              SY=R. mexicanus Meisn. var. s.
                  (Hulten) Boivin
  skottsbergii Deg. & Deg.
  spiralis Small
  stenophyllus Ledeb.
  subalpinus M.E. Jones
  tenuifolius (Wallr.) A. Love
  tomentella Rech. f.
  transitorius Rech. f.
              SY=R. mexicanus Meisn. var. t.
                  (Rech. f.) Boivin
  triangulivalvis (Danser) Rech. f.
        var. oreolapathum Rech. f.
        var. triangulivalvis
              SY=R. mexicanus Meisn.
              SY=R. m. var. angustifolius
                  (Meisn.) Boivin
              SY=R. m. var. strictus M.E. Peck
              SY=R. m. var. subarcticus (Lepage)
                  Boivin
              SY=R. salicifolius Weinm. ssp. t.
                  Danser
              SY=R. salicifolius ssp. t. var.
                  angustivalvis Danser
              SY=R. salicifolius ssp. t. var.
                  mexicanus (Meisn.) C.L.
                  Hitchc.
              SY=R. salicifolius ssp. t. var.
                  montigenitus Jepson
              SY=R. utahensis Rech. f.
  venosus Pursh
  verticillatus L.
  violascens Rech. f.

Stenogonum Nutt.

  flexum (M.E. Jones) Reveal & J.T. Howell
              SY=Eriogonum f. M.E. Jones
              SY=E. f. var. ferronis M.E. Jones
  salsuginosum Nutt.
              SY=Eriogonum s. (Nutt.) Hook.

PONTEDERIACEAE

Eichornia Kunth

  azurea (Sw.) Kunth
  crassipes (Mart.) Solms
              SY=Piaropus c. (Mart.) Britt.
  diversifolius (Vahl) Urban
              SY=Piaropus d. (Vahl) P. Wilson
  paniculata (Spreng.) Solms

Eurystemon Alexander

  mexicanum (S. Wats.) Alexander
              SY=Heteranthera m. S. Wats.

Heteranthera Ruiz & Pavon

  dubia (Jacq.) MacM.
              SY=Zosterella d. (Jacq.) MacM.

  liebmannii (Buch.) Shinners
              SY=Zosterella longituba Alexander
  limosa (Sw.) Willd.
  reniformis Ruiz & Pavon
              SY=H. peduncularis Benth.

Monochoria Presl

  vaginalis (Burm. f.) Kunth
        var. pauciflora (Blume) Merr.
        var. vaginalis

Pontederia L.

  cordata L.
        var. lancifolia (Muhl.) Torr.
              SY=P. c. var. lanceolata (Nutt.)
                  Griseb
              SY=P. lanceolata Nutt.

PORTULACACEAE

Calandrinia H.B.K.

  ambigua (S. Wats.) T.J. Howell
  breweri S. Wats.
  ciliata (Ruiz & Pavon) DC.
              SY=C. c. var. menziesii (Hook.)
                  J.F. Macbr.
  maritima Nutt.

Calyptridium Nutt. ex Torr. & Gray

  monandrum Nutt.
  monospermum Greene
              SY=Spraguea m. (Greene) Rydb.
  parryi Gray
        var. arizonicum J.T. Howell
        var. hesseae Thomas
        var. nevadense J.T. Howell
              SY=C. p. ssp. n. (J.T. Howell)
                  Munz
        var. parryi
  pulchellum (Eastw.) Hoover
  pygmaeum Parish ex Rydb.
  quadripetalum S. Wats.
  roseum S. Wats.
  umbellatum (Torr.) Greene
              SY=C. u. var. caudiciferum (Gray)
                  Jepson
              SY=Spraguea u. Torr.
              SY=S. u. var. c. Gray

Claytonia L.

  acutifolia Pallas ex Roemer & Schultes
  arctica J.E. Adams
  arenicola Henderson
              SY=Montia a. (Henderson) T.J.
                  Howell
  caroliniana Michx.
        var. caroliniana
        var. lewisii McNeill
  cordifolia S. Wats.
              SY=C. sibirica L. var. c. (S.
                  Wats.) R.J. Davis
              SY=Montia c. (S. Wats.) Pax &
                  Hoffman
  eschscholtzii Cham.
              SY=C. acutifolia Pallas ex Roemer
                  & Schultes ssp.
                  graminifolia Hulten
  gypsophiloides Fisch. & Mey.
              SY=Montia g. (Fisch. & Mey.) J.T.

Claytonia (CONT.)
             Howell
    heterophylla (Torr. & Gray) Swanson
             SY=C. sibirica L. var. h. (Torr. &
                Gray) Gray
             SY=Montia h. (Torr. & Gray) Jepson
             SY=M. s. (L.) T.J. Howell var. h.
                (Torr. & Gray) B.L. Robins.
    lanceolata Pursh
        var. chrysantha (Greene) C.L. Hitchc.
             SY=C. c. Greene
             SY=C. l. ssp. c. (Greene) Ferris
        var. flava (A. Nels.) C.L. Hitchc.
             SY=C. f. A. Nels.
        var. idahoensis R.J. Davis
        var. lanceolata
             SY=C. caroliniana Michx. var. l.
                (Pursh) S. Wats.
        var. multiscapa (Rydb.) C.L. Hitchc.
        var. pacifica McNeill
        var. peirsonii Munz & Johnston
        var. sessilifolia (Torr.) A. Nels.
    megarhiza (Gray) Parry ex S. Wats.
        var. bellidifolia (Rydb.) C.L.
             Hitchc.
             SY=C. b. Rydb.
        var. megarhiza
        var. nivalis (English) C.L. Hitchc.
             SY=C. n. English
    nevadensis S. Wats.
    ogilivensis McNeill
    parviflora Dougl. ex Hook.
             SY=C. perfoliata Donn ex Willd.
                var. parviflora (Dougl. ex
                Hook.) Torr.
             SY=Montia perfoliata (Donn) T.J.
                Howell var. parviflora
                (Dougl. ex Hook.) Jepson
    perfoliata Donn ex Willd.
        ssp. perfoliata
        var. nubigena (Greene) v. Poellnitz
             SY=Montia p. (Donn ex Willd.) T.J.
                Howell var. n. (Greene)
                Jepson
        var. perfoliata
             SY=C. p. ssp. p. var. angustifolia
                Greene
             SY=Limnia p. (Donn ex Willd.) Haw.
             SY=Montia p. (Donn ex Willd.) T.J.
                Howell
        var. utahensis (Rydb.) v. Poellnitz
             SY=Limnia u. Rydb.
             SY=Montia p. (Donn ex Willd.) T.J.
                Howell var. u. (Rydb.) Munz
        ssp. viridis (A. Davids.) Fellows
             SY=C. spathulata Dougl. ex Hook.
                var. v. (A. Davids.) Munz
             SY=Montia s. (Dougl. ex Hook.)
                T.J. Howell var. v. A.
                Davids.
    rosea Rydb.
             SY=C. lanceolata Pursh var. r.
                (Rydb.) R.J. Davis
    rubra (T.J. Howell) Tidestrom
             SY=C. perfoliata Donn ex Willd.
                var. depressa (Gray) v.
                Poellnitz
             SY=Montia p. (Donn ex Willd.) T.J.
                Howell ssp. glauca (Nutt. ex
                Torr. & Gray) Ferris
             SY=M. p. var. d. (Gray) Jepson
             SY=M. p. var. g. Nutt. ex Torr. &
                Gray
    sarmentosa C.A. Mey.
             SY=Montia s. (C.A. Mey.) B.L.
                Robins.
    saxosa Brandeg.

             SY=Montia s. (Brandeg.) Brandeg.
    scammaniana Hulten
             SY=Montia s. (Hulten) Welsh
    sibirica L.
        var. bulbillifera Gray
             SY=Montia s. (L.) T.J. Howell var.
                bulbifera (Gray) B.L.
                Robins.
        var. sibirica
             SY=Montia s. (L.) T.J. Howell
    spathulata Dougl. ex Hook.
        var. exigua (Torr. & Gray) Hook. &
             Arn.
             SY=Montia s. (Dougl. ex Hook.)
                T.J. Howell var. e. (Torr. &
                Gray) B.L. Robins.
        var. rosulata (Eastw.) McNeill
             SY=Montia s. (Dougl. ex Hook.)
                T.J. Howell var. r. (Eastw.)
                J.T. Howell
        var. spathulata
             SY=Montia s. (Dougl. ex Hook.)
                T.J. Howell
        var. tenuifolia Torr. & Gray
             SY=Montia s. (Dougl. ex Hook.)
                T.J. Howell var. t. (Torr. &
                Gray) Munz
    tuberosa Pallas ex Willd.
        var. czukczorum (Volk.) Hulten
             SY=C. c. Volk.
        var. tuberosa
             SY=C. caroliniana Michx. var. t.
                (Pallas ex Willd.) Boivin
    umbellata S. Wats.
    virginica L.
        var. acutiflora DC.
             SY=C. simsii Sweet
             SY=C. v. var. s. (Sweet) R.J.
                Davis
        var. virginica
             SY=C. media (DC.) Link
             SY=C. robusta (Somes) Rydb.

Lewisia Pursh

    brachycalyx Engelm. ex Gray
    cantelowii J.T. Howell
    columbiana (T.J. Howell) B.L. Robins.
        var. columbiana
        var. rupicola (English) C.L. Hitchc.
             SY=L. c. ssp. r. (English) Ferris
             SY=L. r. English
        var. wallowensis C.L. Hitchc.
    congdonii (Rydb.) J.T. Howell
             SY=L. columbiana (T.J. Howell)
                B.L. Robins. ssp. congdonii
                (Rydb.) Ferris
    cotyledon (S. Wats.) B.L. Robins.
        var. cotyledon
        var. heckneri (Morton) Munz
        var. howellii (S. Wats.) Jepson
    disepala Rydb.
    kelloggii K. Brandeg.
    leana (Porter) B.L. Robins.
    maguirei N. Holmgren
    nevadensis (Gray) B.L. Robins.
             SY=L. minima A. Nels.
             SY=L. pygmaea (Gray) B.L. Robins.
                var. n. (Gray) Fosberg
    oppositifolia (S. Wats.) B.L. Robins.
    pygmaea (Gray) B.L. Robins.
        ssp. glandulosa (Rydb.) Ferris
        ssp. longipetala (Piper) Ferris
        ssp. pygmaea
    rediviva Pursh
        ssp. minor (Rydb.) N. Holmgren
             SY=L. r. var. m. (Rydb.) Munz

Lewisia (CONT.)
    ssp. rediviva
serrata Heckard & Stebbins
sierrae Ferris
stebbinsii Gankin & Hildreth
triphylla (S. Wats.) B.L. Robins.
tweedyi (Gray) B.L. Robins.

Montia L.

    bostockii (Porsild) Welsh
        SY=Claytonia b. Porsild
    chamissoi (Ledeb. ex Spreng.) Greene
        SY=Claytonia c. Ledeb. ex Spreng.
        SY=Crunocallis c. (Ledeb. ex
          Spreng.) Rydb.
    dichotoma (Nutt.) T.J. Howell
        SY=Claytonia d. Nutt.
    diffusa (Nutt.) Greene
        SY=Claytonia d. Nutt.
    fontana L.
      ssp. amporitana Sennen
        SY=M. f. var. tenerrima (Gray)
          Fern. & Wieg.
      ssp. fontana
        SY=Claytonia f. (L.) R.J. Davis
        SY=M. f. var. lamprosperma (Cham.)
          Fenzl
        SY=M. l. Cham.
      ssp. variabilis S.M. Walters
        SY=M.   funstonii Rydb.
    hallii (Gray) Greene
        SY=Montia dipetala Suksdorf
    howellii S. Wats.
        SY=Claytonia h. (S. Wats.) Piper
    linearis (Dougl. ex Hook.) Greene
        SY=Claytonia l. Dougl. ex Hook.
        SY=Montiastrum l. (Dougl. ex
          Hook.) Rydb.
    minor K.C. Gmel.
        SY=Montia verna Neck.
    parvifolia (Moc. ex DC.) Greene
      ssp. flagellaris (Bong.) Ferris
        SY=Claytonia f. Bong.
        SY=C. p. Moc. ex DC. ssp. f.
          (Bong.) Hulten
        SY=C. p. var. f. (Bong.) R.J.
          Davis
        SY=Montia f. (Bong.) B.L. Robins.
        SY=M. p. var. f. (Bong.) C.L.
          Hitchc.
      var. parvifolia
        SY=Claytonia p. Moc. ex Dc.
    rivularis K.C. Gmel.

Portulaca L.

    caulerpoides Britt. & Wilson
    cyanosperma Egler
    grandiflora Hook.
    halimoides L.
    hawaiiensis Deg.
    lutea Soland. ex Forst. f.
    mundula I.M. Johnston
    neglecta Mackenzie & Bush
    oleracea L.
    parviflorum Nutt.
    parvula Gray
    pilosa L.
    quadrifida L.
    retusa Engelm.
    rubricaulis H.B.K.
        SY=P. phaeosperma Urban
    sclerocarpa Gray
    smallii P. Wilson
    suffrutescens Engelm.
    teretifolia H.B.K.

        SY=P. poliosperma Urban
umbraticola H.B.K.
        SY=P. coronata Small
villosa Cham.

Talinopsis Gray

    frutescens Gray

Talinum Adans.

    angustissimum (Gray) Woot. & Standl.
    appalachianum W. Wolf
    aurantiacum Engelm.
    brevifolium Torr.
    calcaricum Ware
    calycinum Engelm.
    chrysanthum Rose & Standl.
    confertiflorum Greene
    gooddingii P. Wilson
    humile Greene
    longipes Woot. & Standl.
    mengesii W. Wolf
    paniculatum (Jacq.) Gaertn.
      var. paniculatum
        SY=T. reflexum Cav.
      var. sarmentosum (Engelm.) v.
        Poellnitz
    parviflorum Nutt. ex Torr. & Gray
    pulchellum Woot. & Standl.
        SY=T. youngae C.H. Muller
    rugospermum Holz.
    sediforme v. Poellnitz
        SY=T. okanoganense English
    spinescens Torr.
    teretifolium Pursh
    triangulare (Jacq.) Willd.
    validulum Greene

POTAMOGETONACEAE

Phyllospadix Hook.

    scouleri Hook.
    torreyi S. Wats.

Posidonia Koenig

    oceania Koenig

Potamogeton L.

    alpinus Balbis
        SY=P. a. ssp. tenuifolius (Raf.)
          Hulten
        SY=P. a. var. subellipticus
          (Fern.) Ogden
        SY=P. a. var. t. (Raf.) Ogden
    amplifolius Tuckerman
    X argutulus Hagstr. [?alpinus X nodosus]
    bicupulatus Fern.
        SY=P. diversifolius Raf. var.
          trichophyllus Morong
    clystocarpus Fern.
    confervoides Reichenb.
    crispus L.
    diversifolius Raf.
        SY=P. capillaceus Poir. pro parte
        SY=P. c. var. atripes Fern.
        SY=P. d. var. multidenticulatus
          (Morong) Aschers. & Graebn.
    epihydrus Raf.
        SY=P. e. var. nuttallii (Cham. &
          Schlecht.) Fern.

Potamogeton   (CONT.)
          SY=P. e. var. ramosus (Peck) House
     X faxonii Morong [illinoensis X nodosus]
     filiformis Pers.
          SY=P. f. var. alpinus (Blytt)
               Aschers. & Graebn.
          SY=P. f. var. borealis (Raf.) St.
               John
          SY=P. f. var. macounii Morong
          SY=P. f. var. occidentalis (J.W.
               Robbins) Morong
          SY=P. interior Rydb.
     floridanus Small
     foliosus Raf.
          var. fibrillosus (Fern.) Haynes &
               Reveal
               SY=P. fibrillosus Fern.
          var. foliosus
               SY=P. curtissii Morong
               SY=P. f. var. genuinus Fern.
               SY=P. f. var. macellus Fern.
     friesii Rupr.
     gramineus L.
          SY=P. g. var. graminifolius Fries
          SY=P. g. var. maximus Morong ex
               Benn.
          SY=P. g. var. myriophyllus J.W.
               Robbins
          SY=P. g. var. typicus Ogden
     groenlandicus Hagstr.
          SY=P. pusillus L. ssp. g.
               (Hagstr.) Bocher
     X hagstroemii Benn. [gramineus X
          richardsonii]
     hillii Morong
          SY=P. porteri Fern.
     illinoensis Morong
          SY=P. angustifolius Bercht. &
               Presl
          SY=P. heterophyllus Schreb.
          SY=P. lucens sensu auctt. non L.
     insulanus Hagstr.
     lateralis Morong
     latifolius (J.W. Robbins) Morong
          SY=P. filiformis Pers. var. l.
               (J.W. Robbins) Reveal
     longiligulatus Fern.
     methyensis Benn.
     X mysticus Morong [perfoliatus X pusillus
          var. tenuissimus]
     natans L.
     nodosus Poir.
          SY=P. americanus Cham. & Schlecht.
          SY=P. fluitans Roth
     oakesianus J.W. Robbins
     obtusifolius Mert. & Koch
     pectinatus L.
     perfoliatus L.
          SY=P. bupleuroides Fern.
          SY=P. p. ssp. b. (Fern.) Hulten
          SY=P. p. var. b. (Fern.) Farw.
     X perplexus Benn. [natans X nodosus]
     polygonifolius Pourret
          SY=P. oblongus Viviani
     praelongus Wulfen
          SY=P. p. var. angustifolius
               Graebn.
     pulcher Tuckerman
     pusillus L.
          var. gemmiparus J.W. Robbins
               SY=P. g. (J.W. Robbins) Morong
          var. pusillus
               SY=P. panormitanus Biv.
               SY=P. panormitanus var. major G.
                    Fisch.
               SY=P. panormitanus var. minor Biv.
               SY=P. pusillus var. minor (Biv.)

               Fern. & Schub.
          var. tenuissimus Mert. & Koch
               SY=P. berchtoldii Fieber
               SY=P. b. var. acuminatus Fieber
               SY=P. b. var. colpophilus (Fern.)
                    Fern.
               SY=P. b. var. lacunatus (Hagstr.)
                    Fern.
               SY=P. b. var. mucronatus Fieber
               SY=P. b. var. polyphyllus (Morong)
                    Fern.
               SY=P. b. var. t. (Mert. & Koch)
                    Fern.
               SY=P. p. var. m. (Fieber) Graebn.
     X rectifolius Benn. [nodosus X
          richardsonii]
     richardsonii (Benn.) Rydb.
          SY=P. perfoliatus L. ssp. r.
               (Benn.) Hulten
          SY=P. p. var. r. Benn.
     robbinsii Oakes
     X scoliophyllus Hagstr. [amplifolius X
          illinoensis]
     X spathuliformis (J.W. Robbins) Morong
          [?gramineus X nodosus]
          SY=P. varians Morong
     spirillus Tuckerman
          SY=P. dimorphus Raf.
     strictifolius Benn.
          SY=P. pusillus L. var. rutiloides
               (Fern.) Boivin
          SY=P. rutilus sensu auctt. non
               Wolfgang
          SY=P. s. var. rutiloides Fern.
          SY=P. s. var. typicus Fern.
     X subnitens Hagstr. [gramineus X
          perfoliatus]
     X subobtusus Hagstr. [alpinus X nodosus]
     X subsessilis Hagstr. [epihydrus X nodosus]
     subsibiricus Hagstr.
          SY=P. porsildiorum Fern.
     tennesseensis Fern.
     vaginatus Turcz.
          SY=P. interruptus Kit.
     vaseyi J.W. Robbins
     zosteriformis Fern.
          SY=P. compressus sensu auctt. non
               L.

PRIMULACEAE

Anagallis L.

     arvensis L.
          var. arvensis
          var. coerulea (Schreb.) Gren. & Godr.
     minima (L.) Krause
          SY=Centunculus m. L.
     pumila Sw.
          SY=Micropyxis p. (Sw.) Duby
     tenella (L.) L.

Androsace L.

     alaskana Coville & Standl.
     chamaejasme Wulfen
          ssp. andersonii Hulten
          ssp. carinata (Torr.) Hulten
               SY=A. c. Torr.
          ssp. chamaejasme
          ssp. lehmanniana (Spreng.) Hulten
               SY=A. l. Spreng.
     elongata L.
          ssp. acuta (Greene) G.T. Robbins

Androsace (CONT.)
        SY=A. a. Greene
  filiformis Retz.
  occidentalis Pursh
      SY=A. arizonica (Gray) Derganc
      SY=A. o. var. a. (Gray) St. John
      SY=A. o. var. simplex (Rydb.) St.
        John
  septentrionalis L.
    var. glandulosa (Woot. & Standl.) St.
      John
      SY=A. g. Woot. & Standl.
    var. puberulenta (Rydb.) R. Knuth
      SY=A. p. Rydb.
    var. robusta St. John
    var. septentrionalis
    var. subulifera Gray
      SY=A. diffusa Small
      SY=A. septentrionalis var. d.
        (Small) R. Knuth
    var. subumbellata A. Nels.
      SY=A. septentrionalis ssp.
        subumbellata (A. Nels.) G.T.
        Robbins

Dodecatheon L.

  alpinum (Gray) Greene
    ssp. alpinum
    ssp. majus H.J. Thompson
  amethystinum (Fassett) Fassett
  clevelandii Greene
    ssp. clevelandii
    ssp. insulare H.J. Thompson
    ssp. patulum (Greene) H.J. Thompson
      SY=D. p. Greene
    ssp. sanctarum (Greene) Abrams
  conjugens Greene
    var. beamishii Boivin
    var. conjugens
      SY=D. c. ssp. leptophyllum
        (Suksdorf) Piper
      SY=D. cylindrocarpum Rydb.
    var. viscidum (Piper) Mason ex St.
      John
      SY=D. c. ssp. v. (Piper) H.J.
        Thompson
      SY=D. v. Piper
  dentatum Hook.
    ssp. dentatum
    ssp. ellisiae (Standl.) H.J. Thompson
      SY=D. e. Standl.
  frenchii (Vasey) Rydb.
      SY=D. meadia L. ssp. membranaceum
        R. Knuth
      SY=D. m. var. f. Vasey
  frigidum Cham. & Schlecht.
  hansenii (Greene) H.J. Thompson
      SY=D. hendersonii Gray var.
        hansenii Greene
  hendersonii Gray
    ssp. cruciatum (Greene) H.J. Thompson
      SY=D. h. var. c. Greene
    ssp. hendersonii
    ssp. parvifolium (R. Knuth) H.J.
      Thompson
  jeffreyi Van Houtte
    ssp. jeffreyi
      SY=D. j. var. viviparum (Greene)
        Abrams
    ssp. pygmaeum (Hall) H.J. Thompson
  meadia L.
    var. brachycarpum (Small) Fassett
      SY=D. b. Small
      SY=D. b. ssp. b. (Small) R. Knuth
    var. meadia
      SY=D. hugeri Small

      var. obesum Fassett
      var. stanfieldii (Small) Fassett
  poeticum Henderson
  pulchellum (Raf.) Merr.
    var. watsonii (Tidestrom) C.L.
      Hitchc.
      SY=D. pauciflorum (Dur.) Greene
        var. w. (Tidestrom) C.L.
        Hitchc.
      SY=D. pulchellum var. w.
        (Tidestrom) Boivin
      SY=D. radicatum Greene ssp. w.
        (Tidestrom) H.J. Thompson
    ssp. cusickii (Greene) Calder & Taylor
      SY=D. c. Greene
      SY=D. c. var. album Suksdorf
      SY=D. pauciflorum (Dur.) Greene
        var. c. (Greene) Mason ex
        St. John
      SY=D. puberulum (Nutt.) Piper
      SY=D. pulchellum var. album
        (Suksdorf) Boivin
    ssp. macrocarpum (Gray) Taylor &
      MacBryde
      SY=D. pauciflorum (Dur.) Greene
        var. alaskanum (Hulten) C.L.
        Hitchc.
      SY=D. pulchellum ssp. a. (Hulten)
        Hulten
      SY=D. pulchellum ssp. superbum
        (Pennell & Stair) Hulten
      SY=D. pulchellum var. a. (Hulten)
        Boivin
    ssp. monanthum (Greene) H.J. Thompson ex
      Munz
      SY=D. pauciflorum (Dur.) Greene
        var. m. Greene
      SY=D. pulchellum var. m. (Greene)
        C.L. Hitchc.
      SY=D. radicatum Greene ssp. m.
        (Greene) H.J. Thompson
    ssp. pauciflorum (Dur.) Hulten
      SY=D. pauciflorum (Dur.) Greene
    ssp. pulchellum
      SY=D. pauciflorum (Dur.) Greene
        var. salinum (A. Nels.) R.
        Knuth
      SY=D. radiatum Greene
      SY=D. s. A. Nels.
  redolens (Hall) H.J. Thompson
      SY=D. jeffreyi Van Houtte var. r.
        Hall
  subalpinum Eastw.

Douglasia Lindl.

  arctica Hook.
  gormanii Constance
      SY=D. arctica Hook. var. g.
        (Constance) Boivin
  laevigata Gray
    var. ciliolata Constance
      SY=D. l. ssp. c. (Constance)
        Calder & Taylor
    var. laevigata
  montana Gray
      SY=Gregoria m. (Gray) House
  nivalis Lindl.
    var. dentata Gray
    var. nivalis
  ochotensis (Willd.) Hulten

Glaux L.

  maritima L.
    ssp. maritima
    ssp. obtusifolia (Fern.) Boivin

Glaux (CONT.)
                SY=G. m. var. angustifolia Boivin
                SY=G. m. var. macrophylla Boivin
                SY=G. m. var. obtusifolia Fern.

Hottonia L.

    inflata Ell.

Lysimachia L.

    asperulifolia Poir.
    ciliata L.
                SY=Steironema c. (L.) Baudo
                SY=S. pumilum Greene
    clethroides Duby
    X commixta Fern. [terrestris X thyrsiflora]
    daphnoides Hbd.
    filifolia Forbes & Lydgate
    forbesii Rock
    fraseri Duby
    glutinosa Rock
    hillebrandii Hook. f. ex Gray
            var. angustifolia Gray
            var. helleri R. Knuth
            var. hillebrandii
            var. maxima R. Knuth
            var. subherbacea Hbd.
            var. venosa Wawra
    hybrida Michx.
                SY=L. ciliata L. var. validula
                    (Greene ex Woot. & Standl.)
                    Kearney & Peebles
                SY=L. lanceolata Walt. ssp. h.
                    (Michx.) J.D. Ray
                SY=L. l. var. h. (Michx.) Gray
                SY=Steironema h. (Michx.) Raf. ex
                    Jackson
                SY=S. laevigatum T.J. Howell
                SY=S. lanceolatum (Walt.) Gray
                    var. h. (Michx.) Gray
    japonica Thunb.
    kalalauensis Skottsberg
    kipahuluensis St. John
    lanceolata Walt.
                SY=L. graminea (Greene) Hand.-Maz.
                SY=L. l. var. angustifolium (Lam.)
                    Gray
                SY=Steironema heterophyllum
                    (Michx.) Raf.
                SY=S. l. (Walt.) Gray
    loomisii Torr.
    lydgatei Hbd.
    mauritiana Lam.
    nummularia L.
    ovata (Heller) St. John
    X producta (Gray) Fern. [quadrifolia X
        terrestris]
    punctata L.
                SY=L. p. var. verticillata (Bieb.)
                    Klatt
    quadriflora Sims
                SY=L. longifolia Pursh
                SY=Steironema q. (Sims) A.S.
                    Hitchc.
    quadrifolia L.
    X radfordii Ahles [loomisii X quadrifolia]
    radicans Hook.
                SY=Steironema r. (Hook.) Gray
    remyi Hbd.
    terrestris (L.) B.S.P.
                SY=L. t. var. ovata (Rand & Redf.)
                    Fern.
    thyrsiflora L.
                SY=Naumbergia t. (L.) Duby
    tonsa (Wood) R. Knuth
                SY=L. t. var. simplex Kearney

                SY=Steironema intermedium Kearney
                SY=S. t. (Wood) Bickn.
    vulgaris L.

Primula L.

    angustifolia Torr.
    capillaris N. Holmgren & A. Holmgren
    clusiana Tausch
                SY=P. sibirica Jacq.
    cuneifolia Ledeb.
        ssp. cuneifolia
        ssp. saxifragifolia (Lehm.) W.W. Sm. &
            G. Forrest
    cusickiana (Gray) Gray
    egaliksensis Wormskj.
    ellisiae Pollard & Cockerell
    farinosa L.
                SY=P. laurentiana Fern.
    hunnewellii Fern.
    incana M.E. Jones
    maguirei L.O. Williams
    matsumurae Petitm.
    mistassinica Michx.
            var. intercedens (Fern.) Boivin
                SY=P. i. Fern.
            var. macropoda (Fern.) Boivin
            var. mistassinica
                SY=P. m. var. noveboracensis Fern.
    nevadensis N. Holmgren
    parryi Gray
    rusbyi Greene
    specuicola Rydb.
    stricta Hornem.
    suffrutescens Gray
    tenuis Small
                SY=P. borealis Duby
    tschuktschorum Kjellm.
            var. arctica (Koidzumi) Fern.
                SY=P. t. ssp. cairnsiana Porsild
            var. tschuktschorum
    veris L.

Samolus L.

    alyssoides Heller
                SY=S. ebracteatus H.B.K. ssp. a.
                    (Heller) R. Knuth
    cuneatus Small
                SY=S. ebracteatus H.B.K. ssp. c.
                    (Small) R. Knuth
    ebracteatus H.B.K.
                SY=Samodia e. (H.B.K.) Baudo
    vagans Greene
    valerandi L.
        ssp. parviflorus (Raf.) Hulten
                SY=Samolus floribundus H.B.K.
                SY=S. p. Raf.
        ssp. valerandi

Trientalis L.

    borealis Raf.
        ssp. borealis
                SY=T. americana Pursh
        ssp. latifolia (Hook.) Hulten
                SY=T. l. Hook.
    europaea L.
        ssp. arctica (Fisch. ex Hook.) Hulten
                SY=T. a. Fisch. ex Hook.
                SY=T. e. var. a. (Fisch. ex Hook.)
                    Ledeb.
        ssp. europaea

PROTEACEAE

Grevillea R. Br. ex Salisb.

  robusta A. Cunningham

PUNICACEAE

Punica L.

  granatum L.

RAFFLESIACEAE

Pilostyles Guill.

  thurberi Gray

RANUNCULACEAE

Aconitum L.

  X bicolor Schultes [napellus X variegatum]
  columbianum Nutt.
      var. columbianum
        SY=A. c. ssp. pallidum Piper
        SY=A. c. var. bakeri (Greene)
          Harrington
        SY=A. geranioides Greene pro parte
        SY=A. hansenii Greene pro parte
        SY=A. leibergii Greene pro parte
        SY=A. viviparum Greene pro parte
      var. howellii (A. Nels. & J.F.
        Macbr.) C.L. Hitchc.
        SY=A. bulbiferum T.J. Howell
        SY=A. h. A. Nels. & J.F. Macbr.
      var. ochroleucum A. Nels.
  delphiniifolium DC.
    ssp. chamissonianum (Reichenb.) Hulten
    ssp. delphiniifolium
        SY=A. d. ssp. d. var. albiflorum
          Porsild
    ssp. paradoxum (Reichenb.) Hulten
  infectum Greene
  maximum Pallas
  mogollonicum Greene
  napellus L.
  noveboracense Gray
        SY=A. n. var. quasiciliatum
          Fassett
        SY=A. uncinatum L. ssp. n. (Gray)
          Hardin
  reclinatum Gray
  septentrionale Koelle
        SY=A. lycoctonum L.
  uncinatum L.
    ssp. muticum (DC.) Hardin
    ssp. uncinatum
        SY=A. u. var. acutidens Fern.
  variegatum L.

Actaea L.

  X ludovicii Boivin [pachypoda X rubra]
  pachypoda Ell.
        SY=A. alba sensu auctt. non (L.)
          P. Mill.
  rubra (Ait.) Willd.
    ssp. arguta (Nutt.) Hulten
        SY=A. a. Nutt.

        SY=A. a. var. viridiflora (Greene)
          Tidestrom
        SY=A. r. var. a. (Nutt.) Lawson
    ssp. rubra

Adonis L.

  aestivalis L.
        SY=A. a. var. citrina Hoffmann
  annua L.
  vernalis L.

Anemone L.

  berlandieri Pritz.
        SY=A. decapetala sensu auctt. non
          Ard.
        SY=A. d. var. heterophylla (Nutt.
          ex Torr. & Gray) Britt.
        SY=A. h. Nutt. ex Torr. & Gray
  canadensis L.
  caroliniana Walt.
  cylindrica Gray
  deltoidea Hook.
  drummondii S. Wats.
  edwardsiana Tharp
      var. edwardsiana
      var. petraea Correll
  hupehensis (Lem. & Lem. f.) Lem. & Lem. f.
  lancifolia Pursh
        SY=A. trifolia sensu auctt. non L.
  lithophila Rydb.
        SY=A. drummondii S. Wats. var. l.
          (Rydb.) C.L. Hitchc.
  lyallii Britt.
        SY=A. nemorosa L. var. l. (Britt.)
          Ulbr.
        SY=A. quinquefolia L. var. l.
          (Britt.) B.L. Robins.
  minima DC.
  multiceps (Greene) Standl.
  multifida Poir.
      var. globosa (Nutt. ex Pritz.) Torr.
        & Gray
        SY=A. g. Nutt. ex Pritz.
      var. hirsuta C.L. Hitchc.
      var. hudsoniana DC.
        SY=A. h. (DC.) Richards.
      var. multifida
      var. richardsoniana Fern.
      var. saxicola Boivin
  narcissiflora L.
    ssp. alaskana Hulten
        SY=A. n. var. a. (Hulten) Boivin
    ssp. interior Hulten
        SY=A. n. var. i. (Hulten) Boivin
    ssp. sibirica (L.) Hulten
        SY=A. n. var. monantha Schlecht.
    ssp. villosissima (DC.) Hulten
        SY=A. n. var. v. DC.
    ssp. zephyra (A. Nels.) Love, Love &
    Kapoor
        SY=A. z. A. Nels.
  nemorosa L.
      var. bifolia (Farw.) Boivin
      var. nemorosa
  oregana Gray
      var. felix (M.E. Peck) C.L. Hitchc.
        SY=A. f. M.E. Peck
      var. oregana
        SY=A. adamsiana Eastw.
        SY=A. quinquefolia L. var. minor
          (Eastw.) Munz
        SY=A. q. var. o. (Gray) B.L.
          Robins.
  parviflora Michx.
      var. grandiflora Ulbr.

Anemone (CONT.)
         var. parviflora
    piperi Britt.
    quinquefolia L.
         var. grayi (Behr & Kellogg) Jepson
         var. interior Fern.
         var. quinquefolia
             SY=A. nemorosa L. var. q. (L.)
                  Pursh
    ranunculoides L.
    richardsonii Hook.
    riparia Fern.
             SY=A. virginiana L. var. r.
                  (Fern.) Boivin
    stylosa A. Nels.
    tetonensis Porter
             SY=A. multifida Poir. var. t.
                  (Porter) C.L. Hitchc.
    tuberosa Rydb.
    virginiana L.
         var. cylindroidea Boivin
         var. virginiana

Aquilegia L.

    barnebyi Munz
    brevistyla Hook.
    caerulea James
         var. alpina A. Nels.
         var. caerulea
         var. daileyae Eastw.
         var. ochroleuca Hook.
         var. pinetorum (Tidestrom) Payson ex
             Kearney & Peebles
    canadensis L.
         var. australis (Small) Munz
             SY=A. a. Small
         var. canadensis
         var. coccinea (Small) Munz
             SY=A. c. Small
         var. hybrida Hook.
             SY=A. c. var. eminens (Greene)
                  Boivin
         var. latiuscula (Greene) Munz
             SY=A. phoenicantha Cory
    chaplinei Standl. ex Payson
    chrysantha Gray
         var. chrysantha
         var. rydbergii Munz
    desertorum (M.E. Jones) Cockerell ex Heller
    elegantula Greene
    eximia Van Houtte ex Planch.
    flavescens S. Wats.
         var. flavescens
             SY=A. formosa Fisch. var.
                  flavescens (S. Wats.) M.E.
                  Peck
         var. miniata A. Nels. & J.F. Macbr.
    formosa Fisch.
         var. formosa
             SY=A. f. var. communis Boivin
             SY=A. f. var. megalantha Boivin
         var. hypolasia (Greene) Munz
         var. pauciflora (Greene) Boothman
         var. truncata (Fisch. & Mey.) Baker
         var. wawawensis (Payson) St. John
    hinkleyana Munz
    jonesii Parry
         var. elatior Boothman
         var. jonesii
    laramiensis A. Nels.
    longissima Gray
    micrantha Eastw.
         var. mancosana Eastw.
         var. micrantha
    pubescens Coville
    saximontana Rydb.

scopulorum Tidestrom
         var. calcarea (M.E. Jones) Munz
         var. scopulorum
    shockleyi Eastw.
             SY=A. mohavensis Munz
    triternata Payson
    vulgaris L.

Caltha L.

    leptosepala DC.
         ssp. howellii (Huth) P.G. Smith
             SY=C. biflora DC.
             SY=C. b. ssp. h. (Huth) Abrams
             SY=C. h. Huth
             SY=C. l. ssp. b. (DC.) P.G. Smith
             SY=C. l. var. b. (DC.) Lawson
         ssp. leptosepala
             var. leptosepala
                 SY=C. biflora DC. var.
                      rotundifolia (Huth) C.L.
                      Hitchc.
                 SY=C. chelidonii Greene
                 SY=C. l. var. r. Huth
             var. sulfurea C.L. Hitchc.
    natans Pallas
    palustris L.
         var. flabellifolia (Pursh) Torr. &
             Gray
             SY=C. arctica R. Br.
             SY=C. natans Pallas var. a. (R.
                  Br.) Hulten
             SY=C. p. ssp. a. (R. Br.) Hulten
             SY=C. p. var. radicans (Forst.)
                  Hartman
             SY=C. r. Forst.
         var. palustris
             SY=C. asarifolia (DC.) Hulten
             SY=C. natans Pallas var. a. (DC.)
                  Huth
             SY=C. p. ssp. a. (DC.) Hulten
             SY=C. p. var. a. (DC.) Huth

Cimicifuga Wernischeck

    americana Michx.
    arizonica S. Wats.
    elata Nutt.
    laciniata S. Wats.
    racemosa (L.) Nutt.
         var. cordifolia (Pursh) Gray
             SY=C. c. Pursh
         var. racemosa
    rubifolia Kearney

Clematis L.

    addisonii Britt.
             SY=C. viornioides Britt.
             SY=Viorna a. (Britt.) Small
    albicoma Wherry
             SY=Viorna ovata (Pursh) Small
    alpina (L.) P. Mill.
    baldwinii Torr. & Gray
         var. baldwinii
             SY=Viorna b. (Torr. & Gray) Small
         var. latiuscula R.W. Long
    bigelovii Torr.
    catesbyana Pursh
             SY=C. micrantha Small
    coactilis (Fern.) Keener
             SY=C. albicoma Wherry var. c.
                  Fern.
    crispa L.
             SY=Viorna c. (L.) Small
             SY=V. obliqua Small
    dioica L.

Clematis (CONT.)
  drummondii Torr. & Gray
  florida Thunb.
  fremontii S. Wats.
        SY=C. f. var. riehlii Erickson
        SY=Viorna f. (S. Wats.) Heller
  glaucophylla Small
        SY=Viorna g. (Small) Small
  hirsutissima Pursh
        var. arizonica (Heller) Erickson
        var. hirsutissima
  lasiantha Nutt.
  ligusticifolia Nutt.
        var. brevifolia Nutt.
        var. californica S. Wats.
        var. ligusticifolia
        SY=C. suksdorfii B.L. Robins.
  neomexicana Woot. & Standl.
  occidentalis (Hornem.) DC.
        var. dissecta (C.L. Hitchc.) J.
            Pringle
        SY=C. columbiana (Nutt.) Torr. &
            Gray var. d. C.L. Hitchc.
        var. grosseserrata (Rydb.) J. Pringle
        SY=Atragene columbiana Nutt.
        SY=C. c. (Nutt.) Torr. & Gray
        SY=C. c. var. tenuiloba (Gray) J.
            Pringle
        SY=C. o. ssp. g. (Rydb.) Taylor &
            MacBryde
        SY=C. pseudoalpina (Kuntze) A.
            Nels.
        SY=C. t. (Gray) C.L. Hitchc.
        SY=C. verticillaris DC. var. c.
            (Nutt.) Gray
        var. occidentalis
        SY=Atragene americana Sims
        SY=C. verticillaris DC.
        SY=C. v. var. cacuminis Fern.
        SY=C. v. var. grandiflora Boivin
  ochroleuca Ait.
        SY=Viorna o (Ait.) Small
  orientalis L.
        SY=C. aurea A. Nels. & J.F. Macbr.
  palmeri Rose
  pauciflora Nutt.
  pilosa Nutt.
  pitcheri Torr. & Gray
        var. dictyota (Greene) Dennis
        SY=C. d. Greene
        var. filifera (Benth.) B.L. Robins.
            ex Gray
        SY=C. f. Benth.
        var. pitcheri
        SY=Viorna p. (Torr. & Gray) Britt.
  recta L.
  reticulata Walt.
        SY=Viorna r. (Walt.) Small
        SY=V. subreticulata Harbison ex
            Small
  scottii Porter & Coult.
        SY=C. hirsutissima Pursh var. s.
            (Porter & Coult.) Erickson
        SY=Viorna s. (Porter & Coult.)
            Rydb.
  tangutica (Maxim.) Korsh.
        SY=C. orientalis L. var. t. Maxim.
  terniflora DC.
        SY=C. dioscoreifolia Levl. &
            Vaniot
        SY=C. d. var. robusta (Carr.)
            Rehd.
        SY=C. maximowicziana Franch. &
            Sav.
        SY=C. paniculata Thunb.
  texensis Buckl.
  versicolor Small ex Rydb.

        SY=Viorna v. (Small ex Rydb.)
            Small
  viorna L.
        var. flaccida (Small) Erickson
        SY=Viorna f. Small
        var. viorna
        SY=C. beadlei (Small) Erickson
        SY=C. gattingeri Small
        SY=Viorna b. Small
        SY=V. g. (Small) Small
        SY=V. ridgwayi Standl.
        SY=V. v. (L.) Small
  virginiana L.
        SY=C. v. var. missouriensis
            (Rydb.) Palmer & Steyermark
  vitalba L.
  viticaulis Steele
  viticella L.
        SY=Viticella v. (L.) Small

Consolida (DC.) S.F. Gray

  ambigua (L.) Ball & Heywood
        SY=C. ajacis (L.) Schur
        SY=Delphinium ajacis L.
        SY=D. ambigua L.
  orientalis (J. Gay) Schrodinger
        SY=Delphinium o. J. Gay
  regalis S.F. Gray

Coptis Salisb.

  asplenifolia Salisb.
  laciniata Gray
  occidentalis (Nutt.) Torr. & Gray
  trifolia (L.) Salisb.
        ssp. groenlandica (Oeder) Hulten
        SY=C. g. (Oeder) Fern.
        SY=C. t. var. g. (Oeder) Fassett
        ssp. trifolia

Delphinium L.

  alabamicum Kral
  alpestre Rydb.
  andersonii Gray
        ssp. andersonii
        ssp. cognatum (Greene) Ewan
        SY=D. a. var. c. (Greene) R.J.
            Davis
  andesicola Ewan
        ssp. amplum Ewan
        ssp. andesicola
  bakeri Ewan
  barbeyi (Huth) Huth
  bicolor Nutt. & Wyeth
  brachycentrum Ledeb.
  burkei Greene
        SY=D. b. var. distichiflorum Hook.
        SY=D. strictum A. Nels.
        SY=D. s. var. d. (Hook.) St. John
  californicum Torr. & Gray
        ssp. californicum
        ssp. interius (Eastw.) Ewan
  cardinale Hook.
  carolinianum Walt.
        SY=D. azureum Michx.
        SY=D. a. var. nortonianum
            (Mackenzie & Bush) Palmer &
            Steyermark
        SY=D. c. var. crispum Perry
        SY=D. carolinianum var. n.
            (Mackenzie & Bush) Perry
  chamissonis Pritz. ex Walp.
  X confertiflorum Woot. [scaposum X
    virescens ssp. wootonii]
  decorum Fisch. & Mey.

Delphinium (CONT.)
    ssp. decorum
    ssp. tracyi Ewan
  depauperatum Nutt.
            SY=D. cyanoreios Piper
            SY=D. diversifolium Greene ssp.
                harneyense Ewan
            SY=D. diversifolium var. h. (Ewan)
                R.J. Davis
  diversifolium Greene
  elatum L.
  exaltatum Ait.
  geraniifolium Rydb.
  geyeri Greene
  glareosum Greene
    ssp. caprorum (Ewan) Ewan
            SY=D. g. var. c. (Ewan) W.H. Baker
    ssp. glareosum
  glaucescens Rydb.
  glaucum S. Wats.
            SY=D. brownii Rydb.
            SY=D. scopulorum Gray var. g. (S.
                Wats.) Gray
  gracilentum Greene
  grandiflorum L.
  gypsophilum Ewan
    ssp. gypsophilum
    ssp. parviflorum Lewis & Epling
  hansenii (Greene) Greene
    ssp. arcuatum (Greene) Ewan
    ssp. hansenii
    ssp. kernense (A. Davids.) Ewan
  hesperium Gray
    ssp. cuyamacae (Abrams) Lewis & Epling
            SY=D. c. Abrams
    ssp. hesperium
            SY=D. h. var. seditosum Jepson
    ssp. pallescens (Ewan) Lewis & Epling
  hutchinsonae Ewan
  inopinum (Jepson) Lewis & Epling
  kinkiense Munz
  leucophaeum Greene
  luteum Heller
  macrophyllum Woot.
  madrense S. Wats.
  menziesii DC.
    ssp. menziesii
    ssp. pyramidale Ewan
            SY=D. m. var. p. (Ewan) C.L.
                Hitchc.
  multiplex (Ewan) C.L. Hitchc.
  newtonianum D.M. Moore
  novomexicanum Woot.
  nudicaule Torr. & Gray
  X nutans A. Nels. [glaucum X menziesii]
  nuttallianum Pritz. ex Walp.
            var. fulvum C.L. Hitchc.
            var. lineapetalum (Ewan) C.L. Hitchc.
            SY=D. l. Ewan
            var. nuttallianum
            SY=D. nelsonii Greene
            SY=D. nuttallianum ssp. utahense
                (S. Wats.) Ewan
            SY=D. nuttallianum var. laevicaule
                C.L. Hitchc.
  nuttallii Gray
            SY=D. oreganum T.J. Howell
  occidentale (S. Wats.) S. Wats.
    ssp. cucullatum (A. Nels.) Ewan
            SY=D. o. var. c. (A. Nels.) R.J.
                Davis
    ssp. occidentale
            SY=D. scopulorum Gray ssp. o. (S.
                Wats.) Abrams
    ssp. quercicola Ewan
  parishii Gray
    ssp. parishii

            SY=D. amabile Tidestrom
            SY=D. a. ssp. apachense (Eastw.)
                Ewan
    ssp. purpureum Lewis & Epling
    ssp. subglobosum (Wiggins) Lewis &
        Epling
            SY=D. s. Wiggins
  parryi Gray
    ssp. blochmanae (Greene) Lewis & Epling
    ssp. eastwoodae Ewan
    ssp. parryi
    ssp. seditiosum (Jepson) Ewan
  patens Benth.
    ssp. greenei (Eastw.) Ewan
    ssp. hepaticoideum Ewan
    ssp. montanum (Munz) Ewan
    ssp. patens
            SY=D. decorum Fisch. & Mey. var.
                p. (Benth.) Gray
  pavonaceum Ewan
  polycladon Eastw.
            SY=D. scopulorum Gray var. luporum
                (Greene) Jepson
  pratense Eastw.
  purpusii Brandeg.
  ramosum Rydb.
  recurvatum Greene
            SY=D. hesperium Gray var. r.
                (Greene) Davis
  robustum Rydb.
  sapellonis Cockerell
  scaposum Greene
  scopulorum Gray
  sonnei Greene
  stachydeum (Gray) A. Nels. & J.F. Macbr.
            SY=D. umatillense Ewan
  tenuisectum Greene
    ssp. amplibracteatum (Woot.) Ewan
    ssp. tenuisectum
  treleasei Bush ex R.J. Davis
  tricorne Michx.
  trolliifolium Gray
  uliginosum Curran
  umbraculorum Lewis & Epling
  variegatum Torr. & Gray
    ssp. apiculatum (Greene) Ewan
            SY=D. v. var. a. Greene
    ssp. thornei Munz
    ssp. variegatum
  vimineum D. Don
            SY=D. virescens Nutt. var.
                vimineum (D. Don) R.F.
                Martin
  virescens Nutt.
    ssp. penardii (Huth) Ewan
            SY=D. p. Huth
            SY=D. v. var. p. (Huth) Perry
    ssp. virescens
            var. macroceratilis (Rydb.) Cory
            var. virescens
    ssp. wootonii (Rydb.) Ewan
  viridescens Leib.
  xantholeucum Piper

Eranthis Salisb.

  hyemalis (L.) Salisb.

Helleborus L.

  viridis L.

Hepatica P. Mill.

  nobilis P. Mill.
            var. acuta (Pursh) Steyermark
            SY=H. a. (Pursh) Britt.

Hepatica (CONT.)
        SY=H. acutiloba DC.
    var. obtusa (Pursh) Steyermark
        SY=H. americana (DC.) Ker-Gawl.
        SY=H. hepatica (L.) Karst.

Hydrastis Ellis

   canadensis L.

Isopyrum L.

   biternatum (Raf.) Torr. & Gray
       SY=Enemion b. Raf.
   hallii Gray
   occidentale Hook. & Arn.
   savilei Calder & Taylor
   stipitatum Gray

Myosurus L.

   aristatus Benth. ex Hook.
       SY=M. minimus L. var. a. (Benth.
          ex Hook.) Boivin
   cupulatus S. Wats.
   minimus L.
    ssp. apus (Greene) Campbell
      var. apus Greene
      var. filiformis Greene
      var. sessiliflorus (Huth) Campbell
        SY=M. alopecuroides Greene
        SY=M. sessilis S. Wats.
    ssp. major (Greene) Campbell
      var. clavicaulis (M.E. Peck) Campbell
        SY=M. c. M.E. Peck
      var. major (Greene) R.J. Davis
    ssp. minimus
        SY=M. apetalus Gray
        SY=M. m. var. interior Boivin
        SY=M. m. var. lepturus (Gray) T.J.
          Howell
    ssp. montanus Campbell
        SY=M. aristatus Benth. ex Hook.
          ssp. m. (Campbell) Stone
   nitidus Eastw.
       SY=M. egglestonii Woot. & Standl.

Nigella L.

   damascena L.

Pulsatilla P. Mill.

   occidentalis (S. Wats.) Freyn
       SY=Anemone o. S. Wats
   patens (L.) P. Mill.
    ssp. multifida (Pritz.) Zamels
        SY=Anemone ludoviciana Nutt.
        SY=A. nuttalliana DC.
        SY=A. patens L. var. m. Pritz.
        SY=A. p. var. n. (DC.) Gray
        SY=A. p. var. wolfgangiana (Bess.)
          K. Koch
        SY=P. hirsutissima (Pursh) Britt.
        SY=P. l. (Nutt.) Heller
    ssp. patens
        SY=Anemone p. L.

Ranunculus L.

   abortivus L.
    var. abortivus
       SY=R. a. ssp. acrolasius (Fern.)
         Kapoor & Love
       SY=R. abortivus var. acrolasius
         Fern.
       SY=R. abortivus var. typicus Fern.

      var. eucyclus Fern.
      var. indivisus Fern.
   acriformis Gray
    var. acriformis
       SY=R. a. var. typicus L. Benson
    var. aestivalis L. Benson
    var. montanensis (Rydb.) L. Benson
   acris L.
    var. acris
       SY=R. a. ssp. strigulosus (Schur)
         Hyl.
       SY=R. a. var. latisectus G. Beck
       SY=R. a. var. stevenii (Andrz.)
         Lange
       SY=R. a. var. typicus G. Beck
       SY=R. boreanus Jord.
    var. frigidus Regel
       SY=R. grandis Honda
       SY=R. g. var. austrokurilensis
         (Tatew.) Hara
   adoneus Gray
    var. adoneus
       SY=R. a. var. typicus L. Benson
       SY=R. eschscholtzii Schlecht. var.
         a. (Gray) C.L. Hitchc.
    var. alpinus (S. Wats.) L. Benson
       SY=R. eschscholtzii Schlecht. var.
         a. (S. Wats.) C.L. Hitchc.
   alismifolius Geyer ex Benth.
    var. alismellus Gray
    var. alismifolius
       SY=R. a. var. typicus L. Benson
    var. davisii L. Benson
    var. hartwegii (Greene) Jepson
    var. lemmonii (Gray) L. Benson
    var. montanus S. Wats.
   allegheniensis Britt.
   allenii B.L. Robins.
   ambigens S. Wats.
       SY=R. obtusiusculus Raf.
   andersonii Gray
       SY=Beckwithia a. (Gray) Jepson
   aquatilis L.
    var. aquatilis
       SY=R. a. var. codyanus Boivin
       SY=R. a. var. typicus L. Benson
    var. calvescens (W. Drew) L. Benson
       SY=R. confervoides (Fries) Fries
       SY=R. trichophyllus Chaix var.
         calvescens W. Drew
    var. capillaceus (Thuill.) DC.
       SY=Batrachium flaccidum (Pers.)
         Rupr.
       SY=B. trichophyllus (Chaix) F.W.
         Schultz
       SY=R. a. var. t. (Chaix) Gray
       SY=R. t. Chaix
       SY=R. t. var. typicus W. Drew
    var. eradicatus Laestad.
       SY=R. trichophyllus Chaix ssp. e.
        (Laestad.) C.D.K. Cook
       SY=R. t. var. e. (Laestad.) W.
        Drew
    var. harrisii L. Benson
    var. hispidulus E. Drew
       SY=R. trichophyllus Chaix var.
        hispidus (E. Drew) W. Drew
    var. lalondei L. Benson
    var. porteri (Britt.) L. Benson
   arizonicus Lemmon
       SY=R. a. var. typicus L. Benson
   arvensis L.
       SY=R. a. var. tuberculatus DC.
   auricomus L.
    var. auricomus
       SY=R. a. var. typicus G. Beck
    var. glabratus Lynge

Ranunculus (CONT.)
  austro-oreganus L. Benson
  bonariensis Poir.
      var. trisepalus (Gillies) Lourteig
        SY=R. alveolatus Carter
        SY=R. t. Gillies
  bulbosus L.
        SY=R. b. var. dissectus Barbey
        SY=R. b. var. valdepubens (Jord.)
          Briq.
  californicus Benth.
      var. austromontanus L. Benson
      var. californicus
        SY=R. c. var. typicus L. Benson
      var. cuneatus Greene
      var. gratus Jepson
      var. rugulosus (Greene) L. Benson
  canus Benth.
      var. canus
        SY=R. c. var. typicus L. Benson
      var. laetus (Greene) L. Benson
      var. ludovicianus (Greene) L. Benson
  cardiophyllus Hook.
        SY=R. c. var. subsagittatus (Gray)
          L. Benson
        SY=R. c. var. typicus L. Benson
  carolinianus DC.
        SY=R. c. var. villicaulis Shinners
        SY=R. septentrionalis Poir. var.
          pterocarpus L. Benson
  circinatus Sibthorp
      var. circinatus
        SY=Batrachium c. (Sibthorp)
          Reichenb.
      var. subrigidus (W. Drew) L. Benson
        SY=R. amphibius James
        SY=R. aquatilis L. var. s. (W.
          Drew) Breitung
        SY=R. c. var. s. (W. Drew) L.
          Benson
        SY=R. s. W. Drew
  coloradensis (L. Benson) L. Benson
        SY=R. cardiophyllus Hook. var.
          coloradensis L. Benson
  cooleyae Vasey & Rose
  cymbalaria Pursh
      var. alpina Hook.
      var. cymbalaria
        SY=Halerpestes c. (Pursh) Greene
        SY=R. c. var. typicus L. Benson
      var. saximontanus Fern.
        SY=R. c. ssp. s. (Fern.) Thorne
  eastwoodianus L. Benson
  eschscholtzii Schlecht.
      var. eschscholtzii
        SY=R. e. var. typicus L. Benson
        SY=R. nivalis L. var. e.
          (Schlecht.) S. Wats.
      var. eximius (Greene) L. Benson
      var. hultenianus L. Benson
      var. oxynotus (Gray) Jepson
      var. suksdorfii (Gray) L. Benson
      var. trisectus (Eastw.) L. Benson
  falcatus L.
        SY=Ceratocephalus f. (L.) Moench
  fascicularis Muhl. ex Bigelow
      var. apricus (Greene) Fern.
      var. cuneiformis (Small) L. Benson
      var. fascicularis
        SY=R. f. var. typicus L. Benson
  ficaria L.
        SY=Ficaria f. (L.) Karst.
        SY=R. f. var. bulbifera
          Marsden-Jones
  flabellaris Raf.
        SY=R. delphinifolius Torr.
  flammula L.

      var. ?angustifolius Wallr.
      var. filiformis (Michx.) Hook.
        SY=R. reptans L. var. intermedius
          (Hook.) Torr. & Gray
      var. flammula
        SY=R. f. var. genuinus Buch.
      var. ovalis (Bigelow) L. Benson
        SY=R. f. var. samolifolius
          (Greene) L. Benson
        SY=R. reptans L. var. o. (Bigelow)
          Torr. & Gray
  gelidus Kar. & Kir.
        SY=R. g. ssp. g. var.
          shumaginensis Hulten
        SY=R. g. ssp. grayi (Britt.)
          Hulten
        SY=R. grayi Britt.
  glaberrimus Hook.
      var. ellipticus (Greene) Greene
        SY=R. g. var. buddii Boivin
      var. glaberrimus
        SY=R. g. var. typicus L. Benson
      var. reconditus (A. Nels. & J.F.
      Macbr.) L. Benson
        SY=R. r. A. Nels. & J.F. Macbr.
  glacialis L.
      var. chamissonis (Schlecht.) L.
      Benson
        SY=R. g. ssp. c. (Schlecht.)
          Hulten
      var. glacialis
        SY=R. g. var. genuinus L.
  gmelinii DC.
      var. gmelinii
        SY=R. g. var. typicus L. Benson
      var. hookeri (D. Don) L. Benson
        SY=R. g. ssp. purshii (Richards.)
          Hulten
        SY=R. g. var. prolificus (Fern.)
          Hara
        SY=R. g. var. terrestris (Ledeb.)
          L. Benson
        SY=R. purshii Richards.
      var. limosus (Nutt.) Hara
        SY=R. l. Nutt.
  gormanii Greene
  harveyi (Gray) Britt.
      var. ?australis (Brand) L. Benson
      var. harveyi
        SY=R. h. var. pilosus Benke
        SY=R. h. var. typicus L. Benson
  hawaiensis Gray
  hebecarpus Hook. & Arn.
  hederaceus L.
        SY=Batrachium h. (L.) S.F. Gray
  hexasepalus (L. Benson) L. Benson
  hispidus Michx.
      var. eurylobus L. Benson
      var. falsus Fern.
        SY=R. h. var. marilandicus (Poir.)
          L. Benson
        SY=?R. palmatus Ell.
      var. greenmanii L. Benson
      var. hispidus
        SY=R. h. var. typicus L. Benson
  hydrocharoides Gray
      var. hydrocharoides
        SY=R. h. var. typicus L. Benson
      var. ?natans (Nees ex G. Don) L.
      Benson
      var. stolonifer (Hemsl.) L. Benson
  hyperboreus Rottb.
        SY=R. h. ssp. arnellii Scheutz
  hystriculus Gray
  inamoenus Greene
      var. alpeophilus (A. Nels.) L. Benson
      var. inamoenus

Ranunculus   (CONT.)
       SY=R. affinis R. Br. var.
          micropetalus Greene
       SY=R. i. var. typicus L. Benson
    var. subaffinis (Gray) L. Benson
       SY=R. i. ssp. s. (Gray) Taylor &
          MacBryde
jovis A. Nels.
juniperinus M.E. Jones
kamtschaticus DC.
       SY=Oxygraphis glacialis (Fisch.)
          Bunge
lapponicus L.
       SY=Coptidium l. (L.) Gandog.
laxicaulis (Torr. & Gray) Darby
    var. angustifolius (Engelm.) L.
       Benson
    var. laxicaulis
       SY=R. l. var. mississippiensis
          (Small) L. Benson
       SY=R. m. Small
       SY=R. texensis Engelm.
lobbii (Hiern) Gray
longirostre Godr.
       SY=Batrachium l. (Godr.) F.W.
          Schultz
       SY=R. aquatilis L. var. l. (Godr.)
          Lawson
macauleyi Gray
    var. brandegeei L. Benson
    var. macauleyi
       SY=R. m. var. typicus L. Benson
macounii Britt.
       SY=R. m. var. oreganus (Gray)
          Davis
macranthus Scheele
       SY=R. m. var. typicus L. Benson
marginatus d'Urv.
    var. marginatus
    var. trachycarpus (Fisch. & Mey.)
       Azn.
       SY=R. t. Fisch. & Mey.
mauiensis Gray
       SY=R. m. var. longistylus
          Skottsberg
micranthus (Gray) Nutt. ex Torr. & Gray
       SY=R. m. var. cymbalistes (Greene)
          Fern.
       SY=R. m. var. delitescens (Greene)
          Fern.
muricatus L.
natans C.A. Mey.
    var. intertextus (Greene) L. Benson
       SY=R. hyperboreus Rottb. ssp. i.
          (Greene) Kapoor & Love
       SY=R. h. var. i. (Greene) Boivin
    var. natans
       SY=R. n. var. typicus L. Benson
nivalis L.
occidentalis Nutt. ex Torr. & Gray
    var. brevistylis Greene
       SY=R. o. ssp. o. var. b. Greene
    var. dissectus Henderson
    var. eisenii (Kellogg) Gray
    var. howellii (Greene) Greene
    var. nelsonii (DC.) L. Benson
       SY=R. o. ssp. n. (DC.) Hulten
       SY=R. o. ssp. insularis Hulten
    var. occidentalis
       SY=R. o. var. typicus L. Benson
    var. rattanii Gray
    var. ultramontanus Greene
oreogenes Greene
oresterus L. Benson
orthorhynchus Hook.
    var. alaschensis L. Benson
       SY=R. o. ssp. a. (L. Benson)

       Hulten
    var. bloomeri (S. Wats.) L. Benson
       SY=R. b. S. Wats.
    var. hallii Jepson
    var. orthorhynchus
       SY=R. o. var. typicus L. Benson
    var. platyphyllus Gray
       SY=R. o. ssp. p. (Gray) Taylor &
          MacBryde
pacificus (Hulten) L. Benson
pallasii Schlecht.
parviflorus L.
pedatifidus Sm. ex Rees
    var. affinis (R. Br.) L. Benson
       SY=R. a. R. Br.
       SY=R. p. ssp. a. (R. Br.) Hulten
    var. pedatifidus
       SY=R. p. var. leiocarpus (Trautv.)
          Fern.
pensylvanicus L. f.
platensis Spreng.
plebeius R. Br. ex DC.
populago Greene
pusillus Poir.
    var. angustifolius (Engelm.) L.
       Benson
       SY=R. tener C. Mohr
       SY=R. p. var. trachyspermus
          Engelm.
    var. pusillus
       SY=R. lindheimeri Engelm.
       SY=R. oblongifolius Ell.
       SY=R. p. var. l. (Engelm.) Gray
       SY=R. p. var. typicus L. Benson
pygmaeus Wahlenb.
    var. langianus Nathorst
    var. pygmaeus
ranunculinus (Nutt.) Rydb.
       SY=Cyrtorhyncha r. Nutt.
recurvatus Poir.
       SY=R. r. var. adpressipilis
          Weatherby
       SY=R. r. var. typicus L. Benson
repens L.
    var. glabratus DC.
       SY=R. r. var. erectus DC.
    var. pleniflorus Fern.
    var. repens
       SY=R. r. var. linearilobus DC.
       SY=R. r. var. typicus G. Beck
       SY=R. r. var. villosus Lamotte
reptans L.
       SY=R. flammula L. var. r. (L.) E.
          Mey.
rhomboideus Goldie
       SY=R. ovalis Raf.
sabinii R. Br.
       SY=R. pygmaeus Wahlenb. ssp. s.
          (R. Br.) Hulten
sardous Crantz
       SY=R. parvulus L.
sceleratus L.
    var. multifidus Nutt.
       SY=R. s. ssp. m. (Nutt.) Hulten
    var. sceleratus
       SY=R. s. var. typicus L. Benson
septentrionalis Poir.
       SY=R. s. var. caricetorum (Greene)
          Fern.
       SY=R. s. var. nitidus Chapman
       SY=R. siciformis Mackenzie & Bush
          ex Rydb.
subcordatus E.O. Beal
sulphureus Soland. ex Phipps
    var. intercedens Hulten
    var. sulphureus
       SY=R. nivalis L. var. s. (Soland.

Ranunculus   (CONT.)
            ex Phipps) Wahlenb.
    testiculatus Crantz
            SY=Ceratocephalus t. (Crantz) Roth
    trilobus Desf.
    turneri Greene
    uncinatus D. Don
            var. earlei (Greene) L. Benson
            var. parviflorus (Torr.) L. Benson
                SY=R. bongardii Greene
            var. uncinatus
                SY=R. bongardii Greene var.
                    tenellus (Nutt.) Greene
                SY=R. u. var. typicus L. Benson
    verecundus B.L. Robins.

Thalictrum L.

    alpinum L.
            var. alpinum
            var. hebetum Boivin
    arkansanum Boivin
    clavatum DC.
    cooleyi Ahles
    coriaceum (Britt.) Small
            SY=T. caulophylloides Small
    dasycarpum Fisch. & Lall.
            var. dasycarpum
            var. hypoglaucum (Rydb.) Boivin
                SY=T. h. Rydb.
    debile Buckl.
    dioicum L.
    fendleri Engelm. ex Gray
            var. fendleri
                SY=T. f. var. platycarpum Trel.
            var. wrightii (Gray) Trel.
    hultenii Boivin
            SY=T. minus L. ssp. kemense
                (Fries) Hulten
    macrostylum Small & Heller
    minus L.
    mirabile Small
    nigromontanum Boivin
    occidentale Gray
            var. macounii Boivin
            var. occidentale
                SY=T. breitungii Boivin
                SY=T. o. var. megacarpum (Torr.)
                    St. John
            var. palousense St. John
    polycarpum (Torr.) S. Wats.
    pubescens Pursh
            var. hebecarpum (Fern.) Boivin
                SY=T. polygamum Muhl. var. h.
                    Fern.
            var. pubescens
                SY=T. perelegans Greene
                SY=T. polygamum Muhl.
                SY=T. polygamum var. intermedium
                    Boivin
    revolutum DC.
            var. glandulosior Boivin
            var. revolutum
    sparsiflorum Turcz. ex Fisch. & Mey.
            var. richardsonii (Gray) Boivin
            var. saximontanum Boivin
                SY=T. s. var. nevadense Boivin
    steeleanum Boivin
    subrotundum Boivin
    texanum (Gray) Small
    thalictroides (L.) Eames & Boivin
                SY=Anemonella t. (L.) Spach
                SY=Syndesmon t. (L.) Hoffmgg.
    venulosum Trel.
            var. confine (Fern.) Boivin
                SY=T. c. Fern.
                SY=T. c. var. greeneanum Boivin

            var. fissum (Greene) Boivin
            var. lunellii (Greene) Boivin
            var. turneri (Boivin) Boivin
                SY=T. t. Boivin
            var. venulosum
                SY=T. confine Fern. var.
                    columbianum (Rydb.) Boivin
                SY=T. occidentale Gray var.
                    columbianum (Rydb.) M.E.
                    Peck

Trautvetteria Fisch. & Mey.

    caroliniensis (Walt.) Vail
            var. caroliniensis
            var. occidentalis (Gray) C.L. Hitchc.
                SY=T. grandis Nutt.

Trollius L.

    europaeus L.
    laxus Salisb.
        ssp. albiflorus (Gray) Love, Love &
        Kapoor
                SY=T. a. (Gray) Rydb.
                SY=T. l. var. a. Gray
        ssp. laxus
    riederianus Fisch. & Mey.

Xanthorhiza Marsh.

    simplicissima Marsh.

RESEDACEAE

Oligomeris Camb.

    linifolia (Vahl) J.F. Macbr.

Reseda L.

    alba L.
    lutea L.
    luteola L.
    odorata L.

RHAMNACEAE

Adolphia Meisn.

    californica S. Wats.
    infesta (H.B.K.) Meisn.

Alphitonia Reissek

    ponderosa Hbd.
            var. auwahiensis St. John
            var. costata St. John
            var. grandifolia St. John
            var. kauila St. John
            var. lanaiensis St. John
            var. ponderosa

Berchemia Neck. ex DC.

    scandens (Hill) K. Koch

Ceanothus L.

    americanus L.
            var. americanus

Ceanothus (CONT.)
            var. intermedius (Pursh) K. Koch
                SY=C. i. Pursh
            var. pitcheri Torr. & Gray
        arboreus Greene
            var. arboreus
            var. glaber Jepson
    X arcuatus McMinn [cuneatus X fresnensis]
        connivens Greene
        cordulatus Kellogg
        crassifolius Torr.
            var. crassifolius
            var. planus Abrams
        cuneatus (Hook.) Nutt. ex Torr. & Gray
            var. cuneatus
            var. dubius J.T. Howell
            var. submontanus (Rose) McMinn
        cyaneus Eastw.
        dentatus Torr. & Gray
            var. dentatus
            var. floribundus (Hook.) Trel.
        divergens Parry
            ssp. confusus (J.T. Howell) Abrams
            ssp. divergens
        diversifolius Kellogg
        fendleri Gray
                SY=C. f. var. venosus Trel.
                SY=C. f. var. viridis Gray
        ferrisae McMinn
    X flexilis McMinn [cuneatus X prostratus]
        foliosus Parry
            var. foliosus
                SY=C. austromontanus Abrams
            var. medius McMinn
            var. vineatus McMinn
        fresnensis Dudley ex Abrams
        gloriosus J.T. Howell
            var. exaltatus J.T. Howell
            var. gloriosus
            var. porrectus J.T. Howell
        greggii Gray
            var. greggii
            var. orbicularis E.H. Kelso
            var. perplexans (Trel.) Jepson
                SY=C. p. Trel.
            var. vestitus (Greene) McMinn
                SY=C. v. Greene
        griseus (Trel.) McMinn
            var. griseus
            var. horizontalis McMinn
        hearstiorum Hoover & Roof
        herbaceus Raf.
                SY=C. h. var. pubescens (Torr. &
                    Gray) Shinners
                SY=C. ovatus sensu auctt. non
                    Desf.
                SY=C. o. var. p. Torr. & Gray
                SY=C. p. (Torr. & Gray) Rydb.
        impressus Trel.
            var. impressus
            var. nipomensis McMinn
        incanus Torr. & Gray
        insularis Eastw.
                SY=C. megacarpus Nutt. ssp. i.
                    (Eastw.) Raven
                SY=C. m. var. i. (Eastw.) Munz
        integerrimus Hook. & Arn.
            var. californicus (Kellogg) G.T.
                Benson
                SY=C. c. Kellogg
            var. integerrimus
                SY=C. andersonii Parry
            var. macrothyrsus (Torr.) G.T. Benson
            var. puberulus (Greene) Abrams
        jepsonii Greene
            var. albiflorus J.T. Howell
            var. jepsonii

        lemmonii Parry
        leucodermis Greene
    X lobbianus Hook. [dentatus X griseus]
    X lorenzenii (Jepson) McMinn [cordulatus X
        velutinus]
                SY=C. velutinus Dougl. ex Hook.
                    var. l. Jepson
        maritimus Hoover
        martinii M.E. Jones
        masonii McMinn
        megacarpus Nutt.
                SY=C. m. var. pendulus McMinn
        microphyllus Michx.
        oliganthus Nutt.
            var. oliganthus
            var. orcuttii (Parry) Jepson
                SY=C. orcuttii Parry
    X otayensis McMinn [crassifolius X greggii
        var. perplexans]
        palmeri Trel.
        papillosus Torr. & Gray
            var. papillosus
            var. roweanus McMinn
        parryi Trel.
        parvifolius (S. Wats.) Trel.
        pinetorum Coville
        prostratus Benth.
            var. laxus Jepson
            var. occidentalis McMinn
                SY=C. divergens Parry ssp. o.
                    (McMinn) Abrams
            var. prostratus
        pumilus Geene
        purpureus Jepson
                SY=C. jepsonii Greene var. p.
                    (Jepson) Jepson
        ramulosus (Greene) McMinn
            var. fascicularis McMinn
            var. ramulosus
    X regius (Jepson) McMinn [papillosus X
        thyrsiflorus]
                SY=C. papillosus Torr. & Gray ssp.
                    r. Jepson
        rigidus Nutt.
            var. albus Roof
            var. rigidus
        roderickii Knight
    X rugosus Greene [prostratus X velutinus]
        sanguineus Pursh
        serpyllifolius Nutt.
    X serrulatus McMinn [cordulatus X
        prostratus]
        sonomensis J.T. Howell
        sorediatus Hook. & Arn.
        spinosus Nutt.
        thyrsiflorus Eschsch.
            var. repens McMinn
            var. thyrsiflorus
        tomentosus Parry
            var. olivaceus Jepson
                SY=C. t. ssp. o. (Jepson) Munz
            var. tomentosus
    X vanrensselaeri Roof [incanus X
        thyrsiflorus]
    X veitchianus Hook. [griseus X rigidus]
        velutinus Dougl. ex Hook.
            var. hookeri M.C. Johnston
                SY=C. v. var. laevigatus (Hook.)
                    Torr. & Gray
            var. velutinus
        verrucosus Nutt.

Colubrina L.C. Rich. ex Brongn.

        arborescens (P. Mill.) Sarg.
                SY=C. colubrina (Jacq.) Millsp.
        asiatica (L.) Brongn.

Colubrina (CONT.)
 californica I.M. Johnston
 cubensis (Jacq.) Brongn.
        var. floridana M.C. Johnston
 elliptica (Sw.) Briz. & Stern
          SY=C. reclinata (L'Her.) Brongn.
 glandulosa Perkins
        var. antillana (M.C. Johnston) M.C.
            Johnston
          SY=C. rufa (Vell.) Reissek var. a.
                M.C. Johnston
 greggii S. Wats.
        var. greggii
        var. yucatanensis M.C. Johnston
 oppositifolia Brongn. ex Mann
 stricta Engelm. ex M.C. Johnston
 texensis (Torr. & Gray) Gray
        var. pedunculata M.C. Johnston
        var. texensis
 verrucosa (Urban) M.C. Johnston
          SY=C. urbanii M.C. Johnston
          SY=Hybosperma spinosum Urban

Condalia Cav.

 correllii M.C. Johnston
 ericoides (Gray) M.C. Johnston
          SY=Microrhamnus e. Gray
 globosa I.M. Johnston
        var. pubescens I.M. Johnston
 hookeri M.C. Johnston
        var. edwardsiana (Cory) M.C. Johnston
        var. hookeri
          SY=C. obovata Hook.
 mexicana Schlecht.
 spathulata Gray
 viridis I.M. Johnston
 warnockii M.C. Johnston
        var. kearneyana M.C. Johnston
        var. warnockii

Gouania Jacq.

 bishopii Hbd.
 cucullata St. John
 fauriei St. John
 gagnei St. John
 hawaiiensis St. John
 hillebrandii Oliver
 lupuloides (L.) Urban
 lydgatei St. John
 mannii St. John
 meyenii Steud.
 oliveri St. John
 pilata St. John
 polygama (Jacq.) Urban
 remyi St. John
 sandwichiana St. John
 thinophila St. John
 vitifolia Gray

Hovenia Thunb.

 dulcis Thunb.

Karwinskia Zucc.

 humboldtiana (Roemer & Schultes) Zucc.

Krugiodendron Urban

 ferreum (Vahl) Urban

Reynosia Griseb.

 guama Urban
 krugii Urban

septentrionalis Urban
uncinata Urban

Rhamnus L.

 alnifolia L'Her.
 betulifolia Greene
        var. betulifolia
        var. obovata Kearney & Peebles
 californica Eschsch.
   ssp. californica
   ssp. crassifolia (Jepson) C.B. Wolf
   ssp. cuspidata (Greene) C.B. Wolf
   ssp. occidentalis (T.J. Howell) C.B.
      Wolf
          SY=R. c. var. o. (T.J. Howell)
                Jepson
   ssp. tomentella (Benth.) C.B. Wolf
          SY=R. t. Benth.
   ssp. ursina (Greene) C.B. Wolf
          SY=R. c. var. u. (Greene) McMinn
 caroliniana Walt.
        var. caroliniana
        var. mollis Fern.
 cathartica L.
 crocea Nutt.
   ssp. crocea
   ssp. ilicifolia (Kellogg) C.B. Wolf
          SY=R. c. var. i. (Kellogg) Greene
          SY=R. i. Kellogg
   ssp. insula (Kellogg) C.B. Wolf
          SY=R. i. Kellogg
          SY=R. insularis Greene
   ssp. pilosa (Trel.) C.B. Wolf
          SY=R. p. (Trel.) Abrams
 davurica Pallas
        var. nipponica Makino
 frangula L.
          SY=Frangula alnus P. Mill.
          SY=F. a. var. angustifolia Loud.
 lanceolata Pursh
        var. glabrata Gleason
        var. lanceolata
 pirifolia Greene
          SY=R. crocea Nutt. ssp. p.
                (Greene) C.B. Wolf
 purshiana DC.
 rubra Greene
   ssp. modocensis C.B. Wolf
   ssp. nevadensis (A. Nels.) C.B. Wolf
   ssp. obtusissima (Greene) C.B. Wolf
   ssp. rubra
   ssp. yosemitana C.B. Wolf
 serrata Schultes
          SY=R. smithii Greene ssp.
                fasciculata (Greene) C.B.
                Wolf
 smithii Greene
          SY=R. s. ssp. typica C.B. Wolf
 sphaerosperma Sw.

Sageretia Brongn.

 minutiflora (Michx.) Trel.
 wrightii S. Wats.

Ziziphus P. Mill.

 jujuba P. Mill.
          SY=Z. ziziphus (L.) Karst.
 mauritiana Lam.
 obtusifolia (Hook. ex Torr. & Gray) Gray
        var. canescens (Gray) M.C. Johnston
          SY=Condalia lycioides (Gray)
                Weberb. var. c. (Gray) Trel.
          SY=Condaliopsis l. var. c. (Gray)
                Suess.

Ziziphus (CONT.)
        var. obtusifolia
            SY=Condalia lycioides (Gray)
                Weberb.
            SY=Condalia o. (Hook. ex Torr. &
                Gray) Weberb.
            SY=Condaliopsis l. (Gray) Suess.
    parryi Torr.
            SY=Condalia p. (Torr.) Weberb.
            SY=Condaliopsis p. (Torr.) Suess.
    reticulata (Vahl) DC.
            SY=Sarcomphalus r. (Vahl) Urban
    rignonii Del Ponte
            SY=Sarcomphalus domingensis
                (Spreng.) Krug & Urban
    taylori (Britt.) M.C. Johnston
            SY=Sarcomphalus t. Britt.

RHIZOPHORACEAE

Bruguiera Lam.

    gymnorrhiza (L.) Lam.

Cassipourea Aubl.

    guianensis Aubl.
            SY=C. alba Griseb.

Rhizophora L.

    mangle L.

ROSACEAE

Acaena Mutis ex L.

    anserinifolia (J. F. & G. Forst.) Druce
    californica Bitter
    exigua Gray
            var. exigua
            var. glaberrima Bitter
            var. glabriuscula Bitter
            var. subtusstrigulosa Bitter

Adenostoma Hook. & Arn.

    fasciculatum Hook. & Arn.
            var. fasciculatum
            var. obtusifolium S. Wats.
    sparsifolium Torr.

Agrimonia L.

    bicknellii (Kearney) Rydb.
    eupatoria L.
    gryposepala Wallr.
    incisa Torr. & Gray
    microcarpa Wallr.
            SY=A. platycarpa Wallr.
            SY=A. pubescens Wallr. var. m.
                (Wallr.) Ahles
            SY=A. pumila Muhl.
    parviflora Ait.
    pubescens Wallr.
            SY=A. mollis (Torr. & Gray) Britt.
    repens L.
            SY=A. odorata P. Mill.
    rostellata Wallr.
    striata Michx.

Alchemilla L.

    alpina L.
    arvensis (L.) Scop.
            SY=A. cuneifolia Nutt.
            SY=A. occidentalis Nutt.
            SY=Aphanes a. L.
            SY=Aphanes o. (Nutt.) Rydb.
    filicaulis Buser
        ssp. filicaulis
            SY=Alchemilla f. var. denudata
                Buser
            SY=A. vulgaris L. var. f. (Buser)
                Fern. & Wieg.
        ssp. vestita (Buser) M.E. Bradshaw
            SY=Alchemilla minor Huds.
            SY=A. v. (Buser) Raunk.
            SY=A. vulgaris L. var. vestita
                (Buser) Fern. & Wieg.
    glabra Neygenf.
            SY=Alchemilla alpestris F.W.
                Schmidt
            SY=A. vulgaris L. var. grandis
                Blytt
    glaucescens Wallr.
    glomerulans Buser
    lapeyrousii Buser
    microcarpa Boiss. & Reut.
            SY=?Aphanes australis Rydb.
    monticola Opiz
            SY=Alchemilla vulgaris L. var.
                pastoralis (Buser) Boivin
    subcrenata Buser
    venosula Buser
    wichurae (Buser) Stefansson
            SY=Alchemilla vulgaris L. var. w.
                (Buser) Boivin
    xanthochlora Rothm.
            SY=Alchemilla pratensis F.W.
                Schmidt
            SY=A. vulgaris sensu auctt.

Amelanchier Medic.

    alnifolia (Nutt.) Nutt.
            var. alnifolia
            var. cusickii (Fern.) C.L. Hitchc.
            SY=A. c. Fern.
            SY=A. florida Lindl. var. c.
                (Fern.) M.F. Peck
            var. humptulipensis (G.N. Jones) C.L.
                Hitchc.
            SY=A. florida Lindl. var. h. G.N.
                Jones
            var. semiintegrifolia (Hook.) C.L.
                Hitchc.
            SY=A. a. ssp. florida (Lindl.)
                Hulten
            SY=A. f. Lindl.
    arborea (Michx. f.) Fern.
            var. alabamensis (Britt.) G.N. Jones
            SY=A. a. Britt.
            var. arborea
            var. austromontana (Ashe) Ahles
            var. cordifolia (Ashe) Boivin
            var. laevis (Wieg.) Ahles
            SY=A. a. ssp. l. (Wieg.) S. McKay
                ex Landry
            SY=A. X grandiflora Rehd.
            SY=A. l. Wieg.
            SY=A. l. var. nitida (Wieg.) Fern.
    bartramiana (Tausch) M. Roemer
    canadensis (L.) Medic.
            SY=A. c. var. subintegra Fern.
            SY=A. lucida Fern.
            SY=A. oblongifolia (Torr. & Gray)
                M. Roemer

Amelanchier (CONT.)
  denticulata (H.B.K.) W.D.J. Koch
  fernaldii Wieg.
  humilis Wieg.
        SY=A. h. var. campestris Nielsen
        SY=A. h. var. compacta Nielsen
        SY=A. h. var. exserrata Nielsen
        SY=A. mucronata Nielsen
        SY=A. stolonifera Wieg. pro parte
  X intermedia Spach [arborea X canadensis]
  nantucketensis Bickn.
  X neglecta Egglest. [arborea var. laevis X
    bartramiana]
  obovalis (Michx.) Ashe
  pallida Greene
        SY=A. alnifolia (Nutt.) Nutt. var.
            p. (Greene) Jepson
        SY=A. florida Lindl. var. gracilis
            (Heller) M.E. Peck
        SY=A. g. Heller
  pumila Nutt. ex Torr. & Gray
        SY=A. alnifolia (Nutt.) Nutt. var.
            p. (Nutt. ex Torr. & Gray)
            A. Nels.
        SY=A. basalticola Piper
        SY=A. cuneata Piper
        SY=A. glabra Greene
        SY=A. i. polycarpa Greene
  X quinti-martii Louis-Marie [arborea X
    bartramiana]
  sanguinea (Pursh) DC.
        var. gaspensis Wieg.
          SY=A. g. (Wieg.) Fern. & Weath.
        var. sanguinea
          SY=A. amabilis Wieg.
          SY=A. huronensis Wieg.
  stolonifera Wieg.
        SY=A. spicata (Lam.) K. Koch pro
            parte
  utahensis Koehne
    ssp. covillei (Standl.) Clokey
        SY=A. c. Standl.
    ssp. utahensis
        SY=A. alnifolia (Nutt.) Nutt. var.
            oreophila (A. Nels.) R.J.
            Davis
        SY=A. a. var. u. (Koehne) M.E.
            Jones
        SY=A. australis Standl.
        SY=A. bakeri Greene
        SY=A. goldmanii Woot. & Standl.
        SY=A. mormonica Schneid.
        SY=A. oreophila A. Nels.
  X wiegandii Nielsen [arborea X sanguinea]
        SY=A. interior Nielsen

X Amelasorbus Rehd.

    X jackii Rehd. [Amelanchier alnifolia var.
      semiintegrifolia X Sorbus scopulina]

Aronia Medic.

  arbutifolia (L.) Pers.
        SY=Pyrus a. (L.) L. f.
        SY=Sorbus a. (L.) Heynh.
  melanocarpa (Michx.) Ell.
        SY=A. arbutifolia (L.) Pers. var.
            nigra (Willd.) Seymour
        SY=A. nigra (Willd.) Koehne
        SY=Pyrus a. (L.) L. f. var. n.
            Willd.
        SY=P. m. (Michx.) Willd.
        SY=Sorbus m. (Michx.) Schneid.
  prunifolia (Marsh.) Rehd.
        SY=A. arbutifolia (L.) Pers. var.
            atropurpurea (Britt.)

        Seymour
        SY=A. atropurpurea Britt.
        SY=Pyrus arbutifolia (L.) L. f.
            var. atropurpurea (Britt.)
            B.L. Robins.
        SY=P. floribunda Lindl.
        SY=Sorbus arbutifolia (L.) Heynh.
            var. atropurpurea (Britt.)
            Schneid.

Aruncus Schaeffer

  dioicus (Walt.) Fern.
      var. acuminatus (Rydb.) Hara
        SY=A. sylvester Kostel.
        SY=?A. s. var. a. (Dougl.) Jepson
        SY=A. vulgaris Raf.
      var. dioicus
        SY=A. allegheniensis Rydb.
        SY=A. aruncus (L.) Karst.
      var. pubescens (Rydb.) Fern.
        SY=A. p. Rydb.

Cercocarpus H.B.K.

  ledifolius Nutt. ex Torr. & Gray
      var. intricatus (S. Wats.) M.E. Jones
        SY=C. i. S. Wats.
        SY=C. i. var. villosus Schneid.
        SY=C. l. var. hypoleucus (Rydb.)
            M.E. Peck
      var. ledifolius
  montanus Raf.
      var. argenteus (Rydb.) F.L. Martin
        SY=C. a. Rydb.
      var. blancheae (Schneid.) F.L. Martin
        SY=C. alnifolius Rydb.
        SY=C. betuloides Nutt. ex Torr. &
            Gray var. blancheae
            (Schneid.) Little
        SY=C. m. ssp. b. (Schneid.) Thorne
      var. glaber (S. Wats.) F.L. Martin
        SY=C. betuloides Nutt. ex Torr. &
            Gray
        SY=C. b. var. multiflorus Jepson
      var. macrourus (Rydb.) F.L. Martin
        SY=C. betuloides Nutt. ex Torr. &
            Gray var. m. (Rydb.) Jepson
      var. minutiflorus (Abrams) F.L.
        Martin
        SY=C. m. Abrams
      var. montanus
        SY=C. m. var. flabellifolius
            (Rydb.) Kearney & Peebles
      var. paucidentatus (S. Wats.) F.L.
        Martin
        SY=C. breviflorus Gray
        SY=C. b. var. eximius Schneid.
  traskiae Eastw.
        SY=C. betuloides Nutt. ex Torr. &
            Gray var. t. (Eastw.) Dunkle
        SY=C. montanus Raf. var. t.
            (Eastw.) F.L. Martin

Chaenomeles Lindl.

  japonica (Thunb.) Lindl. ex Sapch
  speciosa (Sweet) Nakai
        SY=C. lagenaria (Loisel.) Koidzumi
        SY=Cydonia japonica sensu auctt.
            non (Thunb.) Pers.

Chamaebatia (Benth.) Maxim.

  australis (Brandeg.) Abrams
  foliolosa Benth.

Chamaebatiaria (Porter) Maxim.

    millefolium (Torr.) Maxim.

Chamaerhodos Bunge

    erecta (L.) Bunge
      ssp. erecta
      ssp. nuttallii (Pickering ex Torr. &
        Gray) Hulten
            SY=C. e. var. parviflora (Nutt.)
                C.L. Hitchc.
            SY=C. n. (Pickering ex Torr. &
                Gray) Pickering
            SY=C. n. var. keweenawensis Fern.

Chrysobalanus L.

    icaco L.
            SY=C. i. var. pellocarpus (G.F.W.
                Mey.) DC.
            SY=C. interior Small
            SY=C. p. G.F.W. Mey.

Coleogyne Torr.

    ramosissima Torr.

Cotoneaster Medic.

    acutifolius Turcz.
    franchetii Boiss.
    horizontalis Dcne.
    melanocarpus Lodd.
    microphyllus Wallich ex Lindl.
    pannosus Franch.
    simonsii Baker

Cowania D. Don

    ericifolia Torr.
    mexicana D. Don
        var. dubia Brandeg.
        var. mexicana
        var. stansburiana (Torr.) Jepson
            SY=C. alba Goodding
            SY=C. s. Torr.
    subintegra Kearney

Crataegus L.

    acutiserrata Kruschke
    aemula Beadle
    aestivalis (Walt.) Torr. & Gray
            SY=C. luculenta Sarg.
            SY=C. maloides Sarg.
    ?algens Beadle
    ambitiosa Sarg.
    anamesa Sarg.
            SY=C. antiplasta Sarg.
    ancisa Beadle
    annosa Beadle
    X anomala Sarg. [intricata X mollis]
    apiomorpha Sarg.
            SY=C. a. var. cyanophylla (Sarg.)
                Kruschke
            SY=C. merita Sarg.
            SY=C. vittata Ashe
    arborea Beadle
            SY=C. a. var. ohioensis (Sarg.)
                Kruschke
            SY=C. o. Sarg.
            SY=C. pyracanthoides Beadle var.
                a. (Beadle) Palmer
    arcana Beadle
    ardua Sarg.
    arrogans Beadle

austromontana Beadle
basilica Beadle
        SY=C. alnorum Sarg.
beadlei Ashe
beata Sarg.
        SY=C. formosa Sarg.
berberifolia Torr. & Gray
        SY=C. b. var. edita (Sarg.) Palmer
        SY=C. crocina Beadle
        SY=C. e. Sarg.
        SY=C. edura Beadle
        SY=C. fera Beadle
        SY=C. tersa Beadle
        SY=C. torva Beadle
bicknellii Egglest.
        SY=C. chrysocarpa Ashe var. b.
            (Egglest.) Palmer
biltmoreana Beadle
        SY=C. b. var. stonei (Sarg.)
            Kruschke
        SY=C. s. Sarg.
bisulcata Ashe
brachyacantha Sarg. & Engelm.
brainerdii Sarg.
brazoria Sarg.
        SY=C. dallasiana Sarg.
calpodendron (Ehrh.) Medic.
        SY=C. c. var. gigantea Kruschke
        SY=C. c. var. globosa (Sarg.)
            Palmer
        SY=C. c. var. hispida (Sarg.)
            Palmer
         SY=C. c. var. hispidula (Sarg.)
            Palmer
         SY=C. c. var. microcarpa (Chapman)
            Palmer
        SY=C. c. var. mollicula (Sarg.)
            Palmer
         SY=C. c. var. obesa (Ashe) Palmer
        SY=C. globosa Sarg.
canadensis Sarg.
carrollensis Sarg.
choriophylla Sarg.
chrysocarpa Ashe
        SY=C. c. var. aboriginum (Sarg.)
            Kruschke
        SY=C. c. var. longiacuminata
            Kruschke
        SY=C. c. var. phoenicea Palmer
        SY=C. c. var. piperi (Britt.)
            Kruschke
        SY=C. c. var. rotundifolia
            (Moench) Sarg.
        SY=C. coccinea L. pro parte
        SY=C. columbiana T.J. Howell var.
            piperi (Britt.) Egglest.
        SY=C. jackii Sarg.
        SY=C. mercerensis Sarg.
        SY=C. piperi Britt.
        SY=C. putnamiana Sarg.
        SY=C. rotundifolia Moench pro
            parte
        SY=?C. sicca Sarg.
        SY=?C. s. var. glabrifolia (Sarg.)
            Palmer
chrysophyta Ashe
?cibaria Beadle
coccinioides Ashe
coleae Sarg.
        SY=C. eatoniana Sarg.
        SY=C. macauleyae Sarg.
columbiana T.J. Howell
        SY=C. williamsii Egglest.
compacta Sarg.
compta Sarg.
        SY=C. levis Sarg.
        SY=C. milleri Sarg.

Crataegus (CONT.)
  condigna Beadle
  consanguinea Beadle
  contrita Beadle
  corusca Sarg.
          SY=C. c. var. gigantea Kruschke
          SY=C. c. var. hillii (Sarg.)
                Kruschke
          SY=C. h. Sarg.
  crus-galli L.
          SY=C. acutifolia Sarg.
          SY=C. a. var. insignis (Sarg.)
                Palmer
          SY=C. bushii Sarg.
          SY=C. canbyi Sarg.
          SY=C. cherokeensis Sarg.
          SY=C. cocksii Sarg.
          SY=C. c-g. var. barrettiana
                (Sarg.) Palmer
          SY=C. c-g. var. bellica (Sarg.)
                Palmer
          SY=C. c-g. var. capillata Sarg.
          SY=C. c-g. var. exigua (Sarg.)
                Egglest.
          SY=C. c-g. var. leptophylla
                (Sarg.) Palmer
          SY=C. c-g. var. macra (Beadle)
                Palmer
          SY=C. c-g. var. oblongata Sarg.
          SY=C. c-g. var. pachyphylla
                (Sarg.) Palmer
          SY=C. c-g. var. pyracanthifolia
                Ait.
          SY=C. denaria Beadle
          SY=C. hannibalensis Palmer
          SY=C. mohrii Beadle
          SY=C. palliata Sarg.
          SY=C. palmeri Sarg.
          SY=C. permixta Palmer
          SY=C. pyracanthoides Beadle
          SY=C. regalis Beadle
          SY=C. r. var. paradoxa (Sarg.)
                Palmer
          SY=C. schizophylla Egglest.
          SY=C. signata Beadle
          SY=C. subpilosa Sarg.
          SY=C. tantula Sarg.
          SY=C. uniqua Sarg.
          SY=C. vallicola Sarg.
  curvisepala Lindm.
          SY=C. laevigata (Poir.) DC.
          SY=C. oxyacantha L.
          SY=C. o. var. paulii Rehd.
  X densiflora Sarg. [chrysocarpa X
      flabellata]
          SY=C. flabellata (Bosc) K. Koch
                var. d. (Sarg.) Kruschke
  desueta Sarg.
          SY=C. d. var. wausaukiensis
                Kruschke
  dilitata Sarg.
  dispar Beadle
  disperma Ashe
          SY=C. X collicola Ashe
          SY=C. collina Chapman var.
                collicola (Ashe) Palmer
          SY=C. cuneiformis (Marsh.)
                Egglest.
          SY=C. danielsii Palmer
          SY=C. disperma var. peoriensis
                (Sarg.) Kruschke
          SY=C. X pausiaca Ashe
          SY=C. peoriensis Sarg.
          SY=C. punctata Jacq. var. pausiaca
                (Ashe) Palmer
  dispessa Ashe
  dissona Sarg.

          SY=C. d. var. bellula (Sarg.)
                Kruschke
          SY=C. disjuncta Sarg.
          SY=C. franklinensis Sarg.
          SY=C. incisa Sarg.
          SY=C. pruinosa (Wendl. f.) K. Koch
                var. brachypoda (Sarg.)
                Palmer
          SY=C. p. var. delawarensis (Sarg.)
                Palmer
          SY=C. p. var. dissona (Sarg.)
                Egglest.
  distincta Kruschke
  dodgei Ashe
          SY=C. chrysocarpa Ashe var.
                caesariata (Sarg.) Palmer
          SY=C. d. var. lumaria (Ashe) Sarg.
          SY=C. d. var. rotundata (Sarg.)
                Kruschke
          SY=C. r. Sarg.
  douglasii Lindl.
  engelmannii Sarg.
          SY=C. e. var. sinistra (Beadle)
                Palmer
          SY=C. sublobulata Sarg.
  erythrocarpa Ashe
  erythropoda Ashe
  X evansiana Sarg. [flava X viridis]
  exilis Beadle
  extraria Beadle
  faxonii Sarg.
          SY=C. f. var. durifructa Kruschke
          SY=C. f. var. praecoqua (Sarg.)
                Kruschke
          SY=C. f. var. praetermissa (Sarg.)
                Palmer
  fecunda Sarg.
  flabellata (Bosc) K. Koch
  flava Ait.
          SY=C. aprica Beadle
          SY=C. floridana Sarg.
          SY=C. meridiana Beadle
          SY=C. michauxii Pers.
          SY=C. recurvata Beadle
          SY=C. senta Beadle
  florifera Sarg.
          SY=C. X celsa Sarg.
          SY=C. f. var. c. (Sarg.) Kruschke
          SY=C. f. var. mortonis (Laughlin)
                Kruschke
          SY=C. f. var. shirleyensis (Sarg.)
                Kruschke
          SY=C. f. var. virilis (Sarg.)
                Kruschke
          SY=C. s. Sarg.
          SY=C. silvestris Sarg.
  fontanesiana (Spach) Steud.
          SY=C. tenax Ashe
  fragilis Beadle
  fulleriana Sarg.
          SY=C. confragosa Sarg.
          SY=C. f. var. chippewaensis
                (Sarg.) Kruschke
          SY=C. f. var. gigantea Kruschke
          SY=C. f. var. miranda (Sarg.)
                Kruschke
          SY=C. holmesiana Ashe var. c.
                (Sarg.) Palmer
          SY=C. h. var. magniflora (Sarg.)
                Palmer
          SY=C. X illecebrosa Sarg.
  furtiva Beadle
  georgiana Sarg.
  glareosa Ashe
  grandis Ashe
  grayana Egglest.
          SY=C. flabellata (Bosc.) K. Koch

Crataegus (CONT.)
                    var. g. (Egglest.) Palmer
gregalis Beadle
greggiana Egglest.
X haemocarpa Ashe [flabellata X pruinosa]
harbisonii Beadle
          SY=C. ashei Beadle
harveyana Sarg.
helvina Ashe
holmesiana Ashe
          SY=C. h. var. amicta (Ashe) Palmer
          SY=C. h. var. villipes Ashe
          SY=C. v. (Ashe) Ashe
ideae Sarg.
ignave Beadle
X immanis Ashe [chrysocarpa X pruinosa]
impar Beadle
X improvisa Sarg. [brainerdii X intricata]
inanis Beadle
X incaedua Sarg. [calpodendron X punctata]
incerta Sarg.
indicens Ashe
ingens Beadle
insidiosa Beadle
integra Beadle
intricata Lange
          SY=C. boyntonii Beadle
          SY=C. coccinea L. pro parte
          SY=C. foetida Ashe
          SY=C. fortunata Sarg.
          SY=C. horseyi Palmer
          SY=C. i. var. b. (Beadle) Kruschke
          SY=C. i. var. neobushii (Sarg.)
              Kruschke
          SY=C. i. var. rubella (Beadle)
              Kruschke
          SY=C. i. var. straminea (Beadle)
              Palmer
          SY=C. n. Sarg.
          SY=C. ouachitensis Palmer
          SY=C. o. var. minor Palmer
          SY=C. padifolia Sarg.
          SY=C. p. var. incarnata Sarg.
          SY=C. pallens Beadle
          SY=C. rubella Beadle
          SY=C. straminea Beadle
invicta Beadle
iracunda Beadle
          SY=?C. beckwithae Sarg.
          SY=C. brumalis Ashe
          SY=C. i. var. brumalis (Ashe)
              Kruschke
          SY=C. i. var. diffusa (Sarg.)
              Kruschke
          SY=C. i. var. populnea (Ashe)
              Kruschke
          SY=C. i. var. silvicola (Beadle)
              Palmer
          SY=C. i. var. stolonifera (Sarg.)
              Kruschke
          SY=C. macrosperma Ashe var.
              demissa (Sarg.) Egglest.
          SY=C. p. Ashe
          SY=C. stolonifera Sarg.
irrasa Sarg.
          SY=C. i. var. blanchardii (Sarg.)
              Egglest.
          SY=C. mansfieldensis Sarg.
          SY=C. oakesiana Egglest.
jesupii Sarg.
          SY=C. filipes Ashe
jonesae Sarg.
          SY=C. harryi Sarg.
          SY=C. j. var. brownietta (Sarg.)
              Kruschke
          SY=C. j. var. h. (Sarg.) Kruschke
kelloggii Sarg.

X kennedyi Sarg. [brainerdii X pruinosa]
X kingstonensis Sarg. [brainerdii X
    coccinioides]
knieskerniana Sarg.
lacera Sarg.
lacrimata Small
laetifica Sarg.
          SY=C. nuda Sarg.
          SY=C. X persimilis Sarg.
lanata Beadle
X laneyi Sarg. [brainerdii X succulenta]
          SY=C. X divisa Sarg.
lanuginosa Sarg.
latebrosa Sarg.
laurentiana Sarg.
          SY=C. brunetiana Sarg.
          SY=C. b. var. fernaldii (Sarg.)
              Palmer
          SY=C. l. var. b. (Sarg.) Kruschke
          SY=C. l. var. dissimilifolia
              Kruschke
lemingtonensis Sarg.
leonensis Palmer
lepida Beadle
X lettermanii Sarg. [mollis X punctata]
          SY=C. hirtiflora Sarg.
limata Beadle
limophila Sarg.
X locuples Sarg. [mollis X pruinosa]
          SY=C. coccinioides Ashe var. l.
              (Sarg.) Kruschke
lucorum Sarg.
          SY=C. insolens Sarg.
macrantha Lodd.
          SY=C. divida Sarg.
          SY=C. m. var. colorado (Ashe)
              Kruschke
          SY=C. m. var. d. (Sarg.) Kruschke
          SY=C. m. var. integriloba (Sarg.)
              Kruschke
          SY=C. m. var. occidentalis (Britt.
              & Palmer) Egglest.
          SY=C. m. var. pertomentosa (Ashe)
              Kruschke
          SY=C. X i. Sarg.
          SY=C. p. Ashe
          SY=C. succulenta Schrad. ex Link
              var. i. Sarg.
          SY=C. s. var. m. (Lodd.) Egglest.
          SY=C. s. var. o. Britt. & Palmer
          SY=C. s. var. p. (Ashe) Palmer
macrosperma Ashe
          SY=C. chadsfordiana Sarg.
          SY=C. fretalis Sarg.
          SY=C. m. var. acutiloba (Sarg.)
              Egglest.
          SY=C. m. var. eganii (Ashe)
              Kruschke
          SY=C. m. var. matura (Sarg.)
              Egglest.
          SY=C. m. var. pastorum (Sarg.)
              Egglest.
          SY=C. m. var. pentandra (Sarg.)
              Egglest.
          SY=C. randiana Sarg.
margaretta Ashe
          SY=C. m. var. angustifolia Palmer
          SY=C. m. var. brownii (Britt.)
              Sarg.
          SY=C. m. var. meiophylla (Sarg.)
              Palmer
marshallii Egglest.
media Sarg.
membranacea Sarg.
menandiana Sarg.
mendosa Beadle
meridionalis Sarg.

Crataegus (CONT.)
   micrantha Sarg.
   mira Beadle
   mollis (Torr. & Gray) Scheele
        SY=C. arnoldiana Sarg.
        SY=C. arkansana Sarg.
        SY=C. brachyphylla Sarg.
        SY=C. gravida Beadle
        SY=C. induta Sarg.
        SY=C. invisa Sarg.
        SY=C. limaria Sarg.
        SY=C. m. var. dumetosa (Sarg.)
           Kruschke
        SY=C. m. var. gigantea Kruschke
        SY=C. m. var. incisifolia Kruschke
        SY=C. m. var. sera (Sarg.)
           Egglest.
        SY=C. noelensis Sarg.
   monogyna Jacq.
   munda Beadle
   nitida (Engelm.) Sarg.
   nitidula Sarg.
        SY=C. n. var. limatula (Sarg.)
           Kruschke
        SY=C. n. var. macrocarpa Kruschke
        SY=C. n. var. recedens (Sarg.)
           Kruschke
        SY=C. r. Sarg.
   X notha Sarg. [marshallii X mollis]
   opaca Hook. & Arn.
   operta Ashe
   opulens Sarg.
        SY=C. beata Sarg. var. o. (Sarg.)
           Palmer
   ovata Sarg.
        SY=C. viridis L. var. o. (Sarg.)
           Palmer
   pagensis Sarg.
   panda Beadle
   pearsonii Ashe
   pedicellata Sarg.
        SY=C. albicans Ashe
        SY=C. X aulica Sarg.
        SY=C. coccinea L. pro parte
        SY=C. X ellwangeriana Sarg.
        SY=C. habereri Sarg.
        SY=C. letchworthiana Sarg.
        SY=C. p. var. albicans (Ashe)
           Palmer
        SY=C. p. var. assurgens (Sarg.)
           Palmer
        SY=C. p. var. caesa (Ashe)
           Kruschke
        SY=C. p. var. e. (Sarg.) Egglest.
        SY=C. p. var. robesoniana (Sarg.)
           Palmer
        SY=C. p. var. sertata (Sarg.)
           Kruschke
   penita Beadle
        SY=C. amnicola Beadle
   pennsylvanica Ashe
        SY=C. tatnalliana Sarg.
   phaenopyrum (L. f.) Medic.
        SY=C. populifolia Walt.
        SY=C. youngii Sarg.
   X pilosa Sarg. [intricata X pruinosa]
   pinetorum Beadle
   poliophylla Sarg.
   porrecta Ashe
   pratensis Sarg.
   pringlei Sarg.
        SY=C. p. var. exclusa (Sarg.)
           Egglest.
        SY=C. p. var. lobulata (Sarg.)
           Egglest.
   prona Ashe
        SY=C. gravis Ashe

   pruinosa (Wendl. f.) K. Koch
        SY=C. aspera Sarg.
        SY=C. crawfordiana Sarg.
        SY=C. deltoides Ashe
        SY=C. gattingeri Ashe
        SY=C. g. var. rigida Palmer
        SY=C. gaudens Sarg.
        SY=C. lecta Sarg.
        SY=C. leiophylla Sarg.
        SY=C. littoralis Sarg.
        SY=C. mackenzii Sarg.
        SY=C. m. var. a. (Sarg.) Palmer
        SY=C. m. var. bracteata (Sarg.)
           Palmer
        SY=C. platycarpa Sarg.
        SY=C. porteri Britt.
        SY=C. porteri var. caerulescens
           (Sarg.) Palmer
        SY=C. pruinosa var. grandiflora
           Kruschke
        SY=C. pruinosa var. latisepala
           (Sarg.) Egglest.
        SY=C. pruinosa var. pachypoda
           (Sarg.) Palmer
        SY=C. pruinosa var. rugosa (Ashe)
           Kruschke
        SY=C. pruinosa var. virella (Ashe)
           Kruschke
        SY=C. rugosa Ashe
        SY=C. vicinalis Beadle
        SY=C. virella Ashe
   prunifolia (Poir.) Pers.
   X puberis Sarg. [flabella X punctata]
   pulcherrima Ashe
        SY=C. opima Beadle
        SY=C. robur Beadle
   punctata Jacq.
        SY=C. collina Chapman
        SY=C. c. var. secta (Sarg.) Palmer
        SY=C. c. var. sordida (Sarg.)
           Palmer
        SY=C. c. var. succincta (Sarg.)
           Palmer
        SY=C. fastosa Sarg.
        SY=C. p. var. aurea Ait.
        SY=C. p. var. canescens Britt.
        SY=C. p. var. microphylla Sarg.
        SY=C. verruculosa Sarg.
   putata Sarg.
   raleighensis Ashe
   ravenelii Sarg.
   ravida Ashe
   resima Beadle
   reverchonii Sarg.
        SY=C. discolor Sarg.
        SY=C. r. var. d. (Sarg.) Palmer
        SY=C. r. var. stevensiana (Sarg.)
           Palmer
   rhodella Ashe
   rigens Beadle
   rivularis Nutt.
        SY=C. douglasii Lindl. var. r.
           (Nutt.) Sarg.
   roanensis Ashe
        SY=C. macrosperma Ashe var. r.
           (Ashe) Palmer
        SY=C. r. var. fluviatilis (Sarg.)
           Kruschke
        SY=C. r. var. heidelbergensis
           (Sarg.) Kruschke
   robinsonii Sarg.
   rubrocarnea Sarg.
   sabineana Ashe
   saligna Greene
   sargentii Beadle
        SY=C. venusta Beadle
   scabrida Sarg.

Crataegus   (CONT.)
        SY=C. brainerdii Sarg. var.
           asperifolia (Sarg.) Egglest.
        SY=C. b. var. cyclophylla (Sarg.)
           Palmer
        SY=C. b. var. egglestonii (Sarg.)
           B.L. Robins.
        SY=C. b. var. s. (Sarg.) Egglest.
        SY=C. dunbarii Sarg.
        SY=C. iterata Sarg.
        SY=C. pinguis Sarg.
        SY=C. s. var. a. (Sarg.) Kruschke
        SY=C. s. var. c. (Sarg.) Kruschke
        SY=C. s. var. d. (Sarg.) Kruschke
        SY=C. s. var. e. (Sarg.) Kruschke
        SY=C. s. var. honesta (Sarg.)
           Kruschke
   schuettei Ashe
        SY=C. ferrissii Ashe
        SY=C. s. var. cuneata Kruschke
        SY=C. s. var. ferrissii (Ashe)
           Kruschke
        SY=C. s. var. gigantea Kruschke
        SY=C. tortilis Ashe
   shinnersii Kruschke
   X simulata Sarg. [calpodendron X
     crus-galli]
   spathulata Michx.
   spatiosa Sarg.
   spissa Sarg.
   stenosepala Sarg.
   submollis Sarg.
   suborbiculata Sarg.
        SY=C. durobrivensis Sarg.
        SY=C. kellermanii Sarg.
        SY=C. neobaxteri Sarg.
        SY=C. X silvestris Sarg.
        SY=C. suborbieulata var.
           saundersiana (Sarg.)
           Kruschke
   succulenta Schrad. ex Link
        SY=C. laxiflora Sarg.
        SY=C. neofluvialis Ashe
        SY=C. s. var. gemmosa (Sarg.)
           Kruschke
        SY=C. s. var. laxiflora (Sarg.)
           Kruschke
        SY=C. s. var. michiganensis (Ashe)
           Palmer
        SY=C. s. var. neofluvialis (Ashe)
           Palmer
        SY=C. s. var. pisifera (Sarg.)
           Kruschke
        SY=C. s. var. rutila (Sarg.)
           Kruschke
   suksdorfii (Sarg.) Kruschke
        SY=C. douglasii Lindl. var. s.
           Sarg.
   sutherlandensis Sarg.
   texana Buckl.
   thermopegaea Palmer
   tinctoria Ashe
   tracyi Ashe ex Egglest.
        SY=C. montivaga Sarg.
   triflora Chapman
   tripartita Sarg.
   tristis Beadle
   triumphalis Sarg.
   unifora Muenchh.
   vailiae Britt.
        SY=C. brittonii Egglest.
   valida Beadle
   versuta Beadle
   viburnifolia Sarg.
   viridis L.
        SY=C. abbreviata Sarg.
        SY=C. amicalis Sarg.

        SY=C. atrorubens Ashe
        SY=C. blanda Sarg.
        SY=C. enuculata Sarg.
        SY=C. glabriuscula Sarg.
        SY=C. interior Beadle
        SY=C. velutina Sarg.
        SY=C. viridis var. i. (Beadle)
           Palmer
        SY=C. v. var. lanceolata (Sarg.)
           Palmer
        SY=C. v. var. lutensis (Sarg.)
           Palmer
        SY=C. v. var. velutina (Sarg.)
           Palmer
   visenda Beadle
   vulsa Beadle
   warneri Sarg.
   X websteri Sarg. [brainerdii X
     calpodendron]
   X whittakeri Sarg. [calpodendron X mollis]
   wisconsinensis Kruschke
   wootoniana Egglest.
   xanthophylla Sarg.

Cydonia P. Mill.

   oblonga P. Mill.
        SY=Pyrus cydonia L.

Dalibarda L.

   repens L.

Dryas L.

   drummondii Richards. ex Hook.
      var. drummondii
      var. eglandulosa Porsild
      var. tomentosa (Farr) L.O. Williams
   integrifolia Vahl
     ssp. chamissonis (Spreng.) Scoggan
        SY=D. c. Spreng.
     ssp. crenulata (Juz.) Scoggan
        SY=D. c. Juz.
     ssp. integrifolia
        SY=D. i. var. canescens Simm.
   X lewinii Rouleau [drummondii X
     integrifolia]
   octopetala L.
     ssp. alaskensis (Porsild) Hulten
        SY=D. a. Porsild
     ssp. hookeriana (Juz.) Hulten
        SY=D. h. Juz.
        SY=D. o. var. h. (Juz.) Breitung
     ssp. octopetala
      var. angustifolia C.L. Hitchc.
      var. kamtschatica (Juz.) Hulten
      var. octopetala
        SY=D. o. var. luteola Hulten
     ssp. punctata (Juz.) Hulten
      var. glabrata Hulten
      var. punctata (Juz.) Hulten
        SY=D. o. var. viscida Hulten
        SY=D. p. Juz.
   sylvatica (Hulten) Porsild
        SY=D. integrifolia Vahl ssp. s.
           (Hulten) Hulten

Duchesnea Sm.

   indica (Andr.) Focke
        SY=Fragaria i. Andr.

Exochorda Lindl.

   racemosa (Lindl.) Rehd.
        SY=E. grandiflora (Hook.) Lindl.

Fallugia Endl.

   paradoxa (D. Don) Endl.

Filipendula P. Mill.

   occidentalis (S. Wats.) T.J. Howell
   rubra (Hill) B.L. Robins.
   ulmaria (L.) Maxim.
     ssp. denudata (J. & C. Presl) Hayek
        SY=F. d. (J. & C. Presl) Rydb.
        SY=F. u. var. d. (J. & C. Presl)
          Beck
     ssp. ulmaria
        SY=Spiraea u. L.
   vulgaris Moench
        SY=F. hexapetala Gilib.

Fragaria L.

   X ananassa Duchesne [chiloensis X
     virginiana]
     nm. ananassa
     nm. cuneifolia (Nutt. ex T.J. Howell)
       Staudt
        SY=F. c. Nutt. ex T.J. Howell
        SY=F. grandiflora Ehrh.
   chiloensis (L.) Duchesne
     ssp. chiloensis
        SY=F. c. var. scouleri Hook.
     ssp. lucida (Vilm.) Staudt
     ssp. pacifica Staudt
     ssp. sandwicensis (Duchesne) Staudt
        SY=F. c. ssp. c. var. s.
          (Duchesne) Deg. & Deg.
   crinita Rydb.
        SY=F. vesca L. var. c. (Rydb.)
          C.L. Hitchc.
   vesca L.
     ssp. americana (Porter) Staudt
        SY=F. a. (Porter) Britt.
        SY=F. v. var. a. Porter
     ssp. bracteata (Heller) Staudt
        SY=F. b. Heller
        SY=F. helleri Holz.
        SY=F. v. var. b. (Heller) R.J.
          Davis
     ssp. californica (Cham. & Schlecht.)
       Staudt
        SY=F. c. Cham. & Schlecht.
        SY=F. c. var. franciscana Rydb.
     ssp. vesca
        SY=F. v. var. alba (Ehrh.) Rydb.
   virginiana Duchesne
     ssp. glauca (S. Wats.) Staudt
        SY=F. g. (S. Wats.) Rydb.
        SY=F. pauciflora Rydb.
        SY=F. v. var. g. S. Wats.
        SY=F. v. var. terrae-novae (Rydb.)
          Fern. & Wieg.
     ssp. grayana (Vilm.) Staudt
        SY=F. g. Vilm.
     ssp. platypetala (Rydb.) Staudt
        SY=F. p. Rydb.
        SY=F. sibbaldifolia Rydb.
        SY=F. suksdorfii Rydb.
        SY=F. truncata Rydb.
        SY=F. v. var. illinoensis (Prince)
          Gray
        SY=F. v. var. p. (Rydb.) Hall
     ssp. virginiana
        SY=F. australis (Rydb.) Rydb.
        SY=F. canadensis Michx.
        SY=F. multicipita Fern.
        SY=F. ovalis (Lehm.) Rydb.
        SY=F. v. var. a. Rydb.
        SY=F. v. var. c. (Michx.) Farw.

        SY=F. v. var. o. (Lehm.) R.J.
          Davis

Geum L.

   aleppicum Jacq.
        SY=G. a. ssp. strictum (Ait.)
          Clausen
        SY=G. a. var. s. (Ait.) Fern.
        SY=G. s. Ait.
        SY=G. s. var. decurrens (Rydb.)
          Kearney & Peebles
   X aurantiacum Fries [aleppicum X rivale]
   calthifolium Menzies ex Sm.
        SY=Acomastylis c. (Menzies ex Sm.)
          Bolle
   canadense Jacq.
     var. brevipes Fern.
     var. camporum (Rydb.) Fern. & Weath.
     var. canadense
     var. grimesii Fern. & Weath.
     var. texanum Fern.
   canescens (Greene) Munz
        SY=Sieversia c. (Greene) Rydb.
   geniculatum Michx.
   glaciale J.E. Adams ex Fisch.
   laciniatum Murr.
     var. laciniatum
     var. trichocarpum Fern.
   X macranthum (Kearney) Boivin [calthifolium
     X rossii]
        SY=Acomastylis X m. (Kearney)
          Bolle
        SY=G. X schofieldii Calder &
          Taylor
        SY=Sieversia X m. Kearney
   macrophyllum Willd.
   peckii Pursh
        SY=Sieversia p. (Pursh) Rydb.
   pentapetalum (L.) Makino
        SY=Sieversia p. (L.) Greene
   perincisum Rydb.
     var. intermedium Boivin
     var. perincisum
        SY=G. macrophyllum Willd. ssp. p.
          (Rydb.) Hulten
        SY=G. m. var. p. (Rydb.) Raup
        SY=G. m. var. rydbergii Farw.
        SY=G. oregonense (Scheutz) Rydb.
   X pervale Boivin [perincisum var.
     intermedium X rivale]
   X pulchrum Fern. [macrophyllum X rivale]
   radiatum Michx.
        SY=Sieversia r. (Michx.) Greene
   rivale L.
   rossii (R. Br.) Ser.
     var. depressum (Greene) C.L. Hitchc.
     var. rossii
        SY=Acomastylis r. (R. Br.) Greene
     var. turbinatum (Rydb.) C.L. Hitchc.
        SY=Acomastylis rossii (R. Br.)
          Greene ssp. t. (Rydb.) W.A.
          Weber
        SY=G. gracilipes (Piper) M.E. Peck
        SY=G. t. Rydb.
        SY=Sieversia g. (Piper) Greene
   triflorum Pursh
     var. campanulatum (Greene) C.L.
       Hitchc.
        SY=G. c. (Greene) G.N. Jones
        SY=Sieversia c. (Greene) Rydb.
     var. ciliatum (Pursh) Fassett
        SY=G. c. Pursh
        SY=Sieversia c. (Pursh) G. Don
     var. triflorum
        SY=Erythrocoma t. (Pursh) Greene
        SY=G. ciliatum Pursh var. griseum

Geum (CONT.)
                    (Greene) Kearney & Peebles
                SY=Sieversia t. (Pursh) R. Br.
    urbanum L.
    vernum (Raf.) Torr. & Gray
            SY=Stylipus v. Raf.
    virginianum L.
            SY=G. flavum (Porter) Bickn.
            SY=G. hirsutum Muhl.

Heteromeles M. Roemer

    arbutifolia (Lindl.) M. Roemer
        var. arbutifolia
            SY=Photinia a. Lindl.
        var. cerina Jepson
        var. macrocarpa (Munz) Munz

Hirtella L.

    rugosa Thuill. ex Pers.
            SY=Chrysobalanus r. (Thuill. ex
                Pers.) Maza
            SY=H. portoricensis Humb. & Bonpl.
            SY=Zamzela r. (Thuill. ex Pers.)
                Raf.
    triandra Sw.
            SY=Chrysobalanus t. (Sw.) Morales
            SY=H. multiflora Urban

Holodiscus (K. Koch) Maxim.

    boursieri (Carr.) Rehd.
            SY=H. dumosus (Nutt.) Heller ssp.
                saxicola (Heller) Abrams
    discolor (Pursh) Maxim.
        var. delnortensis Ley
        var. discolor
            SY=H. dumosus (Nutt.) Heller var.
                australis (Heller) Ley
        var. franciscanus (Rydb.) Jepson
            SY=H. d. ssp. f. (Rydb.) Taylor &
                MacBryde
    dumosus (Nutt.) Heller
        var. dumosus
            SY=H. discolor (Pursh) Maxim. var.
                dumosus (Nutt.) Dipple
            SY=H. dumosus var. typicus Ley
            SY=H. glabrescens (Greenm.) Heller
                ex Jepson var. dumosus
                (Nutt.) Maxim. ex Coult.
        var. glabrescens (Greenm.) C.L.
            Hitchc.
            SY=H. g. (Greenm.) Heller ex
                Jepson
            SY=H. microphyllus Rydb.
            SY=H. m. var. sericeus Ley
            SY=H. m. var. typicus Ley

Horkelia Cham. & Schlecht.

    bolanderi Gray
        ssp. bolanderi
        ssp. clevelandii (Greene) Keck
            SY=H. c. (Greene) Rydb.
        ssp. parryi (S. Wats.) Keck
    californica Cham. & Schlecht.
    congesta Dougl. ex Hook.
        ssp. congesta
        ssp. nemorosa Keck
    cuneata Lindl.
        ssp. cuneata
        ssp. puberula (Greene) Keck
        ssp. sericea (Gray) Keck
    daucifolia (Greene) Rydb.
        ssp. daucifolia
        ssp. latior Keck

    elata (Greene) Rydb.
            SY=H. glandulosa Eastw.
    frondosa (Greene) Rydb.
    fusca Lindl.
        ssp. capitata (Lindl.) Keck
            SY=H. f. var. c. (Lindl.) M.E.
                Peck
        ssp. filicoides (Crum) Keck
        ssp. fusca
        ssp. parviflora (Nutt.) Keck
            SY=H. f. var. p. (Nutt.) M.E. Peck
        ssp. pseudocapitata (Rydb.) Keck
            SY=H. f. var. p. (Rydb.) M.E. Peck
        ssp. tenella (S. Wats.) Keck
    hendersonii T.J. Howell
    hispidula Rydb.
    marinensis (Elmer) Crum
    parryi Greene
    sericata S. Wats.
    tenuiloba (Torr.) Gray
    tridentata Torr.
        ssp. flavescens (Rydb.) Keck
        ssp. tridentata
    truncata Rydb.
    tularensis (J.T. Howell) Munz
            SY=Potentilla t. J.T. Howell
    wilderae Parish

Ivesia Torr. & Gray

    aperta (J.T. Howell) Munz
            SY=Potentilla a. J.T. Howell
    argyrocoma (Rydb.) Rydb.
    baileyi S. Wats.
        ssp. baileyi
            SY=Horkelia b. (S. Wats.) Greene
            SY=I. b. ssp. typica Keck
        ssp. setosa (S. Wats.) Keck
            SY=I. b. var. s. S. Wats.
    callida (Hall) Rydb.
    campestris (M.E. Jones) Rydb.
    cryptocaulis (Clokey) Keck
    gordonii (Hook.) Torr. & Gray
            SY=Horkelia g. Hook.
            SY=Potentilla g. (Hook.) Greene
    jaegeri Munz & Johnston
    kingii S. Wats.
            SY=I. eremica Rydb.
    lycopodioides Gray
        ssp. lycopodioides
            SY=I. l. ssp. typica Keck
        ssp. megalopetala (Rydb.) Keck
        ssp. scandularis (Rydb.) Keck
    muirii Gray
    multifoliolata (Torr.) Keck
            SY=Potentilla m. (Torr.) Kearney &
                Peebles
    pickeringii Torr. ex Gray
    purpurascens (S. Wats.) Keck
        ssp. congdonis (Rydb.) Keck
            SY=Horkelia purpurascens S. Wats.
                ssp. c. (Rydb.) Abrams
        ssp. purpurascens
            SY=Horkelia p. S. Wats.
    pygmaea Gray
    rhypara Ertter & Reveal
    sabulosa (M.E. Jones) Keck
            SY=Potentilla s. M.E. Jones
    santolinoides Gray
    sericoleuca (Rydb.) Rydb.
    shockleyi S. Wats.
    tweedyi Rydb.
    unguiculata Gray
    utahensis S. Wats.
    webberi Gray

Kelseya (S. Wats.) Rydb.

Kelseya (CONT.)
   uniflora (S. Wats.) Rydb.

Kerria DC.

   japonica (L.) DC.

Licania Aubl.

   michauxii Prance
        SY=Chrysobalanus incanus Raf.
        SY=C. oblongifolius Michx.
        SY=C. pallidus (Small) L.B. Sm.
        SY=C. retusus Raf.
        SY=Geobalanus o. (Michx.) Small
        SY=G. p. Small

Luetkea Bong.

   pectinata (Pursh) Kuntze
        SY=Saxifraga p. Pursh non Schott,
          Nym. & Kotschy

Lyonothamnus Gray

   floribundus Gray
     ssp. asplenifolius (Greene) Raven
        SY=L. f. var. a. (Greene) Brandeg.
     ssp. floribundus

Malus P. Mill.

   angustifolia (Ait.) Michx.
     var. angustifolia
        SY=Pyrus a. Ait.
     var. puberula Rehd.
        SY=Pyrus a. Ait. var. spinosa
          (Rehd.) Bailey
   X arnoldiana (Rehd.) Rehd. [baccata X
     floribunda]
        SY=Pyrus X a. (Rehd.) Bean
   baccata (L.) Borkh.
        SY=Pyrus b. L.
   coronaria (L.) P. Mill.
     var. coronaria
        SY=M. bracteata Rehd.
        SY=M. c. var. elongata Rehd.
        SY=Pyrus b. (Rehd.) Bailey
        SY=P. c. L.
        SY=P. c. var. e. (Rehd.) Bailey
     var. dasycalyx Rehd.
        SY=Pyrus c. L. var. d. (Rehd.)
          Fern.
        SY=P. c. var. lancifolia (Rehd.)
          Fern.
        SY=P. l. Rehd.
   fusca (Raf.) Schneid.
        SY=M. diversifolia (Bong.) M.
          Roemer
        SY=M. f. var. levipes (Nutt.)
          Schneid.
        SY=Pyrus f. Raf.
        SY=P. rivularis Dougl. ex Hook.
   glabrata Rehd.
   glaucesns Rehd.
        SY=Pyrus g. (Rehd.) Bailey
   ioensis (Wood) Britt.
     var. ioensis
        SY=M. i. var. bushii Rehd.
        SY=M. i. var. palmeri Rehd.
        SY=Pyrus i. (Wood) Bailey
     var. texana Rehd.
        SY=Pyrus i. (Wood) Bailey var. t.
          (Rehd.) Bailey
   X platycarpa Rehd. [coronaria X pumila]
        SY=Pyrus X p. (Rehd.) Bailey
   prunifolia (Willd.) Borkh.

        SY=Pyrus p. Willd.
   pumila P. Mill.
        SY=M. communis Poir.
        SY=M. domestica (Borkh.) Borkh.
        SY=Pyrus p. (P. Mill.) K. Koch
   X soulardii (Bailey) Britt. [ioensis X
     pumila]
        SY=Pyrus X s. Bailey
   sylvestris (L.) P. Mill.
        SY=M. malus (L.) Britt.
        SY=Pyrus m. L.

Neviusia Gray

   alabamensis Gray

Oemleria Reichenb.

   cerasiformis (Torr. & Gray ex Hook. & Arn.)
     Landon
        SY=Osmaronia c. (Torr. & Gray ex
          Hook. & Arn.) Greene

Osteomeles Lindl.

   anthyllidifolia (Sm.) Lindl.

Peraphyllum Nutt.

   ramosissimum Nutt.

Petrophytum (Nutt. ex Torr. & Gray) Rydb.

   acuminatum Rydb.
        SY=P. caespitosum (Nutt.) Rydb.
          ssp. a. (Rydb.) Munz
   caespitosum (Nutt.) Rydb.
     var. caespitosum
        SY=Eriogynia c. (Nutt.) S. Wats.
        SY=Spiraea c. Nutt.
     var. elatius (S. Wats.) Tidestrom
   cinerascens (Piper) Rydb.
   hendersonii (Canby) Rydb.
        SY=Spiraea h. (Canby) Piper

Physocarpus (Camb.) Maxim.

   alternans (M.E. Jones) J.T. Howell
     ssp. alternans
     ssp. annulatus J.T. Howell
     ssp. panamintensis J.T. Howell
   capitatus (Pursh) Kuntze
        SY=P. opulifolius (L.) Maxim. var.
          tomentellus (Ser.) Boivin
   malvaceus (Greene) A. Nels.
   monogynus (Torr.) Coult.
   opulifolius (L.) Maxim.
     var. intermedius (Rydb.) B.L. Robins.
        SY=Opulaster i. Rydb.
        SY=P. i. (Rydb.) Schneid.
     var. opulifolius
        SY=Opulaster alabamensis Rydb.
        SY=O. australis Rydb.
        SY=O. o. (L.) Kuntze
        SY=O. stellatus Rydb.
        SY=Spiraea o. L.

Porteranthus Britt. ex Small

   stipulatus (Muhl. ex Willd.) Britt.
        SY=Gillenia s. (Muhl. ex Willd.)
          Baill.
   trifoliatus (L.) Britt.
        SY=Gillenia t. (L.) Moench

Potentilla L.

Potentilla (CONT.)
   albiflora L.O. Williams
   ambigens Greene
   anglica Laicharding
        SY=P. procumbens Sibthorp
   anserina L.
     ssp. anserina
        SY=Argentina a. (L.) Rydb.
        SY=P. a. var. concolor Ser.
        SY=P. a. var. sericea (L.) Hayne
     ssp. pacifica (T.J. Howell) Rousi
        SY=P. a. var. grandis Torr. & Gray
        SY=P. a. ssp. p. var. lanata
           Boivin
        SY=P. egedei Wormsk. ssp. g.
           (Torr. & Gray) Hulten
        SY=P. e. var. g. (Torr. & Gray)
           J.T. Howell
        SY=P. p. T.J. Howell
   argentea L.
     var. argentea
        SY=Argentina a. (L.) Rydb.
     var. pseudocalabra T. Wolf
   arguta Pursh
     ssp. arguta
        SY=Drymocallis agrimonioides
           (Pursh) Rydb.
        SY=D. arguta (Pursh) Rydb.
     ssp. convallaria (Rydb.) Keck
        SY=P. a. var. c. (Rydb.) T. Wolf
        SY=P. c. Rydb.
   atrosanguinea Lodd. ex D. Don
   biennis Greene
   biflora Willd. ex Schlecht.
   bipinnatifida Dougl.
     var. bipinnatifida
        SY=P. pensylvanica L. var. b.
           (Dougl.) Torr. & Gray
     var. glabrata (Hook.) Kohli & Packer
        SY=P. pensylvanica L. var. g.
           (Hook.) S. Wats.
        SY=P. glabella Rydb.
   brevifolia Nutt. ex Torr. & Gray
   breweri S. Wats.
        SY=P. b. var. expansa S. Wats.
   canadensis L.
     var. canadensis
        SY=P. caroliniana Poir.
        SY=P. pumila Poir.
     var. villosissima Fern.
   collina Wibel
   concinna Richards.
     var. concinna
        SY=P. saximontana Rydb.
     var. divisa Rydb.
        SY=P. c. var. dissecta (S. Wats.)
           Boivin
        SY=P. divisa (Rydb.) Rydb.
     var. macounii (Rydb.) C.L. Hitchc.
     var. rubripes (Rydb.) C.L. Hitchc.
   crinita Gray
     var. crinita
     var. lemmonii (S. Wats.) Kearney &
        Peebles
   diversifolia Lehm.
     var. diversifolia
        SY=P. d. var. glaucophylla (Lehm.)
           S. Wats.
     var. multisecta S. Wats.
        SY=P. m. (S. Wats.) Rydb.
     var. perdissecta (Rydb.) C. L.
        Hitchc.
   drummondii Lehm.
     ssp. bruceae (Rydb.) Keck
     ssp. drummondii
        SY=P. anomalifolia M.E. Peck
   egedei Wormskj.

     ssp. egedei
        var. egedei
           SY=P. anserina L. ssp. e.
              (Wormskj.) Hiitonen
           SY=P. a. var. rolandii (Boivin)
              Boivin
           SY=P. r. Boivin
        var. groenlandica (Tratt.) Polunin
           SY=P. anserina L. var. g. Tratt.
     ssp. yukonensis (Hulten) Hulten
        SY=P. anserina L. var. y. (Hulten)
           Boivin
        SY=P. y. Hulten
   elegans Cham. & Schlecht.
   erecta (L.) Raeusch.
   finitima Kohli & Packer
   fissa Nutt.
        SY=Drymocallis f. (Nutt.) Rydb.
   flabellifolia Hook. ex Torr. & Gray
     var. flabellifolia
     var. grayi (S. Wats.) Jepson
   fruticosa L.
     ssp. floribunda (Pursh) Elkington
        SY=Dasiphora f. sensu auctt. non
           (L.) Rydb.
        SY=Pentaphylloides f. (Pursh) A.
           Love
        SY=Potentilla f. var. tenuifolia
           Lehm.
   glandulosa Lindl.
     ssp. arizonica (Rydb.) Keck
     ssp. ashlandica (Greene) Keck
        SY=Potentilla a. Greene
     ssp. ewanii Keck
     ssp. glabrata (Rydb.) Keck
        SY=Potentilla glandulosa var.
           intermedia (Rydb.) C.L.
           Hitchc.
     ssp. glandulosa
        SY=Drymocallis g. (Lindl.) Rydb.
        SY=Potentilla g. ssp. typica Keck
        SY=P. g. var. campanulata C.L.
           Hitchc.
        SY=P. g. var. incisa Lindl.
        SY=P. rhomboidea Rydb.
     ssp. globosa Keck
     ssp. hansenii (Greene) Keck
     ssp. micropetala (Rydb.) Keck
     ssp. nevadensis (S. Wats.) Keck
        SY=Potentilla g. var. n. S. Wats.
     ssp. pseudorupestris (Rydb.) Keck
        SY=Potentilla g. var. p. (Rydb.)
           Breitung
     ssp. reflexa (Greene) Keck
        SY=Potentilla g. var. r. Greene
   gracilis Dougl. ex Hook.
     var. blasckeana (Turcz.) Jepson
     var. brunnescens (Rydb.) C.L. Hitchc.
        SY=Potentilla b. Rydb.
     var. flabelliformis (Lehm.) Nutt. ex
        Torr. & Gray
        SY=Potentilla f. Lehm.
     var. glabrata (Lehm.) C.L. Hitchc.
        SY=Potentilla angustata Rydb.
        SY=P. camporum Rydb.
        SY=P. etomentosa Rydb.
        SY=P. gracilis ssp. nuttallii
           (Lehm.) Keck
        SY=P. gracilis var. rigida (Nutt.)
           S. Wats.
        SY=P. glomerata A. Nels.
        SY=P. indiges M.E. Peck
        SY=P. jucunda A. Nels.
        SY=P. nuttallii Lehm.
        SY=P. viridescens Rydb.
     var. gracilis
        SY=P. macropetala Rydb.

Potentilla (CONT.)
        var. permollis (Rydb.) C.L. Hitchc.
        var. pulcherrima (Lehm.) Fern.
           SY=Potentilla pulcherrima Lehm.
hickmanii Eastw.
hippiana Lehm.
        var. argyrea (Rydb.) Boivin
           SY=Potentilla a. Rydb.
        var. diffusa Lehm.
           SY=Potentilla leneophylla Torr. &
             James ex Eat.
        var. filicaulis (Nutt.) Boivin
           SY=Potentilla effusa Dougl.
        var. hippiana
hookeriana Lehm.
    ssp. chamissonis (Hulten) Hulten
        SY=Potentilla nivea L. ssp. c.
           (Hulten) Hiitonen
    ssp. hookeriana
        var. furcata (Porsild) Hulten
           SY=Potentilla f. Porsild
        var. hookeriana
           SY=Potentilla nivea L. ssp. h.
           (Lehm.) Hiitonen
        var. umanakensis Hulten
hyparctica Malte
    ssp. hyparctica
        SY=Potentilla emarginata Pursh
        SY=P. flabellifolia Hook. ex Torr.
           & Gray var. e. (Pursh)
           Boivin
    ssp. nana (Willd.) Hulten
        SY=Potentilla h. var. elatior
           (Abrom.) Fern.
inclinata Vill.
        SY=Potentilla canescens Bess.
        SY=P. intermedia L. var. c.
           (Bess.) Rupr.
intermedia L.
millefolia Rydb.
        var. klamathensis (Rydb.) Jepson
           SY=Potentilla k. Rydb.
        var. millefolia
multifida L.
multijuga Lehm.
newberryi Gray
nicolletii (S. Wats.) Sheldon
nivea L.
    ssp. chionodes Hiitonen
        var. chionodes Hiitonen
        var. subguinata (Lange) Hiitonen
           SY=Potentilla n. var. s. Lange
           SY=P. n. var. pentaphylla Lehm.
    ssp. nivea
        var. nivea
           SY=Potentilla n. ssp. fallax
           Porsild
        var. tomentosa Nilss.-Ehle
norvegica L.
    ssp. hirsuta (Michx.) Hyl.
        SY=Potentilla n. var. h. (Michx.)
           Lehm.
        SY=P. n. var. labradorica (Lehm.)
           Fern.
    ssp. monspeliensis (L.) Aschers. &
        Graebn.
        SY=Potentilla m. L.
    ssp. norvegica
oblanceolata Rydb.
ovina Macoun
        var. ovina
           SY=Potentilla versicolor Rydb.
           SY=P. wyomingensis A. Nels.
        var. pinnatisecta S. Wats.
palustris (L.) Scop.
        SY=Comarum p. L.
        SY=Potentilla p. var. parvifolia

           (Raf.) Fern. & Long
        SY=P. palustris var. villosa
           (Pers.) Lehm.
paradoxa Nutt. ex Torr. & Gray
        SY=Potentilla supina L. ssp. p.
           (Nutt. ex Torr. & Gray)
           Sojak
patellifera J.T. Howell
pectinisecta Rydb.
        SY=Potentilla gracilis Dougl. ex
           Hook. var. elmeri (Rydb.)
           Jepson
pensylvanica L.
        var. atrovirens (Rydb.) T. Wolf
           SY=Potentilla a. Rydb.
        var. pectinata (Raf.) Boivin
           SY=Potentilla pectinata Raf.
           SY=P. pensylvanica var. littoralis
           (Rydb.) Boivin
        var. pensylvanica
           SY=Potentilla platyloba Rydb.
        var. strigosa Pursh
           SY=Potentilla s. (Pursh) Pallas ex
           Tratt.
        var. virgulata (A. Nels.) T. Wolf
           SY=Potentilla v. A. Nels.
plattensis Nutt.
pseudosericea Rydb.
pulchella R. Br.
        var. elatior Lange
        var. gracilicaulis Porsild
        var. pulchella
           SY=Potentilla nivea L. var. p. (R.
           Br.) Dur.
           SY=P. usticapensis Fern.
quinquefolia (Rydb.) Rydb.
ranunculus Lange
        SY=Potentilla diversifolia Lehm.
           var. r. (Lange) Boivin
recta L.
        var. obscura (Nestler) W.D.J. Koch
        var. pilosa (Willd.) Ledeb.
        var. recta
        var. sulphurea (Lam. & DC.) Peyr.
reptans L.
rivalis Nutt.
        var. millegrana (Engelm. ex Lehm.) S.
           Wats.
           SY=Potentilla leucocarpa Rydb.
           SY=P. m. Engelm. ex Lehm.
        var. pentandra (Engelm.) S. Wats.
           SY=Potentilla p. Engelm.
        var. rivalis
robbinsiana Oakes
rubella Sorensen
rubricaulis Lehm.
        var. dasyphylla Ledeb.
        var. pedersenii Rydb.
        var. rubricaulis
rupincola Osterhout
saxosa Lemmon ex Greene
    ssp. saxosa
    ssp. sierrae Munz
sierrae-blancae Woot. & Standl.
simplex Michx.
        var. argyrisma Fern.
        var. calvescens Fern.
        var. simplex
           SY=Potentilla s. var. typica Fern.
sterilis (L.) Garcke
stipularis L.
        var. groenlandica Sorensen
        var. stipularis
subjuga Rydb.
subviscosa Greene
        var. ramulosa (Rydb.) Kearney &
           Peebles

Potentilla (CONT.)
    var. subviscosa
  supina L.
  tabernaemontana Aschers.
      SY=Potentilla crantzii (Crantz) G.
        Beck ex Fritsch
      SY=P. c. var. hirta Lange
      SY=P. flabellifolia Hook. ex Torr.
        & Gray var. h. (Lange)
        Boivin
      SY=P. maculata Pourret
      SY=P. verna L. pro parte
  thurberi Gray
    var. atrorubens (Rydb.) Kearney &
      Peebles
    var. sanguinea (Rydb.) Kearney &
      Peebles
    var. thurberi
  thuringiaca Bernh. ex Link
  tridentata (Soland.) Ait.
      SY=Sibbaldiopsis t. (Soland.)
        Rydb.
  vahliana Lehm.
      SY=Potentilla ledebouriana Porsild
      SY=P. uniflora Ledeb.
  villosa Pallas ex Pursh
      SY=Potentilla nivea L. var. v.
        (Pallas ex Pursh) Regel &
        Tiling
      SY=P. v. var. parviflora C.L.
        Hitchc.
  wheeleri S. Wats.
    var. rimicola Munz & Johnston
    var. wheeleri
      SY=Potentilla viscidula Rydb.

Prunus L.

  alleghaniensis Porter
    var. alleghaniensis
    var. davisii W. Wight
  americana Marsh.
  andersonii Gray
  angustifolia Marsh.
    var. angustifolia
      SY=P. a. var. varians W. Wight &
        Hedrick
    var. watsonii (Sarg.) Waugh
  armeniaca L.
  avium (L.) L.
  besseyi Bailey
      SY=P. pumila L. var. b. (Bailey)
        Gleason
  caroliniana (P. Mill.) Ait.
      SY=Laurocerasus c. (P. Mill.) M.
        Roemer
  cerasifera Ehrh.
  cerasus L.
  domestica L.
    ssp. domestica
    ssp. insitita (L.) Schneid.
      SY=P. d. var. i. (L.) Fiori &
        Paol.
      SY=P. X d. L. nm. i. (L.) Boivin
      SY=P. i. L.
  dulcis (P. Mill.) D.A. Webb
      SY=P. amygdalus Batsch
  emarginata (Dougl. ex Hook.) Walp.
      SY=P. e. var. crenulata (Greene)
        Kearney & Peebles
  fasciculata (Torr.) Gray
    var. fasciculata
    var. punctata Jepson
  fremontii S. Wats.
  fruticosa Pallas
  geniculata Harper
  glandulosa Thunb.

gracilis Fngelm. & Gray
gravesii Small
havardii (W. Wight) W. Wight
hortulana Bailey
      SY=P. h. var. mineri Bailey
ilicifolia (Nutt. ex Hook. & Arn.) D.
  Dietr.
injucunda Small
laurocerasus L.
lusitanica L.
lyonii (Eastw.) Sarg.
mahaleb L.
maritima Marsh.
mexicana S. Wats.
      SY=P. m. var. flutonensis (Sarg.)
        Sarg.
      SY=P. m. var. polyandra (Sarg.)
        Sarg.
      SY=P. pensylvanica L. f. var.
        mollis (Dougl.) Boivin
minutiflora Engelm.
munsoniana W. Wight & Hedrick
murrayana Palmer
myrtifolia (L.) Urban
      SY=Laurocerasus m. (L.) Britt.
nigra Ait.
      SY=P. americana Marsh. var. lanata
        Sudworth
      SY=P. a. var. n. (Ait.) Waugh
      SY=P. emarginata (Dougl. ex Hook.)
        Walp. var. mollis (Dougl.)
        Brewer
      SY=P. l. (Sudworth) MacKenzie &
        Bush
occidentalis Sw.
      SY=Laurocerasus o. (Sw.) M. Roemer
X orthosepala Koehne [americana X
  angustifolia]
padus L.
pensylvanica L. f.
    var. pensylvanica
    var. saximontana Rehd.
      SY=P. p. ssp. corymbulosa (Rydb.)
        W. Wight
persica (L.) Batsch
      SY=Amygdalus p. L.
pumila L.
    var. depressa (Pursh) Gleason
      SY=P. d. Pursh
    var. pumila
    var. susquehanae (Willd.) Jaeger
      SY=P. cuneata Raf.
      SY=P. p. var. c. (Raf.) Bailey
      SY=P. s. Willd.
rivularis Scheele
      SY=P. reverchonii Sarg.
serotina Ehrh.
    var. alabamensis (C. Mohr) Little
      SY=Padus a. (C. Mohr) Small
      SY=Padus australis Beadle
      SY=Padus cuthbertii Small
      SY=Prunus alabamensis C. Mohr
      SY=Prunus s. ssp. hirsuta (Ell.)
        McVaugh
    var. eximia (Small) Little
      SY=Prunus s. ssp. e. (Small)
        McVaugh
    var. rufula (Woot. & Standl.) McVaugh
      SY=Prunus s. ssp. virens (Woot. &
        Standl.) McVaugh var. r.
        (Woot. & Standl.) McVaugh
      SY=Prunus s. ssp. v. var. v.
        (Woot. & Standl.) McVaugh
      SY=Prunus v. (Woot. & Standl.)
        Shreve
      SY=Prunus v. (Woot. & Standl.)
        Shreve var. r. (Woot. &

Prunus (CONT.)
    Standl.) Sarg.
   var. serotina
 spinosa L.
 subcordata Benth.
   var. kelloggii Lemmon
   var. oregana (Greene) W. Wight
   var. rubicunda Jepson
   var. subcordata
 texana Dietr.
 tomentosa Thunb.
 umbellata Ell.
   SY=Prunus mitis Beadle
   SY=P. u. var. tarda (Sarg.) W.
     Wight
 virginiana L.
   var. demissa (Nutt.) Torr.
   SY=Prunus d. (Nutt.) Walp.
   SY=P. v. ssp. d. (Nutt.) Taylor &
     MacBryde
   var. melanocarpa (A. Nels.) Sarg.
   SY=Padus m. (A. Nels.) Shafer
   SY=Prunus m. (A. Nels.) Rydb.
   SY=Prunus v. ssp. m. (A. Nels.)
     Taylor & MacBryde
   var. virginana
   SY=Padus nana (Du Roi) M. Roemer
   SY=Padus v. (L.) P. Mill.

Purpusia Brandeg.

 saxosa Brandeg.
   SY=Potentilla osterhoutii (A.
     Nels.) J.T. Howell

Purshia DC. ex Poir.

 glandulosa Curran
   SY=P. tridentata (Pursh) DC. var.
     g. (Curran) M.E. Jones
 tridentata (Pursh) DC.

Pyracantha M. Roemer

 angustifolia (Franch.) Schneid.
 coccinea M. Roemer
   SY=Cotoneaster pyracantha (L.)
     Spach
 fortuneana (Michx.) Li
   SY=P. crenatoserrata (Hance) Rehd.

Pyrus L.

 communis L.
 pyrifolia (Burm. f.) Nakai
   SY=P. serotina Rehd.

Rhodotypos Sieb. & Zucc.

 scandens (Thunb.) Makino

Rosa L.

 acicularis Lindl.
  ssp. acicularis
  ssp. sayi (Schwein.) W.H. Lewis
   SY=R. a. var. bourgeauiana Crepin
   SY=R. a. var. sayiana Erlanson
   SY=R. collaris Rydb.
   SY=R. engelmannii S. Wats.
   SY=R. sayi Schwein.
 arkansana Porter
   var. arkansana
   SY=R. lunellii Greene
   SY=R. rydbergii Greene
   var. suffulta (Greene) Cockerell
   SY=R. alcea Greene

   SY=R. conjuncta Rydb.
   SY=R. lunellii Greene
   SY=R. pratincola Greene
   SY=R. suffulta Greene
   SY=R. s. var. relicta (Erlanson)
     Deam
 blanda Ait.
   var. blanda
   SY=R. rousseauiorum Boivin
   SY=R. williamsii Fern.
   var. carpohispida Schuette
   var. glabra Crepin
   SY=R. johannensis Fern.
   var. glandulosa Schuette
   var. hispida Farw.
 bracteata J.C. Wendl.
 californica Cham. & Schlecht.
   SY=R. aldersonii Greene
 canina L.
   SY=R. c. var. dumetorum Baker
 carolina L.
   var. carolina
   SY=R. c. var. glandulosa (Crepin)
     Farw.
   SY=R. c. var. grandiflora (Baker)
     Rehd.
   SY=R. c. var. obovata (Raf.) Deam
   SY=R. serrulata Raf.
   SY=R. subserrulata Rydb.
   SY=R. texarkana Rydb.
   var. deamii (Erlanson) Deam
   var. sabulosa Erlanson
   var. setigera Crepin
   var. villosa (Best) Rehd.
   SY=R. lyonii Pursh
   SY=R. palmeri Rydb.
 centifolia L.
   var. centifolia
   var. cristata Prev.
   var. muscosa (Ait.) Ser.
 cinnamomea L.
 damascena P. Mill.
 X dulcissima Lunell [blanda X woodsii]
 dumetorum Thuill.
   SY=R. corymbifera Borkh.
 eglanteria L.
   SY=R. rubiginosa L.
 foliolosa Nutt. ex Torr. & Gray
   SY=R. ignota Shinners
 gallica L.
   var. gallica
   var. officinalis Thory
 gymnocarpa Nutt.
   var. gymnocarpa
   var. pubescens S. Wats.
   SY=R. bridgesii Crepin
 hemisphaerica J. Herrm.
 indica L.
 laevigata Michx.
 manca Greene
 micrantha Borrer ex Sm.
 missouriensis Hort. ex Steud.
 moschata J. Herrm.
 multiflora Thunb. ex Murr.
   var. calva author ?
   var. multiflora
 ?myriadenia Greene
 nitida Willd.
 nutkana Presl
   var. hispida Fern.
   SY=R. macdougalii Holz.
   SY=R. spaldingii Crepin
   SY=R. s. var. alta (Suksdorf) G.N.
     Jones
   var. muriculata (Greene) G.N. Jones
   var. nutkana
   SY=R. durandii Crepin

Rosa (CONT.)
    var. setosa G.N. Jones
  obtusiuscula Rydb.
  odorata (Andr.) Sweet
  palustris Marsh.
      SY=R. floridana Rydb.
      SY=R. lancifolia Small
      SY=R. p. var. dasistema (Raf.)
        Palmer & Steyermark
  ?pinetorum Heller
  pisocarpa Gray
      SY=R. p. var. rivalis (Eastw.)
        Jepson
      SY=R. r. Eastw.
  rubrifolia Vill.
      SY=R. glauca Pourret
  rudiuscula Greene
  rugosa Thunb.
  sempervirens L.
  setigera Michx.
    var. setigera
      SY=R. s. var. serena Palmer &
        Steyermark
    var. tomentosa Torr. & Gray
  spinosissima L.
      SY=R. pimpinellifolia L.
  spithamea S. Wats.
    var. sonomensis (Greene) Jepson
      SY=R. s. Greene
    var. spithamea
      SY=R. s. var. solitaria Henderson
  stellata Woot.
    ssp. mirifica (Greene) W.H. Lewis
      var. erlansoniae W.H. Lewis
      var. mirifica (Greene) W.H. Lewis
        SY=R. m. Greene
    ssp. stellata
  tomentosa Sm.
      SY=R. t. var. globulosa Rouy
  virginiana P. Mill.
    var. lamprophylla (Rydb.) Fern.
    var. virginiana
  wichuraiana Crepin
  woodsii Lindl.
    var. glabrata (Parish) Cole
      SY=R. mohavensis Parish
    var. gratissima (Greene) Cole
    var. ultramontana (S. Wats.) Jepson
      SY=R. arizonica Rydb.
      SY=R. a. var. granulifera (Rydb.)
        Kearney & Peebles
      SY=R. covillei Greene
      SY=R. macounii Greene
      SY=R. u. (S. Wats.) Heller
      SY=R. w. ssp. u. (S. Wats.) Taylor
        & MacBryde
    var. woodsii
      SY=R. adenosepala Woot. & Standl.
      SY=R. fendleri Crepin
      SY=R. gratissima Greene
      SY=R. hypoleuca Woot. & Standl.
      SY=R. neomexicana Cockerell
      SY=R. pecosensis Cockerell
      SY=R. standleyi Rydb.
      SY=R. terrens Lunell
      SY=R. w. var. f. (Crepin) Rydb.
  ?yainacensis Greene

Rubus L.

  abactus Bailey
  aborigium Rydb.
      SY=R. almus (Bailey) Bailey
      SY=R. austrinus Bailey
      SY=R. bollianus Bailey
      SY=R. clairbrownii Bailey
      SY=R. foliaceus Bailey

  abundiflorus Bailey
  aculiferus Bailey
  adenocaulis Fern.
  adjacens Fern.
  agilis Bailey
  akermanii Fern.
      SY=R. subinnoxius Fern.
  alaskensis Bailey
      SY=R. pubescens Raf. var. a.
        (Bailey) Boivin
  aliceae Bailey
  allegheniensis Porter ex Bailey
    var. allegheniensis
      SY=R. a. var. plausus Bailey
      SY=R. a. var. populifolius Fern.
      SY=R. auroralis Bailey
      SY=R. fissidens Bailey
      SY=R. longissimus Bailey
      SY=R. nigrobaccus Bailey
      SY=R. nuperus Bailey
      SY=R. pennus Bailey
      SY=R. rappii Bailey
      SY=R. separ Bailey
    var. gravesii Fern.
      SY=R. marilandicus Bailey
      SY=R. tumularis Bailey
      SY=R. uber Bailey
      SY=R. virginianus Bailey
  alter Bailey
  alumnus Bailey
      SY=R. apianus Bailey
      SY=R. barbarus Bailey
      SY=R. campestre Bailey
      SY=R. corei Bailey
      SY=R. facetus Bailey
      SY=R. fernaldianus Bailey
      SY=R. impos Bailey
      SY=R. independens Bailey
      SY=R. licitus Bailey
      SY=R. miriflorus Bailey
      SY=R. paulus Bailey
      SY=R. tennesseeanus Bailey
  ambigens Fern.
  amnicola Blanch.
  andrewsianus Blanch.
  angustifoliatus Bailey
  apogaeus Bailey
      SY=R. exlex Bailey
      SY=R. lassus Bailey
      SY=R. lundelliorum Bailey
      SY=R. uneus Bailey
  aptatus Bailey
  arcticus L.
    ssp. acaulis (Michx.) Focke
      SY=R. a. Michx.
    ssp. arcticus

      var. arcticus
      var. pentaphylloides Hulten
    ssp. stellatus (Sm.) Boivin
      SY=R. a. var. s. (Sm.) Boivin
      SY=R. s. Sm.
  arcuans Fern. & St. John
  arenicolus Blanch.
  argutus Link
      SY=R. incisifrons Bailey
  arizonensis Focke
  arundelanus Blanch.
      SY=R. a. var. jeckylanus (Blanch.)
        Bailey
  arvensis Bailey
      SY=R. saepescandens Bailey
  ascendens Blanch.
      SY=R. ashei Bailey
      SY=R. clausenii Bailey
  ?attractus Bailey
  audax Bailey

Rubus (CONT.)
    baileyanus Britt.
                SY=R. flagellaris Willd. var.
                    humifusus (Torr. & Gray)
                    Boivin
                SY=R. housei Bailey
                SY=R. tenuicaulis Bailey
                SY=R. uniflorifer Bailey
    bartonianus M.E. Peck
    beatus Bailey
    bellobatus Bailey
    betulifolius Small
                SY=R. rhodophyllus Rydb.
    bicknellii Bailey
                SY=R. ellectus Bailey
    biformispinus Blanch.
    bifrons Vest ex Tratt.
    bigelovianus Bailey
    blakei Bailey
    blanchardianus Bailey
    boyntonii Ashe
    burnhamii Bailey
    bushii Bailey
                SY=R. fructifer Bailey
                SY=R. kansanus Bailey
                SY=R. oppisitus Bailey
                SY=R. ozarkensis Bailey
                SY=R. putus Bailey
                SY=R. scibilis Bailey
                SY=R. sertatus Bailey
    cacaponensis Davis & Davis
    caesius L.
    canadensis L.
                SY=R. besseyi Bailey
                SY=R. c. var. imus Bailey
                SY=R. forestalis Bailey
                SY=R. illustris Bailey
                SY=R. irregularis Bailey
                SY=R. laetabilis Bailey
                SY=R. millspaughii Bailey
                SY=R. randii (Bailey) Rydb.
    carolinianus Rydb.
    celer Bailey
    centralis Bailey
    chamaemorus L.
    chapmanii Bailey
                SY=R. cuneifolius Pursh var.
                    angustior Bailey
                SY=R. dixiensis Davis, Fuller &
                    Davis
    clarus Bailey
    conabilis Bailey
    conanictuensis Bailey
    concameratus Davis & Davis
    coronarius Sweet
    cubitans Blanch.
    cuneifolius Pursh
                SY=R. c. var. spiniceps Bailey
                SY=R. c. var. subellipticus Fern.
    curtipes Bailey
    deamii Bailey
                SY=R. gordonii Bailey
                SY=R. rosagnetis Bailey
    decor Bailey
                SY=R. ignarus Bailey
    defectionis Fern.
    deliciosus Torr.
    densissimus Davis & Davis
    depavitus Bailey
    discolor Weihe & Nees
                SY=R. procerus P.J. Muell.
    dissimilis Bailey
                SY=R. apparatus Bailey
                SY=R. perpauper Bailey
    elegantulus Blanch.
                SY=R. adirondackensis Bailey
                SY=R. amabilis Blanch.

                SY=R. amicalis Blanch.
                SY=R. canadensis L. var. e.
                    (Blanch.) Farw.
                SY=R. proprius Bailey
    ellipticus Sm.
        var. ellipticus
        var. obcordatus Focke
    emeritus Bailey
    enslenii Tratt.
                SY=R. bonus Bailey
                SY=R. camurus Bailey
                SY=R. connixus Bailey
                SY=R. cordialis Bailey
                SY=R. dissitiflorus Fern.
                SY=R. longipes Fern.
                SY=R. occultus Bailey
                SY=R. serenus Bailey
                SY=R. tracyi Bailey
    exrubicundus (Woot. & Standl.) Bailey
    exsularis Bailey
    exter Bailey
    fecundus Bailey
    felix Bailey
                SY=R. dives Bailey
    flagellaris Willd.
                SY=R. alacer Bailey
                SY=R. exemptus Bailey
                SY=R. frustratus Bailey
                SY=R. geophilus Blanch.
                SY=R. jaysmithii Bailey var.
                    angustior Bailey
                SY=R. maltei Bailey
                SY=R. neonefrens Bailey
                SY=R. subuniflorus Rydb.
                SY=R. tetricus Bailey
                SY=R. urbanianus Bailey
    flavinanus Blanch.
    floricomus Blanch.
    floridensis Bailey
    floridus Tratt.
    florulentus Focke
    fraternalis Bailey
                SY=R. alius Bailey
                SY=R. fraternus Brainerd & Piet.
                    non Gremli
    frondisentis Blanch.
                SY=R. ravus Bailey
    frondosus Bigelow
                SY=R. brainerdii Fern.
                SY=R. cardianus Bailey
                SY=R. difformis Bailey
                SY=R. eriensis Bailey
                SY=R. heterogeneous Bailey
                SY=R. nescius Bailey
                SY=R. pensilvanicus Poir. var. f.
                    (Bigelow) Boivin
                SY=R. pratensis Bailey
                SY=R. sativus Brainerd
                SY=R. uniquus Bailey
                SY=R. wahlii Bailey
    fryei Davis & Davis
    furtivus Bailey
    fusus Bailey
    glandicaulis Blanch.
                SY=R. acadiensis Bailey
                SY=R. atwoodii Bailey
                SY=R. bracteoliferus Fern.
                SY=R. grandidens Bailey
                SY=R. montpelerensis Blanch.
    glaucifolius Kellogg
        var. ganderi (Bailey) Munz
        var. glaucifolius
    gnarus Bailey
                SY=R. acer Bailey var. subacer
                    Bailey
                SY=R. litoreus Bailey
    grimesii Bailey

Rubus (CONT.)
           SY=R. cathartium Fern.
griseus Bailey
groutianus Blanch.
           SY=R. discretus Bailey
           SY=R. tectus Bailey
gulosus Bailey
hancinianus Bailey
hanesii Bailey
harmonicus Bailey
hawaiiensis Gray
       var. hawaiiensis
       var. inermis Wawra
heterophyllus Willd.
hispidoides Bailey
hispidus L.
           SY=R. h. var. cupulifer Bailey
           SY=R. h. var. obovalis (Michx.)
               Fern.
           SY=R. sempervirens Bigelow
humei Bailey
           SY=R. argutinus Bailey
huttonii Bailey
hypolasius Fern.
ictus Bailey
idaeus L.
    ssp. idaeus
           SY=R. i. var. caudatus (Robins. &
               Schrenk) Fern.
           SY=R. i. var. egglestonii
               (Blanch.) Fern.
           SY=R. i. var. eucyclus Fern. &
               Weath.
           SY=R. i. var. heterolasius Fern.
           SY=R. i. var. peramoenus (Greene)
               Fern.
    ssp. sachalinensis (Levl.) Focke
           SY=R. i. ssp. melanolasius (Dieck)
               Focke
           SY=R. i. ssp. strigosus (Michx.)
               Focke
           SY=R. i. var. aculeatissimus Regel
               & Tiling
           SY=R. i. var. canadensis Richards.
           SY=R. i. var. gracilipes M.E.
               Jones
           SY=R. i. var. melanolasius (Dieck)
               R.J. Davis
           SY=R. i. var. melanotrachys
               (Focke) Fern.
           SY=R. i. var. neglectus Peck
           SY=R. i. var. strigosus (Michx.)
               Maxim.
           SY=R. melanolasius Dieck
           SY=R. X n. Peck
?illecebrosus Focke
immanis Ashe
           SY=R. montnesis Bailey var.
               superior Bailey
impar Bailey
inclinis Bailey
indianensis Bailey
inferior Bailey
iniens Bailey
injunctus Bailey
           SY=R. pluralis Bailey
insons Bailey
           SY=R. honorus Bailey
           SY=R. janssonii Bailey
insulanus Bailey
invisus (Bailey) Britt.
           SY=R. jactus Bailey
           SY=R. macdanielsii Bailey
           SY=R. masseyi Bailey
           SY=R. redundans Bailey
           SY=R. sanfordii Bailey
ithacanus Bailey

           SY=R. densipubens Bailey
           SY=R. fandus Bailey
           SY=R. folioflorus Bailey
           SY=R. pohlii Bailey
           SY=R. satis Bailey
jacens Blanch.
jennisonii Bailey
           SY=R. monongaliensis Bailey
jugosus Bailey
           SY=R. beadlei Bailey
           SY=R. floriger Bailey
junceus Blanch.
kelloggii Blanch.
kennedyanus Fern.
           SY=R. acridens Bailey
           SY=R. lepagei Bailey
           SY=R. ulterior Bailey
kentuckiensis Bailey
laciniatus Willd.
largus Bailey
           SY=R. valentulus Bailey
lasiococcus Gray
latens Bailey
laudatus Berger
           SY=R. ablatus Bailey
           SY=R. condensiflorus Bailey
           SY=R. congruus Bailey
           SY=R. gattingeri Bailey
           SY=R. interioris Bailey
           SY=R. par Bailey
           SY=R. pulchriflorus Bailey
           SY=R. schneckii Bailey
           SY=R. senilis Bailey
           SY=R. subtentus Bailey
lawrencei Bailey
leggii Davis & Davis
leucodermis Dougl. ex Torr. & Gray
       var. bernardinus (Greene) Jepson
           SY=R. l. ssp. b. (Greene) Thorne
       var. leucodermis
           SY=R. occidentalis L. var. l.
               (Dougl. ex Torr. & Gray)
               Focke
       var. trinitatis Berger
leviculus Bailey
           SY=R. census Bailey
libratus Bailey
linkianus Ser.
longii Fern.
louisianus Berger
lucidus Rydb.
           SY=R. nessianus Bailey
macraei Gray
macrophyllus Weihe & Nees
macvaughii Bailey
mananensis Bailey
maniseesensis Bailey
meracus Bailey
           SY=R. tantalus Bailey
michiganensis (Card) Bailey
           SY=R. complex Bailey
           SY=R. cordifrons Bailey
           SY=R. florenceae Bailey
           SY=R. inobvius Bailey
mirus Bailey
           SY=R. magniflorus Bailey
missouricus Bailey
           SY=R. clandestinus Bailey
           SY=R. jejunus Bailey
           SY=R. mediocris Bailey
           SY=R. offectus Bailey
           SY=R. schneideri Bailey
mollior Bailey
moluccanus L.
montensis Bailey
multifer Bailey
           SY=R. jaysmithii Bailey

Rubus (CONT.)
            SY=R. polybotrys Bailey
    multiformis Blanch.
            SY=R. peracer Bailey
    multilicius Bailey
    multispinus Blanch.
    mundus Bailey
    navus Bailey
    nefrens Bailey
    neomexicanus Gray
    nigerrimus (Greene) Rydb.
    nivalis Dougl. ex Hook.
    nocivus Bailey
    notatus Bailey
        var. notatus
            SY=R. boottianus Bailey
            SY=R. n. var. boreus Bailey
        var. ortus Bailey
    novanglicus Bailey
    noveboracus Bailey
    novocaesarius Bailey
    obsessus Bailey
    obvius Bailey
    occidentalis L.
            SY=R. o. var. pallidus Bailey
    odoratus L.
        var. columbianus Millsp.
            SY=R. c. (Millsp.) Rydb.
        var. malachophyllus Fern.
        var. odoratus
            SY=Rubacer o. (L.) Rydb.
    oklahomus Bailey
    onustus Bailey
    orarius Blanch.
    oriens Bailey
    originalis Bailey
    ortivus Bailey
            SY=R. vermontanus Blanch. var. o.
                (Bailey) Boivin
    ostryifolius Rydb.
            SY=R. fatuus Bailey
    paganus Bailey
            SY=R. vagulus Bailey
            SY=R. zaplutus Bailey
    paludivagus Fern.
    X paracaulis Bailey [articus ssp. acaulis X
        pubescens]
            SY=R. pubescens Raf. var.
                paracaulis (Bailey) Boivin
    parcifrondifer Bailey
    parlinii Bailey
    particeps Bailey
            SY=R. eflagellaris Bailey
    particularis Bailey
    parviflorus Nutt.
        var. bifarius Fern.
        var. hypomalacus Fern.
        var. parviflorus
            SY=R. p. var. grandiflorus Farw.
        var. parvifolius (Gray) Fern.
        var. velutinus (Hook. & Arn.) Greene
            SY=R. p. ssp. v. (Hook. & Arn.)
                Taylor & MacBryde
    pascuus Bailey
            SY=R. serissimus Bailey
            SY=R. uliginosus Fern.
    pauxillus Bailey
    pedatus Sm.
    pensilvanicus Poir.
    pergratus Blanch.
        var. pergratus
            SY=R. avipes Bailey
            SY=R. bractealis Bailey
        var. terrae-novae Bailey
    perinvisus Bailey
    permixtus Blanch.
            SY=R. distinctus Bailey

            SY=R. elongatus Brainerd & Piet.
            SY=R. laevior Fern.
            SY=R. sharpii Bailey
            SY=R. vegrandis Bailey
    pernagaeus Bailey
    persistens Rydb.
            SY=R. angustus Bailey
            SY=R. arrectus Bailey
            SY=R. harperi Bailey
            SY=R. penetrans Bailey
            SY=R. zoae Bailey
    perspicuus Bailey
    pervarius Bailey
            SY=R. davisiorum Bailey
    philadelphicus Blanch.
    phoenicolasius Maxim.
    plexus Fern.
    plicatifolius Blanch.
            SY=R. armatus (Fern.) Bailey
            SY=R. botruosus Bailey
            SY=R. bretonis Bailey
            SY=R. coloniatus Bailey
            SY=R. exutus Bailey
            SY=R. obsessus Bailey var.
                unilaris Bailey
            SY=R. pauper Bailey
            SY=R. pololus Bailey
            SY=R. prior Bailey
            SY=R. problematicus Bailey
            SY=R. recurvicaulis Blanch. var.
                armatus Fern.
            SY=R. rhodinsulanus Bailey
            SY=R. rosendahlii Bailey
            SY=R. semierectus Blanch.
            SY=R. victorinii Bailey
    plus Bailey
            SY=R. kalamazoensis Bailey
    porteri Bailey
    positivus Bailey
    praepes Bailey
    prestonensis Davis & Davis
    probabilis Bailey
            SY=R. cuneifolius Pursh var.
                austrifer Bailey
            SY=R. sejunctus Bailey
    probativus Bailey
    probus Bailey
    profusiflorus Bailey
    pronus Bailey
    prosper Bailey
    provincialis Bailey
            SY=R. orbicularis Davis & Davis
    pubescens Raf.
        var. pilosifolius A.F. Hill
        var. pubescens
            SY=R. triflorus Richards.
        var. scius Bailey
    pubifolius Bailey
    pugnax Bailey
            SY=R. allegheniensis Porter ex
                Bailey var. neoscoticus
                (Fern.) Bailey
    quaesitus Bailey
    quebecensis Bailey
    racemiger Bailey
            SY=R. nescius Bailey
            SY=R. reravus Bailey
    randolphiorum Bailey
            SY=R. escatilis Bailey
            SY=R. georgianus Bailey
    recurvans Blanch.
            SY=R. cauliflorus Bailey
            SY=R. limulus Bailey
            SY=R. perfoliosus Bailey
            SY=R. pityophilus S.J. Sm.
            SY=R. r. var. subrecurvans Blanch.
            SY=R. wiegandii Bailey

Rubus (CONT.)
   recurvicaulis Blanch.
        SY=R. r. var. inarmatus Blanch.
   regionalis Bailey
        SY=R. viridifrons Bailey
   ricei Bailey
   riograndis Bailey
        SY=R. duplaris Shinners
        SY=R. trivalis Michx. var. d.
          (Shinners) Mahler
   roribaccus (Bailey) Rydb.
        SY=R. imperiorum Fern.
        SY=R. occidualis Bailey
        SY=R. pauperrimus Bailey
   rosa Bailey

   rosarius Bailey
   rosifolius Sm.
   rossbergianus Blanch.
   russeus Bailey
        SY=R. canaanensis Davis & Davis
   rydbergianus Bailey
   sailorii Bailey
   saltuensis Bailey
   saxatilis L.
   scambens Bailey
   sceleratus Brainerd
   schoolcraftianus Bailey
   segnis Bailey
        SY=R. pudens Bailey
        SY=R. trifrons Blanch. var. p.
          (Bailey) Fern.
   semisetosus Blanch.
        SY=R. benneri Bailey
   setosus Bigelow
        SY=R. condignus Bailey
        SY=R. hispidus L. var. suberectus
          Peck
        SY=R. junior Bailey
        SY=R. nigricans Rydb.
        SY=R. ribes Bailey
        SY=R. significans Bailey
        SY=R. udus Bailey
   severus Brainerd
        SY=R. jacens Blanch. var.
          specialis Bailey
        SY=R. mainensis Bailey
        SY=R. rixosus Bailey
   sewardianus Fern.
   signatus Bailey
   sons Bailey
   spectabilis Pursh
      var. franciscanus (Rydb.) J.T. Howell
        SY=R. s. var. menziesii (Hook.) S.
          Wats.
      var. spectabilis
   spectatus Bailey
   steelei Bailey
        SY=R. currulis Bailey
   stipuatus Bailey
        SY=R. dissensus Bailey
   strigosus Michx.
      var. arizonicus (Greene) Kearney &
        Peebles
      var. strigosus
   subsolanus Bailey
   subtractus Bailey
        SY=R. cupressorum Fern.
   summotus Bailey
   suppar Bailey
   suus Bailey
        SY=R. demareanus Bailey
        SY=R. ramifer Bailey
        SY=R. texanus Bailey
        SY=R. vixargutus Bailey
   tallahasseanus Bailey
   tardatus Blanch.

   temerarius Bailey
   terraltanus Bailey
   tholiformis Fern.
        SY=R. spiculosus Fern.
   tomentosus Borkh.
      var. canescens Wirtg.
      var. tomentosus
   trifrons Blanch.
   ?triphyllus Thunb.
   trivialis Michx.
        SY=R. continentalis (Focke) Bailey
        SY=R. mississipianus Bailey
        SY=R. okeechobeus Bailey
        SY=R. rubrisetus Rydb.
        SY=R. t. var. serosus Bailey
   trux Ashe
   tygartensis Davis & Davis
   ucetanus Bailey
   ulmifolius Schott
      var. inermis (Willd.) Focke
   uniformis Bailey
   ursinus Cham. & Schlecht.
     ssp. macropetalus (Dougl. ex Hook.)
       Taylor & MacBryde
        SY=R. m. Dougl. ex Hook.
        SY=R. u. var. m. (Dougl. ex Hook.)
          R. Br.
     ssp. ursinus
      var. sirbenus (Bailey) J.T. Howell
      var. ursinus
        SY=R. vitifolius Cham. & Schlecht.
          ssp. u. (Cham. & Schlecht.)
          Abrams
   usus Bailey
   uvidus Bailey
        SY=R. associus Hanes
        SY=R. humilior Bailey
        SY=R. licens Bailey
        SY=R. localis Bailey
   vagus Bailey
   variispinus Bailey
   varus Bailey
   velox Bailey
   vermontanus Blanch.
        SY=R. abbrevians Blanch.
        SY=R. deaneanus Bailey
        SY=R. malus Bailey
        SY=R. miscix Bailey
        SY=R. peculiaris Blanch.
        SY=R. perdebilis Bailey
        SY=R. singulus Bailey
        SY=R. superioris Bailey
        SY=R. unanimus Bailey
   vestitus Weihe & Nees
   vigil Bailey
   vigoratus Bailey
   virilis Bailey
   vitifolius Cham. & Schlecht.
      var. eastwoodianus (Rydb.) Munz
      var. titanus (Bailey) Bailey
      var. vitifolius
   vixalacer Bailey
   weatherbyi Bailey
   whartoniae Bailey
   wheeleri (Bailey) Bailey
        SY=R. compos Bailey
        SY=R. fassettii Bailey
        SY=R. fulleri Bailey
        SY=R. potis Bailey
        SY=R. rotundior Bailey
        SY=R. rowleei Bailey
        SY=R. semisetosus Blanch. var. w.
          Bailey
        SY=R. X univocus Bailey
        SY=R. u. Bailey
   wisconsinensis Bailey
        SY=R. latifoliolus Bailey

Rubus   (CONT.)
          SY=R. minnesotanus Bailey
          SY=R. setospinosus Bailey

Sanguisorba L.

   annua (Nutt. ex Hook.) Torr. & Gray
          SY=Poteridium a. (Nutt. ex Hook.)
             Spach
   canadensis L.
          SY=S. c. ssp. latifolia (Hook.)
             Calder & Taylor
          SY=S. c. var. l. Hook.
          SY=S. sitchensis C.A. Mey.
          SY=S. stipulata Raf.
   menziesii Rydb.
   minor Scop.
      ssp. muricata Briq.
          SY=Poterium sanguisorba sensu
             auctt. non L.
   occidentalis Nutt.
   officinalis L.
          SY=S. microcephala Presl
          SY=S. o. ssp. m. (Presl) Calder &
             Taylor
          SY=S. o. var. polygama (Nyl.) Mela
             & Caj.

Sibbaldia L.

   procumbens L.
          SY=Potentilla sibbaldii Haller f.

Sorbaria (Ser.) A. Braun

   sorbifolia (L.) A. Braun
          SY=Schizonotus s. (L.) Lindl.
          SY=Spiraea s. L.

X Sorbaronia Schneid.

   X alpina (Willd.) Schneid. [Aronia
      arbutifolia X Sorbus aria]
          SY=Pyrus X alpina Willd.
          SY=Sorbus X alpina (Willd.) Heynh.
   X arsenii (Britt.) G.N. Jones [Aronia
      arbutifolia X Sorbus decora]
   X fallax Schneid. [Aronia melanocarpa X
      Sorbus aucuparia]
          SY=Pyrus X f. (Schneid.) Fern.
   X hybrida (Moench) Schneid. [Aronia
      arbutifolia X Sorbus aucuparia]
          SY=Pyrus h. Moench
   X jackii Rehd. [Aronia prunifolia X Sorbus
      americana]
          SY=Pyrus X j. (Rehd.) Fern.
   X sorbifolia (Poir.) Schneid. [Aronia
      melanocarpa X Sorbus americana]
          SY=Pyrus mixta Fern.
          SY=Sorbus sargentii Dippel
          SY=Sorbus sorbifolia (Poir.) Hedl.

Sorbus L.

   americana Marsh.
          SY=Pyrus a. (Marsh.) DC.
   aucuparia L.
          SY=Pyrus a. (L.) Gaertn.
   californica Greene
          SY=S. sitchensis M. Roemer ssp. c.
             (Greene) Abrams
   decora (Sarg.) Schneid.
          SY=Pyrus americana (Marsh.) DC.
             var. d. Sarg.
          SY=P. d. (Sarg.) Hyl.
   domestica L.
   dumosa Greene

   groenlandica (Schneid.) Love & Love
          SY=Pyrus decora (Sarg.) Schneid.
             var. g. (Schneid.) Fern.
   hybrida L.
   sambucifolia (Cham. & Schlecht.) M. Roemer
   scopulina Greene
          var. cascadensis (G.N. Jones) C.L.
             Hitchc.
          SY=S. c. G.N. Jones
          var. scopulina
   sitchensis M. Roemer
      ssp. grayii (Wenzig) Calder & Taylor
          SY=S. occidentalis (S. Wats.)
             Greene
          SY=S. s. var. g. (Wenzig) C.L.
             Hitchc.
      ssp. sitchensis
          SY=Pyrus s. (M. Roemer) Piper

Spiraea L.

   alba Du Roi
   betulifolia Pallas
      ssp. corymbosa (Raf.) Taylor & MacBryde
          SY=S. b. var. c. (Raf.) Maxim.
          SY=S. c. Raf.
      ssp. lucida (Dougl. ex Greene) Taylor &
         MacBryde
          SY=S. b. var. lucida (Dougl. ex
             Greene) C.L. Hitchc.
          SY=S. l. Dougl. ex Greene
   X billiardii Herincq. [douglasii X
      salicifolia]
   chamaedryfolia L.
          var. chamaedryfolia
          var. ulmifolia (Scop.) Maxim.
   densiflora Nutt. ex Rydb.
      ssp. densiflora
      ssp. splendens (Baumann ex K. Koch)
         Abrams
          SY=S. d. var. s. (Baumann ex K.
             Koch) C.L. Hitchc.
          SY=S. s. Baumann ex K. Koch
   douglasii Hook.
      ssp. douglasii
          SY=S. d. var. roseata (Rydb.) C.L.
             Hitchc.
      ssp. menziesii (Hook.) Calder & Taylor
          SY=S. d. var. m. (Hook.) Presl
          SY=S. m. Hook.
          SY=S. subvillosa Rydb.
   japonica L. f.
          var. fortunei (Planch.) Rehd.
   latifolia (Ait.) Borkh.
          SY=S. alba Du Roi var. l. (Ait.)
             Dippel
   prunifolia Sieb. & Zucc.
          SY=S. p. var. plena Schneid.
   X pyramidata Greene [betulifolia ssp.
      lucida X douglasii ssp. menziesii]
          SY=S. tomentulosa Rydb.
   salicifolia L.
   septentrionalis (Fern.) Love & Love
          SY=S. alba Du Roi var. s. (Fern.)
             Seymour
          SY=S. latifolia (Ait.) Borkh. var.
             s. Fern.
   stevenii (Schneid.) Rydb.
          SY=S. beauverdiana sensu auctt.
             non Schneid.
   X subcanescens Rydb. [alba X tomentosa]
   thunbergii Sieb. ex Blume
   tomentosa L.
          var. rosea (Raf.) Fern.
          var. tomentosa
   X vanhouttei (Broit) Zabel [cantoniensis X
      trilobata]

Spiraea (CONT.)
 virginiana Britt.
   var. serrulata Rehd.
   var. virginiana

Vauquelinia Correa ex Humb. & Bonpl.

 angustifolia Rydb.
 californica (Torr.) Sarg.
 pauciflora Standl.

Waldsteinia Willd.

 fragarioides (Michx.) Tratt.
  ssp. doniana (Tratt.) Teppner
    SY=W. d. Tratt.
    SY=W. f. var. parviflora (Small)
      Fern.
    SY=W. p. Small
  ssp. fragerioides
 idahoensis Piper
 lobata (Baldw.) Torr. & Gray
 pendula (Urban) Mez

ROXBURGHIACEAE

Croomia Torr. ex Torr. & Gray

 pauciflora (Nutt.) Torr.

RUBIACEAE

Anthocephalus A. Rich.

 chinensis (Lam.) A. Rich. ex Walp.

Antirhea Juss.

 acutata (DC.) Urban
   SY=Stenostomum a. DC.
 coriacea (Vahl) Urban
   SY=Stenostomum c. (Vahl) Griseb.
 lucida (Sw.) Benth. & Hook. f.
   SY=Stenostomum l. (Sw.) Gaertn. f.
 obtusifolia Urban
   SY=Stenostomum o. (Urban) Britt. &
    Wilson
 portoricensis (Britt. & Wilson) Standl.
 sintenisii Urban
   SY=Stenostomum s. (Urban) Britt. &
    Wilson

Asperula L.

 arvensis L.

Bobea Gaud.

 gaudichaudii (Cham. & Schlecht.) St. John &
  Herbst
   SY=B. elatior Gaud.
   SY=?B. e. var. brevipes (Gray)
    Hbd.
   SY=?B. e. var. molokaiensis Rock
 hookeri Hbd.
 mannii Hbd.
 sandwicensis (Gray) Hbd.
 timonioides (Hook. f.) Hbd.

Borreria G.F.W. Mey.

brachysepala Urban
laevis (Lam.) Griseb.
   SY=Spermacoce suffrutescens sensu
    auctt. non Jacq.
ocimoides (Burm. f.) DC.
repens DC.
terminalis Small
verticillata (L.) G.F.W. Mey.
   SY=Spermacoce v. L.

Bouvardia Salisb.

 ternifolia (Cav.) Schlecht.
   SY=B. glaberrima Engelm.

Canthium Lam.

 odoratum (Forst. f.) Seem.

Catesbaea L.

 melanocarpa Krug & Urban
 parviflora Sw.

Cephalanthus L.

 occidentalis L.
   SY=C. o. var. californicus Benth.
   SY=C. o. var. pubescens Raf.
 salicifolius Humb. & Bonpl.

Chiococca P. Br.

 alba (L.) A.S. Hitchc.
   SY=C. racemosa L.
 pinetorum Britt.

Chione DC.

 seminervis Urban & Ekman
 venosa (Sw.) Urban

Coccocypselum P. Br.

 herbaceum Aubl.
   SY=Tontanea h. (Aubl.) Standl.

Coffea L.

 arabica L.
 dewevrei Wildm. & T. Dur.

Coprosma J.R. & G. Forst.

 cymosa Hbd.
 elliptica Oliver
 ernodeoides Gray
  var. ernodeoides
  var. mauiensis St. John
 fauriei Levl.
  var. fauriei
  var. lanaiensis Oliver
  var. oahuensis Oliver
 foliosa Gray
 kauensis (Gray) Heller
 longifolia Gray
  var. longifolia
  var. oppositifolia Fosberg
 menziesii Gray
 molokaiensis St. John
 montana Hbd.
  var. crassa Oliver
  var. montana
  var. orbicularis Oliver
 ochracea Oliver
  var. kaalae St. John
  var. ochracea

Coprosma (CONT.)
        var. rockiana Oliver
    pubens Gray
        var. pubens
        var. sessiliflora Oliver
    repens A. Rich.
    rhynchocarpa Gray
    serrata St. John
    skottsbergiana Oliver
    stephanocarpa Hbd.
    ternata Oliver
    waimeae Wawra

Crucianella L.

    angustifolia L.

Crusea Cham. & Schlecht.

    diversifolia (H.B.K.) W.A. Anderson
        SY=C. wrightii Gray
    subulata (Pavon) Gray

Diodia L.

    apiculata (Willd.) Roemer & Schultes
        SY=Diodella rigida (Willd.) Small
        SY=Diodia r. (Willd.) Cham. &
            Schlecht.
    dasycephala Cham. & Schlecht.
    harperi Small
    hirsuta Pursh
    sarmentosa Sw.
    serrulata (Beauv.) G. Taylor
        SY=Diodia maritima Thonn.
    teres Walt.
        var. angustata Gray
        var. hystricina Fern. & Grisc.
        var. teres
            SY=Diodella t. (Walt.) Small
            SY=Diodia t. var. setifera Fern. &
                Grisc.
    virginiana L.
        var. attenuata Fern.
        var. latifolia Torr. & Gray
        var. virginiana
            SY=Diodia tetragona Walt.

Erithalis P. Br.

    fruticosa L.
    revoluta Urban

Ernodea Sw.

    littoralis Sw.
        var. angusta (Small) R.W. Long
            SY=E. a. Small
        var. littoralis

Exostema (Pers.) L.C. Rich. ex Humb. & Bonpl.

    caribaeum (Jacq.) Roemer & Schultes
    ellipticum Griseb.
    sanctae-luciae (Kentish) Britt.

Faramea Aubl.

    occidentalis (L.) A. Rich.

Galium L.

    album P. Mill.
    ambiguum W. Wight
        ssp. ambiguum
        ssp. siskiyouense (Ferris) Dempster &
            Stebbins

            SY=G. a. var. s. Ferris
    andrewsii Gray
        ssp. andrewsii
        ssp. gatense (Dempster) Dempster &
            Stebbins
            SY=G. a. var. g. Dempster
        ssp. intermedium Dempster & Stebbins
    angustifolium Nutt.
        ssp. angustifolium
            SY=G. a. var. bernardinum Hilend &
                Howell
            SY=G. a. var. diffusum Hilend &
                Howell
            SY=G. a. var. siccatum (W. Wight)
                Hilend & Howell
            SY=G. a. var. typicum Hilend &
                Howell
            SY=G. s. Wight
            SY=G. trichocarpum Nutt.
        ssp. borregoense Dempster & Stebbins
        ssp. foliosum (Hilend & Howell) Dempster
            & Stebbins
            SY=G. a. var. f. Hilend & Howell
        ssp. gabrielense (Munz & Johnston)
            Dempster & Stebbins
            SY=G. g. Munz & Johnston
            SY=G. siccatum Wight var. antinum
                Jepson
        ssp. gracillimum Dempster & Stebbins
        ssp. jacinticum Dempster & Stebbins
        ssp. nudicaule Dempster & Stebbins
        ssp. onycense (Dempster) Dempster &
            Stebbins
            SY=G. a. var. o. Dempster
    aparine L.
            SY=G. agreste var. echinospermum
                Wallr.
            SY=G. aparine var. e. (Wallr.)
                Farw.
            SY=G. aparine var. minor Hook.
            SY=G. aparine var. vaillantii
                (DC.) Koch
            SY=G. spurium L. var. e. (Wallr.)
                Hayek
            SY=G. s. var. v. (DC.) Gren. &
                Godr.
            SY=G. s. var. v. (DC.) G. Beck
            SY=G. v. DC.
    argense Dempster & Ehrend.
    arkansanum Gray
            SY=G. a. var. pubiflorum E.B. Sm.
    asprellum Michx.
    bifolium S. Wats.
    bolanderi Gray
            SY=G. arcuatum Wiegand
            SY=G. culbertsonii Greene
            SY=G. margaricoccum Gray
            SY=G. pubens Gray
            SY=G. p. var. scabridum Gray
            SY=G. subscabridum Wight
    boreale L.
            SY=G. b. ssp. septentrionale
                (Roemer & Schultes) Hara
            SY=G. b. var. hyssopifolium
                (Hoffmann) DC.
            SY=G. b. var. intermedium DC.
            SY=G. b. var. linearifolium Rydb.
            SY=G. b. var. scabrum DC.
            SY=G. b. var. typicum G. Beck
            SY=G. h. Hoffmann
            SY=G. septentrionale Roemer &
                Schultes
            SY=G. strictum Torr.
    brevipes Fern. & Wieg.
    buxifolium Greene
            SY=G. catalinense Gray var. b.
                (Greene) Dempster

Galium (CONT.)
    californicum Hook. & Arn.
      ssp. californicum
      ssp. flaccidum (Greene) Dempster &
          Stebbins
             SY=G. c. var. crebrifolium Nutt.
             SY=G. californicum var. typicum
                Hiland & Howell
             SY=G. f. Greene
             SY=G. occidentale McClatchie
      ssp. luciense Dempster & Stebbins
      ssp. martimum Dempster & Stebbins
      ssp. miguelense (Greene) Dempster &
          Stebbins
             SY=G. c. var. m. (Greene) Jepson
             SY=G. m. Greene
      ssp. primum Dempster & Stebbins
      ssp. sierrae Dempster & Stebbins
    catalinense Gray
      ssp. acrispum Dempster & Stebbins
      ssp. catalinense
    circaezans Michx.
        var. circaezans
            SY=G. bermudense L. pro parte
            SY=G. boreale sensu Walt.
            SY=G. brachiatum sensu Muhl.
            SY=G. circaeoides Roemer &
               Schultes
            SY=G. circaezans var. glabellum
               Britt.
            SY=G. circaezans var. glabrum
               Britt.
            SY=G. c. var. typicum Fern.
            SY=G. rotundifolium var.
               circaezans (Michx.) Kuntze
        var. hypomalacum Fern.
    clementis Eastw.
    cliftonsmithii (Dempster) Dempster &
      Stebbins
          SY=G. nuttallii Gray var. c.
            Dempster
    collomae J.T. Howell
    coloradoense W. Wight
    concinnum Torr. & Gray
    correllii Dempster
    desereticum Dempster & Ehrend.
    emeryense Dempster & Ehrend.
      ssp. emeryense
          SY=G. coloradoense W. Wight var.
            scabriusculum (Ehrend.)
            Dempster & Ehrend.
          SY=G. hypotrichium ssp. s. Ehrend.
          SY=G. s. (Ehrend.) Dempster &
            Ehrend.
      ssp. protoscabriusculum (Dempster &
        Ehrend.) Dempster & Ehrend.
          SY=G. scabriusculum (Ehrend.)
            Dempster & Ehrend. ssp. p.
            Dempster & Ehrend.
    fendleri Gray
    filipes Rydb.
    glabrescens (Ehrend.) Dempster & Ehrend.
      ssp. glabrescens
          SY=G. grayanum Ehrend. ssp.
            glabrescens Ehrend.
      ssp. harticum Dempster & Ehrend.
      ssp. josephinense Dempster & Ehrend.
      ssp. modocense Dempster & Ehrend.
    glaucum L.
          SY=Asperula galioides Bieb.
          SY=A. glauca (L.) Bess.
    grande McClatchie
          SY=G. pubens Gray var. g.
            (McClatchie) Jepson
    grayanum Ehrend.
        var. grayanum
        var. nanum Dempster & Ehrend.

    hallii Munz & Johnston
    hardhamiae Dempster
    hilendiae Dempster & Ehrend.
      ssp. carneum (Hilend & Howell) Dempster
        & Ehrend.
          SY=G. munzii Hilend & Howell var.
            c. Hilend & Howell
      ssp. hilendiae
      ssp. kingstonense (Dempster) Dempster &
        Ehrend.
          SY=G. munzii Hilend & Howell var.
            k. Dempster
    hispidulum Gray
          SY=Bataprine hispidula (Michx.)
            Nieuwl.
          SY=G. carolinianum Dietr.
          SY=G. hispidum Pursh
          SY=G. peregrina (sensu Walt.)
            B.S.P.
          SY=Rubia brownei Michx. non Br.
          SY=R. p. sensu Walt.
          SY=R. walteri DC.
          SY=Valantia hypocarpia L. non Br.
    humifusum Bieb.
          SY=Asperula h. (Bieb.) Bess.
    hypotrichium Gray
      ssp. ebbettsense Dempster & Ehrend.
      ssp. hypotrichium
      ssp. inyoense Dempster & Ehrend.
      ssp. nevadense Dempster & Ehrend.
      ssp. subalpinum (Hilend & Howell)
        Ehrend.
          SY=G. munzii Hilend & Howell var.
            s. Hilend & Howell
      ssp. tomentellum Ehrend.
          SY=G. h. var. t. (Ehrend.)
            Dempster
    jepsonii Hilend & Howell
          SY=G. angustifolium Nutt. var.
            subglabrum Jepson
    johnstonii Dempster & Stebbins
          SY=G. angustifolium Nutt. var.
            pinetorum Munz & Johnston
    kamtschaticum Steller ex Schultes &
      Schultes
    labradoricum (Wieg.) Wieg.
          SY=G. tinctorium L. var. l. Wieg.
    laevipes Opiz
          SY=G. cruciata (L.) Scop.
    lanceolatum Torr.
          SY=?G. circaezans Michx. var. l.
            (Torr.) Torr. & Gray
          SY=G. rotundifolium L. var. l.
            (Torr.) Kuntze
          SY=G. torreyi Bigelow
    latifolium Michx.
          SY=G. l. var. hispidifolium Small
          SY=G. l. var. hispidum Small
    magnifolium (Dempster) Dempster
          SY=G. matthewsii Gray var.
            magnifolium Dempster
    matthewsii Gray
    mexicanum H.B.K.
      ssp. asperrimum (Gray) Dempster
          SY=G. a. Gray
      ssp. asperulum (Gray) Dempster
          SY=G. asperrimum var. asperula
            Gray
          SY=G. asperulum (Gray) Rydb.
          SY=G. m. var. asperulum (Gray)
            Dempster
      ssp. flexicum Dempster
      ssp. mexicanum
    microphyllum Gray
          SY=Relbunium m. (Gray) Hemsl.
    mollugo L.
          SY=G. erectum Huds.

Galium (CONT.)
              SY=G. m. ssp. e. (Huds.) Briq.
              SY=G. m. var. e. (Huds.) Domin
    multiflorum Kellogg
              SY=G. matthewsii Gray var.
                 scabridum Jepson
              SY=G. bloomeri Gray
              SY=G. b. var. hirsutum Gray
              SY=G. multiflorum var. h. (Gray)
                 Jepson
    munzii Hilend & Howell
       ssp. ambivalens Dempster & Ehrend.
       ssp. munzii
    murale (L.) All.
    muricatum W. Wight
              SY=G. chartaceum Wight
    nuttallii Gray
       ssp. insulare Ferris
       ssp. nuttallii
              SY=G. suffruticosum Nutt.
    obtusum Bigelow
       ssp. australe Puff
       ssp. filifolium (Wieg.) Puff
              SY=G. f. (Wieg.) Small
              SY=G. o. var. f. (Wieg.) Fern.
       ssp. obtusum
              SY=G. o. var. ramosum Gleason
              SY=G. trifidum L. var. latifolium
                 Torr.
    odoratum (L.) Scop.
              SY=Asperula o. L.
    oreganum Britt.
              SY=G. kamtschaticum Steller ex
                 Schultes & Schultes var. o.
                 (Britt.) Piper
    palustre L.
    parishii Hilend & Howell
              SY=G. multiflorum Kellogg var.
                 parvifolium Parish
              SY=G. parvifolium (Parish) Jepson
    parisiense L.
       var. leiocarpum Tausch
              SY=G. anglicum Huds.
              SY=G. a. var. parvifolium (Gaud.
                 ex Roemer & Schultes) DC.
              SY=G. divaricatum Lam.
              SY=G. parisiense ssp. anglicum
                 (Huds.) Gaud.
              SY=G. parisiense ssp. parvifolium
                 (Gaud. ex Roemer & Schultes)
                 Gaud.
              SY=G. parisiense var. a. (Huds.)
                 G. Beck
              SY G. parisiense var. divaricatum
                 (Lam.) Davis
              SY=G. parvifolium Gaud. ex Roemer
                 & Schultes
       var. parisiense
              SY=G. litigiosum DC.
              SY=G. l. var. nanum DC.
              SY=G. p. ssp. trichocarpum Tausch
              SY=G. p. var. typicum G. Beck
    pedemontanum (Bellardi) All.
              SY=Cruciata p. (Bellardi) Ehrend.
              SY=Vaillantia p. Bellardi
    pilosum Ait.
       var. laevicaule Weatherby & Blake
       var. pilosum
              SY=G. bermudense L. pro parte
              SY=G. punticulosum Michx. var.
                 pilosum (Ait.) DC.
       var. puncticulosum (Michx.) Torr. &
          Gray
              SY=G. punctatum Pers.
              SY=G. puncticulosum Michx.
              SY=G. purpureum Walt. non L.
              SY=G. walteri J.F. Gmel.

    X pomeranicum Retz. [album X verum]
    porrigens Dempster
       var. porrigens
              SY=G. nuttallii Gray ssp.
                 ovalifolium (Dempster)
                 Dempster & Stebbins
              SY=G. n. var. o. Dempster
       var. tenue (Dempster) Dempster
              SY=G. n. ssp. tenue (Dempster)
                 Dempster & Stebbins
              SY=G. n. var. t. Dempster
    proliferum Gray
              SY=G. p. var. subnudum Greenm.
              SY=G. virgatum Nutt. var. diffusum
                 Gray
    rubioides L.
    saxatile L.
    serpenticum Dempster
       ssp. dayense Dempster & Ehrend.
       ssp. malheurense Dempster & Ehrend.
       ssp. okanoganense Dempster & Ehrend.
       ssp. puberulum (Piper) Dempster &
          Ehrend.
              SY=G. multiflorum Kellogg ssp. p.
                 Piper
              SY=G. m. var. p. (Piper) St. John
              SY=G. watsonii (Gray) Heller ssp.
                 p. (Piper) Ehrend.
       ssp. scabridum (Ehrend.) Dempster &
          Ehrend.
       ssp. scotticum Dempster & Ehrend.
       ssp. serpenticum
       ssp. warnerense Dempster & Ehrend.
       ssp. wenatchicum Dempster & Ehrend.
    sparsiflorum W. Wight
       ssp. glabrius Dempster & Stebbins
       ssp. sparsiflorum
    spurium L.
              SY=G. aparine L. ssp. s. (L.)
                 Simonkai
    stellatum Kellogg
       ssp. eremicum (Hilend & Howell) Ehrend.
              SY=G. s. var. e. Hilend & Howell
       ssp. stellatum
    sylvaticum L.
    texense Gray
              SY=G. californicum Hook. & Arn.
                 var. texanum Torr. & Gray
              SY=G. texanum (Torr. & Gray)
                 Wieg., non Scheele
    tinctorium L.
              SY=G. claytonii Michx.
              SY=G. t. ssp. floridanum (Wieg.)
                 Puff
              SY=G. t. var. diversifolium W.
                 Wight
              SY=G. t. var. f. Wieg.
              SY=G. trifidum L. ssp. tinctorium
                 (L.) Hara
              SY=G. trifidum var. tinctorium
                 (L.) Torr. & Gray
    tricornutum Dandy
              SY=G. tricorne Stokes pro parte
    trifidum L.
       ssp. columbianum (Rydb.) Hulten
              SY=G. columbianum Rydb.
              SY=G. cymosum Wieg.
              SY=G. t. ssp. pacificum (Wieg.)
                 Piper
              SY=G. t. var. p. Wieg.
       ssp. hallophilum (Fern. & Wieg.) Puff
              SY=G. t. var. h. Fern. & Wieg.
       ssp. subbiflorum (Wieg.) Puff
              SY=G. brandegei Gray pro parte
              SY=G. claytonii Michx. var. s.
                 (Wieg.) Wieg.
              SY=G. s. (Wieg.) Rydb.

Galium (CONT.)
            SY=G. tinctorium L. var. s.
                (Wieg.) Fern.
            SY=G. trifidum var. pusillum Gray
            SY=G. trifidum var. s. Wieg.
    ssp. trifidum
            SY=G. brandegei Gray pro parte
    triflorum Michx.
            SY=G. brachiatum Pursh
            SY=G. pennsylvanicum Bart.
            SY=G. t. var. asprelliforme Fern.
            SY=G. t. var. viridiflorum DC.
    uliginosum L.
    uncinulatum Gray
            SY=G. u. var. obstipum (Schlecht.)
                S. Wats.
    uniflorum Michx.
            SY=Bataprine u. (Michx.) Nieuwl.
    verum L.
    virgatum Nutt.
            SY=G. texanum Scheele
            SY=G. v. var. leiocarpum Torr. &
                Gray
    watsonii (Gray) Heller
            SY=G. hypotrichium Gray ssp.
                utahense Ehrend.
            SY=G. multiflorum Kellogg var. w.
                Gray
    wirtgenii F.W. Schultlz
            SY=G. verum L. ssp. w. (F.W.
                Schultlz) Oborny
    wrightii Gray
            SY=G. frankliniense Correll
            SY=G. rothrockii Gray
            SY=G. w. var. r. (Gray) Ehrend. ex
                Ferris

Gardenia Ellis

    brighamii Mann
    jasminoides Ellis
    mannii St. John & Kuykend.
            var. honoluluensis St. John &
                Kuykend.
            var. mannii
    remyi Mann
    weissichii St. John

Genipa L.

    americana L.
    clusiifolia (Jacq.) Griseb.
            SY=Casasia c. (Jacq.) Urban

Geophila D. Don

    repens (L.) I.M. Johnston
            SY=G. herbacea (Jacq.) Schum.

Gonzalagunia Ruiz & Pavon

    spicata (Lam.) Maza
            SY=Duggena hirsuta (Jacq.) Britt.

Gouldia Gray

    hillebrandii Fosberg
            var. hawaiiensis Fosberg
            var. hillebrandii
            var. nodosa Fosberg
    hirtella (Gray) Hbd.
    st.-johnii Fosberg
            var. munroi Fosberg
            var. st.-johnii
    terminalis (Hook. & Arn.) Hbd.
            var. angustifolia Fosberg
            var. antiqua Fosberg

            var. arborescens (Wawra) Fosberg
            var. aspera Fosberg
            var. bobeoides Fosberg
            var. congesta Fosberg
            var. cordata (Wawra) Fosberg
            var. coriacea (Hook. & Arn.) Fosberg
            var. crassicaulis Fosberg
            var. degeneri Fosberg
            var. elongata (Heller) Fosberg
            var. forbesii Fosberg
            var. glabra Fosberg
            var. hathewayi Fosberg
            var. hosakae Fosberg
            var. kaala Fosberg
            var. kapuaensis Fosberg
            var. konaensis Fosberg
            var. lanai Fosberg
            var. macrocarpa (Hbd.) Fosberg
            var. macrothyrsa Fosberg
            var. myrsinoidea Fosberg
            var. osteocarpa Fosberg
            var. ovata (Wawra) Fosberg
            var. parvifolia (Wawra) Fosberg
            var. pedunculata Fosberg
            var. pseudodichotoma Fosberg
            var. pubescens Fosberg
            var. pubistipula Fosberg
            var. purpurea Fosberg
            var. quadrangularis Fosberg
            var. rotundifolia Fosberg
            var. sclerotica Fosberg
            var. skottsbergii Fosberg
            var. stipulacea (Wawra) Fosberg
            var. subcordata Fosberg
            var. tenuicaulis Fosberg
            var. terminalis
            var. wawrana Fosberg

Guettarda L.

    elliptica Sw.
    krugii Urban
    laevis Urban
    ovalifolia Urban
    parviflora Vahl
    pungens Urban
    scabra (L.) Vent.
    valenzuelana A. Rich.

Hamelia Jacq.

    axillaris Sw.
    patens Jacq.
            SY=H. erecta Jacq.

Hedyotis L.

    acerosa Gray
            var. acerosa
                SY=Houstonia a. (Gray) Gray
            var. bigelovii (Greenm.) W.H. Lewis
            var. polypremoides (Gray) W.H. Lewis
                SY=Hedyotis p. (Gray) Shinners
                SY=Houstonia acerosa (Gray) Gray
                    ssp. p. (Gray) Terrell
                SY=Houstonia p. Gray
    acuminata (Cham. & Schlecht.) Steud.
    angusta Fosberg
            var. angusta
            var. koolauensis Fosberg
            var. umbrosa Fosberg
    australis W.H. Lewis & D.M. Moore
            SY=Hedyotis crassifolia Raf. var.
                micrantha Shinners
    boscii DC.
            SY=Oldenlandia b. (DC.) Chapman
    caerulea (L.) Hook.

Hedyotis (CONT.)
                    SY=Houstonia c. L.
                    SY=Houstonia c. var. faxonorum
                        Pease & Moore
    callitrichoides (Griseb.) W.H. Lewis
                    SY=Oldenlandia c. Griseb.
    centranthoides (Hook. & Arn.) Steud.
            var. centranthoides
            var. laevis (Wawra) Fosberg
    cookiana (Cham. & Schlecht.) Steud.
    coriacea Sm.
    correllii W.H. Lewis
    corymbosa (L.) Lam.
                    SY=Oldenlandia c. L.
    crassifolia Raf.
                    SY=Hedyotis caerulea (L.) Hook.
                        var. minima (Beck) Fosberg
                    SY=Hedyotis caerulea var. minor
                        (Michx.) Torr. & Gray
                    SY=Houstonia minima Beck
                    SY=Houstonia patens Ell.
                    SY=Houstonia pusilla Schoepf
    croftiae (Britt. & Rusby) Shinners
    degeneri Fosberg
            var. coprosmifolia Fosberg
            var. degeneri
    elatior (Mann) Fosberg
            var. elatior
            var. ensiformis (Fosberg) Fosberg
            var. herbacea (Levl.) Fosberg
    fluviatilis (Forbes) Fosberg
            var. fluviatilis
            var. hathewayi Fosberg
            var. kamapuaana (Deg.) Fosberg
            var. kauaiensis Fosberg
    foggiana Fosberg
    foliosa (Hbd.) Fosberg
    formosa (Hbd.) Fosberg
    glaucifolia (Gray) Fosberg
            var. glaucifolia
            var. helleri Fosberg
            var. subimpressa Fosberg
            var. waimeae (Wawra) Fosberg
    greenei (Gray) W.H. Lewis
                    SY=Oldenlandia g. Gray
    greenmanii Fosberg ex Shinners
                    SY=Houstonia parvifolia Holz.
    herbacea L.
                    SY=Oldenlandia h. (L.) DC.
    humifusa Gray
                    SY=Houstonia h. (Gray) Gray
    intricata Fosberg
                    SY=Houstonia fasciculata Gray
    knudsenii (Hbd.) Fosberg
    lancifolia (K. Schum.) DC.
                    SY=Oldenlandia l. K. Schum.
    littoralis (Hbd.) Fosberg
    longifolia (Gaertn.) Hook.
                    SY=Hedyotis purpurea (L.) Torr. &
                        Gray var. l. (Gaertn.)
                        Fosberg
                    SY=Houstonia canadensis Willd.
                    SY=Houstonia c. var. ciliolata
                        (Torr.) Boivin
                    SY=Houstonia canadensis var.
                        muscii Boivin
                    SY=Houstonia canadensis var.
                        soperi Boivin
                    SY=Houstonia l. Gaertn.
                    SY=Houstonia setiscaphia L.G. Carr
    mannii Fosberg
            var. cuspidata Fosberg
            var. mannii
            var. munroi Fosberg
            var. scaposa Fosberg
    michauxii Fosberg
                    SY=Houstonia serpyllifolia Michx.

molokaiensis Fosberg
nigricans (Lam.) Fosberg
            var. filifolia (Chapman) Shinners
                    SY=Houstonia f. (Chapman) Small
            var. nigricans
                    SY=Houstonia angustifolia Michx.
                    SY=Houstonia n. (Lam.) Fern.
                    SY=Houstonia pulvinata Small
                    SY=Houstonia salina Heller
            var. parviflora (Gray) W.H. Lewis
            var. rigidiuscula (Gray) Shinners
                    SY=Houstonia r. (Gray) Woot. &
                        Standl.
    nuttalliana Fosberg
                    SY=Houstonia purpurea (L.) Torr. &
                        Gray var. tenuifolia (Nutt.)
                        Fosberg
                    SY=Houstonia t. Nutt.
    ouachitana E.B. Sm.
    parvula (Gray) Fosberg
    procumbens (Walt. ex J.F. Gmel.) Fosberg
            var. hirsuta W.H. Lewis
            var. procumbens
                    SY=Houstonia p. (Walt. ex J.F.
                        Gmel.) Standl.
    purpurea (L.) Torr. & Gray
            var. calycosa (Gray) Fosberg
                    SY=Hedyotis lanceolata Poir.
                    SY=Houstonia l. (Poir.) Britt.
                    SY=Houstonia p. L. var. c. Gray
            var. ciliolata (Torr.) Fosberg
                    SY=Houstonia c. Torr.
            var. montana (Small) Fosberg
                    SY=Houstonia m. Small
            var. purpurea
                    SY=Houstonia p. L.
                    SY=Houstonia purpurea var.
                        pubescens Britt.
    pygmaea Roemer & Schultes
                    SY=Hedyotis wrightii (Gray)
                        Fosberg
                    SY=Houstonia w. Gray
    remyi (Hbd.) Fosberg
            var. nuttallii (Fosberg) Fosberg
                    SY=Hedyotis schlechtendahliana
                        Steud. var. n. Fosberg
            var. plana (Fosberg) Fosberg
                    SY=Hedyotis schlechtendahliana
                        Steud. var. p Fosberg
            var. remyi
                    SY=Hedyotis schlechtendahliana
                        Steud. var. r. (Hbd.)
                        Fosberg
            var. silvicola Fosberg
    rigida Fosberg
            var. rigida
            var. tenuifolia Fosberg
    rosea Raf.
                    SY=Hedyotis taylorae Fosberg
                    SY=Houstonia pygmaea Muell. &
                        Muell.
    rubra (Cav.) Gray
                    SY=Houstonia r. Cav.
    schlechtendahliana Steud.
        ssp. rockii Fosberg
            var. nitens (Wawra) Fosberg
            var. opaca (Wawra) Fosberg
            var. rockii Fosberg
        ssp. schlechtendahliana
            var. glauca (Meyen) Fosberg
            var. schlechtendahliana
            var. secundiflora (Hbd.) Fosberg
        ssp. tenuifolia Fosberg
            var. reticulata Fosberg
            var. tenuifolia Fosberg
    st.-johnii Stone & Lane
    subviscosa (Wright ex Gray) Shinners

Hedyotis (CONT.)
        SY=Oldenlandia s. Wright ex Gray
    thyrsoidea Fosberg
            var. hillebrandii Fosberg
            var. thyrsoidea
    uniflora (L.) Lam.
            var. fasciculata (Bertol.) W.H. Lewis
                SY=H. f. (Bertol.) Small
                SY=Oldenlandia f. (Bertol.) Small
            var. uniflora
                SY=Oldenlandia u. L.

Hemidiodia K. Schum.

    ocymifolia (Willd. ex Roemer & Schultes) K.
        Schum.

Hillia Jacq.

    parasitica Jacq.

Hydrophylax L.f.

    maritima L.f.

Ixora L.

    coccinea L.
    ferrea (Jacq.) Benth.
    grandiflora (Blume) Zoll. & Morr.

Kadua Cham. & Schlecht.

    grandis Gray

Kelloggia Torr. ex Benth. & Hook. f.

    galioides Torr.

Lasianthus Jack

    lanceolatus (Griseb.) Maza
            SY=L. moralesii (Griseb.) C.
                Wright

Lucya DC.

    tetranda (L.) K. Schum.
            SY=Clavenna t. (L.) Standl.

Machaonia Humb. & Bonpl.

    portoricensis Baill.

Mitchella L.

    repens L.

Mitracarpus Zucc.

    breviflorus Gray
    maxwelliae Britt. & Wilson
    polycladus Urban
    portoricensis Urban
    villosus (Sw.) DC.
            SY=M. hirtus (L.) DC.

Morinda L.

    citrifolia L.
    oleifera Lam.
    royoc L.
    sandwicensis Deg.
    trimera Hbd.

Nertera Banks & Soland. ex Gaertn.

granadensis (L.f.) Druce
        var. granadensis
        var. insularis Skottsberg

Paederia L.

    foetida L.

Palicourea Aubl.

    alpina (Sw.) DC.
    barbinervia DC.
    crocea (Sw.) Roemer & Schultes
            SY=P. brevithyrsa Britt. & Standl.
            SY=P. riparia Benth.
    domingensis (Jacq.) DC.

Pentodon Hochst.

    pentandrus (Schum. & Thon.) Vatke
            SY=P. halei (Torr. & Gray) Gray

Phialanthus Griseb.

    grandifolius Alain
    myrtilloides Griseb.

Pinckneya Michx.

    pubens Michx.
            SY=P. bracteata (Bartr.) Raf.

Psychotria L.

    berteroana DC.
    brachiata Sw.
    brownei Spreng.
    fauriei (Levl.) Fosberg
    grandiflora Mann
    grandis Sw.
    guadalupensis (DC.) Howard
            SY=P. grosourdyana (Baill.) Urban
    hathewayi Fosberg
            var. brevipetiolata Fosberg
            var. hathewayi
    hawaiiensis (Gray) Fosberg
            var. glabrithyrsa Fosberg
            var. glomerata (Rock) Fosberg
            var. hawaiiensis
            var. hillebrandii (Rock) Fosberg
            var. molokaiensis (Rock) Fosberg
            var. rotundifolia (Skottsberg)
                Fosberg
            var. scoriacea (Rock) Fosberg
    hexandra Mann
        ssp. hexandra
            var. hexandra Deg. & Fosberg
            var. hirta Wawra
            var. kealiae Fosberg
        ssp. oahuensis Deg. & Fosberg
            var. hosakana Fosberg
            var. oahuensis Deg. & Fosberg
            var. rockii Fosberg
            var. st.-johnii Fosberg
    hobdyi Sohmer
    insularum Gray
            var. paradisii Fosberg
    involucrata A. Rich.
    kaduana (Cham. & Schlecht.) Fosberg
            var. kaduana
                SY=P. leptocarpa (Hbd.) Fosberg
            var. pubiflora (Heller) Fosberg
    ligustrifolia (Northrop) Millsp.
            SY=P. bahamensis Millsp.
    longissima (Rock) St. John
            SY=P. kaduana (Cham. & Schlecht.)
                Fosberg var. 1. (Rock)

Psychotria (CONT.)
              Fosberg
  maleolens Urban
  maricaensis Urban
  mariniana (Cham. & Schlecht.) Fosberg
  mauiensis Fosberg
          var. mauiensis
          var. subcordata (Rock) St. John
  microdon (DC.) Urban
          SY=P. pinularis Sesse & Moc.
  nervosa Sw.
          SY=P. undata Jacq.
  nutans Sw.
  patens Sw.
  portoricensis DC.
  psychotrioides (Heller) Fosberg
          SY=P. greenwelliae Fosberg
  pubescens Sw.
  punctata Vatke
  revoluta DC.
  sulzneri Small
  tenuifolia Sw.
  uliginosa Sw.
  waianensis Fosberg
  wawrae Sohmer

Randia L.

  aculeata L.
          var. aculeata
          var. mitis (L.) Griseb.
            SY=R. m. L.
  formosa (Jacq.) K. Schum.
  portoricensis (Urban) Britt. & Standl.
  rhagocarpa Standl.

Richardia L.

  brasiliensis Gomez
  grandiflora (Cham. & Schlecht.) Steud.
  humistrata (Cham. & Schlecht.) Steud.
  scabra L.
  tricocca (Torr. & Gray) Standl.
          SY=Crusea t. (Torr. & Gray) Heller
          SY=Diodia t. Torr. & Gray

Rondeletia L.

  inermis (Spreng.) Krug & Urban
  pilosa Sw.
  portoricensis Krug & Urban

Rubia L.

  tinctoria L.

Sabicea Aubl.

  cinerea Aubl.
  hirsuta H.B.K.

Schradera Willd.

  vahlii Steyermark
          SY=Urceolaria exotica J.F. Gmel.

Scolosanthus Vahl

  densiflorus Urban
  multiflorus (Sw.) Krug & Urban
          SY=S. grandifolius Krug & Urban
  versicolor Vahl

Sherardia L.

  arvensis L.
  orientalis Boiss. & Hohen.

Spermacoce L.

  confusa Rendle
  glabra Michx.
  tenuior L.
          var. floridana (Urban) R.W. Long
            SY=S. f. Urban
            SY=S. keyensis Small
          var. tenuior
            SY=S. riparia Cham. & Schlecht.
  tetraquetra A. Rich.

Strumpfia Jacq.

  maritima Jacq.

Terebraria Kuntze

  resinosa (Vahl) Sprague
          SY=Laugeria r. Vahl

RUTACEAE

Amyris P. Br.

  balsamifera L.
  elemifera L.
  madrensis S. Wats.
  texana (Buckl.) P. Wilson

Choisya H.B.K.

  arizonica Standl.
  dumosa (Torr.) Gray
  mollis Standl.

Citrus L.

  aurantifolia (Christm.) Swingle
  aurantium L.
  grandis (L.) Osbeck
          SY=C. maxima (Burm.) Merr.
  limon (L.) Burm. f.
  X limonia Osbeck [limon X reticulata]
          SY=C. limonum Risso
  medica L.
  paradisi Macfad.
  reticulata Blanco
  sinensis (L.) Osbeck

Cneoridium Hook. f.

  dumosum (Nutt.) Hook. f.

Esenbeckia H.B.K.

  berlandieri Baill.
          SY=E. runyonii Morton

Fortunella Swingle

  japonica (Thunb.) Swingle
  margarita (Lour.) Swingle

Glycosmis Correa

  parviflora (Sims) Little
          SY=G. citrifolia (Willd.) Lindl.

Helietta Tul.

  parvifolia (Gray) Benth.

Murraya Koenig ex L.

Murraya (CONT.)
   paniculata (L.) Jack
        SY=Chalcas exotica (L.) Millsp.
        SY=C. p. L.

Pelea Gray

  adscendens St. John & Hume
  anapanapaensis St. John
  anisata Mann
     var. anisata
     var. haupuana Stone
  balloui Rock
       SY=P. mannii Hbd. pro parte
  barbigera (Gray) Hbd.
  christophersenii St. John
  cinerea (Gray) Hbd.
     var. cinerea
     var. mauiana Stone
     var. skottsbergii Stone
  cinereops St. John & Hume
  clusiifolia Gray
   ssp. clusiifolia
     var. clusiifolia
     var. crassiloba Stone
     var. cuneata St. John & Hume
     var. ecuneata St. John
     var. fauriei (Levl.) St. John & Hume
     var. minor St. John
     var. pickeringii (St. John) Stone
       SY=P. p. St. John
   ssp. cookeana (Rock) Stone
   ssp. dumosa (Rock) Stone
     var. auriculifolia (Gray) Stone
     var. dumosa (Rock) Stone
   ssp. sapotifolia (Mann) Stone
  cruciata Heller
  degeneri Stone
  descendens St. John
  elliptica (Gray) Hbd.
     var. elliptica
     var. mauiensis St. John
  feddei Levl.
  gayana Rock
  glabra St. John
  grandifolia (Hbd.) St. John & Hume
     var. grandifolia
     var. hualalaiensis (St. John) Stone
     var. lianoides (Rock) Stone
     var. montana (Rock) Stone
     var. ovalifolia (Hbd.) St. John
     var. terminalis (Rock) Stone
  haleakalae Stone
  haupuensis St. John
  hawaiensis Wawra
     var. brighamii (St. John) Stone
     var. gaudichaudii (St. John) Stone
     var. hawaiensis
     var. molokaiana Stone
     var. pilosa St. John
     var. racemiflora (Rock) St. John
     var. remyana Stone
     var. rubra (Rock) Stone
     var. sulfurea (Rock) Stone
  hiiakae Stone
  honoluluensis St. John
  hosakae St. John
  kaalaensis St. John
  kauaensis St. John
  kavaiensis Mann
  kipahuluensis St. John
  knudsenii Hbd.
  lakae Stone
  lanceolata St. John & Hume
  leveillei Faurie ex Levl.
  lohiauana Stone
  lydgatei Hbd.

  macropus Hbd.
  makahae Stone
  molokaiensis Hbd.
       SY=P. stellata St. John
  mucronulata St. John
  multiflora Rock
  munroi St. John
  nealae Stone
  oahuensis Levl.
       SY=P. lucens (Hbd.) St. John
  oblongifolia Gray
     var. manukaensis (St. John) Stone
       SY=P. m. St. John
  obovata St. John
  olowaluensis St. John
       SY=P. mannii Hbd. pro parte
  orbicularis Hbd.
     var. orbicularis
     var. tonsa St. John & Hume
  ovalis St. John
  ovata St. John & Hume
  pallida Hbd.
  paniculata St. John
  parvifolia Hbd.
     var. apoda (St. John) Stone
       SY=P. a. St. John
     var. parvifolia
     var. sessilis (Levl.) Stone
  peduncularis Levl.
     var. cordata Stone
     var. niuensis (St. John) Stone
       SY=P. n. St. John
     var. nummularia Stone
     var. paloloensis (St. John) Stone
       SY=P. paloloensis St. John
     var. pauciflora (St. John) Stone
     var. peduncularis
     var. quadrata Stone
     var. ternifolia (Stone) Stone
  pluvialis St. John
  pseudoanisata Rock
     var. oblanceolata (St. John) Stone
     var. pseudoanisata
  puauluensis St. John
  puberula St. John
  quadrangularis St. John & Hume
  radiata St. John
  recurvata Rock
  reflexa St. John
  rotundifolia Gray
  sandwicensis (Hook. & Arn.) Gray
  semiternata St. John
  st.-johnii Hume
     var. elongata (Hbd.) Stone
       SY=P. e. (Hbd.) St. John
     var. st.-johnii
  storeyana St. John & Hume
  tomentosa St. John & Hume
  volcanica Gray
     var. kohalae Stone
     var. volcanica
  wahiawaensis St. John & Hume
  waialealae Wawra
     var. latior St. John & Hume
     var. pubescens Skottsberg
     var. waialealae
  waimeaensis St. John
  wawraeana Rock
     var. pubens St. John
     var. tenuifolia St. John & Hume
     var. wawraeana
       SY=P. waipioensis St. John
  zahlbruckneri Rock

Phellodendron Rupr.

  amurense Rupr.

Phellodendron (CONT.)
    japonicum Maxim.

Pilocarpus Vahl

    racemosus L.

Platydesma Mann

    cornuta Hbd.
            var. cornuta
            var. decurrens Stone
    remyi (Sherff) Deg., Deg., Sherff & Stone
    rostrata Hbd.
    spathulata (Gray) Stone
            var. pallida (Hbd.) Stone
            var. pubescens (Skottsberg) Stone
            var. spathulata

Poncirus Raf.

    trifoliata (L.) Raf.
            SY=Citrus t. L.

Ptelea L.

    crenulata Greene
    trifoliata L.
        ssp. angustifolia (Benth.) V. Bailey
            var. angustifolia (Benth.) V. Bailey
                SY=P. a. Benth.
            var. persicifolia (Greene) V. Bailey
        ssp. pallida (Greene) V. Bailey
            var. cognata (Greene) Kearney &
                Peebles
                SY=P. angustifolia (Benth.) V.
                    Bailey var. c. (Greene)
                    Kearney & Peebles
                SY=P. t. ssp. a. (Benth.) V.
                    Bailey var. c. (Greene)
                    Kearney & Peebles
            var. confinis (Greene) V. Bailey
            var. lutescens (Greene) V. Bailey
            var. pallida (Greene) V. Bailey
                SY=P. p. Greene
        ssp. polyadenia (Greene) V. Bailey
                SY=P. monticola Greene
        ssp. trifoliata
            var. mollis Torr. & Gray
                SY=P. tomentosa Raf.
            var. trifoliata
                SY=P. baldwinii Torr. & Gray
                SY=P. microcarpa Small
                SY=P. serrata Small
                SY=P. trifoliata var. deamiana
                    Nieuwl.

Ravenia Vell.

    urbanii Engl.

Ruta L.

    chalepensis L.
    graveolens L.

Thamnosma Torr. & Frem.

    montana Torr. & Frem.
    texana (Gray) Torr.

Triphasia Lour.

    trifolia (Burm. f.) P. Wilson

Zanthoxylum L.

    americanum P. Mill.
    bifoliolatum Leonard
    bluettianum Rock
    caribaeum Lam.
    clava-herculis L.
            SY=Z. macrophyllum Nutt.
    coriaceum A. Rich.
    dipetalum Mann
            var. degeneri (Skottsberg) St. John
            var. dipetalum
            var. geminicarpum Rock
            var. hillebrandii (Sherff) St. John
                SY=Z. glandulosum Hbd. non Raf.
            var. mannii (Sherff) St. John
            var. tomentosum Rock
    fagara (L.) Sarg.
            SY=Z. flavum Vahl
    hawaiiense Hbd.
            var. citriodorum Rock
            var. hawaiiense
            var. velutinosum Rock
    hirsutum Buckl.
            SY=Z. clava-herculis L. var.
                fruticosum (Gray) S. Wats.
    kauaense Gray
            var. kauaense
            var. kohalanum (Sherff) St. John
            var. kohuanum (Skottsberg) St. John
            var. tenuifolium (Deg. ex Sherff) St.
            John
    martinicense (Lam.) DC.
    maviense Mann
            var. anceps Rock
            var. cranwelliae (Skottsberg) St.
            John
            var. kaalanum (Sherff) St. John
            var. maviense
            var. rigidum Rock
    monophyllum (Lam.) P. Wilson
    oahuense Hbd.
    parvum Shinners
    punctatum Vahl
    semiarticulatum St. John & Hosaka
            var. semiarticulatum
            var. sessile (Deg.) Fosberg
    skottsbergii (Deg. & Deg.) St. John
    spinifex (Jacq.) DC.
    thomasianum (Krug & Urban) Krug & Urban

SABIACEAE

Meliosma Blume

    herbertii Rolfe
    obstusifolia (Bello) Krug & Urban

SALICACEAE

Populus L.

    X acuminata Rydb. [angustifolia X
        deltoides]
            SY=P. acuminata Rydb. var. rheeri
                Sarg.
    alba L.
            SY=P. a. var. bolleana Lauche
            SY=P. a. var. pyramidalis Bunge
    angustifolia James
            SY=P. balsamifera L. var. a.
                (James) S. Wats.
            SY=P. canadensis Moench var. a.

Populus (CONT.)
                    (James) Wesmael
          SY=P. fortissima A. Nels. & J.F.
             Macbr.
          SY=P. salicifolia Raf.
          SY=P. X sennii Boivin
          SY=P. tweedyi Britt.
    balsamifera L.
       ssp. balsamifera
          SY=P. b. var. candicans (Ait.)
             Gray
          SY=P. b. var. fernaldiana Rouleau
          SY=P. b. var. lanceolata Marsh.
          SY=P. b. var. michauxii (Dode) A.
             Henry
          SY=P. b. var. subcordata Hyl.
          SY=P. c. Ait.
          SY=P. michauxii Dode
          SY=P. tacamahaca P. Mill.
          SY=P. t. var. c. (Ait.) Stout
          SY=P. t. var. l. (Marsh.) Farw.
          SY=P. t. var. m. (Dode) Farw.
       ssp. trichocarpa (Torr. & Gray) Brayshaw
          SY=P. b. var. californica S. Wats.
          SY=P. hastata Dode
          SY=P. t. Torr. & Gray
          SY=P. t. ssp. h. (Dode) Dode
          SY=P. t. var. cupulata S. Wats.
          SY=P. t. var. h. (Dode) A. Henry
          SY=P. t. var. ingrata (Jepson)
             Jepson
    X berolinensis Dippel [laurifolia X nigra]
    X brayshawii Boivin [angustifolia X
       balsamifera]
    X canadensis Moench [deltoides X nigra]
          SY=P. c. var. eugenei
             (Simon-Louis) Schelle
          SY=P. X e. Simon-Louis
    X canescens (Ait.) Sm. [alba X tremula]
          SY=P. alba L. var. c. Ait.
    deltoides Bartr. ex Marsh.
       ssp. deltoides
          SY=P. angulata Ait.
          SY=P. a. var. missouriensis A.
             Henry
          SY=P. balsamifera L. var. m. (A.
             Henry) Rehd.
          SY=?P. b. var. pilosa Sarg.
          SY=P. b. var. virginiana (Foug.)
             Sarg.
          SY=P. canadensis Moench var. v.
             (Foug.) Fiori
          SY=P. d. var. a. (Ait.) Sarg.
          SY=P. d. var. m. (A. Henry) A.
             Henry
          SY=P. d. var. p. (Sarg.) Sudworth
          SY=P. d. var. v. (Foug.) Sudworth
          SY=P. nigra L. var. v. (Foug.)
             Castigl.
          SY=P. palmeri Sarg.
          SY=P. v. Foug.
          SY=P. v. var. pilosa (Sarg.) F.C.
             Gates
       ssp. monilifera (Ait.) Eckenwalder
          SY=Monilistus m. (Ait.) Raf. ex
             Jackson
          SY=P. besseyana Dode
          SY=P. d. var. occidentalis Rydb.
          SY=P. m. Ait.
          SY=P. o. (Rydb.) Britt. ex Rydb.
          SY=P. sargentii Dode
          SY=P. s. var. texana (Sarg.)
             Correll
          SY=P. t. Sarg.
       ssp. wislizenii (S. Wats.) Eckenwalder
          SY=P. d. var. w. S. Wats.
          SY=P. w. (S. Wats.) Sarg.

    fremontii S. Wats.
       ssp. fremontii
          SY=P. arizonica Sarg.
          SY=P. a. var. jonesii Sarg.
          SY=P. canadensis Moench var. f.
             (S. Wats.) Kuntze
          SY=P. f. var. a. (Sarg.) Jepson
          SY=P. f. var. macdougalii (Rose)
             Jepson
          SY=P. f. var. macrodisca Sarg.
          SY=P. f. var. pubescens Sarg.
          SY=P. f. var. thornberi Sarg.
          SY=P. f. var. toumeyi Sarg.
          SY=P. macdougalii Rose
       ssp. mesetae Eckenwalder
          SY=P. f. var. m. (Eckenwalder)
             Little
          SY=P. mexicana sensu auctt. non
             Wesmael
    grandidentata Michx.
          SY=P. g. var. angustata Victorin
          SY=P. g. var. meridionalis
             Tidestrom
          SY=P. g. var. subcordata Victorin
    X heimbergeri Boivin [alba X tremuloides]
    heterophylla L.
          SY=P. argentea Michx. f.
    X hinckleyana Correll [angustifolia X
       fremontii]
    X jackii Sarg. [balsamifera X deltoides]
          SY=P. X andrewsii Sarg.
          SY=P. X bernardii Boivin
          SY=P. X dutillyi Lepage
          SY=P. X gileadensis Rouleau
          SY=P. manitobensis Dode
    nigra L.
          SY=P. dilatata Ait.
          SY=P. italica Moench
          SY=P. n. var. i. Du Roi
    X parryi Sarg. [balsamifera X fremontii]
    X rouleauiana Boivin [alba X grandidentata]
    simonii Carr.
    X smithii Boivin [grandidentata X
       tremuloides]
          SY=P. X barnesii W.H. Wagner
    tremula L.
    tremuloides Michx.
          SY=P. aurea Tidestrom
          SY=P. cercidiphylla Britt.
          SY=P. X polygonifolia Bernard
          SY=P. tremula L. ssp. tremuloides
             (Michx.) Love & Love
          SY=P. tremuloides var. a.
             (Tidestrom) Daniels
          SY=P. tremuloides var. c. (Britt.)
             Sudworth
          SY=P. tremuloides var. intermedia
             Victorin
          SY=P. tremuloides var. magnifica
             Victorin
          SY=P. tremuloides var. rhomboidea
             Victorin
          SY=P. tremuloides var.
             vancouveriana (Trel.) Sarg.
          SY=P. v. Trel.

Salix L.

    alaxensis (Anderss.) Coville
       var. alaxensis
          SY=S. a. var. obovalifolia Ball
          SY=S. speciosa Hook. & Arn., non
             Host
          SY=S. s. var. a. Anderss.
       var. longistylis (Rydb.) Schneid.
          SY=S. a. ssp. l. (Rydb.) Hulten
          SY=S. l. Rydb.

Salix (CONT.)
  alba L.
              SY=S. a. ssp. vitellina (L.)
                 Arcang.
              SY=S. a. var. argentea Wimmer
              SY=S. a. var. sericea Gaud.
              SY=S. a. var. v. (L.) Stokes
              SY=S. v. L.
  X amoena Fern. [glauca X planifolia]
  amygdaloides Anderss.
              SY=S. a. var. wrightii (Anderss.)
                 Schneid.
              SY=S. nigra Marsh. var. a.
                 (Anderss.) Anderss.
              SY=S. w. Anderss.
  arbusculoides Anderss.
              SY=S. humillima Anderss.
              SY=S. h. var. puberula (Anderss.)
                 Anderss.
              SY=S. saskatchevana von Seem.
  arctica Pallas
              SY=S. anglorum Cham.
              SY=S. anglorum var. antiplasta
                 Schneid.
              SY=S. anglorum var. araioclada
                 Schneid.
              SY=S. anglorum var. kophophylla
                 Schneid.
              SY=S. arctica R. Br. ex Richards.
              SY=S. arctica ssp. crassijulis
                 (Trautv.) Skvort.
              SY=S. arctica ssp. petraea
                 (Anderss.) Love, Love &
                 Kapoor
              SY=S. arctica ssp. tortulosa
                 (Trautv.) Hulten
              SY=S. arctica var. antiplasta
                 (Schneid.) Fern.
              SY=S. arctica var. araioclada
                 (Schneid.) Raup
              SY=S. arctica var. brownei
                 Anderss.
              SY=S. arctica var. caespitosa
                 (Kennedy) L. Kelso
              SY=S. arctica var. graminifolia
                 (E.H. Kelso) L. Kelso
              SY=S. arctica var. kophophylla
                 (Schneid.) Polunin
              SY=S. arctica var. pallasii
                 (Anderss.) Kurtz
              SY=S. arctica var. petraea
                 Anderss.
              SY=S. arctica var. petrophila
                 (Rydb.) L. Kelso
              SY=S. arctica var. t. (Trautv.)
                 Raup
              SY=S. b. (Anderss.) Bebb
              SY=S. b. var. petraea (Anderss.)
                 Bebb
              SY=S. caespitosa Kennedy
              SY=S. crassijulis Trautv.
              SY=S. hudsonensis Schneid.
              SY=S. pallasii Anderss.
              SY=S. pallasii var. crassijulis
                 (Trautv.) Anderss.
              SY=S. petrophila Rydb.
              SY=S. petrophila var. caespitosa
                 (Kennedy) Schneid.
              SY=S. t. Trautv.
  arctophila Cockerell ex Heller
              SY=S. arctica Liebm. non Pallas
              SY=S. arctophila var. lejocarpa
                 (Lange) Schneid.
              SY=S. groenlandica Lundstr., non
                 Heer
              SY=S. g. var. l. Lange
              SY=S. waghornei Rydb.

  X argusii Boivin [brachycarpa X candida]
  argyrocarpa Anderss.
              SY=S. a. var. denudata Anderss.
              SY=S. labradorica Rydb. var.
                 pumila Nutt.
  arizonica Dorn
  athabascensis Raup
              SY=S. fallax Raup
              SY=S. pedicellaris Pursh var. a.
                 (Raup) Boivin
  babylonica L.
  ballii Dorn
              SY=S. myrtillifolia Anderss. var.
                 brachypoda Fern.
  barclayi Anderss.
              SY=S. b. var. angustifolia
                 (Anderss.) Anderss. ex
                 Schneid.
              SY=S. b. var. conjuncta (Bebb)
                 Ball ex Schneid.
              SY=S. c. Bebb
              SY=S. hoyeriana Dieck
              SY=S. pyrolaefolia Anderss. var.
                 h. (Dieck) Dippel
              SY=S. regelii Anderss.
              SY=S. rotundifolia Nutt., non
                 Trautv.
  barrattiana Hook.
              SY=S. albertana Rowlee
              SY=S. b. var. angustifolia
                 Anderss.
              SY=S. b. var. latifolia Anderss.
              SY=S. b. var. marcescens Raup
  bebbiana Sarg.
     var. bebbiana
              SY=S. b. var. depilis Raup
              SY=S. b. var. luxurians (Fern.)
                 Fern.
              SY=S. b. var. perrostrata (Rydb.)
                 Schneid.
              SY=S. depressa L. ssp. rostrata
                 (Richards.) Hiitonen
              SY=S. livida Wahlenb. var.
                 occidentalis (Anderss.) Gray
              SY=S. livida var. r. (Richards.)
                 Dippel
              SY=S. p. Rydb.
              SY=S. r. Richards. non Thuill.
              SY=S. r. var. luxurians Fern.
              SY=S. r. var. p. (Rydb.) Fern.
              SY=S. starkeana Willd. ssp. b.
                 (Sarg.) Youngberg
              SY=S. vagans Hook. f. ex Anderss.
                 ssp. cinerascens (Wahlenb.)
                 Anderss. var. o. Anderss.
              SY=S. v. var. r. (Richards.)
                 Anderss.
     var. capreifolia (Fern.) Fern.
              SY=S. rostrata Richards. non
                 Thuill. var. c. Fern.
     var. projecta (Fern.) Schneid.
              SY=S. rostrata Richards. non
                 Thuill. var. p. Fern.
  X beschelii Boivin [bebbiana X discolor]
  X blanda Anderss. [babylonica X fragilis]
              SY=S. X elegantissima K. Koch
  bonplandiana H.B.K.
              SY=S. b. var. toumeyi (Britt. &
                 Shafer) Schneid.
              SY=S. t. Britt. & Shafer
  boothii Dorn
              SY=S. curtiflora sensu auctt. non
                 Anderss.
              SY=S. myrtillifolia sensu auctt.
                 non Anderss.
              SY=S. m. var. c. sensu auctt. non
                 (Anderss.) Bebb ex Rose

Salix (CONT.)
        SY=S. novae-angliae sensu auctt.
            non Anderss.
        SY=S. n-a. ssp. pseudomyrsinites
            (Anderss.) Anderss. var.
            aequalis Anderss.
        SY=S. pseudocordata sensu auctt.
            non (Anderss.) Rydb.
        SY=S. pseudocordata var. a.
            (Anderss.) Ball ex Schneid.
        SY=S. pseudomyrsinites sensu
            auctt. non Anderss.
        SY=S. pseudomyrsinites Anderss.
            var. aequalis (Anderss.)
            Anderss. ex Ball
    brachycarpa Nutt.
        ssp. brachycarpa
            var. brachycarpa
                SY=S. b. var. alticola E.H. Kelso
                SY=S. b. var. antimima (Schneid.)
                    Raup
                SY=S. b. var. glabellicarpa
                    Schneid.
                SY=S. b. var. sansonii Ball
                SY=S. chlorolepis Fern. var.
                    antimima Schneid.
                SY=S. desertorum Richards. var.
                    fruticulosa Anderss.
                SY=S. d. var. stricta Anderss.
                SY=S. stricta (Anderss.) Rydb.
            var. fullertonensis (Schneid.) Argus
                SY=S. b. ssp. f. (Schneid.) Love &
                    Love
                SY=S. f. Schneid.
                SY=S. niphoclada (Rydb.) Argus
                    var. f. (Schneid.) Raup
            var. psammophila Raup
        ssp. niphoclada (Rydb.) Argus
            SY=S. b. var. mexiae Ball
            SY=S. glauca L. var. n. (Rydb.)
                Wiggins
            SY=S. lingulata Anderss.
            SY=S. muriei Hulten
            SY=S. n. Rydb.
            SY=S. n. var. mexiae (Ball) Hulten
            SY=S. n. var. muriei (Hulten) Raup
    X brachypurpurea Boivin [brachycarpa X
        turnorii]
    breweri Bebb
    candida Flugge ex Willd.
        SY=S. c. var. denudata Anderss.
        SY=S. c. var. tomentosa Anderss.
        SY=S. candidula Nieuwl.
    caprea L.
    caroliniana Michx.
        SY=S. amphibia Small
        SY=S. harbisonii Schneid.
        SY=S. longipes Shuttlw. ex
            Anderss.
        SY=S. l. var. pubescens Anderss.
        SY=S. l. var. venulosa (Anderss.)
            Schneid.
        SY=S. l. var. wardii (Bebb)
            Schneid.
        SY=S. marginata Wimmer ex Anderss.
        SY=S. nigra Marsh. var. l.
            (Shuttlw. ex Anderss.) Bebb
        SY=S. n. var. m. (Wimmer ex
            Anderss.) Anderss.
        SY=S. n. var. w. Bebb
        SY=S. occidentalis Bosc ex
            Anderss.
        SY=S. o. var. l. (Shuttlw. ex
            Anderss.) Bebb
        SY=S. w. (Bebb) Bebb
    cascadensis Cockerell
        var. cascadensis

        SY=S. brownii (Anderss.) Bebb var.
            tenera (Anderss.) M.E. Jones
        SY=S. tenera Anderss.
    var. thompsonii Brayshaw
chamissonis Anderss.
chlorolepis Fern.
cinera L.
X clarkei Bebb [candida X petiolaris]
coactilis Fern.
commutata Bebb
        SY=S. barclayi Anderss. var. c.
            (Bebb) L. Kelso
        SY=S. c. var. denudata Bebb
        SY=S. c. var. puberula Bebb
        SY=S. c. var. mixta Piper
        SY=S. c. var. sericea Bebb
cordata Michx.
        SY=S. adenophylla Hook.
        SY=S. c. var. abrasa Fern.
        SY=S. syrticola Fern.
X cryptodonta Fern. [bebbiana X candida]
X delnortensis Schneid. [lasiolepis X
    sitchensis]
        SY=S. breweri Bebb var. d.
            (Schneid.) Jepson
X dieckiana Suksdorf [?geyeriana X
    pedicellaris]
discolor Muhl.
        SY=S. ancorifera Fern.
        SY=?S. conformis Forbes
        SY=S. crassa Barratt
        SY=S. d. var. latifolia Anderss.
        SY=S. d. var. overi Ball
        SY=?S. d. var. prinoides (Pursh)
            Anderss.
        SY=S. fuscata Pursh
        SY=S. p. Pursh
        SY=S. sensitiva Barratt
        SY=S. squamata Rydb.
drummondiana Barratt ex Hook.
        SY=S. bella Piper
        SY=S. chlorophylla Fern. var.
            pellita (Anderss.) Anderss.
            pro parte
        SY=S. covillei Eastw.
        SY=S. d. var. b. (Piper) Ball
        SY=S. d. var. subcoerulea (Piper)
            Ball
        SY=S. pachnophora Rydb.
        SY=S. s. Piper
X dutillyi Lepage [argyrocarpa X
    pedicellaris]
eastwoodiae Cockerell ex Heller
        SY=S. californica Bebb
        SY=S. commutata Bebb var.
            rubicunda Jepson
elaeagnos Scop.
eriocephala Michx.
        SY=S. cordata Michx. var.
            missouriensis (Bebb)
            Mackenzie & Bush
        SY=S. discolor Muhl. var. e.
            (Michx.) Anderss.
        SY=S. m. Bebb
        SY=S. rigida Muhl. var. vestita
            Anderss.
exigua Nutt.
        SY=S. argophylla Nutt.
        SY=S. e. ssp. interior (Rowlee)
            Cronq.
        SY=S. e. ssp. i. var. pedicellata
            (Anderss.) Cronq.
        SY=S. e. var. luteosericea (Rydb.)
            Schneid.
        SY=S. e. var. nevadensis (S.
            Wats.) Schneid.
        SY=S. e. var. stenophylla (Rydb.)

Salix (CONT.)
      Schneid.
    SY=S. e. var. virens Rowlee
    SY=S. fluviatilis Nutt. var. a.
      (Nutt.) Sarg.
    SY=S. f. var. e. (Nutt.) Sarg.
    SY=S. f. var. sericans (Nees)
      Boivin
    SY=S. hindsiana Benth. var.
      tenuifolia Anderss.
    SY=S. interior Rowlee
    SY=S. i. var. angustissima
      (Anderss.) Dayton
    SY=S. i. var. exterior Fern.
    SY=S. i. var. l. (Rydb.) Schneid.
    SY=S. i. var. p. (Anderss.) Ball
    SY=S. i. var. wheeleri Rowlee
    SY=S. linearifolia Rydb.
    SY=S. longifolia Muhl. non Lam.
    SY=S. longifolia var. angustissima
      Anderss.
    SY=S. longifolia var. argophylla
      (Nutt.) Anderss.
    SY=S. longifolia var. exigua
      (Nutt.) Bebb
    SY=S. longifolia var. i. (Rowlee)
      M.E. Jones
    SY=S. longifolia var. opaca
      Anderss.
    SY=S. longifolia var. pedicellata
      Anderss.
    SY=S. longifolia var. wheeleri
      (Rowlee) Schneid.
    SY=S. luteosericea Rydb.
    SY=S. macrostachya Nutt.
    SY=S. malacophylla Nutt. ex Ball
    SY=S. nevadensis S. Wats.
    SY=S. rubra Richards. non Huds.
    SY=S. stenophylla Rydb.
    SY=S. thurberi Rowlee
    SY=S. wheeleri (Rowlee) Rydb.
farriae Ball
    SY=S. f. var. microserrulata Ball
    SY=S. hastata L. var. f. (Ball)
      Hulten
floridana Chapman
    SY=S. astatulana Murrill & Palmer
    SY=S. chapmanii Small
fluviatilis Nutt.
fragilis L.
fuscescens Anderss.
    SY=S. arbutifolia sensu auctt.
    SY=S. f. var. reducta Ball
X gaspeensis Schneid. [brachycarpa X
  chlorolepis]
geyeriana Anderss.
    var. argentea (Bebb) Schneid.
      SY=S. macrocarpa Nutt. var. a.
        Bebb
    var. geyeriana
      SY=S. macrocarpa Nutt.
    var. meleina J.K. Henry
      SY=S. m. (J.K. Henry) G.N. Jones
X glatfelteri Schneid. [amygdaloides X
  nigra]
glauca L.
    ssp. callicarpaea (Trautv.) Bocher
      SY=?S. alpestris Anderss. var.
        americana Anderss.
      SY=S. anamesa Schneid.
      SY=S. atra Rydb.
      SY=?S. barclayi Anderss. var.
        latiuscula Anderss.
      SY=S. callicarpaea Trautv.
      SY=S. cordifolia Pursh
      SY=S. cordifolia ssp. callicarpaea
        (Trautv.) A. Love

      SY=S. cordifolia var. callicarpaea
        (Trautv.) Fern.
      SY=S. cordifolia var. eucycla
        Fern.
      SY=S. cordifolia var. intonsa
        Fern.
      SY=S. cordifolia var. macounii
        (Rydb.) Schneid.
      SY=S. cordifolia var. tonsa Fern.
      SY=S. g. var. m. (Rydb.) Boivin
      SY=S. g. var. stenolepis Polunin
      SY=S. labradorica Rydb.
      SY=S. m. Rydb.
      SY=S. rydbergii Heller
      SY=S. vacciniformis Rydb.
    ssp. glauca
      var. acutifolia (Hook.) Schneid.
        SY=S. g. ssp. a. (Hook.) Hulten
        SY=S. g. var. alicea Ball
        SY=S. g. var. perstipula Raup
        SY=S. g. var. poliophylla
          (Schneid.) Raup
        SY=S. g. var. seemannii (Rydb.)
          Ostenf.
        SY=S. s. Rydb.
        SY=S. villosa Hook. var.
          acutifolia Hook.
      var. glauca
        SY=?S. kobayashii Kimura
      var. villosa (Hook.) Anderss.
        SY=?S. arctica Pallas var.
          subcordata (Anderss.)
          Schneid.
        SY=S. desertorum Richards.
        SY=S. d. var. elata Anderss.
        SY=S. glauca ssp. d. (Richards.)
          Hulten
        SY=S. g. ssp. d. var. sericea
          Hulten
        SY=S. g. ssp. glabrescens
          (Anderss.) Hulten
        SY=S. glauca var. glabrescens
          (Anderss.) Schneid.
        SY=S. glauca var. kenosha (L.
          Kelso) L. Kelso
        SY=S. glauca var. pseudolapponum
          (von Seem.) L. Kelso
        SY=S. glauca var. subincurva (E.H.
          Kelso) L. Kelso
        SY=S. glauca var. villosa (Hook.)
          Anderss.
        SY=S. glaucops Anderss.
        SY=S. glaucops var. glabrescens
          Anderss.
        SY=S. glaucops var. villosa
          (Hook.) Anderss.
        SY=S. nudescens Rydb.
        SY=S. pseudolapponum von Seem.
        SY=S. p. var. k. L. Kelso
        SY=S. p. var. subincurva E.H.
          Kelso
        SY=?S. subcordata Anderss.
        SY=S. villosa Hook.
        SY=S. wolfii Bebb ex Rothrock var.
          p. (von Seem.) M.E. Jones
        SY=S. wyomingensis Rydb.
gooddingii Ball
    SY=S. g. var. variabilis Ball
    SY=S. nigra Marsh. var. vallicola
      Dudley
    SY=S. vallicola (Dudley) Britt. &
      Shafer
X grayi Schneid. [argyrocarpa X planifolia]
X hankensonii Dode [alba var. vitellina X
  nigra]
hastata L.
    SY=S. farriae Ball var. walpolei

Salix (CONT.)

                    Coville & Ball
          SY=S. h. var. subintegrifolia
                    Flod.
          SY=S. w. (Coville & Ball) Ball
herbacea L.
hindsiana Benth.
          SY=S. exigua Nutt. var. parishiana
                    (Rowlee) Jepson
          SY=S. h. var. p. (Rowlee) Ball
          SY=S. p. Rowlee
          SY=S. sessilifolia Nutt. var. h.
                    (Benth.) Anderss.
hookeriana Barratt ex Hook.
          SY=S. amplifolia Coville
          SY=S. h. var. laurifolia J.K.
                    Henry
          SY=S. h. var. tomentosa J.K. Henry
                    ex Schneid.
humboldtiana Willd.
          SY=S. chilensis Molina
          SY=S. h. var. fastigiata Andre
humilis Marsh.
     var. humilis
          SY=S. conifera Muhl.
          SY=S. h. var. angustifolia
                    Anderss.
          SY=S. h. var. hyporhysa Fern.
          SY=S. h. var. keweenawensis Farw.
          SY=S. h. var. rigidiuscula Robins
                    & Fern.
          SY=S. muhlenbergiana Willd.
          SY=?S. recurvata Pursh
     var. microphylla (Anderss.) Fern.
          SY=S. alpina Walt.
          SY=S. h. var. tristis (Ait.)
                    Griggs
          SY=S. longirostris Michx.
          SY=S. t. Ait.
          SY=S. t. var. m. Anderss.
incana Schrank
irrorata Anderss.
X jamesensis Lepage [pedicellaris X
     pellita]
jejuna Fern.
jepsonii Schneid.
          SY=S. covillei Eastw.
          SY=S. sitchensis Sanson ex Bong.
                    var. angustifolia Bebb
          SY=S. s. var. ralphiana Jepson
X jesupii Fern. [alba X lucida]
laevigata Bebb
          SY=S. congesta (Rockroth) T.J.
                    Howell
          SY=S. l. var. angustifolia Bebb ex
                    Rothrock
          SY=S. l. var. araquipa (Jepson)
                    Ball
          SY=S. l. var. c. Bebb ex Rothrock
lanata L.
     ssp. calcicola (Fern. & Wieg.) Hulten
          SY=S. c. Fern. & Wieg.
          SY=S. c. var. glandulosior Boivin
          SY=S. c. var. nicholsiana Polunin
          SY=S. richardsonii Hook. var.
                    macouniana Bebb
     ssp. richardsonii (Hook.) Skvort.
          SY=S. r. Hook.
          SY=S. r. var. mckeandii Polunin
lasiandra Benth.
     var. caudata (Nutt.) Sudworth
          SY=S. c. (Nutt.) Heller
          SY=S. c. var. bryantiana Ball &
                    Bracelin
          SY=S. c. var. parvifolia Ball
          SY=S. fendleriana Anderss.
          SY=S. lasiandra var. f. (Anderss.)

                    Bebb
          SY=S. pentandra L. var. c. Nutt.
     var. lasiandra
          SY=S. arguta Anderss.
          SY=S. a. var. erythrocoma Anderss.
          SY=S. a. var. l. (Benth.) Anderss.
          SY=S. a. var. pallescens
                    (Anderss.) Anderss.
          SY=S. lancifolia Anderss.
          SY=S. lasiandra var. abramsii Ball
          SY=S. lasiandra var. macrophylla
                    (Anderss.) Little
          SY=S. lasiandra var. lancifolia
                    (Anderss.) Bebb
          SY=S. lasiandra var. lyallii Sarg.
          SY=S. lasiandra var. recomponens
                    Raup
          SY=S. lucida Muhl. var. m.
                    Anderss.
          SY=S. lyallii (Sarg.) Heller
          SY=S. p. Anderss.
          SY=S. speciosa Nutt., non Host
lasiolepis Benth.
          SY=S. bakeri von Seem. ex C.F.
                    Baker
          SY=S. bigelovii Torr.
          SY=S. boiseana A. Nels.
          SY=S. franciscana von Seem. ex
                    C.F. Baker
          SY=S. l. var. bakeri (von Seem. ex
                    C.F. Baker) Ball
          SY=S. l. var. bigelovii (Torr.)
                    Bebb
          SY=S. l. var. bracelinae Ball
          SY=S. l. var. sandbergii (Rydb.)
                    Ball
          SY=S. s. Rydb.
          SY=S. suksdorfii Gandog.
X laurentiana Fern. [bebbiana or discolor X
     myricoides]
          SY=S. paraleuca Fern.
lemmonii Bebb
          SY=S. austinae Bebb
          SY=S. l. var. a. (Bebb) Schneid.
          SY=S. l. var. macrostachya Bebb
          SY=S. l. var. melanopsis Bebb
          SY=S. l. var. sphaerostachya Bebb
ligulifolia (Ball) Ball ex Schneid.
          SY=S. cordata var. l. (Ball) L.
                    Kelso
          SY=S. lutea Nutt. var. ligulifolia
                    Ball
lucida Muhl.
          SY=S. arguta Anderss. var.
                    hirtisquama Anderss.
          SY=S. l. var. angustifolia
                    Anderss.
          SY=S. l. var. intonsa Fern.
          SY=S. l. var. latifolia Anderss.
          SY=S. lucida var. ovatifolia
                    Anderss.
          SY=S. pentandra L. var. lucida
                    (Muhl.) Kuntze
lutea Nutt.
          SY=S. cordata Michx. ssp. rigida
                    Muhl. var. vitellina
                    Anderss.
          SY=S. c. var. denveriana L. Kelso
          SY=S. c. var. desolata (F.H.
                    Kelso) L. Kelso
          SY=S. c. var. l. (Nutt.) Bebb
          SY=S. c. var. platyphylla (Ball)
                    L. Kelso
          SY=S. c. var. watsonii Bebb
          SY=S. flava Rydb., non Schoepf
          SY=S. l. var. desolate F.H. Kelso
          SY=S. l. var. famelica Ball

Salix (CONT.)
                    SY=S. l. var. nivaria Jepson
                    SY=S. l. var. p. Ball
                    SY=S. l. var. w. (Bebb) Jepson
                    SY=S. ormsbyensis von Seem.
                    SY=S. rigida Muhl. var. w. (Bebb)
                        Cronq.
                    SY=S. w. (Bebb) Rydb.
    maccalliana Rowlee
    melanopsis Nutt.
                    SY=S. bolanderiana Rowlee
                    SY=S. exigua Nutt. ssp. m. (Nutt.)
                        Cronq.
                    SY=S. e. ssp. m. var. gracilipes
                        (Ball) Cronq.
                    SY=S. e. var. tenerrima
                        (Henderson) Schneid.
                    SY=S. fluviatilis sensu auctt.
                    SY=S. f. var. t. (Henderson) T.J.
                        Howell
                    SY=S. longifolia Muhl. var. t.
                        Henderson
                    SY=S. m. var. b. (Rowlee) Schneid.
                    SY=S. m. var. g. Ball
                    SY=S. m. var. kronkheittii L.
                        Kelso
                    SY=S. m. var. t. (Henderson) Ball
                    SY=S. t. (Henderson) Heller
    monochroma Ball
    monticola Bebb
                    SY=S. amelanchieroides L. Kelso
                    SY=S. barclayi Anderss. var.
                        cochetopiana L. Kelso
                    SY=S. b. var. neomexicana (E.H.
                        Kelso) L. Kelso
                    SY=S. b. var. padophylla (Rydb.)
                        L. Kelso
                    SY=S. b. var. resurrectionis L.
                        Kelso
                    SY=S. b. var. uncompahgre L. Kelso
                    SY=S. b. var. veritomonticola L.
                        Kelso
                    SY=S. cordata Michx. var.
                        crux-aureae L. Kelso
                    SY=S. cordata var. m. (Bebb) L.
                        Kelso
                    SY=S. dissymetrica L. Kelso
                    SY=S. monticola var. n. E.H. Kelso
                    SY=S. padifolia Rydb., non
                        Anderss.
                    SY=S. padophylla Rydb.
                    SY=S. pseudomonticola Ball var.
                        padophylla (Rydb.) Ball
                    SY=S. sawatchicola L. Kelso
    myricoides Muhl.
            var. albovestita (Ball) Dorn
                    SY=S. glaucophylla Bebb var. a.
                        Ball
                    SY=S. glaucophylloides Fern. var.
                        a. (Ball) Fern.
            var. myricoides
                    SY=S. cordata Muhl. var.
                        glaucophylla Bebb
                    SY=S. c. var. myricoides (Muhl.)
                        Carey
                    SY=S. g. (Bebb) Bebb non Bess.,
                        non Anderss.
                    SY=S. g. var. angustifolia Bebb
                    SY=S. g. var. brevifolia Bebb
                    SY=S. g. var. latifolia Bebb
                    SY=S. glaucophylloides Fern.
                    SY=S. glaucophylloides var.
                        brevifolia (Bebb) Ball ex
                        E.G. Voss
                    SY=S. glaucophylloides var.
                        glaucophylla (Bebb) Schneid.
                    SY=S. X myricoides (Muhl.) Carey

    myrsinifolia Salisb.
    myrtillifolia Anderss.
            var. cordata (Anderss.) Dorn
                    SY=S. novae-angliae sensu auctt.
                        non Anderss.
                    SY=S. n-a. Anderss. ssp.
                        pseudo-myrsinites (Anderss.)
                        Anderss. var. c. Anderss.
                    SY=S. pseudocordata (Anderss.)
                        Rydb. var. c. (Anderss.)
                        Ball
                    SY=S. m. var. pseudo-myrsinites
                        (Anderss.) Ball ex Hulten
            var. myrtillifolia
                    SY=S. curtiflora Anderss.
                    SY=S. m. var. c. (Anderss.) Bebb
                        ex Rose
                    SY=S. m. var. lingulata (Anderss.)
                        Ball
                    SY=S. novae-angliae Anderss. ssp.
                        m. (Anderss.) Anderss.
                    SY=S. n-a. ssp. pseudocordata
                        Anderss.
                    SY=S. n-a. ssp. pseudomyrsinites
                        (Anderss.) Anderss.
                    SY=S. n-a. ssp. pseudomyrsinites
                        var. l. Anderss.
                    SY=S. pseudocordata (Anderss.)
                        Rydb.
                    SY=S. pseudo-myrsinites Anderss.
    nigra Marsh.
                    SY=?S. ambigua Pursh
                    SY=?S. denudata Raf.
                    SY=?S. dubia Trautv.
                    SY=S. falcata Pursh
                    SY=S. flavo-virens Hornem.
                    SY=S. houstoniana Pursh
                    SY=S. ligustrina Michx. f.
                    SY=S. n. var. altissima Sarg.
                    SY=S. n. var. falcata (Pursh)
                        Torr.
                    SY=S. n. var. lindheimeri Schneid.
                    SY=S. purshiana Spreng.
    nummularia Anderss.
                    SY=S. n. ssp. tundricola (Schljak)
                        Love & Love
                    SY=S. t. Schljak
    X obtusata Fern. [myricoides X pyrifolia]
    orestera Schneid.
                    SY=S. glauca L. var. o. (Schneid.)
                        Jepson
                    SY=S. g. ssp. o. (Schneid.)
                        Youngberg
    ovalifolia Trautv.
            var. arctolitoralis (Hulten) Argus
                    SY=S. a. Hulten
            var. cyclophylla (Rydb.) Ball
                    SY=S. c. Rydb.
                    SY=S. rotundata Rydb. ex Macoun
            var. glacialis (Anderss.) Argus
                    SY=S. g. Anderss.
            var. ovalifolia
                    SY=S. flagellaris Hulten
                    SY=S. o. var. camdensis Schneid.
    X paraleuca Fern. [discolor X myricoides]
                    SY=S. stenocarpa Fern.
                    SY=?S. unalaschensis Cham.
    X peasei Fern. [herbacea X uva-ursi]
    pedicellaris Pursh
                    SY=?S. dieckiana Suksdorf
                    SY=S. fuscescens Anderss. var.
                        hebecarpa Fern.
                    SY=S. h. (Fern.) Fern.
                    SY=S. myrtilloides L. var. p.
                        Anderss.
                    SY=S. m. var. hypoglauca (Fern.)
                        Ball

Salix (CONT.)

          SY=S. p. var. hypoglauca Fern.
          SY=S. p. var. tenuescens Fern.
pedunculata Fern.
X pellicolor Lepage [discolor X pellita]
pellita Anderss. ex Schneid.
          SY=S. chlorophylla Anderss. var.
              p. (Anderss. ex Schneid.)
              Anderss. pro parte
          SY=?S. obovata Pursh
          SY=S. p. var. psila Schneid.
          SY=?S. seriocarpa Buser
pentandra L.
petiolaris Sm.
          SY=S. gracilis Anderss.
          SY=S. g. var. rosmarinoides
              Anderss.
          SY=S. g. var. textoris Fern.
          SY=S. p. var. angustifolia
              Anderss.
          SY=S. p. var. g. (Anderss.)
              Anderss.
          SY=S. p. var. r. (Anderss.)
              Schneid.
phlebophylla Anderss.
          SY=S. anglorum Cham.
          SY=S. paleoneura Rydb.
X pilosiuscula (Schneid.) Boivin
piperi Bebb
planifolia Pursh
    ssp. planifolia
        var. monica (Bebb) Schneid.
          SY=S. chlorophylla Anderss. var.
              m. (Bebb) Flod.
          SY=S. m. Bebb
          SY=S. phylicifolia L. var. m.
              (Bebb) Jepson
        var. pennata (Ball) Ball ex Dutilly,
            Lapage & Daman
          SY=S. pennata Ball
          SY=S. phylicifolia L. var. pennata
              (Ball) Cronq.
        var. planifolia
          SY=S. chlorophylla Anderss.
          SY=S. c. var. nelsonii (Ball)
              Flod.
          SY=S. c. var. pychnocarpa
              (Anderss.) Anderss.
          SY=S. n. Ball
          SY=S. phylicifolia L. ssp.
              planifolia (Pursh) Hiitonen
          SY=S. planifolia var. n. Ball ex
              E.C. Sm.
          SY=S. pychnocarpa Anderss.
    ssp. pulchra (Cham.) Argus
          SY=S. abusculoides Anderss. var.
              glabra (Anderss.) Anderss.
              ex Schneid.
          SY=S. barclayi Anderss. ssp.
              hebecarpa Anderss.
          SY=S. fulcrata Anderss. var.
              subglauca Anderss.
          SY=S. humillima Anderss. var. g.
              (Anderss.) Anderss.
          SY=S. phylicifolia L. ssp.
              planifolia (Pursh) Hiitonen
              var. s. (Anderss.) Scoggan
          SY=S. phylicifolia var. s.
              (Anderss.) Boivin
          SY=S. phylicoides Anderss.
          SY=S. planifolia ssp. pulchra var.
              yukonensis (Schneid.) Argus
          SY=S. pulchra Cham.
          SY=S. pulchra var. looffiae Ball
          SY=S. pulchra var. palmeri Ball
          SY=S. pulchra var. y. Schneid.
polaris Wahlenb.

          SY=S. polaris ssp. pseudopolaris
              (Flod.) Hulten
          SY=S. polaris var. glabrata Hulten
          SY=S. polaris var. selwynensis
              Raup
          SY=S. pseudopolaris Flod.
X princeps-ourayi L. Kelso [glauca var.
    villosa X monticola]
prolixa Anderss.
          SY=S. cordata Michx. var.
              mackenzieana Hook.
          SY=S. m. (Hook.) Barratt ex
              Anderss.
          SY=S. m. var. macrogemma Ball
          SY=S. rigida Muhl. var.
              mackenzieana (Hook.) Cronq.
          SY=S. r. var. macrogemma (Ball)
              Cronq.
pseudomonticola Ball
          SY=S. barclayi Anderss. var. p.
              (Ball) L. Kelso
purpurea L.
pyrifolia Anderss.
          SY=S. balsamifera Barratt ex Hook.
          SY=S. b. var. alpestris Bebb
          SY=S. b. var. lanceolata Bebb
          SY=S. b. var. vegeta Bebb
          SY=S. cordata Michx. var. b. Hook.
          SY=S. p. var. l. (Bebb) Fern.
raupii Argus
reticulata L.
    ssp. glabellicarpa Argus
    ssp. nivalis (Hook.) Love, Love & Kapoor
          SY=S. n. Hook.
          SY=S. n. var. saximontana (Rydb.)
              Schneid.
          SY=S. r. var. nana Anderss.
          SY=S. r. var. nivalis (Hook.)
              Anderss.
          SY=S. r. var. s. (Rydb.) L. Kelso
          SY=S. s. Rydb.
          SY=S. venusta Anderss.
          SY=S. vestita Pursh var. nana
              Hook.
    ssp. reticulata
          SY=S. orbicularis Anderss.
          SY=S. r. ssp. o. (Anderss.) Flod.
          SY=S. r. var. gigantifolia Ball
          SY=S. r. var. glabra Trautv.
          SY=S. r. var. o. (Anderss.)
              Komarov
          SY=S. r. var. semicalva Fern.
rigida Muhl.
          SY=S. angustata Pursh
          SY=S. cordata Muhl.
          SY=S. c. ssp. a. (Pursh) Anderss.
          SY=S. c. ssp. r. Anderss.
          SY=S. myricoides Muhl. var. a.
              (Pursh) Dippel
          SY=S. m. var. r. (Muhl.) Dippel
          SY=S. r. var. a. (Pursh) Fern.
          SY=S. torreyana Barratt
rotundifolia Trautv.
    ssp. dodgeana (Rydb.) Argus
          SY=S. d. Rydb.
          SY=S. d. var. subrariflora (L.
              Kelso) L. Kelso
    ssp. rotundifolia
          SY=S. behringica von Seem.
          SY=S. polaris Wahlenb. var.
              leiocarpa Cham.
          SY=S. l. (Cham.) Coville
X rubella Bebb [candida X rigida]
X rubens Schrank [alba X fragilis]
X schneideri Boivin [lucida X nigra]
scouleriana Barratt ex Hook.
          SY=S. brachystachys Benth.

Salix (CONT.)
    SY=S. b. ssp. s. var. crassyulis
       Anderss.
    SY=S. b. var. s. (Hook.) Anderss.
    SY=S. capreoides Anderss.
    SY=S. flavescens Nutt.
    SY=S. f. var. capreoides
       (Anderss.) Bebb
    SY=S. f. var. s. (Barratt ex
       Hook.) Bebb
    SY=S. nuttallii Sarg.
    SY=S. n. var. capreoides
       (Anderss.) Sarg.
    SY=S. s. var. b. (Benth.) M.E.
       Jones
    SY=S. s. var. coetanea Ball
    SY=S. s. var. crassijulis
       (Anderss.) Schneid.
    SY=S. s. var. f. (Nutt.) J.K.
       Henry
    SY=S. s. var. poikila Schneid.
    SY=S. s. var. thompsonii Ball
    SY=S. stagnalis Nutt.
sericea Marsh.
    SY=S. grisea Willd.
    SY=S. petiolaris Sm. var. g.
       (Willd.) Torr.
    SY=S. p. var. s. (Marsh.) Anderss.
serissima (Bailey) Fern.
    SY=S. arguta Anderss. var.
       alpigena Anderss.
    SY=S. lucida Muhl. var. serissima
       Bailey
sessilifolia Nutt.
    SY=S. fluviatilis Nutt. var. s.
       (Nutt.) Scoggan
    SY=S. hindsiana Benth. var.
       leucodendroides (Rowlee)
       Ball
    SY=S. longifolia Muhl. var. s.
       (Nutt.) M.E. Jones
    SY=S. macrostachys Nutt.
    SY=S. m. var. cusickii Rowlee
    SY=S. m. var. leucodendroides
       Rowlee
    SY=S. parksiana Ball
    SY=S. s. var. leucodendroides
       (Rowlee) Schneid.
    SY=S. s. var. vancouverensis
       Brayshaw
    SY=S. s. var. villosa Anderss.
setchelliana Ball
    SY=S. aliena Flod.
silicicola Raup
    SY=S. alaxensis (Anderss.) Coville
       var. s. (Raup) Boivin
X simulans Fern. [pedicellaris X sp.]
sitchensis Sanson ex Bong.
    SY=S. coulteri Anderss.
    SY=S. cuneata Nutt.
    SY=S. pellita Anderss. var.
       angustifolia (Bebb) Boivin
    SY=S. s. var. a. Bebb
    SY=S. s. var. congesta Anderss.
    SY=S. s. var. denudata Anderss.
    SY=S. s. var. p. (Schneid.) Jepson
    SY=S. s. var. ralphiana (Jepson)
       Jepson
X smithiana Willd. [caprea X viminalis]
X solheimii E.H. Kelso [reticulata ssp.
   nivalis X rotundifolia ssp. dodgeana]
sphenophylla Skvort.
    SY=S. cuneata Turcz. non Nutt.
    SY=S. s. ssp. pseudotorulosa
       Skvort.
stolonifera Coville
X subsericea (Anderss.) Schneid.

    [petiolaris X sericea]
    SY=S. neo-forbesii Toepffer
    SY=S. petiolaris Sm. var. s.
       Anderss.
    SY=S. sericea Marsh. var.
       subsericea (Anderss.) Rydb.
taxifolia H.B.K.
    SY=S. microphylla Schlecht. &
       Cham.
    SY=S. t. var. lejocarpa Anderss.
    SY=S. t. var. limitanea I.M.
       Johnston
    SY=S. t. var. m. (Schlecht. &
       Cham.) Schneid.
    SY=S. t. var. seriocarpa Anderss.
tracyi Ball
turnorii Raup
    SY=S. lutea Nutt. var. t. (Raup)
       Boivin
tweedyi (Bebb ex Rose) Ball
    SY=S. barrattiana Hook. var. t.
       Bebb ex Rose
    SY=S. rotundifolia Nutt., non
       Trautv.
tyrrellii Raup
X ungavensis Lepage [ballii X glauca ssp.
   callicarpaea]
uva-ursi Pursh
    SY=S. arbuscula L. var.
       labradorica Anderss.
    SY=S. cutleri Tuckerman
    SY=S. c. var. l. (Anderss.)
       Anderss.
    SY=S. irrigtutiana Lundstr.
    SY=S. myrsinites L. var.
       parvifolia Lange
vestita Pursh
    SY=S. fernaldii Blank.
    SY=S. leiolepis Fern.
    SY=S. reticulata L. var. v.
       (Pursh) Anderss.
    SY=S. v. ssp. l. (Fern.) Argus
    SY=S. v. var. erecta Anderss.
    SY=S. v. var. humilior Anderss.
    SY=S. v. var. psilophylla Fern. &
       St. John
viminalis L.
X waghornei Rydb. [arctica X glauca]
X wiegandii Fern. [candida X lanata ssp.
   calcicola]
wolfii Bebb ex Rothrock
    SY=S. idahoensis (Ball) Rydb.
    SY=S. w. var. i. Ball

SANTALACEAE

Buckleya Torr.

   distichophylla (Nutt.) Torr.

Comandra Nutt.

   umbellata (L.) Nutt.
     ssp. californica (Eastw. ex Rydb.) Piehl
       SY=C. c. Eastw. ex Rydb.
       SY=C. u. var. c. (Eastw. ex Rydb.)
        C.L. Hitchc.
     ssp. pallida (A. DC.) Piehl
       SY=C. p. A. DC.
       SY=C. u. var. angustifolia (A.
        DC.) Torr.
       SY=C. u. var. p. (A. DC.) M.E.
        Jones
     ssp. umbellata

Comandra   (CONT.)
            SY=C. richardsiana Fern.

Exocarpos Labill.

    gaudichaudii A. DC.
    luteolus Forbes
    menziesii Stauffer

Geocaulon Fern.

    lividum (Richards.) Fern.
            SY=Comandra l. Richards.

Nestronia Raf.

    umbellula Raf.

Pyrularia Michx.

    pubera Michx.

Santalum L.

    ellipticum Gaud.
            var. ellipticum
            var. latifolium (Gray) Fosberg
            var. littorale (Hbd.) Skottsberg
            var. luteum (Rock) Deg.
    freycinetianum Gaud.
            var. freycinetianum
            var. longifolium (Meurisse) Deg.
    haleakalae Hbd.
    lanaiense (Rock) Rock
    paniculatum Hook. & Arn.
            var. chartaceum Deg. & Deg.
            var. paniculatum
    pyrularium Gray
            var. pyrularium
            var. sphaerolithos Skottsberg
    salicifolium Meurisse

Thesium L.

    linophyllon L.

SAPINDACEAE

Alectryon Gaertn.

    macrococcum Radlk.
    mahoe St. John & Frederick

Allophylus L.

    crassinervis Radlk.
    racemosus Sw.
            SY=A. occidentalis (Sw.) Radlk.

Cardiospermum L.

    corindum L.
            SY=C. keyense Small
    dissectum (S. Wats.) Radlk.
    grandiflorum Sw.
    halicacabum L.
            SY=C. h. var. microcarpum (H.B.K.)
                Blume
            SY=C. m. H.B.K.

Cupania L.

    americana L.
    glabra Sw.

    triquetra A. Rich.

Dodonaea P. Mill.

    eriocarpa Sm.
            var. amphioxea Deg. & Sherff
            var. confertior Sherff
            var. costulata Deg., Deg. & Sherff
            var. degeneri Sherff
            var. eriocarpa
            var. forbesii Sherff
            var. glabrescens Sherff
            var. hillebrandii Sherff
            var. hosakana Sherff
            var. lanaiensis Sherff
            var. molokaiensis Deg. & Sherff
            var. oblonga Sherff
            var. obtusior Sherff
            var. pallida Deg. & Sherff
            var. sherffii Deg. & Deg.
            var. skottsbergii Sherff
            var. vaccinioides Sherff
            var. varians Deg. & Sherff
            var. waimeana Sherff
    sandwicensis Sherff
            var. latifolia Deg. & Sherff
            var. sandwicensis
            var. simulans Sherff
    stenoptera Hbd.
            var. fauriei (Levl.) Sherff
            var. stenoptera
    viscosa (L.) Jacq.
            var. angustifolia (L. f.) Benth.
            var. arborescens (A. Cunningham ex
                Hook.) Sherff
            SY=D. ehrenbergii Schlecht.
            var. linearis (Harvey & Sonder)
                Sherff
            var. viscosa
            SY=D. jamaicensis DC.
            SY=D. microcarya Small

Exothea Macfad.

    paniculata (Juss.) Radlk.

Hypelate P. Br.

    trifoliata Sw.

Koelreuteria Laxm.

    paniculata Laxm.

Matayba Aubl.

    domingensis (DC.) Radlk.
            SY=M. oppositifolia (A. Rich.)
                Britt.

Melicoccus P. Br.

    bijugatus Jacq.
            SY=M. bijuga L.

Paullinia L.

    fuscescens H.B.K.
    pinnata L.
    plumieri Triana & Planch.

Sapindus L.

    drummondii Hook. & Arn.
            SY=S. saponaria L. var. d. (Hook.
                & Arn.) L. Benson
    lonomea St. John

Sapindus (CONT.)
    marginatus Willd.
    oahuensis Hbd.
    saponaria L.
    thurstonii Rock

Serjania P. Mill.

    brachycarpa Gray
    diversifolia (Jacq.) Radlk.
    incisa Torr.
    polyphylla L.) Radlk.

Talisia Aubl.

    pedicellaris Radlk.

Thouinia Poit.

    portoricensis Radlk.
            SY=Thyana p. (Radlk.) Britt.
    striata Radlk.
            SY=Thyana s. (Radlk.) Britt.

Ungnadia Endl.

    speciosa Endl.

Urvillea H.B.K.

    ulmacea H.B.K.

SAPOTACEAE

Bumelia Sw.

    celastrina H.B.K.
        var. angustifolia (Nutt.) R.W. Long
            SY=B. a. Nutt.
        var. celastrina
    krugii Pierre
            SY=B. obovata (Lam.) A. DC. var.
                k. (Pierre) Cronq.
    lanuginosa (Michx.) Pers.
        var. albicans Sarg.
            SY=B. l. ssp. oblongifolia (Nutt.)
                Cronq. var. a. Sarg.
        var. lanuginosa
        var. oblongifolia (Nutt.) R.B. Clark
            SY=B. l. ssp. o. (Nutt.) Cronq.
        var. rigida Gray
            SY=B. l. ssp. r. (Gray) Cronq.
            SY=B. r. (Gray) Small
        var. texana (Buckl.) Cronq.
            SY=B. l. ssp. rigida (Gray) Cronq.
                var. t. (Buckl.) Cronq.
            SY=B. t. Buckl.
    lycioides (L.) Pers.
            SY=B. cassinifolia Small
            SY=B. l. var. ellipsoidalis R.B.
                Clark
            SY=B. l. var. virginiana Fern.
            SY=B. smallii R.B. Clark
    obovata (Lam.) A. DC.
    reclinata (Michx.) Vent.
        var. reclinata
            SY=B. microcarpa Small
        var. rufotomentosa (Small) Cronq.
            SY=B. rufotomentosa Small
    rufa Raf.
    tenax (L.) Willd.
            SY=B. anomala (Sarg.) R.B. Clark
            SY=B. lacuum Small
            SY=B. megacocca Small

thornei Cronq.

Chrysophyllum L.

    argenteum Jacq.
    bicolor Poir.
            SY=C. eggersii Pierre
    cainito L.
    oliviforme L.
    pauciflorum Lam.

Dipholis A. DC.

    bellonis Urban
    cubensis (Griseb.) Pierre
            SY=D. sintenisiana Pierre
    salicifolia (L.) A. DC.

Manilkara Adans.

    albescens (Griseb.) Cronq.
    bahamensis (Baker) Lam. & Meeuse
            SY=Achras emarginata (L.) Little
            SY=Mimusops e. (L.) Britt.
    balata (Aubl.) Dubard
            SY=Achras bidentata A. DC.
            SY=Manilkara bidentata (A. DC.)
                Chev.
            SY=M. nitida (Sesse & Moc.) Dubard
    jaimiqui (C. Wright ex Griseb.) Dubard
    pleeana (Pierre) Cronq.
            SY=Manilkara duplicata (Sesse &
                Moc.) Dubard
    zapota (L.) van Royen
            SY=Achras z. L.
            SY=Sapota achras P. Mill.

Mastichodendron (Engl.) H.J. Lam

    foetidissimum (Jacq.) H.J. Lam
            SY=Sideroxylon f. Jacq.
            SY=S. portoricense Urban

Micropholis (Griseb.) Pierre

    chrysophylloides Pierre
            SY=M. curvata (Pierre) Urban
    garcinifolia Pierre

Nesoluma Baill.

    polynesicum (Hbd.) Baill.

Paralabatia Pierre

    portoricensis Britt. & Wilson

Pouteria Aubl.

    auahiensis (Rock) Fosberg
            SY=Planchonella a. (Rock)
                Skottsberg
    aurantia (Rock) Fosberg
            SY=Planchonella auahiensis (Rock)
                Skottsberg var. aurantia
                (Rock) Skottsberg
    campechiana (H.B.K.) Baehni
            SY=Lucuma nervosa A. DC.
            SY=L. salicifolia H.B.K.
    ceresolii (Rock) Fosberg
            SY=Planchonella c. (Rock) St. John
    dictyoneura (Griseb.) Radlk.
        var. dictyoneura
        var. fuertesii (Urban) Cronq.
    domingensis (Gaertn. f.) Baehni
    hotteana (Urban & Ekman) Baehni
    multiflora (A. DC.) Eyma

Pouteria (CONT.)
        SY=Lucuma m. A. DC.
  rhynchosperma (Rock) Fosberg
        SY=Planchonella r. (Rock) St. John
  sandwicensis (Gray) Baehni & Deg.
        SY=Planchonella puulupensis Baehni
            & Deg.
        SY=Planchonella s. (Gray) Pierre
  spathulata (Hbd.) Fosberg
        SY=Planchonella s. (Hbd.) Pierre
        SY=Planchonella s. var. densiflora
            (Hbd.) St. John
        SY=Planchonella s. var.
            molokaiensis (Levl.) St.
            John

SARRACENIACEAE

Darlingtonia Torr.

  californica Torr.
        SY=Chrysamphora c. (Torr.) Greene

Sarracenia L.

  X ahlesii Bell & Case [alata X rubra]
  alata Wood
        SY=S. sledgei Macfarlane
  X areolata (Macfarlane) Bell [alata X
    leucophylla]
  X catesbaei (Ell.) Bell [flava X purpurea
    ssp. venosa]
        SY=S. c. Ell.
  X chelsonii Masters [purpurea X rubra]
  X courtii Masters [psittacina X purpurea]
  X excellens (Nichols.) Bell [leucophylla X
    minor]
  X exornata (S.G.) Bell [alata X purpurea]
  flava L.
  X formosa Veitch ex Masters [minor X
    psittacina]
  X gilpinii Bell & Case [psittacina X rubra]
  X harperi Bell [flava X minor]
  leucophylla Raf.
        SY=S. drummondii Croom
  minor Walt.
  X mitchelliana (Nichols.) Bell [leucophylla
    X purpurea]
  X mooreana (Veitch) Bell [flava X
    leucophylla]
  oreophila (Kearney) Wherry
  X popei Masters [flava X rubra]
  psittacina Michx.
  purpurea L.
    ssp. purpurea
        SY=S. heterophylla Eat.
        SY=S. p. ssp. gibbosa (Raf.)
            Wherry
        SY=S. p. ssp. h. (Eat.) Torr.
        SY=S. p. var. g. (Raf.) Wherry
        SY=S. p. var. rupicola Boivin
    ssp. venosa (Raf.) Wherry
        SY=S. p. var. v. (Raf.) Fern.
  X readii Bell [leucophylla X rubra]
  X rehderi Bell [minor X rubra]
  rubra Walt.
    ssp. alabamensis (Case & Case) Schnell
        SY=S. a. Case & Case
    ssp. jonesii (Wherry) Wherry
        SY=S. j. Wherry
    ssp. rubra
    ssp. wherryi (Case & Case) Schnell
        SY=S. alabamensis Case & Case ssp.
            w. Case & Case

  X swaniana (Wm. Robins.) Bell [minor X
    purpurea]
  X wrigleyana (S.G.) Bell [leucophylla X
    psittacina]

SAURURACEAE

Anemopsis Hook. & Arn.

  californica (Nutt.) Hook. & Arn.
    var. californica
    var. subglabrata L. Kelso

Saururus L.

  cernuus L.

SAXIFRAGACEAE

Astilbe Buch.-Ham. ex D. Don

  biternata (Vent.) Britt.
  crenatiloba (Britt.) Britt.
  japonica (Morr. & Dcne.) Gray

Bensoniella Morton

  oregona (Abrams & Bacig.) Morton
        SY=Bensonia o. Abrams & Bacig.

Bolandra Gray

  californica Gray
  oregana S. Wats.
        SY=B. o. var. imnahaensis M.E.
            Peck

Boykinia Nutt.

  aconitifolia Nutt.
        SY=Therofon a. (Nutt.) Millsp.
  elata (Nutt.) Greene
        SY=B. occidentalis Torr. & Gray
  major Gray
    var. intermedia Piper
        SY=B. i. (Piper) G.N. Jones
    var. major
  richardsonii (Hook.) Gray
        SY=Therofon r. (Hook.) Kuntze
  rotundifolia Parry

Broussaisia Gaud.

  arguta Gaud.
    var. arguta
    var. pellucida (Gaud.) Fosberg

Carpenteria Torr.

  californica Torr.

Chrysosplenium L.

  americanum Schwein. ex Hook.
  glechomifolium Nutt. ex Torr. & Gray
  iowense Rydb.
        SY=C. alternifolium L. var. i.
            (Rydb.) Boivin
  rosendahlii Packer
        SY=C. alternifolium L. var. r.
            (Packer) Boivin

Chrysosplenium (CONT.)
   tetrandrum (Lund) Th. Fries
        SY=C. alternifolium L. var. t.
          Lund
   wrightii Franch. & Sav.
      var. beringianum (Rose) Hara
      var. wrightii

Conimitella Rydb.

   williamsii (D.C. Eat.) Rydb.

Decumaria L.

   barbara L.

Deutzia Thunb.

   gracilis Sieb. & Zucc.
   scabra Thunb.

Elmera Rydb.

   racemosa (S. Wats.) Rydb.
      var. puberulenta C.L. Hitchc.
      var. racemosa
        SY=Heuchera r. S. Wats.

Fendlera Engelm. & Gray

   linearis Rehd.
   rupicola Gray
      var. falcata (Thornb.) Rehd.
      var. rupicola
   wrightii (Gray) Heller
        SY=F. rupicola Gray var. w. Gray

Fendlerella Heller

   utahensis (S. Wats.) Heller
      var. cymosa (Greene) Kearney &
        Peebles
      var. utahensis

Heuchera L.

   abramsii Rydb.
   alpestris Rosendahl, Butters & Lakela
   americana L.
      var. americana
        SY=H. a. var. brevipetala
          Rosendahl, Butters & Lakela
        SY=H. a. var. calycosa (Small)
          Rosendahl, Butters & Lakela
        SY=H. a. var. heteradenia Fern.
        SY=H. a. var. subtruncata Fern.
        SY=H. calycosa Small
        SY=H. curtisii Torr. & Gray
        SY=H. lancipetala Rydb.
      var. hispida (Pursh) E. Wells
        SY=H. h. Pursh
      var. hirsuticaulis (Wheelock)
        Rosendahl, Butters & Lakela
        SY=H. a. var. interior Rosendahl,
          Butters & Lakela
        SY=H. h. (Wheelock) Rydb.
   bracteata (Torr.) Ser.
   brevistaminea Wiggins
   caespitosa Eastw.
   caroliniana (Rosendahl, Butters & Lakela)
     E. Wells
        SY=H. americana L. var. c.
          Rosendahl, Butters & Lakela
   chlorantha Piper
   X cuneata T.J. Howell [rubescens var.
     truncata X cylindrica var. ovalifolia]
   cylindrica Dougl. ex Hook.

      var. cylindrica
      var. glabella (Torr. & Gray) Wheelock
        SY=H. g. Torr. & Gray
      var. orbicularis (Rosendahl, Butters
        & Lakela) Calder & Saville
        SY=H. ovalifolia Nutt. ex Torr. &
          Gray var. orbicularis
          Rosendahl, Butters & Lakela
      var. ovalifolia (Nutt. ex Torr. &
        Gray) Wheelock
        SY=H. c. var. alpina S. Wats.
        SY=H. o. Nutt. ex Torr. & Gray
        SY=H. o. var. thompsonii
          Rosendahl, Butters & Lakela
      var. septentrionalis Rosendahl,
        Butters & Lakela
   duranii Bacig.
   X easthamii Calder & Savile [chlorantha X
    micrantha var. diversifolia]
   eastwoodiae Rosendahl, Butters & Lakela
   elegans Abrams
   flabellifolia Rydb.
      var. flabellifolia
        SY=H. parvifolia Nutt. ex Torr. &
          Gray var. dissecta M.E.
          Jones
      var. subsecta Rosendahl, Butters &
        Lakela
   glabra Willd. ex Roemer & Schultes
   glomerulata Rosendahl, Butters & Lakela
   grossularifolia Rydb.
      var. grossularifolia
        SY=H. cusickii Rosendahl, Butters
          & Lakela
      var. tenuifolia (Wheelock) C.L.
        Hitchc.
        SY=H. t. (Wheelock) Rydb.
   hallii Gray
   hirsutissima Rosendahl, Butters & Lakela
   leptomeria Greene
      var. leptomeria
        SY=H. versicolor Greene var. l.
          (Greene) Kearney & Peebles
      var. peninsularis Rosendahl, Butters
        & Lakela
   longiflora Rydb.
        SY=H. aceroides Rydb.
        SY=H. l. var. a. (Rydb.)
          Rosendahl, Butters & Lakela
        SY=H. scabra Rydb.
   maxima Greene
   merriamii Eastw.
   micrantha Dougl. ex Lindl.
      var. diversifolia (Rydb.) Rosendahl,
        Butters & Lakela
      var. erubescens (A. Braun & Bouche)
        Rosendahl
      var. hartwegii (S. Wats. ex Wheelock)
        Rosendahl
      var. micrantha
      var. pacifica Rosendahl, Butters &
        Lakela
   missouriensis Rosendahl
   novomexicana Wheelock
   parishii Rydb.
   parviflora Bartl.
        SY=H. p. var. rugelii (Shuttlw.)
          Rosendahl, Butters & Lakela
   parvifolia Nutt. ex Torr. & Gray
      var. arizonica Rosendahl, Butters &
        Lakela
      var. flavescens (Rydb.) Rosendahl,
        Butters & Lakela
      var. major Rosendahl, Butters &
        Lakela
      var. microcarpa Rosendahl, Butters &
        Lakela

Heuchera (CONT.)
    var. nivalis (Rosendahl, Butters &
      Lakela) Love, Love & Kapoor
      SY=H. n. Rosendahl, Butters &
        Lakela
    var. parvifolia
    var. puberula (Mackenzie & Bush) E.
      Wells
      SY=H. p. Mackenzie & Bush
    var. utahensis (Rydb.) Garrett
pilosissima Fisch. & Mey.
    var. hemisphaerica (Rydb.) Rosendahl
    var. pilosissima
pringlei Rydb.
pubescens Pursh
    SY=H. alba Rydb.
    SY=H. pubescens Pursh var.
      brachyandra Rosendahl,
      Butters & Lakela
pulchella Woot. & Standl.
richardsonii R. Br.
    SY=H. hispida sensu auctt. non
      Pursh
    SY=H. r. var. affinis Rosendahl,
      Butters & Lakela
    SY=H. r. var. grayana Rosendahl,
      Butters & Lakela
    SY=H. r. var. hispidior Rosendahl,
      Butters & Lakela
rubescens Torr.
    var. alpicola Jepson
      SY=H. r. var. rydbergiana
        Rosendahl, Butters & Lakela
    var. glandulosa Kellogg
    var. pachypoda (Greene) Rosendahl,
      Butters & Lakela
    var. rubescens
    var. truncata Rosendahl, Butters &
      Lakela
sanguinea Engelm.
    var. pulchra (Rydb.) Rosendahl,
      Butters & Lakela
    var. sanguinea
saxicola E. Nels.
    SY=H. X suksdorfii Rydb.
versicolor Greene
villosa Michx.
    var. arkansana (Rydb.) E.B. Sm.
      SY=H. a. (Rydb.) Rosendahl,
      Butters & Lakela
    var. villosa
      SY=H. crinita Rydb.
      SY=H. macrorhiza Small
      SY=H. v. var. intermedia
        Rosendahl, Butters & Lakela
      SY=H. v. var. m. (Small)
        Rosendahl, Butters & Lakela
wootonii Rydb.

Hydrangea L.

arborescens L.
    ssp. aborescens
      SY=H. a. var. oblonga Torr. & Gray
      SY=H. a. var. sterilis Torr. &
        Gray
    ssp. discolor (Ser. ex DC.) McClintock
      SY=H. a. ssp. deamii (St. John)
        McClintock
      SY=H. a. var. deamii St. John
      SY=H. a. var. discolor Ser. ex DC.
      SY=H. asheii Harbison
      SY=H. cinerea Ser.
    ssp. radiata (Walt.) McClintock
      SY=H. r. Walt.
paniculata Sieb.
quercifolia Bartr.

Itea L.

virginica L.

Jamesia Torr. & Gray

americana Torr. & Gray
    var. americana
    var. californica (Small) Jepson

Jepsonia Small

heterandra Eastw.
malvifolia (Greene) Small
parryi (Torr.) Small

Leptarrhena R. Br.

pyrolifolia (D. Don) R. Br. ex Ser.

Lepuropetalon Ell.

spathulatum (Muhl.) Ell.

Lithophragma (Nutt.) Torr. & Gray

affine Gray
    ssp. affine
    ssp. mixtum R.L. Taylor
      SY=L. tripartita (Greene) Greene
bolanderi Gray
    SY=L. scabrella (Greene) Greene
    SY=L. s. var. peirsonii Jepson
campanulatum T.J. Howell
    SY=L. heterophyllum (Hook. & Arn.)
      Torr. & Gray var. c. (T.J.
      Howell) M.F. Peck
cymbalaria Torr. & Gray
glabrum Nutt.
    SY=L. bulbifera Rydb.
    SY=L. g. var. b. (Rydb.) Jepson
    SY=L. g. var. ramulosum (Suksdorf)
      Boivin
heterophyllum (Hook. & Arn.) Torr. & Gray
maximum Bacig.
parviflorum (Hook.) Nutt. ex Torr. & Gray
tenellum Nutt.
    SY=L. australis Rydb.
    SY=L. breviloba Rydb.
    SY=L. rupicola Greene
    SY=L. t. var. thompsonii (Hoover)
      C.L. Hitchc.
trifoliatum Eastw.

Mitella L.

breweri Gray
    SY=Pectiantia b. (Gray) Rydb.
caulescens Nutt.
    SY=Mitellastra c. (Nutt.) T.J.
      Howell
diphylla L.
    SY=Mitella oppositifolia Rydb.
diversifolia Greene
    SY=Ozomelis d. (Greene) Rydb.
X intermedia Bruhin [diphylla X nuda]
nuda L.
ovalis Greene
    SY=Pectiantia o. (Greene) Rydb.
pentandra Hook.
    SY=Pectiantia p. (Hook.) Rydb.
prostrata Michx.
stauropetala Piper
    var. stauropetala
      SY=Ozomelis s. (Piper) Rydb.
    var. stenopetala (Piper) Rosendahl
      SY=Mitella stenopetala Piper

Mitella (CONT.)
    trifida Graham
        var. trifida
            SY=Ozomelis anomala (Piper) Rydb.
            SY=O. micrantha (Piper) Rydb.
            SY=O. t. (Graham) Rydb.
        var. violacea (Rydb.) Rosendahl
            SY=Mitella v. Rydb.

Parnassia L.

    asarifolia Vent.
    caroliniana Michx.
        SY=P. floridana Rydb.
    cirrata Piper
    fimbriata Koenig
        var. fimbriata
        var. hoodiana C.L. Hitchc.
        var. intermedia (Rydb.) C.L. Hitchc.
            SY=P. i. Rydb.
    glauca Raf.
        SY=P. americana Muhl.
    grandifolia DC.
    kotzebuei Cham. & Schlecht.
        var. kotzebuei
        var. pumila C.L. Hitchc. & Ownbey
    multiseta (Ledeb.) Fern.
        SY=P. palustris L. ssp. neogaea
            (Fern.) Hulten
        SY=P. p. var. n. Fern.
    palustris L.
        var. californica Gray
            SY=P. c. (Gray) Greene
        var. montanensis (Fern. & Rydb.) C.L.
            Hitchc.
            SY=P. m. Fern. & Rydb.
            SY=P. p. ssp. neogaea Fern. &
                Hulten var. m. (Fern. &
                Rydb.) C.L. Hitchc.
        var. palustris
        var. tenuis Wahlenb.
    parviflora DC.
        SY=P. palustris L. var. parviflora
            (DC.) Boivin

Peltiphyllum Engl.

    peltatum (Torr. ex Benth.) Engl.
        SY=Saxifraga p. Torr. ex Benth.

Penthorum L.

    sedoides L.

Philadelphus L.

    argenteus Rydb.
        SY=P. microphyllus Gray ssp. a.
            (Rydb.) C.L. Hitchc.
        SY=P. m. var. a. (Rydb.) Kearney &
            Peebles
    argyrocalyx Woot.
        SY=P. microphyllus Gray ssp. a.
            (Woot.) C.L. Hitchc.
    californicus Benth.
        SY=P. gordonianus Lindl. var.
            columbianus (Koehne) Rehd.
        SY=P. lewisii Pursh ssp.
            californicus (Benth.) Munz
    caucasicus Koehne
    confusus Piper
    cordifolius Lange
    coronarius L.
    crinitus (C.L. Hitchc.) Hu
    ernestii Hu
    floridus Beadle
    gattingeri Hu

    hirsutus Nutt.
        var. hirsutus
        var. intermedius Hu
        var. nanus Hu
    hitchcockianus Hu
    inodorus L.
        var. carolinus Hu
        var. grandiflorus (Willd.) Gray
            SY=P. g. Willd.
            SY=P. gloriosus Beadle
        var. inodorus
        var. laxus (Schrad.) Hu
        var. strigosus Beadle
    insignis Carr.
    intectus Beadle
        var. intectus
            SY=P. pubescens Loisel. var. i.
                (Beadle) A.H. Moore
        var. pubigerus Hu
    lewisii Pursh
        var. angustifolius (Rydb.) Hu
        var. ellipticus Hu
        var. gordonianus (Lindl.) Jepson
            SY=P. g. Lindl.
            SY=P. l. ssp. g. (Lindl.) Munz
        var. helleri (Rydb.) Hu
        var. intermedius (A. Nels.) Hu
        var. lewisii
        var. oblongifolius Hu
        var. parvifolius Hu
        var. platyphyllus (Rydb.) Hu
    maculatus (C.L. Hitchc.) Hu
    madrensis Hemsl.
    mearnsii W.H. Evans ex Koehne
    microphyllus Gray
        var. linearis Hu
        var. microphyllus
            SY=P. m. ssp. typicus C.L. Hitchc.
        var. ovatus Hu
    occidentalis A. Nels.
        var. minutus (Rydb.) Hu
        var. occidentalis
            SY=P. microphyllus Gray ssp. o.
                (A. Nels.) C.L. Hitchc.
    oreganus Nutt. ex Torr. & Gray
    palmeri Rydb.
    pubescens Loisel.
        var. pubescens
            SY=P. latifolius Schrad. ex DC.
        var. verrucosus (Schrad.) Hu
    pumilus Rydb.
        var. ovatus Hu
        var. pumilus
            SY=P. microphyllus Gray ssp. p.
                (Rydb.) C.L. Hitchc.
    serpyllifolius Gray
    sharpianus Hu
        var. parviflorus Hu
        var. sharpianus
    stramineus Rydb.
        SY=P. microphyllus Gray ssp. s.
            (Rydb.) C.L. Hitchc.
    texensis Hu
        var. coryanus Hu
        var. texensis
    tomentosus Wallich
    trichothecus Hu
    wootonii Hu
    zelleri Hu

Ribes L.

    amarum McClatchie
        var. amarum
            SY=Grossularia a. (McClatchie)
                Coville & Britt.
        var. hoffmannii Munz

Ribes (CONT.)
  americanum P. Mill.
  aureum Pursh
        var. aureum
            SY=Chrysobotrya a. (Pursh) Rydb.
        var. gracillimum (Coville & Britt.)
            Jepson
            SY=R. g. Coville & Britt.
  binominatum Heller
        SY=Grossularia b. (Heller) Coville
            & Britt.
  bracteosum Dougl. ex Hook.
  californicum Hook. & Arn.
        var. californicum
            SY=Grossularia c (Hook. & Arn.)
                Coville & Britt.
            SY=R. oligacanthum Eastw.
        var. hesperium (McClatchie) Jepson
            SY=Grossularia h. (McClatchie)
                Coville & Britt.
            SY=R. h. McClatchie
  canthariforme Wiggins
  cereum Dougl.
        var. cereum
            SY=R. reniforme Nutt.
        var. colubrinum C.L. Hitchc.
  cognatum Greene
        SY=Grossularia c. (Greene) Coville
            & Britt.
  coloradense Coville
  cruentum Greene
        var. cruentum
            SY=Grossularia c. (Greene) Coville
                & Britt.
            SY=R. roezlii Regel var. c.
                (Greene) Rehd.
        var. oregonensis Berger
  curvatum Small
        SY=Grossularia c. (Small) Coville
            & Britt.
  cynosbati L.
        var. atrox Fern.
        var. cynosbatii
            SY=Grossularia c. (L.) P. Mill.
        var. glabratum Fern.
  diacanthum Pallas
  divaricatum Dougl.
        var. divaricatum
            SY=Grossularia d. (Dougl.) Coville
                & Britt.
            SY=R. d. var. glabriflorum Koehne
            SY=R. d. var. rigidum M.E. Peck
        var. klamathense (Coville) McMinn
            SY=Grossularia k. Coville
            SY=R. k. (Coville) Fedde
        var. parishii (Heller) Jepson
            SY=Grossularia p. (Heller) Coville
                & Britt.
  echinellum (Coville) Rehd.
        SY=Grossularia e. Coville
  erythrocarpum Coville & Leib.
  glandulosum Grauer
        SY=R. resinosum Pursh
  hendersonii C.L. Hitchc.
  hirtellum Michx.
        var. calcicola (Fern.) Fern.
            SY=R. oxyacanthoides L. var. c.
                Fern.
        var. hirtellum
            SY=Grossularia h. (Michx.) Spach
        var. saxosum (Hook.) Fern.
            SY=R. oxyacanthoides L. var. s.
                (Hook.) Coville
  howellii Greene
        SY=R. acerifolium T.J. Howell
  hudsonianum Richards.
        var. hudsonianum

        var. petiolare (Dougl.) Jancz.
            SY=R. p. Dougl.
  indecorum Eastw.
  inebrians Lindl.
        SY=R. cereum Dougl var. i.
            (Lindl.) C.L. Hitchc.
  inerme Rydb.
        var. inerme
            SY=Grossularia i. (Rydb.) Coville
                & Britt.
            SY=G. neglecta Berger
            SY=R. divaricatum Dougl. var. i.
                (Rydb.) McMinn
        var. pubescens Berger
        var. subarmatum M.E. Peck
  irriguum Dougl.
        SY=Grossularia i. (Dougl.) Coville
            & Britt.
        SY=R. oxyacanthoides L. var. i.
            (Dougl.) Jancz.
  lacustre (Pers.) Poir.
        SY=Limnobotrya l. (Pers.) Rydb.
        SY=R. l. var. parvulum Gray
  lasianthum Greene
        SY=Grossularia l. (Greene) Coville
            & Britt.
  laxiflorum Pursh
  leptanthum Gray
  lobbii Gray
        SY=Grossularia l. (Gray) Coville &
            Britt.
  malvaceum Sm.
        var. malvaceum
            SY=R. m. var. clementinum Dunkle
        var. viridifolium Abrams
  marshallii Greene
        SY=Grossularia m. (Greene) Coville
            & Britt.
  menziesii Pursh
        var. faustum Jepson
        var. hystriculum Jepson
        var. hystrix (Eastw.) Jepson
            SY=Grossularia h. (Eastw.) Coville
                & Britt.
        var. ixoderme Quick
        var. leptosmum (Coville) Jepson
            SY=Grossularia l. Coville
        var. menziesii
            SY=Grossularia m. (Pursh) Coville
                & Britt.
        var. senile (Coville) Jepson
            SY=Grossularia s. Coville
  mescalerium Coville
  missouriense Nutt. ex Torr. & Gray
        var. missouriense
            SY=Grossularia m. (Nutt. ex Torr.
                & Gray) Coville & Britt.
        var. ozarkanum Fassett
  montigenum McClatchie
        SY=Limnobotrya m. (McClatchie)
            Rydb.
  nevadense Kellogg
        var. glaucescens (Eastw.) Berger
        var. jaegeri Berger
        var. nevadense
  nigrum L.
  niveum Lindl.
  odoratum H. Wendl.
  oxyacanthoides L.
        SY=Grossularia o. (L.) P. Mill.
  pinetorum Greene
  quercetorum Greene
        SY=Grossularia q. (Greene) Coville
            & Britt.
  roezlii Regel
        var. amictum (Greene) Jepson
        var. roezlii

Ribes (CONT.)
                SY=Grossularia r. (Regel) Coville
                        & Britt.
        rotundifolium Michx.
                SY=Grossularia r. (Michx.) Coville
                        & Britt.
        rubrum L.
                var. alaskanum (Berger) Boivin
                        SY=R. sativum (Reichenb.) Syme
                        SY=R. sylvestre (Lam.) Mert. &
                                Koch
                        SY=R. vulgare Lam.
        sanguineum Pursh
                var. deductum (Greene) Jepson
                var. glutinosum (Benth.) Loud.
                        SY=R. g. Benth.
                var. melanocarpum (Greene) Jepson
                var. sanguineum
        sericeum Eastw.
                SY=Grossularia s. (Eastw.) Coville
                        & Britt.
        setosum Lindl.
                SY=Grossularia s. (Lindl.) Coville
                        & Britt.
        speciosum Pursh
                SY=Grossularia s. (Pursh) Coville
                        & Britt.
        thacherianum (Jepson) Munz
        triste Pallas
                SY=R. rubrum L. var. propinquum
                        Trautv. & Mey.
        tularense (Coville) Fedde
                SY=Grossularia t. Coville
        uva-crispa L.
                var. sativum DC.
                        SY=Grossularia reclinata (L.) P.
                                Mill.
                        SY=R. grossularia L.
                var. uva-crispa
        velutinum Greene
                var. glanduliferum (Heller) Jepson
                var. gooddingii (M.E. Peck) C.L.
                        Hitchc.
                        SY=R. g. M.E. Peck
                var. velutinum
                        SY=Grossularia v. (Greene) Coville
                                & Britt.
        viburnifolium Gray
        victoris Greene
                var. minus Jancz.
                        SY=Grossularia greeneiana (Heller)
                                Coville & Britt.
                var. victoris
                        SY=Grossularia v. (Greene) Coville
                                & Britt.
        viscosissimum Pursh
                var. hallii Jancz.
                var. viscosissimum
        watsonianum Koehne
                SY=Grossularia w. (Koehne) Coville
                        & Britt.
        wolfii Rothrock
                SY=R. mogollonicum Greene

Saxifraga L.

        adscendens L.
                ssp. oregonensis (Raf.) Bacig.
                        SY=S. a. var. o. (Raf.) Breitung
        aizoides L.
                SY=Chrysobotrya aurea (Pursh)
                        Rydb.
                SY=Leptasea aizoides (L.) Haw.
        aleutica Hulten
        aprica Greene
        bronchialis L.
                ssp. austromontana (Wieg.) Piper

                        SY=S. a. Wieg.
                        SY=S. b. var. a. (Wieg.) G.N.
                                Jones
                ssp. bronchialis
                ssp. cherlerioides (D. Don) Hulten
                        SY=S. b. ssp. b. var. c. (D. Don)
                                Engl.
                        SY=S. c. D. Don
                ssp. funstonii (Small) Hulten
                        SY=S. b. var. purpureo-maculata
                                Hulten
                ssp. vespertina (Small) Piper
                        SY=S. b. var. v. (Small) Rosendahl
                        SY=S. v. (Small) Fedde
        bryophora Gray
        californica Greene
        calycina Sternb.
                ssp. unalaschensis (Sternb.) Hulten
                        SY=S. davurica Willd. var. u.
                                (Sternb.) Engl.
                        SY=S. u. Sternb.
        careyana Gray
                SY=Micranthes c. (Gray) Small
                SY=M. tennesseensis Small
        caroliniana Gray
                SY=Micranthes c. (Gray) Small
        cernua L.
        cespitosa L.
                ssp. cespitosa
                        var. cespitosa
                                SY=Muscaria c. (L.) Haw.
                        var. emarginata (Small) Rosendahl
                ssp. decipiens (Ehrh.) Engl. & Irmsch.
                ssp. delicatula (Small) Porsild
                ssp. exaratoides (Simm.) Engl. & Irmsch.
                        SY=S. c. var. lemmonii Engl. &
                                Irmsch.
                ssp. lauxiuscula Love & Love
                ssp. monticola (Small) Porsild
                        SY=S. c. var. minima Blank.
                ssp. sileneflora (Sternb. ex Cham.)
                        Hulten
                ssp. subgemmifera Engl. & Irmsch.
                        SY=S. c. var. s. (Engl. & Irmsch.)
                                C.L. Hitchc.
                ssp. uniflora (R. Br.) Porsild
        chrysantha Gray
                        SY=S. serphyllifolia Pursh ssp. c.
                                (Gray) W.A. Weber
        davurica Willd.
                ssp. grandipetala (Engl. & Irmsch.)
                        Hulten
                        SY=S. d. var. g. (Engl. & Irmsch.)
                                Boivin
        debilis Engelm. ex Gray
                        SY=S. hyperborea R. Br. ssp. d.
                                (Engelm. ex Gray) Love, Love
                                & Kapoor
        eriophora S. Wats.
        eschscholtzii Sternb.
        fallax Greene
        ferruginea Graham
                var. ferruginea
                var. newcombei Small
                var. vreelandii (Small) Engl. &
                        Irmsch.
                        SY=S. f. var. macounii Engl. &
                                Irmsch.
                        SY=S. v. (Small) Fedde ex Just.
        flagellaris Sternb. & Willd.
                ssp. flagellaris
                ssp. platysepala (Trautv.) Porsild
                        SY=S. f. var. p. Trautv.
                ssp. setigera (Pursh) Tolm.
        foliolosa R. Br.
                var. foliolosa
                        SY=Hydatica f. (R. Br.) Small

Saxifraga (CONT.)
        SY=S. stellaris L. var. comosa
           Poir.
      var. multiflora Hulten
forbesii Vasey
        SY=S. pensylvanica L. var. f.
           (Vasey) Engl. & Irmsch.
fragarioides Greene
X geum L. [hirsuta X umbrosa]
        SY=Micranthes g. (L.) Small
hieracifolia Waldst. & Kit.
      var. angusticapsula Hulten
      var. hieracifolia
      var. rufopilosa Hulten
hirculus L.
   ssp. hirculus
      SY=Leptasea h. (L.) Small
   ssp. propinqua (R. Br.) Love & Love
      SY=S. h. var. p. (R. Br.) Simm.
hirsuta L.
howellii Greene
hyperborea R. Br.
integrifolia Hook.
      var. apetala M.E. Jones
      var. claytoniifolia (Canby) Rosendahl
      SY=S. fragosa Suksdorf
      var. columbiana (Piper) C.L. Hitchc.
      SY=S. c. Piper
      var. integrifolia
      var. leptopetala (Suksdorf) Engl. &
         Irmsch.
      SY=S. plantaginea Small
lyallii Engl.
   ssp. hultenii (Calder & Savile) Calder &
     Taylor
      SY=S. l. var. h. Calder & Savile
   ssp. lyallii
      SY=Micranthes l. (Engl.) Small
      SY=S. l. var. laxa Engl.
marshallii Greene
   ssp. idahoensis (Piper) Krause & Beamish
      SY=S. m. var. i. (Piper) Engl. &
         Irmsch.
      SY=S. occidentalis S. Wats. var.
         i. (Piper) C.L. Hitchc.
   ssp. marshallii
      SY=S. hallii Gray
mertensiana Bong.
      SY=S. m. var. eastwoodiae (Small)
         Engl. & Irmsch.
michauxii Britt.
      SY=Hydatica petiolaris (Raf.)
         Small
      SY=Spathularia m. (Britt.) Small
micranthidifolia (Haw.) Steud.
      SY=Micranthes m. (Haw.) Small
nathorstii (Dusen) Hayek
nelsoniana D. Don
   ssp. carlottae (Calder & Savile) Hulten
      SY=Saxifraga punctata L. ssp. c.
         Calder & Savile
      SY=S. p. var. c. (Calder & Savile)
         Boivin
   ssp. cascadensis (Calder & Savile)
     Hulten
      SY=Saxifraga punctata L. ssp. c.
         Calder & Savile
      SY=S. p. var. c (Calder & Savile)
         Boivin
   ssp. insularis (Hulten) Hulten
      SY=Saxifraga punctata L. ssp. i.
         Hulten
   ssp. nelsoniana
      SY=Saxifraga aestivalis Fisch. &
         Mey.
      SY=S. punctata L. pro parte
      SY=S. p. ssp. n. (D. Don) Hulten

      SY=S. p. var. n. (D. Don) Macoun
   ssp. pacifica (Hulten) Hulten
      SY=Saxifraga punctata L. ssp.
         pacifica Hulten
   ssp. porsildiana (Calder & Savile)
     Hulten
      SY=Saxifraga punctata L. ssp.
         porsildiana Calder & Savile
      SY=S. punctata var. porsildiana
         (Calder & Savile) Boivin
nidifica Greene
nivalis L.
      SY=Micranthes n. (L.) Small
nudicaulis D. Don
nuttallii Small
occidentalis S. Wats.
      var. allenii (Small) C.L. Hitchc.
      var. dentata (Engl. & Irmsch.) C.L.
         Hitchc.
      var. latipetiolata C.L. Hitchc.
      var. occidentalis
      SY=Micranthes o. (S. Wats.) Small
      SY=S. o. var. wallowensis M.E.
         Peck
odontoloma Piper
      SY=Saxifraga arguta sensu auctt.
         non D. Don
      SY=S. punctata L. ssp. a. sensu
         auctt. non (D. Don) Hulten
      SY=S. p. var. a. sensu Engl. &
         Irmsch.
oppositifolia L.
   ssp. glandulisepala Hulten
   ssp. oppositifolia
      SY=Antiphylla o. (L.) Fourr.
   ssp. smalliana (Engl. & Irmsch.) Hulten
oregana T.J. Howell
      var. montanensis (Small) C.L. Hitchc.
      SY=Saxifraga m. Small
      var. oregana
      var. sierrae Coville
      var. subapetala (E. Nels.) C.L.
         Hitchc.
      SY=Saxifraga montanensis Small
            var. s. (E. Nels.) Engl. &
            Irmsch.
palmeri Bush
paniculata P. Mill.
      SY=Chondrosea aizoon (Jacq.) Haw.
      SY=Saxifraga a. Jacq.
      SY=S. a. ssp. neogaea (Butters)
         Love & Love
      SY=S. a. var. n. Butters
pensylvanica L.
   ssp. interior
   ssp. pensylvanica
      SY=Micranthes p. (L.) Haw.
   ssp. tenuirostrata Burns
reflexa Hook.
rhomboidea Greene
      var. franciscana (Small) Kearney &
         Peebles
      var. rhomboidea
      SY=Micranthes r. (Greene) Small
rivularis L.
      var. flexuosa (Sternb.) Engl. &
         Irmsch.
      var. rivularis
rufidula (Small) Macoun
      SY=Saxifraga occidentalis S. Wats.
         ssp. r. (Small) Bacig.
      SY=S. o. var. aequidentata (Small)
         M.E. Peck
      SY=S. o. var. r. (Small) C.L.
         Hitchc.
serphyllifolia Pursh
      SY=Saxifraga s. var. purpurea

Saxifraga (CONT.)
                 Hulten
   sibirica L.
            SY=Saxifraga bracteata D. Don
            SY=S. exilis Steph.
            SY=S. radiata Small
            SY=S. rivularis L. var.
                 laurentiana (Ser.) Engl.
   sibthorpii Boiss.
   spicata D. Don
   stellaris L.
            SY=Hydatica s. (L.) S.F. Gray
   stolonifera Meerb.
            SY=Saxifraga sarmentosa L.
   taylorii Calder & Savile
   tempestiva Elvander & Denton
   tenuis (Wahlenb.) H. Sm.
            SY=Saxifraga gaspensis Fern.
            SY=S. nivalis L. var. g. (Fern.)
                 Boivin
            SY=S. n. var. t. Wahlenb.
   texana Buckl.
            SY=Micranthes t. (Buckl.) Small
   tolmiei Torr. & Gray
         var. ledifolia (Greene) Engl. &
              Irmsch.
         var. tolmiei
   tricuspidata Rottb.
            SY=Leptasea t. (Rottb.) Haw.
   tridactylites L.
   virginiensis Michx.
            var. subintegra Goodman
            var. virginiensis
            SY=Micranthes v. (Michx.) Small

Schizophragma Sieb. & Zucc.

   hydrangeoides Sieb. & Zucc.

Suksdorfia Gray

   ranunculifolia (Hook.) Engl.
            SY=Hemieva r. (Hook.) Raf.
   violacea Gray

Sullivantia Torr. & Gray

   hapemanii (Coult. & Fish.) Coult.
   oregana S. Wats.
   purpusii (Brand) Rosendahl
   renifolia Rosendahl
   sullivantii (Torr. & Gray) Britt.
            SY=S. ohionis Torr. & Gray

Telesonix Raf.

   jamesii (Torr.) Raf.
            var. heucheriformis (Rydb.) Bacig.
            SY=Boykinia h. (Rydb.) A. Nels.
            SY=T. h. Rydb.
            var. jamesii
            SY=Boykinia j. (Torr.) Engl.
            SY=Saxifraga j. Torr.

Tellima R. Br.

   gandiflora (Pursh) Dougl. ex Lindl.
            SY=T. odorata T.J. Howell

Tiarella L.

   cordifolia L.
            var. austrina Lakela
            var. collina Wherry
            SY=T. wherryi Lakela
            var. cordifolia
            SY=T. macrophylla Small pro parte

   laciniata Hook.
            SY=T. californica (Kellogg) Rydb.
            SY=T. trifoliata L. ssp.
                 unifoliata (Hook.) Kern.
                 var. l (Hook.) Wheelock
   trifoliata L.
   unifoliata Hook.
            SY=T. trifoliata L. ssp. u.
                 (Hook.) Kern.

Tolmiea Torr. & Gray

   menziesii (Pursh) Torr. & Gray

Whipplea Torr.

   modesta Torr.

SCHEUCHZERIACEAE

Lilaea Humb. & Bonpl.

   scilloides (Poir.) Hauman
            SY=L. subulata Humb. & Bonpl.

Scheuchzeria L.

   palustris L.
         ssp. americana (Fern.) Hulten
            SY=S. a. (Fern.) G.N. Jones
            SY=S. p. var. a. Fern.

Triglochin L.

   concinna Burtt-Davy
   debilis (M.E. Jones) Love & Love
            SY=T. concinna Burtt-Davy var. d.
                 (M.E. Jones) J.T. Howell
   gaspense Lieth & D. Love
   maritima L.
            SY=T. elata Nutt.
   palustre L.
   striata Ruiz & Pavon

SCHISANDRACEAE

Schisandra Michx.

   coccinea Michx.
            SY=S. glabra (Bickn.) Rehd.

SCROPHULARIACEAE

Agalinis Raf.

   acuta Pennell
            SY=Gerardia a. (Pennell) Pennell
   aphylla (Nutt.) Raf.
            SY=Gerardia a. Nutt.
   aspera (Dougl.) Britt.
            SY=Gerardia a. Dougl.
   besseyana Britt.
            SY=A. tenuifolia (Vahl) Raf. var.
                 macrophylla (Benth.) Blake
   calycina Pennell
   decemloba (Greene) Pennell
            SY=Gerardia d. Greene
   divaricata (Chapman) Pennell

Agalinis (CONT.)
  edwardsiana Pennell
        SY=Gerardia e. (Pennell) Pennell
  fasciculata (Ell.) Raf.
        SY=A. f. var. peninsularis Pennell
        SY=Gerardia f. Ell.
  filifolia (Nutt.) Raf.
        SY=A. filicaulis (Benth.) Pennell
        SY=A. keyensis Pennell
  gattingeri (Small) Small
        SY=Gerardia g. Small
  georgiana (Boynt.) Pennell
  harperi Pennell
  heterophylla (Nutt.) Small
        SY=Gerardia h. Nutt.
  homalantha Pennell
        SY=Gerardia h. (Pennell) Pennell
  linifolia (Nutt.) Britt.
        SY=Gerardia l. Nutt.
  maritima (Raf.) Raf.
    var. grandiflora (Benth.) Shinners
        SY=Gerardia m. Raf. var. g. Benth.
    var. maritima
        SY=Gerardia m. Raf.
  neoscotica (Greene) Fern.
        SY=A. purpurea (L.) Pennell var.
          n. (Greene) Boivin
        SY=Gerardia n. Greene
        SY=G. p. L. var. n. (Greene)
          Gleason
  nuttallii Shinners
        SY=Gerardia longifolia Nutt.
  obtusifolia Raf.
        SY=A. erecta (Walt.) Pennell
        SY=Gerardia o. (Raf.) Pennell
        SY=G. parviflora (Hook.) Chapman
  oligophylla Pennell
  paupercula (Gray) Britt.
    var. borealis Pennell
        SY=Gerardia p. (Gray) Britt. var.
          b. (Pennell) Deam
    var. paupercula
        SY=Gerardia p. (Gray) Britt.
        SY=G. p. var. typica Pennell
  pseudophylla (Pennell) Shinners
        SY=Gerardia p. (Pennell) Pennell
  pulchella Pennell
        SY=A. pinetorum Pennell
        SY=A. pinetorum var. delicatula
          (Pennell) Pennell
  purpurea (L.) Pennell
    var. carteri Pennell
    var. parviflora (Benth.) Boivin
        SY=Gerardia purpurea L. var.
          parviflora Benth.
    var. purpurea
        SY=Gerardia p. L.
  setacea (Walt.) Raf.
        SY=A. laxa Pennell
        SY=A. plukenetii (Ell.) Raf.
        SY=Gerardia gatesii (Benth.)
          Pennell
        SY=G. s. (Walt.) J.F. Gmel.
  skinneriana (Wood) Britt.
        SY=Gerardia s. Wood
  stenophylla Pennell
  strictifolia (Benth.) Pennell
        SY=A. caddoensis Pennell
  tenella Pennell
  tenuifolia (Vahl) Raf.
    var. leucanthera (Raf.) Shinners
        SY=A. t. ssp. l. (Raf.) Pennell
        SY=Gerardia t. Vahl var. l. (Raf.)
          Shinners
    var. parviflora (Nutt.) Pennell
        SY=Gerardia t. Vahl var. p. Nutt.
    var. tenuifolia

        SY=Gerardia t. Vahl
        SY=G. t. var. typica Pennell
  virgata Raf.
        SY=Gerardia purpurea L. var.
          racemulosa (Pennell) Gleason
        SY=G. r. Pennell
  viridis (Small) Pennell
        SY=Gerardia v. Small

Amphianthus Torr.

  pusillus Torr.

Angelonia Humb. & Bonpl.

  angustifolia Benth.
  salicariifolia Humb. & Bonpl.

Antirrhinum L.

  breweri Gray
  cornutum Benth.
    var. cornutum
    var. leptaleum (Gray) Munz
  coulterianum Benth.
    ssp. coulterianum
    ssp. orcuttianum (Gray ex Ives) Pennell
  cyathiferum Benth.
  filipes Gray ex Ives
        SY=Asarina f. (Gray ex Ives)
          Pennell
  kingii S. Wats.
    var. kingii
    var. watsonii (Vasey & Rose) Munz
  majus L.
  multiflorum Pennell
  nuttallianum Benth.
  ovatum Eastw.
  subcordatum Gray
  vexillo-calyculatum Kellogg
  virga Gray

Aureolaria Raf.

  flava (L.) Farw.
    var. flava
        SY=Agalinis f. (L.) Boivin
        SY=Aureolaria calycosa (Mackenzie
          & Bush) Pennell
        SY=Aureolaria f. ssp. reticulata
          (Raf.) Pennell
        SY=Aureolaria f. var. typica
          Pennell
        SY=Dasystoma c. Mackenzie & Bush
        SY=D. f. (L.) Wood
        SY=Gerardia c. (Mackenzie & Bush)
          Fern.
        SY=G. f. L.
        SY=G. f. var. c. (Mackenzie &
          Bush) Steyermark
        SY=G. f. var. reticulata (Raf.)
          Cory
    var. macrantha Pennell
        SY=Agalinis f. (L.) Boivin var. m.
          (Pennell) Boivin
        SY=Gerardia f. L. var. m.
          (Pennell) Fern.
  grandiflora (Benth.) Pennell
    var. cinerea Pennell
        SY=Gerardia g. Benth. var. c.
          (Pennell) Cory
    var. grandiflora
        SY=Dasystoma g. (Benth.) Wood
        SY=Gerardia g. Benth.
    var. pulchra Pennell
        SY=Gerardia g. Benth. var. p.
          (Pennell) Fern.

Aureolaria (CONT.)
        var. serrata (Torr. ex Benth.)
            Pennell
            SY=Dasystoma s. (Torr. ex Benth.)
                Small
            SY=Gerardia g. Benth. var. s.
                (Torr. ex Benth.) B.L.
                Robins.
    laevigata (Raf.) Raf.
            SY=Dasystoma l. (Raf.) Chapman
            SY=Gerardia l. Raf.
    patula (Chapman) Pennell
            SY=Gerardia p. (Chapman) Gray
    pectinata (Nutt.) Pennell
            SY=Aureolaria p. var. eurycarpa
                Pennell
            SY=A. p. var. floridana Pennell
            SY=A. p. var. ozarkensis Pennell
            SY=A. p. var. transcendens Pennell
            SY=A. pedicularia (L.) Raf. var.
                pectinata (Nutt.) Gleason
            SY=Dasystoma pectinata (Nutt.)
                Benth.
            SY=Gerardia pectinata (Nutt.)
                Benth.
            SY=G. pedicularia L. var.
                pectinata Nutt.
    pedicularia (L.) Raf.
        var. ambigens (Fern.) Farw.
            SY=Gerardia p. L. var. a. Fern.
        var. austromontana Pennell
            SY=Gerardia p. L. var. a.
                (Pennell) Fern.
        var. intercedens Pennell
            SY=Gerardia p. L. var. i.
                (Pennell) Fern.
        var. pedicularia
            SY=Agalinis p. (L.) Blake
            SY=Aureolaria p. var. caesariensis
                Blake
            SY=Aureolaria p. var. carolinensis
                Pennell
            SY=Aureolaria p. var. typica
                Pennell
            SY=Dasystoma p. (L.) Benth.
            SY=Gerardia p. L.
    virginica (L.) Pennell
            SY=Agalinis v. (L.) Blake
            SY=Aureolaria dispersa (Small)
                Pennell
            SY=Aureolaria microcarpa Pennell
            SY=Dasystoma v. (L.) Britt.
            SY=Gerardia d. (Small) K. Schum.
            SY=G. v. (L.) B.S.P.

Bacopa Aubl.

    caroliniana (Walt.) B.L. Robins.
            SY=Hydrotrida c. (Walt.) Small
    cyclophylla Fern.
            SY=Herpestis rotundifolia Gaertn.
                f.
    egensis (Poepp.) Pennell
    eisenii (Kellogg) Pennell
    innominata (Maza) Alain
    monnieri (L.) Pennell
            SY=Bramia m. (L.) Drake
    repens (Sw.) Wettst.
            SY=Macuillamia obovata Raf.
            SY=M. r. (Sw.) Pennell
    rotundifolia (Michx.) Wettst.
            SY=Bacopa nobsiana Mason
            SY=Bramia r. (Michx.) Britt.
            SY=Hydranthelium r. (Michx.)
                Pennell
            SY=Macuillamia r. (Michx.) Raf.
    simulans Fern.

        stragula Fern.
        stricta (Schrad.) B.L. Robins.
            SY=Caconapea s. (Schrad.) Britt.

Bartsia L.

    alpina L.

Bellardia All.

    trixago (L.) All.

Besseya Rydb.

    alpina (Gray) Rydb.
            SY=Synthyris a. Gray
    arizonica Pennell
    bullii (Eat.) Rydb.
            SY=Synthyris b. (Eat.) Heller
            SY=Wulfenia b. (Eat.) Barnh.
    plantaginea (James) Rydb.
    ritteriana (Eastw.) Rydb.
            SY=Synthyris r. Eastw.
    rubra (Dougl. ex Hook.) Rydb.
            SY=Synthyris r. (Dougl. ex Hook.)
                Benth.
    wyomingensis (A. Nels.) Rydb.
            SY=B. cinerea (Raf.) Pennell

Brachystigma Pennell

    wrightii (Gray) Pennell

Buchnera L.

    americana L.
            SY=B. breviflora Pennell
            SY=B. floridana Gandog.
    arizonica (Gray) Pennell
    elongata Sw.
            SY=B. longifolia H.B.K.

Capraria L.

    biflora L.

Castilleja Mutis ex L.f.

    affinis Hook. & Arn.
        ssp. affinis
            SY=C. a. ssp. a. var. contentiosa
                (J. F. Macbr.) Bacig.
            SY=C. californica Abrams
            SY=C. douglasii Benth.
        ssp. insularis (Eastw.) Munz
            SY=C. d. ssp. i. (Eastw.) Pennell
    angustifolia (Nutt.) G. Don
    annua Pennell
    applegatei Fern.
        var. applegatei
        var. fragilis (Zeile) N. Holmgren
            SY=C. pinetorum Fern.
            SY=C. wherryana Pennell
        var. pallida (Eastw.) N. Holmgren
            SY=C. breweri Fern.
            SY=C. b. var. p. Eastw.
            SY=C. glandulifera Pennell ssp. p.
                (Eastw.) Pennell
        var. viscida (Rydb.) Ownbey
            SY=C. v. Rydb.
    aquariensis N. Holmgren
    arachnoidea Greenm.
        ssp. arachnoidea
            SY=C. filifolia Eastw.
        ssp. shastensis (Eastw.) Pennell
            SY=C. payneae Eastw.
    arvensis Cham. & Schlecht.

Castilleja (CONT.)
  austromontana Standl. & Blumer
  barnebyana Eastw.
          SY=C. calcicola Pennell ex Edwin
  brevilobata Piper
  cervina Greenm.
  chlorotica Piper
  christii N. Holmgren
  chromosa A. Nels.
          var. chromosa
              SY=C. pyramidalis Edwin
          var. dubia A. Nels.
  chrysantha Greenm.
          SY=C. ownbeyana Pennell
  ciliata Pennell
  cinerea Gray
  coccinea (L.) Spreng.
  covilleana Henderson
  crista-galli Rydb.
  cruenta Standl.
  cryptantha Pennell & G.N. Jones
  culbertsonii Greene
  cusickii Greenm.
          SY=C. lutea Heller
  dissitiflora N. Holmgren
  disticha Eastw.
  elata Piper
          SY=C. miniata Dougl. ex Hook. ssp.
              e. (Piper) Munz
  elmeri Fern.
  elongata Pennell
  exilis A. Nels.
  flava S. Wats.
          SY=C. elkoensis Edwin
          SY=C. linoides Gray
  flavescens Pennell ex Edwin
  foliolosa Hook. & Arn.
  franciscana Pennell
  fraterna Greenm.
  fulva Pennell
  glandulifera Pennell
  gleasonii Elmer
          SY=C. pruinosa Fern. ssp. g.
              (Elmer) Munz
  gracillima Rydb.
  grisea Dunkle
          SY=C. hololeuca Greene ssp. g.
              (Dunkle) Munz
  haydenii (Gray) Cockerell
  hispida Benth. ex Hook.
      ssp. acuta Pennell
              SY=C. h. var. a. (Pennell) Ownbey
              SY=C. taedifera Pennell
      ssp. hispida
              SY=C. h. ssp. abbreviata (Fern.)
                  Pennell
  hololeuca Greene
  hyetophila Pennell
  hyperborea Pennell
          SY=C. villosissima Pennell
  inconstans Standl.
  indivisa Engelm.
  inflata Pennell
          SY=C. wightii Elmer ssp. i.
              (Pennell) Munz
  integra Gray
          var. gloriosa (Britt.) Cockerell
          var. integra
  inverta (A. Nels. & J.F. Macbr.) Pennell &
      Ownbey
          SY=C. pallescens (Nutt.) Greenm.
              var. i. (A. Nels. & J.F.
                  Macbr.) Greenm.
  jepsonii Bacig. & Heckard
  kaibabensis N. Holmgren
  kuschei Eastw.
  lanata Gray

  lapidicola Heller
  lassenensis Eastw.
  latifolia Hook. & Arn.
      ssp. latifolia
      ssp. mendocinensis Eastw.
          SY=C. m. (Eastw.) Pennell
  lemmonii Gray
  leschkeana J.T. Howell
  levisecta Greenm.
  linariifolia Benth. ex DC.
          SY=C. l. var. omnipubescens
              (Pennell) Clokey
          SY=C. trainii Edwin
  lineata Greene
  litoralis Pennell
          SY=C. wightii Elmer ssp. l.
              (Pennell) Munz
  longispica A. Nels.
  ludoviciana Pennell
  lutescens (Greenm.) Rydb.
  martinii Abrams
          var. clokeyi (Pennell) N. Holmgren
              SY=C. c. Pennell
          var. ewanii (Eastw.) N. Holmgren
              SY=C. e. Eastw.
              SY=C. m. ssp. e. (Eastw.) Munz
          var. martinii
              SY=C. gyroloba Pennell
              SY=C. roseana Eastw.
  mexicana (Hemsl.) Gray
          SY=C. tortifolia Pennell
  miniata Dougl. ex Hook.
          var. dixonii (Fern.) A. Nels. & J.F.
              Macbr.
              SY=C. d. Fern.
          var. miniata
              SY=C. chrymactis Pennell
              SY=C. confusa Greene
              SY=C. oblongifolia Gray
              SY=C. peckiana Pennell
  minor (Gray) Gray
  mollis Pennell
  muelleri Pennell
  nana Eastw.
  neglecta Zeile
  nervata Eastw.
          SY=C. latebracteata Pennell
  nivea Pennell & Ownbey
  occidentalis Torr.
          SY=C. parvula Rydb.
  oresbia Greenm.
  organorum Standl.
  pallescens (Nutt.) Greenm.
  pallida (L.) Spreng.
          var. caudata (Pennell) Boivin
              SY=C. c. (Pennell) Rebr.
              SY=C. c. var. auricoma (Pennell)
                  Hulten
              SY=C. p. ssp. c. Pennell
          var. elegans (Malte) Boivin
              SY=C. e. Malte
              SY=C. p. ssp. e. (Malte) Pennell
          var. pallida
  parviflora Bong.
          var. albida (Pennell) Ownbey
              SY=C. oreopola Greenm. ssp. a.
                  Pennell
          var. olympica (G.N. Jones) Ownbey
          var. oreopola (Greenm.) Ownbey
              SY=C. henryae Pennell
              SY=C. o. Greenm.
          var. parviflora
  patriotica Fern.
          SY=C. p. var. blumeri (Standl.)
              Kearney & Peebles
  peirsonii Eastw.
  pilosa (S. Wats.) Rydb.

Castilleja (CONT.)
    ssp. jusselii (Eastw.) Munz
    ssp. pilosa
        SY=C. psittacina (Eastw.) Pennell
        SY=C. rubida Piper var. monoensis
          (Jepson) Edwin
plagiotoma Gray
praeterita Heckard & Bacig.
pruinosa Fern.
puberula Rydb.
pulchella Rydb.
purpurea (Nutt.) G. Don
    var. citrina (Pennell) Shinners
        SY=C. c. Pennell
        SY=C. labiata Pennell
    var. lindheimeri (Gray) Shinners
        SY=C. mearnsii Pennell
        SY=C. williamsii Pennell
    var. purpurea
raupii Pennell
revealii N. Holmgren
rhexifolia Rydb.
        SY=C. lauta A. Nels.
        SY=C. leonardii Rydb.
        SY=C. oregonensis Gandog.
rubida Piper
rupicola Piper
rustica Piper
salsuginosa N. Holmgren
scabrida Eastw.
schizotricha Greenm.
        SY=C. arachnoidea Greenm. ssp. s.
          (Greenm.) Pennell
septentrionalis Lindl.
        SY=C. acuminata sensu auctt. non
          (Pursh) Spreng.
        SY=C. pallida (L.) Spreng. var. s.
          (Lindl.) Gray
sessiliflora Pursh
steenensis Pennell
stenantha Gray
        SY=C. spiralis Jepson
        SY=C. stenantha ssp. spiralis
          (Jepson) Munz
subinclusa Greene
suksdorfii Gray
sulphurea Rydb.
        SY=C. luteovirens Rydb.
        SY=C. mogollonica Pennell
tenuiflora Benth.
        SY=C. laxa Gray
thompsonii Pennell
        SY=C. villicaulis Pennell & Ownbey
uliginosa Eastw.
unalaschcensis (Cham. & Schlecht.) Malte
viscidula Gray
        SY=C. magnistylis Edwin
wallowensis Pennell
wightii Elmer
    ssp. rubra Pennell
    ssp. wightii
        SY=C. w. ssp. anacapensis (Dunkle)
          Pennell
wootonii Standl.
xanthotricha Pennell
yukonis Pennell

Chaenorrhinum (DC.) Reichenb.

  minus (L.) Lange

Chelone L.

  cuthbertii Small
        SY=C. grimesii Weatherby
  glabra L.
    var. dilatata Fern. & Wieg.

    var. elatior Raf.
        SY=C. montana (Raf.) Pennell &
          Wherry
    var. glabra
        SY=C. chlorantha Pennell & Wherry
        SY=C. g. var. c. (Pennell &
          Wherry) Cooperrider
        SY=C. g. var. elongata Pennell &
          Wherry
        SY=C. g. var. ochroleuca Pennell &
          Wherry
        SY=C. g. var. typica Pennell
    var. linifolia Coleman
lyonii Pursh
obliqua L.
    var. erwiniae Pennell & Wherry
    var. obliqua
    var. speciosa Pennell & Wherry

Chionophila Benth.

  jamesii Benth.
  tweedyi (Canby & Rose) Henderson

Collinsia Nutt.

  antonina Hardham
    ssp. antonina
    ssp. purpurea Hardham
  bartsiifolia Benth.
    var. bartsiifolia
    var. davidsonii (Parish) Newsom
    var. hirsuta (Kellogg) Pennell
    var. stricta Newsom
  callosa Parish
  childii Parry ex Gray
  concolor Greene
  corymbosa Herder
  grandiflora Dougl. ex Lindl.
    var. grandiflora
        SY=C. tenella (Pursh) Piper
    var. pusilla Gray
  greenei Gray
  heterophylla Buist ex Graham
    var. austromontana (Newsom) Munz
        SY=C. a. (Newsom) Pennell
    var. heterophylla
  linearis Gray
        SY=C. rattanii Gray var. l. (Gray)
          Newsom
  multicolor Lindl. & Paxton
        SY=C. franciscana Bioletti
  parryi Gray
  parviflora Dougl. ex Lindl.
  rattanii Gray
    ssp. glandulosa (T.J. Howell) Pennell
    ssp. rattanii
  sparsiflora Fisch. & Mey.
    var. arvensis (Greene) Jepson
    var. brucae (M.E. Jones) Newsom
        SY=C. b. M.E. Jones
    var. collina (Jepson) Newsom
        SY=C. solitaria Kellogg
    var. sparsiflora
  tinctoria Hartw. ex Benth.
  torreyi Gray
    var. latifolia Newsom
    var. torreyi
    var. wrightii (S. Wats.) I.M.
      Johnston
        SY=C. w. S. Wats.
  verna Nutt.
  violacea Nutt.

Cordylanthus Nutt. ex Benth.

  bernardinus Munz

Cordylanthus (CONT.)
   capillaris Pennell
   capitatus Nutt. ex Benth.
            SY=C. nevadensis Edwin
   eremicus Coville & Morton
            SY=C. ramosus Nutt. ex Benth. ssp.
               e. (Coville & Morton) Munz
   ferrisianus Pennell
   filifolius Nutt. ex Benth.
   hansenii (Ferris) J.F. Macbr.
   helleri (Ferris) J.F. Macbr.
   kingii S. Wats.
   laxiflorus Gray
   littoralis (Ferris) J.F. Macbr.
      ssp. littoralis
            SY=C. rigidus (Benth.) Jepson ssp.
               l. (Ferris) Jepson
      ssp. platycephalus (Pennell) Munz
            SY=C. p. Pennell
   maritimus Nutt. ex Benth.
      ssp. canescens (Gray) Chuang & Heckard
            SY=C. c. Gray
      ssp. maritimus
      ssp. palustris (Behr) Chuang & Heckard
   mollis Gray
      ssp. hispidus (Pennell) Chuang & Heckard
            SY=C. h. Pennell
      ssp. mollis
   nevinii Gray
   nidularius J.T. Howell
   orcuttianus Greene
   pallescens Pennell
   palmatus (Ferris) J.F. Macbr.
            SY=C. carnulosus Pennell
            SY=C. p. ssp. c. (Pennell) Munz
   parviflorus (Ferris) Wiggins
   pilosus Gray
      ssp. bolanderi (Gray) Munz
            SY=C. b. (Gray) Pennell
      ssp. diffusus (Pennell) Munz
            SY=C. d. Pennell
      ssp. pilosus
   pringlei Gray
   ramosus Nutt. ex Benth.
      ssp. ramosus
            SY=C. r. var. puberulus J.F.
               Macbr.
      ssp. setosus Pennell
   rigidus (Benth.) Jepson
      ssp. brevibracteatus (Gray) Munz
            SY=C. compactus Pennell
      ssp. rigidus
   tecopensis Munz & Roos
   tenuifolius Pennell
   tenuis Gray
      ssp. brunneus (Jepson) Munz
            SY=C. b. (Jepson) Pennell
      ssp. tenuis
   viscidus (T.J. Howell) Pennell
   wrightii Gray ex Torr.
            var. pauciflorus Kearney & Peebles
            var. wrightii

Cymbalaria Hill

   muralis Gaertn., Mey. & Scherb.
            SY=Linaria cymbalaria (L.) P.
               Mill.

Dasystoma Raf.

   macrophylla (Nutt.) Raf.
            SY=Afzelia m. (Nutt.) Kuntze
            SY=Seymeria m. Nutt.

Digitalis L.

   grandiflora P. Mill.
            SY=D. ambigua Murr.
            SY=D. orientalis P. Mill.
   lanata Ehrh.
   lutea L.
   purpurea L.
            var. alba Hort.
            var. purpurea

Dopatrium Buch.-Ham.

   junceum (Roxb.) Buch.-Ham.

Euphrasia L.

   X aequalis Callen [nemorosa X tetraquetra]
   artica Lange ex Rostr.
      ssp. borealis (Townsend) Yeo
            SY=E. b. (Townsend) Wettst.
   bottnica Kihlm.
   disjuncta Fern. & Wieg.
            SY=E. arctica Lange ex Rostr. var.
               d. (Fern. & Wieg.) Cronq.
   frigida Pugsley
            SY=E. arctica sensu auctt. non
               Lange ex Rostr.
            SY=?E. a. var. inundata (F. Jorg.)
               Callen
            SY=E. a. var. obtusata (F. Jorg.)
               Callen
            SY=E. a. var. submollis (F. Jorg.)
               Callen
            SY=E. f. var. obtusata (F. Jorg.)
               Callen
            SY=E. f. var. submollis (F. Jorg.)
               Callen
   hudsoniana Fern. & Wieg.
            var. contracta Sell & Yeo
            var. hudsoniana
            var. ramosior Sell & Yeo
   micrantha Reichenb.
   mollis (Ledeb.) Wettst.
            SY=E. arctica Lange ex Rostr. var.
               m. (Ledeb.) Welsh
   nemorosa (Pers.) Wallr.
            SY=E. americana Wettst.
            SY=E. canadensis sensu auctt. non
               Townsend
            SY=E. curta sensu auctt. pro parte
   oakesii Wettst.
            SY=E. williamsii B.L. Robins.
            SY=E. w. var. vestita Fern. &
               Wieg.
   ostenfeldii (Pugsley) Yeo
            SY=E. cururta sensu auctt. pro
               parte
            SY=E. mollis (Ledeb.) Wettst. var.
               laurentiana Boivin
   randii B.L. Robins.
            var. farlowii B.L. Robins.
            var. randii
            SY=E. purpurea (Raf.) Fern.
            SY=E. r. var. reeksii Fern.
   salisburgensis Funck
            var. hibernica Pugsley
   stricta J.P. Wolff ex J.F. Lehm.
            SY=E. condensata Jord.
            SY=E. officinalis L.
            SY=E. rigidula Jord.
   subarctica Raup
            SY=E. arctica Lange ex Rostr. var.
               dolosa (Boivin) Boivin
            SY=E. disjuncta Fern. & Wieg var.
               dolosa Boivin
            SY=E. pennellii Callen pro parte
   suborbicularis Sell & Yeo
            SY=E. mollis (Ledeb.) Wettst. var.

Euphrasia (CONT.)
            laurentiana Boivin
    tatarica Fisch.
    tetraquetra (Brebiss.) Arrondeau
    vinacea Sell & Yeo

Galvezia Domb. ex Juss.

    speciosa (Nutt.) Gray

Gratiola L.

    aurea Pursh
            SY=G. lutea Raf.
    brevifolia Raf.
    ebracteata Benth.
    flava Leavenworth
    floridana Nutt.
    heterosepala Mason & Bacig.
    neglecta Torr.
            var. glaberrima Fern.
            var. neglecta
    pilosa Michx.
            SY=Sophronanthe p. (Michx.) Small
            SY=Tragiola p. (Michx.) Small &
                Pennell
    ramosa Walt.
    subulata Baldw.
            SY=Sophronanthe hispida Benth.
    virginiana L.
            var. aestuariorum Pennell
            var. virginiana
            SY=G. sphaerocarpa Ell.
    viscidula Pennell
            var. shortii (Durand) Gleason
            SY=G. v. ssp. s. (Durand) Pennell
            var. viscidula
            SY=G. viscosa Schwein.

Hebe Comm. ex Juss.

    X franciscana (Eastw.) Souster
    speciosa (R. Cunningham ex A. Cunningham)
        J.C. Andersen

Keckiella Straw

    antirrhinoides (Benth.) Straw
        ssp. antirrhinoides
            SY=Penstemon a. Benth.
        ssp. microphylla (Gray) Straw
            SY=Penstemon a. Benth. ssp. m.
                (Gray) Keck
            SY=P. m. Gray
    breviflora (Lindl.) Straw
        ssp. breviflora
            SY=Penstemon b. Lindl.
        ssp. glabrisepala (Keck) Straw
            SY=Penstemon b. Lindl. ssp. g.
                Keck
    cordifolia (Benth.) Straw
            SY=Penstemon c. Benth.
    corymbosa (Benth.) Straw
            SY=Penstemon c. Benth.
    lemmonii (Gray) Straw
            SY=Penstemon l. Gray
    rothrockii (Gray) Straw
        ssp. jacintensis (Abrams) Keck
            SY=Penstemon r. Gray ssp. j.
                (Abrams) Keck
        ssp. rothrockii
            SY=Penstemon r. Gray
    ternata (Torr. ex Gray) Straw
        ssp. septentrionalis (Munz & Johnston)
            Straw
            SY=Penstemon t. Torr. ex Gray ssp.
                s (Munz & Johnston) Keck

        ssp. ternata
            SY=Penstemon t. Torr. ex Gray

Kickxia Dumort.

    elatine (L.) Dumort.
    spuria (L.) Dumort.

Lagotis Gaertn.

    glauca Gaertn.
    minor (Willd.) Standl.
            SY=L. glauca Gaertn. ssp. m.
                (Willd.) Hulten
            SY=L. g. var. stelleri (Cham. &
                Schlecht.) Trautv.
            SY=L. s. (Cham. & Schlecht.) Raup

Lendneria Minod

    verticillata (P. Mill.) Britt.

Leucophyllum Humb. & Bonpl.

    candidum I.M. Johnston
            SY=L. violaceum Pennell
    frutescens (Berl.) I.M. Johnston
    minus Gray

Leucospora Nutt.

    multifida (Michx.) Nutt.
            SY=Conobea m. (Michx.) Benth.

Limnophila R. Br.

    indica (L.) Druce
    X ludoviciana Thieret [indica X
        sessiliflora]
    sessiliflora (Vahl) Blume

Limosella L.

    acaulis Sesse & Moc.
    aquatica L.
    australis R. Br.
            SY=L. aquatica L. var. tenuifolia
                Hoffman
            SY=L. subulata Ives
    pubiflora Pennell

Linaria P. Mill.

    aeruginea (Gouan) Cav.
    angustissima (Loisel.) Borbas
            SY=L. italica Trev.
    bipartita (Vent.) Willd.
    canadensis (L.) Dum.-Cours.
    floridana Chapman
    genistifolia (L.) P. Mill.
        ssp. dalmatica (L.) Maire & Petitmengin
            SY=L. d. (L.) P. Mill.
            SY=L. d. var. macedonica Fanzl.
            SY=L. m. (Fanzl.) Griseb.
        ssp. genistifolia
    maroccana Hook. f.
    purpurea (L.) P. Mill.
    repens (L.) P. Mill.
    reticulata (Sm.) Desf.
            SY=L. pinifolia (Poir.) Thellung
    X sepium Allman [repens X vulgaris]
    spartea (L.) Willd.
    supina (L.) Chaz.
    texana Scheele
            SY=L. canadensis (L.) Dum.-Cours.
                var. t. (Scheele) Pennell
    vulgaris P. Mill.

Linaria (CONT.)
            SY=L. linaria (L.) Karst.

Lindernia All.

    brucei Howard
    crustacea (L.) F. Muell.
            var. crustacea
            var. smithii Deg. & Ruhle
    diffusa (L.) Wettst.
            SY=Vandellia d. L.
    dubia (L.) Pennell
            var. anagallidea (Michx.) Cooperrider
                SY=Ilysanthes inequalis (Walt.)
                    Pennell
                SY=L. a. (Michx.) Pennell
            var. dubia
                SY=Ilysanthes attenuata (Muhl.)
                    Small
                SY=I. d. (L.) Barnh.
                SY=L. d. var. riparia (Raf.) Fern.
                SY=L. d. var. major (Pursh)
                    Pennell
                SY=L. d. var. typica Pennell
            var. inundata (Pennell) Pennell
    grandiflora Nutt.
            SY=Ilysanthes g. (Nutt.) Benth.
    monticola Muhl. ex Nutt.
            SY=Ilysanthes m. (Muhl. ex Nutt.)
                Raf.
    procumbens (Krok) Philcox
            SY=L. pyxidaria L. pro parte
    saxicola M.A. Curtis
            SY=Ilysanthes s. (M.A. Curtis)
                Chapman

Macranthera Torr. ex Benth.

    flammea (Bartr.) Pennell

Maurandya Ortega

    acerifolia Pennell
            SY=Asarina a. (Pennell) Pennell
    antirrhiniflora Humb. & Bonpl. ex Willd.
            SY=Asarina a. (Humb. & Bonpl. ex
                Willd.) Pennell
            SY=A. maurandioides Gray
            SY=Maurandella a. (Humb. & Bonpl.
                ex Willd.) Rothm.
    petrophila Coville & Morton
            SY=Asarina p. (Coville & Morton)
                Pennell
            SY=Maurandella p. (Coville &
                Morton) Rothm.
    wislizenii Engelm.
            SY=Asarina w. (Engelm.) Pennell
            SY=Epiziphium w. (Engelm.) Munz

Mazus Lour.

    miguelii Makino
            SY=M. reptans N.E. Br.
    pumilus (Burm. f.) Steenis
            SY=M. japonicus (Thunb.) Kuntze

Mecardonia Ruiz & Pavon

    acuminata (Walt.) Small
            var. acuminata
                SY=Bacopa a. (Walt.) B.L. Robins.
                SY=Pagesia a. (Walt.) Pennell
            var. microphylla (Raf.) Pennell
                SY=Bacopa a. (Walt.) B.L. Robins.
                    var. m. (Raf.) Fern.
            var. peninsularis Pennell
    dianthera (Sw.) Pennell

    tenuis Small
    vandellioides (H.B.K.) Pennell
            SY=M. peduncularis (Benth.) Small
            SY=M. procumbens (P. Mill.) Small

Melampyrum L.

    lineare Desr.
            var. americanum (Michx.) Beauverd
            var. latifolium Bart.
            var. lineare
            var. pectinatum (Pennell) Fern.

Melasma Berg.

    melampyroides (L.C. Rich.) Pennell

Micranthemum Michx.

    glomeratum (Chapman) Shinners
            SY=Hemianthus g. (Chapman) Pennell
    micranthemoides (Nutt.) Wettst.
            SY=Hemianthus m. Nutt.
    umbrosum (Walt.) Blake
            SY=Globifera u. (Walt.) J.F. Gmel.
            SY=Hemianthus callitrichoides
                Griseb.

Mimetanthe Greene

    pilosa (Benth.) Greene
            SY=Mimulus p. (Benth.) S. Wats.

Mimulus L.

    acutidens Greene
            SY=M. inconspicuus Gray var. a.
                (Greene) Gray
    alatus Ait.
    alsinoides Dougl. ex Benth.
    androsaceus Curran ex Greene
            SY=M. palmeri Gray var. a. (Curren
                ex Greene) Gray
    angustatus (Gray) Gray
            SY=Eunanus a. (Gray) Greene
            SY=E. coulteri Harvey & Gray var.
                a. Gray
            SY=M. clarkii Kellogg ex Curran
            SY=M. tricolor Hartw. ex Lindl.
                var. a. Gray
    angustifolius (Greene) A.L. Grant
            SY=Eunanus a. Greene
    arenarius A.L. Grant
    aridus (Abrams) A.L. Grant
            SY=Diplacus a. Abrams
    aurantiacus W. Curtis
        ssp. aurantiacus
            SY=Diplacus a. (W. Curtis) Jepson
            SY=D. glutinosus (J.C. Wendl.)
                Nutt.
            SY=D. latifolius Nutt.
            SY=D. leptanthus Nutt.
            SY=M. g. J.C. Wendl.
            SY=M. g. var. aurantiacus (W.
                Curtis) Lindl.
            SY=M. leptanthus (Nutt.) A.L.
                Grant
            SY=M. viscosus Moench
        ssp. australis (McMinn) Munz
            SY=Diplacus a. McMinn
            SY=D. aurantiacus Jepson ssp.
                australis (McMinn) P.M.
                Beeks
        ssp. lompocensis (McMinn) Munz
            SY=Diplacus l. McMinn
    austinae (Greene) A.L. Grant
    barbatus Greene

Mimulus (CONT.)
    SY=M. deflexus S. Wats.
  bicolor Hartw. ex Benth.
    SY=M. prattenii Dur.
  bifidus Pennell
    ssp. bifidus
      SY=Diplacus glutinosus (J.C.
        Wendl.) Nutt. var.
        grandiflorus Lindl. pro
        parte
      SY=D. grandiflorus (Lindl.)
        Groenl.
    ssp. fasciculatus Pennell
      SY=Diplacus grandiflorus Greene
      SY=D. f. (Pennell) McMinn
  bigelovii (Gray) Gray
    var. bigelovii
      SY=Eunanus b. Gray
    var. cuspidatus A.L. Grant
      SY=M. washoensis Edwin
    var. panamintensis Munz
  biolettii Eastw.
  bolanderi Gray
    SY=Eunanus b. (Gray) Greene
    SY=M. bolanderi var. brachydontus
      A.L. Grant
    SY=M. brevipes Gray
  brachiatus Pennell
  brachystylis Edwin
  brandegei Pennell
  breviflorus Piper
    SY=M. inflatulus Suksdorf
  brevipes Benth.
    SY=Eunanus b. (Benth.) Greene
  breweri (Greene) Coville
    SY=Eunanus b. Greene
    SY=M. rubellus Gray var. b.
      (Greene) Jepson
  cardinalis Dougl. ex Benth.
    SY=Diplacus c. (Dougl. ex Benth.)
      Groenl.
    SY=Erythranthe c. (Dougl. ex
      Benth.) Spach
    SY=M. c. var. exsul Greene
    SY=M. c. var. griseus Greene
    SY=M. c. var. rigens Greene
  cleistogamus J.T. Howell
  clevelandii Brandeg.
    SY=Diplacus c. (Brandeg.) Greene
  clivicola Greenm.
    SY=Eunanus c. (Greenm.) Heller
  coccineus Congd.
    SY=M. stamineus A.L. Grant
    SY=M. wolfii Eastw.
  congdonii B.L. Robins.
    SY=Eunanus c. (B.L. Robins.)
      Greene
    SY=E. douglasii (Benth.) Gray var.
      parviflorus Greene
    SY=E. kelloggii (Curran ex Greene)
      Curran ex Greene var. p.
      (Greene) Jepson
    SY=M. modestus Eastw.
  cusickii (Greene) Piper
    SY=Eunanus c. Greene
    SY=M. bigelovii (Gray) Gray var.
      ovatus Gray
  densus A.L. Grant
  dentatus Nutt. ex Benth.
  dentilobus Robins. & Fern.
    SY=M. parvulus Woot. & Standl.
  diffusus A.L. Grant
    SY=M. grantianus Eastw.
  discolor A.L. Grant
  douglasii (Benth.) Gray
    SY=Eunanus d. Benth.
    SY=E. subuniflorus (Hook. & Arn.)

    Greene
    SY=M. atropurpureus Kellogg
    SY=M. nanus Hook. & Arn. var. s.
      Hook. & Arn.
    SY=M. s. (Greene) Jepson
    SY=M. s. (Hook. & Arn.) Piper
  dudleyi A.L. Grant
  eastwoodiae Rydb.
    SY=M. cardinalis Eastw.
  exiguus Gray
  filicaulis S. Wats.
  flemingii Munz
    SY=Diplacus parviflorus Greene
    SY=M. p. (Greene) A.L. Grant
  floribundus Dougl.
    var. floribundus
      SY=M. deltoides Gandog.
      SY=M. multiflorus Pennell
      SY=M. peduncularis Dougl. ex
        Benth.
      SY=M. pubescens Benth.
      SY=M. serotinus Suksdorf
      SY=M. trisulcatus Pennell
    var. geniculatus (Greene) A.L. Grant
      SY=M. g. Greene
    var. membranaceus (A. Nels.) A.L.
      Grant
      SY=M. m. A. Nels.
    var. subulatus A.L. Grant
      SY=M. s. (A.L. Grant) Pennell
    ssp. ?moorei Iltis
  fremontii (Benth.) Gray
    SY=Eunanus f. Benth.
  gemmiparus W.A. Weber
  glabratus H.B.K.
    var. fremontii (Benth.) A.L. Grant
      SY=M. geyeri Torr.
      SY=M. glabratus var. jamesii Gray
      SY=M. inamoenus Greene
      SY=M. j. Torr. & Gray ex Benth.
      SY=M. j. var. fremontii Benth.
      SY=M. j. var. texensis Gray
      SY=M. reniformis Engelm. ex Benth.
    var. glabratus
      SY=M. g. ssp. typicus Pennell
    var. michiganensis (Pennell) Fassett
      SY=M. g. ssp. m. Pennell
    var. oklahomensis Fassett
    var. utahensis Pennell
  glaucescens Greene
    SY=M. guttatus Fisch. ex DC. var.
      glaucescens (Greene) Jepson
  gracilipes B.L. Robins.
  grayi A.L. Grant
  guttatus Fisch. ex DC.
    ssp. arvensis (Greene) Munz
      SY=M. a. Greene
      SY=M. g. var. a. (Greene) A.L.
        Grant
      SY=M. langsdorfii Donn ex Greene
        var. arvensis (Greene)
        Jepson
    ssp. guttatus
      SY=M. clementinus Greene
      SY=M. decorus (A.L. Grant)
        Suksdorf
      SY=M. equinnus Greene
      SY=M. glabratus Gray var.
        ascendens Gray
      SY=M. grandiflorus J.T. Howell
      SY=M. guttatus var. d. A.L. Grant
      SY=M. guttatus var. depauperatus
        (Gray) A.L. Grant
      SY=M. guttatus var. gracilis
        (Gray) Campbell
      SY=M. guttatus var. grandis Greene
      SY=M. guttatus var. hallii

Mimulus (CONT.)

(Greene) A.L. Grant
   SY=M. guttatus var. laxus (Pennell
     ex M.E. Peck) M.E. Peck
   SY=M. guttatus var. microphyllus
     (Benth.) Pennell ex M.E.
     Peck
   SY=M. guttatus var. puberulus
     (Greene) A.L. Grant
   SY=M. h. Greene
   SY=M. hirsutus J.T. Howell
   SY=M. langsdorfii Donn ex Greene
   SY=M. langsdorfii var. argutus
     Greene
   SY=M. langsdorfii var. guttatus
     (Greene) Jepson
   SY=M. langsdorfii var.
     microphyllus (Benth.) A.
     Nels. & J.F. Macbr.
   SY=M. langsdorfii var. minimus
     Henry
   SY=M. langsdorfii var.
     platyphyllus Greene
   SY=M. laxus Pennell ex M.E. Peck
   SY=M. longulus Greene
   SY=M. luteus L. var. depauperatus
     Gray
   SY=M. maguirei Pennell
   SY=M. microphyllus Benth.
   SY=M. paniculatus Greene
   SY=M. petiolaris Greene
   SY=M. prionophyllus Greene
   SY=M. procerus Greene
   SY=M. puberulus Greene
   SY=M. rivularis Nutt.
   SY=M. tenellus Nutt. ex Gray
   SY=M. thermalis A. Nels.
   SY=M. unimaculatus Pennell
ssp. haidensis Calder & Taylor
ssp. litoralis Pennell
   SY=M. grandis (Greene) Heller
   SY=M. grandis ssp. arenicola
     Pennell
   SY=M. guttatus var. grandis Greene
   SY=M. langsdorfii Donn ex Greene
     var. grandis Greene
ssp. micranthus (Heller) Munz
   SY=M. m. Heller
   SY=M. nasutus Greene var. m.
     (Heller) A.L. Grant
ssp. scouleri (Hook.) Pennell
   SY=M. g. var. s. (Hook.) Pennell
   SY=M. s. Hook.
inconspicuus Gray
jepsonii A.L. Grant
   SY=M. microcarpus Pennell
johnstonii A.L. Grant
jungermannioides Suksdorf
kelloggii (Curran ex Greene) Curran ex Gray
   SY=Eunanus k. Curran ex Greene
laciniatus Gray
   SY=M. eisenii Kellogg
latidens (Gray) Greene
   SY=M. inconspicuus Gray var. l.
     Gray
latifolius Gray
layneae (Greene) Jepson
   SY=Eunanus l. Greene
leptaleus Gray
   SY=Eunanus l. (Gray) Greene
lewisii Pursh
linearis Benth.
   SY=Diplacus l. (Benth.) Greene
   SY=D. longiflorus (Nutt.) A.L.
     Grant var. linearis
     (Benth.) McMinn
   SY=M. glutinosus J.C. Wendl. var.

     linearis (Benth.) Gray
   SY=M. longiflorus (Nutt.) A.L.
     Grant var. linearis (Benth.)
     A.L. Grant
longiflorus (Nutt.) A.L. Grant
ssp. calycinus (Eastw.) Munz
   SY=Diplacus c. Eastw.
   SY=D. longiflorus Nutt. var. c.
     (Eastw.) Jepson
   SY=M. l. var. c. (Eastw.) A.L.
     Grant
ssp. longiflorus
   SY=Diplacus arachnoideus Greene
   SY=D. leptanthus Benth.
   SY=D. longiflorus Nutt.
   SY=D. rutilus (A.L. Grant) McMinn
   SY=D. speciosus Burtt-Davy
   SY=M. glutinosus J.C. Wendl. var.
     brachypus Gray
   SY=M. longiflorus var. rutilus
     A.L. Grant
mephiticus Greene
   SY=Eunanus m. (Greene) Greene
mohavensis Lemmon
   SY=E. m. (Lemmon) Greene
montioides Gray
   SY=M. discolor A.L. Grant
   SY=M. rubellus Gray var.
     latiflorus S. Wats.
moschatus Dougl.
   var. longiflorus Gray
   SY=M. dentatus Nutt. ex Benth.
     var. gracilis Gray
   SY=M. macranthus Pennell
   SY=M. moschatus var. pallidiflorus
     Suksdorf
   var. moniliformis (Greene) Munz
   SY=M. inodorus Greene
   SY=M. leibergii A.L. Grant
   SY=M. moniliformis Greene
   var. moschatus
   var. sessilifolius Gray
nanus Hook. & Arn.
   SY=Eunanus austinae Greene
   SY=E. n. (Hook. & Arn.) Holz.
   SY=E. tolmiei Benth.
   SY=M. a. (Greene) A.L. Greene
   SY=M. microphyton Pennell
   SY=M. t. (Benth.) Rydb.
nasutus Greene
   var. insignis (Greene) A.L. Grant
   SY=M. guttatus Fisch. ex DC. var.
     i. Greene
   SY=M. langsdorfii Donn ex Greene
     var. i. Greene
   var. nasutus
   SY=M. bakeri Gandog.
   SY=M. cordatus Greene
   SY=M. cuspidatus Greene
   SY=M. glareosus Greene
   SY=M. guttatus Fisch. ex DC. var.
     gracilis (Gray) Campbell
   SY=M. guttatus var. lyratus
     (Benth.) Pennell ex M.E.
     Peck
   SY=M. guttatus var. nasutus
     (Greene) Jepson
   SY=M. langsdorfii Donn ex Greene
     var. californicum Jepson
   SY=M. l. var. nasutus (Greene)
     Jepson
   SY=M. luteus L. var. gracilis Gray
   SY=M. lyratus Benth.
   SY=M. marmoratus Greene
   SY=M. minusculus Greene
   SY=M. pardalis Pennell
   SY=M. parishii Gandog.

Mimulus (CONT.)
        SY=M. puberulus Gandog.
        SY=M. puncticalyx Gandog.
        SY=M. subreniformis Greene
nudatus Curran ex Greene
palmeri Gray
parishii Greene
        SY=Eunanus p. (Gray) Greene
parryi Gray
pictus (Curran ex Greene) Gray
        SY=Eunanus p. Curran ex Greene
platycalyx Pennell
platylaemus Pennell
primuloides Benth.
    var. linearifolius A.L. Grant
        SY=M. l. (A.L. Grant) Pennell
        SY=M. linearis sensu auctt. non
            Benth.
        SY=M. p. ssp. linearifolius (A.L.
            Grant) Munz
    var. primuloides
        SY=M. nevadensis Gandog.
        SY=M. pilosellus Greene
        SY=M. primuloides var. minimus
            M.E. Peck
        SY=M. primuloides var. pilosellus
            (Greene) Smiley
pulchellus (E. Drew ex Greene) A.L. Grant
        SY=Eunanus p. E. Drew ex Greene
pulsiferae Gray
puniceus (Nutt.) Steud.
        SY=Diplacus glutinosus (J.C.
            Wendl.) Nutt. var. p.
            (Nutt.) Benth.
        SY=D. puniceus Nutt.
        SY=M. glutinosus J.C. Wendl. var.
            p. (Nutt.) Gray
purpureus A.L. Grant
    var. pauxillus A.L. Grant
    var. purpureus
pygmaeus A.L. Grant
        SY=M. minutissimus Eastw.
rattanii Gray
    ssp. decurtatus (A.L. Grant) Pennell
        SY=M. d. A.L. Grant
        SY=M. r. var. d. (A.L. Grant)
            Pennell
    ssp. rattanii
        SY=Eunanus r. (Gray) Greene
reifschneiderae Edwin
ringens L.
    var. colophilus Fern.
    var. minthodes (Greene) A.L. Grant
        SY=M. m. Greene
    var. ringens
        SY=M. pallidus Salisb.
        SY=M. r. var. congesta Farw.
rubellus Gray
        SY=M. gratioloides Rydb.
rupicola Coville & A.L. Grant
spissus A.L. Grant
    var. lincolensis Edwin
    var. spissus
subsecundus Gray
        SY=Eunanus s. (Gray) Greene
suksdorfii Gray
tilingii Regel
    var. caespitosus (Greene) A.L. Grant
        SY=M. alpinus (Gray) Piper
        SY=M. c. Greene
        SY=M. c. var. implexus (Greene)
            M.E. Peck
        SY=M. i. Greene
        SY=M. luteus L. var. alpinus Gray
        SY=M. scouleri Hook. var. c.
            Greene
    var. corallinus (Greene) A.L. Grant

        SY=M. c. Greene
        SY=M. implicatus Greene
        SY=M. lucens Greene
        SY=M. minusculus Greene
        SY=M. veronicifolius Greene
    var. tilingii
        SY=M. langsdorfii Donn ex Greene
            var. argutus Greene
        SY=M. l. var. minor (A. Nels.)
            Cockerell
        SY=M. l. Donn ex Greene var. t.
            (Regel) Greene
        SY=M. minor A. Nels.
torreyi Gray
        SY=Eunanus t. (Gray) Greene
traskiae A.L. Grant
tricolor Hartw. ex Lindl.
        SY=Eunanus coulteri Harvey & Gray
            ex Benth.
        SY=F. t. (Hartw. ex Lindl.) Greene
verbenaceus Greene
        SY=M. cardinalis Dougl. ex Benth.
            var. v. (Greene) Kearney &
            Peebles
viscidus Congd.
    ssp. constrictus (A.L. Grant) Munz
        SY=M. c. (A.L. Grant) Pennell
        SY=M. subsecundus Gray var. c.
            A.L. Grant
    ssp. viscidus
        SY=M. fremontii (Benth.) Gray var.
            v. (Congd.) Jepson
        SY=M. subsecundus Gray var. v.
            (Congd.) A.L. Grant
washingtonensis Gandog.
        SY=M. ampliatus A.L. Grant
        SY=M. patulus Pennell
        SY=M. peduncularis Gray
whipplei A.L. Grant
whitneyi Gray
        SY=Eunanus bicolor Gray
        SY=M. nanus Hook. & Arn. var. b.
            (Gray) Gray

Misopates Raf.

    orontium (L.) Raf.
        SY=Antirrhinum o. L.

Mohavea Gray

    breviflora Coville
    confertiflora (Benth.) Heller

Neogaerrhinum Rothm.

    kellogii (Greene) Thieret
        SY=Antirrhinum k. Greene
        SY=Asarina stricta (Hook. & Arn.)
            Pennell

Nothochelone (Gray) Straw

    nemorosa (Dougl. ex Lindl.) Straw
        SY=Penstemon n. (Dougl. ex Lindl.)
            Trautv.

Odontites Ludwig

    verna (Bellardi) Dumort.
    ssp. serotina (Dumort.) Corb.
        SY=O. s Dumort.
    ssp. verna

Orthocarpus Nutt.

    attenuatus Gray

Orthocarpus (CONT.)
  barbatus Cotton
  bracteosus Benth.
  brevistylus (Hoover) Hoover
  campestris Benth.
  castillejoides Benth.
      var. castillejoides
      var. humboldtiensis Keck
  copelandii Eastw.
      var. copelandii
      var. cryptanthus (Piper) Keck
  cuspidatus Greene
  densiflorus Benth.
      var. densiflorus
      var. gracilis (Benth.) Keck
      var. obispoensis Keck
  erianthus Benth.
      var. erianthus
      var. gratiosus Jepson & Tracy
      var. micranthus (Gray) Jepson
      var. roseus Gray
  faucibarbatus Gray
      var. albidus (Keck) J.T. Howell
          SY=O. f. ssp. a. Keck
      var. faucibarbatus
  floribundus Benth.
  hispidus Benth.
  imbricatus Torr. ex S. Wats.
  lacerus Benth.
  lasiorhynchus Gray
  linearilobus Benth.
  lithospermoides Benth.
      var. bicolor (Heller) Jepson
      var. lithospermoides
  luteus Nutt.
  pachystachyus Gray
  purpurascens Benth.
      var. latifolius S. Wats.
      var. ornatus Jepson
      var. pallidus Keck
      var. purpurascens
          SY=O. p. var. palmeri Gray
  purpureo-albus Gray
  pusillus Benth.
  succulentus (Hoover) Hoover
          SY=O. campestris Benth. var. s.
              Hoover
  tenuifolius (Pursh) Benth.
  tolmiei Hook. & Arn.

Parentucellia Viviani

  viscosa (L.) Caruel

Pedicularis L.

  albertae Hulten
          SY=P. oederi Vahl var. a. (Hulten)
              Boivin
  angustissima Greene
  attollens Gray
    ssp. attollens
          SY=Elephantella (Gray) Heller
          SY=P. concinna Eastw.
    ssp. protogyna Pennell
  bracteosa Benth. ex Hook.
      var. atrosanguinea (Pennell &
          Thompson) Cronq.
          SY=P. a. Pennell & Thompson
      var. bracteosa
          SY=P. montanensis Rydb.
          SY=P. paddoensis Pennell
          SY=P. thompsonii Pennell
      var. canbyi (Gray) Cronq.
          SY=P. c. Gray
      var. flavida (Pennell) Cronq.
          SY=P. f. Pennell

      var. latifolia (Pennell) Cronq.
          SY=P. l. Pennell
      var. pachyrhiza (Pennell) Cronq.
          SY=P. p. Pennell
      var. paysoniana (Pennell) Cronq.
          SY=P. p. Pennell
      var. siifolia (Rydb.) Cronq.
          SY=P. s. Rydb.
  canadensis L.
    ssp. canadensis
      var. canadensis
          SY=P. gladiata Michx.
      var. dobbsii Fern.
    ssp. fluviatilis (Heller) W.A. Weber
          SY=P. f. Heller
  capitata J.F. Adams
          SY=P. nelsonii R. Br.
  centranthera Gray
      var. centranthera
      var. exulans M.E. Peck
  chamissonis Stev.
          SY=P. romanzovii Cham. ex Spreng.
  contorta Benth. ex Hook.
      var. contorta
      var. ctenophora (Rydb.) A. Nels. &
          J.F. Macbr.
          SY=P. c. Rydb.
          SY=P. lunata Rydb.
  crenulata Benth. ex DC.
          SY=P. albomarginata M.E. Jones
  cystopteridifolia Rydb.
          SY=P. elata Pursh
  densiflora Benth. ex Hook.
    ssp. aurantiaca E. Sprague
    ssp. densiflora
          SY=P. attenuata Benth.
          SY=P. brunnescens Heller
  dudleyi Elmer
  flammea L.
  furbishiae S. Wats.
  groenlandica Retz.
    ssp. groenlandica
          SY=Elephantella g. (Retz.) Rydb.
    ssp. surrecta (Benth. ex Hook.) Piper
          SY=P. g. var. s. (Benth. ex Hook.)
              Gray
  hirsuta L.
  howellii Gray
  labradorica Wirsing
      var. labradorica
          SY=P. euphrasioides Steph. ex
              Willd.
      var. sulphurea Hulten
  lanata Cham. & Schlecht.
          SY=P. kanei Dur.
          SY=P. wildenowii Vved.
  lanceolata Michx.
          SY=P. auriculata Sm.
          SY=P. pallida Nutt.
          SY=P. virginica Poir.
  langsdorfii Fisch. ex Stev.
    ssp. arctica (R. Br.) Pennell
          SY=P. a. R. Br.
          SY=P. hians Eastw.
          SY=P. l. var. a. (R. Br.) Polunin
    ssp. langsdorfii
  lapponica L.
  macrodonta Richards.
          SY=P. parviflora Sm. ex Rees var.
              m. (Richards.) Welsh
  oederi Vahl
          SY=P. versicolor Wahlenb.
  ornithorhyncha Benth.
          SY=P. pedicellata Bunge
          SY=P. subnuda Benth.
  palustris L.
  parryi Gray

Pedicularis (CONT.)
    ssp. mogollonica (Greene) G.D. Carr
        SY=P. m. Greene
    ssp. parryi
    ssp. purpurea (Parry) G.D. Carr
        SY=P. anaticeps Pennell
        SY=P. hallii Rydb.
        SY=P. parryi var. purpurea Parry
  parviflora Sm. ex Rees
    ssp. parviflora
        SY=P. pennellii Hulten var.
            insularis (Calder & Taylor)
            Boivin
    ssp. pennellii (Hulten) Hulten
        SY=P. pennellii Hulten
  procera Gray
        SY=P. grayi A. Nels.
  pulchella Pennell
  racemosa Dougl. ex Hook.
    ssp. alba Pennell
        SY=P. r. var. a. (Pennell) Cronq.
    ssp. racemosa
  rainierensis Pennell & Warren
  semibarbata Gray
      var. charlestonensis Pennell & Clokey
      var. semibarbata
  sudetica Willd.
    ssp. albolabiata Hulten
        SY=P. s. var. bicolor Walp.
    ssp. interioides Hulten
    ssp. interior (Hulten) Hulten
        SY=P. s. var. gymnocephala Trautv.
    ssp. pacifica Hulten
        SY=P. s. var. p. (Hulten) Welsh
    ssp. scopulorum (Gray) Hulten
        SY=P. s. (Gray) Gray
  sylvatica L.
  verticillata L.

Penstemon Mitchell

  abietinus Pennell
  acaulis L.O. Williams
  acuminatus Dougl. ex Lindl.
      var. acuminatus
      var. latebracteatus N. Holmgren
  alamosensis Pennell & Nisbet
  albertinus Greene
  albidus Nutt.
  albomarginatus M.E. Jones
  alluviorum Pennell
        SY=P. laevigatus (L.) Ait. ssp. a.
            (Pennell) Bennett
  alpinus Torr.
    ssp. alpinus
    ssp. brandegei (Porter) Penl.
    ssp. magnus (Pennell) Harrington
  ambiguus Torr.
      var. ambiguus
      var. laevissimus (Keck) N. Holmgren
        SY=P. a. ssp. l. Keck
  anguineus Eastw.
  angustifolius Nutt. ex Pursh
      var. angustifolius
      var. caudatus (Heller) Rydb.
        SY=P. a. ssp. c. (Heller) Keck
        SY=P. c. Heller
      var. venosus (Keck) N. Holmgren
        SY=P. a. ssp. v. Keck
        SY=P. v. (Keck) Reveal
  arenarius Greene
  atenicola A. Nels.
  aridus Rydb.
  arkansanus Pennell
        SY=P. australis Small ssp.
          laxiflorus (Pennell) Bennett
        SY=P. multicaulis Pennell

        SY=P. pallidus Small ssp.
          arkansanus (Pennell) Bennett
        SY=P. wherryi Pennell
  attenuatus Dougl. ex Lindl.
    var. attenuatus
    var. militaris (Greene) Cronq.
    var. palustris (Pennell) Cronq.
        SY=P. a. ssp. p. (Pennell) Keck
    var. pseudoprocerus (Rydb.) Cronq.
        SY=P. a. ssp. p. (Rydb.) Keck
  atwoodii Welsh
  auriberbis Pennell
  australis Small
  azureus Benth.
    ssp. angustissimus (Gray) Keck
    ssp. azureus
        SY=P. a. var. parvulus Gray
  baccharifolius Hook.
  barbatus (Cav.) Roth
    ssp. barbatus
        SY=P. b. var. puberulus Gray
    ssp. torreyi (Benth.) Keck
        SY=P. b. var. t. (Benth.) Gray
        SY=P. t. Benth.
    ssp. trichander (Gray) Keck
        SY=P. b. var. t. Gray
  barnebyi N. Holmgren
  barrettae Gray
  bicolor (Brandeg.) Clokey & Keck
    ssp. bicolor
    ssp. roseus Clokey & Keck
  bracteatus Keck
  bradburii Pursh
        SY=P. grandiflorus Nutt.
  breviculus (Keck) Nisbet & Jackson
        SY=P. jamesii Benth. ssp. b. Keck
  bridgesii Gray
      var. amplexicaulis Monnet
      var. bridgesii
  buckleyi Pennell
  caesius Gray
  caespitosus Nutt. ex Gray
      var. caespitosus
      var. desertipictii (A. Nels.) N.
      Holmgren
        SY=P. c. ssp. desertipictii (A.
          Nels.) Keck
      var. perbrevis (Pennell) N. Holmgren
        SY=P. c. ssp. p. Pennell
  calcareus Brandeg.
  californicus (Munz & Johnston) Keck
  calycosus Small
        SY=P. laevigatus (L.) Ait. ssp. c.
          (Small) Bennett
  campanulatus (Cav.) Willd.
        SY=P. pulchellus Lindl.
  canescens (Britt.) Britt.
        SY=P. brittonorum Pennell
        SY=P. c. var. typicus Pennell
  cardinalis Woot. & Standl.
    ssp. cardinalis
    ssp. regalis (A. Nels.) Nisbet & Jackson
  cardwellii T.J. Howell
  carnosus Pennell
  caryi Pennell
  centranthifolius Benth.
  cinerascens Greene
  cinicola Keck
  cleburnei M.E. Jones
  clevelandii Gray
    ssp. clevelandii
    ssp. connatus (Munz & Johnston) Keck
    ssp. mohavensis Keck
  clutei A. Nels.
  cobaea Nutt.
        SY=P. c. ssp. typicus Pennell
        SY=P. c. var. purpureus Pennell

Penstemon (CONT.)
  comarrhenus Gray
  compactus (Keck) Crosswhite
          SY=P. cyananthus Hook. ssp.
              compactus Keck
  concinnus Keck
  confertus Dougl. ex Lindl.
  confusus M.E. Jones
  crandallii A. Nels.
      ssp. atratus Keck
          SY=P. c. var. a. (Keck) N.
              Holmgren
      ssp. crandallii
      ssp. glabrescens (Pennell) Keck
          var. glabrescens
          var. taosensis (Keck) Nisbet &
              Jackson
      ssp. procumbens (Greene) Keck
  X crideri A. Nels. [eatonii ssp. exsertus X
      pseudospectabilis ssp. connatifolius]
  cusickii Gray
  cyananthus Hook.
  cyaneus Pennell
  cyanocaulis Payson
  cyathophorus Rydb.
  dasyphyllus Gray
  davidsonii Greene
          var. davidsonii
          SY=P. menziesii Hook. ssp. d.
              (Greene) Piper
          SY=P. m. ssp. thompsonii Pennell
          var. menziesii (Keck) Cronq.
          SY=P. m. Hook. pro parte
          var. praeteritus Cronq.
  deamii Pennell
          SY=P. laevigatus (L.) Ait. ssp. d.
              (Pennell) Bennett
  deaveri Crosswhite
          SY=P. virgatus Gray ssp.
              arizonicus (Gray) Keck
  degeneri Crosswhite
  deustus Dougl. ex Lindl.
      ssp. deustus
      ssp. heterander (Torr. & Gray) Pennell &
          Keck
          SY=P. d. var. h. (Torr. & Gray)
              Cronq.
      ssp. variabilis (Suksdorf) Pennell &
          Keck
          SY=P. d. var. v. (Suksdorf) Cronq.
          SY=P. v. Suksdorf
  digitalis Nutt.
          SY=P. laevigatus (L.) Ait. ssp. d.
              (Nutt.) Bennett
          SY=P. l. ssp. d. var. angustus
              Bennett
  diphyllus Rydb.
          SY=P. triphyllus Dougl. ex Lindl.
              ssp. d. (Rydb.) Keck
  discolor Keck
  dissectus Ell.
  dolius M.E. Jones ex Pennell
          var. dolius
          var. duchesnensis N. Holmgren
  eatonii Gray
      ssp. eatonii
      ssp. exertus (A. Nels.) Keck
      ssp. undosus (M.E. Jones) Keck
          SY=P. e. var. u. M.E. Jones
  elegantulus Pennell
  ellipticus Coult. & Fish.
          SY=P. davidsonii Greene var. e.
              (Coult. & Fish.) Boivin
  eriantherus Pursh
          var. argillosus M.E. Jones
          SY=P. whitedii Piper ssp. dayanus
              (T.J. Howell) Keck

          var. eriantherus
          var. redactus Pennell & Keck
          SY=P. whitedii Piper ssp. tristis
              Pennell & Keck
          var. whitedii (Piper) A. Nels.
          SY=P. w. Piper
  euglaucus English
  fendleri Torr. & Gray
  filiformis (Keck) Keck
  flavescens Pennell
  floridus Brandeg.
          var. austinii (Eastw.) N. Holmgren
          SY=P. f. ssp. a. (Eastw.) Keck
          var. floridus
  francisci-pennellii Crosswhite
  fremontii Torr. & Gray
  fruticiformis Coville
      ssp. amargosae Keck
      ssp. fruticiformis
  fruticosus (Pursh) Greene
          var. fruticosus
          var. scouleri (Lindl.) Cronq.
          SY=P. f. ssp. s. (Lindl.) Pennell
              & Keck
          var. serratus (Keck) Cronq.
          SY=P. f. ssp. s. Keck
  gairdneri Hook.
          var. gairdneri
          SY=P. g. ssp. hians (Piper) Keck
          SY=P. g. var. h. Piper
          SY=P. puberulentus Rydb.
          var. oreganus Gray
          SY=P. g. ssp. o. (Gray) Keck
  garrettii Pennell
  glaber Pursh
  glandulosus Dougl. ex Lindl.
          var. chelanensis (Keck) Cronq.
          SY=P. g. ssp. c. Keck
          var. glandulosus
  glaucinus Pennell
  globosus (Piper) Pennell & Keck
  gormanii Greene
  gracilentus Gray
  gracilis Nutt.
          var. gracilis
          var. wisconsinensis (Pennell) Fassett
          SY=P. w. Pennell
  grahamii Keck
  griffinii A. Nels.
  grinnellii Eastw.
      ssp. grinnellii
      ssp. scrophularioides (M.E. Jones) Munz
  guadalupensis Heller
          SY=P. g. var. ernstii (Pennell)
              Cory
  hallii Gray
  harbourii Gray
  harringtonii Penl.
  havardii Gray
  haydenii S. Wats.
  heterodoxus Gray
      ssp. cephalophorus (Greene) Keck
      ssp. heterodoxus
  heterophyllus Lindl.
      ssp. australis (Munz & Johnston) Keck
          SY=P. h. var. a. Munz & Johnston
      ssp. heterophyllus
      ssp. purdyi Keck
  hirsutus (L.) Willd.
          var. hirsutus
          var. minimus Bennett
          var. pygmaeus Bennett
  holmgrenii S.L. Clark
          SY=P. cyananthus Hook. ssp.
              subglaber (Gray) Pennell
  humilis Nutt. ex Gray
          var. brevifolius (Gray) Keck

Penstemon (CONT.)
          var. humilis
              SY=P. cinereus Piper
              SY=P. c. var. foliatus Keck
              SY=P. decurvus Pennell ex
                  Crosswhite
          var. obtusifolius (Pennell) Reveal
              SY=P. h. ssp. o. (Pennell) Keck
    immanifestus N. Holmgren
    incertus Brandeg.
    inflatus Crosswhite
    jamesii Benth.
              SY=P. j. ssp. typicus Keck
    janishiae N. Holmgren
    X jonesii Pennell [eatonii ssp. undosus X
        laevis]
              SY=P. brevibarbatus Crosswhite
    keckii Clokey
    kingii S. Wats.
              SY=P. nyeensis Crosswhite
    labrosus (Gray) Hook. f.
    laetus Gray
        ssp. laetus
        ssp. leptosepalus (Greene ex Gray) Keck
        ssp. roezlii (Regel) Keck
              SY=P. l. var. r. (Regel) Jepson
        ssp. sagittatus Keck
    laevigatus (L.) Ait.
              SY=P. penstemon (L.) Britt.
    laevis Pennell
    lanceolatus Benth.
              SY=P. pauciflorus Greene
              SY=P. ramosus Crosswhite
    laricifolius Hook. & Arn.
        ssp. exilifolius (A. Nels.) Keck
        ssp. laricifolius
    laxiflorus Pennell
              SY=P. australis Small ssp. l.
                  (Pennell) Bennett
    laxus A. Nels.
              SY=P. watsonii Gray ssp. l. (A.
                  Nels.) Keck
    leiophyllus Pennell
    lemhiensis (Keck) Keck & Cronq.
              SY=P. speciosus Dougl. ex Lindl.
                  ssp. l. Keck
    lentus Pennell
          var. albiflorus (Keck) Reveal
              SY=P. l. ssp. a. Keck
          var. lentus
    leonardii Rydb.
    leptanthus Pennell
    linarioides Gray
        ssp. coloradoensis (A. Nels.) Keck
              SY=P. c. A. Nels.
        ssp. compactifolius Keck
        ssp. linarioides
              SY=P. l. ssp. typicus Keck
              SY=P. l. var. viridis Keck
        ssp. maguirei Keck
        ssp. sileri (Gray) Keck
    longiflorus (Pennell) S.L. Clark
              SY=P. cyananthus Hook. ssp. l.
                  Pennell
    lyallii (Gray) Gray
    maguirei Crosswhite
    marcusii (Keck) N. Holmgren
              SY=P. moffattii Eastw. ssp.
                  marcusii Keck
    mensarum Pennell
    X mirus A. Nels. [eatonii ssp. exertus X
        palmeri]
    miser Gray
    moffattii Eastw.
        ssp. moffattii
        ssp. paysonii (Pennell) Keck
    monoensis Heller

    montanus Greene
          var. idahoensis (Pennell & Keck)
              Cronq.
              SY=P. m. ssp. i. Pennell & Keck
          var. montanus
    mucronatus N. Holmgren
    multiflorus Chapman ex Benth.
    murrayanus Hook.
    nanus Keck
    neomexicanus Woot. & Standl.
    neotericus Keck
    newberryi Gray
        ssp. berryi (Eastw.) Keck
        ssp. newberryi
        ssp. sonomensis (Greene) Keck
    nitidus Dougl. ex Benth.
          var. nitidus
          var. polyphyllus (Pennell) Cronq.
    nudiflorus Gray
    oklahomensis Pennell
    oliganthus Woot. & Standl.
    ophianthus Pennell
              SY=P. jamesii Benth. ssp. o.
                  (Pennell) Keck
    osterhoutii Pennell
    ovatus Dougl. ex Hook.
    pachyphyllus Gray ex Rydb.
          var. congestus (M.E. Jones) N.
              Holmgren
              SY=P. c. (M.E. Jones) Pennell
              SY=P. p. ssp. c. (M.E. Jones) Keck
          var. pachyphyllus
    pahutensis N. Holmgren
    pallidus Small
              SY=P. arkansanus Pennell var.
                  pubescens Pennell
              SY=P. brevisepalus Pennell
    palmeri Gray
          var. eglandulosus (Keck) N. Holmgren
              SY=P. p. ssp. e. Keck
          var. macranthus (Eastw.) N. Holmgren
              SY=P. m. Eastw.
          var. palmeri
              SY=P. p. ssp. typicus Keck
    papillatus J.T. Howell
    X parishii Gray [centranthifolius X
        spectabilis]
    parryi (Gray) Gray
    parviflorus Pennell
    parvulus (Gray) Krautter
    parvus Pennell
    patens (M.E. Jones) N. Holmgren
              SY=P. confusus M.E. Jones ssp. p.
                  (M.E. Jones) Keck
    patricus N. Holmgren
    payettensis A. Nels. & J.F. Macbr.
    paysoniorum Keck
    peckii Pennell
    pennellianus Keck
    perpulcher A. Nels.
              SY=P. minidokanus A. Nels. & J.F.
                  Macbr.
    personatus Keck
    petiolatus Brandeg.
    pinifolius Greene
    platyphyllus Rydb.
    pratensis Greene
    procerus Dougl. ex Graham
          var. aberrans (M.E. Jones) A. Nels.
          var. brachyanthus (Pennell) Cronq.
              SY=P. p. ssp. b. (Pennell) Keck
              SY=P. tolmiei Hook. ssp. b.
                  (Pennell) Keck
          var. formosus (A. Nels.) Cronq.
              SY=P. p. ssp. f. (A. Nels.) Keck
              SY=P. tolmiei Hook. ssp. f. (A.
                  Nels.) Keck

Penstemon (CONT.)
  var. modestus (Greene) N. Holmgren
   SY=P. m. Greene
   SY=P. p. ssp. m. (Greene) Keck
  var. procerus
   SY=P. confertus Dougl. ex Lindl.
    ssp. p. (Dougl. ex Graham)
     D.V. Clark
  var. tolmiei (Hook.) Cronq.
   SY=P. t. Hook.
pruinosus Dougl. ex Lindl.
pseudoparvus Crosswhite
pseudospectabilis M.E. Jones
 ssp. connatifolius (A. Nels.) Keck
 ssp. pseudospectabilis
  SY=P. p. ssp. typicus Keck
pudicus Reveal & Beatley
pumilus Nutt.
purpusii Brandeg.
radicosus A. Nels.
rattanii Gray
 ssp. kleei (Greene) Keck
 ssp. rattanii
retrorsus Payson ex Pennell
richardsonii Dougl. ex Lindl.
  var. curtiflorus (Keck) Cronq.
  var. dentatus (Keck) Cronq.
  var. richardsonii
rubicundus Keck
rupicola (Piper) T.J. Howell
rydbergii A. Nels.
  var. aggregatus (Pennell) N. Holmgren
   SY=P. a. Pennell
   SY=P. r. var. a. (Pennell) Keck
  var. rydbergii
  var. varians (A. Nels.) Cronq.
   SY=P. hesperius M.E. Peck
   SY=P. oreocharis Greene
   SY=P. vaseyanus Greene
saxosorum Pennell
scapoides Keck
scariosus Pennell
secundiflorus Benth.
  SY=P. unilateralis Rydb.
seorsus (A. Nels.) Keck
sepalulus A. Nels.
serrulatus Menzies ex Sm.
shastensis Keck
smallii Heller
spathulatus Pennell
speciosus Dougl. ex Lindl.
  SY=P. s. ssp. kennedyi (A. Nels.)
   Keck
spectabilis Thurb. ex Gray
 ssp. spectabilis
 ssp. subviscosus Keck
stenophyllus (Gray) T.J. Howell
stephensii Brandeg.
strictiformis Rydb.
  SY=P. strictus Benth. ssp.
   strictiformis (Rydb.) Keck
strictus Benth.
 ssp. angustus Pennell
 ssp. strictus
subglaber Rydb.
subserratus Pennell
subulatus M.E. Jones
sudans M.E. Jones
  SY=P. deustus Dougl. ex Lindl.
   ssp. s. (M.E. Jones) Pennell
   & Keck
superbus A. Nels.
tenuiflorus Pennell
tenuis Small
teucrioides Greene
thompsoniae (Gray) Rydb.
 ssp. jaegeri Keck

 ssp. thompsoniae
thurberi Torr.
  var. anestius Reveal & Beatley
  var. thurberi
tidestromii Pennell
tracyi Keck
triflorus Heller
 ssp. integrifolius Pennell
  SY=P. brevibarbatus Crosswhite
  SY=P. helleri Small
  SY=P. t. var. i. (Pennell) Cory
 ssp. triflorus
triphyllus Dougl. ex Lindl.
tubiflorus Nutt.
  var. achoreus Fern.
  var. tubiflorus
tusharensis N. Holmgren
  SY=P. caespitosus Nutt. ex Gray
   var. suffruticosus Gray
  SY=P. suffrutescens Rydb.
uintahensis Pennell
utahensis Eastw.
venustus Dougl. ex Lindl.
virens Pennell
virgatus Gray
 ssp. asa-grayi Crosswhite
 ssp. pseudoputus Crosswhite
 ssp. putus (A. Nels.) Crosswhite
 ssp. virgatus
wardii Gray
washingtonensis Keck
watsonii Gray
whippleanus Gray
wilcoxii Rydb.
  SY=P. ovatus Dougl. ex Hook. var.
   pinetorum Piper
wrightii Hook.
yampaensis Penl.

Rhinanthus L.

alectorolophus (Scop.) Pollich
  SY=R. major Ehrh.
arcticus (Sterneck) Pennell
borealis (Sterneck) Chabert
  SY=R. minor L. ssp. b. (Sterneck)
   A. Love
groenlandicus (Ostenf.) Chabert
minor L.
  SY=R. borealis (Sterneck) Chabert
   ssp. kyrolliae (Chabert)
   Pennell
  SY=R. crista-galli L.
  SY=R. c-g. var. fallax (Wimmer &
   Grab.) Druce
  SY=R. k. Chabert
  SY=R. rigidus Chabert
  SY=R. stenophyllus (Schur) Schinz
   & Thellung

Russelia Jacq.

equisetiformis Schlecht. & Cham.
  SY=R. juncea Zucc.

Schistophragma Benth. ex Endl.

intermedia (Gray) Pennell

Schwalbea L.

americana L.
  SY=S. australis Pennell

Scoparia L.

dulcis L.

Scrophularia L.

    atrata Pennell
    californica Cham. & Schlecht.
       ssp. californica
       ssp. floribunda (Greene) R.J. Shaw
             SY=S. c. var. f. Greene
             SY=S. multiflora Pennell
    desertorum (Munz) R.J. Shaw
             SY=S. californica Cham. &
                 Schlecht. var. d. Munz
    laevis Woot. & Standl.
    lanceolata Pursh
    macrantha Greene ex Stiefelhagen
             SY=S. coccinea Gray
             SY=S. neomexicana R.J. Shaw
    marilandica L.
    minutiflora Pennell
    montana Woot.
    nodosa L.
    oregana Pennell
             SY=S. californica Cham. &
                 Schlecht. var. o. (Pennell)
                 Boivin
    parviflora Woot. & Standl.
    villosa Pennell

Seymeria Pursh

    cassioides (Walt.) Blake
             SY=Afzelia c. (Walt.) J.F. Gmel.
    havardii (Pennell) Pennell
    pectinata Pursh
             SY=Afzelia p. (Pursh) Kuntze
    scabra Gray
    texana (Gray) Pennell

Stemodia L.

    durantifolia (L.) Sw.
             SY=S. arizonica Pennell
    maritima L.
    schottii Holz.
    tomentosa (P. Mill.) Greenm. & Thompson

Striga Lour.

    asiatica (L.) Kuntze
             SY=S. lutea Lour.
    gesnerioides (Willd.) Vatke

Synthyris Benth.

    borealis Pennell
    canbyi Pennell
    laciniata (Gray) Rydb.
       ssp. ibapohensis Pennell
       ssp. laciniata
    missurica (Raf.) Pennell
       ssp. hirsuta Pennell
       ssp. missurica
             SY=S. m. ssp. major (Hook.)
                 Pennell
             SY=S. reniformis (Dougl. ex
                 Benth.) Benth. var. major
                 Hook.
             SY=S. stellata Pennell
    pinnatifida S. Wats.
        var. canescens (Pennell) Cronq.
             SY=S. cymopteroides Pennell
             SY=S. hendersonii Pennell
        var. lanuginosa (Piper) Cronq.
             SY=S. l. (Piper) Pennell &
                 Thompson
        var. pinnatifida
             SY=S. paysonii Pennell & Williams
    platycarpa Gail & Pennell

ranunculina Pennell
reniformis (Dougl. ex Benth.) Benth.
    var. cordata Gray
    var. reniformis
schizantha Piper

Tomanthera Raf.

    auriculata (Michx.) Raf.
             SY=Aureolaria a. (Michx.) Farw.
             SY=Gerardia a. Michx.
             SY=Otophylla a. (Michx.) Small
    densiflora (Benth.) Pennell
             SY=Gerardia d. Benth.
             SY=Otophylla d. (Benth.) Small

Tonella Nutt.

    floribunda Gray
    tenella (Benth.) Heller

Verbascum L.

    blattaria L.
    lychnitis L.
    nigrum L.
    phlomoides L.
    phoeniceum L.
    X pterocaulon Franch.
    sinuatum L.
    speciosum Schrad.
    thapsus L.
    virgatum Stokes

Veronica L.

    agrestis L.
    alpina L.
       var. alpina
       var. australis Wahlenb.
          SY=V. alpina ssp. australis
              (Wahlenb.) Love & Love
       var. terrae-novae Fern.
    americana (Raf.) Schwein. ex Benth.
    anagallis-aquatica L.
          SY=V. anagallis L.
    arvensis L.
    austriaca L.
       ssp. teucrium (L.) D.A. Webb
          SY=V. latifolia L.
          SY=V. t. L.
    beccabunga L.
    biloba L.
    catenata Pennell
          SY=V. comosa sensu auctt. non
             Richter
          SY=V. comosa var. glaberrima
             (Pennell) Boivin
          SY=V. comosa var. glandulosa
             (Farw.) Boivin
          SY=V. connata sensu auctt. non
             Raf.
          SY=V. connata Raf. ssp. glaberrima
             Pennell
          SY=V. connata var. glaberrima
             (Pennell) Boivin
          SY=V. connata var. typica Pennell
          SY=V. salina sensu auctt. non
             Schur
    chamaedrys L.
    copelandii Eastw.
    cusickii Gray
    cymbalaria Bodard
          SY=V. glandifera Freyn
    dillenii Crantz
    filiformis Sm.
    fruticans Jacq.

Veronica (CONT.)
    grandiflora Gaertn.
            SY=V. glandifera Pennell
    grandis Fisch. ex Spreng.
    hederifolia L.
    X lackschewitzii Schlenker
        [anagallis-aquatica X catenata]
    longifolia L.
    micromeria Woot. & Standl.
    officinalis L.
            var. officinalis
            var. tournefortii (Vill.) Reichenb.
                SY=V. t. Vill.
    opaca Fries
    peregrina L.
        ssp. peregrina
            SY=V. p. var. typica Pennell
        ssp. xalapensis (H.B.K.) Pennell
            SY=V. p. var. x. (H.B.K.) St. John
                & Warren
            SY=V. sherwoodii M.E. Peck
            SY=V. x. H.B.K.
    persica Poir.
            var. aschersoniana (Lehm.) Boivin
            var. corrensiana (Lehm.) Boivin
            var. persica
    plebia R. Br.
    polita Fries
            SY=V. didyma Ten. pro parte
    scutellata L.
            SY=V. s. var. villosa Schum.
    serpyllifolia L.
        ssp. humifusa (Dickson) Syme
            SY=V. h. Dickson
            SY=V. s. var. borealis Laestad.
            SY=V. s. var. h. (Dickson) Vahl
            SY=V. tenella All.
        ssp. serpyllifolia
    spicata L.
        ssp. incana (L.) Walt.
            SY=V. i. L.
        ssp. spicata
    triphyllos L.
    verna L.
    wormskjoldii Roemer & Schultes
        ssp. alterniflora (Fern.) Pennell
            SY=V. alpina L. var. alterniflora
                Fern.
        ssp. wormskjoldii
            SY=V. alpina L. var. nutans
                (Bong.) Boivin
            SY=V. a. var. unalaschcensis Cham.
                & Schlecht.
            SY=V. stelleri Pallas
            SY=V. s. var. glabrescens Hulten

Veronicastrum Heister ex Fabr.

    virginicum (L.) Farw.
            SY=Leptandra virginica (L.) Nutt.
            SY=Veronica v. L.

SIMAROUBACEAE

Ailanthus Desf.

    altissima (P. Mill.) Swingle
            SY=A. glandulosa Desf.

Alvaradoa Liebm.

    amorphoides Liebm.

Castela Turp.

    erecta Turp.
            SY=Castelaria nicholsonii (Hook.)
                Small
    texana (Torr. & Gray) Rose
            SY=Castela tortuosa Liebm.

Holacantha Gray

    emoryi Gray
            SY=Castela e. (Gray) Moran &
                Felger
    stewartii C.H. Muller

Picramnia Sw.

    pentandra Sw.

Picrasma Blume

    antillana (Eggers) Urban
            SY=Aeschrion a. (Eggers) Small
    excelsa (Sw.) Planch.
            SY=Aeschrion e. (Sw.) Kuntze

Quassia L.

    amara L.

Simarouba Aubl.

    glauca DC.
    tulae Urban

Suriana L.

    maritima L.

SMILACACEAE

Smilax L.

    auriculata Walt.
    biltmoreana (Small) J.B.S. Norton ex
        Pennell
            SY=Nemexia b. Small
            SY=S. ecirrhata (Engelm. ex Kunth)
                S. Wats. var. b. (Small)
                Ahles
    bona-nox L.
            SY=S. b-n. var. exauriculata Fern.
            SY=S. b-n. var. hastata (Willd.)
                A. DC.
            SY=S. b-n. var. hederifolia (Bey.)
                Fern.
    californica (A. DC.) Gray
    coriacea Spreng.
    ecirrhata (Engelm. ex Kunth) S. Wats.
            SY=Nemexia e. (Engelm. ex Kunth)
                Small
    glauca Walt.
            var. glauca
                SY=S. g. var. genuina Blake
            var. leurophylla Blake
    havanensis Jacq.
    herbacea L.
            SY=Nemexia h. (L.) Small
    hispida Muhl.
            SY=S. h. var. australis J.B.S.
                Norton
            SY=S. h. var. montana Coker
            SY=S. tamnoides L. var. h. (Muhl.)
                Fern.
    hugeri (Small) J.B.S. Norton ex Pennell
            SY=Nemexia h. Small

Smilax (CONT.)
          SY=S. ecirrhata (Engelm. ex Kunth)
          S. Wats. var. h. (Small)
          Ahles
    illinoiensis Mangaly
    jamesii G. Wallace
    lasioneuron Hook.
          SY=Nemexia l. (Hook.) Rydb.
          SY=S. herbacea L. var. l. (Hook.)
          A. DC.
    laurifolia L.
          SY=S. megacarpa Morong pro parte
    melastomifolia Sm.
          var. melastomifolia
          var. subinermis Hbd.
    pseudo-china L.
          SY=Nemexia leptanthera (Pennell)
          Small
          SY=N. tamnifolia (Michx.) Small
          SY=S. l. Pennell
          SY=S. t. Michx.
    pulverulenta Michx.
          SY=Nemexia p. (Michx.) Small
          SY=S. herbacea L. var. p. (Michx.)
          Gray
    pumila Walt.
    renifolia Small
    rotundifolia L.
          SY=S. r. var. crenulata Small &
          Heller
          SY=S. r. var. quadrangularis
          (Muhl.) Wood
    sandwicensis Kunth
          var. crassifolia Hbd.
          var. sandwicensis
    smallii Morong
          SY=S. domingensis Willd.
          SY=S. lanceolata L.
    tamnoides L.
    walteri Pursh
          SY=S. megacarpa Morong pro parte

SOLANACEAE

Acnistus Schott

    arborescens (L.) Schlecht.

Atropa L.

    belladonna L.

Browallia L.

    americana L.

Brugmansia Pers.

    arborea (L.) Steud.
    X candida (Pers.) Safford [aurea X
      versicolor]
    suaveolens (Humb. & Bonpl. ex Willd.)
      Bercht. & Presl
          SY=Datura s. Humb. & Bonpl. ex
                Willd.

Brunfelsia L.

    americana L.
    densifolia Krug & Urban
    lactea Krug & Urban
    portoricensis Krug & Urban

Capsicum L.

    annuum L.
        var. annuum
        var. glabriusculum (Dunal) Heiser &
          Pickersgill
          SY=C. a. var. aviculare (Dierbach)
          D'Arcy & Eschsch.
          SY=C. a. var. minimum (P. Mill.)
          Shinners
          SY=C. baccatum sensu American
          auth., non L.
    frutescens L.

Cestrum L.

    alternifolium (Jacq.) O.E. Schulz
    diurnum L.
        var. diurnum
        var. portoricense O.E. Schulz
    fasciculatum (Schlecht.) Miers
    macrophyllum Vent.
          SY=C. laurifolium L'Her.
    nocturnum L.
    parqui L'Her.

Chamaesaracha (Gray) Benth. & Hook.

    coniodes (Moric. ex Dunal) Britt.
    coronopus (Dunal) Gray
    crenata Rydb.
    edwardsiana Averett
    pallida Averett
    sordida (Dunal) Gray
    villosa Rydb.

Datura L.

    discolor Bernh.
    ferox L.
    innoxia P. Mill.
          SY=D. meteloides DC.
          SY=D. wrightii Regel
    metel L.
          SY=D. fastuosa L.
    quercifolia H.B.K.
    stramonium L.
        var. stramonium
        var. tatula (L.) Torr.
          SY=D. t. L.

Goetzea Wydler

    elegans Wydler

Hunzikera D'Arcy

    texana (Torr.) D'Arcy
          SY=Browalia t. Torr.
          SY=Leptoglossis t. (Torr.) Gray
          SY=Nierembergia viscosa Torr.

Hyoscyamus L.

    albus L.
    niger L.

Jaltomata Schlecht.

    antillana (Krug & Urban) D'Arcy ined.
          SY=Saracha a. Krug & Urban
    procumbens (Cav.) J.L. Gentry
          SY=Saracha p. (Cav.) Ruiz & Pavon

Leucophysalis Rydb.

    grandiflora (Hook.) Rydb.
          SY=Chamaesaracha g. (Hook.) Fern.
          SY=Physalis g. Hook.

Leucophysalis (CONT.)
  nana (Gray) Averett
          SY=Chamaesaracha n. (Gray) Gray

Lycianthes Hassler

  virgata (Lam.) Bitter

Lycium L.

  andersonii Gray
          var. andersonii
          var. deserticola (C.L. Hitchc. ex
              Munz) Jepson
          var. wrightii Gray
  barbarum L.
          SY=L. halimifolium P. Mill.
  berberiodes Correll
  berlandieri Dunal
          var. berlandieri
          var. longistylum C.L. Hitchc.
          var. parviflorum (Gray) Terracc.
              SY=L. berlandieri var. brevilobum
                  C.L. Hitchc.
  californicum Nutt.
  carolinianum Walt.
          var. carolinianum
          var. quadrifidum (Dunal) C.L. Hitchc.
  chinense P. Mill.
  cooperi Gray
  exsertum Gray
  fremontii Gray
  macrodon Gray
  pallidum Miers
          var. oligospermum C.L. Hitchc.
          var. pallidum
  parishii Gray
  puberulum Gray
  richii Gray
          var. hassei (Greene) I.M. Johnston
              SY=L. brevipes Benth. var. h.
                  (Greene) C.L. Hitchc.
              SY=L. h. Greene
          var. richii
              SY=L. brevipes Benth.
  rickardii C.H. Muller
  sandwicense Gray
          SY=L. carolinianum Walt. var. s.
              (Gray) C.L. Hitchc.
  texanum Correll
  torreyi Gray
  tweedianum Griseb.
          var. chrysocarpum (Urban & Ekman)
              C.L. Hitchc.
              SY=L. americanum Jacq.
          var. tweedianum
  verrucosum Eastw.

Lycopersicon P. Mill.

  esculentum P. Mill.
          var. esculentum
              SY=L. lycopersicon (L.) Karst. ex
                  Farw.
              SY=Solanum l. L.
          var. leptophyllum (Dunal) D'Arcy
              SY=L. cerasiforme Dunal
              SY=L. e. var. c. (Dunal) Gray

Margaranthus Schlecht.

  solanaceus Schlecht.
          SY=M. lemmonii Gray
          SY=M. purpurascens Rydb.

Nectouxia H.B.K.

  formosa H.B.K.

Nicandra Adans.

  physalodes (L.) Gaertn.
          SY=Physalodes p. (L.) Britt.

Nicotiana L.

  acuminata (Graham) Hook.
          var. acuminata
          var. multiflora (Phil.) Reiche
  alata Link & Otto
          SY=N. affinis T. Moore
  attenuata Torr. ex S. Wats.
  bigelovii (Torr.) S. Wats.
          var. bigelovii
              SY=N. quadrivalvis Pursh var. b.
                  (Torr.) DeWolf
          var. multivalvis (Lindl.) Fast
              SY=N. m. Lindl.
              SY=N. quadrivalvis Pursh var. m.
                  (Lindl.) Mansf.
          var. quadrivalvis (Pursh) Fast
              SY=N. q. Pursh
          var. wallacei Gray
              SY=N. quadrivalvis Pursh var. w.
                  (Gray) Mansf.
  clevelandii Gray
  glauca Graham
  longiflora Cav.
  palmeri Gray
  plumbaginifolia Viviani
  repanda Willd. ex Lehm.
  rustica L.
  sylvestris Speg. & Comes
  tabacum L.
  trigonophylla Dunal

Nierembergia Ruiz & Pavon

  hippomanica Miers
          var. coerulea (Miers) Millan
          var. hippomanica
  scoparia Sendtner
          SY=N. frutescens Durieu

Nothocestrum Gray

  breviflorum Gray
          var. breviflorum
          var. longipes Hbd.
  latifolium Gray
  longifolium Gray
          var. brevifolium Hbd.
          var. longifolium
          var. rufipilosum B.C. Stone
  peltatum Skottsberg
  subcordatum Mann

Oryctes S. Wats.

  nevadensis S. Wats.

Petunia Juss.

  parviflora Juss.
          SY=P. integrifolia (Hook.) Schinz
              & Thellung
  violacea Lindl.
          SY=P. X atkinsiana D. Don ex Loud.
          SY=P. axillaris (Lam.) B.S.P.
          SY=P. X hybrida Vilm.

Physalis L.

  acutifolia (Miers) Sandw.

Physalis (CONT.)
          SY=P. wrightii Gray
   alkekengii L.
   angulata L.
          var. angulata
          var. lanceifolia (Nees) Waterfall
               SY=P. l. Nees
          var. pendula (Rydb.) Waterfall
               SY=P. p. Rydb.
   angustifolia Nutt.
   arenicola Kearney
          var. arenicola
          var. ciliosa (Rydb.) Waterfall
               SY=P. c. Rydb.
   carpenteri Riddell ex Rydb.
   caudella Standl.
   cordata P. Mill.
          SY=P. barbadensis Jacq. var.
               glabra (Michx.) Fern.
          SY=P. b. var. obscura (Michx.)
               Rydb.
          SY=P. pubescens L. var. g.
               (Michx.) Waterfall
   crassifolia Benth.
          var. crassifolia
               SY=P. c. var. cardiophylla (Torr.)
               Gray
          var. versicolor (Rydb.) Waterfall
               SY=P. v. Rydb.
   eggersii O.E. Schulz
   fuscomaculata de Rouv.
   greenei Vasey & Rose
   hederifolia Gray
          var. comata (Rydb.) Waterfall
               SY=P. c. Rydb.
               SY=P. rotundata Rydb.
          var. cordifolia (Gray) Waterfall
               SY=P. fendleri Gray
               SY=P. f. var. c. Gray
          var. hederifolia
               SY=P. h. var. puberula Gray
   heterophylla Nees
          var. clavipes Fern.
          var. heterophylla
               SY=P. ambigua (Gray) Britt.
               SY=P. h. var. a. (Gray) Rydb.
               SY=P. h. var. nyctaginea (Dunal)
               Rydb.
               SY=P. n. Dunal
               SY=P. sinuata Rydb.
          var. villosa Waterfall
   ixocarpa Brot. ex Hornem.
   lagascae Roemer & Schultes
   latiphylla Waterfall
   peruviana L.
   philadelphica Lam.
          var. immaculata Waterfall
          var. philadephica
   pubescens L.
          var. grisea Waterfall
          var. integrifolia (Dunal) Waterfall
               SY=P. pruinosa sensu auctt. non L.
          var. missouriensis (Mackenzie & Bush)
               Waterfall
               SY=P. m. Mackenzie & Bush
          var. pubescens
               SY=P. barbadensis Jacq.
               SY=P. floridana Rydb.
   pumila Nutt.
      ssp. hispida (Waterfall) Hinton
               SY=P. lanceolata Michx.
               SY=P. longifolia Nutt. var.
                    hispida (Waterfall)
                    Steyermark
               SY=P. virginiana P. Mill. var.
                    hispida Waterfall
      ssp. pumila

   subglabrata Mackenzie & Bush
          SY=P. heterophylla Nees var. s.
               (Mackenzie & Bush) Waterfall
          SY=P. longifolia Nutt. var. s.
               (Mackenzie & Bush) Cronq.
          SY=P. macrophysa Rydb.
          SY=P. virginiana P. Mill. var. s.
               (Mackenzie & Bush) Waterfall
   subulata Rydb.
          var. neomexicana (Rydb.) Waterfall
               SY=P. foetens Poir. var. n.
                    (Rydb.) Waterfall
               SY=P. n. Rydb.
          var. subulata
               SY=?P. foetens Poir.
   turbinata Medic.
   variovestita Waterfall
   virginiana P. Mill.
          var. campaniforma Waterfall
          var. polyphylla (Greene) Waterfall
          var. sonorae (Torr.) Waterfall
               SY=P. heterophylla Nees var. s.
                    (Torr.) Waterfall
               SY=P. longifolia (Nutt.) Trel.
               SY=P. l. var. s. (Torr.) Waterfall
               SY=P. rigida Pollard & Ball
          var. texana (Rydb.) Waterfall
          var. virginiana
               SY=P. intermedia Rydb.
               SY=P. monticola C. Mohr
   viscosa L.
      ssp. maritima (M.A. Curtis) Waterfall
          var. elliottii (Kunze) Waterfall
               SY=P. e. Kunze
          var. maritima (M.A. Curtis) Waterfall
               SY=P. m. M.A. Curtis
          var. spathulifolia (Torr.) Gray
      ssp. mollis (Nutt.) Waterfall
          var. cinerascens (Dunal) Waterfall
          var. mollis
               SY=P. m. Nutt.
      ssp. viscosa

Quincula Raf.

   lobata (Torr.) Raf.
          SY=Physalis l. Torr.

Salpichroa Miers

   origanifolia (Lam.) Baill.
          SY=Perizoma rhomboidea (Gillies &
               Hook.) Small
          SY=S. r. (Gillies & Hook.) Miers

Salpiglossis Ruiz & Pavon

   erecta (DC.) D'Arcy
          SY=Bouchetia e. DC.

Schizanthus Ruiz & Pavon

   pinnatus Ruiz & Pavon

Scopolia Jacq.

   carniolica Jacq.

Solandra Sw.

   grandiflora Sw.
          SY=Swartzia g. (Sw.) J.F. Gmel.

Solanum L.

   americanum P. Mill.
          var. americanum

Solanum (CONT.)
        SY=S. nigrum L. var. a. (P. Mill.)
           O.E. Schulz
      var. patulum (L.) Edmonds
antillarum O.E. Schulz
aviculare Forst. f.
bahamense L.
      var. bahamense
      var. luxurians D'Arcy
      var. rugelii D'Arcy
bulbocastanum Dunal
campechiense L.
        SY=S. guanicense Urban
capsicastrum Link ex Schauer
capsicoides All.
        SY=S. aculeatissimum Jacq.
        SY=S. ciliatum Lam.
carolinense L.
      var. carolinense
      var. floridanum (Dunal) Chapman
        SY=S. f. Dunal
        SY=S. godfreyi Shinners
      var. hirsutum (Nutt.) Gray
citrullifolium A. Braun
clokeyi Munz
        SY=S. wallacei (Gray) Parish var.
           c. (Munz) McMinn
conocarpum Dunal
cornutum Lam.
        SY=Androcera rostrata (Dunal)
           Rydb.
        SY=S. r. Dunal
davisense M.D. Whalen
deflexum Greenm.
dimidiatum Raf.
        SY=S. perplexum Small
        SY=S. torreyi Gray
diphyllum L.
donianum Walp.
        SY=S. blodgettii Chapman
douglasii Dunal
        SY=S. nigrum L. var. d. (Dunal)
           Gray
drymophilum O.E. Schulz
dulcamara L.
      var. dulcamara
      var. villosissimum Desv.
elaeagnifolium Cav.
erianthum D. Don
        SY=S. e. var. adulterinum (Ham. ex
           G. Don) Baker & Simmonds
        SY=S. verbascifolium sensu auctt.
           non L.
fendleri Gray
      var. fendleri
      var. texense Correll
ficifolium Ortega
        SY=S. torvum Sw.
furcatum Dunal
gayanum (Remy) Phil. f.
glaucophyllum Desf.
        SY=S. glaucum Dunal
gracilius Herter
        SY=S. gracile Otto
haleakalaense St. John
heterodoxum Dunal
      var. heterodoxum
      var. novomexicanum Bartlett
      var. setigeroides M.D. Whalen
hillebrandii St. John
incompletum Dunal
      var. glabratum Hbd.
      var. incompletum
      var. mauiensis Hbd.
interius Rydb.
jamaicense P. Mill.
jamesii Torr.

kauaiense Hbd.
laciniatum Ait.
lanceifolium Jacq.
lanceolatum Cav.
leptosepalum Correll
lumholtzianum Bartlett
luteum P. Mill.
   ssp. alatum (Moench) Dostal
        SY=S. a. Moench
   ssp. luteum
        SY=S. nigrum L. var. villosum L.
        SY=S. v. (L.) P. Mill.
mammosum L.
marginatum L. f.
mauritianum Scop.
        SY=S. auriculatum Ait.
melongena L.
?mucronatum O.E. Schulz
nelsonii Dunal
      var. nelsonii
      var. thomasiifolium Seem.
nigrescens Mart. & Gal.
nigrum L.
      var. nigrum
      var. virginicum L.
nodiflorum Jacq.
        SY=S. americanum P. Mill. var. n.
           (Jacq.) Edmonds
parishii Heller
persicifolium Dunal
pinnatisectum Dunal
polygamum Vahl
pseudocapsicum L.
pseudogracile Heiser
puberulum Nutt. ex Seem.
quitoense Lam.
racemosum Jacq.
riedlei Dunal
rugosum Dunal
sandwicense Hook. & Arn.
sarachoides Sendtner
seaforthianum Andr.
sisymbriifolium Lam.
sodomeum L.
sublobatum Willd. ex Roemer & Schultes
        SY=S. gracile sensu auctt. non
           Link
        SY=S. ottonis Hyl.
tenuilobatum Parish
tenuipes Bartlett
      var. latisectum M.D. Whalen
      var. tenuipes
triflorum Nutt.
triquetrum Cav.
tuberosum L.
umbelliferum Eschsch.
      var. glabrescens Torr.
      var. incanum Torr.
      var. umbelliferum
wallacei (Gray) Parish
woodburyi Howard
xantii Gray
      var. glabrescens Parish
      var. hoffmannii Munz
      var. intermedium Parish
      var. montanum Munz
      var. obispoense (Eastw.) Wiggins
      var. xantii

SPARGANIACEAE

Sparganium L.

   americanum Nutt.

Sparganium (CONT.)
    androcladum (Engelm.) Morong
    chlorocarpum Rydb.
            SY=S. acaule (Beeby) Rydb.
            SY=S. c. var. a. (Beeby) Fern.
    emersum Rehmann
            SY=S. angustifolium Michx.
            SY=S. e. var. a. (Michx.) Taylor &
                MacBryde
            SY=S. e. var. multipedunculatum
                (Morong) Reveal
            SY=S. m. (Morong) Rydb.
            SY=S. simplex Huds.
    eurycarpum Engelm.
            SY=S. californicum Greene
            SY=S. e. var. greenei (Morong)
                Graebn.
            SY=S. g. Morong
    fluctuans (Morong) B.L. Robins.
    glomeratum Laestad. ex Beurling
    hyperboreum Laestad. ex Beurling
    minimum (Hartman) Fries

STAPHYLEACEAE

Staphylea L.

    bolanderi Gray
    trifolia L.

Turpinia Raf.

    paniculata Vent.

STERCULIACEAE

Ayenia L.

    ardua Cristobal
            SY=?A. insularis Cristobal
            SY=A. pusilla sensu auctt. non L.
    compacta Rose
            SY=A. californica Jepson
    euphrasiifolia Griseb.
    filiformis S. Wats.
    insulaecola Cristobal
    limitaris Cristobal
    microphylla Gray
    pilosa Cristobal

Firmiana Marsili

    simplex (L.) W. Wight
            SY=F. platanifolia (L. f.) Schott
                & Endl.

Fremontodendron Coville

    californicum (Torr.) Coville
        ssp. californicum
            SY=Fremontia californica Torr.
            SY=Fremontia c. var. diegensis M.
                Harvey
            SY=Fremontia c. var. integra M.
                Harvey
            SY=Fremontia c. var. typica M.
                Harvey
            SY=Fremontia c. var. viridis M.
                Harvey
        ssp. crassifolium (Eastw.) Thomas
            SY=Fremontia californica Torr.

                var. crassifolia (Eastw.)
                    Abrams
                SY=Fremontia crassifolia Eastw.
        ssp. decumbens (R. Lloyd) Munz
            SY=Fremontodendron d. R. Lloyd
        ssp. napense (Eastw.) Munz
            SY=Fremontia californica Torr.
                var. n. (Eastw.) McMinn
            SY=Fremontia n. Eastw.
        ssp. obispoense (Eastw.) Munz
            SY=Fremontia californica Torr.
                ssp. o. (Eastw.) Munz
            SY=Fremontia o. Eastw.
    mexicanum A. Davids.
            SY=Fremontia m. (A. Davids.) J.F.
                Macbr.

Guazuma P. Mill.

    ulmifolia Lam.
            SY=G. guazuma (L.) Cockerell

Helicteres L.

    jamaicensis Jacq.

Hermannia L.

    pauciflora S. Wats.
    texana Gray

Melochia L.

    corchorifolia L.
    nodiflora Sw.
    pyramidata L.
            SY=Moluchia p. (L.) Britt.
    tomentosa L.
        var. frutescens (Jacq.) DC.
        var. tomentosa
            SY=Moluchia t. (L.) Britt.
    umbellata (Houtt.) Stapf
            SY=Melochia indica Kurz
    villosa (P. Mill.) Fawcett & Rendle
            SY=Melochia hirsuta Cav.
            SY=M. h. var. glabrescens Gray
            SY=Riedlea h. (Cav.) DC.

Nephropetalum Robins. & Greenm.

    pringlei Robins. & Greenm.

Pentapetes L.

    phoenicea L.

Theobroma L.

    cacao L.

Waltheria L.

    calcicola Urban
    indica L.
        var. americana (L.) R. Br. ex Hosaka
            SY=W. a. L.
            SY=W. pryolifolia Gray

STYRACACEAE

Halesia Ellis ex L.

    carolina L.
            SY=H. parviflora Michx.

Halesia (CONT.)
  diptera Ellis
          var. diptera
          var. magniflora Godfrey
  tetraptera Ellis
          var. monticola (Rehd.) Reveal &
              Seldin
          SY=H. carolina L. var. m. Rehd.
          SY=H. m. (Rehd.) Sarg.
          SY=H. m. var. vestita Sarg.
          var. tetraptera

Styrax L.

  americana Lam.
          var. americana
          var. pulverulenta (Michx.) Rehd.
          SY=S. p. Michx.
  grandifolia Ait.
  officinalis L.
          var. fulvescens (Eastw.) Munz &
              Johnston
          SY=S. o. ssp. f. (Eastw.)
              Beauchamp
          var. officinalis
          var. redivia (Torr.) Howard
          SY=S. californica Torr.
          SY=S. officinalis L. var. c.
              (Torr.) Rehd.
  platanifolia Engelm. ex Torr.
          var. platanifolia
          var. stellata Cory
  portoricensis Krug & Urban
  texana Cory
  youngae Cory

SYMPLOCACEAE

Symplocos Jacq.

  lanata Krug & Urban
  martinicensis Jacq.
          SY=S. polyantha Krug & Urban
  micrantha Krug & Urban
  tinctoria (L.) L'Her.
          SY=S. t. var. ashei Harbison
          SY=S. t. var. pygmaea Fern.

TACCACEAE

Tacca J.R. & G. Forst.

  leontopetaloides (L.) Kuntze

TAMARICACEAE

Tamarix L.

  africana Poir.
  aphylla (L.) Karst.
          SY=T. articulata Vahl
  aralensis Bunge
  canariensis Willd.
  chinensis Lour.
          SY=T. pentandra Pallas
  gallica L.
  parviflora DC.
          SY=T. tetrandra Pallas sensu

          auctt.
  ramosissima L.

THEACEAE

Cleyera Thunb.

  albopunctata (Griseb.) Krug & Urban
          SY=Eroteum a. (Griseb.) Britt.

Eurya Thunb.

  sandwicensis Gray
          var. grandifolia Wawra
          var. sandwicensis

Franklinia Bartr. ex Marsh.

  alatamaha Bartr. ex Marsh.

Gordonia Ellis

  lasianthus (L.) Ellis

Laplacea H.B.K.

  portoricensis (Krug & Urban) Dyer
          SY=Haemocharis p. Krug & Urban

Stewartia L.

  malachodendron L.
  ovata (Cav.) Weatherby
          SY=Malachodendron pentagynum
              (L'Her.) Small
          SY=S. p. L'Her.

Ternstroemia Mutis ex L. f.

  heptasepala Krug & Urban
          SY=Taonabo h. (Krug & Urban)
              Britt.
  luquillensis Krug & Urban
          SY=Taonabo l. (Krug & Urban)
              Britt.
  peduncularis DC.
          SY=Taonabo p. (DC.) Britt.
  stahlii Krug & Urban
          SY=Taonabo pachyphylla (Krug &
              Urban) Britt.
          SY=Taonabo s. (Krug & Urban)
              Britt.
  subsessilis (Britt.) Kobuski
          SY=Taonabo s. Britt.

THEOPHRASTACEAE

Jacquinia L.

  arborea Vahl
          SY=J. barbasco (Loefl.) Mez pro
              parte
  berterii Spreng.
  keyensis Mez
  revoluta Jacq.
  stenophylla Urban
  umbellata DC.

THYMELIACEAE

Daphne L.

    laureola L.
    mezereum L.

Daphnopsis Mart.

    americana (P. Mill.) J.R. Johnston
      ssp. caribaea (Griseb.) Nevl.
            SY=D. c. Griseb.
    helleriana Urban
    philippiana Krug & Urban

Dirca L.

    occidentalis Gray
    palustris L.

Edgeworthia Meisn.

    papyrifera Sieb. & Zucc.
         SY=E. chrysantha Lindl.

Passerina L.

    annua Wikstr.
         SY=Thymelaea passerina (L.) Coss.
            & Germ.

Wikstroemia Endl.

    basicordata Skottsberg
    bicornuta Hbd.
      ssp. bicornuta
      ssp. montis-eke Skottsberg
    caumii Skottsberg
    degeneri Skottsberg
    elongata Gray
    eugenioides Skottsberg
    forbesii Skottsberg
    furcata (Hbd.) Rock
    haleakalensis Skottsberg
    hanalei Wawra
    isae Skottsberg
    lanaiensis Skottsberg
        var. acutifolia Skottsberg
        var. lanaiensis
    leptantha Skottsberg
    macrosiphon Skottsberg
    monticola Skottsberg
        var. monticola
        var. occidentalis Skottsberg
    oahuensis (Gray) Rock
    palustris Hochr.
        var. major Skottsberg
        var. palustris
    perdita Deg. & Deg.
    phillyreifolia Gray
        var. buxifolia (Gray) Skottsberg
        var. phillyreifolia
        var. rigida Gray
    pulcherrima Skottsberg
    recurva (Hbd.) Skottsberg
        var. neriifolia Skottsberg
        var. recurva
    sandwicensis Meisn.
    sellingii Skottsberg
    skottsbergiana Sparre
    uva-ursi Gray
        var. kauaiensis Skottsberg
        var. uva-ursi
    vacciniifolia Skottsberg
    villosa Hbd.

TILIACEAE

Corchorus L.

    aestuans L.
         SY=C. acutangulus Lam.
    hirsutus L.
    hirtus L.
        var. glabellus Gray
        var. hirtus
        var. orinocensis (H.B.K.) K. Schum.
            SY=C. o. H.B.K.
    siliquosus L.

Tilia L.

    americana L.
         SY=T. a. var. neglecta (Spach)
            Fosberg
         SY=T. glabra Vent.
         SY=T. n. Spach
         SY=T. palmeri F.C. Gates
         SY=T. truncata Spach
         SY=T. venulosa Sarg.
    caroliniana P. Mill.
         SY=T. australis Small
         SY=T. c. var. rhoophila Sarg.
         SY=T. floridana (V. Engl.) Small
         SY=T. f. var. hypoleuca Sarg.
         SY=T. f. var. oblongifolia Sarg.
         SY=T. georgiana Sarg.
         SY=T. leptophylla (Vent.) Small
         SY=T. leucocarpa Ashe
         SY=T. leucocarpa var.
            brevipedunculata (Sarg.)
            Ashe
         SY=T. littoralis Sarg.
         SY=T. porracea Ashe
         SY=T. pubescens Ait.
    cordata P. Mill.
    heterophylla Vent.
         SY=T. eburnea Ashe
         SY=T. h. var. michauxii (Nutt.)
            Sarg.
         SY=T. lasioclada Sarg.
         SY=T. m. Nutt.
         SY=T. monticola Sarg.
    petiolaris DC.
    platyphyllos Scop.
    X vulgaris Hayne [cordata X platyphyllos]
         SY=T. europaea L. pro parte

Triumfetta L.

    bogotensis DC.
    lappula L.
    pentandra A. Rich.
    rhomboidea Jacq.
         SY=T. bartramia L.
         SY=T. excisa Urban
    semitriloba Jacq.
    velutina Vahl

TRAPACEAE

Trapa L.

    natans L.

TROPAEOLACEAE

Tropaeolum L.

Tropaeolum (CONT.)
    majus L.

TURNERACEAE

Piriqueta Aubl.

    caroliniana (Walt.) Urban
            var. caroliniana
            var. exasperata Urban
                SY=P. tracyi Gandog.
            var. glabra (DC.) Urban
                SY=P. glabrescens Small
            var. tomentosa Urban
                SY=P. t. sensu Small, non H.B.K.
            var. viridis (Small) G.S. Torr.
                SY=P. v. Small
    cistoides (L.) Griseb.
    ovata (Bello) Urban
    viscosa Griseb.

Turnera L.

    diffusa Willd.
            var. aphrodisiaca (G.H. Ward) Urban
            var. diffusa
    pumilea L.
    ulmifolia L.

TYPHACEAE

Typha L.

    angustata Bory & Chaubard
    angustifolia L.
            SY=T. a. var. calumetensis Peattie
    domingensis Pers.
    X glauca Godr. [angustifolia X latifolia &
        domingensis X latifolia]
    latifolia L.

ULMACEAE

Celtis L.

    iguanaea (Jacq.) Sarg.
            SY=Momisia i. (Jacq.) Rose &
                    Standl.
    laevigata Willd.
            var. laevigata
                SY=C. l. var. anomala Sarg.
                SY=C. l. var. brachyphylla Sarg.
                SY=C. mississippiensis Bosc
            var. smallii (Beadle) Sarg.
                SY=C. s. Beadle
            var. texana Sarg.
    lindheimeri Engelm. ex K. Koch
    occidentalis L.
            var. canina (Raf.) Sarg.
            var. occidentalis
                SY=C. o. var. crassifolia (Lam.)
                    Gray
            var. pumila (Pursh) Gray
                SY=C. p. Pursh
                SY=C. p. var. deamii Sarg.
    pallida Torr.
            SY=C. spinosa Spreng. var. p.
                    (Torr.) M.C. Johnston

            SY=Momisia p. (Torr.) Planch.
    reticulata Torr.
            SY=C. brevipes S. Wats.
            SY=C. douglasii Planch.
            SY=C. occidentalis L. var. r.
                    (Torr.) Sarg.
            SY=C. r. var. vestita Sarg.
    tenuifolia Nutt.
            var. georgiana (Small) Fern. & Schub.
                SY=C. g. Small
                SY=C. occidentalis L. var. g.
                    (Small) Ahles
                SY=C. pumila (Muhl.) Pursh var. g.
                    (Small) Sarg.
            var. soperi Boivin
            var. tenuifolia
    trinervia Lam.

Planera J.F. Gmel.

    aquatica (Walt.) J.F. Gmel.

Trema Lour.

    cannabina Lour.
    lamarckiana (Roemer & Schultes) Blume
    micrantha (L.) Blume
            SY=T. floridana Britt.

Ulmus L.

    alata Michx.
    americana L.
            SY=U. a. var. floridana (Chapman)
                    Little
            SY=U. f. Chapman
    crassifolia Nutt.
    glabra Huds.
            SY=U. campestris L. pro parte
    minor P. Mill.
            SY=U. carpinifolia Ruppius ex G.
                    Suckow
    parvifolia Jacq.
            SY=U. chinensis Pers.
    procera Salisb.
            SY=U. campestris L. pro parte
    pumila L.
    rubra Muhl.
            SY=U. fulva Michx.
    serotina Sarg.
    thomasii Sarg.

Zelkova Spach

    serrata (Thunb.) Makino

URTICACEAE

Boehmeria Jacq.

    cylindrica (L.) Sw.
            var. cylindrica
            var. drummondiana (Weddell) Weddell
                SY=B. c. var. scabra Porter
                SY=B. decurrens Small
                SY=B. drummondiana Weddell
                SY=B. s. (Porter) Small
    grandis (Hook. & Arn.) Heller
            var. cuneata Skottsberg
            var. grandis
            var. kauaiensis Skottsberg
    nivea (L.) Gaud.
            SY=Ramium n. (L.) Small
    repens (Griseb.) Weddell

Cypholophus Weddell

    macrocephalus Weddell

Hesperocnide Torr.

    sandwicensis Weddell
    tenella Torr.

Laportea Gaud.

    aestuans (L.) Chew
        SY=Fleurya a. (L.) Gaud.
    canadensis (L.) Weddell
        SY=Urticastrum divaricatum (L.)
            Kuntze
    interrupta (L.) Chew

Neraudia Gaud.

    angulata Cowan
        var. angulata
        var. dentata Deg. & Cowan
    cookii author?
    kahoolawensis Hbd.
    kauaiensis (Hbd.) Cowan
        var. helleri Cowan
        var. kauaiensis
    melastomifolia Gaud.
        var. gaudichaudii Cowan
        var. melastomifolia
        var. pallida Cowan
        var. parvifolia (Wawra) Hbd.
        var. pubescens Cowan
        var. uncinata Cowan
    ovata Gaud.
    sericea Gaud.

Parietaria L

    floridana Nutt.
        SY=P. nummularia Small
    hespera Hinton
        var. californica Hinton
        var. hespera
    judaica L.
    obtusa Rydb. ex Small
        SY=P. pensylvanica Muhl. ex Willd.
            var. o. (Rydb. ex Small)
              Shinners
    officinalis L.
    pensylvanica Muhl. ex Willd.
    praetermissa Hinton

Pilea Lindl.

    fontana (Lunell) Rydb.
        SY=P. opaca (Lunell) Rydb.
    herniarioides (Sw.) Weddell
    inaequalis (Juss. ex Poir.) Weddell
    krugii Urban
    leptophylla Urban
    margarettae Britt.
    microphylla (L.) Liebm.
    multicaulis Urban
    nummularifolia (Sw.) Weddell
    obtusata Liebm.
    parietaria (L.) Blume
    peploides (Gaud.) Hook. & Arn.
        var. major Hbd.
        var. peploides
    pumila (L.) Gray
        var. deamii (Lunell) Fern.
        var. pumila
        SY=Adicea p. (L.) Raf.
    repens (Sw.) Weddell
    richardii Urban

sanctae-crucis Liebm.
    semidentata (Juss.) Weddell
    serpyllifolia (Poir.) Weddell
    tenerrima Miq.
    yunquensis (Urban) Britt. & Wilson

Pipturus Weddell

    albidus (Hook. & Arn.) Gray
    brighamii Skottsberg
    forbesii Krajina
    gaudichaudianus Weddell
        var. asperrimus Skottsberg
        var. gaudichaudianus
        var. hualalaiensis Skottsberg
    hawaiensis Levl.
        var. eriocarpus (Skottsberg)
            Skottsberg
        var. hawaiensis
        var. integrifolius Deg. & Deg.
        var. molokaiensis (Skottsberg) Deg. &
            Deg.
    helleri Skottsberg
    kauaiensis Heller
    oahuensis Skottsberg
    pachyphyllus Skottsberg
    pterocarpus Skottsberg
    rockii Skottsberg
    ruber Heller
    skottsbergii Krajina

Pouzolzia Gaud.

    occidentalis Weddell

Rousselia Gaud.

    humilis (Sw.) Urban

Soleirola Gaud.

    soleirolii (Req.) Dandy
        SY=Helxine s. Req.

Urera Gaud.

    baccifera (L.) Weddell
    caracasana Gaud.
    chlorocarpa Urban
    glabra (Hook. & Arn.) Weddell
    kaalae Wawra
    konaensis St. John
    lobata L.
    sandvicensis Weddell
        var. kauaiensis Rock
        var. mollis Weddell
        var. sandvicensis

Urtica L.

    chamidryoides Pursh
        SY=U. c. var. runyonii Correll
    dioica L.
      ssp. dioica
      ssp. gracilis (Ait.) Seland.
        var. angustifolia Schlecht.
          SY=U. breweri S. Wats.
          SY=U. serra Blume
        var. californica (Greene) C.L.
          Hitchc.
          SY=U. c. Greene
          SY=U. lyallii S. Wats. var. c.
            (Greene) Jepson
        var. gracilis (Ait.) Taylor &
          MacBryde
          SY=U. g. Ait.
        var. holosericea (Nutt.) C.L. Hitchc.

Urtica (CONT.)
          SY=U. h. Nutt.
          var. lyallii (S. Wats.) C.L. Hitchc.
               SY=U. l. S. Wats.
          var. procera (Muhl.) Weddell
               SY=U. p. Muhl.
               SY=U. viridis Rydb.
     gracilenta Greene
     urens L.

VALERIANACEAE

Centranthus Neck. ex Lam. & DC.

     ruber (L.) DC.
               SY=Kentranthus r. (L.) Druce

Plectritis (Lindl.) DC.

     ciliosa (Greene) Jepson
        ssp. ciliosa
               SY=P. californica (Suksdorf) Dyal
        ssp. insignis (Suksdorf) Morey
               SY=P. californica (Suksdorf) Dyal
                    var. rubens (Suksdorf) Dyal
               SY=P. ciliosa var. davyana
                    (Jepson) Dyal
               SY=P. macroptera (Suksdorf) Rydb.
                    var. patelliformis
                    (Suksdorf) Dyal
     congesta (Lindl.) DC.
        ssp. brachystemon (Fisch. & Mey.) Morey
               SY=P. anomala (Gray) Suksdorf
               SY=P. a. var. gibbosa (Suksdorf)
                    Dyal
               SY=P. aphanoptera (Gray) Suksdorf
               SY=P. c. var. major (Fisch. &
                    Mey.) Dyal
               SY=P. magna (Greene) Suksdorf
               SY=P. samolifolia (DC.) Hoeck
               SY=P. s. var. involuta (Suksdorf)
                    Dyal
        ssp. congesta
        ssp. nitida (Heller) Morey
               SY=P. magna (Greene) Suksdorf var.
                    n. (Heller) Dyal
     macrocera Torr. & Gray
        ssp. grayii (Suksdorf) Morey
               SY=P. m. var. g. (Suksdorf) Dyal
               SY=P. m. var. mamillata (Suksdorf)
                    Dyal
        ssp. macrocera
               SY=P. eichleriana (Suksdorf)
                    Heller
               SY=P. jepsonii (Suksdorf)
                    Burtt-Davy
               SY=P. m. var. collina (Heller)
                    Dyal
               SY=P. m. var. macroptera Suksdorf
               SY=P. macroptera (Suksdorf) Rydb.

Valeriana L.

     acutiloba Rydb.
               SY=V. capitata Pallas ex Link ssp.
                    a. (Rydb.) F.G. Mey.
     arizonica Gray
               SY=V. acutiloba Rydb. var. ovata
                    (Rydb.) A. Nels.
     californica Heller
               SY=V. capitata Pallas ex Link ssp.
                    californica (Heller) F.G.
                    Mey.
     capitata Pallas ex Link

     columbiana Piper
     dioica L.
        ssp. sylvatica (Soland. ex Richards.)
             F.G. Mey.
               SY=V. d. var. s. (Soland. ex
                    Richards.) S. Wats.
               SY=V. septentrionalis Rydb.
     edulis Nutt. ex Torr. & Gray
        ssp. ciliata (Torr. & Gray) F.G. Mey.
               SY=V. c. Torr. & Gray
               SY=V. e. var. c. (Torr. & Gray)
                    Cronq.
        ssp. edulis
     occidentalis Heller
     officinalis L.
     pauciflora Michx.
     pubicarpa Rydb.
               SY=V. acutiloba Rydb. var. p.
                    (Rydb.) Cronq.
               SY=V. capitata Pallas ex Link ssp.
                    p. (Rydb.) F.G. Mey.
     scandens L.
               var. canolleana (Gard.) C.A. Muell.
               var. scandens
     scouleri Rydb.
               SY=V. sitchensis Bong. ssp.
                    scouleri (Rydb.) F.G. Mey.
               SY=V. sitchensis var. scouleri
                    (Rydb.) M.E. Jones
     sitchensis Bong.
        ssp. sitchensis
        ssp. uliginosa (Torr. & Gray) F.G. Mey.
               SY=V. s. var. u. (Torr. & Gray)
                    Boivin
               SY=V. u. (Torr. & Gray) Rydb.
     sorbifolia H.B.K.
     texana Steyermark

Valerianella P. Mill.

     amarella (Lindheimer ex Engelm.) Krok
     carinata Loisel.
     chenopodiifolia (Pursh) DC.
     dentata (L.) Pollich
     florifera Shinners
     locusta (L.) Betcke
               SY=V. olitoria Pollich
     longiflora (Torr. & Gray) Walp.
     nuttallii (Torr. & Gray) Walp.
     ozarkana Dyal
               SY=V. bushii Dyal
     palmeri Dyal
     radiata (L.) Dufr.
               SY=V. r. var. fernaldii Dyal
               SY=V. r. var. missouriensis Dyal
               SY=V. stenocarpa (Engelm. ex Gray)
                    Krok var. parviflora Dyal
     stenocarpa (Engelm. ex Gray) Krok
     texana Dyal
     umbilicata (Sullivant) Wood
               SY=V. intermedia Dyal
               SY=V. patellaria (Sullivant) Wood
               SY=V. radiata (L.) Dufr. var. i.
                    (Dyal) Gleason
     woodsiana (Torr. & Gray) Walp.

VERBENACEAE

Aegiphila Jacq.

     elata Sw.
     martinicensis Jacq.
          var. martinicensis
          var. oligoneura (Urban) Moldenke

Aloysia Ortega & Palau ex Pers.

    gratissima (Gillies & Hook.) Troncoso
        var. gratissima
            SY=A. lycioides Cham.
        var. schulzae (Standl.) Moldenke
            SY=A. lycioides Cham. var. s.
               (Standl.) Moldenke
    macrostachya (Torr.) Moldenke
    triphylla (L'Her.) Britt.
            SY=A. citriodora (Cav.) Ortega
            SY=Lippia c. (Ortega) H.B.K.
    wrightii (Gray) Heller
            SY=Lippia w. Gray

Avicennia L.

    germinans (L.) L.
            SY=A. nitida Jacq.

Bouchea Cham.

    linifolia Gray
    prismatica (L.) Kuntze
        var. brevirostra Grenz.
        var. longirostra Grenz.
        var. prismatica
    spathulata Torr.

Callicarpa L.

    americana L.
        var. americana
        var. lactea F.J. Mull.
    ampla Schauer
    dichotoma (Lour.) K. Koch
        SY=C. purpurea Juss.

Citharexylum L.

    berlandieri B.L. Robins.
    brachyanthum (Gray) Gray
    caudatum L.
    fruticosum L.
        var. fruticosum
        var. subvillosum Moldenke
        var. villosum (Jacq.) O.E. Schulz
    pentandrum Vent.
    X perkinsii Moldenke [caudatum X spinosum]
    spathulatum Moldenke & Lundell
    spinosum L.

Clerodendrum L.

    aculeatum (L.) Schlecht.
        SY=Volkameria a. L.
    bungei Steud.
        SY=C. foetidum Bunge
    fragrans (Vent.) Willd.
        var. fragrans
            SY=C. philippinum Schauer pro
               parte
        var. multiplex (Sweet) Moldenke
    glabrum E. Mey.
    indicum (L.) Kuntze
        SY=Siphonanthus i. L.
    inerme (L.) Gaertn.
    japonicum (Thunb.) Sweet
    speciosissimum Van Geert

Cornutia L.

    obovata Urban
    pyramidata L.

Duranta L.

repens L.
    var. alba (Masters) Bailey
    var. microphylla (Desf.) Moldenke
    var. repens

Glandularia J.F. Gmelin

    bipinnatifida (Nutt.) Nutt.
        var. bipinnatifida
            SY=Verbena ambrosifolia Rydb. ex
               Small
            SY=V. b. Nutt.
            SY=V. b. var. latilobata Perry
            SY=V. ciliata Benth.
            SY=V. c. var. longidentata Perry
            SY=V. c. var. pubera (Greene)
               Perry
        var. brevispicata Umber
    canadensis (L.) Nutt.
        SY=G. drummondii (Lindl.) Small
        SY=G. lambertii (Sims) Small
        SY=Verbena c. (L.) Britt.
        SY=V. c. var. atroviolacea Dermen
        SY=V. c. var. compacta Dermen
        SY=V. c. var. d. (Lindl.) E.M.
            Baxter
        SY=V. c. var. grandiflora (Haage &
            Schmidt) Moldenke
        SY=V. c. var. l. (Sims) Thellung
        SY=V. X oklahomensis Moldenke
    chiricahensis Umber
    delticola (Small) Umber
        SY=Verbena d. Small
    elegans (H.B.K.) Umber
        var. elegans
            SY=Verbena e. H.B.K.
        var. asperata (Perry) Umber
            SY=Verbena e. var. a. Perry
    goodingii (Briq.) Solbrig
        SY=Verbena g. Briq.
        SY=V. g. var. nepetifolia
            Tidestrom
    maritima (Small) Small
        SY=Verbina m. Small
    peruviana (L.) Small
        SY=G. chamaedrifolia Juss.
        SY=Verbena p. (L.) Britt.
    polyantha Umber
    pulchella (Sweet) Troncoso
    pumila (Rydb.) Umber
        SY=Verbena p. Rydb.
    quandrangulata (Heller) Umber
        SY=Verbena q. Heller
    racemosa (Eggert) Umber
        SY=Verbena r. Eggert
    tampensis (Nash) Small
        SY=Verbena t. Nash
    ?tenuisecta (Briq.) Small
        SY=Verbena t. Briq.
    tumidula (Perry) Umber
        SY=Verbena t. Perry
    vercunda Umber
    wrightii (Gray) Umber
        SY=Verbena w. Gray

Ghinia Schreb.

    boxiana Moldenke
        SY=G. spinosa (Sw.) Britt. &
            Wilson

Lantana L.

    achyranthifolia Desf.
    arida Britt.
    camara L.
        var. aculeata (L.) Moldenke

Lantana (CONT.)
        SY=L. a. L.
    var. camara
    var. flava (Medic.) Moldenke
    var. hybrida (Neubert) Moldenke
    var. mista (L.) Bailey
    var. mutabilis (Hook.) Bailey
    var. sanguinea (Medic.) Bailey
depressa Small
        SY=L. ovatifolia Britt. var.
            reclinata R.W. Long
frutilla Moldenke
fucata Lindl.
    var. antillana Moldenke
horrida H.B.K.
insularis Moldenke
involucrata L.
    var. involucrata
    var. odorata (L.) Moldenke
macropoda Torr.
microcephala A. Rich.
        SY=Goniostachyum citrosum Small
montevidensis (Spreng.) Briq.
        SY=L. sellowiana Link & Otto
ovatifolia Britt.
reticulata Pers.
strigosa (Griseb.) Urban
tiliifolia Cham.
trifolia L.
urticifolia P. Mill.
urticoides Hayek
    var. hispidula Moldenke
    var. urticoides
velutina Mart. & Gal.

Lippia L.

alba (P. Mill.) N. E. Br.
        SY=L. geminata H.B.K.
graveolens H.B.K.
micromera Schauer
    var. helleri (Britt.) Moldenke
        SY=L. h. Britt.

Petitia Jacq.

domingensis Jacq.

Petrea L.

kohautiana Presl
volubilis L.

Phryma L.

leptostachya L.
        SY=P. l. var. confertifolia Fern.

Phyla Lour.

cuneifolia (Torr.) Greene
        SY=Lippia c. (Torr.) Steud.
lanceolata (Michx.) Greene
        SY=Lippia l. Michx.
        SY=L. l. var. recognita Fern. &
            Grisc.
        SY=P. l. var. r. (Fern. & Grisc.)
            Soper
nodiflora (L.) Greene
    var. antillana Moldenke
    var. canescens (H.B.K.) Moldenke
        SY=Lippia c. H.B.K.
        SY=L. n. (L.) Michx. var. c.
            (H.B.K.) Kuntze
    var. incisa (Small) Moldenke
        SY=Lippia i. (Small) Tidestrom
        SY=P. i. Small

    var. longifolia Moldenke
    var. nodiflora
        SY=Lippia n. (L.) Michx.
    var. reptans (Spreng.) Moldenke
        SY=Lippia n. (L.) Michx. var. r.
            (Spreng.) Kuntze
        SY=L. r. Spreng.
        SY=P. n. var. repens (Spreng.)
            Moldenke
    var. rosea (D. Don) Moldenke
        SY=Lippia n. (L.) Michx. var. r.
            (D. Don) Munz
    var. texensis Moldenke
scaberrima (Juss.) Moldenke
stoechadifolia (L.) Small
        SY=Lippia s. (L.) H.B.K.
strigulosa (Mart. & Gal.) Moldenke
    var. sericea (Kuntze) Moldenke
        SY=P. s. var. parvifolia
            (Moldenke) Moldenke
    var. strigulosa

Priva Adans.

lappulacea (L.) Pers.
portoricensis Urban

Stachytarpheta Vahl

cayennensis (L.C. Rich.) Vahl
    var. albiflora Moldenke
    var. cayennensis
        SY=Valerianoides c. (L.C. Rich.)
            Kuntze
dichotoma (Ruiz & Pavon) Vahl
        SY=S. australis Moldenke
X hybrida Moldenke [jamaicensis X strigosa]
X intercedens Danser [dichotoma X
    jamaicensis]
jamaicensis (L.) Vahl
        SY=Valerianoides j. (L.) Kuntze
mutabilis (Jacq.) Vahl
speciosa Pohl ex Schauer ex A. DC.
strigosa Vahl
        SY=Valerianoides s. (Vahl) Britt.
X trimenii Rech. [indica X mutabilis]
urticifolia (Salisb.) Sims

Stylodon Raf.

carneus (Medic.) Moldenke
        SY=S. carolinensis (Walt.) Small
        SY=Verbena carnea Medic.
        SY=V. carolinensis (Walt.) J.G.
            Gmel.

Tetraclea Gray

coulteri Gray
    var. angustifolia (Woot. & Standl.)
        A. Nels. & J.F. Macbr.
    var. coulteri
viscida Lundell

Verbena L.

abramsii Moldenke
        SY=V. lasiostachys Link var. a.
            (Moldenke) Jepson
X blanchardii Moldenke [hastata X simplex]
bonariensis L.
    var. conglomerata Briq.
bracteata Lag. & Rodr.
        SY=V. bracteosa Michx.
        SY=V. imbricata Woot. & Standl.
brasiliensis Vell.
californica Moldenke

Verbena (CONT.)
  cameronensis L.I. Davis
        SY=V. lundelliorum Moldenke
  canescens H.B.K.
      var. canescens
      var. roemeriana (Scheele) Perry
  carolina L.
        SY=Glandularia peruviana (L.)
           Small
  clemensorum Moldenke
  cloverae Moldenke
        SY=V. c. var. lilacina Moldenke
  X deamii Moldenke [bracteata X stricta]
        SY=V. X dodgei Boivin
  ehrenbergiana Schauer
  X engelmannii Moldenke [hastata X
    urticifolia]
  gracilis Desf.
  halei Small
  hastata L.
      var. hastata
      var. scabra Moldenke
  hybrida Voss ex Rumpl.
  X illicita Moldenke [stricta X urticifolia]
  lasiostachys Link
      var. lasiostachys
        SY=V. prostrata R. Br. non Savi
      var. septentrionalis Moldenke
  litoralis H.B.K.
  macdougalii Heller
  menthifolia Benth.
  X moechina Moldenke [simplex X stricta]
  neomexicana (Gray) Small
      var. hirtella Perry
      var. neomexicana
      var. xylopoda Perry
  officinalis L.
  perennis Woot.
  X perriana Moldenke [bracteata X
    urticifolia]
  pinetorum Moldenke
  plicata Greene
      var. degeneri Moldenke
      var. plicata
  rigida Spreng.
  riparia Raf. ex Small & Heller
  robusta Greene
  runyonii Moldenke
  X rydbergii Moldenke [hastata X stricta]
        SY=V. X paniculatistricta Engelm.
  scabra Vahl
  simplex Lehm.
        SY=V. angustifolia Michx.
  stricta Vent.
  tenera Spreng.
  urticifolia L.
      var. incarnata (Raf.) Moldenke
      var. leiocarpa Perry & Fern.
      var. simplex Farw.
      var. urticifolia
  xutha Lehm.

Vitex L.

  agnus-castus L.
      var. caerulea Rehd.
  divaricata Sw.
  glabrata R. Br.
  negundo L.
      var. heterophylla (Franch.) Rehd.
      var. intermedia (P'ei) Moldenke
      var. negundo
  parviflora Juss.
  trifolia L.
      var. simplicifolia Cham.
        SY=V. ovata Thunb.
      var. subtrisecta (Kuntze) Moldenke

      var. trifolia
      var. variegata Moldenke

VIOLACEAE

Calceolaria Loefl. ex Britt. & Brown

  chelidonioides H.B.K.
        SY=C. scabiosifolia Roemer &
           Schultes

Hybanthus Jacq.

  attenuatus (Humb. & Bonpl.) G.K. Schulze
  concolor (T.F. Forst.) Spreng.
        SY=Cubelium c. (T.F. Forst.) Raf.
        SY=Viola c. T.F. Forst.
  linearifolius (Vahl) Urban
        SY=Ionidium l. (Vahl) Britt.
  portoricensis Urban
        SY=Ionidium p. (Urban) Krug &
           Urban
  verticillatus (Ortega) Baill.
      var. platyphyllus (Gray) Cory & Parks
      var. verticillatus
        SY=H. linearis (Torr.) Shinners
        SY=Calceolaria v. (Ortega) Kuntze

Isodendrion Gray

  forbesii St. John
  hawaiiense St. John
  hillebrandii St. John
  hosakae St. John
  lanaiense St. John
  laurifolium Gray
  longifolium Gray
  lydgatei St. John
  maculatum St. John
  molokaiense St. John
  pyrifolium Gray
  remyi St. John
  subsessilifolium Heller
  waianaeense St. John

Viola L.

  X aberrans Greene [fimbriatula X sororia]
        SY=V. X fernaldii House
  X abundans House [fimbriatula X sagittata]
        SY=V. X erratica House
  adunca Sm.
      var. adunca
        SY=?V. a. ssp. ashtonae M.S. Baker
        SY=?V. adunca ssp. radicosa M.S.
           Baker
        SY=V. adunca ssp. typica M.S.
           Baker
        SY=V. montanensis Rydb.
        SY=V. subvestita Greene
      var. bellidifolia (Greene) Harrington
        SY=V. b. Greene
      var. cascadensis (M.S. Baker) C.L.
        Hitchc.
        SY=V. c. M.S. Baker
      var. kirkii Duran
      var. oxyceras (S. Wats.) Jepson
        SY=V. a. ssp. o. (S. Wats.) Piper
      var. uncinulata (Greene) C.L. Hitchc.
        SY=V. a. ssp. u. (Greene)
           Applegate
  affinis Le Conte
        SY=V. latiuscula Greene
  X angellae Pollard [palmata X triloba]

Viola (CONT.)
  appalachiensis Henry
  arvensis Murr.
  bakeri Greene
    ssp. bakeri
        SY=V. nuttallii Pursh var. b.
            (Greene) C.L. Hitchc.
    ssp. grandis M.S. Baker & Clausen
    ssp. shastensis M.S. Baker
  beckwithii Torr. & Gray
    ssp. beckwithii
    ssp. glabrata M.S. Baker
  X bernardii Greene [pedatifida X sororia]
  biflora L.
    ssp. biflora
    ssp. carlottae Calder & Taylor
        SY=V. b. var. c. (Calder & Taylor)
            Boivin
  X bissellii House [obliqua X sororia]
        SY=V. X conturbata House
  brittoniana Pollard
      var. brittoniana
      var. pectinata (Bickn.) Alexander
        SY=V. p. Bickn.
  X caesariensis House [sagittata X triloba]
  californica M.S. Baker
  canadensis L.
      var. canadensis
      var. corymbosa Nutt. ex Torr. & Gray
        SY=V. canadensis ssp. rydbergii
            (Greene) House
        SY=V. canadensis var. rugulosa
            (Greene) C.L. Hitchc.
        SY=V. rugulosa Greene
      var. scariosa Porter
      var. scopulorum Gray
  canina L.
      var. canina
      var. montana (L.) Lange
        SY=V. c. ssp. m. (L.) Hartman
  chamissoniana Gingins
  X champlainensis House [affinis X
      septentrionalis]
  charlestonensis M.S. Baker & Clausen
        SY=V. purpurea Kellogg var. c.
            (M.S. Baker & Clausen) Welsh
            & Reveal
  X consobrina House [affinis X hirsutula]
        SY=V. X dowelliana House
  X consocia House [affinis X obliqua]
  conspersa Reichenb.
  X convicta House [fimbriatula X palmata]
  X cordifolia (Nutt.) Schwein. [hirsutula X
      sororia]
  cuneata S. Wats.
  X davisii House [affinis X brittoniana]
  X dimissa House [hirsutula X sagittata]
  X discors House [affinis X palmata]
  X dissensa House [affinis X sagittata]
  X dissita House [hirsutula X triloba]
  douglasii Steud.
  X eamesii House [brittoniana X palmata]
  egglestonii Brainerd
  epipsila Ledeb.
    ssp. repens (Turcz.) Becker
  esculenta Ell.
  X festata House [obliqua X sagittata]
  X filicetorum Greene [affinis X sororia]
        SY=V. X columbiana House
        SY=V. X consona House
  fimbriatula Sm.
        SY=V. sagittata Ait. var. ovata
            (Nutt.) Torr. & Gray
  flettii Piper
  floridana Brainerd
        SY=V. chalcosperma Brainerd
  glabella Nutt.

  X greenei House [sagittata X sororia]
  X greenmanii House [obliqua X triloba]
  hallii Gray
  hastata Michx.
  helena Forbes & Lydgate
      var. helena
      var. lanaiensis Rock
  hirsutula Brainerd
  X hollickii House [affinis X fimbriatula]
  howellii Gray
  incognita Brainerd
        SY=V. i. var. forbesii Brainerd
  X insessa House [nephrophylla X obliqua]
  X insolita House [brittoniana X sororia]
  kauaensis Gray
      var. kauaensis
      var. wahiawaensis Forbes
  labradorica Schrank
        SY=V. adunca Sm. var. minor
            (Hook.) Fern.
  lanceolata L.
    ssp. lanceolata
    ssp. occidentalis (Gray) Russell
        SY=V. o. (Gray) T.J. Howell
    ssp. vittata (Greene) Russell
        SY=V. l. var. v. (Greene)
            Weatherby & Grisc.
        SY=V. v. Greene
  langloisii Greene
        SY=V. l. var. pedatiloba Brainerd
  langsdorfii (Regel) Fisch.
  lobata Benth.
      var. integrifolia S. Wats.
      var. lobata
  lovelliana Brainerd
  X luciae Skottsberg [maviensis X robusta]
  maccabeiana M.S. Baker
  macloskeyi Lloyd
    ssp. macloskeyi
    ssp. pallens (Banks ex DC.) M.S. Baker
        SY=V. blanda Willd.
        SY=V. m. var. p. (Banks ex DC.)
            C.L. Hitchc.
        SY=V. p. (Banks ex DC.) Brainerd
        SY=V. p. var. subreptans Rouss.
  X malteana House [conspersa X rostrata]
  X marylandica House [brittoniana X
      sagittata]
        SY=V. X holmiana House
  maviensis Mann
      var. kohalana Rock
      var. maviensis
  X melissifolia Greene [obliqua X
      septentrionalis]
  missouriensis Greene
  X mistura House [palmata X sagittata]
  X modesta House [lanceolata X primulifolia]
  X mollicula House [macloskeyi ssp. pallens
      X primulifolia]
  X montivaga House [septentrionalis X
      sororia]
  X mulfordae Pollard [brittoniana X
      fimbriatula]
  X napae House [pratincola X sororia]
  nephrophylla Greene
      var. arizonica (Greene) Kearney &
          Peebles
      var. cognata (Greene) C.L. Hitchc.
      var. nephrophylla
  X notabilis Bickn. [brittoniana X obliqua]
  novae-angliae House
  nuttallii Pursh
    ssp. nuttallii
    ssp. vallicola (A. Nels.) Taylor &
        MacBryde
        SY=V. n. var. v. (A. Nels.) St.
            John

Viola (CONT.)
            SY=V. v. A. Nels.
    oahuensis Forbes
    obliqua Hill
            SY=V. cucullata Ait.
            SY=V. c. var. microtitis Brainerd
    ocellata Torr. & Gray
    odorata L.
    orbiculata Geyer ex Hook.
            SY=V. sempervirens Greene var.
                orbiculoides M.S. Baker
    palmata L.
    palustris L.
        var. brevipes (M.S. Baker) R.J. Davis
        var. palustris
    X parca House [fimbriatula X
        septentrionalis]
    X peckiana House [palmata X sororia]
    pedata L.
            SY=V. p. var. concolor Holm
            SY=V. p. var. lineariloba DC.
            SY=V. p. var. ranunculifolia DC.
    pedatifida G. Don
    pedunculata Torr. & Gray
        ssp. pedunculata
        ssp. tenuifolia M.S. Baker & Clausen
    X perpera House [hirsutula X sororia]
    X pinetorum Greene [purpurea ssp. mesophyta
        X purpurea ssp. xerophyta]
            SY=V. purpurea Kellogg var.
                pinetorum (Greene) Greene
    X populifolia Greene [sororia X triloba]
            SY=V. variabilis Greene
    X porteriana Pollard [fimbriatula X
        obliqua]
    praemorsa Dougl. ex Lindl.
            SY=V. linguifolia Nutt.
            SY=V. nuttallii Pursh var. l.
                (Nutt.) Jepson
            SY=V. n. var. major Hook.
            SY=V. n. ssp. p. (Dougl. ex
                Lindl.) Piper
            SY=V. n. var. p. (Dougl. ex
                Lindl.) S. Wats.
            SY=V. p. ssp. arida M.S. Baker
            SY=V. p. ssp. l. (Nutt.) M.S.
                Baker & Clausen ex M.E. Peck
            SY=V. p. ssp. major (Hook.) M.S.
                Baker
            SY=V. p. ssp. oregona M.S. Baker &
                Clausen ex M.E. Peck
            SY=V. p. var. l. (Nutt.) M.E. Peck
            SY=V. p. var. major (Hook.) M.E.
                Peck
    pratincola Greene
            SY=V. retusa Greene
    priceana Pollard
            SY=V. papilionacea Pursh var.
                priceana (Pollard) Alexander
    primulifolia L.
        var. acuta (Bigelow) Torr. & Gray
        var. primulifolia
        var. villosa Eat.
            SY=V. p. ssp. v. (Eat.) Russell
    psychodes Greene
            SY=V. lobata Benth. ssp. p.
                (Greene) Munz
    pubescens Ait.
        var. eriocarpa (Schwein.) Russell
            SY=V. e. Schwein.
            SY=V. pensylvanica Michx.
        var. leiocarpa (Fern. & Wieg.)
            Seymour
            SY=V. eriocarpa Schwein. var. l.
                Fern. & Wieg.
            SY=V. pensylvanica Michx. var. l.
                (Fern. & Wieg.) Fern.

        var. peckii House
        var. pubescens
    purpurea Kellogg
        ssp. atriplicifolia (Greene) M.S. Baker
            & Clausen
            SY=V. p. var. a. (Greene) M.E.
                Peck
        ssp. aurea (Kellogg) J. Clausen
            SY=V. a. Kellogg
            SY=V. purpurea var. a. (Kellogg)
                M.S. Baker
        ssp. dimorpha M.S. Baker & Clausen
        ssp. geophyta M.S. Baker & Clausen
        ssp. integrifolia M.S. Baker & Clausen
        ssp. mesophyta M.S. Baker & Clausen
        ssp. mohavensis (M.S. Baker & Clausen ex
            Blake) Clausen
            SY=V. aurea Kellogg ssp.
                arizonensis M.S. Baker &
                Clausen
            SY=V. aurea ssp. m. M.S. Baker &
                Clausen ex Blake
        ssp. purpurea
        ssp. venosa (S. Wats.) M.S. Baker &
            Clausen
            SY=V. p. var. v. (S. Wats.)
                Brainerd
        ssp. xerophyta M.S. Baker & Clausen
    quercetorum M.S. Baker & Clausen
    rafinesquii Greene
            SY=V. bicolor Pursh
            SY=V. kitaibeliana Roemer &
                Schultes pro parte
            SY=V. kitaibeliana Roemer &
                Schultes var. r. (Greene)
                Fern.
    X ravida House [hirsutula X palmata]
    X redacta House [fimbriatula X hirsutula]
    renifolia Gray
        var. brainerdii (Greene) Fern.
        var. renifolia
    X robinsoniana House [fimbriatula X
        triloba]
    robusta Hbd.
    rostrata Pursh
    rotundifolia Michx.
    X ryoniae House [obliqua X palmata]
    sagittata Ait.
        var. sagittata
            SY=V. X cestrica House
            SY=V. emarginata (Nutt.) Le Conte
            SY=V. e. var. acutiloba Brainerd
        var. subsagittata (Greene) Pollard
    selkirkii Pursh ex Goldie
    sempervirens Greene
            SY=V. sarmentosa Dougl. ex Hook.
    septemloba Le Conte
    septentrionalis Greene
        var. grisea Fern.
        var. septentrionalis
    sheltonii Torr.
    X slavinii House [affinis X triloba]
            SY=V. X milleri Moldenke
    sororia Willd.
            SY=V. palmata L. var. s. (Willd.)
                Pollard
            SY=V. papilionacea Pursh pro parte
            SY=V. rosacea Brainerd
    stoneana House
    striata Ait.
    X subaffinis House [affinis X nephrophylla]
    X sublanceolata House [lanceolata X
        macloskyi ssp. pallens]
    tomentosa M.S. Baker & Clausen
    tracheliifolia Gingins
        var. olokelensis (Skottsberg)
            Skottsberg

Viola (CONT.)
        var. populifolia Skottsberg
        var. tracheliifolia
tricolor L.
triloba Schwein.
        var. dilatata (Ell.) Brainerd
            SY=V. falcata Greene
        var. triloba
            SY=V. palmata L. var. t.
                (Schwein.) Gingins ex DC.
trinervata T.J. Howell
tripartita Ell.
            SY=V. t. var. glaberrima (DC.)
                Harper
umbraticola H.B.K.
        var. glaberrima Becker
        var. umbraticola
utahensis M.S. Baker & Clausen
X variabilis Greene [sororia X triloba]
viarum Pollard
villosa Walt.
            SY=V. rugosa Small
wailenalenae (Rock) Skottsberg
walteri House

VITACEAE

Ampelopsis Michx.

    arborea (L.) Koehne
    brevipedunculata (Maxim.) Trautv.
            SY=A. heterophylla (Thunb.) Sieb.
                & Zucc.
    cordata Michx.
            SY=Cissus ampelopsis Pers.
    mexicana Rose

Cissus L.

    caustica Tussac
    erosa L.C. Rich.
    incisa (Nutt.) Des Moulins
    intermedia A. Rich.
    obovata Vahl
    sicyoides L.
    trifoliata (L.) L.
    tuberculata Jacq.

Parthenocissus Planch.

    heptaphylla (Buckl.) Small
    inserta (Kern.) Fritsch
            SY=P. vitacea (Knerr) A.S. Hitchc.
    quinquefolia (L.) Planch.
        var. murorum (Focke) Rehd.
        var. quinquefolia
            SY=P. hirsuta (Pursh) Graebn.
            SY=P. q. var. h. (Pursh) Planch.
            SY=P. q. var. saint-paulii (Koehne
                & Gray) Rehd.
    tricuspidata (Sieb. & Zucc.) Planch.
            SY=Ampelopsis t. Sieb. & Zucc.

Vitis Adans.

    acerifolia Raf.
            SY=V. doaniana Munson ex Viala
            SY=V. longii Prince
            SY=V. l. var. microsperma Bailey
    aestivalis Michx.
            SY=V. a. var. argentifolia
                (Munson) Fern.
            SY=V. aestivalis var. bicolor (Le
                Conte) Britt. & Brown

            SY=V. argentifolia Munson
            SY=V. baileyana Munson pro parte
            SY=V. bicolor Le Conte
            SY=V. lecontiana House
            SY=V. lincecumii Buckl.
            SY=V. lincecumii var. glauca
                Munson
            SY=V. lincecumii var. lactea Small
            SY=V. rufotomentosa Small
            SY=V. simpsonii Munson
    arizonica Engelm.
        var. arizonica
            SY=V. a. var. galvinii Munson
        var. glabra Munson
            SY=V. treleasei Munson ex Bailey
    berlandieri Planch.
            SY=V. helleri (Bailey) Small
    bourquina Munson ex Viala
    californica Benth.
    champinii Planch.
    cinerea Engelm. ex Millard
        var. canescens Bailey
        var. cinerea
        var. floridana Munson
    girdiana Munson
    illex Bailey
    labrusca L.
            SY=V. l. var. subedentata Fern.
            SY=V. labruscana Bailey
    monticola Buckl.
            SY=V. texana Munson
    mustangensis Buckl.
        var. diversa (Bailey) Shinners
            SY=V. candicans Engelm. var. d.
                Bailey
        var. mustangensis
            SY=V. candicans Engelm.
    novae-angliae Fern.
    palmata Vahl
    riparia Michx.
            SY=V. r. var. praecox Engelm.
            SY=V. r. var. syrticola (Fern. &
                Wieg.) Fern.
            SY=V. vulpina L. var. p. (Engelm.)
                Bailey
            SY=V. v. var. s. Fern. & Wieg.
    rotundifolia Michx.
            SY=Muscadinia munsoniana (Simpson
                ex Munson) Small
            SY=M. r. (Michx.) Small
            SY=V. m. Simpson ex Munson
    rupestris Scheele
            SY=V. r. var. dissecta Eggert ex
                Bailey
    shuttleworthii House
            SY=V. coriacea Shuttlw.
    smalliana Bailey
    sola Bailey
    tiliifolia Humb. & Bonpl. ex Willd.
    vinifera L.
    vulpina L.
            SY=V. baileyana Munson pro parte
            SY=V. cordifolia Lam.
            SY=V. c. var. foetida Engelm.
            SY=V. c. var. sempervirens Munson

XYRIDACEAE

Xyris L

    ambigua Bey. ex Kunth
    baldwiniana Schultes
            SY=X. b. var. tenuifolia (Chapman)
                Malme

Xyris (CONT.)
  brevifolia Michx.
  caroliniana Walt.
          SY=X. arenicola Small
          SY=X. flexuosa Muhl. ex Ell
          SY=X. pallescens (C. Mohr) Small
  complanata R. Br.
  difformis Chapman
          var. curtissii (Malme) Kral
          SY=X. bayardii Fern.
          SY=X. c. Malme
          SY=X. neglecta Small
          var. difformis
          SY=?X. elata Chapman
          var. floridana Kral
  drummondii Malme
  elliottii Chapman
  fimbriata Ell.
  flabelliformis Chapman
  isoetifolia Kral
  jupicai L.C. Rich.
          SY=X. communis Kunth
  laxifolia Mart.
          var. iridifolia (Chapman) Kral ined.
          SY=X. i. Chapman
          var. laxifolia
  longisepala Kral
  montana Ries
  platylepis Chapman
  scabrifolia Harper
  serotina Chapman
  smalliana Nash
          SY=X. congdonii Small
          SY=X. s. var. olneyi (Wood)
              Gleason
  stricta Chapman
  tennesseensis Kral
  torta Sm.
          SY=X. t. var. macropoda Fern.
          SY=X. t. var. occidentalis Malme

ZANNICHELLIACEAE

Cymodocea Koenig

  filiformis (Kuetz.) Correll
          SY=C. manatorum Aschers.
          SY=Syringodium f. Kuetz.

Halodule Endl.

  beaudettei (den Hartog) den Hartog
          SY=Diplanthera b. den Hartog
          SY=D. wrightii (Aschers.) Aschers.
          SY=H. w. Aschers.

Zannichellia L.

  palustris L.
          SY=Z. p. var. major (Boenn.)
              W.D.J. Koch
          SY=Z. p. var. stenophylla Aschers.
              & Graebn.

ZINGIBERACEAE

Alpinia Roxb.

  zerumbet (Pers.) Burtt & R.M. Sm.
          SY=A. speciosa (Wendl.) K. Schum.
          SY=Catimbium s. (Wendl.) Holttum

          SY=Languas s. (Wendl.) Merr.

Costus L.

  guanaiensis Rusby
          var. macrostrobilus (K. Schum.) Maas
          SY=C. m. K. Schum.
  spicatus (Jacq.) Sw.
          SY=Alpinia s. Jacq.
          SY=C. cylindricus Jacq.

Curcuma L.

  longa L.

Hedychium Koenig

  coronarium Koenig
  flavescens Carey ex Roscoe

Renealmia L. f.

  alpina (Rottb.) Maas
          SY=Alpinia exaltata (L. f.) Roemer
              & Schultes
          SY=R. e. L. f.
  jamaicensis (Gaertn.) Horan.
          SY=Alpinia antillara Roemer &
              Schultes
          SY=A. j. Gaertn.
          SY=R. a. (Roemer & Schultes)
              Gagnepain
  occidentalis (Sw.) Sweet
          SY=Alpinia aromatica Aubl.
          SY=A. o. Sw.
          SY=Renealmia a. (Aubl.) Griseb.

Zingiber Boehmer

  officinalis Roscoe
          SY=Amomum zingiber L.
          SY=Z. z. (L.) Karst.
  purpureum Roscoe
          SY=Z. cassumunar Roxb.
  zerumbet (L.) Roscoe
          SY=Amomum z. L.

ZOSTERACEAE

Ruppia L.

  anomala Ostenf.
  cirrhosa (Petag.) Grande
          SY=R. spiralis L. ex Dumort.
  maritima L.
          var. brevirostris Agardh
          var. exigua Fern. & Wieg.
          var. longipes Hagstr.
          SY=R. pectinata Rydb.
          var. maritima
          SY=R. m. var. intermedia (Thed.)
              Aschers. & Graebn.
          var. occidentalis (S. Wats.) Graebn.
          SY=R. o. S. Wats.
          var. pacifica St. John & Fosberg
          var. rostrata Agardh
          var. spiralis Morris
          SY=R. m. var. obliqua (Schur)
              Aschers. & Graebn.
          var. subcapitata Fern. & Wieg.

Zostera L.

  marina L.

Zostera (CONT.)
        var. latifolia Morong
        var. marina
        var. stenophylla Aschers. & Graebn.
    nana Roth

ZYGOPHYLLACEAE

Fagonia L.

    californica Benth.
    laevis Standl.
            SY=F. californica Benth. ssp. l.
                (Standl.) Wiggins
    longipes Standl.
    pachyacantha Rydb.
            SY=F. californica Benth. var.
                glutinosa Vail

Guaiacum L.

    angustifolium Engelm.
            SY=Porlieria a. (Engelm.) Gray
    officinale L.
    sanctum L.

Kallstroemia Scop.

    californica (S. Wats.) Vail
    grandiflora Torr. ex Gray
    hirsutissima Vail
    maxima (L.) Hook. & Arn.
    parviflora J.B.S. Norton
            SY=K. intermedia Rydb.
    perennans B.L. Turner
    pubescens (G. Don) Dandy
            SY=K. caribaea Rydb.

Larrea Cav.

    tridentata (Sesse & Moc. ex DC.) Coville
            SY=L. divaricata sensu auctt. non
                Cav.
            SY=L. glutinosa Engelm.

Peganum L.

    harmala L.
    mexicanum Gray

Tribulus L.

    cistoides L.
    terrestris L.

Zygophyllum L.

    fabago L.
        var. brachycarpum Boiss.

Errata

Page	Column	Line	
XIV		8	Flora of Utah and Nevada (1925)
1	1	4	daniifolium
1	2	11	SY = C. pyramidalis
2	2	9	Coult.)
12	1	68	ssp. asiatica (Makino) A. Love
13	1	57	SY = L. c. var. **megastachyon**
13	2	5	SY = Huperzia **d.**
19	2	38	clausa (Chapman
21	2	27	SY = Calophanes h. **(Michx.)**
26	2	22	var. graminifolium
26	2	26	ADD **SY** = **A. gramineum var. wahlenbergii (Holmb.) Raymond & Kucyniak**
26	2	38	ADD **SY** = **A. p-a. var. michaletti (Aschers. & Graebn.) Buch.**
27	1	13	var. calycina
27	2	58	echinocephala
28	1	47	non (Gray) Benth. ex **S.** Wats.
44	1	33	ADD **SY** = **A. a. (Ashe) Bickn.**
46	1	18	SY = Lachnostoma a.
52	2	39	SY = A. campest**ris**
61	1	3	X cubana Britt. & Blake [arborescens
66	2	2	SY = Cirsium **muticum**
67	2	53	sophiifolia
68	2	4	(DC.)
70	1	27	SY = R. l. var. **helleri**
76	1	72	SY = Ageratina altissima **(L.) King & H.E. Robins.**
83	2	33	SY = C. o. var. **rudis**
83	2	55	SY = H. **scabra**
84	1	75	SY = H. X dorei Lepage
85	2	32	var. canotomentosus
86	2	16	ambrosiifolia
95	1	27	exilis **(Gray) Gray**
95	1	33	SY = P. e. var. a. **(Gray) Gray**
95	2	20	SY = Laphamia halimifolia
96	1	42	var. alpina **K**oidzumi
100	1	29	cannabinifolius
101	1	66	var. pseudaureus
102	1	43	SY = S. **simpsonii** Greene
111	2	26	SY = B. X **caerulea-grandis**
115	1	48	SY = **C.** j.
115	1	56	SY = **C.** j.
122	1	51	Reichenb.
125	1	46	SY = D. **pinnata** var. **paysonii**
134	1	55	SY = T. **fendleri** Gray
145	1	6	ADD **SY** = **L. gaudichaudii DC. var. g-m. (Rock) St. John & Hosaka**
146	2	15	SY = I. **arborea**
146	2	16	SY = I. **arborea**
149	1	5	var. alabamense
149	1	52	var. cephaloidea
149	2	1	SY = A. **congesta**
151	1	64	SY = **Arenaria** d. Fenzl ex Torr. & Gray var. e.
151	2	2	SY = Arenaria **rossii**
151	2	44	SY = Arenaria **stricta** Michx.
154	2	17	SY = Melandrium **affine (J. Vahl ex Fries) J. Vahl**
156	2	40	SY = A. **stricta**
159	2	49	SY = C. humile sensu auctt. non **Hook.**
162	1	51	var. depauperata Hodgdon
162	1	52	SY = L. minor L. var. d. (Hodgdon)
175	1	70	SY = C. **flava**
176	2	59	SY = C. **complanata Torr. & Hook.**
178	1	59	Mackenzie
187	2	20	SY = S. **cyperinus**
187	2	61	SY = Bolboschoenus paludosus **(A. Nels.) Soo**
189	2	23	syriaca
192	2	63	SY = C. u. var. **cisatlantica**
197	2	30	SY = A. l. var. major Pax & Hoffmann
200	1	34	SY = Euphorbia **multiformis**
200	1	35	**manoana**
202	2	56	SY = **Tithymalopsis** m.
217	2	66	SY = D. a. var. sintenisii
219	1	44	SY = Micropteryx **c-g.**
222	2	37	neoincanus
224	1	27	var. kiellmannii
225	1	37	SY = L. **leucopsis J.G. Agardh**
226	1	47	SY = L. **densiflorus**
226	1	48	SY = L. **densiflorus**
226	1	49	SY = L. **densiflorus**
226	1	50	SY = L. **densiflorus**
226	1	51	SY = L. **densiflorus**

Page	Column	Line	
226	1	52	SY = L. **densiflorus**
227	1	22	ssp. dudleyi (C. P. Sm.) Kenney & **D.** Dunn
227	1	63	(Henderson) Kenney & **D.** Dunn
228	1	42	ssp. k. (C. P. Sm.)
228	1	43	**SY = L. s.** var. k.
232	1	10	riparia
232	1	37	var. viscida
238	1	36	SY = V. **americana**
238	1	37	SY = V. **americana**
244	2	46	macounii
251	2	25	phacelioides Nutt. ex **W.** Bart.
257	2	67	SY = J. **arcticus**
259	1	9	X nodosiformis Fern [**a**lpinus
260	1	33	var. kjellman**ioides**
260	1	34	SY = L. kjellmann**iana**
260	1	36	SY = L. m. var. kjellmann**iana**
276	1	60	pygmaea
278	1	24	ADD **SY = S. liliacea (Greene) Wynd**
279	2	41	SY = Calliprora ixioide**s**
281	2	10	SY = L. lewisii ssp. lepag**ei**
286	2	49	pictum (Gillies
294	1	44	pomifera (Raf. **ex Sarg.**)
294	2	9	Skottsberg
297	2	9	SY = N. **marina**
305	1	11	ssp. tracyi (Jep**son**)
305	2	61	SY = E. **leptocarpum Hausskn.**
314	2	41	SY = **Platanthera**
318	2	22	SY = P. odoratissimus L. f. var. **oahuensis**
320	2	45	SY = P. **macounii**
327	1	66	SY = A. **hiemalis (Walt.) B.S.P.**
329	2	60	decipiens
331	2	29	SY = Brom**o**psis
336	1	75	leibergii (Vasey) Freckman**n**
364	2	43	var. fernandina
370	1	45	twisselmann**ii**
370	1	64	var. **dichro**cephalum
375	2	58	Hoffmann
378	1	25	SY = P. **gramineus**
378	1	27	SY = P. **gramineus**
378	1	29	SY = P. **gramineus**
379	1	54	SY = D. **meadia**
379	1	73	SY = D. **m.**
394	1	61	dilatata
397	1	38	SY = C. suborbi**c**ulata
401	2	24	SY = Dasiphora **fruticosa**
401	2	26	SY = Pentaphylloides **floribunda**
401	2	28	SY = Potentilla **fruticosa**
405	2	55	ADD **SY = R. arcticus ssp. arcticus var. acaulis (Michx.) Boivin**
407	1	58	SY = R. **montensis**
408	1	44	SY = R. zapultus
408	2	66	SY = R. **nocivus**
412	2	18	SY = G. s. **W.** Wright
413	1	8	Hilend
413	1	33	SY = G. **circaezans**
414	1	70	SY = G. puncticulosum
414	2	9	SY = G. **nuttallii Gray** ssp. **t.**
415	1	31	(Schu**lt**z)
428	2	3	SY = S. petiolaris Sm. var. **subsericea**
430	1	10	polycephala (L.)
433	1	61	ssp. arborescens
435	1	43	var. cynosbati
437	2	23	SY = **Saxifraga**
438	1	57	grandiflora
438	2	29	ADD **var. elata (Nutt.) Gray**
438	2	30	SY = T. **e.**
438	2	30	ADD **var. maritima**
443	2	51	SY = E. **curta**
444	2	34	Hoffmann
445	2	32	SY = M. palmeri Gray var. a. (Curran
448	1	62	var. lincol**n**ensis
451	1	69	SY = P. **eatonii**
456	1	16	pseud**o**china
466	2	5	Glandularia J.F. Gmel.
467	2	19	SY = P. **strigulosa**

DATE			